ENCYCLOPEDIA OF MODERN PHYSICS

EXECUTIVE ADVISORY BOARD

ENCYCLOPEDIA OF MODERN PHYSICS

Robert A. Meyers, Editor

TRW, Inc.

STEVEN N. SHORE, Scientific Consultant

Computer Sciences Corporation

NASA Goddard Space Flight Center

and DEMIRM, Observatoire de Meudon

Academic Press, Inc.

Harcourt Brace Jovanovich, Publishers

San Diego   New York   Berkeley   Boston

London   Sydney   Tokyo   Toronto

Academic Press, Inc.
San Diego, California 92101

United Kingdom Edition published by
Academic Press Limited
24–28 Oval Road, London NW1 7DX

Library of Congress Cataloging-in-Publication Data

Encyclopedia of modern physics / Robert A. Meyers, editor;
Steven N. Shore, scientific consultant.
     p. cm.
   ISBN 0-12-226692-7 (alk. paper)
   1. Physics--Dictionaries.   I. Meyers, Robert A. (Robert Allen),
   Date. II. Shore, Steven N.
  QC5.E544   1989
  530'.03--dc20   89-17886
                                      CIP

Printed in the United States of America
89 90 91 92   9 8 7 6 5 4 3 2 1

EDITORIAL ADVISORY BOARD

CONTENTS

Contributors ix
Foreword xi
Preface xiii

Amorphous Semiconductors 1
David Adler

Atomic and Molecular Collisions 15
Robert E. Johnson

Atomic Physics 43
Francis M. Pipkin

Bonding and Structure in Solids 61
J. C. Phillips

Chaos 69
Roderick V. Jensen

Chemical Physics 97
Richard Bersohn and Bruce J. Berne

CP Violation 109
John F. Donoghue and Barry R. Holstein

Dense Matter Physics 115
George Y. C. Leung

Electrodynamics, Quantum 149
W. P. Healy

Elementary Particle Physics 171
Timothy Barklow and Martin Perl

Energy Transfer, Intramolecular 205
Paul W. Brumer

Excitons, Semiconductor 231
Donald C. Reynolds and Thomas C. Collins

Ferromagnetism 251
H. R. Khan

Holography 261
Clark C. Guest

Laser Cooling and
Trapping of Atoms.........275
John E. Bjorkholm

Lasers...............................285
William T. Silfvast

Molecular Optical
Spectroscopy..................303
Thomas M. Dunn

Multiphoton
Spectroscopy..................325
Y. Fujimura and S. H. Lin

Neutron Scattering........353
William B. Yelon

Nonlinear Optical
Processes........................361
John F. Reintjes

Nuclear Physics..............415
Lawrence Wilets

Optical Diffraction........437
Salvatore Solimeno

Plasma Science
and Engineering.............449
J. L. Shohet

Positron Microscopy.....473
Arthur Rich and James Van House

Quantum Mechanics.....481
Albert Thomas Fromhold, Jr.

Quantum Optics............519
J. H. Eberly and P. W. Milonni

Quasicrystals..................551
Marko V. Jarić

Radiation Physics...........569
Andrew Holmes-Siedle

Relativity, General.........589
James L. Anderson

Relativity, Special..........613
John D. McGervey

Statistical Mechanics....637
W. A. Wassam, Jr.

Superconductivity
Mechanisms....................679
J. Spalek

Ultrasonics......................717
John Szilard

Unified Field
Theories..........................737
L. Dolan

Index...................................749

CONTRIBUTORS

Adler, David. *Massachusetts Institute of Technology.* Amorphous Semiconductors.

Anderson, James L. *Stevens Institute of Technology.* Relativity, General.

Barklow, Timothy. *Stanford University.* Elementary Particle Physics (in part)

Berne, Bruce J. *Columbia University.* Chemical Physics (in part).

Bersohn, Richard. *Columbia University.* Chemical Physics (in part).

Bjorkholm, John E. *AT&T Bell Laboratories.* Laser Cooling and Trapping of Atoms.

Brumer, Paul W. *University of Toronto.* Energy Transfer, Intramolecular.

Collins, Thomas C. *University of Tennessee.* Excitons, Semiconductor (in part).

Dolan, L. *Rockefeller University.* Unified Field Theories.

Donoghue, John F. *University of Massachusetts.* CP Violation (in part).

Dunn, Thomas M. *University of Michigan.* Molecular Optical Spectroscopy.

Eberly, J. H. *University of Rochester.* Quantum Optics (in part).

Fromhold, Albert Thomas, Jr. *Auburn University.* Quantum Mechanics.

Fujimura, Y. *Tohoku University.* Multiphoton Spectroscopy (in part).

Guest, Clark C. *University of California, San Diego.* Holography.

Healy, W. P. *University College London and Royal Melbourne Institute of Technology.* Electrodynamics, Quantum.

Holmes-Siedle, Andrew. *Fulmer Research Laboratories.* Radiation Physics.

Holstein, Barry R. *University of Massachusetts.* CP Violation (in part).

Jarić, Marko V. *Texas A&M University.* Quasicrystals.

Jensen, Roderick V. *Yale University.* Chaos.

Johnson, Robert E. *University of Virginia.* Atomic and Molecular Collisions.

Khan, H. R. *Forschungsinstitut für Edelmetalle und Metallchemie and University of Tennessee.* Ferromagnetism.

Leung, George Y. C. *Southeastern Massachusetts University.* Dense Matter Physics.

Lin, S. H. *Arizona State University.* Multiphoton Spectroscopy (in part).

McGervey, John D. *Case Western Reserve University.* Relativity, Special.

Milonni, P. W. *Los Alamos National Laboratory.* Quantum Optics (in part).

Perl, Martin. *Stanford University.* Elementary Particle Physics (in part).

Phillips, J. C. *AT&T Bell Laboratories.* Bonding and Structure in Solids.

Pipkin, Francis M. *Harvard University.* Atomic Physics.

Reintjes, John F. *Naval Research Laboratory.* Nonlinear Optical Processes.

Reynolds, Donald C. *Wright Patterson Air Force Base.* Excitons, Semiconductor (in part).

Rich, Arthur. *The University of Michigan.* Positron Microscopy (in part).

Shohet, J. L. *University of Wisconsin.* Plasma Science and Engineering.

Silfvast, William T. *AT&T Bell Laboratories.* Lasers.

Solimeno, Salvatore. *University of Naples.* Optical Diffraction.

Spalek, J. *Purdue University.* Superconductivity Mechanisms.

Szilard, John. *Loughborough University.* Ultrasonics.

Van House, James. *The University of Michigan.* Positron Microscopy (in part).

Wassam, W. A., Jr. *Centro de Investigación y de Estudios Avanzados del IPN, Mexico.* Statistical Mechanics.

Wilets, Lawrence. *University of Washington.* Nuclear Physics.

Yelon, William B. *University of Missouri Research Reactor.* Neutron Scattering.

FOREWORD

Modern physics, quantum mechanics and relativity, is a product of the twentieth century. It is founded firmly in the last century's attempt to address the microstructure of matter and to come to grips with the concepts of action-at-a-distance, electromagnetism, and heat radiation. Such diverse physical phenomena as atomic and molecular spectral lines, the gas laws, phase transformations, and interference of light have now become routine subject matter for unified approaches.

The development of modern physics, like the physics of the nineteenth century, is also intimately connected with the improvement in technology that has both led and followed the theoretical and laboratory investigations. The development of the Crookes tube led to the discovery of X rays at the close of the last century and provided the tool for evaluating the inner structures of atoms and the details of the structure of condensed matter. The work on phosphors led directly to the ability to measure the impact of charged particles on a luminescent screen, permitting the measurement of properties of electrons and protons. Developments in chemistry made the separation of radioactive elements feasible and served as the tool for the discovery of fission. Improvements in gratings and optics provided new spectroscopic tools for the study of fine and hyperfine structure of spectral lines, and for the exploration of new wavelength regions like the infrared and ultraviolet. Lasers, masers, and quantum optics are only a few of the fertile outgrowths of this continued exploration.

The discovery of the Zeeman effect was a product of the last century and paved the way for electron theory. The explanation of the so-called anomalous effect in the alkaline rare earths, which required the introduction of spin as a quantum number, opened a whole new area of both theory and practice. Through the development of quantum statistics and the exclusion principle, both atomic structure and, eventually, the behavior of electrons in solids could be explained. The transistor, the cornerstone of modern technology, can be seen as the ultimate practical product of this chain of theoretical development.

The discovery of superconductivity and superfluidity required the development of vacuum technology and cryogenics, both of which followed directly from nineteenth-century studies of thermodynamics and the properties of gases and liquids at low temperature. The same theories that were developed to cope with atomic and molecular structure were directed toward these new phenomena and resulted in a coherent theory to explain the quantum behavior of the condensed state, the prediction of the Josephson effect, the development of SIS junctions, and the dramatic new discoveries concerning high-temperature superconductivity. The study of plasma phenomena has found applications in solid-state physics as well as many other areas.

Mathematical studies of the orbits of satellites began the development of the tools now used to study turbulence and nonlinear dynamics. Fractals and stochastic theory have merged with these tools to provide powerful descriptions of nonlinear phenomena and have opened an entirely new area of study, chaos, which crosses virtually all disciplinary lines. Developments in computing have spurred this study and have permanently altered the way in which both physical and technological questions are addressed.

Simulation of physical environments and phenomena inaccessible in the laboratory has become possible on scales ranging from the many-body interactions of quarks to the large-scale dynamics of galaxies.

Biologists and geologists of the last century were stymied by their inability to reconcile the time scales required for evolution and the classical limits of the age of the earth and the sun. The twentieth century has been able not only to provide a time scale through the application of nuclear physics to stellar interiors, but also to extend the theory to the understanding of the origin of the elements through nucleosynthesis in stars and in the early universe. It can truly be stated that evolution, the product of zoological systematics in the last century, has found its driving mechanisms in the physical theory of the twentieth century.

The unification of relativity and quantum theory, begun in the attempt to understand the interaction of light with matter, has developed into the rich and wide-ranging areas of field theory, whose subject matter ranges from lasers to the ultimate unification of all the forces of nature. The weak interaction and electromagnetism are now seen as manifestations of a unified field; theories have been developed for the incorporation of the strong or nuclear force into a grand unified theory; and gravitation may yet yield to the theoretical onslaught and be included in an ultimate "theory of everything." This century has seen the scale of energy probed in the laboratory increase by more than a factor of one hundred thousand as particle accelerators have been developed that permit the probing of the deep structure of the proton. By the end of this century the superconducting supercollider, the most technologically complex particle accelerator ever conceived, will be on-line and should be able to address many of the predictions of the unified field theories.

The rapid development of particle physics has also found an expression in the cosmic context. Theoretical investigations of the early universe, collapsed states of self-gravitating matter, and the prediction and discovery of black holes are all the result of the attempts to combine general relativity theory with quantum mechanics. It is now possible to glimpse the earliest moments of the universe and to begin the discussion of its end states, a dream of the last century.

Steven N. Shore

PREFACE

The *Encyclopedia of Modern Physics* is intended to provide a survey of the most rapidly advancing fields of theoretical and applied physics for scientists and engineers in all fields of physical, or for that matter biological, sciences who have a basic knowledge of chemistry, physics, and calculus. Thus, the contents should be accessible to upper division and graduate university students, faculty and professional scientists, and engineers in all fields of endeavor.

This *Encyclopedia* is unique in that it is essentially a condensation of 34 textbooks that could be written to comprehensively detail the theory and application of recent advances in physics. Each author and article was selected by this Editor-in-Chief in consultation with our advisory board. The article content was specified to include (1) a glossary of terms useful to the typical scientifically educated reader not expert in the particular field of the article, (2) a concise definition, (3) a first-principal based treatment of the subject, and (4) a bibliography.

While there must be some degree of subjectivity involved in selection of topics to be classified as "modern physics," after selection of the articles it was noted that basically all areas of physics are represented in the *Encyclopedia*, indicating that there have been major recent advances across the board in the field of physics.

Of course, inevitably, even as the articles were being selected, written, revised, and set in print, additional advances in physics were being made, and thus a book on modern physics must necessarily be somewhat incomplete.

The classification of some of the articles is certainly open to discussion; however, it is worthwhile to attempt a grouping and organization of a field as diverse as modern physics. Here, then, is a listing of the articles by classification.

Atomic and molecular physics: Atomic and Molecular Collisions; Atomic Physics; Chemical Physics; Energy Transfer, Intramolecular; Molecular Optical Spectroscopy; Multiphoton Spectroscopy; Neutron Scattering.

Phenomenology: Ferromagnetism; Nonlinear Optical Processes; Holography; Laser Cooling and Trapping of Atoms; Lasers; Ultrasonics.

Quantum physics, mechanics, and fields: Electrodynamics, Quantum; Quantum Mechanics; Quantum Optics.

Condensed matter: Bonding and Structure in Solids; Dense Matter Physics; Optical Diffraction; Amorphous Semiconductors; Excitons, Semiconductor; Positron Microcopy; Superconductivity Mechanisms; Quasicrystals.

Plasmas: Plasma Science and Engineering.

Nuclear physics and physics of elementary particles and fields: Nuclear Physics; Radiation Physics; CP Violation; Elementary Particle Physics.

Relativity and gravitation: Relativity, General; Relativity, Special; Unified Field Theories.

Statistical physics and thermodynamics: Chaos; Statistical Mechanics.

Many of these article titles bring to mind recent advances in physics that have received attention in the scientific and trade press including, Chaos, Amorphous Semiconductors, Superconductivity Mechanisms, Positron Microscopy, Laser Cooling and Trapping of Atoms, and Dense Matter Physics.

Robert A. Meyers

AMORPHOUS SEMICONDUCTORS

David Adler   *Massachusetts Institute of Technology*

I.   Classification
II. Structure
III. Electronic Structure
IV. Defects
V.   Doping
VI. Chemical Modification
VII. Transient Effects
VIII. Applications

GLOSSARY

Conduction band: Lowest-energy unfilled set of states available to an electron in a semiconductor.

Defect center: Atom with a coordination that is not chemically optimal for the particular solid under consideration.

Density of states: Number of states per unit volume per unit energy range available to an electron as it moves through the solid.

Doping: Process of introducing an impurity in small concentrations into a semiconductor, thereby creating a defect center that results in a significant change in the energy necessary to create a free carrier of electricity.

Field effect: Process of modulating the electrical conductivity of a semiconductor by inducing positive or negative charge as a result of applying a voltage across a structure in which an insulator separates the semiconductor from a metallic electrode (called the *gate*).

Glass transition temperature: Softening point of a glass, above which the viscosity rapidly decreases by many orders of magnitude and crystallization becomes possible.

Hopping: Mechanism of electrical conduction in which an electron moves between two localized states, usually with the assistance of the energy of thermal atomic vibrations.

Localized state: Electronic state in which the electron is confined to a particular region of space, as opposed to an *extended state*, in which it can be located throughout the solid.

Metastable state: Phase in which the solid does not have the lowest possible free energy for its particular temperature and pressure, but additional energy is necessary to attain equilibrium. The retarding forces are said to impose a *potential barrier* impeding atomic motion.

Valence band: Highest-energy filled set of states available to an electron in a semiconductor.

Amorphous semiconductors are solids whose atoms do not exhibit any long-range periodic structure but are, nevertheless, capable of conducting electricity when thermal, optical, or chemical energy is supplied to their electrons. Although insulators and metals have been well known since the discovery of electricity, semiconductors are a much more recent discovery both scientifically and technologically. Since the first semiconductors investigated were crystalline and the current solid-state theory was restricted to periodic solids, it was believed for many years that amorphous semiconductors did not exist. Consequently, it was a surprise when it was reported in the 1950s that many chalcogenide glasses, that is, amorphous alloys containing significant concentrations of tellurium, selenium, or sulfur and prepared by rapid cooling of the liquid, were semiconducting. The field of amorphous semiconductors attracted little scientific interest before 1968, when Ovshinsky reported the existence of reversible switching effects in thin films of chalcogenide glasses and proposed a wide array of potential applications. This work spurred intensive investigations into the electronic properties of chalcogenides and amorphous analogues of the conventional, tetra-hedrally coordinated, crystalline semiconductors (e.g., silicon and germanium). In fact, many amorphous-based semiconductor devices, from computer memories and television pickup tubes to solar cells, are now commercially available.

I. Classification

The classification scheme that is most useful for the analysis of electronic behavior is based on predominant atomic coordination. As is the case for crystalline solids, semiconducting behavior is generally associated with covalent s–p bonding, and the maximum coordination number under ordinary circumstances is 4. When sp^3 bonding predominates, the solid is tetrahedral. Examples are amorphous-silicon-based alloys, amorphous germanium, and Group III–V alloys (e.g., amorphous GaAs). When p^3 bonding predominates because of the presence of large concentrations of Group VA atoms (i.e., pnictogens), the amorphous solids are usually called pnictide glasses. When p^2 bonding predominates as a result of the presence of large concentrations of Group VIA atoms or chalcogens, the amorphous solids are called chalcogenide glasses. Although other amorphous semiconductors exist, these three groups have been by far the most intensively studied.

II. Structure

Before any analysis of the physical properties of an amorphous solid can be undertaken, a knowledge of its structure is essential. Because of the lack of long-range periodicity in amorphous materials, it is much more difficult to determine their structure than those of crystalline solids. In addition, the preparation methods necessary to obtain the material in the amorphous phase require the consideration of several issues that are not usually contemplated when crystals are investigated. First, there is the question of composition. It is not always evident what is actually in the solid. If the material has been prepared by cooling from the liquid, there are usually not many compositional uncertainties; however, many of the most important amorphous semiconductors cannot be fabricated in this way, but rather are deposited as thin films directly from the vapor phase, and even relatively easily formed glassy solids are often prepared in thin-film form for convenience or for

commercial purposes. In such cases, the composition must be carefully determined, and preferential deposition and unsuspected impurities are the rule rather than the exception. The composition can be a sensitive function not only of the preparation technique, but also of the many deposition parameters and of the geometry of the system. The development of modern techniques, for example, secondary-ion mass spectroscopy (SIMS), has been an immense aid in determining the composition, but most important is the necessity of such a determination.

Once the composition is known, two other investigations are essential. First, it is important to determine the homogeneity of the material. Identification of microscopic phase separation over regions greater than 100 nm is relatively straightforward, but inhomogeneities on a smaller scale may be much more subtle. Aggregation of one of the species can be important on a scale of only ~1 nm. The second consideration is of possible anisotropy. It is evident that all thin films have at least two interface regions, that is, one near the substrate and the other contacting a solid, liquid, or gas, and these interfaces probably are significantly different from the bulk. In addition to these lateral effects, anisotropy perpendicular to the substrate (e.g., columnar growth) is often observed.

Amorphous solids do not exhibit the long-range order of crystals, and this results in a freedom from periodic constraints that allows a wide range of potential compositions and structures. In this sense, an amorphous solid can be considered a giant chemical molecule. Since small molecules can have many different structures with very different properties, it is clear that amorphous solids with the same composition are far from unique. Under ordinary conditions, a great deal of short-range order is expected; that is, each atom is most likely to have a local coordination consistent with its electronic structure. Thus, neutral atoms in Groups I to IV should have one to four nearest neighbors, respectively, whereas neutral atoms in Groups V, VI, and VII should have coordination numbers of 3, 2, and 1, respectively, in accordance with the $8 - N$ rule of chemical bonding. This arises because any deviation from the optimal coordination would ordinarily cause a large increase in total energy and thus would be suppressed if at all possible. [See BONDING AND STRUCTURE IN SOLIDS.]

It is important to consider the ideal structure of an amorphous semiconductor. In amorphous alloys, different chemical bonds are possible, and the strongest bonds are favored. Each type of bond has an optimal length, and the large values of the bond-stretching frequencies indicate the presence of strong forces, which tend to suppress significant bond-length deviations. Although considerably weaker, the bond-bending frequencies are usually sufficiently large that there are, at most, small bond-angle distortions under ordinary circumstances. The forces tending to impose third-neighbor (i.e., dihedral-angle) constraints are generally very much weaker. It has been suggested that the optimal average coordination for an amorphous alloy is the one in which the number of constraints introduced by fixed bond lengths and bond angles is equal to the number of degrees of freedom, namely, 3. If the average coordination number is m, then the average number of bonds per atom is $m/2$. Since the average number of bond angles is $m(m - 1)/2$, the criterion suggests that the optimal coordination for glass forming, m_0, is given by

$$(m_0/2) + [m_0(m_0 - 1)/2] = 3$$

or

$$m_0 = \sqrt{6} \simeq 2.4 \qquad (1)$$

This is an oversimplification, particularly if some of the bonds are predominately ionic, in which case the bond-bending frequencies are relatively small. Nevertheless, most glass-forming alloys have average coordinations between 2 and 3, and amorphous solids with higher coordinations tend to be overconstrained. For such materials, there is competition between formation of the optimal number of bonds and relief of the concomitant strains thereby introduced.

The total energy of amorphous alloys containing one or more elements from Groups I to III with one or more from Groups V and VI can be reduced by the formation of dative bonds. This tends to increase the average coordination at the expense of additional strains.

Therefore, the lowest-energy structure of a particular amorphous semiconductor is that which maximizes the number of chemical bonds consistent with the given composition, maximizes the strength of the possible chemical bonds, optimize the bond lengths, and optimizes the bond angles. If the average coordination is $\simeq$ 2.4, these criteria can be satisfied, at least in principle, without undue strain; however, if the average coordination is >2.4, some strains are introduced. Most of these strains result in bond-angle distortions, the minimal cost in energy.

However, particularly in the case of tetrahedral materials, very strained regions will almost certainly be characterized by deviations from the optimal coordination.

In reality, the ideal structure is never attained, even for glasses with low average coordinations. This is because of the preparation techniques, which ordinarily require rapid thermal quenching to prevent crystallization. Even when bulk glasses are made by cooling from the liquid phase, significant atomic motion does not occur below a well-defined glass transition temperature T_g in the vicinity of which the viscosity rapidly increases with decreasing temperature from values characteristic of a liquid to those of a solid. Since T_g is a finite temperature, thermodynamics suggest that the equilibrium phase is not the one with the lowest energy; rather it is the one with the lowest free energy. For example, if a particular defect, that is, nonoptimally coordinated, configuration with a relatively low creation energy ΔE_d exists, then a minimum concentration of these defects, which is given by

$$N_d = N_0 \exp[-\Delta E_d/kT_g] \qquad (2)$$

where N_0 is the total concentration of atoms in the glass, is thermodynamically quenched during preparation. Less careful fabrication techniques can increase the defect concentrations well beyond that given by Eq. (2). Defect configurations ordinarily control the electrical properties of amorphous semiconductors.

III. Electronic Structure

As indicated, the quantum theory of solids, as originally formulated, was based entirely on crystalline periodicity. In the absence of such periodicity, the problem of electronic structure becomes mathematically much more complex, and additional approximations are essential. The electronic structure of amorphous semiconductors is generally regarded as at least qualitatively understood.

The main results of the quantum theory of crystalline solids are that the electronic states lie in densely packed regions of energy called bands, which are separated by regions called gaps; the electronic density of states $g(E)$ decreases near a band edge E_0 as

$$g(E) = A|E - E_0|^{1/2} \qquad (3)$$

where A is a constant; all states within a band are extended (i.e., the occupying electron has equal probability of being located near any equivalent atom in the periodic solid); and localized states can exist only within gaps and result only from the presence of defects or surfaces. The extended nature of the vast majority of electronic states in crystals result in high carrier mobilities and ultimately in metallic behavior. The gaps are responsible for the existence of insulators and semiconductors. Since amorphous metals exist, extended states must also be present in noncrystalline solids. Also, since amorphous insulators exist, gaps must be present even in the absence of periodicity; however, Eq. (3) is not valid for noncrystalline materials. Instead of sharp band edges, the breakdown of long-range periodicity introduces a tailing of states from the bands into the gaps; these states are appropriately called band tails. From the previous discussion, it is clear that these tails arise primarily from the bond-angle distortions that take up the strains in materials with $m > 2.4$. When $m \approx 2.4$, the principal cause may be dihedral-angle variations, and it might be expected that the tails are less extensive. Theoretical calculations ordinarily suggest a Gaussian band tail, although exponential tails have also been derived; the experimental evidence thus far is in support of the latter. The necessary presence of these tails effectively eliminates the band gap and emphasizes the question of how an amorphous semiconductor can exist at all.

This problem is exacerbated by the breakdown of Bloch's theorem in disordered systems. In perfectly crystalline solids, this theorem requires that all electronic states be extended throughout the solid, a fact, as noted above, that is generally believed to produce the high mobilities of free carriers in semiconductors.

In the absence of long-range periodicity, the existence of extended states has never been rigorously demonstrated. In fact, in one dimension, it has been shown that all states are localized, and there is some indication that a similar result holds in two dimensions. The current belief is that in three dimensions well-defined energy levels exist at which points the carrier mobility $\mu(E)$ sharply increases from essentially zero to finite values; these energy levels are called *mobility edges*. Although the concept of mobility edges is extremely useful, we must bear in mind that the actual mobility of a carrier at any energy E depends on the occupation numbers of all the other states in the band, so that $\mu(E)$ is not a well-defined quantity. Neverthe-

less, if an energy range Δ exists within which all states are essentially equivalent, the idea of a mobility edge can be heuristically justified. For example, if $g(E)$ is the density of states in the conduction band tail, the average separation between equivalent localized states is then

$$d = [g(E)\Delta]^{-1/3}$$

If the localized wave function falls off over a characteristic distance r_0, then the carrier mobility should be proportional to the probability of quantum-mechanical tunneling between two adjacent localized states. Thus, we can write

$$\mu(E) = \mu_0 \exp(-2d/r_0)$$
$$= \mu_0 \exp\{-2/[\Delta g(E)]^{1/3} r_0\} \quad (4)$$

If the density of states in the conduction band tail falls off exponentially below a particular energy E_0, as is typically observed, then

$$g(E) = A^3 \exp[-(E_0 - E)/kT_0] \quad (5)$$

where T_0 is a parameter describing the extent of the tail. The substitution of Eq. (5) into Eq. (4) yields

$$\mu(E) = \mu_0 \exp\left[-\frac{2}{Ar_0\Delta^{1/3}} \exp\left(\frac{E_0 - E}{3kT_0}\right)\right] \quad (6)$$

Equation (6) is essentially a step function, which justifies the concept of a mobility edge. The position of the edge E_c is given by the value of E for which the magnitude of the term in square brackets in Eq. (6) is unity. Thus,

$$E_c = E_0 - 3kT_0 \ln(Ar_0\Delta^{1/3}/2) \quad (7)$$

If Δ is determined by the broadening due to phonons, we can take $\Delta \approx kT$. In this case, the conduction-band mobility edge decreases logarithmically with increasing temperature. Substitution of Eq. (7) into Eq. (5) yields the critical density of states, g_c, that defines the position of the mobility edge:

$$g_c = 8/\Delta r_0^3 \quad (8)$$

We should expect that the wave function falls off rather slowly near the mobility edge. If we take $r_0 \approx 100$ Å and $\Delta = 0.025$ eV (kT at 300 K), then $g_c \approx 3 \times 10^{20}$ cm^{-3} eV^{-1}, a reasonable value.

If the density of states in the band tail decreases in a Gaussian rather than an exponential manner, similar results apply. In either case, the valence-band mobility edge has the inverse behavior and might be expected to increase logarithmically with increasing temperature. The difference between the two mobility edges is often called the *mobility gap* E_g. Although it is often remarked that the mobility gap plays the same role in a disordered semiconductor as the band gap does in a crystalline semiconductor, this is somewhat misleading. In crystals, the band gap determines the position of the optical absorption edge and the concentration of intrinsic carriers; on the other hand, in disordered materials, optical absorption can involve localized states in the band tails and these states can also control the position of the Fermi level in some cases.

Cohen, Fritzsche, and Ovshinsky suggested that, for multicomponent amorphous solids, the valence and conduction-band tails could overlap in the center of the gap, thus yielding a finite density of states at the Fermi energy $g(\varepsilon_f)$. They also assumed that all the atoms locally satisfied their chemical valence requirements, thus precluding the existence of sharp bumps within the mobility gap, as occurs in doped crystalline semiconductors. Their model, often called the CFO model (Fig. 1b), provides a useful structure for the analysis of transport and optical data; however, it is important to ask how the assump-

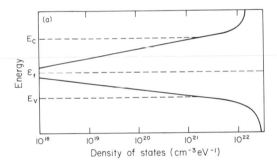

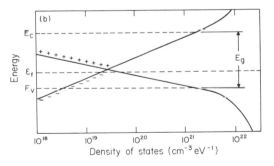

FIG. 1. Mott–CFO model for the density of states of covalent amorphous semiconductors. (a) Elemental amorphous semiconductors, for example, amorphous silicon; (b) multicomponent amorphous alloys. Here E_v and E_c represent the valence and conduction-band mobility edges, respectively, E_g is the mobility gap, and ε_f indicates the position of the Fermi energy.

tions of the model might be tested by such experiments.

Extensive band tails should be evident in ordinary optical-absorption experiments. In crystalline semiconductors, the material is essentially transparent to light of frequency below that of the energy gap, but the optical-absorption coefficient rapidly increases with photon frequency above the gap. It might be expected that, if extensive band tails exist, the optical-absorption coefficient should begin to increase with photon frequencies well below the mobility gap, and this increase should be much less sharp than in the case of crystals with direct edges.

The most obvious manifestation of a sharp mobility edge would be an electrical conductivity that varies as

$$\sigma = N_c e \mu_e \exp[-(E_c - \varepsilon_f)kT]$$
$$+ N_v e \mu_h \exp[-(\varepsilon_f - E_v)/kT] \qquad (9)$$

where N_c and N_v are the effective densities of states in the conduction and valence bands, respectively, μ_e and μ_h the electron and hole mobilities beyond the mobility edges, respectively, and E_c and E_v are the positions of the conduction-band and valence-band mobility edges, respectively. Equation (9) would yield a log σ versus T^{-1} plot that is either a single straight line or two intersecting straight lines. On the other hand, if an exponential density of states exists, as indicated by Eq. (5), the use of Eq. (6) indicates that hopping of carriers through the localized states in the band tail dominates the equilibrium transport for $T < T_0$. A log σ versus T^{-1} plot is still quite linear at ordinary temperatures, with an activation energy $E_a = E_0 - E_f$ somewhat greater than the expected $E_c - E_f$. If the data are extended to temperatures comparable with T_0, some upward curvature of the log σ versus T^{-1} plot should become evident.

If the CFO suggestion that a finite $g(\varepsilon_f)$ exists is correct, the Fermi energy would then appear to be pinned, in the sense that conductivity would be insensitive to substitutional doping or to the injection of excess carriers. In addition, the redistribution of electrons, which should leave high-energy valence-band states above ε_f (see Fig. 1), should create large densities of unpaired spins; these would be expected to be observable in both electron spin resonance (ESR) and magnetic-susceptibility experiments.

Mott pointed out that, for a sufficiently large $g(\varepsilon_f)$, phonon-assisted hopping among localized states near ε_f might dominate bandlike conduction beyond the mobility edges at sufficiently low temperatures. It was further suggested that at very low temperatures, when very few energetic phonons are present, such hopping conduction would take place preferentially to farther rather than to nearest-neighboring localized states in order to reduce the energy that must be obtained from the phonons. For this mechanism of variable-range hopping conduction, the conductivity varies as

$$\sigma = \sigma_0 \exp[-(T_0/T)^{1/4}] \qquad (10)$$

where σ_0 and T_0 are constants. When Eq. (10) is obeyed, the plot log σ versus T^{-1} is concave upward, but the plot log σ versus $T^{-1/4}$ is linear.

The possibility of variable-range hopping somewhat muddles the test for the existence of sharp mobility edges. A linear plot of log σ versus T^{-1} could reflect either the absence of a sharp mobility edge or the predominance of variable-range hopping conduction.

In principle, field-effect measurements provide an excellent tool for determining the density of states near the Fermi energy. This technique involves inducing either excess negative or excess positive charge in the semiconductor. The resultant increase in the electron concentration should then cause ε_f to shift upward or downward, respectively. The extent of this shift can be determined from the corresponding change in conductivity and it can then be related directly to $g(E)$. In principle, the field-effect technique can also be used to determine whether $g(E)$ is smooth or exhibits structure, the latter indicating a significant lack of local valence satisfaction.

Thus, the CFO model suggests that semiconductors (e.g., amorphous silicon, germanium, and selenium) should exhibit a large field effect, the possibility of substitutional doping, a small density of unpaired spins, no variable-range hopping, and a sharp absorption edge. In contrast, a multicomponent amorphous alloy should be characterized by a small field effect, no substitutional doping, a large unpaired-spin density, variable-range hopping, and a diffuse absorption edge.

When the experiments were carried out, the results were surprising. For example, pure amorphous silicon or germanium deposited on a room-temperature substrate exhibits none of the predicted properties: There is no observable field effect, no doping, a large unpaired-spin density, variable-range hopping conduction, and a diffuse absorption edge. Many of these properties, however, are not intrinsic and anneal away

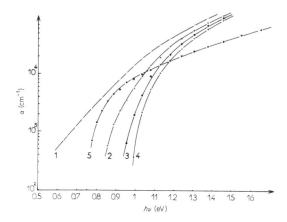

FIG. 2.   Absorption coefficient α as a function of photon energy $h\nu$ for amorphous germanium film. 1, Deposited; 2, annealed at 200°C; 3, annealed at 300°C; 4, annealed at 400°C; 5, crystallized. [From Theye, M. L. (1971). *Mat. Res. Bull.* **6,** 103.]

at temperatures below the crystallization point. Figure 2 indicates that the diffuse absorption edge in deposited amorphous germanium films evolves on annealing into an edge no more diffuse than in the crystallized film. The strong correlation between the unpaired-spin density, the electrical conductivity, and the optical edge of amorphous silicon as functions of annealing is evident in Fig. 3.

All of these properties arise because of the overconstrained nature of the fourfold-coordinated tetrahedral network. Strains introduced on deposition lead to distorted bond angles and

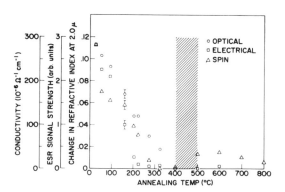

FIG. 3.   Correlation between room-temperature electrical conductivity, ESR signal strength, and change in refractive index for amorphous silicon films as a function of annealing (S/cm = Ω^{-1} cm^{-1}). The shaded area represents the region in which the films crystallize. $\bigcirc$, Optical; $\square$, electrical, $\triangle$, spin. [From Brodsky, M. H., Title, R. S., Weiser, K., and Petit, G. D. (1970). *Phys. Rev. B* **1,** 2632.]

well-defined chemical defects (e.g., dangling bonds). These defects can be reduced by depositing the film on a high-temperature substrate or by annealing it after deposition, but such procedures are limited by the crystallization temperature. The behavior predicted by the CFO model is never observed.

Alternatively, when amorphous silicon is hydrogenated during deposition or afterward, the expected behavior is observed in each case. This was originally discovered accidentally when silane gas, SiH_4, was decomposed by a radio-frequency glow discharge in an attempt to obtain pure amorphous silicon. The resulting material, particularly when deposited on a substrate at $\approx 300°C$, exhibits a field effect indicative of a very low ($\sim 10^{17}$ cm^{-3} eV^{-1}) density of states near ε_f, as evident from Fig. 4. The density of defects is sufficiently low that p and n doping of the films is easily achieved by mixing small concentrations of either B_2H_6 or PH_3 gas with the SiH_4 before decomposition. The results are shown in Fig. 5. All silane-decomposed films contain 5–30 at % residual hydrogen, which bonds only monovalently (ignoring the possibilities of three-center and hydrogen bonding), thus

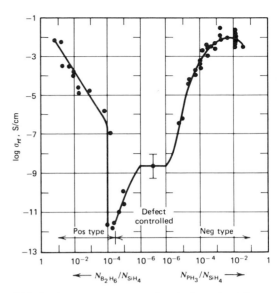

FIG. 5. Room-temperature electrical conductivity as a function of phosphorus and boron doping in a-Si:H films. [From Spear, W. E., and Le Comber, P. G. (1976). *Philos. Mag.* **33**, 935.]

decreasing the average coordination number and vastly reducing the strains. The reason for the optimal substrate temperature near 300°C appears to be the thermodynamic requirement that hydrogen effuse from the network at high temperatures. Thus, the minimum defect density very likely represents a balance between the beneficial effects of annealing and the deleterious effects of hydrogen effusion, as is evident from the temperature dependence of the unpaired-spin density shown in Fig. 6.

Other alloys besides amorphous silicon–hydrogen (a-Si:H) exhibit excellent electronic properties. One particularly promising material is a-Si:F:H, which is fabricated by the glow-discharge decomposition of SiF_4–H_2 mixtures. This alloy appears to have lower defect densities than a-Si:H, most likely because of the extra flexibility of the primarily ionic Si—F bond. The Si—F bond is also stronger than the Si—H bond, resulting in a harder material that retains its properties to higher annealing temperatures.

Despite these wide variations in behavior, the properties of tetrahedral amorphous semiconductors can be understood by extending one or the other of the models sketched in Fig. 1. However, the same is not the case for chalcogenide alloys. In almost all of these materials, the Fermi energy appears to be very strongly pinned, as evidenced by the absence of extrinsic

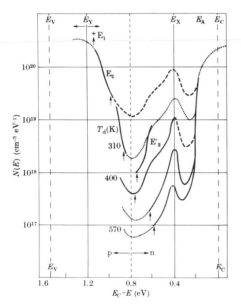

FIG. 4. Density of states as a function of energy for evaporated films of amorphous silicon and glow-discharge films of a-Si:H deposited at various substrate temperatures T_d in kelvins. The position of the Fermi energy is indicated by the arrow. The full lines represent the results deduced from field-effect measurements. [From Madan, A., Le Comber, P. G., and Spear, W. E. (1976). *J. Non-Cryst. Solids* **20**, 239.]

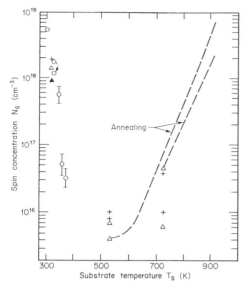

FIG. 6.   Unpaired-spin concentration as a function of substrate temperature and annealing for a-Si:H films. Inductive system: $\bigcirc$; diode system: $\triangle$ (A), $\blacktriangle$ (A*), + (C). [From Fritzsche, H., Tsai, C. C., and Persans, P. (1978). *Solid State Technol.* **21**, 55.]

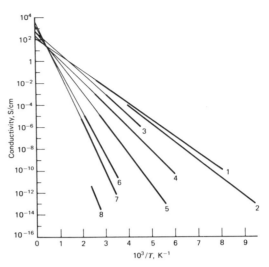

FIG. 7.   Electrical conductivity as a function of temperature for several amorphous chalcogenide alloys (S/cm = Ω^{-1} cm^{-1}). [From Mott, N. F., and Davis, E. A. (1979). "Electronic Processes in Non-Crystalline Materials," 2nd ed. Clarendon, Oxford.]

conduction, doping, or an observable field effect under ordinary conditions. The extremely linear relation between log σ and T^{-1} over many decades in conductivity for many different amorphous chalcogenides is shown in Fig. 7; no extrinsic region is evident down to conductivities of 10^{14} S/cm (= Ω^{-1} cm^{-1}). On the other hand, no variable-range hopping or any unpaired-spin density has been observed except under the most nonequilibrium conditions, and the absorption edge is no more diffuse in multicomponent glasses than in amorphous selenium or amorphous tellurium, as is clear from Fig. 8.

Relatively large field effects have been observed in multicomponent glasses containing tellurium and arsenic, but these are transient and decay with time. Thus, there is evidence both for and against a large density of states at the

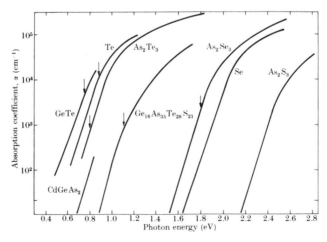

FIG. 8.   Absorption coefficient α as a function of photon energy for several amorphous chalcogenide glasses. [From Mott, N. F., and Davis, E. A. (1979). "Electronic Processes in Non-Crystalline Materials," 2nd ed. Clarendon, Oxford.]

Fermi energy, and the experimental results appear to be contradictory.

The resolution of these and other problems that have emerged with regard to the electronic structure of amorphous semiconductors lies in the recognition that well-defined defects exist in these materials. As indicated previously, these defects control the transport properties of the semiconductors.

IV. Defects

Although many of the defects common in crystalline solids, such as, vacancies, interstitials, dislocations, and grain boundaries, are not likely to occur in the absence of periodicity, chemical defects, such as undercoordinated or overcoordinated atoms and wrong bonds, are possible. There is a great deal of evidence for their existence in all classes of amorphous semiconductors. One of the simplest examples is the apparent substitutional doping of amorphous silicon alloys. If phosphorus and boron do enter these materials in tetrahedral sites, such centers then have the wrong chemical coordination and represent defects. In the CFO model, the valence requirements of each atom are locally satisfied, and this type of doping would then be impossible. Even in undoped samples of a-Si:H, the existence of defects can be inferred unambiguously. For example, it is clear from the results in Fig. 6 that the spin density decreases with increasing substrate temperature until $\approx 300°C$. However, as samples are annealed above 300°C, hydrogen effuses and the spin density is observed by ESR measurements, which indicates a common center for all spins. Clearly, samples prepared at low substrate temperatures must contain some of the same defect centers as those created when hydrogen effuses. It can be concluded that well-defined defects are present in tetrahedrally coordinated amorphous solids.

There is also a great deal of evidence for defects in amorphous chalcogenides. Perhaps the simplest demonstration comes from the photoluminescence of crystalline and amorphous SiO_2. Amorphous SiO_2 has an absorption edge at ≈ 10 eV; however, the photoluminescence excitation spectrum peaks near 7.6 eV and is clearly separated from the continuum, as is evident from Fig. 9. Also shown in Fig. 9 are the photoluminescence, excitation, and absorption spectra for amorphous As_2Se_3 and amorphous SiO_2. The similarities in data for two such different glasses

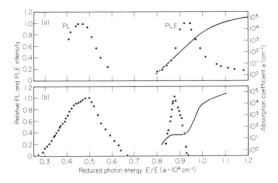

FIG. 9. Photoluminescence (PL), excitation (PLE), and absorption spectra for amorphous As_2Se_3 and amorphous SiO_2, normalized to the energy at which the absorption coefficient α is 10^4 cm^{-1}. (a) For As_2S_3, $E(\alpha = 10^4$ cm$^{-1}) = 2.55$ eV; (b) for SiO_2, $E(\alpha = 10^4$ cm$^{-1}) = 8.6$ eV. [From Gee, C. M., and Kastner, M. (1979). *Phys. Rev. Lett.* **42**, 1765.]

are strong evidence in favor of a similar mechanism for the luminescence.

The photoluminescence and excitation spectra for both neutron-irradiated crystalline and amorphous SiO_2 are essentially the same; however, whereas the temperature dependence of the luminescence intensities of unirradiated amorphous As_2Se_3 and amorphous SiO_2 are identical when the temperature is scaled to the glass transition temperature, irradiated amorphous SiO_2 has a luminescence 20 times stronger. Irradiated samples of crystalline SiO_2 have a strong absorption peak at 7.6 eV, which can result from nothing but a defect center. After irradiation, the excitation spectrum still peaks near 7.6 eV and the luminescence still peaks at 4.5 to 5.0 eV, but the luminescence is greatly enhanced. Irradiation simply increases the density of the defects that are responsible for the luminescence. In addition, the similarities in the scaled temperature quenching of the photoluminescence and the scaled energy dependence of luminescence and excitation spectra for several different glasses suggest a common mechanism for the entire process. Thus, it can be concluded that well-defined defects similar to those in the corresponding crystals are responsible for photoluminescence in chalcogenide glasses.

The recognition that well-defined defects exist in amorphous silicon alloys provides a simple explanation of the diverse electronic properties of pure and hydrogenated amorphous silicon. Pure amorphous silicon has an average coordination near 4, and the resulting strains lead to high defect concentrations. These defects yield localized states in the gap, which pin the Fermi energy, produce large spin densities, and lead to the predominance of variable-range hopping conduction; however, a-Si:H has a much lower average coordination, which relieves the vast majority of the strains and removes almost all of the localized states from the gap. This is clear from the field-effect results shown in Fig. 4. The structure evident in the data from Fig. 4 is usually assumed to result from dangling bonds on silicon atoms, although the situation is undoubtedly much more complex.

The resolution of the anomalous behavior in amorphous chalcogenides is much more subtle. These materials are not overconstrained and strain-produced defects are not expected; however, a particularly low energy defect, which has been called a valence-alternation pair (VAP), exists in these materials. A typical VAP, sketched in Fig. 10, consists of a positively charged threefold-coordinated chalcogen and a negatively charged singly coordinated chalcogen. In the usual notation, in which C refers to a chalcogen atom, P to a pnictogen, and T to a tathogen (Group IV), such a VAP is represented by C_3^+–C_1^-, where the subscript gives the coordination number and the superscript gives the charge state of the center. The creation energy of this VAP can be estimated to be approximately equal to a quantity U, the additional Coulomb repulsion that results from the presence of the extra electron on the C_1^- site. The quantity U represents the difference between the ionization potential of the chalcogen atom and its electron affinity, screened by the dielectric response of the material. Recent experimental results suggest that U is 1.0 eV for chalcogenides; therefore, Eq. (2) predicts that there are ~10^{16} cm^{-3} or more VAPs in chalcogenide glasses under ordinary conditions.

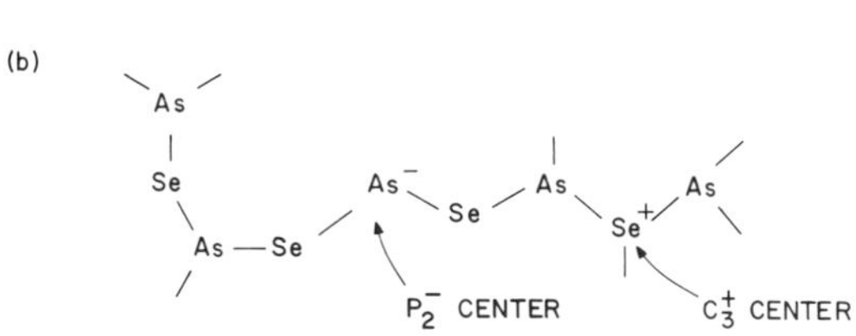

The reason for the low creation energy of a VAP is that the total number of bonds in a C_3^+–C_1^- pair is four, exactly the same as in a pair of chalcogen atoms in their ground states (C_2^0). Furthermore, both C_3^+ and C_1^- centers have low energies: A positively charged chalcogen is isoelectronic with a pnictogen, and therefore it is optimally threefold-coordinated; a negatively charged chalcogen is isoelectronic with a halogen and thus is optimally singly coordinated.

In addition to their low creation energy, VAPs also have a very unusual property that resolves the question of why the Fermi energy is so strongly pinned in chalcogenides, whereas there is no measurable unpaired-spin density and variable-range hopping conduction is not observed in any temperature range. The origin of this is the fact that the two defect centers, C_1 and C_3, have coordination numbers separated by two. This provides the possibility of converting a C_3 to a C_1 center simply by the breaking of a bond, since breaking of a two-electron bond lowers the total coordination of the atoms in a network by 2. It is this possibility, and the fact that a positively charged C_3 center and a negatively charged C_1 center have a lower total energy than two neutral centers with either the C_1 or C_3 configuration, that leads to Fermi energy pinning. The Fermi energy is the average energy necessary to add a few electrons to the material. However, if VAPs are present, adding two electrons simply converts a C_3^+ to a C_1^- center, independent of the charge state of the material, until no more C_3^+ centers are left. This occurs in two steps. The first electron is attracted to the C_3^+ center, converting it to C_3^0. The second electron then moves to any one of the three nearest neighboring sites of the C_3^0, and the bond between that site and the C_3^0 breaks. This returns the C_3^0 center to a C_2^0 ground state, leaving the neighbor as a C_1^- center. Since C_3^0 contains an antibonding electron but C_1^- does not, the second electron enters with a lower energy than the first. This means that excess electrons tend to enter the material in relatively localized pairs; since each pair has the same total energy, ε_f is pinned. The presence of VAPs also explains a

FIG. 10. Examples of valence-alternation pairs. (a) Amorphous selenium; (b) amorphous As$_2$Se$_3$. C, Chalcogen; P, pnictogens.

great deal of other unusual characteristics of chalcogenide glasses including the complete absence of spins, that is, neither a C_3^+ nor a C_1^- center contains any unpaired spins, no observed variable-range hopping (two electrons or holes must hop simultaneously to interconvert C_3^+ and C_1^- centers, and the resulting electrostatic relaxations require too much energy), and the details of the photoconductivity, photoluminescence, and photostructural effects.

Under the assumption that heteropolar bonding predominates in binary alloys, the lowest-energy VAPs in Group V and VI alloys are expected to be $C_3^+-P_2^-$ pairs, where P represents a pnictogen atom; similarly, in Group IV and VI alloys, the lowest-energy VAP is a $C_3^+-T_3^-$ pair, where T represents a tathogen atom. However, in both of these cases, the electronegativity differences are such that they have larger creation energies than a $C_3^+-C_3^-$ pair in a pure chalcogen material and thus appear with considerably lower concentrations. In addition, the $C_3^+-T_3^-$ VAP only pins the Fermi energy for excess positive charge because additional bond formation is impossible for tathogens. Although a $P_4^+-T_3^-$ pair has a low creation energy in Group IV and V alloys, it is not a VAP because of steric constraints. Although a P_4^0 center can in principle convert to a T_3^0 center by breaking a bond, the resulting relaxation would require both the tathogen and the pnictogen atoms to move toward each other after the bond breaks, which could be an energetically unfavorable motion. In pure pnictogen materials, the creation energy of a $P_2^--P_2^-$ VAP would be expected to be large relative to those discussed previously because of the additional $s-p$ promotion necessary to form the P_4^+ center, and there are some steric constraints retarding interconvertibility. In general, pnictogens exhibit behavior intermediate between chalcogenides and tetrahedral solids. In pure tathogens, VAPs are impossible because 4 is the maximal coordination with only s and p orbitals. However, charged defect centers are still likely and can be induced thermodynamically.

When an amorphous solid is not overconstrained, it would be expected that the lowest-energy defect predominates; however, there are several complications. First, it is possible that minimizing the free energy at T_g requires the presence of states in which the oppositely charged centers are located near each other, that is, intimate pairs (IVAPs), in which case a distribution of such pairs with various separations is present in the glass. Second, in nonstoichiometric alloys, homopolar bonds must exist, and these can yield localized gap states. Third, many other defects besides single overcoordination or undercoordination can exist. In particular, tathogens often bond with twofold coordination (T_2^0), thereby saving the $s-p$ promotion energy, and these might be expected in tetrahedral alloys. Other unusual configurations (e.g., three-center bonds) could also be present. None of these defects would ordinarily be expected to exist in sufficiently large concentrations to be observable in direct structural studies, but they could be of the utmost importance in transport behavior.

V. Doping

The fabrication of conventional semiconductor devices requires the ability to decouple the electrical activation energy from the optical gap. In crystalline semiconductors, this is accomplished by substitutional doping, in which small concentrations of an impurity atom with a different valence from the host are introduced into the crystal. The periodic constraints then force a defect configuration, which leads to a large change in ε_f. This would appear to be impossible in an amorphous semiconductor, in which no crystalline constraints exist; however, a-Si:H and a-Si:F:H alloys can be routinely doped with either phosphorus or boron. It is thus important to determine why phosphorus and boron enter the amorphous network tetrahedrally instead of in their ground-state (P_3^0 or B_3^0) configurations. One possibility is that phosphorus or boron acts as a nucleation center for the growth of doped microcrystalline silicon, and there is evidence that this occurs in some very heavily doped samples; however, there is no direct structural evidence for any crystallites in lightly or moderately doped amorphous silicon alloys.

Both a-Si:H and a-Si:F:H have sufficiently low defect concentrations that doping would be straightforward if P_4 and B_4 centers could form. If the Si–P bond is sufficiently strong compared to the Si–Si bond, the creation energy of a $P_4^+-T_3^-$ pair (relative to the ground-state pair, $P_3^0-T_4^0$) might not be very large (since the total number of bonds is preserved). In that case, doping can be understood on thermodynamic grounds, a given concentration of $P_4^+-T_3^-$ pairs being introduced during growth, with ε_f increasing to the midpoint of the P_4^+ and T_3^- levels. For moderately heaving doping, the presence of the band tail or of twofold-coordinated silicon may be essential. Boron doping can be explained analogously by the creation of $T_3^+-B_4^-$ pairs; however, the unusual chemistry of boron can complicate the situation considerably.

VI. Chemical Modification

When low-energy VAPs exist, the Fermi energy is effectively pinned and ordinary doping is impossible. Even if an impurity atom could be positioned in a nonoptimal chemical environment, the resulting release of excess electrons, for example, would convert only some of the C_3^+ centers to C_1^- without moving ε_f. To overcome this problem, Ovshinsky introduced an ingenious technique for decoupling ε_f, even in the presence of large VAP densities, thereby making possible the use of chalcogenides in conventional semiconductor devices.

The technique of chemical modification requires the use of large concentrations of one or more modifying elements to overcome the large VAP density and the introduction of the modifier in a nonequilibrium manner to preclude its incorporation in the network. A simple example is the postdiffusion of indium into a chalcogenide glass. Because indium has an odd valence, it cannot enter the formed matrix in a fully bonded position. A low-energy state for indium is one in which it gives up an electron and forms two ordinary and two dative bonds. For moderate indium concentrations, the excess electrons in pairs convert C_3^+ to C_1^- centers, and the conductivity is unaffected; however, eventually, the C_3^+ centers are completely depleted and ε_f begins to increase, thus reducing the electrical activation energy. Typical results are shown in Fig. 11.

Alkali atoms and transition metals can also be used as chemical modifiers, and co-sputtering and ion-implantation techniques can be effective under certain conditions. Pure materials and alloys from Groups III to VI are amenable to chemical modification. Because of the nonequilibrium preparation methods, the chemistry of the phenomena can be extremely complicated.

VII. Transient Effects

Virtually all of the present electronic uses of amorphous semiconductors involve nonequilibrium processes introduced by the application of light or applied electric fields. Under these conditions, the nature and concentration of the

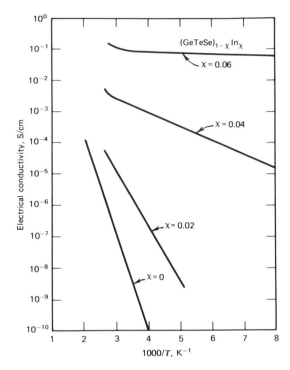

FIG. 11. Electrical conductivity as a function of temperature for pure amorphous GeTeSe and films chemically modified with indium (S/cm = Ω^{-1} cm^{-1}).

defects are of the utmost importance, because they ordinarily completely control the kinetics of the return to equilibrium.

When VAPs are present, the positively and negatively charged centers represent traps with large cross sections for electrons and holes, respectively. However, because of the interconvertibility of the charge centers, equilibration requires atomic relaxations (e.g., bond breaking). The complete return to equilibrium requires at least three steps. For example, in a chalcogen, these are trapping of an electron by the positively charged center,

$$C_3^+ + e^- \rightarrow C_3^0 \qquad (11)$$

a relaxation involving, in this case, bond breaking,

$$C_3^0 \rightarrow C_1^0 \qquad (12)$$

and trapping of a second electron by the neutral center,

$$C_1^0 + e^- \rightarrow C_1^- \qquad (13)$$

In field-effect observations, for example, excess positive or negative charge is induced in the semiconductor by applying a voltage to a gate electrode. If a negative charge is induced, processes (11)–(13) ensure that ε_f is pinned and the field effect is unobservable. However, this pinning takes place only if the time scale of the experiment is long compared with the longest of the three processes. This can be the resolution of the question of why field effects are routinely observed in arsenic–tellurium glasses. If any one of the rates controlling processes (11)–(13) is high, a bottleneck can retard equilibrium. The most likely origin of such a bottleneck is process (12), since probably both C_3^0 and C_1^0 are local minima in the potential energy of the network; thus, a barrier should exist between the two defect centers, and an appropriate Boltzmann factor is included in the equilibration kinetics. For very short times, C_3^+ centers trap electrons and C_1^- centers trap holes, but the two neutral centers do not interconvert. Thus, the apparent density of states is that of a doped, compensated semiconductor. The Fermi energy is not pinned in this time scale, and a transient field effect should be observable; however, this field effect should decay with time, with the time constant yielding the height of the potential barrier. Such transient effects have been observed in tellurium–arsenic glasses and have been quantitatively analyzed. The relaxation time for process (12) is given by

$$\tau \simeq \tau_{ph} \exp(\Delta E/kT) \qquad (14)$$

where ΔE is the height of the barrier and τ_{ph} a phonon time of $\sim 10^{-12}$ sec. Thus, at room temperature, with a ΔE of 0.5 eV, τ is ~ 100 μsec; however, for a ΔE of 1.0 eV, τ is several hours. These transient effects can persist for a very long time, and this is the case for glasses in the tellurium–arsenic system.

Despite the sharp contrasts in steady-state behavior between amorphous silicon alloys and chalcogenides, many of the unusual transient phenomena observed in chalcogenide glasses have also been observed in a-Si:H. These include dispersive transport, photoluminescence fatigue, light-induced creation of unpaired spins, and apparent photostructural changes. These and other observations provide evidence that charged centers also exist in a-Si:H; however, since unpaired spins always occur in the dark, in a-Si:H, T_3^0 states almost certainly exist. Nevertheless, there is both theoretical and experimental evidence that T_3^+–T_3^- pairs have lower energy than two $2T_3^0$ centers, *provided* that complete local relaxation exists. This arises because threefold coordination is *optimal* for both positively and negatively charged tathogens, but not for the neutral atoms. But a T_3^+ center optimally

bonds sp^2 with a 120° bond angle, while a T_3^- center primarily bonds p^3 with bond angles in the 95° to 105° range. Thus, it is possible that T_3^0 centers are stable in locally strained regions because the requisite atomic relaxations necessary to achieve low-energy T_3^+–T_3^- pairs cannot easily take place.

Such a model explains many of the unusual properties of a-Si:H, including the existence of long-time transient phenomena. Because of the different bond angles surrounding T_3^+ and T_3^- centers, the two neutral centers that result from electron and hole capture, respectively, are interconvertible but not identical. Therefore, a potential barrier can retard the kinetics of equilibration just as in chalcogenides, and this can account for the many similarities in transient phenomena between these two classes of materials, which otherwise exhibit very different behavior.

A metastable state can be induced in a-Si:H after exposure to intense light for a long time. Whereas deposited films exhibit room-temperature conductivity of $\sim 10^{-6}$ S/cm (= Ω^{-1} cm^{-1}) and activation energy of 0.57 eV, after the film is exposed to light the conductivity decreases to $\sim 10^{-10}$ S/cm and the activation energy increases to 0.87 eV. The original state can be restored by annealing the film for 2 hr at 150°C. The relaxation time for recovering the deposited state has an activation energy of ≈ 1 eV, so that the room-temperature relaxation time is of the order of days.

The existence of charged defect centers provides a natural explanation of such effects, since electrons can be trapped at the positive centers to form a metastable state. If the potential barrier retarding equilibration is about 1 eV, the very long time relaxations can be understood.

VIII. Applications

The initial commercial excitement that followed the demonstration of the existence of amorphous semiconductors arose primarily from their low preparation costs and their resistance to deterioration under severe conditions, rather than from their unique physical properties. Chalcogenide glasses, with their pinned Fermi energy and efficient electron and hole traps, initially found applications arising from their high ratios of photoconductivity to dark conductivity in electrophotographic copying machines and television pickup tubes. When a sufficient electric field is applied to neutralize the charged traps, the resulting high conducting state is exploited in computer logic applications. The ease of crystallization of some of these glasses forms the basis of many memory systems, including high-bit-density optical memories.

In contrast, amorphous silicon-based alloys behave as conventional semiconductors. Their first application was as photovoltaic devices, which have been uniformly deposited in a roll-to-roll process over wide areas. This has also spurred interest in the use of these materials as thin-film semiconductor devices such as thin-film transistors, phototransistors, or even as ordinary thin-film diodes.

Ultimately, however, it may be the possibility of molecular engineering, as exemplified by such techniques as chemical modification and layering on a 10-Å scale, that will have the greatest impact, since in these ways truly unique materials can be designed and prepared.

BIBLIOGRAPHY

Adler, D. (1978). *Phys. Rev. Lett.* **41,** 1755.
Adler, D. (1985). *In* "Physical Properties of Amorphous Materials" (D. Adler, B. B. Schwartz, and M. C. Steele, eds.), p. 5. Plenum, New York.
Adler, D., Shur, M. S., Silver, M., and Ovshinsky, S. R. (1980). *J. Appl. Phys.* **51,** 3289.
Cohen, M. H., Fritzsche, H., and Ovshinsky, S. R. (1969). *Phys. Rev. Lett.* **22,** 1065.
Kastner, M. A. (1985). *In* "Physical Properties of Amorphous Materials" (D. Adler, B. B. Schwartz, and M. C. Steele, eds.), p. 381. Plenum, New York.
Kastner, M., Adler, D., and Fritzsche, H. (1976). *Phys. Rev. Lett.* **37,** 1504.
Mott, N. F. (1980). *J. Phys. C* **31,** 5433.
Mott, N. F., and Davis, E. A. (1979). "Electronic Processes in Non-Crystalline Materials," 2nd ed. Oxford Univ. Press (Clarendon), London and New York.
Ovshinsky, S. R. (1968). *Phys. Rev. Lett.* **21,** 1450.
Ovshinsky, S. R. (1985). *In* "Physical Properties of Amorphous Materials" (D. Adler, B. B. Schwartz, and M. C. Steele, eds.), p. 105. Plenum, New York.
Ovskinsky, S. R., and Madan, A. (1978). *Nature (London)* **276,** 482.
Phillips, J. C. (1979). *J. Non-Cryst. Solids* **34,** 153.
Scher, H., and Montroll, E. W. (1975). *Phys. Rev. B: Solid State* [3] **12,** 2455.
Spear, W. E., and LeComber, P. G. (1976). *Philos. Mag.* [8] **33,** 935.

ATOMIC AND MOLECULAR COLLISIONS

Robert E. Johnson *University of Virginia*

I.   Introduction
II.   Impact Parameter Cross Sections
III. Elastic Scattering
IV.   Wave Mechanics of Scattering
V.   Interaction Potentials
VI.   Inelastic Collisions

GLOSSARY

Born approximation: First-order estimate of the collision cross sections.

Born–Oppenheimer approximation: Separation of the electron and nuclear motion. The latter is often treated classically.

Charge exchange: Process of transferring an electron from one of the colliding particles to the other; usually from a neutral to an ion.

Cross section: Probability of an interaction between two colliding particles expressed as an area.

Differential cross sections: Cross sections that are functions of one of the collision results (e.g., angle scattered; energy transfer). Summing over all possible results gives the cross section.

Elastic collision: Collision involving a deflection of the colliding particles but no change in their internal state.

Impact parameter: Defines the closeness of a collision as that component of the distance between the two colliding particles that is perpendicular to their relative velocity.

Inelastic collision: Collision resulting in a change in the internal states of the colliding particles.

Interaction potentials: Net change in potential energy of the two colliding particles. A potential exists for each set of initial states of the particles.

Mean free path: Average distance between encounters for a particle traversing a material.

Semiclassical method: Calculation of wave-mechanical effects using classical quantities.

Stopping cross section: Product of cross section and the energy loss in an inelastic process, summed over all such processes.

Stopping power: Stopping cross section times the material number density. Gives the energy loss per unit path length for a particle traversing a material.

The field of atomic and molecular collisions involves the study of the effects produced by the motion of atomic and molecular particles when they approach each other. The effects produced are generally described in terms of the amount of energy transferred between the colliding particles. This may be simply an elastic energy transfer corresponding to a deflection or an inelastic energy transfer producing changes in the internal states of the particles. The study of these changes is used to determine the details of the forces of interaction between atomic and molecular particles. In addition, knowledge of these effects is used to describe phenomena in which collisions play an important role, such as the behavior of gases and plasmas and the modification of solids by ion bombardment.

I. Introduction

A. OVERVIEW

The need to understand the behavior of colliding atoms and molecules is self-evident as we live in a world constructed from atomic building blocks. It is a dynamic construction of moving particles governed by a few fundamental forces. The interaction between moving atoms or molecules is thought of as a collision, and the effects produced by these collisions are a primary concern of this chapter. In addition, ever since Rutherford's discovery of the nucleus, collisions between atoms have provided a means of determining atomic structure and the forces of interaction. Therefore the field of atomic and molecular collisions has been sustained both by investigations into the nature of the interactions and by application of the results to help understand our atomic and molecular environment.

Collision events are correctly described via quantum mechanics (wave mechanics), but it has become customary when discussing such events to employ classical notions. In some cases this simplifies the understanding of the physics, in which case wave-mechanical effects, such as interference and diffraction, can then be incorporated as corrections. Such an approach is referred to as a semiclassical method, of which there are a variety. Basically they all have the same justification: that is, they are employed when the quantum-mechanical wavelength associated with the collision is small compared to the dimensions of the system. This is the same basis for using geometric optics to approximate the passage of light through a medium. When discussing collisions, the incident "radiation" is a beam of particles and the medium is the field of a "target" atom or molecule. The wavelength is then given by $\lambda = h/p$ where h is Planck's constant and p is the momentum of the particles. Comparing λ to an atomic radius (e.g., a_0, the Bohr radius), we establish a rough criteria for the usefulness of semiclassical methods: collision energies much greater than a Rydberg (27.2 eV) for incident electrons and much greater than hundredths of an electron-volt for incident ions, atoms, or molecules. For the heavy particles, therefore, this criterion is satisfied for most energies of interest. Care must be taken, however, in making such a statement. Diffraction regions (i.e., scattering at small angles) are *always* dominated by wave-mechanical effects, as are regions in which transitions take place. In many cases such regions determine the nature of the collision process. Therefore, the above, very useful criterion must be applied cautiously.

B. Cross Section Defined

The effect of atomic particles on each other is generally described via an interaction cross section. The conceptually simplest cross section, the total collision cross section, is obtained from an experiment like that in Fig. 1. A beam of

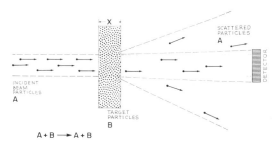

FIG. 1. Beam experiment to obtain the scattering cross section of A by particles B. [From Johnson, R. E. (1982). *Introduction to Atomic and Molecular Collisions*. Plenum Press, New York.]

particles is incident on a target containing atoms, and the change in intensity of the beam is monitored. If the target is "thin" so that an incident particle is only likely to make a single collision, then the change in intensity, ΔI, for a small change in thickness, Δx, is written

$$\Delta I = -\sigma n_B I \, \Delta x \qquad (1)$$

where n is the density of target atoms and I is the measured intensity (particles/cm^2/sec). In Eq. (1) the proportionality constant σ is the cross section, indicating, roughly, the range of the interaction between the colliding particles. Integrating Eq. (1), the intensity of unscattered particles versus thickness x can be found for thick samples:

$$I = I_0 e^{-n_B \sigma x} \qquad (2)$$

Differentiating I in Eq. (2), the quantity $dI/I_0 = (n_B \sigma)e^{-n_B \sigma x} \, dx$ is the Poisson probability of the first collision occurring between x and $x + dx$, and $(n_B \sigma)^{-1}$ is called the mean free path between collisions.

One of the first results found from such measurements is that the total cross section varies slowly with velocity as shown in Fig. 2. That is, atoms have diffuse boundaries and the effective range of the interaction region changes with velocity. This dependence is wave mechanical in nature, as the cross section is determined by the amount of scattering at small angles, which is the diffraction region mentioned above. Total cross sections that are calculated strictly classically from *realistic* potentials are always infinite. Hence, we see in this simplest of examples the need for cross-section approximations that incorporate wave-mechanical effects.

A second set of experiments for studying

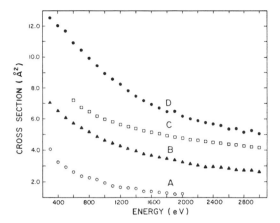

FIG. 2.   He + He. Detector apertures: A 0.57°, B 0.26°, C 0.11°, D 0.056°. [From W. J. Savola, Jr., F. J. Erikssen, and E. Pollack (1973). *Phys. Rev.* **A7**, 932.]

atomic interactions requires detection of those particles that are scattered rather than those not scattered. For a beam of atoms A incident on a target containing atoms B, the angular differential cross section

$$d\sigma = (d\sigma/d\Omega)_A \, d\Omega$$

relates the number of incident particles per unit time scattered into a region of solid angle $d\Omega$ to the incident flux. Similarly,

$$d\sigma = (d\sigma/d\Omega)_B \, d\Omega$$

relates the number of target particles per unit time ejected into a region of solid angle to the flux of incident particles. Of course, collecting all particles A scattered into all solid angles is equivalent to detecting all scattered particles. Hence, integrating $d\sigma$ over all angles yields the total scattering cross section σ described above. In the experiment described, one might also discriminate between any internal changes that have occurred (e.g., changes in mass, charge, energy, etc.) The cross sections determined from such an experiment, therefore, would indicate both the nature of the interaction between A and B and the likelihood of an internal change (transition) in either A or B. These two aspects of the cross section will be brought out more clearly in the subsequent discussion. When no internal changes occur the collisions are called elastic; when such changes occur we refer to them as inelastic.

In this chapter we first elaborate on the nature of the cross section using a semiclassical description referred to as the impact parameter method. We then discuss the forces between atomic particles and subsequently use those forces to calculate cross sections and transition probabilities. Such calculations are divided into those methods useful for transitions in fast collisions and those useful in slow collisions. For incident ions and atoms, collisions are fast or slow depending on whether the ratio τ_c/τ_0 is less than or greater than 1. Here τ_c is the collision time ($\sim d/v$, where d is a characteristic dimension associated with A and B, e.g., $d \approx a_0$ for outer shell electrons, and v is the relative speed) and τ_0 is the characteristic period of the system of particles. For ionization of outer-shell electrons, the characteristic period is $\tau_0 \approx 10^{-16}$–10^{-17} sec, whereas for molecular processes, $\tau_0 \approx 10^{-14}$ sec for vibrational motion and $\tau_0 \approx 10^{-13}$ for rotational motion. Setting $\tau_c \approx \tau_0$, such characteristic times translate into incident-particle energies of the order 1–100 keV/amu, 0.1 eV/amu, and 0.001 eV/amu, where amu is the atomic mass unit. When $\tau_c/\tau_0 \approx 1$, the collision time is about the same size as the characteristic period and the transition probabilities and cross sections are large. At much larger or much smaller times, the cross sections are found to decrease.

As the interaction between even the simplest atoms can be quite complex, many of the details of the following discussion are treated qualitatively. The purpose of the presentation following is to let the reader have a "feel" for the complexities and yet acquire the ability to understand the nature of the approximate expressions and formulas used by many workers in the field of atomic and molecular collisions. The discussion starts using the classical impact parameter concept to formulate cross sections of various types so the reader has a clear idea of the definitions of the quantities calculated and used later.

II. Impact Parameter Cross Sections

A. FORMULATION

The trajectory of an incident particle A interacting repulsively with an initially *stationary* target particle B is shown in Fig. 3. The quantity b in that figure indicates the closeness of approach and is referred to as the impact parameter. It is the perpendicular distance between the incident velocity vector **v** and the position **R** of particle A measured from B. The impact parameter also indicates the angular momemtum of the colliding particles (i.e., $|L| = |M_A \mathbf{R} \times \mathbf{v}| = M_A v b$). The

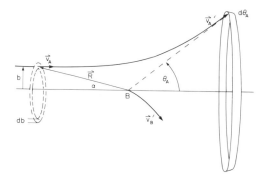

FIG. 3. Scattering of A by B: b is impact parameter, **R** shows initial position of A with respect to B, $\mathbf{v}_A$ is the initial velocity, $\mathbf{v}'_A$ and $\mathbf{v}'_B$ are the final velocities, and θ_A is the scattering angle for A: $\mathbf{v}_A = \mathbf{v}$ here.

impact parameter (or angular momentum) and the relative velocity v (or energy) are sufficient to characterize collisions between spherically symmetric particles. As the impact parameter between two colliding particles determines the likelihood of a deflection, we assign a collision probability $P_c(b, v)$ for each impact parameter. If the incident and/or target particle has a spin or is a molecule, then initial orientations with respect to the collision axis have to be specified, for example, $P_c(b, \phi, \omega, v)$ where ϕ is the azimuthal angle of approach and ω is an orientation angle. However, if there are no preferential aligning fields (e.g., outside fields or target B in a crystalline lattice), then the incident and target particles are presumed to be randomly oriented and $P_c(b, v)$ is an average over collisions involving all possible orientations.

Based on the experimental definition in Eq. (1), the interaction cross section between A and B can now be written

$$\sigma(v) = 2\pi \int_0^\infty P_c(b, v) b \, db \qquad (3)$$

That is, the particles passing through the ring of area $2\pi b \, db$ about a single target atom, (e.g., Fig. 3) will be scattered from the beam with a probability P_c and, hence, contribute to the observed change in intensity of the beam, ΔI in Eq. (1). In classical mechanics, processes are deterministic and hence, P_c is either 0 or 1. For a finite-range interaction (e.g., collision between two spheres of radius r_A and r_B), $P_c = 0$ for $b > r_A + r_B$ and the cross section is finite, $\sigma = \pi(r_A + r_B)^2$. For interactions between atomic or molecular particles the forces are infinite in range (i.e., $P_c = 1$ for all b) yielding a classical cross section which is infinite. Quantum me-

chanics, on the other hand, gives finite cross sections for most realistic interactions, like the measured values in Fig. 2. That is, $P_c(b, v) \xrightarrow{b \to \infty} 0$ in a quantum-mechanical calculation for interactions that decay rapidly at large separation. This deficiency in the classical estimate of the cross section may not be of practical importance in many applications. That is, one generally wants to know whether a particle experiences a deflection or loses an amount of energy larger than some prescribed minimum size and *not* the likelihood of being deflected even at infinitesimally small angles (i.e., extremely large b).

From Fig. 3 it is seen that if the forces between the particles are independent of their orientations (i.e., no azimuthal dependence), then those particles passing through the ring of area $2\pi b \, db$ will be scattered into an angular region $2\pi \sin \theta_A \, d\theta_A$. Therefore

$$d\sigma = (d\sigma/d\Omega)_A 2\pi \sin \theta_A \, d\theta_A = 2\pi b \, db$$

so that the differential cross section discussed earlier is

$$\left(\frac{d\sigma}{d\Omega}\right)_A \equiv \sigma(\theta_A) = \left| \frac{b \, db}{\sin \theta_A \, d\theta_A} \right| \qquad (4a)$$

where the simplified notation for azimuthally symmetric cross section, $\sigma(\theta_A)$, is often used instead of $(d\sigma/d\Omega)_A$. Similarly, the cross section for scattering of target particles is

$$\left(\frac{d\sigma}{d\Omega}\right)_B \equiv \sigma(\theta_B) = \left| \frac{b \, db}{\sin \theta_B \, d\theta_B} \right| \qquad (4b)$$

For an elastic binary collision of the type we have been discussing, the energy and momentum of each particle after the collision can be related to the incident energy and the scattering angle using the conservation of energy and momentum. The scattering angle can be expressed either in the laboratory or in the center of mass (CM) system. In a practical problem the former is more useful, but in describing the relationship between the atomic interactions and the resulting deflections the latter is much more convenient. In Table I the relationships between laboratory and CM quantities are summarized for the case we have been discussing, a moving particle A incident on a stationary particle B. We write the CM deflection angle versus b as $\chi(b)$ (the deflection function) and, therefore, by analogy with the above discussion, the classical differential cross section in the CM system is

$$\sigma(\chi) = \left| \frac{b \, db}{\sin \chi \, d\chi} \right| \qquad (5)$$

TABLE I.   Relationship between Laboratory and CM Variables

Velocity of CM	$\mathbf{V_c} = (M_A\mathbf{v}_A + M_B\mathbf{v}_B)/(M_A + M_B) \equiv \dot{\mathbf{R}}_c$
Relative velocity in CM	$\mathbf{v} = (\mathbf{v}_A - \mathbf{v}_B) \equiv \dot{\mathbf{R}}$

Total laboratory quantities	CM quantities
$M = M_A + M_B$	$m = M_A M_B/(M_A + M_B)$
$E_A + E_B = \frac{1}{2}MV_c^2 + E$	$E = \frac{1}{2}mv^2$
$\mathscr{P} = M\mathbf{V}_c$	$\mathbf{P} = 0$
$\mathscr{L} = M\mathbf{R}_c \times \mathbf{V}_c + \mathbf{L}$	$\mathbf{L} = m\mathbf{R} \times \mathbf{v}$

Transformations (for v_B initially zero and elastic collisions):

$$\theta_B = \tfrac{1}{2}(\pi - \chi)$$
$$\tan \theta_A = \mu \sin \chi/(1 + \mu \cos \chi); \qquad \mu = M_B/M_A$$
$$T = \gamma E_A \sin^2(\chi/2); \qquad \gamma = 4M_B M_A/(M_A + M_B)^2$$
$$(d\sigma/d\Omega)_A = \sigma(\chi) \left| \frac{d\cos \chi}{d\cos \theta_A} \right| = \sigma(\chi) \frac{(\mu^2 + 2\mu \cos \chi + 1)^{3/2}}{\mu^2|\mu + \cos \chi|}$$
$$(d\sigma/d\Omega)_B = \sigma(\chi) \left| \frac{d\cos \chi}{d\cos \theta_B} \right| = \sigma(\chi)|4 \sin(\chi/2)|$$
$$\frac{d\sigma}{dT} = \frac{4\pi}{\gamma E_A} \sigma(\chi)$$

This can be transformed (see Table I) to give the laboratory scattering cross sections in Eqs. (4) for the incident or target particle.

B. Energy Loss Cross Sections

For elastic collisions the energy transfer to B is simply related to χ when B is initially stopped (see Table I) and, therefore, it is often useful to consider the elastic energy transfer cross section instead of $\sigma(\chi)$. Calling T the energy transfer to particle B, one writes

$$\frac{d\sigma}{dT} = \frac{4\pi}{\gamma E_A} \sigma(\chi) \qquad (6)$$

where γ is given in Table I. The reader will find both $\sigma(\chi)$ and $d\sigma/dT$ used in the literature and, therefore, should keep in mind that they are related. In addition to transferring kinetic energy T to particle B, there is also a probability that one or both particles will experience a change in internal energy (a transition), which we label $P_{0\to f}$ (b, v). The subscripts indicate the initial (0) and final (f) states of the colliding particles, and those collisions for which $P_{0\to f} \neq 0$ are inelastic. As in Eq. (3), an *inelastic* cross section can be written

$$\sigma_{0\to f}(v) = 2\pi \int_0^\infty P_{0\to f}(b, v)b \, db \qquad (7)$$

As each inelastic process is associated with an internal energy change $Q_{0\to f}$, the average inelastic energy lost to such processes is

$$\bar{Q}(b) = \sum_f Q_{0\to f} P_{0\to f}(b, v) \qquad (8)$$

In writing Eq. (8) we assumed the sum of probabilities over all processes, including no internal energy change, is unity [$\sum_f P_{0\to f} = P_c(b, v) = 1$].

We can use the above to define the average energy loss per unit path length for particle A traversing a material consisting of particles B. In passing through a target, A loses energy both to electronic excitations and to kinetic energy transfer to target nuclei B. Noting that $(n_B\sigma_{0\to f})^{-1}$ is the mean free path for the occurrence of an energy loss $Q_{0\to f}$, the averaged electronic energy loss per unit path length in the target is

$$\left(\frac{dE}{dx}\right)_e = \sum_f Q_{0\to f} (n_B\sigma_{0\to f}) \qquad (9)$$

Similarly, if $[n_B(d\sigma/dT) \, dT]^{-1}$ is the mean free path for losing an amount of kinetic energy between T and $T + dT$ to a nucleus B, then the averaged elastic loss of A *per unit path length* in

the target is

$$\left(\frac{dE}{dx}\right)_n = \int_0^\infty T[n_B(d\sigma/dT)dT] \qquad (10)$$

To the extent that the excitations and collisional energy transfers are independent, the average net energy loss is the sum of these. Because they both depend on the material number density, this is generally written

$$\left(\frac{dE}{dx}\right) = n_B(S_e + S_n) \qquad (11)$$

with

$$S_e = \sum Q_{0 \to f} \, \sigma_{0 \to f} = 2\pi \int_0^\infty Q(b)b \, db \qquad (12a)$$

$$S_n = \int_0^\infty T\left(\frac{d\sigma}{dT}\right) dT = 2\pi \int_0^\infty T(b)b \, db \qquad (12b)$$

The quantity (dE/dx), which has been measured extensively, is referred to as the stopping power of the material, and the quantities S_e and S_n are the electronic and nuclear elastic stopping cross sections. These are quantities determined only by the single collision interaction of A with B and not properties of the target as a whole. Results for S_e are tabulated, and simple, generally useful expressions for S_n are available, for example, see Table II.

The nuclear stopping cross section S_n is related to another useful quantity, the diffusion cross section σ_d. The change in the parallel com-

---

TABLE II.   Elastic Collision Expressions

$$\chi(b) = \pi - 2b \int_{R_0}^\infty \frac{dR}{R^2}\left(1 - \frac{b^2}{R^2} - \frac{V}{E}\right)^{-1/2}$$

$$\approx -\left(\frac{1}{2E}\right)\frac{d}{db}\int_{-\infty}^\infty V(R)\, dZ$$

$$\eta^{sc}(b) = \frac{p_0}{\hbar}\left[\int_{R_0}^\infty \left(1 - \frac{b^2}{R^2} - \frac{V}{E}\right)^{1/2} dR\right.$$

$$\left. - \int_b^\infty \left(1 - \frac{b^2}{R^2}\right)^{1/2} dR\right] \approx \frac{-1}{2\hbar v}\int_{-\infty}^\infty V(R)\, dZ$$

$$L\chi(b) = (2\hbar b)\frac{\partial \eta^{sc}(b)}{\partial b}, \qquad L = p_0 b = m v b$$

$$\sigma(\chi) = \frac{\gamma E_A}{4\pi}\frac{d\sigma}{dT} \quad \text{(B initially stopped)}, \qquad S_n = \frac{\gamma E_A}{2}\sigma_d$$

$$\rho = \chi \sin \chi \sigma(\chi), \qquad \iota = \chi E, \quad E - \frac{M_B}{M_A + M_B}E_A$$

Quadratures (m = number of integration points)

$$g(x) \equiv \frac{b}{R_0}\left[\frac{1 - x^2}{1 - (b/R_0)^2 x^2 - \dfrac{V(R_0/x)}{E}}\right]^{1/2}$$

$$\chi(b) \approx \pi\left[1 - \frac{1}{m}\sum_{i=1}^m g(x_i)\right],$$

$$x_i = \cos[(2i - 1)\pi/4m]$$

$$\eta^{sc}(b) \approx \frac{2L}{\hbar}\left\{\frac{\pi}{(2m + 1)}\sum_{j=1}^m \sin^2\left(\frac{j\pi}{2m + 1}\right)[g(x_j)^{-1} - 1]\right\},$$

$$x_j = \cos\left[\frac{j\pi}{2m + 1}\right]$$

(continues)

Impulse estimates ($V/E \ll 1$)

for $V = C_n/R^n$, $\chi(b) \approx a_n V(b)/E$,

$$a_n = \pi^{1/2}\Gamma\left(\frac{n+1}{2}\right) / \Gamma\left(\frac{n}{2}\right) \xrightarrow{n \to \infty} (\pi n/2)^{1/2}$$

$$\eta^{sc}(b) \approx \frac{-L\chi(b)}{2\hbar(n-1)}$$

$[\Gamma(x + 1) = x\Gamma(x), \Gamma(1) = 1, \Gamma(\tfrac{1}{2}) = \pi^{1/2}$;

gamma function tabulated]

$\rho \approx n^{-1}(a_n C_n/\tau)^{2/n}$

$$\frac{d\sigma}{dT} \approx \frac{\pi}{n} \mathscr{A}_n^2/(\gamma E_A)^{1/n} T^{1+1/n}; \quad \mathscr{A}_n = \left(\frac{2M_A}{M_A + M_B} a_n C_n\right)^{1/n}$$

$$S_n \approx \frac{\pi}{n-1} \mathscr{A}_n^2 (\gamma E_A)^{1-2/n}, \, n > 2$$

for $V = (Z_A Z_B e^2) \exp(-\beta R)/R$

$$\chi(b) \approx \frac{(Z_A Z_B e^2)\beta}{E} K_1(\beta b) \xrightarrow{b \to \infty} \left(\frac{\pi \beta b}{2}\right)^{1/2} \frac{V(b)}{E}$$

$$\eta^{sc}(b) \approx -\frac{(Z_A Z_B e^2)}{\hbar v} K_0(\beta b) \xrightarrow{b \to \infty} \frac{-L\chi(b)}{2\hbar(\beta b)}$$

(K_n are the modified Bessel functions; tabulated)

Massey–Mohr [Eq. (31)]

$$\sigma \approx 2\pi(\bar{b})^2 \left(1 + \frac{1}{2n-4}\right) \qquad (n > 2)$$

$$\bar{b} \approx \left[\frac{2a_n C_n}{(n-1)\hbar v}\right]^{1/(n-1)}$$

Born approximation

for $V = (Z_A Z_B e^2) \exp(-\beta R)/R$

$$\frac{d\sigma}{dT} \approx \pi\mathscr{A}^2/(\gamma E_A)(T + T_0)^2$$

$$\mathscr{A} = \left(\frac{2M_A}{M_A + M_B} Z_A Z_B e^2\right)$$

$$T_0 = \frac{(\hbar\beta)^2}{2M_B}$$

$$\sigma \approx \pi\mathscr{A}^2/T_0(\gamma E_A + T_0)$$

$$S_n \approx \frac{\pi\mathscr{A}^2}{(\gamma E_A)} \left[\ln\left(\frac{\gamma E_A}{T_0} + 1\right) - \frac{\gamma E_A}{(\gamma E_A + T_0)}\right]$$

Thomas–Fermi (Lindhard)

Scaling quantities

$a_{TF} = 0.8853 a_0/(Z_A^{2/3} + Z_B^{2/3})^{1/2}$

$\varepsilon = (\gamma E_A)a_{TF}/2\mathscr{A}, \qquad \mathscr{A}$ as above

$t = \varepsilon^2 T/(\gamma E_A)$

$$\frac{d\sigma}{dt} \approx \pi a_{TF}^2 \xi/4t^{4/3}[1 + \xi^{2/3}t^{4/9}]^{3/2}, \qquad \xi \approx 2.62$$

$$S_n \approx \left(\frac{9\pi}{2}\right) \frac{\mathscr{A}^2}{(\gamma E_A)} \{\ln[\mathscr{B} + (1 + \mathscr{B}^2)^{1/2}] - \mathscr{B}/(1 + \mathscr{B}^2)^{1/2}\}$$

$$\mathscr{B} = \xi^{1/3}\varepsilon^{4/9}$$

(continues)

TABLE II (*Continued*)

Lenz–Jensen potential

$$V = \frac{Z_A Z_B e^2}{R} \, \Phi(R/a_{TF})$$

$$\Phi = (1 + b_1 y + b_2 y^2 + b_3 y^3 + b_4 y^4) e^{-y}$$

$$y = (9.67 R/a_{TF})^{1/2}, \qquad b_1 = 1, \qquad b_2 = 0.3344$$

$$b_3 = 0.0485, \qquad b_4 = 0.002647$$

ponent of momentum of a very slow particle A due to a collision with B involving no electronic excitations is written $(\Delta \mathbf{p}_a)_{//} = p_0 (1 - \cos \chi)$. Therefore, the drag force on a particle A traversing a target containing particles B is the collision rate times this momentum loss,

$$(\Delta \mathbf{F}_A)_{//} = \int (\Delta \mathbf{p}_a)_{//} \, v n_B \, (2\pi b \, db)$$

This can be written as $(\Delta \mathbf{F}_A)_{//} = n_B p_0 v \sigma_d$, where

$$\sigma_d = 2\pi \int_0^\infty [1 - \cos \chi(b)] b \, db$$

$$= 2\pi \int_{-1}^1 (1 - \cos \chi) \sigma(\chi) \, d(\cos \chi) \quad (13)$$

Using the expression for T given in Table I we see that S_n and σ_d are related:

$$S_n = \frac{\gamma E_A}{2} \, \sigma_d \tag{14}$$

That is, the drag force on a particle is related to the stopping power, i.e., $(\Delta \mathbf{F}_A)_{//} = (1 + \mu) (dE/dx)_n$, where $\mu = M_B/M_A$. Because the factor $(1 - \cos \chi)$ goes to zero as χ goes to zero, both S_n and σ_d, unlike σ, can be calculated classically, as we will see later.

We have defined a number of cross sections in the above discussion. These are all described without references to the forces between the particles. These forces produce both the deflections and the transitions (inelastic processes) discussed above. However, it has become customary to separate the discussion of these two effects. We follow this tradition here and consider first the calculation of deflections for two spherically symmetric colliding particles.

III. Elastic Scattering

A. Classical Deflection Function

The determination of $\chi(b)$, the CM deflection function, begins with angular momentum conservation. In the CM system the two colliding particles follow trajectories that are equivalent and the collision can be described as the scattering of a particle of reduced mass m by a stationary center of force. Using the coordinates given in Fig. 3. we write the angular momentum as

$$L = mvb = mR^2 \dot{\alpha} \tag{15}$$

Rearranging Eq. (15) and integrating over time, we obtain

$$\chi(b) = \pi - \int_{-\infty}^\infty \dot{\alpha} \, dt = \pi - vb \int_{-\infty}^\infty \frac{dt}{R^2}$$

Finally, assuming an interaction potential $V(R)$ that depends only on the separation R, energy conservation is expressed as

$$\frac{1}{2} m\dot{R}^2 + \frac{L^2}{2mR^2} + V(R) = E$$

Employing Eq. (15), this expression can be written as a radial velocity,

$$\dot{R} = \pm v \left[1 - \frac{b^2}{R^2} - \frac{V}{E} \right]^{1/2} \tag{16}$$

At the start of the collision the radial velocity is $(-v)$, that is, the atoms approach each other at a speed v. As R decreases, $\dot{R}$ approaches zero. R then begins to increase, as the particles recede, until it becomes $(+v)$ at large separations. Using Eq. (16) the deflection function above can be written

$$\chi(b) = \pi - 2b \int_{R_0}^\infty \frac{dR}{R^2} \left[1 - \frac{b^2}{R^2} - \frac{V}{E} \right]^{-1/2} \tag{17}$$

where R_0 is the distance of closest approach at which $\dot{R} = 0$. This expression can be integrated analytically for a few potentials of the form $V(R) = C_n/R^n$ but otherwise is treated by a simple numerical procedure given in Table II.

B. Impulse Approximation

For fast incident ions [ratio V/E small for all R in Eq. (17)] the deflections are small and it is useful to replace the above expressions for $\chi(b)$

by an impulse approximation. That is,

$$\chi(b) = \frac{(\Delta p)_\perp}{p_0} \approx \frac{\int_{-\infty}^{\infty} F_\perp \, dt}{p_0}$$

$$= -\frac{d}{db}\left[\frac{1}{2E}\int_{-\infty}^{\infty} V \, dZ\right] \qquad (18)$$

In this equation we assumed a straight line trajectory, i.e., $R^2 \approx b^2 + Z^2$, with $Z = vt$. For the general class of power-law potentials ($V = C_n/R^n$), $\chi(b)$ calculated from Eq. (18) is given in Table II. It is seen, that the deflection is determined primarily by the nature of the potential *near the distance of closest approach* of the colliding particles ($R_0 \approx b$). Therefore, it is *not* necessary to know the potential accurately at all R to obtain a reasonable estimate of the deflection function for collisions between ions and atoms as long as $V/E \ll 1$. Since V is of the order of electron-volts, this criterion is satisfied for a large number of problems involving incident ions, atoms, or molecules.

From Eq. (18) the quantity $\tau = \chi E$ (which in the laboratory frame is $\tau \approx \theta_A E_A$ for small angles) is seen to depend only on the impact parameter. Also, the modified cross section $\rho = \chi (\sin \chi) \sigma (\chi)$ [in the lab frame, $\theta_A \sin \theta_A \sigma (\theta_A)$] obtained using Eq. (18) in Eq. (5), is *independent of the energy E* for small angles (Table II). Based on these results, experimental measure-

ments of ρ versus τ can be directly converted into power-law potentials that are applicable over limited ranges of R. It is seen in Fig. 4 that for colliding atoms, $n \to 1$ as $R_0 \to 0$ (i.e., τ gets large). That is, the interaction is the repulsive interaction between the nuclei. As R_0 increases (τ decreases), n increases, reflecting the electron screening of the repulsive interaction. Because of the simplicity of the power potential results, they are used frequently to calculate effects of energetic charged particles incident on solids for which knowledge of the potential is only needed over a limited range of interaction radii.

Before discussing the determination of interaction potentials for specific collision pairs, we examine the nature of differential cross section for two forms of the potential, a purely repulsive and a long-range attractive plus short-range repulsive potential. The latter potential is characteristic of the interaction of an ion with an atom or molecule in its ground state. In Fig. 5 are shown the deflection functions and ρ versus τ plots. Based on Table II it is clear that the deflection function should approximately follow the form of the potential. Therefore, for the repulsive potential shown, χ is always positive, attaining a maximum value of π for head-on collisions [$b = 0$, Eq. (17)]. For the other potential, χ is negative at large b, eventually becomes positive, and again reaches π at $b = 0$. However, χ goes through a minimum, χ_r, at the impact parameter labeled b_r. At this minimum, $d\chi/db = 0$, and, therefore, the classical calculation of cross section in Eq. (5) becomes *infinite*. This large

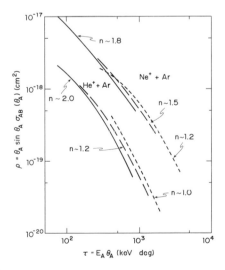

FIG. 4. Values of ρ versus τ experiment for three values of E_A: ——— 25 keV; ——— 50 keV; ——— 100 keV. Approximate power laws indicated by n. [Data from E. N. Fuls, P. R. Jones, F. P. Ziemba, and E. Everhart (1957). *Phys. Rev.* **107**, 104. Figure from R. E. Johnson (1982). "Introduction to Atomic and Molecular Collisions." Plenum Press, New York.]

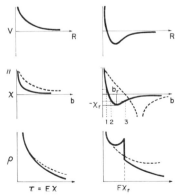

FIG. 5. Values of χ versus b and ρ versus τ for a repulsive and an attractive potential. Rainbow angle and impact parameter are χ_r and b_r. Three impact parameter's contribution for $|\chi| < |\chi_r|$ are labeled. Solid line higher energy, dashed line low energy. [From R. E. Johnson (1982). "Introduction to Atomic and Molecular Collisions," p. 53. Plenum Press, New York.]

enhancement in the scattering probability is similar to the effect that produces rainbows in the scattering of light from water droplets; hence, χ_r is called the rainbow angle. When $\chi_r < \pi$, then for angles greater than χ_r, only one impact parameter contributes to the cross section. However, for angles less than χ_r, three impact parameters contribute that have the same value of $\cos \chi$.

Using a Lennard–Jones form for the long-range attractive plus short-range repulsive potential, $V = C_{2n}/R^{2n} - C_n/R^n$, and for the power-law results in Table II the rainbow angle (for $V \ll E$) is

$$\chi_r \approx \frac{a_n^2}{a_{2n}} \left(\frac{V_{min}}{E} \right) \qquad (19)$$

That is, the rainbow angle is determined by the depth of the potential minimum V_{min} (which is $V_{min} = -C_n^2/4C_{2n}$ for the Lennard–Jones potential), and the shape of the potential via a_n and a_{2n}. Because the ratio a_n^2/a_{2n} changes slowly with n, measuring the rainbow angle can directly give an estimate of potential well depth.

As the collision energy decreases, χ_r can become much larger than π, implying that the two particles may orbit each other before separating. In fact, for very small velocities there is a particular impact parameter for which the particles can be trapped in orbit (i.e., $\ddot{R} = 0$ at $\dot{R} = 0$). Using only the attractive part of potential, along with Eq. (16) and the power-low results in Table II, the orbiting impact parameter is

$$b_0 \approx \left[\left(\frac{n}{2} \right) \frac{C_n}{E} \right]^{1/n} \bigg/ \left(\frac{n-2}{n} \right)^{(n-2)/2n} \qquad (20)$$

Therefore, for $b \leq b_0$, the interaction times can be very long and the particles approach each other closely, whereas for $b > b_0$ the particles simply scatter. This fact has been used extensively in estimating ion–molecule interaction cross sections [see Section VI,H].

Near the rainbow angle (or when orbiting occurs), a number of impact parameters contribute to the scattered flux at a given observation angle; hence, interference phenomena occur and classical cross-section estimates are not valid. However, if the angular resolution is not large the simple addition of the contribution to the scattered flux from each impact parameter gives an adequate representation of the cross section. This was used to give ρ versus τ in Fig. (5). That is,

$$\sigma(\chi) \approx \sum_i \left| \frac{b \, db}{\sin \chi \, d\chi} \right|_{b=b_i} \qquad (21)$$

where the b_i are all those impact parameters giving the same value of $\cos \chi$ [i.e., $\chi \to \pm(\chi + 2\pi m)$, m an integer]. As the interference phenomenon can give important additional information about the *details* of the interaction potentials, wave-mechanical calculations are useful. In addition, the effects at very small angles and at low energies can only be described via wave mechanics. In the following, therefore, we briefly review that wave-mechanical description of the elastic scattering cross section that parallels the above discussion. We thereby obtain results for diffraction (e.g., total cross sections σ) and interference (e.g., rainbow) phenomena. The results have parallels in light scattering.

IV. Wave Mechanics of Scattering

A. THE SCATTERING AMPLITUDE

In wave mechanics the scattering of the particle is replaced by scattering of a wave, but here again we can describe this scattering in the CM system, as the transformation between the laboratory and CM quantities is the same as in Table I. The beam of particles incident on a target is replaced by a plane wave, $\exp(i\mathbf{K} \cdot \mathbf{R} - i\omega t)$, where $\hbar\mathbf{K}$ replaces the momentum and $\hbar\omega = \hbar^2 K^2/2m$ is the energy per particle in the beam. K^{-1} is often written as $\lambdabar$ where $\lambdabar = \lambda/2\pi$, with λ the wavelength. Describing the beam of particles as a plane wave implies that the particles are nonlocalized, that is, it is equally probable to find a particle at any point in the beam. When this wave is scattered, as in light scattering, the outgoing wave at very large distances from the scattering center has the form

$$\left[\exp(i\mathbf{K} \cdot \mathbf{R}) + f(\chi) \frac{e^{iKR}}{R} \right] e^{-i\omega t} \qquad (22)$$

where we have assumed, as in the above discussions, spherical scattering centers (i.e., no azimuthal dependence). The magnitude of the scattered wave in an angular region about χ is given by $|f(\chi)|^2/R^2$ in Eq. (22), where $f(\chi)$ is referred to as the scattering amplitude and R is the distance from the scattering center. The plane wave in Eq. (22) is the unscattered portion of the wave, which is assigned unit amplitude. This implies that the scattering potential is only a small disturbance. The differential cross section (probability of scattering into a unit solid angle about χ) is simply

$$\sigma(\chi) = |f(\chi)|^2 \qquad (23)$$

Therefore, the scattering problem reduces to solving the Schroedinger wave equation subject to the boundary condition at large R, which is expressed by the form of the wave function in Eq. (22).

To draw analogies with the classical calculation of cross section, it is customary to express the plane wave in terms of multipole moments,

$$\exp(i\mathbf{K} \cdot \mathbf{R}) = \sum_{l=0}^{\infty} i^l(2l + 1)P_l(\cos \chi)j_l(KR)$$

The $j_l(KR)$ are the spherical Bessel functions, which have the asymptotic form, as $R \to \infty$

$$j_l(KR) \quad \to \quad \frac{\sin(KR - l\pi/2)}{KR}$$

The P_l are the Legendre polynomials, where l labels the various moments ($l = 0$, spherical; $l = 1$, dipole; $l = 2$, quadrupole; etc.). In the present problem, however, l also is the angular momentum index $[L^2 = l(l + 1)\hbar^2]$. Therefore, l replaces the impact parameter b, which we used in the classical description

$$b \to \sqrt{l(l + 1)} \ \hbar/mv \approx (l + \tfrac{1}{2})/K = (l + \tfrac{1}{2})\lambdabar$$

$$(24)$$

Upon intersecting a scattering center, each spherical wave j_l experiences a phase shift (η_l), becoming $\sin(KR - l\pi/2 + \eta_l)/KR$ in the asymptotic region. Therefore, one can use these expressions to write

$$f(\chi) = \frac{1}{2Ki}$$

$$\times \sum_{l=0}^{\infty} (2l + 1) [\exp(2i\eta_l) - 1] P_l(\cos \chi) \quad (25)$$

Substituting Eq. (25) into Eq. (23) indicates that the *determination of the scattering cross section* reduces to the *determination of the phase shifts*. Before calculating η_l we use the form for $f(\chi)$ in Eq. (25).

The total cross section, obtained by integrating over the angle in Eq. (23), becomes

$$\sigma = 2\pi \int_{-1}^{1} |f(\chi)|^2 \, d(\cos \chi)$$

$$= \frac{8\pi}{K^2} \sum_{l=0}^{\infty} (l + \tfrac{1}{2}) \sin^2 \eta_l \quad (26)$$

If the phase shifts are zero, no scattering has occurred and the cross section is zero. Further, for large-momentum collisions (small wavelengths) the sum in l above can be approximated by an integral in b using Eq. (24). The expres-

sion for σ can then be written in the same form as the classical cross section in Eq. (3) if we define $P_c(b) \approx 4 \sin^2 \eta(b)$. (Here the dependence of the phase shift η on l is written as a dependence on b.) This expression for P_c clearly demonstrates that at large impact parameters (or l) when $\eta(b)$ goes to zero the wave-mechanical scattering probability $P_c(b)$ goes to zero.

Comparing Eqs. (25) and (26), it is also seen that

$$\sigma = \frac{4\pi}{K} \operatorname{Im} f(0) \quad (27)$$

This result is referred to as the optical theorem, based on analogy with light scattering, and it shows directly what we stated much earlier. The cross section σ is determined by scattering at zero degrees, which for wave scattering is a diffraction region. Note also that the probability P_c above has an average value of 2 at small b and not unity as it is classically. These facts both emphasize that the total scattering cross section σ is a nonclassical quantity. Note that, although the sum in Eq. (25) can also be replaced by an integral over b at small wavelengths, the expression for differential cross section so obtained is quite *different* from the classical expression in Eq. (21). That is, in wave mechanics the *scattering amplitudes* for each impact parameter (or l) that contribute to scattering at angle χ are added, whereas classically the *cross sections* that contribute at angle χ are added [e.g., Eq. (21)]. Hence, the wave-mechanical cross section can exhibit an *oscillatory* behavior.

B. SEMICLASSICAL CROSS SECTION

The relationship between the classical and wave-mechanical cross sections becomes clearer if the sum in Eq. (25) is examined in more detail. At those impact parameters b_i producing classical scattering into the angular region χ, *constructive interference* occurs in the sum in Eq. (25). Therefore, the classical trajectories are like the light rays in geometrical optics. The angular differential cross section

$$\sigma(\chi) = |f(\chi)|^2$$

$$\approx \left| \sum_i \sigma[\chi(b_i)]^{1/2} \exp[i\alpha(b_i)] \right|^2 \quad (28)$$

replaces the classical expression in Eq. (21), where the $\sigma[\chi(b_i)]$ are the classical cross sections, $|b \, db/\sin \chi \, d\chi|_{b=b_i}$, used in Eq. (21). Equation (28) directly exhibits the interference between contributions from different impact parameters. In this expression the phase factor is $\alpha(b) = A/\hbar \pm \varepsilon$, where ε is a multiple of $\pi/4$. Here A is the difference in the classical action between an undeflected particle and a scattered particle, $A = 2\hbar\eta^{sc}(b) - L\chi$ with

$$\hbar\eta^{sc}(b) = \int_{R_0}^{\infty} p(R)\,dR - \int_{b}^{\infty} p_0(R)\,dR$$

The regions of constructive interference are those for which the *classical action is a minimum* ($\partial A/\partial b = 0$), so that

$$\chi = \left(\frac{2}{K}\right)\frac{\partial\eta^{sc}(b)}{\partial b} \qquad (29)$$

On substituting the form for $\eta^{sc}(b)$ given above, Eq. (29) becomes equivalent to the expression for $\chi(b)$ in Eq. (17). Therefore, each region of constructive interference corresponds to a classical trajectory that contributes at angle χ. If only one impact parameter contributes in Eq. (28), the semiclassical result for $\sigma(\chi)$ is *identical* to the classical result. When more than one impact parameter contributes at a given scattering angle, $\sigma(\chi)$ is oscillatory. For rainbow scattering, which we discussed earlier, a schematic diagram of the differential cross section is shown in Fig. 6 indicating the difference between the wave-mechanical and classical behavior. That is, at large τ (close collisions) one impact parameter contributes and the semiclassical cross section follows the classical calculation. At small τ interferences occur and the semiclassical cross section oscillates about the classical result.

C. OTHER APPROXIMATIONS

Using the semiclassical expression for $\eta(b)$ given above and the impulse approximation, a simple estimate of $\eta^{sc}(b)$ that is applicable when (V/E) is small, is given in Table II. Using this in Eq. (25) *and* an approximation to $P_l(\cos\chi)$ valid at small angles, one obtains an estimate of $f(\chi)$ valid at small angles,

$$f(\chi) \simeq -\frac{m}{2\pi\hbar^2}\int d^3R\, V(R)e^{-i\,\Delta\mathbf{p}\cdot\mathbf{R}/\hbar} \qquad (30)$$

This is referred to as the first Born approximation, and results for $\sigma(\chi)$ using Eq. (30) are given in Table II.

As $\eta(b) \xrightarrow{b\to 0} 0$ (see Table II) for potentials that decrease faster than $1/R$, the impulse approximation to $\eta(b)$ can be used to estimate the integrated cross section σ. That is, we replace $\sin^2\eta(b)$ by $\frac{1}{2}$ in Eq. (26) out to some large impact parameter $\bar{b}$ beyond which $\eta(b)$ is always small. Then $\sin^2\eta(b) \approx \eta^2(b)$ at larger b, and σ is written

$$\sigma \simeq 2\pi\bar{b}^2 + 8\pi\int_{\bar{b}}^{\infty}\eta^2(b)b\,db \qquad (31)$$

This is referred to as the Massey–Mohr approximation, and the result for power-law potentials is given in Table II. The expression given is finite for $n > 2$, unlike the classical result, which is never finite for a potential of infinite range. The cross section is also seen to decrease monotonically with increasing energy as shown earlier in Fig. (2). Using the optical theorem [Eq. (27)] in reverse, we see that, if σ is finite, then the

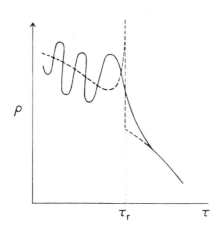

FIG. 6. Schematic diagram of angular differential cross section shown as ρ versus τ plot. τ_r indicates rainbow value for $E\chi_r$. Dashed curved, classical cross section; oscillation occurs where more than one trajectory contributes. [From R. E. Johnson (1982). "Introduction to Atomic and Molecular Collisions," p. 88. Plenum Press, New York.]

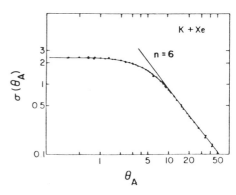

FIG. 7. Elastic scattering cross section (arbitrary units) versus laboratory scattering angle. Line indicates classical calculation using vander Waals potential. [Data from Helbing and H. Pauly (1964). *Z. Physik* **179**, 16. Figure from R. E. Johnson (1982). "Introduction to Atomic and Molecular Collisions," p. 163. Plenum Press, New York.]

differential cross section must be finite as $\chi \to 0$ (see Fig. 7), unlike the classical result. By contrast, a similar calculation of the diffusion cross section σ_d (or S_n) gives the *same form* as the classical expression [Eq. (13)] at small wavelengths. This comes about because scattering at $\chi - 0$ is excluded in the expression for σ_d. Therefore, classical expressions of the diffusion cross, and hence the nuclear stopping cross section, are accurate over a broad range of incident ion energies, allowing extensive use of classical estimates for ion penetration of gases and solids.

In the opposite extreme, when the wavelength is *large* compared to the dimensions of the scattering center ($K^{-1} \gg d$), then no classical analogies exist. However, only the lowest l values contribute in Eq. (25); hence, only one term is kept in the sum, so that

$$f(\chi) \simeq \frac{1}{K} \sin \eta_0 \exp(i\eta_0) \qquad \text{and}$$

$$\sigma(\chi) = \frac{\sin^2 \eta_0}{K^2} \quad (32)$$

(Note that if $\eta_0 \approx n\pi$ then higher terms are needed: this is referred to as the Ramsauer–Townsend effect.) The expression for the cross section in Eq. (32) represents the *same region of impact parameter* as the semiclassical result. However, at very low velocities, it is seen from Eq. (24) that an increase of l from 0 to 1 can make b very large. When b is much greater than the size of the colliding particles, d, then all interactions are encompassed, which at low velocities may only require a single value of l. In this limit it is seen that the differential cross section is *independent of angle for any potential*, a result obtained classically only for the collision of spheres. Because of its simplicity, an isotropic scattering cross section is often used in low-energy collisions, even when the large-wavelength criterion does not rigorously apply. At large wavelengths, the integrated cross section in Eq. (26) is simply written as

$$\sigma \approx \frac{4\pi}{K^2} \sin^2 \eta_0$$

which is also equal to the diffusion cross section σ_d at low velocities.

In the above we have considered very general properties of cross sections. In wave mechanics, as in the scattering of light, the cross sections will exhibit oscillatory behavior and/or resonances for a variety of potential forms. For example, when the long-range potential is attractive, a forward glory is seen in the integrated cross section. In the following sections we will briefly review the nature of interaction potentials, but in doing so we will also consider inelastic effects such as excitations and ionizations.

V. Interaction Potentials

A. OVERVIEW

The primary force determining the behavior of colliding atoms or molecules is the Coulomb interaction. This force acts between each of the constituent electrons and nuclei. Using Coulomb potentials the complete interaction potential for all of the particles can be immediately written down. However, each of the constituents moves relative to the center of mass of its parent molecule. As this motion is superimposed on the overall collisional motion, the description of a collision can be quite complex even for the simplest atoms. Rather than solve the wave equation for the complete, many-body system, one introduces the concept of single-interaction potentials averaged over the relative motion of the constituent particles. In using this concept we again exploit the huge mass difference between electrons and nuclei. This mass difference allows us, with reasonable accuracy, to separate the motion of the electrons and nuclei, a procedure referred to as the Born–Oppenheimer separation.

The behavior of the electrons during a collision is clearly going to depend on the relative motion of the nuclei. Therefore, interaction potentials are generally calculated in two limits. If the collisions are fast relative to the internal motion of the electrons ($v \gg v_e$), then during the collision the electronic distribution is static, except for abrupt changes (transitions) that occur when the particles are at their closest approach. The transitions reflect the ability of the molecules to absorb (emit) energy when exposed to the time-varying field of the passing particle in the same way that these molecules absorb or emit photons. Before and after the transition the potentials are determined from the separated charge distributions.

In the opposite extreme (slow collisions, $v \ll \bar{v}_e$) the electrons adjust continuously and smoothly to the nuclear motion, returning to their initial state at the end of the collision. This collision process is called adiabatic as the electrons do not gain or lose energy. That is, even though the molecules may be deflected and change kinetic energy, their initial and final *elec-

tronic states remain the same. The electronic distribution evolves from a distribution in which electrons are attached to separate centers at large R to a distribution in which the electrons are shared by the two centers at small R, a covalent distribution. Therefore, for every possible initial state there is a corresponding adiabatic potential, resulting in rather complex potential diagrams. Such potentials also determine the ability of the two particles to bind together to form a molecule.

As molecules in a collision are not moving infinitely slowly, the motion of the nuclei can induce transitions between states. For near-adiabatic collisions these transitions occur at well-defined internuclear separations—for example, at those internuclear separations at which the atomic character of the wave function gives way to the molecular, covalent character. Because a large computational effort is required to obtain the set of potential curves, and an additional large computational effort is required to describe the collisions when transitions occur, simplifying procedures are very attractive and approximate potentials are often constructed.

As the electrons in different shells have very different velocities, the above separation into fast and slow collisions allows us to treat the orbitals separately. For instance, when a collision is fast with respect to the outer-shell electrons, it may be adiabatic with respect to the inner-shell electrons. Therefore, the inner-shell electrons return to their initial state. Their effect is to only screen the nuclear interactions and, hence, they play a passive role in the collision. Alternatively, when inner-shell excitations occur the outer-shell electrons can be considered to be static during the collision. As a point of reference, it is useful to remember that a nucleus with a speed equivalent to an electron in the ground state of a hydrogen atom has an energy of about 25 keV/amu. Further, the orbital speed of an electron in an atom can be scaled to the speed of an electron in the ground state of hydrogen using the effective nuclear charge.

The interaction potential between two atoms can be written as a sum of the nuclear repulsion and the averaged electronic energy $\varepsilon_j(R)$,

$$V_j(R) = \frac{Z_A Z_B e^2}{R} + [\varepsilon_j(R) - \varepsilon_j] \quad (33)$$

In this expression j labels the electronic state, Z_A and Z_B are the nuclear charges, and ε_j is the total electronic energy of the colliding atoms, $\varepsilon_j(R)$, as $R \to \infty$. Each state j is associated with a pair of separated atomic states at large R. In the *electrostatic limit* ($v \gg \bar{v}_e$), the electronic energy $\varepsilon_j(R)$ is a sum of the electronic energies of the separated atoms, $(\varepsilon_{Aj} + \varepsilon_{Bj}) = \varepsilon_j$, *plus* the averaged interaction of the electrons on each atom with the electrons and nucleus of the other atom, which we call V_j^e. The quantity V_j^e is written

$$
\begin{aligned}
V_j^e(R) = & -Z_B e^2 \int \frac{\rho_{Aj}(\mathbf{r}_A)}{|\mathbf{R} - \mathbf{r}_A|} d^3 r_A \\
& - Z_A e^2 \int \frac{\rho_{Bj}(\mathbf{r}_B)}{|\mathbf{R} - \mathbf{r}_B|} d^3 r_B \\
& + e^2 \int \frac{\rho_{Aj}(\mathbf{r}_A)\, \rho_{Bj}(\mathbf{r}_B)}{|\mathbf{R} - \mathbf{r}_A + \mathbf{r}_B|} d^3 r_A\, d^3 r_B
\end{aligned}
$$
$$(34)$$

where the ρ_{Aj} and ρ_{Bj} are the atomic densities for the electrons on atoms A and B. V_j^e is the classical electrostatic interaction and is evaluated in many texts for various form for ρ.

In the adiabatic approximation, $\varepsilon_j(R)$ in Eq. (33) is calculated from the full electronic wave equation at each R. An approximation that is particularly useful for light atoms is to estimate $\varepsilon_j(R)$ via the molecular-orbital method used by chemists. For heavy atoms, on the other hand, the Thomas–Fermi method is often employed to estimate $\varepsilon_j(R)$. In this model the electrons are treated as a gas subject to the Pauli principle. In the following, rather than calculate, $V_j(R)$ for all R, we will consider its behavior for various regions of R. Such an approach is reasonable as it is seen from Table II that the deflection function is determined primarily by a narrow region of R about the distance of closest approach.

B. Short-Range Potentials

For close collisions ($R \ll \bar{r}_A, \bar{r}_B$ where $\bar{r}_A$ and $\bar{r}_B$ are the mean atomic radii), the nuclear repulsion dominates. The electrons essentially screen the nuclear repulsive interaction; hence, one often approximates $V_j(R)$ by

$$V_j(R) = \frac{Z_A Z_B e^2}{R} \Phi\left(\frac{R}{a_j}\right) \quad (35)$$

where a_j is a screening length and Φ the screening function. Considerable effort has been expended determining Φ and a_j for many-electron atoms and often Φ is written as $\exp(-R/a_j)$. At small distances of closest approach which are usually associated with fast collision, an electrostatic calculation [Eq. (34)] can be used to estimate Φ and a_j. In fact, the electrons in different shells have different screening lengths. Therefore, for a bare ion colliding with an atom, a potential of the form

$$V_j(R) = \frac{Z_A e^2}{R} \sum_i N_{B_i} e^{-R/a_i}$$

is used, where N_{B_i} is the number of electrons in the ith shell on B. A good approximation for light atoms is to assume that the ith shell has a screening constant determined by the ionization energy I_i, $a_i \approx a_0/Z_i'$ with the effective charge $Z_i' = (I_i/I_0)^{1/2}$, where I_0 is the ground-state, hydrogen atom ionization potential.

For collisions between large atoms ($Z \gtrsim 10$), the Thomas–Fermi screening constant used by Lindhard,

$$a_{TF} - 0.8853 a_0 (Z_A^{2/3} + Z_B^{2/3})^{-1/2} \qquad (36)$$

has been shown to be generally applicable and more elaborate functional forms for Φ are employed. Expressions for cross section and stopping cross section have been given in Table II for the potential in Eq. (35) using the exponential screening function. Results obtained by Lindhard for the Thomas–Fermi screening function are also given. As the short-range collision region has been extensively studied, one should be careful to use experimentally determined parameters when available.

C. LONG-RANGE INTERACTIONS

For slow collisions even small disturbances can lead to deflections; therefore the interaction potential at long range (large separations between colliding partners, $R >> \bar{r}_A$, $\bar{r}_B$) is of interest. As the electronic distributions of each of the colliding particles are distorted very little by the presence of the other, the interaction potential is written in powers of R,

$$V_j(R) = \sum C_n/R^n \qquad (37)$$

and, generally, only the largest term (lowest power of n) is kept in a calculation. Therefore, the power-law expression for cross section that we developed earlier can be used for weak-interaction, long-range collisions. The lead terms in Eq. (37) can be obtained from the electrostatic interaction in Eq. (34) by expanding the denominators in powers of R. For ion–ion collisions the lead term in Eq. (37) is $n = 1$, $C_1 = \bar{Z}_A \bar{Z}_B e^2$, which is the coulomb interaction with $\bar{Z}_A = Z_A - N_A$ (the nuclear charge of A minus the number of electrons on A or the net charge). For

ion collisions with a neutral molecule having an electric dipole moment μ_B (e.g., water molecule), the lead term is $n = 2$, $C_2 = \bar{Z}_A \mu_B e \cos \theta$ where $\cos \theta$ is the angle between the internuclear axis and the dipole moment. For neutral molecules that do not have dipole moments (e.g., O_2, N_2), the ion–quadrupole interaction dominates ($n = 3$). For two neutral molecules having dipole moments, the dipole–dipole interaction ($n = 3$) dominates, and so on.

For colliding atomic particles the electrostatic multipole moments of the charge distribution are zero and for some molecules (e.g., H_2) the multipole moments are small. However, the field of the other particle induces moments in the separated charge distribution. The lead term for ion-neutrals is the ion-induced dipole interaction ($n = 4$),

$$C_4 = -\frac{\alpha_B}{2} (\bar{Z}_A e)^2$$

where α_B is the polarizability of the neutral particle B. For collisions between two neutrals there is not average field at B due to A or vice versa. However, instantaneous fluctuations in charge density on either atom lead to short-lived fields that induce moments in the other. Therefore, for neutrals the lead term is the induced-dipole–induced-dipole interaction, which is the well-known van der Waals potential used to describe the behavior of realistic gases. This interaction ($n = 6$) is always attractive (i.e., C_6 negative) and the coefficients have been extensively evaluated from collision experiments.

D. INTERMEDIATE RANGE POTENTIALS AND CHARGE EXCHANGE

When R is the order of the atomic radii ($\bar{r}_A$, $\bar{r}_B$), the electron clouds on the two interacting particles overlap. In this region the distortion of the charge clouds on each center becomes too large to treat as simply a perturbative polarization of the separated electronic distributions. This is an especially important region for determining transitions between colliding particles but is also important in molecular structure determinations. The emphasis here is on those aspects important in collisions.

The overlap of charge on the two centers allows for the possibility of charge-exchange collisions (e.g., $H^+ + O_2 \rightarrow H + O_2^+$). When charge exchange occurs the potentials before and after the collisions can be drastically different. For

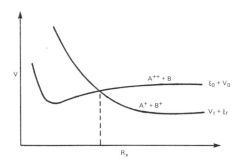

FIG. 8. Interaction potentials versus internuclear separations; curve crossing for the $A^{2+} + B \rightarrow A^+ + B^+$ collisions.

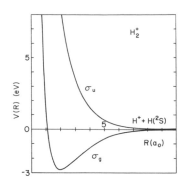

FIG. 9. Interaction potential energy curves versus internuclear separation for ground state H_2^+: nuclear repulsive plus electronic (ε_g or ε_u).

example, for $O^{2+} + S \rightarrow O^+ + S^+$, a process of interest in the Jovian magnetosphere plasma, the initial long-range interaction is attractive and is determined by the polarizability of S. On the other hand, the final interaction is clearly repulsive. Depending on the final states of O^+ and S^+, the initial and final potential curves may cross as shown in Fig. 8. Therefore, electron exchange between O^{2+} and S can occur with *no net change* in the total electronic energy *at* the crossing point. Of course, the change in state gradually results in a net change in the *final* electronic energy as the particles separate. Therefore, such crossings indicate transition regions and the likelihood of charge exchange is determined by the overlap of the charge distributions at the crossing point.

For this region of intermediate R the covalent (electron exchange) interaction can be examined by considering the $H^+ + H$ (i.e., H_2^+) system. In such a system, the exchange $H^+ + H \rightarrow H + H^+$ occurs *without* a net change in internal energy, unlike the case discussed above. The two identical states at large R ($H + H^+$ and $H^+ + H$) split at smaller R, due to electron sharing, forming both attractive and repulsive potentials, as shown in Fig. 9. If the wave functions placing the electron on centers A and B in identical states of the hydrogen atom are ϕ_A and ϕ_B, then the linear combination appropriate to the H_2^+ molecule are $\psi_{g,u} \cong (1/\sqrt{2})(\phi_A \pm \phi_B)$. The labels g and u refer to symmetric and antisymmetric (gerade and ungerade) states of H_2^+. In the scattering of protons by hydrogen, as the initial state is *either* ϕ_A or ϕ_B, half the collisions will be along repulsive potentials (u state) and half along an attractive potential (g state). The difference in energy between these states is referred to as the exchange energy and will determine the

behavior of charge-exchange collisions. This exchange energy decays as the overlap of the wave functions on A and B ($<\phi_A|\phi_B> \propto e^{-R/\bar{a}}$), which is an exponential function. Therefore, this exchange interaction is eventually dominated at very large R by the long-range, power-law potentials discussed earlier.

Molecular orbital potentials for H_2^+ can also be constructed for each excited state of H, or, more usefully, they can be constructed for one electron in the field to two identical nuclei *and* other electrons. In Fig. 10 we give a diagram showing the general behavior of these one-elec-

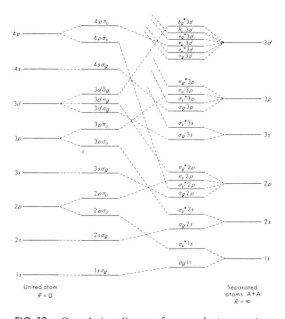

FIG. 10. Correlation diagram for one electron on two identical centers with effective charge Z'_A. [From R. E. Johnson (1982). "Introduction to Atomic and Molecular Collisions," p. 124. Plenum, New York.]

tron, molecular orbital binding energies $\varepsilon_j(R)$. As this quantity does not include the nuclear repulsive term in Eq. (33), we can follow the behavior of these states down to $R = 0$. This diagram correlates the one-electron states of the separated centers ($R \rightarrow \infty$) with those of the united atom ($R \rightarrow 0$), indicating how the electronic binding energy changes versus R. The states are labeled by their symmetry (g or u) under inversion and by the component of electronic angular momentum along the internuclear axis ($|m_l| = 0,1,2,...$ as $\sigma,\pi,\delta,...$). The correlation between states at large and small R is determined by the fact that states of the same symmetry [rotation $|m_l|$ and inversion (g, u)] do not cross (i.e., do not become degenerate in energy). Two sets of notation are used for these states in order to indicate which are the corresponding atomic states at either $R \rightarrow 0$ or $R \rightarrow \infty$. For example, the two lowest states we considered above for H_2^+ are labeled $\sigma_g 1s$ and $\sigma_u 1s$ at large R, indicating they originate from the ground state (e.g., 1s state of H). At small R they are labeled $1s\sigma_g$ and $2p\sigma_u$, indicating they correlate with the 1s and 2p states of the united atom (e.g., He^+ for the H_2^+ molecule).

To obtain interaction potentials for a particular pair of atoms the net binding energy of each electron, determined from the correlation diagram and the Pauli principle, is combined with the nuclear repulsive potential in Eq. (33). Therefore, two hydrogen atoms (H_2) interact via a singlet state $(\sigma_g 1s)^2$ $^1\Sigma_g^+$ which is *attractive* at intermediate R, and a triplet state $(\sigma_g 1s)(\sigma_u 1s)$ $^3\Sigma_u^+$. The latter is a net *repulsive* potential at intermediate and small R (i.e., ignoring the polarization at very large R). Therefore, for the collision between two H atoms $\frac{3}{4}$ of the collisions are repulsive and $\frac{1}{4}$ attractive. Two helium atoms (He_2) interact via a $(\sigma_g 1s)^2$ $^1\Sigma_g^+$ ground repulsive state. The lowest state for two oxygen atoms is $(\sigma_g 1s)^2$ $(\sigma_g 1s)^2$ $(\sigma_g 2p)^2$ $(\pi_u 2p)^4$ $(\pi_g 2p)^4$ $(\sigma_g 2s)^2$ $^3\Sigma_g^+$. In the above the $\Sigma,\Pi,\Delta,...$ represent the *net* angular momentum along the internuclear axis ($|\Sigma m_{li}| = 0,1,2,...$, where i implies a sum over orbitals). These states are all doubly degenerate (i.e., the angular momentum vector can be pointed either way) except the Σ states, which are, therefore, also labeled according to their behavior under reflection ($+$ or $-$) about a plane containing the nuclei.

The molecular orbital diagrams are extremely useful for determining which transitions are likely to occur. Following the molecular orbital states at large R into small R will result in a number of curve crossings as seen in the potential diagram for certain states of He_2^{2+} in Fig. 11. In a full adiabatic calculation of the electronic energies, for which the electrons *totally* adjust to one another, crossings are avoided for states of the same total symmetry. Therefore the crossings in the molecular orbital diagram indicate that value of R at which the character of the wave function is changing and hence indicate the transition regions. The $(\sigma_u 1s)^2$ $^1\Sigma_g^+$ state of He_2^{2+} is seen to make a series of crossings as both of the electrons are promoted to (2p) electrons in the united atom limit. Therefore, half of the collisions for He^{2+} on He will proceed along an attractive $^1\Sigma_u^+$ for which transitions *are unlikely* at low velocities and half along a strongly repulsive $^1\Sigma_g^+$ state for which transitions are *very likely* to occur if the distance of closest approach is small enough. As pointed out by Lichten and Fano, such electron promotion occurs in all colliding systems as there are two atomic orbital states at large R for every atomic orbital state at small R (see Figs. 9 and 10).

A molecular orbital correlation diagram can also be constructed for an electron in the field of two positive centers having different effective charges Z_A and Z_B. As the inversion symmetry is broken, the correlations are determined using $|m_l|$ only, thereby lessening the promotion effect. The importance of promotion, and hence transition, depends in part, therefore, on how similar the effective charges are on each center. For instance, for a proton interacting with an argon atom (ArH^+ potential), the ground-state potentials at intermediate R are well described by a single electron shared by an Ar^+ core and a proton. As these have binding energies $I_{Ar} = 15.2$ eV and $I_H = 13.6$ eV, the effective charges are very similar ($Z'_{Ar} = 1.06$, $Z'_H = 1$). Therefore, the lowest Σ states of ArH^+ are similar to those of H_2^+ at intermediate R. Stated another way, states of the same symmetry that are close in energy at large R are strongly coupled; therefore, they tend to "repel" (diverge) from each other. The covalent nature of these states is associated with an exchange interaction similar to that in H_2^+, which we write in a general form as

$$\Delta\varepsilon = A \exp(-R/\bar{a}) \qquad (38)$$

where the effective radius is $\bar{a} = 2a_0/(Z'_A + Z'_B)$.

This confusion of potentials presents problems for the user. For example, although the $O^+ + O$ (O_2^+ system) has a lowest-lying attractive (bound) state $^2\Pi_g$, in a collision of a ground-state oxygen ion with a ground-state oxygen

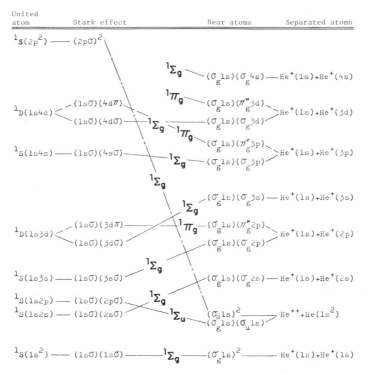

FIG. 11. Correlation diagram for certain states of He_2^{2+}. He^{2+} + He gives a $^1\Sigma_g$ that crosses many states and a $^1\Sigma_u$ that does not. [From R. E. Johnson (1982). "Introduction to Atomic and Molecular Collisions," p. 123. Plenum, New York.]

atom seven different potential curves evolve, each with a different multiplicity. Therefore, simplifying procedures are desirable. Although it is not strictly correct to use a single potential, the repulsive curves often dominate and a simple repulsive potential is often used to approximate the interaction at intermediate R. Based on the form for electron exchange interaction discussed above, this potential is generally represented as an exponential, $V = Ae^{-R/\bar{a}}$. Such a potential is called a Moliere potential and is commonly used in particle penetration calculations. It can be thought of as an extension of the short-range, repulsive interactions discussed earlier. In fact, a universal potential form intended to cover both ranges of R for heavy elements is the Lenz–Jensen form for Φ in Eq. (35), which we give in Table II.

VI. Inelastic Collisions

A. OVERVIEW

The calculation of the changes in internal motion of a molecular system of particles induced by a collision with another atom or molecule is a complicated many-body problem for which a number of approximate models have been developed. When the interacting particles are moving, as in a collision, then the interactions described in the preceding sections become dynamic. In addition to the overall deflections, calculated by the potentials described, the time-dependent fields produce changes in the internal state of the molecule. The motion of each of the constituent particles of the molecules can be characterized by a frequency ω and a mean radial extent from the center of mass, $\bar{r}$. The interaction with a passing particle of velocity v at a distance b from the center of a target atom or molecule is said to be nearly adiabatic if $\tau_c \gg \omega^{-1}$ where $\tau_c \approx b/v$. For nonadiabatic collisions, transitions become likely, as discussed in Section I,B.

In most instances, the approximations used involve only two states. Such models can be divided into two categories: strong interactions, for which the evolution of the electronic states during the collision is important (e.g., curve-crossing transitions) and weak interactions, for which the initial charge distributions can be considered static. The former case generally applies to incident ion velocities comparable to or smaller than the velocities of the constituent particles of the target and the latter to large velocities. If, in addition, $b \ll \bar{r}$, then the collision is a close collision and the incident particle can be thought of as interacting with each of the constituents of the target molecule separately. This is referred to as the binary-encounter limit. On the other hand, if $b \gg \bar{r}$, a distant collision, the target atom or molecule must be viewed as a whole. Classically the constituent particles are often treated as bound oscillators of frequency ω. These oscillators are then excited by the time-dependent field of the passing particle. (Of course, in wave mechanics the impact parameter is not a well-defined concept and the close and distant criterion is based, more correctly, on whether the momentum transfer to the constituent particles is large or small.) In the following we calculate, classically, the large-momentum transfer (close collision) and small-momentum transfer (distant collision) contributions to the energy-loss cross section. These results are synthesized by considering the Bethe–Born approximation to the cross section. Following this we consider models for strong interactions such as charge exchange.

B. BINARY-ENCOUNTER APPROXIMATION

In the often-used binary-encounter approximation (BEA), each target constituent is assumed to interact separately with the incident particle. For example, the stopping power [Eq. (11)] can be written as a sum of contributions from each of the constituents:

$$S = S_e + S_n \approx \sum_j N_j \int Q \, d\bar{\sigma}_j(Q)$$
$$+ \sum_k \int T \, d\sigma_k(T) \qquad (39)$$

The differential cross section $d\bar{\sigma}_j(Q)$ is the energy-transfer cross sections to the electrons in orbit j with population N_j and $d\sigma_k(T)$ is the energy-transfer cross section to nucleus k in the target molecule. The latter contribution we considered earlier when describing elastic collisions. Results are given in Table II for $d\sigma(T)$ $[\equiv (d\sigma/dT) \, dT]$ and S_n for a single nucleus using a number of potentials. We note in passing that the total S_n can be rewritten for a molecular target having more than one nucleus in such a way that the energy transfer to the center of mass of the target molecule and that energy going into vibrational and rotational excitation of the mole-

cule are separated. The latter contributions are, in fact, inelastic effects and, of course, sufficient collisional energy transfer can lead to dissociation of a molecular target. If M is the total molecular mass and D_k the removal energy for nucleus k of mass M_k, then the BEA estimate for the collisional dissociation cross section is

$$\sigma_D \approx \sum_k \int_{MD_k/(M-M_k)}^{\gamma_{Ak} E_A} d\sigma_k(T) \qquad (40)$$

This is easily calculated using the results in Table II.

The cross section $d\bar{\sigma}_j(Q)$ in the BEA is an "elastic" cross section for a collision between the incident particle and a free electron leading to a change Q in the electron's energy. The bar implies an averaging over the initial orbital velocity distribution of the electron. In general, any appropriate cross section can be used to describe the target electron–incident particle collision. If the incident particle is a bare ion of charge Z_A, then the collision with the electrons is coulombic. If, further, the ion's velocity is large compared to the mean orbital velocity of the electrons, then the averaging is not important and collision is described by the Rutherford cross section

$$d\bar{\sigma}_j(Q) \simeq \frac{2\pi}{m_e v^2} \left(\frac{Z_A e^2}{Q} \right)^2 dQ \qquad (41a)$$

(e.g., use results in Table II with $M_B \to m_e$, $n = 1$, $c_1 = Z_A e^2$, and $T \to Q$). Using this cross section in Eq. (39), the reader can verify that electronic stopping cross section [e.g., Eq. (12a)] can be written

$$S_e \approx \frac{2\pi}{m_e v^2} (Z_A e^2)^2 \sum_j N_j \ln \frac{2m_e v^2}{(Q_j)_{min}} \qquad (41b)$$

where $(Q_j)_{min}$ and $2m_e v^2$ are the lower and upper limits on the energy transfer to the "stationary" electrons. Similarly, the BEA approximation to the total ionization cross section for fast ions, referred to as the Thompson cross section, is

$$\sigma_I \approx \frac{2\pi}{m_e v^2} (Z_A e^2)^2 \sum_j N_j \left(\frac{1}{I_j} - \frac{1}{2m_e v^2} \right) \qquad (41c)$$

where I_j is the removal energy for electrons in the jth orbit. The BEA ionization and stopping cross sections for incident ions vary as $1/E_A$ and $(\ln E_A)/E_A$ respectively at high energies. For an incident neutral a screened Coulomb cross section should be used for $d\bar{\sigma}_j(Q)$; for example, the Born approximation for the screened Coulomb interaction is given in Table II with $M_B \to m_e$ and $T \to Q$. Using these the *form* for S_e and σ_I

will closely resemble that of S_n and σ_D calculated using a screened Coulomb potential.

C. CLASSICAL OSCILLATOR

For distant-collisions, and low-momentum transfer collisions, we can treat the bound electrons and/or nuclei as classical oscillators that are excited by the time-varying field of the passing particle. The motion of an electron oscillator in a field $\mathcal{E}_j(t)$ is

$$m_e\ddot{\mathbf{r}}_j + \Gamma_j\dot{\mathbf{r}}_j + m_e\omega_j^2\mathbf{r}_j = -e\mathcal{E}_j(t) \quad (42)$$

where ω_j is the binding frequency and Γ_j is a damping constant. Writing the energy transfer as

$$Q_j = \int_{-\infty}^{\infty} \dot{\mathbf{r}}_j \cdot [-e\mathcal{E}_j(t)]\, dt$$

then as $\Gamma_j \to 0$ it is straightforward to show that

$$Q_j \to \frac{\pi}{m_e}|e\mathcal{E}_j(\omega)|^2$$

where $\mathcal{E}_j(\omega)$ is the Fourier transform of $\mathcal{E}_j(t)$, that is,

$$\mathcal{E}_j(\omega) = \frac{1}{\sqrt{2\pi_i}}\int_{-\infty}^{\infty} \mathcal{E}_j(t)\, \exp(-i\omega t)\, dt$$

Note that $|e\mathcal{E}_j(\omega)|$ has units of momentum, indicating the net impulse to the oscillator.

Describing $\mathcal{E}_j$ as the field associated with a screened coulomb potential, $V = (Z_A e^2/R)e^{-\beta R}$ and assuming straight line trajectories ($R^2 = b^2 + v^2 t^2$), the energy transfer is

$$Q_j = \frac{2(Z_A e^2)^2}{m_e v^2}\left(\frac{1}{b^2}\right)\Big[(\beta_j' b)^2 K_1^2(\beta_j' b)$$

$$+ \left(\frac{\omega_j b}{v}\right)^2 K_0^2(\beta_j' b)\Big] \quad (43a)$$

where $\beta'^2 = \beta^2 + (\omega_j/v)^2$ and K_1 and K_0 are modified Bessel functions. This energy transfer behaves asymptotically as

$$Q_j \to \frac{2(Z_A e^2)^2}{m_e v^2}\left(\frac{1}{b^2}\right)\begin{cases} 1 & \text{for} \quad b\beta \ll 1 \\[2mm] \left.\begin{array}{l}\dfrac{\pi}{2}(\beta b)\exp(-2\beta b) \\[2mm] \text{for} \quad b\beta \gg 1\end{array}\right\}\dfrac{\omega_j b}{v} \ll 1 \\[4mm] \dfrac{\pi}{2}\left(\dfrac{\omega_j b}{v}\right)\exp\left(\dfrac{-2\omega_j b}{v}\right); \\[3mm] \dfrac{\omega_j b}{v} \gg 1, \dfrac{\omega_j}{v} \gg \beta \end{cases}$$

$$(43b)$$

It is important to note in these expressions that screening of the collision is a result *both* of the direct screening in the potential via β [$\beta \to a_j^{-1}$ in Eq. (35)] and that due to the motion of the electron via ω_j. In Eq. (43b), ($\omega_j b/v \ll 1$) is the classical BEA limit for a screened coulomb potential, in which the incident ion is fast compared to the bound electron motion; Q decreases rapidly for $b > \beta^{-1}$. Further, ($\omega_j b/v \gg 1$) is the adiabatic limit as described earlier; Q is seen to decrease rapidly for $b > v/\omega_j$. Writing the electronic stopping power as

$$S_e = \sum_j N_j 2\pi \int_{(b_{\min})_j}^{\infty} Q_j(b)b\, db$$

where N_j is the number of electrons of frequency ω_j and using Eq. (43a) the expression for S_e is exactly integrable. Setting $b_{\min}$ equal to the wavelength in the center of mass of the electron–incident–ion system ($\hbar/m_e v$), then

$$S_e \simeq 4\pi \frac{(Z_A e^2)^2}{m_e v^2}\sum_j N_j \Big[\ln\Big(\frac{1.123 m_e v}{\beta_j'\hbar}\Big)$$

$$- \frac{1}{2}\Big(\frac{\beta}{\beta_j'}\Big)^2\Big] \quad (44a)$$

When the direct screening dominates (i.e., $\beta_j' = \beta$ for large v), then Eq. (44a) has the form of Eq. (41b) with $Q_{\min} \propto \beta^2\hbar^2/(2m_e)$. That is, the screening constant β determines the low energy cutoff. On the other hand, when $\beta = 0$ (i.e., no direct screening), then

$$S_e \simeq 4\pi \frac{(Z_A e^2)^2}{m_e v^2}\sum_j N_j\Big(\ln\frac{1.123 m_e v^2}{\hbar\omega_j}\Big) \quad (44b)$$

This is roughly twice the BEA result given in Eq. (41b). The origin of this factor of two will be clear from the subsequent discussion. Here we only point out that this classical oscillator model very nicely describes both distant and close collisions when calculating the energy loss cross section. However, it is much less useful for determining, for instance, the ionization cross section.

D. BETHE–BORN

The first-order estimate of the inelastic cross section in quantum mechanics is given by the Born approximation. The Born scattering amplitude for a transition from an initial state (0) to a final state (f) is given by

$$f_{0\to f} \approx \frac{-m}{2\pi\hbar^2}\int e^{-i\mathbf{K}_f\cdot\mathbf{R}}V_{f0}(R)e^{i\mathbf{K}_0\cdot\mathbf{R}}\, d^3R \quad (45a)$$

This leads to a cross section, as in Eq. (23), of

the form

$$\sigma_{0\to f}(\chi) = \frac{K_f}{K_0}|f_{0\to f}|^2 \qquad (45b)$$

where $\hbar K_f$ and $\hbar K_0$ are the final and initial momentum and $\cos \chi = \hat{K}_f \cdot \hat{K}_0$. In Eq. (45a) the exponentials represent incoming and outgoing plane waves and $V_{f0}(R)$ is the interaction potential averaged over the final and initial states. Equation (45) can be related to Eq. (30), the elastic scattering result. For elastic scattering, the final and initial electronic states are identical; therefore K_j and K_0 are the same size but differ only in direction. In describing elastic scattering, the potential $V_{00}(R)$ is simply the electrostatic potential for the ground state that we called $V_0(R)$ in Eq. (33). In addition, $e^{i(K_0-K_f)\cdot R} = e^{-i\Delta p \cdot R/\hbar}$, yielding the result in Eq. (30).

For a fast collision of an incident ion with a neutral, Bethe approximated the cross section above as

$$d\sigma_{0\to f} \approx \frac{2\pi(Z_A e^2)^2}{m_e v^2} Z_B \frac{dQ}{Q^2} |F_{0\to f}(Q)|^2 \quad (46)$$

where $Q = \Delta p^2/2m_e$ and the quantity $F_{0\to f}(Q)$ is the interaction matrix element of the target atom. Note that if $|F_{0\to f}|^2 = 1$, this expression becomes identical to the Rutherford cross section above. Equation (46) differs from the BEA in that Δp is *not* the momentum transfer to a single electron but to the system as a whole. Hence, even for very small net change in momentum, electronic transitions, which require significant internal energy changes, *can* occur. The interaction matrix element is a weighting factor, which has the property that

$$\sum_f (\varepsilon_f - \varepsilon_0)|F_{0\to f}(Q)|^2 = Q$$

so that the *average* effect of all the electronic transitions is an energy change equal to that used in the BEA. $F_{0\to f}(Q)$ contains the screening effect of the moving electron, so that for large-momentum transfers, and hence large Q,

$$|F_{0\to f}(Q)|^2 \approx \begin{cases} 0 & \varepsilon_f - \varepsilon_0 \neq Q \\ 1 & \varepsilon_f - \varepsilon_0 = Q \end{cases}$$

On substitution into Eq. (46) this gives the classical BEA result, Eq. (41a), discussed earlier. This means that for large energy transfers, the momentum transfer is, with a *high probability*, equal to that transferred to a single electron raising it to an excited state. At small momentum transfers, hence low Q,

$$Z_B|F_{0\to f}(Q)|^2 \approx \left(\frac{Q}{\varepsilon_f - \varepsilon_0}\right) f_{0f}$$

This is the limit in which the excitations are predominantly dipole excitations and f_{0f} is the dipole oscillator strength, which is also used to determine the polarizability of an atom in the field an ion. [*See* ATOMIC PHYSICS.]

Using these expressions for the generalized oscillator strength to calculate the stopping power, one finds the remarkable result that the *form* of the stopping power is the same at high and low Q (or Δp), so that the result is *not* sensitive to the value of Q dividing these regions. The combined stopping power can be written

$$S_e \equiv \sum_f \int (\varepsilon_f - \varepsilon_0)\, d\sigma_{0\to f}$$

$$\approx \frac{4\pi(Z_A e^2)^2}{m_e v^2} Z_R \left(\ln \frac{2m_e v^2}{I} - 1 - C\right) \quad (47)$$

In this expression the sum over states is absorbed in the values of I and C. Here C is a factor associated with the shell structure of the target atom or molecule and I is an average ionization potential. Values of I and C have been extracted from experiment. This expression for S_e has the same form as that in Eq. (44b) for the coulomb interaction with an oscillator and, therefore, is also twice the BEA result at high velocities. The value of 2 indicates that the distant *and* close collisions described here contribute roughly equally to S_e.

The quantities above can also be used to determine the leading terms at large v for total ionization cross section σ_I and the straggling cross section $S_e^{(2)}$. The lead term at large v is

$$\sigma_I \equiv \sum \int d\sigma_{0\to f} \approx \frac{4\pi(Z_A e^2)^2}{m_e v^2} \left(\frac{2m_e \overline{r_0^2}}{3\hbar^2}\right) \ln\left(\frac{2mv_e^2}{I}\right)$$

$$(48a)$$

where f implies all final states leading to an ionization and $\overline{r_0^2}$ is the mean-squared radius of the initial state. Similarly,

$$S_e^{(2)} \equiv \sum \int (\varepsilon_f - \varepsilon_0)^2\, d\sigma_{0\to f} \approx 4\pi(Z_A e^2)^2 Z_B$$

$$(48b)$$

which is summed over all final states, ionizations, and excitations. The latter quantity, which weights large energy transfers heavily, is equal to the BEA calculation at high velocities as close collisions (large momentum transfers) dominate. On the other hand, the ionization cross section, which unlike S_e and $S_e^{(2)}$ is not weighted by the energy transfer, is determined primarily by *dipole* excitations (small momentum transfers). The ionization cross, therefore, is *not* very well represented by Thompson expression in Eq. (41c); rather, it has a similar energy dependence, for fast incident ions, to that of the stopping cross section, i.e., $(\ln E_A)/E_A$. This commonality in energy dependence has the very useful consequence that the *number of ionizations* per unit path length produced by a fast incident ion stopping in a gas or solid is roughly proportional to S_e.

E. TWO-STATE MODELS: CHARGE EXCHANGE

For collision velocities comparable to or smaller than the speed of the constituents of the target particles, the details of the interaction potentials described earlier can control the transition probability. When the coupling between neighboring states is strong and the existence of other states can be treated as a weak perturbation, a number of two-state models can be used for calculating these probabilities. The models are based on the impact parameter cross section in Eq. (7), and forms for the transition probabilities $P_{0 \to f}$ are given in Table III. A requirement for strong coupling is that the energy difference between the states at any point in the collision is small compared to the uncertainty in the energy of the states during the collision. This uncertainty is estimated as $\Delta E \approx \hbar(\Delta R_x/v)$ where ΔR_x is the extent of the transition region. The general nature of these two-state inelastic cross sections was described earlier. At low velocities (ΔE above much less than the state spacing, $\varepsilon_f - \varepsilon_0$), the levels are well defined and transitions are not likely. At high velocities the uncertainty ΔE becomes large so that the states effectively overlap. However, at high velocities, the time for a transition to occur becomes short so that transition probabilities again become small and the Born approximation described above should be used. When the spacing between states during the collision is comparable to ΔE, then the cross section is large (i.e., $\sim \pi a_0^2$ for transition involving outer-shell electrons).

TABLE III. Inelastic Collision Expressions

Impact parameter results

$$\left(\sigma_{0 \to f} = 2\pi \int_0^\infty P_{0 \to f}(b) b \, db, \quad \varepsilon_f - \varepsilon_0 \equiv \hbar \omega_{f0} \equiv Q_{0 \to f}\right)$$

First order (Born)

$$P_{0 \to f}(b) \approx \left| \frac{1}{i\hbar} \int_{-\infty}^{\infty} \Delta\varepsilon_{0f}(R) \exp(i\omega_{f0}t) \, dt \right|^2$$

for $\Delta\varepsilon_{0f} = V_0 e^{-\beta R}$

$$P_{0 \to f}(b) \approx \left(\frac{2V_0 b}{\hbar v}\right)^2 \left(\frac{\beta}{\beta'}\right)^2 K_1^2(b\beta')$$

$$\beta'^2 = \beta^2 + \frac{\omega_{f0}^2}{v^2}$$

$$\sigma_{0 \to f} \approx \frac{4\pi}{3} \left(\frac{2V_0}{\hbar\omega_{f0}\beta}\right)^2 \frac{(\omega_{f0}/v\beta)^2}{[1 + (\omega_{f0}/v\beta)^2]^3}$$

Landau–Zener–Stueckleberg

(curve crossing, $\varepsilon_f + V_f = \varepsilon_0 + V_0$ at $R = R_x$)

$$P_{0 \to f}(b) \approx 2\overline{P}_{0 \to f} \sin^2\left[\frac{1}{\hbar}\int_0^{t_x} (\varepsilon_f - \varepsilon_0 + V_f - V_0) \, dt\right]$$

$$\overline{P}_{0 \to f} = 2p_{0f}(1 - p_{0f}), \qquad p_{0f} = 1 - \exp(-\delta)$$

$$\delta = \left| \frac{2\pi}{\hbar} \Delta\varepsilon_{0f}^2 \middle/ \left(\frac{dR}{dt}\right) \frac{d}{dR}(V_f - V_0) \right|_{R=R_x}$$

$$\sigma_{0 \to f} \approx \pi R_x^2 \, \overline{P}_{0 \to f}$$

(continues)

TABLE III (*Continued*)

Landau–Zener–Stueckleberg (*continued*)

Useful form: $\Delta\varepsilon_{0f} \approx V_0(R/\bar{a}) \exp(-0.86R/\bar{a})$, $V_0 = (I_A I_B)^{1/2}$,

$$\bar{a} = a_0/(Z'_A + Z'_B), \qquad Z'_A \equiv \left(\frac{2a_0}{e^2} I_A\right)^{1/2}$$

Rosen–Zener (noncrossing)

$$P_{0\to f}(b) \approx \overline{P}_{0\to f} \sin^2\left[\frac{1}{\hbar}\int_{-\infty}^{\infty} \Delta\varepsilon_{0f} \, dt\right]$$

$$\overline{P}_{0\to f} = \frac{1}{2}\left|\frac{\int_{-\infty}^{\infty} \Delta\varepsilon_{0f} \exp(i\omega_{f0}t) \, dt}{\int_{-\infty}^{\infty} \Delta\varepsilon_{0f} \, dt}\right|^2$$

Demkov (exponential coupling, $\Delta\varepsilon_{0f} = A e^{-R/\bar{a}}$)

$$|\Delta\varepsilon_{0f}|_{R=R_x} = \frac{1}{2}|\varepsilon_f - \varepsilon_0 + V_f - V_0|_{R=R_x}, \qquad \text{defines } R_x$$

$$P_{0\to f}(b) \approx \overline{P}_{0\to f} \sin^2\left[\frac{1}{\hbar}\int_0^{t_x} (\varepsilon_f - \varepsilon_0 + V_f - V_0) \, dt\right]$$

$$\overline{P}_{0\to f}(b) \approx \frac{1}{2} \operatorname{sech}^2\left[\frac{\pi\bar{a}}{2\hbar}(\varepsilon_f - \varepsilon_0 + V_f - V_0)/\left(\frac{dR}{dt}\right)\right]_{R=R_x}$$

$$\sigma_{0\to f} \approx \pi R_x^2 \overline{P}_{0\to f}(b)$$

Resonant charge exchange

$v \lesssim \bar{v}_e$ (Firsov)

$$P_{ct}(b) = \sin^2\left[\frac{1}{2\hbar}\int_{-\infty}^{\infty} \Delta\varepsilon(R) \, dt\right]$$

$$\sigma_{ct} \approx \frac{1}{2}\pi b_x^2, \qquad \left[\frac{1}{2\hbar}\int_{-\infty}^{\infty} \Delta\varepsilon \, dt\right]_{b_x} \approx \pi^{-1}$$

$v \gg \bar{v}_e$ for (H$^+$ + H)

$$P_{ct} \to \frac{64\pi(b/a_0)^3}{(v/v_0)^7} \exp\left(-\frac{bv}{a_0 v_e}\right), \qquad \sigma_{ct} \propto v^{-12}$$

Langevin

$$\sigma_{0\to f} = \overline{P}_{0\to f}\pi b_0^2 \qquad [b_0 \text{ in Eq. (20)}]$$

For ion–molecule reaction:

$$b_0^2 = \frac{2}{v}\left(\frac{\alpha_B Z_A e^2}{m}\right)^{1/2}, \qquad \alpha_B = \text{polarizability}$$

Born approximation

$$\sigma_{0\to f}(\chi) = \frac{K_f}{K_0}\left[\frac{m}{2\pi\hbar^2}\int d^3R \, V_{f0}e^{i(\mathbf{K}_0 - \mathbf{K}_f)\cdot\mathbf{R}}\right]^2$$

using $V_{f0} = V_0 \exp(-\beta R)$

$$\sigma_{0\to f}(\chi) = \frac{K_f}{K_0}\left[\frac{4m\beta V_0}{\hbar^2}\right]^2 \frac{1}{[\beta^2 + |\mathbf{K}_0 - \mathbf{K}_f|^2]^4}$$

$$\sigma_{0\to f} = \frac{\pi}{3K_0^2}\left[\frac{4m\beta V_0}{\hbar^2}\right]^2 \left\{\frac{1}{[\beta^2 + (K_0 - K_f)^2]^3} - \frac{1}{[\beta^2 + (K_0 - K_f)^2]^3}\right\}$$

Approximate models for the two-state impact parameter cross section in Eq. (7) can be written in the form

$$\sigma_{0 \to f} \approx \overline{P}_{0 \to f} \pi b_x^2 \qquad (49)$$

where b_x is that impact parameter giving an onset for transitions and $\overline{P}_{0 \to f}$ is an averaged transition probability. For symmetric resonant charge exchange (e.g., $H^+ + H \to H + H^+$) the probability of charge exchange is an oscillatory function at low v (Fig. 12 and Table III). The average probability of a transition (exchange) is roughly $\frac{1}{2}$ as the initial and final states are identical (i.e., $\varepsilon_0 = \varepsilon_f$). As $\Delta\varepsilon(R)$ from Eq. (38) divided by $\hbar$ is roughly the rate of electron transfer, then at large R, when the time for sharing of an electron becomes longer than the collision time, the cross section goes to zero. Firov, therefore, estimated the size of the cross section by finding the largest value of b for which the number of transfers is small,

$$\frac{1}{2\hbar} \int_{-\infty}^{\infty} \Delta\varepsilon \; dt \approx \pi^{-1} \qquad (50a)$$

This is similar to the Massey–Mohr procedure used for estimating the total elastic cross section in Eq. (31). As the exchange energy at large R decreases exponentially, Eq. (50a) reduces to the well-known result

$$b_x = b_{ct} \approx \overline{a}(B - \ln v) \qquad (50b)$$

where B is a very slowly varying function of v

and $\overline{a}$ is the mean radius of the atomic electron cloud. Using Eq. (50b) with Eq. (49), it is seen that the charge exchange cross section increases slowly with decreasing velocity at intermediate velocities.

At very high velocities, exchange of an electron from a stationary atom to a fast ion requires a significant change in momentum. Therefore, $\overline{P}_{0 \to f}$ goes to zero rapidly ($\sim v^{-12}$ for $H^+ + H$). This begins to occur for $v > \overline{v}_e$. Therefore, at high velocities the ionization cross sections dominate over the charge transfer cross sections. At very low velocities the ion and neutral can orbit due to the long-range polarization potential. Orbiting occurs at an impact parameter determined in Eq. (20), which for the polarization interaction ($n = 4$) changes as $v^{-1/2}$. As this value of b increases more rapidly with decreasing v than b_{ct} in Eq. (50b), orbiting can dominate as $v \to 0$. A schematic diagram of the net cross section over many order of magnitude of v is given in Fig. 13.

It is seen in Fig. 13 that the cross section grows even at very low velocities when $\varepsilon_f = \varepsilon_0$. On the other hand, charge exchange between nonsymmetric systems (e.g., $S^+ + O$ or $H^+ + O$) requires a small change in energy. Therefore, the cross section will exhibit a maximum as discussed above and indicated in Fig. 13. When ΔE is much greater than the state separa-

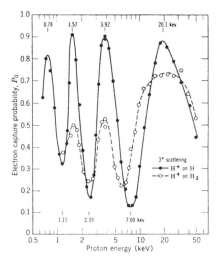

FIG. 12.   Charge-exchange probability at fixed scattering angle versus incident proton energy. Peaks nearly equally spaced in v^{-1}. [From G. J. Lockwood and E. Everhart (1962). *Phys. Rev.* **125**, 567.]

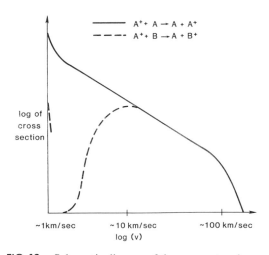

FIG. 13.   Schematic diagram of the resonant and nonresonant charge exchange cross sections: decrease at very high energy due to momentum required; cross section increases with decreasing velocity at very low energies due to orbiting, for nonresonant only if exothermic.

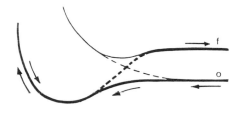

R → 0   R → ∞

FIG. 14. Effective potentials for the two trajectories leading to a transition. Dashed lines are approximate diabatic potential curves that cross as in Fig. 8. [From R. E. Johnson (1982). "Introduction to Atomic and Molecular Collisions," p. 139. Plenum Press, New York.]

tion*$|\varepsilon_f - \varepsilon_0|$, then the cross section behaves like the symmetric resonant collision [i.e., b_x given by Eq. (50) and $\overline{P}_{0\to f} \approx \frac{1}{2}$]. At velocities for which ΔE is comparable to or smaller than $|\varepsilon_f - \varepsilon_0|$, transitions generally occur in a narrow range of internuclear separations ΔR_x about a particular value R_x. In this case the transition probability is a rapidly varying function of v. Such transitions are divided into two classes, those for which a curve crossing exists (e.g., Fig. 8), and those for which the states do not cross. Therefore, knowledge of the potentials discussed earlier is required.

For the curve crossing case, b_x in Eq. (49) is set equal to the crossing point R_x. For impact parameters less than b_x, the colliding particles pass through the point R_x twice, as indicated in Fig. 14. At each passage we assign an average transition probability p_{0f} between the states. After the collision a *net* change in state will have occurred if a transition occurred on the first passage but not the second [$p_{0f}(1 - p_{0f})$] and vice versa [$(1 - p_{0f})p_{0f}$], yielding $\overline{P}_{0f} = 2p_{0f}(1 - p_{0f})$. Using this, Eq. (49) becomes

$$\sigma_{0f}^{LZS} \approx 2p_{0f}(1 - p_{0f})\pi R_x^2 \qquad (51)$$

The fact that two different pathways lead to the same result implies inference occurs; hence, the differential cross section is oscillatory. The Landau–Zener–Stueckleberg expression for p_{0f} is given in Table III. These expressions apply up to velocities where ΔE becomes greater than the energy splitting $|\varepsilon_f - \varepsilon_0|$, at which point the cross section behaves like the resonant case of Eq. (50). Figure 15 shows the results for three collision processes.

Demkov and Rosen and Zener have considered the case of the noncrossing interaction potentials, for example, $A^* + B \to A + B^*$. The transition region R_x is defined as that point at which the exchange interaction between the initial and final states [e.g., Eq. (38)] $\Delta\varepsilon_{0f}(R)$ is approximately equal to half the spacing between

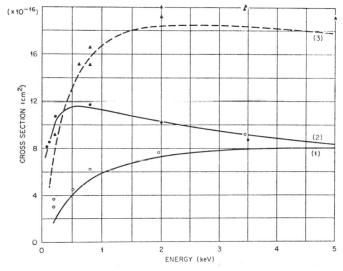

FIG. 15. Single-electron capture by doubly charged ions. Curves, LZS calculation; points are data: (1, ○) $Ar^{+2} + Ne \to Ar^+ + Ne^+$; (2, ●) $N^{+2} + He \to N^+ + He^+$; (3, △) $Ne^{+2} + Ne \to Ne^+ + Ne^+$. [From R. A. Mapleton (1972). *Theory of Charge Exchange*, p. 212. Wiley, New York.]

states $|\varepsilon_f - \varepsilon_0 + V_f - V_0|$, where V_f and V_0 are the interaction potentials for these states [e.g., Eq. (33)]. Demkov developed a very simple and useful expression for $\bar{P}_{0 \to f}$ given in Table III. The cross section so calculated would again join smoothly onto the symmetric-resonant-like result at higher velocities where the energy difference $|\varepsilon_f - \varepsilon_0|$ becomes unimportant.

At very low velocities, the cross sections for both the crossing and noncrossing cases may again increase *if the transitions are exothermic*. That is, even though the transition probability is small for any one pass through the transition region, if orbiting occurs [see Eq. (20)] then after many passes the cumulative effect can lead to a significant probability for a change in the electronic state as suggested in Fig. 13. As the collision energy is low this can only occur when internal energy is *released* due to an exothermic change in the internal state. In such a case $b_x \approx b_0$ of Eq. (20) and $\bar{P}_{0 \to f}$ in Eq. (49) is equal to the statistical probability of populating any of the exothermic states.

Although we have emphasized charge-exchange collisions and, further, only collisions involving ions and atoms, these procedures apply generally to molecular collisions and to all varieties of internal change in state including molecular reactions. For example, the above discussion of orbiting applies to low-energy ion–molecule reactions yielding the often-used Langevin cross section (see Table III). In all cases, including the processes discussed here, readers are urged to refer to the specific literature for accurate measured and calculated cross sections and use the approximations here as a guide or for rough estimates when more accurate results are not available.

F. DETAILED BALANCE

We briefly consider a property of the inelastic collision cross sections that can be quite usefully exploited in some cases. The equations of motion, both quantum-mechanical and classical, are such that

$$|f_{0 \to f}(\chi)|^2 = |f_{f \to 0}(\chi)|^2$$

$$\text{or} \qquad P_{0 \to f}(b) = P_{f \to 0}(b) \qquad (52a)$$

This is due to the time-reversal symmetry of the collision process. Using Eq. (45b) with Eq. (52a), the differential cross sections for the forward and reverse reactions can be related. Further, integrating over angle, the integrated in-

elastic cross sections are related by

$$p_0^2 \sigma_{0 \to f}(p_0) = p_f^2 \sigma_{f \to 0}(p_f) \qquad (52b)$$

where p_0 and p_f are the initial and final momenta. In the semiclassical region $p_0 \approx p_f$, and therefore the forward and reverse reactions have roughly the same cross section. That is, in this region endothermic and exothermic processes behave similarly. At low velocities, near threshold for the endothermic process, the forward and reverse reactions can differ markedly as pointed out when discussing orbiting in Section VI,E. However, these processes are simply related by Eq. (52b). This relationship is a statement of the principle of detailed balance used when describing equilibrium in statistical mechanics. Based on the notions of statistical mechanics, Eq. (52b) can be extended to cases where there are a number of equivalent initial states ξ_0 and/or final states ξ_f (e.g., spin or angular momentum states),

$$p_0^2 \xi_0 \sigma_{0 \to f}(p_0) = p_f^2 \xi_f \sigma_{f \to 0}(p_f)$$

Such a relationship allows one to determine, for instance, deexcitation cross sections from data on excitation cross sections and provides a constraint when calculating cross sections by approximate methods.

G. STOPPING CROSS SECTION SUMMARY

In the above we have considered the individual collisional processes that will determine the behavior of a fast particle penetrating a gas, liquid, or solid. The relative importance of these processes in determining the overall energy loss rate of the fast particle is illustrated in Fig. 16 for protons losing energy to H_2O. The quantity shown is the total stopping cross section $S \equiv n^{-1}(dE/dx)$ [see Eq. (11)]. First, it has been demonstrated that the molecular effects are small in the total energy loss process, so that S for H_2O is very nearly a sum of S for two H atoms and an O atom. In fact the most important difference would be that the ionization potential is lower for H_2O than for either H or O. Second, it is seen from Fig. 16 that the separate contributions, S_e and S_n, dominate the total stopping in very different velocity regions. At high velocities S_e and S_n decay as $[(\ln E_A)/E_A]$ as described above, but their magnitudes differ by the large difference in the mass of the electron versus that of the nucleus (i.e., m_e versus M_B in the denominators of the expressions for S_e and S_n; see Eq. (41b) and Table II).

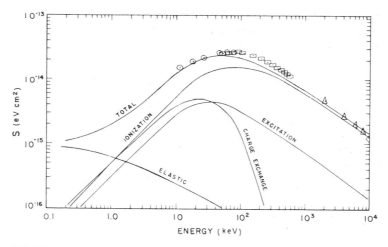

FIG. 16. Stopping cross section versus proton energy from protons incident on H_2O. Contribution from various processes (lines); experiment (points). [From J. H. Miller and A. E. S. Green (1973) *Radiat. Res.* **54**, 354.]

In the region where the electronic stopping power is large, it is seen in Fig. 16 that ionization dominates at high velocities, with a roughly 10% contribution from excitations, but charge exchange dominates at lower velocities. In this description, energy loss due to charge exchange is in fact a two-step process differing, therefore, from the other single-collision contributions shown. That is, on traversing a material, ions capture an electron, becoming neutral, and later ionize again. This process of capture and loss at low velocities can be described in the solid state as a drag force on the ion. This force comes about as the electron cloud is distorted (polarized) by the passing ion, thereby increasing the electron density in the vicinity of the ion (i.e., the covalent effect discussed earlier). The drag force produced by this distortion of the electron cloud changes roughly as $(dE/dx)_e \propto v$. Lindhard has given a very simple expression for the stopping power based on the Thomas–Fermi model of the atom that is accurate for large ions and atoms but is also reasonable for light ions and atoms for $v < \bar{v}_e$:

$$ S_e \approx \xi_e 8\pi e^2 a_0 \frac{Z_A Z_B}{Z} \left(\frac{v}{v_0}\right) \qquad \xi_e \approx Z_A^{1/6} $$

In this expression v_0 is the velocity of an electron on the hydrogen atom and $Z = (Z_A^{2/3} + Z_B^{2/3})^{3/2}$.

Because the expressions given in available tables are *measurements* of S, the charge-exchange cycle is always included as the materials

are macroscopically thick. This means that these tabulated values of S are *not* necessarily those of the initial charge state of the ion as it enters the solid. Because we imagine the incident ion to be in a variety of charge states during its passage through even a relatively thin material, these stopping cross sections are referred to as equilibrium-charge-state stopping cross sections. At low velocities the particle quickly becomes neutralized in the material, and at high velocities it eventually becomes fully stripped. The depth in the material at which the equilibrium charge state is achieved is related to the atomic density of the material and the average charge exchange and electron loss cross sections, all quantities we described above. This depth corresponds to hundreds of monolayers of material at high velocity and tens of monolayers at low velocities.

Transmission of ions through thin films allows us to estimate the equilibrium charge state Z_{eff}. A number of semiempirical expressions have been developed since Z_{eff} depends primarily on the ion speed. For example,

$$ Z_{eff} \approx Z_A (1 + y^{-1/0.6})^{-0.6} $$
$$ y \equiv 3.86\sqrt{E_A/M_A}/Z_A^{0.45} $$

with E_A in mega electron volts and M_A in atomic mass units. Such expressions for the effective charge can be used to estimate S_e by substituting Z_{eff} for Z_A in expressions for a bare incident ion, for example in Eq. (47). In fact, since I in Eq.

(47) depends primarily on the target properties, heavy ion values of S_e can be estimated from proton values of S_e by multiplying by the Z_{eff}^2 above.

BIBLIOGRAPHY

Bernstein, R. B. (1979). "Atom-Molecule Collision Theory." Plenum, New York.

Bates, D. R., and Bederson, B. (1965–1982). "Advances in Atomic and Molecular Physics," Vols. 1–18. Academic Press, New York.

Child, M. S. (1974). "Molecular Collision Theory." Academic Press, New York.

Hasted, J. B. (1972). "Physics of Atomic Collisions," 2nd ed. Am. Elsevier, New York.

Hirschfelder, J. O., Curtiss, F., and Bird, R. B. (1964). "Molecular Theory of Gases and Liquids." Wiley, New York.

Johnson, R. E. (1982). "Introduction to Atomic and Molecular Collisions." Plenum, New York.

McDowell, M. R. C., and Coleman, J. P. (1970). "Introduction to the Theory of Ion–Atom Collisions." North-Holland Publ., Amsterdam.

Massey, H. S. W. (1979). "Atomic and Molecular Collisions." Halsted Press, New York.

Massey, H. S. W., Burhop, E. H. S., and Gilbody, H. G. (1970–1974). "Electronic Ionic Impact Phenomena," 2nd ed., Vol. 1–5. Oxford Univ. Press, London and New York.

Mott, N. F., and Massey, H. S. W. (1965). "The Theory of Atomic Collisions," 3rd ed. Oxford Univ. Press, London and New York.

ATOMIC PHYSICS

Francis M. Pipkin   *Harvard University*

I.   Historical Development
II. Constituents of the Atom
III. The Hydrogen Atom
IV. The Helium Atom
V.   Complex Atoms
VI. Interaction of Atoms with Radiation
VII. Atomic Collisions

GLOSSARY

Atomic number: Positive charge on the nucleus of an atom in units of the electron charge. It is given the symbol Z.

Atomic orbital: Wave function describing the distribution of an electron in one quantum state of an atom.

Bohr magneton: Natural unit in which to measure the magnetic moment of an electron or atom. It is equal to $e\hbar/2m$. It is represented by the symbol μ_B.

Electric dipole radiation: Principal form in which electromagnetic radiation is emitted by atoms. It requires a change in parity and $\Delta j = 0, \pm 1$ where j is the angular momentum.

Electron volt: Energy acquired by an electron when it is accelerated by a potential difference of 1 V. It is represented by the symbol eV and is equal to 1.602×10^{-19} J.

Fine structure constant: Measure of the strength of the interaction between a charged particle and the electromagnetic field. It is given by $\alpha = e^2/(4\pi\varepsilon_0 c\hbar)$ and has a value of 1/137.03604.

g-Value: Ratio of the magnetic moment of an atom or an electron to the product of the largest component of the angular momentum and the Bohr magneton. For nuclei, the nuclear magneton is used instead of the Bohr magneton.

Hartree: Unit of energy for atomic units. It is $2hcR_\infty$ and is equal to 27.211608 eV.

Ionization potential: Smallest energy required to remove a bound electron from an atom.

Isotope: Elements with the same atomic number Z but with different mass number A. They have the same number of protons in the nucleus but different numbers of neutrons.

j–j Coupling: Angular momentum coupling scheme for atoms in which for each electron the spin angular momentum is tightly coupled to its own orbital angular momentum. The total angular momentum is obtained by first adding for each electron its spin angular momentum and orbital angular momentum and then adding the angular momenta of the individual electrons.

Mass number: Total number of protons and neutrons in the nucleus of an atom. It is given the symbol A.

Nucleus: Heavy central part of an atom, containing most of the mass and all of the positive charge. It is composed of neutrons and protons.

Neutron: Neutral elementary particle with approximately the same mass as the hydrogen atom. It is one of the constituents of the atomic nucleus.

Parity: Behavior of the wave function when the coordinate system is inverted and for each electron (x_i, y_i, z_i) is replaced by $(-x_i, -y_i, -z_i)$. The parity is even or odd depending on whether or not the wave function changes sign.

Russell Saunders coupling: Angular momentum coupling scheme for an atom in which the spin and orbital angular momentum of individual electrons are independently added together to form the total spin and total orbital angular momentum and then these angular momenta are added together to obtain the total angular momentum.

Spontaneous emission: Radiation of electromagnetic energy by an atom when there are no photons present.

Stimulated emission: Radiation of electromagnetic energy by an atom caused by the presence of photons.

Atomic physics is that branch of physics that deals with the structure of atoms, with the structure of ions formed by addition or removal of electrons from neutral atoms, with the interactions of atoms and ions with one another, and with their interaction with the electromagnetic field and free electrons.

I. Historical Development

The idea that all matter consists of indivisible particles called atoms can be traced to the third to fifth century B.C. Greek philosophers Leusippus, Democritus, and Epicurus, who speculated that everything was composed of indivisible particles of a small number of simpler substances in different combinations and proportions. The first century B.C. poem entitled De Rerum Natura by the Latin poet Lucretius provides an elegant summary of early atomism. This idea was attractive and provided encouragement for the alchemists in the Middle Ages who were trying to convert cheaper metals such as iron and lead to the noble metals silver and gold by changing the relative proportions of a few basic constituents.

The modern theory of atoms was founded by the English chemist John Dalton at the beginning of the nineteenth century. Dalton realized that when the chemical elements were combined to form compounds, the weights with which they combined always occurred in well-defined proportions. A good example is provided by the combination of nitrogen and oxygen to form the several oxides of nitrogen. Fourteen parts by weight of nitrogen combine with 8, 16, 24, 32, or 40 parts by weight of oxygen to form the five common oxides of nitrogen. Dalton interpreted these proportions in terms of the combination of different numbers of atoms, each of which has a fixed weight. He was thus able to understand the oxides of nitrogen in terms of the combination of nitrogen atoms (N) with a weight of 14 units with oxygen atoms (O) with a weight of 16 units in the proportions N_2O, NO, N_2O_3, NO_2, and N_2O_5, where the subscripts denote the number of atoms of each species in the smallest units of the compounds of nitrogen and oxygen. Dalton had no way of knowing the weight or size of one atom, but this gave him a way of knowing the ratio of the weights of the atoms. Through a study of the available chemical reactions, chemists were able to develop a table of combining weights for the elements.

Studies of gases by Amedeo Avogadro and others showed that common gases such as hydrogen, nitrogen, and oxygen consisted of molecules each of which was composed of two atoms, and that equal volumes of gases at the same temperature and pressure contained the same number of molecules. Rare gases such as helium, neon, and argon, which are chemically inert and occur in the gas phase as single atoms, were not clearly identified until the last decade of the nineteenth century.

In the first few decades of the nineteenth century, when scientists began to study the behavior of solutions when an electric current was passed through them, they realized that in solutions atoms could become positively or negatively charged. These positively or negatively charged atoms were called ions. In 1833 Michael Faraday found that the passage of a fixed quantity of electricity through a solution containing a compound of hydrogen would always cause the appearance of the same amount of hydrogen gas at the negative terminal, irrespective of the kind of hydrogen compound that had been dissolved and irrespective of the strength of the solution. More generally, it was found that the weights of the material given off or deposited at both the positive and negative terminals by the same quantity of electricity were proportional to the chemical combining weights. The electrical behavior of ions in solutions lead to the speculation that all electrical currents were formed by the motion of charged particles. The measurements on solutions were used to determine the ratio of the charge to the mass (e/m) for the ions in solution. The largest ratio was obtained for the positive hydrogen ion.

In the middle decades of the nineteenth century, physicists developed kinetic theory. This theory describes the macroscopic behavior of gases in terms of the behavior of independent atoms and molecules that move according to Newton's laws of motion and occasionally collide elastically with one another. The temperature of the gas is related to the average kinetic energy of the constituents; the pressure is related to the number of constituents per second striking the walls of the container. For a typical gas at room temperature, the average velocity is 500 m/sec. The average distance a constituent travels before it strikes another constituent is called the mean free path and it depends upon the diameter of the constituents and the number of constituents per unit volume. Kinetic theory relates the viscosity, the rate of diffusion, and the thermal conductivity of gases to the diameter and average velocity of the constituents.

It was also discovered that when atoms were excited in a flame or an electrical discharge, they emitted light at discrete wavelengths or frequencies that were characteristic of the particular elements. The observed frequencies did not display the characteristic harmonic structure found in mechanical vibrations. All attempts to understand the pattern of observed frequencies in terms of vibrating mechanical structures failed. In 1885 a Swiss school teacher, Johann Jacob Balmer, discovered that the frequencies of the known spectrum lines of hydrogen could be represented by a simple empirical formula, which with hindsight can be written in the form

$$\nu = Rc \left(\frac{1}{2^2} - \frac{1}{n^2} \right)$$

where ν is the frequency, c the speed of light, R an empirical constant, and $n = 3, 4, 5, \ldots, 16$. The constant R is now called the Rydberg constant in honor of J. R. Rydberg, a Swedish physicist who made major contributions to our understanding of the spectra of atoms.

In the last third of the nineteenth century, physicists studying the conduction of gases in discharge tubes at low pressure identified cathode rays, which moved toward the negative terminal (cathode), and canal rays, which moved toward the positive terminal (anode). The two components were studied by using cathodes and anodes with holes in them to prepare beams of the rays. Initially it was not clear whether they were waves or particles.

In 1897, Joseph John Thomson, an English physicist, showed that the cathode rays were negative particles whose properties did not depend on the nature of the gas. They were called electrons—a name first suggested in 1891 by G. J. Stoney for the natural unit of electricity. Thomson found that the charge-to-mass ratio of the electron was roughly 1836 times larger than that found for the hydrogen ion in solutions.

In 1898, Wilhelm Wien showed that the canal rays have ratios of charge to mass that depend on the gas and are similar to those found for ions in solution. It was later shown that for dis-

charges in hydrogen gas there are canal rays with the same charge-to-mass ratio as that found for the hydrogen ion in solutions. This particle, the hydrogen ion, is now called the proton. The study of canal rays led Thomson to the development of mass-spectrometric methods for measuring precisely the masses of the elements.

The discovery of the electron lead to a model for the atom as a structure made up of positively charged protons and negatively charged electrons. The charge of the electron was subsequently measured by observing the influence of an electric field on the motion of droplets that were charged due to the removal or addition of one or more electrons. The most extensive set of measurements was made by Robert A. Millikan, using oil drops. Millikan's experiments also showed that the charge of the proton had the same magnitude as the charge of the electron.

In 1900 Max Planck, a German physicist, found that he could predict the distribution of electromagnetic radiation emitted by a black body—a body that absorbs all the radiation incident upon it—by postulating that the electromagnetic radiation was emitted and absorbed in characteristic units called quanta whose energy E is related to the frequency ν of the electromagnetic radiation by the equation

$$E = h\nu$$

where h is the Planck constant. Einstein later used the Planck hypotheses to explain the manner in which light caused metals such as sodium to emit electrons.

In 1911, Ernest Rutherford and his co-workers discovered through experiments in which alpha particles were scattered from thin gold foils that all the positive charge in an atom was concentrated in a very small region of space much smaller than the size of the atom determined from studies of gases, liquids, and solids. This led to the picture of the atom as a miniature solar system consisting of a heavy positively charged nucleus about which a group of electrons moved. Like the solar system, most of the atom was empty space.

Niels Bohr, a Danish physicist, subsequently combined the insight provided by the Rutherford experiment with the quantum hypothesis of Planck and Einstein to make the first quantitative model of the atom. Bohr pictured the hydrogen atom as a single positively charged proton with a size of roughly 10^{-15} m with the electron moving around the proton in a circular orbit with a diameter of 10^{-10} m. To determine the structure, Bohr introduced three new hypotheses. He assumed that not all circular orbits for the electron were possible but that the only allowed orbits were those for which the orbital angular momentum of the electron was an integer multiple of the Planck constant divided by 2π. The orbital angular momentum of an object is a measure of its rotational inertia and for a particle in a circular orbit is equal to the product of the mass, the speed, and the radius of the orbit. It is represented by a vector perpendicular to the plane of the orbit with a magnitude equal to mvr. Bohr also assumed, in contradiction to the predictions of classical electromagnetic theory, that while the electron was moving around the proton in one of the allowed orbits, its total energy was constant and it did not radiate electromagnetic energy. Bohr postulated instead that the electron radiated electromagnetic energy only when it passed from one allowed circular orbit to a second allowed circular orbit with lower energy, and that the frequency of the radiation was given by the equation

$$\nu = \frac{E_1 - E_2}{h}$$

Here E_1 and E_2 are the total energies when the electron is in each of the allowed orbits and h is the Planck constant. By using classical mechanics to calculate the total energy in each of the orbits, Bohr obtained for the frequency of the radiation the formula

$$\nu = (2\pi^2 m_e/h^3)(e^2/4\pi\varepsilon_0)^2(1/n_2^2 - 1/n_1^2)$$
$$= Rc(1/n_2^2 - 1/n_1^2)$$

where m_e is the mass of the electron, e the charge of the electron, ε_0 the permittivity of empty space, R the Rydberg constant, and n_1 and n_2 are integers that characterize the angular momentum of the two allowed orbits. This expression is the same as that given by the Balmer formula, with the Rydberg constant given in terms of independently known quantities. This theory was very successful and predicted not only the Balmer formula but three other spectral series in hydrogen that were independently observed and the spectral series for ionized helium. The theory was subsequently refined by Sommerfeld and Wilson in order to treat properly the three-dimensional nature of the problem and to take account of special relativity.

The refinements of the theory showed that the angular momentum of the atom was space-quantized. The angular momentum for an electron

moving about the nucleus can be represented as a vector perpendicular to the plane of the orbit. Classically, the angular momentum vector points in a fixed direction in an inertial coordinate system, and it can assume any orientation with respect to the reference axis. The quantization of the angular momentum introduced in the Bohr theory predicts that the angular momentum vector can only assume a small number of orientations with respect to the reference axis. This quantization of the angular momentum was confirmed by Stern and Gerlach through experiments in which an inhomogeneous magnetic field was used to deflect a beam of silver atoms. They found that the deflection of the beam were quantized. [See QUANTUM MECHANICS.]

In 1924 Louis de Broglie introduced the hypothesis that the electron behaved like a wave with a wavelength λ that depended on its momentum p through the relationship

$$\lambda = \frac{h}{p}$$

De Broglie found that he could understand the Bohr orbits in terms of circles, each of whose circumference was an integral number of wavelengths. This hypothesis lead Erwin Schroedinger to the invention in 1926 of a partial differential equation, now called the Schroedinger equation, for the description of the wave behavior of the electron. Schroedinger found that those solutions for which the amplitude of the wave function remained large around the nucleus for all time predicted for the electron a discrete set of possible energies, which were the same as those predicted by the Bohr theory. The solutions of the Schroedinger equation are now called the wave functions for the electron. The value of the wave function at any point is in general a complex number, and its absolute value squared gives the relative probability of the electron being found at a given point in space and time if a measurement is made.

More precise measurements of the radiation emitted by atoms showed that many of the spectral lines were not one line but several lines close together. The additional structure is called the fine structure. It was found that this structure could be understood if the electron were not a simple pointlike particle but a small spinning sphere with an angular momentum that could only have two directions with respect to a given reference axis. The electron could either have a component of angular momentum $\frac{1}{2}(h/2\pi)$ or $-\frac{1}{2}(h/2\pi)$ along the reference axis. This internal angular momentum is now referred to as the spin angular momentum and an electron is said to have spin $\frac{1}{2}$. Since the electron is charged, the rotation produces a current, and this current results in a magnetic moment. The electron magnetic moment is twice as large as one would expect classically for a rotating uniformily charged sphere. The moment is said to be anomalous. The interaction of the magnetic moment with the magnetic field in the rest frame of the electron—produced by the motion of the electron in the electric field of the nucleus—causes the fine structure of the spectral lines. The energy of the electron is different when the spin points in the same direction as the orbital angular momentum than when it points in the opposite direction.

In 1928, P. A. M. Dirac discovered a generalization of the Schroedinger equation that takes proper account of special relativity. Dirac found that this equation required that the electron have spin $\frac{1}{2}$ and predicted that the magnetic moment would be exactly twice as large as one would expect classically for a rotating uniformily charged sphere. This theory also predicted that there was a positive partner to the electron that had the same mass and spin angular momentum. This particle, which is now called the positron, was discovered in cosmic rays by Carl Anderson in 1932. The Dirac theory was a major triumph for relativistic quantum mechanics.

The other characteristic property of the electron is that no two electrons can occupy the same quantum state. This was first hypothesized by Wolfgang Pauli and is now called the Pauli exclusion principle. It is a characteristic of all particles with intrinsic spin $\frac{1}{2}$. The Pauli principle is the major ingredient for explaining the periodic chart of the elements.

The alpha-particle scattering experiments of Rutherford and his collaborators and the measurements of the characteristic X rays of the elements by Moseley, which were carried out in the early part of the twentieth century, showed that each element could be characterized by the charge Z of the nucleus and that the number of electrons in the outer orbits was equal to the charge of the nucleus. Measurements of the mass of the nuclei, using mass spectrometers invented by Thomson and refined by Aston, Dempster, and others, showed that for each element the mass A of the nucleus was roughly twice the mass of the protons in the nucleus and that there were generally several nuclei with the same Z and a different total mass. Nuclei with the same Z and different masses are called isotopes. It was first hypothesized that the nucleus contained electrons that compensated the charge from some of the protons. This hypothesis was contradicted by quantum mechanics and by the total angular momentum observed experimentally for some nuclei. It was speculated by Rutherford that there was a heavy neutral particle called the neutron. In 1932 James Chadwick discovered the neutron and showed that it had a mass that was very slightly greater than that of the proton. [*See* NUCLEAR PHYSICS.]

II. Constituents of the Atom

The three basic constituents required to understand the structure of atoms are the proton, the neutron, and the electron. The properties of these three particles are summarized in Table I. The nucleus of a typical atom consists of Z protons and N neutrons; it is assigned the atomic number Z and mass number $A = Z + N$. In general, for any given Z there are several isotopes with different values of N and A. To a good approximation the nucleus is a spherical object with a radius equal to $1.2A^{1/3} \times 10^{-15}$ m. The electrons occupy the region around the nucleus and are contained in a sphere with a diameter of $(1-3) \times 10^{-10}$ m, which depends on Z.

TABLE I. Properties of the Constituents of the Atom

Property	Measured value
Electron	
Spin	$\frac{1}{2}$
Charge	$-1.6021892(46) \times 10^{-19}$ C
Mass	$9.109534(47) \times 10^{-31}$ kg
Magnetic moment	$9.284851(65) \times 10^{-24}$ J/T
Proton	
Spin	$\frac{1}{2}$
Charge	$1.6021892(46) \times 10^{-19}$ C
Mass	$1.672614(11) \times 10^{-27}$ kg
Magnetic moment	$1.4106171(30) \times 10^{-26}$ J/T
Neutron	
Spin	$\frac{1}{2}$
Charge	0
Mass	$1.674920(11) \times 10^{-27}$ kg
Magnetic moment	$-0.9662437(17) \times 10^{-26}$ J/T

III. The Hydrogen Atom

The simplest atom is the hydrogen atom. It consists of a single proton and electron. Figure 1 shows the energy levels for the hydrogen atom predicted by the Schroedinger equation with the addition of the electron magnetic moment and the inclusion of special relativity to second order. The energy levels are characterized by the energy E, the principal quantum number n, the orbital angular momentum l, the spin angular momentum s, and the total angular momentum j. It is convenient to express the angular momentum in units of $\hbar$ where $\hbar = h/(2\pi)$; we will use this unit throughout the rest of this article. The wave function for a particular state is sometimes referred to as the atomic orbital for that state. Lower-case letters are used to designate the orbitals for a single electron, and upper-case letters are used to designate the angular momentum states of the whole atom, which may in general be due to the sum of the orbital and spin angular momenta of several electrons. The following notation is used to characterize the angular momentum of each quantum state:

$$^{2S+1}L_J$$

L is the orbital angular momentum, and it has the possible values 0, 1, 2, 3, ... It is customary to use the letters S, P, D, F, G, etc. to designate the respective angular momenta. For each angular momentum L there are $2L + 1$ substrates, which are characterized by the projection M of the angular momentum along the z axis. For angular momentum L, M can assume the values

$$-L, -(L - 1), \cdots, 0, \cdots, L - 1, L$$

In the absence of a magnetic field, these levels all have the same energy and are said to be degenerate. Quantum-mechanically, the angular momentum can be pictured as a vector of length $\sqrt{l(l + 1)}$, which has a component m along the z axis and lies in a cone with its axis along the z axis. The relationship of the classical angular momentum and the quantum-mechanical representation is shown in Fig. 2. The cone represents the equally probable orientations of the angular momentum, and the angular momentum vector is at rest at an indeterminate position in the conical surface. This is shown graphically for $l = 2$ in Fig. 3. For a single electron, the spin angular momentum has a value of $\frac{1}{2}$. The quantum-mechanical magnitude is $\sqrt{3}/2$ and it can have a component of $\frac{1}{2}$ or $-\frac{1}{2}$ along the z axis. In general the total spin S is the vector sum of the spins of several electrons. For each spin S there are $2S + 1$ sublevels. The multiplicity $2S + 1$ is written as a superscript in front of the letter designating the orbital angular momentum. To obtain the total angular momentum, one adds quantum-mechanically the spin angular momen-

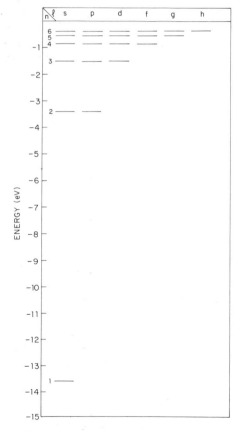

FIG. 1.   The energy levels of the hydrogen atom.

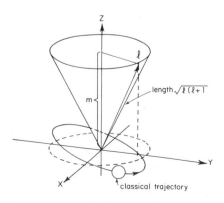

FIG. 2.   The relationship between the classical angular momentum and the quantum-mechanical representation.

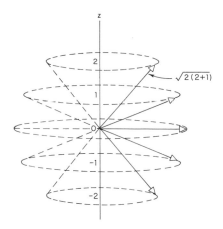

FIG. 3. The allowed orientations for angular momentum $l = 2$.

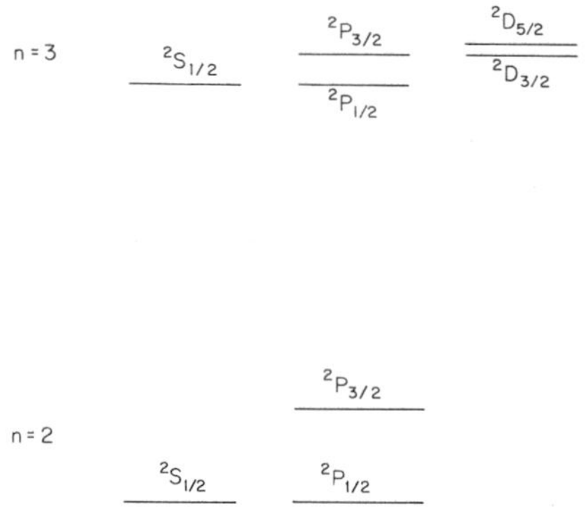

FIG. 4. The energy levels for the $n = 2$ and $n = 3$ states of hydrogen predicted by the Dirac theory and the Schroedinger theory with inclusion of the electron magnetic moment and special relativity to $(v/c)^4$.

tum and the orbital angular momentum. That is,

$$\mathbf{J} = \mathbf{L} + \mathbf{S}$$

The possible values for the total angular momentum are

$$|L - S|, \ldots, |L + S|$$

The total angular momentum can have both half-integral and integral values. For an electron in the 2p state, $L = 1$, $S = \frac{1}{2}$, $2S + 1 = 2$, and

$$1 + \frac{1}{2} = \frac{3}{2}$$
$$\nearrow$$
$$J$$
$$\searrow$$
$$1 - \frac{1}{2} = \frac{1}{2}$$

Thus the 2p electron gives rise to the $(2p)^2P_{1/2}$ and $(2p)^2P_{3/2}$ states. Due to the interaction of the magnetic moment of the electron with the magnetic field produced by the motion of the electron about the proton, the energy of the $^2P_{1/2}$ and $^2P_{3/2}$ states will be different. The resultant energy levels for the $n = 2$ and $n = 3$ states of hydrogen are shown in Fig. 4. The degeneracy of the states with the same total angular momentum is true for both the Schroedinger theory and the Dirac theory. It is due to the inverse square law behavior of the coulomb force law.

The wave function of an atom can also be characterized by its parity. The parity is defined by the change in the sign of the wave function when the coordinate system is inverted and (x, y, z) is replaced by $(-x, -y, -z)$. That is,

$$\psi(-x, -y, -z) = \pm\psi(x, y, z)$$

If the multiplying factor is $+1$ or -1, respectively, the parity of the wave function is said to be even or odd. For a given angular momentum

state, all the substrates have the same parity; the hydrogen atom states with even angular momentum have even parity, and states with odd angular momentum have odd parity.

Higher-order electrodynamic corrections cause a decrease of the effective coulomb force when the electron is near the proton and result in a smaller binding energy for the S states. This shift in the energy levels is called the Lamb shift. It is shown for the $n = 2$ and $n = 3$ levels in Fig. 5.

The square of the absolute value of the normalized wave function for each of the quantum states gives the probability per unit volume for finding the electron at each point in space. Figure 6 shows the spatial probability for the 1s, 2s, $2p_0$, $2p_{+1}$, and $2p_{-1}$ orbitals, where the subscript

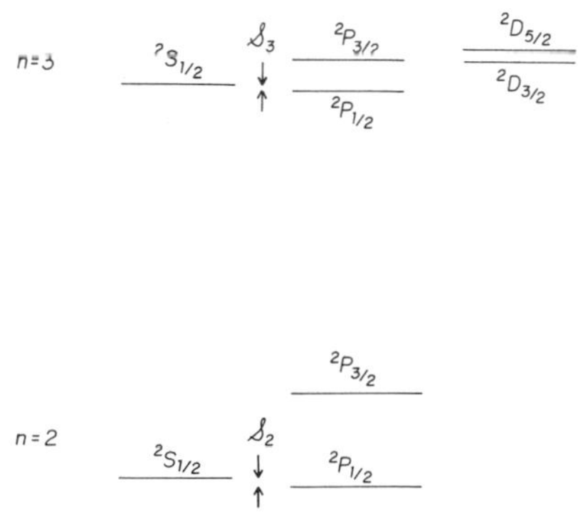

FIG. 5. The energy levels for the $n = 2$ and $n = 3$ states of hydrogen with the inclusion of the Lamb shift. The Lamb shift also makes the binding energy for the $3^2P_{3/2}$ level slightly smaller than that for the $3^2D_{3/2}$ level, so these energy levels are no longer degenerate.

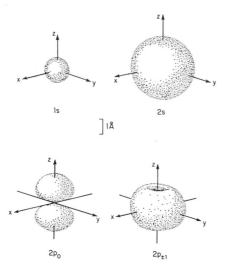

FIG. 6. Boundary surface plots of the probability density for the 1s, 2s, $2p_0$, and $2p_{\pm1}$ states of the hydrogen atom. The boundary surface excludes all space in which $|\psi|^2$ is less than one-tenth of its maximum value. [Adapted from Frank J. Bockoff, F. J. (1969). Elements of Quantum Theory." Addison-Wesley, Reading, Mass.]

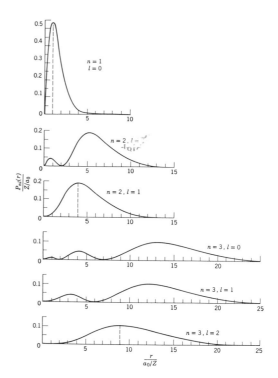

FIG. 7. The radial probability density for the electron in a one-electron atom for $n = 1$, 2, 3 and all the possible values of l. The dashed lines shows the radius of the corresponding circular orbit in the Bohr theory. [Adapted from Eisberg, R., and Resnick, R. (1985). "Quantum Physics of Atoms, Molecules, Solids, Nuclei, and Particles." John Wiley and Sons, New York.]

denotes the z component of the angular momentum. The s orbitals are spherically symmetric with the greatest probability for the electron to be close to the nucleus. The p orbitals have an anisotropic distribution. The distributions for the p_{+1} and p_{-1} orbitals are consistent with what one would expect for an electron moving in a circular orbit in the x–y plane taking into account the fact that the angular momentum vector lies in a cone with a half-angle of 45°. The distribution for the p_0 orbital is that expected for an electron that is in a circular orbit in a plane containing the z axis with all possible orientations of the plane equally probable. The radial probability density P_{nl} for the electron in the nl atomic orbital to be in a spherical shell with radius between r and $r + dr$ is given by integrating

the probability per unit volume over the volume enclosed between spheres of radii r and $r + dr$. Figure 7 shows the radial probability density for the $n = 1$, 2, and 3 states of hydrogen. The dashed line shows the radius of the corresponding orbit in the Bohr theory. In dealing with atoms it is advantageous to use atomic units that are based on the dimensions and energy of the hydrogen atom. Table II summarizes the atomic units.

TABLE II. Atomic Units

Quantity	Description	Symbol	Value
Fine structure constant	$e^2/4\pi\varepsilon_0 ch$	α	$1/137.03604(11)$
Mass	Mass of the electron	m_e	$9.109534(47) \times 10^{-31}$ kg
Length	Radius of hydrogen atom	a_0	$0.52917706(44) \times 10^{-10}$ m
Energy	Hartree = $2hcR_\infty$	H	$27.211608(72)$ eV
Velocity	Velocity in first Bohr orbit	αc	2.187691×10^6 m/sec
Magnetic moment	Bohr magneton = $e\hbar/2m$	μ_B	$9.2740775(152) \times 10^{-24}$ J/T

IV. The Helium Atom

The nucleus of the helium atom consists of two protons and two neutrons. It is sometimes referred to as the alpha particle and was first observed in the decay of heavy radioactive nuclei. The energy levels for He$^+$, which is a helium nucleus with only one bound electron, are similar to those for hydrogen, with a larger binding energy due to the higher Z for the nucleus. To a good approximation the energy levels are given by

$$E_{nl} = RcZ^2/n^2$$

To obtain the energy levels for He one adds a second electron to the He$^+$ ion. Since the second electron cannot be in the same quantum state and thus have the same quantum numbers as the first electron, it must be either in a different orbital state or in a different spin state. Since the spin has little effect upon the energy levels, a lower energy is obtained by putting the second electron in the same orbital state as the first electron but with a different value for the projection of the spin angular momentum along the z axis. Thus the quantum numbers (n, l, m_l, m_s) for the first electron will be $(1, 0, 0, \frac{1}{2})$ and the quantum numbers for the second electron will be $(1, 0, 0, -\frac{1}{2})$. The z component of the total spin angular momentum, which is the sum of the z components of the spin angular momenta of the two electrons, will be zero; it can be shown that the total spin angular momentum will also be zero. The binding energy for the second electron will be smaller than that for the first electron because the first electron tends to shield the nucleus and thus reduce the effective coloumb field seen by the second electron.

In building up the states for the helium atom, we spoke of the electrons as if they were distinguishable and there were a first electron and a second electron. In reality the electrons are indistinguishable and we cannot determine which is the first electron and which is the second. The Pauli exclusion principle, which requires that no two electrons occupy the same quantum state, can be reformulated in terms of a requirement that the wave function describing the electrons in any quantum-mechanical system be antisymmetric with respect to the exchange of any two electrons. The corresponding wave function for the ground state of helium is

$$\psi(\mathbf{r}_1, \mathbf{r}_2) = \psi_{10}(\mathbf{r}_1)\psi_{10}(\mathbf{r}_2)\sqrt{\tfrac{1}{2}}[\chi_1(\tfrac{1}{2})\chi_2(-\tfrac{1}{2})$$
$$- \chi_1(-\tfrac{1}{2})\chi_2(\tfrac{1}{2})]$$

Here $\psi_{10}(\mathbf{r}_i)$ is the orbital wave function of the ith electron in the ground state and $\chi_i(m_s)$ is the spin wave function for the ith electron with component m_s along the z axis. This wave function retains the indistinguishability of the two electrons. It says that the electrons occupy the same orbital state and that one electron has spin component $+\frac{1}{2}$ along the z axis and the other electron has spin component $-\frac{1}{2}$ along the z axis. It does not say which electron has which spin component.

To obtain the first excited state of helium one can add the second electron in either a 2s or 2p state of the helium ion with the spin vectors of the two electrons adding to either 0 or 1. Since the two electrons now occupy different orbital states, there is no restriction on the spin states: There are four possible states for the helium atom:

$$(1s)(2s)^1S \qquad (1s)(2s)^3S$$
$$(1s)(2p)^1P \qquad (1s)(2p)^3P$$

where the S and P refer, respectively, to the total orbital angular momentum of the two electrons and the leading superscript 1 or 3 uses the multiplicity $2S + 1$ to designate the total spin. The states are referred to, respectively, as singlet and triplet states. For each of these states one can further add the spin and orbital angular momenta to obtain the total angular momentum. The complete list of possible states is

$$(1s)(2s)^1S_0 \qquad (1s)(2s)^3S_1$$
$$(1s)(2p)^1P_1 \qquad (1s)(2p)^3P_0, {}^3P_1, {}^3P_2$$

For the singlet states the spin function is antisymmetric under the interchange of the two electrons and the orbital wave function is symmetric; for the triplet states the spin wave function is symmetric under the interchange of the two electrons and the orbital wave function is antisymmetric. The net result is that in the singlet state the electrons are in general closer together and the coulomb repulsion is larger. This results in a smaller binding energy for the singlet states. The electrons act as if they were subject to a force that depends on the relative orientation of the spins. This quantum-mechanical effect is sometimes referred to as due to the exchange force; it has no classical analog. The electrons are said to be correlated, and the difference in energy of the singlet and triplet states is due to the difference in the coulomb repulsion between the two electrons due to the electron correlation.

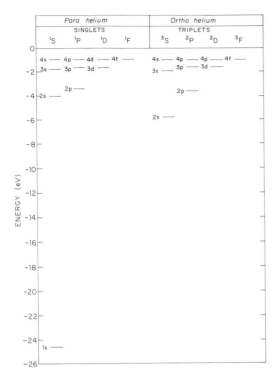

FIG. 8.   The energy levels for the helium atom.

It requires a detailed calculation to predict the order of the energy levels. Figure 8 shows the observed energy-level diagram for helium. The parity of the wave function is determined by inverting the coordinates of both electrons.

The interaction that causes a helium atom in an excited state to decay through the emission of photons is due primarily to the interaction between the electromagnetic field and the charge of the electron. The interaction of the electromagnetic field with the spins is much smaller. The net result is that there is little coupling between the singlet and triplet states. Helium acts as if there were two different species, one with spin 0 and one with spin 1. They are sometimes referred to, respectively, as para- and orthohelium. A similar behavior is observed for molecular hydrogen, which has energy levels very similar to those for helium.

A further result of this weak coupling is that the $(1s)(2s)^3S_1$ state of helium has a very small probability for decay. It is said to be metastable. In the absence of external perturbations, it decays by single-photon magnetic dipole radiation with a lifetime of 0.841×10^4 sec. This is to be compared with a lifetime of 20 msec for the $(1s)(2s)^1S_0$ state, which decays by the emission of two electric dipole photons.

V. Complex Atoms

The scheme used to construct the energy levels of helium can be generalized and used to construct the energy levels of more complex atoms with many electrons. One adds electrons one at a time, with each electron placed in the unoccupied orbital with the lowest energy in accord with the Pauli exclusion principle. To first order it can be assumed that the electrons are independent and that the order for filling the levels is the same as that for hydrogen. When all the orbitals for a given angular momentum with different z component of angular momentum are filled one obtains a spherically symmetric configuration with total angular momentum equal to zero, which is referred to as a closed shell. Helium, which has two electrons in the 1s orbital, is the first atom with a closed shell. The ground state of helium has total orbital angular momentum equal to zero, total spin angular momentum equal to zero, and total angular momentum equal to zero. The second closed shell is for two electrons in the 2s orbital; this is the beryllium atom. The third closed shell is for six electrons in the 2p orbital; this is the neon atom. The complete configuration of neon is $(1s)^2(2s)^2(2p)^6$. The closed shells that correspond to the filling of all the levels with the same principle quantum number in hydrogen are major closed shells that correspond to the rare gases. This accounts for their inertness and failure to be chemically active.

As Z increases, the order of filling is modified from that for the energy levels of hydrogen due to the coulomb interaction between the electrons. Figure 9 shows the empirical order of filling of the levels. Table III gives the ground-state electronic configuration and first ionization potentials for all the elements. The first ionization potential is the energy required to remove the most weakly bound ground-state electron.

It has not been possible to solve exactly the Schroedinger equation for a many-electron system. Techniques such as the Hartree consistent field method have been developed to take into account the interaction between the electrons. The Hartree method uses a wave function that is the product of single-particle wave functions. A guess is made of the wave function for each electron, and these wave functions are used to calculate the electrostatic potential due to the charge distribution of the electrons. These potentials are averaged over angles and summed over all the electrons but one. This summed po-

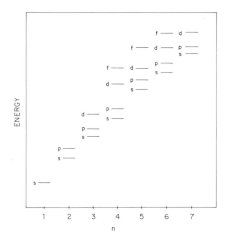

FIG. 9. The order of filling of the levels for many-electron atoms.

tential is used in the Schroedinger equation for the remaining electron to solve for the wavefunction for that electron. This procedure is repeated for each electron, and a new wave function for each electon is obtained. The calculation is iterated until there is no change in the wavefunctions. Figure 10 shows the calculated wavefunctions for the argon atom. Figure 11 shows the total radial probability distribution and the average charge $Z(r)$ when the effective potential is expressed in the form

$$V(r) = -Z(r)e^2/4\pi\varepsilon_0 r$$

A more elaborate procedure called the Hartree–

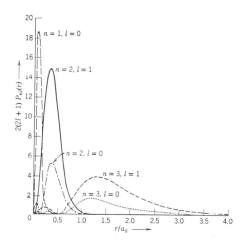

FIG. 10. The Hartress theory radial probability densities for the filled quantum states of the argon atom. [Adapated from Eisberg, R, and Resnick, R. (1985). "Quantum Physics of Atoms, Molecules, Solids, Nuclei, and Particles." John Wiley and Sons, New York.]

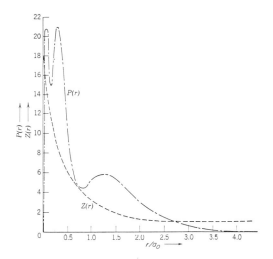

FIG. 11. The total radial probability density $P(r)$ of the argon atom, and the quantity $Z(r)$ that specifies its net potential. [Adapted from Eisberg, R., and Resnick, R. (1985). "Quantum Physics of Atoms, Molecules, Solids, Nuclei, and Particles." John Wiley and Sons, New York.]

Fock method has been developed to take proper account of the Pauli principle.

To obtain the angular momentum wave function of a complex atom, one must add together the spin and orbital angular momenta of the individual electrons outside the closed shells to obtain the total angular momentum of the atom. Two approximate coupling schemes are used, Russell–Saunders (or L–S) coupling and j–j coupling. In Russell–Saunders coupling, the orbital angular momenta of the individual electrons are added together to obtain the total orbital angular momentum of the complex atom, and the spin angular momenta of the individual electrons are added together to form the total spin angular momentum. The total orbital angular momentum is then added to the total spin angular momentum to obtain the total angular momentum. According to Hund's rule, the state with the highest multiplicity is the lowest energy state. In j–j coupling, for each electron the orbital and the spin angular momentum are added together to form the total angular momentum for that electron. The total angular momenta of the individual electrons are then added together to obtain the total angular momentum for the atom. Low-Z atoms are best described by Russell–Saunders coupling, and high-Z atoms by j–j coupling. Helium provides a good example of Russell–Saunders coupling. The parity of the wave

TABLE III.   Electronic Configuration and Ionization Energies for the Elements[a]

Z	Atom	Orbital electronic configuration	Ionization potential (eV)
1	H	$(1s)^1$	13.598
2	He	$(1s)^2$	24.587
3	Li	$(He)(2s)^1$	5.392
4	Be	$(He)(2s)^2$	9.322
5	B	$(He)(2s)^2(2p)^1$	8.298
6	C	$(He)(2s)^2(2p)^2$	11.260
7	N	$(He)(2s)^2(2p)^3$	14.534
8	O	$(He)(2s)^2(2p)^4$	13.618
9	F	$(He)(2s)^2(2p)^5$	17.422
10	Ne	$(He)(2s)^2(2p)^6$	21.564
11	Na	$(Ne)(3s)^1$	5.139
12	Mg	$(Ne)(3s)^2$	7.646
13	Al	$(Ne)(3s)^2(3p)^1$	5.986
14	Si	$(Ne)(3s)^2(3p)^2$	8.151
15	P	$(Ne)(3s)^2(3p)^3$	10.486
16	S	$(Ne)(3s)^2(3p)^4$	10.360
17	Cl	$(Ne)(3s)^2(3p)^5$	12.967
18	Ar	$(Ne)(3s)^2(3p)^6$	15.759
19	K	$(Ar)(4s)^1$	4.341
20	Ca	$(Ar)(4s)^2$	6.113
21	Sc	$(Ar)(4s)^2(3d)^1$	6.54
22	Ti	$(Ar)(4s)^2(3d)^2$	6.82
23	V	$(Ar)(4s)^2(3d)^3$	6.74
24	Cr	$(Ar)(4s)^1(3d)^5$	6.766
25	Mn	$(Ar)(4s)^2(3d)^5$	7.435
26	Fc	$(Ar)(4s)^2(3d)^6$	7.870
27	Co	$(Ar)(4s)^2(3d)^7$	7.86
28	Ni	$(Ar)(4s)^2(3d)^8$	7.635
29	Cu	$(Ar)(4s)^1(3d)^{10}$	7.726
30	Zn	$(Ar)(4s)^2(3d)^{10}$	9.394
31	Ga	$(Ar)(4s)^2(3d)^{10}(4p)^1$	5.999
32	Ge	$(Ar)(4s)^2(3d)^{10}(4p)^2$	7.899
33	As	$(Ar)(4s)^2(3d)^{10}(4p)^3$	9.81
34	Se	$(Ar)(4s)^2(3d)^{10}(4p)^4$	9.752
35	Br	$(Ar)(4s)^2(3d)^{10}(4p)^5$	11.814
36	Kr	$(Ar)(4s)^2(3d)^{10}(4p)^6$	13.999
37	Rb	$(Kr)(5s)^1$	4.177
38	Sr	$(Kr)(5s)^2$	5.695
39	Y	$(Kr)(5s)^2(4d)^1$	6.38
40	Zr	$(Kr)(5s)^2(4d)^2$	6.84
41	Nb	$(Kr)(5s)^1(4d)^4$	6.88
42	Mo	$(Kr)(5s)^1(4d)^5$	7.099
43	Tc	$(Kr)(5s)^2(4d)^5$	7.28
44	Ru	$(Kr)(5s)^1(4d)^7$	7.37
45	Rh	$(Kr)(5s)^1(4d)^8$	7.46
46	Pd	$(Kr)(4d)^{10}$	8.34
47	Ag	$(Kr)(5s)^1(4d)^{10}$	7.576
48	Cd	$(Kr)(5s)^2(4d)^{10}$	8.993
49	In	$(Kr)(5s)^2(4d)^{10}(5p)^1$	5.786
50	Sn	$(Kr)(5s)^2(4d)^{10}(5p)^2$	7.344
51	Sb	$(Kr)(5s)^2(4d)^{10}(5p)^3$	8.641
52	Te	$(Kr)(5s)^2(4d)^{10}(5p)^4$	9.009
53	I	$(Kr)(5s)^2(4d)^{10}(5p)^5$	10.451
54	Xe	$(Kr)(5s)^2(4d)^{10}(5p)^6$	12.130

(*continues*)

TABLE III (*Continued*)

Z	Atom	Orbital electronic configuration	Ionization potential (eV)
55	Cs	$(Xe)(6s)^1$	3.894
56	Ba	$(Xe)(6s)^2$	5.212
57	La	$(Xe)(6s)^2(5d)^1$	5.577
58	Ce	$(Xe)(6s)^2(4f)^1(5d)^1$	5.47
59	Pr	$(Xe)(6s)^2(4f)^3$	5.42
60	Nd	$(Xe)(6s)^2(4f)^4$	5.49
61	Pm	$(Xe)(6s)^2(4f)^5$	5.55
62	Sm	$(Xe)(6s)^2(4f)^6$	5.63
63	Eu	$(Xe)(6s)^2(4f)^7$	5.67
64	Gd	$(Xe)(6s)^2(4f)^7(5d)^1$	6.14
65	Tb	$(Xe)(6s)^2(4f)^9$	5.85
66	Dy	$(Xe)(6s)^2(4f)^{10}$	5.93
67	Ho	$(Xe)(6s)^2(4f)^{11}$	6.02
68	Er	$(Xe)(6s)^2(4f)^{12}$	6.10
69	Tm	$(Xe)(6s)^2(4f)^{13}$	6.18
70	Yb	$(Xe)(6s)^2(4f)^{14}$	6.254
71	Lu	$(Xe)(6s)^2(4f)^{14}(5d)^1$	5.426
72	Hf	$(Xe)(6s)^2(4f)^{14}(5d)^2$	7.0
73	Ta	$(Xe)(6s)^2(4f)^{14}(5d)^3$	7.89
74	W	$(Xe)(6s)^2(4f)^{14}(4d)^4$	7.98
75	Re	$(Xe)(6s)^2(4f)^{14}(5d)^5$	7.88
76	Os	$(Xe)(6s)^2(4f)^{14}(5d)^6$	8.7
77	Ir	$(Xe)(6s)^2(4f)^{14}(5d)^7$	9.1
78	Pt	$(Xe)(6s)^1(4f)^{14}(5d)^9$	9.0
79	Au	$(Xe)(6s)^1(4f)^{14}(5d)^{10}$	9.225
80	Hg	$(Xe)(6s)^2(4f)^{14}(5d)^{10}$	10.437
81	Tl	$(Xe)(6s)^2(4f)^{14}(5d)^{10}(6p)^1$	6.108
82	Pb	$(Xe)(6s)^2(4f)^{14}(5d)^{10}(6p)^2$	7.416
83	Bi	$(Xe)(6s)^2(4f)^{14}(5d)^{10}(6p)^3$	7.289
84	Po	$(Xe)(6s)^2(4f)^{14}(5d)^{10}(6p)^4$	8.42
85	At	$(Xe)(6s)^2(4f)^{14}(5d)^{10}(6p)^5$	—
86	Rn	$(Xe)(6s)^2(4f)^{14}(5d)^{10}(6p)^6$	10.748
87	Fr	$(Rn)(7s)^1$	—
88	Ra	$(Rn)(7s)^2$	5.279
89	Ac	$(Rn)(7s)^2(6d)^1$	6.9
90	Th	$(Rn)(7s)^2(6d)^2$	—
91	Pa	$(Rn)(7s)^2(5f)^2(6d)^1$	—
92	U	$(Rn)(7s)^2(5f)^3(6d)^1$	—
93	Np	$(Rn)(7s)^2(5f)^4(6d)^1$	—
94	Pu	$(Rn)(7s)^2(5f)^6$	5.8
95	Am	$(Rn)(7s)^2(5f)^7$	6.0
96	Cm	$(Rn)(7s)^2(5f)^7(6d)^1$	—
97	Bk	$(Rn)(7s)^2(5f)^9$	—
98	Cf	$(Rn)(7s)^2(5f)^{10}$	—
99	Es	$(Rn)(7s)^2(5f)^{11}$	—
100	Fm	$(Rn)(7s)^2(5f)^{12}$	—
101	Md	$(Rn)(7s)^2(5f)^{13}$	—
102	No	$(Rn)(7s)^2(5f)^{14}$	—
103	Lr	$(Rn)(7s)^2(5f)^{14}(6d)^1$	—

[a] Adapted from "Gray, H. B. "Chemical Bonds: An Introduction to Atomic and Molecular Structure." Benjamin/Cummings, Publ., Menlo Park, California.

function for a many-electron atom is determined by inverting the coordinates of all the electrons.

There is additional structure in the energy levels of many atoms due to the interaction between the electrons and the magnetic and quadrupole moments of the nucleus. For these cases the nucleus has angular momentum and one must add the nuclear angular momentum to the electronic angular momentum to obtain the total angular momentum. Measurements of the hyperfine structure are used to determine the magnetic and quadrupole moments of nuclei.

VI. Interaction of Atoms with Radiation

The interaction of an atom with an external electromagnetic field results in the emission and absorption of radiation. An atom in an excited state will decay exponentially with a mean life that depends on the available energy and the character of the initial and final states. The dominant form of radiation is electric dipole radiation; it takes place only between states with opposite parity.

In order to describe the interaction of atoms with the radiation field, it is necessary to quantize the electromagnetic field. The universe is pictured as a large cubic box with reflecting walls, and the set of solutions used to describe the electromagnetic field is taken to be the set of plane waves that satisfy the boundary conditions for a box with perfectly reflecting walls. The electromagnetic field is then expanded in a Fourier series in terms of these plane waves. The field can be specified by giving the amplitude of each of the Fourier components. For each mode there are two polarizations with the polarization vectors perpendicular to the direction of propagation of the plane wave. Due to the form of Maxwell's equations, the time dependence of the amplitude of these individual modes is the same as that for a simple harmonic oscillator whose frequency is the frequency of the Fourier component. A familiar example of a simple harmonic oscillator is a weight at the end of a spring where the restoring force due to the spring is proportional to it's extension from the equilibrium configuration. To quantize the electromagnetic field, one uses the same technique as is used to quantize the simple harmonic oscillator. This quantization gives for each mode a set of equally spaced energy levels whose separation is $h\nu$, where ν is the frequency of the

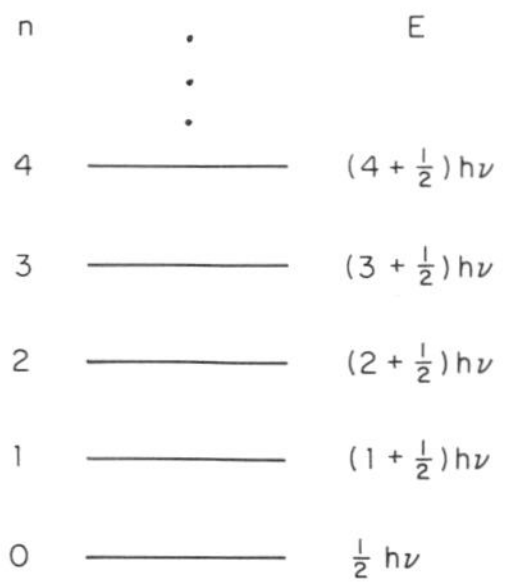

FIG. 12. The energy levels for the simple harmonic oscillator.

Fourier component. These energy levels are depicted in Fig. 12. Each excitation corresponds to one photon. If for the oscillator with frequency, ν direction of propagation $\mathbf{k}$, and polarization $\mathbf{e}$, the nth energy level is occupied, then there are n photons with frequency ν and polarization $\mathbf{e}$ propagating in direction $\mathbf{k}$.

In the electric dipole approximation the interaction between the electrons in the atom and the electromagnetic field has the form

$$\mathcal{H}_I = -\sum_i e\mathbf{r}_i \cdot \mathbf{E}(\mathbf{r}_i)$$

where $\mathbf{r}_i$ is the coordinate of the ith electron and $\mathbf{E}(\mathbf{r}_i)$ is the external electromagnetic field at the position of the ith electron. The interaction of an atom with the radiation field results in the emission and absorption of radiation. If an atom is in an excited state and there are no photons present, the interaction will cause the atom to decay to a lower state, with a characteristic mean life dependent on the energy difference between the levels and the wave functions describing the initial and final states of the atom. This emission of radiation in the absence of external radiation is called spontaneous emission.

Electric dipole transitions take place only between states of opposite parity, and only certain changes in the quantum numbers between the initial and final states are allowed. The selection rules are summarized in Table IV. The probability for the atom to be in the upper state decreases exponentially with time. The radiation emitted by an atom is not monochromatic but has a frequency distribution given by the equation

$$I(v) = I_0 \left(\frac{\gamma}{2\pi}\right) \frac{1}{(v - v_0)^2 + (\gamma/2)^2}$$

This distribution was first discussed by Lorentz,

TABLE IV.   Selection Rules for Electric
Dipole Radiation

For a single electron
$\Delta l = \pm 1$
$\Delta m = 0, \pm 1$
$\Delta j = 0, \pm 1$
$\Delta m_j = 0, \pm 1$
Not $j = 0$ to $j = 0$
For $\Delta j = 0$ not $m_j = 0$ to $m_j = 0$

For a many-electron configuration
$\Delta L = 0, \pm 1$ but not $L = 0$ to $L = 0$
$\Delta M_L = 0, \pm 1$
$\Delta J = 0, \pm 1$ but not $J = 0$ to $J = 0$
$\Delta M_J = 0, \pm 1$ but for $J \to J$ not $M_J = 0$
  to $M_J = 0$

and it is called the Lorentzian line profile. The
mean life τ for decay is related to the width of
the distribution by the equation

$$\tau = \frac{1}{\gamma}$$

The angular distribution and polarization of
the radiation depends on the change in the z
component of the angular momentum when the
atom goes from the initial state to the final state.
The radiation pattern for the $\Delta m = 0$ transitions
is the same as that for an electron that is oscillating harmonically with frequency v about the origin along the z axis. It has the normalized angular distribution

$$(3/8\pi) \sin^2 \theta$$

where θ is the angle between the direction of
propagation of the radiation and the z axis. The
radiation pattern for $\Delta m = \pm 1$ transitions is the
same as that for an electron which moves in the
xy plane in a circular orbit about the origin with
frequency v. For $\Delta m = -1$ the motion is coun-
terclockwise from x to y; for $\Delta m = +1$ the motion is clockwise from x to $-y$. It has the normalized angular distribution

$$(3/16\pi)(1 + \cos^2 \theta)$$

These radiation patterns and the polarization of
the radiation are depicted in Fig. 13.

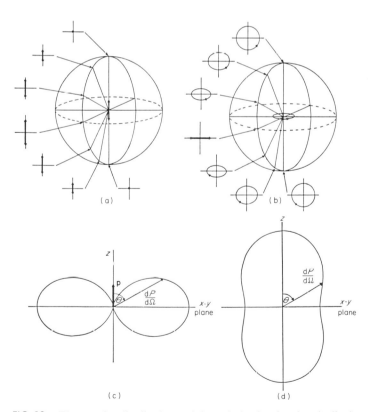

FIG. 13.   The angular distribution and the polarization for electric dipole
radiation. [Adapted from Corney, A. (1977). ''Atomic and Laser Spectroscopy.'' Oxford University Press, Oxford.]

If an atom is struck by a plane electromagnetic wave whose frequency v is related to the difference in energy of the two levels E_1 and E_2 through the Bohr relation

$$v = (E_1 - E_2)/h$$

and there is an allowed dipole transition connecting the two levels, then an atom in the lower state can be excited into the higher state with the absorption of a photon or an atom in the upper state can be stimulated to make a transition to the lower state with the emission of a photon with the same frequency, polarization, and direction as the incident electromagnetic wave. These two processes are referred to, respectively, as absorption and stimulated emission. It is the latter process that makes possible the laser.

The interaction of the atom with the electromagnetic field also produces corrections to the energy levels. These corrections arise through higher-order processes in which virtual photons are emitted and reabsorbed or a photon changes for a short time into a virtual electron–positron pair. The understanding and calculation of these energy shifts is one of the major triumphs of quantum electrodynamics. The first convincing experimental observation of the need for such corrections was the measurement of the Lamb shift in hydrogen. According to the Dirac theory, in the $n = 2$ state of hydrogen the $(2s)^2S_{1/2}$ and $(2p)^2P_{1/2}$ state have exactly the same energy. In an experiment carried out at Columbia University by Willis Lamb and his co-workers, it was found that the $(2s)^2S_{1/2}$ state was less bound than the $(2p)^2P_{1/2}$ state by $\frac{1}{10}$ of the fine structure splitting between the $(2p)^2P_{1/2}$ and $(2p)^2P_{3/2}$ states. It was shown later that this shift agrees with the predictions of quantum electrodynamics. The electron in the $(2s)^2S_{1/2}$ state spends more time near the proton than an electron in the $(2p)^2P_{1/2}$ state. Due to the emission and reabsorption of virtual photons, the electron is pushed away from the proton so that its interaction with the coulomb field due to the proton is weaker. This decreases the binding energy for the $(2s)^2S_{1/2}$ state more than for the $(2p)^2P_{1/2}$ state and results in a shift of the $(2s)^2S_{1/2}$ state upward from the $(2p)^2P_{1/2}$ state. The Lamb shift has now been measured to 10 ppm with satisfactory agreement between experiment and theory.

The most precisely measured higher-order effect due to the electromagnetic field is the correction to the magnitude moment of the electron. The magnetic moment of the electron differs by a small amount from the value predicted by the Dirac theory due to the interaction with the electromagnetic field. The calculated value of the ratio g of the electron magnetic moment to the Bohr magneton is

$$\frac{g - 2}{2} = [1\ 159\ 652\ 460\ (137)] \times 10^{-12}$$

The experimental value is

$$\frac{g - 2}{2} = [1\ 159\ 652\ 200\ (40)] \times 10^{-12}$$

The agreement is spectacular.

VII. Atomic Collisions

A major aspect of atomic physics is the description and understanding of collisions between electrons, photons, ions, and atoms. Atomic collisions are ubiquitous phenomena that play a major role in a wide area of science and technology ranging from light sources and lasers to fusion reactors. The collisions can be divided into two major categories: elastic collisions in which the total kinetic energy of the colliding atoms is the same after the collision as before, and inelastic collisions in which some of the kinetic energy goes into the excitation (or ionization) of the colliding atoms.

A. Cross Section

The scattering reaction is described in terms of the cross section. For a collision in which particle A strikes target particle B, which is assumed to be at rest, the total cross section σ is defined as the planar area centered on B which particle A must strike for the two particles to interact with one another. The cross section is expressed in square centimeters and for an incident flux of 1 particle/cm^2 gives the probability for the two particles to interact. The differential cross section $d\sigma/d\Omega$ is defined as the area for the incident particle to be deflected through an angle θ with respect to the incident direction into an area $d\Omega$ on a sphere of unit radius centered on the target particle. The particle is said to be scattered into the solid angle $d\Omega$. For a collision in which the internal state of one or both of the colliding atoms is changed, one must also specify the internal state of A and B before and after the collision. Figure 14 illustrates the definition of the total and differential cross sections.

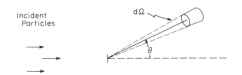

FIG. 14. Diagram illustrating the definition of the differential cross section.

B. Elastic Collisions

The study of the elastic scattering of atoms first arose in the development of kinetic theory, where the goal was to understand the transport of energy and momentum by gases. In a typical gas at room temperature, the atoms are moving with a Maxwellian distribution of velocities with an average kinetic energy of 0.038 eV, which is much smaller than the typical energy of 2–3 eV required to excite an atom from the ground state into the lowest excited state. The collisions are elastic and result in a change in the directions of motion and relative kinetic energies of the colliding atoms. The collision cross section determines the mean free path, which is the average distance an atom travels before it interacts with another atom, the viscosity of the gas, which is the resistance to shearing motion, the rate for diffusion of one gas within another, and the transport of heat in the gas. The cross section can be determined experimentally either through measurements of the viscosity and the rate of diffusion, or through experiments in which an atomic beam is incident on a target gas or one atomic beam is passed through a second beam.

Similar cross sections can be defined for low-energy ions in a gaseous environment. For ions, an important quantity is the mobility, which is a measure of the rate for an ion to move through a gas under the influence of an electric field.

A special category of elastic collisions is reactions of the form

$$A + A^* \rightarrow A^* + A$$

in which there is an interchange of the excitation of the colliding atoms. These are sometimes referred to as resonant collisions.

As the energy for the colliding atoms increases, the relative probability for elastic collisions decreases and the cross section for inelastic collisions in which one or both of the colliding partners is excited or ionized dominates.

C. Inelastic Collisions

One of the most important classes of reactions is those in which an electron collides with an atom and either raises the atom to an excited state or ionizes the atom. This is a dominant process in gaseous discharges. The first studies of this process were carried out by Franck and Hertz and were used to verify the quantization of the energy levels predicted by the Bohr theory. Figure 15 shows a diagram of the apparatus used by Franck and Hertz. Electrons emitted by a hot filament were accelerated by an electric field produced by an external voltage source, and they passed through a gas of atoms such as mercury. A second grid was located in front of the collecting electrode and put at a small positive voltage with respect to the collecting plate in order to prevent electrons with very low energy from reaching the collecting plate. Franck and Hertz observed the number of electrons transmitted as a function of the accelerating voltage and found that for certain voltages the transmission was smaller. These voltages corresponded to those for which the acceleration gave the electrons sufficient energy to excite the mercury atoms from the ground state up into one of the excited states. As the voltage was increased they excited successive states with higher energy. This experiment confirmed the quantization predicted by the Bohr theory and showed that the cross section for excitation had a peak near threshold.

The excitation of the target particles by the electron is due to the electric field produced by the electron, and a similar excitation takes place when any charged particle, such as a proton, passes through a gas. This excitation is the dominant mechanism for the loss of energy by charged particles passing through matter and provides a mechanism for the detection and measurement of charged particles.

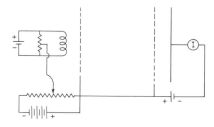

FIG. 15. Schematic diagram of the Franck–Hertz experiment.

D. Charge-Exchange Collisions

Another important class of collisions is those of the form

$$A^+ + B \rightarrow A + B^+$$

where an electron passes from the target particle to the projectile. This process is called charge exchange and is particularly important in fusion research plasmas where there are multiply charged ions. The cross section is at a maximum when the relative velocity of the atoms is the same as the velocity of the electron in the outer Bohr orbit and decreases as the energy increases. Charge exchange is one form of a larger class of rearrangement collisions in which particles are interchanged between the projectile and target.

E. Reactive Collisions

An important class of collisions that forms a major part of chemistry is reactive collisions, in which the projectile and target combine to form molecules. Some simple examples are

$$H + H \rightarrow H_2$$
$$H + D_2 \rightarrow HD + D$$

The first collision requires the presence of a third body to conserve energy and momentum.

The second collision is an example of a rearrangement collision involving a simple molecule.

BIBLIOGRAPHY

Atkins, P. W. (1974). "Quanta: A Handbook of Concepts." Oxford Univ. Press, London and New. York.

Bates, D. R., and Bederson, B. (1975–1985). "Advances in Atomic and Molecular Physics," Vols. 11–20. Academic Press, New York.

Bethe, H. A., and Salpeter, E. (1957). "Quantum Mechanics of One- and Two-Electron Atoms." Springer-Verlag, Berlin and New York.

Corney, A. (1977). "Atomic and Laser Spectroscopy." Oxford Univ. Press, London and New York.

Drukarev, G. F. (1985). "Collisions of Electrons with Atoms and Molecules." Plenum, New York.

Eisberg, R., and Resnick, R. (1985). "Quantum Physics of Atoms, Molecules, Solids, Nuclei, and Particles." Wiley, New York.

Lindgren, I., and Morrison, J. (1982). "Atomic Many-Body Theory." Springer-Verlag, Berlin and New York.

Mott, N. E., and Massey, H. S. W. (1965). "Theory of Atomic Collisions." Oxford Univ. Press, London and New York.

Van Dyck, R. S., Jr., and Fortson, E. N., eds. (1984) "Atomic Physics," Vol. 9. World Scientific Publ. Co., Singapore.

BONDING AND STRUCTURE IN SOLIDS

J. C. Phillips *AT&T Bell Laboratories*

I. Introduction: Molecules and Solids
II. Molecular Crystals
III. Ionic Crystals and Electronegativity
IV. Covalent Crystals and Directed Valence Bonds
V. Mixed Covalent and Ionic Bonding
VI. Metallic Bonding
VII. Quantum Structural Diagrams
VIII. Complete Quantum Structure Analysis

GLOSSARY

Atom: Smallest unit of an element.

Bond: Electronic configuration that binds atoms together.

Covalent bond: Chemical bond formed by electron sharing.

Crystals: Solids in which the atoms are arranged in periodic fashion.

Electronegativity: Measure of the ability of an atom to attract electrons.

Glass: Solid in which the atoms are not arranged in periodic fashion and which melts into a supercooled liquid when heated rapidly.

Ionic bond: Chemical bond caused by charge transfer.

Metallic: Material with high electrical conductivity at low frequency.

Molecule: Bonded atoms in a gas.

Valence: Number of electrons used by an atom to form chemical bonds.

The relative positions of atoms in molecules and solids are described and explained in terms of the arrangements of their nearer neighbors. Together with the chemical valences of the atoms as given by the periodic table, these arrangements of the bonding determine the structure and physical properties of solids. Both structure and properties can be used to separate solids into various classes where further quantitative trends can be systemically described by structural diagrams.

I. Introduction: Molecules and Solids

The combinations of atoms found in the vapor phase are called molecules. Molecules containing a small number of atoms have been studied accurately and extensively. Most of our knowledge of chemical bonding between atoms comes from these studies. When atoms are condensed to form solids, the atomic density is much greater, as reflected by the number of atoms that are nearest neighbors of any given atom. This number is called the coordination number. An example is the molecule NaCl, in which each atom has one nearest neighbor. In solid NaCl each atom has six nearest neighbors.

Solids in their pure forms are nearly always crystalline. A crystal is a periodic arrangement of atoms along lines, which in turn is repeated periodically along planes. Finally, the planes are repeated periodically to form the crystal lattice.

Most of our knowledge of crystal structures comes from the diffraction of waves of photons, electrons, or neutrons by lattice planes. Usually all the atomic positions in the crystal can be determined this way. By comparing chemical trends in bond lengths in crystals with those in molecules, one can often infer the nature of the electronic charge distribution responsible for chemical bonding in the crystal. From this it may be possible to predict the nature of chemical bonding at crystalline defects or even in noncrystalline solids, which are amorphous or glassy.

The structures of millions of solids are known by diffraction. To understand these structures one begins by studying the simplest cases and

TABLE I.   Electronegativity Table of the Elements According to Pauling

Li	Be	B											C	N	O	F
1.0	1.5	2.0											2.5	3.0	3.5	4.0
Na	Mg	Al											Si	P	S	Cl
0.9	1.2	1.5											1.8	2.1	2.5	3.0
K	Ca	Sc	Ti	V	Cr	Mn	Fe	Co	Ni	Cu	Zn	Ga	Ge	As	Se	Br
0.8	1.0	1.3	1.5	1.6	1.6	1.5	1.8	1.8	1.8	1.9	1.6	1.6	1.8	2.0	2.4	2.8
Rb	Sr	Y	Zr	Nb	Mo	Tc	Ru	Rh	Pd	Ag	Cd	In	Sn	Sb	Te	I
0.8	1.0	1.2	1.4	1.6	1.8	1.9	2.2	2.2	2.2	1.9	1.7	1.7	1.8	1.9	2.1	2.5
Cs	Ba	La–Lu	Hf	Ta	W	Re	Os	Ir	Pt	Au	Hg	Tl	Pb	Bi	Po	At
0.7	0.9	1.1–1.2	1.3	1.5	1.7	1.9	2.2	2.2	2.2	2.4	1.9	1.8	1.8	1.9	2.0	2.2
Fr	Ra	Ac	Th	Pa	U	Np–No										
0.7	0.9	1.1	1.3	1.5	1.7	1.3										

classifying them into groups. The main groups are characterized as molecular, metallic, ionic, and covalent. In most solids the actual bonding is a mixture to some degree of these different kinds of chemical interaction. While most solids are complex, the inorganic solids, which are best understood because they have had the widest technological applications, are usually either simple examples from a main group or are closely related to them. In contrast, organic and biologically important molecules may be quite complex. The chemical and structural simplicity of technologically important inorganic solids stems from the requirement of availability of techniques for production in bulk.

Certain general techniques are widely used for describing bonding and structure in solids. Tables of atomic radii are available for ionic, covalent, and metallic bonding. Deviations of bond lengths from values predicted by these radii of order 1 to 3% often reveal critical structural features of importance to material fabrication and properties. The cohesion of solids can be connected to the cohesion of the elements. A binary solid A_mB_n is said to have heat of formation ΔH_f, which is the difference between m times the cohesive energy of A plus n times that of B minus the cohesive energy of A_mB_n. This heat of formation can be estimated with often remarkable accuracy from Pauling's table of elemental electronegativities $X(A)$. This is probably the most widely used table in science apart from the periodic table of the elements, and it is shown here as Table I.

II. Molecular Crystals

We now turn to the differently bonded main groups of solids. The molecular crystals are the simplest case, because the intermolecular forces are typically much weaker than the intramolecular ones. As a result the structure of the molecules, as reflected, for example, by bond lengths and vibration frequencies, is almost the same in the solid as in the gas phase. Some examples of materials that form molecular solids are the inert gases, diatomic halogens, closed-shell molecules such as methane, and many planar aromatic molecules such as benzene. Typically in molecular crystals the heat of fusion per molecule per bond is at least 10 times smaller than the bond dissociation energy.

The binding forces that hold molecular crystals together may arise from electric dipoles if the molecules carry permanent dipole moments (e.g., HCl). When the molecules have no permanent moment, binding arises from mutually induced dipole moments (van der Waals interactions).

The structures of molecular crystals are determined primarily by packing considerations and thus vary from material to material according to molecular shape. Molecular solids are generally poor conductors of electricity, and even the photoconductivity is generally small unless metallic impurities are added to "sensitize" the material.

III. Ionic Crystals and Electronegativity

Before discussing the structure of ionic crystals in detail, we shall familiarize ourselves with the concept of electronegativity, defined by Pauling as "the ability of atoms in the bonded state to attract electrons to themselves." Atoms in solids are in a variety of bonded states, and it is due to Pauling's insight that we have come to realize that the atomic electronegativity that he defined in terms of heat of formation (Section I) is indeed nearly constant for each element. His idea is that in solids charge flows from cations with smaller electronegativity to anions with greater electronegativity and that the heat of formation resulting from this charge flow is proportional to $(X_c - X_a)^2$, where X_c and X_a are the cation and anion electronegativities, respectively.

Ionic crystals are composed of cations and anions with very large electronegativity differences, such as alkali metals and halides, columns I and VII of the periodic table, respectively. In this case the charge transfer of valence electrons is almost complete, so that the core configurations become isoelectronic to those of inert-gas atoms (e.g., Na^+ to Ne, Cl^- to Ar). While some energy is required to ionize the cations and transfer electrons to the anions, this energy is more than recovered thanks to the larger electronegativity of the anions and the mutual attraction of cations by their anion neighbors. In the case of the alkali halides, the cohesive energies can be estimated within a few percentage points by assuming complete charge transfer and evaluating the electrostatic energies (including ion polarization energies). A core–core repulsive energy, required by the exclusion principle, completes the calculation, which was first sketched around 1910.

As one might expect, the overall features of the crystal structures of ionic crystals are given quite well by packing spherical cations and anions in the appropriate proportions indicated by their chemical formulas. However, the ions are not quite the incompressible spheres suggested by their isoelectronic analogy to inert-gas atoms. If they were, one could use simple geometrical arguments (originating around 1930) to predict a coordination number of 8 (CsCl structure), 6 (NaCl structure), or 4 (ZnS structure). These correspond to packing cations and anions of nearly equal size (CsCl structure), and then successively larger anion/cation size ratios lead to increasing anion–anion contacts, thus reducing coordination numbers. These "radius ratio" rules do not actually describe the crystal structures, as shown in Fig. 1. What this means is that the ions should not be regarded as hard spheres, but rather as centers of quantum mechanically determined electronic charge distributions. Additional evidence for the breakdown of classical electrostatic models is contained in the elastic constants of the alkali halides. If these models were correct, the elastic constants would satisfy certain relations (the Cauchy rela-

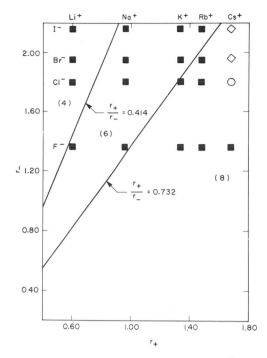

FIG. 1. The structures of the alkali halides M^+X^- as functions of classical ionic radii r_+ and r_-, respectively. (Coordination numbers in parentheses are those predicted by the classical ionic model). In the upper left corner, for example, Li^+I^- is predicted by the classical model to have coordination number 4, but the symbol indicates it is actually six-fold coordinated. Key: ■, sixfold coordinated; ◇, eightfold coordinated; ○, six- or eightfold coordinated. [From Phillips, J. C. (1974). *In* "Solid State Chemistry" (N. B. Hannay, ed.), Vol. 1. Plenum, New York.]

tions) valid for central force interactions. These relations are not satisfied for most of the alkali halides, another indication of quantum mechanical interactions. Some simplified modern treatments of these and related problems are discussed in the following sections.

IV. Covalent Crystals and Directed Valence Bonds

Whereas ionic crystals can be (at least roughly) described in classical terms, structure and bonding in covalent crystals can be understood only in terms of quantum mechanical electron orbital wave functions. Prototypical covalent crystals have the diamond structure. Many technologically important semiconductors such as silicon and germanium have this structure or a closely related one, the zinc blende or wurtzite structure. In these structures each atom is tetrahedrally coordinated (Fig. 2).

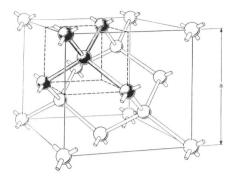

FIG. 2. The tetrahedrally coordinated diamond structure, which describes many technologically important semiconductors such as silicon. [From Phillips, J. C. (1970). *Phys. Today* **23** (Feb.), p. 23.]

The structure shown in Fig. 2 can be explained simply in terms of directed valence electron orbitals. The valence configuration of the atom is $ns^2 np^2$, with $n = 3$ for silicon. In the crystal this becomes $ns\,np^3$ so that (counting electron spin twofold degeneracy) both the ns and np levels are half-filled. These four states can be combined to form four directed valence orbitals with tetrahedral geometry. The wave functions on nearest neighbors can be combined in phase to form bonding states or out of phase to form antibonding states. Then wave-function overlap produces an energy gap between these states (Fig. 3). This energy gap is the basis of the technologically important electronic and optical properties of semiconductors.

The covalent energy gained by wave-function overlap or interference is much more sensitive to structural perfection than is the energy associated with classical ionic interactions. A very important consequence of this sensitivity is that it is possible to produce semiconductor crystals such as silicon in far purer and far more structurally perfect states than has been possible with any other solid. It is possible to add impurities with designed concentrations and locations to tailor the chemical and mechanical design of the solid with far greater precision and ease than for any other solid. Thus the quantum mechanical nature of structure and bonding in covalent silicon is the key to its technological significance.

V. Mixed Covalent and Ionic Bonding

Most semiconductors and insulators have neither purely covalent nor purely ionic bonding, but their bonding is described as a mixture of covalent and ionic effects. The way in which the mixture occurs is of great importance both scientifically and technologically. We shall discuss several important examples.

The simplest case occurs for the tetrahedrally coordinated covalent structure shown in Fig. 2. This structure contains two kinds of atomic sites: site A with only B neighbors, and vice versa. In silicon and gemanium both sites are occupied by the same atom, which has four valence electrons. However, one can occupy the two sites with different atoms, such that the total number of valence electrons is eight per atom pair (formally represented by $A^N B^{8-N}$). Many compounds of this type with $N = 3$ and $N = 2$

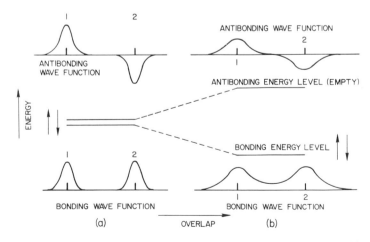

FIG. 3. Sketch of electronic interactions between directed valence orbitals that produce an energy gap between bonding states (electrons shared between nearest neighbors) and antibonding states (no electron sharing). [From Phillips, J. C. (1970). *Phys. Today* **23** (Feb.), p. 23.]

are known. One example with $N = 3$ is GaAs, which is important technologically in optoelectronic semiconductor applications.

The next interesting case is the triatomic material SiO_2 (silica). The electronegativity difference between silicon and oxygen is large, so the bonding here contains a large ionic component. At the same time each oxygen atom contains six valence electrons while silicon has four, so the total number of valence electrons per molecular unit is 16. This favors covalent bonding. In the solid each silicon atom has tetrahedral oxygen neighbors, while each oxygen atom has two silicon neighbors, which is again the coordination characteristic of covalent bonding. Silica is chemically stable and can be made very pure, for much the same reasons that silicon can. This high purity is essential to technological applications in the context of optical fibers for communications.

Another feature of silica is that it can easily be cooled into a solid state that is not crystalline but more like a frozen liquid. The state is called a glass. The ductility of glasses at high temperatures is essential to the manufacture of optical fibers. However, glasses are also ductile on a molecular scale and so do not form molecular "cracks," which would be arrays of broken bonds that might be electrically active and destructive to the electronic capabilities of semiconductor devices such as transistors. It is one of nature's most fortunate accidents that silicon electronic devices can be packaged by simply oxidizing the surfaces of solid silicon to form a protective coating of silica, SiO_2, which is chemically stable because of its covalent bonding and which is mechanically stable because SiO_2 is not only the oldest known but also the best glass.

The last case is primarily ionic materials with a covalent component. Many oxides in which the oxygen atoms are three- to six-fold coordinated fall in this class, and this includes many ceramic materials. These materials can have high melting points and good chemical stability, but they are brittle and for this reason their range of technological applications is limited.

VI. Metallic Bonding

Broadly speaking three kinds of elements are found in metals. They are the simple $s–p$ metallic elements from the left-hand side of the periodic table, such as lithium, aluminum, and lead; the rare-earth and transition elements with f and d valence electrons, such as titanium, iron, and nickel; and metalloid elements, such as carbon, silicon, and phosphorus, which may also in certain combinations form covalent solids. In metals the coordination number (or number of nearest neighbors) is much larger (usually twice as large or more) than the number of valence electrons. This means that the directed valence bonds found in molecules or in covalent crystals are much weaker (although not completely absent) in metals.

The high electrical and thermal conductivity of metals is a result of the absence of a gap in the energy spectrum between filled and empty electronic states. This high electrical conductivity in turn reduces the contribution to cohesion associated with charge transfer because the internal electric fields are limited by electronic redistribution or charge flow on an atomic scale. Thus ionic interactions are reduced in metals compared with ionic crystals.

The reduction of covalent or molecular bonding as well as ionic bonding in metals presents a paradox. If neither of these bonding mechanisms is fully effective, to what forces do metals owe their cohesion? Modern quantum theory shows a complex correlation of the motion of metallic valence electrons, which reduces the Coulomb repulsive energy between these like charges while leaving almost unchanged the attractive Coulomb interaction between negatively charged electrons and positively charged atom cores. It is this correlation energy that is primarily responsible for metallic cohesion.

From studies of the structure and cohesion of metals it appears that d valence electrons (as in the transition metals) contribute almost as effectively to metallic cohesion as s and p electrons. The f electrons in rare earth metals, on the other hand, play a minor role in metallic cohesion but occasionally have magnetic properties. Transition metals are notable for their strong magnetic properties (iron, cobalt, and nickel), as well as their high melting points and refractory properties, which result from the large number of combined s, p, and d valence electrons. The compound with the highest known melting point is tungsten carbide (WC), an interesting combination of a transition element whose d levels are half-full with a metalloid element whose s and p valence levels are half-full. Also, here tungsten is very large and carbon is very small, which makes possible an ionic contribution to the cohesive and refractory properties.

VII. Quantum Structural Diagrams

The description of structure and bonding in solids given in the preceding sections is largely qualitative, but it is a fair (although abbreviated) account of most of what was generally known as a result of quantum mechanical analysis in the period from 1930 to 1960. Starting in 1960 a more quantitative description was developed that enables us to inspect systematic trends in structure and bonding with the aid of quantum structural diagrams.

With a structural diagram one assigns to each element certain characteristics and then treats these characteristics as configuration coordinates, which are used to construct structural maps. The natural classical configuration coordinates are atomic size and electronegativity, as defined by Pauling (see Table I). To these we may add the number of valence electrons per atom. One then takes a class of binary compounds, say $A^N B^M$, with the same value of $P = N + M$ and uses size differences (or ratios), as well as electronegativity differences, as Cartesian coordinates. If the characteristics or configuration coordinates have genuine value for describing structure and bonding, compounds composed of different elements A and B, but with similar values of their Cartesian coordinates, should have the same crystal structure. Put somewhat differently, the structural map should separate into simple regions, with each region containing compounds with the same crystal structure.

Early attempts to construct structural maps of this kind using classical coordinates were only partially successful; as many as 10 or 20% or more of the compounds were misplaced. From this failure most workers concluded that the problem of structure and bonding in solids, and especially in metals where the number of known compounds exceeds 10^4, was simply too complex to solve in any simple way. Finding a solution was left to the indefinite future, when computers became large enough and quantum mechanical methods accurate enough to predict structures on a case-by-case basis.

Recent research has shown that the idea of structural diagrams is itself valid but that previous failures arose from the use of largely classical coordinates. In addition to the number of valence electrons per atom (a quantum con-

cept), one must also use other quantum variables to replace the classical variables of atomic size and electronegativity. This has been done in several ways, which are substantially equivalent. The simplest case is $A^N B^M$ compounds where A and B have only s and p valence electrons and $N + M = P = 8$, which means that the s and p valence levels are half-full. In this case one can separate ionic and covalent crystal structures by separating the average energy gap between occupied and empty electronic states into ionic and covalent components, represented by C and E_h, respectively. Both NaCl (ionic) and diamond, silicon, and germanium (covalent) crystals (Fig. 2) belong in this group, with $C/E_h = 0$ in the latter and C/E_h large in the former. The quantum structural diagram for $A^N B^{8-N}$ non-transition-metal compounds shown

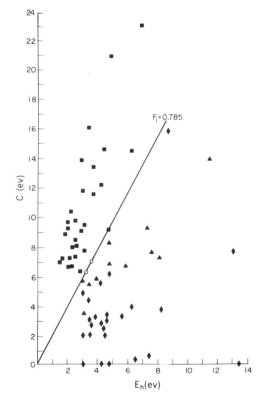

FIG. 4. The separation of the energy gap shown in Fig. 3 into covalent and ionic components (E_h and C, respectively) generates a structural map that separates fourfold- and sixfold-coordinated $A^N B^{8-N}$ crystals perfectly (no transition or rare-earth elements). The structures and coordination numbers (in parentheses) are as follows: ◆, diamond, zincblend (4); ▲, wurtzite (4); ■, rock salt (6); ○, rock salt/wurtzite (6, 4). [From Cohen, M. L., Heine, V., and Phillips, J. C. (1982). Sci. Am. **246** (6), 82.]

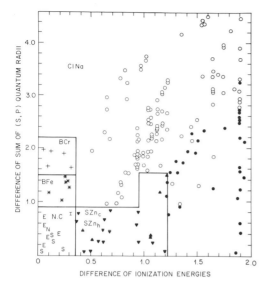

FIG. 5. A general separation of $A^N B^{8-N}$ crystal structures utilizes quantum coordinates defined for all elements including rare-earth and transition metals. Compounds containing the latter are indicated by open symbols. [From Villars, P. (1983). *J. Less-Common Met.* **92**, 215.]

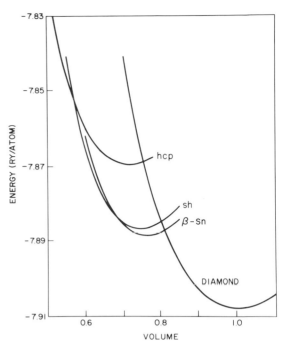

FIG. 6. A plot of the total energy of silicon crystals in different crystal structures as a function of atomic volume. At atmospheric pressure the diamond structure has the lowest energy, but at pressure of hundreds of thousands of atmospheres silicon is more stable in other structures. Such high pressures can be produced in the laboratory and they are also found at great depths below the earth's surface. [From Chang, K. J., and Cohen, M. L. (1984). *Phys. Rev. B* **30**, 5376.]

in Fig. 4 not only is a huge improvement on the classical structure diagram shown in Fig. 1 but also is an exact separation of covalent and ionic crystal structures.

To extend this analysis to transition and rare-earth metals as well as compounds in which the valence shell is not exactly half-full is a monumental task that includes ~1000 AB compounds, ~1000 AB_2 compounds, and more than 1000 AB_3 and A_3B_5 compounds, as well as more than 7000 ternary compounds. The correct quantum coordinates for these 10,000 compounds have been identified from a field of 182 candidate coordinates, some classical and some quantum coordinates. All the best coordinates are found to be quantum coordinates, and these turn out to be the atomic ionization potential and a suitably defined quantum core size. The result for $A^N B^{8-N}$ compounds (where A or B or both may be transition or rare-earth elements) is shown in Fig. 5. It is representative of the best global analysis of structure and bonding in solids available in 1985. This structural map is 97% successful.

VIII. Complete Quantum Structure Analysis

On a case-by-case basis a full discussion of structure and bonding in a given solid can be achieved using the most advanced computational techniques combined with the most sophisticated computers. Work with sufficient precision and flexibility to describe the structure of solid surfaces, point defects, and solid transitions under high pressures became available in selected cases in the 1980s. An excellent example is shown in Fig. 6, which gives the total energy of crystalline silicon in different crystal structures as a function of volume. From these curves transition pressures and volumes can be obtained from the tie-line (common tangent) construction due (~100 years ago) to Gibbs. It is interesting that all the results shown in Figs. 4, 5, and 6 are based on a particular approach to the quantum structure of solids that is known as the pseudopotential method.

BIBLIOGRAPHY

Adams, D. M. (1974). "Inorganic Solids." Wiley, New York.

Chang, K. J., and Cohen, M. L. (1984). *Phys. Rev. B30*, 5376.

Cohen, M. L., Heine V., and Phillips, J. C. (1982). *Sci. Am.* **246**(6), 82.

Pauling, L. (1960). "Nature of the Chemical Bond." Cornell University Press, Ithaca.

Phillips, J. C. (1970). The chemical bond and solid state physics, *Phys. Today,* **23** (February), p. 23.

Phillips, J. C. (1974). *in* "Solid State Chemistry" (N. B. Hannay, ed.), Vol. 1: The Chemical Structure of Solids. Plenum, New York.

Tosi, M. P. (1964). *Solid-State Phys.* **16,** 1.

Villars, P. (1983). *J. Less-Common Met.* **92,** 215.

Villars, P. (1985). *J. Less-Common Met.* **109,** 93.

Wigner, E. P., and Seitz, F. (1955). *Solid-State Phys.* **1,** 1.

CHAOS

Roderick V. Jensen   *Yale University*

I.   Introduction
II. Classical Chaos
III. Dissipative Dynamical Systems
IV. Hamiltonian Systems
V.   Quantum Chaos

GLOSSARY

Cantor set: Simple example of a fractal set of points with a noninteger dimension.

Deterministic equations: Equations of motion with no random elements for which formal existence and uniqueness theorems guarantee that once the necessary initial and boundary conditions are specified the solutions in the past and future are uniquely determined.

Dissipative system: Dynamical system in which frictional or dissipative effects cause volumes in the phase space to contract and the long-time motion to approach an attractor consisting of a fixed point, a periodic cycle, or a strange attractor.

Dynamical system: System of equations describing the time evolution of one or more dependent variables. The equations of motion may be difference equations if the time is measured in discrete units, a set of ordinary differential equations, or a set of partial differential equations.

Ergodic theory: Branch of mathematics that introduces statistical concepts to describe average properties of deterministic dynamical systems.

Fractal: Geometrical structure with self-similar structure on all scales that may have a noninteger dimension, such as the outline of a cloud, a coastline, or a snowflake.

Hamiltonian system: Dynamical system that conserves volumes in phase space, such as a mechanical oscillator moving without friction, the motion of a planet, or a particle in an accelerator.

KAM theorem: The Kolmogorov–Arnold–Moser theorem proves that when a small, nonlinear perturbation is applied to an integrable Hamiltonian system it remains nearly integrable if the perturbation is sufficiently small.

Kicker rotor: Simple model of a Hamiltonian dynamical system that is exactly described by the classical standard map and the quantum standard map.

Kolmogorov–Sinai entropy: Measure of the rate of mixing in a chaotic dynamical system that is closely related to the average Lyapunov exponent, which measures the exponential rate of divergence of nearby trajectories.

Localization: Quantum interference effect, introduced by Anderson in solid-state physics, which inhibits the transport of electrons in disordered or chaotic dynamical systems as in the conduction of electronics in disordered media or the microwave excitation and ionization of highly excited hydrogen atoms.

Mixing: Technical term from ergodic theory that refers to dynamical behavior that resembles the evolution of cream poured in a stirred cup of coffee.

Phase space: Mathematical space spanned by the dependent variables of the dynamical system. For example, a mechanical oscillator moving in one-dimension has a two-dimensional phase space spanned by the position and momentum variables.

Poincaré section: Stroboscopic picture of the evolution of a dynamical system in which the values of two dependent variables are plotted as points in a plane at regular intervals of time.

Random matrix theory: Theory introduced to describe the statistical fluctuations of the spacings of nuclear energy levels based on the statistical properties of the eigenvalues of matrices with random elements.

Resonance overlap criterion: Simple analytical estimate of the conditions for breakup of KAM surfaces leading to wide-spread, global chaos.

Strange attractor: Aperiodic attracting set with a fractal structure that often characterizes the long-time dynamics of chaotic dissipative systems.

A wide variety of natural phenomena exhibit complex, irregular behavior. In the past, many of these phenomena were considered to be too difficult to analyze; however, the advent of high-speed digital computers coupled with new mathematical and physical insight has led to the development of a new interdisciplinary field of science called nonlinear dynamics, which has been very successful in finding some underlying order concealed in nature's complexity. In particular, research in the physical sciences and in engineering has revealed how very simple, deterministic mathematical models of the motion of asteroids, the vibrations of stressed beams, of electrical and chemical oscillators, and turbulent fluid flows can exhibit behavior as complex as any found in nature. The apparently random behavior of these deterministic, nonlinear dynamical systems is called chaos.

Since many different fields of science and engineering are confronted with difficult problems involving nonlinear equations, the field of nonlinear dynamics has evolved in a highly interdisciplinary manner, with important contributions coming from biologists, mathematicians, engineers, and physicists. In the physical sciences, important advances have been made in our understanding of complex processes and patterns in dissipative systems, such as damped, driven, nonlinear oscillators and turbulent fluids, and in the derivation of statistical descriptions of Hamiltonian systems, such as the motion of celestial bodies and the motion of charged particles in accelerators and plasmas. Moreover, the predictions of chaotic behavior in simple mechanical systems have led to the investigation of the manifestations of chaos in the corresponding quantum systems, such as atoms and molecules in very strong fields.

This article attempts to describe some of the fundamental ideas; to highlight a few of the important advances in the study of chaos in classical, dissipative, and Hamiltonian systems; and to indicate some of the implications for quantum systems.

I. Introduction

In the last 10 or 15 years, the word *chaos* has emerged as a technical term to refer to the complex, irregular, and apparently random behavior of a wide variety of physical phenomena, such as turbulent fluid flow, oscillating chemical reactions, vibrating structures, the behavior of nonlinear electrical circuits, the motion of charged particles in accelerators and fusion devices, the orbits of asteroids, and the dynamics of atoms and molecules in strong fields. In the past, these complex phenomena were often referred to as random or stochastic, which meant that researchers gave up all hope of providing a detailed microscopic description of these phenomena and restricted themselves to statistical descriptions alone. What distinguishes chaos from these older terms is the recognition that many complex physical phenomena are actually described by deterministic equations, such as the Navier–Stokes equations of fluid mechanics, Newton's equations of classical mechanics, or Schrödinger's equation of quantum mechanics, and the important discovery that even very simple, deterministic equations of motion can exhibit exceedingly complex behavior and structure that is indistinguishable from an idealized random process. Consequently, a new term was required to describe the irregular behavior of these deterministic dynamical systems that reflected the new found hope for a deeper understanding of these various physical phenomena. These realizations also led to the rapid development of a new, highly interdisciplinary field of scientific research called nonlinear dynamics, which is devoted to the description of complex, but deterministic, behavior and to the search for "order in chaos."

The rise of nonlinear dynamics was stimulated by the combination of some old and often obscure mathematics from the early part of the 20th century that were preserved and developed by isolated mathematicians in the United States, the Soviet Union, and Europe; the deep natural insight of a number of pioneering researchers in meteorology, biology, and physics; and by the widespread availability of high-speed digital computers with high-resolution computer graphics. The mathematicians constructed simple, but abstract, dynamical systems that could generate complex behavior and geometrical patterns. Then, early researchers studying the nonlinear evolution of weather patterns and the fluctuations of biological populations realized that their crude approximations to the full mathematical equations, in the form of a single difference equation or a few ordinary differential equations, could also exhibit behavior as complex and seemingly random as the natural phenomena. Finally, high-speed computers provided a means for detailed computer experiments on these simple mathematical models with complex behavior. In particular, high-resolution computer graphics have enabled experimental mathematicians to search for order in chaos that would otherwise be buried in reams of computer output. This rich interplay of mathematical theory, physical insight, and computer experimentation, which characterizes the study of chaos and the field of nonlinear dynamics, will be clearly illustrated in each of the examples discussed in this article.

Chaos research in the physical sciences and engineering can be divided into three distinct areas relating to the study of nonlinear dynamical systems that correspond to (1) classical dissipative systems, such as turbulent flows or mechanical, electrical, and chemical oscillators; (2) classical Hamiltonian systems, where dissipative processes can be neglected, such as charged particles in accelerators and magnetic confinement fusion devices or the orbits of asteroids and planets; and (3) quantum systems, such as atoms and molecules in strong static or intense electromagnetic fields.

The study of chaos in classical systems (both dissipative and Hamiltonian) is now a fairly well-developed field that has been described in great detail in a number of popular and technical books. In particular, the term *chaos* has a very precise mathematical definition for classical nonlinear systems, and many of the characteristic features of chaotic motion, such as the extreme sensitivity to initial conditions, the appearance of strange attractors with noninteger, fractal dimensions, and the period-doubling route to chaos, have been cataloged in a large number of examples and applications, and new discoveries continue to fill technical journals. In Section II, we will begin with a precise definition of chaos for classical systems and present a very simple mathematical example that illustrates the origin of this complex, apparently random motion in simple deterministic dynamical systems. In Section II, we will also consider additional examples to illustrate some of the other important general features of chaotic classical systems, such as the notion of geometric structures with noninteger dimensions.

Some of the principal accomplishments of the application of these new ideas to dissipative systems include the discovery of a universal theory for the transition from regular, periodic behavior to chaos via a sequence of period-doubling bifurcations, which provides quantitative predictions for a wide variety of physical systems, and the discoveries that mathematical models of turbulence with as few as three nonlinear differential equations can exhibit chaotic behavior that is governed by a strange attractor. In particular, the realization that only a few degrees of freedom are necessary to generate dynamics as complex as that observed in fully developed turbulence has provided great hope that many irreguar phenomena, even in systems with very large numbers of degrees of freedom, might be adequately described by the dynamics on a low-dimensional strange attractor. Although research along these lines has yet to provide a comprehensive theory of turbulence, which has remained an outstanding problem of theoretical physics and engineering for over 100 years, the ideas and analytical methods introduced by the simple models of nonlinear dynamics have provided important analogies and metaphors for describing complex natural phenomena that should ultimately pave the way for a better theoretical understanding. Section III

will be devoted to a detailed discussion of several models of dissipative systems with important applications in the description of turbulence and the onset of chaotic behavior in a variety of nonlinear oscillators.

In the realm of Hamiltonian systems, the exact nonlinear equations for the motion of particles in accelerators and fusion devices and of celestial bodies are simple enough to be analyzed using the analytical and numerical methods of nonlinear dynamics without any gross approximations. Consequently, accurate quantitative predictions of the conditions for the onset of chaotic behavior that play significant roles in the design of accelerators and fusion devices and in understanding the irregular dynamics of asteroids can be made. Moreover, the important realization that only a few interacting particles, representing a small number of degrees of freedom, can exhibit motion that is sufficiently chaotic to permit a statistical description has greatly enhanced our understanding of the microscopic foundations of statistical mechanics, which have also remained an outstanding problem of theoretical physics for over a century. Section IV will examine several simple mathematical models of Hamiltonian systems with applications to the motion of particles in accelerators and fusion devices and to the motion of celestial bodies.

Finally, in Section V, we will discuss the more recent and more controversial studies of the quantum behavior of strongly coupled and strongly perturbed Hamiltonian systems, which are classically chaotic. In contrast to the theory of classical chaos, there is not yet a consensus on the definition of quantum chaos because the Schrödinger equation is a linear equation for the deterministic evolution of the quantum wave function, which is incapable of exhibiting the strong dynamical instability that defines chaos in nonlinear classical systems. Nevertheless, both numerical studies of model problems and real experiments on atoms and molecules reveal that quantum systems can exhibit behavior that resembles classical chaos for long times. In addition, considerable research has been devoted to identifying the distinct signatures or symptoms of the quantum behavior of classically chaotic systems. At present, the principal contributions of these studies has been the demonstration that atomic and molecular physics of strongly perturbed and strongly coupled systems can be very different from that predicted by the traditional perturbative methods of quantum mechanics. For example, experiments with highly excited hydrogen atoms in strong microwave fields have revealed a novel ionization mechanism that depends strongly on the intensity of the radiation but only weakly on the frequency. This dependence is just the opposite of the quantum photoelectric effect, but the sharp onset of ionization in the experiments is very well described by the onset of chaos in the corresponding classical system.

II. Classical Chaos

This section provides a summary of the fundamental ideas that underlie the discussion of chaos in all classical dynamical systems. It begins with a precise definition of chaos and illustrates the important features of the definition using some very simple mathematical models. These examples are also used to exhibit some important properties of chaotic dynamical systems, such as extreme sensitivity to initial conditions, the unpredictability of the long-time dynamics, and the possibility of geometric structures corresponding to strange attractors with noninteger, fractal dimensions. The manifestations of these fundamental concepts in more realistic examples of dissipative and Hamiltonian systems will be provided in Sections III and IV.

A. THE DEFINITION OF CHAOS

The word *chaos* describes the irregular, unpredictable, and apparently random behavior of nonlinear dynamical systems that are described mathematically by the deterministic iteration of nonlinear difference equations or the evolution of systems of nonlinear ordinary or partial differential equations. The precise mathematical definition of chaos requires that the dynamical system exhibit mixing behavior with positive Kolmogorov–Sinai entropy (or positive average Lyapunov exponent). This definition of chaos invokes a number of concepts from ergodic theory, which is a branch of mathematics that arose in response to attempts to reconcile statistical mechanics with the deterministic equations of classical mechanics. Although the equations that describe the evolution of chaotic dynamical systems are fully deterministic (no averages over random forces or initial conditions are involved), the complexity of the dynamics invites a statistical description. Consequently, the statistical concepts of ergodic theory provide a natural language to define and characterize chaotic behavior.

1. Ergodicity

One concept that is probably familiar to most physical scientists is the notion of ergodicity. Roughly speaking, a dynamical system is ergodic if the system comes arbitrarily close to every possible point (or state) in the accessible phase space over time. In this case, the celebrated ergodic theorem guarantees that long-time averages of any physical quantity can be determined by performing averages over phase space with respect to a probability distribution. However, although there has been considerable confusion in the physical literature, ergodicity alone is not sufficiently irregular to account for the complex behavior of turbulent flows or interacting many-body systems. A simple mathematical example clearly reveals these limitations.

Consider the dynamical system described by the difference equation

$$x_{n+1} = x_n + a, \qquad \text{Mod } 1 \qquad (1)$$

which takes a real number x_n between 0 and 1, adds another real number a, and subtracts the integer part of the sum (Mod 1) to return a value of x_{n+1} on the unit interval $[0, 1]$. The sequence of numbers, $\{x_n\}_{n=0,1,2,3,\ldots}$, generated by iterating this one-dimensional map describes the time history of the dynamical variable x_n (where time is measured in discrete units labeled by n). If $a = p/q$ is a rational number (where p and q are integers), then starting with any initial x_0, this dynamical system generates a time sequence of $\{x_n\}$ that returns to x_0 after q iterations since $x_q = x_0 + p$ (Mod 1) $= x_0$ (Mod 1). In this case, the long-time behavior is described by a periodic cycle of period q that visits only q different values of x on the unit interval $[0, 1]$. Since this time sequence does not come arbitrarily close to every point in the unit interval (which is the phase or state space of this dynamical system), this map is not ergodic for rational values of a.

However, if a is an irrational number, the time sequence never repeats and x_n will come arbitrarily close to every point in the unit interval. Moreover, since the time sequence visits every region of the unit interval with equal probability, the long-time averages of any functions of the dynamical variable x can be replaced by spatial averages with respect to the uniform probability distribution $P(x) = 1$ for x in $[0, 1]$. Therefore, for irrational values of a, this dynamical system, described by a single, deterministic difference equation, is an ergodic system.

Unfortunately, the time sequence generated by this map is much too regular to be chaotic. For example, if we initially colored all the points in the phase space between 0 and $\frac{1}{4}$ red and iterated the map, then the red points would remain clumped together in a continuous interval (Mod 1) for all time. But, if we pour a little cream in a stirred cup of coffee or release a dyed gas in the corner of the room, the different particles of the cream or the colored gas quickly spread uniformly over the accessible phase space.

2. Mixing

A stronger notion of statistical behavior is required to describe turbulent flows and the approach to equilibrium in many-body systems. In ergodic theory, this property is naturally called mixing. Roughly speaking, a dynamical system described by deterministic difference or differential equations is said to be a mixing system if sets of initial conditions that cover limited regions of the phase space spread throughout the accessible phase space and evolve in time like the particles of cream in coffee. Once again a simple difference equation serves to illustrate this concept.

Consider the shift map:

$$x_{n+1} = 2x_n, \qquad \text{Mod } 1 \qquad (2)$$

which takes x_n on the unit interval, multiplies it by 2, and subtracts the integer part to return a value of x_{n+1} on the unit interval. If we take almost any initial condition, x_0, then the this deterministic map generates a time sequence $\{x_n\}$ that never repeats and for long times is indistinguishable from a random process. Since the successive iterates wander over the entire unit interval and come arbitrarily close to every point in the phase space, this map is ergodic. Moreover, like Eq. (1), the long-time averages of any function of the $\{x_n\}$ can be replaced by the spatial average with respect to the uniform probability distribution $P(x) = 1$.

However, the dynamics of each individual trajectory is much more irregular than that generated by Eq. (1). If we were to start with a set of red initial conditions on the interval $[0, \frac{1}{4}]$, then it is easy to see that these points would be uniformly dispersed on the unit interval after only 2 iterations of the map. Therefore, we call this dynamical system a mixing system. (Of course, if we were to choose very special initial conditions, such as $x_0 = 0$ or $x_0 = p/2^m$, where p and m are positive integers, then the time sequence would still be periodic. However, in the set of all possible initial conditions, these exceptional initial conditions are very rare. Mathematically, they comprise a set of zero measure, which means the chance of choosing one of these special initial conditions by accident is nil.)

It is very easy to see that the time sequences generated by the vast majority of possible initial conditions is as random as the time sequence generated by flipping a coin. Simply write the initial condition in binary representation, that is, $x_0 = 0.0110011011100011010. \ldots$. Multiplication by 2 corresponds to a register shift that moves the binary point to the right (just like multiplying a decimal number by 10). Therefore, when we iterate Eq. (2),

we read off successive digits in the initial condition. If the leading digit to the left of the binary point is a one, then the Mod 1 replaces it by a 0. Since a theorem by Martin–Löf guarantees that the binary digits of almost every real number are a random sequence with no apparent order, the time sequence $\{x_n\}$ generated by iterating this map will also be random. In particular, if we call out heads whenever the leading digit is a 1 (which means that x_n lies on the interval $[\frac{1}{2}, 1]$) and tails whenever the leading digit is a 0 (which means that x_n lies on the interval $[0, \frac{1}{2}]$), then the time sequence $\{x_n\}$ generated by this deterministic difference equation will jump back and forth between the left and right halves of the unit interval in a process that is indistinguishable from that generated by a series of coin flips.

The technical definition of chaos refers to the behavior of the time sequence generated by a mixing system, such as the shift map defined by Eq. (2). This simple, deterministic dynamical system with random behavior is the simplest chaotic system, and it serves as the paradigm for all chaotic systems.

3. Extreme Sensitivity to Initial Conditions

One of the essential characteristics of chaotic systems is that they exhibit extreme sensitivity to initial conditions. This means that two trajectories with initial conditions that are arbitrarily close will diverge at an exponential rate. The exponential rate of divergence in mixing systems is related to a positive Kolmogorov–Sinai entropy. For simple systems, such as the one-dimensional maps defined by Eqs. (1) and (2), this local instability is characterized by the average Lyapunov exponent, which in practice is much easier to evaluate than the Kolmogorov–Sinai entropy.

It is easy to see that Eq. (2) exhibits extreme sensitivity to initial conditions with a positive average Lyapunov exponent, while Eq. (1) does not. If we consider two nearby initial conditions x_0 and y_0, which are $d_0 = |x_0 - y_0|$ apart, then after one iteration of a map, $x_{n+1} = F(x_n)$ of the form of Eqs. (1) or (2), the two trajectories will be approximately separated by a distance $d_1 = |(dF/dx)(x_0)| \, d_0$. Clearly, if $|dF/dx| < 1$, the distance between the two points decreases, if $|dF/dx| > 1$, the distance increases, while if $|dF/dx| = 1$, the two trajectories remain approximately the same distance apart. We can easily see by differentiating the map or looking at the slopes of the graphs of the return maps in Figs. 1 and 2 that $|dF/dx| = 1$ for Eq. (1), while $|dF/dx| = 2$ for Eq. (2). Therefore, after many iterations of Eq. (1), nearby initial conditions will generate trajectories that stay close together (the red

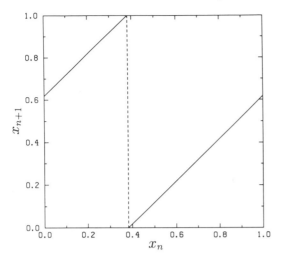

FIG. 1.   A graph of the return map defined by Eq. (1) for $a = (\sqrt{5} - 1)/2 \simeq 0.618$. The successive values of the time sequence $\{x_n\}_{n=1,2,3,...}$ are simply determined by taking the old values of x_n and reading off the new values x_{n+1} from the graph.

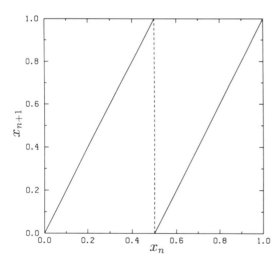

FIG. 2.   A graph of the return map defined by Eq. (2). For values of x_n between 0 and 0.5, the map increases linearly with slope 2; but for x_n larger than 0.5, the action of the Mod 1 requires that the line reenter the unit square at 0.5 and rise again to 1.0.

points remained clumped), while the trajectories generated by Eq. (2) diverge at an exponential rate (the red points mix throughout the phase space). Moreover, the average Lyapunov exponent for these one-dimensional maps, defined by

$$\lambda = \lim_{N \to \infty} \frac{1}{N} \sum_{n=0}^{N} \ln \left| \frac{dF}{dx}(x_n) \right| \qquad (3)$$

provides a direct measure of the exponential rate of divergence of nearby trajectories. Since the slope of the return maps for Eqs. (1) and (2) are the same for almost all values of x_n, the average Lyapunov exponents can be easily evaluated. For Eq. (1), we get $\lambda = 0$, while Eq. (2) gives $\lambda = \log 2 > 0$.

However, it is important to note that all trajectories generated by Eq. (2) do not diverge exponentially. As mentioned earlier, the set of rational x_0's with even denominators generate regular periodic orbits. Although these points are a set of measure zero compared with all of the real numbers on the unit interval, they are dense, which means that in every subinterval, no matter how small, we can always find one of these periodic orbits. The significance of these special trajectories is that, figuratively speaking, they play the role of rocks or obstructions in a rushing stream around which the other trajectories must wander. If this dense set of periodic points were not present in the phase space, then the extreme sensitivity to initial conditions alone would not be sufficient to guarantee mixing behavior. For example, if we iterated Eq. (2) without the Mod 1, then all trajectories would diverge exponentially, but the red points would never be able to spread throughout the accessible space that would consist of the entire positive real axis. In this case, the dynamical system is simply unstable, not chaotic.

4. Unpredictability

One important consequence of the extreme sensitivity to initial conditions is that long-term prediction of the evolution of chaotic dynamical systems, like predicting the weather, is a practical impossibility. Although chaotic dynamical systems are fully deterministic, which means that once the initial conditions are specified the solution of the differential or difference equations are uniquely determined for all time, this does not mean that it is humanly possible to find the solution for all time. If nearby initial conditions diverge exponentially, then any errors in specifying the initial conditions, no matter how small, will also grow exponentially. For example, if we can only specify the initial condition in Eq. (2) to an accuracy of one part in a thousand, then the uncertainty in predicting x_n will double each time step. After only 10 time steps, the uncertainty will be as large as the entire phase space, so that even approximate predictions will be impossible. [If we can specify the initial conditions to double precision accuracy on a digital computer (1 part in 10^{18}), then we can only provide approximate predictions of the future values of x_n for 60 time steps before the error spans the entire unit interval.] In contrast, if we specify the initial condition for Eq. (1) to an accuracy of 10^{-3}, then we can always predict the future values of x_n to the same accuracy. [Of course, errors in the time evolution can also arise from uncertainties in the parameters in the equations of evolution. However, if we can only specify the parameter a in Eq. (1) to an accuracy of 10^{-3}, we could still make approximate predictions for the time sequence for as many as 10^3 iterations before the uncertainty becomes as large as the unit interval.]

B. FRACTALS

Another common feature of chaotic dynamical systems is the natural appearance of geometrical structures with noninteger dimensions. For example, in dissipative dynamical systems, described by systems of differential equations, the presence of dissipation (friction) causes the long-time behavior to converge to a geometrical structure in the phase space called an attractor. The attractor may consist of a single fixed point with dimension 0, a periodic limit cycle described by a closed curve with dimension 1, or if the long-time dynamics is chaotic, the attracting set may resemble a curve with an infinite number of twists, turns, and folds that never closes on itself. This strange attractor is more than a simple curve with dimension 1, but it may fail to completely cover an area of dimension 2. In addition, strange attractors are found to exhibit the same level of structure on all scales. If we look at the complex structure through a microscope, it does not look any simpler no matter how much we increase the magnifications. The term *fractal* was coined by Benoit Mandelbrot to describe these complex geometrical objects. Like the shapes of snowflakes and clouds and the outlines of coastlines and mountain ranges, these fractal objects are best characterized by a noninteger dimension.

The simplest geometrical object with a noninteger dimension is the middle-thirds Cantor set. If we take the unit interval [0, 1] and remove all of the points in the middle third, then we will be left with a set consisting of two pieces $[0, \frac{1}{3}]$ and $[\frac{2}{3}, 1]$ each of length $\frac{1}{3}$. If we remove the middle thirds of these remaining pieces, we get a set consisting of four pieces of length $\frac{1}{9}$. By repeating this construction *ad infinitum*, we end up with a strange set of points called a Cantor set. Although it consists of points, none are isolated. In fact, if we magnify any interval containing elements of the set, for example, the segment contained on the interval $[0, \frac{1}{3}^n]$ for any positive n, then the magnified interval will look the same as the complete set (see Fig. 3).

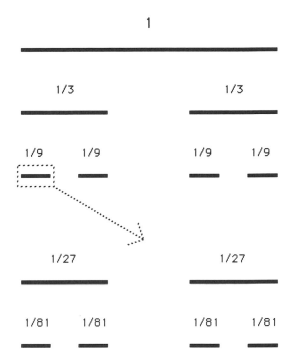

FIG. 3. The middle-thirds Cantor set is constructed by first removing the points in the middle third of the unit interval and then successively removing the middle thirds of the remaining intervals *ad infinitum*. This figure shows the first four stages of the Cantor set construction. After the first two steps, the first segment is magnified to illustrate the self-similar structure of the set.

In order to calculate a dimension for this set, we must first provide a mathematical definition of dimension that agrees with our natural intuition for geometrical objects with integer dimension. Although there are a variety of different definitions of dimension corresponding to different levels of mathematical rigor, one definition that serves our purpose is to define the dimension of a geometrical object in terms of the number of boxes of uniform size required to cover the object. For example, if we consider two-dimensional boxes with sides of length L (for example, $L = 1$ cm), then the number of boxes required to cover a two-dimensional object, $N(L)$, will be approximately equal to the area measured in units of L^2 (that is, cm^2). Now, if we decrease the size of the boxes to L', then the number of boxes $N(L')$ will increase approximately as $(L/L')^2$. [If $L' = 1$ mm, then $N(L') \approx 100N(L)$.] Similarly, if we try to cover a one-dimensional object, such as a closed curve, with these boxes, the number of boxes will increase only as (L/L'). (For $L = 1$ cm and $L' = 1$ mm, $N(L)$ will be approximately equal to the length of the curve in centimeters, while $N(L')$ will be approximately the length of the curve in millimeters.) In general,

$$N(L) \propto L^{-d} \tag{4}$$

where d is the dimension of the object. Therefore, one natural mathematical definition of dimension is provided by the equation

$$d = \lim_{L \to 0} \log N(L)/\log(1/L) \tag{5}$$

obtained by taking the logarithm of both sides of Eq. (4). For common geometrical objects, such as a point, a simple curve, an area, or a volume, this definition yields the usual integer dimensions 0, 1, 2, and 3, respectively. However, for the strange fractal sets associated with many chaotic dynamical systems, this definition allows for the possibility of noninteger values. For example, if we count the number of boxes required to cover the middle-thirds Cantor set at each level of construction, we find that we can always cover every point in the set using 2^n boxes of length $(\frac{1}{3})^n$ (that is, 2 boxes of length $\frac{1}{3}$, 4 boxes of length $\frac{1}{9}$, etc.). Therefore, Eq. (5) yields a dimension of $d = \log 2/\log 3 = 0.63093 \ldots$, which reflects the intricate self-similar structure of this set.

III. Dissipative Dynamical Systems

In this section, we will examine three important examples of simple mathematical models that can exhibit chaotic behavior and that arise in applications to problems in science and engineering. Each represents a dynamical system with dissipation so that the long-time behavior converges to an attractor in the phase space. The examples increase in complexity from a single difference equation, such as Eqs. (1) and (2), to a system of two coupled difference equations and then to a system of three coupled ordinary differential equations. Each example illustrates the characteristic properties of chaos in dissipative dynamical systems with irregular, unpredictable behavior that exhibits extreme sensitivity to initial conditions and fractal attractors.

A. THE LOGISTIC MAP

The first example is a one-dimensional difference equation, like Eqs. (1) and (2), called the logistic map, which is defined by

$$x_{n+1} = ax_n(1 - x_n) \equiv F(x_n). \tag{6}$$

For values of the control parameter a between 0 and 4, this nonlinear difference equation also takes values of x_n between 0 and 1 and returns a value x_{n+1} on the unit interval. However, as a is varied, the time sequences $\{x_n\}$ generated by this map exhibit extraordinary transitions from regular behavior, such as that generated by Eq. (1), to chaos, such as that generated by Eq. (2). Although this mathematical model is too simple to be directly applicable to problems in physics and engineering, which are usually described by differential equations, Mitchell Feigenbaum has shown that the transition from order to chaos in dissipative dynamical systems exhibits universal characteristics (to be discussed later), so that the logistic map is representative of a large class of dissipative dynamical systems. Moreover, since the analysis of this deceptively simple difference equation involves a number of standard techniques used in the study of nonlinear dynamical systems, we will examine it in considerable detail.

As noted in the seminal review article in 1974 by Robert May, a biologist who considered the logistic map as a model for annual variations of insect populations, the time evolution generated by the map can be easily studied using a graphical analysis of the return maps displayed in Fig. 4. Equation (6) describes an inverted parabola that intercepts the $x_{n+1} = 0$ axis at $x_n = 0$ and 1, with a maximum of $x_{n+1} = a/4$ at $x_n = 0.5$. Although this map can be easily iterated using a short computer program, the qualitative behavior of the time sequence $\{x_n\}$ generated by any initial x_0 can be examined by simply tracing lines on the graph of the return map with a pencil as illustrated in Fig. 4.

For values of $a < 1$, almost every initial condition is attracted to $x = 0$ as shown in Fig. 4 for $a = 0.95$. Clearly, $x = 0$ is a fixed point of the nonlinear map. If we start with $x_0 = 0$, then the logistic map returns the value $x_n = 0$ for all future iterations. Moreover, a simple linear analysis, such as that used to define the Lyapunov exponent in Section II, shows that for $a < 1$ this fixed point is stable. (Initial conditions that are slightly displaced from the origin will be attracted back since $|(dF/dx)(0)| = a < 1$.)

However, when the control parameter is increased to $a > 1$, this fixed point becomes unstable and the long-time behavior is attracted to a new fixed point, as shown in Fig. 4 for $a = 2.9$, which lies at the other intersection of the 45° line and the graph of the return map. In this case, the dynamical system approaches an equilibrium with a nonzero value of the dependent variable x. Elementary algebra shows that this point corresponds to the nonzero root of the quadratic equation $x = ax(1 - x)$ given by $x^* =$

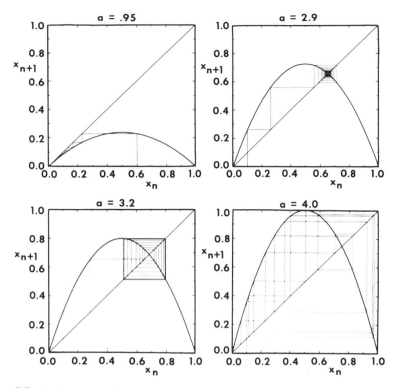

FIG. 4. Return maps for the logistic map, Eq. (6), are shown for 4 different values of the control parameter a. These figures illustrate how pencil and paper can be used to compute the time evolution of the map. For example, if we start our pencil at an initial value of $x_0 = 0.6$ for $a = 0.95$, then the new value of x_1 is determined by tracing vertically to the graph of the inverted parabola. Then, to get x_2, we could return to the horizontal axis and repeat this procedure, but it is easier to simply reflect off the 45° line and return to the parabola. Successive iterations of this procedure give rapid convergence to the stable, fixed point at $x = 0$. However, if we start at $x_0 = 0.1$ for $a = 2.9$, our pencil computer diverges from $x = 0$ and eventually settles down to a stable, fixed point at the intersection of the parabola and the 45° line. Then, when we increase $a > 3$, this fixed point repels the trace of the trajectory, which settles into either a periodic cycle, such as the period 2 cycle for $a = 3.2$, or a chaotic orbit, such as that for $a = 4.0$.

$(a - 1)/a$. Again, a simple linear analysis of small displacements from this fixed point reveals that it remains stable for values of a between 1 and 3. When a becomes larger than 3, this fixed point also becomes unstable and the long-time behavior becomes more complicated, as shown in Fig. 4.

1. Period Doubling

For values of a slightly bigger than 3, empirical observations of the time sequences for this nonlinear dynamical system generated by using a hand calculator, a digital computer, or our "pencil computer" reveals that the long-time behavior approaches a periodic cycle of period 2, which alternates between two different values of x. Because of the large nonlinearity in the difference equation, this periodic behavior could not be deduced from any analytical arguments based on exact solutions or from perturbation theory. However, as typically occurs in the field of nonlinear dynamics, the empirical observations provide us with clues to new analytical procedures for describing and understanding the dynamics. Once again, the graphical analysis provides an easy way of understanding the origin of the period 2 cycle.

Consider a new map,

$$x_{n+2} = F^{(2)}(x_n) = F[F(x_n)]$$
$$= a^2(x_n - x_n^2) - a^3(x_n^2 - 2x_n^3 + x_n^4) \qquad (7)$$

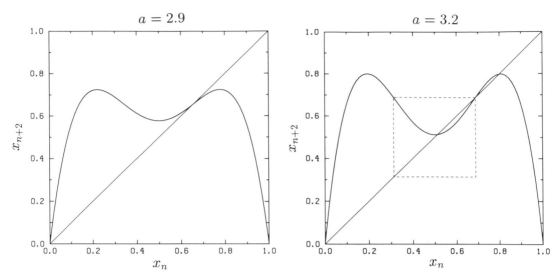

FIG. 5. The return maps are shown for the second iterate of the logistic map, $F^{(2)}$, defined by Eq. (7). The fixed points at the intersection of the 45° line and the map correspond to values of x that repeat every 2 periods. For $a = 2.9$, the 2 intersections are just the period-1 fixed points at 0 and a^*, which repeat every period and therefore every other period as well. However, when a is increased to 3.2, the peaks and valleys of the return map become more pronounced and pass through the 45° line and 2 new fixed points appear. Both of the old, fixed points are now unstable because the absolute value of the slope of the return map is larger than 1, but the new points are stable, and they correspond to the two elements of the period 2 cycle displayed in Fig. 4. Moreover, because the portion of the return map contained in the dashed box resembles an inverted image of the original logistic map, one might expect that the same bifurcation process will be repeated for each of these period 2 points as a is increased further.

constructed by composing the logistic map with itself. The graph of the corresponding return map, which gives the values of x_n every other iteration of the logistic map, is displayed in Fig. 5. If we use the same methods of analysis as we applied to Eq. (6), we find that there can be at most 4 fixed points that correspond to the intersection of the graph of the quartic return map with the 45° line. Because the fixed points of Eq. (4) are values of x that return every other iteration, these points must be members of the period 2 cycles of the original logistic map. However, since the period 1 fixed points of the logistic map at $x = 0$ and x^* are automatically period 2 points, 2 of the fixed points of Eq. (7) must be $x = 0$, x^*. When $1 < a < 3$, these are the only 2 fixed points of Eq. (7), as shown in Fig. 5 for $a = 2.9$. However, when a is increased above 3, 2 new fixed points of Eq. (7) appear, as shown in Fig. 5 for $a = 3.2$, on either side of the fixed point at $x = x^*$, which has just become unstable.

Therefore, when the stable period 1 point at x^* becomes unstable, it gives birth to a pair of fixed points, $x^{(1)}$, $x^{(2)}$ of Eq. (7), which form the elements of the period 2 cycle found empirically for the logistic map. This process is called a pitchfork bifurcation. For values of a just above 3, these new fixed

points are stable and the long-time dynamics of the second iterate of the logistic map, $F^{(2)}$, is attracted to one or the other of these fixed points. However, as a increases, the new fixed points move away from x^*, the graphs of the return maps for Eq. (7) get steeper and steeper, and when $|dF^{(2)}/dx|_{x^{(1)},x^{(2)}} > 1$, the period 2 cycle also becomes unstable. [A simple application of the chain rule of differential calculus shows that both periodic points destabilize at the same value of a, since $F(x^{(1),(2)}) = x^{(2),(1)}$ and $(dF^{(2)}/dx)(x^{(1)}) = (dF/dx)(x^{(2)})(dF/dx)(x^{(1)}) = (dF^{(2)}/dx)(x^{(1)})$.]

Once again, empirical observations of the long-time behavior of the iterates of the map reveal that when the period 2 cycle becomes unstable it gives birth to a stable period 4 cycle. Then as a increases, the period 4 cycle becomes unstable and undergoes a pitchfork bifurcation to a period 16 cycle, then a period 32 cycle, and so on. Since the successive period-doubling bifurcations require smaller and smaller changes in the control parameter, this bifurcation sequence rapidly accumulates to a period cycle of infinite period at $a_x = 3.57\ldots$.

This sequence of pitchfork bifurcations is clearly displayed in the bifurcation diagram shown in Fig. 6. This graph is generated by iterating the map for sev-

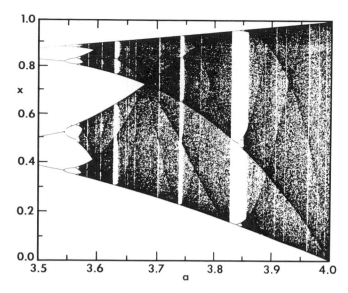

FIG. 6. A bifurcation diagram illustrates the variety of long-time behavior exhibited by the logistic map as the control parameter a is increased from 3.5 to 4.0. The sequences of period-doubling bifurcations from the period 4 to the period 8 to the period 16 are clearly visible in addition to ranges of a in which the orbits appear to wander over continuous intervals and ranges of a in which periodic orbits, including odd periods, appear to emerge from the chaos.

eral hundred time steps for successive values of a. For each value of a, we plot only the last hundred values of x_n to display the long-tme behavior. For $a < 3$, all of these points land close to the fixed point at a^*; for $a > 3$, these points alternate between the 2 period 2 points, then between the 4 period 4 points, and so on.

The origin of each of these new periodic cycles can be qualitatively understood by applying the same analysis that we used to explain the birth of the period 2 cycle from the period 1. For the period 4 cycle, we consider the second iterate of the period 2 map,

$$x_{n+4} = F^{(4)}(x_n) = F^{(2)}[F^{(2)}(x_n)] \\ = F\{F[F(F(x_n))]\} \tag{8}$$

In this case, the return map is described by polynomial of degree 16 that can have as many as 16 fixed points that correspond to intersections of the 45° line with the graph of the return map. Two of these period 4 points correspond to the period 1 fixed points at 0 and x^*, and for $a > 3, 2$ correspond to the period 2 points at $x^{(1)}$ and $x^{(2)}$. The remaining 12 period 4 points can form 3 different period 4 cycles that appear for different values of a. Figure 7 shows a graph of $F^{(4)}(x_n)$ for $a = 3.2$, where the period 2 cycle is still stable, and for $a = 3.5$, where the unstable period 2 cycle has bifurcated into a period 4 cycle.

(The other 2 period 4 cycles are only briefly stable for other values of $a > a_*$.)

We could repeat the same arguments to describe the origin of the period 8; however, now the graph of the return map of the corresponding polynomial of degree 32 would begin to tax the abilities of our graphics display terminal as well as our eyes. Fortunately, the "slaving" of the stability properties of each periodic point via the chain rule argument (described previously for the period 2 cycle) means that we only have to focus on the behavior of the successive iterates of the map in the vicinity of the periodic point closest to $x = 0.5$. In fact, a close examination of Figs. 4, 5, and 7 reveals that the bifurcation process for each $F^{(n)}$ is simply a miniature replica of the original period-doubling bifurcation from the period 1 cycle to the period 2 cycle. In each case, the return map is locally described by a parabolic curve (although it is not exactly a parabola beyond the first iteration and the curve is flipped over for every other $F^{(N)}$).

Because each successive period-doubling bifurcation is described by the fixed points of a return map $x_{n+N} = F^{(N)}(x_n)$ with ever greater oscillations on the unit interval, the amount the parameter a must increase before the next bifurcation decreases rapidly, as shown in the bifurcation diagram in Fig. 6. The

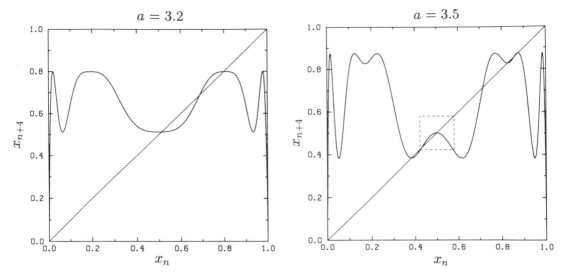

FIG. 7.   The appearance of the period 4 cycle as a is increased from 3.2 to 3.5 is illustrated by these graphs of the return maps for the fourth iterate of the logistic map, $F^{(4)}$. For $a = 3.2$, there are only 4 period 4 fixed points that correspond to the 2 unstable period 1 points and the 2 stable period 2 points. However, when a is increased to 3.5, the same process that led to the birth of the period 2 fixed points is repeated again in miniature. Moreover, the similarity of the portion of the map near $x_n = 0.5$ to the original map indicates how this same bifurcation process occurs again as a is increased.

differences in the changes in the control parameter for each succeeding bifurcation, $a_{n+1} - a_n$, decreases at a geometric rate that is found to rapidly converge to a value of

$$\delta = \frac{a_n - a_{n-1}}{a_{n+1} - a_n} = 4.6692016 \ldots \quad (9)$$

In addition, the maximum separation of the stable daughter cycles of each pitchfork bifurcation also decreases rapidly, as shown in Fig. 6, by a geometric factor that rapidly converges to

$$\alpha = 2.502907875 \ldots \quad (10)$$

2. Universality

The fact that each successive period doubling is controlled by the behavior of the iterates of the map, $F^{(N)}(x)$, near $x = 0.5$, lies at the root of a very significant property of nonlinear dynamical systems that exhibit sequences of period-doubling bifurcations called universality. In the process of developing a quantitative description of the period doubling in the logistic map, Mitchell Feigenbaum discovered that the precise functional form of the map did not seem to matter. For example, he found that a map on the unit interval described by $F(x) = a \sin \pi x$ gave a similar sequence of period-doubling bifurcations.

Although the values of the control parameter a at which each period-doubling bifurcation occurs are different, he found that both the ratios of the changes in the control parameter and the separations of the stable daughter cycles decreased at the same geometrical rates δ and α as the logistic map.

This observation ultimately led to a rigorous proof, using the mathematical methods of the renormalization group borrowed from the theory of critical phenomena, that these geometrical ratios were universal numbers that would apply to the quantitative description of any period-doubling sequence generated by nonlinear maps with a single quadratic extremum. The logistic map and the sine map are just two examples of this large universality class. The great significance of this result is that the global details of the dynamical system do not matter. A thorough understanding of the simple logistic map is sufficient for describing both qualitatively and, to a large extent, quantitatively the period-doubling route to chaos in a wide variety of nonlinear dynamical systems. In fact, we will see that this universality class extends beyond one-dimensional maps to nonlinear dynamical systems described by more realistic physical models corresponding to two-dimensional maps, systems of ordinary differential equations, and even partial differential equations.

3. Chaos

Of course, these stable periodic cycles, described by Feigenbaum's universal theory, are not chaotic. Even the cycle with an infinite period at the period-doubling accumulation point $a_\times$ has a zero average Lyapunov exponent. However, for many values of a above $a_\times$, the time sequences generated by the logistic map have a positive average Lyapunov exponent and therefore satisfy the definition of chaos. Figure 8 plots the average Lyapunov exponent computed numerically using Eq. (3) for the same range of values of a, as displayed in the bifurcation diagram in Fig. 6.

Wherever the trajectory appears to wander chaotically over continuous intervals, the average Lyapunov exponent is positive. However, embedded in the chaos for $a_\times < a < 4$ we see stable period attractors in the bifurcation diagram with sharply negative average Lyapunov exponents. The most prominent periodic cycle is the period 3 cycle, which appears near $a_3 = 3.83 \ldots$. In fact, between $a_\times$ and a_3, there is a range of values of a for which cycles of every odd and even period are stable. However, the intervals for the longer cycles are too small to discern in Fig. 6. The period 5 cycle near $a = 3.74$ and the period 6 cycles near $a = 3.63$ and $a = 3.85$ are the most readily apparent in both the bifurcation diagram and the graph of the average Lyapunov exponent.

Although these stable periodic cycles are mathematically dense over this range of control parameters, the values of a where the dynamics are truly chaotic can be mathematically proven to be a significant set with nonzero measure. The proof of the positivity of the average Lyapunov exponent is much more difficult for the logistic map than for Eq. (2) since $\log |(dF/dx)(x_n)|$ can take on both negative and positive values depending on whether x_n is close to $\frac{1}{2}$ or to 0 or 1. However, one simple case for which the logistic map is easily proven to be chaotic is for $a = 4$. In this case, the time sequence appears to wander over the entire unit interval in the bifurcation diagram, and the numerically computed average Lyapunov exponent is positive. If we simply change variables from x_n to $y_n = (2/\pi) \sin^{-1} \sqrt{x_n}$, then the logistic map for $a = 4$ transforms to the tent map:

$$y_{n+1} = \begin{cases} 2y_n & 0 \le y_n \le 0.5 \\ 2(1 - y_n) & 0.5 \le y_n \le 1 \end{cases} \quad (11)$$

which is closely related to the shift map, Eq. (2). In particular, since $|dF/dy| = 2$, the average Lyapunov exponent is found to be $\lambda = \log 2 \approx 0.693$, which is the same as the numerical value for the logistic map.

B. THE HÉNON MAP

Most nonlinear dynamical systems that arise in physical applications involve more than one dependent variable. For example, the dynamical description of any mechanical oscillator requires at least two variables—a position and a momentum variable. One of the simplest dissipative dynamical systems that describes the coupled evolution of two variables was introduced by Michel Hénon in 1976. It is defined by taking a one-dimensional quadratic map for x_{n+1} similar to the logistic map and coupling it to a second linear map for y_{n+1}:

$$x_{n+1} = 1 - ax_n^2 + y_n \quad (12a)$$

$$y_{n+1} = bx_n \quad (12b)$$

This pair of difference equations takes points in the $x–y$ plane with coordinates (x_n, y_n) and maps them to new points (x_{n+1}, y_{n+1}). The behavior of the sequence of points generated by successive iterates of this two-dimensional map from an initial point (x_0, y_0) is determined by the values of two control parameters a and b. If a and b are both 0, then Eqs. (12) map every point in the plane to the attracting fixed point at $(1, 0)$ after at most two iterations.

If $b = 0$ but a is nonzero, then the Hénon map reduces to a one-dimensional quadratic map that can be transformed into the logistic map by shifting the variable x. Therefore, for $b = 0$ and even for b

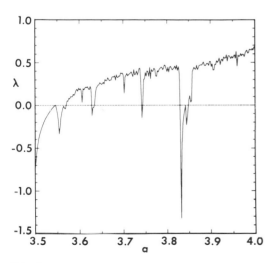

FIG. 8.   The values of the average Lyapunov exponent, computed numerically using Eq. (3), are displayed for the same values of a shown in Fig. 6. Positive values of λ correspond to chaotic dynamics, while negative values represent regular, periodic motion.

small, the behavior of the time sequence of points generated by the Hénon map closely resembles the behavior of the logistic map. For small values of a, the long-time iterates are attracted to stable periodic orbits that exhibit a sequence of period-doubling bifurcations to chaos as the nonlinear control parameter a is increased. For small but nonzero b, the main difference from the one-dimensional maps is that these regular orbits of period N are described by N points in the $(x-y)$ plane rather than points on the unit interval. (In addition, the basin of attraction for these periodic cycles consists of a finite region in the plane rather than the unit interval alone. Just as in the one-dimensional logistic map, if a point lies outside this basin of attraction, then the successive iterates diverge to ∞.)

The Hénon map remains a dissipative map with time sequences that converge to a finite attractor as long as b is less than 1. This is easy to understand if we think of the action of the map as a coordinate transformation in the plane from the variables (x_n, y_n) to (x_{n+1}, y_{n+1}). From elementary calculus, we know that the Jacobian of this coordinate transformation, which is given by the determinant of the matrix

$$M = \begin{pmatrix} -2ax & 1 \\ b & 0 \end{pmatrix} \tag{13}$$

describes how the area covered by any set of points increases or decreases under the coordinate transformation. In this case, $J = \text{Det } M = -b$. When $|J| > 1$, areas grow larger and sets of initial conditions disperse throughout the $x-y$ plane under the iteration of the map. But when $|J| < 1$, the areas decrease under each iteration, so areas must contract to sets of points that correspond to the attractors.

1. Strange Attractors

However, these attracting sets need not be a simple fixed point or a finite number of points forming a periodic cycle. In fact, when the parameters a and b have values that give rise to chaotic dynamics, the attractors can be exceedingly complex, composed of an uncountable set of points that form intricate patterns in the plane. These strange attractors are best characterized as fractal objects with noninteger dimensions.

Figure 9 displays 10,000 iterates of the Hénon map for $a = 1.4$ and $b = 0.3$. [In this case, the initial point was chosen to be $(x_0, y_0) = (0, 0)$, but any initial point in the basin of attraction would give similar results.] Because $b < 1$, the successive iterates rapidly converge to an intricate geometrical

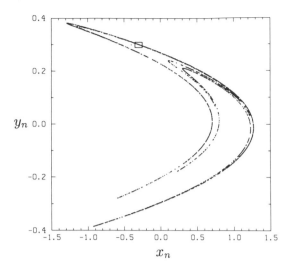

FIG. 9. The first 10,000 iterates of the two-dimensional Hénon map trace the outlines of a strange attractor in the x_n-y_n plane. The parameters were chosen to be $a = 1.4$ and $b = 0.3$ and the initial point was $(0, 0)$.

structure that looks like a line that is folded on itself an infinite number of times. The magnifications of the sections of the attractor shown in Fig. 10 display the detailed self-similar structure. The cross sections of the folded line resemble the Cantor set described in Section II, B. Therefore, since the attractor is more than a line but less than an area (since there are always gaps between the strands at every magnification), we might expect it to be characterized by a fractal dimension that lies between 1 and 2. In fact, an application of the box-counting definition of fractal dimension given by Eq. (5) yields a fractal dimension of $d = 1.26 \ldots$.

Moreover, if you were to watch a computer screen while these points are plotted, you would see that they wander about the screen in a very irregular manner, slowly revealing this complex structure. Numerical measurements of the sensitivity to initial conditions and of the average Lyapunov exponents (which are more difficult to compute than for one-dimensional maps) indicate that the dynamics on this strange attractor are indeed chaotic.

C. The Lorenz Attractor

The study of chaos is not restricted to nonlinear difference equations such as the logistic map and the Hénon map. Systems of coupled nonlinear differential equations also exhibit the rich variety of behavior that we have already seen in the simplest nonlinear dynamical systems described by maps. A classic ex-

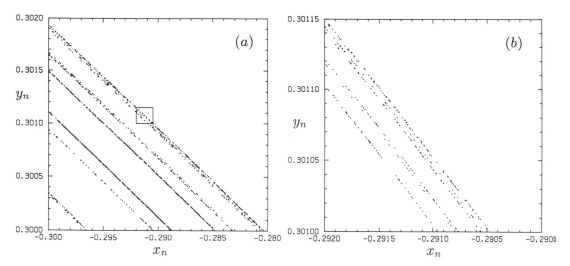

FIG. 10.   To see how strange the attractor displayed in Fig. 9 really is, we show two successive magnifications of a strand of the attractor contained in the box in Fig. 9. Here, (a) shows that the single strand in Fig. 9 breaks up into several distinct bands, which shows even finer structure in (b) when the map is iterated 10,000,000 time steps.

ample is provided by the Lorenz model described by three coupled nonlinear differential equations:

$$dx/dt = -\sigma x + \sigma y \qquad (14a)$$

$$dy/dt = -xz + rx - y \qquad (14b)$$

$$dz/dt = xy - bz \qquad (14c)$$

These equations were introduced in 1963 by Edward Lorenz, a meteorologist, as a severe truncation of the Navier–Stokes equations describing Rayleigh–Benard convection in a fluid (like Earth's atmosphere), which is heated from below in a gravitational field. The dependent variable x represents a single Fourier mode of the stream function for the velocity flow and the variables y and z represent two Fourier components of the temperature field, and the constants r, σ, and b are the Rayleigh number, the Prandtl number, and a geometrical factor, respectively.

The Lorenz equations provide our first example of a model dynamical system that is reasonably close to a real physical system. (The same equations provide an even better description of optical instabilities in lasers, and similar equations have been introduced to describe chemical oscillators.) Numerical studies of the solutions of these equations, starting with Lorenz's own pioneering work using primitive digital computers in 1963, have revealed the same complexity as the Hénon map. In fact, Hénon originally introduced Eqs. (12) as a simple model that exhibits the essential properties of the Lorenz equations.

A linear analysis of the evolution of small volumes in the three-dimensional phase space spanned by the dependent variables x, y, and z shows that this dissipative dynamical system rapidly contracts sets of initial conditions to an attractor. When the Rayleigh number r is less than 1, the point $(x, y, z) = (0, 0, 0)$ is an attracting fixed point. But when $r > 1$, a wide variety of different attractors that depend in a complicated way on all three parameters r, σ, and b are possible. Like the Hénon map, the long-time behavior of the solutions of these differential equations can be attracted to fixed points, to periodic cycles, which are described by limit cycles consisting of closed curves in the three-dimensional phase space, and to strange attractors, which are described by a fractal structure in phase space. In the first two cases, the dynamics is regular and predictable, but the dynamics on strange attractors is chaotic and unpredictable (as unpredictable as the weather).

The possibility of strange attractors for three or more autonomous differential equations, such as the Lorenz model, was established mathematically by Ruelle and Takens. Figure 11 shows a three-dimensional graph of the famous strange attractor for the Lorenz equations corresponding to the values of the parameters $r = 28$, $\sigma = 10$, and $b = \frac{8}{3}$, which provides a graphic illustration of the consequences of their theorem. The initial conditions were chosen to be $(1, 1, 1)$. The trajectory appears to loop around on two surfaces that resemble the wings of a butterfly, jumping from one wing to the other in an irregular manner. However, a close inspection of these surfaces reveals that under successive magnification they exhibit the same kind of intricate, self-similar structure as the striations of the Hénon attractor.

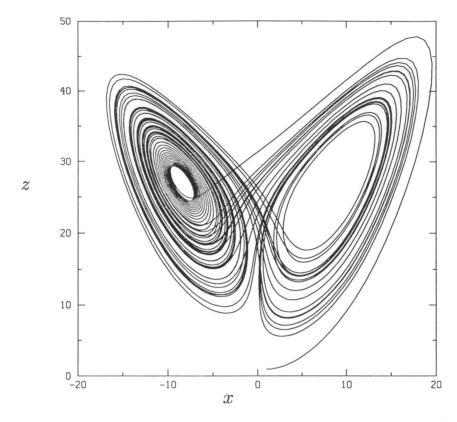

FIG. 11. The solution of the Lorenz equations for the parameters $r = 28$, $\sigma = 10$, and $b = \frac{8}{3}$ rapidly converges to a strange attractor. This figure shows a projection of this three-dimensional attractor onto the $x-z$ plane, which is traced out by approximately 100 turns of the orbit.

This detailed structure is best revealed by a so-called Poincaré section of continuous dynamics, shown in Fig. 12, which was generated by plotting a point in the $x-z$ plane every time the orbit passes from negative y to positive y.

Since we could imagine that this Poincaré section was generated by iterating a pair of nonlinear difference equations, such as the Hénon map, it is easy to understand, by analogy with the analysis described in Section III, B, how the time evolution can be chaotic with extreme sensitivity to initial conditions and how this cross section of the Lorenz attractor, as well as the Lorenz attractor itself, can have a noninteger, fractal dimension.

D. APPLICATIONS

Perhaps the most significant conclusion that can be drawn from these three examples of dissipative dynamical systems that exhibit chaotic behavior is that the essential features of the behavior of the more realistic Lorenz model are well described by the prop-

erties of the much simpler Hénon map and to a large extent by the logistic map. These observations provide strong motivation to hope that simple nonlinear systems will also capture the essential properties of even more complex dynamical systems that describe a wide variety of physical phenomena with irregular behavior. In fact, the great advances of nonlinear dynamics and the study of chaos in the last 10 years can be attributed to the fulfillment of this hope in both numerical studies of more complicated mathematical models and experimental studies of a variety of complicated natural phenomena.

The successes of this program of reducing the essential features of complicated dynamical processes to simple nonlinear maps or to a few coupled, nonlinear differential equations have been well documented in a number of conference proceedings and textbooks. For example, the universality of the period-doubling route to chaos and the appearance of strange attractors have been demonstrated in numerical studies of a wide variety of nonlinear maps, systems of nonlinear, ordinary, and partial differen-

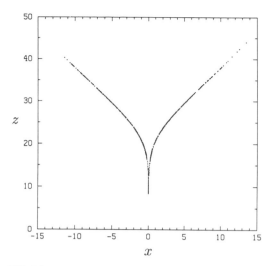

FIG. 12. A Poincaré section of the Lorenz attractor is constructed by plotting a cross section of the butterfly wings. This graph is generated by plotting the point in the x–z plane each time the orbit displayed in Fig. 11 passes through $y = 0$. This view of the strange attractor is analogous to that displayed in Fig. 9 for the Hénon map. This figure appears to consist of only a single strand, but this is because of the large contraction rate of the Lorenz model. Successive magnifications would reveal a fine-scale structure similar to that shown in Fig. 10 for the Hénon map.

tial equations. Even more importantly, Feigenbaum's universal constants δ and α, which characterize the quantitative scaling properties of the period-doubling sequence, have been measured to good accuracy in a number of careful experiments on Rayleigh–Bénard convection and nonlinear electrical circuits and in oscillating chemical reactions (such as the Belousov–Zabotinsky reaction), laser oscillators, acoustical oscillators, and even the response of heart cells to electrical stimuli. In addition, a large number of papers have been devoted to the measurement of the fractal dimensions of strange attractors that may govern the irregular, chaotic behavior of chemical reactions, turbulent flows, climatic changes, and brain-wave patterns.

However, the complete description of turbulence remains an outstanding unsolved problem. Nevertheless, the results of studying these simple nonlinear dynamical systems as paradigms for fully developed turbulence have introduced promising new tools and a fresh, new way of approaching this very old problem. Perhaps the most important lesson of nonlinear dynamics has been the realization that complex behavior need not have complex causes and that many aspects of irregular, unpredictable phenomena may be understood in terms of simple nonlinear models.

However, the study of chaos also teaches us that despite an underlying simplicity and order we will never be able to describe the precise behavior of chaotic systems analytically nor will we succeed in making accurate long-term predictions no matter how much computer power is available. In fact, the problem of turbulence may never be solved in the sense that physical scientists usually mean. At the very best, we may hope to discern some of this underlying order in an effort to develop reliable statistical methods for making predictions for average properties of chaotic systems.

IV. Hamiltonian Systems

Although most physics textbooks on classical mechanics are largely devoted to the description of Hamiltonian systems in which dissipative, frictional forces can be neglected, such systems are rare in nature. The most important examples arise in celestial mechanics, which describes the motions of planets and stars; accelerator design, which deals with tenuous beams of high-energy charged particles moving in guiding magnetic fields; and the physics of magnetically confined plasmas, which is primarily concerned with the dynamics of trapped electrons and ions in high-temperature fusion devices. Although few in number, these examples are very important.

In this section, we will examine three simple examples of classical Hamiltonian systems that exhibit chaotic behavior. The first example is the well-known baker's transformation, which clearly illustrates the fundamental concepts of chaotic behavior in Hamiltonian systems. Although it has no direct applications to physical problems, the baker's transformation, like the logistic map, serves as a paradigm for all chaotic Hamiltonian systems. The second example is the standard map, which has direct applications in the description of the behavior of a wide variety of periodically perturbed nonlinear oscillators ranging from particle motion in accelerators and plasma fusion devices to the irregular rotation of Hyperion, one of the moons of Saturn. Finally, we will consider the Hénon–Heiles model, which corresponds to an autonomous Hamiltonian system with two degrees of freedom, describing, for example, the motion of a particle in a nonaxisymmetric, two-dimensional potential well or the interaction of three nonlinear oscillators (the 3-body problem).

A. The Baker's Transformation

The description of a Hamiltonian system, like a frictionless mechanical oscillator, requires at least two dependent variables that usually correspond to a generalized position variable and a generalized momentum variable. These variables define a phase space for the mechanical system, and the solutions of the equations of motion describe the motion of a point in the phase space. Starting from the initial conditions specified by an initial point in the 2–d plane, the time evolution generated by the equations of motion trace out a trajectory or orbit.

The distinctive feature of Hamiltonian systems is that the areas or volumes of small sets of initial conditions are preserved under the time evolution, in contrast to the dissipative systems, such as the Hénon map or the Lorenz model, where phase-space volumes are contracted. Therefore, Hamiltonian systems are not characterized by attractors, either regular or strange, but the dynamics can nevertheless exhibit the same rich variety of behavior with regular periodic and quasi-periodic cycles and chaos.

The simplest Hamiltonian systems correspond to area-preserving maps on the x–y plane. One well-studied example is the so-called baker's transformation, defined by a pair of difference equations:

$$x_{n+1} = 2x_n, \qquad \text{Mod } 1 \qquad (15a)$$

$$y_{n+1} = \begin{cases} 0.5y_n & 0 \leq x_n \leq 0.5 \\ 0.5(y_n + 1) & 0.5 \leq x_n \leq 1 \end{cases} \qquad (15b)$$

The action of this map is easy to describe by using the analogy of how a baker kneads dough (hence, the origin of the name of the map). If we take a set of points (x_n, y_n) covering the unit square ($0 \leq x_n \leq 1$ and $0 \leq y_n \leq 1$), Eq. (15a) requires that each value of x_n be doubled so that the square (or dough) is stretched out in the x direction to twice its original length. Then, Eq. (15b) reduces the values of y_n by a factor of two and simultaneously cuts the resulting rectangular set of points (or dough) in half at $x = 1$ and places one piece on top of the other, which returns the dough to its original shape, as shown in Fig. 13. Then, this dynamical process (or kneading) is repeated over and over again.

Since area is preserved under each iteration, this dynamical system is Hamiltonian. This can be easily seen mathematically if we think of the successive iterates of the baker's transformation as changes of coordinates from x_n, y_n to x_{n+1}, y_{n+1}. As in the case of the Hénon map, we can analyze the effects of this transformation by evaluating the Jacobian of the coordinate transformation that is the determinant of the matrix

$$M = \begin{pmatrix} 2 & 0 \\ 0 & \tfrac{1}{2} \end{pmatrix} \qquad (16)$$

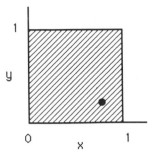

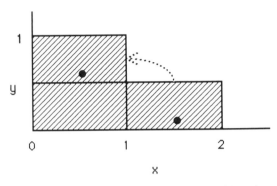

FIG. 13.   The baker's transformation takes all of the points in the unit square (the dough), compresses them vertically by a factor of $\tfrac{1}{2}$, and stretches them out horizontally by a factor of 2. Then, this rectangular set of points is cut at $x = 1$, the two resulting rectangles are stacked one on top of the other to return the shape to the unit square, and the transformation is repeated over again. In the process, a "raisin," indicated schematically by the black dot, wanders chaotically around the unit square.

Since $J = \text{Det } M = 1$, we know from elementary integral calculus that volumes are preserved by this change of variables.

1. Chaotic Mixing

Starting from a single initial condition (x_0, y_0), the time evolution will be described by a sequence of points in the plane. [To return to the baking analogy we could imagine that (x_0, y_0) specifies the initial coordinate of a raisin in the dough.] For almost all initial conditions, the trajectories generated by this simple, deterministic map will be chaotic. Because the evolution of the x coordinate is completely determined by the one-dimensional, chaotic shift map, Eq. (2), the trajectory will move from the right half to the left half of the unit square in a sequence that is indistinguishable from the sequence of heads and tails generated by flipping a coin. Moreover, since the location of the orbit in the upper half or lower half of the unit square is determined by the same random sequence, the successive iterates of the initial point (the raisin) will wander around the unit square in a chaotic fashion.

In this simple model, it is easy to see that the mechanism responsible for the chaotic dynamics is the process of stretching and folding of the phase space. In fact, this same stretching and folding lies at the root of all chaotic behavior in both dissipative and Hamiltonian systems. The stretching is responsible for the exponential divergence of nearby trajectories, which is the cause of the extreme sensitivity to initial conditions that characterizes chaotic dynamics. The folding ensures that trajectories return to the initial region of phase space so that the unstable system does not simply explode.

Since the stretching only occurs in the x direction for the baker's transformation, we can easily compute the value of the exponential divergence of nearby trajectories, which is simply the logarithm of the largest eigenvalue of the matrix M. Therefore, the baker's transformation has a positive Kolmogorov–Sinai entropy, $\lambda = \log 2$, so that the dynamics satisfy our definition of chaos.

B. The Standard Map

Our second example of a simple Hamiltonian system that exhibits chaotic behavior is the standard map described by the pair of nonlinear difference equations

$$x_{n+1} = x_n + y_{n+1}, \qquad \text{Mod } 2\pi \qquad (17a)$$

and

$$y_{n+1} = y_n + k \sin x_n \qquad \text{Mod } 2\pi \qquad (17b)$$

Starting from an initial point (x_0, y_0) on the "2π square," Eq. (17b) determines the new value for y_1 and Eq. (17a) gives the new value of x_1. The behavior of the trajectory generated by successive iterates is determined by the control parameter k, which measures the strength of the nonlinearity.

The standard map provides a remarkably good model of a wide variety of physical phenomena that are properly described by systems of nonlinear differential equations (hence the name standard map). In particular, it serves as a paradigm for the response of all nonlinear oscillators to periodic perturbations. For example, it provides an approximate description of a particle interacting with a broad spectrum of traveling waves, an electron moving in the imperfect magnetic fields of magnetic bottles used to confine fusion plasmas, and the motion of an electron in a highly excited hydrogen atom in the presence of intense electromagnetic fields. In each case, x_n and y_n correspond to the values of the generalized position and momentum variables, respectively, at discrete times n. Since this model exhibits most of the generic features of Hamiltonian systems that exhibit a transition from regular behavior to chaos, we will examine this example in detail.

The standard map actually provides the exact mathematical description for one physical system called the "kicked rotor." Consider a rigid rotor in the absence of any gravitational or frictional forces that is subject to periodic kicks every unit of time $n = 1$, $2, 3 \ldots$. Then, x_n and y_n describe the angle and the angular velocity (angular momentum) just before the nth kick. The rotor can be kicked either forward or backward depending on the sign of $\sin x_n$, and the strengths of the kicks are determined by the value of k. As the nonlinear parameter k is increased, the trajectories generated by this map exhibit a dramatic transition from regular, ordered behavior to chaos. This remarkable transformation is illustrated in Fig. 14, where a number of trajectories are plotted for four different values of k.

When $k = 0$, the value of y remains constant at y_0 and the value of x_n increases each iteration by the amount y_0 (Mod 2π, which means that if x_n does not lie on the interval $[0, 2\pi]$ we add or subtract 2π until it does). In this case, the motion is regular and the trajectories trace out straight lines in the phase space. The rotor rotates continuously at the constant angular velocity y_0. If y_0 is a rational multiple of 2π, then Eq. (17a), like Eq. (1), exhibits a periodic cycle. However, if y_0 is an irrational multiple of 2π, then the dynamics is quasi-periodic for almost all initial values of x_0 and the points describing the orbit gradually trace out a solid horizontal line in the phase space.

1. Resonance Islands

As k is increased to $k = 0.5$, most of the orbits remain regular and lie on smooth curves in the phase space; however, elliptical islands begin to appear around the point $(\pi, 0) = (\pi, 2\pi)$. (Remember, the intrinsic periodicity of the map implies that the top of the 2π square is connected to the bottom and the right-hand side to the left.) These islands correspond to a resonance between the weak periodic kicks and the rotational frequency of the rotor. Consequently, when the kicks and the rotations are synchronous, the rotor is accelerated. However, because it is a nonlinear oscillator (as opposed to a linear, harmonic oscillator), the rotation frequency changes as the

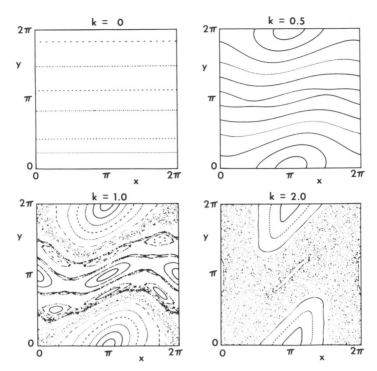

FIG. 14. Successive iterates of the two-dimensional standard map for a number of different initial conditions are displayed for 4 values of the control parameter k. For small values of k, the orbits trace out smooth, regular curves in the two-dimensional phase space that become more distorted by resonance effects as k increases. For $k = 1$, the interaction of these resonances has generated visible regions of chaos in which individual trajectories wander over large regions of the phase space. However, for $k = 1$ and $k = 2$, the chaotic regions coexist with regular islets of stability associated with strong nonlinear resonances. The boundaries of the chaotic regions are defined by residual KAM curves. For still larger values of k (not shown), these regular islands shrink until they are no longer visible in these figures.

velocity increases so that the motion goes out of resonance and therefore the kicks retard the motion and the velocity decreases until the rotation velocity returns to resonance, then this pattern is repeated. The orbits associated with these quasi-periodic cycles of increasing and decreasing angular velocity trace out elliptical paths in the phase space, as shown in Fig. 14.

The center of the island, $(\pi, 0)$, corresponds to a period 1 point of the standard map. [This is easy to check by simply plugging $(\pi, 0)$ into the right-hand side of Eqs. (17).] Figure 14 also shows indications of a smaller island centered at (π, π). Again, it is easy to verify that this point is a member of a period 2 cycle [the other element is the point $(2\pi, \pi) = (0, \pi)$]. In fact, there are resonance islands described by chains of ellipses throughout the phase space as-

sociated with periodic orbits of all orders. Howeve most of these islands are much too small to show u in the graphs displayed in Fig. 14.

As the strength of the kicks increases, these islands increase in size and become more prominent. For $k = 1$, several different resonance island chains are clearly visible corresponding to the period 1, period 2, period 3, and period 7 cycles. However, as the resonance regions increase in size and they begin to overlap, individual trajectories between the resonance regions become confused, and the motion becomes chaotic. These chaotic orbits no longer lie on smooth curves in the phase space but begin to wander about larger and larger areas of the phase space as k is increased. For $k = 2$, a single orbit wanders over more than half of the phase space, and for $k = 5$ (not shown), a single orbit would appear to uniformly cover the entire 2π square (although a microscopic examination would always reveal small regular regions near some periodic points).

2. The Kolmogorov–Arnold–Moser Theorem

Since the periodic orbits and the associated resonance regions are mathematically dense in the phase space (though a set of measure zero), there are always small regions of chaos for any nonzero value of k. However, for small values of k, an important mathematical theorem, called the Kolmogorov–Arnold–Moser (KAM) theorem, guarantees that if the perturbation applied to the integrable Hamiltonian system is sufficiently small, then most of the trajectories will lie on smooth curves, such as those displayed in Fig. 14 for $k = 0$ and $k = 0.5$. However, Fig. 14 clearly shows that some of these so-called KAM curves (also called KAM surfaces or invariant tori in higher dimensions) persist for relatively large values of $k \sim 1$.

The significance of these KAM surfaces is that they form barriers in the phase space. Although these barriers can be circumvented by the slow process of Arnold diffusion in four or more dimensions, they are strictly confining in the two-dimensional phase space of the standard map. This means that orbits starting on one side cannot cross to the other side, and the chaotic regions will be confined by these curves. However, as resonance regions grow with increasing k and begin to overlap, these KAM curves are destroyed and the chaos spreads, as shown in Fig. 14.

The critical k_c for this onset of global chaos can be estimated analytically using Chirikov's resonance overlap criteria, which yields an approximate value of $k_c \approx 2$. However, a more precise value of k_c can be determined by a detailed examination of the breakup of the last confining KAM curve. Since the resonance regions associated with low-order periodic orbits are the largest, the last KAM curve to survive is the one furthest from a periodic orbit. This corresponds to an orbit with the most irrational value of average rotation frequency, which is the golden mean $- (\sqrt{5} - 1)/2$. Careful numerical studies of the standard map show that the golden mean KAM curve, which is the last smooth curve to divide the top of the phase space from the bottom, is destroyed for $k \simeq 1$ (more precisely for $k_c = 0.971635406$). For $k > k_c$, Mackay, Meiss, and Percival have shown that this last confining curve breaks up into a so-called cantorus, which is a curve filled with gaps resembling a Cantor set. These gaps allow chaotic trajectories to leak through so that single orbits can wander throughout large regions of the phase space, as shown in Fig. 14 for $k = 2$.

3. Chaotic Diffusion

Because of the intrinsic nonlinearity of Eq. (17b), the restriction of the map to the 2π square was only a graphical convenience that exploited the natural periodicities of the map. However, in reality, both the angle variable and the angular velocity of a real physical system described by the standard map can take on all real values. In particular, when the golden mean KAM torus is destroyed, the angular velocity associated with the chaotic orbits can wander to arbitrarily large positive and negative values.

Because the chaotic evolution of both the angle and angular velocity appears to execute a random walk in the phase space, it is natural to attempt to describe the dynamics using a statistical description despite the fact that the underlying dynamical equations are fully deterministic. In fact, when $k \gg k_c$, careful numerical studies show that the evolution of an ensemble of initial conditions can be well described by a diffusion equation. Consequently, this simple deterministic dynamical system provides an interesting model for studying the problem of the microscopic foundations of statistical mechanics, which is concerned with the question of how the reversible and deterministic equations of classical mechanics can give rise to the irreversible and statistical equations of classical statistical mechanics and thermodynamics.

C. The Hénon–Heiles Model

Our third example of a Hamiltonian system that exhibits a transition from regular behavior to chaos is described by a system of four coupled, nonlinear differential equations. It was originally introduced by Michel Hénon and Carl Heiles in 1964 as a model of the motion of a star in a nonaxisymmetric, two-dimensional potential corresponding to the mean gravitational field in a galaxy. The equations of motion for the two components of the position and momentum,

$$dx/dt = p_x \tag{18a}$$

$$dy/dt = p_y \tag{18b}$$

$$dp_x/dt = -x - 2xy \tag{18a}$$

$$dp_y/dt = -y + y^2 - x^2 \tag{18b}$$

are generated by the Hamiltonian

$$H(x, y, p_x, p_y)$$
$$= \frac{p_x^2}{2} + \frac{p_y^2}{2} + \frac{1}{2}(x^2 + y^2) + x^2 y - \frac{1}{3} y^3 \tag{19}$$

where the mass is taken to be unity. Equation 19 corresponds to the Hamiltonian of two uncoupled harmonic oscillators $H_0 = (p_x^2/2) + (p_y^2/2) + \frac{1}{2}(x^2 + y^2)$ (consisting of the sum of the kinetic and a quadratic potential energy) plus a cubic perturbation $H_1 = x^2 y - \frac{1}{3}y^3$, which provides a nonlinear coupling for the two linear oscillators.

Since the Hamiltonian is independent of time, it is a constant of motion that corresponds to the total energy of the system $E = H(x, y, p_x, p_y)$. When E is small, both the values of the momenta (p_x, p_y) and the positions (x, y) must remain small. Therefore, in the limit $E \ll 1$, the cubic perturbation can be neglected and the motion will be approximately described by the equations of motion for the unperturbed Hamiltonian, which are easily integrated analytically. Moreover, the application of the KAM theorem to this problem guarantees that as long as E is sufficiently small the motion will remain regular. However, as E is increased, the solutions of the equations of motion, like the orbits generated by the standard map, will become increasingly complicated. First, nonlinear resonances will appear from the coupling of the motions in the x and the y directions. As the energy increases, the effect of the nonlinear coupling grows, the sizes of the resonances grow, and when they begin to overlap, the orbits begin to exhibit chaotic motion.

1. Poincaré Sections

Although Eqs. (18) can be easily integrated numerically for any value of E, it is difficult to graphically display the transition from regular behavior to chaos because the resulting trajectories move in a four-dimensional phase space spanned by x, y, p_x, and p_y. Although we can use the constancy of the energy to reduce the dimension of the accessible phase space to three, the graphs of the resulting three-dimensional trajectories would be even less revealing than the three-dimensional graphs of the Lorenz attractor since there is no attractor to consolidate the dynamics. However, we can simplify the display of the trajectories by exploiting the same device used to relate the Hénon map to the Lorenz model. If we plot the value of p_x versus x every time the orbit passes through $y = 0$, then we can construct a Poincaré section of the trajectory that provides a very clear display of the transition from regular behavior to chaos.

Figure 15 displays these Poincaré sections for a number of different initial conditions corresponding to 3 different energies, $E = \frac{1}{12}$, $\frac{1}{8}$, and $\frac{1}{6}$. For very small E, most of the trajectories lie on an ellipsoid in the four-dimensional phase space, so the intersection of the orbits with the p_x–x plane traces out simple ellipses centered at $(x, p_x) = (0, 0)$. For $E = \frac{1}{12}$, these ellipses are distorted and island chains associ-

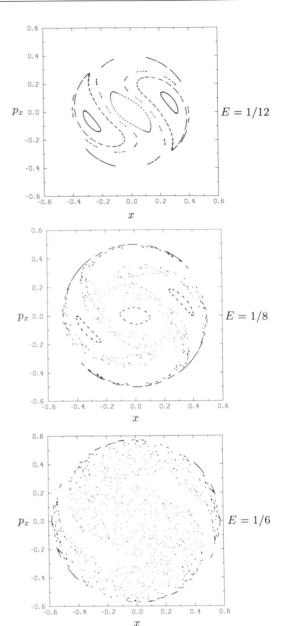

FIG. 15. Poincaré sections for a number of different orbits generated by the Hénon–Heiles equations are plotted for 3 different values of the energy E. These figures were created by plotting the position of the orbit in the x–p_x plane each time the solutions of the Hénon–Heiles equations passed through $y = 0$ with positive p_y. For $E = \frac{1}{12}$, the effect of the perturbation is small and the orbits resemble the smooth but distorted curves observed in the standard map for small k, with resonance islands associated with coupling of the x and y oscillations. However, as the energy increases and the effects of the nonlinearities become more pronounced, large regions of chaotic dynamics become visible and grow until most of the accessible phase space appears to be chaotic for $E = \frac{1}{6}$. (These figures can be compared with the less symmetrical Poincaré sections plotted in the y–p_y plane that usually appear in the literature.)

ated with the nonlinear resonances between the coupled motions appear; however, most orbits appear to remain on smooth, regular curves. Finally, as E is increased to $\frac{1}{8}$ and $\frac{1}{6}$, the Poincaré sections reveal a transition from ordered motion to chaos, similar to that observed in the standard map.

In particular, when $E = \frac{1}{6}$, a single orbit appears to uniformly cover most of the accessible phase space defined by the surface of constant energy in the full four-dimensional phase space. Although the dynamics of individual trajectories is very complicated in this case, the average properties of an ensemble of trajectories generated by this deterministic but chaotic dynamical system should be well described using the standard methods of statistical mechanics. For example, we may not be able to predict when a star will move chaotically into a particular region of the Galaxy, but the average time that the star spends in that region can be computed by simply measuring the relative volume of the corresponding region of the phase space.

D. APPLICATIONS

The earliest applications of the modern ideas of nonlinear dynamics and chaos to Hamiltonian systems were in the field of accelerator design starting in the late 1950s. In order to maintain a beam of charged particles in an accelerator or storage ring, it is important to understand the dynamics of the corresponding Hamiltonian equations of motion for very long times (in some cases, for more than 10^8 revolutions). For example, the nonlinear resonances associated with the coupling of the radial and vertical oscillations of the beam can be described by models similar to the Hénon–Heiles equations, and the coupling to field oscillations around the accelerator can be approximated by models related to the standard map. In both cases, if the nonlinear coupling or perturbations are too large, the chaotic orbits can cause the beam to defocus and run into the wall.

Similar problems arise in the description of magnetically confined electrons and ions in plasma fusion devices. The densities of these thermonuclear plasmas are sufficiently low that the individual particle motions are effectively collisionless on the time scales of the experiments, so dissipation can be neglected. Again, the nonlinear equations describing the motion of the plasma particles can exhibit chaotic behavior that allows the particles to escape from the confining fields. For example, electrons circulating along the guiding magnetic field lines in a toroidal confinement device called a TOKAMAK will feel a periodic perturbation because of slight variations in the magnetic fields, which can be described by a model similar to the standard map. When this perturbation is sufficiently large, electron orbits can become chaotic, which leads to an anomalous loss of plasma confinement that poses a serious impediment to the successful design of a fusion reactor.

The fact that a high-temperature plasma is effectively collisionless also raises another problem in which chaos actually plays a beneficial role and which goes right to the root of a fundamental problem of the microscopic foundations of statistical mechanics. The problem is how do you heat a collisionless plasma? How do you make an irreversible transfer of energy from an external source, such as the injection of a high-energy particle beam or high-intensity electromagnetic radiation, to a reversible, Hamiltonian system? The answer is chaos. For example, the application of intense radio-frequency radiation induces a strong periodic perturbation on the natural oscillatory motion of the plasma particles. Then, if the perturbation is strong enough, the particle motion will become chaotic. Although the motion remains deterministic and reversible, the chaotic trajectories associated with the ensemble of particles can wander over a large region of the phase space, in particular to higher and lower velocities. Since the temperature is a measure of the range of possible velocities, this process causes the plasma temperature to increase.

Progress in the understanding of chaotic behavior has also caused a revival of interest in a number of problems related to celestial mechanics. In addition to Hénon and Heiles' work on stellar dynamics described previously, Jack Wisdom at MIT has recently solved several old puzzles relating to the origin of meteorites and the presence of gaps in the asteroid belt by invoking chaos. Each time an asteroid, which initially lies in an orbit between Mars and Jupiter, passes the massive planet Jupiter, it feels a gravitational tug. This periodic perturbation on small orbiting asteroids results in a strong resonant interaction when the two frequencies are related by low-order rational numbers. As in the standard map and the Hénon–Heiles model, if this resonant interaction is sufficiently strong, the asteroid motion can become chaotic. The ideal Kepler ellipses begin to precess and elongate until their orbits cross the orbit of Earth. Then, we see them as meteors and meteorites, and the depletion of the asteroid belts leaves gaps that correspond to the observations.

The study of chaotic behavior in Hamiltonian systems has also found many recent applications in physical chemistry. Many models similar to the Hénon–Heiles model have been proposed for the description of the interaction of coupled nonlinear oscillators that correspond to atoms in a molecule. The interesting questions here relate to how energy is transferred from one part of the molecule to the other. If the classical dynamics of the interacting atoms is regular, then the transfer of energy is impeded by KAM surfaces, such as those in Figs. 14 and 15. However, if the classical dynamics is fully chaotic, then the molecule may exhibit equipartition of energy as predicted by statistical theories. Even more interesting is the common case where some regions of the phase space are chaotic and some are regular. Since most realistic, classical models of molecules involve more than two degrees of freedom, the unraveling of this complex phase space structure in six or more dimensions remains a challenging problem.

Finally, most recently there has been considerable interest in the classical Hamiltonian dynamics of electrons in highly excited atoms in the presence of strong magnetic fields and intense electromagnetic radiation. The studies of the regular and chaotic dynamics of these strongly perturbed systems have provided a new understanding of the atomic physics in a realm in which conventional methods of quantum perturbation theory fail. However, these studies of chaos in microscopic systems, like those of molecules, have also raised profound, new questions relating to whether the effects of classical chaos can survive in the quantum world. These issues will be discussed in Section V.

V. Quantum Chaos

The discovery that simple nonlinear models of classical dynamical systems can exhibit behavior that is indistinguishable from a random process has naturally raised the question of whether this behavior persists in the quantum realm where the classical nonlinear equations of motion are replaced by the linear Schrödinger equation. This is currently a lively area of research. Although there is general consensus on the key problems, the solutions remain a subject of controversy. In contrast to the subject of classical chaos, there is not even agreement on the definition of quantum chaos. There is only a list of possible symptoms for this poorly characterized disease. In this section, we will briefly discuss the problem of quantum chaos and describe some of the characteristic features of quantum systems that correspond to classically chaotic Hamiltonian systems. Some of these features will be illustrated using a simple model that corresponds to the quantized description of the kicked rotor described in Section IV, B. Then, we will conclude with a description of the comparison of classical and quantum theory with real experiments on highly excited atoms in strong fields.

A. The Problem of Quantum Chaos

Guided by Bohr's correspondence principle, it might be natural to conclude that quantum mechanics should agree with the predictions of classical chaos for macroscopic systems. In addition, because chaos has played a fundamental role in improving our understanding of the microscopic foundations of classical statistical mechanics, one would hope that it would play a similar role in shoring up the foundations of quantum statistical mechanics. Unfortunately, quantum mechanics appears to be incapable of exhibiting the strong local instability that defines classical chaos as a mixing system with positive Kolmogorof–Sinai entropy.

One way of seeing this difficulty is to note that the Schrödinger equation is a linear equation for the wave function, and neither the wave function nor any observable quantities (determined by taking expectation values of self-adjoint operators) can exhibit extreme sensitivity to initial conditions. In fact, if the Hamiltonian system is bounded (like the Hénon–Heiles Model), then the quantum mechanical energy spectrum is discrete and the time evolution of all quantum mechanical quantities is doomed to quasi-periodic behavior [such as that of Eq. (1)].

Although the question of the existence of quantum chaos remains a controversial topic, nearly everyone agrees that the most important questions relate to how quantum systems behave when the corresponding classical Hamiltonian systems exhibit chaotic behavior. For example, how does the wave function behave for strongly perturbed oscillators, such as those modeled by the classical standard map, and what are the characteristics of the energy levels for a system of strongly coupled oscillators, such as those described by the Hénon–Heiles model?

B. Symptoms of Quantum Chaos

Even though the Schrödinger equation is a linear equation, the essential nonintegrability of chaotic Hamiltonian systems carries over to the quantum domain. There are no known examples of chaotic classical systems for which the corresponding wave equations can be solved analytically. Consequently, theoretical searches for quantum chaos have also relied heavily on numerical solutions. These detailed numerical studies by physical chemists and physicists studying the dynamics of molecules and the excitation and ionization of atoms in strong fields have led to the identification of several characteristic features of the quantum wave functions and energy levels that reveal the manifestation of chaos in the corresponding classical systems.

One of the most studied characteristics of nonintegrable quantum systems that correspond to classically chaotic Hamiltonian systems is the appearance of irregular energy spectra. The energy levels in the hydrogen atom, which is described classically by regular, elliptical Kepler orbits, form an orderly sequence, $E_n = -1/(2n^2)$, where $n = 1, 2, 3, \ldots$ is the principal quantum number. However, the energy levels of chaotic systems, such as the quantum Hénon–Heiles model, do not appear to have any simple order at large energies that can be expressed in terms of well-defined quantum numbers. This correspondence makes sense since the quantum numbers that define the energy levels of integrable systems are associated with the classical constants of motion (such as angular momentum), which are destroyed by the nonintegrable perturbation. For example, Fig. 16 displays the calculated energy levels for a hydrogen atom in a magnetic field that shows the transition from the regular spectrum at low magnetic fields to an irregular spectrum ("spaghetti") at high fields in which the magnetic forces are comparable to the Coulomb binding fields.

This irregular spacing of the quantum energy levels can be conveniently characterized in terms of the statistics of the energy level spacings. For example,

Fig. 17 shows a histogram of the energy level spacings, $s = E_{i+1} - E_i$, for the hydrogen atom in a magnetic field that is strong enough to make most of the classical electron orbits chaotic. Remarkably, this distribution of energy level spacings, $P(s)$, is identical to that found for a much more complicated quantum system with irregular spectra–compound nuclei. Moreover, both distributions are well described by the predictions of random matrix theory, which simply replaces the nonintegrable (or unknown) quantum Hamiltonian with an ensemble of large matrices with random values for the matrix elements. In particular, this distribution of energy level spacings is expected to be given by the Wigner-Dyson distribution, $P(s) \sim s \exp(-s^2)$, displayed in Fig. 17. Although these random matrices cannot predict the location of specific energy levels, they do account for many of the statistical features relating to the fluctuations in the energy level spacings.

Despite the apparent statistical character of the quantum energy levels for classically chaotic systems, these level spacings are not completely random. If they were completely uncorrelated, then the spacings statistics would obey a Poisson distribution $P(s) \sim \exp(-s)$, which would predict a much higher probability of nearly degenerate energy levels. The absence of degeneracies in chaotic systems is easily understood because the interaction of all the quantum states induced by the nonintegrable perturbation leads to a repulsion of nearby levels. In addition, the energy levels exhibit an important long-range

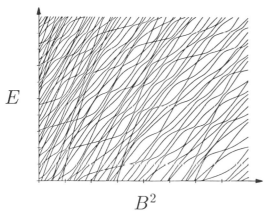

FIG. 16.   The quantum mechanical energy levels for a highly excited hydrogen atom in a strong magnetic field are highly irregular. This figure shows the numerically calculated energy levels as a function of the square of the magnetic field for a range of energies corresponding to quantum states with principal quantum numbers $n \approx 40$–50. Because the magnetic field breaks the natural spherical and Coulomb symmetries of the hydrogen atom, the energy levels and associated quantum states exhibit a jumble of multiple avoided crossings caused by level repulsion, which is a common symptom of quantum systems that are classically chaotic. [From Delande, D. (1988). Ph.D. Thesis, Université Pierre & Marie Curie, Paris.]

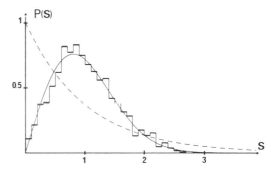

FIG. 17.   The repulsion of the quantum mechanical energy levels displayed in Fig. 16 results in a distribution of energy level spacings, $P(s)$, in which accidental degeneracies ($s = 0$) are extremely rare. This figure displays a histogram of the energy level spacings for 1295 levels, such as those in Fig. 16. This distribution compares very well with the Wigner–Dyson distribution (solid curve), which is predicted for the energy level spacing for random matrices. If the energy levels were uncorrelated random numbers, then they would be expected to have a Poisson distribution indicated by the dashed curve. [From Delande, D., and Gay, J. C. (1986). *Phys. Rev. Lett.* **57,** 2006.]

correlation called spectral rigidity, which means that fluctuations about the average level spacing are relatively small over a wide energy range. Recently, Michael Berry has traced this spectral rigidity in the spectra of simple chaotic Hamiltonians to the persistence of regular (but not necessarily stable) periodic orbits in the classical phase space. Remarkably, these sets of measure-zero classical orbits appear to have a dominant influence on the characteristics of the quantum energy levels and quantum states.

Recent experimental studies of the energy levels of Rydberg atoms in strong magnetic fields by Karl Welge and collaborators at the University of Bielefeld appear to have confirmed many of these theoretical and numerical predictions. Unfortunately, the experiments can only resolve a limited range of energy levels, which makes the confirmation of statistical predictions difficult. However, the experimental observations of this symptom of quantum chaos are very suggestive. In addition, the experiments have provided very striking evidence for the important role of classical regular orbits embedded in the chaotic sea of trajectories in determining gross features in the fluctuations in the irregular spectrum. In particular, there appears to be a one-to-one correspondence between regular oscillations in the spectrum and the periods of the shortest periodic orbits in the classical Hamiltonian system. Although the corresponding classical dynamics of these simple systems is fully chaotic, the quantum mechanics appears to cling to the remnants of regularity.

Another symptom of quantum chaos that is more direct is to simply look for quantum behavior that resembles the predictions of classical chaos. In the cases of atoms or molecules in strong electromagnetic fields where classical chaos predicts ionization or dissociation, this symptom is unambiguous. (The patient dies.) However, quantum systems appear to be only capable of mimicking classical chaotic behavior for finite times determined by the density of quantum states (or the size of the quantum numbers). In the case of as few as 50 interacting particles, this break time may exceed the age of the universe, however, for small quantum systems, such as those described by the simple models of Hamiltonian chaos, this time scale, where the Bohr correspondence principle for chaotic systems breaks down, may be accessible to experimental measurements.

C. The Quantum Standard Map

One model system that has greatly enhanced our understanding of the quantum behavior of classically chaotic systems is the quantum standard map, which was first introduced by Casati, Chirikov, Israelev,

and Ford in 1979. The Schrödinger equation for the kicked rotor described in Section IV, B also reduces to a map that describes how the wave function (expressed in terms of the unperturbed quantum eigenstates of the rotor) spreads at each kick. Although this map is formally described by an infinite system of linear difference equations, these equations can be solved numerically to good approximation by truncating the set of equations to a large but finite number (typically ≤ 1000 states).

The comparison of the results of these quantum calculations with the classical results for the evolution of the standard map over a wide range of parameters has revealed a number of striking features. For short times, the quantum evolution resembles the classical dynamics generated by evolving an ensemble of initial conditions with the same initial energy or angular momenta but different initial angles. In particular, when the classical dynamics is chaotic, the quantum mechanical average of the kinetic energy also increases linearly up to a break time where the classical dynamics continue to diffuse in angular velocity but the quantum evolution freezes and eventually exhibits quasi-periodic recurrences to the initial state. Moreover, when the classical mechanics is regular the quantum wave function is also confined by the KAM surfaces for short times but may eventually "tunnel" or leak through.

This relatively simple example shows that quantum mechanics is capable of stabilizing the dynamics of the classically chaotic systems and destabilizing the regular classical dynamics, depending on the system parameters. In addition, this dramatic quantum suppression of classical chaos in the quantum standard map has been related to the phenomenon of Anderson localization in solid-state physics where an electron in a disordered lattice will remain localized (will not conduct electricity) through destructive quantum interference effects. Although there is no random disorder in the quantum standard map, the classical chaos appears to play the same role.

D. Microwave Ionization of Highly Excited Hydrogen Atoms

As a consequence of these suggestive results for the quantum standard map, there has been a considerable effort to see whether the manifestations of classical chaos and its suppression by quantum interference effects could be observed experimentally in a real quantum system consisting of a hydrogen atom prepared in a highly excited state that is then exposed to intense microwave fields.

Since the experiments can be performed with atoms prepared in states with principal quantum num-

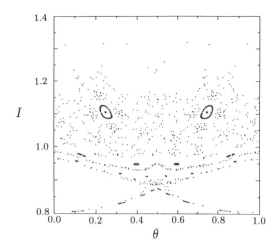

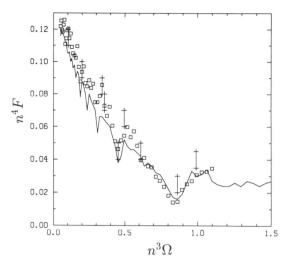

FIG. 18. This Poincaré section of the classical dynamics of a one-dimensional hydrogen atom in a strong oscillating electric field was generated by plotting the value of the classical action I and angle θ once every period of the perturbation with strength $I^4F = 0.03$ and frequency $I^3\Omega = 1.5$. In the absence of the perturbations, the action (which corresponds to principal quantum number n by the Bohr–Sommerfeld quantization rule) is a constant of motion. In this case, different initial conditions (corresponding to different quantum states of the hydrogen atom) would trace out horizontal lines in the phase space, such as those in Fig. 14, for the standard map at $k = 0$. Since the Coulomb binding field decreases as $1/I^4$ (or $1/n^4$), the relative strength of the perturbation increases with I. For a fixed value of the perturbing field F, the classical dynamics is regular for small values of I with a prominent nonlinear resonance below $I = 1.0$. A prominent pair of islands also appears near $I = 1.1$, but it is surrounded by a chaotic sea. Since the chaotic orbits can wander to arbitrarily high values of the action, they ultimately lead to ionization of the atom.

FIG. 19. A comparison of the threshold field strengths for the onset of microwave ionization predicted by the classical theory for the onset of chaos (solid curve) with the results of experimental measurements on real hydrogen atoms with $n = 32$ to 90 (open squares) and with estimates from the numerical solution of the corresponding Schrödinger equation (crosses). The threshold field strengths are conveniently plotted in terms of the scaled variable $n^4F = I^4F$, which is the ratio of the perturbing field F to the Coulomb binding field $1/n^4$ versus the scaled frequency $n^3\Omega = I^3\Omega$, which is the ratio of the microwave frequency Ω to the Kepler orbital frequency $1/n^3$. The prominent features near rational values of the scaled frequency, $n^3\Omega = 1, \frac{1}{2}, \frac{1}{3}$, and $\frac{1}{4}$, which appear in both the classical and quantum calculations as well as the experimental measurements, are associated with the presence of nonlinear resonances in the classical phase space.

bers as high as $n = 100$, one could hope that the dynamics of this electron with a 0.5-μm Bohr radius would be well described by classical dynamics. In the presence of an intense oscillating field, this classical nonlinear oscillator is expected to exhibit a transition to global chaos such as that exhibited by the classical standard map at $k \approx 1$. For example, Fig. 18 shows a Poincaré section of the classical action-angle phase space for a one-dimensional model of a hydrogen atom in an oscillating field for parameters that correspond closely to those of the experiments. For small values of the classical action I, which correspond to low quantum numbers by the Bohr–Somerfeld quantization rule, the perturbing field is much weaker than the Coulomb binding fields and the orbits lie on smooth curves that are bounded by invariant KAM tori. However, for larger values of I, the relative size of the perturbation increases and the orbits become chaotic, filling large regions of phase

space and wandering to arbitrarily large values of the action and ionizing. Since these chaotic orbits ionize, the classical theory predicts an ionization mechanism that depends strongly on the intensity of the radiation and only weakly on the frequency, which is just the opposite of the dependence of the traditional photoelectric effect.

In fact, this chaotic ionization mechanism was first experimentally observed in the pioneering experiments of Jim Bayfield and Peter Koch in 1974, who observed the sharp onset of ionization in atoms prepared in the $n \approx 66$ state, when a 10-GHz microwave field exceeded a critical threshold. Subsequently, the agreement of the predictions of classical chaos on the quantum measurements has been confirmed for a wide range of parameters corresponding to principal quantum numbers from $n = 32$ to 90. Figure 19 shows the comparison of the measured thresholds for the onset of ionization with the theoretical predictions for the onset of classical chaos in a one-dimensional model of the experiment.

Moreover, detailed numerical studies of the solution of the Schrödinger equation for the one-dimensional model have revealed that the quantum mechanism that mimics the onset of classical chaos is the abrupt delocalization of the evolving wave packet when the perturbation exceeds a critical threshold. However, these quantum calculations also showed that in a parameter range just beyond that studied in the original experiments the threshold fields for quantum delocalization would become larger than the classical predictions for the onset of chaotic ionization. This quantum suppression of the classical chaos would be analogous to that observed in the quantum standard map. Very recently, the experiments in this new regime have been performed, and the experimental evidence supports the theoretical prediction for quantum suppression of classical chaos, although the detailed mechanisms remain a topic of controversy.

These experiments and the associated classical and quantum theories are parts of the exploration of the frontiers of a new regime of atomic and molecular physics for strongly interacting and strongly perturbed systems. As our understanding of the dynamics of the simplest quantum systems improves, these studies promise a number of important applications to problems in atomic and molecular physics, physical chemistry, solid-state physics, and nuclear physics.

BIBLIOGRAPHY

Berry, M. V. (1983). Semi-classical mechanics of regular and irregular motion. In "Chaotic Behavior of Deterministic Systems" (G. Iooss, R. H. G. Helleman, and R. H. G. Stora, eds.), p. 171. North-Holland Publ., Amsterdam.

Berry, M. V. (1985). Semi-classical theory of spectral rigidity. *Proc. R. Soc. London, A* **400**, 229.

Campbell, D., ed. (1983). "Order in Chaos, Physica 7D." Plenum, New York.

Casati, G., ed. (1985). "Chaotic Behavior in Quantum Systems." Plenum, New York.

Casati, G., Chirikov, B. V., Shepelyansky, D. L., and Guarneri, I. (1987). Relevance of classical chaos in quantum mechanics: the hydrogen atom in a monochromatic field. *Phys. Rep.* **154**, 77.

Crutchfield, J. P., Farmer, J. D., Packard, N. H., and Shaw, R. S. (1986). Chaos. *Sci. Am.* **255**, 46.

Cvitanovic, P., ed. (1984). "Universality in Chaos." Adam Hilger, Bristol. (This volume contains a collection of the seminal articles by M. Feigenbaum, E. Lorenz, R. M. May, and D. Ruelle, as well as an excellent review by R. H. G. Helleman.)

Ford, J. (1983). How random is a coin toss? *Phys. Today* **36**, 40.

Gleick, J. (1987). "Chaos: Making of a New Science." Viking, New York.

Jensen, R. V. (1987). Classical chaos. *Am. Sci.* **75** 166.

Jensen, R. V. (1987). Chaos in atomic physics. *In* "Atomic Physics 10" (H. Narami and I. Shimimura, eds.), p. 319. North-Holland, Amsterdam.

Jensen, R. V. (1988). Chaos in atomic physics. *Phys. Today* **41** S-30.

Lichtenberg, A. J., and Lieberman, M. A. (1983). "Regular and Stochastic Motion." Springer, New York.

MacKay, R. S., and Meiss, J. D., eds. (1987). "Hamiltonian Dynamical Systems." Adam Hilger, Bristol.

Mandelbrot, B. B. (1982). "The Fractal Geometry of Nature." Freeman, San Francisco.

Ott, E. (1981). Strange attractors and chaotic motions of dynamical systems. *Rev. Mod. Phys.* **53**, 655.

Physics Today (1985). Chaotic orbits and spins in the solar system. *Phys. Today* **38**, 17.

Schuster, H. G. (1984). "Deterministic Chaos." Physik-Verlag, Weinheim, Federal Republic of Germany.

CHEMICAL PHYSICS

Richard Bersohn and Bruce J. Berne *Columbia University*

I. Properties of Individual and Pairs of Molecules
II. Collective Properties

GLOSSARY

Born–Oppenheimer approximation: A quantum mechanical explanation for the approximate separation of molecular energy into electronic, vibrational, and rotational energies.

Electric Multipole moment: If the charge density of a system is $\rho(r, \theta, \phi)$ where r, θ, ϕ are spherical polar coordinates, then the lth multipole moments are the set of averages $\int \rho(r, \theta, \phi) r^l Y_l^m(\theta, \phi) dV$. The moments of a spherically symmetric charge distribution are zero.

Green–Kubo relations: Expressions for transport coefficients like viscosity, thermal conductivity, and rate constants in terms of time correlation functions.

Molecular dynamics method: A method for simulating the properties of many-body systems based on solving classical equations of motion.

Monte Carlo method: A method for simulating the equilibrium properties of many-body systems based on random walks.

Normal coordinates: The coordinates of a vibrating system that oscillate with a single frequency.

Partition function: A sum over quantum states used to determine thermodynamic properties from the quantum mechanical energy levels.

Path integral methods: A formulation of quantum mechanics and quantum statistical mechanics developed by Feynman.

Radial distribution function: The average density of fluid atoms as a function of distance from a given fluid atom.

Raman scattering: An inelastic scattering of a photon by a molecule; the difference in energy between the incident and scattered photon is a difference of molecular energy levels.

Spectroscopy: The measurement of energy levels.

Statistical mechanics: A general theory of many particle systems that relates bulk properties to microscopic properties.

Time correlation functions: A function that describes the correlation between properties of a system at different times.

Chemical physics is the physics of the individual and collective properties of molecules. However, the distinction between chemistry and chemical physics is largely a matter of emphasis. The approach of the chemical physicist is theoretical. He searches for underlying theoretical principles and the molecules that he uses are often a means to an end, whereas the synthetic chemist usually considers the molecules that he synthesizes and their reactions as ends in themselves. This article on chemical physics is divided into two sections, one on phenomena which depend primarily on the properties of individual and pairs of molecules and the other on phenomena which are primarily collective.

I. Properties of Individual and Pairs of Molecules

Studies in chemical physics can be loosely classified as spectroscopic, structural, and dynamic. Spectroscopy is concerned with the determination of molecular energy levels. Structural studies are aimed at finding the distribution of particles within a molecule and molecules within a liquid or solid. The location of the nuclei defines the structure of the molecule, that is, the distances between nuclei and the angles between internuclear vectors. The distribution of electrons is intimately connected with the forces that hold the atoms together. Dynamics involves the relation of the rate of molecular transformations and changes of state caused by collision to the intra- and intermolecular forces.

A. MOLECULAR SPECTROSCOPY

Spectroscopy is the measurement of energy level differences. This is most usually accomplished by measuring the frequencies of light which are absorbed or emitted by a molecule, but it is sometimes done by measurements of the energy of an incident photon or particle together with a measurement of a scattered photon or particle. For example, the frequency of scattered light may differ from that of the incident light. The absolute value of the frequency difference is a difference of energy levels of the molecule divided by Planck's constant. This phenomenon called Raman scattering has many analogs. Electron loss spectroscopy is extensively used to measure vibrational frequencies of surfaces. The difference in energy between incident and scattered electrons is, in general, a quantized energy left in the solid. When very slow ("cold") neutrons are scattered by a warm liquid or solid, the scattered neutrons move faster than those in the incident beam.

In some spectroscopies the scattered particle whose energy is measured is not the same as the incident particle. For example, in photoelectron spectroscopy an incident photon with known energy whose wavelength is in the XUV (<100 nm) or X-ray region produces a photoelectron. Careful measurements of the kinetic energy of the electron yield energy level spacings in the positive ion. In photodissociation spectroscopy when a molecule is dissociated into fragments by light, measurement of the kinetic energy of a fragment yields the internal energy of the fragments.

Before discussing spectroscopy, it might be appropriate to warn and console the reader about the multitude of units of energy which are used to describe spectroscopic phenomena. The SI unit of energy is the joule (1 J = 10^7 ergs) but this unit is rarely used by spectroscopists. Just as the nuclear spectroscopist uses one-million electron volts (1 MeV = 9.65×10^{10} J) as a unit of energy, the X-ray spectroscopist uses electron volts (1 eV = 9.65×10^4 J). The ultraviolet,

visible and infrared spectroscopists use the cm^{-1} unit which is not an energy unit but a reciprocal of a wavelength (1 $cm^{-1} \times h/c = 11.96$ J). The microwave and radiofrequency spectroscopists use the megahertz (1 MHz $\times h = 3.99 \times 10^{-4}$ J) and the NMR spectroscopists often use a dimensionless quantity, the relative frequency shift in parts per million. Some spectroscopists use units of energy, others units of reciprocal wavelength, and still others frequency units. Indeed the reference to spectral features by their wavelength is still a very common practice, although it is to be deplored. There are natural reasons for these different choices of energy units but they are bewildering to the beginner. In a way, these units are like the light year, a nonstandard unit convenient to the astronomer but to nobody else.

Molecular spectroscopy is generally divided into four branches, each corresponding to a different type of motion and, in general, to a different frequency range. Electronic spectroscopy is the study of the differences in electronic energy levels which occur in atoms and molecules; the corresponding frequencies are 10^{14}–10^{17} Hz. Vibrational spectroscopy is the study of molecular vibrations, whose frequencies are typically of the order of 10^{12}–10^{14} Hz. Rotational spectroscopy is the study of the rotational energy levels of molecules; corresponding rotational frequencies are typically in the range of microwaves, 10^{10}–10^{12} Hz. Finally nuclear magnetic resonance spectroscopy is the study of the magnetic fields acting on a nucleus.

The given divisions are not quite as sharp as represented. Frequency ranges overlap and electronic spectroscopy is often a source of information on vibrations and rotations as well as electronic motion. Nevertheless, this classification of spectra is of fundamental importance. The theoretical justification for this classification is the Born–Oppenheimer approximation. This is an approximation that exploits the fact that the electron is 10^4–10^5 times lighter than a typical nucleus. Classically speaking, the electron revolves around the molecule so rapidly that during an electron period the nuclei do not have time to react to the different positions of the electron and hardly move. Thus electronic energies can be calculated with the assumption that the nuclei are stationary. They are not stationary of course but the electronic energy levels can be (in principle) calculated for any given arrangement of the nuclei. The arrangement or configuration that produces the least electronic energy is the equilibrium structure of the molecule. [*See* MOLECULAR OPTICAL SPECTROSCOPY; MULTIPHOTON SPECTROSCOPY.]

Let us pause to consider the coordinates used to describe the shape and location of an isolated set of N atoms. The configuration of a molecule with N atoms is determined by $3N$ coordinates. Of these, six are external coordinates in that they describe the location of the center of mass and the spatial orientation of the molecule. However, the electronic energy is independent of these coordinates, for example, the position and orientation of the molecule. Thus there are $3N - 6$ internal coordinates ($3N - 5$ for a linear molecule) which are, in general, vibrations.

The electronic energy, which is considered as a function of the nuclear coordinates, behaves as a potential energy function for the nuclear motion. In a stable molecule the atoms execute small vibrations about the position of the minimum of electronic energy. This statement implies that, in general, for each electronic state the molecule will have a different equilibrium structure. In mathematical terms, the potential in the nth electronic state can be expanded in powers of the displacement about the equilibrium position

$$V_n(\mathbf{Q}) = V_n(\mathbf{Q}_0) + \sum_i \left(\frac{\partial V}{\partial Q_i}\right)_0 (Q_i - Q_{i0})$$

$$+ \frac{1}{2} \sum_i \sum_j \left(\frac{\partial^2 V_n}{\partial Q_i \partial Q_j}\right)_0$$

$$\times (Q_i - Q_{i0})(Q_j - Q_{j0})$$

where $\mathbf{Q}$ is the set of vibrational coordinates and $\mathbf{Q}_0$ is their equilibrium value. If $\mathbf{Q}_0$ is truly the equilibrium position, then all the first derivatives in the preceding equation will vanish. If the Q_i are chosen to be normal coordinates, then all the cross derivatives vanish as well and the following equation is obtained.

$$V_n(\mathbf{Q}) = V_n(\mathbf{Q}_0) + \frac{1}{2} \sum_{i=1}^{3N-6} \frac{\partial^2 V_n}{\partial Q_{i0}^2}$$

$$\times (Q_i - Q_{i0})^2 + \cdots$$

A normal coordinate is a coordinate that classically will vary sinusoidally in time with a single frequency. In contrast, an arbitrary nonnormal coordinate will vary in time with a number (up to $3N - 6$) of frequencies. The quantities $(\partial^2 V_n/$

$\partial Q_i^2)_0$ are the force constants, the curvatures of the potential with respect to the normal coordinates i. From this definition, it is clear that the force constant may be different in different electronic states and indeed even the normal coordinates may differ. The simplest example is a diatomic molecule where the equilibrium internuclear distance is different in two different electronic states.

1. Electronic Spectroscopy

The electronic spectroscopy of atoms is clearly separated by energy domains into transitions of valence electrons and transitions of inner shell electrons. When two or more atoms combine to form a molecule, to a good approximation, core electrons remain core electrons but valence electrons occupy three types of states: bonding, nonbonding, and antibonding. For example, in the water molecule there are two core electrons in the oxygen inner shell, two pairs of bonding electrons and two pairs of nonbonding electrons. Lower energy electronic transitions are often explained by promotion of electrons from nonbonding to antibonding states. At somewhat higher energies, electrons in bonding states may be promoted to antibonding states. At still higher energies, one has Rydberg series, which are a set of atomiclike transitions in which valence electrons are excited to progressively higher-energy states where the electron circumnavigates the ion core at, typically, a considerable distance. Excitations of inner shell electrons require photons with energies in the X-ray region.

Electronic spectra are identified not only by their energies but also by two closely related quantities: their intensity, and their polarization. Electromagnetic fields cause transition in molecules by exciting resonances in the electronic or nuclear motion. The interaction energy is almost invariably expanded in multipoles, that is, in powers of the size of the molecule divided by the wavelength of light. This ratio is so small that for practical purposes only the electric and magnetic dipoles interacting with the external electric and magnetic fields, respectively, are considered. There is a fairly sharp distinction between the domain of spectroscopy in which the electric and magnetic dipole moments are important in causing transitions. The electric dipole moment matrix elements are typically of the order of ea_0, the electronic charge multiplied by the Bohr radius. The magnetic dipole moment matrix elements are typically of the order of $e\hbar/mc$, the electronic charge multiplied by the Compton wavelength. However, $\hbar/mc$ is a factor of $e^2/\hbar c \simeq \frac{1}{137}$ smaller than $a_0 = \hbar^2/me^2$. Transition probabilities are classically proportional to the square of the time derivative of the appropriate dipole moment, and quantum mechanically to the square of the matrix element between the initial and final states of that dipole moment operator. Hence, magnetic dipole transition probabilities are typically four to five orders of magnitude weaker than electric dipole transitions. Magnetic dipole transitions are therefore rarely observed in molecular spectroscopy except in paramagnetic systems where only the magnetic dipole interaction can accomplish magnetic moment reorientation.

The polarization of an electric dipole transition is the direction along which the external time dependent electric field must lie in order to cause the transition. Transitions in an atom placed in a potential, independent of angle, would be unpolarized in the sense that the transition probability would be independent of direction of the time dependent optical field. In all molecules except those with tetrahedral or octahedral symmetry, the absorption is anisotropic; in some directions there will be no absorption. The vector μ_{if}, the dipole moment matrix element between the initial and final states is often called the transition dipole. In a linear molecule the transition dipole must lie either parallel or perpendicular to the molecular axis. In a planar molecule, typical of the majority of dyes, the transition dipole must lie either in the plane in some particular direction or perpendicular to the plane. The polarization of an absorption is most often determined by measuring the absorption of polarized light by an oriented set of molecules as in a single crystal.

2. Vibrational Spectroscopy

Infrared and Raman spectroscopy are used to determine the $3N - 6$ vibrational frequencies of a molecule with N atoms ($3N - 5$ if the molecule is linear). There are both experimental and theoretical problems. On the experimental side the intensity of a vibrational transition depends on the square of the rate of change of the dipole moment during the vibration. However, certain modes have no accompanying change in dipole moment and therefore electric dipole transitions are forbidden; examples of the modes are the

symmetric stretch of CO_2 or the vibration of N_2. Raman spectroscopy often can supply the missing information. On the theoretical side, the $3N - 6$ vibrational frequencies once determined must be interpreted in terms of $(3N - 6)/(3N - 7)/2$ force constants, that is, second derivatives of the potential function. One instructive point has emerged: vibrational frequencies of most molecules display very little differences in the condensed phase and in the gas phase. Obvious exceptions to this rule are vibrations in such strongly interacting molecules as the water molecule.

One of the frontiers of chemical physics is the vibrationally highly excited "hot" molecule. Vibrations are no longer neatly divided into a set of harmonic normal mode oscillations. Vibrations are strongly anharmonic. Vibrations and rotations are no longer clearly separated and the nature of the states are difficult to describe. Yet it is precisely these energy-rich molecules that are the precursors to chemical reactions, whether they are unimolecular or bimolecular.

3. Radiofrequency Spectroscopy

Radio-frequency spectroscopy comprises many branches including nuclear magnetic resonance, nuclear quadrupole resonance, electron spin resonance, atomic beam resonance, optical pumping of atoms, and microwave spectroscopy of rotating molecules. The results of the last technique are described in the section on molecular structure. Nuclear magnetic resonance has contributed vastly more to chemistry than the other techniques.

The development of nuclear magnetic resonance (NMR) is an extraordinary chapter in the history of science in which the original goal, while eventually achieved, was dwarfed by an unforseen discovery. Nuclei with odd numbers of neutrons or protons have a net quantized angular momentum $I\hbar$ where $I = \frac{1}{2}, 1, \frac{3}{2}, \ldots$. The nuclear angular momentum implies a rotation of positive charge which in turn produces a magnetic moment. The existence of nuclear magnetic moments was manifested by the hyperfine structure of atoms, an interaction energy between the nuclear magnetic moment and a magnetic field generated by the electrons. Nuclear magnetic moments had also been detected by atomic beam experiments in which atomic beams were deflected in different directions by inhomogeneous magnetic fields; radio-frequency transitions changed the relative intensities. However, the nuclear magnetic moment could be determined only if the effective magnetic field of the electron could be calculated; in the late 1940s, when NMR was developed, computers adequate to calculate these fields did not exist. Nuclear magnetic resonance measurements on solid or liquid samples seemed like the perfect solution. The moments μ could be derived from the simple equation $h\nu = \mu B$ where h is Planck's constant, B is the magnetic induction and ν the frequency used to excite the transition between the $2I + 1$ magnetic sublevels of the nucleus.

In practice it was found that the apparent magnetic moment varied with the chemical environment of the nuclei. The effect was due to a local magnetic susceptibility, which, in general, increased with the number of atomic electrons. For the particular case of the proton ^1_1H and $^{13}_6C$, resonance fields (or frequencies) differed at different nonequivalent positions and this fact serves as a decisive structural tool for organic molecules. As important as the structure determination is, the really unique feature is the ability to measure exchange rates between different environments. For example, cyclohexane has six axial and six equatorial hydrogen atoms that interchange as the molecule converts from one chair form through the boat form to a second equivalent chair form. At room temperature, only a single resonance line is seen because the two sites interchange at a rate much faster than the frequency difference between them. When the liquid is cooled, the rate decreases and two lines are seen. The contributions of NMR to solid-state physics, to all branches of chemistry, to biochemistry and even medicine through the imaging of internal organs, have been overwhelming. The original goal targets, the magnetic moments of the stable nuclei, have all been measured and tabulated in reference books but a theory has not been developed to explain them.

B. Molecular Structures

Molecular structure is most often determined by scattering methods. A beam of photons or material particles (chosen so that their wavelengths are comparable to close internuclear distances) is directed at the sample. At particular scattering angles, constructive interference will occur between the waves scattered by pairs of atoms or, in a crystal, planes of atoms. By measuring the relative scattered intensities at the various angles at which constructive interference occurs, a structure may be inferred. The structure of a condensed phase is probed by X-rays or neutrons; the structure of a gas molecule or a surface is commonly determined by electron scattering.

In several types of spectroscopy, molecular rotational energies are determined from which the three moments of inertia can be deduced. By measuring the spectra of the same molecule with different nuclear isotopes, additional sets of moments of inertia can be obtained.

Molecular structure is the distribution of nuclei and electrons within a molecule. The location of the nuclei, often just called the structure, is the molecular architecture. Closely connected to this architecture is the distribution of electrons circumnavigating the heavy nuclei. The techniques for determining the nuclear positions are based on interferences in scattering (diffraction) or spectroscopic measurements of moments of inertia.

1. Diffraction Methods

When an incident wave is scattered by a molecule, the scattered wave consists of a sum of contributions, large and small, from each of the electrons and nuclei in the molecule. The scattered intensity, which is the square of the scattered amplitude, will consist of a sum of squares of the individual amplitudes plus a sum of products of scattered amplitudes originating from different particles. The former sum is independent of the distances between particles, whereas the latter is structure dependent. The constructive interference between the different amplitudes gives rise to significant diffraction peaks when the interparticle distances are comparable to the wavelength of the incident wave. In practice, this means that the wavelength must be of the order of an angstrom, ($1 \text{ Å} = 10^{-10}$ m). The most common "particles" used for scattering have been the X-ray ($\lambda \sim 1–2$ Å) and the velocity selected slow neutron ($\lambda \sim 1$ Å). In a few cases, especially for diffraction from surfaces, hydrogen atoms and helium atoms have been used.

X-ray diffraction from single molecular crystals has been our most important source of knowledge of molecular structure. The assumption, in most cases solidly confirmed by experiment, is that the forces that hold molecules together in the solid are much weaker than the interatomic forces that determine the molecular structure. Therefore, the structure in the solid will be negligibly different in the solid, liquid and gas phases. There are, of course, exceptions to this rule such as PCl_5 which is $PCl_4^+ PCl_6^-$ in the solid state but the rule is generally valid.

Nuclei scatter X-rays to a negligible extent because they are so heavy that they do not react to the time varying electric field of the photon. The electrons will contribute a total scattering amplitude proportional to

$$f = \int \rho(\mathbf{r})e^{i\mathbf{s}\cdot\mathbf{r}} \, dV$$

where $\mathbf{r}$ is a vector whose origin is at the nucleus, $'\rho(\mathbf{r})$ is the electron density at the point $\mathbf{r}$, $\mathbf{s} = 2\pi(\mathbf{k} - \mathbf{k}_0)$ and dV is a volume element. The wave vectors $\mathbf{k}$ and $\mathbf{k}_0$ are of the scattered incident X-rays, respectively. The atomic scattering factor f varies with the atom, being larger for atoms with more electrons. The amplitude for scattering from an entire unit cell of a crystal is represented (approximately) as a sum over atomic scattering amplitudes

$$F = \sum_j f_j e^{i\mathbf{s}\cdot\mathbf{R}_j}$$

where f_j is the scattering amplitude of the jth atoms whose mean position is at the point $\mathbf{R}_j$ within the unit cell. There will be constructive interference, that is, Bragg peaks whenever $\mathbf{s} \cdot (\mathbf{R}_j - \mathbf{R}_k)$ is an integral multiple of 2π. There will be a large number of planes of atoms in the crystal for which this Bragg condition will be satisfied and therefore the crystal structure is in principle overdetermined.

As is shown by the previous equation, the atomic scattering amplitude is roughly proportional to the number of electrons in the atom. Thus, it is particularly difficult to observe scattering interferences from hydrogen atoms. Neutron scattering amplitudes depend on nuclear cross sections which are of comparable magnitudes for all nuclei. An identical expression for F is obtained except that the f's are now nuclear scattering amplitudes. Because these amplitudes are small, neutron diffraction is carried out with condensed phases and not with gases.

Electron diffraction has also been used for determination of molecular structure. Electrons interact far more strongly with matter than X-rays and can penetrate only a few hundred angstroms into a crystal. They have been used, therefore, for structural measurements of solid surfaces and gaseous molecules. The fact that the gaseous molecules exhibit all orientations to an incoming electron beam, greatly reduces the information content of the experiment. In other words, only systems with a small number of structural parameters can be effectively studied. The scattering amplitudes for individual atoms include a nuclear as well as an electronic term but the essence of the structurally dependent interference remains the same.

2. Structure Inferred from Moments of Inertia

At very high resolution, electronic spectra exhibit rotational structure from which the moments of inertia of both the ground and excited electronic states can be extracted. Similarly vibrational spectra taken at very high resolution will yield the moments of inertia in the ground and the excited vibrational states. Rotational Raman spectra have yielded moments of inertia of small symmetric molecules. The most extensive and accurate source by far of moment of inertia data is microwave spectroscopy. This technique is applicable to any polar gaseous molecule.

Frequencies in microwave spectroscopy that are directly proportional to inverse moments of inertia are usually measured to five or six significant figures. It would thus appear that bond distances are determined to better than 0.001 Å. In reality there are usually more structural parameters required than measured moments of inertia. For example, in CH_3I, a pyramid-shaped molecule the structure is determined by two bond distances and a bond angle. Because of the threefold symmetry axis only two moments of inertia are measurable. The usual solution is an isotopic substitution, for example, deuterium for hydrogen. It is often found, however, that a unique structure cannot be found consistent with all the moments of inertia. The reason is that the different vibrational wave functions in the isotopically substituted molecules yield slightly different average structures.

C. Dynamics of Molecular Processes

A system at equilibrium is characterized by a balance of opposing forces and reactions so that its macroscopic properties are independent of time. Time-dependent processes are caused by deviations from thermal equilibrium. For example, if there is a spatial gradient in temperature, electric potential, momentum or concentration, equilibrium is restored by the transport of energy, charge, momentum or identity, respectively. The ratios between the fluxes of these properties and the corresponding gradient are called the transport coefficients, namely, the coefficients of thermal conductivity, electrical conductivity, viscosity and diffusion, respectively.

Even in a spatially uniform system there can be substantial departures from equilibrium. The system may not be in an equilibrium distribution over its internal states and often approaches equilibrium ("relaxes") exponentially with a rate characteristic of each degree of freedom, which is the reciprocal of the relaxation time. For example, there are electronic relaxation times, vibrational relaxation times, rotational relaxation times and even nuclear or electronic spin lattice relaxation times. There are great variations in the rate of approach to equilibrium of these various degrees of freedom varying from 10^{12} sec^{-1} to 10^{-3} sec^{-1}.

When the chemical composition of a system changes with time, it is less clear how to express this change mathematically. When dealing with a reaction, for example, of the form

$$A + B \rightleftarrows X_1 \rightleftarrows X_2 \cdots \rightleftarrows X_n \rightarrow C + D$$

where A,B are reactants, C,D are products and the X_i are intermediates, known or unknown, it can often fit the observations to an empirical formula with one or more terms of the following form:

$$d(C)/dt = k(A)^a(B)^b(C)^c(D)^d$$

where a, b, c, and d are not necessarily integer constants and the parentheses indicate concentrations of the enclosed species. The constant k is called a rate constant. The complexity of this equation is due to the complexity of the reaction mechanism. If the reaction were really elementary, that is, if there were no intermediates, then the equation

$$d(C)/dt = k(A)(B)$$

would express the fact that the reaction takes place as a consequence of collisions of A and B molecules.

Except for radiative and nonradiative decay of electronically excited molecules the approach to equilibrium of a set of molecules is always accomplished by intermolecular interactions of some sort. In a low pressure gas the central event is a bimolecular collision. A collision changes the states of the pair of molecules involved in it. The most gentle change of state is the elastic scattering that involves only a rotation of the relative momentum of the pair. The probability of such an event is expressed by a differential cross section, $d\sigma/d\Omega$ which gives the number of the pairs whose relative momentum is scattered into $d\Omega$ (at a certain orientation, Ω) per unit time, per unit incident flux. The integral of this differential cross section over all solid angles is called the total scattering cross section.

In perhaps the most famous experiment of atomic physics, Rutherford demonstrated that the deflection of α particles could be connected with the Coulomb law of force between nuclei and α particles. If the potential energy was known in advance classical or quantum equations of motion could be used to determine the differential scattering cross section; in principle the process can be reversed, that is, the theoretical potential energy acting between the two scatterers can be deduced from the experimental differential cross section. Indeed in principle if the scatterers are small enough, preferably atomic in nature, the potential energy can be calculated using the Schrödinger equation of quantum mechanics. A large amount of research in atomic scattering may be briefly summarized as follows. Atoms interact with a very steeply rising short range potential and a relatively weak long range attraction. The long range attraction gives rise to very large scattering at small angles superimposed on a much less intense isotropic scattering due to the short range repulsion. [See ATOMIC PHYSICS.]

Elastic collisions can relax translational energies producing a Boltzmann distribution of translational energies. However relaxations of internal degrees of freedom can only be accomplished by inelastic collisions in which energy is exchanged between internal degrees of freedom and translation or between internal degrees of freedom of both molecules. The efficiency of these inelastic collisions expressed by the magnitudes of the inelastic cross sections depends on the nature and strength of the interaction between the molecules. Specifically, the intermolecular potential must depend on the molecular orientations in order to cause changes of rotational state and on an internal normal coordinate in order to cause changes in vibrational states. As a general rule, rotational equilibrium is usually achieved with only five to ten collisions but vibrational equilibrium requires hundreds to thousands of collisions. The reason is simply that intermolecular potential energies are much more dependent on molecular orientation than on the relatively small vibrational amplitudes. A more subtle point is the following. The probability of an inelastic collision transferring an amount of energy $\hbar\omega$ between a more or less random translation and an internal degree of freedom is proportional to the spectral density of the random interaction at the frequency ω. The typical frequency associated with translational motion during a collision would be the reciprocal of a collision duration. This frequency is just right for rotations, too small for vibrations and too large for nuclear magnetic energy level separations. Because V → T (vibration to translation) relaxation is slow, another form of vibrational relaxation occurs earlier which is called V → V′. These V → V′ transitions between vibrational states of nearly the same energy, for example,

$$2HBr(v = 1) = HBr(v = 0) + HBr(v = 2)$$

have much larger cross sections than a V → T process such as

$$HBr(v = 1) + M = HBr(v = 0) + M$$

where M is another molecule.

There is a hierarchy of relaxation problems in dilute gases. One begins with macroscopic transport processes whose transport coefficients can be expressed in terms of averages of scattering cross sections. These in turn can, in principle, be calculated from an intermolecular potential energy. Spectroscopically the evolution of an initially nonequilibrium distribution over internal states can be followed toward a Boltzmann distribution. The relaxation rate can be expressed in terms of inelastic scattering cross sections that can, again in principle, be calculated from the intermolecular potential energy function.

The chemical reaction itself can be treated as a collision problem in mechanics. The measured temperature dependent rate constant for a chemical reaction $k(T)$ is an average over the Boltzmann distributions for both translation and internal energies. There is not enough information in the function $k(T)$ to permit a potential energy function, a surface to be extracted by an inversion. An area of research, chemical dynamics is the study of reactions in which some form of state selection of the reactants or the products has been carried out. The state selection can take many forms. If the reactant molecules are formed into molecular beams, the direction and sometimes also the magnitude of the relative velocity will be known and the differential reaction cross section can be determined. A reactant can be excited to a specific vibrational state and the effect on the cross section can be measured. Similarly the vibrational and rotational state distribution of product molecules can sometimes be determined by laser induced fluorescence, multiphoton ionization or infrared emission. When sufficient information is obtained, it should ultimately be possible to extract a potential surface on which the reactants move and are transformed into products. The assumption here is that the set of reactant molecules and the set of product molecules correspond to different potential minima on the same surface. If this is not the case, the reaction is said to be nonadiabatic. The theoretical problem is more complicated; two potential surfaces must be extracted as well as a perturbing potential that couples them.

II. Collective Properties

Liquids and solids are systems that contain on the order of Avagodro's number (6×10^{23}) of strongly interacting molecules. How do the macroscopic properties of such systems depend on the properties of the constituent isolated molecules? The properties of collective systems can be roughly divided into structural properties, and dynamic properties. The theoretical study of such collective properties falls within the province of statistical mechanics.

The macroscopic properties of condensed systems are of central importance to almost all fields of science. These properties are described by phenomenological theories such as thermodynamics, hydrodynamics, electrodynamics, and chemical kinetics. Nevertheless, it should be clear that the laws of thermodynamics, the thermodynamic equation of state for given materials, the laws of hydrodynamics, and electrodynamics, and the properties germane to these as well as all of chemical kinetics should be derivable from the underlying classical or quantum mechanical theory of atoms and molecules (in fact the whole of chemistry should be predictable from mechanics.)

A. STATISTICAL MECHANICS

Statistical mechanics is a molecular theory of macroscopic systems. It provides the bridge between the microscopic world of nuclei and electrons and the macroscopic world of the phenomenological theories. Starting with the Hamiltonian of the system, the laws of equilibrium thermodynamics, hydrodynamics, electrodynamics and chemical kinetics can be derived. The Hamiltonian H is the classical energy as a function of the momenta and positions of all the particles in the system. It consists of the sum of the kinetic and the potential energies. [See STATISTICAL MECHANICS.]

Statistical mechanics can be subdivided into equilibrium and nonequilibrium statistical mechanics. The former deals with systems in thermodynamic equilibrium, whereas the latter deals with the time evolution of macroscopic systems.

1. Equilibrium Statistical Mechanics

The thermodynamic equation of state of a system can be derived from an expression that relates the Helmholtz free energy $A_N(T,V)$ to the microscopic properties of the system,

$$A_N(T,V) = -kT \ln Q_N(T,V)$$

where T is the thermodynamic temperature (degrees Kelvin) and V is the volume, and where

$$Q_N(T,V) = \sum_n \exp(-E_n/kT)$$

is the canonical partition function. Thus to derive the thermodynamic equation of state; that is, the relationship between the free energy, the number of particles, the temperature T, and the volume V, $Q_N(T,V)$ must be calculated. In these equations k is Boltzmann's constant, the sum goes over all quantum mechanical states of the system and $\{E_n\}$ is the set of energy eigenvalues corresponding to these states. These energy levels are found by solving the Schrödinger equation of quantum mechanics, namely,

$$H\Phi_n(q_1, ..., q_N) = E_n\Phi_n(q_1, ..., q_N)$$

where $\Phi_n(q_1, ..., q_N)$ is an energy eigenfunction of the system and $\{q_1, ..., q_N\}$ are the coordinates specifying the positions of the particles. Thus to predict the thermodynamic properties the allowed energies for the system at the given volume have to be determined. In very dilute gases it is possible to calculate the thermodynamic properties very accurately. This is already an impressive achievement, but liquids are more difficult to treat. In quantum mechanics every particle has wavelike properties. It follows that the atoms and molecules that liquids are composed of occupy a region in space characterized by a diameter proportional to the particle's thermal deBroglie wavelength, a quantity inversely proportional to $(mT)^{1/2}$. Only at very low temperatures does the matter wave of one molecule interfere with the matter wave of another molecule. This interference can give rise to astounding properties as it does in liquid He which displays superfluidity (a purely quantum mechanical phenomenon). Because most liquids exist at high temperatures it is possible to ignore these quantum effects, and to treat them using classical statistical mechanics; that is, statistical mechanics based on classical mechanics. Although the treatment of strongly interacting systems is still very difficult in classical statistical mechanics, progress has been made on a variety of fronts. Statistical mechanical perturbation theory allows the properties of fluids to be calculated using the exact results for a simple reference system such as the hard sphere fluid. This is a fluid consisting of spheres that are not allowed to interpenetrate. The structure and thermodynamic properties of fluids can be determined by integral equations. The behavior of systems undergoing phase transformations can be determined by renormalization group techniques. The properties of real imperfect gases can be determined by diagrammatic techniques.

What is meant by the term "the structure of liquids"? Imagine an instantaneous photograph of a liquid. The atoms are packed together in a noncrystalline arrangement; that is, there is no long range order. Nevertheless, the positions of nearby atoms are correlated. Because the atoms cannot overlap, the minimum distance from one atomic center to another is the atomic diameter. Thus around an atom there will be an exclusion sphere outside of which near neighbors are found, but these neighbors once again define a region of space which excludes next nearest neighbors and so on. The radial distribution function gives the average density of atoms as a function of distance from a given atom in the system. Due to the previously stated packing effects, this function will exhibit peaks and troughs as a function of distance. At very large distances it will be equal to the bulk density because the presence of an atom at one place will not be felt by another atom very far from it.

These functions can be measured by diffraction methods such as neutron scattering and X-ray scattering. They can also be determined from statistical mechanics.

2. Nonequilibrium Statistical Mechanics

In hydrodynamics, the time evolution of fluid flow depends not only on thermodynamic properties, but also on such transport properties as shear and bulk viscosities, thermal conductivity, mutual diffusion coefficients, etc. These characterise the transport of momentum, heat, and mass. In electrodynamics the response of systems to the imposition of electric and magnetic fields depends on dielectric response functions of the system, and in chemical kinetics the approach to equilibrium is described by empirical rate laws and chemical reaction rate constants. Expressions for the transport coefficients in terms of the microscopic dynamics of the quantum or classical systems can be derived by modern statistical mechanics. Moreover it is now possible to derive the equations of macroscopic rate processes such as hydrodynamics and chemical kinetics starting with the Hamiltonian of the system and using the methods of nonequilibrium statistical mechanics.

Transport coefficients can be expressed in terms of time correlation functions or covariance functions of spontaneous fluctuations in equilibrium systems. For example, the self diffusion coefficient can be expressed as,

$$D = \tfrac{1}{3} \int_0^\infty dt \, \langle \mathbf{v}(0) \cdot \mathbf{v}(t) \rangle$$

where $\langle \mathbf{v}(0) \cdot \mathbf{v}(t) \rangle$ is the autocorrelation function of the velocity of a labeled particle, $\mathbf{v}$. Such expressions for transport coefficients are called Green–Kubo relations. All transport coefficients and chemical rate constants can be expressed as time integrals of time correlation functions; that is, as Green–Kubo relations. In addition, spectroscopic band shapes, NMR and ESR line shapes and differential cross-sections for the scattering of light and thermal neutrons can be expressed as space-time Fourier transforms of appropriate time correlation functions. The theory of such processes is called linear response theory. To treat nonlinear processes it is necessary to use mode-mode coupling theory.

B. Numerical Statistical Mechanics

One of the great developments in statistical mechanics during the past several decades is the development of methods for simulating strongly interacting systems such as liquids, liquid crystals, solids and glasses on computers. The formalism of statistical mechanics provides exact analytical expressions. Because it is often impossible to evaluate these expressions analytically numerical methods must be used. In fact, prior to the development of computers very little progress was made in understanding liquids and amorphous solids. There are two major techniques used in computer simulation; molecular dynamics (MD) and Monte Carlo (MC), and these have contributed enormously to the theory of condensed systems.

1. Molecular Dynamics

A starting point for the simulation of a classical liquid is the Hamiltonian of the system. If there are N particles in the system $6N$ equations of motion must be solved: one for each position coordinate and one for each momentum coordinate starting with a given set of positions and momenta these equations are solved by finite difference techniques on a computer. At the end of each time interval the new positions and momenta of all the particles are recorded. After very many time steps this gives a trajectory of the whole many-body system. Thermodynamic properties are found by time averaging dynamical properties over the trajectory. Likewise time correlation functions and thereby transport coefficients, rate constants, differential scattering cross-sections, spectral line shapes, etc. can be determined from the trajectory. This method is called molecular dynamics. Recently, it has been possible to simulate nonequilibrium systems by clever applications of constraints and boundary conditions. Because simulations are done on finite systems, periodic boundary conditions are usually adopted to bypass unusual surface effects. Nevertheless, it is also possible to simulate surface properties. In order to use molecular dynamics, there must be knowledge of the forces that exist between the molecules. Only in the simplest systems is it possible to determine the forces between two molecules (either by experiment or by ab-initio quantum chemical calculations). Thus molecular modeling is required before molecular dynamics can be applied.

2. Monte Carlo Simulations

If there is interest only in equilibrium properties, an alternative to molecular dynamics is the

Monte Carlo technique. This ingenious method is based on the theory of finite Markov chains. In classical statistical mechanics the probability distribution for finding the particles at a given position in configuration space is proportional to the Boltzmann factor, $\exp(-U/kT)$ where U is the potential energy of the system in that configuration. In Monte Carlo an initial configuration is chosen. A particle is next moved to a new position. This move is either accepted or rejected based on a certain criterion. This process is repeated for each particle. The criterion for acceptance or rejection is based on a random number generator in such a way that the configurations thus generated are distributed according to the Boltzmann factor. These sampled configurations give a trajectory in configuration space that looks like a random walk. Averages of position dependent properties over this random walk trajectory then give thermodynamic properties.

Simulation techniques have been applied to a wide variety of many-body systems. For example, hydrogen bonding in water and aqueous solutions including protein solutions has been studied in this way. Such methods are often used as experimental tests of theoretical assumptions. In many cases totally new phenomena were first observed using computer simulations and only later observed in real systems. The methods of statistical mechanics and numerical statistical mechanics are applicable to many fields outside of chemistry. For example the question of quark confinement in high energy physics can be formulated as a problem in statistical mechanics that can be treated by Monte Carlo techniques using the Feynman path integral representation. Another example involves the fragmentation of nuclei that can be regarded as a liquid droplet of nucleons.

With the advent of supercomputers these methods will continue to play a very large role in the study of collective phenomena. One day very complex reactions will be simulated on computers.

BIBLIOGRAPHY

Berne, B. J., and Pecora, R. (1976). "Dynamic Light Scattering: With Applications to Chemistry, Biology and Physics." Wiley (Interscience), New York.

Berne, B. J. (ed.) (1976). "Modern Theoretical Chemistry," Vol. 6, Part A and B. Plenum, New York.

McQuarrie, Donald A. (1975). "Statistical Mechanics." Harper and Row, New York.

CP VIOLATION

John F. Donoghue and Barry R. Holstein   *University of Massachusetts*

I. CP Symmetry
II. Observation of CP Violation
III. Future Areas of Research

GLOSSARY

Charge conjugation invariance: Invariance of physical laws under the process of interchanging particle and antiparticle.

Decay: Elementary particles can transform into other combinations of particles as long as energy, momentum, charge, etc., are conserved. The process of an isolated particle transforming into several lighter particles is called decay.

Electric dipole moment: Classically, an electric dipole is a separation of charges so that, although the whole system is electrically neutral, the distribution of charge has a region of positive charge and a region of negative charge separated along some axis.

Kaon: Elementary particle with a mass of about one-half of that of the proton. The kaon is the lightest of the particles with a quantum number called "strangeness."

Parity invariance: Invariance of physical laws under the process of reversing spatial coordinates but not time. If accompanied by a 180° rotation, this is equivalent to a reflection in a mirror.

CP violation is said to occur when two processes, which differ by the combined action of charge conjugation and parity reversal, do not occur at the same rate. This phenomenon is rare, but it has been observed in the decay of the neutral K meson system. The origin of this slightly broken symmetry is not presently understood, and it may tell us more about the structure of fundamental interactions.

I. CP Symmetry

From ancient times, the concept of symmetry has commanded a powerful influence upon our view of the universe. However, many such symmetries are only approximate, and the way in which they are broken can reveal much about the underlying dynamics of physical law. Perhaps the earliest example of a broken symmetry was the required modification of the presumed perfect circular orbits of the outer planets by epicycles—circles on circles—in order to explain the observation that occasionally the trajectories of these planets through the sky double back on themselves (see Fig. 1). This breaking of perfect symmetry, although small, forced scientists to search more deeply into the basic forces responsible for celestial orbits, leading ultimately to Newton's law of universal gravitation.

More recently, in the mid-1950s, the concept of parity invariance—left–right symmetry—was found to be violated by the weak interaction, that is, the force responsible for such processes as nuclear beta decay. The concept of right or left in such a process is realized by particles whose direction of spin is respectively parallel or antiparallel to the particle momentum. Wrapping one's hand about the momentum vector with fingers pointing in the direction of rotation, as in Fig. 2, the particle is said to be left/right handed if the left/right thumb points in the direction of the momentum vector. Parity invariance would require the absence of handedness, that is, the emission of equal numbers of left and right handed particles in a decay process. Beta decay processes occur with both electron (e^-) and positron (e^+) emission:

$$A \rightarrow B + e^- + \bar{\nu}_e$$

$$B' \rightarrow A' + e^+ + \nu_e$$

where ν_e ($\bar{\nu}_e$) is an accompanying neutrino (antineutrino). In such decays, the neutrinos (antineutrinos) are found to be completely left- (right-) handed indicating a maximal violation of parity, due to the left-handed character of the particles that mediate the weak interaction, the W and Z bosons.

Even after the overthrow of parity in 1957, it was

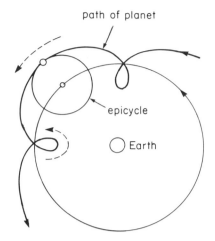

FIG. 1.   In order to account for the fact that the outer planets occasionally double back on themselves during their traversal of the heavens, the early Greek astronomer Hipparchus postulated the idea that instead of perfect circles the planetary orbits were better described by epicycles, or circles on circles.

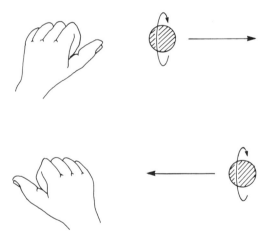

FIG. 2.   A spinning particle is described as either left- or right-handed depending upon which hand, when wrapped around the direction of motion with fingers pointing in the direction of the spin, has the thumb pointing in the direction of travel. Thus, the top figure indicates left-handed motion and the bottom figure shows right-handed motion.

believed that a modified remnant of the symmetry remained, that of CP. Here, P designates the parity operation, while C signifies charge conjugation, which interchanges particle and antiparticle. Thus, CP invariance requires equality of the rates for

$$A \rightarrow B + e^- + \bar{\nu}_e \text{ (right)}$$

and

$$\bar{A} \rightarrow \bar{B} + e^+ + \nu_e \text{ (left)}$$

Such CP invariance occurs naturally in the theories that have been developed to explain beta decay. It would then also extend to other weak interaction processes, such as the decays of other elementary particles.

There is also a related symmetry called time reversal, T. In this case, the symmetry corresponds to the replacement of the time t by $-t$ in all physical laws (plus the technical addition of using the complex conjugate of transition amplitudes). Pictorially, this corresponds to the equivalent of taking a film of a scattering process $A + B \rightarrow C + D$ and then running the film backward to obtain $C + D \rightarrow A + B$ (see Fig. 3). Time reversal invariance requires that the two processes occur with equal probability. In addition, there is a very powerful theorem, called the CPT theorem, that asserts that in all of the presently possible theories the combined action of CP and T transformations must be a symmetry. Of course, this CPT invariance is also being subjected to experimental scrutiny, but the theorem is so powerful that for all of the theories now being applied to nature it does hold. In this case, any violation of CP invariance

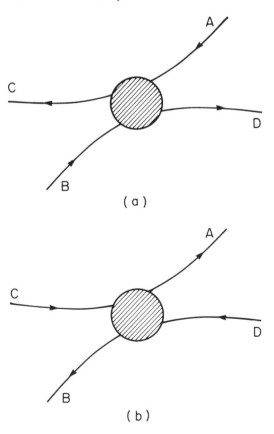

FIG. 3.   According to time reversal symmetry, the reaction $A + B \rightarrow C + D$ indicated in (a) must be identical to the reaction $C + D \rightarrow A + B$ shown in (b).

must also be accompanied by a corresponding violation of T symmetry. To the level that they have been tested in the beta decay processes just described, both CP and T appear to be true symmetries.

II. Observation of CP Violation

In 1964, however, a small breaking of CP symmetry was found in other weak reactions, as we shall describe. In order to understand how this phenomenon was observed, we need to know that under a left–right transformation a particle can have an intrinsic eigenvalue, either $+1$ or -1, since under two successive transformations one returns to the original state. Pi mesons, for example, are spinless particles with a mass of about 140 MeV/c^2. They exist in three different charge states, positive, negative, and neutral, and are found to transform with a negative sign under the parity operation. Under charge conjugation, the three pions transform into one another:

$$C\begin{pmatrix} \pi^+ \\ \pi^0 \\ \pi^- \end{pmatrix} \rightarrow \begin{pmatrix} \pi^- \\ \pi^0 \\ \pi^+ \end{pmatrix}$$

(Note that the π^0 is its own antiparticle.) Under CP, then, a neutral pion is negative:

$$CP\,|\pi^0\rangle = -|\pi^0\rangle$$

so that a state consisting of two (three) neutral pions is respectively even (odd) under CP. The argument is somewhat more subtle for charged pions, but it is found that spinless states $|\pi^+\pi^-\rangle$ and a symmetric combination of $|\pi^+\pi^-\pi^0\rangle$ are also even and odd, respectively, under the CP operation.

Before 1964, it was believed that the world was CP invariant, which has interesting implications for the system of K mesons (spinless negative parity particles with a mass of about 498 MeV/c^2) which decay into 2π and 3π states by means of the weak interaction. There exist two neutral K species, particle and antiparticle, with opposite strangeness quantum numbers:

$$K^0 \qquad m = 497.7 \text{ MeV} \qquad S = +1$$
$$\overline{K}^0 \qquad m = 497.7 \text{ MeV} \qquad S = -1$$

In terms of the framework of the quark model, the K^0 is a bound state of a down quark and a strange antiquark, while the $\overline{K}^0$ contains a down antiquark and a strange quark. Under CP, we have

$$CP\,|K^0\rangle = |\overline{K}^0\rangle$$
$$CP\,|\overline{K}^0\rangle = |K^0\rangle$$

If the world were to be CP invariant, then the particle that decays to a two-pion final state must itself be an eigenstate of CP with CP $= +1$, while that which decays to a three-pion final state must have CP $= -1$. Although neither K^0 nor $\overline{K}^0$ are CP eigenstates, one can form linear combinations that are:

$$|K_1^0\rangle = \sqrt{\frac{1}{2}}\,(|K^0\rangle + |\overline{K}^0\rangle) \qquad CP = +1$$

$$|K_2^0\rangle = \sqrt{\frac{1}{2}}\,(|K^0\rangle - |\overline{K}^0\rangle) \qquad CP = -1$$

and by CP symmetry, it is these superpositions of K^0, $\overline{K}^0$ that decay into 2π, 3π channels:

$$K_1^0 \rightarrow \pi^+\pi^-\,,\; \pi^0\pi^0$$
$$K_2^0 \rightarrow \pi^+\pi^-\pi^0,\; \pi^0\pi^0\pi^0$$

In fact, precisely this phenomenon is observed. The neutral kaons that decay weakly into these pionic channels are different particles with different lifetimes:

$$\tau_1 \sim 10^{-10} \text{ sec}$$
$$\tau_2 \sim 10^{-8} \text{ sec}$$

but without definite strangeness. One can even observe strangeness oscillations in the time development of the neutral kaon system, which is a fascinating verification of the superposition principle at work at the microscopic level, but it is not our purpose to study this phenomenon here.

Rather, we return to the year 1964 when an experiment at Brookhaven National Laboratory by Christenson, Cronin, Fitch, and Turlay observed that the same particle—the long-lived ($\sim 10^{-8}$ sec) kaon could decay into both 2π and 3π channels. The effect was not large—for every 300 or so $K \rightarrow 3\pi$ decays, a single $K \rightarrow 2\pi$ was detected—but it was definitely present. Since these channels possess opposite CP eigenvalues, it was clear that CP symmetry had been observed.

What are the implications of this breaking of CP symmetry? Unfortunately, the answer is not fully known, which in turn is the reason why studies of CP violation are so important at present. The breaking may be caused by a new, previously unknown interaction. Indeed, one of the first theories on the subject, the superweak model of Wolfenstein, proposed that there is a new, very weak force that can mix K_1 and K_2. Because it is so weak, it is very unlikely to be seen in any other reactions outside of the K_1, K_2 system. However, this model does yield clear predictions in kaon decay, because all possible signals of CP violation are attributable to a single parameter governing the mixing of K_1 and K_2. It is also possible, however, that CP violation can be accommodated without the introduction of any new forces, as was first demonstrated by Kobayashi and Maskawa (KM). They noted that the weak interactions of the various quarks do not always have to have the same complex phase. Normally, the phase ϕ of an amplitude $A = |A|e^{i\phi}$ is not observable, since the decay probability is given by only the absolute value of the amplitude, $|A|^2$. However, phases can be observed. For example, if the mass eigenstates are K_1 and K_2 as described previously, then the relative phase of $K_0 \rightarrow \pi\pi$ and $\overline{K}_0 \rightarrow \pi\pi$ would be observable.

$$A(K_0 \rightarrow \pi\pi) = |A|e^{i\phi}$$

$$A(\overline{K}_0 \rightarrow \pi\pi) = |A|e^{-i\phi}$$

$$A(K_1 \rightarrow \pi\pi) = \sqrt{2}|A| \cos \phi$$

$$A(K_2 \rightarrow \pi\pi) = \sqrt{2}|A| \sin \phi$$

In the KM scheme, this phase can be generated by interactions with heavy quark intermediate states and is naturally small. The origin of this phase in the heavy quark couplings is not well understood, but its existence is compatible with the theory. There are, in addition, several other theories that can explain the observed CP violation, all of which predict new physics beyond the present standard model. The breaking of CP symmetry may thus turn out to be the key for new directions in particle physics.

The two generic mechanisms for finding CP violation in kaons were mentioned previously. One involves a mixing between K_1 and K_2 so that the physical mass eigenstate contains a mixture of both CP eigenvalues. For the long-lived kaon, this means that we can write

$$|K_L\rangle = \frac{1}{(1 + |\varepsilon|^2)^{1/2}} (|K_2^0\rangle + \varepsilon|K_1^0\rangle)$$

where ε is the mixing parameter that characterizes CP violation. From experiment, we know that ε is quite small:

$$|\varepsilon| \approx 2 \times 10^{-3}$$

The other general mechanism involves the direct generation of CP violating phases in the $K \rightarrow 2\pi$ amplitude. In the first case, all kaon CP violation is governed by the same number ε, and both the rates for $K_L \rightarrow \pi^+\pi^-$ and $K_L \rightarrow \pi^0\pi^0$ are predicted to be exactly proportional to the corresponding CP conserving $K_s \rightarrow 2\pi$ rates. However, the direct weak interactions of $K^0 \rightarrow \pi^+\pi^-$ and $K^0 \rightarrow \pi^0\pi^0$ are not necessarily the same, and to the extent that they are

different, the weak phases of the two amplitudes may differ. Roughly,

$$\text{Amp}(K_L \rightarrow \pi^+\pi^-) - \text{Amp}(K_L \rightarrow \pi^0\pi^0)$$
$$\approx \left[\frac{|\text{Amp}(K^0 \rightarrow \pi^+\pi^-)| - |\text{Amp}(K^0 \rightarrow \pi^0\pi^0)|}{|\text{Amp}(K^0 \rightarrow \pi^+\pi^-)| + |\text{Amp}(K^0 \rightarrow \pi^0\pi^0)|} \right]$$
$$\times |A| \sin \phi$$

The factor in parenthesis above is known experimentally to be equal to $\frac{1}{20}$, so that this difference in rates will not be large. Conventionally, we describe the differing direct CP violating $K_L \rightarrow 2\pi$ amplitudes by a parameter ε', such that

$$\frac{\text{Amp}(K_L \rightarrow \pi^+\pi^-)}{\text{Amp}(K_S \rightarrow \pi^+\pi^-)} = \varepsilon + \varepsilon'$$

$$\frac{\text{Amp}(K_L \rightarrow \pi^0\pi^0)}{\text{Amp}(K_S \rightarrow \pi^0\pi^0)} = \varepsilon - 2\varepsilon'$$

The KM model predicts $\varepsilon'/\varepsilon \sim 0.005$ (with significant theoretical uncertainties), and attempts to measure ε' have lead to some extremely beautiful experiments over the past 25 years. We are now seeing a hint of a nonzero value of ε' in the result of an experiment at CERN (Centre Européenne pour la Recherche Nucléaire) in Geneva, which reports

$$\frac{\varepsilon'}{\varepsilon} = 0.0033 \pm 0.0011$$

while an experiment at Fermilab (near Chicago) at present has only a bound

$$\frac{\varepsilon'}{\varepsilon} = 0.0032 \pm 0.0028 \pm 0.0012$$

Present results are thus consistent with the predictions of the KM model and seem to rule out the superweak theory, which would require $\varepsilon'/\varepsilon = 0$. Nevertheless, both experiments are being repeated and soon expect to have a more precise measurement, with an uncertainty of only ± 0.0005. Because of the difficult nature of experiments at this sensitivity, one is probably best advised to wait until the new experiments confirm the CERN value before trusting in a nonzero ε'. However, the present situation makes one hopeful that we may soon be able to rule out large classes of models and to focus our attention on more realistic theories.

III. Future Areas of Research

Clearly, in order to further distinguish between models and thus to pinpoint the origin of the CP nonconservation, additional observations of CP violation

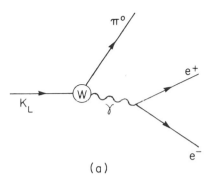

(a)

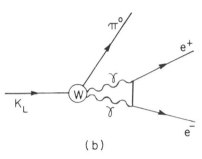

(b)

FIG. 4. The reaction $K_L \to \pi^0 e^+ e^-$ can occur via either one- or two-photon intermediate states. The former, shown in (a), must be CP violating, while the latter, shown in (b), is CP conserving.

are required. At the present time, despite nearly a quarter century of effort, the only experimental observation of CP violation is within the neutral kaon system described previously. Nevertheless, work is underway around the world that may help to clarify this situation. We mention three such efforts.

Before leaving the neutral kaon system, consider the rare decay mode

$$K_L \to \pi^0 e^+ e^-$$

Such a process can arise to lowest order via either single- or double-photon exchange diagrams, as shown in Fig. 4, and it is found that the former or latter is, respectively, odd or even under CP symmetry. The two-photon diagram is suppressed with respect to its single-photon analog by a power of the fine structure constant

$$\frac{\alpha}{\pi} = \frac{e^2}{4\pi^2 \hbar c} \approx \frac{1}{137\pi}$$

This implies that the CP violating and CP conserving components of the amplitude should contribute nearly at the same level:

Amp (CP conserving) $\sim$ Amp (2 photon)

$$\sim \text{Amp (1 photon)} \times \frac{\alpha}{\pi}$$

Amp (CP violating) $\sim$ Amp (1 photon) $\times \varepsilon$

If this is the case, such opposite CP amplitudes should give rise to a strong interference, which can be detected, for example, as an asymmetry in the energy spectrum of the outgoing electron and position. This would be easily detectable except for one fact: compared to the $K_L \to 3\pi$ case, this amplitude is of order

$$\frac{\text{Amp}(K_L \to \pi^0 e^+ e^-)}{\text{Amp}(K_L \to 3\pi)} \sim \frac{\alpha}{\pi} \varepsilon \sim 10^{-6}$$

Thus, the branching ratio, the fraction of $K_L \to \pi^0 e^+ e^-$ that decays for each K_L decay, is only $O(10^{-12})$! Nevertheless, experiments are being planned at present to detect such rare transitions and, provided they can reach the desired sensitivity, should be able to provide a new probe of the CP violating interaction. Again, the superweak model predicts CP violation to arise only of size ε, while within the KM picture, effects should be significantly larger.

A second important area of current and future research is that of measurement of electric dipole moments of particles, that is, seeking a term in the interaction between a particle and an applied electric field of the form

$$H \propto \mathbf{S} \cdot \mathbf{E}$$

where $\mathbf{S}$ is the spin and $\mathbf{E}$ is the electric field. Imagining a spinning particle to be placed in an electric field set up by two oppositely charged capacitor plates, it is easy to see that if the flow of time is reversed, the spin reverses but not the electric field, so that an interaction of the form $\mathbf{S} \cdot \mathbf{E}$ violates time reversal (see Fig. 5) and hence, by the CPT theorem, also violates CP. Experiments attempting to find a

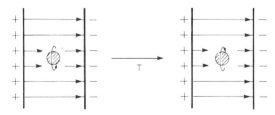

FIG. 5. Under time reversal, a spinning particle placed between capacitor plates, as shown, will reverse its direction of spin, but the direction of the electric field stays the same. Thus, an interaction of the form $\mathbf{S} \cdot \mathbf{E}$ violates time reversal invariance.

possible electric dipole moment of the neutron have been underway for many years both in France (at the Institute Laue Langevin at Grenoble) and the Soviet Union (at the Leningrad Nuclear Physics Institute). (Use of a neutral particle is a necessity since a charged particle would accelerate out of the experimental region under the influence of the electric field.) At the present time, both groups report the absence of an observed dipole moment at the level 10^{-25} e-cm. This is an incredibly sensitive level. Thus, imagine a neutron expanded to the size of the earth. The above limit then corresponds to a charge separation of only about 10 microns! Similar searches for a nonzero dipole moment have been performed within atoms. However, interpretation of these measurements is clouded by the shielding of the nucleus from the full effect of the electric field because of the motion of the electron cloud.

There exist important implications of this result for theoretical models of CP violation. In the case of the superweak and KM theories, one would not expect to have observed any effect at the present experimental level, since in the former case a $\Delta S = 2$ interaction is involved, while in the latter, one requires the heavy quark sector for a nonzero effect. However, calculations using many other models predict an effect at about the present upper limit, so it is important to push the experiments even further in order to confirm or refute these predictions.

The third area of active experimental interest has to do with heavy quark, c or b, manifestations of quark–antiquark mixing, such as those that gave rise to the CP violation observed in the kaon sector. In the case of the charmed, c, quark, there are strong theoretical reasons to suspect that such mixing will not be large and this expectation is borne out by experiment. However, in the bottom, b, quark sector, substantial mixing is anticipated and has recently been observed at a level even larger than had been predicted. Observation of this mixing is a first step, but it will be considerably more difficult to detect the CP violation itself, which will, as in the strange quark case, be only a small modification superim-posed upon the basic decay process. It is estimated that in order to successfully observe this phenomenon using b quarks nearly one hundred million such decays will be required. This is far beyond the ability of present accelerators, but there is considerable interest in generating this capability, and the next few years may well see a good deal of activity in this area.

Finally, it should be noted that while we have emphasized that the neutral kaon system is the only one in which the experimental observation of CP violation has thus far been detected, there exists strong evidence of such nonconservation in one additional area—that of cosmology, specifically the observation that the universe consists primarily of protons, neutrons, and electrons, with very few of their corresponding antiparticles. Since it is natural to assume that during the big bang equal numbers of particles and antiparticles were produced, one requires a mechanism whereby the interactions of such particles after production creates a slight predominance of particles over antiparticles. Then, when particles and antiparticles annihilate into photons, only a small particle excess will remain, as observed. It can be shown that this scenario requires the existence of CP violating interactions, and thus, the phenomenon of CP nonconservation can be said to be of cosmic significance.

BIBLIOGRAPHY

Georgi, H. (1984). "Weak Interactions and Modern Particle Theory." Benjamin/Cummings, Reading, Massachusetts.

Gottfried, K., and Weisskopf, V. F. (1984). "Concepts of Particle Physics," Vol. 1. Oxford Univ. Press, London and New York.

Halzen, F., and Martin, A. D. (1984). "Quarks and Leptons: An Introductory Course in Modern Particle Physics." Wiley, New York.

Kane, G. (1987). "Modern Elementary Particle Physics." Addison-Wesley, Reading, Massachusetts.

Perkins, D. H. (1982). "Introduction to High Energy Physics." 2nd ed. Addison-Wesley, Reading, Massachusetts.

DENSE MATTER PHYSICS

George Y. C. Leung   *Southeastern Massachusetts University*

I. Background and Scope
II. Basic Theoretical Method
III. Composition of Dense Matter
IV. Equation of State of Dense Matter
V. Transport Properties of Dense Matter
VI. Neutrino Emissivity and Opacity

GLOSSARY

Adiabat: Equation of state of matter that relates the pressure to the density of the system under a constant entropy.

Baryons: Elementary particles belonging to a type of fermions that includes the nucleons, hyperons, delta particles, and others. Each baryon is associated with a baryon number of one, which is a quantity conserved in all nuclear reactions.

Bosons: Elementary particles are divided into two classes called bosons and fermions. The bosons include the photons, phonons, and mesons. At thermal equilibrium, the energy distribution of identical bosons follows the Bose–Einstein distribution.

Degenerate electrons: System of electrons that occupy the lowest allowable momentum states of the system, thus constituting the absolute ground state of such a system.

Fermions: Class of elementary particles that includes the electrons, neutrinos, nucleons, and other baryons. Identical fermions obey Pauli's exclusion principle and follow the Fermi–Dirac distribution at thermal equilibrium.

Isotherm: Equation of state of matter that relates the pressure to the density of the system at constant temperature.

Neutrinos: Neutral, massless fermions that interact with matter through the weak interaction. Neutrinos are produced, for example, in the decay of the neutrons.

Neutronization: Form of nuclear reaction in which the neutron content of the reaction product is always higher than that of the reaction ingredient. It occurs in dense matter as its density increases from 10^7 to 10^{12} g/cm^3.

Nuclear matter: Matter substance forming the interior of a nucleus. Its density is approximately 2.8×10^{12} g/cm^3, which is relatively independent of the nuclear species. It is composed of nearly half neutrons and half protons.

Phonons: Lattice vibrations of a solid may be decomposed into a set of vibrational modes of definite frequencies. Each frequency mode is composed of an integral number of quanta of definite energy and momentum. These quanta are called phonons. They are classified as bosons.

Photons: Particle–wave duality is an important concept of quantum theory. In quantum theory, electromagnetic radiation may be treated as a system of photons endowed with particle properties such as energy and momentum. A photon is a massless boson of unit spin.

Quarks: Subparticle units that form the elementary particles. There are several species of quarks, each of which possesses, in addition to mass and electric charge, other fundamental attributes such as c-charge (color) and f-charge (flavor).

Superconductivity: Electrical resistance of a superconductor disappears completely when it is cooled below the critical temperature. The phenomenon is explained by the fact that due to the presence of an energy gap in the charge carriers' (electrons or protons) energy spectrum, the carriers cannot be scattered very easily, and the absence of scattering leads to superconductivity.

Superfluidity: Superfluidity is the complete absence of viscosity. The conditions leading to superconductivity also lead to superfluidity in the proton or electron components of the substance. In the case of neutron matter, the neutron component may turn superfluid due to the absence of scattering. The critical temperatures for the proton and neutron components in neutron matter need not be the same.

Dense matter physics is the study of the physical properties of material substance compressed to high density. The density range begins with hundreds of grams per cubic centimeter and extends to values ten to fifteen orders of magnitudes higher. Although such dense matter does not occur terrestrially, it exists inside stellar objects such as the white dwarf stars, neutron stars, and black holes, and possibly existed during the early phase of the universe. Dense matter physics therefore provides the scientific basis for the investigation of these objects.

I. Background and Scope

Matter is the substance of which all physical objects are composed. The density of matter is the ratio of its mass to volume and is a measure of the composition of matter and the compactness of the constituent entities in it. In units of grams per cubic centimeter (g/cm^3) the density of water is 1.0 g/cm^3, and the densities of all macroscopic objects on earth do not exceed roughly 20 g/cm^3. However, some stellar objects are believed to be formed of matter with much higher densities. In the 1920s, a star called Sirius B, a binary companion of the star Sirius, was found to be a highly compact object having the mass of our sun but the size of a planet, and thus must be composed of matter of very high density, estimated to reach millions of g/cm^3. Sirius B is now known to belong to a class of stellar objects called the white dwarf stars. Dense matter physics began as an effort to understand the structure of the white dwarf stars. It matured into a branch of science devoted to the study of the physical properties of dense matter of all types that may be of interest to astrophysical and cosmological investigations.

Since the type of matter under study cannot be found terrestrially, it is impossible to subject it to direct laboratory examination. Hence the study of dense matter physics is mainly theoretical in nature. In the 1920s the emergence of quantum mechanics was making a strong impact on physics, and a theory of dense matter based on the quantum mechanical behavior of electrons at high density was constructed. It marked the dawn of dense matter physics, and this theory remains valid today for the study of white dwarf stars. The subsequent identification of other compact stellar objects such as the neutron stars and black holes greatly intensified the study of dense matter physics. We survey here what can be expected theoretically from dense matter and the implications of present theories on the structure of these compact stellar objects.

On the experimental side, the study is benefited by the fact that if the concept of matter density is extended to include microscopic bodies such as the atomic nuclei, then a substance called nuclear matter, which possesses extremely high density, may be identified. Through nuclear physics study, it is then possible to subject matter with such high density to laboratory examinations. Such experimental information provides an invaluable guide to the study of matter forming the neutron stars. [*See* NUCLEAR PHYSICS.]

Compact stellar objects are mainly the remains of stars whose nuclear fuels have been exhausted and are drained of the nuclear energy needed to resist the pull of the gravitational force. As the gravitational force contracts the stellar body, it also grows in strength. This unstable situation is described as gravitational collapse, which continues until a new source of reaction strong enough to oppose the gravitational force becomes available. The search for the physical properties of dense matter responsible for resistances to gravitational collapse is an important aspect in dense matter physics since the results have important astrophysical implications.

The structure and stability of a compact stellar object depend on its composition and the equation of state of the form of matter that it is composed of. The equation of state expresses the pressure generated by the matter substance as a function of its density and temperature. The determination of the composition and the equation of state of dense matter is a prime objective in dense matter studies. These topics are discussed in Sections III and IV after a brief introduction of the basic theoretical method involved is presented in Section II.

Compact stellar objects perform rotations, pulsations, and emissions, and to understand these processes we would need to know, in addition to the equation of state, the properties of dense matter under nonequilibrium conditions. These are called the transport properties, which include the electrical and thermal conductivity and viscosity. These intrinsic properties of dense matter are discussed in Section V. The effects of a strong magnetic field on the transport properties, however, are not included in this writing. Properties related to radiative transfer, such as emissivity and opacity, are discussed in Section VI. However, radiative transfer by photons in dense matter is completely superceded by conductive transfer, and since it does not play an important role, the photon emissivity and opacity in dense matter will not be discussed. Instead, Section VI concentrates on the much more interesting topic of neutrino emissivity and opacity in dense matter.

The properties of dense matter will be discussed in several separate density domains, each of which is characterized by typical physical properties. In the first density domain, from 10^2 to 10^7 gm/cm^3, the physical properties are determined to a large extent by the electrons among the constituent atoms. The electrons obey an important quantum mechanical principle called Pauli's exclusion principle which forbids two electrons to occupy the same quantum state in a system. All electrons must take up quantum states that differ in energy, and as the electron density is increased, more and more of the electrons are forced to take on new high-energy quantum states. Consequently, the total energy of the electrons represents by far the largest share of energy in the matter system. It is also responsible for the generation of an internal pressure in the system. All white dwarf stars are believed to be composed of matter with densities falling in this domain, which is known to sustain stable stellar configurations. The physical mechanism mentioned here for electrons is central to establishing the physical properties of dense matter at all density domains, and for this reason it is first introduced in Section II.

The second density domain ranges from 10^7 to 10^{12} g/cm^3, where nuclear physics plays a key role. Above 10^7 g/cm^3 the constituent atomic nuclei of the dense matter experiences nuclear transmutations. In general, an increase in density above this point leads to the appearance of nuclei that are richer in neutron content than those occurring before. This process, called neutronization, continues throughout the entire density domain. The process also suppresses the increase in electron number with increase in matter density and thus deprives the matter system of its major source of energy. Matter with densities belonging to this density domain experiences a gradual reduction in compressibility with increasing density and is no longer able to sustain stable stellar configurations after its density exceeds 10^8 g/cm^3.

As matter density approaches 10^{12} g/cm^3, some nuclei become so rich in neutrons that they cease to bind the excess neutrons; nuclei now appear to be immersed in a sea of neutrons. The onset of such a phenomenon is called neutron drip, a term suggesting that neutrons are dripping out of the nuclei. This leads to the third density domain ranging from 10^{12} to 10^{15} g/cm^3. A rapid increase in neutron density accompanying an increasing matter density leads to the production of energetic neutrons, since neutrons (like electrons) obey Pauli's exclusion principle. Hence the same quantum mechanical mechanism characterizing the first density domain becomes operative here. As soon as neutrons were discovered experimentally in the 1930s, this mechanism was invoked to suggest the possible existence of stable neutron stars, long before neutron stars were actually identified in astronomical observations. Unlike the electrons, however, neutrons interact among themselves with nuclear forces that are comparatively strong and must be handled with great care. The average density of atomic nuclei, or nuclear matter density, is of the order of 10^{14} g/cm^3. Much of the needed physics in understanding matter with density in this density domain must come from nuclear physics.

Our understanding of matter with densities above 10^{15} g/cm^3 is very tentative; for this reason we shall, for discussion, assign matter with densities above 10^{15} g/cm^3 into the fourth and last density domain. In this area we shall discuss the physical basis for topics such as hyperonic matter, pion condensation, and quark matter.

Since the study of dense matter is highly theoretical in nature, we begin our discussion with an introduction to the basic theoretical method needed for such an investigation in establishing the composition of dense matter and its equation of state.

II. Basic Theoretical Method

Matter may first be considered as a homogeneous system of atoms without any particular structure. Such a system may be a finite portion in an infinite body of such substance, so that boundary effects on the system are minimized. The system in the chosen volume possesses a fixed number of atoms. At densities of interest all atoms in it are crushed, and the substance in the system is best described as a plasma of atomic nuclei and electrons. We begin by studying pure substances, each formed by nuclei belonging to a single nuclear species, or nuclide. The admixture of other nuclides in a pure substance can be accounted for and will be considered after the general method of investigation is introduced. [*See*, PLASMA SCIENCE AND ENGINEERING.]

The physical properties of the system do not depend on the size of the volume, since it is chosen arbitrarily, but depend on parameters such as density, which is obtained by dividing the total mass of the system by its volume. The concept of density will be enlarged by introducing a host of densitylike parameters, all of which are obtained by dividing the total quantities of these items in the system by its volume. The term density will be qualified by calling it mass density whenever necessary. In addition, parameters such as the electron number density and the nuclei number density will be introduced.

Each nuclide is specified by its atomic number Z and mass number A. Z is also the number of protons in the nucleus and A the total number of protons and neutrons there. Protons and neutrons have very nearly the same mass and also very similar nuclear properties; they are often referred to collectively as nucleons. Each nuclide will be designated by placing its mass number as a left-hand superscript to its chemical symbol, such as ^{4}He and ^{56}Fe.

Let the system consist of N atoms in a volume V. The nuclei number density is

$$n_A = N/V \qquad (1)$$

and the electron number density is

$$n_e = NZ/V \qquad (2)$$

which is the same as the proton number density, because the system is electrically neutral. The mass of a nuclide is given in nuclear mass tables to high accuracy. To determine the mass density of a matter system we usually do not need to know the nuclear mass to such accuracy and may simply approximate it by the quantity Am_p, where $m_p = 1.67 \times 10^{-24}$ g is the proton mass. The actual mass of a nucleus should be slightly less than that because some of the nuclear mass, less than one percent, is converted into the binding energy of the nucleus. The mass density ρ of the matter system is given simply by

$$\rho = NAm_p/V \qquad (3)$$

For example, for a system composed of electrons and helium nuclei, for which $Z = 2$ and $A = 4$, a mass density of 100 g/cm^3 corresponds to a nuclei number density of $n_A = 4 \times 10^{24}$ cm^{-3} and an electron number density $n_e = 2n_A$.

Like all material substances, dense matter possesses an internal pressure that resists compression, and there is a definite relation between the density of a substance and the pressure it generates. The functional relationship between pressure and density is the equation of state of the substance and is a very important physical property for astrophysical study of stellar objects. The structure of a stellar object is determined mainly by the equation of state of the stellar substance. In this section, the method for the derivation of the equation of state will be illustrated.

Unlike a body of low-density gas, whose pressure is due to the thermal motions of its constituent atoms and is therefore directly related to the temperature of the gas, dense matter derives its pressure from the electrons in the system. When matter density is high and its electron number density is correspondingly high, an important physical phenomenon is brought into play, which determines many physical properties of the system. We shall illustrate the application of this phenomenon to the study of dense matter with densities lying in the first density domain, 10^2–10^7 g/cm^3.

A. Pauli's Exclusion Principle

It is a fundamental physical principle that all elementary particles, such as electrons, protons, neutrons, photons, and mesons, can be classified into two major categories called fermions and bosons. The fermions include electrons, protons, and neutrons, while the bosons include photons and mesons. Atomic nuclei are regarded as composite systems and need not fall into these classes. In this study, one fermionic property, Pauli's exclusion principle, plays a particularly important role. The principle states that *no two identical fermions should occupy the same quantum state in a system*. Let us first explain the meaning of identical fermions. Two fermions are identical if they are of the same type and have the same spin orientation. The first requirement is clear, but the second deserves further elaboration. A fermion possesses intrinsic angular momentum called spin with magnitude equal to $\frac{1}{2}, \frac{3}{2}, \frac{5}{2}, \ldots$ basic units of the angular momentum (each basic unit equals Planck's constant divided by 2π). The electron spin is equal to $\frac{1}{2}$ of the basic unit, and electrons are also referred to as spin-half particles. Spin is a vector quantity and is associated with a direction. In the case of electron spin, its direction may be specified by declaring whether it is oriented parallel to or antiparallel to a chosen direction. The fact that there are no other orientations besides these two is a result of nature's quantal manifestation, the recognition of which laid the foundation of modern atomic physics. These two spin orientations are simply referred to as spin-up and spin-down. All spin-up electrons in a system are identical to each other, as are all spin-down electrons. The term identical electrons is thus defined. Normally, these two types of electrons are evenly distributed, since the system does not prefer one type of orientation over the other.

We come now to the meaning of a quantum state. Each electron in a dynamical system is assigned a quantum state. The quantum state occupied by an electron may be specified by the electron momentum, denoted by $\mathbf{p}$, a vector that is expressable in Cartesian coordinates as p_x, p_y, p_z. When an electron's momentum is changed to $\mathbf{p}'$ in a dynamical process, it moves to a new quantum state. The momentum associated with the new quantum state $\mathbf{p}'$ must, however, differ from $\mathbf{p}$ by a finite amount. For a system of uniformly distributed electrons in a volume V, this amount may be specified in the following way. If each of the three momentum components are compared, such as p_x and p_x', they must differ by an amount equal to h/L, where h is Planck's constant and $L - V^{1/3}$. In other words, in the space of all possible momenta for the electrons, each quantum state occupies a size of h^3/V, which is the product of the momentum differences in all three components.

B. Fermi Gas Method

According to Pauli's exclusion principle, all electrons of the same spin orientation in a system must possess momenta that differ from each other by the amount specified above. The momenta of all electrons are completely fixed if it is further required that the system exist in its ground state, which is the lowest possible energy state for the system because there is just one way that this can be accomplished, which is

for the electrons to occupy all the low-momentum quantum states before they occupy any of the high-momentum quantum states. In other words, there exists a fixed momentum magnitude p_F; and in the ground-state configuration, all states with momentum magnitudes below p_F are occupied while the others with momentum magnitudes above p_F are unoccupied. Graphically, the momenta of the occupied states plotted in a three-dimensional momentum space appear like a solid sphere centered about the point of zero momentum. The summation of all occupied quantum states is equivalent to finding the volume of the sphere, since all states are equally spaced from each other inside the sphere. Let the radius of the sphere be given by p_F, called the Fermi momentum, then the integral for the volume of the sphere with the momentum variable expressed in the spherical polar coordinates is given by

$$\int_0^{p_F} 4\pi \, p^2 \, dp = \left(\frac{4\pi}{3}\right) p_F^3 \qquad (4)$$

where p is the magnitude of $\mathbf{p}$, $p = |\mathbf{p}|$. The number of electrons accommodated in this situation is obtained by dividing the above momentum volume by h^3/V and then multiplying by 2, which is to account for electrons with two different spin orientations. The result is proportional to the volume V which appears because the total number of electrons in the system is sought. The volume V is divided out if the electron number density is evaluated, which is given by

$$n_e = (8\pi/3h^3) \, p_F^3 \qquad (5)$$

Equation (5) may be viewed as a relation connecting the electron number density n_e to the Fermi momentum p_F. Henceforth, p_F will be employed as an independent variable in establishing the physical properties of the system.

The total kinetic energy density (energy per unit volume) of the electrons can be found by summing the kinetic energies of all occupied states and then dividing by the volume. Each state of momentum $\mathbf{p}$ possesses a kinetic energy of $p^2/2m_e$, where m_e is the electron mass, and the expression for the electron kinetic energy density is

$$\varepsilon_e = \frac{2}{h^3} \int_0^{p_F} 4\pi p^2 \, dp \, \frac{p^2}{2m_e} = \frac{2\pi}{5h^3 m_e} p_F^5 \qquad (6)$$

The average kinetic energy per electron is obtained by dividing ε_e by n_e:

$$\varepsilon_e/n_e = 0.6(p_F^2/2m_e) \qquad (7)$$

where $p_F^2/2m_e$ is the kinetic energy associated with the most energetic electrons in the system, and here we see that the average kinetic energy of all electrons in the system is just six-tenths of it. This is an important result to bear in mind. It tells us that p_F is a representative parameter for the system and may be used for order of magnitude estimates of many of the energy related quantities. The method described here for finding the energy of the system of electrons may be applied to other fermions such as protons and neutrons and is referred to generally as the Fermi gas method.

It is now instructive to see what would be the average electron kinetic energy in a matter system that is composed of ^{4}He nuclei at a density of 100 g/cm^3. From the electron number density that we have computed before, $n_e = 8 \times 10^{24}$ cm^{-3}, we find that

$$\frac{\varepsilon_e}{n_e} = \frac{0.3h^2c^2}{m_ec^2}\left(\frac{3}{4\pi}n_e\right)^{2/3} \approx 300 \text{ eV} \qquad (8)$$

where eV stands for electron volts. In evaluating expressions such as Eq. (8), we shall follow a scheme that reduces all units needed in the problem to two by inserting factors of h or c at appropriate places. These two units are picked to be electron volts (eV) for energy and centimeters (cm) for length. Thus, instead of expressing the electron mass in grams it is converted into energy units by multiplying by a factor of c^2, and $m_ec^2 = 0.511 \times 10^6$ eV. The units for h can also be simplified if they are combined with a factor of c, since $hc = 1.24 \times 10^{-4}$ eV cm. This scheme is employed in the evaluation of Eq. (8) as the insertions of the c factors are explicitly displayed.

The average kinetic energy per electron given by Eq. (8) is already quite formidable, and it increases at a rate proportional to the two-thirds power of the matter density. For comparison, we estimate the average thermal energy per particle in a system, which may be approximated by the expression k_BT, where k_B is Boltzmann's constant and T the temperature of the system in degrees Kelvin (K). The average thermal energy per particle at a temperature of 10^6 K is about 100 eV. Consequently, in the study of compact stellar objects whose core densities may reach millions of g/cm^3 while the temperature is only 10^7 K, the thermal energy of the particles may justifiably be ignored.

Electrons belonging to the ground state of a system are called degenerate electrons, a term that signifies the fact that all electron states

whose momenta have magnitudes below the Fermi momentum p_F are totally occupied. Thus, whenever electrons are excited dynamically or thermally to states with momenta above p_F, leaving states with momenta below p_F empty, such a distribution of occupied states is called partially degenerate. Partially degenerate electrons will be discussed later in connection with systems at high temperatures.

C. Pressure

The internal pressure P of a system of particles is given by the thermodynamic expression

$$P = -\left.\frac{dE}{dV}\right|_N \qquad (9)$$

where E is the total energy of the system and N its particle number in the volume V. For example, in the case of the electron gas, $E = \varepsilon_eV$ and $N = n_eV$. The evaluation of the derivative in Eq. (9) is best carried out by using p_F as an independent variable and converting the derivative into a ratio of the partial derivatives of E and V with respect to p_F, while keeping N fixed. For a noninteracting degenerate electron system, the pressure can also be derived from simple kinematical considerations, and the result is expressed as

$$P_e = \frac{2}{h^3}\int_0^{p_F} 4\pi p^2\, dp\, \frac{pv}{3} \qquad (10)$$

where v is the velocity of the electron, $v = p/m_e$. By either method, the electron pressure is evaluated to be

$$P_e = \frac{8\pi}{15m_eh^3}p_F^5 \qquad (11)$$

Pressure due to a degenerate Fermi system is called degenerate pressure.

Pressure generated by the nuclei may be added directly to the electron pressure, treating both as partial pressures. They contribute additively to the total pressure of the system. Since nuclei are not fermions (with the sole exception of the hydrogen nuclei, which are protons), they do not possess degenerate pressure but only thermal pressure, which is nonexistent at zero temperature. In the case of the hydrogen nuclei, the degenerate pressure they generate may also be neglected when compared with the electron pressure, since the degenerate pressure is inversely proportional to the particle mass; and the proton mass being 2000 times larger than the electron mass makes the proton pressure 2000 times smaller than the electron pressure (bearing in mind also that the proton number density is the same as the electron number density). Thus the pressure from the system P is due entirely to the electron pressure:

$$P = P_e \qquad (12)$$

D. RELATIVISTIC ELECTRONS

Because the electrons have very small mass, they turn relativistic at fairly low kinetic energies. Relativistic kinematics must be employed for the electrons when the kinetic energy approaches the electron rest energy $m_e c^2$. For a noninteracting degenerate electron system, its most energetic electrons should reach this energy threshold when the electron number density is $n_e = 2 \times 10^{30}$ cm^{-3}, which translates into a helium matter density of $\rho = 3 \times 10^7$ g/cm^3.

In the relativistic formalism, the evaluation of the electron number density of the system remains unchanged since it depends only on p_F. The evaluation of the electron energy density must, however, be modified by replacing the individual electron kinetic energy from the expression $p^2/2m_e$ to the relativistic expression

$$e_k = (p^2 c^3 + m_e^2 c^4)^{1/2} - m_e c^2 \qquad (13)$$

so that

$$\varepsilon_e = \frac{2}{h^3} \int_0^{p_F} 4\pi p^2 \, dp \, e_k \qquad (14)$$

The electron pressure may also be found from Eq. (9) without modification, but if Eq. (10) is used, the electron velocity must be modified to the relativistic form, which is

$$v = \frac{p}{(p^2 c^2 + m_e^2 c^4)^{1/2}} \qquad (15)$$

In summary, the physical quantities of a noninteracting degenerate electron system are

$$n_e = \frac{8\pi}{3\lambda_e^3} t^3 \qquad (16a)$$

$$\varepsilon_e = \frac{\pi m_e c^2}{\lambda_e^3} \{ t(2t^2 + 1)(t^2 + 1)^{1/2} $$
$$- \ln[t + (t^2 + 1)^{1/2}] \} \qquad (16b)$$

$$P_e = \frac{\pi m_e c^2}{\lambda_e^3} \left\{ \frac{t}{3} (2t^2 - 3)(t^2 + 1)^{1/2} \right.$$
$$\left. + \ln[t + (t^2 + 1)^{1/2}] \right\} \qquad (16c)$$

where $\lambda_e = h/m_e c$ is the electron Compton wavelength having the dimension of a length and $t = p_F/m_e c$, which is dimensionless.

E. EQUATION OF STATE

The mass density of dense matter given by Eq. (3) may be expressed in terms of the electron number density as

$$\rho = m_p n_e / Y_e \qquad (17)$$

where Y_e, called the electron fraction, is the ratio of the number of electrons to nucleons in the system. For a pure substance, it is just $Y_e = Z/A$, but for a mixed substance, where a variety of nuclides are present, it is given by

$$Y_e = \sum_i x_i \left(\frac{Z_i}{A_i} \right) \qquad (18)$$

where the subscript i designates the nuclide type and x_i denotes the relative abundance of that nuclide in the system. For example, $Y_e = 1.0$ for a pure hydrogen system, $Y_e - 0.5$ for a pure ^{4}He system, and $Y_e = \frac{26}{56} = 0.464$ for a pure ^{56}Fe system. All pure substances that are composed of nuclides lying between He and Fe in the periodic table have electron fractions bounded by the values 0.464 and 0.5. Since pressure depends on n_e of a system while its mass density depends on n_e/Y_e, it is clear that matter systems with the same Y_e would have the same equation of state.

In Fig. 1, typical equations of state of dense matter at zero temperature are plotted for the cases of $Y_e = 1.0$ and $Y_e = 0.5$. The pressure is

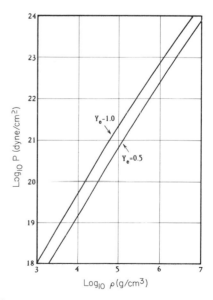

FIG. 1.   Two equations of state at zero temperature, one for matter composed of helium nuclei ($Y_e - 0.5$) and the other for matter composed of hydrogen nuclei ($Y_e = 1.0$), computed by the Fermi gas method.

due entirely to that of a noninteracting degenerate electron system and is computed from Eq. (16c). Pressure is expressed in units of dynes per square centimeter (dyn/cm²). From the approximate linearity of these curves on a log–log plot, it is apparent that the pressure and density are related very nearly by a power law. It is therefore both convenient and useful to express the equation of state in the form

$$P \propto \rho^{\Gamma} \tag{19}$$

where

$$\Gamma = \frac{d(\ln P)}{d(\ln \rho)} \tag{20}$$

is called the adiabatic index of the equation of state. Γ should be approximately constant in the density range belonging to the first density domain. We find $\Gamma = \frac{5}{3}$ at the low-density end of the density domain, and it decreases slowly to values approaching $\Gamma = \frac{4}{3}$ at the high-density end of the regime. The magnitude of Γ is a measure of the ''stiffness'' of the equation of state, in the language of astrophysics. A high Γ means the equation of state is stiff (the substance is hard to compress). Such information is of course of utmost importance to the study of stellar structure.

The numerical values of the equation of state of dense matter composed of helium nuclei at zero temperature are listed in Table I. In spite of the fact that this equation of state is obtained by neglecting electrostatic interaction, it should still be applicable to the study of white dwarf stars which are expected to be composed predominantly of dense helium matter. As we shall see below, the electrostatic interaction is not going to be important at high densities. The exact composition of the star is also not important in establishing a representative equation of state for the substance forming the white dwarf star, since all nuclides from He to Fe yield very similar Y_e factors, therefore any mixture of these nuclides would give very nearly the same equation of state. The maximum core density of the largest white dwarf star should not exceed 10^7 g/cm³.

F. ELECTROSTATIC INTERACTION

In a plasma of electrons and nuclei, the most important form of interaction is the electrostatic interaction among the charged particles, which could modify the total energy of the system. Several estimates of the electrostatic interaction energy will be given here beginning with the crudest.

Each electron in the plasma experiences attraction from the surrounding nuclei and repulsion by the other electrons. Its electrostatic interaction does not depend on the overall size of the system chosen for the investigation, since it is electrically neutral and should depend only on the local distribution of the electric charges. The positive charges carried by the protons are packed together in units of Z in the nucleus, while the negative charges carried by the electrons are distributed in the surrounding space. To investigate the electrostatic energy due to such a charge distribution, let us imagine that the system can be isolated into units, each occupied by a single nucleus of charge Z and its ac-

TABLE I. Equation of State of Matter Composed of Helium Nuclei at Zero Temperature

t	ρ (g/cm³)	P (dyn/cm²)	$\mathrm{Log}_{10}\,\rho$	$\mathrm{Log}_{10}\,P$	Γ
0.037	1.00×10^2	6.70×10^{15}	2.0	15.826	$\frac{5}{3}$
0.054	3.16×10^2	4.58×10^{16}	2.5	16.661	$\frac{5}{3}$
0.080	1.00×10^3	3.12×10^{17}	3.0	17.494	$\frac{5}{3}$
0.117	3.16×10^3	2.11×10^{18}	3.5	18.324	1.66
0.172	1.00×10^4	1.43×10^{19}	4.0	19.156	1.66
0.252	3.16×10^4	9.63×10^{19}	4.5	19.984	1.65
0.371	1.00×10^5	6.41×10^{20}	5.0	20.807	1.64
0.544	3.16×10^5	4.16×10^{21}	5.5	21.619	1.61
0.798	1.00×10^6	2.59×10^{22}	6.0	22.414	1.57
1.172	3.16×10^6	1.52×10^{23}	6.5	23.182	1.51
1.720	1.00×10^7	8.36×10^{23}	7.0	23.922	1.45
2.524	3.16×10^7	4.31×10^{24}	7.5	24.634	1.40
3.706	1.00×10^8	2.13×10^{25}	8.0	25.328	1.37

companying Z electrons. The volume of the unit is given by Z/n_e, since $1/n_e$ is the volume occupied by each electron and there are Z electrons in the unit. The shape of the unit may be approximated by a spherical cell with the nucleus residing at its center. In each spherical cell, the electrons are distributed in a spherically symmetric way about the center. In other words, if a set of spherical polar coordinates were introduced to describe this cell, with the coordinate origin at the nucleus, the electron distribution can only be a function of the radial variable r and not of the polar angle variables. With such a charge distribution, the cell appears electrically neutral to charges outside the cell, and all electrostatic energy of such a unit must be due to interactions within the cell. The radius of the cell r_s is given by the relation $(4\pi/3)r_s^3 = Z/n_e$. A more accurate treatment would call for the replacement of the spherical cell by polygonal cells of definite shapes by assuming that the nuclei in the system are organized into a lattice structure, but the energy corrections due to such considerations belong to higher orders and may be ignored for the time being.

The determination of the radial distribution of the electrons within the cell is complicated by the fact that it cannot be established by classical electrostatic methods, because in this case with cell size in atomic dimensions, Pauli's exclusion principle plays a major role. As a first estimate we may assume the electrons are distributed uniformly within the cell. The interaction energy due to electron–nucleus interaction is

$$E_{N-e} = -3/2(Ze)^2/r_s \tag{21}$$

and that due to electron–electron interaction is

$$E_{e-e} = 3/5(Ze)^2/r_s \tag{22}$$

giving a total electrostatic energy per cell of

$$E_s = E_{N-e} + E_{e-e} = -9/10(Ze)^2/r_s \tag{23}$$

In these expressions, the electric charge is expressed in the Gaussian (c.g.s.) units, so the atomic fine-structure constant is given by $\alpha = 2\pi e^2/hc = (137.04)^{-1}$. Dividing Eq. (23) by Z gives us the average electrostatic energy per electron, which, if expressed as a function of the electron number density of the system, is

$$\varepsilon_s = E_s/Z = -(9e^2/10)(4\pi Z^2 \, n_e/3)^{1/3} \tag{24}$$

Note that the interaction energy is negative and thus tends to reduce the electron pressure in the system. Quantitatively, for an electron number density of $n_e = 8 \times 10^{24}$ cm^{-3} and a corresponding helium matter density of 100 g/cm^3, we find $\varepsilon_s = 65$ eV, which is about 20% of the average electron kinetic energy. At higher densities, the relative importance of the electrostatic interaction energy to the total energy of the system is actually reduced.

G. Thomas–Fermi Method

The assumption of uniform electron distribution in a cell, though expedient, is certainly unjustified, since the electrodes are attracted to the nucleus; and thus the electron distribution in the neighborhood of the nucleus should be higher. The exact determination of the electron distribution in a cell is a quantum mechanical problem of high degree of difficulty. The best known results on this problem are derived from the Thomas–Fermi method, which is an approximate method for the solution of the underlying quantum mechanical problem. The method assumes the following:

1. It is meaningful to introduce a radially dependent Fermi momentum $p_F(r)$ which determines the electron number density, as in Eq. (5), so that

$$n_e(r) = 8\pi/3h^3 \, [p_F(r)]^3 \tag{25}$$

2. The maximum electron kinetic energy, given by $p_F^2/2m_e$, now varies radially from point to point; but for a system in equilibrium, electrons at all points must reach the same maximum total energy, kinetic and potential, and thus

$$[p_F(r)]^2/2m_e - e \, \Phi(r) = \mu \tag{26}$$

where $\Phi(r)$ is the electrostatic potential energy, and μ the chemical potential, a constant independent of r (μ replaces p_F in establishing the electron number density in the cell).

3. The electrostatic potential obeys Gauss' law

$$\nabla^2\Phi = -4\pi[Ze \, \delta(r) - en_e(r)] \tag{27}$$

where $\delta(r)$ is the mathematical delta function. From these three relations the electron distribution may be solved. Since the first relation is based on Eq. (5), which is derived from a system of uniformly distributed electron gas, its accuracy depends on the actual degree of variation in electron distribution in the cell. The second relation can be improved by adding to the electrostatic potential a term called the exchange energy, which is a quantum mechanical result. The improved method which includes the exchange energy is called the Thomas–Fermi–Dirac method.

The solution of these three relations for $n_e(r)$ is quite laborious, but it has been done for most of the stable nuclides. After it is found, the electron pressure in the system can be evaluated by employing Eq. (16c), setting p_F in the expression equal to $p_F(r_s)$, which is the Fermi momentum of the electrons at the cell boundary. There is no net force acting at the cell boundary, and the pressure there is due entirely to the kinetic energy of the electrons. Thus, the electron pressure can be evaluated by Eq. (16c). Note that, this condition would not be fulfilled by an interior point in the cell. Denoting the degenerate electron pressure evaluated for noninteracting electrons by P_e, the pressure P of the matter system corrected for electrostatic interaction by means of the Thomas–Fermi–Dirac method may be expressed as

$$P = F(\xi)P_e \qquad (28)$$

where ξ is a dimensionless parameter related to r_s and Z.

$$\xi = (0.62)Z^{-1/3}(a_0/r_s) \qquad (29)$$

where $a_0 = (h/2\pi)^2/e^2m_e = 5.292 \times 10^{-9}$ cm is the Bohr radius and $F(\xi)$ a rather complicated function that is displayed graphically in Fig. 2. The dashed lines represent the correction factor derived from the Thomas–Fermi method, and it

holds for all nuclides. The correction factor derived from the Thomas–Fermi–Dirac method depends on Z, and curves for two extreme cases are shown. Curves for nuclides of intermediate Z should fall between these two curves. The correction curve based on the assumption of uniform electron distribution is shown by the dotted curve for comparison. It is computed for the case of Fe matter. All curves tend to converge for large values of ξ. The relation between matter density and ξ is given by

$$\rho \approx 12AZ\,\xi^3 \quad \text{g/cm}^3 \qquad (30)$$

The values of $F(\xi)$ for all the curves shown in Fig. 2 are less than unity, and the corrected pressure is always less than that given by the noninteracting degenerate electron system. As ξ increases, $F(\xi)$ tends toward unity, implying that the electrostatic correction on the pressure becomes less and less significant with increasing density. At $\xi = 1$, the pressure of the system is about 20% less than that given by the electron degenerate pressure. For that value of ξ, the density of He matter is about 10^2 g/cm^3, while the density of Fe matter is already 10^4 g/cm^3. Thus, the proper treatment of electrostatic interaction is more relevant to the high-Z nuclei than to the low-Z nuclei. None of the curves shown in Fig. 2 is applicable to dense He matter, whose ξ values would be larger than unity. For He matter, electrostatic correction based on uniform electron distribution should be applicable. At a He matter density of 10^4 g/cm^3, the electrostatic correction accounts for a 5% reduction in pressure. Electrostatic correction based on uniform electron distribution may be applied to all matter system whenever the ξ values exceed unity.

Even though the Thomas–Fermi–Dirac method works quite satisfactorily for dense matter systems, the reader should, however, be reminded that it is not suitable for studying common metals whose densities are rather low. In these cases the cells are so large that the electrons fall into orbital motions around the nucleus and must be handled differently.

H. HIGH TEMPERATURES

At high temperatures, of the order of millions of degrees Kelvin, the thermal energy of the particles in the system becomes comparable to the average kinetic energy of the electrons, in which case the zero-temperature equation of state must be corrected for thermal effects. We use the term finite temperature to denote a situation

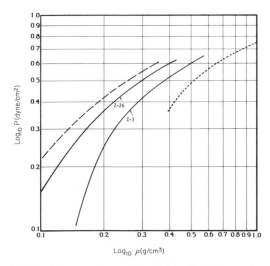

FIG. 2. Pressure corrections factor $F(\zeta)$ due to electrostatic interaction determined from the Thomas–Fermi method (dashed line), the Thomas–Fermi–Dirac method (solid lines), and by the assumption of uniform electron distribution (dotted line).

where temperature shows a perceptible effect on the energy state of the system. The proper treatment of finite temperature will first be discussed by ignoring the electrostatic interaction, which will be discussed later.

Without electrostatic interaction, the nuclei behave like a classical ideal gas, and the pressure P_A they contribute to the system is given by the classical ideal gas law:

$$P_A = \rho/Am_p \, k_B T \qquad (31)$$

The electrons, on the other hand, must be handled by the Fermi gas method. At finite temperatures electrons no longer occupy all the low-energy states; instead, some electrons are thermally excited to high-energy states. Through quantum statistics it is found that the probability for occupation of a quantum state of momentum $\mathbf{p}$ is given by the function

$$f(\mathbf{p}) = \{1 + \exp[(\varepsilon(\mathbf{p}) - \mu)/k_B T]\}^{-1} \quad (32)$$

where $\varepsilon(\mathbf{p}) = p^2/2m_e$ and μ is the chemical potential, which in the present noninteracting case is given by the Fermi energy ε_F:

$$\mu = \varepsilon_F = p_F^2/2m_e \qquad (33)$$

It accounts for the energy needed to introduce an addition particle into the system. The Fermi–Dirac distribution $f(\mathbf{p})$ is the distribution of occupied quantum states in an identical fermion system in thermal equilibrium. Since the distribution that we shall employ depends on only the magnitude of the momentum and not its direction, it shall henceforth be written as $f(p)$. The finite temperature parameters of a noninteracting electron system are to be evaluated from

$$n_e = \frac{2}{h^3} \int 4\pi p^2 \, dp \, f(p) \qquad (34a)$$

$$\varepsilon_e = \frac{2}{h^3} \int 4\pi p^2 \, dp \, f(p) \, \varepsilon(p) \qquad (34b)$$

$$P_e = \frac{2}{h^3} \int 4\pi p^2 \, dp \, f(p) \, \frac{pv}{3} \qquad (34c)$$

where the range of the p integration extends from zero to infinity. Note that, for relativistic electrons, $\varepsilon(p)$ in Eq. (34b) should be replaced by e_k of Eq. (13) and v in Eq. (34c) should be changed from $v = p/m_e$ to that given in Eq. (15). The integrals in these equations cannot be evaluated analytically, and tabulated results of these integrals are available.

It is also useful to evaluate an entropy density (entropy per unit volume) for the system. The entropy density due to the electrons may be expressed in terms of quantities already evaluated as

$$s_e = (P_e + \varepsilon_e - n_e\mu)/T \qquad (35)$$

The entropy density due to the nuclei is

$$s_A = (P_A + \varepsilon_A)/T = (5/3)P_A/T \qquad (36)$$

where ε_A is the energy density of the nuclei. The entropy density has the same dimensions as Boltznman's constant k_B.

The equation of state relates the pressure of the system to its density and temperature. Since the pressure is determined by two independent parameters, it is hard to display the result. Special cases are obtained by holding either the temperature constant or the entropy of the system constant as the density is varied. An equation of state that describes a thermodynamic process in which the temperature remains constant is called an isotherm, and an equation of state that describes an adiabatic process for which the entropy of the system remains constant is called an adiabat. In the case where the total number of particles in the system remains unchanged as its volume is varied, constant entropy means the entropy per particle remains a constant, which is obtained by dividing the entropy density given by either Eq. (35) or Eq. (36) by the particle number density. In Fig. 3, typical isotherms and adiabats for a system of (noninteracting) neu-

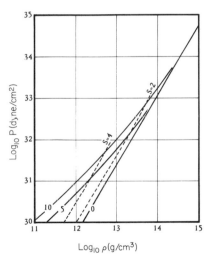

FIG. 3. Isotherms (solid lines) and adiabats (dashed lines) for a dense system of neutrons computed by the Fermi gas method with complete neglect of nuclear interaction. Isotherms shown are computed for temperatures equivalent to $k_B T = 0$, 5, and 10 MeV, and the adiabats for entropy per neutron equal to $s/k_B = 2$ and 4.

trons at high densities are shown. Such a system of neutrons is called neutron matter, which is an important form of dense matter and will be discussed in detail later. Neutrons are fermions, and the method of Fermi gas employed for the study of an electron system may be applied in the same manner to the neutron system. In Fig. 3 it is evident that the adiabats have much steeper rises than the isotherms, and this feature holds for all matter systems in general.

The inclusion of electrostatic interaction to a finite temperature system can be dealt with by extending the Thomas–Fermi–Dirac method described before. At a finite temperature the electron distribution in the system's quantum states must be modified from that of degenerate electrons to that of partially degenerate electrons as determined by the Fermi–Dirac distribution of Eq. (32). Results of this type have been obtained, but they are too elaborate to be summarized here. Interested readers are referred to references listed in the bibliography, where reference to the original articles on this topic can be found.

This section concentrates mainly on simple matter systems with densities lying in the first density domain, but the concepts discussed are applicable to the study of dense matter at all densities. At higher densities, the electrostatic interaction is replaced by nuclear interaction as the dominant form of interaction. The discussion of nuclear interaction will be deferred to the next section.

III. Composition of Dense Matter

The composition of dense matter will be discussed separately for the four density domains delineated in Section I. In the first density domain where density is 10^2–10^7 g/cm^3, dense matter can in principle exist in any composition, especially at low temperatures. The constituent nuclei of the matter system may belong to any nuclide or combination of nuclides listed in the nuclear table. A sufficient number of electrons must be present to render the system electrically neutral. Stated in such general terms, not much more needs to be added. However, since dense matter is normally found in the interior of stellar objects, its composition naturally depends on nucleosynthesis related to stellar evolution. A review of the stellar evolution process will suggest the prevalent compositions of dense matter in this density domain and the conditions under which they exist.

A star derives its energy through the fusion of hydrogen nuclei to form ^{4}He. It is called hydrogen burning, and its ignition temperature is about 10^7 K. The resulting helium nuclei are collected at the core of the star. As the helium core becomes hot enough and dense enough (estimated to be 1.5×10^8 K and 5×10^4 g/cm^3), nuclear reactions based on ^{4}He occur, initiating the helium flash. Helium burning generates the following sequence of relatively stable products: ^{12}C $\rightarrow$ ^{16}O $\rightarrow$ ^{20}Ne $\rightarrow$ ^{24}Mg. These ^{4}He-related processes account for the fact that after ^{1}H and ^{4}He, which are respectively the most and second most abundant nuclear species in stellar systems, ^{16}O is the third most abundant, ^{12}C the fourth, and ^{20}Ne the fifth. Both hydrogen burning and helium burning proceed under high temperatures and high densities. The reason is nuclei are electrically charged and do not normally come close enough for reactions to take place unless their thermal energy is large enough to overcome the repulsive electrical force. The repulsive force increases with the electric charge Z of the nuclei, which is why reactions of this type begin with low-Z nuclei and advance to high-Z nuclei as core temperature rises and density increases. These are exothermic reactions in which energy is generated.

As helium burning proceeds, the helium-exhausted core contracts sufficiently to initiate ^{12}C + ^{12}C and ^{16}O + ^{16}O reactions, called carbon burning and oxygen burning, respectively. The final products of these processes are ^{24}Mg, ^{28}Si, and ^{32}S. Upon conclusion of oxygen burning, the remanent nuclei may not wait for the further contraction of the core to bring about reactions such as ^{24}Mg + ^{24}Mg, since the electrical repulsion between them is very large and the temperature and density needed to initiate such reactions are correspondingly high. Instead, these nuclei react with photons to transform themselves in successive steps to the most tightly bound nuclide in the nuclear table, which is ^{56}Fe, in a process called photodisintegration rearrangement. Therefore, the most likely composition of matter with a density reaching the high end of the first density domain is of electrons and ^{56}Fe nuclei.

A. NEUTRONIZATION

As matter density advances into the second density domain, 10^7–10^{12} g/cm^3, the ground-state composition of the dense matter system changes with density. For example, if a dense matter system composed of ^{56}Fe nuclei at 10^7 g/cm^3 is compressed, then as density increases the constituent nuclei transmute in a sequence like Fe → Ni → Se → Ge and so on. Transmutations occur because the availability of high-energy electrons in a dense system is making such reactions possible, and transmutations result if the transmuted nuclei lower the total energy of the matter system. This process is called neutronization, because the resulting nuclide from such a reaction is always richer in neutron content than that entering into the reaction.

The physical principle involved can be seen from the simplest of such reactions, in which an electron e is captured by a proton p to produce a neutron n and a neutrino ν:

$$e + p \rightarrow n + \nu \qquad (37)$$

This is an endothermic reaction. For it to proceed, the electron must carry with it substantial amount of kinetic energy, the reason being that there is a mass difference between neutron and proton that, expressed in energy units, is $(m_n - m_p)c^2 = 1.294$ MeV, where 1 MeV = 10^6 eV. However, when matter density exceeds 10^7 g/cm^3, electrons with kinetic energies comparable to the mass difference become plentiful, and it is indeed energetically favorable for the system to have some of the high-energy electrons captured by protons to form neutrons. The neutrinos from the reactions in general escape from the system since they interact very weakly with matter at such densities. Reactions based on the same principle as Eq. (37) are mainly responsible for the neutronization of the nuclei in the matter system.

Let us denote the mass of a nucleus belonging to a certain nuclide of atomic number Z and mass number A by $M(Z, A)$ and compare the energy contents of different pure substances, each of which is composed of nuclei of a single species. Each system of such a pure substance is characterized by a nuclei number density n_A. Since n_A is different for different systems, it is best to introduce instead a nucleon number density n_B. All systems with the same density have the same n_B, where the subscript B refers to baryons, a generic term including nucleons and other nucleonlike particles. Each baryon is assigned a unit baryon number, which is a quantity conserved in all particle reactions. Thus, n_B will not be changed by the reaction shown in Eq. (37) or the neutronization process in general, which makes it a useful parameter. It is related to the

nuclei number density by $n_B = An_A$ and to the electron number density by $n_e = n_B Z/A$.

At zero temperature, the energy density of such a system is given by

$$\varepsilon = (n_B/A)M(Z, A)c^2 + \varepsilon_s + \varepsilon_e \qquad (38)$$

where ε_s is the electrostatic interaction energy density and ε_e the degenerate electron energy density, which is to be calculated from Eq. (16b) for relativistic electrons. The masses of most stable nuclides and their isotopes have been determined quite accurately and are listed in the nuclear table. One may therefore compute ε according to Eq. (38) for all candidate nuclides using listed masses from nuclear table. As a rule, only even–even nuclides, which are nuclides possessing even numbers of protons and even numbers of neutrons, are needed for consideration. These nuclides are particularly stable and are capable of bringing the dense matter system to a low energy state.

The electrostatic interaction energy density ε_s may be estimated from methods discussed in Section II. Even though electrostatic interaction energy plays a minor role in determining the pressure of the system once the density is high, it has an effect on the composition of the system, which depends on relatively small amount of energy difference. The ε_s may be estimated by evaluating the lattice energy due to the electrostatic interaction energy of a system of nuclei that form into a lattice, while the electrons are assumed to be distributed uniformly in space. The lattice giving the maximum binding is found to be the body-centered-cubic lattice, and the corresponding lattice energy is

$$\varepsilon_L = -0.89593(Ze)^2/r_s \qquad (39)$$

where r_s is the cell radius defined for Eq. (21) and is related to n_B by

$$r_s = (4\pi n_B/3A)^{-1/3}$$

For each nuclide, the energy density of the system evaluated from Eq. (38) is a function of n_B. Next, plot the quantity ε/n_B versus $1/n_B$ as shown in Fig. 4, where the curves corresponding to nuclides ^{62}Fe, ^{62}Ni, and ^{64}Ni are drawn. From the curves indicated by ^{62}Fe and ^{62}Ni, it is obvious that a matter system composed of ^{62}Ni nuclei would have an energy lower than that composed of ^{62}Fe nuclei at densities shown. The ground-state composition of dense matter is determined by nuclides whose curves form the envelope to the left of all the curves, as illustrated, for example, by the curves for ^{62}Ni and ^{64}Ni in

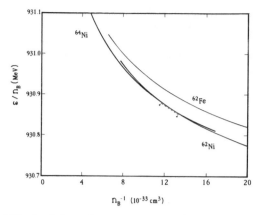

FIG. 4. Schematic energy curves for matter composed of nuclides ^{62}Fe, ^{62}Ni, and ^{64}Ni illustrating the tangent construction method for establishing the domain of a first-order phase transition in matter. The short dashed line is a line tangent to the ^{62}Ni and ^{64}Ni curves. The triangular pointers indicate the points of tangency. [After Y. C. Leung, "Physics of Dense Matter." Science Press, Beijing, China, 1984, by permission of Science Press, Beijing, China.]

Fig. 4. Two of these curves on the envelope cross at a certain density. This means that a change in composition occurs in the neighborhood of that density. Such a change, called a first-order phase transition, begins at a density below the density where the curves cross, when some of the ^{62}Ni nuclei are being rearranged into ^{64}Ni, and which ends at a density above that intersection when all of the ^{62}Ni nuclei are changed into ^{64}Ni. Throughout a first-order phase transition, the pressure of the system remains constant, which is a requirement of thermal equilibrium. The exact density at which a phase transition begins and the density at which it ends can be found from the tangent construction method shown in Fig. 4. By this method one draws a straight line tangent to the convex curves on the envelope as shown by the dotted lines. The values of n_B at the points of tangency correspond to the onset and termination of phase transition. All points on the tangent exhibit the same pressure since the expression for pressure, Eq. (9), may be converted to read

$$P = -\frac{\partial(\varepsilon/n_B)}{\partial(1/n_B)} \qquad (40)$$

Within the range of phase transition, the tangent now replaces the envelope in representing the energy of the matter system.

The sequence of nuclides constituting the zero-temperature ground state of dense matter

in the density range 10^7–10^{11} g/cm^3 obtained by this method with ε_s replaced by ε_L is shown in Table II. As indicated, neutronization begins at a density of 8.1×10^6 g/cm^3, which is quite close to 10^7 g/cm^3, and that is the reason why we choose 10^7 g/cm^3 to mark the begining of the second density domain. Note that the sequence of nuclides in the table shows a relative increase in neutron content with density, which is indicated by the diminishing Z/A ratios.

The lattice structure can be destroyed by thermal agitation. At each density, there is a corresponding melting temperature T_m that can be estimated by comparing the lattice energy with its thermal energy. It is usually taken to be

$$k_B T_m = -c_m^{-1} (Ze)^2/r_s \qquad (41)$$

where $c_m \approx 50$–100. Thus, for Fe matter at $\rho = 10^8$ g/cm^3, the melting temperature is about $T_m \approx 2 \times 10^8$ K.

B. Nuclear Semiempirical Mass Formula

As matter density exceeds 4×10^{11} g/cm^3 or so, the ground-state composition of the matter system may prefer nuclides that are so rich in neutrons that these nuclides do not exist under normal laboratory conditions, and their masses would not be listed in nuclear tables. To understand these nuclides, theoretical models of the nucleus must be constructed to deduce their masses and other properties. This is a difficult

TABLE II.   Ground-State Composition of Dense Matter in the Second Density Domain[a]

Nuclide	Z	Z/A	ρ_{max} (g/cm^3)
^{56}Fe	26	0.4643	8.1×10^6
^{62}Ni	28	0.4516	2.7×10^8
^{64}Ni	28	0.4375	1.2×10^9
^{84}Se	34	0.4048	8.2×10^9
^{82}Ge	32	0.3902	2.2×10^{10}
^{80}Zn	84	0.3750	4.8×10^{10}
^{78}Ni	28	0.3590	1.6×10^{11}
^{76}Fe	26	0.3421	1.8×10^{11}
^{124}Mo	42	0.3387	1.9×10^{11}
^{122}Zr	40	0.3279	2.7×10^{11}
^{120}Sr	38	0.3166	3.7×10^{11}
^{118}Kr	36	0.3051	4.3×10^{11}

[a] ρ_{max} is the maximum density at which the nuclide is present. [From G. Baym, C. Pethick, and P. Sutherland, *Astrophysical J.* **170**, 299 (1971). Reprinted with permission by *The Astrophysical Journal;* published by the University of Chicago Press; © 1971 The American Astronomical Society.]

task since nuclear forces are complicated, and the problem is involved. A preliminary investigation of this problem should rely on as many empirical facts about the nucleus as possible. For the present purpose, the nuclear semiempirical mass formula, which interpolates all known nuclear masses into a single expression, becomes a useful tool for suggesting the possible masses of these nuclides.

The nuclear mass is given by the expression

$$M(Z, A) = (m_p + m_e)Z$$
$$+ m_n(A - Z) - B(Z, A) \quad (42)$$

where m_p, m_e, and m_n denote the proton, electron, and neutron mass, respectively, and $B(Z, A)$ the binding energy of the nucleus. The nuclear semiempirical mass formula for even–even nuclides expresses the binding energy in the following form:

$$B(Z, A) = a_V A - a_S A^{2/3} - a_C Z^2 A^{-1/3}$$
$$- \frac{a_A(A - 2Z)^2}{A} + a_P A^{-3/4} \quad (43)$$

where the coefficients have been determined to be $a_V = 15.75$ MeV, $a_S = 17.8$ MeV, $a_C = 0.710$ MeV, $a_A = 23.7$ MeV, and $a_P = 34$ MeV. This mass formula is not just a best-fit formula since it has incorporated many theoretical elements in its construction. This feature, hopefully, will make it suitable for extension to cover unusual nuclides.

Applying Eqs. (42) and (43) to the computation of the matter energy density given by Eq. (38), one finds matter composition with density approaching 4×10^{11} g/cm^3. The general result is that as matter density increases, the constituent nuclei of the system become more and more massive and at the same time become more and more neutron rich. In general, nuclei become massive so as to minimize the surface energy which is given by the term proportional to a_S in Eq. (43), and they turn neutron rich so as to minimize the electrostatic interaction energy within the nuclei, given by the term proportional to a_C in Eq. (43). In general, the nuclear semiempirical mass formula is not believed to be sufficiently accurate for application to nuclides whose Z/A ratios are much below $Z/A \approx 0.3$. For these nuclides, we must turn to more elaborate theoretical models of the nucleus.

C. NEUTRON DRIP

When the nuclei are getting very large and very rich in neutron content, some of the neutrons become very loosely bound to the nuclei; as density is further increased, unbound neutrons begin to escape from the nuclei. The nuclei appear to be immersed in a sea of neutrons. This situation is called neutron drip, a term suggesting that neutrons are dripping out of the giant nuclei to form a surrounding neutron sea. At zero temperature, this occurs at a density of about 4×10^{11} g/cm^3. When it occurs, matter becomes a two-phase system, with one phase consisting of the nuclei and the other consisting of neutrons (with possibly a relatively small admixture of protons). The nuclei may be visualized as liquid drops suspended in a gas consisting of neutrons. These two phases coexist in phase equilibrium. The energy densities of these two phases are to be investigated separately. If the surface effects around the nuclei may be neglected, then these two phases are assumed to be uniform systems, each of which exhibits a pressure and whose neutron and proton components reach certain Fermi energies [cf. Eq. (26)]. At phase equilibrium, the pressures of these two phases must be identical, and the Fermi energies of the respective neutron and proton components of these two phases must also be identical. The composition of matter in the density domain of neutron drip is established by the system that is in phase equilibrium and at the same time reaching the lowest possible energy state.

The evaluation of the Fermi energies for the neutron and proton components depends on the nature of the nuclear interaction. Since the nuclear interaction is not as well known as the electrostatic interaction, the results derived for post-neutron drip densities are not as well established as those found for the first density domain. At the present time, there are several forms of effective nuclear interactions constructed specifically to explain nuclear phenomenology, which are quite helpful for this purpose. Effective nuclear interaction is to be distinguished from realistic nuclear interaction which is derived from nuclear scattering data and is regarded as a more fundamental form of nuclear interaction.

D. LIQUID DROP MODEL

A nucleus in many respects resembles a liquid drop. It possesses a relatively constant density over its entire volume except near the surface, and the average interior density is the same for nuclei of all sizes. A model of the nucleus that takes advantage of these features is the liquid drop model. It considers the total energy of the nucleus to be the additive sum of its bulk energy, surface energy, electrostatic energy, and translational energy. The bulk energy is given by multiplying the volume of the nucleus by the energy density of a uniform nuclear matter system with nuclear interaction included. The surface energy is usually taken to be a semiempirical quantity based on calculations for nuclei of finite sizes and on experimental results on laboratory nuclei. In a two-phase situation the surface energy must be corrected for nucleon concentration outside the nucleus. The electrostatic energy involves the interaction among all charged particles inside and outside the nucleus. The position of the nuclei may again be assumed to form a body-centered-cubic lattice. Both electrostatic lattice energy and electrostatic exchange energy contribute to it. The translational energy is due to the motion of the nucleus.

At high temperatures, the bulk energy must be computed according to a system of partially degenerate nucleons. This is usually done by employing effective nuclear interaction, in which case the nucleons are assumed to be uncorrelated, and the Fermi–Dirac distribution of Eq. (32) may be directly applied to the nucleons. The problem is much more complicated if the realistic nuclear interaction is employed, in which case particle correlations must be included, and for this reason the computation is much more elaborate. The surface energy shows a reduction with temperature. This result can be extracted from finite nuclei calculations. The lattice energy also shows temperature modification, since nuclei at the lattice points agitate with thermal motions. The same thermal motion also contributes to the translational energy of the nuclei, which may be assumed to possess thermal velocities given by the Boltzmann distribution.

The results of a study based on the liquid drop model of the nucleus is depicted in Fig. 5, where the variation of the size of a nucleus (its A number) with matter density and temperature is shown. The study is done with an effective nuclear interaction called the Skyrme interaction and for a matter system having an overall electron fraction $Y_e = 0.25$. A wide range of temperatures, expressed in terms of $k_B T$ in units of MeV, is included. Note that the temperature corresponding to $k_B T = 1$ MeV is $T = 1.16 \times 10^{10}$ K. The dashed lines indicate the A numbers of the nuclei at various densities and temperatures. The solid line forms the boundary separating a one-phase system from a two-phase system. The conditions for a two-phase system are

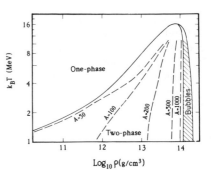

FIG. 5. Variations in composition of hot dense matter with $Y_e = 0.25$. A two-phase system exists in the region under the uppermost solid line and a one-phase system above it. The dashed lines indicate the types of equilibrium nuclides participating in the two-phase system. [After Lamb, D. Q., Lattimer, J. M., Pethick, C. M., and Ravendall, D. G. *Phys. Rev.* **41,** 1623 (1978), by permission of the authors.]

included in the plot under the solid line. The two-phase condition disappears completely when matter density exceeds 2×10^{14} g/cm^3. Just before that density, there exists a density range where the nuclei in the system merge and trap the surrounding neutrons into a form of bubbles. This range is indicated by the cross-hatched area in the plot and is labeled bubble.

E. NEUTRON MATTER

In general, when matter density exceeds 2×10^{14} g/cm^3, all nuclei are dissolved into a homogeneous system, and for a range of densities above that, matter is composed almost entirely of neutrons. The admixture of protons is negligibly small (about one percent), because all protons must be accompanied by an equal number of electrons which, being very light, contribute a large amount of kinetic energy to the system. Hence, all ground-state systems tend to avoid the presence of electrons. Such a nearly pure neutron system is called neutron matter.

The average density of the atomic nucleus, quite independent of its species, is approximately 2.8×10^{14} g/cm^3, and therefore many of the methods employed for the study of the nucleus may be applied to the study of neutron matter. As we have mentioned before, the distribution of mass density of a nucleus has been found to be quite uniform over its entire volume, and that uniform mass density is very nearly the same from one nuclide to the other. For a heavy nucleus, its volume is extended so as to maintain a mass density common to nuclei of all sizes. This remarkable fact is described as nuclear saturation. Nuclear substance possessing this mass density is called nuclear matter, and its mass density, called nuclear density, is found to be 2.8×10^{14} g/cm^3.

Neutron matter, however, is not nuclear matter. Neutron matter consists nearly entirely of neutrons, whereas nuclear matter consists of neutrons and protons in roughly equal fractions; and while nuclear matter forms bound units or nuclei, neutron matter is an unbound system. The ground-state composition of an extended system is not given by that of nuclear matter but neutron matter. Neutron matter does not exist terrestrially; its existence is only inferred theoretically. Neutron matter may be studied in analogy to nuclear matter. To accomplish this there are methods based on realistic nuclear interaction expressed in the form of nuclear potentials, the best known of which are the Brueckner–Bethe–Goldstone method and the constrained variational method. There are also methods based on the use of effective nuclear interaction. These methods combine the Hartree–Fock method with some form of effective nuclear potential. Though they are considered less fundamental than the two methods mentioned before, they usually yield more direct and accurate results at nuclear density where they are designed to perform. Since neutron matter exists for a range of densities, the choice of an applicable method depends also on how well these methods may be extended to cover such a wide range of densities. In this respect, a phenomenological model called the relativistic mean field model seems very attractive. We shall return to these methods in the next section when the equation of state of neutron matter is discussed.

It has been seriously proposed that known nuclear interaction forces would make neutron matter superfluid and the proton component in it superconductive when its temperature is below a critical temperature T_c which is estimated to be as high as 10^9–10^{10} K. These phenomena would have profound effects on the transport properties of neutron matter and will be discussed in more detail in Sections V and VI.

Since neutrons, like electrons, are fermions and obey Pauli's exclusion principle, many features of a dense degenerate electron system are exhibited by neutron matter. In spite of having a strong nuclear interaction, these features still play dominant roles in the neutron system.

Thus, results based on a dense degenerate neutron system serve as useful guides to judge the properties of neutron matter over a wide range of densities. Such guidance is particularly needed when hyperons and other massive particles begin to make their appearance in dense matter, because the precise nature of their interactions is still poorly known.

F. BARYONIC MATTER

As matter density advances beyond 10^{15} g/cm^3 and into the fourth and last density domain, many new and unexpected possibilities may arise. It is fairly sure that light hyperons appear first. Hyperons are like nucleons in every aspect except that they are slightly more massive and carry a nonzero attribute called hypercharge, or strangeness. They are $\Lambda(1115)$ and $\Sigma(1190)$, where the numbers in the parentheses give their approximate masses, expressed as mc^2 in units of MeV. Λ is electrically neutral, but there are three species of Σ with charges equal to $+1$, 0, and -1 of the proton charge. More exact values of their masses may be found in a table of elementary particles. These particles are produced in reactions such as

$$p + e \rightarrow \Lambda + \nu \qquad (44a)$$

$$p + e \rightarrow \Sigma^0 + \nu \qquad (44b)$$

$$n + e \rightarrow \Sigma^- + \nu \qquad (44c)$$

The hyperons appear as soon as the kinetic energies of the neutrons exceed the mass difference between neutron and these particles, in a manner similar to the neutronization process.

In order of ascending mass, the next group of nucleonlike particles to appear are the $\Delta(1232)$, which come in four species with electric charges equal to $+2$, $+1$, 0, and -1 of the proton charge. Δ are produced as excited states of the nucleons. The nucleons, hyperons, and Δ fall into the general classification of baryons. All baryons are assigned the baryon number $+1$ and antibaryons the number of -1. The baryon number is conserved additively in all particle reactions. Matter systems composed of nucleons, hyperons, Δ, and possibly other baryons are called baryonic matter.

G. PION CONDENSATION

Pions, $\pi(140)$, fall into a class of elementary particles called mesons which are quite different from the baryons. They participate in nuclear reactions with the baryons, but they are bosons and not restricted by Pauli's exclusion principle. They are produced in a large variety of particle interactions. They do not normally appear in a matter system, because energy is needed to create them, and their presence would mean an increase in the total energy of the matter system. They do appear, however, if their interaction with the nucleons creates an effective mass for the nucleon that is lower than its actual mass by an amount comparable to the mass of the pion. This occurs when the baryonic matter reaches some critical density, estimated to be of the order of 10^{15} g/cm^3. This result is not definite since the elementary particle interaction is still far from being understood. Different estimates yield rather different results. When the pions do appear, they may possess very similar momenta as required by the interaction which is momentum dependent. Such a state of a boson system is called a condensate. The appearance of a pion condensate in dense matter is called pion condensation.

H. Quark Matter

With the advent of the quark theory some of the traditional notions about elementary particles must be revised. The baryons and mesons, which have been traditionally called elementary particles, must be viewed as composite states of quarks. Current quark models have achieved such success that it is hardly in doubt that quark theory must play a key role in the understanding of the subparticle world. Quarks are spin-half fermions possessing fractional electric charges like $\frac{2}{3}$, $\frac{1}{3}$, and $-\frac{1}{3}$ of the proton charge. Besides mass, spin, and electric charge, they possess additional attributes such as c-charge (also called color) and f-charge (also called flavor). The electric charge will be called q-charge in the present terminology. In a quark model, the baryons mentioned before are the bound states of three quarks, and the mesons are the bound states of quark and antiquark pairs. Multiquark states involving more than three quarks are also possible. The interactions of the quarks are governed by quantum chromodynamics (QCD) through their c-charges, quantum flavodynamics (QFD) through their f-charges, and quantum electrodynamics (QED) through their q-charges. The bound-state configurations of the quarks are described by QCD which is a highly nonlinear theory, and for this reason the transition from one form of manifestation to another can occur abruptly.

Such an abrupt transition is believed to be responsible for the distinct boundary of a nucleon, for example. Inside the nucleon boundary, where three quarks are in close proximity to each other, the effective quark interaction is very weak, but once any one of them reaches the boundary, the interaction turns strong so rapidly that the quark does not have enough energy to penetrate the boundary; thus, in effect, the quarks are confined inside the boundary. Each nucleon, or baryon, therefore occupies a volume.

What if the baryon density inside a matter system is so high that the volumes occupied by them are crushed? Naturally, the boundaries will be gone, and all quarks merge to form a uniform system. Such a state of dense matter is called quark matter. The transition to quark matter has been estimated to occur at densities as low as 2 to 4 × 10^{15} g/cm^3 at zero temperature and at even lower densities at high temperatures. Current estimates of the average quark-confining energy is about 200 MeV per quark; therefore if the temperature is above 2 × 10^{12} K or a corresponding thermal energy of $k_B T = 200$ MeV, the nucleons are vaporized and the quarks set free, forming a sort of quark gas. These results, we hasten to add, are only tentative.

There is very little confidence in postulating anything beyond the quark matter, and for this reason we shall end our discussion at this point.

IV. Equation of State of Dense Matter

The equation of state of a substance is a functional relation between the pressure generated by the substance and its density and temperature. It reflects the composition and internal structure of the substance. The equation of state of dense matter is of prime importance to astrophysical study of compact stellar objects like white dwarf stars and neutron stars. It will be discussed in detail for the first three density domains. Our knowledge about dense matter in the fourth density domain is insufficient to provide accurate quantitative evaluation of its properties at this time.

A. The First Density Domain

For the first density domain (10^2–10^7 g/cm^3), we shall concentrate our discussion on a matter system composed of Fe^{56} nuclei, which represents the most stable form of matter system in

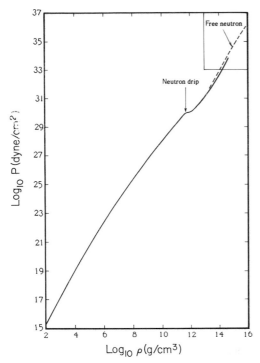

FIG. 6. The ground-state equation of state of dense matter. The numerical values of the solid line are listed in Table III under the heading BPS. The onset of neutron drip is indicated. The different versions of the equation of state in the high-density region (outlined by rectangle) is given in Fig. 7. For correlation purposes, the equation of state of a noninteracting neutron system, indicated by "free neutron," is drawn here.

that domain and shall be referred to as Fe matter. The equation of state of Fe matter at zero temperature in this density range is determined basically by the degenerate electrons in the system. Electrostatic interaction among the charged particles plays a relatively minor role, which has been demonstrated in Section II. Nevertheless, the degenerate electron pressure must be corrected for electrostatic effects. For Fe matter with densities below 10^4 g/cm³, electrostatic interaction must be computed on the basis of the Thomas–Fermi–Dirac method which is discussed in Section II, and for Fe matter with densities above 10^4 g/cm³, the electrostatic interaction may be computed on the assumption that electrons are distributed uniformly around the nuclei.

Figure 6 shows the ground-state equation of state of dense matter at zero temperature in a log–log plot over the first three density domains. The portion below 10^7 g/cm³ belongs to the

equation of state for Fe matter in the first density domain. The numerical values of this equation of state are given in Table III under the heading BPS. The adiabatic indices are listed under Γ. The equation of state in this density domain should be quite accurate since its composition is well established.

At high temperatures, degenerate electrons become partially degenerate in a way described

TABLE III.   Ground-State Equation of State of Dense Matter

	BPS			
ρ (g/cm³)	P (dyn/cm²)	Z	A	Γ
2.12×10^2	5.82×10^{15}	26	56	—
1.044×10^4	9.744×10^{18}	26	56	1.796
2.622×10^4	4.968×10^{19}	26	56	1.744
1.654×10^5	1.151×10^{21}	26	56	1.670
4.156×10^5	5.266×10^{21}	26	56	1.631
1.044×10^6	2.318×10^{22}	26	56	1.586
2.622×10^6	9.755×10^{22}	26	56	1.534
1.655×10^7	1.435×10^{24}	28	62	1.437
3.302×10^7	3.833×10^{24}	28	62	1.408
1.315×10^8	2.604×10^{25}	28	62	1.369
3.304×10^8	8.738×10^{25}	28	64	1.355
1.045×10^9	4.129×10^{26}	28	64	1.344
2.626×10^9	1.272×10^{27}	34	84	1.340
1.046×10^{10}	7.702×10^{27}	32	82	1.336
2.631×10^{10}	2.503×10^{28}	30	80	1.335
6.617×10^{10}	8.089×10^{28}	28	78	1.334
1.049×10^{11}	1.495×10^{29}	28	78	1.334
2.096×10^{11}	3.290×10^{29}	40	122	1.334
4.188×10^{11}	7.538×10^{29}	36	118	1.334
4.460×10^{11}	7.890×10^{29}	40	126	0.40
6.610×10^{11}	9.098×10^{29}	40	130	0.40
1.196×10^{12}	1.218×10^{30}	42	137	0.63
2.202×10^{12}	1.950×10^{30}	43	146	0.93
6.248×10^{12}	6.481×10^{30}	48	170	1.31
1.246×10^{13}	1.695×10^{31}	52	200	1.43
2.210×10^{13}	3.931×10^{31}	58	241	1.47
6.193×10^{13}	1.882×10^{32}	79	435	1.54
1.262×10^{14}	5.861×10^{32}	117	947	1.65
2.761×10^{14}	2.242×10^{33}	—	—	1.82
5.094×10^{14}	7.391×10^{33}	—	—	2.05

FP[b]		SKM	
ρ (g/cm³)	P (dyn/cm²)	ρ (g/cm³)	P (dyn/cm²)
4.11×10^{13}	1.01×10^{32}	4.11×10^{13}	8.00×10^{31}
5.64×10^{13}	1.52×10^{32}	5.64×10^{13}	1.42×10^{32}
9.75×10^{13}	3.36×10^{32}	9.75×10^{13}	4.34×10^{32}
1.55×10^{14}	8.60×10^{32}	1.55×10^{14}	1.23×10^{33}
2.31×10^{14}	2.30×10^{33}	2.31×10^{14}	3.21×10^{33}
3.21×10^{14}	5.71×10^{33}	3.29×10^{14}	7.64×10^{33}
4.97×10^{14}	1.82×10^{34}	4.51×10^{14}	1.67×10^{34}
7.72×10^{14}	6.22×10^{34}		
1.23×10^{14}	2.21×10^{35}		

(continues)

TABLE III (*Continued*)

RMF	
ρ (g/cm^3)	P (dyn/cm^2)
3.36×10^{14}	9.69×10^{33}
4.60×10^{14}	3.39×10^{34}
6.25×10^{14}	9.04×10^{34}
8.46×10^{14}	1.99×10^{35}
1.14×10^{15}	3.73×10^{35}
2.25×10^{15}	1.12×10^{36}
7.42×10^{15}	5.03×10^{36}

[a] From G. Baym, C. Pethick, and P. Sutherland, *Astrophysical Journal* **170**, 199 (1971). Reprinted with permission of *The Astrophysical Journal,* published by the University of Chicago Press; © 1971 The American Astronomical Society.
[b] From B. Friedman and V. R. Pandharipande, *Nuclear Physics* **A361**, 502 (1981). Reprinted with permission of *Nuclear Physics;* © 1981 North-Holland Publishing Co., Amsterdam.

by Eq. (32). Finite temperature equations of state will not be shown here. General behaviors of the isotherms and adiabats should be similar to those shown in Fig. 3. At densities below 10^4 g/cm^3, where the energy correction due to electrostatic interaction is important, the Thomas–Fermi–Dirac method must be extended to include thermal effects. Interested readers are referred to references listed in the Bibliography for further details.

B. THE SECOND DENSITY DOMAIN

The ground-state composition of matter in the second density domain (10^7–10^{12} g/cm^3) is described in Section III. The composition varies with density due to neutronization of the constituent nuclei. As the composition changes from one form to another, a first-order phase transition is involved and, over the density range where phase transition occurs, the pressure remains constant. Therefore, in a detailed plot of the equation of state in this density domain, pressure rises with density except at regions of phase transition, where it remains constant. The equation of state appears to rise in steps with increasing density, but since the steps are quite narrow, the equation of state in this density domain may be approximated by a smooth curve.

The pressure of the matter system is due entirely to the degenerate electrons. In establishing the ground-state composition of the matter system, electrostatic interaction energy in the form of lattice energy has been included, but it is quite negligible as far as the pressure of the system is concerned. The pressure in this density domain does not rise as rapidly with density as it does in the first density domain because the ground-state composition of the system shows a gradual decline in the Z/A values with density. The composition is quite well established up to a density of 4×10^{11} g/cm^3, which marks the onset of neutron drip. The composition consists of nuclei found under normal laboratory conditions or their nearby isotopes, and their masses can be either measured or extrapolated from known masses with reasonable certainty. The ground-state equation of state at zero temperature in this density range is shown in Fig. 6. The numerical values of the equation of state before neutron drip, together with the atomic number Z and mass number A of the constituent nuclei at these densities, are listed in Table III under the heading BBP.

C. THE THIRD DENSITY DOMAIN

The third density domain (10^{12}–10^{15} g/cm^3) begins properly with the onset of neutron drip, which occurs at a density of 4×10^{11} g/cm^3. With the onset of neutron drip, matter is composed of giant nuclei immersed in a sea of neutrons. Nucleons inside the nuclei coexist with nucleons outside the nuclei forming a two-phase system. Theoretical studies of such a system rely heavily on the nuclear liquid drop model which gives a proper account of the different energy components in a nucleus. The nuclear liquid drop model is described in Section III. At each density, matter is composed of a particular species of nuclei characterized by certain Z and A numbers, which exist in phase equilibrium with the neutron sea outside which has a much lower Z/A ratio. Such a two-phase system constitutes the ground-state composition of matter in this density range.

Once the ground-state composition is established, the equation of state can be found as before by establishing the electron fraction Y_e of the system and proceeding to evaluate the electron pressure. The electron pressure remains the main pressure component of the system until neutron pressure takes over at higher densities. Electron number density, however, is kept fairly constant throughout the neutron drip region. It is the neutron number density and not the electron number density that is rising with increasing matter density. The suppression in electron number density with increasing density has kept the system pressure relatively constant in the density range between 4×10^{11}–10^{12} g/cm^3. This is shown in Fig. 6. Eventually, enough neutrons are produced, and the system pressure is taken over entirely by the neutron pressure. Then the equation of state shows a rapid rise subsequent to 10^{12} g/cm^3.

The numerical values of this portion of the equation of state are listed in Table III under the heading BBP. They are computed by means of the so-called Reid soft-core potential, which is determined phenomenologically by fitting nucleon–nucleon scattering data at energies below 300 MeV, as well as the properties of the deuteron. It is considered one of the best realistic nuclear potentials applicable to nuclear problems. The computation is done in the elaborate pair approximation which includes correlation effects between pairs of nucleons. In practice, this usually means solving the Brueckner–Bethe–Goldstone equations for the nucleon energy of each quantum state occupied by the nucleons. The summation of the nucleon energies for all occupied states yields the energy density of the system. Computations with the realistic nuclear potential are very involved, and several additional corrections are needed to achieve agreement with empirical results. Extension of the method to include finite temperature calculations has not been attempted.

A second form of approach is described as the independent particle approximation, in which case all particles are uncorrelated and move in the system without experiencing the presence of the others except through an overall nuclear potential. The success of the method depends on the adequacy of the effective potential that is prescribed for interaction between each pair of nucleons. A large variety of nuclear effective potentials has been devised. Potentials are usually expressed in functional forms depending on the separation between the interacting pair of nucleons. These potentials depend not only on the particle distance but also on their spin orientations. Some even prescribe dependence on the relative velocity between the nucleons. One effective nuclear potential deserving special mention is the Skyrme potential which, like the Reid soft-core potential, belongs to the class of velocity-dependent potentials. Computations based on effective nuclear potentials seem to constitute the only viable method in dealing with the neutron drip problem at finite temperatures. Some results of finite temperature equations of state in the neutron drip region have been obtained by means of the Skyrme potential. The difficulty in working with a neutron drip system lies in the treatment of phase equilibrium, which must be handled with delicate care. The nuclear liquid drop model serves to reduce much of that work to detailed algebraic manipulations. Still, quantitative results of finite temperature equations of state in the neutron drip region are scarce.

D. Neutron Matter Region

As matter density increases towards 10^{12} g/cm^3, the constituent nuclei become so large and their Z/A ratio so low that they become merged with the neutron sea at a density of approximately 2×10^{14} g/cm^3, where the phenomenon of neutron drip terminates, and the ground state of the matter system is represented by nearly pure neutron matter. Since neutron matter is so similar to nuclear matter, all successful theories that describe nuclear matter properties have been applied to predict the properties of neutron matter. The ground-state equation of neutron matter that is determined by pair approximation is listed in Table II under BBP. It terminates at a density of 5×10^{14} g/cm^3, which is the upper limit of its applicability. There are also results obtained by the constrained variational method and the independent particle method with the Skyrme potential. These differential equations of state are compared in Fig. 7. It gives us some idea as to how unsettled the issue remains at the present time.

The numerical results of the ground-state equation of state obtained by the constrained variational method are listed in Table III under the heading FP. They are evaluated from a form of realistic nuclear potential by a method that solves the nuclear many-body problem by means of a variational technique. The finite temperature equation of state evaluated by this method is also available.

The numerical results of the ground-state equation of state evaluated from the Skyrme potential are listed in Table III under the heading SKM. Being an effective potential, the Skyrme potential contains adjustable parameters that are established by fitting nuclear properties. As the potential is tried out by different investigators, new sets of potential parameters are being proposed. The results given here are based on a recent set of parameters designated as SKM in the literature.

There are also attempts to formulate the nuclear interaction problem in a relativistic formal-

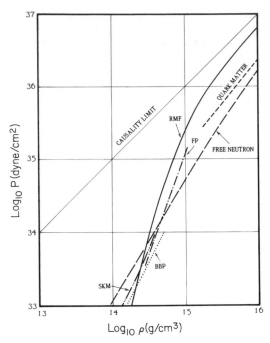

FIG. 7. Different versions of the equation of state for ground-state neutron matter. Numerical values of curves indicated by BBP, FP, SKM, and RMF are listed in Table III under the same headings. See text for discussion of free neutron, quark matter, and causality limit.

ism. This seems to be quite necessary if the method is to be applicable to matter density in the 10^{15} g/cm^3 region. A simplified nuclear interaction model based on the relativistic quantum field theory seems quite attractive. It is called the relativistic mean field model. In this model, nuclear interaction is described by the exchange of mesons. In its simplest form, only two types of mesons, a scalar meson and a vector meson, both electrically neutral, are called upon to describe nuclear interaction at short range. The scalar meson is called the σ-meson and the vector meson the ω-meson. Employed on a phenomenological level, the model needs only two adjustable parameters to complete its description, and when it is applied to study the nuclear matter problem, it successfully predicts nuclear saturation. These two parameters may then be adjusted so that saturation occurs at the right density and with the correct binding energy. The model is then completely prescribed and may then be extended to study the neutron matter problem. The equation of state thus obtained for neutron matter at zero temperature is plotted in Fig. 7, together with others for comparison. The

numerical results are listed in Table III under the heading RMF.

The relativistic mean field model shows great promise as a viable model for dense matter study. In the present form it has some defects. For example, it over-predicts the value of the compression modulus of nuclear matter. Also, the inclusion of a few other types of mesons may be needed to improve on the description of nuclear interaction. In particular, a charge vector meson that plays no role in the nuclear matter system contributes nevertheless to the neutron matter system and should be included. Since empirical results on dense matter are limited, it is difficult to fix the added parameters due to the new mesons in a phenomenological way. Future improvements of this model may have to be based on new results coming from experimental heavy ion collision results.

The diagonal line labeled causality limit in Figure 7 represents an equation of state given by $P = c\rho$. Any substance whose equation of state extends above this line would give rise to a sound speed exceeding the speed of light, which is not allowed because it would violate causality; thus the causality limit defines an upper limit to all equations of state. It is drawn here to guide the eye.

The pressure of neutron matter rises rapidly with increasing density. Such an equation of state is called stiff. This feature is already exhibited by a free neutron system that is devoid of interaction. The equation of state of a free neutron system is plotted alongside the others in Fig. 7 for comparison. It is labeled free neutron. The stiffness of the free neutron equation of state had led scientists in the 1930s to suggest the possible existence of neutron stars in the stellar systems almost as soon as neutrons were discovered. Neutron stars were detected a quarter of century later.

E. Pion Condensation Region

The stiffness of the equation of neutron matter could be greatly modified should pions be found to form a condensate in neutron matter. A pion condensate occurs in a matter system only if the interaction between the nucleons and pions lower the interaction energy sufficiently to account for the added mass and kinetic energy of the pions. The possibility of pion condensation has been suspected for a long time since pion–nucleon scattering experiments have revealed a strong attractive interaction between pions and nucleons, and this attractive mode could produce the pion condensation phenomenon. However, a quantitatively reliable solution of this problem is yet to be established. Nuclear interactions are hard to deal with, and most theoretical results are not considered trustworthy unless they can be collaborated by some empirical facts. In the pion condensation problem, no empirical verification is available, and many of the estimates about pion condensation, though plausible, cannot be easily accepted.

The effect of pion condensation on the equation of state of neutron matter is to lower the pressure of the system, because it adds mass to the system without contributing the comparable amount of kinetic energy. The softening of the equation of state is a reflection of the presence of an attractive interaction that brings about pion condensation and serves to lower the pressure. Some estimates have suggested that pion condensation should occur at a neutron matter density of around 5×10^{15} g/cm^3. No quantitative results, however, will be presented here for the pion condensation situation.

F. Baryonic Matter

As the density of neutron matter increases, the Fermi energy of the neutron system increases accordingly. If we neglect interactions among the particles, then as soon as the neutron Fermi energy reaches the energy corresponding to the mass difference of the neutron and Λ, the ground state of the matter system favors the replacement of the high-energy neutrons by Λ. Λ is the least massive baryon heavier than the nucleons. Based on an estimate in which interactions are neglected, Λ should appear in neutron matter at a density of 1.6×10^{15} g/cm^3. On a scale of ascending mass, the next group of baryons after the nucleons and Λ is the Σ, which consists of three charged species, and after that is the Δ, which consists of four charged species. Similar estimates put the appearance of Σ at 2.9 $\times$ 10^{15} g/cm^3 and the appearance of Δ at 3.7×10^{15} g/cm^3.

The above estimates emphasize the mass of the baryons. It turns out that the electric charge of the baryon is also important. For example, when a Σ^- appears, it replaces not only an energetic neutron; because of its charge it replaces an energetic electron as well. Again, based on estimates in which interactions are neglected, Σ^- appears at a density of 10^{15} g/cm^3 at Δ^- at 1.3

$\times 10^{15}$ g/cm^3, which are lower values than those given above.

To study baryonic matter properly, particle interactions must be included. Unfortunately, our knowledge of baryon interaction is still quite limited, and the interaction strengths among the different species of baryons are not yet established with sufficient accuracy. Therefore, much of our understanding of the baryonic matter is based on theoretical conjectures of the nature of baryon interaction.

One of the best methods appropriate to the study of baryonic matter is the relativistic mean field model mentioned before. The model considers all particle interactions to be mediated by two mesons, the scalar σ-meson and the vector ω-meson, both electrically neutral. Particle interaction is established once the coupling constants of these mesons with different baryons are known. There are theoretical justifications to believe that the ω-meson interacts with all baryons at equal strength, in the same manner that the electric field interacts with all electric charges equally no matter which particles carry them. In analogy with the electric charge, each baryon is associated with a unit of baryonic charge and its antiparticle with a negative unit of baryonic charge. The baryonic force mediated by the ω-meson behaves very much like the electric force in the sense that like charges repel and unlike charges attract. It is repulsive between baryons but attractive between baryons and antibaryons. If this is the case, the coupling constants of the ω-meson with all baryons in the relativistic mean field model may be taken to be the same as that adopted for the nucleons. One possible fallacy in this reasoning is that the coupling constants in the relativistic mean field model represent effective coupling constants that incorporate modifications due to the presence of neighboring particles. Therefore, even though the concept of a baryonic charge to which the ω-meson couples is correct, the effective ω-meson coupling constants with different baryons in the relativistic mean field model need not be the same.

For the σ-meson, the best estimate of its coupling constants is probably by means of the quark model which assigns definite quark contents to the baryons and σ-meson. If the quarks are assumed to interact equally among themselves, and given the fact that particle attributes such as electric charge and hypercharge must be conserved in the interaction process, it is possible to deduce that the σ-meson couples equally between nucleons and Δ, and also equally between Λ and Σ, but its coupling with Λ and Σ is only two-thirds of the coupling with the nucleon and Δ.

Incorporating these couplings in the relativistic mean field model, we can deduce the equation of state for baryonic matter. It shows a slight decrease in pressure compared with that due to neutron matter when matter density exceeds 10^{15} g/cm³. The effect of admixing other baryons in a neutron matter system is relatively minor on the system's equation of state. On the other hand, the effect of particle interaction on the equation of state is quite significant. Therefore, a proper understanding of the equation of state of matter in the 10^{15} g/cm³ range awaits better knowledge of particle interactions as well as methods in dealing with a many-body system.

G. Quark Matter

In a quark model of the elementary particles, the nucleon is viewed as the bound state of three quarks, which interact via the exchange of gluons as determined by quantum chromodynamics. The size of the nucleon is therefore determined by the confining radius of the quark interaction. When matter density reaches the point that the average separation of the nucleons is less than its confining radius, the individual identity of the nucleon is lost, and the matter system turns into a uniform system of quarks forming quark matter.

Quarks are fermions and like electrons and neutrons obey Pauli's exclusion principle. There are different species of quarks, and the known quarks are given the names of up quark (u), down quark (d), strange quark (s), and charm quark (c). There may be others. Inside a nucleon, their effective masses are estimated to be (expressed in mc^2) under 100 MeV for the u and d quarks, approximately 100–200 MeV for the s quark, and approximately 2000 MeV for the c quark. If quark interaction is neglected, the equation of state of quark matter may be deduced in the same manner as it is for a degenerate electron system. The treatment of quark interaction turns out to be not as difficult as it is for neutrons. In a quark matter, the effective quark interaction is relatively weak, and perturbative treatment of the interaction is possible. This interesting feature of quantum chromodynamics prescribes that the effective interaction of the quarks should decrease in strength as the quarks interact in close proximity to each

other, and this has been verified to be true in experiments. The equation of state of quark matter at zero temperature computed with quark interaction has been obtained. One version of it is plotted in Fig. 6 for illustration.

The transition from baryonic matter to quark matter is a first-order phase transition that may be established by a tangent construction method on the energy density of the baryonic matter and that of the quark matter in a manner similar to that discussed in Section III for the neutronization process and illustrated in Fig. 4. The onset of transition from neutron matter to quark matter has been estimated to begin at densities around 1.5 to 4×10^{15} g/cm³.

In Table III, only the basic ground-state equation of state of dense matter is presented. It serves to suggest the possible behavior of matter compressed to high densities. It would be too elaborate to detail all aspects of the equation of state at finite temperatures. At the present time, the study of dense matter is being actively pursued.

V. Transport Properties of Dense Matter

Properties of dense matter under nonequilibrium conditions, such as during the transfer of mass and conduction of heat and electricity, are of much physical interest. These are the transport properties, and knowledge of them is important in understanding stellar structure and stellar evolution. We discuss here the following transport properties of dense matter: electrical conductivity, heat conductivity, and shear viscosity.

A. Electrical Conductivity

The electrical conductivity of a substance is given by the ratio of the induced electric current density to the applied electric field. In the case of a metal, the electric current is due to the flow of conduction electrons. If all conduction electrons are given an average drift velocity v_d, then the current density j_e is given by

$$j_e = nev_d \qquad (45)$$

where n is the number density of the conduction electrons and e the electric charge of the electrons. The electrons acquire the drift velocity as they are accelerated by the electric field. When an electron collides with either the lattice or another electron, its drift velocity is redirected randomly, and the subsequent velocity is averaged to zero statistically. Thus, after each collision the electron may be assumed to be restored to thermal equilibrium and begins anew with zero drift velocity. The average drift velocity is therefore determined by the mean time between collisions τ, during which an electron is accelerated to its drift velocity by the electric field $\mathscr{E}$.

$$v_d = (e\mathscr{E}/m_e)\tau \qquad (46)$$

τ is also called the relaxation time. Putting these two expressions together gives the electrical conductivity:

$$\sigma = (ne^2/m_e)\tau \qquad (47)$$

The electrical conductivity is in units of inverse seconds, or $\sec^{-1}$, in the cgs system of units and is related to the mks system of units by $\sec^{-1} = (9 \times 10^9 \, \Omega \, \text{meter})^{-1}$. When the electrons in the system are degenerate, the electron mass in Eq. (47) is replaced by p_F/c, where p_F is the Fermi momentum of the degenerate electron system. The relaxation time is the most crucial parameter in this investigation, which must be related to the electron density, the average electron speed, the number and types of scatterers in the system, and the scattering cross sections of the electrons with different types of scatterers.

When the temperature is below the melting temperature T_m given by Eq. (41), matter in the density range of 10^2–10^{14} g/cm^3 is in a solid state possessing a crystalline structure. The constituent nuclei are organized into a lattice while the electrons are distributed more or less uniformly in the space between. In the neutron drip region, 10^{12}–10^{14} g/cm^3, neutrons are also outside the nuclei. The relaxation time of the electrons is determined by the frequency of scatterings with the lattice nuclei and with the other electrons and neutrons. When there are several scattering mechanisms present, the relaxation times due to different mechanisms are found by the inverse sum of their reciprocals:

$$\tau^{-1} = \tau_1^{-1} + \tau_2^{-1} + \cdots \qquad (48)$$

Upon electron scattering, the lattice vibrates and the vibration propagates collectively like sound waves through the lattice. In quantized form the sound waves behave like particles, called phonons. Electron scattering from the lattice is usually described as electron–phonon scattering. Phonons increase rapidly in number with temperature. We therefore expect the electric conductivity due to electron–phonon scattering to decrease with increasing temperature.

Electron–electron scatterings have minor effects on the electric conductivity, because in dense matter all the electrons are not bound to the nuclei. Electrons lose energy only if they are scattered from bound electrons. Elastic scattering of electrons does not alter the current being transported. Hence this form of scattering does not affect electric conductivity and may be ignored.

Let us imagine that the dense matter in consideration is formed from a dynamical process, as in the formation of a neutron star, in which case the material substance is adjusting to reach the proper density while it is being cooled. If the solidification rate is faster than the nuclear equilibrium rate, there will be large admixtures of other nuclei with the equilibrium nuclei. The nonequilibrium nuclei act as impurities in the system. Also, there may be defects in the system due to rapid rates of cooling and rotation. Electrons and phonons are scattered by these impurities and defects which also limit the transport of electric charge by electrons.

There are estimates of the electric conductivity in dense matter that consider all the above discussed features. Typical results for a wide range of densities at a temperature of 10^8 K are shown in Fig. 8. Electric conductivities due to different scattering mechanisms are shown separately. The total conductivity should be determined by the lowest lying portions of the curves, since the total conductivity, like the relaxation time, is found by taking the inverse sum of the reciprocals of the conductivities due to different scattering mechanisms.

Electron–phonon scattering should be the dominant mechanism. The conductivity due to electron–phonon scattering has a temperature dependence of approximately T^{-1}. Electron–impurities scattering depends on the impurity concentration and the square of the charge difference between the impurity charge and the equilibrium nuclide charge. Let us define a concentration factor

$$x_{\text{imp}} = \sum \left(\frac{n_i}{n_A}\right)(Z_i - Z)^2 \qquad (49)$$

where the summation is over all impurity species designated by subscript i; n_i and Z_i are respectively the number density and charge of each impurity species. The conductivity curve due to impurity scattering σ_{imp} is drawn in Fig. 8

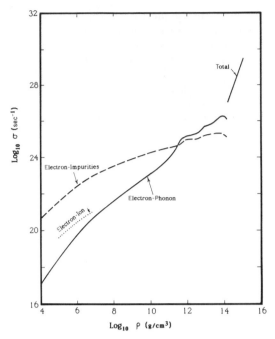

FIG. 8.   The electrical conductivity of dense matter at a temperature of $T = 10^8$ K. For densities below 2×10^{12} g/cm³, the solid line corresponds to the electrical conductivity limited by the mechanism of electron–phonon scattering while the system is in a solid state. Possible melting occurs in the low-density region at this temperature, and electron–phonon scattering is replaced by electron–ion scattering. The electrical conductivity due to electron–ion scattering is drawn as a dotted line. The electrical conductivity due to electron–impurities scattering is drawn as a dashed line, taking an arbitrary impurity concentration factor of $x_{imp} = 0.01$. For densities above 2×10^{12} g/cm³, the electrical conductivity is limited mainly by the mechanism of electron–proton scattering, and it is drawn as a solid line and labeled "total." [From Elliott, F., and Itoh, N. (1979). *Astrophys. J.* **206,** 218 (1976); **230,** 847 (1979). Reprinted with permission of *The Astrophysical Journal,* published by the University of Chicago Press; © 1976 and 1979 The American Astronomical Society.]

with an assumed $x_{imp} = 0.01$. It is independent of T.

The electron–phonon curve would not be correct at the low-density end where the melting temperature is below 10^8 K. Above melting temperature, the lattice structure would not be there, and electrons would not be impeded by scattering from phonons but instead from the nuclei. Electric conductivity due to electron–nuclei scattering is shown in the possible melting region.

In the neutron matter region, 10^{14}–10^{15} g/cm³, the system is composed basically of three types of particles, neutrons, protons, and electrons, of which both protons and electrons act as carriers for electrical conduction. The main difficulty in dealing with this situation is to take into full account the interactions among the particles. With the presence of a strong attractive interaction among the nucleons, it is quite likely that the protons will be paired to turn the system into a superconducting state when the temperature falls below the critical temperature, in which case the electric conductivity becomes infinite. The critical temperature T_c is estimated to be as high as 10^9–10^{10} K. Such a temperature, though it appears high, corresponds actually to a thermal energy $k_B T$ that is small compared with the Fermi energies of the particles present.

If the system is not in a superconducting state, or in other words it is in a normal state, its electrical conductivity is due mainly to electrons as carriers. The electron relaxation time is determined by scattering by protons and has been evaluated. The electrical conductivity of neutron matter in the normal state is plotted in Fig. 8 in the density range 10^{14}–10^{15} g/cm³.

B. THERMAL CONDUCTIVITY

The thermal conductivity of a substance is given by the ratio of the amount of heat transferred per unit area per unit time to the temperature gradient. Kinetic theory of dilute gas yields the following expression for thermal conductivity:

$$k = (1/3)nc_s\bar{v}^2\tau \qquad (50)$$

where n is the carrier number density, c_s its specific heat, $\bar{v}$ the average thermal velocity, and τ the relaxation time of the carriers. The thermal conductivity is expressed in units of erg/cm K sec. When there are several types of carriers participating in the transport of heat, the final thermal conductivity is the sum of individual conductivities due to different types of carriers. In the case of a solid, the important carriers are the electrons and phonons, but in general the phonon contribution to thermal conduction is negligible compared with the electron contribution.

In thermal conduction by electrons, no electric current is generated while energy is being transported. The type of electron–phonon scattering important for thermal conduction is different from that of electrical conduction. At each point in the system, the electron number density obeys the Fermi–Dirac distribution of Eq. (32), which assigns a higher probability of occupation of high-energy states when the temperature is high than when the temperature is low. When a thermal gradient exists in the system, neighboring points have different electron distributions. Electrons moving from a high-temperature point to a low-temperature point must lose some of their energy to satisfy the distribution requirement. If this can be accomplished over a short distance, the thermal resistivity of the substance is high, or its thermal conductivity is low. The most important mechanism of energy loss is through inelastic scatterings of electrons by phonons at small angles. Such scatterings constitute the major source of thermal resistivity. On the other hand, elastic scatterings of electrons do not lead to energy loss and aid in thermal conduction. The frequency of inelastic scattering to that of elastic scattering depends on the thermal distribution of phonons. The temperature dependence of thermal conductivity due to electron–phonon scatterings is therefore complicated. For matter density below 10^8 g/cm^3, it is relatively independent of temperature, but above that it shows a decrease with increasing temperature.

Although electron–electron scatterings do not contribute to the electric conductivity, they contribute to the thermal conductivity by redistributing the electron energies. The thermal conductivity due to impurity scattering k_{imp} is directly related to the electric conductivity due to impurity scattering σ_{imp} by the Wiedemann–Franz rule:

$$k_{imp} = \pi^2 T/3(k_B/e)^2 \sigma_{imp} \quad (51)$$

The thermal conductivities due to different mechanisms at 10^8 K are shown in Fig. 9. The thermal conductivity due to impurity scattering is drawn with $x_{imp} = 0.01$, and it has a temperature dependence linear in T. The thermal conductivity due to electron–electron scattering is expected to have a temperature dependence of T^{-1}. The thermal conductivity due to electron–ion scattering is shown for the region where 10^8 K is expected to be above the melting temperature. The total thermal conductivity of dense matter is determined by the mechanism that yields the lowest thermal conductivity at that density, since the total thermal conductivity, like the relaxation time, is found from the inverse sum of the reciprocals of the conductivities due to different mechanisms.

In the neutron matter region, 10^{14}–10^{15} g/cm^3,

all three types of particles, neutrons, protons, and electrons, contribute to the thermal conductivity

$$k = k_e + k_n + k_p$$

where the subscripts e, n, and p denote contributions to the thermal conductivity by electrons, neutrons, and protons, respectively. When the particles are in a normal state (i.e., not in a superfluid or superconducting state), the thermal conductivity is determined primarily by the highly mobile electrons, whose motion is impeded largely by scatterings with the protons and other electrons, and much less by scatterings with the neutrons. The neutron contribution to the thermal conductivity is substantial because of its high number density. Neutrons encounter neutron–proton scattering and neutron–neutron scattering in the process. The proton contribution to the thermal conductivity is small because the proton number density is low, but otherwise the protons contribute in a manner similar to the neutrons. The thermal conductivities due to these three components for a system in the normal state are shown in short dashed lines in Fig. 9. The total conductivity is drawn as a solid line there.

The system may also become superfluid when its temperature falls below the critical temperature of $T_c \approx 10^9$–10^{10} K. The critical temperature of the protons is in general different from that of the neutrons, and therefore it is possible that while one turns superfluid, the other remains normal. Also, when the temperature falls below the critical temperature of a certain type of particles, say the neutrons, there remains a normal component of neutrons in the system. This situation is usually described by a two-fluid model that consists of both the superfluid and normal fluid components. In general, scattering of particles off the superfluid component is negligible for transport purposes.

If only the protons turn superfluid (and superconducting) while the neutrons remain normal, the thermal conductivity due to the superfluid protons vanishes. The thermal conductivity found for the system in the normal state is basically unaltered, because the protons give a very small contribution to the thermal conductivity, as shown in Fig. 9.

If the neutrons turn superfluid while the protons remain normal, the superfluid component of the neutrons gives vanishing thermal conductivity, and the contribution by the normal component of the neutrons to the thermal conduc-

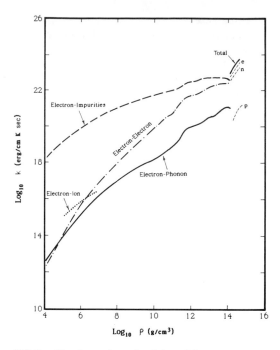

FIG. 9.   The thermal conductivity of dense matter at a temperature of $T = 10^8$ K. For densities below 2×10^{12} g/cm³, the solid line corresponds to the thermal conductivity limited by the mechanism of electron–phonon scattering. If the lattice is melted, the solid line should be replaced by the dotted line, which is the thermal conductivity due to electron–ion scattering. The thermal conductivity due to electron–electron scattering is given by dot-dashed lines, and that due to electron–impurities scattering is given by dashed lines, where the impurity concentration factor is assumed to be $x_{imp} = 0.01$. For densities above 2×10^{12} g/cm³, the thermal conductivity received contributions from the electrons, neutrons, and protons, which are drawn in thin dashed lines and marked by e, n, and p, respectively. The total thermal conductivity from these carriers is drawn in a solid line and labeled "total." [From Elliott, F., and Itoh, N. *Astrophys. J.* **206**, 218 (1976); **230**, 847 (1979). Reprinted with permission of *The Astrophysical Journal,* published by the University of Chicago Press; © 1976 and 1979 The American Astronomical Society.]

tivity diminishes rapidly with decreasing temperature below the critical temperature, because the number density of the normal neutrons decreases rapidly with decreasing temperature. The thermal conductivity in this case is therefore determined entirely by the electron contribution to the thermal conductivity, which is modified slightly from the normal matter case due to the absence of scattering by superfluid. The net result is that the thermal conductivity is only slightly reduced from the normal case

shown in Fig. 9. However, if both the neutrons and protons turn superfluid, the thermal conductivity is due entirely to electron–electron scattering, and the general result is indicated by the extension of the electron–electron curve for densities below 10^{14} g/cm³.

C. Shear Viscosity

When a velocity gradient exists in a fluid, a shearing stress is developed between two layers of fluid with differential velocities. The shear viscosity is given by the ratio of the shearing stress to the transverse velocity gradient. Elementary kinetic theory suggests that the shear viscosity of a dilute gas is given by

$$\eta = \frac{1}{3} nm\bar{v}^2\tau \tag{52}$$

where n is the molecular density, m the mass of each molecule, $\bar{v}$ the average thermal velocity of the molecules, and τ the relaxation time. Viscosity is expressed in units of g/cm sec, which is also called poise. In appearance it is similar to the expression for thermal conductivity with the exception that the specific heat per particle in the thermal conductivity is being replaced by the particle mass. Consequently, we may suspect that the electron component of the viscosity should behave similarly to the thermal conductivity. This, however, is not true due to the fact that the relaxation time involved relates to different aspects of the scattering mechanism. Also, there is an additional component to the viscosity. The solid lattice can make a great contribution to the total viscosity. Unfortunately, the determination of the lattice viscosity is very difficult, and no adequate work has been performed to determine its properties at the present time. The following discussion relates only to the electron viscosity.

The electron relaxation time is determined by electron scatterings by phonons, impurities, electrons, and nuclei. Shearing stress is developed when electrons belonging to fluid layers of different velocities are exchanged. Thus, viscosity is related to mass transfer or the transfer of electrons. This is similar to electric conduction where the transfer of electrons gives rise to charge transfer and is different from heat conduction which involves the adjustment of electron energy distributions. The evaluation of the relaxation times for the viscosity due to different scattering mechanisms is similar to that for the electrical conductivity.

The viscosity of dense matter at 10^8 K due to different scattering mechanisms is shown in the Fig. 10. The temperature dependence of the viscosity due to electron–phonon scattering is approximately T^{-1}, as in the case of the electrical conductivity. This is also true of the viscosity due to electron–impurities scattering, which is independent of temperature as in the case of the electrical conductivity. While electron–electron scatterings do not contribute to the electrical conductivity, they play a role in viscosity giving rise to a temperature dependence of T^{-2}.

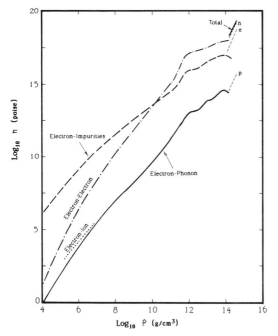

FIG. 10. The viscosity of dense matter at a temperature of $T = 10^8$ K. For densities below 2×10^{12} g/cm³, the viscosity due to electron–phonon scattering is drawn as a solid line. If the lattice is melted, the viscosity is due to electron–ion scattering, which is drawn as a dotted line. The viscosity due to electron–electron scattering is drawn as dot-dashed lines, and the viscosity due to electron–impurities scattering is drawn as dashed lines, where the impurity concentration factor is taken to be $x_{imp} = 0.01$. For densities above 2×10^{12} g/cm³, the viscosity receives contributions from neutrons, electrons, and protons, and they are shown as thin dashed lines and labeled n, e, and p, respectively. The total viscosity from these three components is drawn as a solid line and labeled "total." [From Elliott, F., and Itoh, N. *Astrophys. J.* **206**, 218 (1976); **230**, 847 (1979). Reprinted with permission of *The Astrophysical Journal*, published by the University of Chicago Press; © 1976 and 1979 The American Astronomical Society.]

In the neutron matter region, 10^{14}–10^{15} g/cm³, all three types of particles, neutrons, protons, and electrons, contribute to the viscosity. The contributions from neutrons, protons, and electrons to the viscosity are shown separately in Fig. 10 in this density range by dashed lines. The total viscosity is drawn as a solid line. They all have a temperature dependence of T^{-2}.

When the temperature drops below the critical temperature T_c, superfluid proton and/or neutron components appear. The behavior of the viscosity in the superfluid state is very similar to the thermal conductivity. If the protons turn superfluid, the viscosity is basically unaltered from the normal viscosity, because the proton contribution is small. If the neutrons turn superfluid, the superfluid component of the neutrons has vanishing viscosity. The viscosity is dominated by the electron contribution which is determined by the electron–electron scattering and electron–proton scattering mechanisms. When both protons and neutrons turn superfluid, then the viscosity is determined entirely by electron–electron scattering, and the general result is indicated by the extension of the electron–electron curve for densities below 10^{14} g/cm³.

VI. Neutrino Emissivity and Opacity

In a nonequilibrium situation where a temperature gradient exists in a substance, energy is transported not only by means of thermal conduction, as discussed in the last section, but also by radiation. Parameters of radiative transfer intrinsic to the matter system are its emissivity and opacity. There are two major forms of radiation. In one form, the radiative energy is transmitted by photons and in the other by neutrinos.

In a dense matter system whose constituent electrons are highly degenerate (i.e., when the electron Fermi energy is high compared with the thermal energy $k_B T$), the degenerate electrons cannot avail themselves as effective scatterers for the passage of energy carriers created by the thermal gradient; and the thermal conductivity is correspondingly high. Therefore, energy transport is far more effective through heat conduction than it would be for radiative transfer. For this reason, the problem of photon emissivity and opacity in dense matter receives very little attention. Most astrophysical studies of photon emissivity and opacity are performed for relatively low-density substances, such as those forming the interiors of luminous stars. Since our subject is dense matter, these parameters will not be discussed here. Interested readers are referred to astrophysical texts suggested in the bibliography. More relevant to dense matter physics is the topic of neutrino emissivity and opacity.

A. Neutrino Emissivity

The reaction described by Eq. (37) is a sample reaction in which a neutrino is produced. Indeed, neutrinos are produced throughout the second density domain whenever neutronization occurs, and in these processes neutrinos are produced even at zero temperature. At high temperature there are other reactions that are more effective in neutrino production. Reactions involving neutrinos belong to a class of interaction called the weak interaction. The coupling constant for the weak interaction is much smaller than that of the electromagnetic interaction. Neutrino interaction rates are in general slower than comparable photon interaction rates by a factor of at least 10^{20}. Neutrinos can pass through thick layers of substance and experience no interaction. For example, the neutrino mean free path through matter with density similar to our sun is about a billion solar radii. Hence, once they are produced in a star they are lost into space, and they serve as an efficient cooling mechanism for hot and dense stellar objects.

During a supernova process, the collapse of the stellar core raises the core temperature to as high as 10^{11} K. It quickly cools to a temperature of 10^9 K through neutrino emission as the core stabilizes into a neutron star. Neutrino emission dominates photon emission until the temperature drops to 10^8 K. It is estimated that neutrino cooling dominates for at least the first few thousand years of a neutron star after its formation, and therefore the neutrino cooling mechanisms deserve attention. We are primarily interested in those neutrino emission mechanisms that supersede photon emission under similar conditions. They determine therefore the cooling rate of neutron stars in the early period, and they also play a major role in the dynamics of a supernova process.

An important neutrino production process is the modified Urca reactions which involve neutrons, protons, and electrons:

$$n + n \rightarrow n + p + e + \bar{\nu} \qquad (53a)$$

$$n + p + e \rightarrow n + n + \nu \qquad (53b)$$

where ν and $\bar{\nu}$ denote neutrino and antineutrino, respectively. In the following we shall not distinguish neutrinos from antineutrinos and address them collectively as neutrinos. These two reactions are very nearly the inverse of each other, and when they are occurring at equal rates, the numbers of neutrons, protons, and electrons in the system is unaltered, while neutrinos are being produced continuously. Urca is the name of a casino in Rio de Janeiro. Early pioneers of neutrino physics saw a parallel between nature's way of extracting energy from the stellar systems and the casino's way of extracting money from its customers, so they named the reactions after the casino. Reactions (53a,b) are modifications of the original Urca reactions by adding an extra neutron to the reaction. This increases the energy range over which neutrinos may be produced and thus improves the production rate.

Our current understanding of the weak interaction theory is provided by the Weinberg–Salam–Glashow theory. Even though most of the neutrino reactions to be discussed in this section have never been verified under laboratory conditions, they are nevertheless believed to be correct, and quantitative estimates of their reaction rates are reliable. When reaction (53a) occurs, a transition is made from a state of two neutrons to a state of a neutron, proton, electron, and neutrino. The theory evaluates the transition probability from an initial state of two neutrons that occupy quantum states of definite momenta to a final state of four particles of definite momenta. Whenever a transition is made, a neutrino of a specific energy is produced. The total neutrino energy emitted from the system per unit time is found by summing the neutrino energies of all allowed transitions multiplied by their respective transition probabilities. The neutrino emissivity is the total neutrino energy emitted per unit time per unit volume of the substance.

Consider a neutron matter system in the density range of 10^{14}–10^{15} g/cm^3 that is composed mainly of neutrons with a small admixture (about 1%) of protons and electrons. Quantum states occupied by the particles are given by the Fermi–Dirac distribution f_j of Eq. (32). Note that the temperature dependence of the final result comes from the temperature factor in the distribution. Designating the initial-state neutrons by subscripts 1 and 2, the final-state neutron by 3, and the proton, electron, and neutrino by p, e, and ν, respectively, we find that the emissivity of the neutron matter system is given by the following equation:

$$E_{\text{Urca(a)}} = \int [d^3n_1 f_1][d^3n_2 f_2][d^3n_3(1 - f_3)]$$
$$\times [d^3n_p(1 - f_p)][d^3n_e(1 - f_e)]$$
$$\times [d^3n(1 - f_\nu)]W(i \rightarrow f)e_\nu \quad (54)$$

where the summations over quantum states are represented by integrations over $d^3n_j = V(d^3p_j/h^3)$, $e_\nu = c|\mathbf{p}_\nu|$ is the neutrino energy, and $W(i \rightarrow f)$ is the transition probability from initial state i to final state f per unit time and per unit volume of the system. It has the following structure:

$$W(i \rightarrow f) = (2\pi)^4 \, \delta(\mathbf{p}_i - \mathbf{p}_f) \, \delta(E_i - E_f)$$
$$\times \sum |H(i \rightarrow f)|^2 \quad (55)$$

where $\mathbf{p}_i$ and $\mathbf{p}_f$ denote the total momenta of the initial and final states, respectively, E_i and E_f the total energies of the initial and final states, respectively, and $H(i \rightarrow f)$ the matrix element of the Hamiltonian describing the interaction; the summation symbol $\sum$ indicates the summations over all spin orientations of the particles. The mathematical delta functions are here to insure that only energy–momentum-conserving initial and final states are included in this evaluation. In Eq. (54), the initial state particles are assigned distributions f_j while the final state particles are assigned distributions $(1 - f_j)$ because the particles produced in the reaction must be excluded from states that are already occupied, and therefore they must take up states that are not occupied, which are expressed by $(1 - f_j)$. The chemical potentials in the distributions f_j determine the particle numbers in the system. They are related by $\mu_1 = \mu_2 = \mu_3$ and $\mu_p = \mu_e$, while in most cases $\mu_\nu = 0$. The quantity μ_ν will be different from zero only in circumstance where the neutrino opacity is so high that the neutrinos are trapped momentarily and become partially degenerate.

If the neutrons and protons in the system are assumed to be noninteracting, the emissivity due to the reaction (53a) can be easily evaluated, and the result may be expressed conveniently as

$$E_{\text{Urca(a)}} = (6.1 \times 10^{19} \text{ erg/cm}^3 \text{ sec})(\rho/\rho_0)^{2/3}(T_9)^8$$
$$(56)$$

where $\rho_0 = 2.8 \times 10^{14}$ g/cm^3 is the density of nuclear matter, and the symbol T_9 stands for temperature in units of 10^9 K. Here the emissivity (in units of ergs per cubic centimeter per second) is evaluated for a dense system of neutron matter whose proton and electron contents are determined by the ground-state requirement un-

der reaction Eq. (37). It is expressed in this form because neutron matter exists only with densities comparable to nuclear matter density, and T_9 is a typical temperature scale for neutron star and supernova problems.

An interesting point to note is that this emissivity is given by eight powers of temperature. This comes about for the following reason. For the range of temperature considered, the thermal energy is still small compared with the degenerate or Fermi energy ε_F of the fermions in the system (except neutrinos), and most of the easily accessible quantum states are already occupied, leaving only a small fraction of quantum states in each species contributing to the reaction (of the order of $k_B T/\varepsilon_F$ per species). Since there are two fermion species in the initial state and three (not counting neutrinos) in the final state, a factor of T^5 is introduced. The allowed neutrino states are restricted only by energy conservation, and their number is given by the integration over $d^3n_\nu \, \delta(E_i - E_f)$, which is proportional to the square of the neutrino energy e_ν^2. This, together with the neutrino energy term in Eq. (54), gives e_ν^3. Since e_ν must be related to $k_B T$, which is the only energy variable in the problem, all together they give the T^8 dependence to the expression. The above deduction has general applicability, and if there were one less fermion in both the initial and final states, then the emissivity from such a process should be proportional to T^6.

The emissivity due to reaction (53b), $E_{\text{Urca(b)}}$, can be shown to be of comparable magnitude to that evaluated above, and the total emissivity due to the modified Urca process is simply twice that of Eq. (56).

$$E_{\text{Urca}} = (1.2 \times 10^{20} \text{ erg/cm}^3 \text{ sec})(\rho/\rho_0)^{2/3}(T_9)^8$$
$$(57)$$

At high density, muons would appear alongside electrons, and neutrino production reactions similar to those of Eq. (53a) and (53b) become operable, but with the electrons there replaced by muons. Similar results are obtained, but muons do not appear in dense matter until its density exceeds 8×10^{14} g/cm^3, and the emissivity due to the additional muon processes is only a minor correction.

A more significant consideration is the inclusion of nuclear interaction, which has been neglected in the above evaluation. Nuclear interaction appears in the evaluation of the matrix element $H(i \rightarrow f)$. When nuclear interaction is included, the total emissivity due to modified

Urca process is changed to

$$E_{\text{Urca}} = (7.4 \times 10^{20} \text{ erg/cm}^3 \text{ sec})(\rho/\rho_0)^{2/3}(T_9)^8$$

(58)

which is a factor of six higher than that evaluated without the inclusion of nuclear interaction.

Other important neutrino production mechanisms in neutron matter are

$$n + n \rightarrow n + n + \nu + \bar{\nu}$$ (59a)

$$n + p \rightarrow n + p + \nu + \bar{\nu}$$ (59b)

in each of which a pair of neutrinos is produced as the nucleons scatter from each other. Neutrino emissivities for these processes are evaluated to be

$$E_{\text{nn}} = (1.8 \times 10^{19} \text{ erg/cm}^3 \text{ sec})(\rho/\rho_0)^{1/3}(T_9)^8$$

(60a)

$$E_{\text{np}} = (2.0 \times 10^{19} \text{ erg/cm}^3 \text{ sec})(\rho/\rho_0)^{2/3}(T_9)^8$$

(60b)

where the subscript nn denotes reaction (59a) and np denotes (59b). These emissivities are evaluated with nuclear interaction taken into consideration. The processes are, however, not as effective as the modified Urca processes in neutrino production.

If neutron matter turns superfluid after its temperature falls below the critical temperature T_c, then the neutrino production rates evaluated above must be reduced; T_c should be in the range 10^9–10^{10} K. Superfluidity is explained by the fact that an energy gap appears in the energy spectrum of the particle, just above its Fermi energy. What that means is that a group of quantum states whose energies fall in the energy gap are excluded from the system. Normally, in an inelastic scattering process, the neutrons or protons are scattered into these states; but since the states are absent, they must be excited into much higher energy states, and thus scattering becomes difficult and less likely to occur. The consequence is that neutrons and protons may move about relatively freely without being impeded by scatterings. This is the explanation of superfluidity. By the same token, the above neutrino production mechanisms that depend on the scattering of neutrons and protons are similarly reduced. Qualitatively, if superfluidity occurs in both the neutron and proton components, E_{Urca} and E_{np} are reduced by a factor of $\exp[-(\Delta_n + \Delta_p)/k_B T]$, where Δ_n and Δ_p are the widths of the energy gaps for neutron and proton superfluidity, respectively. For E_{nn}, the reduction is $\exp[-2\Delta_n/k_B T]$. If superfluidity occurs in just

one component, the reductions are obtained by setting the energy gap of the normal component in the above expressions to zero.

In the case of pion condensation in neutron matter, interactions involving pions and nucleons also modify the neutrino production rate. Some estimates have shown that the neutrino production rates thus modified can be significant. However, due to the uncertainty in our present understanding of the pion condensation problem, no emissivity for this situation will be quoted here.

There are also neutrino production mechanisms not involving the direct interaction of two nucleons, such as the following:

1. Pair annihilation

$$e + e^+ \rightarrow \nu + \bar{\nu}$$ (61a)

2. Plasmon decay

$$\text{plasmon} \rightarrow \nu + \bar{\nu}$$ (61b)

3. Photoannihilation

$$e + \gamma \rightarrow e + \nu + \bar{\nu}$$ (61c)

4. Bremsstrahlung

$$e + (Z, A) \rightarrow (Z, A) + e + \nu + \bar{\nu}$$ (61d)

5. Neutronization

$$e + (Z, A) \rightarrow (Z - 1, A) + \nu$$ (61e)

where e^+ denotes a positron, plasmon a photon propagating inside a plasma, and (Z, A) a nucleus of atomic number Z and mass number A. A photon in free space cannot decay into a neutrino pair, because energy and momentum cannot be conserved simultaneously in the process. However, when a photon propagates inside a plasma, its relation between energy and momentum is changed in such a way that the decay becomes possible. Quanta of the electromagnetic wave in a plasma are called plasmons. The neutrino production rates for these processes have been found to be relatively insignificant at typical neutron star densities and temperatures and will not be listed here.

B. NEUTRINO OPACITY

When a radiation beam of intensity I (erg/cm^2 sec) is incident on a substance of density ρ (g/cm^3), the amount of energy absorbed from the beam per unit volume per unit time E (erg/cm^3 sec) is proportional to the opacity $\kappa = E/\rho I$. The opacity is expressed in units of cm^2/g. Each neutrino emission process has an inverse process corresponding to absorption. In addition to absorption, scattering can also impede the passage of neutrinos through the medium. Both absorption and scattering contribute to the opacity. Some of the more important processes are listed below.

1. Scattering by neutrons

$$\nu + n \rightarrow \nu + n \qquad (62a)$$

2. Scattering by protons

$$\nu + p \rightarrow \nu + p \qquad (62b)$$

3. Scattering by electrons

$$\nu + e \rightarrow \nu + e \qquad (62c)$$

4. Scattering by nuclei

$$\nu + (Z, A) \rightarrow \nu + (Z, A) \qquad (62d)$$

5. Absorption by nucleons,

$$\nu + n \rightarrow p + e \qquad (62e)$$

6. Absorption by nuclei

$$\nu + (Z, A) \rightarrow (Z + 1, A) + e \qquad (62f)$$

Similar processes occur for antineutrinos, which shall not be displayed here.

For each of these reactions, a reaction cross section is evaluated from the Weinberg–Salam–Glashow theory of weak interactions. The cross section represents an area effective in obstructing the incident beam of neutrinos. The opacity may be expressed in terms of the reaction cross sections as follows:

$$\kappa = \rho^{-1} \sum n_j \sigma_j \qquad (63)$$

where n_j denotes the number density of the particles that react with the neutrino and σ_j their reaction cross sections, and the summation Σ is over all the reactions listed above. Often, neutrino opacity is expressed by a neutrino mean free path, which is defined as

$$\lambda = (\rho \kappa)^{-1} \qquad (64)$$

Among these reactions, the contribution of reaction (62f) to the opacity of dense matter is quite negligible because it entails the production of electrons; and since electrons in the system are already highly degenerate, it is difficult to accommodate the newly produced electrons.

The most important reaction in this regard is reaction (62d) where the neutrino is scattered by the nuclei. This is the result of coherent scattering, in which all nucleons in a nucleus partici-

pate as a single entity in the process. The cross section of a coherent process involving A nucleons is proportional to A^2 times the cross section of scattering from a single nucleon. This reaction therefore dominates all others when the matter system is composed of giant nuclei, which is the case when matter density is below nuclear density, that is, before nuclei dissolve into neutron matter.

For neutron matter the important reactions are scatterings by neutrons, protons, and electrons as indicated by reactions (62a), (62b), and (62c). Neutrinos are scattered elastically by the nucleons and nuclei since the scatterers are massive. The neutrinos may change directions after scattering but do not lose their energies to the scatterers. They lose energy only if they are scattered by electrons. Electron scattering is therefore an important process in lowering the energies of the high-energy neutrinos, bringing them into thermal equilibrium with all neutrinos should the neutrinos be trapped in the system for a duration long enough for this to happen. Even though neutrinos interact very weakly and are therefore very difficult to confine, neutrino trappings is in fact believed to occur at the moment when the collapsing stellar core reaches the point of rebound initiating the explosive supernova process. Therefore a great deal of attention has been given to the problem of neutrino opacity and the issue of neutrino trapping.

The cross sections for these processes in the reference frame of the matter system are evaluated to be as follows:

1. Neutrino–electron scattering,

$$\sigma_e \approx (1/4)\sigma_0(e_\nu/m_e c^2)(\varepsilon_F/mc),$$

$$e_\nu \ll \varepsilon_F \qquad (65)$$

2. Neutrino–nucleon scattering,

$$\sigma_N \approx (1/4)\sigma_0(e_\nu/m_e c^2)^2,$$

$$e_\nu \ll m_n c^2 \qquad (66)$$

3. Neutrino–nucleus scattering,

$$\sigma_A \approx (1/16)\sigma_0(e_\nu/m_e c^2)^2 A^2[1 - (Z/A)],$$

$$e_\nu \ll 300 A^{-1/3} \text{ MeV} \qquad (67)$$

where $\sigma_0 = 1.76 \times 10^{-44}$ cm^2 is a typical weak interaction cross section, e_ν the neutrino energy, and ε_F the Fermi energy of the electrons. The total neutron opacity of the substance is given by

$$\kappa = \rho^{-1}[n_e \sigma_e + n_N \sigma_N + n_A \sigma_A] \qquad (68)$$

According to the Weinberg–Salam–Glashow theory, neutrinos also interact with quarks, and therefore dense quark matter also emits and absorbs neutrinos. However, since our understanding of quark matter is still far from complete, no results related to neutrino emissivity and opacity in quark matter will be quoted at this time.

BIBLIOGRAPHY

Chandrasekhar, S. (1939). "An Introduction to the Study of Stellar Structure." Dover, New York.

Clayton, D. D. (1968). "Principles of Stellar Evolution and Nucleosynthesis." McGraw-Hill, New York.

Gasiorowicz, S. (1979). "The Structure of Matter: A Survey of Modern Physics." Addison-Wesley, Reading, Massachusetts.

Leung, Y. C. (1984). "Physics of Dense Matter." Science Press, Beijing, China.

Schwarzschild, M. (1958). "Structure and Evolution of the Stars." Dover, New York.

Shapiro, S. L., and Teukolsky, S. A. "Black Holes, White Dwarfs, and Neutron Stars, The Physics of Compact Objects." Wiley, New York.

Zeldovich, Ya. B., and Novikov, I. D. (1971). "Relativistic Astrophysics." Chicago University Press, Chicago.

ELECTRODYNAMICS, QUANTUM

W. P. Healy   *University College London and Royal Melbourne Institute of Technology*

I. Introduction
II. Nonrelativistic Quantum Electrodynamics
III. Relativistic Quantum Electrodynamics

GLOSSARY

Anomalous magnetic moment: Difference between the intrinsic magnetic moment of a charged spin-$\frac{1}{2}$ particle and that predicted by the single-particle Dirac theory.

Coherent state: State of the quantized radiation field in which the average electric and magnetic fields and the average energy are the same as the corresponding quantities for a state of the classical electromagnetic field.

C, P, and T symmetries: Invariance of quantum electrodynamics under the operations of charge conjugation (or matter–antimatter interchange), parity (or left–right interchange), and time reversal, respectively.

Dirac equation: Four-component relativistic quantum mechanical wave equation for a spin-$\frac{1}{2}$ particle.

Einstein's A and B coefficients: Factors determining the rates of spontaneous emission, induced emission, and absorption of radiation by atoms.

Feynman diagram: Pictorial representation of a process in quantum electrodynamics in which states of particles or atoms are depicted as lines and their interactions as vertices where two or more lines meet.

Gauge invariance: Independence of a quantity of the choice of potentials used to represent the electromagnetic field.

Lamb shift: Change in atomic energy levels (from the values predicted by the single-particle Dirac theory) caused by electromagnetic interactions, or the splitting of spectral lines due to this change.

Leptons: Spin-$\frac{1}{2}$ particles subject to the weak and, if charged, the electromagnetic force, but not subject to the strong nuclear force, and including electrons, muons, tauons, neutrinos and their antiparticles.

Maxwell's equations: General fundamental equations for the electromagnetic field, summarizing the basic laws of electromagnetism.

Occupation-number state: State of a quantized field (such as the Maxwell or Dirac field) that has a definite number of particles or quanta in each field mode.

Photon: Particle or quantum of the electromagnetic field that travels at the speed of light, has no charge or rest mass, and has intrinsic spin 1.

Renormalization: Elimination of unobservable mass and charge of bare particles in favor of observed mass and charge of physical particles.

S-matrix element: Probability amplitude for a scattering process in which the incoming and outgoing particles are specified by their momenta and polarization or spin states.

Quantum electrodynamics is the fundamental theory of electromagnetic radiation and its interaction with microscopic charged particles, particularly electrons and positrons. In its most accurate form, the theory combines the methods of quantum mechanics with the principles of special relativity; often, however, it is sufficient to treat the charged particles in nonrelativistic approximation. Each part of the complete dynamical system of radiation and charges displays a characteristic wave–particle duality. Thus, electrons behave in many circumstances as particles, but they can also exhibit wave properties such as interference and diffraction. Similarly electromagnetic radiation, which was considered classically as a wave field, may have particle properties ascribed to it under suitable conditions, e.g., in scattering experiments. The particles or quanta associated with the electromagnetic field are called photons. Quantum electrodynamics is a highly successful theory, despite certain mathematical and interpretational difficulties inherent in its formulation. Its success is due in part to the weakness of the coupling between the radiation and the charges, which makes possible a perturbative treatment of the interaction of the two parts of the system. The theory accounts for many phenomena, including the emission or absorption of radiation by atoms or molecules, the scattering of photons or electrons, and the creation or annihilation of electron–positron pairs. Its most famous predictions concern the electromagnetic shift of energy levels observed in atomic spectra and the anomalous magnetic moment of the electron; both of these predictions are in good agreement with experimental results. Quantum electrodynamics also includes the interaction of photons with muons and tauons (which differ from electrons only in mass) and their antiparticles. The validity of the theory has been tested in high-energy collision experiments involving these particles down to distances less than 10^{-15} cm.

I. Introduction

A. Early Theories of Light

Since quantum electrodynamics is the modern theory of electromagnetic radiation, including visible light, it is instructive to begin with a brief historical review of previous theories. The nature of light has long been a subject of interest to philosophers and scientists. In the fifth century B.C., Empedocles of Acragas held that light takes time to travel from one place to another but that we cannot perceive its motion. He knew that the moon shines by light reflected from the sun and was also aware of the cause of solar eclipses. Heron of Alexandria, who is thought to have lived in the first or second century A.D., discussed the rectilinear propagation properties of light. In his book *Catoptrica* he derived the law of reflection using a principle of minimal distance. The law of refraction was not formulated until 1621, when it was discovered experimentally by Snell. Snell's law was later derived theoretically from Fermat's celebrated principle of least time.

From about the middle of the seventeenth century to the end of the nineteenth century there were two competing, and mutually contra-

dictory, theories of light. The wave theory was initiated by Hooke and Huygens following the first observations of interference and diffraction. Huygens enunciated a principle, based on the wave theory, from which he derived the laws of reflection and refraction. He also discovered the polarization properties of light. These properties, as well as the law of rectilinear propagation, were difficult to explain by the wave theory, which at that time dealt only with longitudinal waves in a hypothetical "aether," analogous to sound waves in air. These difficulties led Newton to propose a corpuscular theory, according to which light is emitted from luminous bodies in a stream of small particles or corpuscles. Newton's views inhibited any further advances in the wave theory until about the beginning of the nineteenth century. In the meantime, the fact that light has a finite speed was confirmed by Römer through observations of eclipses of the moons of Jupiter. This occurred in 1675, more than two millenia after the time of Empedocles. (The speed of light in empty space is denoted by c and is approximately 2.998×10^{10} cm/sec in cgs units.)

B. Classical Electrodynamics

The revival of the wave theory began with Young's interpretation of interference experiments. In particular, the destructive interference of two light beams at certain points in space seemed totally inexplicable on the corpuscular hypothesis but was readily accounted for by the wave theory. Young also suggested that light waves execute transverse rather than longitudinal vibrations, as this could then explain the observed polarization properties. The wave theory was further developed by Fresnel, who applied it to phenomena involving diffraction, interference of polarized light, and crystal optics. An important test of the theory was provided by the comparison of the speeds of light in media with different refractive indexes. According to the wave theory light travels slower in an optically denser medium, but according to the corpuscular theory it travels faster. The results of experiments carried out in 1850 agreed with the predictions of the wave theory.

The wave theory was in a certain sense completed when Maxwell established his equations for the electromagnetic field and showed that they have solutions corresponding to transverse electromagnetic waves in which both the electric and magnetic induction field vectors oscillate perpendicularly to the direction of propagation. The speed of these waves in empty space could be calculated from constants (the permittivity and permeability of the vacuum) obtained by purely electric and magnetic measurements and was found to be the speed of light. This conclusion became the basis of the electromagnetic theory of light. It was subsequently found that the frequencies of visible light form only a small part of the complete spectrum of electromagnetic radiation, which also includes radio waves, microwaves and infrared radiation on the low-frequency side, and ultraviolet radiation, X-rays, and gamma rays on the high-frequency side.

C. Photons

Despite the success of classical electromagnetic theory in dealing with the propagation, interference, and scattering of light, experiments carried out about the end of the nineteenth century and the beginning of the twentieth century led to the reintroduction of the corpuscular theory, though in a form different to that proposed by Newton. The departure from classical concepts began in 1900 when Planck published his law of black-body radiation. In this law the quantum of action h (approximately 6.626×10^{-27} erg sec), now known as Planck's constant, made its first appearance in physics. Planck's law for the variation with frequency of the energy in black-body radiation at a given temperature is closely related to the existence of discrete energy levels for the electromagnetic field, even though Planck, in his original derivation of the law, did not consider the field itself to be quantized. A black body is one that absorbs all the electromagnetic energy incident on it. It was shown by Kirchoff in 1860 that when such a body is heated, the emitted radiation does not depend on the detailed composition of the body but only on its absolute temperature. Radiation confined in a state of thermal equilibrium in a cavity with perfectly reflecting walls behaves as black-body radiation. According to classical electromagnetic theory, the cavity radiation can undergo simple harmonic motion at a number of certain allowed or characteristic frequencies ν, the values of which depend on the shape and size of the enclosure. These so-called radiation oscillators may be quantized, as in ordinary quantum mechanics. Then for each nonnegative integer n, an oscillator with frequency ν has a nondegenerate stationary state with energy $nh\nu$ above the ground-state energy (see Fig. 1). The possible values of the energy at this frequency thus form a discrete set $0, h\nu, 2h\nu, 3h\nu, \ldots$ instead of a continuum. It can be shown that quantization of the oscillators in this way for all the allowed frequencies leads directly to Planck's law.

In 1905, Einstein made use of the idea of light quanta in order to explain the photoelectric effect and later applied it to the emission as well as the absorption of radiation by atoms. The light quantum hypothesis states not only that the energy of monochromatic radiation of frequency ν is made up of integral multiples of the quantum $h\nu$, but also that the momentum is made up of integral multiples of the quantum h/λ, where λ is the wavelength of the radiation (ν and λ are related by the equation $\nu\lambda = c$). This contrasts sharply with the classical picture in which the energy and momentum are regarded as continuously variable. The existence of discrete light quanta, or photons, is not immediately evident on a macroscopic scale, however. Due to the smallness of Planck's constant, even in a weak electromagnetic field there is an enormous number of photons, provided the frequency is not too high. For example, black-body radiation at a

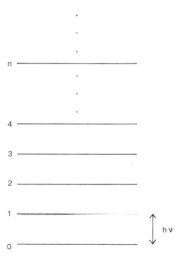

FIG. 1.   The horizontal lines represent the discrete energy levels of a quantized harmonic oscillator of frequency ν. The energy of the ground state is taken to be 0 and the equally spaced levels $0, h\nu, 2h\nu, \ldots, nh\nu, \ldots$ are labelled by the quantum numbers $0, 1, 2, \ldots, n, \ldots$, respectively. Excitation of the radiation field at frequency ν to level $nh\nu$ corresponds to the addition of n photons, each with energy $h\nu$, to the field.

temperature of 300 K (room temperature) contains about 5.5×10^8 photons/cm³, most of which correspond to frequencies in the infrared part of the electromagnetic spectrum. At a temperature of 6000 K (roughly that at the surface of the sun), the bulk of the radiation has frequencies in the visible spectrum, and there are about 4.4×10^{12} photons/cm³. (The total number of photons in black-body radiation is proportional to the cube of the absolute temperature.)

Individual photons manifest themselves only through their interaction with atomic systems. According to Einstein's treatment of the absorption and emission of radiation, for example, an atom in a stationary state can make a transition to a lower or a higher energy level accompanied by the creation or annihilation, respectively, of a photon. If the atomic energies are E_r and E_s, where $E_r > E_s$, then the energy $h\nu$ of the photon must equal the difference $E_r - E_s$ (see Fig. 2). This is called Bohr's frequency condition and is equivalent to the law of conservation of energy applied to the complete system of atom and radiation; any energy lost or gained by the atom is given up to or abstracted from the radiation field in the form of photons. It should be noted that the number of photons in the radiation field need not be constant—photons can be created or annihilated through the interaction of the field with atoms.

The scattering of X rays by free electrons also furnishes direct evidence for the corpuscular properties of radiation. In 1922, Compton discovered that when X rays of wavelength λ are incident on a graphite target, the scattered X rays have intensity peaks at two wavelengths, λ and λ', where $\lambda' > \lambda$. The shift in wavelength given by $\Delta\lambda = \lambda' - \lambda$ is a function of the angle of scattering (i.e., the angle between the direction of the incident and scattered X rays) but is independent of wavelength and the target material. The X rays with unchanged wavelength λ were understood to have been elastically scattered by atoms, which suffer no appreciable recoil, and they could readily be accounted for on the basis of classical electrodynamics. The scattered X rays with shifted wavelength λ', however, required a new interpretation. If it is assumed that the incident X rays consist of photons, then these may collide with essentially free electrons in the target. In this case a photon gives up some of its energy $h\nu$ to an electron and is scattered with a lower frequency ν' and a longer wavelength λ', where $\nu'\lambda' = c$.

The wavelength shift $\Delta\lambda$ can be calculated as a function of scattering angle by using the laws of conservation of energy and momentum. By treating the problem relativistically and taking the electron to be at rest initially (see Fig. 3), it is not difficult to show that $\Delta\lambda$ depends on the scattering angle θ alone through the formula

$$\Delta\lambda = \lambda_c(1 - \cos\theta)$$

where the constant λ_c is the Compton wavelength given by

$$\lambda_c = h/mc \simeq 2.43 \times 10^{-10} \text{ cm}$$

Here m is the rest mass of the electron (approximately 9.11×10^{-28} g). This formula was verified experimentally. The energy and distribution of the recoil electrons and scattered X rays were also in accord with the predictions of the photon theory. [*See* OPTICAL DIFFRACTION; RELATIVITY, SPECIAL.]

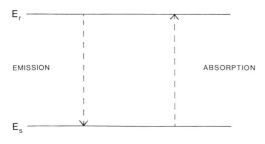

FIG. 2. An atom can make a transition from a higher energy level E_r to a lower level E_s while emitting a photon of frequency ν, where $h\nu = E_r - E_s$. The emission may be spontaneous or induced by radiation. The atom can make the upward transition from level E_s to level E_r by absorbing a photon of frequency ν.

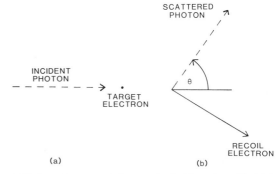

FIG. 3. Photon and electron in the Compton effect (a) before collision and (b) after collision. The scattering angle θ is the angle between the initial and final directions of the photon.

D. Quantum Electrodynamics

The use of the photon concept to explain certain phenomena does not imply a return to a naive classical particle view of light and other forms of electromagnetic radiation. A proper account must also be given of the wave properties of radiation, such as interference and diffraction. Indeed, the formulas for the energy and momentum of the photons—$h\nu$ and h/λ—are based on the assumption that the photons are associated with waves of definite frequency and wavelength. The nature of electromagnetic radiation is such that it appears, *under different experimental conditions*, sometimes to have particle properties and sometimes to have wave properties—these two aspects are said to be complementary. A single coherent theory that encompassed the dual nature of radiation, and with it settled the age-old controversy between the wave theory and the corpuscular theory, was made possible only by the development of quantum mechanics in the mid-1920s. In 1927, Dirac used the new methods of quantization, which had been successfully applied to atomic systems, to quantize the radiation field enclosed in a cavity, and was thus able to give a fully dynamical treatment of the emission and absorption of light by atoms. The beginning of quantum electrodynamics may be taken to date from this time.

The wave properties of radiation can be adequately described by using Maxwell's equations for the electromagnetic field, and these are retained as operator equations in the quantum theory of radiation. Suppose, for example, that the electromagnetic field has a node (i.e., a point where the field amplitudes always vanish) due to interference at P. Then an atom placed at P has, in so far as it can be regarded as a geometrical point, zero probability of absorbing a photon from the field. The field amplitudes are, however, subject to uncertainty relations, involving Planck's constant h, which are analogous to the Heisenberg uncertainty relations for the position and momentum of a particle in ordinary quantum mechanics. The origin of the uncertainty relations for the fields may be understood by considering a simple example.

Let $\bar{\mathscr{E}}$ denote the average value of a component of the electric field over a volume V and a time interval T. (Since a field component at a definite point in space and a definite instant of time appears an abstraction from physical reality, only such average values need be consid-

ered.) Now $\bar{\mathscr{E}}$ may be found by measuring the change produced by the field in the momentum of a charged test body occupying the volume during this time. Although the position and momentum of the test body are uncertain by amounts Δq and Δp that satisfy the Heisenberg uncertainty relation $\Delta q\,\Delta p \sim h$, it can be shown that this does not impair the accuracy of the field measurement, provided a sufficiently massive and highly charged body (which is therefore part of a macroscopic measuring instrument) is used. The charge Q on the body must be such that the product $Q\Delta q$ is large; $\bar{\mathscr{E}}$ can then be measured to any desired accuracy $\Delta\bar{\mathscr{E}}$. However, in the measurement of two average field strengths $\bar{\mathscr{E}}$ and $\bar{\mathscr{E}}'$ (taken over two space regions V and V' during two times intervals T and T', respectively), it may not be possible to make both $\Delta\bar{\mathscr{E}}$ and $\Delta\bar{\mathscr{E}}'$ as small as is desired. If the separation distance L between V and V' (see Fig. 4) is such that most light signals emitted from V during the time interval T will reach V' during the time interval T', then the measurement of $\bar{\mathscr{E}}$ will influence that of $\bar{\mathscr{E}}'$ in a way that is to some extent unknown. For the field produced by the test body used to measure $\bar{\mathscr{E}}$ is superimposed on $\bar{\mathscr{E}}'$ and cannot be fully subtracted out, as its value is somewhat uncertain (due to the uncertainty Δq in the position of the test body). This field, and hence $\Delta\bar{\mathscr{E}}'$, can indeed be made as small as is desired by making the product $Q\Delta q$ sufficiently small, but then $\Delta\bar{\mathscr{E}}$ becomes relatively large. The experimental conditions for measurements of $\bar{\mathscr{E}}$ and $\bar{\mathscr{E}}'$ are complementary—those that serve to measure $\bar{\mathscr{E}}$ more precisely will measure $\bar{\mathscr{E}}'$ less precisely, and vice versa. The order of magnitude of the uncertainty product is given by

$$\Delta\bar{\mathscr{E}}\,\Delta\bar{\mathscr{E}}' \sim h/L^3T$$

and is independent of both Q and Δq. Thus, only for well-separated regions or over long intervals of time can both averages be measured with unlimited accuracy.

FIG. 4. Regions V and V' in which the average electric fields $\bar{\mathscr{E}}$ and $\bar{\mathscr{E}}'$ are measured during time intervals T and T', respectively. The separation distance L is such that most light signals emitted from V during the interval T will reach V' during the interval T'.

E. Electrons and Positrons

Dirac's original radiation theory had to be modified to bring it into line with the special theory of relativity. This was true particularly of the treatment of the charged particles with which the electromagnetic field interacts. In 1928, Dirac had developed a one-particle relativistic wave equation for the electron that automatically accounted for the observed electron spin and predicted values for the fine structure of the energy levels of the hydrogen atom and of hydrogen-like ions that were in agreement with the experimental data of that time. The Dirac equation, however, also has extraneous solutions corresponding to negative-energy states. To eliminate these, Dirac introduced in 1930 the so-called hole theory, according to which most of the negative-energy states are occupied, each having one electron. Any unoccupied states, or holes, may be interpreted as particles with positive energy and positive charge. These particles were at first thought by Dirac to be protons, but were later identified with positrons, or antiparticles of electrons.

The experimental discovery of the positron by Anderson in 1932 lent support to Dirac's hole theory. Nevertheless, difficulties remained, such as the infinite (but unobservable!) charge density associated with the "sea" of negative-energy electrons. These difficulties can be removed, however, by treating Dirac's one-particle wave equation for the electron as a field equation and subjecting it to a process of quantization, similar in some respects to the quantization of the classical electromagnetic field. This method, which is often referred to as second quantization, was applied to the Dirac equation by Heisenberg and others and resulted in the appearance of electrons and positrons, on an equal footing, as quanta of the Dirac field, just as photons appear as quanta of the Maxwell field. There are, however, some differences between the methods of second quantization used for the Dirac and Maxwell fields, which stem from the different characteristics of the associated particles or quanta. Photons have zero rest mass (but have nonzero momentum because they travel at the speed of light), are electrically neutral, have spin 1 (in units of $\hbar$, which is Planck's constant divided by 2π), and are bosons (i.e., any number of photons can occupy a given state). Electrons and positrons have the same nonzero rest mass, carry equal but oppositely signed charges (by convention, this is negative for the electron and positive for the positron), have spin $\frac{1}{2}$, and are fermions (i.e., not more than one electron or positron can occupy a given state). It was shown by Pauli that there is a connection between the spin of a particle and its so-called statistics— particles with integer spin are bosons and are not subject to the exclusion principle, whereas particles with half odd-integer spin are fermions and are subject to the exclusion principle. It is necessary for photons to be bosons in order that the quantized electromagnetic field may have a classical counterpart, which is realized in the limit of large photon occupation numbers. The quantized Dirac field, on the other hand, does not have a physically realizable classical limit.

F. Divergences and Renormalization

In quantum electrodynamics, as in classical electrodynamics, there are no known exact solutions to the equations for the complete dynamical system of radiation and charges. Indeed, from a purely mathematical viewpoint, the question of even the existence of such solutions is still an open one. Approximate solutions may be found by assuming that the coupling between the two parts of the system is weak and using perturbation theory. This is justified by the smallness of the fine-structure constant α, which gives a measure of the strength of the coupling:

$$\alpha = e^2/4\pi\hbar c \approx \tfrac{1}{137}$$

where e is the magnitude of the charge on the electron and rationalized cgs units are being used ($e \approx 1.355 \times 10^{-10}$ g$^{1/2}$ cm$^{3/2}$/sec).

It was found in the 1930s and 1940s that the calculations for many processes, when taken beyond the first approximation, gave divergent results. Some divergences (the so-called infrared divergences) were due to deficiencies in the approximation method itself. Others (ultraviolet divergences) were associated with the problem of the structure and self-energy of the electron and other elementary particles. This problem had also arisen in classical electrodynamics, where the electron was assumed to have a structure-dependent electromagnetic contribution included in its inertial mass. In quantum electrodynamics, however, there occurred additional divergences of a radically different nature, due to effects that have no classical analogues. For example, the possibility of electron–positron pair creation gave rise to an infinite vacuum polarization in an external field and also implied an infinite self-energy for the photon.

The need to extract finite results from the formalism became acute when refinements in experimental technique revealed small discrepancies between the observed fine structure of the energy levels of atomic hydrogen and that given by Dirac's one-particle relativistic wave equation. These differences, whose existence had been suspected for some time, were measured accurately by Lamb and Retherford in 1947. In the same year, Kusch and Foley found that the value of the intrinsic magnetic moment of an electron in an atom also differs slightly from that predicted by the Dirac theory. To show that these discrepancies could be explained as radiative effects in quantum electrodynamics, it was necessary first to recognize that the mass and charge of the bare electrons and positrons that appear in the formalism cannot have their experimentally measured values. Since the electromagnetic field that accompanies an electron, for example, can never be "switched off," the inertia associated with this field contributes to the observed mass of the electron; the bare mechanical mass itself is unobservable. Similarly, an electromagnetic field is always accompanied by a current of electrons and positrons whose influence on the field contributes to the measured values of charges. The parameters of mass and charge, therefore, had to be renormalized to express the theory in terms of observable quantities. The results for the shift of energy levels (now known as the Lamb shift) and the anomalous magnetic moment of the electron then turned out to be finite and were, moreover, in good agreement with experiment. The use of explicitly relativistic methods of calculation, developed by Tomonaga and Schwinger, was essential in avoiding possible ambiguities in this procedure. Further important contributions were made by Dyson, who showed that the renormalized theory gave finite results for interaction processes of arbitrary order, corresponding to arbitrary powers of the coupling constant e, and by Feynman, who introduced a diagrammatic representation of the mathematical expressions for these processes, which are often of considerable complexity.

The Feynman-diagram technique and Dyson's perturbation theory are now part of the standard formulation of quantum electrodynamics. This formulation and some of its applications will be outlined in Sections II and III. In this article only the electromagnetic interactions of electrons and positrons (or, more generally, of charged leptons, which include muons and tauons and their antiparticles) are considered. These particles also participate in the so-called weak interaction (and, of course, in the much weaker gravitational interaction). A unified theory of the electromagnetic and weak interactions has been developed in recent years. Many elementary processes, however, are dominated by electromagnetic effects, and these alone form the subject matter of quantum electrodynamics. [*See* UNIFIED FIELD THEORIES.]

II. Nonrelativistic Quantum Electrodynamics

A. APPROXIMATIONS

Any treatment of the pure radiation field based on Maxwell's equations in empty space must satisfy the principles of the special theory of relativity, even though it might not be expressed in a form that makes this evident. Quantum electrodynamics has, however, a well-defined nonrelativistic limit in so far as the motion of the charged particles with which the electromagnetic field interacts is concerned. The nonrelativistic theory is of an approximate character, but it involves a much simpler mathematical formalism than its more exact relativistic counterpart. Moreover, it can be applied to a wide range of problems in physics and chemistry, particularly in the areas of atomic spectroscopy, intermolecular forces, laser physics, and quantum optics. [*See* QUANTUM OPTICS.]

Nonrelativistic quantum electrodynamics provides an accurate description of phenomena when the following two conditions are satisfied:

1. The charged particles move at such slow speeds (in the inertial frame of a given observer) that their masses can be considered constant and equal to their rest masses. Since the relativistic mass of a particle with speed v and rest mass m is $m/\sqrt{(1 - v^2/c^2)}$, this requires that $v/c \ll 1$. Now this inequality generally holds for the constituent particles of atoms under normal laboratory conditions. For example, the root mean square speed $\bar{v}$ (relative to the supposedly slowly moving nucleus) of the electron of a hydrogen-like ion in a state with principal quantum number n is $Ze^2/(2nh)$ where Ze is the nuclear charge. If $Z = 1$ and $n = 1$ (the hydrogen atom in its ground state), then $\bar{v}/c$ equals the fine-structure constant α (approximately $\frac{1}{137}$) and the corresponding fractional increase in mass (over and above the rest mass) is only about 3 parts in 10^5.

This ratio is larger for higher values of Z but smaller for higher values of n. The variation of mass with velocity is, therefore, expected to be appreciable only for the inner-shell electrons of the heavier elements.

2.   The *number* of each type of charged particle (electron, proton, etc.) is *conserved*; that is, such particles are neither created nor destroyed in any process. This assumption imposes a restriction on the frequency ν of the radiation with which the particles may interact, since photons of sufficiently high energy are capable of creating particle–antiparticle pairs. This possibility requires an energy of order mc^2 (where m is the rest mass of the *lightest* charged particle, namely the electron) and will therefore be excluded if $\nu \ll \nu_c$, where ν_c is defined by $h\nu_c = mc^2$ and is about 10^{20} Hz. (Here ν_c is the frequency associated with the Compton wavelength of the electron given by $\lambda_c = h/mc$.) It follows that hard X rays and high-energy gamma rays are to be omitted from consideration in this section.

B. An Assembly of Photons

The classical electromagnetic field in empty space is equivalent to an infinite number of one-dimensional simple harmonic oscillators. One oscillator is associated with each *plane wave* component of the field, specified by its frequency ν, wave vector $\mathbf{k}$ (where $|\mathbf{k}| = 2\pi\nu/c$), and unit polarization vector $\hat{\mathbf{e}}$. The waves are transverse waves, which implies that the polarization vector is perpendicular to the direction of propagation $\hat{\mathbf{k}}$ (see Fig. 5). Hence, for each propagation direction, there are two independent polarization vectors $\hat{\mathbf{e}}^{(\lambda)}$ ($\lambda = 1, 2$). A radiation oscillator may therefore be labelled by the pair $(\mathbf{k}, \lambda)$, which specifies the frequency, propagation direction, and polarization for the corresponding mode of the field.

A mathematical description of an assembly of noninteracting photons is obtained when each of the radiation oscillators is treated as a quantum mechanical system. This involves little more than the use of the matrix theory of the harmonic oscillator developed in elementary quantum mechanics but extended to cover the case of a set of independent oscillators. The result of this quantization of the electromagnetic field can be briefly summarized. States of the complete system are represented by vectors in a generalized (in fact, infinite-dimensional) vector space and dynamical variables (such as energy and momentum) by linear operators, which act on the vectors to produce other vectors of the same kind. The vacuum state is that for which every oscillator has its lowest energy. It can be assumed, for convenience, that the energy of the vacuum state is zero. The so-called zero-point energy $\frac{1}{2}h\nu$ of an oscillator with frequency ν is therefore discarded, but this amounts merely to a shift in the datum point for measuring energies. (Nevertheless, changes in the zero-point

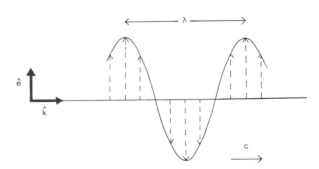

FIG. 5.   A linearly polarized electromagnetic wave of wavelength λ propagating in empty space with speed c in the direction $\hat{\mathbf{k}}$. The (real) unit polarization vector $\hat{\mathbf{e}}$ and $\hat{\mathbf{k}}$ together determine the plane of polarization, at any point of which the magnetic induction vector $\mathscr{B}$ (broken arrows) is parallel to $\hat{\mathbf{e}}$ and oscillates in simple harmonic motion of frequency ν, where $\nu\lambda = c$. The electric vector $\mathscr{E}$ (not shown) also oscillates with frequency ν and in phase with $\mathscr{B}$ but is perpendicular to the plane of polarization.

energy can give rise to measureable forces, for example, between conducting plates. This is called the Casimir effect.)

If a radiation oscillator of mode $(\mathbf{k}, \lambda)$ is excited to its nth stationary state, with energy $nh\nu$, then this is taken to correspond physically to the presence of n photons, each with energy $h\nu$, for that mode of the field. An *occupation-number state* is one with a specified number of photons in each mode. The number $n_{\mathbf{k}\lambda}$ of photons in mode $(\mathbf{k}, \lambda)$ is then called the *occupation number* for that mode. Only a finite number of occupation numbers can be nonzero and the total numbers of photons is $\Sigma\, n_{\mathbf{k}\lambda}$ with the summation extending over occupied modes. Similarly, the total energy E of the photons in an occupation-number state is $\Sigma(n_{\mathbf{k}\lambda}\, h\nu)$. The vacuum state can be thought of as an occupation-number state for which every occupation number is zero.

The operator that represents the total energy of the radiation field is called the *Hamiltonian* operator and is denoted by H_{RAD}. It can be expressed in terms of photon *annihilation* and *creation* operators. The annihilation operator for mode $(\mathbf{k}, \lambda)$, when acting on an occupation-number state vector, reduces the number of photons for that mode by one, and when acting on the vacuum-state vector gives the zero vector. Similarly the creation operator increases the number of photons by one. (In the context of the elementary theory of the harmonic oscillator, these operators are usually called *lowering* and *raising* operators, respectively.)

A general state of the radiation field at time t is represented by a state vector Ψ of unit length that is a linear combination of the occupation-number state vectors, with coefficients $c(\ldots, n_{\mathbf{k}\lambda}, \ldots; t)$ depending on the occupation numbers and the time. The square of the magnitude of $c(\ldots, n_{\mathbf{k}\lambda}, \ldots; t)$ is the *probability* that if a measurement of the occupation numbers is carried out at time t, then these will be found to have precisely the values $\ldots, n_{\mathbf{k}\lambda}, \ldots$. So long as no measurements are made on the system, the time evolution of the state vector is governed by *Schrödinger's equation*

$$i\hbar\, \frac{\partial \Psi}{\partial t} = H_{\mathrm{RAD}}\Psi$$

The (unit) length of the state vector does not change with time and so the dynamical behavior of the system may be said to correspond to a *pure rotation* in the generalized vector space. Indeed, the individual probabilities $|c|^2$ do not change with time, since the probability amplitudes c change only through a phase factor

$\exp(-iEt/\hbar)$. This is consistent with the fact that for the free field, photons are neither created nor destroyed.

C. The Quantized Electromagnetic Field

While the treatment of the radiation field as an assembly of photons may seem to emphasize its corpuscular aspects, the wave properties are, nevertheless, also contained in the formalism. In particular, *Maxwell's equations in empty space remain valid,* although they now appear as operator equations rather than as equations for classical fields. Both the electric field $\mathscr{E}$ and the magnetic induction field $\mathscr{B}$ become operators that can be expressed as linear combinations of the photon annihilation and creation operators. If these expressions are inserted into the classical formula for the field energy, then the expansion of the Hamiltonian operator in terms of the annihilation and creation operators is recovered. Thus,

$$H_{\mathrm{RAD}} = \frac{1}{2} \iiint (\mathscr{E}^2 + \mathscr{B}^2)\, dV$$

(It is true that an infinite zero-point energy also appears. This energy may, however, be discarded, as were the zero-point energies of the individual oscillators.) Similarly, the classical expression

$$\frac{1}{c} \iiint \mathscr{E} \times \mathscr{B}\, dV$$

for the field momentum, obtained from Poynting's theorem, implies, when reinterpreted in terms of annihilation and creation operators, that a momentum $\hbar\mathbf{k}$ is to be ascribed to each photon of mode $(\mathbf{k}, \lambda)$. This is in agreement with Einstein's hypothesis, since $\hbar|\mathbf{k}| = h/\lambda$.

Another consequence of the quantization of the field is the occurrence of *uncertainties* or *fluctuations* that have no counterpart in the classical theory. In the vacuum state, for example, the mean value of the electric field is zero but its *root mean square deviation from the mean,* $\Delta\mathscr{E}$, is nonzero. The fluctuation $\Delta\mathscr{E}$ arises from the collective *zero-point motions* of the radiation oscillators and, if calculated for a nonzero volume with linear dimensions of order L and a nonzero time interval with length of order T, assumes a value whose order of magnitude is given by

$$\Delta\mathscr{E} \sim \begin{cases} \dfrac{\sqrt{\hbar c}}{L^2}, & \text{if } L \geq cT \\[2ex] \dfrac{\sqrt{\hbar c}}{L(cT)}, & \text{if } L \leq cT \end{cases}$$

In any other of the occupation-number states, for which the mean values are also zero, the field fluctuations are of greater magnitude than in the vacuum state. Now the occupation-number states resemble incoherent superpositions of classical plane-wave states, since they are associated with definite wave vectors and polarization vectors but do not have well-defined phases (in the classical sense), whereas a classical plane-wave field has a simple harmonic time dependence. There exist other states of the quantized field, however, called coherent or quasi-classical states, in which the phase is more well defined but the number of photons, and hence the energy and momentum, is less sharp than for the occupation-number states. To each state of the classical field there corresponds a unique coherent state of the quantized field such that (a) the mean values of the quantized field components are equal to the classical field components and (b) the mean value of the quantized energy is equal to the classical energy. The coherent states are also remarkable in the following respect: *the field fluctuations for these states are exactly the same as those for the vacuum state.*

The operators representing the components of the electric field $\mathscr{E}$ and the magnetic induction field $\mathscr{B}$ satisfy certain commutation relations, which may be derived from those for the photon annihilation and creation operators. Just as in ordinary quantum mechanics the commutation relation

$$qp - pq = i\hbar$$

between the position operator q and the momentum operator p of a particle leads to the Heisenberg uncertainty relation

$$\Delta q \, \Delta p \sim \hbar$$

so the commutation relations between the components of $\mathscr{E}$ and $\mathscr{B}$ lead to uncertainty relations for the electromagnetic field strengths. These uncertainty relations are in agreement with the way in which the field strengths can, at least in principle, be measured by means of macroscopic test bodies. This was shown in detail by Bohr and Rosenfeld in 1933.

D. Interactions of Photons and Atoms

The quantized radiation field has so far been considered as a system by itself. A set of nonrelativistic charged particles, interacting through instantaneous Coulomb forces and also, perhaps, acted on by prescribed external static electric or magnetic fields, can also be considered as a system by itself, as in ordinary quantum mechanics. This system will, for convenience, be referred to as an atom, although it may really be a molecule, an ion, or a collection of atoms, molecules, or ions. It is assumed that there are N charged particles with masses m_1, $m_2, ..., m_N$, charges $e_1, e_2, ..., e_N$, position operators $\mathbf{q}_1, \mathbf{q}_2, ..., \mathbf{q}_N$, and momentum operators $\mathbf{p}_1, \mathbf{p}_2, ..., \mathbf{p}_N$. The Hamiltonian operator for this system is given by

$$H_{\text{ATOM}} = \sum_{\alpha=1}^{N} \frac{1}{2m_\alpha} \mathbf{p}_\alpha^2 + U$$

where the first term represents the kinetic energy and the second the potential energy. The potential energy U depends on the positions and momenta of the particles, their charges, and the external fields, if any are present.

It is often a good approximation to treat the nuclei as fixed and to regard the coordinates and momenta of the electrons alone as dynamical variables. This is possible because of the large mass of the protons and neutrons compared with that of the electrons (proton mass $\approx 1836 \times$ electron mass). The fixed-nuclei approximation involves, among other things, the neglect of the recoil of the atoms which should accompany the absorption or emission of photons. The recoil velocity is, however, normally very small. For example, the speed imparted to a hydrogen atom by a photon with a frequency in the visible spectrum is of the order of 10^{-8} times the speed of light *in vacuo*. Such a speed results in only a very slight Doppler shift in the frequency of the emitted radiation. In the fixed-nuclei approximation, the Hamiltonian operator H_{ATOM} has, in general, a discrete set of energy levels $E_r, E_s, ...$ corresponding to bound states as well as a continuous set of energy levels E corresponding to ionized states. Here $r, s, ...$ are shorthand notations for sets of quantum numbers sufficient to specify the states completely.

A state vector for the complete system consisting of the atom and the radiation field is obtained by multiplying a state vector for the field directly into a state vector for the atom. For example, there are states for which the photon occupation numbers have definite values and the atom is in a stationary state with a definite energy. The general state of the complete system is, at any instant, a superposition of such product states.

The Hamiltonian H for the complete system is not simply the sum of the radiation and atomic

Hamiltonians given previously. This sum must be supplemented by an interaction term H_{INT}:

$$H = H_{RAD} + H_{ATOM} + H_{INT}$$

The inclusion of the interaction term is essential if the operator equations of motion are to reproduce (a) Maxwell's equations for $\mathscr{E}$ and $\mathscr{B}$ with the charges and currents as sources and (b) the Lorentz-force law for the charged particles when acted on by $\mathscr{E}$ and $\mathscr{B}$, that is, the expected equations of motion for the interacting systems. The interaction Hamiltonian H_{INT} contains some operators that refer to the field and some that refer to the particles and, hence, is responsible for the coupling between the two parts of the complete system. In the absence of H_{INT}, the product vectors of the type mentioned above represent stationary states in which the photon occupation numbers are constant and the atom has a fixed energy. Due to the presence of H_{INT}, however, transitions between these states can occur, in which, for example, the atom loses or gains energy and the number of photons is correspondingly increased or decreased. The interaction Hamiltonian may be expressed as the sum of two parts, one proportional to e and the other to e^2:

$$H_{INT} = eH_1 + e^2H_2$$

where H_1 is linear and H_2 is quadratic in the photon annihilation and creaton operators. As a consequence, these two terms give rise to processes in which the number of photons changes by one or two, respectively.

The use of the so-called Coulomb gauge is very convenient in nonrelativistic theory. In this gauge only the transverse electromagnetic field, which is a superposition of modes with transverse polarization vectors ($\hat{\mathbf{e}}^{(\lambda)} \cdot \hat{\mathbf{k}} = 0$), is quantized. The effect of the longitudinal field, responsible for the instantaneous Coulomb interaction between the charges, is treated as a potential as in ordinary quantum mechanics and is included in the expression for H_{ATOM}.

The time evolution of the complete system is governed by Schrödinger's equation

$$i\hbar \frac{\partial \Psi}{\partial t} = H\Psi$$

where now H is the total Hamiltonian and Ψ represents the state of both the field and the atom at time t. No exact solutions of this equation are known. Fortunately, however, H_{INT} is of order e and, hence, can be regarded as a small perturbation to the unperturbed Hamiltonian

$H_{RAD} + H_{ATOM}$. Time-dependent perturbation theory can then be used to calculate approximately the probabilities for transitions between unperturbed states. The total energy is always exactly conserved in transitions between initial and final states. Since the perturbation is small, the unperturbed energy is approximately conserved in such transitions.

E. APPLICATIONS

Applications of the theory to the emission and absorption of photons by atoms and the scattering of photons by free electrons will now be considered.

1. Spontaneous Emission— Einstein's A Coefficient

If initially (a) the atom is in an excited state r with energy E_r and (b) the radiation field is in the vacuum state, then there is a probability that after a time t a photon of mode $(\mathbf{k}, \lambda)$ has been created and the atom has made a transition to a state s with lower energy E_s, where

$$h\nu \approx E_r - E_s$$

Since there are no photons at all present initially, this process is known as spontaneous emission. It is represented graphically by the Feynman diagram in Fig. 6. Single-photon spontaneous emission involves, in the lowest order of perturbation theory, only that term in the interaction Hamiltonian that is proportional to e.

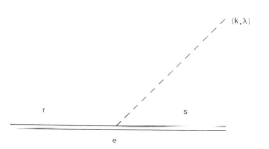

FIG. 6. Feynman diagram for spontaneous emission. The left-hand and right-hand portions of the parallel horizontal lines represent the initial and final atomic states r and s, respectively. (Double lines are used to indicate that the electrons are not free but are bound to the atomic nucleus.) The dotted line represents the emitted photon of mode $(\mathbf{k}, \lambda)$. This is created when the atom undergoes the transition $r \to s$. The vertex labelled e corresponds to the first-order term in the interaction Hamiltonian, which is responsible for this process in the lowest order of perturbation theory.

Furthermore, the so-called dipole approximation can be used for optical or lower frequencies and bound states of atoms or small molecules, since then the wavelength of the emitted photon is much larger than the dimensions of the region in which the atomic wave functions differ significantly from zero. The emission probability can sometimes be expressed in terms of a constant transition rate (that is, a probability per unit time for the transition to occur) known as Einstein's A coefficient. The total transition rate for emission of the photon in any direction and with any polarization is given in dipole approximation by

$$A_s^r = (16\pi^3\nu^3/3hc^3)|\boldsymbol{\mu}^{rs}|^2$$

where $\boldsymbol{\mu}^{rs}$ denotes the dipole transition moment, which can be calculated once the wave functions for the atomic states r and s are known. Thus, in dipole approximation, Einstein's A coefficient is proportional to the cube of the transition frequency and the square of the length of the dipole transition moment.

The reciprocal of A_s^r is the average lifetime of the upper state r with respect to the lower state s. For example, for optical transitions with a photon wavelength of order 5000 Å (1 Å = 10^{-8} cm) and a dipole transition moment of order ea_0 (where a_0 is the Bohr radius of hydrogen, approximately 0.53 Å), the lifetime is of order 10^{-8} sec. The transition probability is proportional to the time t so long as t is large compared with the atomic period $1/\nu$ and small compared with the lifetime. Since, with the above assumptions, the period is of order 10^{-15} sec, there is indeed a range of values of t that satisfy both conditions. The detection of the emitted photons must take place at times t lying in this range, or else the emission rate is not approximately constant.

2. Absorption and Stimulated Emission—Einstein's B Coefficients

If the atom is initially at the lower level E_s, but there is radiation already present, it may make a transition to the higher level E_r by absorbing a photon with energy approximately equal to $E_r - E_s$. The Feynman diagram for absorption is shown in Fig. 7. The transition rate for this process is proportional to the photon occupation number $n_{k\lambda}$ and hence to the intensity I (erg cm^{-3} Hz^{-1}) of the incident radiation in the spectral region from which the photon is absorbed. If the atom is bathed in isotropic unpolarized radiation (so that I is independent of $\hat{k}$ and λ), the total absorption rate is B_r^sI where B_r^s is Ein-

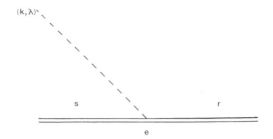

FIG. 7.   Feynman diagram for absorption. In the initial state (left of diagram), the atom has energy E_s and there is a photon of mode (k, λ) present, whereas in the final state (right of diagram), the atom has higher energy E_r and the photon has been annihilated.

stein's B coefficient for absorption, given in dipole approximation as

$$B_r^s = (2\pi^2/3h^2)|\boldsymbol{\mu}^{rs}|^2$$

The upper limit on the time for the validity of this transition rate is now much less than the reciprocal of B_r^sI. Times less than this upper limit but much greater than the period can be found, provided the intensity of the radiation is not too high.

For an atom initially at the upper level E_r with radiation present as before, there is a probability for a transition to the lower level E_s accompanied by the emission of a photon with the same characteristics as some of those in the incident beam. This emission may, of course, occur spontaneously, that is, even when all the photon occupation numbers are zero. There is in addition, however, emission that is stimulated or induced by the incident radiation, at a rate proportional to its intensity. For isotropic unpolarized radiation, the stimulated emission rate is B_s^rI, where the B coefficient for emission $r \rightarrow s$ is the same as that for absorption $s \rightarrow r$, that is, $B_s^r = B_r^s$.

3. Thomson Scattering

The nonrelativistic limit of the Compton scattering of photons by free electrons is known as Thomson scattering. This limit applies when both the electron and photon momenta have magnitudes small compared to mc. If $\mathbf{p}$ and $\mathbf{p}'$ denote the initial and final momenta of the electron and $(\mathbf{k}, \lambda)$ and $(\mathbf{k}', \lambda')$ denote the wave vectors and polarizations of the incident and scattered photons, then it follows from the laws of conservation of energy and momentum that $k' \approx k$ and hence that $p' \approx p$. Thus, the magnitudes of the momenta are effectively unaltered, although, in general, their directions change. In particular, for the limit considered, there is no shift in the frequency of the scattered photon.

The Feynman diagrams that give the leading contributions to Thomson scattering are shown in Fig. 8. Each diagram depicts the incident photon and initial electron arriving from the left, and the scattered photon and recoil electron disappearing to the right. The contribution of Fig. 8a arises from the e^2 term in the interaction Hamiltonian; it may be said that here the incident photon is annihilated and the scattered photon simultaneously created. In Fig. 8b and c, on the other hand, the annihilation and creation are represented by two different one-photon vertices, each arising from the e term in the interaction Hamiltonian. These diagrams differ in the order in which the creation and annihilation take place. Overall momentum is conserved at every vertex.

It must be emphasized, however, that the contributions of Fig. 8a–c cannot be physically separated, since only the initial and final states are observed. For this reason, the intermediate states are often referred to as virtual states. Indeed, it may be shown that the contributions of Fig. 8b and c effectively cancel, as their sum is of order $\hbar k/(mc)$ times that of Fig. 8a.

In scattering experiments the measured quantity is the cross section (having dimensions cm^2), defined as the number of scattered particles per unit time divided by the number of incident particles per unit area per unit time. For Thomson scattering, the differential cross section per unit solid angle Ω for scattering the photon with polarization λ' and direction within $d\Omega$ of $\hat{\mathbf{k}}'$ is given, from the contribution of Fig. 8a, by

$$\frac{d\sigma}{d\Omega} = r_0^2 |\hat{\mathbf{e}}^{(\lambda)} \cdot \hat{\mathbf{e}}^{(\lambda')}|^2 = r_0^2 \cos^2 \Theta$$

where (a) it is assumed that the polarization vectors are real, (b) Θ is the angle between the directions of polarization of the incoming and outgoing photons, and (c)

$$r_0 = e^2/4\pi mc^2 \approx 2.82 \times 10^{-13} \text{ cm}$$

and is the so-called classical electron radius. (This is the radius that an electron of uniformly distributed charge must have if its electrostatic energy is to equal its rest energy.) It should be noted in particular that, with the approximations indicated, the cross section is independent of frequency and vanishes if the incident and scattered polarization vectors are perpendicular.

If the incident photons are randomly polarized and the polarization of the scattered photons is not observed, then the cross section should be averaged over initial polarization indexes and summed over final polarization indexes. The resulting differential cross section per unit solid angle depends only on the scattering angle θ (where $\cos \theta = \hat{\mathbf{k}} \cdot \hat{\mathbf{k}}'$):

$$d\sigma/d\Omega = \tfrac{1}{2}r_0^2(1 + \cos^2 \theta).$$

The total unpolarized cross section, obtained by integrating this over all solid angles, is given by

$$\sigma_{\text{tot}} = (8\pi/3)r_0^2 \approx 6.65 \times 10^{-25} \text{ cm}^2.$$

4. Other Applications

The field of application of nonrelativistic quantum electrodynamics has expanded considerably in recent years due to the development of lasers and their use as spectroscopic tools for investigating a variety of physical and chemical systems. Lasers are sources of highly coherent and very intense beams of light. In the subject of quantum optics, the quantum statistical properties, such as the degree of coherence, of the light beam itself may be the object of investigation. For example, the quasi-classical states of the radiation field referred to earlier exhibit a higher

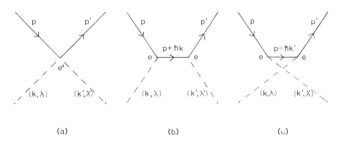

FIG. 8. Feynman diagrams for Thomson scattering. Dotted lines represent photons and solid lines represent free electrons.

degree of coherence than the occupation number states, and these in turn are less chaotic than fields in thermal equilibrium, in which the distribution of photons follows Planck's radiation law. [*See* QUANTUM OPTICS.]

The high intensity of the light beams that may be achieved by using laser sources can also give rise to nonlinear effects that are unobservable at lower intensities. A typical example of a nonlinear process is third-harmonic generation, that is, the absorption of three photons of frequency ν by an atom and the emission of a single photon of frequency 3ν (the third harmonic of the incident frequency). The rate for this process (after which the atom returns to its initial state and overall energy is consequently conserved) is proportional to the cube of the intensity of the incident beam. This should be contrasted with a linear process such as single-photon absorption for which the transition rate is proportional to the intensity itself, the factor of proportionality being Einstein's B coefficient.

III. Relativistic Quantum Electrodynamics

A. RELATIVISTIC THEORY

Relativistic quantum electrodynamics is formed by the union of the special theory of relativity, characterized by the speed of light, and quantum mechanics, characterized by Planck's constant. In discussing the relativistic theory it is useful and customary to employ the natural system of units, in which speeds are measured as multiples of c and angular momenta are measured as multiples of $\hbar$. Since no natural length appears in the theory, lengths continue to be measured in centimeters. The expression for any quantity in (rationalized) natural units is obtained from the corresponding expression in (rationalized) cgs units simply by setting $c = 1$ and $\hbar = 1$. For example, the cgs expressions $\hbar \mathbf{k}$, mc^2, and $e^2/(4\pi\hbar c)$ for the momentum of a photon, the rest energy of an electron and the fine structure constant, respectively, become $\mathbf{k}$, m and $e^2/(4\pi)$, respectively, in natural units. It is also easy to convert from natural units to cgs units, by inserting appropriate factors of $\hbar$ and c.

If quantum electrodynamics is to satisfy the principles of the special theory of relativity, its equations must be covariant under Lorentz transformations. Lorentz transformations relate the space–time coordinates x, y, z, and t of events as seen by observers using inertial frames of reference moving with uniform velocity relative to each other. (The coordinates x, y, z, and t are the components of a four-dimensional vector, to be denoted simply by x.) A covariant equation has the same form for two such observers. The fact that physical laws are expressible as covariant equations means that these laws are the same for all observers using inertial reference frames.

The state of a quantum-mechanical system (for example, the electromagnetic field *in vacuo*) is specified in relativistic theory on a three-dimensional spacelike hyperplane. In a given inertial frame, this consists of either all the points in three-dimensional space at a particular instant of time or all the events on a two-dimensional plane moving for all time perpendicularly to itself with constant speed greater than that of light. Two distinct events on a spacelike hyperplane cannot be connected by signals travelling with speed less than or equal to the speed of light, and so two measurements made in the vicinity of the corresponding space–time points will not interfere. This is known as microscopic causality. The whole of the four-dimensional space–time manifold is filled with a set of parallel spacelike hyperplanes, which may be labelled by an invariant timelike parameter τ, with τ ranging from $-\infty$ to ∞. The evolution of the system is described by specifying the state for each hyperplane τ and is determined dynamically, through Schrödinger's equation, on the interval $[\tau_1, \tau_2]$, if the state is specified at τ_1 and no measurements are made until τ_2.

B. ELECTRONS AND POSITRONS

The one-particle relativistic theory of the electron is based on the Dirac equation. This is a differential equation, with matrix coefficients, for a spinor wave function $\psi(x)$ having four components $\psi^\mu(x)$ ($\mu = 0, 1, 2, 3$). The requirement that the Dirac equation be covariant determines the behavior of ψ under Lorentz transformations. Since the electron is now described by a four-component spinor rather than by a one-component scalar wave function, it has extra degrees of freedom, over and above those allowed by the Schrödinger theory. These correspond to the spin or intrinsic angular momentum of magnitude $\frac{1}{2}$ (in natural units). Hence the spin appears automatically in the Dirac theory of the electron and does not have to be added on in an *ad hoc* fashion, as it does in nonrelativistic quantum mechanics.

The difficulties of interpretation associated with the negative-energy solutions of the Dirac equation have already been mentioned. These difficulties disappear in the second-quantized version of the theory, in which electrons and positrons are treated on an equal footing and all have positive energy. Moreover, this version provides a calculus for processes involving annihilation and creation of electrons and positrons—in the high-energy regime, the number of these particles is no longer conserved.

The spinor ψ and its related adjoint spinor $\bar{\psi}$ are first expressed in terms of plane-wave solutions of the free-particle equation. These solutions correspond to particles with energy E, momentum $\mathbf{p}$, and rest mass m satisfying the relativistic energy–momentum relation

$$E^2 = |\mathbf{p}|^2 + m^2$$

The coefficients in the expansions of ψ and $\bar{\psi}$ may then be interpreted as annihilation and creation operators for electrons and positrons in definite momentum and spin states. The spin states can be chosen to be helicity states of the electrons and positrons, that is, states in which the component of spin in the direction of motion is either $\frac{1}{2}$ (right-hand helicity) or $-\frac{1}{2}$ (left-hand helicity). It is only the component of spin in the direction of motion that is invariant under Lorentz transformations.

The algebra of the creation and annihilation operators for electrons and positrons differs from that of the creation and annihilation operators for photons. (Technically, it involves anticommutation instead of commutation relations.) The difference arises from the fact that, whereas photons are bosons, electrons and positrons are fermions, and are subject to the exclusion principle. The only possible occupation numbers for electrons or positrons are therefore 0 or 1. This is in agreement with Pauli's spin-statistics theorem, since the spin of an electron or positron is half an odd integer. It can be shown that quantizing the Dirac field by using commutation relations instead of anticommutation relations leads to a Hamiltonian operator with energy levels that are not bounded below and, hence, one for which no stable vacuum (ground) state exists. (Similarly, quantizing the Maxwell field, which has intrinsic spin 1, by using anticommutation relations instead of commutation relations leads to a breakdown of microscopic causality.)

C. Covariant Quantization of the Electromagnetic Field

The quantization of the electromagnetic field in the Coulomb gauge, though very useful for dealing with bound systems in nonrelativistic approximation, is not a manifestly covariant procedure. In the Coulomb-gauge formalism, only the transverse field is quantized, while the longitudinal field gives rise to an instantaneous interaction between the charges. The division of the field into transverse and longitudinal components, however, is not Lorentz covariant—these components do not transform separately on going from one inertial frame to another. To exhibit the covariance of the theory, it is necessary to use a gauge condition on the electromagnetic potentials that is itself covariant. The most convenient such condition is the so-called Lorentz condition. This condition also leads to certain difficulties which are, however, overcome in the formalism developed by Gupta and Bleuler.

The electromagnetic field is, in the first instance, quantized in a covariant way without reference to the Lorentz gauge condition. In contrast to the noncovariant treatment, there are now, for each wave vector $\mathbf{k}$, four types of photon, corresponding to timelike and longitudinal as well as two transverse polarization vectors, which are, in addition, four-dimensional rather than three-dimensional vectors. Moreover, the inner (or scalar) product of the infinite-dimensional vector space on which the photon creation and annihilation operators act is not positive definite; that is, there exist nonzero vectors in this space the square of whose length is zero or negative. (This is due to the metric of the space–time continuum, which distinguishes timelike from spacelike directions. Thus, the four-dimensional vector x is spacelike, lightlike, or timelike, relative to the origin, according as $x^2 + y^2 + z^2 - t^2$ is positive, zero, or negative.) This constitutes a serious difficulty, since the quantum mechanical statistical interpretation requires a positive-definite inner product. For the resolution of this problem, the use of the Lorentz gauge condition, which has yet to be imposed, is of decisive importance.

It may be shown that neither the Lorentz condition nor Maxwell's equations are satisfied as operator equations in the covariant theory, because they are incompatible with the commutation relations. They are, however, satisfied as equations for expectation values (and hence are satisfied in the classical limit), provided a subsidiary condition is imposed on those state vectors that are to represent physically realizable states. The effect of the subsidiary condition is to make the timelike and longitudinal photons unobservable in real states of the system. These states have either no timelike or longitudinal photons at all or only certain allowed admixtures of them. Moreover, changing the allowed admixtures is merely equivalent to carrying out a gauge transformation that maintains the Lorentz condition. The allowed admixtures are always such that the contributions of timelike and longitudinal photons to, for example, the energy and momentum, cancel out, and only the contributions of the transverse, observable photons remain. Similarly, the statistical interpretation of the theory is consistent, when this is restricted to the calculation of probabilities for physically realizable states.

Despite the fact that timelike and longitudinal photons are unobservable in real states of the system, their presence is important and cannot be neglected in intermediate or virtual states. For example, the Coulomb interaction may be described in terms of the virtual exchange of timelike and longitudinal photons by charged particles. The appearance of these photons in the formalism is also required, of course, if the theory is to be manifestly Lorentz covariant.

D. SYMMETRIES AND CONSERVATION LAWS

The coupling between the quantized Maxwell and Dirac fields is represented by a Lorentz-invariant interaction Hamiltonian density $\mathcal{H}_{\mathrm{INT}}$ that links the scalar and vector potentials to the charge and current densities. Here $\mathcal{H}_{\mathrm{INT}}$ is linear in the electromagnetic potentials and bilinear in the spinor fields ψ and $\bar{\psi}$ and is also of order e—there is no e^2 term as in nonrelativistic theory. The interaction Hamiltonian density may be derived from a Lagrangian density known as the minimal-coupling Lagrangian density.

It is interesting to note that certain continuous symmetries of the coupled systems are reflected in the structure of the complete Lagrangian density $\mathcal{L}$, which is Lorentz invariant and gauge invariant. According to a theorem of Noether, these symmetries must lead to conservation laws. For example, the invariance of $\mathcal{L}$ under time displacements implies the conservation of energy; its invariance under space displace-

ments and rotations implies the conservation of linear and angular momentum, respectively; and its invariance under gauge transformations implies the conservation of charge.

The complete system also has three discrete symmetries. It is invariant under (a) charge conjugation C, that is, the interchange of particles and antiparticles (which affects only electrons and positrons, since the photon is its own antiparticle); (b) the parity operation P, that is, space inversion or the interchange of left and right; and (c) time reversal T. This invariance under C, P and T is not shared by all the laws of nature. The nonconservation of parity in the weak interaction, which is responsible for the dynamics of beta emission, was suggested by Lee and Yang in 1956 and subsequently confirmed experimentally. That the combined transformation of charge conjugation and parity is also not a symmetry follows from the decay of the long-lived neutral K meson into two charges pions, a decay that is forbidden by CP conservation. Invariance under the combined CPT transformation, established on very general assumptions (Lorentz covariance and locality), then implies that time reversal is also not a symmetry of the physical world. Hence, the separate conservation of C, P, and T is only an approximation which is, however, valid for phenomena that are adequately described by electrodynamics alone.

E. THE S MATRIX AND FEYNMAN DIAGRAMS

The S matrix in quantum electrodynamics is used to calculate probability amplitudes for processes in which particles (electrons, positrons or photons) that are initially free are allowed to interact and scatter. In the so-called interaction picture of the motion, the state vector Ψ for the complete system evolves under the influence of the interaction Hamiltonian $\mathcal{H}_{\mathrm{INT}}$ alone, and the S operator (or scattering operator) maps the state vector on the hyperplane $\tau = -\infty$ (that is, long before the interaction takes place) onto the state vector on the hyperplane $\tau = \infty$ (that is, long after the interaction has ceased):

$$\Psi(\infty) = S\Psi(-\infty)$$

The S operator can be developed as a power series in the coupling constant e. With the help of a theorem due to Wick, the structure of the nth-order contribution, corresponding to the nth power of e in the expansion, may be systematically analysed and represented by Feynman diagrams. It is usually convenient to use Feynman diagrams in energy–momentum space. These represent all possible virtual processes that can take place for given initial and final momentum and polarization or spin states of the particles. The Feynman rules enable expressions for the probability amplitude or S-matrix element S_{fi} for the process $i \rightarrow f$ to be written down directly from the diagrams. From this the cross section for the process may be calculated to a given order in e and compared with the experimentally obtained value.

The lowest order of perturbation theory ($n = 1$) involves only the first power of the interaction Hamiltonian $\mathscr{H}_{INT}$, which is linear in the photon annihilation and creation operators and bilinear in the fermion (electron or positron) annihilation and creation operators. This gives rise to processes such as those depicted in the Feynman diagrams of Fig. 9. These diagrams are called basic vertex diagrams. There are in all eight such diagrams, corresponding to processes in which a photon is either annihilated or created and two fermions are annihilated or created or one is annihilated and the other created.

Every Feynman diagram is a combination of some or all of the eight basic vertex diagrams—an nth-order diagram contains n vertices. Energy and momentum (which together form a four-dimensional vector) are conserved at every vertex. (This is in contrast to the nonrelativistic theory, in which momentum but not energy is conserved in virtual processes.) However, the relativistic relation between energy and momentum need not be satisfied for virtual particles. Now this relation cannot be satisfied by all the particles participating in a basic vertex

process, which must therefore be a virtual rather than a real process. For example, electron–positron annihilation with the production of a single photon is forbidden by energy–momentum conservation, even though it is allowed by charge conservation. Hence the basic vertex diagrams can appear only as parts of larger Feynman diagrams depicting processes for which overall energy and momentum are conserved and the relativistic energy–momentum relation is satisfied by the (real) particles in the initial and final states.

As an example of a real process, consider the Compton scattering of photons by electrons. This is allowed in the second order of perturbation theory, and the Feynman diagrams, each containing two vertices, are shown in Fig. 10. The corresponding polarized cross section for the laboratory reference system, in which the target electron is initially at rest, is given by the Klein–Nishina formula:

$$\frac{d\sigma}{d\Omega} = \frac{\alpha^2}{4m^2} \left(\frac{\nu'}{\nu}\right)^2 \left[\frac{\nu}{\nu'} + \frac{\nu'}{\nu} + 4(\varepsilon \cdot \varepsilon)^2 - 2\right]$$

Here ν and ν' are the frequencies of the incident and scattered photons, respectively, and ε and ε' are their (four-dimensional) transverse polarization vectors, which in this formula are assumed to be real (so that the photons are linearly polarized). In the low-energy limit ($\nu \ll m$ and $\nu' \approx \nu$), this reduces to the Thomson cross section derived from the nonrelativistic theory. (Note that $r_0 = \alpha/m$.) The unpolarized cross section, obtained by averaging over initial and summing over final polarizations, is given by

$$\frac{d\sigma}{d\Omega} = \frac{\alpha^2}{2m^2} \left(\frac{\nu'}{\nu}\right)^2 \left[\frac{\nu}{\nu'} + \frac{\nu'}{\nu} - \sin^2 \theta\right]$$

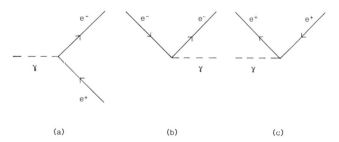

(a)   (b)   (c)

FIG. 9. Virtual processes depicted by basic vertex diagrams (to be viewed from left to right). (a) Photon (γ) annihilation and electron–positron (e^-e^+) pair production. (b) Electron scattering and photon creation. (c) Photon annihilation and positron scattering. Note the convention for the sense of the arrows used on the fermion lines to distinguish electrons and positrons.

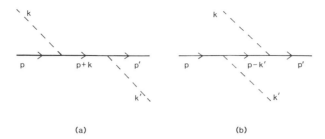

(a)   (b)

FIG. 10.   Feynman diagrams for Compton scattering. The lines
are labelled by the four-dimensional energy–momentum vectors of the particles. Energy and momentum are conserved
overall and at every vertex. Polarization and spin labels have
been suppressed. Diagrams (a) and (b) differ in the order in
which the incident photon is annihilated and the scattered photon created.

where θ is the angle of scattering, as in Fig. 3. This reduces to the unpolarized Thomson cross section in the low-energy limit.

F. Radiative Corrections

The first approximation to the S-matrix element for a given process may be improved by adding contributions from higher-order perturbation theory. These contributions, known as radiative corrections, often, though not always, involve integrals with ultraviolet divergences (that is, the integrals tend to infinity as the upper limits on the momenta of the virtual photons or fermions involved tend to infinity). For example, radiative corrections of second order in e (or of first order in α) relative to the lowest-order term are expected when one of the modifications shown in Fig. 11 is made in a Feynman diagram.

Each of the integrals corresponding to the modified diagrams, however, has an ultraviolet divergence.

The divergence difficulties of relativistic quantum electrodynamics may be overcome by first regularizing the theory, that is, by altering it so that all the integrals converge. In the method of dimensional regularization, for example, this is achieved by replacing (in a well-defined sense) divergent four-dimensional expressions by convergent $(4 - \varepsilon)$-dimensional expressions, where $\varepsilon > 0$. This may be described as reducing the dimensions of energy-momentum space from 4 to $4 - \varepsilon$. The regularized theory is not equivalent to quantum electrodynamics, which is restored only in the limit as $\varepsilon \to 0$, in which limit the divergences reappear.

The mass and charge of the fermions are then renormalized; that is, the predictions of the

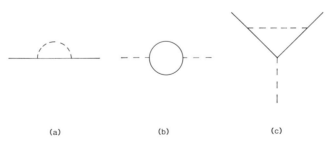

(a)   (b)   (c)

FIG. 11.   Modifications of a fermion line, a photon line, and a basic vertex part leading to second-order radiative corrections. (a) Fermion self-energy arising from emission and reabsorption of virtual photons. (b) Photon self-energy (or vacuum polarization) arising from virtual pair creation and annihilation. (c) Vertex modification arising from virtual photon exchange.

regularized theory are expressed in terms of the observed mass and charge of the physical particles rather than the unobservable mass and charge of the bare particles. In this connection, certain relations between fermion self-energy and vertex-modification contributions (see Fig. 11a and c), known as Ward's identities, allow a great simplification to be made. In particular, they imply that charge renormalization arises solely from vacuum polarization effects (see Fig. 11b). It is important to note also that mass and charge renormalization would have to be carried out even if no divergences appeared in the formalism.

Finally, quantum electrodynamics is recovered by removing the regularization. If the method of dimensional regularization is used, this means taking the limit as $\varepsilon \to 0$. In this limit, infinities reappear in the relations between the observed and bare masses and charges. These relations, however, are not susceptible to experimental verification, as the bare masses and charges themselves are unobservable. Moreover, as $\varepsilon \to 0$, the physical predictions of the theory (for example, radiative corrections to scattering cross sections or electromagnetic shifts of energy levels) are finite in all orders of perturbation theory and are expressed in terms of the observed masses and charges. (For this reason quantum electrodynamics is said to be a renormalizable quantum field theory.) These predictions can therefore be tested against experimental results.

1. The Lamb Shift

The nonrelativistic Lamb shift for atomic hydrogen was first calculated by Bethe in 1947, following the experiments of Lamb and Retherford. Whereas Dirac's one-particle relativistic theory predicts that the $2S_{1/2}$ and $2P_{1/2}$ states of the hydrogen atom have the same energy, Lamb and Retherford showed that the $2S_{1/2}$ level is actually higher and found a difference in energy corresponding to a frequency of about 1000 MHz. In Bethe's treatment, the effect was interpreted as a difference between the electron's self-energy when free (Fig. 11a) and when bound to the proton (Fig. 12a); a cutoff of order mc was used for the momenta of the virtual photons emitted in each case. The calculation gave no shift for the $2P_{1/2}$ level but did give an upward shift of about 1040 MHz for the $2S_{1/2}$ level and was therefore, in view of the nonrelativistic treatment and the approximations made, in good

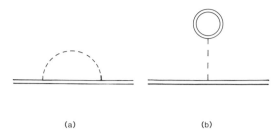

(a)   (b)

FIG. 12. Feynman diagrams for second-order radiative corrections to atomic energy levels (Lamb shift). (a) Electron self-energy and (b) vacuum polarization.

agreement with the observed value of the separation. This agreement has been enhanced by subsequent refinements in both theory and experiment. [*See* ATOMIC PHYSICS.]

The relativistic treatment of the Lamb shift, which is a bound-state problem, requires a more elaborate formalism (involving the so-called bound interaction picture) than the S-matrix theory outlined above. In addition to electron self-energy effects, there are also vacuum polarization effects (see Fig. 12) and effects due to the finite mass and nonzero size of the nucleus. (The proton is not a pointlike object but has an effective radius of about 0.86×10^{-13} cm.) Vacuum polarization gives a downward shift of about 27 MHz to the $2S_{1/2}$ level. There are still difficulties in calculating higher-order binding corrections to the Lamb shift and these have resulted in discrepancies of the order of 0.045 MHz between different calculated values. The discrepancies are small but nevertheless larger than the experimental uncertainties.

Recent theoretical values for the $2S_{1/2} - 2P_{1/2}$ level splitting in atomic hydrogen given by Borie are 1057.843(15) MHz and 1057.888(15) MHz where the figures in brackets represent uncertainties in the last two digits quoted. These values are based on earlier calculations by Mohr and Erickson, respectively, but take previously neglected nuclear-size effects into account. Recent experimental values are 1057.826(20) MHz, obtained by Andrews and Newton in 1976, and 1057.845(9) MHz, obtained by Lundeen and Pipkin in 1981. Despite the discrepancies between the theoretical values, the agreement between theory and experiment is impressive. The uncertainties, both theoretical and experimental, in the Lamb-shift frequency are now of the order of 10^4 Hz. This should be compared with a frequency of order 10^9 Hz for the Lamb shift itself and a frequency of order 10^{15} Hz for an optical transition.

2. The Anomalous Magnetic Moment of the Electron

The comparison of the measured and calculated values of the anomalous magnetic moment of the electron is regarded as an important test of quantum electrodynamics. The anomalous moment arises from small deviations of the electron's gyromagnetic ratio from 2, which is the value predicted by the Dirac theory. The gyromagnetic ratio g_{e^-} is defined through the relation between the intrinsic magnetic moment $\mathbf{M}$ of the electron and its spin angular momentum $\mathbf{S}$, namely

$$\mathbf{M} = -g_{e^-}\left(\frac{e}{2mc}\right)\mathbf{S}$$

The directly measured quantity is not the gyromagnetic ratio (or g-factor, as it is also called) itself but the electron anomaly a_{e^-}, which is the difference between this ratio and 2, all divided by 2. Thus,

$$a_{e^-} = \frac{g_{e^-} - 2}{2}$$

The electron anomaly is, like the g-factor, a dimensionless constant. Its value is approximately one-tenth of one percent and has been both measured and calculated with great precision. The experimental and theoretical values of a_{e^-} agree to nine places of decimals:

$$a_{e^-} = 0.001159652$$

The value of g_{e^-} is obtained from this by simple arithmetic:

$$g_{e^-} = 2.002319304$$

In experiments carried out at the University of Washington in Seattle, the accuracy of the measurement has been greatly increased. In these experiments, electrons are held in a configuration of static electric and magnetic fields in a cavity with linear dimensions of order 1 cm. The arrangement is known as a Penning trap. Even single electrons can be held in it for weeks at a time. The electrons in a Penning trap have a discrete set of energy levels and are sometimes considered as part of an atom with a nucleus of macroscopic size, namely the experimental apparatus or, indeed, the earth on which it rests; the atom is called geonium.

The most recent measured value of a_{e^-} was reported by Van Dyck, Schwinberg, and Dehmelt in 1985 as

$$a_e^{\text{exp}} = 0.001159652193(4)$$

where again the figure in brackets represents the probable uncertainty in the last digit. The electron anomaly, or the g-factor, is the most accurately known of all physical constants. Its measurement does not depend on a knowledge of either the values of other physical constants or the strength of the magnetic field involved in the experiment.

The theoretical value of a_{e^-} is obtained by considering the scattering of electrons by an external (prescribed) field. Feynman diagrams for the lowest-order contribution to this process and a radiative correction of order α are shown in Fig. 13. The change in the momentum of the scattered electron is supplied by the external field. The lowest-order contribution to the electron anomaly is known exactly (its value $\alpha/2\pi$ was calculated by Schwinger in 1948), as is the contribution of order α^2. Further contributions of order α^3 and α^4 have also been calculated, partly analytically and partly numerically. (The contribution of order α^4 arises from 891 different Feynman diagrams!) A recent theoretical value, due to Kinoshita and Lindquist, is given by

$$a_{e^-}^{\text{th}} = 0.001159652460(127)\,(075)$$

where the first error stems from experimental uncertainties in the value of the fine-structure constant α and the second from computational and other theoretical uncertainties. The slight inconsistency, in the last three digits, between the experimental and theoretical values has not yet been resolved.

The positron anomaly a_{e^+} was measured by Schwinberg, Van Dyck, and Dehmelt in 1981 using the geonium experiment. They concluded

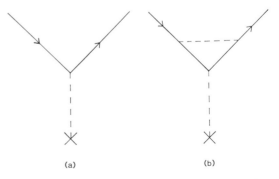

(a)　　　　　　　　(b)

FIG. 13.　Feynman diagrams for electron scattering by an external field (Denoted by X). (a) Zeroth-order contribution yielding a g-factor of 2 or an electron anomaly of 0. (b) Radiative correction of order α yielding a g-factor of $2 + (\alpha/\pi)$ or an electron anomaly of $\alpha/2\pi$.

that any difference between the ratio of the positron g-factor to the electron g-factor and unity must be less than 10^{-10}. This conclusion is strong evidence for the validity of the *CPT* theorem. Any departure of g_{e+}/g_{e-} from 1 would signal a breakdown of the combined charge conjugation, parity and time-reversal transformation as a symmetry of nature.

G. Interaction of Photons and Leptons

Relativistic quantum electrodynamics may readily be extended to include the interaction of photons with certain other charged particles besides electrons and positrons. These are the muon (symbol μ^-) and the tauon (symbol τ^-) and their antiparticles μ^+ and τ^+. Muons and tauons have, within experimental accuracy, the same charge $(-e)$ and spin $(\frac{1}{2})$ as electrons, but different masses. While the rest energy (measured in electron volts) of the electron is about 0.511 MeV, that of the muon is about 105.659 MeV and that of the tauon is (with a possible error of about 3 MeV) about 1784 MeV. The fact that the electron, muon, and tauon seem to have identical characteristics (apart from mass) is known as $e-\mu-\tau$ universality. The muon and the tauon have lifetimes of order 10^{-6} sec and 10^{-13} sec, respectively. Electrons, muons, and tauons are all called leptons (as are neutrinos, which are, however, uncharged); they do not, in contrast to hadrons, experience the strong (nuclear) force. On the other hand, they do participate in the weak and gravitational interactions as well as in the electromagnetic interaction.

An example of a scattering process that involves more than one kind of lepton is muon pair production in electron–positron collisions. The Feynman diagram for the lowest-order contribution to this is shown in Fig. 14. An electron and a positron are annihilated and a virtual photon created; this in turn is annihilated and a muon and an antimuon created. For the process to occur, the electron and positron together must have at least the threshold energy equal to twice the rest energy of the muon (about 211 MeV). It should be noted that in Fig. 14 the lepton number, defined as the number of leptons minus the number of antileptons, is conserved at each vertex for both electrons and muons. This is true generally (and for tauons as well) and arises from the form of the interaction Hamiltonian. Each basic vertex involves only one type of lepton or antilepton. There are no vertices involv-

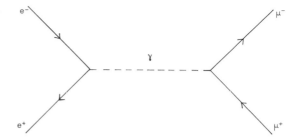

FIG. 14. Muon pair production through electron–positron annihilation. The lepton number (in this case 0) is conserved at each vertex for each kind of lepton separately.

ing, for example, the annihilation of an electron and the simultaneous creation of a muon.

In the center-of-mass reference system (in which the electrons and positrons collide head-on with the same energy E), the threshold energy is reached when the particles have been accelerated to speeds about 0.999988 times the speed of light. For energies much greater than the threshold energy (or speeds even closer to that of light), the unpolarized differential and total cross sections for muon pair production in the center-of-mass system reduce to

$$\frac{d\sigma}{d\Omega} = \frac{\alpha^2}{16E^2}(1 + \cos^2\theta)$$

and

$$\sigma_{\text{tot}} = \frac{\pi\alpha^2}{3E^2}$$

where θ is the angle between the incoming electron and outgoing muon (or between the incoming positron and outgoing antimuon). The second of these formulas has been verified in experiments using total center-of-mass energies of the order of 30 GeV.

The results of the high-energy experiments can be used to set bounds on possible deviations from exact quantum electrodynamics. The existence of heavy photons, for example, would modify the structure of the theory. (Thus, in the static limit, the Coulomb potential would no longer have a simple inverse distance dependence). To good approximation, the cross section for muon pair production would be altered according to

$$\sigma_{\text{tot}} \to \sigma_{\text{tot}}\left(1 \pm \frac{4E^2}{\Lambda_\pm^2}\right)$$

where $\Lambda_\pm$ are cutoff parameters with the dimensions of energy. For consistency with the experimental results $\Lambda_{\pm}$ must be at least of the order of 150 GeV. (Recent results obtained at the Stanford Linear Accelerator Center suggest that $\Lambda_{\pm} > 172$ GeV.) This corresponds to a test of the pointlike nature of photon–lepton interactions down to distances of the order of $1/\Lambda_{\pm}$, that is, less than 10^{-15} cm.

BIBLIOGRAPHY

Barut, A. O. (Ed.) (1984). ''Quantum Electrodynamics and Quantum Optics,'' N.A.T.O. A.S.I. Series Vol. 110. Plenum, New York.

Berestetskii, V. B., Lifshitz, E. M., and Pitaevskii, L. P. (1982). ''Quantum Electrodynamics,'' Course of Theoretical Physics Vol. 4, 2nd ed. Translated from the Russian by J. B. Sykes and J. S. Bell. Pergamon, Oxford.

Craig, D. P., and Thirunamachandran, T. (1984). ''Molecular Quantum Electrodynamics: an Introduction to Radiation–Molecule Interactions.'' Academic Press, Orlando.

Gräff, G., Klempt, E., and Werth, G. (Eds.) (1981). ''Present Status and Aims of Quantum Electrodynamics,'' Lecture Notes in Physics Vol. 143, Springer-Verlag, Berlin and New York.

Healy, W. P. (1982). Non-Relativistic Quantum Electrodynamics.'' Academic Press, New York.

Mandl, F., and Shaw, G. (1984). ''Quantum Field Theory.'' Wiley, New York.

ELEMENTARY PARTICLE PHYSICS

Timothy Barklow and Martin Perl   *Stanford University*

I.   The Nature of Elementary Particles
II. Basic Forces and Gauge Bosons
III. Lepton Family of Elementary Particles
IV. Quark Family of Elementary Particles
V.   Quantitative Description of Particle Interactions
VI. Experimental Techniques
VII. Quark Model of Hadrons and Quantum Chromodynamics
VIII. Electromagnetic and Weak Interactions
IX. Current Issues in Elementary Particle Physics

GLOSSARY

Antiparticle: Partner of a particle, and identical to the particle except that all chargelike properties (electric charge, strangeness, charm, etc.) are opposite. When a particle and its antiparticle meet, these properties cancel in a process called annihilation. The particle and antiparticle then disappear, and other particles are produced.

Asymptotic freedom: Concept that the strong force between quarks gets weaker as the quarks get very close together.

Baryon: Type of hadron. The baryon family includes the proton, neutron, and those other particles whose eventual decay products include the proton. Baryons are composed of 3-quark combinations.

Color: Property of quarks and gluons, analogous to electric charge, which describes how the strong force acts on a quark or gluon.

Electromagnetic force or interaction: Long-range force and interaction associated with the electric and magnetic properties of particles. This force is intermediate in strength between the weak and strong force. The carrier of the electromagnetic force is the photon.

Electroweak force or interaction: Force and interaction that represent the unification of the electromagnetic force and the weak force.

Flavor: General name for the various kinds of quarks, such as up, down, and strange. Also sometimes applied to the various kinds of leptons.

Generation: Classification of the leptons and quarks into families according to a mass progression. The first generation consists of the electron and its neutrino and of the up and down quarks. The second generation consists of the muon and its neutrino and of the charm and strange quarks. The third generation consists of the tau and its neutrino and of the bottom and (expected) top quarks.

Gluon: Massless particle that carries the strong force.

Hadron: Subnuclear, but not elementary, particle composed of quarks. The hadron family of particles consists of baryons and mesons. These particles can interact with each other via the strong force.

Lepton: Member of the family of weakly interacting particles, which includes the electron, muon, tau, and their associated neutrinos and antiparticles. Leptons are acted upon not by the strong force but by the electroweak and gravitational forces.

Luminosity: Measure of the rate at which particles in a collider interact. The larger the luminosity the greater the rate of interaction.

Meson: Any strongly interacting particle that is not a baryon. Mesons are composed of quark–antiquark combinations.

Photon: Massless particle that carries the electromagnetic force.

Quantum chromodynamics (QCD): Theory that describes the strong force among quarks in a manner similar to the description of the electromagnetic force by quantum electrodynamics.

Quantum electrodynamics (QED): Theory that describes the electromagnetic interaction in the framework of quantum mechanics. The quantum of the electromagnetic force is the photon.

Quarks: Family of elementary particles that make up the hadrons. The quarks are acted upon by the strong, electroweak, and gravitational forces. Five quarks are known, called up, down, strange, charm, and bottom. A sixth, called top, is expected to exist.

Standard model: Collection of established experimental knowledge and theories in particle physics which summarizes our present (1986) picture of that field. It includes the three generations of quarks and leptons, the electroweak theory of the weak and electromagnetic forces, and the quantum chromodynamic theory of the strong force.

Strong force or interaction: Short-range force and interaction between quarks, which is carried by the gluon. The strong force also dominates the behavior of interacting mesons and baryons and accounts for the strong binding among nucleons.

W: Charged particle that carries the weak force, also called an intermediate vector boson. Its mass is about 90 times the proton mass.

Weak force or interaction: Force and interaction that is much weaker than the strong force but stronger than gravity. It causes the decay of many particles and nuclei. It is carried by the W and Z particles.

Z: Neutral particle that carries the weak force, also called an intermediate vector boson. It is slightly heavier than the W particle, with a mass about 100 times the proton mass.

This article has two parts. The first part provides a general introduction to elementary particle physics: the nature of elementary particles; the basic forces and the gauge bosons which carry the forces; the lepton family of particles; the quark family of particles; the interaction of particles in collisions and decays; and experimental techniques. The second part, beginning with Section VII, summarizes on a more detailed and technical level: the quark model of hadrons and quantum chromodynamics, the electromagnetic and weak interactions, and current issues in elementary particle physics.

An elementary particle is the simplest and most basic form of matter; it is much smaller than atoms or nuclei. Three kinds of elementary particles are known: leptons, quarks, and force-carrying particles also called gauge bosons. The best known example of an elementary particle is the electron, which is a lepton. The best known example of a force-carrying particle or gauge boson is the photon, which carries the electromagnetic field.

I. The Nature of Elementary Particles

A. DEFINITION OF AN ELEMENTARY PARTICLE

A piece of matter is called an elementary particle when it has no other kinds of particles inside of it and no parts that can be identified. How does one know whether a particle is elementary? Only by experimenting with it to see if it can be broken up or by studying it to determine if it has an internal structure or parts. This is illustrated in Fig. 1. Molecules are not elemen-

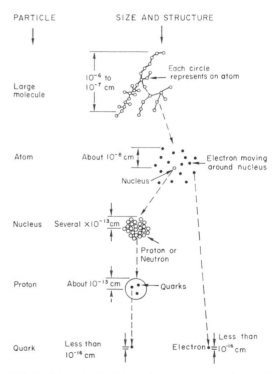

FIG. 1.   Many basic objects in nature are made up of yet simpler objects. For example, molecules are made up of atoms, and atoms are made up of electrons moving around a nucleus. To the best of our present knowledge, the elementary particles, electrons and quarks, are not made up of simpler particles.

tary because they can be broken into atoms by chemical reactions, heating, or other means. Nor are atoms elementary: They can be broken into electrons and nuclei by bombarding the atom with other atoms or with light rays. Nor is the nucleus elementary: If we bombard nuclei with high-energy particles or high-energy light rays called gamma rays, the nucleus can be broken into protons and neutrons. For about fifty years physicists considered the neutron and proton to be elementary, but in the last two decades it was found that they are made of simpler particles called quarks. Hence the neutron and proton are not elementary particles, but to the best of our knowledge the quarks are elementary. With respect to the other constituent of ordinary matter, the electron, no experiment has succeeded in breaking it up or in finding an internal structure for it. Hence the electron is an elementary particle.

B. Size of Elementary Particles

As one proceeds down through the sequence of molecule, atom, nucleus, proton and neutron, and finally quark and electron, the sizes of the particles get smaller. The size of atoms is of the order of 10^{-8} cm (0.00000001 cm). This one hundred millionth of a centimeter is very small by everyday standards. Molecules are larger, their size depending in a rough way on the number of atoms in the molecule. Molecules containing hundreds of atoms, such as organic molecules, can be examined by electron microscopy and thus can almost be seen in the ordinary sense of the word.

But once one goes below the atomic level to nuclei, there is no way to "look" at these particles with any sort of microscope. The nuclei consist of neutrons and protons packed closely together. The proton and neutron are both about 10^{-13} cm, or about 1/100,000 of the size of an atom. Nuclei are a few times larger than a neutron or proton, depending upon how many of these particles they contain; but they are still not much larger than 10^{-13} cm. Nuclei, neutrons, and protons are so small they must be measured by indirect methods.

When we come to an elementary particle such as a quark or an electron, there is a yet smaller scale. By indirect means the sizes of quarks and electrons are known to be less than 10^{-16} cm, less than 1/1000 of the size of neutron or proton. Indeed, there is no evidence that these particles have any size at all, they may be thought of as

points of matter occupying no space. Thus elementary particle physics is also the physics of the smallest objects in nature (Fig. 2). [See ATOMIC PHYSICS; NUCLEAR PHYSICS.]

C. High-Energy and Elementary Particles

Elementary particle physics, the physics of the very small, is also called high-energy physics. The term high-energy refers to the energies of the particles used to produce particle reactions; high-energy means that the kinetic energy (energy of motion) of a particle is much higher than its rest mass energy. To experiment with elementary particles high-energy is needed for two reasons. First, kinetic energy can be converted into mass, and mass can be converted into kinetic energy according to the Einstein equation

$$E = Mc^2 \qquad (1)$$

Here E is the kinetic energy, which can be converted into mass M, and c the velocity of light. Therefore, to produce heavier new particles, larger amounts of energy E are needed.

The second reason for needing high-energy particles is that one investigates the size and structure of a particle by bombarding it with other particles. And the deeper one wishes to penetrate into a particle, the higher must be the energy of the bombarding particles.

The Heisenberg uncertainty principle also leads to the conclusion that the investigation of small distances requires high-energies. To measure small distances very precisely, there must be a large uncertainty in the momentum associated with that measurement. A large uncertainty in momentum can only be accommodated by a large initial momentum, which requires high energies.

The principal way in which we give high energy to a particle is to accelerate it by the force an electric field exerts on the particle's charge, in a large device called an accelerator (Section VI). A convenient unit for measuring both energy and mass is the electron-volt (eV). This is the energy acquired by an electron or proton passing through an electric potential with a total voltage of 1 V. Larger units are

MeV = 10^{+6} eV = 1 million electron-volts

GeV = 10^{+9} eV = 1 billion electron-volts

TeV = 10^{+12} eV = 1 trillion electron-volts

The significance of these energy units can be appreciated by looking at some particle masses

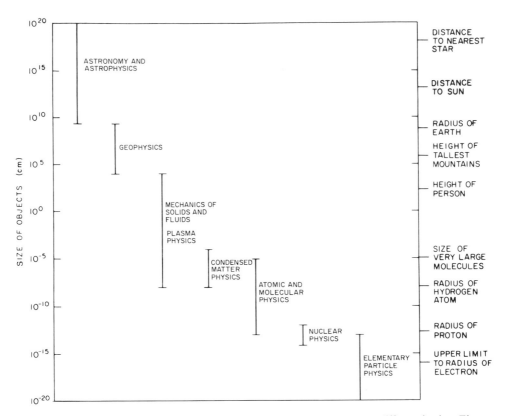

FIG. 2.   The different subfields of physics study parts of nature that are very different in size. Elementary particle physics studies the smallest objects in nature, objects that are smaller than 10^{-13} cm.

expressed in electron-volts:

1.   The electron mass is about 0.5 MeV.
2.   The proton mass is about 1 GeV.
3.   The heaviest known particle, the Z^0, has a mass of about 100 GeV = 0.1 TeV. The highest effective particle energies produced by existing accelerators are in the range of several hundred GeV.

In this article, to simplify the units, we express mass in terms of the equivalent from Eq. 1, rather than use the conventional unit eV/c^2.

D. Properties and Types of Elementary Particles

Each kind of elementary particle is distinguished by the intrinsic properties of the particle, mass and spin, and by how the particle connects with the basic forces.

Table I gives the masses and spins of the well-established elementary particles. The masses have a vast range of values: The photon's mass is 0, and the Z^0's mass is about 93 GeV. More massive elementary particles may exist; there have not yet been comprehensive searches for these more massive particles because of the energy limitations of existing accelerators.

Some particles have a permanent angular momentum called spin angular momentum. In classical physics, one can think of spin angular momentum as being due to a particle permanently rotating or spinning about an axis through its center. However this picture cannot be used quantitatively in elementary particle physics because such particles may have zero size. The spin angular momentum can be expressed as $sh/2\pi$, where h is Planck's constant ($h = 6.63 \times 10^{-27}$ erg-sec) and s the spin of the particle. General quantum mechanical principles limit s to the values 0, 1/2, 1, 3/2, 2, ... (Particles having half-integer spin 1/2, 3/2, ... are called fermions; those with integer spin 0, 1, ... are bosons.) The known elementary particles have spin 1/2 or 1.

Besides its mass and spin, each kind of particle is distinguished by how it connects with the four basic forces: gravitation, electromagnetism, strong force, and weak force. (These

TABLE I.   Properties of the Established Elementary Particles

Family	Name	Symbol	Mass	Spin	Electric charge[b]	Antiparticle symbol
Leptons	Electron	e^- or e	0.511 MeV	1/2	−1	e^+ or $\bar{e}$
	Muon	μ^- or μ	105.7 MeV	1/2	−1	μ^+ or $\bar{\mu}$
	Tau	τ^- or τ	1784. ± 3. MeV	1/2	−1	τ^+ or $\bar{\tau}$
	Electron neutrino	ν_e	<0.000046 MeV	1/2	0	$\bar{\nu}_e$
	Muon neutrino	ν_μ	<0.25 MeV	1/2	0	$\bar{\nu}_\mu$
	Tau neutrino	ν_τ	<70. MeV	1/2	0	$\bar{\nu}_\tau$
Quarks[a]	Up	u	~350 MeV	1/2	+2/3	$\bar{u}$
	Down	d	~350 MeV	1/2	−1/3	$\bar{d}$
	Strange	s	~500 MeV	1/2	−1/3	$\bar{s}$
	Charm	c	~1500 MeV	1/2	+2/3	$\bar{c}$
	Bottom or beauty	b	~5000 MeV	1/2	−1/3	$\bar{b}$
Gauge bosons	Photon	γ	0.0	1	0	γ
	W^+ boson	W^+	81.8 ± 1.5 GeV	1	+1	W^-
	W^- boson	W^-	81.8 ± 1.5 GeV	1	−1	W^+
	Z^0 boson	Z^0	92.6 ± 1.7 GeV	1	0	Z^0
	Gluon	g	Assumed 0	1	0	g

[a] The constituent quark masses are calculated from hadron masses.
[b] The electric charge is expressed in units of the magnitude of the electron's charge, 1.602×10^{-19} C.

forces are described in the next section.) For example, particles can interact with the electromagnetic force through their electric charge. Table I gives the electric charge in units of the magnitude of the charge of the electron $q = 1.6 \times 10^{-19}$ C. The photon has zero electric charge, but it carries the electromagnetic field.

All the established elementary particles fall into three classes according to how they connect with the basic forces:

1. leptons, which do not connect with the strong force but connect with the three other forces,

2. quarks, which connect with all four forces, and

3. force-carrying particles, which carry the forces.

Other properties of a particle such as its lifetime are determined by its mass, its spin, and how it connects with the forces. Most nonzero-mass particles are unstable: They spontaneously decay according to the experimental law,

$$P = e^{-t/\tau}$$

Here P is the probability that a particle still exists a time t after it was formed, and τ the particle lifetime.

Roman or Greek letters are used as symbols for particles, as shown in Table I.

E. Particles and Antiparticles

Consider first the charged particles. The electron, which has negative charge, has a related particle, called the antielectron or positron, that has positive charge. This same relation applies to the quarks. For example, the bottom quark has a charge of −1/3; the bottom antiquark has the same mass but has a charge of +1/3. Hadrons also have their antiparticles. The most famous example is the antiproton, the antiparticle of the proton. Relativistic quantum theory predicts that every charged particle must have an antiparticle. An antiparticle is indicated by a bar over the particle symbol. Thus

e = electron, $\bar{e}$ = positron

b = bottom quark, $\bar{b}$ = bottom antiquark

p = proton, $\bar{p}$ = antiproton

Sometimes e^- is used for the electron, e^+ is used for the positron, and so forth.

With respect to neutral particles, theory and experiment show that a particle can have a different neutral antiparticle or it can be its own antiparticle. For example, each neutrino has an antiparticle that is different from itself, but the photon is its own antiparticle. The operation that transforms a particle into its antiparticle and vice-versa is called charge conjugation and is denoted by C. Thus, $Ce = \bar{e}$, $Cp = \bar{p}$, and so on.

F. Hadrons

Hadrons are subnuclear particles, but they are not elementary particles. To the best of our knowledge, hadrons (Section VII) are made either of three quarks or of one quark and one antiquark bound together by the strong force. Table II lists a few of the known hadrons. The first hadrons to be discovered were the proton and neutron; now more than a hundred types are known.

Although hadrons are not elementary particles, they are nevertheless very important in elementary particle physics research. First, we do not know how to isolate quarks, so to do experiments on quarks we must use the quarks in hadrons. Second, hadrons are fascinating forms of matter, and it is interesting to study them in their own right.

G. Reactions of Elementary Particles and Hadrons

Reactions involving elementary particles or hadrons are represented in the same fashion as chemical reactions. For example,

$$e^- + p \rightarrow \nu_e + n$$

means that an electron e^- interacts with a proton p to form an electron neutrino ν_e and a neutron n. In

$$e^- + p \rightarrow e^- + \pi^+ + \pi^- + p$$

an electron and proton interact to form four particles. When the two particles in the final state are the same as those in the initial state,

$$e^- + p \rightarrow e^- + p$$

and

$$e^- + \mu^- \rightarrow e^- + \mu^-$$

for example, it is called an elastic reaction or elastic scattering. All other reactions are called inelastic.

The decay of a single particle is similarly represented:

$$\mu^- \rightarrow \nu_\mu + e^- + \bar{\nu}_e$$

shows how the muon decays, and

$$\pi^- \rightarrow \mu^- + \bar{\nu}_\mu$$

is the most frequent way the π^- hadron decays.

Reaction and decay processes occur through the various basic forces discussed in Section II. The reaction equation does not directly indicate which forces are taking part in the reaction.

H. Conservation Laws

All reactions and decays of elementary particles and hadrons are described by a set of laws called conservation laws. Here the term conservation means that some quantity remains unchanged (that is, is conserved) in a reaction. The simplest example of a conserved quantity is the total energy.

In some reaction, say,

$$a + b \rightarrow c + d + e + f$$

the initial total energy is the sum of the kinetic and the rest energies of a and b. (The rest energy of a particle is mc^2, where m is the particle mass.) The final total energy is the sum of the kinetic energies and rest energies of particles c through f. The law of conservation of energy states that the final total energy is equal to the initial total energy. If particles c through f have a total mass greater than the masses of particles a plus b, then some initial kinetic energy is converted into mass in the reaction.

In some decay, say,

$$a \rightarrow b + c + d$$

some of the mass of a goes to produce the masses of the particles b through d, and the rest gives kinetic energy to the particles.

Two other conserved quantities are the total linear momentum and the total angular momentum. The conservation of total angular momentum leads to the rule that the sum of the spins of

TABLE II.   Properties of Some Hadrons

Name	Symbol	Mass (GeV)	Charge[a]	Quarks in the hadrons
Proton	p	0.938	+1	uud
Antiproton	p̄	0.938	−1	ūūd̄
Neutron	n	0.940	0	udd
Antineutron	n̄	0.940	0	ūd̄d̄
Positive pion	π^+	0.140	+1	ud̄
Negative pion	π^-	0.140	−1	ūd
Neutral pion	π^0	0.135	0	uū, dd̄
Positive kaon	K$^+$	0.494	+1	us̄
Negative kaon	K$^-$	0.494	−1	ūs
Neutral kaon	K^0	0.498	0	ds̄, d̄s
D$^+$ charm meson	D$^+$	1.869	+1	cd̄
D$^-$ charm meson	D$^-$	1.869	−1	c̄d
D^0 charm meson	D^0	1.865	0	cū, c̄u
Psi or J	ψ/J	3.097	0	cc̄
Upsilon	Υ	9.460	0	bb̄

[a] Expressed in units of the magnitude of the electron's charge, 1.602×10^{-19} C.

the final particles must be integer or half integer according to whether the sum of the spins of the initial particles is integer or half integer.

The total electric charge is also conserved. For example

$$\pi^- + p \rightarrow \pi^- + \pi^+ + p$$

does not occur. Other conservation laws, such as lepton conservation, are discussed later in this article.

II. Basic Forces and Gauge Bosons

A. THE FOUR BASIC FORCES

Elementary particles interact with each other through four basic forces (Table III). Two of these have been known for hundreds of years: the forces of electromagnetism and gravitation. The other two forces were discovered in the twentieth century: the strong or nuclear force that holds the atomic nucleus together and the weak force that operates in many forms of radio-activity. The forces are described and distinguished from each other by their strengths, by the distance or range over which they act, and by the precise manner in which they act on particles. Here their strength and range properties are described; the precise manner in which the forces act is discussed later in the sections devoted to each force.

The most powerful of the four forces is the strong force (Section VII). It is not felt directly in everyday phenomena because its range does not extend beyond a distance of about 10^{-13} cm from an elementary particle. The strong force is so powerful that it holds the quarks together, forming hadrons. The proton and neutron are the best known examples. The range of the strong force determines the size of the protons and neutrons, whose radii are about 10^{-13} cm. Protons and neutrons in turn are held together by residual effects of the strong force, making nuclei. Thus the strong force is sometimes called the nuclear force, but it is more useful to think of the nuclear force as being a manifestation of the strong force. Since atoms and molecules are 10^{-8} cm or larger in size and thus are not within the range of the strong force, and since electrons are not affected by the strong force, there is no direct effect of the strong force at the level of atomic or molecular physics. There are some small, indirect effects due to the sizes and structures of nuclei. The present theory of the strong force is called quantum chromodynamics.

The electromagnetic force (Section VIII) between elementary particles follows the same laws as it does when it is used in motors, generators, and electronic equipment. The elementary particles behave as very small particles with electric charge. If the particle has nonzero spin, it also behaves as a very small magnet in relation to the electromagnetic force. (The strength of the equivalent magnetic properties of a particle is expressed by its magnetic moment.) Unlike the strong force, the electromagnetic force is effective to very large distances. The strength of the electromagnetic coulomb force F_{coul} exerted by a particle of charge Q weakens as the distance from the particle increases.

$$F_{coul} = \frac{Q^2}{R^2}$$

But there is no sharp cutoff as in the strong force. Hence the range of the electromagnetic force is said to be infinite. The electromagnetic

TABLE III.   The Four Basic Forces

Type of force	Strong or nuclear	Electromagnetic	Weak	Gravitational
Behavior over distance	Limited to less than about 10^{-13} cm	Extends to very large distances	Limited to less than about 10^{-16} cm	Extends to very large distances
Strength relative to strong force at a distance of 10^{-13} cm	1	10^{-2}	10^{-13}	10^{-38}
Particle that carries the force	Gluon, identified indirectly, but not (and perhaps cannot be) isolated	Photon	W^+, W^-, and Z^0 intermediate bosons	Not discovered, but called graviton

force determines the structures and behaviors of atoms and molecules. The theory of the electromagnetic force is called quantum electrodynamics.

The weak force (Section VIII) acts over distances of less than about 10^{-16} cm and is much less powerful than the strong force. Yet the weak force is not negligible. In a certain sense it is more pervasive than the strong force. Some elementary particles, such as leptons and the $W^{\pm}$ and Z^0, are not affected by the strong force but are affected by the weak force. The radioactive decay of the neutron and of nuclei, as well as the decays of many of the elementary particles, occur through the weak force. Since the range of the weak force is so small, the force is not seen directly in atomic or human-scale phenomena.

The gravitational force is important in human-scale and astronomical phenomena because of the immense mass of the earth, planets, and stars; but the effect of the gravitational force exerted by one elementary particle is very small compared with that of the three other forces exerted by that particle. Experimental methods in particle physics are not at present sufficiently sensitive to detect the gravitational force of one elementary particle.

The strengths of the electromagnetic, strong, and weak forces are described by a quantity called the coupling constant, defined later.

Since the 1920s, physicists have speculated about the possibility that the theories of the various forces can be unified into one general theory. The question is, are the four seemingly different forces simply different manifestations of one force? First thoughts were about unifying the gravitational and electromagnetic forces; however, that has not been done, and we do not know if it can be done. But within the last 15 years, a significant unification of the electromagnetic and weak forces has been made (Section VIII) and has been experimentally verified. [See UNIFIED FIELD THEORIES.]

B. Force-Carrying Particles and Gauge Bosons

It is a basic principle of quantum mechanics that a force has a dual nature: It can be transmitted by means of either a wave or a particle. The clearest example is the electromagnetic force, which can be treated in some situations as being carried by an electromagnetic wave (radio waves, light waves, etc.) and in other situations

as being carried by a particle (the photon). The question then arises whether the other forces also obey quantum mechanics in this sense and thus can be thought of as being carried by particles. Table III summarizes our present knowledge.

As already noted, the electromagnetic force is carried by the photon. Photons are the particles in a ray of visible light or in an x-ray beam. At high energies they are sometimes called gamma rays, hence the usual symbol for a photon is the Greek letter gamma (γ). The photon has zero mass (or close to zero mass), it has spin 1, it has zero electric charge, and it is stable. Photons comprising an electromagnetic wave of frequency f Hz have an energy per photon of

$$E_\gamma = hf$$

where h is Planck's constant. The photon does not interact in any direct way with the weak or strong force. It does interact with the gravitational force according to general relativity, the force being proportional to E_γ.

The weak force is carried by three different, recently discovered, particles called the W^+, W^-, and Z^0. They all have masses close to 90 GeV (Table I), and they have spin 1. The W^+ and W^- each have one unit of positive or negative electric charge, hence they interact with the electromagnetic force; the Z^0, being electrically neutral, does not interact electrically. The W^+, W^-, and Z^0 interact with the gravitational force but not with the strong force.

Experiment and theory indicate that the strong force is carried by particles called gluons (Section VII) and symbolized by g. However, unlike the photon, $W^{\pm}$, and Z^0, the gluon has not been isolated experimentally and directly studied. Indeed, most forms of the current theory of the strong force, (quantum chromodynamics) state that gluons cannot be isolated. Gluons, like quarks, are said to be confined within hadrons (a subject discussed later). Indirect experimental evidence and current theory give the gluon a zero mass, spin 1, and zero electric charge. It does not interact with the electromagnetic or weak forces. Inside a hadron, the gluon carries some part of the hadron's energy E_g, and the gravitational force interacts with the gluon in proportion to E_g.

The particle conjectured to carry the gravitational force has been called the graviton and ascribed a spin of 2; but such a particle has not yet been discovered, and there is no experimental evidence for its existence. Because of the feebleness of the gravitational interaction among elementary particles, its detection would be extraordinarily difficult. Furthermore, there is no successful application of quantum theory to general relativity at present. Therefore the nature of the particle carrying the gravitational force (if there is such a particle) is an open question.

The mathematical theories that describe the strong, electromagnetic, and weak forces obey a general principle called a gauge symmetry, hence they are called gauge theories (Section VII). The gluon, photon, $W^{\pm}$, and Z^0 particles are intrinsic to those theories and are called gauge bosons, the boson term indicating that their spins are integers.

III. Lepton Family of Elementary Particles

A. DEFINITION OF A LEPTON

The lepton family of elementary particles is defined by two properties:

1. Leptons are affected by the gravitational, electromagnetic, and weak forces, but not by the strong force.

2. Leptons cannot be arbitrarily created or destroyed; all reactions involving leptons follow a principle called lepton generation conservation, or lepton conservation.

Table IV shows the six known leptons. The tau neutrino ν_τ has not been detected, but there is a great deal of indirect evidence for its existence and properties. The leptons come in pairs formed according to the lepton conservation principle, each pair consisting of one charged

lepton and one neutral lepton called a neutrino. Each pair is called a generation, and in each generation the mass of the neutrino is much less than the mass of the charged lepton. The generation pairs are formed according to the lepton conservation principle. The usual representation for the lepton pairs is

$$\begin{pmatrix} \nu_e \\ e^- \end{pmatrix} , \quad \begin{pmatrix} \nu_\mu \\ \mu^- \end{pmatrix} , \quad \begin{pmatrix} \nu_\tau \\ \tau^- \end{pmatrix}$$

Each pair has an antiparticle pair, respectively,

$$\begin{pmatrix} \bar{\nu}_e \\ e^+ \end{pmatrix} , \quad \begin{pmatrix} \bar{\nu}_\mu \\ \mu^+ \end{pmatrix} , \quad \begin{pmatrix} \bar{\nu}_\tau \\ \tau^+ \end{pmatrix}$$

B. LEPTON CONSERVATION

The charged lepton and neutrino in each pair show a unique property, called lepton number and given the symbols n_e, n_μ, n_τ. For example the τ^- and ν_τ have

$$n_e = 0, \qquad n_\mu = 0, \qquad n_\tau = 1$$

Their antiparticles τ^+ and $\bar{\nu}_\tau$ have

$$n_e = 0, \qquad n_\mu = 0, \qquad n_\tau = -1$$

Table V gives the complete scheme. All quarks and hadrons have n equal to zero.

All experiments so far show that a reaction involving leptons can occur only if the separate sums of n_e, n_μ, and n_τ do not change in a reaction. For example, the reaction

$$e^- + p \rightarrow \nu_e + p$$

occurs since $\Sigma n_e = 1$, $\Sigma n_\mu = 0$, and $\Sigma n_\tau = 0$ before and after the reaction. But the reactions

$$e^- + p \rightarrow \mu^- + p \quad \text{or} \quad e^- + p \rightarrow \nu_\mu + n$$

TABLE IV.   Properties of the Known Leptons

Lepton generation	Lepton	Symbol	Mass (MeV)	Charge	Lifetime (sec)	Major decay modes	
1	Electron	e^-	0.511	-1	Stable[a]	None	
1	Electron neutrino	ν_e	<0.00003	0	Stable[a]	None	
2	Muon	μ^-	105.7	-1	2.197×10^{-6}	$\nu_\mu + e^- + \bar{\nu}_e$	
2	Muon neutrino	ν_μ	<0.25	0	Stable[a]	None	
3	Tau	τ^-	$1784. \pm 3$	-1	3.4×10^{-13}		
						$\nu_\tau + e^- + \bar{\nu}_e$	16.5%
						$\nu_\tau + \mu^- + \bar{\nu}_\mu$	18.5%
						$\nu_\tau + \pi^-$	10.3%
						$\nu_\tau + \rho^-$	22.1%
						$\nu_\tau + \rho^0 + \pi^-$	5.4%
3	Tau neutrino	ν_τ	<70	0	Stable[a]	None	

[a] Stability based partly on measurement and partly on theoretical expectations.

TABLE V. Lepton Conservation Numbers

Particle	n_e	n_μ	n_τ
e^-, ν_e	+1	0	0
e^+, $\bar{\nu}_e$	−1	0	0
μ^-, ν_μ	0	+1	0
μ^+, $\bar{\nu}_\mu$	0	−1	0
τ^-, ν_τ	0	0	+1
τ^+, $\bar{\nu}_\tau$	0	0	−1
All other particles and all hadrons	0	0	0

do not occur since Σn_e and Σn_μ change during the reaction. In reactions with several leptons,

$$e^- + e^+ \rightarrow \mu^- + \mu^+ \quad \text{and} \quad e^- + e^+ \rightarrow \tau^+ + \tau^-$$

occurs since $\Sigma n_e = 0$, $\Sigma n_\mu = 0$, and $\Sigma n_\tau = 0$ before and after. But

$$e^- + e^+ \rightarrow \mu^- + e^+$$

does not occur.

The lepton conservation law holds for the three forces with which the leptons interact: the electromagnetic, the weak, and the gravitational. Present theory does not explain the nature of the property in each lepton pair that gives it a unique lepton number, nor does present theory explain why the sums of the lepton numbers cannot change in a reaction. In the last few years there has been speculation, but as yet no evidence, that the proton might very rarely decay to a lepton plus hadrons. If that turns out to be true, lepton conservation would not hold in the process.

C. Muon and Tau Lepton Decays

The muon and the tau lepton are unstable, but their decay process are limited by lepton conservation. Indeed, studies of these decay processes give one of the proofs of the lepton conservation

law. The muon has only one decay mode, for the μ^-,

$$\mu^- \rightarrow \nu_\mu + e^- + \bar{\nu}_e$$

and for the μ^+,

$$\mu^+ \rightarrow \bar{\nu}_\mu + e^+ + \nu_e$$

Since the second process involves only the antiparticles of the first process, there is no need to write the second decay.

The tau lepton has much more mass than the muon, hence it can decay in more ways. Table IV lists the main decay modes. Note that the τ^- always decays to a ν_τ plus other particles, an example of tau lepton conservation.

IV. Quark Family of Elementary Particles

A. Definition of a Quark

The quark family of elementary particles (Table VI) is defined by two properties:

1. Quarks are affected by all four basic forces. Because they are affected by the strong force (Section VII), quarks act very differently from leptons in many situations. In particular, it is either impossible or very difficult to isolate quarks, whereas leptons can easily be isolated.

2. Quarks obey a conservation law similar to, but more complicated than, the lepton conservation law.

A very peculiar property of the quarks is that they have electric charges of 2/3 or 1/3 of the unit of electric charge carried by the electron. All other particles, elementary or not, have either zero or integral charges. This fractional charge property of the quarks is an intrinsic part of the present standard theory of how hadrons are composed of quarks. For example, the proton with charge +1 is composed of 2 up quarks

TABLE VI. Properties of the Established Quarks

Quark generation	Quark	Symbol	Mass (GeV)	Charge	Known decay modes of heavier quarks
1	Up	u	~350	$+\frac{2}{3}$	
1	Down	d	~350	$-\frac{1}{3}$	
2	Charm	c	~1500	$+\frac{2}{3}$	$c \rightarrow s$ + other particles
					$c \rightarrow d$ + other particles
2	Strange	s	~500	$-\frac{1}{3}$	$s \rightarrow u$ + other particles
3	Bottom or beauty	b	~5000	$-\frac{1}{3}$	$b \rightarrow c$ + other particles

with charge $+2/3$ each and of one down quark with charge $-1/3$ (Table II). The total quark charge is

$$\frac{2}{3} + \frac{2}{3} - \frac{1}{3} = +1$$

On the other hand, the neutron is composed of one up and two down quarks, giving it a total charge

$$\frac{2}{3} - \frac{1}{3} - \frac{1}{3} = 0$$

The four smallest mass quarks are arranged in generation pairs

$$\begin{pmatrix} u \\ d \end{pmatrix} , \begin{pmatrix} c \\ s \end{pmatrix}$$

Each pair has a $+2/3$ charge quark and a $-1/3$ charge quark.

The existence of five quarks has been well established. Most particle physicists believe that there is a sixth quark with a $+2/3$ charge, called the t or top quark, which will complete the third generation pair:

$$\begin{pmatrix} t \\ b \end{pmatrix}$$

Like the leptons, each quark has an antiquark of opposite electric charge.

There is an important unanswered question concerning quarks. Can one quark be isolated from all other matter so that it exists by itself as a free particle? We know from experiment that all the leptons can exist as free particles. But can the quarks be free? At present most physicists believe that quarks are always confined inside the more complicated hadrons. This belief is based on the failure of almost all experiments that have tried to make or find free quarks. We say *almost all* because one series of experiments has indicated that free quarks might exist. In the end, this is a question that can only be resolved with more experiments. In the meantime, most versions of the theory of how quarks interact with the strong force assume that the strong force always confines the quarks inside hadrons. The same theories also assure that the gluons, which carry the strong force, are also confined to make hadrons.

B. Quark Conservation Laws

Quark conservation laws are more complicated than the lepton conservation law for two reasons: (a) There is an additional force, the strong force, and (b) quark types are almost but not exactly conserved when they interact with the weak force.

Under the strong, electromagnetic, or gravitational force, each quark type (u, d, c, s, b, t) can be created or destroyed only when its antiparticle is created or destroyed. For example, through the electromagnetic force,

$$e^+ + e^- \rightarrow u + \bar{u}$$

can occur but

$$e^+ + e^- \rightarrow u + \bar{c}$$

cannot. The strong force process of proton–antiproton annihilation can occur because the u and d quarks in the proton can combine with and annihilate the $\bar{u}$ and $\bar{d}$ quarks, respectively, in the antiproton.

Under the weak force, each generation pair is mostly conserved within its own generation. For example, the u and d quarks, when interacting through the weak force, mostly interact with each other. The major decay modes of the c quark through the weak force are

$$c \rightarrow s + u + \bar{d}$$
$$c \rightarrow s + l^+ + \nu_l$$

Here l^+ means e^+, μ^+, or τ^+. But about 5% of the time the c decays to the d quark, which is in a different generation. Then

$$c \rightarrow d + u + \bar{d}$$
$$c \rightarrow d + l^+ + \nu_l$$

This is discussed quantitatively in Section VIII.

V. Quantitative Description of Particle Interactions

A. Reactions and Feynman Diagrams

Elementary particles and hadrons are too small to be directly studied. We study them indirectly by colliding two particles together and determining what particles come out of the reactions. Each time there is a collision, a number of different reactions can happen. This is illustrated in Fig. 3, a time sequence of two protons colliding. In (a) the protons are about to collide, and in (b) they have just collided to form a complicated concentration of mass and energy. This concentration of mass and energy is unstable and can change again into particles in many different ways. Thus in (c) two protons may come out again, or a large number of hadrons may

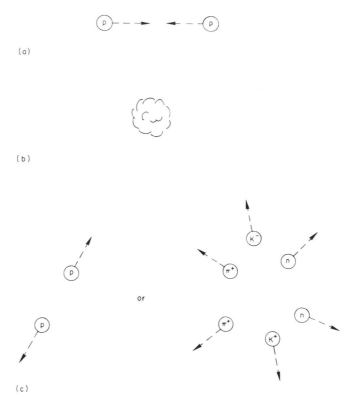

(a)

(b)

(c)

FIG. 3. In (a) two protons are about to collide head-on. When they collide in (b), their mass and energy are concentrated in a small region of space. That concentration of mass and energy is unstable and very quickly breaks up into new particles as in (c). Sometimes just two particles come out of the collision. But at high energies many particles usually come out of the collisions, and none of them need be the original protons.

come out of the collision, or other kinds of particles not shown here may be produced. There are many possibilities. By studying the different reactions and the different particles produced, we can explore the behavior of the basic forces and elementary particles and can search for new particles or new basic forces.

It is clumsy to draw time sequences such as Fig. 3. Instead, we use a kind of shorthand in which the collision process is pictured in a single diagram. Thus Fig. 4 shows the collision diagram for two protons going into two protons and also for two protons going into many hadrons, the same two processes shown in Fig. 3. (This article uses the convention in which time advances from left to right.)

The concentration of mass and energy in Figs. 3 and 4 represents the crux of how particles interact through the basic forces. Often we know enough about that concentration region to ex-

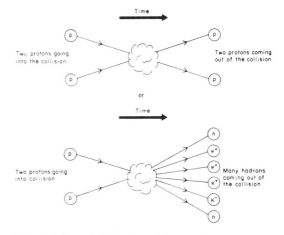

FIG. 4. The collisions of particles can be represented succinctly in a diagram in which time advances from left to right. For example, the lower figure shows two protons going into a collision producing a concentration of mass and energy, which then breaks up into six particles.

plain it in simple terms. For example, the reaction

$$e^+ + e^- \rightarrow \mu^+ + \mu^-$$

occurs as diagrammed in Fig. 5(b). These pictures are of great assistance in making calculations, and in this context they are called Feynman diagrams after their inventor. As time advances (i.e., moving to the right in the figure), the electron and positron collide; then the collision annihilates the electron and positron and produces a highly excited photon, which contains all of the collision energy. This photon is extremely unstable and quickly decays into a muon and an antimuon. Alternatively, the photon could produce a quark–antiquark pair or another electron–positron pair.

Figure 6(a) shows another example, the reaction

$$e^- + p \rightarrow \nu_e + n$$

which occurs through the weak force and the exchange of a W^- particle. Decays can also be represented by Feynman diagrams; Fig. 6(b) shows the decay

$$\mu^- \rightarrow \nu_\mu + e^- + \bar{\nu}_e$$

which occurs through the weak force.

In calculations, the mathematical expression corresponding to the exchanged force-carrying

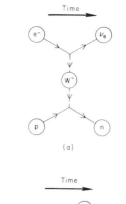

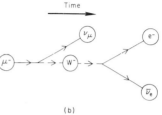

FIG. 6.   Feynman diagrams for (a) $e^- + p \rightarrow \nu_e + n$ and (b) $\mu^- \rightarrow \nu_\mu + e^- + \bar{\nu}_e$.

particle is called the propagator. In the simplest situations the propagator is given by

$$\frac{1}{q^2 - M^2}$$

Here M is the mass of the force-carrying practice and

$$q^2 = E^2 - p^2$$

where E and p are, respectively, the energy and momentum transferred by the force-carrying particle.

B. COUPLING CONSTANTS

In Figs. 5 and 6 the diagrams have parts called vertices in which two leptons or two quarks or two hadrons join with one force-carrying particle such as a photon or a W. In the quantum mechanical picture these vertices represent the emission or absorption of a force-carrying particle by a lepton or quark. The strength with which the force interacts with the leptons or quarks is represented by a parameter called the coupling constant. The coupling constant provides a measure of the strength of the force, but it is not the complete measure of that strength. The strength may also depend on the energy of the particle involved in the vertex (discussed later). The strong force has the largest coupling constant by far.

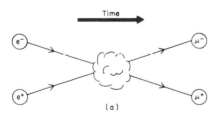

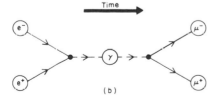

FIG. 5.   The collision of an electron and a positron can lead to the production of a positive muon and a negative muon. The electron and positron actually disappear; the technical term is that they annihilate each other. A sketch of how that interaction occurs is shown in (a). A more detailed description of the interaction is given in (b).

C. Reaction Cross Section

In some reaction, say,

$$a + b \rightarrow c + d + e$$

there are many properties of the final particles that can be studied. For example, the angles at which the particles emerge from the reaction and how they divide the available kinetic energy. One reaction property is of particular importance: The cross section σ measures the probability that particles $c + d + e$ are produced when a and b collide. The units of σ are area, such as cm^2. This can be understood through a classical picture: The larger the area occupied by a and b the greater the probability of collision and hence reaction.

The strong force has the largest cross sections:

$$\sigma_{strong} \lesssim 10^{-24}\ cm^2$$

($\lesssim$ means approximately equal to or less than.) The weak force has the smallest cross sections:

$$\sigma_{weak} \lesssim 10^{-37}\ cm^2$$

The electromagnetic force has in-between cross sections:

$$\sigma_{electromagnetic} \lesssim 10^{-28}\ cm^2$$

But all three forces can also produce rare processes with very small cross sections; there is no lower limit to the size of a cross section, although at present it is very difficult to study reactions whose cross sections are less than 10^{-40} cm^2.

The small sizes of cross sections when expressed in cm^2 has led to the unit

$$1\ barn = 10^{-24}\ cm^2$$

The usual metric prefixes, milli-, micro-, nano-, and pico- denote smaller cross-section units. For example, 1 microbarn $= 10^{-30}\ cm^2$.

D. Higher-Order Feynman Diagrams

So far we have discussed only the simplest Feynman diagram for a reaction. However, as illustrated in Fig. 7, for the reaction

$$e^+ + e^- \rightarrow \mu^+ + \mu^-$$

there are other diagrams—Figs. 7(b) and 7(c)—in addition to the simplest diagram. The more complicated diagrams have additional photons being exchanged by the leptons; hence they have more vertices. The number of vertices is called the order of the diagram, Fig. 7(a) is order 2, Fig. 7(b) is order 4, and Fig. 7(c) is order 6.

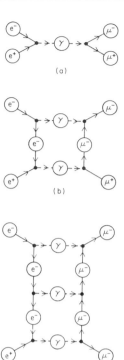

FIG. 7.   Feynman diagrams for $e^+ + e^- \rightarrow \mu^+ + \mu^-$ of order: (a) two, (b) four, and (c) six.

At each of these vertices there is a coupling constant for the electromagnetic force, thus higher-order diagrams have more coupling constants. The size of the reaction cross section for a diagram depends in part on the product of all the coupling constants. Since the coupling constant for the electromagnetic force is much smaller than 1, the higher-order diagrams have much smaller reaction cross sections than the simplest diagram, Fig. 7(a). Therefore, to calculate reaction cross sections for processes carried out through the electromagnetic force, the simplest diagram is usually sufficiently accurate. If more accuracy is required, then the next highest-order diagrams are used to calculate the additions to the reaction cross section from the diagrams.

The use of the simplest diagram, and the next higher-order diagram when necessary, is called a perturbation calculation. Such calculations can also be used for reactions involving the weak force, since the weak force coupling constant is also much smaller than 1. The relative simplicity of perturbation calculations has allowed complete comparisons of experiment with theory in reactions involving the electromagnetic and weak forces. This has allowed the physicist to gain a deep understanding of these forces.

Perturbation calculations cannot be used for most reactions involving the strong force, because the strong force coupling constant is larger than 1; hence higher-order diagrams are as important as lower-order diagrams, and a very large number of diagrams contribute to a reaction cross section. At present we know how to use strong force perturbation calculations in a limited range of strong force reactions. These are reactions in which large amounts of momentum are transferred to or from a quark, as discussed in Section VII.

VI. Experimental Techniques

A. INTRODUCTION

The purposes of experiments in elementary particle physics is to study the behavior of the basic forces and elementary particles and to look for new types of particles and forces. But few of these studies and searches can be carried out with the apparatus found in the ordinary physics laboratory. For example, elementary particles are too small to be seen with visible light microscopes or even electron microscopes. Furthermore, many elementary particles have very short lifetimes and do not exist for a long enough time to be studied directly. Also, the search for new particles usually requires that other particles collide at high energies to produce the new particles.

The primary experimental method in elementary particle physics involves the collision of two particles at high energy and the subsequent study of the reactions that occur and the particles produced. There are two methods, as shown in Fig. 8: fixed-target experiments at ac-

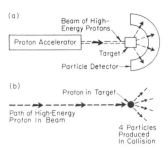

FIG. 9.   In a fixed-target experiment a beam of high-energy particles (e.g., protons) is produced by an accelerator. The beam of particles interacts with the target, producing new particles. The particles are detected and their properties studied using a particle detector. In (a) the entire experiment is sketched. In (b) the interaction of the particle itself is shown: A proton in the beam interacts with a proton in the target and produces four particles.

celerators and experiments with colliding-beam accelerators.

An example of a fixed-target experiment using an accelerator is shown in Fig. 9. A beam of protons, accelerated to high energy by a proton accelerator, leaves the accelerator and passes into a mass of material, called a target, that is fixed in position. The collisions occur between the protons in the beam and the material in the target. The simplest material to use for the target is hydrogen, because the hydrogen atom consists of a single electron moving around the single proton that forms the nucleus of the hydrogen atom, but other materials such as deuterium and heavy elements are used. To determine what has happened one needs an apparatus that can detect the particles coming out of the collision, called a particle detector. Particle detectors cannot see particles directly, but they can determine the energies, directions of motion, and nature of the particles.

In fixed-target experiments, the useful energy for the collision does not increase nearly as fast as the energy of the primary beam increases. Quantitatively the useful energy is the total energy in the center-of-mass of the colliding particles:

$$E_{cm} \approx \sqrt{2E_{beam}\,M_{target}}$$

In this high-energy approximation, E_{beam} is the kinetic energy of a particle in the accelerator beam and M the mass of a particle in the target. The alternative is to collide two beams of particles moving in opposite directions, as shown in Figs. 8(b) and 10. In this case the useful energy

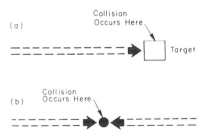

FIG. 8.   (a) In fixed-target experiments, a beam of high-energy particles collides with particles at rest in a target. (b) In colliding-beam experiments, two beams of high-energy particles collide head-on.

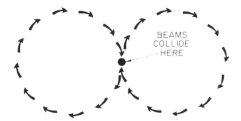

FIG. 10. In the simplest form of colliding-beam facilities, two beams of particles travel in the same direction in circles that are tangent at just one point. The beams collide at that point.

is actually the sum of the energies of the two beams (if the two beam energies are equal):

$$E_{cm} = 2E_{beam}$$

Particle colliders now produce the highest useful energy of any of our machines. Since there is no fixed target in a particle collider, the particle detector must look directly at the region where the opposing beams of particles collide. Figure 10 shows how this is done in a circular collider, where the beams of particles move in opposite directions around two circles. In this simple example the beams collide at just one point. In a real collider the beams would be arranged to collide at several different points, providing the opportunity to carry out several experiments at once.

Some experiments in elementary particle physics are carried out without using accelerators. Some use particles from fission reactors or from cosmic rays. Others look for new particles, such as free quarks or magnetic monopoles, in ordinary matter. Still others study with great precision the properties of the stable or almost stable particles, testing, for example, the equality of the sizes of the electric charge of the electron and the proton.

B. Particle Accelerators and Colliders

Figure 11 shows how an accelerator works. A bunch of electrically charged particles, either electrons or protons, passes through an electric field. The particles gain energy because they are accelerated by the electric field, hence the name accelerator. The energy gained by each particle is given by the voltage across the electric field. Thus an electron passing through a potential of 1 volt gains an energy of 1 electron-volt (1 eV). And an electron passing through 1 million volts gains an energy of 1 million electron-volts (1 MeV). Since protons have the same electric

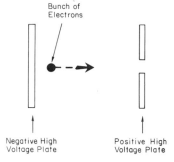

FIG. 11. Accelerators work by exerting an electric force on a charged particle. In the example here a negative plate repels the bunch of electrons and a positive plate attracts them. The electrons thus gain energy in moving from the negative plate to the positive plate. By the time they reach the positive plate, they are traveling so quickly that they pass through the hole in the plate.

charge as electrons, a proton passing through a million volts also gains an energy of 1 MeV.

Accelerators are either linear or circular (Fig. 12). In the linear accelerator, the particle is propelled by very strong electromagnetic fields, gaining all of its energy in one pass through the machine. In the circular accelerator, the particles are magnetically constrained to circulate many times around a closed path or orbit, and the particle energy is increased on each successive orbit by an accelerating electric field. High-

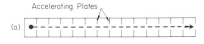

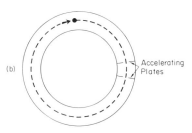

FIG. 12. Very high energies cannot be obtained by using just one pair of plates as in Fig. 11. There are two ways to solve this problem. In (a), a linear accelerator, many pairs of plates are lined up, and the particles being accelerated are given more and more energy as they pass through each pair of plates. In (b), a circular accelerator, only one pair of plates is used, but the particles are made to travel in a circle, thus passing through that pair of plates again and again. Each time they pass through the pair of plates they gain more energy.

energy accelerators and colliders are large and expensive machines. Thus few are built, and these are used as intensively as possible.

The machines under construction are all colliders, because they can achieve the largest useful (center-of-mass) energy. Another simplified example of a particle collider is shown in Fig. 13(a). In a circular machine, a bunch of electrons and a bunch of positrons circulate in opposite directions, the particle bunches being held in the machine by a magnetic guide field. (These machines are also called storage rings.) At two opposite places in the machine, the bunches meet head on. Most of the particles in one bunch simply pass through the other bunch without actually colliding and continue to circulate around the storage ring, for hours or even days, making thousands or even millions of rotations per second.

The following combinations of particles are now used or will be used in colliders: $e^+ + e^-$, $p + p$, $p + \bar{p}$, $e^- + p$, and $e^+ + p$. A critical property of colliders is the luminosity L, which is a measure of the rate at which particle collisions occur. Since particle collisions are the essence of particle experiments, the more collisions per second the more useful the collider. Quantitatively,

$$\text{collisions per second} = \sigma L$$

where σ is the cross section. Existing colliders have luminosities in the range of 10^{29} to 10^{32} $cm^{-2}\ sec^{-1}$.

An alternative to storage rings for particle colliders is the use of colliding beams produced by linear accelerators, Fig. 13(b), a linear particle collider. The colliding bunches of particles pass through each other just once. Much denser bunches must be used to compensate for the absence of repeated passes.

C. PARTICLE DETECTORS

Electrically charged high-energy particles such as electrons and pions are detected by means of their ionization of the material through which they pass. This ionization leaves free electrons and ions along the particle's path. There are various ways to find the positions of the ions or electrons and therefore to indirectly detect the high-energy particle.

The bubble chamber is the classic example. The liquid in a bubble chamber is heated above its boiling point, but it is prevented from boiling by high pressure in the chamber. If that pressure is released for a very short time and then reapplied, the liquid still does not boil. However, if a charged particle passes through the chamber while the pressure is released, the resulting ionization leads to the formation of a string of bubbles along the path of a particle. This string of bubbles is then photographed to produce a picture of the paths or tracks of the charged particles in their passage through the chamber.

The drift chamber is the most common way to detect the tracks of charged particles. Thin metal wires are arranged in parallel rows in a chamber containing a gas such as argon. The wires are at permanent high voltage, adjacent wires having opposite voltages. When a high-energy particle passes through the gas, the electrons from the ionization path drift to the positive voltage wires, inducing electrical signals in the wires nearest the particle path. These signals are then processed electronically, and the path of the particle is determined.

Ionization produced by a charged particle is detected in other ways by other types of particle detectors. In a semiconductor detector, the charged particle ionizes the semiconductor; in a

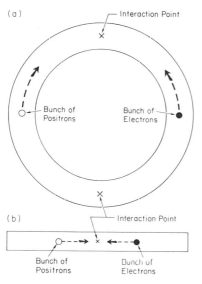

FIG. 13.   (a) A colliding-beam storage ring accelerator has two bunches of particles moving in opposite directions. The bunches move through each other at the two interaction points, and a small number of particles collide; but most of the particles continue to circulate again and again around the orbits, with some collisions taking place at each pass. (b) In a linear colliding-beam accelerator the two bunches of particles meet only once. To make best use of that single pass, the bunches have to be much denser than those in a circular collider.

scintillator the ionization produces visible light which is detected by a phototube. A few charged-particle detectors, such as Cerenkov radiation detectors, do not use ionization.

Electrically neutral high-energy particles are detected by their interaction with matter through the strong or electromagnetic forces. Through the strong force, a neutron can produce protons or charged pions, which in turn are detected by their ionization. Photons interact with matter through the electromagnetic force, producing electrons and positrons which produce ionization. The point at which the ionization is detected is the point where the photon interacted.

The energy of a high-energy particle is measured in two principal ways. If the particle is charged, it will have a circular path in a magnetic field of radius r. The particle's momentum p is

$$p = qBr$$

where B is the magnetic field strength and q the particle charge. For a particle of mass m the energy is

$$E = \sqrt{p^2 + m^2}$$

The other way to measure a particle's energy is to allow it to interact in matter so that all its energy is converted into ionization energy, in a device called a calorimeter.

VII. Quark Model of Hadrons and Quantum Chromodynamics

A. EVIDENCE FOR THE QUARK MODEL

The concept that hadrons are the bound states of more fundamental particles of fractional charge and half-integer spin was invented by M. Gell-Mann and G. Zweig in 1964. Gell-Mann and Zweig proposed that hadrons were made of either a quark and an antiquark (a meson) or three quarks (a baryon). Their quark model explained why hadrons with certain combinations of quantum numbers were observed while hadrons with other combinations were not. For example, mesons in S-states (0 orbital angular momentum) were observed with total spin J, parity P, and charge conjugation C in the combinations $J^{PC} = 0(^{-+})$, $1(^{--})$ but were not observed in the combinations $J^{PC} = 0(^{--})$, $1(^{-+})$. The first set of quantum numbers is allowed when combining a spin 1/2 particle and a spin 1/2 antiparticle, but the second is not allowed.

At the time the quark model was proposed, it could account for the quantum numbers of the known hadrons, but it could do little else. In fact, there appeared to be a glaring discrepancy with experiment in that no fractionally charged particles had ever been observed. Despite the fact that there still has been no independently confirmed observation of free quarks, the evidence for the existence of quarks today is incontrovertible.

The evidence for quarks since 1964 has taken several forms:

1. As new hadrons have been discovered, they have continued to fit into the quark model scheme of hadron spectroscopy. The measurements of the mass spectra of mesons made of a charm quark and a charm antiquark (charmonium states) and of a bottom quark and a bottom antiquark (bottomonium states) have been particularly revealing. Energy level diagrams of these states look very much like the energy level diagram for positronium (a bound state of an electron and a positron). Indeed, the charmonium and bottomonium states can be quantitatively accounted for by nonrelativistic potential models of bound fermion–antifermion pairs.

2. Hadron–hadron, lepton–hadron, and lepton–lepton scattering experiments have all been consistent with, and have lent great support to, the quark model of hadrons. One of the most striking pieces of evidence for the existence of quarks comes from e^+e^- annihilation into hadrons. The process proceeds through the Feynman graph shown in Fig. 14(b). The electron and positron annihilate through a virtual photon that couples to the charges of the final state quark–antiquark pair. As the quark and antiquark move away from each other, they exert an attractive force on each other and thereby build up potential energy. Eventually, it becomes energetically favorable to create additional quark–antiquark pairs, and the quarks and antiquarks so produced combine with each other and with the original quarks to form hadrons. What one sees in the laboratory are two back-to-back jets of hadrons, oriented in the direction of the original quark–antiquark pair [Fig. 14(a)]. The angular distribution of the jets is precisely that expected from pointlike objects with spin 1/2.

B. THE COLOR DEGREE OF FREEDOM

Quarks have a degree of freedom not shared by the other family of fundamental fermions, the leptons. This degree of freedom, called color, is

MARK II – SLC/PEP

(a)

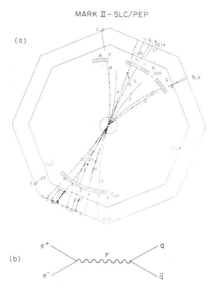

(b)

FIG. 14. e^+e^- annihilation into two jets of hadrons: (a) as observed in a particle detector and (b) the Feynman diagram for the underlying fundamental process $e^+ + e^- \to q + \bar{q}$.

baryon wave function can be totally antisymmetric.

Quarks come in three colors, denoted R, Y, and B for red, yellow, and blue. Antiquarks come in three anticolors: $\bar{R}$, $\bar{Y}$, and $\bar{B}$ for antired, antiyellow, and antiblue. All hadrons are colorless. If the c quark in a D^+ pseudoscalar meson has the color red, then the $\bar{d}$ quark has the color antired. A baryon always contains one red quark, one yellow quark, and one blue quark. It is said to be colorless because it is invariant under unitary transformations in the space spanned by the three color eigenstates.

The rate of production for the process

$$e^+ + e^- \to \text{hadrons}$$

corresponds to what one would expect from quark pair-production, once the color degree of freedom is included. This is illustrated in Fig. 15 where the value

$$R = \frac{\sigma(e^+ + e^- \to \text{hadrons})}{\sigma(e^+ + e^- \to \mu^+ + \mu^-)} \qquad (2)$$

is plotted as a function of e^+e^- center-of-mass energy. The rate for the process

$$e^+ + e^- \to f + \bar{f}$$

is proportional to the square of the fermion's charge. Thus,

$$R = \frac{4}{9} \quad \text{(red up quark only)}$$

$$R = \frac{4}{9} + \frac{4}{9} + \frac{4}{9} = \frac{4}{3} \quad \text{(all up quarks)}$$

$$R = \sum_q 3 \cdot e_q^2 \quad \text{(all quarks q above threshold)}$$

$$(3)$$

intimately associated with the strong force felt by quarks. The color degree of freedom was first proposed as a means to make the wave function of baryons antisymmetric under the interchange of any two constituent quarks, as required by the Pauli exclusion principle. It is easiest to see the need for an extra degree of freedom in the case of the doubly charged baryon Δ^{++} with spin $J = 3/2$. The state consists of three up quarks with their spins aligned. The total wave function of such a state is symmetric under the interchange of any two quarks. If quarks are given an additional tri valued degree of freedom, then the

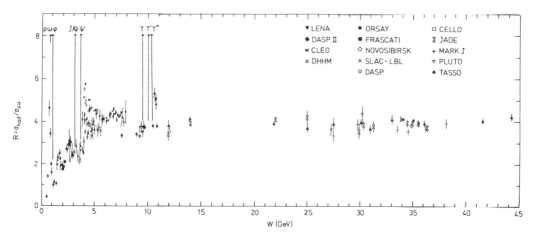

FIG. 15. $R = [\sigma(e^+e^- \to \text{hadrons}]/[\sigma(e^+e^- \to \mu^+\mu^-)]$ as a function of e^+e^- center-of-mass energy.

where e_q is the charge of quark q. For center-of-mass energies above 11 GeV (above bb threshold),

$$R = 3 \cdot \left(\frac{4}{9} + \frac{1}{9} + \frac{1}{9} + \frac{4}{9} + \frac{1}{9} \right) = \frac{11}{3} \qquad (4)$$

Among the fundamental fermions, only quarks feel the strong force. It is no coincidence that they are also the only fundamental fermions with the color degree of freedom. Color is the "electric charge" of the strong force. To understand how color generates the strong force, some background in a class of quantum field theories called quantum gauge fields theories is necessary.

C. GAUGE FIELD THEORIES

The theories that describe the strong, electromagnetic, and weak interactions are all gauge field theories. In gauge field theories, interactions are a consequence of the requirement that the Lagrangian for free matter fields be invariant under a set of transformations called local gauge transformations. The Lagrangian for a free fermion field has the form

$$\mathscr{L} = \bar{\psi}(i\gamma^{\mu}\partial_{\mu} - m)\psi \qquad (5)$$

where ψ is the fermion field (an electron, for example), m the mass of the fermion, and γ^{μ} are matrices. The term $\gamma^{\mu}\partial_{\mu}$ represents the sum

$$\sum_{\mu=0}^{3} \gamma^{\mu}\partial_{\mu}$$

Repeated indices will always indicate a sum, unless otherwise indicated.

The first known quantum gauge field theory was quantum electrodynamics (QED), the fundamental theory of electromagnetic interactions. In QED, the local gauge transformation is a phase rotation of the fermion field:

$$\psi(x) \rightarrow \exp[iq\theta(x)]\psi(x) \qquad (6)$$

The phase transformation varies from point to point in space-time x. This is what is meant by a local gauge transformation. If the same phase rotation is applied at all points in space-time, then the transformation is called a global gauge transformation.

If the Lagrangian describing the fermion field ψ is to be invariant under arbitrary local phase rotations, then the derivative ∂_{μ} must be redefined to be

$$\partial_{\mu} \rightarrow D_{\mu} \equiv \partial_{\mu} + iqA_{\mu}(x) \qquad (7)$$

where A_{μ} is a new field called the gauge field. The gauge field is in fact the scalar and vector potentials of classical electrodynamics (i.e., the electromagnetic field), and the constant q is the charge of the fermion field ψ. If the gauge field A_{μ} undergoes the transformation

$$A_{\mu}(x) \rightarrow A_{\mu}(x) - \partial_{\mu}\theta(x) \qquad (8)$$

when the fermion undergoes the transformation (6), then the Lagrangian in Eq. (5) with the redefined derivative in Eq. (7) is invariant under local phase rotations. Gauge transformations of the form of Eq. (8) are familiar from classical electrodynamics.

The set of phase rotations in Eq. (6) forms the mathematical group $U(1)$. A gauge group is said to be Abelian when the product of two successive gauge transformations is independent of the order in which the transformations are made. The $U(1)$ gauge group is Abelian.

D. QUANTUM CHROMODYNAMICS

The strong force results from requiring that the free-quark Lagrangian be invariant under local rotations in the space spanned by the three color eigenstates. The resulting quantum gauge field theory is called quantum chromodynamics (QCD). The gauge group is $SU(3)_{\text{color}}$, where the subscript is a reminder that the gauge transformations are rotations in the space spanned by the three color eigenstates of the quarks. $SU(3)$ is a non-Abelian group, so QCD is a non-Abelian gauge field theory. In QCD (or strictly speaking, perturbative QCD) massless vector gauge bosons called gluons are exchanged by colored quarks, much as massless photons are exchanged by charged particles in QED.

QCD has a richer structure than QED due to the fact that it is a non-Abelian gauge theory. There are eight kinds of gluons, compared with only one photon. Furthermore, the gluons themselves are colored (i.e., color charged) whereas the photon carries no electric charge. A red quark might interact with a yellow quark, for example, by exchanging a red–antiyellow gluon, as shown in Fig. 16(a). Because the gluons are themselves colored, they can interact with each other, and Feynman graphs such as those shown in Fig. 17 are possible. Such interactions are not found in QED. Self-interacting gauge boson fields are a feature of non-Abelian gauge field theories.

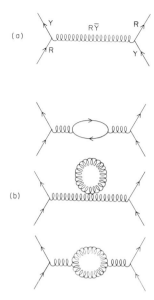

FIG. 16. Feynman diagrams for gluon exchange by two quarks: (a) the Born diagram and (b) higher-order diagrams.

E. THE QCD RUNNING COUPLING CONSTANT

An important property of QCD is the running coupling constant of perturbative QCD. The lowest-order Feynman diagram for the scattering of two quarks through virtual gluon exchange is shown in Fig. 16(a). Such a diagram is called a Born diagram. When higher-order diagrams such as those shown in Fig. 16(b) are calculated, the results can be expressed as a modification to the coupling constant used in the Born diagram calculation. The modified coupling constant in general depends on the momentum transferred between the two quarks. The momentum transfer is usually expressed in terms of q^2, the square of the invariant mass of the virtual gluon exchanged by the two quarks. The dependence of α_S, the strong force coupling constant, on q^2 has the form

$$\alpha_S(q^2) = \alpha_S(q_0^2) \left[1 + \frac{\alpha_S(q_0^2)}{12\pi} \right.$$
$$\left. \times \log\left(\frac{-q^2}{q_0^2}\right)(2n_f - 33) + \cdots \right] \quad (9)$$

where $\alpha_S(q_0^2)$ is the strong force coupling constant at a reference value of $q^2 = q_0^2$ and n_f the number of quarks with masses $M^2 < q^2$.

Running coupling constants are common to all quantum field theories. Higher-order corrections to the QED Born diagram, for example, lead to a q^2 dependent QED coupling constant of the form

$$\alpha(q^2) = \alpha(q_0^2) \left[1 + \frac{\alpha(q_0^2)}{3\pi} \log\left(\frac{-q^2}{q_0^2}\right) + \cdots \right]$$
$$(10)$$

where again $\alpha(q_0^2)$ is the coupling constant at a reference q^2. The importance of the QCD running coupling constant lies in the coefficient $(2n_f - 33)$ of the logarithm term. As Fig. 18 illustrates, the running coupling constant $\alpha_S(q^2)$ decreases with increasing q^2 (or, equivalently, with smaller distances), so that for large enough q^2, $\alpha_S \ll 1$. This behavior is known as asymptotic freedom. It implies that perturbative QCD calculations will be meaningful for large enough q^2.

Asymptotic freedom also implies that as q^2 decreases (i.e., as larger distances are probed) α_S grows. For small enough q^2, α_S exceeds unity, so perturbative QCD is no longer valid. The growth of the strong coupling constant with distance provides hope that QCD can someday explain the apparent confinement of color (i.e., the confinement of quarks and gluons inside hadrons).

Asymptotic freedom plays an important role in the interpretation of high-momentum transfer collisions between leptons and hadrons. When a

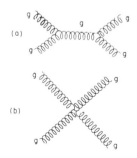

FIG. 17. Feynman diagrams for gluon self-interactions.

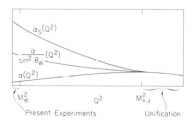

FIG. 18. Evolution of the electromagnetic, weak, and strong coupling constants (α, $\alpha/\sin^2 \theta_W$, and α_s, respectively); q^2 is the square of the energy scale at which a particular interaction takes place.

lepton scatters off a hadron by transferring a large amount of momentum to the hadron, the process is termed deep inelastic lepton–hadron scattering. The leptons can be electrons, muons, or neutrinos. The hadrons are typically protons and neutrons. During deep inelastic lepton–hadron collisions, the leptons appear to scatter off pointlike, quasi-free, approximately massless spin 1/2 objects. These objects are, of course, the quarks, but not until the discovery of asymptotic freedom was the quasi-free nature of the quarks inside hadrons understood. If quarks felt the same strong force at high q^2 as they do at low q^2, then the quark constituents of hadrons would not appear to be quasi-free. They would instead appear to be rigidly attached to the hadrons during deep inelastic collisions, and the energy–angular distributions of the scattered leptons would be modified accordingly. The discovery of the q^2 dependence of α_S given by Eq. (9) helped resolve the apparent paradox that while the quarks acted like quasi-free objects during deep inelastic collisions, they were nevertheless permanently confined inside hadrons.

F. EXPERIMENTAL EVIDENCE FOR GLUONS

Have gluons ever been observed? Like quarks, gluons have never been observed as free particles. Gluons carry color, and the only free objects observed in high-energy physics laboratories so far have been colorless objects. There is nevertheless strong experimental support for the existence of gluons.

One of the things experimentalists can do in deep inelastic lepton–nucleon scattering is to measure the momentum distribution of quarks inside the nucleon. Both electron–nucleon and neutrino–nucleon scattering give the same quark momentum distributions. (This is yet another piece of evidence for the quark model of hadrons.) Both types of deep inelastic scattering experiments indicate that quarks can account for only about half of the nucleon's momentum. The other half of the nucleon's momentum is carried by constituents that are blind to the electromagnetic and weak forces. These extra constituents are the gluons that bind the quarks together inside the nucleon. The gluons also produce virtual quark–antiquark pairs, so there is an observable antiquark component in nucleons.

While lepton–hadron scattering experiments provided the first experimental evidence for gluons, the most dramatic and convincing piece of evidence for gluons has come from e^+e^- annihilation. Recall that when an electron and positron annihilate and form two back-to-back hadron jets, the angular distribution and rate strongly reflect the underlying fundamental process,

$$e^+ + e^- \rightarrow q + \bar{q}$$

For large enough e^+e^- center-of-mass energies (greater than about 25 GeV), the momenta of the quark and antiquark are large enough that perturbative QCD is applicable. Perturbative QCD predicts that one of the quarks will sometimes radiate a gluon, with a probability proportional to the strong coupling constant [Fig. 19(b)]. If the gluon is radiated at a large angle with respect to the quark, then three jets of hadrons will be formed. Such three-jet events are actually observed [Fig. 19(a)].

The rate of three-jet events produced in e^+e^- annihilation is a measure of the strong coupling constant α_S. However, a measurement of α_S based on the rate of three-jet events produced in e^+e^- annihilation is difficult to make because the fragmentation of quarks and gluons into jets of hadrons cannot be calculated from first principles and must be modeled. Nevertheless, we

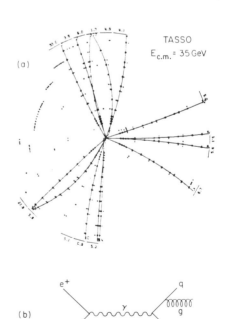

FIG. 19. e^+e^- annihilation into three jets of hadrons: (a) as observed in a particle detector and (b) the Feynman diagram for the underlying fundamental process $e^+e^- \rightarrow q + \bar{q} + g$.

know that at center-of-mass energies of about 30 GeV, the value of α_S appears to be in the neighborhood of 0.2.

VIII. Electromagnetic and Weak Interactions

One of the great achievements of twentieth-century physics has been the understanding that the electromagnetic and weak forces are manifestations of the same force, now called the electroweak force. In this chapter we present an overview of the fundamental $SU(2) \times U(1)$ gauge theory of electroweak interactions and discuss electroweak phenomenology and experimental evidence for the electroweak theory.

A. Left-Handed and Right-Handed Fermions

A prerequisite for a discussion of the $SU(2) \times U(1)$ theory is the concept of left-handed and right-handed fermions. Relativistic fermions are governed by the Dirac equation. A Dirac fermion (sometimes called a Dirac spinor) contains four components:

$$\psi(x) - \begin{pmatrix} \psi_1(x) \\ \psi_2(x) \\ \psi_3(x) \\ \psi_4(x) \end{pmatrix} \quad (11)$$

The Dirac equation is actually a set of four coupled differential equations that can be represented in matrix notation by

$$(i\gamma^\mu \partial_\mu - m)\psi = 0 \quad (12)$$

where γ^μ, $\mu = 0, 1, 2, 3$, are 4×4 matrices. (Recall from Section VII,C that repeated indices indicate a sum over the indices.) These matrices can take the form

$$\gamma^0 = \begin{pmatrix} 1 & 0 \\ 0 & -1 \end{pmatrix}, \quad \gamma^i = \begin{pmatrix} 0 & \sigma^i \\ -\sigma^i & 0 \end{pmatrix} \quad (13)$$

where the 2×2 σ^i matrices are the Pauli matrices familiar from nonrelativistic quantum mechanics:

$$\sigma^1 = \begin{pmatrix} 0 & 1 \\ 1 & 0 \end{pmatrix}, \quad \sigma^2 = \begin{pmatrix} 0 & -i \\ i & 0 \end{pmatrix},$$

$$\sigma^3 = \begin{pmatrix} 1 & 0 \\ 0 & -1 \end{pmatrix} \quad (14)$$

The spin degree of freedom is contained in the Dirac fermion wave function. An electron at rest

with its spin oriented in the $+z$ direction would have the wave function

$$\psi = \exp(-imt) \begin{pmatrix} 1 \\ 0 \\ 0 \\ 0 \end{pmatrix} \quad (15)$$

An electron at rest with its spin pointing in the $-z$ direction would have the wave function

$$\psi = \exp(-imt) \begin{pmatrix} 0 \\ 1 \\ 0 \\ 0 \end{pmatrix} \quad (16)$$

An alternate quantization axis for the fermion's spin is the direction of motion of the fermion. The spin eigenstates along such a quantization axis are called helicity eigenstates. A fermion in a helicity eigenstate with the spin pointing parallel (antiparallel) to the direction of motion is said to be right-handed (left-handed).

For a Dirac fermion state ψ, the right-handed and left-handed states ψ_R and ψ_L are defined by

$$\psi_R \equiv \frac{1}{2}(1 + \gamma_5)\psi$$

$$\psi_L \equiv \frac{1}{2}(1 - \gamma_5)\psi \qquad \gamma_5 \equiv i\gamma^0\gamma^1\gamma^2\gamma^3 \quad (17)$$

The states ψ_R and ψ_L are so named because they are right-handed and left-handed helicity eigenstates in the limit of zero fermion mass.

The decomposition of fermions into right-handed and left-handed components is important in the $SU(2) \times U(1)$ theory because the left- and right-handed components transform differently under $SU(2) \times U(1)$ gauge transformations. The difference between the transformation properties of right-handed and left-handed fermions leads ultimately to the phenomenon of parity violation in the weak interactions. A parity transformation inverts the x, y, z cartesian coordinates for a system. This makes the left-handed component of a fermion right-handed and vice versa. Parity violation occurs when a system is not invariant under parity transformations.

B. Introduction to the $SU(2) \times U(1)$ Gauge Theory

The $SU(2) \times U(1)$ electroweak theory is a quantum gauge field theory, just as are QED and QCD. [QED is actually contained in the $SU(2) \times$

$U(1)$ theory, as we shall see.] An $SU(2) \times U(1)$ local gauge transformation is a product of a local rotation in weak isospin space and a local phase rotation with a constant of proportionality given by a quantity called the weak hypercharge. The group of rotations in weak isospin space is $SU(2)$, and the group of phase rotations is $U(1)$; hence the electroweak gauge group is $SU(2) \times U(1)$.

Weak isospin rotations, when applied to leptons, mix left-handed charged leptons with the left-handed neutrino of the same generation. Weak isospin rotations have no effect on right-handed leptons. The left-handed leptons therefore form weak isospin doublets:

$$\begin{pmatrix} \nu_e \\ e^- \end{pmatrix}_L, \qquad \begin{pmatrix} \nu_\mu \\ \mu^- \end{pmatrix}_L, \qquad \begin{pmatrix} \nu_\tau \\ \tau^- \end{pmatrix}_L \qquad (18)$$

The right-handed charged leptons form weak isospin singlets:

$$e_R^-, \qquad \mu_R^-, \qquad \tau_R^- \qquad (19)$$

If a neutrino is massless (Table IV), then the left-handed neutrino is the entire neutrino spinor, and the right-handed neutrino does not exist. We shall assume that the three neutrinos ν_e, ν_μ, ν_τ are exactly massless, so that there are no right-handed neutrino weak isospin singlets. The left- and right-handed leptons are eigenstates of T^3, the third component of weak isospin, with eigenvalues given in Table VII.

The weak hypercharge Y of a particle is given by its charge Q in units of the magnitude of the electron charge and its third component of weak isospin T^3:

$$Y = 2(Q - T^3) \qquad (20)$$

Note that because T^3 is different for the left- and right-handed components of a fermion, the hypercharge will also be different for the two components. The weak hypercharges for the left- and right-handed components of leptons are summarized in Table VII.

The $SU(2) \times U(1)$ local gauge invariant Lagrangian for leptons is given by

$$\mathscr{L} = i\bar{\psi}\gamma^\mu D_\mu \psi - \frac{1}{4} F^i_{\eta\rho} F^{i\eta\rho} \qquad (21)$$

where D_μ is the gauge covariant derivative

$$D_\mu = \partial_\mu + ig W^i_\mu T^i + ig' B_\mu \frac{Y}{2} \qquad (22)$$

and $-(1/4)F^i_{\eta\rho}F^{i\eta\rho}$ is a kinetic energy term for the gauge boson fields. The operators T^i, called the generators of $SU(2)$, are given by

$$T^i = \frac{1}{2} \sigma^i \qquad i = 1, 2, 3 \qquad (23)$$

in the case of $SU(2)$ doublets and

$$T^i = 0 \qquad i = 1, 2, 3 \qquad (24)$$

for $SU(2)$ singlets. T^3 is the third component of weak isospin operator. The fields W^i_μ, $i = 1, 2, 3$, are the gauge boson fields associated with $SU(2)$ transformations, and B_μ is the gauge boson field associated with weak hypercharge $U(1)$ transformations.

We now study the lepton interaction term

$$\bar{\psi}i\gamma^\mu(D_\mu - \partial_\mu)\psi = \bar{\psi}_L i\gamma^\mu(D_\mu - \partial_\mu)\psi_L$$
$$+ \bar{\psi}_R i\gamma^\mu(D_\mu - \partial_\mu)\psi_R \qquad (25)$$

For the case

$$\psi_L = \begin{pmatrix} \nu_e \\ e \end{pmatrix}_L \quad \text{and} \quad \psi_R = e_R \qquad (26)$$

TABLE VII.   Third Component of Weak Isospin, Weak Hypercharge, and Z^0 Coupling of the First Generation Quarks and Leptons[a]

Fermion	Third component of weak isospin T^3	Weak hypercharge Y	Z^0 coupling $\lvert T^3 - Q \sin^2 \theta_W \rvert$
ν_e	1/2	−1	0.5
e_L	−1/2	−1	0.28
e_R	0	−2	0.22
u_L	1/2	1/3	0.35
d_L	−1/2	1/3	0.43
u_R	0	4/3	0.15
d_R	0	−2/3	0.07

[a] The values for the 2nd and 3rd generation quarks and leptons are the same as for the corresponding 1st generation fermions.

the lepton interaction term is given explicitly by

$$\bar{\psi}i\gamma^{\mu}(D_{\mu} - \partial_{\mu})\psi = -\bar{e}_{L}\frac{g}{2}(W_{\mu}^{1} + iW_{\mu}^{2})\nu_{e}$$

$$- \bar{\nu}_{e}\frac{g}{2}(W_{\mu}^{1} - iW_{\mu}^{2})e_{L}$$

$$- \bar{\nu}_{e}\left(g\gamma^{\mu}W_{\mu}^{3}T^{3} + g'\gamma^{\mu}B_{\mu}\frac{Y}{2}\right)\nu_{e}$$

$$- \bar{e}_{L}\left(g\gamma^{\mu}W_{\mu}^{3}T^{3} + g'\gamma^{\mu}B_{\mu}\frac{Y}{2}\right)e_{L}$$

$$- \bar{e}_{R}g'B_{\mu}\frac{Y}{2}e_{R} \tag{27}$$

1. Charged Current Interactions

The terms

$$\bar{e}_{L}\frac{g}{2}(W_{\mu}^{1} + iW_{\mu}^{2})\nu_{e} \quad \text{and}$$

$$\bar{\nu}_{e}\frac{g}{2}(W_{\mu}^{1} - iW_{\mu}^{2})e_{L} \tag{28}$$

represent the absorption of a positively and negatively charged gauge boson, respectively, an example of which is shown in Fig. 6b. We write

$$W_{\mu}^{\pm} = \frac{W_{\mu}^{1} \mp iW_{\mu}^{2}}{\sqrt{2}} \tag{29}$$

and call $W^{\pm}$ the charged W gauge bosons. Interactions mediated by the charged W bosons are called charged current interactions. The parameter g is the coupling constant for charged current interactions. Note that the W bosons couple only to left-handed leptons, so that charged current interactions are manifestly parity-violating in the $SU(2) \times U(1)$ theory.

2. Neutral Current Interactions

The terms

$$\bar{\nu}_{e}\left(g\gamma^{\mu}W_{\mu}^{3}T^{3} + g'\gamma^{\mu}B_{\mu}\frac{Y}{2}\right)\nu_{e}$$

$$\bar{e}_{L}\left(g\gamma^{\mu}W_{\mu}^{3}T^{3} + g'\gamma^{\mu}B_{\mu}\frac{Y}{2}\right)e_{L}$$

and

$$\bar{e}_{R}g'B_{\mu}\frac{Y}{2}e_{R} \tag{30}$$

represent neutral current interactions. Notice that there are two gauge boson fields involved here: W_{μ}^{3} and B_{μ}. Some linear combination of these fields should be the photon A_{μ}, and so we

write

$$A_{\mu} = \sin\theta_{W}W_{\mu}^{3} + \cos\theta_{W}B_{\mu} \tag{31}$$

and call the orthogonal combination Z_{μ}:

$$Z_{\mu} = \cos\theta_{W}W_{\mu}^{3} - \sin\theta_{W}B_{\mu} \tag{32}$$

Rewriting the neutral current terms [Eq. (30)] in terms of the fields A_{μ} Z_{μ},

$$-\bar{\nu}_{e}\left[\gamma^{\mu}A_{\mu}\left(g\sin\theta_{W}T^{3} + \frac{1}{2}g'\cos\theta_{W}Y\right)\right.$$

$$+ \gamma^{\mu}Z_{\mu}\left(g\cos\theta_{W}T^{3} - \frac{1}{2}g'\sin\theta_{W}Y\right)\right]\nu_{e}$$

$$-\bar{e}_{L}\left[\gamma^{\mu}A_{\mu}\left(g\sin\theta_{W}T^{3} + \frac{1}{2}g'\cos\theta_{W}Y\right)\right.$$

$$+ \gamma^{\mu}Z_{\mu}\left(g\cos\theta_{W}T^{3} - \frac{1}{2}g'\sin\theta_{W}Y\right)\right]e_{L}$$

$$-\bar{e}_{R}\left(\gamma^{\mu}A_{\mu}\frac{g'}{2}\cos\theta_{W}Y - \frac{1}{2}\gamma^{\mu}Z_{\mu}g'\sin\theta_{W}Y\right)e_{R} \tag{33}$$

we can identify A_{μ} with the photon if

$$g\sin\theta_{W} = g'\cos\theta_{W} = e \tag{34}$$

where e is the magnitude of the electron's charge. The angle θ_{W} that determines the mixing of the W_{μ}^{3} and B_{μ} fields is called the Weinberg angle. Equation (34) gives a relation between the strength of the electromagnetic force and the strength of the charged weak force.

The gauge boson Z_{μ} is called the Z^{0} gauge boson. The coupling of the Z^{0} to fermions can be read from Eq. (33) and is

$$g\cos\theta_{W}T^{3} - \frac{1}{2}g'\cos\theta_{W}Y \tag{35}$$

Expressing g and g' in terms of e and θ_{W}, the Z^{0} coupling can be written

$$\frac{e}{\sin\theta_{W}\cos\theta_{W}}[T^{3} - Q\sin^{2}\theta_{W}] \tag{36}$$

Values for $T^{3} - Q\sin^{2}\theta_{W}$ are summarized in Table VII. Processes mediated by the Z^{0} gauge boson are called weak neutral current interactions.

Although the neutrinos are electrically neutral and therefore do not couple to the photon, they nevertheless experience neutral current interactions through their nonzero coupling to the Z^{0}.

$$\nu_{\mu} + e \rightarrow \nu_{\mu} + e$$

is an example of a reaction that can only be mediated by the Z^{0} (Fig. 20).

FIG. 20.   Feynman diagram for the weak neutral current process $\nu_\mu + e^- \to \mu_\mu + e^-$.

The Z^0 couples to both left and right-handed charged leptons, as does the photon. Unlike the photon, the Z^0 couples with different strengths to right-handed and left-handed charged leptons, and the neutral current interactions mediated by the Z^0 are therefore parity violating.

C. THE HIGGS MECHANISM

Although our development of the $SU(2) \times U(1)$ theory has so far been correct, it nonetheless has been incomplete. The problem is that, in the theory presented so far, the gauge bosons are massless. Because of the short-range nature of the weak force, we know that the $W^\pm$ and Z^0 gauge bosons cannot be massless. Therefore, to be phenomenologically successful, something in the $SU(2) \times U(1)$ theory must give a lot of mass to the $W^\pm$ and Z^0. Furthermore, for technical reasons mass must be given to the $W^\pm$ and Z^0 without breaking the local $SU(2) \times U(1)$ gauge invariance of the Lagrangian.

An explicit mass term in a Lagrangian for a gauge boson X_μ has the form

$$\frac{M^2}{2} X_\mu X^\mu \qquad (37)$$

where M is the mass of the gauge boson. An explicit mass term for a fermion f takes the form

$$m(\bar{f}_R f_L + \bar{f}_L f_R) \qquad (38)$$

where m is the fermion mass. Explicit gauge boson mass terms break the $SU(2) \times U(1)$ local gauge symmetry and must therefore be absent from the Lagrangian.

Scalar (spin 0) fields can be added to the $SU(2) \times U(1)$ locally gauge invariant Lagrangian, Eq. (21), without spoiling the gauge invariance. If we add a pair of complex fields ϕ^+ and ϕ^0 and put them in a weak isospin doublet,

$$\Phi \equiv \begin{pmatrix} \phi^+ \\ \phi^0 \end{pmatrix} \qquad (39)$$

then the following terms can be added to Eq. (21) without destroying the gauge invariance:

$$(D^\mu \Phi)^\dagger (D_\mu \Phi) - V(\Phi^\dagger \Phi)$$
$$- G[\psi_R(\Phi^\dagger \psi_L) + (\bar{\psi}_L \Phi)\psi_R] \qquad (40)$$

The superscripts of ϕ^+ and ϕ^0 indicate the charge of the scalar fields. G is an arbitrary constant called a Yukawa coupling constant. The only restriction on the potential V is that it not contain powers of $\Phi^\dagger \Phi > 2$. We let the potential $V(\Phi^\dagger \Phi)$ take the form

$$V(\Phi^\dagger \Phi) = \mu^2(\Phi^\dagger \Phi) + |\gamma|(\Phi^\dagger \Phi)^2 \qquad (41)$$

where μ and λ are arbitrary constants. The potential $V(\Phi^\dagger \Phi)$ is plotted in Fig. 21 for the cases $\mu^2 > 0$ and $\mu^2 < 0$.

The Lagrangian

$$\mathcal{L} = \mathcal{L}_{\text{lepton}} + \mathcal{L}_{\text{scalar}} \qquad (42)$$

with the terms $\mathcal{L}_{\text{lepton}}$ and $\mathcal{L}_{\text{scalar}}$ given by Eqs. (21) and (40), respectively, is the complete electroweak $SU(2) \times U(1)$ locally gauge invariant Lagrangian, which today so successfully describes electroweak phenomena. However, the true nature of the Lagrangian is hidden from us when it is expressed in such a form. To ascertain the true nature of the Lagrangian in Eq. (42) we must study the lowest energy state of the system, called the vacuum state. The field Φ corresponding to the lowest energy state is called the vacuum expectation value of Φ, denoted by $\langle \Phi \rangle_0$. From Fig. 21 we see that if $\mu^2 > 0$, the vacuum expectation value of Φ is simply $\langle \Phi \rangle_0 = 0$. However if $\mu^2 < 0$, the lowest energy of the

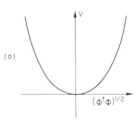

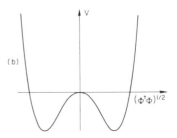

FIG. 21.   The Higgs scalar potential $V(\Phi^\dagger \Phi) = \mu^2(\Phi^\dagger \Phi) + |\lambda|(\Phi^\dagger \Phi)^2$: (a) $\mu^2 > 0$ and (b) $\mu^2 < 0$.

system corresponds to

$$V = -\frac{|\mu|^4}{4|\lambda|} \tag{43}$$

The vacuum expectation value of Φ is then any state $\langle\Phi\rangle_0$ with

$$\langle\Phi\rangle_0^\dagger\langle\Phi\rangle_0 = -\frac{\mu^2}{2|\lambda|} \tag{44}$$

It turns out that nature has chosen $\mu^2 < 0$.

The true behavior of the Lagrangian (42) is revealed when the scalar field Φ is assumed to deviate only slightly from its nonzero vacuum expectation value $\langle\Phi\rangle_0$. To this end we choose

$$\langle\Phi\rangle_0 = \begin{pmatrix} 0 \\ v/\sqrt{2} \end{pmatrix}, \quad v = \sqrt{\frac{-\mu^2}{|\lambda|}} \tag{45}$$

and write

$$\Phi(x) = \langle\Phi\rangle_0 + \begin{pmatrix} 0 \\ \eta(x)/\sqrt{2} \end{pmatrix} + \sum_{j=1}^{3} i\frac{\xi^j(x)}{v}(T^j\langle\Phi\rangle_0) \tag{46}$$

where ξ^j and η are small-valued functions of the space-time coordinates x. After a carefully chosen $SU(2) \times U(1)$ local gauge transformation, the scalar term of the Lagrangian (42) will then contain the terms

$$\frac{1}{2}(\partial^\mu\eta)(\partial_\mu\eta) - \mu^2\eta^2$$

$$+ \frac{g^2v^2}{8}[|W_\mu^+|^2 + |W_\mu^-|^2]$$

$$+ \frac{v^2}{8}[(gW_\mu^3 - g'B_\mu^3)^2]$$

$$- \frac{Gv}{\sqrt{2}}(\bar{e}_R e_L + \bar{e}_L e_R) \tag{47}$$

where we have taken ψ_R and ψ_L to be

$$\psi_R = e_R \quad \text{and} \quad \psi_L = \begin{pmatrix} \nu_e \\ e \end{pmatrix}_L \tag{48}$$

The term

$$\frac{g^2v^2}{8}[|W_\mu^+|^2 + |W_\mu^-|^2] \tag{49}$$

looks like a mass term for the $W^\pm$ gauge bosons. The term

$$\frac{v^2}{8}[(gW_\mu^3 - g'B_\mu^3)^2]$$

$$= \frac{v^2e^2}{8\sin^2\theta_W\cos^2\theta_W}$$

$$[(\cos\theta_W W_\mu^3 - \sin\theta_W B_\mu^3)^2] \tag{50}$$

is a mass term for the linear combination of W_μ^3 and B_μ corresponding to the Z^0 gauge boson, Eq. (32). There is no mass term for the orthogonal combination corresponding to the photon A_μ.

When the field Φ is expanded about a nonzero vacuum expectation value $\langle\Phi\rangle_0$, and only low-order terms are kept, the Lagrangian is no longer gauge-invariant. The symmetry is said to be spontaneously broken by the nonzero vacuum expectation value. The method by which the gauge bosons of gauge field theories acquire mass through spontaneous symmetry breaking is called the Higgs mechanism. S. Weinberg and A. Salam were the first to realize that the Higgs mechanism could be applied to a gauge theory of the electromagnetic and weak interactions to give mass to the weak force gauge bosons.

The term

$$\frac{1}{2}(\partial^\mu\eta)(\partial_\mu\eta) - \mu^2\eta^2 \tag{51}$$

in Eq. (47) can be identified with the kinetic energy term of a scalar particle of mass $\sqrt{2}|\mu|$. Of the four real fields,

$$\xi^1, \quad \xi^2, \quad \xi^3, \quad \eta \tag{52}$$

in our expression (46) for the weak isospin doublet Φ, three have disappeared as three gauge bosons have acquired mass, while one remains. The remaining scalar particle has 0 charge, is called the neutral Higgs boson, and is denoted by H^0. The mass of the H^0 is a free parameter of the $SU(2) \times U(1)$ theory. The H^0 particle has never been observed.

D. ELECTROWEAK COUPLING OF QUARKS

Like the leptons, the left-handed components of quarks mix under weak isospin rotations, and the right-handed components of quarks remain unaltered. Left-handed quarks form the following weak isospin doublets:

$$L_u \equiv \begin{pmatrix} u \\ d' \end{pmatrix}_L, \quad L_c \equiv \begin{pmatrix} c \\ s' \end{pmatrix}_L, \quad L_t \equiv \begin{pmatrix} t \\ b' \end{pmatrix}_L \tag{53}$$

where the primed quantities d', s', and b' are mixtures of the left-handed components of d, s, and b quarks:

$$
\begin{pmatrix} d'_L \\ s'_L \\ b'_L \end{pmatrix}
$$

$$
= \begin{pmatrix} c_1 & -s_1 c_3 & -s_1 s_3 \\ s_1 c_2 & c_1 c_2 c_3 - s_2 s_3 e^{i\delta} & c_1 c_2 c_3 + s_2 c_3 e^{i\delta} \\ s_1 s_2 & c_1 s_2 c_3 + c_2 s_3 e^{i\delta} & c_1 s_2 s_3 - c_2 c_3 e^{i\delta} \end{pmatrix}
$$

$$
\times \begin{pmatrix} d_L \\ s_L \\ b_L \end{pmatrix} \quad (54)
$$

where $c_i = \cos \theta_i$ and $s_i = \sin \theta_i$ for $i = 1, 2$, and 3. The matrix (54) is called the Kobayashi–Maskawa or K–M matrix. The angles θ_1, θ_2, θ_3, and δ are called the Kobayashi–Maskawa angles. The components of the matrix have the values

$$
\begin{pmatrix} 0.9705 \text{ to } 0.9770 & 0.21 \text{ to } 0.24 \\ 0.21 \text{ to } 0.24 & 0.971 \text{ to } 0.973 \\ 0. \text{ to } 0.024 & 0.036 \text{ to } 0.069 \end{pmatrix}
$$

$$
\left.\begin{matrix} 0. \text{ to } 0.014 \\ 0.036 \text{ to } 0.070 \\ 0.997 \text{ to } 0.999 \end{matrix}\right) \quad (55)
$$

so the d' state is mostly d_L, the s' state is mostly s_L, and the b' state is mostly b_L. The right-handed quark states are each weak isospin singlets:

$$
u_R, \, d_R, \, s_R, \, c_R, \, b_R, \, t_R \quad (56)
$$

The weak isospin eigenvalues T^3 and hypercharges Y for left-handed and right-handed quarks are given in Table VII.

The mixing of different generation quarks in the same weak isospin doublet comes about because nature has taken full advantage of the most general $SU(2) \times U(1)$ locally gauge invariant coupling between the Higgs scalar Φ and fermions. This coupling can be written as

$$
\begin{aligned}
& -G_1[(\bar{L}_u \bar{\Phi}) u_R + \bar{u}_R (\bar{\Phi}^\dagger L_u)] \\
& -G_2[(\bar{L}_u \Phi) d_R + \bar{d}_R (\Phi^\dagger L_u)] \\
& -G_3[(\bar{L}_u \Phi) s_R + \bar{s}_R (\Phi^\dagger L_u)] \\
& -G_4[(\bar{L}_c \bar{\Phi}) c_R + \bar{c}_R (\bar{\Phi}^\dagger L_c)] \\
& -G_5[(\bar{L}_c \Phi) d_R + \bar{d}_R (\Phi^\dagger L_c)] \\
& -G_6[(\bar{L}_c \Phi) s_R + \bar{s}_R (\Phi^\dagger L_c)]
\end{aligned} \quad (57)
$$

plus analogous terms involving L_t, b_R, and t_R. The G_i are Yukawa coupling constants. The same Higgs mechanism that gave mass to the $W^\pm$ and Z^0 gauge bosons will also lead to the

terms

$$
-\frac{G_1 v}{\sqrt{2}} (\bar{u}_R u_L + \bar{u}_L u_R)
$$

$$
-\left(\frac{G_2 v}{\sqrt{2}} \cos \theta_1 + \frac{G_5 v}{\sqrt{2}} \sin \theta_1 \cos \theta_2 \right)
$$

$$
\times (\bar{d}_R d_L + \bar{d}_L d_R)
$$

$$
-\left(\frac{G_6 v}{\sqrt{2}} \left(\cos \theta_1 \cos \theta_2 \cos \theta_3 - \sin \theta_2 \sin \theta_3 e^{i\delta} \right) \right.
$$

$$
\left. -\frac{G_3 v}{\sqrt{2}} \sin \theta_1 \cos \theta_3 \right)(\bar{s}_R s_L + \bar{s}_L s_R)
$$

$$
-\frac{G_4 v}{\sqrt{2}} (\bar{c}_R c_L + \bar{c}_L c_R)
$$

$$
+ \text{ terms involving } b_L, b_R, t_L, \text{ and } t_R \quad (58)
$$

where v is related to the Higgs vacuum expectation value Eq. (45). These terms look like fermion mass terms (recall Eq. 38). The d quark, for example, has a mass

$$
\frac{G_2 v}{\sqrt{2}} \cos \theta_1 + \frac{G_5 v}{\sqrt{2}} \sin \theta_1 \cos \theta_2 \quad (59)
$$

If $G_3 = G_5 = 0$, then the weak eigenstates d' and s' are the left-handed components of the mass eigenstates, and there is no mixing of weak eigenstates and mass eigenstates. If G_3 and G_5 are nonzero, then there will be mixing.

Weak Neutral Coupling of Quarks

The term in the $SU(2) \times U(1)$ Lagrangian that corresponds to the weak neutral current coupling of quarks is

$$
\frac{e}{\sin \theta_W \cos \theta_W} [\bar{L}_u \gamma^\mu Z_\mu (T_3 - Q \sin^2 \theta_W) L_u
$$

$$
+ \bar{L}_c \gamma^\mu Z_\mu (T_3 - Q \sin^2 \theta_W) L_c
$$

$$
- \sin^2 \theta_W (\bar{u}_R \gamma^\mu Z_\mu Q u_R + \bar{d}_R \gamma^\mu Z_\mu Q d_R
$$

$$
+ \bar{s}_R \gamma^\mu Z_\mu Q s_R + \bar{c}_R \gamma^\mu Z_\mu Q c_R)] \quad (60)
$$

where we have neglected b and t quarks. Expanding the left-handed coupling, we have

$$
\bar{L}_\mu \Gamma L_u + \bar{L}_c \Gamma L_c = \bar{u}_L \Gamma u_L + \bar{d}'_L \Gamma d'_L
$$

$$
+ \bar{s}'_L \Gamma s'_L + \bar{c}_L \Gamma c_L \quad (61)
$$

where

$$
\Gamma \equiv \gamma^\mu Z_\mu (T^3 - Q \sin^2 \theta_W) \quad (62)
$$

Expanding the terms for d_L and s_L, we have

$$
\bar{d}'_L \Gamma d'_L = \cos^2 \theta_1 \bar{d}_L \Gamma d_L + \sin^2 \theta_1 \bar{s}_L \Gamma s_L
$$

$$
- \sin \theta_1 \cos \theta_1 (\bar{s}_L \Gamma d_L + \bar{d}_L \Gamma s_L) \quad (63)
$$

and

$$\bar{s}'_L \Gamma s'_L = \sin^2 \theta_1 \bar{d}_L \Gamma d_L + \cos^2 \theta_1 \bar{s}_L \Gamma s_L$$
$$+ \sin \theta_1 \cos \theta_1 (\bar{s}_L \Gamma d_L + \bar{d}_L \Gamma s_L) \quad (64)$$

where we have set $\theta_2 = \theta_3 = 0$.

At the time Weinberg and Salam proposed their $SU(2) \times U(1)$ theory, the charm quark had not been discovered, so L_c and hence the term $\bar{s}'_L \Gamma s'_L$ in Eq. (61) did not exist. This meant that the term

$$- \sin \theta_1 \cos \theta_1 (\bar{s}_L \Gamma d_L + \bar{d}_L \Gamma s_L) \quad (65)$$

in the expression (63) for $\bar{d}'_L \Gamma d'_L$ was left uncanceled. Such an interaction term would lead to decay processes such as

$$K^+ \rightarrow \pi^+ + \nu + \bar{\nu} \quad (66)$$

which, in the quark model view, is the decay

$$\bar{s} \rightarrow \bar{d} + \nu + \bar{\nu} \quad (67)$$

The predicted rate for such a strangeness-changing neutral current decay was eight orders of magnitude larger than the measured limit. S. Glashow, J. Iliopoulos, and L. Maiani demonstrated in 1970 that a fourth quark would lead to the cancellation of terms in the Lagrangian responsible for strangeness-changing neutral currents. Higher-order diagrams would still lead to strangeness-changing neutral currents, even with a fourth (charm) quark present, and to suppress again the decay rate of Eq. (66) it was predicted that the charm quark mass must be less than a few GeV. In 1974 the J/Ψ particle, a meson with a mass of about 3 GeV consisting of a charm quark and a charm antiquark, was discovered simultaneously in hadron–hadron collisions at Brookhaven and in e^+e^- collisions at SLAC. This gave a charm quark mass of about 1.5 GeV. The discovery of the J/Ψ meson therefore provided support, not only for the quark model of hadrons but also for the $SU(2) \times U(1)$ electroweak theory.

E. ELECTROWEAK PHENOMENOLOGY

Muon decay [Fig. 6(b)] proceeds through the charged vector boson W^-. The amplitude for the decay rate is proportional to

$$\frac{g^2}{M_W^2} \quad (68)$$

This is because each vertex in Fig. 6(b) contributes a factor g, while the W^- propagator contributes a factor

$$\frac{1}{(q^2 - M_W^2)} \quad (69)$$

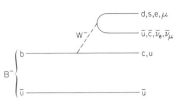

FIG. 22.   Diagram for the decay of the B$^-$ meson.

where q^2 is the square of the momentum transferred by the W^- in the decay. For muon decay, $|q|^2 \ll M_W^2$, so that the muon decay amplitude is simply proportional to Eq. (68).

Hadrons can also decay through the $W^\pm$. For example, a meson with a $\bar{u}$ and b quark that cannot decay by any other means decays eventually through the charged weak coupling (Fig. 22). The decay rate amplitude for a particular final state is proportional to the product of g^2/M_W^2 and the two K–M matrix elements associated with the two vertices in Fig. 22. By comparing the charged weak decays of many particles (both leptons and hadrons), we can measure the mixing angles of the Kobayashi–Maskawa matrix (54).

Three independent parameters determine the coupling of the $SU(2) \times U(1)$ gauge bosons to quarks and leptons. These can be taken to be the magnitude of the charge of the electron e, the Weinberg angle θ_W, and a quantity called the Fermi constant G_F. G_F is defined to be

$$G_F = \frac{g^2}{4\sqrt{2}M_W^2} \quad (70)$$

Notice that experiments involving low q^2 $W^\pm$ exchange, such as muon decay and charged current neutrino scattering experiments, measure G_F, but they do not measure g and M_W separately. G_F is measured to be $G_F = 1.16632 \pm 0.00004 \times 10^{-5}$ GeV^{-2}.

From Eq. (34) we have the relation

$$g = \frac{e}{\sin \theta_W} \quad (71)$$

so the mass of the $W^\pm$ can be expressed as

$$M_W^2 = \frac{e^2}{4\sqrt{2}G_F \sin^2 \theta_W} \quad (72)$$

According to Eq. (50) the Z^0 mass is

$$M_Z^2 = \frac{e^2 v^2}{4 \sin^2 \theta_W \cos^2 \theta_W} \quad (73)$$

where v is, from Eq. (49),

$$v^2 = \frac{4M_W^2}{g^2} = \frac{1}{\sqrt{2}G_F} \quad (74)$$

so that

$$M_Z^2 = \frac{M_W^2}{\cos^2 \theta_W} \tag{75}$$

When Weinberg and Salam proposed their $SU(2) \times U(1)$ theory in 1967, the charge of the electron e was, of course, known, and the Fermi coupling constant G_F had also been measured. However, the Weinberg angle θ_W was an unknown quantity, and hence there was no estimate of the $W^\pm$ and Z masses.

When neutral weak current events were observed in the early 1970s, physicists got their first look at the magnitude of θ_W. Figure 20 shows the Born diagram for the neutral weak current process

$$\nu_\mu + e \rightarrow \nu_\mu + e$$

As usual, we represent the amplitude for this process by multiplying together the couplings at the two vertices and the propagator. The neutrino vertex is

$$\frac{e}{2 \sin \theta_W \cos \theta_W} \tag{76}$$

and the propagator contributes

$$\frac{1}{M_Z^2} \tag{77}$$

and we represent the electron vertex as the square root of the sum of squares of the left- and right-handed contributions:

$$\left(\frac{e}{2 \sin \theta_W \cos \theta_W}\right)\left(\frac{1}{4} - \sin^2 \theta_W + 2 \sin^4 \theta_W\right)^{1/2} \tag{78}$$

Combining the three terms, the amplitude for elastic muon–neutrino electron scattering is proportional to

$$\frac{e^2}{4 \sin^2 \theta_W \cos^2 \theta_W} \frac{1}{M_Z^2}$$
$$\times \left(\frac{1}{4} - \sin^2 \theta_W + 2 \sin^4 \theta_W\right)^{1/2} \tag{79}$$

We have neglected effects due to angular momentum conservation and the relative spins of the neutrino and the electron. When these are included, the relative contribution of the right-handed electron is reduced by 66%. The important point is that the neutral current cross section is sensitive to θ_W. Once θ_W had been measured, in experiments such as muon–neutrino electron, electron–neutrino electron, and neutrino–hadron scattering, the mass of the $W^\pm$ and Z^0 gauge bosons could be predicted.

The measured value of θ_W is usually expressed in terms of $\sin^2 \theta_W$, which is about 0.2. Thus the mass of the $W^\pm$ boson should be about 84 GeV, and the mass of the Z^0 should be about 94 GeV. Direct observations of the $W^\pm$ and Z^0 bosons did not take place until 1983 when the $W^\pm$ and Z^0 bosons were produced in proton–antiproton collisions at 540 GeV center-of-mass energy at CERN. Recalling the quark composition of the proton (Table II), we find that the fundamental processes leading to $W^\pm$ production are

$$u + \bar{d} \rightarrow W^+$$
$$d + \bar{u} \rightarrow W^- \tag{80}$$

and the fundamental processes leading to Z^0 production are

$$u + \bar{u} \rightarrow Z^0$$
$$d + \bar{d} \rightarrow Z^0 \tag{81}$$

The $W^\pm$ are observed through their decay to $\nu_e + e^+$ or $\bar{\nu}_e + e^-$, and the Z^0 is observed through its decay to an e^+e^- or $\mu^+\mu^-$ pair. The measured masses of the $W^\pm$ and Z^0 gauge bosons (Table I) are as predicted by Eqs. (72) and (75).

IX. Current Issues in Elementary Particle Physics

Research efforts in elementary particle physics today can be broadly classified into two categories: continued tests of the standard model and experimental searches for, and theoretical models of, physics beyond the standard model. By standard model we mean the description of elementary particle physics presented in this article: three generations of quarks and leptons which interact through the $SU(2) \times U(1)$ theory of the electroweak force and the $SU(3)_{color}$ theory of the strong force.

A. CONTINUED TESTS OF THE STANDARD MODEL

All high-energy physics phenomena appear to fit the standard model. There are, however, many aspects of the standard model that have yet to be tested. First, almost all experiments sensitive to the electroweak force have involved the exchange of gauge bosons with invariant-mass squared values of $|q^2| \ll M_W^2, M_Z^2$. The one (very notable) exception is the observation of W and Z gauge bosons at the CERN proton–antiproton collider. Will the standard model continue to stand up under close scrutiny at $|q^2| > M_W^2, M_Z^2$? Second, most experiments involving the strong force have taken place at energies where the strong coupling constant α_S was greater than 1. It is very difficult to test quantum chromodynamics quantitatively under conditions where perturbative calculations cannot be made. It is hoped that as experiments are performed at higher energies, there will be a greater opportunity to compare QCD predictions with experiment. Third, two particles are predicted by the standard model that have yet to be seen: the top quark and the neutral Higgs boson. The standard model does not predict the masses of these two particles, but it does predict that they must exist at some mass.

The standard model parameters that experimental physicists measure are

1. the mass of the Z^0 gauge boson, M_Z,
2. the mass of the W gauge boson, M_W,
3. the mixing angles of the Kobayashi–Maskawa matrix, θ_1, θ_2, and θ_3,
4. upper limits on the masses of the electron–neutrino, muon–neutrino, and tau–neutrino,
5. The QCD coupling constant α_S, evaluated at a particular q^2, and
6. the couplings of the quarks and leptons to the γ, W, and Z gauge bosons.

All of these quantities have been measured to some degree of accuracy already. Many of them will be measured to a greater accuracy at $q^2 = M_Z^2$ in the late 1980s at the SLC e^+e^- linear collider at SLAC and at the LEP e^+e^- storage ring at CERN. The cross section for

$$e^+ + e^- \rightarrow Z^0 \rightarrow \text{anything}$$

has a sharp peak at center-of-mass energies equal to the mass of the Z^0, so many Z^0 bosons will be produced at SLC and LEP, even if the luminosities at these machines are in a certain sense average or below. Other experiments at lower q^2 will continue to provide better values for the mixing angles of the Kobayashi–Maskawa matrix and better upper limits on the masses of the neutrinos.

Some standard model parameters are not known at all. These are the top quark mass, the neutral Higgs boson mass, and the phase angle δ of the Kobayashi–Maskawa matrix. Actually, we do know a little about these parameters. We know, since hadrons containing top quarks have not been produced at the PETRA e^+e^- collider at DESY, that the top quark mass must be greater than 22.5 GeV. The neutral Higgs boson mass is constrained through theoretical considerations to be between 7 and 1000 GeV.

B. CP Violation

The phase angle δ in Eq. (54) is of great interest because a nonzero δ can produce CP violation in weak interactions. CP represents the combined operations of parity transformation P and charge conjugation C. CP violation occurs when the operation CP is not a symmetry of the Lagrangian. The mass eigenstates of a system that violates CP are not eigenstates of CP.

CP violation shows up experimentally in the decays of neutral K mesons. The neutral K meson mass eigenstates K_S^0 and K_L^0 are linear combinations of the strange quark eigenstates $|K^0> = |\bar{d}s>$ and $|\bar{K}^0> = |d\bar{s}>$:

$$|K_S^0> = a_{11}|K^0> + a_{12}|\bar{K}^0>$$
$$|K_L^0> = a_{21}|K^0> + a_{22}|\bar{K}^0>$$

$$\begin{pmatrix} a_{11} & a_{12} \\ a_{21} & a_{22} \end{pmatrix} \approx \begin{pmatrix} \dfrac{1}{\sqrt{2}} & \dfrac{1}{\sqrt{2}} \\ \dfrac{1}{\sqrt{2}} & -\dfrac{1}{\sqrt{2}} \end{pmatrix}$$

It was once thought that $|K_S^0>$ and $|K_L^0>$ were eigenstates of CP; but in 1964, J. Cronin and V. Fitch discovered a decay mode of the $|K_L^0>$ state that indicated that $|K_S^0>$ and $|K_L^0>$ were not quite eigenstates of CP.

The standard $SU(2) \times U(1)$ weak interaction with only two generations of fermions will not have a complex phase $e^{i\delta}$ (the K–M matrix is parametrized in such a case by only one angle). Such a Lagrangian will violate parity and charge conjugation separately but will conserve the combined operation of CP. In 1973, M. Kobayashi and T. Maskawa demonstrated that, in the standard $SU(2) \times U(1)$ theory with three generations of quark doublets, a complex phase $e^{i\delta}$ arises in the matrix describing the mixing of weak and mass quark eigenstates. This phase will lead to a violation of CP in the $SU(2) \times U(1)$ Lagrangian.

Although we know today that there are three generations of fundamental fermions, we do not know if the complex phase in the Kobayashi–Maskawa matrix can explain the observed CP violation of the $K^0\bar{K}^0$ system. The amount by which CP is violated is very small, and quantitative predictions of $K^0\bar{K}^0$ CP violation in terms of δ, θ_1, θ_2, θ_3, and the quark masses are difficult to make because nonperturbative QCD effects are very important. Nevertheless, thanks to recent improved experimental measurements of θ_1, θ_2, and θ_3, physicists have been able to place tight limits on the allowed values of the phase angle δ. New, more sensitive $K^0\bar{K}^0$ CP violation experiments are planned for the mid to late 1980s, and these may tell us whether or not the standard model can account for $K^0 \bar{K}^0$ CP violation.

There is a general result from quantum field theory that states that the combined operations of charge conjugation, parity transformation, and time reversal (CPT) should be a symmetry of any Lagrangian. CP violation therefore implies time reversal violation. Time reversal violation should show up as a small electric dipole moment in the neutron. Experimental limits on the electric dipole moment of the neutron continue to improve; if ever a nonzero value is found, it will add substantially to the understanding of CP violation.

C. Physics Beyond the Standard Model

Even if all predictions of the standard model were to be confirmed to the umpteenth decimal place, few physicists would be satisfied with the standard model. Perhaps the biggest problem with the standard model is the multitude of fundamental constants and fundamental fields. The standard model contains 18 fundamental parameters: 3 coupling constants (α, α_S, $\sin^2 \theta_W$), the neutral Higgs boson mass, the Higgs vacuum expectation value, and 13 Yukawa couplings, corresponding to 4 Kobayashi–Maskawa angles, 3 charged lepton masses, and 6 quark masses. The standard model has nothing to say about these parameters. Should the fundamental theory of elementary particle physics contain so many arbitrary constants? The number of fermion generations (three) is also arbitrary (there may even be more generations at higher masses—we shall have to determine this from experiment).

Other questions left unanswered by the standard model are the following. Why are fermions in the representations they are in: left-handed quarks and leptons in $SU(2)$ doublets, right-handed quarks and leptons in $SU(2)$ singlets, all leptons in $SU(3)_{color}$ singlets, and all quarks in $SU(3)_{color}$ triplets? Why do the fundamental fermions have the hypercharges and hence the electric charges that they have? (Why is the charge of the electron precisely three times the charge of the down quark?) Why are neutrinos massless or at least nearly massless? Where does the form of the Higgs scalar potential come from, and why do the constants describing this potential have the values that they have? Is the Higgs scalar a fundamental particle or is it com-

posite? Are quarks and leptons composite objects? (The repetition of quark–lepton generations with higher-mass fermions in successive generations certainly suggests such a possibility.)

There are many ideas for physics beyond the standard model, each of which attempts to deal with one or more of the above questions. These ideas include

1. grand unified gauge theories of the electroweak and strong forces,
2. dynamical symmetry breaking, and
3. supersymmetric gauge theories.

1. Grand Unified Gauge Theories

Grand unified gauge theories attempt to combine the $SU(3)_{color}$ and $SU(2) \times U(1)$ gauge theories into a single gauge theory with one gauge group and one coupling constant. Such theories reduce the number of fundamental coupling constants from three to one and explain why leptons and quarks have the electric charge ratios that they do. The development of grand unified gauge theories has been motivated by the success of the $SU(3)_{color}$ and $SU(2) \times U(1)$ gauge theories, the striking similarity between quarks and leptons, and the possibility that the running coupling constants of the electromagnetic, weak, and strong forces might converge at some very large energy (Fig. 18).

The simplest grand unified gauge theory is based on the group $SU(5)$. In this theory all the quarks and leptons in one generation are placed in the same $SU(5)$ multiplet. Such an assignment produces simple algebraic relations between the electric charges of quarks and leptons that explain the observed ratios of quark and lepton charges and the fact that the sum of electrical charges over fermions in the same generation is zero.

Another consequence of placing quarks and leptons in the same gauge group multiplet is that there will be gauge bosons, called X and Y gauge bosons, that mediate transitions between quarks and leptons and that therefore induce interactions that violate baryon number and lepton number conservation. It is presumed that some Higgs mechanism breaks the $SU(5)$ symmetry to $SU(3)_{color} \times SU(2) \times U(1)$ and in the process imparts very large masses to the X and Y gauge bosons. This first level of spontaneous symmetry breakdown leaves the $SU(3) \times SU(2) \times U(1)$ gauge bosons massless, and the $W^{\pm}$ and Z^0 gauge bosons then acquire mass in a second

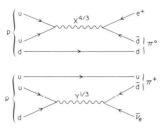

FIG. 23. Diagrams for the decay of the proton through the $SU(5)$ X and Y gauge bosons.

spontaneous symmetry breakdown in a manner analogous to the process described in Section VIII,C.

The X and Y gauge bosons induce proton decay (Fig. 23); and for the proton lifetime to be extended to values greater than the current lower limit of 10^{32} years, the masses of the X and Y gauge bosons must be greater than about 10^{15} GeV. The X and Y gauge boson masses are further constrained by the requirement that they be roughly equal to the energy at which the running weak and strong coupling constants converge. Applying the boundary condition that these two running coupling constants are equal at some unification energy, we can obtain relations between the electromagnetic and strong coupling constants. If we substitute known low-energy values for these coupling constants, the unification energy is found to be about 10^{15} GeV. The current limits on proton decay are therefore very close to the values predicted by some grand unified theories; in fact, the $SU(5)$ grand unified theory appears to be excluded by present experimental limits.

Although proton decay is not observed at the rate predicted by the $SU(5)$ grand unified theory, the theory does do an admirable job of predicting $\sin^2 \theta_W$ at present-day energies. Requiring that the weak and strong coupling constants be equal at a unification energy leads to an expression for $\sin^2 \theta_W (q^2)$ in terms of $\alpha(q^2)$ and the unification energy. With a known low-energy value for α, and 10^{15} GeV for the unification energy, $\sin^2 \theta_W$ is predicted to be 0.21 at 100 GeV.

2. Dynamical Symmetry Breaking

The standard model doublet of complex scalar fields Φ is a set of four fundamental scalar fields. Some physicists find fundamental scalar fields objectionable. The problem is that the mass of the neutral Higgs boson (or the mass of any other fundamental scalar field) is overly sensi-

tive to details of the electroweak theory at very high energies. It is often said that there is no "natural" way for the Higgs boson mass to be less than 1000 GeV, because small variations in parameters describing the electroweak theory at large energies cause enormous variations in the Higgs boson mass. Fundamental fermions do not have this problem.

Models have been proposed in which the standard model fundamental scalar fields are replaced by composite scalar fields made of new fundamental fermions. Such models are called dynamical symmetry breaking models.

One well-known dynamical symmetry-breaking model is technicolor. In technicolor there are new fundamental fermions called techniquarks that interact and form bound states of technihadrons through a new force called technicolor. Techniquarks and the technicolor force are entirely analogous to quarks and the $SU(3)_{\text{color}}$ force; the major difference between the two systems is that the technicolor mass scale is 1 TeV, compared with 1 GeV for ordinary $SU(3)_{\text{color}}$.

Technicolor is relevant to the question of fundamental scalar fields, because the pions of the technicolor sector (technipions) can be used instead of fundamental scalar fields to give mass to the $W^{\pm}$ and Z^0 gauge bosons. When the $SU(2) \times U(1)$ symmetry is broken by technipions, the $W^{\pm}$ and Z^0 gauge bosons acquire masses that are in the same ratio as in the standard model. A technicolor mass scale of 1 TeV will produce $W^{\pm}$ and Z^0 masses on the order of 100 GeV.

The problem with dynamical symmetry breaking is that there is no known way to construct a phenomenologically acceptable model that gives mass to the quarks and leptons. Recall that the single complex Higgs doublet of the standard model gave mass not only to the $W^{\pm}$ and Z^0 gauge bosons but also to quarks and leptons.

3. Supersymmetric Gauge Field Theories

All quarks and leptons have spin $\frac{1}{2}$, all gauge bosons (photon, gluon $W^{\pm}$ and Z^0) have spin 1, and the neutral Higgs scalar has spin 0. Some physicists believe that there exist as yet unseen quarks and leptons with spin 0, and gauge bosons and Higgs scalars with spin $\frac{1}{2}$. Such particles are predicted by a class of gauge field theories called supersymmetric gauge field theories. Supersymmetric theories predict that for every fundamental particle with spin s, there is another particle, called the supersymmetric partner, with a spin that differs from s by $\frac{1}{2}$ unit.

The transformation that changes a particle's spin by $\frac{1}{2}$ unit is called a supersymmetry transformation. The Lagrangians of supersymmetric theories are invariant under supersymmetry transformations.

Supersymmetric theories are considered attractive for a number of reasons. First, in supersymmetric theories the masses of fundamental scalar fields are no longer overly sensitive to details of the theory at very high energies. Second, theorists find that supersymmetric quantum field theories are "better behaved" than other quantum field theories. Third, in locally supersymmetric theories there is a connection between supersymmetry transformations and space-time transformations. This implies that it may be possible to use locally supersymmetric theories to unify the gravitational force with the strong and electroweak forces.

To date there is no experimental evidence for supersymmetry. If there are supersymmetric partners of the known quarks, leptons, and gauge bosons, then their masses must be greater than about 20 GeV.

BIBLIOGRAPHY

Georgi, H. (1984). "Weak Interactions and Modern Particle Theory." Benjamin/Cummings, Menlo Park, CA.

Gottfried, K., and Weisskopf, V. F. (1984). "Concepts of Particle Physics." Clarendon Univ. Press, Oxford.

Hazlen, F., and Martin, A. D. (1984). "Quarks and Leptons: An Introductory Course in Modern Particle Physics." Wiley, New York.

Particle Data Group. (1984). *Review of Modern Physics* **56**(2), Part II, April.

Perkins, D. H. (1982). "Introduction to High-Energy Physics." 2nd ed. Addison-Wesley, Reading, MA.

Quigg, C. (1983). "Theories of the Strong, Weak, and Electromagnetic Interactions." Benjamin/Cummings, Menlo Park, CA.

ENERGY TRANSFER, INTRAMOLECULAR*

Paul W. Brumer   *University of Toronto*

I.   Introduction
II. Qualitative Dynamics
III. Dynamics: Theory
IV. Statistical Approximations and Dynamics
V.   Experimental Studies
VI. Summary

GLOSSARY

Adiabatic: Process during which there is no change in the electronic configuration of the molecule.

Constant of the motion: Property of a system, expressible as a smooth function of the coordinates and momenta, whose numerical value remains unchanged during the course of the system dynamics.

Ergodic: Classical mechanical system in which a trajectory uniformly cover a specific surface in phase space. The physics literature utilizes this term to imply uniform coverage of the surface in phase space defined by fixed total energy.

Integrable: Also termed regular or quasiperiodic. Classical mechanical system characterized by the existence of a set of independent constants of the motion equal in number to the number of degrees of freedom of the system. Specific attributes of such systems are discussed in the text.

Mixed state: State of a system in which there is some information missing relative to a complete specification of the state.

Mixing: Classical mechanical system that is ergodic and possesses additional properties associated with relaxation.

* We are grateful to the donors of the Petroleum Research Fund, administered by the American Chemical Society, for partial support of the research on which this overview is based.

Pure state: State of a system in which all information about the state is known and specified.

Relaxation: Tendency of a system to evolve from a specific time-dependent initial state to a time-invariant final state.

Statistical: Qualitative term signifying models or dynamics with characteristics of ergodic and mixing behavior.

Unimolecular decay: Process in which an energized molecule breaks up into smaller molecular constituents. The molecular analog of nuclear fission.

Intramolecular dynamics is the time evolution of rotational, vibrational, and electronic degrees of freedom of isolated individual molecules, that is, molecules in a collision-free environment. As in any mechanical system, one can consider the time evolution of any of a variety of system properties. Intramolecular energy transfer focuses on the dynamical flow of energy among subcomponents of the molecule as the primary property of interest.

I. Introduction

Naturally occurring systems in the gaseous or liquid phase at typical temperatures are comprised of vast numbers of atoms and molecules in constant motion. The well-defined system temperature is a reflection of the stored energy content, both in the form of *inter*molecular features such as molecule–molecule interactions and translational motion and in the form of motion internal to individual molecules. The latter, the motion of *isolated* molecules, is termed *intra*molecular dynamics. Included within this definition are both dynamics at energies below the dissociation energy of the molecule, in which case it remains perpetually bound, and dynamics at energies above dissociation, in which case the molecule can break up into different chemical products. [*See* CHEMICAL PHYSICS.]

Intramolecular energy transfer is the subset of intramolecular dynamics in which the focus of attention is on the flow of energy within the isolated molecule. As the simplest of models, one may imagine a linear molecule A—B—C as being three mass points coupled by springs. Initiating an oscillation of the model by stretching the A—B spring, and subsequently observing the time-dependent alteration of lengths of the A—B and B—C springs corresponds to a simple model of an experiment on vibrational energy transfer. Indeed, modern experimental techniques allow for the creation of collision-free environments in which such isolated intramolecular dynamics may be studied and possibly externally influenced.

The isolated molecule, comprised of a set of N atoms bound by interatomic forces, is a complex physical system. It possesses $3N$ degrees of freedom related to nuclear motion, three being associated with center-of-mass translation, three with rotation (or two if the molecule is linear), and the vast remainder with vibration. The molecule also has electronic degrees of freedom associated with the configuration of its electrons. Additional degrees of freedom, related to internal nuclear composition are not readily altered at the energies of interest in molecular chemistry and physics and can therefore be neglected. Thus, even the simplest description of the dynamics of a typical small molecule such as benzene (C_6H_6), which regards it as being composed of 12 atoms, involves the complex motion of 33 interacting degrees of freedom. [*See* STATISTICAL MECHANICS.]

It is convenient to visualize intramolecular dynamics in terms of three steps, not necessarily independent. First, the molecule is prepared in a time-dependent state by any of a variety of means (e.g., collisions, chemical reaction, laser excitation). Second, the molecule evolves in time in accordance with quantum mechanics. Third, the time-evolved state of the system is analyzed, that is, measured. Explicit emphasis in a measurement (and sometimes preparation) on the flow of energy among subcomponents of the molecule (e.g., chemical bonds or rotational degrees of freedom) constitutes the study of intramolecular energy transfer.

It is advantageous to distinguish two qualitatively different types of intramolecular energy flow, which we shall term reversible and irreversible. In the former, energy flows from one part of the molecule to another, but then subsequently returns, reforming the initial state. That is, the system shows no long-term trend toward a final redistribution of energy among subcomponents of the molecule. In the latter, there is a transfer of energy within the molecule, with a general trend toward a stationary final state. The latter behavior is that associated with relaxation, or statistical, dynamics and has historically been assumed to occur in highly excited molecules. Conditions under which molecular systems appear to display reversible versus irreversible energy transfer are discussed in detail later below.

Intramolecular dynamics and intramolecular energy transfer have been, and continue to be, areas of intense scientific interest. Such interest falls into two categories, loosely termed "practical" and "fundamental." From the practical viewpoint, note that the outcome of a molecular process is heavily linked to the flow of energy in the molecular participants. That is, an understanding of intramolecular energy transfer proves central to the interpretation of chemical processes and their dependence on system conditions. For example, unimolecular decay, in which an isolated energized molecule dissociates into chemical products (e.g, ABC → A + BC, where ABC, A, and BC are arbitrary molecules), occurs via the concentration of sufficient energy in the A—BC bond. Further practical interest arises from important developments in laser technology that permit the introduction of energy into molecules in a variety of controlled ways. This opens the possibility of externally influencing the outcome of a molecular process by varying the initial mechanism of preparation (e.g., producing AB + C, rather than A + BC, from ABC by judicious preparation of the initial state). Whether such selectivity can be achieved in practice is the subject of intense current theoretical and experimental interest.

From a fundamental viewpoint, intramolecular energy transfer links to three basic scientific issues: (1) reversible versus relaxation phenomena, (2) quantum versus classical chaos, and (3) quantum/classical correspondence. These are briefly introduced in Section II, where it becomes clear that intramolecular energy-transfer studies provide a useful laboratory for the study of fundamental questions in these areas.

Experimental and theoretical studies of intramolecular energy transfer and intramolecular dynamics have a long history. This article, designed as an introduction, rather than as a survey, focuses on recent directions in this field of research. Of specific interest are insights gained into general rules for understanding and measuring intramolecular dynamics. For this reason we cite only a few sample computations to illustrate relevant general features and provide only a brief qualitative discussion of experimental methods. References to more historical interests in the field are provided at the end of this article.

The organization of this article is as follows. Section II is designed to provide qualitative insight into intramolecular energy transfer via two subsections. The first, Section II,A, discusses selected computational results on two molecular models, and the second, Section II,B, qualitatively introduces the fundamental problems alluded to above. The reader who is interested in a qualitative picture is urged to first focus on these sections. Section III contains a description of isolated molecule dynamics from both the quantum and classical viewpoints. Emphasis here is on several general features of classical and quantum intramolecular dynamics. Finally, two brief sections, Sections IV and V, discuss statistical approximations to intramolecular energy transfer and the nature of modern experiments designed to probe molecular motion.

Space limitations, coupled with the author's intention to provide a useful introductory treatment, have led to restrictions on the material that can be covered. Thus, we focus, throughout this article, on adiabatic processes, that is, dynamics that take place without change in the electronic configuration of the molecule. When this is not the case, a remark to this effect is made. In addition, we assume throughout that the radiation field, be it associated with radiative absorption or emission, is sufficiently weak to be treatable as a perturbation. Further, the field of intramolecular dynamics is replete with model approaches that, albeit reasonable, have not been justified either theoretically or experimentally—the latter due principally to technological limitations. The modern focus on accurate dynamics is emphasized in this article, with the consequence that such simple models are necessarily slighted.

II. Qualitative Dynamics

Information regarding intramolecular dynamics is available from three sources: experimental studies on specific molecular systems, theoretical computations on specific systems or models, and formal studies of typical ("generic") systems. At present, the latter two provide considerably greater detail than the first and involve two separate steps. In the first step, the forces between the atoms in the molecule are determined or modeled, while in the second step one considers the dynamics determined by these forces. This dynamics is done either quantum-mechanically, which is correct but difficult, or via classical mechanics, which is often a good approximation to the quantum result. In either case, the forces describing the dynamics are sufficiently complex to necessitate numerical computer solutions. As an introduction to the nature of intramolecular dynamics and to issues of interest in this area, we consider two examples.

A. Two Model Calculations

As a first example, consider nuclear motion in a four-atom system, A—B—C—D. Two qualitatively different energy ranges are possible. In the first, the system is provided with sufficient energy to induce vibrational and rotational motion but insufficient energy to break any of the bonds. This is bound-state intramolecular dynamics. In the second regime, there is sufficient energy to allow molecular dissociation to one or more of the molecular products (e.g., A—B + C—D). Typically, each of the interatomic bonds will have a different dissociation energy, the energy required to break the bond. Thus, several dissociation "channels" are possible, such as

$$A—B—C—D \rightarrow A—B + C—D \quad E_1$$
$$A—B—C—D \rightarrow A + B—C—D \quad E_2$$
$$A—B—C—D \rightarrow A + B + C + D \quad E_3$$

where the lowest energy required for each of the particular processes is arbitrarily labeled E_1, E_2, etc.

Consider the specific case of NaBrKCl for which theoretical, classical dynamics studies are available at energies where two dissociative channels are energetically accessible. (Questions as to the validity of the classical picture are relegated to later sections.) This system possesses attractive forces between the atoms such that the bound NaBrKCl species lies at an energy of approximately 40 kcal/mol below NaBr + KCl or NaCl + KBr. Specifically, consider the case where energized NaBrKCl is formed by the collision

$$NaBr + KCl \rightarrow NaBrKCl$$

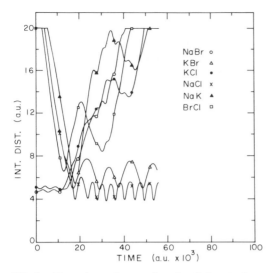

FIG. 1. Time dependence of each of the six interatomic distances during a trajectory describing the collision of NaBr with KCl. System energy is $E = 0.0912$ a.u. [From Brumer, P. (1972). Ph.D. dissertation, Harvard University.]

with sufficient energy to dissociate as

$$NaBrKCl \rightarrow NaCl + KBr$$

$$NaBrKCl \rightarrow NaBr + KCl$$

The initial collision between NaBr and KCl is here regarded as the preparatory step to the subsequent intramolecular dynamics and decay of the intermediate NaBrKCl species. An analogy to the collision, bound-molecule dynamics, and subsequent decay is a system of two balls at-

tached by a spring, colliding with two other such balls on a billiard table containing a deep hole in the center. The springs are breakable and endowed with the ability to exchange between pairs of particles.

The basic dynamics in classical mechanics is embodied in trajectories, that is, the dynamics following precise specification of the system initial conditions. Comparison with experiment then involves averaging results over a set of trajectories consistent with the specific experimental conditions. Two such trajectories are considered in Figs. 1 and 2 where we show the time dependence of the distances between all the atoms during the collision. In one case the initial conditions lead to a long-lived intermediate and in the other case they lead to a short-lived species. Consider the shorter trajectory (Fig. 1) first. A careful examination of the figure indicates the initial oscillations of the bound NaBr and KCl, with the other atomic distances shrinking in time as the two diatomics approach one another. The collision between them occurs at approximately 12,000 atomic time units (denoted atu, where an atomic time unit is 2.4×10^{-17} sec) and is promptly followed by decay to NaCl + KBr. Thus, for these particular initial conditions, the intermediate species NaClKBr is only a fleeting phase in the collision. Results in Fig. 2 are in sharp contrast, showing a long-lived intermediate that displays complex dynamics in a bound energized molecule. Here the system forms a bound four-body molecule at approximately 38,000 atu that persists until $t = 240,000$ atu. During this time the various bond distances

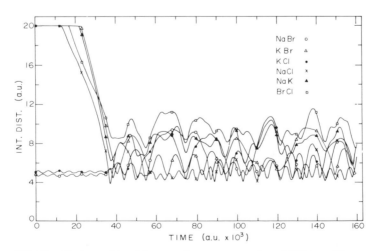

FIG. 2. Similar to Fig. 1 but at a lower energy, $E = 0.0227$ a.u. [From Brumer, P. (1972). Ph.D. dissertation, Harvard University.] (*Figure continues.*)

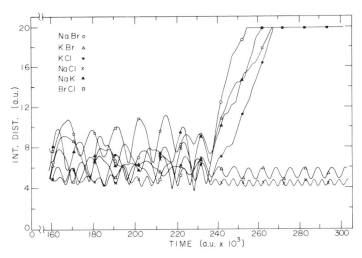

FIG. 2.   (*Continued*)

go through a variety of values between 4 and 11 Bohr radii, a sort of vibrant dance of four bound particles. Decay follows thereafter to the product diatomics.

In a classical picture, knowledge of the distances and momenta of the atoms as a function of time permits complete knowledge of all properties. Figure 3 shows, for example, the calculated energies in each of the alkali halide bonds during the course of the dynamics for the same collision as seen in Fig. 2. Figure 4 contains an analogous picture of the rotational energy of the four-atom system. The dynamics is clearly marked by extensive energy exchange between the bonds and among rotation and vibration.

These examples provide a picture of the complexity of individual trajectories in bound intramolecular dynamics. Such a trajectory emerges from a precise specification of initial particle momenta and coordinates. A comparison with real phenomena requires, however, averaging over all initial conditions not precisely specified in the given experiment. For example, the experiment may only have initially fixed the trans-

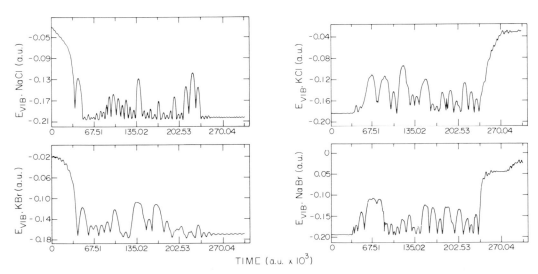

FIG. 3.   Vibrational energy in each of the four alkali–halide bonds, as a function of time, for the trajectory shown in Fig. 2. [From Brumer, P. (1972). Ph.D. dissertation, Harvard University.]

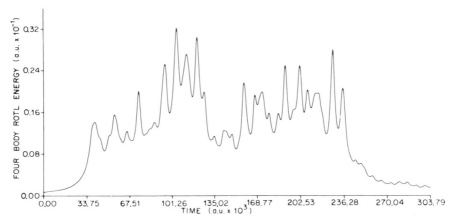

FIG. 4.   Total rotational energy associated with the NaClKBr system during the trajectory shown in Fig. 2. [From Brumer, P. (1972). Ph.D. dissertation, Harvard University.]

lational and internal energies of the colliding partners. Figure 5 provides a typical example of the result of averaging over a set of trajectories where the trajectory in Fig. 2 is one participant. Here we show a typical measureable in the decay of NaClKBr, that is, the probability of independently finding the products NaCl and KBr with particular vibrational energies. Note that the results are simpler than the complex underlying trajectories. This is a result of both the comparatively simpler question being asked and the averaging implicit in the computation.

The details of the dynamics ongoing during the course of the collision also often simplify as

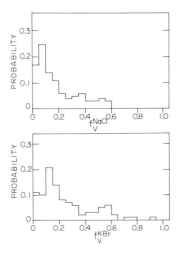

FIG. 5.   Probability of producing a specific vibrational state of NaCl or of KBr from collisions of NaBr with KCl at $E = 0.0411$ a.u. [From Wardlaw, D. (1982). Ph.D. dissertation, University of Toronto.]

a result of averaging over a set of trajectories. Consider, for example, Fig. 6, which shows the *average* vibrational energy in each of the four alkali halide bonds during the collision; in this case the calculation only shows trajectories prior to dissociation. The results are to be compared to Fig. 3, although the overall energies in the two figures are somewhat different. One sees that each of the bond energies changes rapidly over the initial time period but then tends to level out. This is an example of apparent intramolecular energy *relaxation* during the course of the dynamics. That is, the system seems to reach, at least with respect to these variables, a relatively invariant state reminiscent of the behavior of macroscopic systems relaxing to equilibrium. The observation of relaxation behavior is, of course, not ubiquitous and is dependent on gross system conditions (e.g., total energy, total angular momentum, etc.). Understanding conditions under which intramolecular dynamics may be approximated by a statistical model is indeed one of the main themes in intramolecular energy transfer. Such statistical models have both advantages and disadvantages. They are advantageous in that they considerably simplify the description of the dynamics. They are disadvantageous in that they imply that the final state of the system is relatively insensitive to the initial state. That is, the outcome of a chemical event is not readily influenced by altering initial conditions.

The example discussed here displays a number of relevant features. Clearly noticeable is the complexity of individual trajectories modeling long-lived dynamics as well as the possibility of

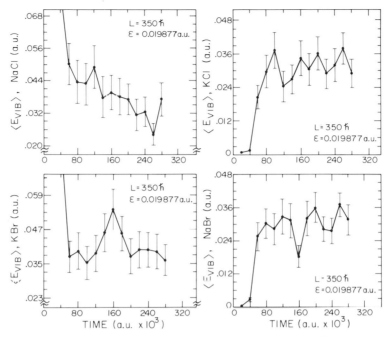

FIG. 6. Average energy in each of the four alkali–halide bonds as a function of time. Here the average values are obtained from a set of NaBr + KCl trajectories. "Error bars" show the range of values associated with the set of trajectories incorporated in the calculation. [From Brumer, P. (1972). Ph.D. dissertation, Harvard University.]

simplifications that result if a statistical description that includes relaxation applies. The participation of rotations as well as vibrations in the dynamics is also evident. In addition it makes clear the important role of the relative *rates* of intramolecular energy transfer and other competitive processes. That is, the degree of intramolecular energy flow within the molecule depends on the length of time the system exists as a bound entity, as well as on some (as yet undefined) "rate of intramolecular energy flow." In cases where competitive processes such as dissociation (or the ever-present radiative emission) are possible, effective intramolecular energy flow requires a larger rate of energy transfer than of competitive processes. (Note, however, that in the case of competitive dissociative or other nuclear rearrangement processes, both energy transfer and the competitive process are governed by the same set of underlying dynamical equations for the nuclear motion. This leads to a natural difficulty associated with attempting to partition the dynamics into distinct parts labeled intramolecular dynamics and dissociation.)

The computation described above is com-

pletely classical: the nuclear motion is assumed to be well described by Newton's equations. The extent to which classical mechanics provides a useful description of intramolecular energy flow is another focus of current research in this area. As one example of the validity of classical mechanics, consider the bound-state dynamics of a three-atom system confined to a line, that is, A—B—C. Computations on the case where the A—B and B—C bonds are anharmonic have been performed using both classical and quantum mechanics, where the initial state is a mixed state (see Section III,C). One useful measure of system dynamics is the probability of the system returning to the state from which it started. Figure 7 shows the time dependence of this probability for one case, where the quantum and classical results are seen to be in excellent agreement. In sharp contrast is the comparison shown in Fig. 8, which corresponds to the probability of return to another, higher-energy initial mixed state where classical–quantum disagreement is substantial. The effect has also been seen for the higher-energy decay of A—B—C to AB + C as shown in Figs. 9 and 10. Specifically, the probability of the system re-

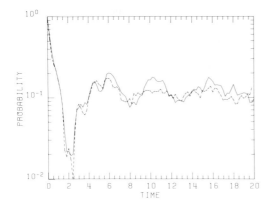

FIG. 7.   Classical (dashed) and quantum (solid) probability, as a function of time, of a coupled Morse oscillator system remaining in its original state. [From Kay, K. (1980). *J. Chem. Phys.* **72**, 5955.]

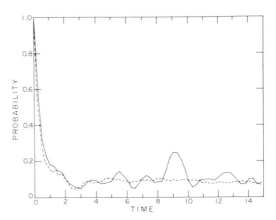

FIG. 9.   Classical (dashed) and quantum (solid) probability of a coupled Morse–harmonic oscillator system remaining in its initial state. The energy is sufficient to allow dissociation. [From Kay, K. (1984). *J. Chem. Phys.* **80**, 4973.]

maining in its initial state is shown in Figs. 9 and 10 for dynamics initiated in two different mixed states. Once again, agreement is achieved between classical and quantum mechanics in one case but there is substantial disagreement in the other, with the classical being more statistical. The origin of this difference lies, in this case, in the existence of so-called quantum trapping states, which lead to classical results that behave more statistically than the quantum. Such states occur when there is a very asymmetric distribution of energy among the bonds in the molecule. Although here classical mechanics is more statistical than is quantum mechanics, this is not always the case.

The computations described above are useful in that they provide some qualitative insight into the nature of intramolecular dynamics and en-

ergy flow in particular systems. They are, in essence, theoretical *experiments* on given systems in that they provide only hints of the general rules that govern rates, nature, and degree of intramolecular energy interchange. In the sections that follow, we describe the current state of understanding on the general principles that underlie these processes. Prior to doing so we emphasize, in the next section, some of the fundamental issues related to intramolecular energy transfer that have been alluded to above.

B. Fundamental Issues: Qualitative Overview

The goal of science is to provide a qualitative and quantitative description of natural phenom-

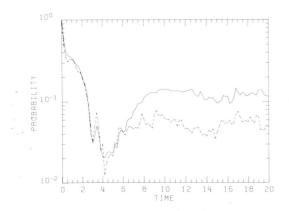

FIG. 8.   Same as Fig. 7, but for a different initial state. [From Kay, K. (1980). *J. Chem. Phys.* **72**, 5955.]

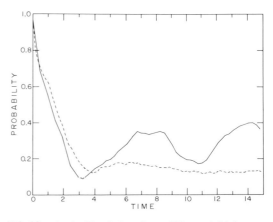

FIG. 10.   As in Fig. 9, but for a different initial state. [From Kay, K. (1984). *J. Chem. Phys.* **80**, 4973.]

ena. Such a description is most useful if it is as simple as possible. For example, there is little reason to invoke relativistic quantum mechanics to describe planetary motion; classical mechanics will normally suffice. Furthermore, such a scientific description is most useful if it is computationally tractable. Thus, for example, thermodynamics often provides a more useful route to disallowing certain processes in a bulk system than does a full dynamics calculation involving the underlying Avogadro number of molecules. It is in this spirit that the utility of two *approximate* descriptions—classical mechanics in lieu of quantum mechanics, and statistical approaches in lieu of full dynamics calculations— is a central theme in contemporary intramolecular dynamics. Indeed, they intertwine in an interesting fashion.

Consider first the essential difference between a dynamics description and a statistical description of a system. By the former we mean a description based on Newton's equations in classical mechanics or Schrodinger's equation in quantum mechanics. By the latter we mean descriptions characteristic of nonequilibrium statistical mechanics, such as the Boltzmann or Fokker–Planck equation. Dynamics (either quantum or classical) is based on a set of basic equations that are time-reversal invariant. This property means that the final state of a process may be time-reversed to recover the initial state. It implies that the final state of dynamical evolution contains all the information associated with the initial state. As a consequence, relaxation to a final equilibrium state, independent of the fine details of the initial state, does not occur, and hence the final state may be a sensitive function of the initial state. Relaxation to equilibrium is, however, a familiar feature in macroscopic systems and the equations of statistical mechanics are designed to provide a nonmicroscopic description that encompasses the relaxation process. The link between the underlying time-reversible equations of motion and the macroscopic irreversible equations is not well established and has been the subject of extensive, long-standing discussions on the basic equations of nature. Typical questions include: are time-reversible equations more fundamental than statistical relaxation equations, or do they have equal, but independent, roles as models of nature? Is observed relaxation a consequence of coarse graining associated with macroscopic measurements on intrinsically time reversible systems?

Isolated-molecule dynamics is expected to be a sufficiently elementary process to permit observation of microscopic reversibility in the dynamics and hence to display a dependence of the outcome of dynamics on initial conditions. This dependence is desirable since the ability to retain information about initial conditions is necessary in order to achieve the technologically desirable goal of externally influencing chemical reactions. However, a great many experiments, perhaps with insufficiently well-characterized preparation and measurement, have indicated that time-irreversible relaxation is a useful model for many intramolecular processes. Thus, isolated-molecule intramolecular dynamics serves as a laboratory for the study of the interrelationship between irreversible relaxation behavior in systems that are fundamentally describable by time reversible equations of motion. It also presents an experimental challenge to prepare sufficiently well-characterized states to observe time-reversibility and sensitivity to initial conditions.

Two further issues of fundamental importance, that of quantum–classical correspondence and that of "quantum chaos," are intimately linked to studies in intramolecular dynamics. Extensive theoretical studies, beginning in the early 1960s, showed that a great many phenomena involving the dynamics of atoms and molecules are well described by classical mechanics. That is, atomic and molecular dynamics are at the borderline between classical and quantum mechanics, in that dynamic phenomena in molecules are often well approximated by classical dynamics. This is not always the case, as shown in the example in Section II,A, and so studies in intramolecular dynamics provide insight into the utility of classical mechanics as an approximation to quantum mechanics. Phrased in this manner, this appears to be a question of relative accuracy, with the expectation that the process is qualitatively similar in both mechanics. Recently, however, major *qualitative* distinctions between classical and quantum mechanics have been noted. Specifically, it is now known that even small classical mechanical systems (e.g., two degrees of freedom) can display highly statistical behavior, termed chaos. One can show formally, however (as will be described further below), that such chaotic behavior is not possible in bound state quantum mechanics. Further, typical semiclassical schemes that base quantization on the classical system motion (e.g., Einstein–Brillouin–

Keller quantization) do not hold in this classically chaotic regime.

Computational studies have indicated that chaotic behavior is expected in classical mechanical descriptions of the motion of highly excited molecules. As a consequence, intramolecular dynamics relates directly to the fundamental issues of quantum versus classical chaos and semiclassical quantization. Practical implications are also clear: if classical mechanics is a useful description of intramolecular dynamics, it suggests that isolated-molecule dynamics is sufficiently complex to allow a statistical-type description in the chaotic regime, with associated relaxation to equilibrium, and a concomitant loss of controlled reaction selectivity.

III. Dynamics: Theory

This section provides an introduction to the theory of classical and quantum intramolecular dynamics, with emphasis on general principles.

A. The Hamiltonian

A complete description of the dynamics of any molecular system is contained in the Hamiltonian H, which is the energy operator in quantum mechanics or energy function in classical mechanics. In general, the Hamiltonian is a function of the electronic and nuclear degrees of freedom, as is the description of the system dynamics. This complex problem simplifies through the adoption of the Born–Oppenheimer approximation, which is the assumption that nuclear and electronic motion are independent due to their substantially different time scales and masses. This assumption allows one to first solve for the dynamics of the electrons and then obtain the forces experienced by the nuclei as determined by this fixed-electron configuration.

Within the Born–Oppenheimer approximation, the nuclear Hamiltonian may be written in the form

$$H(\mathbf{q}, \mathbf{p}) = T(\mathbf{p}) + V(\mathbf{q}) \tag{1}$$

where $T(\mathbf{p})$ is the nuclear kinetic energy and $V(\mathbf{q})$ is the nuclear potential energy. Classically, $\mathbf{p}$ is the momentum of the nuclei, whereas in quantum mechanics it is the momentum operator. In general $\mathbf{q}$ has $3N$ components for an N-atom system. However, it is always possible to eliminate three coordinates corresponding to the center of mass of the system, and we shall assume this reduction to the internal molecular co-

ordinate system has been carried out; thus $\mathbf{q}$ denotes $3(N - 1)$ coordinates.

Central to the *nature* of dynamics is the notion of coupled versus uncoupled degrees of freedom. Consider, for example, a system whose Hamiltonian is the sum of two terms, such that

$$H(\mathbf{q}, \mathbf{p}) = h_1 + h_2 \tag{2}$$

where h_1 depends on a different set of coordinates and momenta than does h_2. Under these circumstances, the dynamics of the degrees of freedom in h_1 are independent of those in h_2, that is, the total system is composed of two independent subsystems. Introduction of a coupling term, to give

$$H = h_1 + h_2 + V(1, 2) \tag{3}$$

where $V(1, 2)$ is a term dependent on the coordinates and momenta of both h_1 and h_2, induces energy exchange and interrelated dynamics among the entire system. Major qualitative changes in the dynamics can result from small perturbations or couplings.

It is important to note that although Eq. (3) is conceptually pleasing, there is no unique division into subsystems and into intersubsystem coupling for a given physical system. Rather, such a division must be motivated by an experimental or theoretical interest in a particular property, such as the energy flow between subsystems h_1 and h_2. Central to the nature of intramolecular energy transfer is an understanding of the effect of the coupling $V(1, 2)$ on the energy flow among the subsystems represented by h_1 and h_2.

B. Classical Mechanics

There are a number of formulations of classical mechanics, each providing different insights into its nature. For example, Hamilton's method, used here, describes dynamics in terms of trajectories in generalized coordinates and momenta. Consider an M degrees of freedom system with system Hamiltonian $H(\mathbf{q}, \mathbf{p})$, where $(\mathbf{q}, \mathbf{p})$ is a complete set of M conjugate generalized coordinates and momenta. The time evolution of the system is given by Hamilton's equations,

$$dq_i/dt = \partial H/\partial p_i \qquad dp_i/dt = -\partial H/\partial q_i \tag{4}$$

with $i = 1, \ldots, M$. Specifying the state of the system at $t = 0$ via the initial conditions $\mathbf{q}(t = 0) = \mathbf{q}_0$, $\mathbf{p}(t = 0) = \mathbf{p}_0$ then leads to a solution, a trajectory $\mathbf{p}(t)$, $\mathbf{q}(t)$, from which the time dependence of all system properties along that trajectory can be computed. The motion may be best visualized as taking place in a $2M$-dimensional space, termed phase space, with $(\mathbf{q}, \mathbf{p})$ as coordinates. A trajectory is a curve in this space parametrized by the index t.

Modeling a realistic system necessitates producing a set of trajectories with varying initial conditions and looking at averages of system properties over this set. It is this trajectory technique, where $V(\mathbf{q})$ is a model or realistic potential and where Hamilton's equations are solved numerically on a computer, that has been extensively used to study intramolecular dynamics in small molecules. Results typical of those obtained were shown in Section II,A.

The flexibility of Hamilton's approach lies in the appearance of generalized canonical coordinates $(\mathbf{p}, \mathbf{q})$. As a consequence of their generality, one may seek out the set of coordinates and momenta within which the dynamics is most easily performed and understood. For example, a conservative Hamiltonian system may have, along any trajectory, a constant value of total angular momentum $\mathbf{J}$. Such a quantity is said to be a constant of the motion and can prove useful as a momentum, since the equation of motion for $\mathbf{J}$ is particularly simple, $d\mathbf{J}/dt = 0$. Indeed, the idea of seeking constants of motion for use as coordinates or momenta is the central goal of the Hamilton–Jacobi approach to classical dynamics. The essential approach is simple. One seeks a set of M constants of the motion $\mathbf{A}$ of the system via a systematic procedure. Once found, these momenta are known to be constant along any trajectory, and the time dependence of the conjugate coordinates is generally simple. With the constants of the motion known, along with the procedure for generating them, it is a relatively straightforward algebraic problem to express the desired $\mathbf{q}(t)$, $\mathbf{p}(t)$ in terms of these constants and their conjugate coordinates. That is, we have

$$\mathbf{p}(t) = \mathbf{A} = \text{const } 1$$
$$\mathbf{q}(t) = \mathbf{q}(t, \mathbf{A}) \tag{5}$$

Although we shall not deal with Hamilton–Jacobi theory here, the concept of a set of constants of the motion is vital to an understanding of the issue of intramolecular energy flow and statistical versus nonstatistical behavior. In essence, the number of global constants of the motion provides a method for grouping systems into general categories.

Consider an M degrees of freedom system. How many constants of the motion can we identify for a given trajectory? The answer is clearly $2M$, with the simplest set consisting of the initial values of the coordinates and momenta that define the trajectory and that surely constitute a set of $2M$ equations for $\mathbf{q}(t)$, $\mathbf{p}(t)$ in terms of constants, that is,

$$\mathbf{q}(t) = \mathbf{q}(\mathbf{q}_0, \mathbf{p}_0; t) \qquad \mathbf{p}(t) = \mathbf{p}(\mathbf{q}_0, \mathbf{p}_0; t)$$

These constants of the motion, however, are of little interest. Although knowing them does identify the trajectory uniquely, it provides no reduction of the complexity of solving the full problem for $\mathbf{q}(t)$, $\mathbf{p}(t)$. Clearly, what we seek is a set of global constraints that define smooth surfaces in phase space on which the trajectories lie. Members of such a set of functions are called global constants of the motion. The number of global constants of motion that a system possesses provides a means of generally classifying the system behavior.

1. Integrable Systems

Consider first the case in which an M degrees of freedom system possesses M independent global constants of the motion, denoted $\mathbf{K}$. Selecting these as the M generalized momenta $\mathbf{K}$, with conjugate momenta $\mathbf{Q}$, allows us to write the Hamiltonian as $H(\mathbf{K})$. Hamilton's equations then become

$$dK_i/dt = -\partial H/\partial Q_i = 0$$
$$dQ_i/dt = \partial H/\partial K_i = \omega_i(\mathbf{K}) \tag{6}$$

where the last equation defines the frequencies $\omega_i(\mathbf{K})$. Systems allowing this description are termed *integrable*, or *regular*, and possess a number of important properties:

1. Trajectories lie on M-dimensional surfaces in phase space whose topology is that of a torus. The tori are labeled by the values of $\mathbf{K}$. For example, in a two-degrees-of-freedom system, the motion lies on the surface of a doughnut (see Fig. 11).

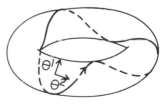

FIG. 11. Sample quasiperiodic trajectory in a two-degrees-of-freedom system as it moves on the surface of a torus in phase space. The trajectory shown is actually periodic; in general, the trajectory will fill the entire torus surface.

2. The frequencies of motion about the independent directions on the torus are given by $\omega(\mathbf{K})$.

3. The time dependence of the coordinates and momenta for a given trajectory are given by the Fourier *series* (where $\mathbf{n}$ is a vector of M integers)

$$\mathbf{q}(t) = \sum_{\mathbf{n}} \mathbf{q_n}(\mathbf{K})\exp[i\mathbf{n} \cdot \omega(\mathbf{K})t]$$

$$\mathbf{p}(t) = \sum_{\mathbf{n}} \mathbf{p_n}(\mathbf{K})\exp[i\mathbf{n} \cdot \omega(\mathbf{K})t] \tag{7}$$

Note then that trajectories in such systems come arbitrarily close, during the course of their dynamics, to their original starting position. For this reason, such dynamics is termed *quasiperiodic*. The vast majority of "textbook" problems dealt with in elementary and advanced analytical mechanics treatises are of this type. Examples include the hydrogen atom or the small-vibrations Hamiltonian, although these systems tend to be, in addition, separable, that is, of the form $H = H(K_1) + H(K_2) + \cdots$.

Trajectories in integrable systems are stable with respect to small changes in initial conditions. In particular, consider a trajectory [$\mathbf{q}(t)$, $\mathbf{p}(t)$] emanating from initial conditions $\mathbf{q}(0) = \mathbf{q}_0$, $\mathbf{p}(0) = \mathbf{p}_0$ and an initially close trajectory [$\mathbf{q}'(t)$, $\mathbf{p}'(t)$] originating from $\mathbf{q}'(0) = \mathbf{q}_0 + \delta Q$, $\mathbf{p}'(0) = \mathbf{p}_0 + \delta P$ with δQ and δP very small. Then define $d(t)$ as the time-dependent "distance in phase space" between these two trajectories,

$$d(t) = [p_1(t) - p_1'(t)]^2 + [p_2(t) - p_2'(t)]^2$$
$$+ \cdots + [q_1(t) - q_1'(t)]^2 + \cdots]^{1/2} \tag{8}$$

This quantity measures the rate at which two nearby trajectories separate as a function of time. Then:

4. For regular systems, $d(t)$ grows linearly with t, a relatively slow rate of separation characteristic of stability.

All these features are, in a qualitative sense, indicative of essentially predictable motion in integrable systems. The quasiperiodicity assures that, in some sense, system behavior is repetitious, or at least describable by a simple set of frequencies. The linear growth of $d(t)$ suggests that knowledge of the behavior of a single trajectory allows prediction, over a fair length of time, of the behavior of its initially nearby neighbors.

As a simple example of regular dynamics, to be embellished later below, consider a system of two uncoupled oscillators,

$$H = H_1(p_1, q_1) + H_2(p_2, q_2) \tag{9}$$

where the individual oscillators, with Hamiltonians $H_i(q_i, p_i) = T(p_i) + V(q_i)$, have potentials terms of the Morse form,

$$V(q_i) = D_i[\exp(-q_i/a_i) - 1]^2 \tag{10}$$

Here D_i is the oscillator dissociation energy and a_i is a system parameter. Equation (10) might, for example, model three atoms on a line where the coupling potential between the oscillators has been eliminated. The individual Hamiltonians H_1 and H_2 are then the conserved integrals, and their numerical values remain constant at their initial values throughout the dynamics. Since this is a two-degrees-of-freedom system and since two constants of the motion exist, the system is integrable.

Consider a measurement on the dynamics of this system. If an experiment were able to begin with an initial state consisting of a single trajectory, then the time evolution is on a torus in phase space and there is no energy transfer between the subsystems H_1 and H_2. Sufficiently long-time observations on the system would reveal repetitive quasiperiodic motion. Two remarks regarding such measurements are, however, important. First, if one is interested in the dynamics of a subcomponent other than H_1 and H_2—say one interrogates the time dependence of the harmonic oscillator $p_1^2/2m_1 + q_1^2$—then the measurement would reveal energy flow into and out of this subsystem. Quasiperiodicity would still be evident, however, after a suitable time. Second, if the time scale of the experiment is short compared to the relevant system frequencies, then this quasiperiodicity will not be manifest. The essential point then is that the nature of the measurement of interest determines whether the regular system behavior is fundamental to, or observable in, the particular experimental study.

There is another feature of integrable systems that is important. Specifically, consider the concept of statistical behavior in dynamics. The fundamental features of such behavior are that a trajectory at energy E fills the entire volume of phase space associated with that energy E, that a set of trajectories relaxes to a long time limit that no longer varies with time and that the final state is dependent solely on the energy of the system. It is clear that the first of these properties is not satisfied by a regular system, since a trajectory lies on the surface of a torus of dimensionality M whereas the constraint to constant energy would confine dynamics to a larger surface of dimension $2M - 1$. It would also appear, from the list of properties above, that a regular system does not relax. This is, in fact, not the case. That is, properties (a)–(d) constitute features of the *trajectories* of a regular system. As already remarked, however, typical comparisons with physical systems require information on the average behavior of the time development of a collection, or ensemble, of trajectories. It is therefore important to note that despite the quasiperiodic behavior of integrable systems, an ensemble of trajectories in a regular system can relax to a long-time, stationary distribution. The final relaxed state of the system is, however, intimately related to the initial conditions of the dynamics. This is clear from the simplest of considerations. That is, each of the trajectories in the set of trajectories retains its original values of the conserved quantities. Thus the final state of the system will depend on more than just the total overall energy of the system.

Each of properties (a)–(d) gives rise to useful computational tools for the theoretical identification of integrable behavior in models of molecular motion. Relationships to actual experimental techniques and measurements are, however, not well formulated.

The regular system constitutes one major category of observable dynamical behavior. Systems that are integrable are well known and have been experimentally observed, as discussed later. A second major category of Hamiltonian systems emerges from formal ergodic theory, which defines a set of increasingly idealized statistical systems. Such systems, (in terms of increasing statistical characteristics) are termed ergodic, mixing, K-systems, and Bernoulli systems. Each category imposes additional conditions, leading to requirements difficult to verify for realistic systems. Thus they are to be regarded as idealized models of statistical motion.

2. Ergodic

Consider first the integrable system where each trajectory lies on the surface of a torus. Two conditions are possible. In the first, the trajectory wraps about the torus and closes on itself without covering the torus completely. An example is shown in Fig. 11, where it is clear that this property arises if the frequencies of motion about the torus are related to one another by the relation $n_1\omega_1 + n_2\omega_2$. Such a set of frequencies is said to be rationally related and results in the trajectory returning exactly to its original position. On the other hand, the frequencies on the torus may not be rationally related, in which case the trajectory fills the entire surface of the torus. Under such conditions the dynamics is said to be ergodic on the torus. This formal terminology does not correspond to the historical use of the term ergodic as found in the physics literature. There, ergodic tends to mean ergodic on the energy hypersurface, that is, on the $(2M - 1)$-dimensional surface in $2M$-dimensional phase space that results from constraining the system to constant energy. For clarity we shall term this E-ergodic.

Thus, the characteristic of an E-ergodic system is the existence of a single trajectory at each energy E that comes arbitrarily close to all points on the energy hypersurface. It is important to note, however, that this property does not ensure that the system displays irreversible relaxation during the course of the dynamics. A pictorial analog of possible motion of an ergodic system is provided by imagining a speck of carbon in a continuously stirred fluid. The carbon speck, representative of the system in phase space, moves throughout the fluid without constraint, but does not settle down to some long-time stationary state.

3. Mixing

A system that is ergodic but has the rudimentary properties associated with statistical irreversible behavior is the mixing system. Such a system displays the following properties.

Consider the system at energy E. Denote the average over the energy surface by the expression $\langle f \rangle$, where $f(\mathbf{q}, \mathbf{p})$ is any dynamical property. Then

1. $\lim f[\mathbf{q}(t), \mathbf{p}(t)] = \langle f \rangle \qquad t \to \infty$
2. The correlation between any two dynamical properties, that is,

$$\langle g[\mathbf{q}(t), \mathbf{p}(t)], f[\mathbf{q}(0), \mathbf{p}(0)] \rangle$$
$$- \langle g[\mathbf{q}(t), \mathbf{p}(t)] \rangle \langle f[\mathbf{q}(0), \mathbf{p}(0)] \rangle$$

goes to zero as $t \to \infty$.
3. Subdivide the total phase space into regular regions of particular volume. Then the probability of going from region i to region j in the long-time limit depends only on the size of the phase space regions i and j.

Thus, a mixing system satisfies a number of simple properties that are in qualitative agreement with statistical relaxation dynamics. A particle of soluble colored material stirred into water provides a pictorial analog of mixing dynamics. Once again, the fluid models the phase space. The system evolves over time to reach a final macroscopically invariant distribution of uniformly colored fluid throughout the container.

It is unfortunate that the formal definitions of ergodic, mixing, etc. systems involve the infinite time limit. As a consequence, a system may, for example, still be mixing even if relaxation is not observed in the *finite* time associated with a realistic measurement. This limitation significantly reduces the practical utility of formal concepts such as mixing behavior.

A host of other formal systems with additional, and hence stricter, requirements have been defined. Here we only mention the *C*-system, which is ergodic and mixing and which possesses the important characteristic that the distance $d(t)$ between any two initially close trajectories in phase space grows exponentially in time. This trajectory instability leads to the rapid parting of trajectories from one another and hence the inability to predict the dynamics of trajectories, even for a relatively short time period, from knowledge of the dynamics of their neighbors.

A system that displays characteristics of mixing as well as exponential divergence of adjacent trajectories is termed irregular or chaotic. In contrast with the characteristic properties of a regular system, an irregular system displays

(1)   trajectories that lie upon the $(2M - 1)$-dimensional energy hypersurface in phase space (additional simple constants of the motion such as angular momentum may also be incorporated),

(2) and (3)   trajectory dynamics that can not be written in terms of a Fourier series involving a simple set of discrete frequencies and their overtones and combinations, and

(4)   a distance $d(t)$ between nearby trajectories that grows exponentially, that is, $d(t) = d(0)\exp(kt)$, indicative of trajectory instability.

This set of properties gives rise to useful theoretical indicators of irregular motion, but connections with actual experimental observables are not well established.

From the viewpoint of measurement, were one able to prepare a single trajectory as the initial state of an irregular system, then the subsequent measurement of any property, other than energy, would show continual variation with time. The trajectory would, in addition, display no tendency to return to the original state over any finite time. If one prepared an ensemble of trajectories it would approach a long-time stationary distribution dependent solely on energy.

4. Typical Molecular Systems

Both regular and irregular motion are extremes of behavior, and their relation to the dynamics of realistic systems has principally been established through numerical computer studies. These studies indicate that many, but certainly not all, molecular systems display behavior characterizable as regular at low energies and irregular at higher energies. The example of carbonyl sulfide, OCS, is shown in Fig. 12, where the percent of phase space not showing exponential divergence is shown. The system is seen to display a transition to chaotic motion at an energy of approximately 14,000 cm^{-1}. By 20,000 cm^{-1}, close to dissociation, almost all of the phase space is irregular.

To appreciate the origin of the regular behavior at low energies, we note two common approximations in low-energy molecular motion. The first, rotation–vibration decoupling, assumes that the rotational and vibrational motion are essentially uncoupled at low energies, that is, that the Hamiltonian is the sum of vibrational and rotational parts:

$$H = H_{\text{vib}} + H_{\text{rot}} \qquad (11)$$

Second, we recall the standard normal-mode procedure for small-amplitude vibrational mo-

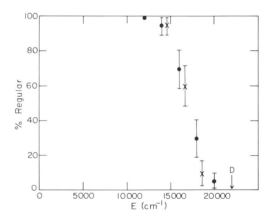

FIG. 12.   Percent regular trajectories as a function of energy for model OCS. The symbol D denotes the molecular dissociation energy. [From Carter, D., and Brumer, P. (1982). *J. Chem. Phys.* **77**, 4208.]

tion wherein, at sufficiently small energy, the vibrational Hamiltonian is of the form

$$H_{vib} = H_0$$

$$= \frac{1}{2}\left[\sum_i (P_i^2 + \lambda_i Q_i^2)\right] + V(\mathbf{P},\mathbf{Q}) \quad (12)$$

with $V(\mathbf{P},\mathbf{Q})$ sufficiently small to be negligible. The $\mathbf{Q}$, $\mathbf{P}$ are called normal coordinates and momenta. Thus, low-energy vibration is well approximated by a sum of M harmonic-oscillator Hamiltonians. In some instances an alternative separable Hamiltonian, composed of the sum of bond Hamiltonians, provides a superior separable representation. In either case, the low-energy vibrational motion is regular and separable.

The situation changes dramatically with increasing energy as $V(\mathbf{P},\mathbf{Q})$ becomes larger and the system begins to exchange energy between the decoupled harmonic oscillators. The subsequent dynamics, as observed in the measurement of the energy in a normal mode, depends intimately on the nature of the coupling, which is typically expandable in the form

$$V(\mathbf{P},\mathbf{Q}) = \sum_n V_n(\mathbf{P},\mathbf{Q}) \quad (13)$$

where $V_n(\mathbf{P},\mathbf{Q})$ denotes polynomial terms of the form $Q_i^k P_j^m$ with $k + m = n$.

As a simple example of the effect of coupling, consider a two-degrees-of-freedom system with

$$V(\mathbf{P},\mathbf{Q}) = V_2(\mathbf{P},\mathbf{Q}) = AQ_1Q_2 \quad (14)$$

It is convenient to first identify the constants of the motion in the harmonic Hamiltonian and use them as the new momenta. Consider then the momenta

$$I_i = (4\lambda_i)^{-1/2}[P_i^2 + \lambda_i Q_i^2] \quad (15)$$

and conjugate coordinates

$$\theta_i = \cot^{-1}[-\lambda_i^{1/2}Q_i/P_i] \quad (16)$$

In these coordinates, H_0 assumes the form

$$H_0 = \omega_1 I_1 + \omega_2 I_2 \quad (17)$$

where $\omega_i = \lambda_i^{1/2}$. We shall assume ω_1 and ω_2 to be unequal. These specific type of momenta and coordinates $\mathbf{I}$ are termed action-angle variables.

The relationships in Eqs. (15)–(16) allow us to rewrite $V_2(\mathbf{P},\mathbf{Q})$ as

$$V_2 = A(I_1I_2/\omega_1\omega_2)^{1/2}[\cos(\theta_1 - \theta_2) \\ - \cos(\theta_1 + \theta_2)] \quad (18)$$

where A is a constant.

Hamilton's equations of motion [Eq. (4)] then provide expressions for dI_i/dt that are nonzero due to the coupling V_2. In the event that the coupling is small, one may approximate the solution for the time dependence of the angles as that of the time dependence in the absence of the perturbation. This approach, a classical perturbation theory, gives the following result for the time dependence of $I_1(t)$:

$$I_1(t) = I_1(0) - A\left(\frac{I_1I_2}{\omega_1\omega_2}\right)^{1/2}$$

$$\cos[(\omega_1 - \omega_2)t + \theta_1^0 - \theta_2^0]/(\omega_1 - \omega_2)$$

$$- \cos[(\omega_1 + \omega_2)t + \theta_1^0 + \theta_2^0]/(\omega_1 + \omega_2)$$

$$(19)$$

where θ_1^0, θ_2^0 are the initial values of the angles. The quantity $I_2(t)$ is similar, but out of phase. Thus, the action variables oscillate about their unperturbed values with frequencies $(\omega_1 - \omega_2)$ and $(\omega_1 + \omega_2)$. Since $\omega_1 - \omega_2$ is assumed large, the total variation of I_1 and I_2 as a function of time is small.

The result is quite different if the system is resonant, that is, $\omega_1 = \omega_2$. In this case, the effect of the perturbation is more drastic, and energy can be exchanged completely, albeit periodically, between the two harmonic oscillators.

There are several reasons why the example treated above is a gross oversimplification of the situation in molecules. First, the unperturbed system is assumed harmonic, that is, linear in $\mathbf{I}$. Second, the perturbation has been assumed to be comprised of a single term; third, only one type of perturbation has been included. We now qualitatively examine the important effects associated with the breakdown of these simplifying assumptions.

1. Anharmonicity of $\mathbf{H_0}$: In general, H_0 is not harmonic, but is rather of the general anharmonic form $H(\mathbf{I})$. As a result, the zero-order system frequencies $\omega_i(\mathbf{I}) = \partial H_0/\partial I_i$ are no longer independent of the actions $\mathbf{I}$; that is, the frequencies depend on the energy content of the oscillators. As a consequence, anharmonic systems will display regions of the I_1, I_2 space where $\omega_1(\mathbf{I})$ and $\omega_2(\mathbf{I})$ are resonant as well as other regions where they are not. Thus, in the course of the dynamics, the zero-order system can go into and out of resonance as the energy of the oscillator varies. Regions of $\mathbf{I}$ space where the system is resonant are called resonance zones. Note that despite the coupling, the dynamics within this resonance region is regular.

A well-known example provides a picture of the resonance zone and the "trapping effect" associated with a nonlinear resonance. Consider a child's swing being pushed at fixed frequency. The nonlinear swing, well approximated by a pendulum, will gain energy from the "pusher" until the system is well out of resonance. At this stage the swing loses energy until it once again comes into resonance with the driving frequency. The system is therefore effectively trapped in a range of swing energies determined by the resonance zone associated with this driven pendulum. A similar effect is associated with the single resonance region associated with $\omega_1(\mathbf{I}) = \omega_2(\mathbf{I})$ in the example above, the system being essentially trapped in the region about the resonance center if the dynamics is initiated in that region.

2. Other Coupling Contributions: The above discussion emphasizes the $\omega_1 = \omega_2$ resonance, which results from the assumed form of the coupling in Eq. (14). In general, the coupling is more complicated, but is still expected to be expandable in the form

$$V(\mathbf{I},\theta) = \sum_{mn} V_{m,n}(I_1,I_2)\exp(in\theta_1 + im\theta_2) \quad (20)$$

The V_2 coupling term in Eq. (14) is an example of $|n| = 1$, $|m| = 1$ contributions to this expansion and leads to the $\omega_1 = \omega_2$ condition for resonance. Similarly, the n, m term in this series leads to an "n, m resonance" at action variables satisfying $n\omega_1(\mathbf{I}) = m\omega_2(\mathbf{I})$. Once again, within the neighborhood of this single resonance, the system displays regular energy transfer between the zero-order oscillators. We note that the size of the resonance zones tends to decrease with increasing n, m.

3. Overlapping Resonances: When a few terms in Eq. (20) contribute to the coupling, there is little reason to expect that specific regions of $\mathbf{I}$ space are influenced by solely one resonance. Under rather general conditions, the resonance regions in phase space arising from different terms in the coupling expansion [Eq. (20)] may overlap. Numerical studies have shown that energy flow between the zero-order oscillators assumes chaotic characteristics in regions of overlapping resonances.

As an example, consider the resonance and resonance-overlap structure associated with a collinear molecule A—B—C where H_0 is the sum of two Morse oscillators [Eq. (10)] corresponding to the bond potentials and the coupling term is of the form Ap_1p_2. Here p_j is the momentum associated with the jth bond. With E_i defined as the energy of the ith bond and E the total energy of the system, quantitative application of resonance theory allows for the explicit determination of E_i, E regions dominated by either a single resonance or by overlapping resonances. Sample results are shown in Fig. 13, where the solid shading indicates regions dominated by a single resonance and the cross-hatched areas are those dominated by overlapping resonances. In the case shown, there is a general trend toward overlapping resonances as the energy increases, consistent with the observation of increasing chaotic behavior with increasing energy. For the particular parameters shown, however, the system, even at energies near dissociation ($E = 1.5D_c$), displays regions of regular behavior dominated by a single resonance. Alternate system parameters can result in larger or smaller contributions from overlapping resonances.

In summary, the picture that emerges with respect to energy transfer between specified zero-order oscillators is qualitatively straightforward. The coupling between the specified oscillators induces nonresonant energy transfer between the oscillators if the system is initiated, and remains, within a nonresonant region of $\mathbf{I}$ values.

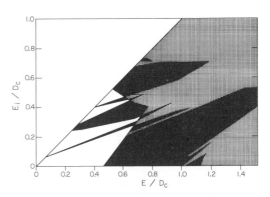

FIG. 13. Resonance structure of a model system A—B—C where each atom has a mass equal to that of carbon. The A—B bond has frequency 1000 cm⁻¹ and dissociation energy D_c whereas the B—C bond has corresponding parameters 1300 cm⁻¹, $1.5D_c$. Black areas denote single-resonance regions, and cross-hatched areas denote regions of overlapping resonances. [From Oxtoby, D., and Rice, S. A. (1976). *J. Chem. Phys.* **65**, 1676.]

Resonant energy transfer results if the system begins within a resonance zone, or enters the resonance zone, during the dynamics. In both cases, energy transfer has well-defined pathways: the energy transfer is well described in terms of the time-dependent energy content of the zero-order oscillators. Finally, chaotic energy transfer between the zero-order oscillators is expected in the I-space regime dominated by overlapping resonances. Computational results on bound molecules indicates that the the volume of resonance regions increases with increasing system energy.

One important aspect of this discussion is worthy of emphasis. Specifically, the subdivision of the system into zero-order oscillators and coupling terms, and the subsequent expansion of the coupling term, must be motivated by the experimentally measured quantities. Specifically, one may see apparent chaotic motion between particular zero-order oscillators even if the system is regular. This would be the case if the time scale of measurement is short and the observed oscillators are not those directly related to the conserved integrals of motion. This feature also makes clear that overlapping resonances do not necessarily ensure true irregular motion.

Detailed studies on the dynamics of realistic molecular systems are just becoming available. As a consequence, it is unclear whether the vast majority of highly excited molecules are weakly coupled with few overlapping resonances or are strongly chaotic. As a specific example of resonant coupling with weak coupling characteristics, and hence a specific energy-transfer pathway, we discuss below the study of overtones in the benzene molecule. As an example of chaotic energy transfer, we call attention to the NaBrKCl example discussed in Section II.

Recent experimental studies on benzene have shown that the absorption spectrum contains local mode features, that is, evidence of local isolated bond motions. In the benzene case, the C—H bonds, if they contain sufficient energy, appear directly in the spectrum, as if they were decoupled from the remainder of the molecular framework. In particular, one sees evidence of excitation to the overtones of C—H stretch, that is, five, six, seven, etc. quanta of energy in the bond. The experimental results further indicate that if energy is deposited in these bonds, it would transfer to the remainder of the benzene nuclear framework within about 10^{-13} sec. Although apparently rapid, this rate of energy

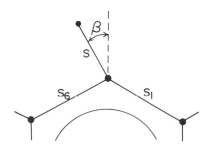

FIG. 14. Coordinates defining the C—H bond distance (s), C—C bond distances s_6, s_1, and wag angle β in benzene. [From Sibert, E. L., Hynes, J. T., and Reinhardt, W. P. (1984). *J. Chem. Phys.* **81**, 1135.]

transfer is substantially slower than that expected from a C—H bond democratically linked to all the degrees of freedom in the ring.

Detailed quantum and classical studies of the dynamics of benzene indicate the following picture. Consider first the immediate local environment of a C—H attached to the ring (Fig. 14). Two C—C bond distance are labeled s_1 and s_6, with the C—H bond distance being labeled s. Also shown is the angle β associated with the wag motion of the C—H relative to the ring. The picture that emerges from numerical studies is that with increasing excitation of the C—H bond, the C—H bond frequency comes into resonance with the wag, where the resonance is characterized as $n = 2$, $m = 1$. Energy transfer from the C—H bond first occurs as resonant energy transfer to the wag. Energy is subsequently transferred from the CCH wag to the remainder of the modes of the molecule. This is pictorially shown in Fig. 15. Trajectory calcula-

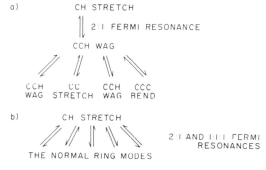

FIG. 15. A schematic of the coupling scheme linking the C—H stretch to the various modes of the benzene ring. Schemes (a) and (b) represent the same coupling schemes described in two different zero-order mode languages. [From Sibert, E. L., Hynes, J. T., and Reinhardt, W. P. (1984). *J. Chem. Phys.* **81**, 1135.]

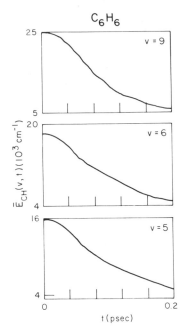

tions of the time dependence of the flow of energy out of the excited C—H mode, for various degrees of excitation, are shown in Fig. 16. A complementary picture of the growth of energy in the ring modes of the benzene framework and into the lower-lying states of C—H on the benzene ring is shown in Figs. 17 and 18. On the time scale shown, the energy flow out of the C—H bond is irreversible. Agreement with experiment is good, providing evidence that energy flow in this case occurs through a well-defined pathway of resonances. The comparative quantum calculations are discussed later.

Further experimental and theoretical efforts are underway to establish the extent to which energy-transfer mechanisms in molecules are either chaotic or rather specific in their nature.

FIG. 16.   Time dependence of the average energy in the C—H oscillator for three cases: excitation in the ninth C—H vibrational level ($v = 9$), $v = 6$, and $v = 5$. [From Sibert, E. L., Hynes, J. T., and Reinhardt, W. P. (1984). *J. Chem. Phys.* **81**, 1135.]

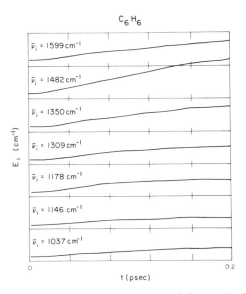

FIG. 17.   Growth of energy content of ring modes in benzene associated with the $v = 6$ case in Fig. 16. Note that only a few ring modes, labeled by their frequency, are shown. [From Sibert, E. L., Hynes, J. T., and Reinhardt, W. P. (1984). *J. Chem. Phys.* **81**, 1135.]

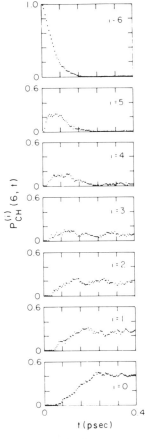

FIG. 18.   Probability of finding i quanta in the C—H oscillator as a function of time for the case of initial v = 6. [From Sibert, E. L., Hynes, J. T., and Reinhardt, W. P. (1984). *J. Chem. Phys.* **81**, 1135.]

C. Quantum Dynamics

Molecules are, of course, properly described by quantum mechanics, and classical mechanics is recognized as a particular approximation. Nonetheless, an introductory description of intramolecular energy transfer via classical mechanics has proven useful since it contains few concepts that are truly unfamiliar to the macroscopic world. Although it is possible to cast both quantum and classical mechanics in a similar formal language (i.e., distributions in phase space and a Liouville propagator), standard quantum mechanics is based on a mathematical structure that is substantially different from that of classical Hamiltonian mechanics. We first provide a brief qualitative summary of some results of quantum investigations, and then we present details that can be best appreciated by the reader who is well versed in quantum mechanics.

First and foremost, we note that classical mechanics does not allow a number of phenomena that occur in nature. A familiar example is tunneling, in which a system has finite probability of being in a region of phase space where it is not permitted classically. The simplest example of tunneling occurs in a system consisting of a particle moving in a potential that has two minima with a potential barrier between them. Classically, a particle initiated in one of the wells with energy below the barrier height is confined to that well forever. In the quantum case, however, the system flows between the two wells: it "tunnels" through the potential barrier. Tunneling effects are most certainly important in the intramolecular dynamics of systems at energies below such potential barriers. Less well known are symmetry effects, resonances, etc. that can play important roles in intramolecular dynamics. Space limitations prevent a discussion of these quantum phenomena, and the interested reader is referred to the bibliography for details. We focus rather on the dynamical consequences of energy quantization in quantum mechanics. This property means that a system can only exist at specific energy values, a property shared by other observables as well. Energy is, however, intimately linked to dynamics, since the Hamiltonian determines system time propagation, as discussed later.

One important remark is in order. That is, although quantum phenomena have been observed in molecular systems, we possess only the very qualitative "traditional" rules regarding conditions under which quantum effects predominate. Specifically, if the initial state involves large classical actions and the initial state is one that is allowed classically, then quantum effects tend to be small. Considerably more work is necessary, however, before more quantitative, predictive statements can be made and before our understanding of classical/quantum correspondence in bound molecular systems is complete.

Considerations of the quantum dynamics of bound molecules shows that, in the absence of the emission of radiation from energized molecules, all dynamics is quasiperiodic and regular. That is, quantum mechanics does not admit the possibility of long-time relaxation to a time-independent stationary state, a property that characterizes a classical mixing system. This property creates a number of difficulties in understanding the formal relationship between classical and quantum mechanics, particularly for energized molecules that display classically chaotic behavior. Practically, however, one finds that if the system is close to the classical limit, then quantum and classical dynamics agree over a significant time scale. This time scale is expected, in the vast majority of typical chemical experiments, to be in excess of the time of interest for the process. Under these circumstances, the formal discrepancy between classical and quantum mechanics is irrelevant to the specific chemical problem. Nevertheless, since quantum mechanics does not admit anything other than quasiperiodic behavior, attention has recently been focused on other quantities that might provide the qualitative distinction between quantum systems that display, more or less, statistical behavior.

Quantum mechanics describes system behavior in terms of operators that represent measurable quantities, their eigenfunctions, which describe possible states of the system, and their eigenvalues, which correspond to allowable values of the measurable. If the system is presumed best described in terms of a specification of the system energy, then one looks for states $|\psi\rangle$ that are eigenfunctions of the molecular Hamiltonian operator $\mathbf{H}$. That is, one solves the problem

$$\mathbf{H}|\psi(t)\rangle = -i\hbar \, \partial|\psi(t)\rangle/\partial t \qquad (21)$$

where t is the time. The nature of this equation is such as to admit the solutions

$$|\psi_j(t)\rangle = |\psi_j\rangle \exp(-iE_j t/\hbar) \qquad (22)$$

where $|\psi_j\rangle$ is the solution to the eigenvalue problem

$$\mathbf{H}|\psi_j\rangle = E_j|\psi_j\rangle \qquad (23)$$

Here the Hilbert-space vector $|\psi_j\rangle$ has a coordinate space representation $\psi_j(\mathbf{q}) = \langle\mathbf{q}|\psi_j\rangle$ and $|\psi_j(\mathbf{q})|^2$ is the probability of observing a given value of $\mathbf{q}$ when the system is in a state defined by energy E_j. In general, the system may be degenerate, in which case several E_j may have the same numerical value.

This treatment and the one that follows provide an idealized picture in which the molecule is entirely isolated from external influences. Such an ideal picture cannot, in fact, apply. Specifically, although one may be able to experimentally isolate the molecule from interactions with other molecules (e.g., via high-vacuum techniques), the molecule will always interact with the background radiation field to radiate energy. In this discussion we regard this emission as a small perturbation that can be introduced as part of the measurement process.

Consider time dependence in quantum mechanics, with the experimentally prepared initial state assumed completely specified as the Hilbert space vector $|\phi(0)\rangle$. That is, the system is initially in a pure state and may be expanded in a linear combination of energy eigenstates $|\psi_j\rangle$ as

$$|\phi(t = 0)\rangle = \sum_j c_j|\psi_j\rangle \qquad c_j = \langle\psi_j|\phi(0)\rangle \quad (24)$$

Then, from Eq. (22), the subsequent time evolution is given by

$$|\phi(t)\rangle = \sum c_j|\psi_j\rangle\exp(-iE_jt/\hbar) \qquad (25)$$

In the alternative case, the so-called mixed state, one does not have a complete specification of the initial state; that is, the initial state cannot be described as a single Hilbert space vector, nor can it be written in terms of a linear combination of $|\psi_j\rangle$. It can, however, be written in terms of a density matrix,

$$\rho(0) = \sum_n w_n|\chi_n\rangle \langle\chi_n| \qquad (26)$$

where the states $|\chi_n\rangle$ can be written in terms of a linear combination of $|\psi_j\rangle$ and where $\sum_n w_n = 1$, with w_n equal to or greater than 0. The essential feature of such states is that $\rho(0)$ is lacking information on the relative phases of the participating eigenstates $|\psi_j\rangle$.

We focus on the pure state, although it is the exception, rather than the rule, in experimentally prepared systems. In accordance with quantum mechanics, a measurement of a particular property during the course of the system time evolution consists of evaluating the average value of the corresponding operator. For example, if one measures the property described by the quantity F, then the average value of F as a function of time is given by

$$\langle F(t)\rangle = \langle\phi(t)|F|\phi(t)\rangle$$
$$= \sum_{i,j} d_{i,j} \exp(i\omega_{i,j}) \qquad (27)$$
$$d_{i,j} = \langle\psi_j|F|\psi_i\rangle c_j^* c_i$$

where $\omega_{i,j} = (E_i - E_j)/h$. Thus $\langle F(t)\rangle$ may be written as a linear combination of a terms involving a discrete set of frequencies $\omega_{i,j}$. By analogy with the discussion of classical systems, this sum is seen to be quasiperiodic.

The number of terms contributing to the sum in Eq. (27) and the relationship between the frequencies $\omega_{i,j}$ determine the kind of qualitative behavior observed. In the case where only a few terms contribute, the dynamics is almost periodic. Such behavior is observable as, for example, a periodic modulation of the fluorescence emitted from a molecule prepared in a linear combination of a few states, a phenomenon known as quantum beats. An example of beats in SO_2 is shown in Fig. 19, where the fluorescence reflects the interference between two contributing levels. Such simple behavior emerges only when the initially created state is comprised of a few levels. This is seldom the case, as

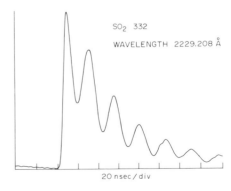

FIG. 19.   The intensity of fluorescence, as a function of time, from SO_2 created in a superposition state composed of two levels. [From Ivanco, M., Hager, J., Sharfin, W., and Wallace, S. C. (1983). *J. Chem. Phys.* **78**, 6531.]

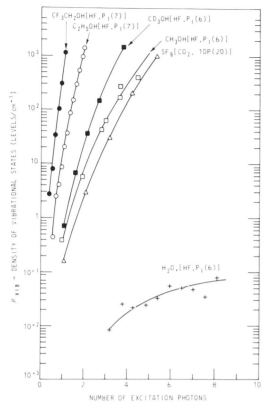

FIG. 20.   Density of states (i.e., number of states per unit energy interval) as a function of energy for a number of molecules. The abcissa is in number of photons, rather than energy, where the type of laser photon used depends on the particular molecule. For example, the state density for CF₃CH₂OH is plotted versus photons, from an HF laser, associated with the P₁(7) line. The energy of each of these photons is 0.01660 a.u. Other photons used are the P₁(6) line with a photon energy of 0.01683 a.u. and the CO₂10P(20) line with energy of 0.00430 a.u. per photon. [From McAlpine, R. D., Evans, D. K., and McClusky, F. K. (1980). *J. Chem. Phys.* **73**, 1153.]

seen from Fig. 20, which shows the density of states (i.e., the number of states per unit energy interval) $D(E)$ for some typical molecules. The quantity $D(E)$ is seen to be an increasing function of the size of the molecule and, for even small molecules such as SF_6, can reach very large values (e.g., $\geq 10^3/cm^{-1}$). As a result, chemical experiments, whose energy resolution is often not sharp, will typically involve an initial state comprised of many many levels. The subsequent dynamics emerges through the simultaneous interference of a multitude of terms, and the resultant behavior is difficult to qualitatively

extract from a formal sum [Eq. (27)] of contributions from individual levels.

For example, the system can display short-time behavior reminiscent of relaxation. Consider the case where the coefficients $d_{i,j}$ are distributed in a smooth fashion about a particular frequency value. Then one can show that for short times compared to the density of frequencies, $\langle F(t) \rangle$ decays smoothly as a function of time, with a time scale governed by the inverse of the frequency width of the coefficients $d_{i,j}$. This behavior is termed dephasing, to distinguish it from irreversible relaxation of the initial state. That is, despite this decay, the system will eventually reassemble to form the initial state, although this time scale may be exceedingly long. Thus, an experiment measuring $\langle F(t) \rangle$ over a time scale short compared to the recurrence time will show apparent relaxation of $\langle F(t) \rangle$. Nonetheless, formally, the system is quasiperiodic. The facts that quantum bound-state dynamics is quasiperiodic, classical mechanics can be mixing, and the latter is expected to approximate the former, make up the essence of the important unresolved problem "what is quantum chaos?"

The quantum picture of bound-state dynamics calls attention to an important aspect of intramolecular dynamics. Specifically, the state $|\phi(t)\rangle$ is comprised of a linear combination of states $|\psi_j\rangle$. The probability of observing the system in an exact eigenstate $|\psi_j\rangle$ at time t is given by $|c_j \exp(-E_j t/\hbar)|^2 = |c_j|^2$. Thus, the population of each exact eigenstate does not change as a function of time. If, in fact, one were solely interested in the population of these exact levels, then there is no such thing as time dependence in the dynamics of bound molecules (other than radiative emission)! Clearly, the focus of intramolecular dynamics and energy transfer is on attributes other than exact eigenstates populations.

To appreciate the desired description of energy flow in chemistry, recall the historical origin of the interest in intramolecular energy flow. The most prominant case is that of unimolecular decay, in which a molecule, sufficiently energized, breaks into a variety of products, (e.g., ABC → A + BC). In this case the focus of attention, and therefore of the measurement, is on the energy content of the A—B bond. This is typical of chemical descriptions in which the analysis is in terms of subunits of the molecule that are not, in themselves, naturally distinct subcomponents of the molecule. Such a description results when a zero-order basis set is used. Specifically, consider the Hamiltonian for a two-degrees-of-freedom system written, as in the classical case, in the form

$$H = H_0 + V \qquad H_0 = H_1 + H_2 \qquad (28)$$

where H_1 and H_2 describe two distinct subcomponents of interest in a particular experiment and the eigenfunctions of H_i are denoted by $|\chi_j^i\rangle$, where

$$H_i|\chi_j^i\rangle = \varepsilon_j^i|\chi_j^i\rangle \qquad i = 1, 2 \qquad (29)$$

The perturbation V couples the zero-order states so that exact-energy eigenstates $|\psi_k\rangle$ are linear combinations of these zero-order states or vice versa. That is,

$$|\chi_i^1\rangle|\chi_j^2\rangle = \sum_k b_{i,j}^k|\psi_k\rangle \qquad (30)$$

Using this expression and Eq. (22) gives the following form for the time evolution of these zero-order basis-set states;

$$|\chi_i^1\rangle|\chi_j^2\rangle(t) = \sum_k b_{i,j}^k|\psi_k\rangle\exp(-iE_kt/\hbar) \qquad (31)$$

If the initial state consists of a linear combination of the zero-order states, then the populations of the zero-order states are seen to be time dependent.

The degree to which the zero-order states enter into the exact eigenstates [i.e., the nature of the sum in Eq. (30)] is a measure of the strength of the coupling and provides a insight into the nature of the exact eigenstates *from the viewpoint of this particular zero-order basis*. It essentially provides the time-independent picture of the possible zero-order states that can be coupled during the dynamical evolution of the system. Equation (28), treated quantum mechanically, admits the same kind of perturbation treatment as in the classical case, with a similar emphasis on isolated resonances and overlapping resonances emerging.

As in the classical case, the question of the nature of intramolecular energy flow—whether it is statistical or whether it displays a specific pathway—is of interest. Unfortunately, few quantum calculations on realistic molecular systems have been performed. The example of the dynamics of benzene, initially prepared in an excited state of the C—H bond, has, however, been treated quantum-mechanically. Here the relevant zero-order Hamiltonian is of the form

$$H = H_L + H_N + H_{LN} \qquad (32)$$

where H_L is the Hamiltonian for the local C—H vibrational motion, H_N is the Hamiltonian for the remainder of the molecule, and H_{LN} is the coupling between them. Computations have been carried out, including the coupling between the C—H vibration and the CCH wag motion, via terms in H_{LH}. In the adopted model, H_N is essentially harmonic so that the zero-order states are of the form $|v_L,k_N\rangle = |v_L\rangle|k_N\rangle$, denoting v quanta in the C—H stretch and k quanta in the ring modes.

As in the classical treatment in Section II, interest is in the dynamics of benzene in the energy range where the C—H bond is prepared with considerable energy. Figure 21 shows some of the many zero-order energy eigenstates in the energy regime associated with the zero-order state $|6_L, 0_N\rangle$. Not shown are states with less than four quanta in H_L, which, although coupled to the $|6_L, 0_N\rangle$ state, were too numerous to be included in the computation. For convenience, the levels are stacked in ladders, or "tiers," with the ladder labeled by the quanta of energy in the C—H oscillator. Calculations show that the exact system eigenstates are a strongly coupled mixture of the zero-order states.

Figure 22 shows the result of a calculation on the quantum dynamics of benzene initially ex-

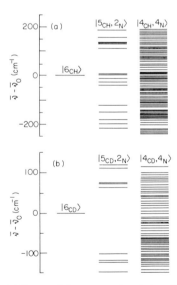

FIG. 21. (a) Some of the zero-order states of benzene in the energy neighborhood near that of excitation to the sixth vibrational level of C—H. States are organized, for clarity, in ladders associated with the number of quanta in the C—H mode. (b) Levels for the analogous monodeuterated benzene case. [From Sibert, E. L., Reinhardt, W. P., and Hynes, J. T. (1984). *J. Chem. Phys.* **81**, 1115.]

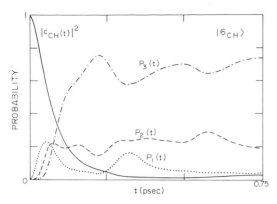

FIG. 22. The time dependence of various quantities associated with the quantum dynamics of benzene initially excited to the sixth level of C—H bond excitation. Labels are defined in the text, other than $P_i(t)$, which denotes the probability of being in the ith tier of energy levels. Only three tiers were included in this calculation. [From Sibert, E. L., Reinhardt, W. P., and Hynes, J. T. (1984). *J. Chem. Phys.* **81**, 1115.]

cited to the sixth vibrational level of C—H. The population of the sixth level (denoted $|c_{CH}(t)|^2$) is seen to decay rapidly in time, with concomitant growth of population in the various tiers. The essential features of this quantum calculation are in accord with the classical results by the same authors and describe the dynamics of benzene, initially energized in the C—H vibration, as decay via a specific pathway of resonances between zero-order modes.

Although specific examples of this kind are of considerable importance, general rules regarding quantum intramolecular dynamics would be far more useful. Indeed, this was the motivation for investigating ergodic, mixing, etc. systems classically. Similarly, it is the reason for the general interest in the question of "what is quantum chaos?" We comment briefly (and only qualitatively), on recent developments in this area, noting first that an appropriate definition of chaos can only involve properties of the central participants in the time evolution, that is, the energy eigenstates and eigenvalues.

As a starting point, note that chaotic classical systems possess a number of prominent features. First, the system relaxes to a long-time stationary distribution. Second, the long-time limit is sensitive to only a few simple constants of the motion (e.g., the total energy). Third, as a direct consequence, the computation of system properties can be done by replacing the actual dynamics by simplified statistical models. The first property is clearly not satisfied by a quan-

tum bound system, which has been formally shown to be quasiperiodic. Nonetheless, one possibility is that the system approach a long-time value about which fluctuations are small and that this long-time value be sensitive to only a few simple properties. Indeed, if all wave functions in an energy interval are basically the same (i.e. have similar properties), then the system will evolve in a fashion that is relatively insensitive to the nature of the preparation. Further, if there are no integrals of motion other than the total energy, then one might expect the energy eigenvalues to display rather simple properties reflecting this characteristic. These remarks motivate one current view on the nature of quantum chaos. Specifically, in such a system the eigenfunctions are proposed to be similar to one another in character as the energy varies, and the probability of observing a particular level spacing is expected to be of a specific form (the Wigner adjacent-level distribution). The former requirement, coupled with the condition that all eigenfunctions be orthogonal, that is,

$$\langle \psi_i | \psi_j \rangle = 0 \qquad (33)$$

hints at the nature of these wavefunctions—that is, they display erratic nodal patterns. This is clearly not the case for the vast majority of typical systems studied in elementary quantum mechanics (e.g., the H atom, the particle in a box), which are, in fact, integrable and separable in classical mechanics as well.

Note that the condition of erratic nodal patterns makes sense with respect to the view resulting from an analysis in terms of a zero-order basis set. In particular, expanding such exact wave functions in any arbitrary basis is expected to yield populations spread over all zero-order wave functions in the energy neighborhood. Thus there is a form of statistical coupling between all zero-order basis functions.

Numerical computations on model systems have, in fact, revealed erratic wavefunctions for some systems that are classically mixing. An example is shown in Fig. 23, where the system is a particle confined, by infinite potential walls, to a region that is the shape of a racetrack (the so-called stadium system). The nodal patterns are clearly highly disordered and in marked contrast with the simple nodal lines associated with, for example, the separable particle in a rectangle case. Unfortunately, there is no one-to-one correspondence between systems that display chaotic classical behavior and the observation of erratic wave-function nodal patterns.

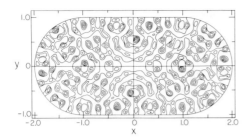

FIG. 23. Contour plot of $\psi(q)$ associated with a typical state of the particle in a stadium potential. Only positive contours are shown. Note the highly erratic pattern of contour lines proposed to be a characteristic of wavefunctions associated with chaotic classical systems. [From Brumer, P., and Taylor, R. D. (1983). *Faraday Disc. Chem. Soc.* **75**, 171.]

IV. Statistical Approximations and Dynamics

Accurate dynamical studies, such as those discussed above, are limited to small molecular systems and relatively short times (e.g., typically 100 vibrational periods). These restrictions stem from a number of sources. First, reliable computations of forces between atoms in large molecular systems are extremely difficult, and few quantitatively accurate models of interatomic forces exist. Second, even if such potentials are available, large-molecule motion treated quantum-mechanically involves huge numbers of participating energy eigenstates. Techniques for efficiently computing large numbers of eigenstates in systems with significant numbers of degrees of freedom are only now being developed. Alternative techniques, which rely on the direct, numerical, temporal propagation of initial states [i.e., via Eq. (21)] do not utilize eigenstates but suffer inaccuracies in long-time applications. An analogous difficulty exists in attempting numerical studies of dynamics using classical mechanics; in this case, the exponential growth of distances in phase space (see Section III) translates into the rapid growth, with time, of numerical errors.

Since the vast majority of interesting molecules have many degrees of freedom, the need for models that simplify the dynamics is evident. The most popular of such models relies on a statistical assumption of relaxation during the course of the dynamics. That is, one assumes that after preparation of the energized molecule, the system relaxes to a well-defined state that is dependent on only gross features (e.g., the total energy) of the preparation. Assumptions of this

kind predate detailed dynamical studies of molecular dynamics.

To appreciate the simplifications resulting from such models, consider the paradigm case of unimolecular decomposition ($A \rightarrow B + C$, where A, B, and C are molecules), where this assumption leads to statistical theories of the rate of unimolecular decay. Here A, sufficiently energized to dissociate, is prepared by any of a variety of means (e.g., laser excitation, collisions, or as the product of a chemical reaction). A detailed computation of the dynamics of this process for a realistic molecule, an intractable computational feat, would entail the following steps. One first specifies the exact nature of the molecular potential, the nature of the process that prepares the excited molecule, and the state of the molecule prior to preparation. Second, the exact dynamics of the evolution of the molecule, from preparation to decay, is computed. Such a computation must be repeated for each and every type of initial state of the molecule and each and every type of state preparation. By contrast, the formulation of a typical statistical model proceeds as follows. First, define the region of phase space in which the molecule A is to be regarded as being bound and a complementary region in which it is characterizable as B + C. Then assume that the rate of dissociation of the excited molecule depends primarily on the magnitude of simple known constants of the motion, that is, energy E and total angular momentum $\mathbf{J}$. Assume further that any initially prepared state rapidly relaxes to a uniform distribution over the surface in phase space characterized by fixed E, $\mathbf{J}$ and within the region where A is regarded as bound. The computation of the rate of decay of such a system then entails an analysis of the rate at which the relaxed uniform distribution crosses over into the dissociation region. Such a computation is straightforward in both classical and quantum mechanics if a number of approximations on the structure of the molecule are made, and leads to statistical theories, such as RRKM theory, discussed in the references.

The essence of such theories is, then, that the system undergoes memory erosion during the dynamics so that information on initial conditions is lost; only knowledge of E and $\mathbf{J}$ remains. Similar theories have been proposed for a wide variety of processes that involve the participation of long-lived molecular intermediates. These include chemical reactions that proceed via collision complexes, photodissociation where molecular preparation is through controlled laser excitation, molecules adhering to surfaces where detachment is induced via a variety of means, etc. Such theories have the advantage of yielding rather general results, which are amenable to both theoretical and experimental analysis. For example, in the case of unimolecular decay, the rate constant for dissociation is found to increase with increasing energy and decrease with the number of participating degrees of freedom in the system.

Experimental studies on the validity of such theories have been ongoing for many years, as discussed briefly in the next section. The theoretical examination of the validity of such approaches is more recent and links directly to the issues discussed in Section III. At present, classical mechanical studies have shown the possibility of both statistical and nonstatistical decays, depending on the degree and extent of exponential divergence of trajectories in phase space. The greater the degree and extent of exponential separation, the closer the agreement with statistical approaches. Similarly, quantum-mechanical studies have shown that model systems can display unimolecular rate constants whose energy dependence is inconsistent with that predicted by simple statistical theories. It is fair to say, however, that a clear understanding of the interrelationship between molecular properties and the validity of statistical theories is in its early stages of development.

V. Experimental Studies

The ideal experiment on intramolecular energy transfer, as yet unachieved, entails a number of simple features. Specifically, the molecule is prepared in a well-defined and well-characterized state and evolves for a known time interval, after which the state of the system is precisely determined via a high-finesse experimental probe. The importance of each of these components to the resolution of even the most qualitative of questions (e.g., is the energy transfer statistical or not) should be clear. If, for example, as is the case in some of the available techniques, the initial state of the molecule is in itself highly randomized energetically, then a subsequent measurement that shows that the energy is statistically distributed among the system modes is of little consequence. That is, this particular observation would be a consequence of the preparation, as distinct from the evolution, of the molecule.

The dynamics of molecules with low energy content is usually well described by a Hamiltonian that is the sum of independent degrees of freedom, and absorption spectroscopy has provided considerable information on the nature of such systems. The situation with respect to energized molecules is far more complex. A number of experimental tools have been utilized over the past 30 years to examine the nature of energy transfer in such systems, as discussed next. Most suffer from the inaccurate knowledge of the initial system state. In our brief description of several experimental methods it will become clear that experimental tools have been rapidly developing over the past few years and that an explosion of highly informative experimental data is expected over the next decade.

First and foremost, note that interest is in the nature of intramolecular dynamics of molecules in isolation. That is, observations must be made over a time scale where the molecule does not collide with others in the reaction vessel. Modern techniques allow very low pressures under which such measurements can be made. Most desirable among these methods are beam techniques in which molecules are studied in a low-density beam produced, for example, by vaporizing molecules in an oven. Experiments prior to this "beam age" (circa 1960) often inferred information about intramolecular dynamics from bulk data, which contained effects due to collisions, with resultant loss in accuracy.

Measurement techniques in typical experiments can be subdivided into two categories; the first and most modern, probes the bound-molecular dynamics directly, for example by observing radiation emitted or absorbed during the course of the dynamics. The second infers information about the nature of bound-molecular dynamics by indirect means, typically by analyzing the *outcome* of a process that involves the molecule of interest as a long-lived intermediate. These latter types of measurement are readily clarified by considering the NaCl + KBr example discussed in Section II. Specifically, this particular reaction proceeds via the bound NaClKBr intermediate to yield two different sets of products, either NaCl + KBr or NaBr + KCl. The probability of observing these two product "channels" depends on the nature of the NaClKBr dynamics. That is, the product ratio will provide insight into whether intramolecular energy transfer in NaClKBr is rapid or not. More extensive measurements will entail analysis of the internal states of the product molecules (see Fig. 5), as well as of the relative velocities of the products. Thus, these kinds of measurements allow one to infer information on the characteristics of the dynamics of the intermediate without directly observing it. Other examples of this kind include measurements on the products of unimolecular decay, photodissociation, etc.

Similarly, most experiments on intramolecular energy transfer fall into one of two categories with respect to preparation of the molecule. In the crudest, the system is prepared by a "coarse" technique where little detail about the initial molecular state is available. Such experiments include preparation via collision—that is, where the molecule of interest, A, collides with, and absorbs energy from, another molecule B— and preparation via reaction, where the molecule A is the product of a "precursor" reaction. Although the bulk of early work on intramolecular dynamics was carried out with these techniques, far greater insight emerges from modern experiments in which the molecule A is prepared by the absorption of radiation.

The optimum experiment would therefore proceed by preparing the molecules in a precise state using beam methods and laser excitation, followed by measurement of the radiative emission as well as other properties of the bound molecule. Such experiments are now underway in a number of laboratories throughout the world. Also in a stage of rapid development are pump-probe techniques in which two lasers are utilized, one to prepare the molecule in a desired initial state and the second to probe the state of the evolving molecule. Such techniques are just on the brink of providing concise information on isolated molecular evolution and are discussed in the references. Along with these experimental developments there is a need for reliable theories to understand the interrelationship between observed features and the nature of the dynamics. Such developments are also in progress.

VI. Summary

Understanding the nature of intramolecular energy flow in isolated molecules is of great practical and fundamental interest. Early developments, both theoretical and experimental, were hampered by a number of technological problems now being overcome. As a consequence, general features that determine the rate and extent of intramolecular energy transfer are slowly emerging. These include generic features of classical Hamiltonian systems and the way in which coupling terms influence the nature of the dynamics, the dependence of observed energy transfer on the zero-order system, the interaction between state preparation and state measurement on the qualitative interpretation of the dynamics, and the differences and similarities between the quantum and classical views of dynamics. Nonetheless, general rules regarding the rates and extent of intramolecular energy flow have yet to be established. Similarly, a number of fundamental issues arising in the study of intramolecular energy flow have yet to be resolved. Rapid technological developments in computational and experimental tools hold great promise for substantial developments over the next decade.

BIBLIOGRAPHY

Berry, M. W. (1978). *In* "Topics in Nonlinear Mechanics" (S. Jorna, ed.), Amer. Inst. Phys., New York.

Bondybey, V. E. (1984). *Annu. Rev. Phys. Chem.* **35,** 591.

Brumer, P. (1981). *Adv. Chem. Phys.* **47,** 201.

Brumer, P. and Shapiro, M. (1987). *Adv. Chem. Phys.* (in press).

Faraday Discuss. Chem. Soc. (1983). No. 75.

Forst, W. (1973). "Theory of Unimolecular Reactions." Academic Press, New York.

Levine, R. D. (1969). "Quantum Mechanics of Molecular Rate Processes." Oxford Univ. Press, London.

McDonald, J. D. (1979). *Annu. Rev. Phys. Chem.* **30,** 29.

Noid, D. W., Koszykowski, M. L., and Marcus, R. A. (1981). *Annu. Rev. Phys. Chem.* **32,** 267.

Rice, S. A. (1975). *In* "Excited States" (E. C. Lim, ed.), vol. 2. Academic Press, New York.

Rice, S. A. (1981). *Adv. Chem. Phys.* **47,** 117.

Stechel, E. B. and Heller, E. J. (1984). *Annu. Rev. Phys. Chem.* **35,** 563.

EXCITONS, SEMICONDUCTOR

Donald C. Reynolds *Wright Patterson Air Force Base*
Thomas C. Collins *University of Tennessee*

I. Intrinsic Exciton Characteristics
II. Extrinsic Exciton Characteristics
III. Interaction of Excitons with Other Systems
IV. Special Properties of Excitons

GLOSSARY

Band gap: Energy difference between two allowed bands of electron energy in a solid.

Bound exciton: Exciton localized at an impurity or defect in the crystal.

Brillouin zones: Those volumes in K space bounded by intersecting surfaces defined by points in K space where the energy is discontinuous and Bragg reflection occurs.

Central-cell correction: Differences in the binding energies of different chemical impurity atoms (donors or acceptors) in a host lattice resulting from different core configurations.

Degenerate semiconductor: Semiconductor in which a band has orbital as well as spin degeneracy.

Direct semiconductor: Semiconductor in which the minimum in the conduction band and the maximum in the valence band occur at the same wave vector.

Hamiltonian: Total energy operator of the system, kinetic plus potential energy operator.

Indirect semiconductor: Semiconductor in which the minimum in the conduction band and the maximum in the valence band occur at different wave vectors.

Nondegenerate semiconductor: Semiconductor in which the bands have only spin degeneracy.

Oscillator strength: Measure of the intensity of a particular energy transition.

Phonon: Expression for a quantized lattice vibration. Acoustic phonons refer to in-phase motion of neighboring ions in the same cell in a lattice vibration; optical phonons refer to out-of-phase motion of the same ions.

The exciton is a quantum of electronic excitation produced in a periodic structure such as an insulating or semiconducting solid. This quantum of energy has motion, and the motion is characterized by a wave vector. Frenkel was the first to treat the theory of optical absorption in a solid as a quantum process consisting of atomic excitations. The excitation process implies that the excited electron does not leave the cell from which it was excited. In his attempt to gain insight into the transformation of light into heat in solids, Frenkel was able to explain the transformation by first-order perturbation theory of a system of N atoms having one valence electron per atom with the following properties:

(a) The coupling between different atoms in a crystal is small compared with the forces holding the electron within the separate atoms.

(b) The Born–Oppenheimer approximation is valid.

(c) The total wave function is a product of one-electron functions.

Thus the Frenkel exciton is a tight-binding description of an electron and a hole bound at a single site such that their separate identities are not lost. This model of the exciton emerges as a limiting case of the general theory of excitons and is applicable to insulating crystals. In the case of semiconductors, nonequilibrium electrons and holes are bound in excitonic states at low temperatures by Coulomb attraction. Semiconducting crystals are characterized by large dielectric constants and small effective masses; therefore the electrons and holes may be treated in a good approximation as completely independent particles, despite the Coulomb interaction. This results because the dielectric constant reduces the Coulomb interaction between the hole and electron to the extent that it produces a weakly bound pair of particles that still retain much of their free character. The exciton represents a state of slightly lower energy than the unbound hole–electron. The effective mass theory used to describe such weakly bound particles was developed by Wannier. These weakly bound excitons are appropriately described using the one-band electronic structure picture by adding the Coulomb interaction between the hole and electron. Semiconductor materials are the heart of most modern optical and electronic devices; as a result, the dominant technological interest is focused on these materials. In view of this interest, this article emphasizes the Wannier excitons, which are appropriate for these materials. [See BONDING AND STRUCTURE IN SOLIDS.]

I. Intrinsic Exciton Characteristics

A. Introduction

The intrinsic fundamental-gap exciton in semiconductors is a hydrogenically bound hole–electron pair, the hole being derived from the top valence band and the electron from the bottom conduction band. It is a normal mode of the crystal created by an optical excitation wave, and its wave functions are analogous to those of the Block wave states of free electrons and holes. When most semiconductors are optically excited at low temperatures, it is the intrinsic excitons that are excited. The energies of the ground and excited states of the exciton lie below the band-gap energy of the semiconductor. Hence, the exciton structure must first be determined in order to determine the band-gap energy. The exciton binding energy can be determined from spectral analysis of its hydrogenic ground and excited state transitions (this also gives central-cell corrections). Precise band-gap energies can be determined by adding the exciton binding energy to the experimentally measured photon energy of the ground-state transition.

Both direct and indirect exciton formation occurs in semiconductors, depending on the band structure. The former is characteristic of many of the II–VI and III–V compounds, and the latter is characteristic of germanium and silicon. For indirect optical transitions, momentum is conserved by the emission or absorption of phonons. The detailed nature of the valence-band structure of degenerate and nondegenerate semiconductors is elucidated by understanding the intrinsic-exciton structure of these semiconductors.

B. Excitation of the *n*-Particle System

Excitons are excited states of the system in which the number of electrons does not change. Ordinarily in solid-state physics one thinks of calculations of one-body approximations as excitation of the system. However, this only works in metals near the Fermi surface or where one adds an electron or takes one away from the system. In the case of metals, the electrons and holes are very diffuse and the interaction between the excited "particle" and the other "particles" is very small. When an electron is added to the system, one obtains electron affinities or $N + 1$ solutions, and when an electron is removed one obtains ionization energies or $N - 1$ solutions. All solutions to one-body calculations such as the one-particle Green's function method, Hartree–Fock calculations, and similar approximations (unless specifically added) do not contain the interaction of the excited "particle" with the other "particles." It is necessary to calculate solutions to the two-body Green's function (or some approximation to the two-body problem) in order to have an exciton. [See QUANTUM MECHANICS.]

Another approach is to calculate the many-body excited states by construction of excited wave functions that are orthogonal to the approximate ground-state wave function. The excited-state approximate total energy is calculated, from which is subtracted the approximate total ground-state energy. The formalism gives an effective operator that looks like a scaled or screened hydrogen Hamiltonian. This approach was used by both Frenkel and Wannier. In the case of Frenkel, the excited states were required to be a linear combination of very local excitations. In particular, he required the atoms in his model to all be in the ground state except one. This one was excited, but the excitation was localized to the one atomic site. The many-body solution using these trial wave functions was identical to the normal modes of vibration of the phonon spectra. The solution is also similar to spin waves where one starts with either an up or down spin on each lattice site.

Wannier, on the other hand, let the excitation trial wave functions be more diffuse than one lattice site. This representation he called "exciton waves," which could vary from very local excitations (Wannier-like functions) to very diffuse (Bloch-like functions). His effective excitation operator was a hydrogenic operator with multipole corrections. The many-body effects were approximated through screening of the Coulomb potential between electrons and the hole left by the excitation. This formalism also has the advantage that when small perturbing fields are turned on, the resulting effective operator adds the perturbing operator, similar to what one finds for the hydrogen atom or molecule case with screened potentials.

C. SYSTEMS OF EXCITONS IN VARIOUS CRYSTAL SYMMETRIES

1. Direct Nondegenerate Semiconductors

Nondegenerate semiconductors are typified by those materials that belong to the wurtzite crystal structure. This is a uniaxial structure having sixfold rotational symmetry and belongs to the C_{6V} crystal point group. In this structure the degeneracy in the valence band is removed by crystal-field interactions.

The tight-binding approximation in conjunction with group theory was first used to describe the irreducible representations, band symmetries, and selection rules for the wurtzite structure. If one considers the absorption (emission) of electromagnetic radiation by atoms, the probability of the occurrence of a transition between two unperturbed states ψ_i and ψ_f as caused by the interaction of an electromagnetic radiation field and a crystal is dependent on the matrix element

$$\int \psi_f^* H_{\text{int}} \psi_i \, dr \tag{1}$$

where

$$H_{\text{int}} = \frac{e\hbar}{imc} \mathbf{A} \cdot \nabla \tag{2}$$

where $\mathbf{A}$ is the vector potential of the radiation field and has the form

$$\mathbf{A} = \mathbf{n}|A_0|e^{i(\mathbf{q}\cdot\mathbf{r}-\omega t)}$$

where e is the electron charge, m is the electron mass, c is the velocity of light, $\mathbf{n}$ is a unit vector

in the direction of polarization, and $\mathbf{q}$ is the wave vector. Expanding the spatial part of $\mathbf{A}$ in a series gives

$$H_{\text{int}} = \sum_{j=0}^{\infty} H_{\text{int}}^j$$

where

$$H_{\text{int}}^j \approx (\mathbf{q} \cdot \mathbf{r})^j (\mathbf{n} \cdot \nabla)$$

and the dipole term is then the first term ($j = 0$). The matrix element in Eq. (1) is now expressed as a series, and for an electric dipole transition to be allowed, the matrix element between the initial and final states must be nonzero.

In the case of transitions between two states of an atom (that is in a crystalline field), the initial and final states of the atom are characterized by irreducible representations of the point group of the crystal field. Also, the dipole moment operation must transform like one of the irreducible representations of the group. If one denotes the representations that correspond to the initial and final states of the transition and to the multipole radiation of order j ($j = 0$ for electric dipole radiation) by Γ_i, Γ_f, and $\Gamma_r^{(j)}$, respectively, at $\mathbf{k} = 0$, then the matrix element in Eq. (1) transforms under rotations like the triple direct product

$$\Gamma_f \times \Gamma_r^{(j)} \times \Gamma_i \tag{3}$$

The selection rules are then determined by which of the triple-direct-product matrix elements in question do not vanish.

The dipole moment operator for electric dipole radiation transforms like x, y, or z, depending on the polarization. When the electric vector $\mathbf{E}$ of the incident light is parallel to the crystal axis, the operator corresponds to the Γ_1 representation. When it is perpendicular to the crystal axis, the operator corresponds to the Γ_5 representation.

Since the crystal has a principal axis, the crystal field removes part of the degeneracy of the p levels. Thus, disregarding spin–orbit coupling, the following decomposition at the center of the Brillouin zone is obtained:

$$\text{conduction band} \qquad S \rightarrow \Gamma_1$$
$$\text{valence band} \qquad \begin{array}{l} P_x, P_y \rightarrow \Gamma_5 \\ P_z \rightarrow \Gamma_1 \end{array}$$

Introducing the spin doubles the number of levels. The splitting caused by the presence of spin

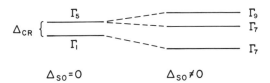

FIG. 1. Band structure and band symmetries for the wurtzite structure.

is represented by the inner products

$$\Gamma_5 \times D_{1/2} \rightarrow \Gamma_7 + \Gamma_9$$

$$\Gamma_1 \times D_{1/2} \rightarrow \Gamma_7$$

and the band structure at $\mathbf{k} = 0$ along with the band symmetries is shown in Fig. 1.

2. Direct Degenerate Semiconductors

Materials that crystallize in the diamond or the zinc-blende structures are representative of degenerate semiconductors. Two materials that have been extensively investigated and that are characteristic of direct degenerate semiconductors are GaAs and InP. These materials crystallize in the zinc-blende structure, which has T_d point-group symmetry.

The dipole momement operator for electric dipole radiation in zinc-blende structures transforms like Γ_5. The conduction band is s-like, while the valence band is p-like. This structure does not have a principal axis; therefore, the crystal-field energy is zero and the full degeneracy of the p levels is retained. Thus, disregarding spin–orbit coupling, the following decomposition at the center of the Brillouin zone is obtained:

conduction band $S \rightarrow \Gamma_1$

valence band $P \rightarrow \Gamma_4$

Introducing the spin doubles the number of levels. Consider the Γ_1 s-like conduction band and the triply degenerate p-like valence band. The states at the center of the Brillouin zone, which belong to Γ_1 and Γ_4 representations of the single group, are shown in Fig. 2. The splitting caused by the presence of spin is represented by

the inner product as follows:

$$\Gamma_1 \times D_{1/2} \rightarrow \Gamma_6$$

$$\Gamma_4 \times D_{3/2} \rightarrow \Gamma_7 + \Gamma_8$$

Physically this result means that the six valence band states, consisting of the three p-like states, each associated with one or the other of the two spin states, and that are degenerate in the absence of spin–orbit interaction, now split into two levels, one having Γ_7 symmetry and the other having Γ_8 symmetry. The Γ_8 level is fourfold degenerate, while the Γ_7 level is twofold degenerate.

The energy of the exciton is described by the Hamiltonian

$$H_{ex} = H_e + H_h + H_{eh} \qquad (4)$$

where H_e and H_h are the Hamiltonians for the electron and the hole and H_{eh} is the interaction Hamiltonian between the electron and the hole.

3. Indirect Transitions

Two of the most extensively studied indirect materials are the elemental semiconductors Si and Ge. Both of these materials have indirect band gaps, and therefore the lowest energy electronic state is an indirect exciton. For this lowest energy state to be optically excited, momentum must be conserved; thus, additional momentum must be supplied by the creation or annihilation of an appropriate phonon. These materials crystallize in the diamond structure and belong to the O_h point-group symmetry.

The band structures of Si and Ge are similar; as a result, Si will be used as the example in this discussion to describe the indirect exciton. The

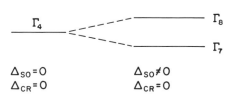

FIG. 2. Band structure and band symmetries for the zinc-blende structure.

band structure of Si is shown in Fig. 3. The conduction-band minimum occurs as a Δ_1 symmetry approximately 85% of the way to the zone boundary in the $\langle 100 \rangle$ direction. The corresponding valence band symmetry is Δ_5. Using the group character tables at $\Gamma = (0, 0, 0)$ and $\Delta = (k, 0, 0)$,

$$\Delta_1 \times \Delta_5 = \Gamma_{15}^+ + \Gamma_{25}^+ + \Gamma_{15}^- + \Gamma_{25}^- \quad (5)$$

The presence of the Γ_{15}^- symmetry means that this is an allowed transition. The maximum in the valence band occurs at $\mathbf{k} = 0$ having Γ_{25}^+ symmetry. The $\Gamma_{25}^+ \rightarrow \Gamma_{15}^-$ transition is allowed as

$$\Gamma_{25}^+ \times \Gamma_{15}^- = \Gamma_2^- + \Gamma_{12}^- + \Gamma_{25}^- + \Gamma_{15}^- \quad (6)$$

The momentum-conserving phonon in the valence band from Δ_5 to Γ_{25}^+ is

$$\Gamma_{25}^+ \times \Delta_5 = (\Delta_5) + (\Delta_1 + \Delta_2') + \Delta_1' + \Delta_2$$

$$= (TO + TA) + (LA + LO) \quad (7)$$

where TO is transverse optical, TA is transverse acoustic, and L indicates longitudinal. All phonons are allowed, as seen from Fig. 4. In this case, the energy denominator in the intermediate state is fairly large where the radiative transition occurs first; therefore the transitions may be weak.

The more important phonon for conserving momentum is in the conduction band from Δ_1 to Γ_{15}^-,

$$\Gamma_{15}^- \times \Delta_1 = \Delta_1 + \Delta_5 = LA + (TO + TA) \quad (8)$$

Here only the LO phonon is forbidden.

The same momentum considerations apply to the exciton as apply to the bands, since an exciton that is constructed from bands whose extrema differ by an amount $\mathbf{k}\Delta$ will have a momentum $\mathbf{k}\Delta$, which must be supplied by the phonon field during an optical transition.

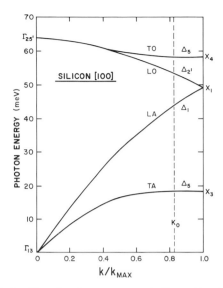

FIG. 4.   The vibration spectrum of silicon. [As determined by Brockhouse (1959). *Phys. Rev. Lett.* **2**, 256.]

D. PERTURBATIONS

1. Magnetic Fields

When the nondegenerate crystal is placed in a uniform magnetic field, there are several new terms in the Hamiltonian, which will be described below. In this description, the band-gap extrema are at $\mathbf{k} = 0$ with their shape parabolic at least to second order in $\mathbf{k}$ and with only double spin degeneracy. One may write the exciton equation as a simple hydrogen Schrodinger equation including mass and dielectric anisotropy. In general one finds the mass anisotropy is small, allowing first-order perturbation calculations to be made for the energy states as well as for the magnetic-field effects.

Since in this model the valence and conduction band extrema are at $\mathbf{k} = 0$, the wave vector of the light that creates the exciton $\mathbf{k}$ will also represent the position of the exciton in $\mathbf{k}$ space. If one divides the momentum and space coordinates into the center-of-mass coordinates and the internal coordinates, the exciton Hamiltonian can be divided into seven terms as follows:

$$H = H_1 + H_2 + H_3 + H_4 + H_{k1}$$
$$+ H_{k2} + H_{k3}; \quad (9)$$

$$H_1 = \frac{-\hbar}{2m} \left[\frac{1}{\mu_x} \left(\frac{\partial^2}{\partial x^2} + \frac{\partial^2}{\partial y^2} \right) + \frac{1}{\mu_z} \left(\frac{\partial^2}{\partial z^2} \right) \right]$$
$$- \frac{e^2}{\varepsilon \eta^{1/2}} (x^2 + y^2 + \eta^{-1} z^2)^{-1/2} \quad (10)$$

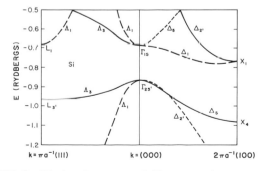

FIG. 3.   The band structure of silicon near the energy gap. [As computed by Phillips (1958). *Phys. Rev.* **112**, 685.]

$$H_2 = -2i\zeta \left[\frac{A_x}{\Delta_x} \left(\frac{\partial}{\partial x} \right) + \frac{A_y}{\Delta_y} \left(\frac{\partial}{\partial y} \right) + \frac{A_z}{\Delta_z} \frac{\partial}{\partial z} \right] \tag{11}$$

$$H_3 = \frac{e^2}{2mc^2} \left\{ \frac{A_x{}^2}{\mu_x} + \frac{A_y{}^2}{\mu_y} + \frac{A_z^2}{\mu_z} \right\} \tag{12}$$

$$H_4 = \frac{\zeta}{2} \sum_{\gamma = x,y,z} (g_{e\gamma} S_{e\gamma} + g_{h\gamma} S_{h\gamma}) H_\gamma \tag{13}$$

$$H_{k1} = -\frac{i\hbar}{2m} \left[\frac{K_x}{\Delta_x} \left(\frac{\partial}{\partial x} \right) + \frac{K_y}{\Delta_y} \left(\frac{\partial}{\partial y} \right) + \frac{K_z}{\Delta_z} \left(\frac{\partial}{\partial z} \right) \right] \tag{14}$$

$$H_{k2} = \zeta \left(\frac{K_x}{\mu_x} A_x + \frac{K_y}{\mu_y} A_y + \frac{K_z}{\mu_z} A_z \right) \tag{15}$$

$$H_{k3} = \frac{\hbar^2}{8m} \left\{ \frac{K_x{}^2}{\mu_x} + \frac{K_y{}^2}{\mu_y} + \frac{K_z^2}{\mu_z} \right\}, \tag{16}$$

where m is the free-electron mass and μ_γ is the reduced effective mass of the exciton in the direction γ. Also,

$$\zeta = \frac{e\hbar}{2mc}, \qquad \eta = \frac{\varepsilon_z}{\varepsilon_x} \tag{17}$$

$$A = \tfrac{1}{2}(\mathbf{H} \times \mathbf{r}) \tag{18}$$

$$\frac{1}{\Delta_\gamma} = \left(\frac{m}{m_{e\gamma}^*} - \frac{m}{m_{h\gamma}^*} \right) \tag{19}$$

in the wurtzite structure $\mu_x = \mu_y$.

The first term is the Hamiltonian for a hydrogenic system in the absence of external fields. This term has the possibility of including the mass and dielectric anisotropies. The second term is an $\mathbf{A} \cdot \mathbf{p}$ term, which leads to the linear (Zeeman) magnetic field term. In this term, the momentum operator $\mathbf{p}_i$ becomes $\mathbf{p}_i - (2e/c)\mathbf{A}_i$, where $\mathbf{A}_i = \tfrac{1}{2}\mathbf{H} \times \mathbf{r}_i$ is the vector potential, $\mathbf{H}$ is the magnetic field, and $\mathbf{r}_i$ is the coordinate of the ith electron. The A^2 term is the diamagnetic field term proportional to $|\mathbf{H}|^2$. The fourth term is the linear interaction of the magnetic field with the spin of the electron and hole. If one has small effective reduced mass for the electron and a large dielectric constant, the radii of the exciton states are much larger than the corresponding hydrogen-state radii. Hence, since the spin–orbit coupling is proportional to r^{-3} and thus quite small, it is legitimate to write the magnetic-field perturbations in the Paschen–Back limit as done above.

The last three terms are the $\mathbf{K} \cdot \mathbf{P}$, $\mathbf{K} \cdot \mathbf{A}$, and K^2 terms; $\mathbf{K}$ is the center-of-mass momentum. Treating the $\mathbf{K} \cdot \mathbf{P}$ to second order and adding the K^2 term, one can obtain an energy term that appears like the center-of-mass kinetic energy. The $\mathbf{K} \cdot \mathbf{A}$ term has little effect upon the energy; however, it has very interesting properties. This term represents the quasi-electric field that an observer riding with the center-of-mass of the exciton would experience because of the magnetic field in the laboratory. This quasi-field would produce a stark effect linear in $\mathbf{H}$, and this would give rise to a maximum splitting interpretable as a "g value."

2. Strain Field

A detailed study has been made of the stress-induced splitting of the exciton states in both wurtzite and zinc-blende structures. The band symmetries for the structures are shown in Figs. 1 and 2, respectively. In these materials the conduction band is s-like and the valence bands are p-like. The theory was developed around the effective Hamiltonian

$$H = H_{v-0} + H_{v-p} + H_c + H_{ex}, \tag{20}$$

where H_{v-0} describes the zero pressure mixing of the three p-like valence bands, H_{v-p} describes the strain-dependent mixing of the valence bands, H_c describes the conduction-band energy, and H_{ex} describes the valence-hole–conduction-electron interaction. The final term in the Hamiltonian is where the spin-exchange interaction is introduced. This term is written as

$$H_{ex} = BE + J\boldsymbol{\sigma}_h \cdot \boldsymbol{\sigma}_e, \tag{21}$$

where BE is the exciton binding energy and the last term is the exchange interaction term. The exchange constant J can be calculated from known band properties. The Hamiltonians have been formalized for both zinc-blende and wurtzite structures. Solutions of the Hamiltonians lead to the matrix elements for the allowed optical transitions $\langle 0 | \boldsymbol{\epsilon} \cdot \boldsymbol{\nabla} | \psi \rangle$, where $\boldsymbol{\epsilon}$ gives the polarization of the light and $|0\rangle$ corresponds to a filled valence band and empty conduction band.

Experimentally observed splittings of exciton lines under uniaxial stress were observed in wurtzite-type II–VI compounds. In wurtzite structures, all of the orbital degeneracies of the valence band are removed by the spin–orbit interaction and by the trigonal crystal field. It was evident that this splitting could not be accounted for by the usual deformation-potential theory based on one-electron energy bands. It was determined that the observed splitting could be attributed to the decomposition of the degenerate Γ_5 exciton state by the deformation of the wurtzite lattice. It is the combined effect of stress and exchange coupling that gives rise to the splitting. The inclusion of the spin-exchange interaction also allows one to bridge the gap between the description of excitons in the J–J coupling scheme and the L–S coupling scheme.

II. Extrinsic Exciton Characteristics

A. INTRODUCTION

The intrinsic exciton may bind to various impurities, defects, and complexes, and the subsequent decay from the bound state yields information concerning the center to which it was bound. Bound-exciton complexes are extrinsic properties of materials. These complexes are observed as sharp-line optical transitions in both photoluminescence and absorption. The binding energy of the exciton to the impurity or defect is generally weak compared to the free-exciton binding energy. The resulting complex is molecular-like (analogous to the hydrogen molecule or molecule–ion) and has spectral properties that are analogous to those of simple diatomic molecules. The emission or absorption energies of these bound-exciton transitions are always below those of the corresponding free-exciton transitions, due to the molecular binding energy.

Bound excitons were first reported in the indirect semiconductor silicon. Here it was found that when group V elements were added to silicon, sharp photoluminescent lines were produced, and these lines were displaced in energy in a regular way. The binding energies of exciton complexes produced by adding different group V donors were described by the linear relation

$$E = 0.1E_i \qquad (22)$$

where E is the binding energy of the exciton and E_i is the ionization energy of the donor. The small differences in ionization energies for different-effective-mass chemical donors result from central cell corrections. A similar relationship was found when the group III acceptors were added to silicon. A modified linear relationship has been found for donors and acceptors in compound semiconductors.

The sharp spectral lines of bound exciton complexes can be very intense (large oscillator strength). The line intensities will, in general, depend on the concentrations of impurities and/or defects present in the sample.

If the absorption transition occurs at $\mathbf{k} = 0$ and if the discrete level associated with the impurity approaches the conduction band, the intensity of the absorption line increases. The explanation offered for this intensity behavior is that the optical excitation is not localized in the impurity but encompasses a number of neighboring lattice points of the host crystal. Hence, in the absorption process, light is absorbed by the entire region of the crystal consisting of the impurity and its surroundings.

The oscillator strength of the bound exciton, F_d, relative to that of the free exciton f_{ex} can be expressed as

$$F_d = (E_0/|E|)^{3/2} f_{ex} \qquad (23)$$

where $E_0 = (2\hbar^2/m)(\pi/\Omega_0)^{2/3}$, E is the binding energy of the exciton to the impurity, m is the effective mass of the intrinsic exciton, and Ω_0 is the volume of the unit cell.

It has been shown in some materials that F_d exceeds f_{ex} by more than four orders of magnitude. An inspection of Eq. (23) reveals that, as the intrinsic exciton becomes more tightly bound to the associated center, the oscillator strength, and hence the intensity of the exciton-complex line, should decrease as $(1/E)^{3/2}$.

In magnetic fields, bound excitons have unique Zeeman spectral characteristics, from which it is possible to identify the types of centers to which the free excitons are bound. Bound-exciton spectroscopy is a very powerful analytical tool for the study and identification of impurities and defects in semiconductor materials.

B. BOUND-EXCITON COMPLEXES IN DIFFERENT SYMMETRIES

1. Degenerate Semiconductors

The model of the donor–bound-exciton complex is used to describe bound-exciton complexes in zinc-blende structures. The Hamiltonian of the system may be written

$$H = H_{ex} + H_d + H_{dex} \qquad (24)$$

where H_{ex} and H_d are the exciton and donor Hamiltonians and H_{dex} describes the interaction between the exciton and donor. The Hamiltonian for the exciton is given in Eq. (4). In this equation, H_{ch} is the interaction Hamiltonian between the electron and hole,

$$H_{eh} = -e^2/\varepsilon |\mathbf{r}_e - \mathbf{r}_h| + H_{exch} \qquad (25)$$

Here ε is the dielectric constant and H_{exch} is the electron–hole exchange Hamiltonian. The exchange Hamiltonian H_{exch} is

$$H_{exch} = A_1 \boldsymbol{\sigma} \cdot \mathbf{J} + A_2(\sigma_x J_x^3 + \sigma_y J_y^3 + \sigma_z J_z^3) \tag{26}$$

where $\boldsymbol{\sigma}$ and $\mathbf{J}$ are the operators for electron spin and effective hole spin, respectively, and A_1 and A_2 are parameters describing the exchange energy.

The model of the exciton bound to a neutral donor is shown in Fig. 5. In the initial state the two electrons pair to form a bonding state, leaving an unpaired hole. When the exciton collapses from this state, the final state may consist of the donor in the ground state or in an excited state. The donor may pick up energy from the exciton recombination, thus leaving the donor in an excited state. The model of the exciton bound to the acceptor is more complicated than the donor. The initial state of the neutral acceptor-bound exciton consists of two $J = \frac{3}{2}$ holes and one $J = \frac{1}{2}$ electron as shown in Fig. 6. The two $J = \frac{3}{2}$ holes combine to give a $J = 0$ and a $J = 2$ state. The interaction of the electron spin operator $\boldsymbol{\sigma}$ with the effective hole spin operator $\mathbf{J}$ results in three $\mathbf{J} + \boldsymbol{\sigma}$ states, $\frac{1}{2}$, $\frac{3}{2}$, and $\frac{5}{2}$. As in the case of the donor, when the exciton collapses, the final state will consist of the neutral acceptor in the ground or an excited state.

The photoluminescent spectrum for GaAs is shown in Fig. 7, where X is the free exciton transition; the $D°X$ lines are associated with the neutral donor-bound excitons, and the $A°X$ lines are associated with the neutral acceptor-bound excitons. The $J = \frac{5}{2}$ and $J = \frac{3}{2}$ neutral acceptor-

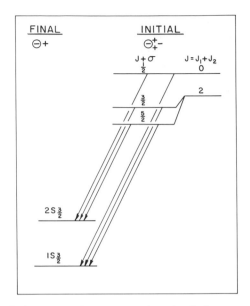

FIG. 6.   Schematic representation of radiative recombination of an exciton bound to a neutral acceptor, where the final state is the acceptor in the ground or in the excited configuration.

bound exciton transitions are clearly observed. The $(D°X)$ $n = 2$ lines are associated with the collapse of the exciton from the neutral donor-bound exciton state, leaving the donor in an excited state. The energy of the transition is expressed as

$$E_T = E_{ex} - E_{Dex} - E_D^* \tag{27}$$

where E_{ex} is the free exciton energy, E_{Dex} is the binding energy of the exciton to the donor, and E_D^* is the energy required to place the neutral donor in an excited state. The analogous neutral acceptor-bound exciton transitions in which the

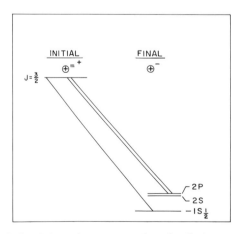

FIG. 5.   Schematic representation of radiative recombination of an exciton bound to a neutral donor, where the final state is the donor in the ground or in the excited configuration.

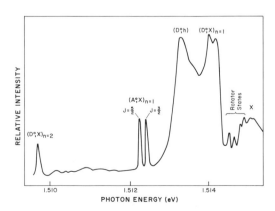

FIG. 7.   Photoluminescence spectrum of GaAs in the near bandgap region.

final state of the acceptor is left in an excited state are not shown. These transitions occur at appreciably lower energies, due to the larger binding energies of the acceptors.

2. Nondegenerate Semiconductors

The theory for the nondegenerate case is based on the wurtzite structure, with the salient factors of the band structure such as band symmetries and selection rules being derived from group theory.

Consider any simple optical transition in which an electron bound to an impurity is taken from one band to another. Suppose the initial and final states can be approximately assigned effective mass wave functions. The initial-state wave function can then be written as

$$f_i(r) U_{i0}(r) \qquad (28)$$

where $f_i(r)$ is a slowly varying function of r and U_{i0} is the periodic part of the Bloch function of band i for wave vector zero. Similarly, the final-state wave function can be written

$$f_f(r) U_{f0}(r) \qquad (29)$$

Since f is a slowly varying function, the optical matrix element

$$\int f_i(r) U_{i0}(r) \mathbf{P} f_f^*(r) U_{f0}^*(r) \, d^3 r \qquad (30)$$

can be approximately written as

$$\left[\int f_i(r) f_f^*(r) \, d^3 r \right] \frac{1}{\Omega} \left[\int U_{i0}(r) \mathbf{P} U_{f0}^*(r) \, d\tau \right] \qquad (31)$$

where the second integration is carried out over the unit cell, whose volume is Ω. In this approximation, the only large optical-matrix elements will arise when the analogous band-to-band transition is allowed.

A similar argument can be made to show that in this effective-mass approximation, large g values can be expected only when the parent energy-band wave functions exhibit large g values.

In the case of weakly bound states at substitutional impurities and energy bands at $\mathbf{k} = 0$ in the wurtzite structure, it is reasonable to describe the states as though they belonged to the point group of the crystal rather than to the group of the impurity. Such a description gives the degeneracy of the states correctly. This description neglects certain optical transitions that are technically allowed, but that are weak in the effective mass approximation and will set equal to zero certain g values that should be much smaller than usual g values. The advantage of the description is that it neglects these small ef-

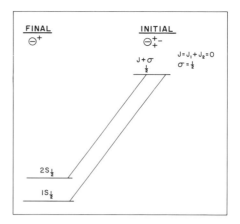

FIG. 8. Model of the exciton bound to the neutral acceptor for a nondegenerate semiconductor.

fects and thus permits the full use of group theory without the clutter of what should be small perturbations.

The electron g value, g_e, should be very nearly isotropic, since the conduction band is simple and the g shift of the free electron is small, only weakly dependent on the state of binding of the electron. The hole g value, g_h, should be completely anisotropic with g_h equal to zero (for the top Γ_9) for magnetic fields perpendicular to the hexagonal axis. It is to be expected that the hole g value will be sensitive to its state of binding, since the different valence bands will be strongly mixed in bound-hole states. The model of the exciton bound to a neutral donor for the nondegenerate case is very similar to that for the degenerate case, shown in Fig. 5, for zero applied magnetic field. The unpaired hole for the nondegenerate semiconductor is twofold degenerate, as compared to the fourfold degeneracy for the degenerate semiconductor. The model of the exciton bound to the neutral acceptor for the nondegenerate case is less complicated than for the degenerate case. The initial state of the complex consists of paired doubly degenerate holes, leaving an unpaired electron as shown in Fig. 8. The final state consists of the acceptor, either in the ground state or an excited state. In the absence of an applied magnetic field, a single optical transition is observed.

C. Perturbations

1. Magnetic Field

When a magnetic field is applied to the donor-bound exciton complexes in a degenerate semiconductor, the line splitting due to the presence of the magnetic field can be predicted from Fig. 5. In the initial state, the $J = \frac{3}{2}$ unpaired hole will split into a quartet while the 1s final state will split into a doublet. This splitting will result in six allowed transitions. When the final state consists of the excited $n = 2$ states, the splitting is much more complicated. The 2s state is doubly degenerate, while the 2p state is sixfold degenerate. In addition to the increased multiplicity of lines, rather large diamagnetic shifts are also observed. The energies of these transitions in a magnetic field have been calculated. In general it is not easy to solve the Hamiltonian for the donor-bound exciton complex in a magnetic field. In the low-field regime, the exciton Hamiltonian of Eq. (4) can be separated into two parts: the spherical symmetric (s-wave-like) and asymmetric (d-wave-like) parts. For the perturbation calculation, in the low-field regime one can treat the s-wave-like part as an unperturbed Hamiltonian and the d-wave-like part as a perturbed Hamiltonian.

In the high-field regime—i.e., when the magnetic energy is much greater than the Coulomb energy—the solution of Eq. (4) may be obtained by an adiabatic method which can be written as

$$E_{ij} = L_{ij}\gamma \qquad (32)$$

where the L_{ij} are the linear coefficients for the Landau-type solutions. They turn out to be of the order of $L_{ij} = 0.01$. These energies shift much more rapidly with magnetic field than is experimentally observed in the intermediate-field region.

In the intermediate-field regime where the magnetic energy is of the order of the Coulomb energy, the solution of Eq. (4) is not easily obtained. A phenomenological scheme for solution in this region was used to bridge the gap between the solutions in the low- and high-field regimes. In the intermediate-field region, a variety of functional forms for eigenvalues can be constructed. In the framework of infinite-order perturbation calculations, one may conclude that the dominant correction term will be an even function of an applied magnetic field, provided the linear Zeeman energy term is absorbed in the unperturbed Hamiltonian. For simplicity, the following form for the eigenvalues for all fields was chosen:

$$E_{ij} = E_B + (G_{ij}\gamma + D_{ij}\gamma^2 + \beta_{ij}L_{ij}\gamma^3)/(1 + \beta_{ij}\gamma^2) \qquad (33)$$

In the low-field regime, the above reduces to the perturbation scheme, and in the high-field regime, it reduces to the adiabatic scheme, that is, the Landau-level-type solutions.

The magnetic field splitting of the acceptor bound exciton is quite complicated. It can be seen from Fig. 6 that the three initial states will split into a total of 12 states, and the final 1s state will split into a quartet. These transitions have been observed experimentally; however, the energies of the transitions have not been calculated due to the complexity of the problem, which involves the degenerate valence band.

The theory of bound excitons in nondegenerate semiconductors is based on the wurtzite structure. In considering transitions involving bound excitons formed from holes in the top valence band, the g value of the electron is isotropic. The g value of the hole has the form $g_h = g_{h\parallel}\cos\theta$, where θ is the angle between the c axis of the crystal and the magnetic field direction. The symbols $\oplus$ and $\ominus$ refer to ionized donors and acceptors, respectively, and $+$ and $-$ refer to electrons and holes. The neutral donor-bound exciton is very similar to that for the degenerate case in Fig. 5; however, in the initial state, the unpaired hole is only doubly degenerate. Therefore, a total of four transitions is observed in the presence of a magnetic field. For the orientation $\mathbf{C} \perp \mathbf{H}$, only two transitions will be observed, since the hole g value goes to zero for this orientation. The neutral acceptor-bound exciton will also exhibit a four-line transition in the presence of a magnetic field. In the initial state, the unpaired electron is doubly degenerate, while the final state consisting of the unpaired hole is also doubly degenerate. In this case also, the hole splitting goes to zero for the orientation $\mathbf{C} \perp \mathbf{H}$, resulting in a two-line transition.

2. Stress Field

In zinc-blende-type semiconductors, the uniaxial strain patterns and electric dipole selection rules have been derived for lines arising from weakly bound exciton complexes. The effect of stress on excitons bound to shallow neutral acceptors in zinc-blende structures has been rather thoroughly investigated.

In the unstrained crystal, a hole from the $J = \frac{3}{2}$ (Γ_8) valence band in combination with an electron from the $J = \frac{1}{2}$ (Γ_6) conduction band gives rise to the ground-state exciton. Uniaxial stress splits the $J = \frac{3}{2}$ degenerate valence band into two bands, one with $M_j = \pm\frac{1}{2}$, the other with $M_j = \pm\frac{3}{2}$. This splitting is reflected in optical transitions involving holes from the valence band. The shallow acceptor removes an energy state from the valence band and establishes it as a quantum state of lower energy in the gap region. This state is made up of valence-band wave functions and therefore will also reflect valence-band splittings.

When an exciton is captured by the shallow acceptor, an acceptor-bound exciton complex (A^0X) is formed that consists of two holes and one electron weakly bound to a negative acceptor ion. In the absence of stress, three transitions are observed, as shown in Fig. 6.

When a uniaxial stress is applied to the (A^0X) complex described above, the degeneracy of the states is lifted due to the splitting of the Γ_8 hole states. A schematic plot of the resulting energies is shown in Fig. 9. The lower part of the figure shows the splitting of the final (one-hole) state after the collapse of the exciton. The upper part shows the splitting and shifts of the initial energy states prior to exciton decay. The energies of these states have been calculated and compared with experimental observations, as shown in Fig. 10. The lines in Fig. 10 show the predicted energy levels (in the absence of a crystal field) for transitions between the initial and final states, with the center of gravity shift included.

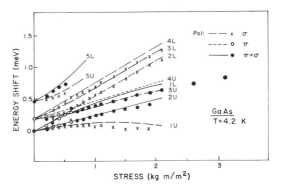

FIG. 10.   Experimental (points) and calculated (lines) line shifts under increasing uniaxial stress. The stress is applied in a [100] direction. [From Schmidt, M., Morgan, T. N., and Shairer, W. (1975). *Phys. Rev.* **B11,** 5002.]

The σ lines are for transitions polarized perpendicular to the applied stress, while the π lines show transitions polarized parallel to the applied stress. The agreement between theory and experiment is very good.

D. Multibound Excitons

Sharp photoluminescent lines have been observed at energies less than the energy of the line associated with an exciton bound to a neutral donor in silicon, germanium, and silicon carbide. Similar lines have also been observed that are associated with acceptors in silicon and gallium arsenide. The energies and widths of these lines were such that they could not be explained in terms of any recombination mechanism involving just a single exciton bound to a neutral shallow impurity center. A model involving a multiexciton complex bound to a donor (acceptor) was invoked in which each line was associated with radiative recombinations of an exciton in the bound multiexciton complex.

A series of emission lines was observed in silicon crystals also lightly doped with boron or phosphorus. The series began with the bound-exciton line and converged toward the energetic position of the maximum of emission of the condensed electron–hole state. The emission series is shown for both boron and phosphorus dopants in Fig. 11. The impurities can bind a series of intermediate "multiple-exciton states" containing the single bound exciton and electron–hole droplet state.

A model was proposed in which the multiple-exciton complex is built up by successive capture of free excitons at neutral impurity centers.

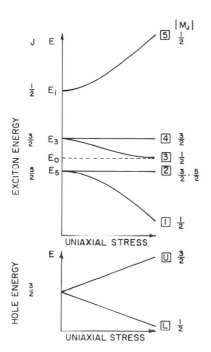

FIG. 9.   Schematic plot of the splitting of the (A^0X) initial and ground states under uniaxial stress. [From Schmidt, M., Morgan, T. N., and Shairer, W. (1975). *Phys. Rev.* **B11,** 5002.]

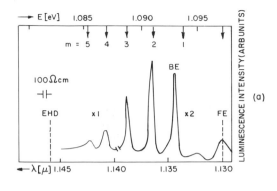

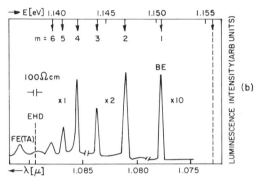

FIG. 11. Emission spectra of (a) Si : B with TO phonon and (b) Si : P without phonon assistance. Excitation intensity 7.5 W cm^{-2}, $T = 2$ K. The dashed lines indicate the positions of the FEs and the maxima of EHD emission. In Si : P(NP), the FE does not really appear; however, its position is known from the phonon-assisted FE spectrum. The EHD emission in the NP spectrum only appears at higher doping levels and under high excitation. The arrows mark the calculated values $h\nu_m^*$ given by $h\nu_m^* = -18.5[1 - \exp(-0.21m)]$ meV for Si : B and $h\nu_m^* = -18.5[1 - \exp(-0.32m)]$ meV for Si : P. [From Sauer, R. (1973). *Phys. Rev. Lett.* **31**, 376.]

A multiple-exciton complex having index m can capture another free exciton and then have the index $m + 1$; the decay of an exciton would decrease the index to $m - 1$. The observed photon energy $h\nu_m$ is the difference between the energies of the initial and final states,

$$h\nu_m = E_g - E_{FE} - E_m = h\nu_{FE} - E_m \quad (34)$$

where E_{FE} is the binding energy of the free exciton and E_m that of an exciton in the m complex. The energy difference between the mth line $h\nu_m$ and the free-exciton line is a measure of the binding energy E_m.

The model was successful in obtaining an empirical fit to the series of emission lines with the

series formulas

$$h\nu_m^* = -18.5[1 - \exp(-0.21m)] \quad \text{meV} \quad (35)$$

for Si : B (except for the bound exciton line) and

$$h\nu_m^* = -18.5[1 - \exp(-0.32m)] \quad \text{meV} \quad (36)$$

for Si : P. The calculated line positions are shown in Fig. 11; $h\nu_0^* = 0$ corresponds to the free-exciton line. Other models have been proposed, including a shell model in which all of the electrons and all of the holes in the bound multiexciton complex are assumed to be equivalent and therefore must conform to the Pauli principle. The complex is then built up along the lines of a shell model similar to what has been used to study nuclei and many-electron atoms.

The behavior of these new lines in the presence of magnetic and stress fields helped to establish the viability of the bound multiexciton complex model.

E. DONOR–ACCEPTOR PAIRS

Donor–acceptor pairs introduce transitions in the bound-exciton region whose behavior is quite different from excitons bound to foreign impurities or defects. The pairs can produce bound states distributed in energy. The range of energies results both from the possible impurities or defects interacting as pairs and from a dependence on pair separation. Discrete pair spectra were first observed as a very complicated spectra in GaP consisting of very many sharp lines. The donors and acceptors will occupy substitutional or interstitial sites. In the case of substitutional sites, both the donor and acceptor can occupy sites on the same sublattice for a compound material such as GaP, or they may be on opposite sublattice sites. Another arrangement is with one impurity at an interstitial site and the other at a particular lattice site. All of these arrangements have been observed.

The energy required to bring a hole and an electron from infinity to an ionized donor–acceptor pair separated by a distance R may be written as

$$E(R) = E_g - E_A - E_D + e^2/\varepsilon R \quad (37)$$

In this expression $E(R)$ is the energy of the pair recombination line, E_g the band gap of the semiconductor, E_A and E_D the acceptor and donor binding energies, respectively, R the donor–acceptor separation, and ε the low-frequency dielectric constant. When the donor–acceptor distances become small [$R < R_0 =$

(donor–acceptor concentration)$^{1/3}$], a van der Waals attractive term may become important, and Eq. (37) becomes

$$E(R) = E_g - E_A - E_D + e^2/\varepsilon R - (e/K)(a/R)^6$$
(38)

In the case of random pair distribution, it would be expected that over a small range of R, the line intensity would reflect the statistical probability of a specific pair occurring. In considering GaP, which has the zinc-blende structure, and assuming that both the donors and acceptors result from substitutional impurities and that both occupy sites on the same sublattice, it is possible to relate R to a given observed line. For the preceding case, $R_m = a_0(\frac{1}{2}m)^{1/2}$, where a_0 is the GaP lattice constant and R_m the distance to the mth nearest neighbor on the radius of the mth shell. The donors and acceptors occupy face-centered cubic sites, and the number of pairs for a given m can be tabulated. The variation in number of pairs allows a correlation with observed spectra. For the case when the donors and acceptors occupy opposite sublattice sites, $R_m = a_0(\frac{1}{2}m - \frac{5}{16})^{1/2}$ and $N(R) > 0$ for all m.

The value of $E(R)$ and R are determined from experiment; therefore Eqs. (37) and (38) can be helpful in identifying the donors and acceptors involved in donor–acceptor pair recombination.

III. Interaction of Excitons with Other Systems

A. PHONONS

Emission from bound exciton complexes has been observed in many materials. These are very sharp transitions which in many cases are replicated by emission lines that are separated in energy from the parent transition by an optical phonon energy for the particular lattice in question. In crystals having the wurtzite symmetry there is a Γ_1 and a Γ_5 longitudinal optical–transverse optical (LO–TO) splitting due to long-range electrostatic forces as well as a Γ_1–Γ_5 LO–LO and TO–TO splitting due to anisotropic short-range interatomic forces. The Γ_1–Γ_5 LO–LO splitting has been observed on the phonon sidebands, due to the interaction with the macroscopic longitudinal optical phonon electric field, in both CdS and ZnO. In CdS, the phonon-assisted line results from the collapse of the exciton bound to a neutral acceptor with the creation of an LO phonon. Both the Γ_1 and the Γ_5 LO phonons are created in this process. These two phonons differ in energy by 2.4 cm^{-1}. This small difference in energy is clearly resolved, showing that the phonon-assisted transitions are not appreciably broadened by the phonon interaction.

In the case of CdS, the exciton is rather weakly bound (17 meV) to the acceptor. This results in a localized state in **K** space. The phonon energies show that it is localized near **K** = 0. The phonon dispersion curves were calculated for CdS using a mixed binding model. In this model the potential contains a short-range part corresponding to covalent bonding and a long-range part due to Coulomb interactions between point ion charges. The calculations show that the LO phonon dispersion curves are quite flat in the vicinity of **K** = 0. This would account for the very small line broadening observed in the phonon interaction.

A similar interaction between the LO phonon and an exciton bound to a neutral donor was observed in ZnO. In this material, the energy separation between the Γ_1 LO phonon and the Γ_5 LO phonon is larger (11 cm^{-1}).

1. Exciton–Bound-Phonon Quasi-Particle

Optical transitions have been observed in a number of ionic crystals in which the energy separating the parent transition and its LO-phonon sideband is less than the LO-phonon energy $\hbar\omega_0$ by approximately 10%. In absorption spectra of AgBr:I, transitions associated with the bound exciton occur at energy separations approximately 30% less than $\hbar\omega_0$. These results can be explained in terms of a bound-phonon quasi-particle model. The calculated binding energies and oscillator strengths for this new quasiparticle can account for phonon interactions whose energies are less than that of the LO phonon $\hbar\omega_0$. LO phonons bound to neutral donors have been observed in both Raman-scattered and luminescence spectra of GaP crystals that were doped with S, Te, Si, and Sn. These results were interpreted as impurity modes associated with dielectric effects of the neutral donors, rather than as local modes associated with mass defects of the substituents.

The spectra due to Raman scattering from the neutral donor LO-phonon bound states are shown in Fig. 12. The binding energies associated with the exciton–bound-phonon quasi-particle for several different donors are given in Table I.

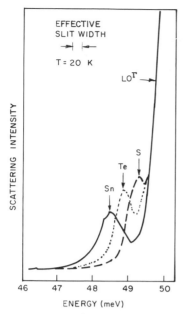

FIG. 12.   Raman scattering of 5145-Å Ar⁺ laser light from GaP containing ~10^{15} cm⁻³ neutral Sn, Te, or S donors, recorded just below the $k = 0$, LO phonon at 50.2 meV, showing the new local modes. These modes can be seen easily at donor concentrations as low as 10^{17} cm⁻³, although their strengths relative to the LO Γ lattice normal mode decrease in proportion to the neutral donor concentrations. [From Dean P. J., Manchon, Jr., D. D., and Hopfield, J. J. (1970). *Phys. Rev. Lett.* **25**, 1027.]

The virtual process

donor + LO phonon **k** → excited donor →

donor + LO phonon **k′**

produces the interaction of an LO phonon with a donor site. The effective scattering matrix element $H_{\mathbf{kk'}}$ is proportional to

$$\sum_j \frac{(E_j - E_0))\langle 0|e^{i\mathbf{k}\cdot\mathbf{r}}|j\rangle\langle j|e^{i\mathbf{k'}\cdot\mathbf{r}}|0\rangle}{[(E_j - E_0)^2 - (\hbar\omega)^2]KK'} \qquad (39)$$

where $|0\rangle$ is the donor ground-state wave function and the sum is over excited donor states j. The interaction is attractive when $\hbar\omega$ is less than the excitation energy of the donor. For a spherical approximation, the interaction will produce bound states for each angular momentum of the LO phonon around the donor. When the first excited state of the donor is comparable with the phonon frequency, an approximation in which only the lowest-donor excited state is kept will give a reasonable lower bound to the binding energy for $H_{\mathbf{KK'}}$.

Using this approximation, one obtains the following wave functions and energies for the s and p states. For s states,

$$E_B = \frac{32}{729}\hbar\omega_0\left(\frac{\varepsilon_0}{\varepsilon_\infty} - 1\right)\frac{e^2}{2a\varepsilon_0}$$

$$\frac{E_{2s} - E_{1s}}{(E_{2s} - E_{1s})^2 - (\hbar\omega_0)^2} \qquad (40)$$

$$\Psi_{\mathbf{K}} \propto \frac{K}{[(\tfrac{3}{2}a)^2 + K^2]^3} \qquad (41)$$

TABLE I.   Experimental and Calculated Binding Energies E_B of LO Phonons Localized at Neutral Donors in GaP Observed through Raman Scattering $(E_B)^p$, and as Sidebands in the Luminescence of Excitons Bound to Neutral Donors $(E_B)^{s\,a}$

Donor	(meV)	$E_{1s}(A_1) \to E_{2s}(A_1)$ (meV)	$E_{1s}(A_1) \to E_{2p}^b$ (meV)	$(E_B)_{calc}^s$ (meV)	$(E_B)_{exp}^s$ (meV)	$(E_B)_{calc}^p$ (meV)	$(E_B)_{exp}^p$ (meV)
S	104.2^c	82.6^c	94.0^g	0.98	1.2 ± 0.2	0.58	0.8 ± 0.2
Te	89.8^d	68.2^d	79.9^d	1.40	1.9 ± 0.2	0.70	1.2 ± 0.2
Si	82.5^c	70.2^f	72.3^c	1.16	?	0.82	1.3 ± 0.2
Sn	65.5^e	53.2^e	58.5^e	[3.6]^h	?	[1.35]^h	1.6 ± 0.2

^{a} From Dean, P. J. *et al.* (1970). *Phys. Rev. Lett.* **25**, 1027.
^{b} Calculated as weighted mean of $E_{1s}(A_1) \to E_{2p0}$ and $E_{1s}(A_1) \to E_{2p\pm}$.
^{c} From A. Onton (1969). *Phys. Rev.* **186**, 786.
^{d} From A. Onton and R. C. Raylor (1970). *Phys. Rev. B* **1**, 2587.
^{e} From P. J. Dean, R. A. Faulkner, and E. G. Schonherr (1970). *Proc. Int. Conf. Phys. Semicond., 10th*, Cambridge, Mass., p. 286.
^{f} Assuming the $E_{2s}(A_1)$ is the same as for Sn.
^{g} Assuming the E_{2p0} and $E_{2p\pm}$ are the same for S and Te.
^{h} Using degenerate perturbation theory.

For p states,

$$E_B = \frac{224}{6561} \hbar\omega_0\left(\frac{\varepsilon_0}{\varepsilon_\infty} - 1\right)\frac{e^2}{2a\varepsilon_0}$$

$$\frac{E_{2p} - E_{1s}}{(E_{2p} - E_{1s})^2 - (\hbar\omega_0)^2} \quad (42)$$

$$\Psi_K \alpha \frac{\cos\theta_b}{[(\frac{3}{2}a)^2 K^2]^3} \quad (43)$$

In these expressions, ε_0 and ε_∞ are the static and high-frequency dielectric constants of the Frohlich electron–phonon interaction. The donor Bohr radius is a, and the donor binding energy is $e^2/2a\varepsilon_0$ for a hydrogenic donor. The theoretical binding energies from Eqs. (40) and (42) are included in Table I.

B. PHOTONS

1. Spatial Resonance Dispersion

For those states in a crystal where the photon wave vector and the exciton wave vector are essentially equal, the energy denominator for exciton–photon mixing is small and the mixing becomes large. These states are not to be considered as pure photon states or pure exciton states, but rather mixed states. Such a mixed state has been called a polariton. When there is a dispersion of the dielectric constant, spatial dispersion has been invoked to explain certain optical effects of crystals. It was originally thought that it would introduce only small corrections to such things as the index of refraction, until it was demonstrated that if there was more than one energy transport mechanism, as in the case of excitons, this was not true. Spatial dispersion addresses the possibility that two different kinds of waves of the same energy and same polarization can exist in a crystal differing only in wave vector. The one with an anomalously large wave vector is an anomalous wave. In the treatment of dispersion by exciton theory, it was shown that if the normal modes of the system were allowed to depend on the wave vector, a much higher-order equation for the index of refraction would result. The new solutions occur whenever there is any curvature of the ordinary exciton band in the region of large exciton–photon coupling. These results apply to the Lorentz model as well as to quantum-mechanical models whenever there is a dependence of frequency on wave vector.

It was pointed out early in the investigation of spatial dispersion that the specific dipole moment of polarization of a crystal and the electric field intensity are not in direct proportion. It was found that the two were related by a differential equation that resulted in giving Maxwell equations of higher order. This led to the existence of several waves of the same frequency, polarization, and direction but with different indices of refraction. Subsequent studies of the reflectivity of CdS demonstrated the effects of spatial dispersion. Extensive calculations resulted in the following expression for the index of refraction:

$$n^2 = \frac{c^2 k^2}{\omega^2} = \varepsilon$$
$$+ \sum_j \frac{4\pi(\alpha_{0j} + \alpha_{2j}k^2)\omega_{0j}^2}{\omega_{0j}^2 + (\hbar\omega_{0j}k^2/m_j^*) - \omega^2 - i\omega\Gamma_j} \quad (44)$$

In this equation the sum over j is to include the excitons in the frequency region of interest, and the contributions from other oscillators are included in a background dielectric constant ε. In Eq. (44) one has expanded both the numerator and denominator in powers of k, keeping terms to order of k^2, m^* is the sum of the effective masses of the hole and electron that comprise the exciton, and ω_{0j} is the frequency of the jth oscillator at $\mathbf{k} = 0$. Eliminating k^2 from Eq. (44) and neglecting the line width of the oscillators, the working equation is

$$n^2 = \varepsilon + \sum_j \frac{4\pi(\alpha_{0j} + \alpha_{2j}\omega^2 n^2/c^2)\omega_{0j}^2}{\omega_{0j}^2 + (\hbar\omega_{0j}\omega^2 n^2/c^2 m_j^*) - \omega^2} \quad (45)$$

The sum is over excitons from the top two valence bands, where the "allowed" excitons have been included with $\alpha_{0j} \neq 0$, $\alpha_{2j} = 0$ while the "forbidden" (which are seen only because $\mathbf{k} \neq 0$) excitons are included with $\alpha_{0j} = 0$, $\alpha_{2j} \neq 0$. The above equation reduces to a polynomial in n^2 whose roots give the wavelengths of the various "normal modes" for transfer of energy within the crystal.

In the classical case, $\alpha_{2j} = \hbar\omega_{0j}/m_j^* = 0$. For a given frequency, the two roots of n^2 for Eq. (45) are $-n$ and $+n$. Thus in the classical case, for a given principle polarization, frequency, and direction of propagation, only one transverse mode exists.

2. Two-Photon Processes

The concept of two-photon processes dates back more than 40 years, and was first treated theoretically. The observation of two-photon transitions occurred approximately 30 years later. The two-photon transition is a nonlinear process, and as such its full potential as a tool for investigating material parameters was not realized until the advent of laser sources.

The process is one in which two quanta are simultaneously absorbed in an electronic transition. The energy sum of the two quanta must be equal to the energy of the electronic transition. In one-photon absorption spectra, the absorption coefficient as a function of photon energy is obtained. In two-photon transitions, the absorption is dependent on two frequencies; therefore, instead of a plane curve, the spectrum is a two-dimensional surface. The two processes are complementary; however, they yield different information. The single-photon process is allowed between states of different parity, while the two-photon transitions are allowed between states of the same parity. The polarization dependence of two-photon absorption is more complicated than it is for one-photon processes. Even for isotropic materials, the mutual orientation of the electric vectors of the absorbed fields is important. Thus, the two-photon spectrum contains more information than the one-photon spectrum. All two-quantum processes require an intermediate state in which one photon is absorbed or emitted and the atom is left in an excited state.

The fine structure of the 2P states of the exciton from the top valence level in CdS has been studied. In this experiment, a visible dye laser and a CO_2 laser were used. From the polarization selection rules, it was shown that the visible photon created the virtual 1s exciton, and the absorption of the infrared photon brings it to the final p state. Using these results, the two-photon absorption coefficient was calculated and compared with the experimentally measured results. The agreement between theory and experiment is very good.

IV. Special Properties of Excitons

A. Introduction

The excitonic properties of the semiconductor are key to the development of a large number of devices. In this section five of the properties that have direct application will be outlined. The first is the exciton states in quantum wells. The binding energy of the exciton can be adjusted due to the thickness of the layers of the different semiconductor materials. This leads to having different energy responses, as well as many other electrical and optical properties.

The second area is hole–electron droplets,

which are extremely useful in the study of many-body effects and the phase transition from a gas into a liquid. The role of excitons in developing a high-temperature superconductor is given in Section IV,D. When this property is exploited, it could have a major impact on modern electronic devices. The last two areas outlined are lasing transitions and optical bistability. These two areas lead to totally optical switching devices, which again may lead to many useful devices such as the optical computer.

B. Excitons in Quantum Wells

Superlattice structures have generated considerable interest for more than a decade because of the novel transport phenomena predicted for such structures. The superlattice is a multilayered periodic structure having dimensions varying from a few angstroms to hundreds of angstroms. Carriers in semiconductor superlattices may be confined to certain layers by the superlattice potential variations, resulting in new conductivity properties. The allowed carrier energy levels are determined by quantization effects when the confined regions are sufficiently small. The confined regions produce new optical effects as well as electrical effects. Superlattices emerged as practical structures when the metal organic chemical vapor deposition (MOCVD) and molecular beam epitaxy (MBE) crystal growth techniques evolved, making high-quality structures feasible. Very thin layers with smooth surface morphology can be grown by these techniques. One of the very common heterostructures produced by these techniques is $GaAs/Al_xGa_{1-x}As$. By cladding the GaAs layer with GaAlAs barriers, the electrons and holes are confined within the GaAs well, resulting in a modification of their energy levels in the well. Repeating the growth of these layers results in a multiquantum well (MQW) structure, the number of wells being equal to the number of repeated cycles. When the layer thicknesses are small, the electrons are confined as quantized electron waves. If the barrier-layer thicknesses are large enough, tunneling between wells does not occur. The confinement of carriers within the GaAs well results in an effective increase in the bandgap. The low-temperature bandgap of bulk GaAs is 1.5196 eV. $Al_xGa_{1-x}As$ has a direct bandgap for $x < 0.45$, with a bulk bandgap which is $1.25\,x$ eV greater than GaAs. In the quantum-well structure, the difference in bandgap is divided between the conduction band

FIG. 13. Quantum-state energy levels in GaAs/ $Al_x Ga_{1-x} As$ quantum well. CB and VB refer to conduction and valence bands, respectively.

and the valence band. The percentage contribution to each band is a measure of the confining barrier for that band. Well-size quantization results in a shift of the allowed energies for electrons. If infinite confining barriers are assumed, the allowed minimum energies for electrons are given by

$$E_n = h^2 n^2 / 8 m^* L^2 \qquad (46)$$

where h is Planck's constant, n is an integer marking the number of half-wavelengths of the confined electron, m^* is the effective mass, and L is the well thickness. These energy shifts for electrons in the conduction band are shown in Fig. 13.

The energy levels in the valence band of the quantum well are also modified. In bulk GaAs, the light and heavy hole valence bands are degenerate at $K = 0$. From Eq. (46) it is seen that the energy of the confined particles is different for different masses. The layered structure has reduced the cubic symmetry of bulk GaAs to uniaxial symmetry. The optical transitions thus become nondegenerate. This feature is shown in the valence band of Fig. 13.

The line shape of the absorption and photoluminescence bands of GaAs (MQW) is excitonic. Verification of these bands as being assigned to light and heavy hole free-exciton transitions has been made from polarization measurements. These measurements include optical spin-orientation measurements and linear polarization measurements of emission emanating from a cleaved edge of the (MQW).

C. HOLE–ELECTRON DROPLETS

Nonequilibrium electrons and holes in semiconductors are bound in excitons at low temperatures by Coulomb attraction. The exciton

forms because it represents a state of slightly lower energy than the unbound hole–electron. At high exciting intensities, the density of hole–electron pairs is increased and excitons are formed at a higher rate. At high concentrations the interaction among electrons becomes very important, and when a certain threshold density is reached liquid droplets are formed. These collective droplets consist of nonequilibrium electrons and holes; therefore, when electron–hole recombination occurs, specific radiation is emitted.

This intense radiation was first observed in Si. The spectra contained, in addition to the well-known peaks due to the annihilation of free excitons with appropriate phonons, some broad bands of radiation shifted toward lower energies.

The formation of a condensed phase of nonequilibrium carriers was considered, to account for the new radiation peaks. If the collective interaction predominates in the condensed phase, survival of excitons as quasi-particles is doubtful. It was shown that the condensed phase consists of a degenerate electron–hole plasma characterized by metallic properties.

The most convincing evidence that the new substance is in a separate phase was obtained from light-scattering experiments. The effect is analogous to the scattering of light by drops dispersed in a fog. The range of angles through which the radiation is scattered depends on the size of the particles, the larger scattering angles resulting from smaller particles.

The far-infrared absorption of the condensed phase in Ge was investigated. The results were explained by assuming that the absorption is caused by the excitation of plasmons in the drops of the condensed phase. From these experiments, the drop radius was estimated to be $R \approx 10^{-3}$ to 10^{-4} cm.

When the excitons enter the liquid state (electron–hole drops), the electron and hole give up their exclusive association and enter a sea of particles in which they are bound equally to all of the other charge carriers in the droplet. The droplet therefore is made up of independent electrons and holes. Since the density in the droplet is greater than in the exciton gas, it can be qualitatively understood why droplets form, by considering the relation of energy to charge carrier distance. Free electrons and holes that have the greatest separation recombine with the highest energy. Excitons form only when the electron and hole are coupled, resulting in an appropriate Bohr radius for the particular material being considered (in the case of Ge it is approximately 115 Å). The energy of the resulting recombination radiation from the excitons is less than the energy of the recombination radiation of the free electrons and holes, reflecting the exciton binding energy. In the case of the liquid droplet, the electron–hole distance is still further reduced (100 Å for Ge). Therefore the liquid state forms because it is a reduced energy state of the system.

The liquid is made up of independent electrons and holes, which gives a metallic character to the liquid, whereas the exciton gas is an insulator. In the exciton gas, the particles are far enough apart to behave as a classical gas; they move independently and their velocities are determined by random processes. The probability of finding a particle with a given energy falls off exponentially as the energy is increased. The exciton gas obeys Maxwell–Boltzmann statistics. The combination of the statistical distribution in the gas with the density of states gives the shape of the luminescent line as shown in Fig. 14. In the liquid droplet, the electrons and holes are close enough together so that the droplet must be considered as a single system. The availability of states for occupation in such a system is determined by the Pauli exclusion principle. The probability of the charge carriers

occupying the available states is determined by Fermi–Dirac statistics. The uppermost filled state at the absolute zero of temperature can be considered the Fermi level. Here again, the statistical distribution coupled with the density of states determines the shape of the luminescent peak. Here it is assumed that any electron is equally likely to recombine with any hole. Recombination in which both particles have very high or very low energies is unlikely. The maximum intensity will occur when the difference in energy of the two particles is near the medium value as shown in Fig. 14. The luminescence spectrum of free excitons and electron–hole drops in Ge is also shown in Fig. 14. The line shapes agree reasonably well with theory.

D. Exciton Mechanism in Superconductivity

There are four well-defined experiments that have not been explained to date without invoking a superconducting state in which the phonon-mediated electron–electron interaction is replaced by excitons. As in the case of phonons, the excitons cause the electrons to attract, forming bound states known as Cooper pairs.

The first experiment is one in which an attempt was made to measure the effects of gravity on the position of an electron in a copper tube. A large temperature-dependent transition in the magnitude of the ambient axial electric field inside the vertical copper tube was found. Above a temperature of 4.5 K, the ambient field was 3×10^{-7} V/m or greater. Below 4.5 K the magnitude of the ambient field drops very rapidly, to 5×10^{-11} V/m at 4.2 K. This was measured by time-of-flight spectra of an electron travelling in the center of the copper tube, and the field effects on the tube were screened out. The equation for the time of flight spectra is

$$T = \left(\frac{m}{2}\right)^{1/2} \int_0^h$$

$$[W - ezE_{amb}(z) - ezE_{app} - mgz]^{-1/2} \, dz \quad (47)$$

where W is the related kinetic-energy term of the electron, h is the length of the tube, mgz is the gravitational term, E_{amb} is the effective ambient electric field [E_{amb} is assumed to consist of a constant term due to gravitationally induced distortions of the tube and a term due to the patch (roughness of the surface) effect with a complicated z dependence], and E_{app} is the uniform applied field. It is believed that this ambient field is screened out by the superconducting electrons

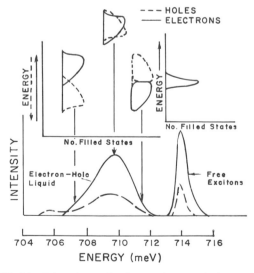

FIG. 14.   Line shapes for free excitons and electron–hole drops in Ge——, the combination of the statistical distribution with the density states – – –, and experimental spectra. [From Lo, T. K. (1974). *Solid State Commun.* **15**, 1231.]

that are in the oxide layer that forms on the copper tube.

The next two experiments were measurements of CuCl that revealed large changes in electrical conductivity and magnetic susceptibility. It was found that when polycrystalline samples of CuCl under hydrostatic pressure of approximately 5 kbar were rapidly cooled (20 K/min), they went through repeated transitions from a state of weak diamagnetism to a state of strong diamagnetism. The diamagnetic susceptibility (χ) varied between -10^{-5} and approximately -1. When $\chi \approx -1$, the Meissner effect, in which the magnetic field is totally excluded from the sample, was observed. This phenomena is characteristic of superconducting materials. The strong diamagnetic state was accompanied by a sharp increase in electrical conductivity.

Similar experiments were performed on the same material in carefully controlled environments. In these experiments the diamagnetic anomaly was observed above 90 K over a temperature range of 10–20 K accompanied by a sharp increase in electrical conductivity.

The model used to explain the above experiment is based on the calculated band structure of CuCl. In zero applied pressure, CuCl is a direct-gap material with minimum energy at the Γ point. When pressure is introduced into the band calculation by reducing the lattice constant, it is found that the conduction band at the X point moves down in energy relative to the conduction band at the Γ point. The conduction band at X becomes degenerate with the conduction band at Γ for a lattice constant reduction of 0.2%, which is consistent with the pressures used in the experiments. Another important point is that a fair amount of oxygen was present in all of the CuCl samples. The calculations reveal that the energy of the oxygen bound electron is less than the binding energy of the exciton, and oxygen will give up part of its electrons to the conduction band at the experimental temperature. An upper limit on the critical temperature T_C of an electron–exciton coupled superconductor can be obtained assuming the static limit of the coupling of the electron and exciton and assuming the T_C can be approximated by

$$T_C = 1.14 \frac{\hbar \omega_{ex}}{K_B} \exp[-1/uD(E_F)] \qquad (48)$$

Here ω_{ex} is the energy of the exciton, u is the coupling coefficient of the exciton, K_B is Boltzmann's constant, and $D(E_F)$ is the density of electrons at the Fermi surface. When the conduction bands at Γ and X are degenerate, $D(E_F)$ is a maximum. It was found that for carrier densities $N(E) = 10^{-1}$ e/unit cell, T_C for a nondegenerate conduction band was 38 K, while it was 1745 K for the degenerate case. For $N(E) = 10^{-2}$ e/unit cell, the corresponding numbers are 10^{-2} K and 55 K, respectively. Therefore it is possible that with 10^{-2} e/unit cell, superconductivity close to or above liquid-nitrogen temperatures can be achieved.

In the final experiment, studies of CdS samples that had been pressure-quenched at 77 K showed strong diamagnetic effects. A super paramagnetic effect was also detected. In this experiment one must explain the pressure effects, the super paramagnetic effect, the diamagnetic anomaly, and the fact that these are only observed in selected samples. As in the case of CuCl, these effects are interpreted in terms of a superconducting state induced by the interaction of the band structure with applied pressure and specific impurity effects.

E. Lasing Transitions

Semiconductor lasers were first reported in 1962, the first being the GaAs injection laser. Since that time many semiconductor lasers have been produced from III–V compounds, and they cover an appreciable portion of the spectrum from 0.65 to 8.5 μm.

Shortly thereafter, rapid developments were made in the area of II–VI compound lasers. The first report of high-efficiency laser action was in electron-beam-pumped CdS. Even higher efficiencies were later achieved in electron-beam-pumped CdSe. The spontaneous line, centered at 6800 Å for CdSe, corresponds to an emission line that has been observed in photoluminescence experiments and has been attributed to an exciton bound to an acceptor. The spontaneous line in CdS at 4.2 K is the 4888-Å line and was also associated with an exciton bound to a neutral acceptor site.

The recombination radiation from highly excited CdS crystals was investigated. The experimental technique allowed the determination of the spectral dependence of the optical gain. From these investigations, it was concluded that at least three different processes can contribute to laser action. A low-gain process results from the annihilation of a free exciton and the emission of a photon and an LO phonon. A medium-gain process is due to an exciton–exciton interaction, and a high-gain process involves an exciton–electron interaction.

The interpretation of the excitation dependence of the spontaneous emission and of the gain for the free exciton-related processes are as follows:

1. For the excitation intensity $J < 1$ A/cm^2, only the E_x–LO process yielded some gain.

2. For 1 A/cm$^2 < J < 3$ A/cm^2, the low-energy tail resulting from electron–exciton interaction line dominates.

3. For $J > 3$ A/cm^2, the low-energy tail resulting from electron–exciton interaction is the dominant gain process. CdS, CdS : Se, and CdSe lasers provide a tunability from 0.5 to 0.7 m. In CdS, mode-locked pulses shorter than 4 psec have been obtained.

F. Optical Bistability

Optical bistability can be defined as any optical system possessing two different steady-state transmissions for the same input intensity. To achieve optical bistability, the optical device must have feedback. This implies that the transmission intensity must have some dependence on the output intensity. Many of the bistable devices have been Fabry–Perot etalons containing materials having nonlinear indices of refraction at high input light intensities. In this type of device, the cavity is tuned so that a transmission maximum lies close to the laser frequency, but still having low transmission at low input intensities. As the input intensity is increased, the light penetrating the cavity will be sufficient to cause the nonlinear index material to tune the cavity toward the laser frequency. This has been termed intrinsic dispersive optical bistability— intrinsic because the nonlinear index material provides the feedback, and dispersive because the reflective or real part of the nonlinear susceptibility is more important than the imaginary or absorption part.

For practical optical bistable devices, attention has been focused on semiconductor materials. Semiconductors have high absorption coefficients, particularly at resonant excitonic transitions. These materials will produce absorptive and dispersive bistable devices. Useful absorption in these materials is achieved in very short transversing paths, making very fast switching devices achievable. Both the free- and bound-exciton transitions in semiconductors have shown promise for high-speed, low-power switching devices. These characteristics make possible fast, all-optical, signal-processing devices.

Bibliography

Cardona, M. (1969). "Modulation Spectroscopy", ed. by F. Seitz and D. Turnbull and E. Ehrenreich. *Solid State Physics, Suppl. 11*. Academic Press.

Craig, D. P. and Walmsley (1968). "Excitons in Molecular Crystals", Benjamin.

Dexter, D. L. and Knox, R. S. (1965). "Excitons", Interscience Publishers.

Dimmock, J. O. (1967). "Theory of Exciton States", *Semicond. and Semimetals*, 3, Chap. 7, Academic Press, New York.

Gossard, A. C. (1982). "Treatise on Materials Science and Technology," 24, 13.

Knox, R. S. (1963). "Theory of Excitons," *Solid State Phys*. Suppl. 5, Academic Press, New York.

Rashba, E. I., and Sturge, M. D., (ed.) (1982). "Excitons", North Holland Publishing Co., Amsterdam.

Reynolds, D. C., and Collins, T. C. (1981). "Excitons, Their Properties and Uses", Academic Press, New York.

Thomas, D. G., and Hopfield, J. J. (1959). *Phys. Rev.* **116**, 573.

Thomas, D. G., and Hopfield, J. J. (1962). *Phys. Rev.* **128**, 2135.

Wheeler, R. G., and Dimmock, J. O. (1962). *Phys. Rev.* **125**, 1805.

FERROMAGNETISM

H. R. Khan   *Forschungsinstitut für Edelmetalle und Metallchemie and University of Tennessee*

I. Basic Concept of Magnetism
II. Origin of Magnetism
III. Magnetization Curves
IV. The Hysteresis Loop
V. Anisotropic Magnetization
VI. Magnetic Order
VII. Ferromagnetic Domains
VIII. Magnetostriction
IX. Magnons
X. Ferromagnetism and Superconductivity
XI. Ferromagnetic Materials and Their Applications

GLOSSARY

Coercive force (H_c):   Negative value of the magnetizing field H that makes the magnetic induction B of a ferromagnetic material zero.

Curie temperature (T_C):   Temperature above which the spontaneous magnetization of a ferromagnetic material vanishes.

Magnetic flux density: In the international system of units (SI), the magnetic field intensity H (A/m) and the magnetization M (A/m) are related to the magnetic induction or the magnetic flux density B (Wb/m^2) through the relation

$$B = \mu_0(H + M)$$

where μ_0 is the permeability of the free space and has the value 12.57×10^{-7} Wb/A m).

Magnetic moment: The magnetic moment of a small plane coil is a product of the current I flowing in the coil and the area of the coil A, IA (A m^2). The magnetic moment of a small magnet is equal to the magnetic moment of a small coil that would experience the same torque when placed in the same orientation at the same location in the same magnetic field.

Magnetic susceptibility (χ):   The ratio M/H is called the magnetic susceptibility of a mate-rial and may be expressed in mass, volume, or molar units.

Magnetization (M):   Magnetic moment per unit volume.

Neel temperature (T_N):   Temperature below which the interaction between the atomic moments affecting antiparallel orientation surmounts the thermal agitation. At Neel temperature T_N, the susceptibility of a material has its maximum.

Permeability (μ):   The ratio B/H is the permeability of a material.

Remanence: If the magnetic flux density does not fall to zero upon reducing the magnetic field on the specimen to zero, then this remained magnetic flux density in the specimen is called the remanence.

A special arrangement of electrons in the atoms causes a material to become a ferromagnetic material. For example, the incompletely filled M shells of iron, cobalt, and nickel atoms are responsible for the ferromagnetism in these metals. Atoms behave as small magnets ordered in parallel arrangement in ferromagnetic materials. The magnetization curve and the hysteresis loop determine whether it is a hard or soft ferromagnetic. The parameters determined from the hysteresis loop are the permeability, coercivity, remanence, and the area of the loop itself. The area of the loop gives the energy loss per unit volume of the specimen per cycle and is dissipated as heat energy called as hysteresis loss. The hysteresis loop and the parameter derived from it determine the suitability of a material in a particular application.

I. Basic Concept of Magnetism

All the materials occurring in nature are magnetic. They may be paramagnetic, diamagnetic, ferromagnetic, antiferromagnetic, or ferrimagnetic. The magnetic behavior of a material depends on its electronic structure. For example, the ferromagnetism of iron, cobalt, and nickel in the periodic system is due to the incompletely filled M shells of their atoms. Due to these incompletely filled shells, the atoms behave as magnets ordered in parallel arrangement in ferromagnetic materials. In antiferromagnetic materials, the atomic magnets are ordered in antiparallel arrangement. Ferrimagnetic materials are a special case of ferromagnetic materials. The neighboring atoms interact with each other in a material, and this interaction force is dependent on the distance between neighboring atoms and the diameter of the atomic shell responsible for the atomic magnetic moment. The sign and magnitude of this interaction force cause a material to show different magnetic behavior. The usefulness of a ferromagnetic material is shown by its magnetization curve and hysteresis loop. The hysteresis loop provides information about the "permeability" and "coercivity" of a ferromagnetic material. For example, a soft magnetic material to be used as a transformer core should have a high value of permeability, whereas a hard or permanent magnetic material should have a high value of coercivity. The physical condition, purity, and composition of a material control the useful magnetic properties like permeability and coercivity, and they can be modified by controlling different parameters of a material. By rapid solidification of materials from the melt, very soft magnetic and hard magnetic materials can be produced. Both the hard magnetic and soft magnetic materials find applications in the electrical and electronic industry. Some examples of their uses are transformers, motors, generators, relays, telephone cables, audio and video recording and replaying, and data memory systems and computers.

A magnetic field gradient is generated by a magnet consisting of a flat north pole and a pointed south pole, as shown in Fig. 1. The magnetic field is stronger near the pointed pole. If a piece of material in the cylindrical form is suspended with a string between the poles, then a magnetic force is generated on this material. The kind of magnetism on the material is determined from the direction of the magnetic force on this material. If the material is strongly attracted toward the pointed pole, then it is ferromagnetic. A paramagnetic material is weakly attracted toward the pointed pole, and a diamagnetic material is repelled by the pointed pole. For example,

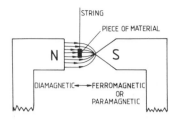

FIG. 1. An experimental set-up to check the paramagnetism, diamagnetism, and ferromagnetism of a material.

iron is ferromagnetic, aluminum is paramagnetic, and bismuth is diamagnetic. Materials can be synthesized that show ferromagnetism, although the constituents may be paramagnetic or diamagnetic. Ferrimagnetism and antiferromagnetism are closely related to ferromagnetism. Many compounds are ferrimagnetic. Ferrimagnetic materials in general are oxides of the ferromagnetic metals.

II. Origin of Magnetism

In the classical picture, an atom consists of a nucleus surrounded by a number of electrons that depends on the element and its position in the periodic system. The electrons are distributed in different shells, named K, L, M, and N, outward from the center. Each shell can accommodate only a certain number of electrons, the maximum number being $2n^2$, where n is the number of the shell. The innermost shell K is complete with $2 \times 1^2 = 2$ electrons, the second shell L is complete with $2 \times 2^2 = 8$, the third shell M with $2 \times 3^2 = 18$ electrons, etc. For example, the ferromagnetic element iron has its M shell incompletely filled and contains only 14 electrons. An electron carries a negative charge, and its motion in an orbit gives rise to an electric current. The orbital motion of the electron is equivalent to a thin magnet and produces a magnetic field. Besides the orbital motion, the electron also spins around its own axis. The spinning negative charge also gives rise to an electric current and behaves like a small magnet. The completely filled shells are magnetically neutral, because an equal number of electrons spin in clockwise and anticlockwise directions. As mentioned earlier, the M shell of iron contains only 14 electrons instead of 18, and nine electrons spin in clockwise and five in anticlockwise directions. The uncompensated four electrons

IRON   COBALT   NICKEL

FIG. 2.   The crystal structures of the ferromagnetic metals iron, cobalt, and nickel.

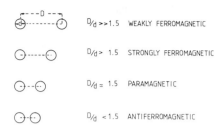

$D/d \gg 1.5$ WEAKLY FERROMAGNETIC

$D/d > 1.5$ STRONGLY FERROMAGNETIC

$D/d \approx 1.5$ PARAMAGNETIC

$D/d < 1.5$ ANTIFERROMAGNETIC

FIG. 4.   Ratios of the interatomic distance D to the unfilled atomic shell diameter d causing different kinds of magnetism.

produce a magnetic field at a distance equal to four electrons. Therefore iron has a magnetic moment of 4 units. The other ferromagnetic element, cobalt, has only 15 electrons in the M shell and carries a magnetic moment of 3 units. Nickel is also ferromagnetic because its atom in the M shell has only 16 electrons and carries a magnetic moment of 2 units. Thus we have seen that ferromagnetism originates from the unfilled M shells in iron, cobalt, and nickel atoms.

The free isolated atoms of iron, cobalt, and nickel have magnetic moments equivalent to 4, 3, and 2 units. In metals, the atoms are not isolated but are packed together. This packing of atoms influences the distribution of electrons in the M shell, because this is outside of an atom. Therefore the experimentally measured values of the magnetic moments of iron, cobalt, and nickel are 2.22, 1.71, and 0.606 units, and are lower compared to the isolated free atoms.

Each element in the periodic system has a definite crystal structure. In iron and nickel atoms, the atoms occupy the body-centered cubic (b.c.c.) and face-centered cubic (f.c.c.) lattice sites, respectively, whereas in cobalt the atoms occupy the hexagonal lattice sites as shown in Fig. 2. In ferromagnetic elements, each atom carries a magnetic moment and a magnetic axis, and even in the absence of an externally applied magnetic field these atomic magnets point in one direction, as shown in Fig. 3. The internal magnetic field required to order the atomic magnets in one direction is called a Weiss molecular field. For example in the case of iron, its intensity is 5.5×10^6 Å/cm.

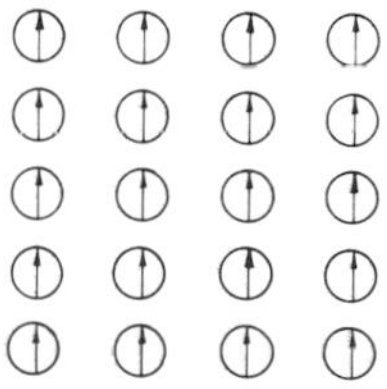

The interaction of neighboring atoms causes the alignment of atomic magnets in the same direction (ferromagnets) or in opposite directions (antiferromagnets), as seen from the quantum mechanics point of view. The force of interaction is a function of the ratio of the distance D between the neighboring atoms and the diameter d of the atomic shell responsible for the atomic magnetic moment. The force of interaction changing from positive to negative value causes a material to become ferromagnetic, weakly ferromagnetic, paramagnetic, or antiferromagnetic as shown in Figs. 4 and 5. The iron, cobalt, and nickel are strongly ferromagnetic because the ratio D/d is larger than 1.5, whereas gadolinium is weakly ferromagnetic because the value of D/d is about 3.1.

When a ferromagnetic material with its atomic magnets pointed in one direction is heated, the thermal vibrations of the atoms become stronger with the increase in temperature. When the temperature reaches a value called a magnetic change point or Curie temperature, the atomic magnets orient themselves randomly and the ferromagnetic material transforms to a paramag-

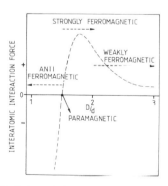

FIG. 5.   Variation of the interatomic interaction force with the ratio D/d of the interatomic distance D to the diameter d of the unfilled atomic shell.

FIG. 3.   Arrangement of the atomic magnets in a ferromagnetic material.

netic state. The Curie temperature of iron is 770°C, 1130°C for cobalt, and 360°C for nickel.

III. Magnetization Curves

The magnetization M of a material is defined as the magnetic moment per unit volume. The practical usefulness of a ferromagnetic material is determined from its magnetization curve. The experimental set-up for plotting the magnetization curve is shown in Fig. 6. A thin torroidal ring of the ferromagnetic material of cross section A is wound with N turns per meter. A current of I amperes flowing through the winding generates a magnetic flux $\Phi = BA$ in the ring. A flux meter connected with a secondary coil of few turns measures this flux Φ. The flux density $B = \Phi/A$ is composed of two parts: one arising from the external current flowing in the winding, and the second arising from the internal current associated with the motion of electrons in the ferromagnetic material.

The flux density arising from the external current NI per meter of the winding is $\mu_0 NI$, where μ_0 is a constant. The magnetization M arising from the internal currents in the ferromagnetic material may be considered constant across the cross section. The magnetic field set up by this magnetization is $\mu_0 M$. The magnetic intensity in the core is the sum of these two contributions:

$$B_{\text{core}} = \mu_0 NI + \mu_0 M \tag{1}$$

In the absence of a ferromagnetic core,

$$B_{\text{no core}} = \mu_0 NI \tag{2}$$

The ratio $B_{\text{core}}/B_{\text{no core}}$ is defined as the permeability of a ferromagnetic material. The permeability is the ratio of the magnetic intensity in a toroidal core to the intensity that the same current in the same winding would produce in the absence of a ferromagnetic core, and it is dimensionless. The permeability of a material depends on its history and is very high (~ 1000) for the soft magnetic materials like iron.

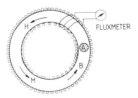

FIG. 6.   Experimental set-up for plotting the magnetization curve.

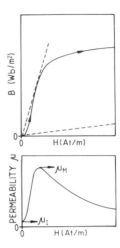

FIG. 7.   Plot of the magnetic induction B as a function of the magnetizing field H of a ferromagnetic material, and plot of the permeability μ as a function of the magnetizing field H (ampere turns per meter).

The value of the magnetizing field $H = NI$ is increased by increasing the current I in the winding, which increases the value of B measured by the flux meter. A plot of B versus H is called a magnetization curve (B–H curve) and is shown in Fig. 7. The term μ_I is the initial permeability of the material obtained from initial slope. When the magnetizing field H is increased, the ratio B/H also increases until a maximum value is reached. The slope at this point, $\mu_m = B/H$, is called the maximum permeability. The term B also achieves its maximum value, called the saturation magnetic intensity B_s.

The variation of the initial and maximum permeability as a function of the magnetizing field of a ferromagnetic material is shown in Fig. 7. All ferromagnetic materials show this kind of B–H and μ–H behavior, but the magnitudes of the permeability and the scales of the B and H are different for different materials. Many practical uses of ferromagnetic materials require them to possess high values of magnetizing fields. Typical examples are the core of low-current transformers, low-current relays, inductive loading of telephone cables, and the sensitive detectors of small field changes. The best ferromagnetic materials for these applications are those with highest μ_I and μ_m values.

IV. The Hysteresis Loop

The initial magnetization as shown in Fig. 7 is not reversible. When H slowly increases, the

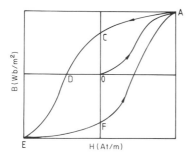

FIG. 8.   Plot of the magnetic induction B as a function of magnetizing field H, hysteresis loop.

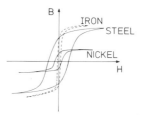

FIG. 9.   Hysteresis loops of some ferromagnetic materials.

value of B also increases until it reaches its maximum value at point A as shown in Fig. 8. Thus 0A represents the initial magnetization curve. When the magnetic field is slowly decreased, the flux density B follows the curve AC. The magnetic field is zero at C, but at this point a flux density equal to 0C remains in the ferromagnetic material. If the magnetic field is reversed to 0D, the flux density is completely removed from the material, and further reducing the field in the negative direction brings the flux density to point E. The flux density follows the curve EFA if the direction of the magnetic field is changed and slowly increased. If the cycle is repeated few times, it brings the material in the cyclic state. The loop ACEFA is called the *hysteresis loop*. The flux density at the point C is called *remanence*, and the reverse field at point D is called *coercive force*.

The total energy required to magnetize a unit volume of the specimen from 0 to A on the initial curve is given by

$$W = \int_0^A H \, dB$$

If the magnetic field is reversed to zero, then the returned path on the hysteresis loop is AC and the total energy taken from the magnetizing field H is the area 0AC. If the curve 0A were traced back to the original path, then the energy taken from the magnetizing field H would have been returned to it and there would have been no loss of energy. But in the case of the hysteresis loop shown in Fig. 8, there is a loss of energy. The total energy per unit volume of the specimen taken from the magnetizing field H for one complete cycle of the hysteresis loop is the area of the hysteresis loop, which is $\int_0^A H \, dB$. This energy is dissipated as heat energy and is called hysteresis loss.

Hysteresis loops of small areas are observed for the soft ferromagnetic materials, whereas the hard ferromagnetic materials show large areas. The hysteresis loop of an ideal soft ferromagnet should show just a line, but in practice such materials do not exist. Some amorphous iron-based materials show hysteresis loops of very small area. A typical example of a hard ferromagnetic with a relatively large hysteresis loop area is carbon steel. The hysteresis loop area is also an important parameter in determining the application of a ferromagnetic material. The hysteresis loops of some ferromagnetic materials are shown in Fig. 9.

Good permanent magnets possess high values of residual flux and coercive force. These ferromagnetic materials with high residual flux and coercive force can not be used in motors or transformers, because the flux changes continuously and there is an energy loss in each cycle. The energy loss is proportional to the area of the hysteresis loop. Therefore a ferromagnetic material subjected to cyclic magnetization, as in motors and transformers, should have as narrow a hysteresis loop as possible.

V. Anisotropic Magnetization

The energy in single crystals of ferromagnetic material that governs the magnetization along the crystallographic axes is called the magneto-crystalline or anisotropy energy. It is easy to magnetize an iron crystal along the cubic-edge directions rather than along directions of other crystal axes. However, a nickel crystal is easily magnetized along the long diagonal axis compared to the cubic-edge directions. Cobalt with its hexagonal crystal structure can be easily magnetized in the direction of the hexagonal axis. These crystallographic axes of a ferromagnetic crystal along which they are easily magnetized are called directions of easy magnetization.

VI. Magnetic Order

The magnetic susceptibility χ per unit volume of a magnetic material is defined as the ratio of the magnetization M to the macroscopic magnetizing field intensity B: $\chi = M/B$. The magnetic susceptibility may be defined in terms of the unit mass, "mass susceptibility," or mole (molar susceptibility) of a material, or unit volume, called volume susceptibility. The variation of χ with temperature of a paramagnetic material is shown in Fig. 9 and is related to the temperature as $\chi = C/T$, called the Curie law. Here the constant C is called the Curie constant.

A. Ferromagnetic Order

In a ferromagnetic material, the individual magnetic moments are ordered in parallel arrangement, as shown in Fig. 3. A ferromagnet possess a magnetic moment even in the absence of an externally applied magnetic field, and this spontaneous magnetic moment is also called the *saturation moment*. When a ferromagnet is heated, the parallel arrangement disappears above the Curie temperature. The magnetic susceptibility of a ferromagnetic material at temperatures close to the Curie temperature is related to the temperature by

$$\chi = \frac{C}{(T - T_C)}$$

which is also shown in Fig. 9.

B. Antiferromagnetic Order

The atomic magnetic moments order in an antiparallel arrangement in an antiferromagnetic material is shown in Fig. 10. The resultant moment is zero below the ordering or *Neel temperature* T_N. The variation of the magnetic susceptibility χ of an antiferromagnetic material with

FIG. 11.   Temperature-dependent magnetic susceptibility χ behavior of a paramagnetic, ferromagnetic material.

temperature is shown in Fig. 11 and is given by the relation

$$\chi = \frac{C}{(T + \theta)}$$

where θ is the Neel temperature T_N.

C. Ferrimagnetic Order

In some ferromagnetic materials, the saturation magnetization does not correspond to the parallel alignment of the individual atomic magnetic moments. These materials are the magnetic oxides with the general chemical formula $MO \cdot Fe_2O_3$, where M may be the metals, for example zinc, cadmium, iron, nickel, copper, cobalt, or magnesium. These ferrites have the spinel crystal structure. There are eight occupied tetrahedral sites and 16 occupied octahedral sites in this crystal structure. In these ferrites, iron has two different ionic states, ferrous (Fe^{2+}) and ferric (Fe^{3+}). The eight tetrahedral sites in the cubic spinel structure are occupied by Fe^{3+} ions, whereas half of the 16 octahedral sites are occupied by Fe^{3+} and the rest by Fe^{2+} ions. The magnetic moments of eight Fe^{3+} ions on the tetrahedral and octahedral sites cancel each other, leaving only the magnetic moments of the eight Fe^{2+} ions, as shown in Fig. 12.

VII. Ferromagnetic Domains

At temperatures below the Curie point, the magnetic moment may be much less than the saturation moment of a ferromagnetic material.

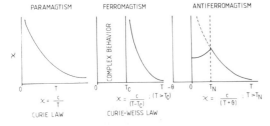

FIG. 10.   The arrangement of the atomic magnets in an antiferromagnetic material.

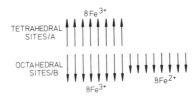

FIG. 12.   The arrangement of the atomic magnets in a ferrite of cubic spinel structure.

The polycrystalline as well as the single-crystal specimen consist of small regions called *domains,* within each of which the local magnetization is saturated. The magnetic axes of these domains may point in different directions, and it is possible that for a certain arrangement it might give a zero resultant magnetic moment of the specimen. The application of an external magnetic field saturates the specimen, because the external field causes the orientation of the domain magnetization in the direction of the applied magnetic field. Small cylindrical magnetic domains may be stabilized in a thin crystal of uniaxial material by applying a bias magnetic field. The bubble diameter is on the order of 10 μm. These magnetic bubbles are of interest in high-density memory-storage devices.

As discussed in Section III, the permeability and coercivity are important parameters that control the practical application of a ferromagnetic material. The domain structure of a ferromagnetic material affects both of these parameters. A pure, well oriented, and homogenous material facilitates the domain boundary displacement and possess high permeability. On the other hand, an inhomogenous material consisting of multiple phases suppresses the boundary displacement and possess high "coercivity."

F. Bitter developed a simple method to observe the domain boundaries. A drop of a colloidal suspension of a finely divided ferromagnetic material such as magnetite is placed on the surface of a ferromagnetic material. The colloidal

particles in the suspension concentrate strongly on the boundaries between the domains where the strong local magnetic fields exist that attract the magnetic particles. A simple domain structure in a silicon–iron single crystal is shown in Fig. 13.

VIII. Magnetostriction

When a ferromagnetic material is magnetized, small changes in the physical dimensions of the specimen take place, and this effect is called *magnetostriction.* This magnetostriction of a material is defined as the increase in length per unit length in the direction of the magnetization. Magnetostriction is different for different axes of a ferromagnetic single crystal. The useful parameter "permeability" of a ferromagnetic material is related to the magnetostriction.

IX. Magnons

A ferromagnetic material in the ground state has all its spins arranged parallel in one direction, as shown in Fig. 14(a). The excited state is obtained if the spins are reversed. Figure 14(b) shows the excited state where one spin is antiparallel. The elementary excitations are the spin waves, and one wavelength is shown in Fig. 14(c). These elementary excitations are called *magnons* and are analogous to the lattice vibrations or phonons. Spin waves are the relative orientation of the spins on a lattice, whereas the lattice vibrations are the oscillations in the relative position of the atoms on a lattice. Spin waves have been observed by neutron scattering experiments near the Curie temperature or even above the Curie temperature.

X. Ferromagnetism and Superconductivity

Both ferromagnetism and superconductivity involve spin ordering. The difference is that in a ferromagnet the spins order parallel, whereas in

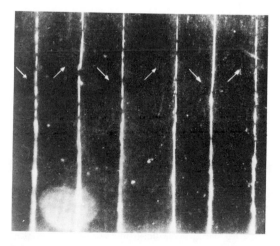

FIG. 13.   Simple domain structure in silicon–iron single crystal. [After Williams, Bozorth, and Shockley (1949). *Phys. Rev.* **75,** 155.]

FIG. 14.   (a) The arrangement of spins in a ferromagnetic material. (b) The elementary excitation occurs when one spin is antiparallel. (c) One wavelength of the spin wave.

a superconductor they order antiparallel below the superconducting transition temperature and form the "Cooper pairs." A possibility of coexistence of superconductivity and ferromagnetism in the same material was proposed. To observe this coexistence, the ferromagnetic impurities were dissolved in superconducting materials, for example, gadolinium (ferromagnetic) in lanthanum (superconducting). The lanthanum–gadolinium compounds were superconducting up to 1 at. % gadolinium and became ferromagnetic for the concentrations of gadolinium above 2.5 at. %.

Recently some ternary compounds of the formula MRh_4B_4, with B = as thorium, yttrium, neodymium, samarium, gadolinium, terbium, dysprosium, holmium, erbium, or lutetium, having $CeCo_4B_4$ structure, have been discovered. A typical example of a compound showing both ferromagnetism and superconductivity is $Er_4Rh_4B_4$, which is superconducting at 8.7 K and ferromagnetic at 0.9 K.

XI. Ferromagnetic Materials and Their Applications

The hysteresis loop of a ferromagnetic material provides information about its usefulness in technical applications, as discussed in Section IV. The hysteresis loop depends on the physical condition, composition, and purity of a specimen. Depending on the application of a ferromagnetic material, the important properties are the "permeability" and "coercivity."

When a strain-free material is cold-worked, the permeability of the material is reduced and the hysteresis loss is increased. The strain-relieving heat treatment of the cold-worked specimen again brings the original magnetic properties back: for example, the permeability is increased and the hysteresis loss reduced. In general, the strain-free crystals show the minimum hysteresis loss.

The presence of the impurities as carbon, oxygen, nitrogen, sulfur, etc. affects the permeability and hysteresis loss of a ferromagnetic material. In general, the materials with high permeability and low hysteresis loss are pure materials.

The composition of a ferromagnetic material also influences its magnetic properties. The addition of silicon to iron increases the permeability and reduces the hysteresis loss. However, high concentrations of silicon decrease the saturation magnetization. Therefore, the iron–silicon alloys with low concentrations of silicon are desirable in applications like the cores of transformers and in electric motors and generators.

The iron–nickel alloys possess high values of initial and maximum permeability and very low hysteresis loss compared to the iron–silicon alloys. An alloy of composition with 78.5% nickel and 21.5% iron is called *permalloy* and has an initial relative permeability of ~10,000 compared to 250 for the pure iron. These alloys are in general used for the magnetic screening of the electronic equipment. Small additions of the metals chromium or molybdenum further modify the magnetic properties of these materials to be used as cores in transformers or inductors working at the audio or higher frequencies. For example, the magnetic cores of inductors and transformers working at radiofrequencies (~100 Mc/sec) show high eddy current losses. Used here are the ferrites, which have high resistivity (~10^6 times that of metals) and high permeability. In other applications, the ferromagnetic materials with a high value of "coercivity" and large area of hysteresis loop are required. These materials possess hard magnetism compared to the already discussed soft magnetism. The addition of carbon to iron increases the hysteresis loss. The carbon steel was used as a material for permanent magnets in earlier days. However, aging degenerates the magnetic properties of carbon-steel magnets. Addition of metals as cobalt, chromium, or tungsten improves the magnetic properties, and these materials are less susceptible to aging. A large number of alloys composed of iron, nickel, cobalt, aluminum, copper, platinum, manganese, and oxides of iron and rare-earth metals have been developed that show high values of coercivity and are suitable for permanent magnets.

Some oxides like $\gamma\text{-}Fe_2O_3$ and CrO_2 possess high coercivity and are used as recording tapes in the form of this layers of fine powders. Permalloys are used to construct the magnetic heads to write signals as residual magnetization on the tapes or to reproduce the electrical signals from the magnetized tapes. Magnetic discs or drums are made for the memory systems in computers.

Development of modifying the magnetic properties by rapid solidification of alloys from the melt has created a new field. By rapid solidification, the microstructure can be affected—in some cases the phases may be finely dispersed, and in other the alloys may become noncrystalline or amorphous. The amorphous alloys have no crystal lattice and no magnetic anisotrophy.

There are no extended defects that would otherwise interact strongly with the domain walls in these noncrystalline materials. In certain cobalt-based noncrystalline alloys, the magnetostriction can be adjusted to zero such that the internal and applied stresses have minimal effect on the magnetic properties. Amorphous magnetic alloys have high hardness and yield strength and are magnetically soft. In particular, cobalt-containing alloys have vanishingly small magnetostriction. Combining the good mechanical and magnetic properties, they are very useful materials. They can be strained elastically over wide limits, are insensitive to irreversible magnetic damage, and are very suitable materials where the elastic deformability is desired. They are also useful materials for making recording heads in audio, video, and data recording systems, due to their good high-frequency response and wear resistance.

Soft magnetic and highly elastic mechanical behavior has led to the development of flexible magnetic shielding.

The amorphous magnetic materials consist of two main groups. In one group, the materials are composed of transition metals and metalloids; in the second group, they are composed of only different metals. Some amorphous materials used as soft magnetic materials are $Fe_{81}(Si, B, C)_{19}$; $(FeNi)_{78}(Mo, Si, B)_{22}$; $(Co, Fe)_{70...76}(Mo, Si, B)_{30...24}$; and $(Co, Mn)_{70...76}(Mo, Si, B)_{30...24}$.

Hard magnetic materials can also be produced by rapid solidification techniques. For example, a magnetic material of composition $Fe_{14}Nd_2B$ produced by rapid solidification is superior to the Co–Sm material. Other hard magnetic materials (some transition and rare earth metals, and boron), have also been produced by rapid cooling. The rapid cooling technique is also less expensive compared to the conventional methods.

BIBLIOGRAPHY

Brailsford, F. (1968). "An Introduction to the Magnetic Properties of Materials." Longmans Green and Co. Ltd., London.

Chikazumi, S. (1964). "Physics of Magnetism." John Wiley and Sons, New York.

Craik, D. J., and Tebble, R. S. (1966). "Ferromagnetism and Ferromagnetic Domains." North-Holland Publ., Amsterdam

Della Torre, E., and Bobeck, A. H. (1974). "Magnetic Bubbles." North-Holland Publ., Amsterdam.

Kittel, Charles. (1979). "Introduction to Solid State Physics." John Wiley and Sons, New York.

Morrish, A. H. (1965). "Physical Principles of Magnetism." John Wiley and Sons, New York.

Standley, K. J. (1972). "Oxide Magnetic Materials," 2nd ed. Oxford Univ. Press, London.

Steeb, S., and Warlimont, H. (Eds.) (1985). "Rapidly Quenched Metals," Vol. II. North-Holland Publ., Amsterdam.

Vonsovskii, S. V. (1975). "Magnetism." Halsted Press, New York.

Wohlfarth, E. P. (1980–1982). "Ferromagnetic Materials," Vol. I (1980), Vol. II (1980), Vol. III (1982). North-Holland Publ., Amsterdam.

Zeiger, H. J. (1973). "Magnetic Interaction in Solids." Oxford Univ. Press, London.

HOLOGRAPHY

Clark C. Guest *University of California, San Diego*

I. Introduction
II. Basic Principles
III. Classification of Holograms
IV. Recording Materials
V. Applications
VI. History of Holography

GLOSSARY

Diffraction: Property exhibited by waves, including optical waves. When part of a wavefront is blocked, the remaining portion spreads to fill in the space behind the obstacle.

Diffraction orders: When a wave passes through regularly spaced openings, such as a recorded holographic interference pattern, the diffracted waves combine to form several beams at different angles. These beams are called diffraction orders.

Fringes: Regular pattern of bright and dark lines produced by the interference of optical waves.

Hologram: Physical record of an interference pattern. It contains phase and amplitude information about the wavefronts that produced it.

Holograph: Although this is an English word, meaning signature, it is often improperly used as a synonym for hologram.

Holography: Process of recording holograms and reproducing wavefronts from them.

Index of refraction: Property of transparent materials related to their polarizability at optical frequencies. The speed of light in a vacuum divided by the speed of light in a material gives the index of refraction for the material.

Interference: When two waves are brought together they may be in phase, in which case their amplitudes add, or they may be out of phase, in which case their amplitudes can-

cel. The reinforcement and cancellation of wavefronts due to their relative phase is called interference. Waves that do not have the same phase structure will add in some regions of space and cancel in others. The resulting regions of high and low amplitude form an interference pattern.

Parallax: Difference, due to perspective, in a scene viewed from different locations.

Planewave: Wave configuration in which surfaces of constant phase form parallel flat planes. All the light in a planewave is travelling the same direction, perpendicular to the surface of the planes.

Reconstructed beam: Light diffracted by a hologram that reproduces a recorded wavefront.

Reconstructing beam: Beam that is incident on a hologram to provide light for the reconstructed beam.

Reconstruction: Either the process of reading out a recorded hologram, or the wavefront produced by reading out a hologram.

Refractive index: See "index of refraction."

Spherical wave: Wave configuration in which surfaces of constant phase form concentric spherical shells or segments of shells. Light in an expanding spherical wave is propagating radially outward from a point, and light in an converging spherical wave is propagating radially inward toward a point.

Surface relief pattern: Ridges and valleys on the surface of a material.

Wavefront: Surface of constant phase in a propagating wave.

Holography is the technology of recording wavefront information and producing reconstructed wavefronts from those recordings. The record of the wavefront information is called a hologram. Any propagating wave phenomenon such as microwaves or acoustic waves is a candidate for application of the principles of holography, but most interest in this field has centered on waves in the visible portion of the electromagnetic spectrum. Therefore, this article will concentrate on optical holography.

I. Introduction

Although holography has many applications, it is best known for its ability to produce three-dimensional images. A hologram captures the perspective of a scene in a way that no simple photograph can. For instance, when viewing a hologram it is possible by moving one's head to look around objects in the foreground and see what is behind them. Yet holograms can be recorded on the same photographic film used for photographs.

Two questions naturally occur: What makes holograms different from photographs, and how can a flat piece of film store a three-dimensional scene? Answering these questions must begin with a review of the properties of light. As an electromagnetic wave, light possesses several characteristics: amplitude, phase, polarization, color, and direction of travel. When the conditions for recording a hologram are met, there is a very close relationship between phase and direction of travel. The key to the answers to both our questions is that photographs record only amplitude information (actually, they record intensity, which is proportional to the square of the amplitude), and holograms record both amplitude and phase information. How ordinary photographic film can be used to record both amplitude and phase information is described in Section II of this article.

Another interesting property of holograms is that they look nothing like the scene they have recorded. Usually a hologram appears to be a fairly uniform gray blur, with perhaps a few visible ring and line patterns randomly placed on it. In fact, all the visible patterns on a hologram are useless information, or noise. The useful information in a hologram is recorded in patterns that are too small to see with the unaided eye; features in these patterns are about the size of a wavelength of light, one two-thousandth of a millimeter.

One useful way to think of a hologram is as a special kind of window. Light is reflected off the objects behind the window. Some of the light passes through the window, and with that light we see the objects. At the moment we record the hologram, the window "remembers" the amplitude and direction of all the light that is passing through it. When the hologram is used to play back (reconstruct) the three-dimensional scene, it uses this recorded information to reproduce the the original pattern of light amplitude and direction that was passing through it. The light reaching our eye from the holographic window is the same as when we were viewing the objects themselves. We can move our heads around and view different parts of the scene just as if we were viewing the objects through the window. If part of the hologram is covered up, or cut off, the entire scene can still be viewed, but through a restricted part of the window.

There are actually many different types of holograms. Although photographic film is the most widely used recording material, several other recording materials are available. The properties of a hologram are governed by the thickness of the recording material and the configuration of the recording beams. The various classifications of holograms will be discussed in Section III. Holograms can be produced in materials that record the light intensity through alterations in their optical absorption, their index of refraction, or both. Materials commonly used for recording holograms are discussed in Section IV.

Holograms have many uses besides the display of three-dimensional images. Applications include industrial testing, precise measurements, optical data storage, and pattern recognition. A presentation of the applications of holography is given in Section V.

II. Basic Principles

Photographs record light-intensity information. When a photograph is made, precautions must be taken to ensure that the intensities in the scene are suitable; they must be neither too dim nor too bright. Holograms record intensity and phase information. In addition to the limits placed on the intensity, the phase of light used to record holograms must meet certain conditions as well. These phase conditions require that the light is coherent. There are two types of coherence, temporal and spatial; the light used for recording holograms must have both types of coherence. Temporal coherence is related to the colors in the light. Temporally coherent light contains only one color: it is monochromatic. Each color of light has a phase associated with it; multicolored light cannot be used to record a hologram because there is no one specific phase to record. Spatial coherence is related to the direction of light. At any given point in space, spatially coherent light is always travelling in one direction, and that direction does not change with time. Light that constantly changes its direction also constantly changes its relative phase in space, and so is unsuitable for recording holograms.

Temporal and spatial coherence are graded quantities. We cannot say that light is definitely coherent or definitely incoherent; we can only say that a given source of light is temporally and spatially coherent by a certain amount. Ordinary light, from a light bulb for example, is temporally and spatially very incoherent. It contains many different colors, and at any point in space it is changing directions so rapidly that our eyes cannot keep up; we just see that on average it appears to come from many different directions. The temporal coherence of ordinary light can be improved by passing it through a filter that lets only a narrow band of colors pass. Spatial coherence can be improved by using light coming from a very small source, such as a pinhole in an opaque mask. Then we know that light at any point in space has to be coming from the direction of the pinhole. Ordinary light that has been properly filtered for color and passed through a small aperture can be used to record holograms. However, light from a laser is naturally very temporally and spatially coherent. For this reason, practically all holograms are recorded using laser light. [See LASERS.]

The simplest possible hologram results from bringing together two linearly polarized optical planewaves. Imagine that a planewave is incident at an angle θ_1 on a flat surface. At a particular moment in time we can plot the electric field of the planewave at positions on that surface. This is done in Fig. 1(a). If a second planewave is incident on the same surface at a different angle θ_2, its electric field can also be plotted. This, along with the combined field from both planewaves is plotted in Fig. 1(b). Both planewaves incident on the surface are, of course, travelling forward at the speed of light. In Fig. 1, parts (c)–(e) show the electric field at the observation surface for each planewave and for the combined field when the waves have each travelled forward by one-fourth, one-half, and three-quarters of a wavelength, respectively. The interesting thing to notice is that the locations on the observation plane where the total electric field is zero remain fixed as the waves travel. Locations midway between zero electric field lo-

cations experience an oscillating electric field. To an observer, locations with constant zero electric field appear dark, and locations with oscillating electric field appear bright. These alternating bright and dark lines, called fringes, form the interference pattern produced by the planewaves. Likewise, locations with zero electric field will leave photographic film unexposed, and locations with oscillating electric field will expose film. Thus, the interference pattern can be recorded.

The interference fringes resulting from two planewaves will appear as straight lines. Wavefronts do not have to be planewaves to produce an interference pattern, nor do the wavefronts have to match each other in shape. Interference between arbitrary wavefronts can appear as concentric circles or ellipses, or as wavy lines.

The distance L from one dark interference fringe to the next (or from one bright fringe to the next) depends on the wavelength λ of the light and the angle θ between the directions of propagation for the wavefronts,

$$L = \lambda/[2\sin(\theta/2)] \qquad (1)$$

For reasonable angles, this fringe spacing is about the size of a wavelength of light, around one two-thousandth of a millimeter. This explains why holograms appear to be a rather uniform gray blur: the useful information is recorded in interference fringes that are too small for the eye to see.

Variations in the amplitudes of the recording beams are also recorded in the hologram and contribute to accurate reproduction of the reconstructed wavefront. During hologram recording, locations of zero electric field will occur only if the two beams are equal in amplitude. If one beam is stronger than the other, complete cancelation of their electric fields is impossible, and the depth of modulation, or contrast, of the recorded fringes is decreased. In practical terms, this means that all of the hologram is exposed to some extent.

There are two steps to the use of ordinary photographs, taking (and developing) the photograph, and viewing the photograph. Similarly, there are two steps to using a hologram: recording (and developing) the hologram, and reconstructing the holographic image. As we have just seen, recording a hologram amounts to recording the interference pattern produced by two coherent beams of light. Reconstructing the holographic image is usually accomplished by shining one of those two beams through the de-

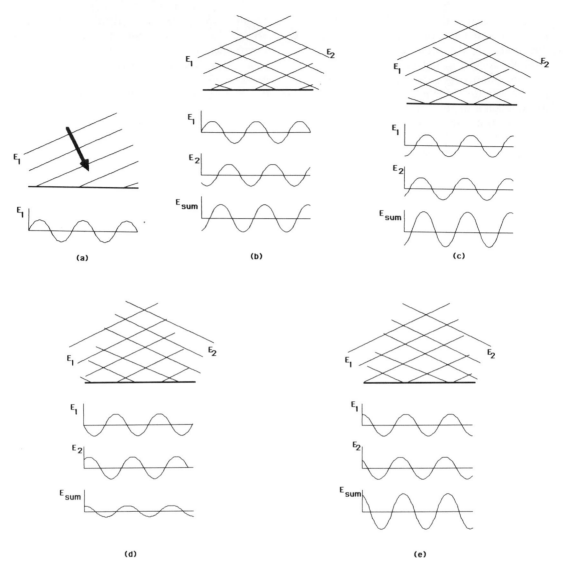

FIG. 1. (a) The electric field of a planewave incident on a surface. The electric field due to two planewaves incident on a surface at different angles as the waves are (b) initially, (c) after one-quarter wavelength of propagation, (d) after one-half wavelength of propagation, and (e) after three-quarters wavelength of propagation.

veloped hologram. Through a wave phenomenon known as diffraction, the recorded interference fringes redirect some of the light in the reconstructing beam to form a replica of the second recording beam. This replica, or reconstructed, beam travels away from the hologram with the same variation in phase and amplitude that the original beam had. Thus, for the eye, or for common image recording instruments such as a photographic camera or a video camera, the reconstructed wavefront is indistinguishable from the original wavefront and therefore pos-

sesses all the visual properties of the original wavefront, including the three-dimensional aspects of the scene. [See OPTICAL DIFFRACTION.]

The interference fringe spacing that is recorded depends on the angle between the recorded beams. During reconstruction, this fringe spacing is translated, through diffraction, back into the proper angle between the reconstructing and the reconstructed beams. During reconstruction, locations on the hologram with high fringe contrast divert more optical energy into the reconstructed beam than locations with low fringe contrast, thereby reproducing the amplitude distribution of the original beam.

For most holograms, even if the contrast of the recorded fringes is large, not all of the light in the reconstructing beam can be diverted into the reconstructed image. The ratio of the amount of optical power in the reconstructed image to the total optical power in the reconstructing beam is the diffraction efficiency of the hologram. For common holograms recorded on photographic film, the maximum diffraction efficiency is limited to 6.25%. The rest of the light power, 93.75%, passes through the hologram unaffected, or ends up in beams that are called higher diffraction orders. These higher diffraction orders leave the hologram at different angles and are generally not useful. As will be noted in Section III, certain types of holograms, notably thick-phase holograms, are able to eliminate the undiffracted light and the higher diffraction orders; they can produce a diffraction efficiency close to 100%.

Certain assumptions that are made in the explanation given above should now be discussed. First, it is assumed that both beams used for recording are the same optical wavelength, that is, the same color. This is necessary to produce a stationary interference pattern to record. It is also assumed that the wavelength of the reconstructing beam is the same as that of the recording beams. This is often the case, but is not necessary. Using a reconstructing beam of a different wavelength changes the size of the reconstructed image: a shorter wavelength produces a smaller image, and a longer wavelength produces a larger image. Oddly, the dimensions of the image parallel to the plane of the hologram scale linearly with wavelength, but the depth dimension of the image scales proportional to the square of the wavelength, so three-dimensional images reconstructed with a different wavelength will appear distorted. Also, if the change in wavelength is large, distortions will occur in the other dimensions of the image, and the diffraction efficiency will decrease.

Another assumption is that the two recording beams have the same polarization. If beams with different polarizations or unpolarized beams are used, the contrast of the fringes, and therefore the diffraction efficiency of the hologram, is decreased. Beams that are linearly polarized in perpendicular directions cannot interfere and so cannot record a hologram. The polarization of the reconstructing beam usually matters very little in the quality of the reconstruction. An exception to this is when the hologram is recorded as an induced birefringence pattern in a material. Then the polarization of the reconstructing beam should be aligned with the maximum variation in the recording material index of refraction.

It is also possible that the reconstructing beam is not the same as one of the recording beams, either in its phase variation, its amplitude variation, or both. If the structure of the reconstructing beam differs significantly from both of the recording beams, the reconstructed image is usually garbled. One particular case where the image is not garbled is if the reconstructing beam has the same relative phase structure as one of the recording beams, but approaches the hologram at a different angle. In this case, provided the angle is not too large and that the recording behaves as a thin hologram (see Section III for an explanation of thin and thick holograms), the reconstructed image is produced.

Photographic film is assumed to be the recording material used in the example above. Many other materials can be used to record holograms, and these are the subject of Section IV. Photographic film gives a particular type of recording, classified as a thin absorption hologram. The meaning and properties of different classifications of holograms are dealt with in Section III.

III. Classification of Holograms

Many different types of holograms are possible. Holograms are classified according to the material property in which the interference pattern is recorded, the diffraction characteristics of the hologram, the orientation of the recording beams with respect to the hologram, and the optical system configuration used for recording and reconstructing the hologram.

Holograms are recorded by exposing an optically sensitive material to light in the interference pattern produced by optical beams. For example, exposing photographic film to light triggers a chemical reaction that, after development, produces a variation in the optical absorption of the film. Portions of the film exposed to high optical intensity become absorbing, and unexposed portions of the film remain transparent. Other materials also exhibit this characteristic of changing their optical absorption in response to exposure to light. Holograms that result from interference patterns recorded as variations in material absorption are known as amplitude holograms.

There are also materials whose index of refraction changes in response to exposure to light. These materials are usually quite transparent, but the index of refraction of the material increases or decreases slightly where it is exposed to light. Holograms that result from interference patterns recorded as index-of-refraction variations are known as phase holograms. During reconstruction, light encountering regions with a higher index of refraction travels more slowly than light passing through lower-index regions. Thus the phase of the light is modified in relation to the recorded interference pattern.

It is not correct to assume that amplitude holograms can reconstruct wavefronts with only amplitude variations, and phase holograms can reconstruct wavefronts with only phase variations. In Section II it was explained that wavefront direction (i.e., phase) is recorded by interference fringe spacing, and wavefront amplitude is recorded by interference fringe contrast. In reality, both amplitude and phase types of holograms are capable of recording wavefront amplitude and phase information.

Holograms are also classified as being "thin" or "thick." These terms are related to the diffraction characteristics of the hologram. A thin hologram is expected to produce multiple diffraction orders. That is, although only two beams may have been used for recording, a single reconstructing beam will give rise to several reconstructed beams, called diffraction orders. Another property associated with thin holograms is that if the angle at which the reconstructing beam approaches the hologram is changed, the hologram continues to diffract light, with little change in diffraction efficiency. The diffraction orders will rotate in angle as the reconstructing beam is rotated. Thick holograms, on the other hand, produce only a single diffracted beam; a portion of the reconstructing beam may continue through the hologram in its original direction as well. Also, noticeable diffraction efficiency for thick holograms occurs only if the reconstructing beam is incident on the hologram from one of a discrete set of directions, called the Bragg angles. If the beam is not at a Bragg angle, it passes through the hologram and no diffracted beam is produced. The property of thick holograms that diffraction efficiency falls off if the reconstructing beam is not at a Bragg angle is called angular selectivity.

Many thick holograms can be recorded in the same material and reconstructed separately by arranging for their Bragg angles to be different.

The terms thin and thick were originally applied to holograms based solely on the thickness of the recording material. The situation is, in fact, more complicated. Whether a particular hologram displays the characteristics associated with being thick or thin depends not only on the thickness of the recording material, but also on the relative sizes of the optical wavelength and the interference fringe spacing, and on the strength of the change produced in the absorption or refractive index of the material.

The next category of hologram classification has to do with the arrangement of the recording beams (and therefore the reconstructing and reconstructed beams) with respect to the recording material. When two planewave beams produce interference fringes, the fringes form a set of planes in space. The planes lie parallel to the bisector of the angle between the beams, as shown in Fig. 2(a). If the recording material is arranged so that both recording beams approach it from the same side, fringes are generally perpendicular to the material surfaces, as shown in Fig. 2(b), and a transmission-type hologram is formed. During readout of the transmission hologram, the reconstructing and the recon-

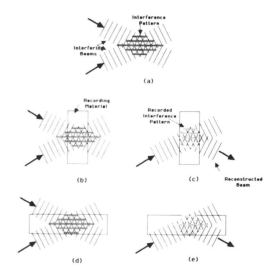

FIG. 2. (a) The interference produced by two planewave beams. (b) The orientation of the recording material for a transmission hologram. (c) The configuration used for reconstruction with a transmission hologram. (d) The orientation of the recording material for a reflection hologram. (e) The configuration used for reconstruction with a reflection hologram.

structed beams lie on opposite sides of the holo-gram, as in Fig. 2(c). Alternatively, the recording material can be arranged so that the recording beams approach it from opposite sides. In this case, the fringes lie parallel to the surfaces of the material, as shown in Fig. 2(d), and a reflection-type hologram is formed. For a reflection hologram, the reconstructing and the reconstructed beams lie on the same side of the hologram, portrayed in Fig. 2(e)

The three hologram classification criteria dis-cussed so far—phase or absorption, thick or thin, and transmission or reflection—play a role in determining the maximum possible diffraction efficiency of the hologram. Table I summarizes the diffraction efficiencies for holograms pos-sessing various combinations of these character-istics. Because the fringes for reflection holo-grams lie parallel to the surface, there must be an appreciable material thickness to record the fringes; therefore, thin reflection holograms are not possible and are absent from the table. (Of-ten, holograms use reflected light but have fringes perpendicular to the material surface and are properly classified as transmission holo-grams.) Notice that phase holograms are gener-ally more efficient than absorption holograms, and thick phase holograms are able to achieve perfect efficiency; all of the reconstructing beam power is coupled into the reconstructed beam. Keep in mind that the figures in the table repre-sent the absolute highest diffraction efficiencies that can be achieved. In practice, hologram dif-fraction efficiencies are often substantially lower.

The final classification criterion to be dis-cussed has more to do with the configuration of the optical system used for recording than with the hologram itself. For recording a hologram, coherent light is reflected from or transmitted through an object and propagates to the position of the recording material. A second beam of co-herent light produces interference with light from the object, and the interference pattern is recorded in the material. If the object is very close to the recording material or is imaged onto the recording material, then an image-plane ho-logram is formed. If the separation between the object and the recording material is a few times larger than the size of the object or the material, a Fresnel hologram is formed. If the object is very far from the recording material, a Fraunho-fer hologram is recorded. Another possible con-figuration is to place a lens between the object and the recording material such that the distance between the lens and the material is equal to the focal length of the lens. This arrangement pro-duces a Fourier transform hologram, so called because during reconstruction, the Fourier transform of the reconstructed wave must be taken (with a lens) to reproduce the recorded image. These previous configurations use a planewave as the second recording beam. If an expanding spherical wave is used instead, with the source of the spherical wave near the loca-tion of the object, a quasi-Fourier-transform ho-logram, or lensless Fourier transform hologram, is formed. This type of hologram shares many of the properties of the true Fourier transform ho-logram.

Other classes of holograms also exist, such as polarization holograms, rainbow holograms, synthetic holograms, and computer generated holograms. However, these are specialized top-ics, best dealt with separately from the central concepts of holography.

IV. Recording Materials

There are many materials that can be used for recording holographic interference patterns. Some materials record the pattern as a change in their optical absorption; this yields absorption holograms. Other materials record the patterns as changes in their index of refraction or as a relief pattern on their surface; these materials produce phase holograms. The thickness of the recording layer is also important to the charac-teristics of the hologram, as discussed in Sec-tion III.

Practical concerns related to holographic re-cording materials include the recording resolu-tion, material sensitivity as a function of optical wavelength, and the processing steps required to develop the hologram. The spacing of inter-

TABLE I. Maximum Diffraction Efficiencies of Holo-gram Classes

Thickness	Modulation	Configuration	Maximum efficiency (%)
Thin	Absorption	Transmission	6.25
Thin	Phase	Transmission	33.90
Thick	Absorption	Transmission	3.70
Thick	Absorption	Reflection	7.20
Thick	Phase	Transmission	100.00
Thick	Phase	Reflection	100.00

ference fringes can be adjusted by changing the angle between the beams: a small angle gives large fringes, and a large angle gives fringes as small as one-half the wavelength of the light used. An ideal recording material would have a resolution of at least 5000 fringes per millimeter.

Some materials are sensitive to all visible wavelengths, and others to only a portion of the spectrum. The wavelength sensitivity of the recording material must be matched to the light source used. Sensitive recording materials are generally desirable, since high sensitivity reduces the recording exposure time and the amount of optical power required of the source. Long exposures are undesirable because of the increased chance that a random disturbance will disrupt the coherence of the recording beams.

Many recording materials require some chemical processing after exposure to develop the holographic pattern. Some materials develop when heat is applied, and a few materials require no processing at all: the hologram is immediately available. Of course, the need for developing complicates the system and introduces a delay between recording the hologram and being able to use it. Important characteristics of the most common hologram recording materials are summarized in Table II.

Silver halide photographic emulsions are the most common recording material used for holograms. They are a mature and commercially available technology. The emulsion may be on a flexible acetate film or, for greater precision, a flat glass plate. Photographic emulsions have a very high sensitivity and respond to a broad spectral range. The ordinary development procedure for photographic emulsions causes them to become absorptive at the locations that have been exposed to light. Thus, an absorption hologram is produced. Alternate developing proce-

dures employing bleaches leave the emulsion transparent but modulate the index of refraction or the surface relief of the emulsion. These processes lead to phase holograms. Many photographic emulsions are thick enough to produce holograms with some characteristics of a thick grating.

Dichromated gelatin is one of the most popular materials for recording thick phase holograms. Exposure to light causes the gelatin molecules to crosslink. The gelatin is then washed, followed by dehydration in alcohol. Dehydration causes the gelatin to shrink, causing cracks and tears to occur in the regions of the gelatin that are not crosslinked. The cracks and tears produce a phase change in light passing through those regions. Phase holograms recorded in dichromated gelatin are capable of achieving diffraction efficiencies of 90% or better with very little optical noise. The primary limitations of dichromated gelatin are its very low sensitivity and the undesirable effects of shrinkage during development.

Photoresists are commonly used in lithography for fabrication of integrated circuits, but can be utilized for holography too. Negative and positive photoresists are available. During developing, negative photoresists dissolve away in locations that have not been exposed to light, and positive photoresists dissolve where there has been exposure. In either case, a surface relief recording of the interference pattern is produced. This surface relief pattern can be used as a phase hologram either by passing light through it or by coating its surface with metal and reflecting light off it. The photoresist can also be electroplated with nickel, which is then used as the master for embossing plastic softened by heat. The embossing process can be done rapidly and inexpensively, and so is useful for mass produc-

TABLE II.   Holographic Recording Materials

Material	Modulation	Sensitivity (J/cm^2)	Resolution (line pairs/mm)	Thickness (μm)
Photographic emulsion	Absorption or phase	$\sim 5 \times 10^{-5}$	~ 5000	< 17
Dichromated gelatin	Phase	$\sim 7 \times 10^{-2}$	> 3000	12
Photoresist	Phase	$\sim 1 \times 10^{-2}$	~ 1000	> 1
Photopolymer	Phase	$\sim 1 \times 10^{-2}$	3000	3–150
Photoplastic	Phase	$\sim 5 \times 10^{-5}$	> 4100	1–3
Photochromic	Absorption	~ 2	> 2000	100–1000
Photorefractive	Phase	~ 3	> 1000	5000

ing holograms for use in magazines and other large-quantity applications.

Photopolymers behave in a fashion similar to photoresists, but instead of dissolving away during development, exposure of a photopolymer to light induces a chemical reaction in the material that changes its index of refraction or modulates its surface relief. Some photopolymers require no development processing, and others must be heated or exposed to ultraviolet light.

Photoplastics are noted for their ability to record and erase different holographic patterns through many cycles. The photoplastics are actually multilayer structures. A glass substrate plate is coated with a conductive metal film. On top of this is deposited a photoconductor. The final layer is a thermoplastic material. For recording, a uniform static electric charge is applied to the surface of the thermoplastic. The voltage drop due to the charge is divided between the photoconductor and the thermoplastic. The structure is then exposed to the holographic interference pattern. The voltage across the illuminated portions of the photoconductor is discharged. Charge is then applied a second time to the surface of the device. This time excess charge accumulates in the regions of lowered voltage. The device is now heated until the thermoplastic softens. The electrostatic attraction between the charge distribution on the surface of the thermoplastic and the conductive metal film deforms the plastic surface into a surface relief phase hologram. Cooling the plastic then fixes it in this pattern. The hologram may be erased by heating the plastic to a higher temperature so that it becomes conductive and discharges its surface.

Photochromics are materials that change their color when exposed to light. For example, the material may change from transparent to absorbing for a certain wavelength. This effect can be used to record absorption holograms. Furthermore, the recording process can be reversed by heating or exposure to a different wavelength. This allows patterns to be recorded and erased. These materials, however, have very low sensitivity.

Photorefractive materials alter their refractive index in response to light. These materials can be used to record thick phase holograms with very high diffraction efficiency. This recording process can also be reversed, either by uniform exposure to light or by heating. These materials, too, have rather low sensitivity, but research is continuing to produce improvements.

V. Applications

Holography is best known for its ability to reproduce three-dimensional images, but it has many other applications as well. Holographic nondestructive testing is the largest commercial application of holography. Holography can also be used for storage of digital data and images, precise interferometric measurements, pattern recognition, image processing, and holographic optical elements. These applications are treated in detail in this section.

An image reconstructed from a hologram possesses all the three-dimensional characteristics of the original scene. The hologram can be considered a window through which the scene is viewed. As described in Section II, a hologram records information about the intensity and direction of the light that forms it. These two quantities (along with color) are all that the eye uses to perceive a scene. The light in the image reconstructed by the hologram has the same intensity and direction properties as the light from the original scene, so the eye sees an image that is nearly indistinguishable from the original. There are two important aspects in which the holographic reconstruction of a scene differs from the original: color and speckle. Most holograms are recorded using a single wavelength of light. The image is reconstructed with this same wavelength, so all objects in the scene have the same color. Also, because of the coherence properties of laser light, holographic images do not appear smooth: they are grainy, consisting of many closely spaced random spots of light called speckles. Attempts to produce full-color holograms and eliminate speckle will be described in this section.

The simplest method of recording an image hologram is shown in Fig. 3(a) Light reflecting off the object forms one recording beam, and light incident on the film directly from the laser is the other beam needed to produce the interference pattern. The exposed film is developed and then placed back in its original position in the system. The object that has been recorded is removed from the system, and the laser is turned on. Light falling on the developed hologram reconstructs a virtual image of the object that can be viewed by looking through the hologram toward the original position of the object, as shown in Fig. 3(b).

Many variations on the arrangement described above are possible. Often, the portion of the beam directed toward the object is split into

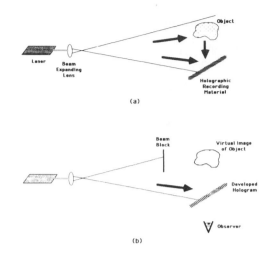

FIG. 3. A simple optical system for (a) recording and (b) viewing a transmission hologram.

several beams that are arranged to fall on the object from different directions so that it is uniformly illuminated. Attention to lighting is especially important when the scene to be recorded consists of several separate objects that are different distances from the film. Another recording arrangement is to have the laser beam pass through the film plane before reflecting from the object, as shown in Fig. 4(a). A reflection hologram is formed with this arrangement, and can be viewed as shown in Fig. 4(b). An important advantage of the reflection hologram is that a laser is not needed to view it: a point source of white light will work quite well. Use of white light to reconstruct a hologram is not only more convenient, but eliminates speckle too. This is possible because the reflection hologram also acts as a color filter, efficiently reflecting light of only the proper wavelength. It is important, however, that source is very small (as compared to its distance from the hologram), otherwise the reconstructed image will be blurred.

A hologram that reconstructs an image containing all the colors of the original scene would, obviously, be a desirable accomplishment. The principle obstacle in achieving this goal is that holographic interference patterns result only when very monochromatic light is used. Recording a hologram with a single color and reconstructing the image with several colors does not work. Reconstructing a hologram with a wavelength other than the one used to record it changes the size and direction of the reconstructed image, so the variously colored images produced with white-light illumination are not aligned. It is possible to record three different

holograms of the same object on one piece of film, using a red, a blue, and a green laser. The three holograms can be reconstructed simultaneously with the appropriate colors, and a reasonable representation of the colors of the original object is produced. The problem, however, is that each reconstructing beam illuminates not only the hologram it is intended for, but the other two holograms as well. This leads to numerous false-color images mingled with the desired image. The false images can be eliminated with a thick hologram recording material by using the angular selectivity of thick holograms.

Another approach to color holography is the stereogram. Sets of three separate holograms (one each for red, green, and blue) are recorded for light coming from the scene at various angles. A projection system using a special screen is used for viewing. Light from different angular perspectives is directed at each eye of the viewer, thus providing the parallax information needed to yield a three dimensional image. This system requires that the viewer is positioned rather exactly with respect to the screen.

Interferometry is a means of making precise measurements by observing the interference of optical wavefronts. Since holograms record phase information about the light from an object, they are useful in making before and after comparisons of the deformation of objects in response to some change in their environment. A typical arrangement for holographic interferometry is shown in Fig. 5. The first step is to record and develop a hologram of the object. The hologram is replaced in the system, and the image of the object is reconstructed from it. The object

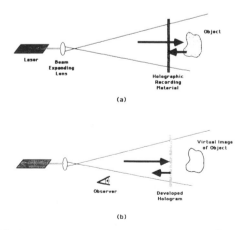

FIG. 4. A simple optical system for (a) recording and (b) viewing a reflection hologram.

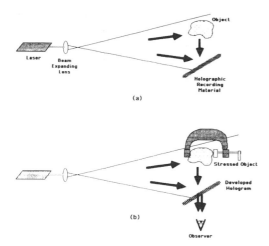

FIG. 5. (a) For real-time holographic interferometry, a hologram of the object is recorded. (b) Then the stressed object is viewed through the developed hologram.

itself is subjected to some change and illuminated as it was to record the hologram. When viewing the object through the hologram, light from the object and from the holographic reconstruction of the object will interfere. If the surface of the object has deformed, bright and dark fringes will appear on the surface of the object. Each fringe corresponds to a displacement in the surface of the object by one-quarter of the optical wavelength. Thus, this is a very sensitive measurement technique.

The same principles apply to transparent objects that undergo an index of refraction change. Changes in index of refraction alter the optical path length for light passing through the object. This leads to interference fringes when the object is viewed through a holographic recording of itself. Such index of refraction changes can occur in air, for example, in response to heating or aerodynamic flows.

A slight modification to the technique described above can provide a permanent record of the interference pattern. A hologram of the original object is recorded but not developed. The object is subjected to some change and a second recording is made on the same film. The film is now developed and contains superimposed images of both the original and the changed object. During reconstruction, wavefronts from the two images will interfere and show the desired fringe pattern.

Another variation of holographic interferometry is useful for visualizing periodic vibrations of objects. Typically, the object is subjected to some excitation that causes its surface to vibrate. A hologram of the vibrating object is made with an exposure time that is longer than the period of the vibration. A time integration of the lightwave phases from locations on the object is recorded on the film. Portions of the object that do not move, vibrational nodes, contribute the same lightwave phase throughout the exposure. This produces constructive interference, and these locations appear bright in the reconstruction. Bright and dark fringes occur elsewhere on the object, with the number of fringes between a particular location and a vibrational node indicating the amplitude of the vibration.

Deformations on the surface of an object in response to applied pressure or heating can often be used to make determinations concerning changes within the volume of the object under the surface. For this reason, holographic interferometry is useful for holographic nondestructive testing; characteristics of the interior of an object can be determined without cutting the object apart. If interference fringes are concentrated on one area of a stressed object, that portion of the object is probably bearing a greater portion of the stress than other locations. Or a pattern of fringes may indicate a void in the interior of the object, or a location where layers in a laminated structure are not adhering.

Holography plays an important role in the field of optical data processing. A common optical system using a holographic filter is shown in Fig. 6. One application of this system is pattern recognition through image correlation. This is useful for detecting the presence of a reference object in a larger scene. The first step is to record a Fourier hologram of the object that is to be detected. A transparency of the object is placed in the input plane and is holographically recorded onto photographic film in the filter

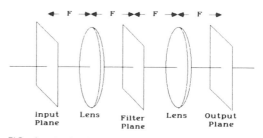

FIG. 6. A simple, yet versatile, optical system for using holographic filters for image processing.

plane; a reference beam is also incident on the film. After exposure, the film is removed from the system, developed, and then replaced in the filter plane. The scene believed to contain occurrences of the reference object is placed in the input plane and illuminated. A portion of the light distribution in the output plane will represent the correlation of the input image with the reference image. This means that wherever the reference image occurs in the input scene, a bright spot of light will be present at the corresponding position on the output plane. As a specific example, the reference object might be a printed word. A hologram of the word would be placed in the filter plane. Input to the system could be pages of printed text. For each occurrence of the selected word within the text, a bright spot would be present at the output plane. In the simple form described, the system can detect occurrences of the reference object that are very close to the same size and rotational orientation as the object used to record the hologram. However, research is being conducted to produce recognition results without regard to scaling or rotation of the object, and has already met with considerable success.

Other image-processing operations can be performed using the same optical system. The hologram is recorded to represent a frequency-domain filter function rather than a reference object. For example, blurring of an image produced by an unfocused optical system can be modeled as a filtering of the spatial frequencies in the image. Much of the detail can be restored to a blurred image by placing a hologram representing the inverse of the blurring function in the filter plane and a transparency of the blurred image in the input plane. Holographic filters can also be used to produce other useful image-processing operations, such as edge enhancement.

Holography is also attractive for data storage. Because holograms record information in a distributed fashion, with no one point on the hologram corresponding to a particular part of the image, data recorded in holographic form are resistant to errors caused by defects in the recording material. Returning to the analogy of the hologram as a window through which to view an image, if part of the hologram is obscured or destroyed, it is still possible to recover all the data simply by looking through the remaining usable portion of the window. Obviously, if part of an ordinary photograph containing data is destroyed, the data on that portion of the photograph is irrevocably lost. The data stored in a holographic system can be pages of text, images, or digitally encoded computer data. If thick recording materials are used, many pages of data can be stored in one piece of material by utilizing the angular selectivity of thick holograms.

Holograms can be recorded to provide the same functions as refractive optical elements such as lenses and prisms. For example, if a hologram is recorded as the interference of a planewave and a converging spherical wave, when the developed hologram is illuminated with a planewave it will produce a converging spherical wave, just as a positive lens would. Holographic optical elements (HOEs), as they are called, have two principle disadvantages with respect to the more common refractive elements. HOEs work as designed only for one wavelength of light, and because they usually have a diffraction efficiency less than 100%, not all of the light is redirected as desired. However, HOEs also have advantages over refractive elements in certain applications. First, the optical function of a HOE is not linked to its physical shape. A HOE may be placed on a curved or angled surface. For example, HOEs are used on the surface of the visor on pilots' helmets to serve as a projection lens for instrumentation displays. Also, HOEs can be created that provide the function of refractive elements that would be very difficult to fabricate, such as an off-axis segment of a lens. Several HOEs can be recorded on the same piece of material to give the function of multiple refractive elements located at the same spatial position. Another popular configuration is to record a HOE on the surface of an existing lens. The lens takes care of most of the refraction required by an optical system, and the HOE provides small additional corrections to reduce aberrations. A second advantage of HOEs is their small physical volume and weight. A HOE recorded on film can provide the same function as a thick, heavy piece of glass. In particular, lenses with a large aperture and short focal length can be quite massive when fabricated in glass, while those characteristics are readily produced using a HOE. Finally, there are applications in which the limited range of working wavelengths of HOEs is desirable. Reflective holograms are often used in these applications. Information presented in the intended color can be reflectively imaged toward a viewer while the scene behind the hologram, containing mainly other colors, is also visible through the hologram without distortion.

The applications just cited are only a sampling of the uses of holography. It is a field that is the subject of ongoing research and that enjoys the continuing discovery of new forms and applications.

VI. History of Holography

The fundamental principle of holography, recording the phase of a wavefront as an intensity pattern by using interference, was devised by Dennis Gabor in 1948 to solve a problem with aberrations in electron microscopy. He also coined the word hologram from Greek roots meaning whole record. The results of Gabor's work of translating electron diffraction patterns into optical diffraction patterns and then removing aberrations from the resulting image were less than satisfactory. Other researchers following his lead were not significantly more successful, so this first application of holography soon faded from the scene.

Other researchers experimented with optical holography for its own sake during the early and mid-1950s, but results were generally disappointing. At that time, holography suffered from two important disadvantages. First, no truly interesting application had been found for it. The rather poor-quality images it produced were laboratory curiosities. Second, coherent light to record and view holograms was not readily available. The laser was not available until the early 1960s. Coherent light had to be produced through spectral and spatial filtering of incoherent light.

In 1955 an interesting, and eventually very successful, application for holography was uncovered. Leith and Upatnieks, working at the University of Michigan's Radar Laboratory, discovered that holography could be used to reconstruct images obtained from radar signals. Their analysis of this technique led them to experiment with holography in 1960, and by 1962 they had introduced an important improvement to the field. Gabor's holograms had used a single beam to produce holograms, recording the interference of light coming only from various locations on the object. This led to reconstructed images that were seriously degraded by the presence of higher diffraction orders and undiffracted light. Leith and Upatnieks introduced a second beam for recording. The presence of this second beam separated the reconstructed image from other undesired light, thus significantly improving the image quality.

Also in the early 1960s, Denisyuk was producing thick reflection holograms in photographic emulsions. These holograms have the advantage that they can be viewed with a point source of white (incoherent) light, since the reflection hologram acts also as a color filter.

Advances in practical applications of holography began to occur in the mid 1960s. In 1963, Vander Lugt introduced the form of holographic matched filter that is still used today for pattern recognition. Powell and Stetson discovered holographic interferometry in 1965. Other groups working independently introduced other useful forms of this important technique.

The first form of thin transmission hologram that can be viewed in white light, the rainbow hologram, was developed by Benton in 1969. Another form of hologram that can be viewed in white light, the multiplex hologram, was produced by Cross in 1977.

BIBLIOGRAPHY

Caulfield, H. J., (ed.) (1979). "Handbook of Optical Holography." Academic Press, New York.
Goodman, J. W. (1968). "Introduction to Fourier Optics." McGraw-Hill, New York.

LASER COOLING AND TRAPPING OF ATOMS

John E. Bjorkholm   *AT&T Bell Laboratories*

I. Introduction
II. Radiation Pressure Forces
III. The Single-Beam, Dipole-Force Optical Trap
IV. Laser Cooling of Atoms
V. Optical Trapping of Atoms
VI. Conclusion

GLOSSARY

Dipole force: Force exerted on an atom due to the stimulated scattering of light by the atom. Also called the stimulated force or the gradient force.

Laser cooling: Using laser radiation pressure forces to extract kinetic energy from atomic motion.

Laser radiation pressure: Radiation pressure exerted by a laser beam. Because laser beams are very bright, laser radiation pressure forces are generally much larger than the radiation pressure forces exerted by beams of incoherent light.

Optical molasses: Three-dimensional configuration of laser beams used to cool atoms to very low temperatures (on the order of 10^{-4} K).

Optical trapping: Using laser radiation pressure forces to confine an atom, or a group of atoms, within a small region of space.

Quantum heating: Heating of atomic motion caused by the fluctuations of the radiation pressure forces around their average values.

Radiation pressure: Forces exerted on a body due to the generalized scattering of light by that body.

Resonance-radiation pressure: Radiation pressure exerted on an atom by light tuned close to the frequency of one of the atom's resonance transitions.

Spontaneous force: Force exerted on an atom due to the spontaneous scattering of light by the atom. Also called the scattering force.

Laser light can exert significant forces on a free atom when the light frequency is tuned to be close to, or equal to, the frequency of one of the atom's resonance-absorption lines. These resonance-radiation pressure forces have found unique application with atomic beams. Most noteworthy, gaseous collections of atoms in vacuum have been cooled to ultralow temperatures (as low as 10^{-4} K); also, such ultracold atoms have been optically trapped within a small volume of space for appreciable periods of time (many seconds). When viewed in hindsight, 1987 may well be seen as the year when the study of laser radiation pressure on atoms came of age. In previous years efforts were mainly devoted to understanding the forces in detail and to learning how to use them. In the years ahead most of the effort will be focused on using these techniques to help in carrying out measurements of a more fundamental nature. The purpose of this article is to discuss the basic forces exerted on atoms by resonance-radiation pressure and to describe some of the new optical techniques that can be used to manipulate atoms in ways not heretofore possible.

I. Introduction

It is well known that a beam of light carries momentum and that it will exert forces on objects that it illuminates. These forces arise out of the generalized scattering of the light by the object. Before the invention of the laser, however, applications of radiation pressure were virtually nonexistent. This was because the light emitted by conventional incoherent light sources is neither intense enough nor spectrally narrow enough to cause large effects. Nonetheless, in 1933 O. Frisch was able to demonstrate the transverse deflection of some of the atoms in an atomic beam of sodium caused by the light from a resonance lamp. The deflection observed was very small, about 3×10^{-5} rad, and was caused by the absorption and re-emission of a single photon by each deflected atom! The invention of the laser, with its intense and highly directional light beams, dramatically changed the situation. Now it was possible for an atom or other body to interact with a large number of photons in a short period of time. This was made clear in 1970 when A. Ashkin used a laser beam to accelerate, trap, and manipulate micrometer-sized, transparent dielectric spheres suspended in water. He further pointed out that significant radiation pressure forces similarly could be exerted on an atom if the frequency of the light was tuned near the frequency of one of the atomic resonance transitions. This was an attractive idea since neutral atoms are not easily manipulated using conventional techniques. Ashkin's seminal paper started worldwide thinking about laser radiation pressure on atoms. Experiments demonstrating these forces followed less rapidly. As laser radiation pressure became better understood throughout the 1970s, new applications became apparent. In 1975 it was realized that light pressure could be used to cool atoms to very low temperatures and in 1978 Ashkin proposed a particularly simple optical trap for atoms. It was not until the mid-1980s that some of these interesting possibilities were actually demonstrated experimentally. The delay between initial conception and eventual realization was caused by the need to develop the complex techniques and equipment required for the experiments. The first demonstrations of the basic forces and of the optical cooling and trapping of atoms utilized sodium atoms and precisely tuneable, cw dye lasers operating near the 589-nm resonance line of sodium. The lasers and the associated apparatus were quite complex and expensive. Recent advances have led to great simplification. In particular, since 1986 relatively simple and inexpensive cw GaAs diode lasers have been used to cool and trap cesium atoms, using the cesium resonance line at 852 nm.

II. Radiation Pressure Forces

The forces of laser radiation pressure are of two types. The first is usually referred to as the spontaneous force, but is sometimes called the scattering force. The second is variously referred to as the dipole force, the stimulated force, or the gradient force. A complete description of these forces is complex since it must account for the quantized nature of light. That is, an atom scatters only one photon at a time and this fact leads to statistical fluctuations of the forces in direction and in time. These quantum fluctuations tend to heat the atomic motion and are a limiting factor in many applications of radiation pressure. A complete description of these fluctuations is beyond the scope of this article, but some of the consequences of these fluctuations will be discussed.

A. SPONTANEOUS FORCE

The spontaneous force of radiation pressure arises because of the spontaneous scattering of nearly resonant light by the atom. It is most easily understood using the photon picture of light. Consider Fig. 1 which shows an atom illuminated by a traveling-wave light beam of frequency $\nu = c/\lambda$, where c is the speed of light and λ is the light wavelength. Because the light is tuned to be nearly resonant with the atom, the atom occasionally absorbs a photon from the beam and makes the transition to its excited state. In absorbing the photon the atom picks up the photon momentum h/λ. After a short time the atom decays back to its ground state and, in the process, re-emits a photon of frequency ν in some random direction, as shown by the out-going wavey lines in the figure. When averaged over many scattering events, the out-going photons carry away no momentum since the scattering distribution is symmetric. Thus F_s, the average force exerted on the atom due to spontaneous scattering, is in the direction of the light propagation. Its magnitude is the rate at which momentum is absorbed from the incoming photons and is given by

$$F_s = (h/\lambda)(f/\tau)$$

where f is the probability that the atom is in its excited state and τ is the excited-state lifetime. To proceed further, we make the simplifying assumption that the atom can be described by the two-level model. This model assumes that the atom has only two energy levels, those of the ground and excited states. While this model is usually overly simplistic, its use makes a discussion of the basic physics much

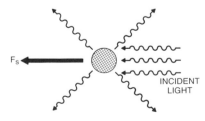

FIG. 1.   Schematic diagram showing an atom being illuminated by a beam of nearly resonant light. The out-going wavey lines indicate photons spontaneously scattered by the atom and the vector F_S denotes the resulting average spontaneous force exerted on the atom.

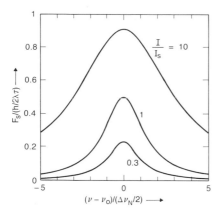

FIG. 2.   The normalized spontaneous force as a function of the normalized light frequency for three values of I/I_S.

easier. The model is widely used, but rarely can it be applied to actual situations without at least some modification. For the two-level atom,

$$f = p/2(1 + p)$$

where p is the saturation parameter and is given by

$$p = (I/I_s)/(1 + q^2)$$

In this equation, I is the light intensity, I_s the saturation intensity for the transition, and q the normalized detuning given by $q = 2(\nu - \nu_0)/\Delta\nu_N$ where ν_0 is the resonant frequency of the transition and $\Delta\nu_N = 1/2\pi\tau$ its natural linewidth (FWHM). Figure 2 shows the dependence of F_S on detuning of the light from resonance for several values of I/I_s. The frequency dependence of the force reflects the atomic absorption line shape. For $p << 1$ the line shape is Lorentzian. The natural linewidth is usually quite narrow, $\Delta\nu_N = 10$ MHz for the sodium atom and 5.2 MHz for cesium; this is why precisely tuneable lasers are required to exert significant spontaneous forces on atoms. When $p >> 1$ the line exhibits broadening and the maximum force saturates. The maximum force is small but nonetheless significant; for sodium and cesium, the corresponding accelerations amount to roughly 10^5 and 6×10^4 times the acceleration of gravity, respectively. [See LASERS.]

A number of experiments have demonstrated or have used the spontaneous force on neutral atoms of sodium and cesium, which are alkali metals. Alkali atoms are particularly useful since they have a single electron in their outermost shell and a correspondingly simple atomic energy level structure. Atoms with more than one outermost electron have more complicated level structure and this leads to experimental difficulties. Nonetheless, spontaneous forces

have been exerted on excited neon atoms in which a metastable excited level functioned as an effective ground state. Spontaneous forces have also been utilized on singly ionized atoms of magnesium and barium contained in ion traps. These atoms have two electrons in their outermost shell, so the singly ionized species have a reasonably simple "alkali-metal-like" level structure.

B. FLUCTUATIONS OF THE SPONTANEOUS FORCE

The force F_s is the net force averaged over many spontaneous scattering events. Because an atom scatters only one photon at a time, and because each photon is scattered into a random direction, there are fluctuations of the instantaneous force around the average. These fluctuations cause the atomic motion to contain a significant random component, which grows with time. This heating of the atomic motion is described by a so-called momentum diffusion coefficient, D_p. The rate at which W, the kinetic energy associated with the random motion, grows is given by $dW/dt = D_p/m$, where m is the atomic mass. For a uniform, traveling, plane wave and in the limit $p << 1$, this expression becomes

$$dW/dt = (m/2)(h/m\lambda)^2(p/\tau)$$

The quantity $h/m\lambda$ is the speed of atomic recoil due to the absorption of a single photon; for the sodium atom it is about 3 cm/sec and for cesium it is 0.35 cm/sec. For situations in which p is large or when the light field contains intensity gradients the expression for D_p becomes much more complex. Counteracting quantum heating is crucial for the cooling and trapping of atoms.

C. DIPOLE FORCE

The dipole force of resonance-radiation pressure is most easily understood using the wave picture of light. In this picture it is simply the force exerted on an induced dipole situated in an electric field gradient. It can also be viewed as arising from the stimulated scattering of light by the atom. The average dipole force can be written as

$$\overline{F}_d = (4\pi/c)\alpha \, \overline{\nabla}I$$

where α is the atomic polarizability and I is the light intensity. For the idealized two-level atom,

$$\alpha = -\frac{1}{2}\frac{(\lambda/2\pi)^3 q}{(1 + q^2)(1 + p)}$$

Several characteristics of this force should be stressed. First, it exists only when there is a gradient of the light intensity. Second, the dipole force has no

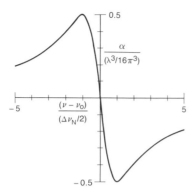

FIG. 3. The normalized atomic polarizability as a function of the normalized light frequency for $I/I_S \ll 1$.

upper limit; importantly, it can be much larger than the spontaneous force. Finally, the frequency dependence of this force is dispersive in character, as shown in Fig. 3. The force is zero for $\nu = \nu_0$; for $\nu < \nu_0$, the direction of the force is such as to pull the atom into the high-intensity regions of the light beam; for $\nu > \nu_0$, the atom is pushed away from the intense regions. Notice also that the force can be large even for $q \gg 1$.

It is often convenient to think of the dipole force as being derivable from a conservative optical potential $U(\bar{r})$, which is given by

$$U(\bar{r}) = (h\,\Delta\nu_N/4)q\,\ln[1 + p(\bar{r})]$$

This potential energy is the same as the shift in energy of the atomic ground state caused by the optical Stark effect. Thus, whenever dipole forces are exerted on an atom, there will also be optical Stark shifts of the atomic energy levels. Because these level shifts can often be large compared with $\Delta\nu_N$, it is usually difficult to effectively apply the spontaneous force to an atom that is simultaneously subjected to a dipole force.

D. Fluctuations of the Dipole Force

Due to the quantum nature of light there are also random fluctuations of the dipole force about its average. These fluctuations are much more difficult to describe and understand than are those of the spontaneous force. We will not consider them here, but it must be realized that the quantum heating caused by these fluctuations can be very large, often very much greater than that due to spontaneous fluctuations. The dipole force and its fluctuations are proportional to the gradient of the light intensity. Thus dipole force quantum heating can be exceptionally large in light fields that have large intensity gradients, as in a standing-wave field.

E. Early Demonstrations of the Forces

Early demonstrations of the spontaneous force were made by illuminating a sodium atomic beam at normal incidence with the light from a cw dye laser tuned onto the atomic resonance. The spontaneous force caused substantial deflection of the atoms. Deflection angles as large as 5×10^{-3} rad were observed, corresponding to the scattering of about 200 photons and an acquired transverse velocity of 600 cm/sec. The maximum deflections obtained were limited by the Doppler shifts associated with the transverse speed. That is, as the atoms acquired transverse speed, they were Doppler shifted out of resonance with the light and the force was greatly diminished. The transverse speed of 600 cm/sec corresponds to a Doppler shift of 10 MHz, the full natural linewidth of sodium.

The first demonstration of the dipole force was made around 1980 by superimposing a copropagating cw dye laser beam on top of an atomic beam of sodium atoms. The laser beam had a Gaussian intensity profile (TEM$_{00}$ mode) and was tuned several gigahertz away from the atomic resonance. Because of the intensity gradients and the cylindrical symmetry of the illumination, transverse dipole forces were exerted on the atoms. For tunings below resonance the forces were such as to pull the atoms to the axis of the laser beam; in other words, the light exerted focusing, confining forces on the atoms. For tunings above resonance the light forces were opposite and brought about defocusing effects on the atoms. The dramatic changes in the atomic beam profile caused by these forces are shown in Fig. 4. Because the light was tuned far from resonance, the effects caused by the average spontaneous force were small. Significantly, however, it was demonstrated that the size of the spot to which the atomic beam could be focused was determined by the transverse heating of the atomic motion caused by the fluctuations of the spontaneous force. While not yet demonstrated, it should be possible to focus atomic beams to very small spot sizes by using a TEM$_{01}$, or donut-mode, laser beam tuned above the atomic resonance. In this case the atoms would tend to be concentrated on the laser beam axis where the light intensity is lowest and where the spontaneous heating is minimized.

In a generalized sense these experiments demonstrated that laser beams can be used to manipulate atomic motion in useful ways. It is expected that the use of lasers to modify and control atomic motion

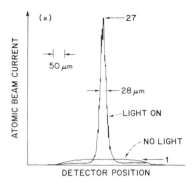

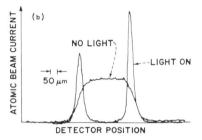

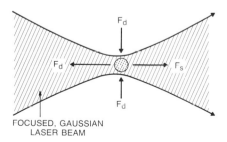

FIG. 4. Focusing and defocusing of an atomic beam by a superimposed, Gaussian laser beam. The figures show the atomic beam current density as a function of position, with the light on and off. In (a) the light frequency is less than the atomic resonance frequency and the atomic beam is focused by the light. In (b) the light is tuned above the atomic resonance and defocusing takes place.

will become a useful technique in atomic beams work.

III. The Single-Beam, Dipole-Force Optical Trap

The first optical trap to be demonstrated is deceptively simple to describe. As shown in Fig. 5, this trap is formed by a sharply focused Gaussian-mode

FIG. 5. Schematic diagram of the single-beam, dipole-force optical trap. There is a point of stable equilibrium for the atom just beyond the focus of the light beam. The dipole and spontaneous forces exerted on the atom are shown.

laser beam tuned far below the atomic resonance ($q << -1$). The intensity of a Gaussian (or TEM$_{00}$-mode) laser beam propagating along the z axis (the longitudinal direction) and focused at $z = 0$ is given by

$$I(r, z) = (2P/\pi w^2) \exp(-2r^2/w^2)$$

where $w(z) = w_0[1 + (\lambda z/\pi w_0^2)^2]^{1/2}$, w_0 is the focal spot size at $z = 0$, and P is the power in the laser beam. As shown in Fig. 5, there is a point of stable equilibrium for the atom just beyond the focus. Transverse confinement of the atom is provided by the transverse dipole forces. Longitudinal confinement is brought about by balancing the longitudinal spontaneous force with the longitudinal dipole force that exists because of the strong longitudinal gradients of the intensity (strong focusing). It is useful to consider the depth of the trap for parameters comparable to those used in the first demonstration of optical trapping. In those experiments, which trapped the sodium atom, the laser power P was 200 mW, the focal spot size w_0 was 10 μm, and the longitudinal trap depth was maximized by detuning the laser -150 GHz from the resonance (corresponding to $q = -3 \times 10^4$). For these parameters, and approximating the sodium atom with the two-level model, the transverse well depth is found to be equivalent to an atomic temperature of 25 mK and the longitudinal trap depth equivalent to 15 mK. For reasons to be explained later, the trap depths for the actual experiment were only about 40% of the above values.

From this example it now can be readily understood why it took a number of years to develop the techniques required to demonstrate an optical trap. The problems that needed to be confronted and solved were as follows. First, since the trap is very shallow, a collection of ultracold atoms ($T << 1$ mK) is needed to load it. No source of such cold atoms existed prior to 1985. Second, the trap volume is very small, on the order of 10^{-7} cm^3. Thus, to efficiently load the trap, the collection of cold atoms must also be dense. Finally, even if cold atoms could be loaded into the trap, the quantum fluctuations of the forces would rapidly heat them. Without an effective means for counteracting this heating, the initially cold atoms would "boil" out of the trap about 10 msec after being loaded. Thus, before an optical trap could be demonstrated it was necessary to develop techniques for cooling atoms to ultralow temperatures and for keeping them cold in the presence of quantum heating. As will shortly be described, the solution to these problems is a technique called "optical molasses," which was first demonstrated in 1985.

IV. Laser Cooling of Atoms

In this section we will discuss the techniques that have been used to cool atoms to temperatures low enough for placing them in optical traps.

A. ATOMIC BEAM SLOWING

The seemingly most straightforward way to slow, or cool, the atoms in an atomic beam is to let them propagate against a light beam tuned to the atomic resonance. The deceleration corresponding to the maximum spontaneous force is $h/2m\lambda\tau$; for sodium and cesium, this amounts to -9.15×10^7 cm/sec^2 and -5.7×10^6 cm/sec^2, respectively. Consider a typical thermal atomic beam of sodium atoms, for which the longitudinal velocity distribution peaks at roughly 10^5 cm/sec. In principle, atoms moving at this speed could be brought to rest in about 1 msec, during which time the atoms would travel over a distance of about 46 cm while scattering about 3×10^4 photons. Unfortunately, such efficient deceleration is not straightforward to achieve because of the Doppler shifts of the light frequency that occur as the atom slows down. For instance, a velocity change of only 10^3 cm/sec causes a Doppler shift of about 17 MHz, which is larger than the 10-MHz linewidth of the sodium absorption line. In other words, unless something is done, the atoms quickly shift out of resonance with the light and the deceleration is greatly reduced.

Several experimental techniques have been devised to counteract these Doppler shifts and to keep the atoms in resonance with the light as they slow down. The first technique uses spatially dependent Zeeman shifts of the atomic energy levels to compensate for the changing Doppler shifts. In this method, the atomic beam is directed down the bore of a solenoid having a tapered longitudinal magnetic field. The field varies in such a way that the Zeeman shifts of the atomic energy levels compensate for the changing Doppler shifts as the atoms slow down. In the second technique the frequency of the laser is directly "chirped," or swept in frequency, at a rate appropriate for keeping the atoms in resonance with the light. Both techniques work well and both have been used to bring the mean longitudinal velocity of some of the atoms in an atomic beam to zero.

These "stopped" atoms, however, are not cold enough for use in optical trap experiments because of the quantum heating that occurs as they are slowed down. Considering sodium once again, an atom starting with a longitudinal velocity of 10^5 cm/sec scatters 3×10^4 photons in being stopped. Because of the random scattering of the emitted photons, there is a spread in the velocity distribution along each axis given approximately by $(h/m\lambda)\sqrt{N/3}$, where N is the number of photons scattered in bringing the atom to rest. For sodium this amounts to roughly 300 cm/sec; the corresponding atomic temperature is roughly 75 mK. In order to achieve optical trapping, even colder atoms are required.

Atoms having temperatures of this magnitude have found other applications. For example, the first trapping of neutral atoms was demonstrated in 1985 using a magnetic trap that had a potential well depth on the order of 5 K. This trap was filled using cold atoms obtained using the Zeeman slowing technique. Techniques similar to the chirping technique have also been used to cool hot atoms confined in ion traps that have well depths on the order of 10 eV (corresponding to a temperature of about 2×10^5 K). Ions in traps have been cooled to temperatures of about 10 mK.

B. OPTICAL MOLASSES

"Optical molasses" is a technique that uses the spontaneous force to rapidly cool already cold atoms to much lower temperatures.

The basic idea behind optical molasses is easily described for one dimension. Consider an atom illuminated from opposite directions by two laser beams of equal intensity traveling along the x axis. The light frequency is tuned to be slightly below the atomic resonance. When the light intensity is low, the net force acting on the atom is simply the sum of the forces of each beam acting alone. When the atom is at rest it sees no net spontaneous force since the forces exerted by the two light beams are equal and opposite. Now let the atom have a velocity component along the x axis. In this situation the frequency of the beam propagating against the atomic motion is Doppler shifted closer to resonance and the force it exerts on the atom increases. The opposite holds for the copropagating light beam. As a result, there is a net average force that opposes the atomic velocity and that is proportional to it. For small velocities it is given by

$$F_x = (8\pi h/\lambda^2)(I/I_s)(q/1 + q^2)v_x = -\beta v_x$$

where I is the intensity of each beam. This expression is valid as long as the Doppler shifts are small compared with $\Delta\nu_N$; for sodium this corresponds to velocities less than 150 cm/sec. This force appears as a viscous damping force to the atom. Damping is maximized for $q = -1$, which corresponds to a detuning a $\Delta\nu_N/2$ below the atomic resonance. An initial velocity exponentially damps to zero with a decay time of m/β. For the sodium atom, $q = -1$, and $I/I_s = 0.1$, this decay time is 16 μsec; thus the damping is seen to be quite strong.

Optical molasses is the extension of the above ideas to three dimensions, using three pairs of oppositely propagating laser beams along the three orthogonal axes. A slowly moving atom situated in the mutual intersection of the six laser beams experiences strong three-dimensional viscous damping and its average velocity is rapidly reduced to zero. Because of the quantum fluctuations of the optical forces, the atoms do not actually come to rest. As will be discussed, the atoms execute a random walk motion. Consequently, atoms find it difficult to escape from the optical molasses and they are confined within it for a long time. In spite of containment times approaching 1 sec that have been achieved with sodium, optical molasses is not an optical trap since there are no restoring forces exerted on the atoms.

The long time required for an atom to escape from optical molasses is easily understood in terms of the quantum fluctuations. An atom in optical molasses with an average velocity of zero experiences a velocity "kick" of $h/m\lambda$ in a random direction each time it scatters a photon. Each velocity kick is damped out and the net result is that the atom executes a random walk motion in three dimensions with a step size of $(\lambda/2\pi)(I/I_s)^{-1}$. After a time t, the mean-square deviation of the atom from its starting point is

$$\langle r^2 \rangle = (\lambda/2\pi)^2 (I/I_s)^{-2} N$$

where N is the number of scattered photons and is roughly given by $(6I/I_s)(t/\tau)$, for $I/I_s \ll 1$. Confinement times of about 0.5 sec are obtained using optical molasses with a diameter of 1 cm. More careful analysis yields somewhat shorter confinement times.

While the average atomic velocity in optical molasses is zero, the mean-square atomic velocity is determined by the interplay of the heating caused by the fluctuations of the optical forces and the cooling caused by the average spontaneous force. Equilibrium is established when the rate of heating equals the average cooling rate. A careful analysis, which includes the dipole heating rate due to the standing-wave nature of the light beams, yields an equilibrium temperature for the collection of atoms given by

$$kT_{eq} = m\langle v^2 \rangle = h\Delta\nu_N/2$$

For the sodium atom $T_{eq} = 240$ μK and for cesium it is 100 μK.

Optical molasses was first demonstrated in 1985 using the sodium atom and cw dye lasers. The experiments were carried out using a pulsed atomic beam. Atoms initially traveling with a speed of 2×10^4 cm/sec were decelerated to 2×10^3 cm/sec using a chirped slowing laser beam. These slow atoms were then allowed to drift into the optical molasses region, which was roughly spherical in shape and about 0.5 cm^3 in volume, where final cooling and retention took place. Atomic densities of 10^6 cm^{-3} and retention times of several tenths of a second were obtained. A direct measurement of the temperature of the atoms in optical molasses was made using a time-of-flight technique. It was found that T_{eq} was 240 μK, in agreement with the quantum heating prediction. During 1987 similar, but experimentally simpler, optical molasses experiments were carried out using cw laser diodes and the cesium atom; the atomic temperature achieved was the predicted limit of 100 μK.

The ideas discussed in this section apply only in the low intensity limit, $I < I_s$. In the high intensity limit, $I \gg I_s$, the behavior of the optical forces becomes much more complicated and, in some situations, can confound our physical understanding. As an example, for low intensities optical molasses provides damping of the atomic motion for tunings below the atomic resonance and heating for tunings above resonance. In the high intensity limit, the situation is reversed! In this case it is the dipole forces, which exist because of the standing waves in the optical molasses configuration, that do the cooling. This situation is usually referred to as "stimulated molasses." The damping rate can far exceed that for the usual optical molasses, but so does the heating. The equilibrium temperature achieved with stimulated molasses is not as low as that achieved with spontaneous molasses.

V. Optical Trapping of Atoms

A. SINGLE-BEAM, DIPOLE-FORCE TRAP

The successful demonstration of optical molasses in 1985 provided experimentalists with the remaining tools needed to accomplish optical trapping. Optical trapping of atoms was first carried out in 1986 using sodium atoms, cw dye lasers, and the single-beam, dipole-force optical trap described in Section III.

The experiment was carried out by first injecting sodium atoms into optical molasses. After a delay of several milliseconds to allow the atoms to reach equilibrium, the optical trap beam was introduced into the interior of the optical molasses, as shown sche-

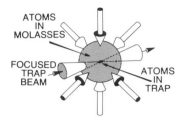

ATOMS
IN
MOLASSES

FOCUSED
TRAP
BEAM

ATOMS
IN
TRAP

FIG. 6.   Schematic diagram of the interaction region used for trapping atoms. The broad arrows represent the collimated laser beams that intersect to form "optical molasses." The shaded sphere represents the fluorescence emitted by the collection of ultracold atoms contained and executing random-walk motion within the optical molasses. The optical trap is formed just beyond the focus of the trap laser beam, which is also shown. The black spot represents the intense fluorescence emitted by the dense collection of atoms confined within the optical trap.

matically in Fig. 6. Trapping was observed as the buildup of a small, but intense, spot of fluorescence situated within the much larger and much weaker cloud of fluorescence from the atoms in the optical molasses (see Fig. 7). The brightness of the small spot indicated that the density of trapped atoms was much higher than the density of atoms in optical mo-

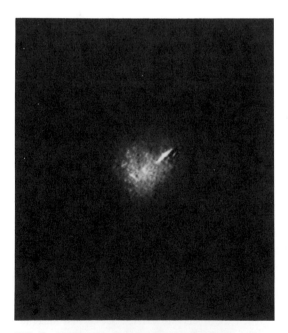

FIG. 7.   Photograph of trapped atoms. The large, diffuse cloud is fluorescence emitted by atoms contained within optical molasses. The small, bright spot is the fluorescence from the much higher density of atoms confined within the optical trap.

lasses. Recall that optical molasses is needed to keep the trapped atoms cool. However, optical molasses is not effective when the trapping light is present because of the large optical Stark shifts associated with the potential well of the trap. Thus, in this experiment it was necessary to alternatively chop on and off the trap and optical molasses beams. Furthermore, it was important that the chopping be done rapidly enough; if too slow, the oscillation of the atoms within the optical potential well becomes unstable and trapping does not occur.

The optical trap was observed to behave as expected from calculations in all respects but one. The one discrepancy was the observed trap lifetime; it was found only to be about 1 sec, about four orders if magnitude less than expected from simple calculations. It was surmised, and later confirmed, that the short lifetime was determined by the imperfect vacuum in the vacuum chamber (about 10^{-9} Torr). Trapped atoms were ejected from the trap by even very weak collisions with the residual gas atoms, which were at a temperature of about 300 K.

A summary of the operating parameters for the original optical trap and the results obtained with it are as follows. The trap beam was focused within optical molasses to a spot size of 10 μm, its power was 200 mW, and it was tuned approximately 150 GHz below the sodium D1 resonance line. Good results were obtained for chopping periods ranging from about 0.4 to 10 μsec. Under best conditions about 10^3 atoms were trapped within a volume of 10^{-9} cm^3 at a density of 10^{12} cm^{-3}. The atomic temperature was inferred to be about 400 μK.

B. SPONTANEOUS FORCE OPTICAL TRAPS

In the early 1980s there was a good deal of speculation about the various forms that optical traps might take. This speculation led to what has come to be called the optical Earnshaw's theorem. Simply put, the theorem states that it is impossible to form a dc optical trap using only forces that are directly proportional to the light intensity. Since the spontaneous force is proportional to I in the low intensity limit, this theorem initially was interpreted as stating that optical traps could not be constructed using only the spontaneous force. However, it is now realized that there are at least two ways to avoid this conclusion. The first is to allow for time dependence of the forces. Thus spontaneous force traps can be constructed in which at least some of the optical beams exhibit time dependence. These traps are sometimes referred to as "optodynamic" traps. Another way around the theorem is to break the assumption that the forces are linearly related to the optical intensity.

Thus saturation of the forces due to high intensities, the optical pumping of real multilevel atoms, and the application of other force fields can all be used to allow dc, spontaneous force traps to be constructed. Similar techniques have been used in ion traps to avoid the consequences of the electrical Earnshaw's theorem.

The first spontaneous force optical trap was demonstrated during 1987 and is referred to as the "magnetic molasses" trap. This trap is constituted by superimposing a simple spherical quadrupole magnetic field on an optical molasses formed using circularly polarized light beams. The magnetic field causes spatially dependent Zeeman shifts of the atomic energy levels. For sodium these Zeeman shifts result in restoring forces that are proportional to the displacement of an atom from the origin (defined by the magnetic field). This trap is much larger (several millimeters diameter) and much deeper (about 500 mK) than the single-beam, dipole-force trap. As a result, it is easier to fill and many more atoms could be trapped. Trap lifetimes were increased by reducing the chamber background pressure. Figure 8 shows an example of the exponential decay of the trap population exhibiting a 1/e-lifetime of 65 sec. Trap lifetimes as long as 100 sec were obtained at a background pressure of 2×10^{-10} Torr. As many as 10^7 sodium atoms were confined in the magnetic molasses trap at densities as high as 10^{11} cm^{-3}.

When the magnetic molasses trap was loaded with a high initial atomic density, the decay with time of the number of trapped atoms was observed to be nonexponential and faster than the exponential decay observed with low initial fills. An example of such an observation is shown in Fig. 9. It was found that the departure from exponential decay was proportional to the square of the density of trapped atoms, indi-

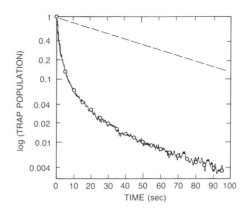

FIG. 9. A logarithmic plot of the nonexponential decay of the atomic population in a magnetic molasses trap observed for a high initial atomic density (about 10^{11} cm^{-3}). The initial decay is 150 times faster than the decay that would be obtained in the absence of density-dependent losses, as shown by the dashed straight line.

cating that the additional losses were caused by collisions between the ultracold trapped atoms. This finding is of interest for several reasons. First, density-dependent losses such as these are detrimental since they limit the maximum atomic densities that can be achieved in an optical atom trap. This will make it difficult to push a great deal further into a new regime of gas physics in which high atomic densities and ultralow temperatures are simultaneously achieved. On the other hand, it is now possible to observe the effects of the relatively infrequent collisions between slow atoms. In work with thermal atomic beams or gas cells such collisions are swamped out by the much more frequent collisions between fast atoms. Collisions between slow atoms are not well understood and are only beginning to be studied. Recent observations of associative ionization between slow atoms in an optical trap have indicated that the cross section for the process is several orders of magnitude larger than for atoms traveling at typical thermal velocities.

VI. Conclusion

The year 1987 is somewhat of a "watershed" year in that the nature of work on laser radiation pressure on atoms is undergoing a subtle change. In the past, efforts were directed towards demonstrating and understanding the basic forces and the ways in which they can be utilized to affect atomic trajectories and to cool and trap atoms. In the future, the emphasis will be placed on using these forces and techniques as experimental tools that will make it possible to

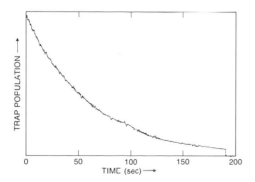

FIG. 8. Time dependence of the atomic population in a "magnetic molasses" optical trap as a function of time. The decay is well described by an exponential curve having a 1/e-lifetime of 65 sec.

carry out measurements of a fundamental nature that had not heretofore been possible. Areas of interest include the study of atomic collisions at slow speeds, collisions between slow atoms and surfaces, the possible observation of collective quantum effects between cold atoms, possible extension of laser radiation pressure to molecules, and precision spectroscopic measurements. Because of these tantalizing prospects, the number of scientists interested in applying laser radiation pressure on atoms is undergoing rapid growth.

BIBLIOGRAPHY

Theoretical discussions:

Cook, R. J. (1979). *Phys. Rev. A* **20,** 224.
Dalibard, J., and Cohen-Tannoudji, C. (1985). *J. Opt. Soc. Am. B* **2,** 1707.
Gordon, J. P., and Ashkin, A. (1980). *Phys. Rev. A* **21,** 1606.

Experimental demonstrations:

Bjorkholm, J. E., Freeman, R. R., Ashkin, A., and Pearson, D. B. (1980). *Opt. Lett.* **5,** 111.

Bjorkholm, J. E., Chu, S., Ashkin, A., and Cable, A. (1987). *In* "Advances in Laser Science—II" (Lapp, M., Stwalley, W. C., and Kenney-Wallace, G. A., eds.), *Am. Inst. Phys. Conf. Proc.* **160,** 319. Am. Inst. Phys. New York.
Chu, S., Hollberg, L. W., Bjorkholm, J. E., Cable, A., and Ashkin, A. (1985) *Phys. Rev. Lett.* **55,** 48.
Chu, S., Bjorkholm, J. E., Ashkin, A., and Cable, A. (1986). *Phys. Rev. Lett.* **57,** 314.
Ertmer, W., Blatt, R., Hall, J. L., and Zhu, M. (1985). *Phys. Rev. Lett.* **54,** 996.
Migdall, A., Prodan, J., Phillips, W. D., Bergeman, T., and Metcalf, H. (1985). *Phys. Rev. Lett.* **54,** 2596.
Phillips, W. D., and Metcalf, H. J. (1982). *Phys. Rev. Lett.* **48,** 596.
Phillips, W. D., and Metcalf, H. J. (1987). *Sci. Am.,* March, p. 50.
Prodan, J., Migdall, A., Phillips, W. D., So, I., Metcalf, H., and Dalibard, J. (1985). *Phys. Rev. Lett.* **54,** 992.
Raab, E. L., Prentiss, M. G., Cable, A., Chu, S., and Pritchard, D. E. (1987). *Phys. Rev. Lett.* **59,** 2631.
Watts, R. N., and Wieman, C. E. (1986). *Opt. Lett.* **11,** 291.
Wineland, D. J., and Itano, W. M. (1987). *Phys. Today,* June, p. 34.

LASERS

William T. Silfvast *AT&T Bell Laboratories*

I. Laser History
II. Laser Gain Media
III. Laser Beam Properties
IV. Laser Linewidth
V. Laser Wavelengths
VI. Types of Lasers
VII. Applications of Lasers

GLOSSARY

Absorption: Extinction of a photon of light when it collides with an atom and excites an internal energy state of that atom.

Emission: Radiation produced by an atomic species when an electron moves from a higher energy level to a lower one.

Frequency: Reciprocal of the time it takes for a lightwave to oscillate through a full cycle.

Gain: Condition that causes a beam of light to be intensified when it passes through a specially prepared medium.

Linewidth: Frequency or wavelength spread over which emission, absorption, and gain occur in the amplifier.

Mode: Single frequency beam of light that follows a unique path as it grows in the amplifier and emerges as a beam.

Photon: Discrete quantum of light of an exact energy or wavelength.

Population inversion: Condition in which more atoms exist in a higher energy state than a lower one, leading to amplification, or gain, at a wavelength determined by electron transitions between those states.

Wavelength: Distance over which light travels during a complete cycle of oscillation.

A laser is a device that amplifies, or increases, the intensity of light, producing a strong, highly directional or parallel beam of light of a specific wavelength. The word "laser" is an acronym for "light amplification by stimulated emission of radiation." Stimulated emission is a natural process, first recognized by Einstein, that occurs when a beam of light passes through a medium and initiates or stimulates atoms of that medium to radiate more light in the same direction and at the same wavelength as that of the original beam. A specific laser device (see Fig. 1) consists of (1) an amplifying, or gain, medium that produces an increase in the intensity of a light beam and (2) an optical resonator or mirror arrangement that provides feedback of the amplified beam into the gain medium thereby producing the beamlike and ultrapure frequency or coherent properties of the laser. The optical cavity, or resonator, which typically consists of two highly reflecting mirrors arranged at opposite ends of an elongated amplifier, allows a strong beam of light to develop due to multiple reflections that produce growth of the beam as it bounces back and forth through the amplifier. A useful beam emerges from a laser either by making one of the laser mirrors partially transmitting or by using a mirror with a small hole in it. Some lasers have such a high gain that the intensity increase is large enough, after only a single pass of the beam through the amplifier, that mirrors are not necessary to achieve a strong beam. Such amplifiers produce a directional beam by having the gain medium arranged in a very elongated shape, causing the beam to grow and emerge from the amplifier in only the elongated direction.

Some of the unique properties of lasers include high beam power for welding or cutting, ultrapure frequency for communications and holography, and an ultraparallel beam for long-distance propagation and extremely tight focusing. Laser wavelengths cover the far infrared to the near infrared, the visible to the ultraviolet, and the vacuum ultraviolet to the soft X-ray region. Laser sizes range from small semiconductor lasers the size of a grain of salt, for use in optical

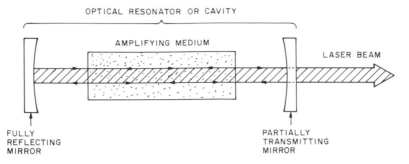

FIG. 1.   Laser components including amplifying medium, optical cavity, and laser beam.

communications, to large solid-state and gas lasers the size of a large building, for use in laser fusion programs.

I. Laser History

Charles Townes was the first person to take advantage of the stimulated emission process as an amplifier by conceiving and constructing the first maser (an acronym for "microwave amplification by stimulated emission of radiation"). The maser produced a pure beam of microwaves that were anticipated to be useful for communications in a similar way to that of a klystron or a traveling-wave tube. The first maser was produced in ammonia vapor and the inversion occurred between two energy levels that produced gain at a wavelength of 1.25 cm. In the maser, the radiation wavelengths are comparable to the size of the device and therefore the means of producing and extracting the beam could not obviously be extrapolated to the optical spectrum in which the wavelengths of light are of the order of 100th the size of a human hair.

In 1958, Townes along with Arthur Schawlow began thinking about extending the maser principle to optical wavelengths. At that time they developed the concept of a laser amplifier and an optical mirror cavity to provide the multiple reflections thought to be necessary for rapid growth of the light signal into an intense visible beam. Townes later (1964) shared the Nobel prize in physics with A. Prokhorov and N. Basov of the Soviet Union for the development of the maser–laser principle.

In 1960, Theodore Maiman of the Hughes Research Laboratories produced the first laser using a ruby crystal as the amplifier and a flash lamp as the energy source. The helical flash lamp surrounded a rod-shaped ruby crystal and the optical cavity was formed by coating the flattened ends of the ruby rod with a highly reflecting material. In operation, an intense red beam emerged from the ends of the rod when the flash lamp was initiated.

Shortly after the ruby laser came the first gas laser, developed in 1961 in a mixture of helium and neon gases by A. Javan, W. Bennett, and D. Herriott of Bell Laboratories. At the same laboratories, L. F. Johnson and K. Nassau first demonstrated the now well-known and high-power neodymium laser. This was followed in 1962 by the first semiconductor laser demonstrated by R. Hall at the General Electric Research Laboratories. In 1963, C. K. N. Patel of Bell Laboratories discovered the infrared carbon dioxide laser which later became one of the most powerful lasers. Later that year A. Bloom and E. Bell of Spectra-Physics discovered the first ion laser, in mercury vapor. This was followed in 1964 by the argon ion laser developed by W. Bridges of Hughes Research Laboratories and in 1966 the blue helium–cadmium metal vapor ion laser discovered by W. T. Silfvast, G. R. Fowles, and B. D. Hopkins at the University of Utah. The first liquid laser in the form of a fluorescent dye was discovered that same year by P. P. Sorokin and J. R. Lankard of the IBM Research Laboratories leading to the development of broadly tunable lasers. The first of the now well-known rare-gas–halide excimer lasers was first observed in xenon fluoride by J. J. Ewing and C. Brau of the Avco–Everett Research Laboratory in 1975. In 1976 J. M. J. Madey and co-workers at Stanford University developed the first free-electron laser amplifier operating at the infrared carbon dioxide laser wavelength. In 1985 the first soft X-ray laser was successfully demonstrated in a highly ionized selenium plasma by D. Matthews and a large number of co-workers at the Lawrence Livermore Laboratories.

II. Laser Gain Media

A. ENERGY LEVELS AND THE EMISSION AND ABSORPTION OF LIGHT

All lasers (except the free-electron laser) result from electron energy changes among discrete energy levels of atomic species, including: (1) individual atoms or ions, (2) small uniquely bonded groups of atoms (molecules), (3) periodically arranged groups of atoms (semiconductors or crystalline solids), or (4) randomly arranged groups of atoms (liquids and amorphous solid structures). All of these species contain a lowest energy level (ground state) in which the electrons reside at low temperatures, and a spectrum of higher lying levels that are occupied when energy is pumped into the species either by heating or irradiating it with light or energetic particles such as fast electrons or fast atomic particles. Light originates from these species when electrons jump or decay from some of these high lying, or excited, energy levels to lower lying energy levels. The energy that is lost by the particular atomic material when the electron decays is given up in the form of a photon, a discrete particle of light. In order to satisfy the law of conservation of energy, the emitted photon must be of the exact energy corresponding to the energy difference between the higher lying level and the lower lying level, and the wavelength or frequency of the emitted photon is associated with that energy difference. Many photons together can form a beam or a wave of light.

Light, emitted when an electron decays, can occur either spontaneously, due to inherent interactions of the atomic structure, or by stimulated emission whereby the electron is forced or driven to radiate by an approaching photon of the appropriate energy or wavelength (see Fig 2). Absorption, the opposite process of stimulated emission, occurs when an atom having an electron in a low lying energy level absorbs and thereby eliminates an approaching photon, using the absorbed energy to boost that electron to a higher lying energy level.

B. POPULATION INVERSIONS

A laser amplifier is produced when conditions are created within the amplifying medium such that there are more atoms having electrons at a higher energy level than at a lower energy level for a specific pair of levels. This condition is known as a population inversion since it is the

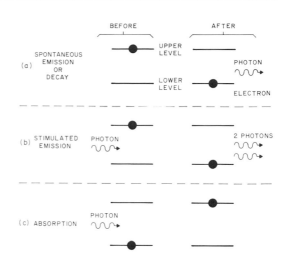

FIG. 2. Electronic transitions between energy levels depicting (a) spontaneous emission, (b) stimulated emission, and (c) absorption.

opposite or inverse of almost all physical situations at or near thermal equilibrium in which there are more atoms with electrons in lower energy levels than at higher levels. Under the conditions of a population inversion, when a beam of light passes through the amplifier more photons will be stimulated than absorbed thereby resulting in a net increase in the number of photons or an amplification of the beam.

C. EXCITATION MECHANISMS IN LASERS (ENERGY SOURCES)

Since all laser emission involves radiation from excited states of atoms, the energy must be fed to those atoms to produce the excited states. This energy is provided in the form of highly energetic electrons (moving at rapid speeds of the order of 10^8 to 10^9 cm/sec), energetic heavier particles such as protons, neutrons, or even other atoms, or electromagnetic radiation (light) in the form of (1) a broad frequency spectrum of emission such as a flash lamp or (2) a narrow frequency spectrum provided by another laser. The most common excitation source is that provided by energetic electrons since they are easily accelerated by applying an electric field or voltage drop to an amplifier. Electrons are typically used in most gas and semiconductor lasers whereas light is most often used in liquid (dye) lasers and crystalline solid-state lasers. Electron excitation sources tend to be the most efficient since flash lamps are themselves generally excited by electrons before they are used to pump lasers. Heavier particles are generally less efficient as pumping sources since they are much more difficult to energize than either electrons or lamps.

D. INVERSIONS IN GASES, METAL VAPORS, AND PLASMAS

Inversions in gases, metal vapors, and plasmas are generally produced by applying a voltage drop across the elongated gain region thereby producing an electric field that accelerates the electrons. These rapidly moving electrons then collide with the gas atoms and excite them to a wide range of excited energy levels. Some of those levels decay faster than others (primarily by spontaneous emission) leaving population inversions with higher levels. If the populations in the inverted levels are high enough, then the gain may be sufficient to make a laser. Typically if 1 in 10^5 or 10^6 atoms is in a specific upper laser level, that will be a sufficient fraction to produce enough gain to make a laser. Most gas lasers have relatively low gains and therefore amplifier lengths of the order of 25 to 100 cm are necessary. Since spontaneous emission rates are much faster for shorter wavelength transitions, power input for short-wavelength lasers is significantly higher than for visible and infrared lasers. Typical gas pressures for gas lasers range from 1/1000th to 1/100th of an atmosphere although there are some gas lasers that operate at atmospheric pressure and above and require very closely spaced electrodes (transverse excitation) in order to produce the necessary excited state populations.

E. INVERSIONS IN LIQUIDS (DYES)

Most excited states of liquids decay so rapidly by collisions with surrounding atoms or molecules (10^{-13} sec) that it is difficult to accumulate enough population in an upper laser level to make significant gain. Also, since it is difficult to use electron excitation in liquids, the primary energy source is optical excitation, either by flash lamps or by other lasers. Fluorescing dyes are the best liquid media for lasers. The fact that dyes fluoresce suggests that their excited energy levels stay populated long enough to lose their energy by radiation of light rather than by collisions with surrounding atoms or electrons. To help establish the population inversion, the lower laser level of a dye decays very rapidly by collisions.

F. INVERSIONS IN SEMICONDUCTORS

Inversions in semiconductors are produced when a p–n junction is created by joining two slightly different semiconducting materials (in a similar way to that of a transistor). The n-type material has an excess of electrons and the p-type material has an excess of holes (missing electrons). When they are joined, the excess electrons of the n-type material are pulled over into the p region and vice versa by their charge attraction, causing the electrons and holes in that region to recombine and emit recombination radiation. This neutralizes the junction region, leaving a small, inherent electric field to prevent further recombination of the remaining electrons and holes near the junction. If an external electric field is applied in the appropriate direction, by applying a voltage across the junction, more electrons and holes can be pulled together, causing them to recombine and emit more radiation but also to produce an inversion. This inversion occurs on transitions originating from energy levels located just above the inherent energy band gap of the material. Semiconductor lasers operate in a similar way to light-emitting diodes except that the requirements for constructing the lasers are much more restrictive than the diodes, due primarily to the necessity for higher electric current densities that are essential to produce large gains, and also due to the need for better heat-dissipation capabilities to remove the heat produced by the higher current densities.

G. INVERSIONS IN SOLIDS

Inversions in most solid-state lasers are obtained by implanting impurities (the laser species) within a host material such as a crystal or a glass in a proportion ranging from approximately 1 part in 100 to 1 part in 10,000. In most solid-state lasers the impurities are the form of ions in which the energy states are screened from the surrounding atoms so the energy levels are narrow, like those of isolated atoms or ions, rather than broad like those of liquids. In color center lasers the impurities are crystal defects produced by irradiating the crystal with X rays. In solids, as in liquids, electrons cannot easily be accelerated by electric fields to excite the laser energy levels of the impurity species so the energy must be fed to the medium via flash lamps or other lasers. The input lamp energy occurs over a broad wavelength region to a large number of excited energy levels. These levels then decay to the upper laser level which acts much like a temporary storage reservoir, collecting enough population to make a large inversion with respect to lower lying levels that have a rapid decay to the ground state.

H. Bandwidth of Gain Media

The frequency or wavelength spectrum (gain bandwidth) over which gain occurs in a laser amplifier is determined by a number of factors. The minimum width is the combined width of the energy levels involved in the laser transition. This width is due primarily to the uncertainty of the natural radiative decay time of the laser levels. This width can be increased by collisions of electrons with the laser states in high-pressure gas lasers, by interaction of nearby atoms and bonding electrons in liquids and solids, and by Doppler shifted frequencies in most gas lasers. This gain linewidth of the laser amplifier is only part of the contribution to the linewidth of the laser beam. Significant line narrowing due to optical cavity effects will be described in the next section.

III. Laser Beam Properties

A. Beam Properties

The beam properties of a laser, such as the direction and divergence of the beam and the wavelength or frequency characteristics that are not related to the bandwidth of the laser gain medium, are determined largely by the laser structure. The features of the structure affecting the beam properties include the width and length of the gain medium, the location, separation and reflectivity of the mirrors of the optical cavity (if the gain is low and the gain duration long enough to make use of mirrors), and the presence of losses in the beam path within the cavity. These features determine unique properties of the laser beam referred to as laser modes.

B. Shape of the Gain Medium

If a laser gain medium were in the shape of a round ball, then stimulated emission would occur equally in all directions and the only result might be a slight increase in the intensity of the light, for a slightly shorter duration than would occur if gain were not present, but the effect would probably not be noticeable to an observer. The goal of a laser designer is to cause most of the laser photons to be stimulated in a specific direction in order to produce a highly directional beam, at the expense of allowing those same photons to radiate in random directions by either stimulated or spontaneous emission (as was the case for the round ball). This is achieved by making the gain medium significantly longer in one dimension than in the other two.

C. Growth of the Beam and Saturation

When gain is produced in the amplifier and spontaneously emitted photons begin to be amplified by stimulated emission, photons that are emitted in directions other than the elongated direction of the amplifier soon reach the walls of the medium and die out. The photons that are emitted in the elongated direction continue to grow by stimulating other atoms to emit additional photons in the same direction until all of those photons reach the end of the amplifier. They then arrive at the mirror and are reflected back through the amplifier where they continue to grow. Finally, after a number of round trips, a beam begins to evolve. If the duration of the gain is long enough, amplification will lead to more photons in the beam than there are atoms available to stimulate, and growth can therefore no longer occur. The beam is then said to be saturated and the beam power reaches a steady value, determined by the amount of energy being fed into the upper laser level by the pumping source. If the population inversion can be maintained on a continuous basis in the amplifier, the laser-beam output becomes steady and the laser is referred to as a continuous wave, or cw, laser. If the gain only lasts for a short duration, the laser output occurs as a burst of light and the laser is referred to as a pulsed laser.

D. Optical Cavity (Optical Resonator)

The optical cavity or resonator, typically comprised of a mirror at each end of the elongated gain region (see Fig. 1), allows the rapidly growing beam to bounce back and forth or resonate between the mirrors. Although the first laser used flat mirrors, as suggested in the original Schalow–Townes paper, in 1961 Fox and Li suggested the use of slightly curved mirrors, especially for cavities where the amplifier consisted of a long narrow tube, in order to reconcentrate the beam in the center of the gain medium after each reflection from the mirrors thereby reducing the diffraction losses of the narrow tube. Stable modes of lower loss are possible for curved mirrors than for flat mirrors if the separation between the mirrors is less than twice their radius of curvature. Mirror reflectivities of 99.9% at the laser wavelength, using dielectric layered coatings, make possible laser operation under conditions of very low gain.

E. Stable and Unstable Resonators

The term resonator implies a wave that is in harmony or resonance with the device that is generating the wave, whether it be an organ pipe, a flute, or a microwave cavity. An optical wave can also have this property. The term "resonance" suggests that an exact integral number of wavelengths (a mode) of the wave fit between the mirrors of the resonator. A stable resonator refers to a mirror arrangement (usually one with a mirror at each end of the elongated cavity), producing modes that are continually reproducible during the duration that gain occurs in the amplifier (which could be a thousandth of a second or many days or longer). An unstable resonator is a mirror arrangement that is used to obtain modes when the amplifier gain is high and has a short duration (less than a millionth of a second) such that a normal mode would not have time to evolve. In that situation, the energy is extracted by using a mirror arrangement in which the beam begins to resonate in a small unstable region of the amplifier. Part of this beam is leaked into the larger portion of the amplifier where it rapidly grows and extracts most of the amplifier energy in a few passes.

F. Laser Modes

Laser modes are wavelike properties relating to the oscillating character of a light beam as the beam passes back and forth through the amplifier and grows at the expense of existing losses. The development of modes involves an attempt by competing light beams of similar wavelengths to fit an exact number of their waves into the optical cavity with the constraint that the oscillating electric field of the light beam is zero at each of the mirrors. This is much like a vibrating guitar string which is constrained at each end by the bridge and a fret, but is free to vibrate with as many nodes and antinodes in the region in between as it chooses. As an example, a laser mode of green light having a wavelength of exactly 5.0×10^{-5} cm will fit exactly 1,000,000 full cycles of oscillation between laser cavity mirrors separated by a distance of exactly 50 cm. Most lasers have a number of modes operating simultaneously, in the form of both longitudinal and transverse modes, which give rise to a complex frequency and spatial structure within the beam in what might otherwise appear as a relatively simple, pencil-like beam of light.

G. Longitudinal Modes

Each longitudinal mode is a separate light beam traveling along a distinct path between the mirrors and having an exact integral number of wavelengths along that path. In the example of green light mentioned previously, three different longitudinal modes would have very slightly different wavelengths of green light (indistinguishable in color to the eye) undergoing respectively 1,000,000, 1,000,001, and 1,000,002 full cycles of oscillation between the mirrors while traveling exactly the same path back and forth through the amplifier (see Fig. 3). In this situation each mode would differ in frequency by exactly 300 MHz as determined by the velocity of light (3 ×

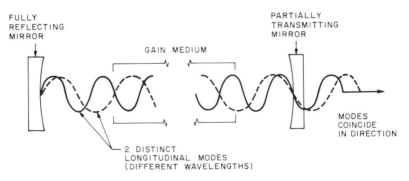

FIG. 3.   Two distinct longitudinal modes occupying the same spatial region of the laser optical cavity.

10^{10} cm/sec) divided by twice the cavity length (2 × 50 cm). A gas laser amplifier having a relatively narrow gain width of 3 GHz could fit 10 longitudinal modes within the gain bandwidth whereas a liquid (dye) laser having a bandwidth covering up to one-fifth of the visible spectrum of light could have as many as 100,000 longitudinal modes all oscillating simultaneously.

H. Transverse Modes

Whereas longitudinal modes involve many light beams traveling exactly the same path through the amplifier, but differing in wavelength by an amount determined by the total number of wave cycles that fit between the mirrors, different transverse modes are represented by slightly different optical paths as they travel through the amplifier (Fig. 4). Thus each transverse mode traveling over its unique path could consist of several longitudinal modes oscillating along that path. In most instances, closely located transverse modes differ in frequency by a smaller value than do adjacent longitudinal modes that follow the same path through the amplifier.

IV. Laser Linewidth

A. Homogeneous Linewidth of Gain Media

All laser amplifiers have a finite frequency width or wavelength width over which gain can occur. This width is related to the widths of the energy levels involved in the population inversion. Single atoms or ions have the narrowest widths, determined by the very narrow energy levels inherent in the atomic structure. The gain linewidth for these amplifiers is the sum of the linewidths of the upper and lower laser levels.

The linewidth is determined by both a homogeneous component and an inhomogeneous component. The homogeneous component includes all mechanisms that involve every atom in both the upper and lower laser levels of the gain medium in identically the same way. These mechanisms include the natural radiative lifetime due to spontaneous emission, broadening due to collisions with other particles such as free electrons, protons, neutrons, or other atoms or ions, or power broadening in which a high intensity laser beam rapidly cycles an atom between the upper and lower levels at a rate faster than the normal lifetime of the state. In each of these effects, the rate at which the process occurs determines the number of frequency components or the bandwidth required to describe the process, with faster processes needing broader linewidths. These rates can vary anywhere from 10^6 Hz in the infrared to 10^{10} Hz in the soft X-ray spectral region for spontaneous emission and up to 10^{12} Hz or more or rapid collisional broadening.

B. Inhomogeneous Linewidth of Gain Media

The inhomogeneous component of the broadening results from processes that affect different atoms in the upper and lower laser levels in different ways depending on unique characteristics of those atoms. The most common example of inhomogeneous broadening is Doppler broadening or motional broadening that results from the random thermal motion of the atoms due to their finite temperature. In this effect, atoms traveling in one direction would see an approaching lightwave as having a specific frequency and atoms traveling in the opposite direction would see that same lightwave as having a lower fre-

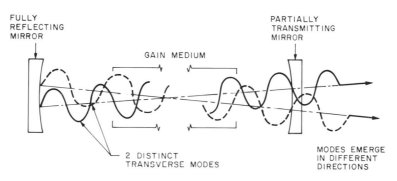

FIG. 4. Two distinct transverse modes oscillating over different spatial regions of the laser optical cavity.

quency. In a similar way, light emitted from each of those atoms would be seen by an observer as having different frequencies or wavelengths even though each atom emitted the same frequency. This Doppler effect is similar to that when a train approaches with its horn blowing and the pitch of the sound changes from a high tone to a lower tone as the train first approaches the listener and then moves away, even though the frequencies emitted from the horn remain the same.

C. Linewidth in Laser Amplifiers

The dominant broadening effect for most gas laser amplifiers is Doppler broadening which produces a linewidth of the order of 10^9 to 10^{10} Hz for visible and shorter wavelength amplifiers. Molecules have linewidths that are determined not only by the jumping of electrons between energy levels, as is the case of individual atoms, but also by the rotation of the atoms around a common center of gravity and by the vibrations of the atoms as though they were tied together by various springs. These rotations and vibrations produce a series of emissions of light over a relatively narrow range of wavelengths and in some instances these emissions can overlap in frequency to produce broad emission lines. If no overlap occurs, then the emission consists of an equally spaced series of narrow lines.

Liquids have their energy levels broadened due to the interaction of the closely packed atoms of the liquid. Such dense packing leads to a very rapid collisional interruption of the electronic states causing them to smear out into very broad energy bands thereby producing emission spectrum widths of the order of 10^{13} to 10^{14} Hz.

Solids, in addition to having closely located atoms (only slightly more dense than liquids) in many cases have a periodic structure due to the regularity of the atomic distribution (much as in the way eggs are lined up in an egg carton). This produces in many species a large energy gap between the lowest energy level and the first excited energy level. Emission from this excited level can have a very broad frequency spectrum similar to that of liquids; however, there also exist discrete levels within the bandgap, known as exciton states, that have a much narrower emission spectrum (of the order of 10^{12} Hz), due to electron–hole pairs bonding together to form atomiclike particles. Such exciton states are involved in some semiconductor lasers. Another class of emission and broadening in solids occurs when an atom like chromium is embedded in another solid material such as aluminum oxide (sapphire) to make a ruby crystal. In such a crystal, the outer electrons of the chromium atoms are shared with the surrounding aluminum oxide atoms while the inner electrons are screened from the neighboring atoms causing the atom to behave more like an isolated atom having narrow energy levels. These crystalline solids, containing specific impurities, can have narrow emission lines of the order of 10^{11} to 10^{12} Hz. The first laser was made in ruby using such levels.

D. Line Narrowing Due to Cavity Effects

A laser cavity tends to select specific wavelengths (Fig. 5) within the normal gain bandwidth of the gain medium as determined by the wavelengths of light that have an exact integral number of waves that fit between the mirrors (modes). These modes, which are equally spaced in frequency or wavelength, tend to be amplified at the expense of other wavelengths that suffer losses by not exactly "fitting" between the mirrors. In principal these modes can "narrow up" to widths of the order of a few hertz, but cavity stability problems tend to keep them from going much below 1000 Hz unless extremely stable environments and rigid cavity structures are available.

V. Laser Wavelengths

A. Range of Wavelengths

Wavelengths of electromagnetic radiation covering the spectral region where lasers occur are referred to in terms of fractions of meters. The two most common units are the micrometer (μm) or 10^{-6} m and the nanometer (nm) 10^{-9} m. Micrometers are used to designate the infrared region ranging from 0.7 μm to approximately 1,000 μm. Nanometers are used to cover the spectral range from the visible at 700 nm (0.7 μm) down to approximately 10 nm in the soft X-ray region. Each spectral region covers a specific wavelength range although the boundaries are not always exact. The infrared is broken up into three regions. The far infrared ranges from about 15 to approximately 1,000 μm (approaching the microwave region). The middle infrared covers from 2 to 15 μm and the near infrared ranges from 0.7 to 2 μm. The visible region in-

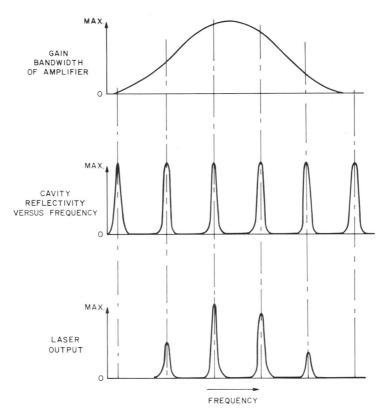

GAIN BANDWIDTH OF AMPLIFIER

CAVITY REFLECTIVITY VERSUS FREQUENCY

LASER OUTPUT

FREQUENCY

FIG. 5. Laser output resulting from effects of laser amplifier gain bandwidth and laser optical cavity modes.

cludes the rainbow spectrum ranging from the violet at 400 to 440 nm, the blue from 440 to 490 nm, the green from 490 to 550 nm, the yellow from 550 to 580 nm, the orange from 580 to 610 nm, and the red from 610 to 700 nm. The ultraviolet ranges from 400 to 200 nm with the 400- to 300-nm region termed the near ultraviolet and the 200- to 300-nm region the far ultraviolet. The vacuum ultraviolet covers the range from 100 to 200 nm, since it is the region where radiation can no longer transmit through the air (thus requiring a vacuum) because of absorption due to air molecules (mostly oxygen) and because transmission optics can still be used in this region. The shortest region is the extreme ultraviolet, which extends into the soft X-ray region, covering the range from 10 to 100 nm. In this region only reflection optics can be used and even then the reflectivities of the best materials are quite low (especially below 40 nm) when compared to those available in the visible. Free-electron lasers have the potential for operating at all wave-

lengths and will therefore not be singled out for any specific wavelength in this section.

B. INFRARED LASERS

The most powerful lasers occur in the middle and near infrared, but a large number of lasers have been produced in the far infrared. These include the discharge excited water-vapor lasers with wavelengths ranging from 17–200 μm and the cyanide laser operating at 337 μm. The other principle far infrared lasers are the laser-pumped gases of methyl fluoride (450–550 μm) and ammonia (81 μm). In the middle infrared, the carbon dioxide laser is the most significant laser operating primarily at 9.6 and 10.6 μm. The other strong laser is the carbon monoxide laser which emits in the 5–6-μm region. There is also an important chemical laser, the hydrogen–fluoride laser, which operates at 3.7 μm. In the near infrared, many more lasers are available. The most significant laser is the neodymium laser, usually available in a yttrium–aluminum–garnet (YAG) host crystal referred to as the Nd:YAG laser, or in a glass host called a Nd:glass laser, both operating at 1.06 μm. Semiconductor lasers operate over the range from 0.7 to approximately 1.70 μm; their wavelengths being extended in both directions with continuing research. Color center lasers are tunable from 0.8 to 4.0 μm and the alexandrite laser is tunable from 0.7 to 0.8 μm. Tunable dye lasers also extend into the infrared to about 1.5 μm.

C. Visible Lasers

Visible lasers are primarily dominated by gas lasers and tunable dye lasers. The helium–neon laser at 633 nm is probably the most commonly seen laser. The argon ion laser covers the blue and green spectrum with the most prominent wavelengths at 488 and 515 nm. The krypton ion laser covers the spectrum from green to red with some of the most prominent wavelengths at 521, 568, and 647 nm. The helium–cadmium laser operates in the blue at 442 nm. Tunable dye lasers operate over the entire visible spectrum, generally pumped by a slightly shorter wavelengh laser such as the argon ion laser or the frequency doubled or tripled Nd:YAG laser. The rhodamine 6G dye has the lowest pump threshold and operates as a laser over a wavelength range from 570 to 630 nm.

D. Ultraviolet Lasers

There are fewer lasers available in the ultraviolet primarily because pump thresholds become much higher than for visible lasers and because cavity mirror reflectivities are lower and less durable. The two prominent continuous lasers are the argon ion laser, operating primarily at 351 nm and the helium cadmium laser at 325 nm. The rare-gas–halide excimer lasers produce high pulsed powers at 351 nm for xenon–fluoride, 308 nm for xenon chloride, and 248 nm for krypton fluoride. The pulsed nitrogen laser operating at 331 nm was extensively used for pumping dye lasers but this has now largely been replaced with the frequency doubled and tripled Nd:YAG laser and the excimer lasers. Tunable dye lasers as short as 310 nm are available but they are not easily pumped by commercial lasers and therefore frequency mixing and doubling of visible and infrared dye lasers is the most common technique for producing tunable laser radiation in this region.

E. Vacuum Ultraviolet Lasers

The molecular hydrogen laser was the first laser developed in this spectral region, operating in the 120- and 160-nm ranges but it never became a useful device. There are now only two readily available vacuum ultraviolet lasers. These are the argon–fluoride excimer laser operating at 193 nm and the fluorine excimer laser emitting at 153 nm. Much of the research requiring lasers in this spectral region is accomplished by frequency summing and mixing of various visible and ultraviolet lasers to produce coherent output, but such techniques are not yet readily available to the general laser community.

F. Extreme Ultraviolet and Soft X-Ray Lasers

Until 1984 there were no lasers to report in this spectral region. At that time stimulated emission was reported in krypton at 93 nm. In 1985 several highly ionized atoms provided laser output in the soft X-ray end of this spectrum, including twenty-fourth ionized selenium at 21 nm, fifth ionized carbon at 18 nm, twenty-ninth ionized yttrium at 15 nm and thirty-second ionized molybdenum at 13 nm. These lasers are so new that no experiments have yet been carried out with them. They are initiated with powerful neodymium and carbon dioxide lasers the size of large buildings and are therefore not easily duplicated in other laboratories.

VI. Types of Lasers

A. Gas Lasers

The most common types of gas lasers are the helium–neon laser, the argon and krypton ion lasers, the carbon dioxide laser, the rare-gas–halide excimer lasers, and the chemical lasers, most notably the hydrogen–flouride laser. Metal vapor lasers also fit into this category but are treated separately in Section V. With few exceptions these lasers receive their energy input via collisions of gas atoms with high-energy electrons. This energy is provided by applying a high voltage between electrodes located within the gaseous medium in order to accelerate the electrons to the necessary high energies. In some instances the electrons first excite a storage level in a separate species within the gaseous medium rather than directly pumping the laser state. The energy is subsequently transferred from that storage level to the laser level of the lasing species by direct collisional exchange of energy.

1. Helium–Neon Laser

The helium–neon laser was the first gas laser. The original laser transitions were in the near infrared but the most commonly used transition is the red laser at a wavelength of 632.8 nm. This laser, which has more units in use than any other laser, is available in sizes ranging from approximately 10 cm in length to over 100. It has continuous power outputs ranging from less than a milliwatt to over 100 mW and has a lifetime approaching 50,000 h for some commercial units. The excitation mechanism involves electrons colliding with helium atoms to produce helium metastable atoms, which then transfer their energy to neon laser levels. This laser is used in surveying, construction, supermarket checkout scanners, printers, and many other applications.

2. Argon and Krypton Ion Lasers

Argon and krypton ion lasers were discovered shortly after helium–neon lasers. They were the first lasers to operate in the green and blue regions of the spectrum and some versions provide ultraviolet output. These lasers have the capability of producing more than 20 W of continuous power for the largest versions. The size of the laser tubes range from 50 to 200 cm in length with a separate power supply. They are relatively inefficient and consequently require high input power and water cooling for most units. The high power requirements, which put great demands on the strength of the laser-discharge region, limit the lifetime of the high-power versions of these lasers. Some smaller, lower power versions of the argon ion laser are air cooled and offer lifetimes of 5000 h. The excitation mechanisms for these lasers involve electrons collisions first populating the ion ground states of the argon and krypton species with subsequent electron excitation to the upper laser level. Applications include phototherapy of the eye, pumping dye lasers, printing, and lithography.

3. Carbon Dioxide Laser

The carbon dioxide lasers are some of the most powerful lasers, operating primarily in the middle infrared spectral region at a wavelength of 10.6 μm. They range from small versions with a few milliwatts of continuous power to large pulsed versions the size of large buildings producing over 10,000 J of energy. They are among the most efficient lasers (up to 30%) and can produce continuous output powers of over 100 kW in room-size versions. Small versions of these lasers are referred to as waveguide lasers because the excitation region is of a cylindrical shape small enough to guide the beam down the bore in a waveguide type of mode. They can produce continuous power outputs of up to 100 W from a device smaller than a shoe box (with a separate power supply). The lasers typically operate in a mixture of carbon dioxide, nitrogen, and helium gases. Electron collisions excite metastable (storage) levels in nitrogen molecules with subsequent transfer of that energy to carbon dioxide laser levels. The helium gas acts to keep the average electron energy high in the gas-discharge region. This laser is used for a wide variety of applications including eye and tissue surgery, welding, cutting, and heat treatment of materials, laser fusion, and beam weapons. Figure 6 shows a carbon dioxide laser being used to drill a hole in a turbine blade. The white streamers are hot metal particles being ablated from the region of the hole.

4. Rare-Gas–Halide Excimer Lasers

The rare-gas–halide excimer lasers, relative newcomers to the laser industry, operate primarily in the ultraviolet spectral region in mixtures of rare gases, such as argon, krypton, or xenon with halide molecules such as chlorine and fluorine. They include the argon–fluoride laser at 193 nm, the krypton–fluoride laser at 248 nm, the xenon–chloride laser at 308 nm, and the xenon–flouride laser at 351 nm. These lasers typically produce short pulses of energy ranging from tens of millijoules to thousands of joules in pulse durations of 10 to 50 nsec and repetition rates of up to 1000 pulses per second. They range in size from an enclosure that would fit on a kitchen table top to lasers the size of a very large room. They are relatively efficient (1–5%) and provide useful energy in a wavelength region that has not had powerful lasers available previously. Their operating lifetimes are related to the development of discharge tubes, storage regions, and gas pumps that can tolerate the corrosive halogen molecules that are circulated rapidly through the gain region. Typical lifetimes are of the order of the time it takes to produce 10^6 to 10^7 pulses. The laser species are mixed with helium gas to provide a total pressure of 2 atms. Excitation occurs by electron dissociation

FIG. 6. Carbon dioxide laser used to drill holes in a turbine blade. White streamers are hot metal particles being ejected from the hole region.

and ionization of the rare gas molecule to produce Ar^+, Kr^+, or Xe^+ ions. These ions then react with the halide molecules pulling off one of the atoms of that molecule to create an excited-state dimer (abbreviated as excimer) molecule. This excimer molecule then radiates rapidly to an unstable (rapidly dissociating) lower laser level. The very high operating pressures cause the molecules to react rapidly in order to produce the upper laser levels at a rate that can compete with the rapid decay of those levels. Applications include laser surgery, pumping of dye lasers, and lithography.

5. Chemical Lasers

In these lasers the molecules undergo a chemical reaction that leaves the molecule in an excited state that has a population inversion with respect to a lower lying state. An example of this type of laser is the hydrogen–fluoride laser in which molecular hydrogen and molecular fluorine react to produce hydrogen–fluoride molecules in their excited state resulting in stimulated emission primarily at 2.8 μm. There are no commercially available chemical lasers but they have undergone extensive development for military applications.

B. METAL VAPOR LASERS

Metal vapor lasers are actually a type of gaseous laser since the laser action occurs in the atomic or molecular vapor phase of the species at relatively low pressures, but the lasers have peculiar problems associated with vapors, such as having to vaporize a solid or liquid into the gaseous state either before or during the excitation and lasing process. These problems, along with the problems associated with controlling the condensed vapors after they diffuse out of the hot region in some designs, and with minimizing the corrosive effects of hot metal atoms and ions in other designs, have caused them to be classified in a separate category. The two most well-known types of metal vapors lasers are the helium–cadmium ion laser and the pulsed copper vapor laser.

1. Helium–Cadmium Laser

The helium–cadmium laser operates primarily at two wavelengths, in the blue at 441.6 nm and in the ultraviolet at 325.0 nm. It produces continuous power outputs of the order of 5–100 mW for the blue and 1–25 mW in the ultraviolet, in sizes ranging from 50–200-cm long. The cadmium vapor is obtained by heating the cadium metal in a reservoir located near the helium discharge. The cadmium vapor diffuses into the excited helium gas where it is ionized and the cataphoresis force on the cadmium ions causes them to move towards the negative potential of the cathode thereby distributing the metal relatively uniformly in the discharge region to produce a uniform gain. Typical operating life for the laser is of the order of 5000 h with the limiting factors being the control of the vapor and the loss of helium gas via diffusion through the glass tube walls. The laser levels are excited by collisions with helium metastable atoms, by electron collisions, and by photoionization resulting from radiating helium atoms. Applications include printing, lithography, and fluorescence analysis.

2. Copper Vapor Laser

This laser operates in the green at 510.5 nm and in the yellow at 578.2 nm. It efficiently (2%) produces short laser pulses (10–20-nsec duration) of 1 mJ of energy at repetition rates of up to 20,000 times per second yielding average powers of up to 20 W. The size is similar to that of an excimer laser. Commercial versions of this laser are designed to heat the metallic copper up to temperatures of the order of 1600°C in order to provide enough copper vapor to produce laser action. The lifetime associated with operating these lasers at such high temperatures has been limited to a few hundred hours before servicing is required. The excitation mechanism is primarily by electron collisions with ground-state copper atoms to produce upper laser states. Applications of these lasers include uranium isotope enrichment, large-screen optical imaging, and pumping of dye lasers. A gold vapor laser that is similar to the copper laser, but emits red light at a wavelength of 624.0 nm, is used for cancer phototherapy.

C. SOLID-STATE LASERS

These lasers generally consist of transparent crystals or glasses as "hosts" within which ionic species of laser atoms are interspersed or "doped." Typical host materials include aluminum oxide (sapphire), garnets, and various forms of glasses with the most common lasing species being neodymium ions and ruby ions (the first laser). In color center lasers the host is typically an alkali–halide crystal and the laser species is an electron trapping defect in the crystal. The energy input in all of these lasers is provided by a light source that is focused into the crystal to excite the upper laser levels. The light source is typically a pulsed or continuously operating flash lamp, but efficient diode lasers are also being used to pump small versions of neodymium lasers and argon ion lasers are used to pump color center lasers.

1. Neodymium Lasers

Neodymium atoms are implanted primarily in host materials such as yttrium–aluminum–garnet (YAG) crystals or various forms of glasses in quantities of approximately one part per hundred. When they are implanted in YAG crystals, the laser emits in the near infrared at 1.06 μm with continuous powers of up to 250 W and with pulsed powers as high as several megawatts. The YAG crystal growth difficulties limit the size of the laser rods to approximately one centimeter in diameter. The YAG host material, however, has the advantage of having a relatively high thermal conductivity to remove wasted heat, thus allowing these crystals to be operated at high repetition rates of the order of many pulses per second. Glass hosts also produce Nd

lasers in the 1.06-μm wavelength region but with a somewhat broader bandwidth than YAG. They can also be grown in much larger sizes than YAG, thereby allowing the construction of very large amplifiers, but glasses have a much lower thermal conductivity, thus requiring operation at much lower repetition rates (of the order of one pulse every few minutes or less). Thus Nd:YAG is used for continuous lasers and relatively low-energy pulsed lasers (1 J per pulse) operating at up to 10 pulses per second whereas glass lasers exist in sizes up to hundreds of centimeters in diameter, occupy large buildings and are capable of energies as high as 100 kJ per pulse for laser fusion applications. Neodymium lasers typically have very long lifetimes before servicing is required, with the typical failure mode being the replacement of flash lamps. Neodymium lasers are used for surgery applications, for pumping dye lasers (after doubling and tripling their frequencies with nonlinear optical techniques), as military range finders, for drilling holes in solid materials, and for producing X-ray plasmas for X-ray light sources, and for laser fusion and for making X-ray lasers.

2. Ruby Laser

The ruby laser, the first laser discovered, is produced by implanting chromium ions into an aluminum oxide crystal host and then irradiating the crystal with a flash lamp to excite the laser levels. Although ruby lasers were frequently used during the early days of the laser, the difficulties associated with growing the crystals, compared with the ease of making neodymium lasers, has led to their being used much less often in recent times.

3. Color Center Lasers

Color center lasers use a different form of impurity species implanted in a host material in quantities of one part per ten thousand. In such lasers the laser species is generally produced by irradiating the crystal with X rays to produce defects that attract electrons. These defect centers produce energy levels that absorb and emit light and are capable of being inverted to produce gain. Color center lasers typically operate in the infrared from 0.8–4 μm and are tunable within that range by using different crystals having different emission wavelengths. Their tunability makes them attractive lasers for doing spectroscopy.

D. Semiconductor Lasers

Semiconductor or diode lasers, typically about the size of a grain of salt, are the smallest lasers yet devised. They consist of a *p–n* junction formed in an elongated gain region, typically in a gallium–arsenide crystal, with parallel faces at the ends to serve as partially reflecting mirrors. They operate with milliamps of current at a voltage of only a few volts. The entire laser package is very small and could be incorporated into an integrated circuit board if required. Heterostructure lasers, a more recently developed type of diode laser, include additional layers of different materials of similar electronic configuration, such as aluminum, indium and phosphorous on the sides of the junction to help confine the electronic current to the junction region in order to minimize current and heat dissipation requirements. Semiconductor lasers range in wavelengh from 0.7 to 1.8 μm with typical continuous output powers of up to 10 mW. By constructing a row of *p–n* junctions next to each other, all of the separate gain media can be forced to emit together in a phased array to produce an effective combined power output of over one watt. Applications for semiconductor lasers are primarily in the communications field in which the near-infrared beams can be transmitted over long distances through low-loss fibers. In addition, they have recently found a large market as the reading device for compact disc players. Figure 7 shows a diode laser array, consisting of 10 diode lasers, recording at a high data rate onto a multitrack optical disk.

E. Liquid (Dye) Lasers

Dye lasers are similar to solid-state lasers in that they consist of a host material (in this case a solvent such as alcohol) in which the laser (dye) molecules are dissolved at a concentration of the order of one part in ten thousand. Different dyes have different emission spectra or colors thus allowing dye lasers to cover a broad wavelength range from the ultraviolet (320 nm) to the infrared at about 1500 nm. A unique property of dye lasers is the broad emission spectrum (typically 30–60 nm) over which gain occurs. When this broad gain spectrum is combined with a diffraction grating or a prism as one of the cavity mirrors, the dye laser output can be a very narrow frequency beam (10 GHz or smaller) tunable over a frequency range of 10^{13} Hz. Frequency tuning over even larger ranges is accomplished by inserting different dyes into the laser cavity.

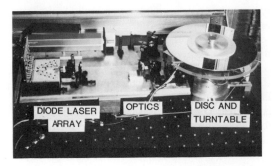

FIG. 7. High-data-rate multichannel optical recording using an array of addressable diode lasers. [Courtesy of RCA Laboratories.]

Dye lasers are available in either pulsed (up to 50–100 mJ) or continuous output (up to a few watts) in table-top systems that are pumped by either flash lamps or by other lasers such as frequency-doubled or -tripled YAG lasers or argon ion lasers. Most dye lasers are arranged to have the dye and its solvent circulated by a pump into the gain region from a much larger reservoir, since the dye degrades slightly during the excitation process. Dyes typically last for 3 to 6 months in systems where they are circulated. Dye lasers are used mostly for applications where tunability of the laser frequency is required, either for selecting a specific frequency that is not available from one of the solid-state or gas lasers, or for studying the properties of a material when the laser frequency is varied over a wide range. Most of these applications are in the area of scientific experiments. Another large application of dye lasers is for producing ultra-short optical pulses by a technique known as mode locking. In this process, all of the longitudinal modes of a dye laser (as many as 10,000) are made to oscillate together (in phase), causing individual pulses as short as 50 fsec (5×10^{-14} sec) to emerge from the laser, spaced at intervals of the order of 20 nsec. These short pulses are of interest in studying very fast processes in solids and liquids and may have applications for optical communications.

F. Free-Electron Lasers

Free-electron lasers are significantly different from any other type of laser in that the laser output does not occur from discrete transitions in atoms or molecules of gases, liquids, or solids. Instead, a high-energy (of the order of one million electron volts) beam of electrons is directed to pass through a spatially varying magnetic field that causes the electrons to oscillate back and forth in a direction transverse to their beam direction, at a frequency related to the magnet separation in the transverse direction and also to the energy of the electron beam. This oscillation causes the electrons to radiate at the oscillation frequency and to stimulate other electrons to oscillate and thereby radiate at the same frequency, in phase with the original oscillating electrons, thereby producing an intense beam of light emerging from the end of the device. Mirrors can be placed at the ends of the magnet region to feed the optical beam back through the amplifier to stimulate more radiation and cause the beam to grow. The free-electron laser, although still more of a laboratory curiosity than a useful device, offers to produce very high power output over a wide range of wavelengths from the far infrared to the vacuum ultraviolet.

VII. Applications of Lasers

A. LASER PROPERTIES ASSOCIATED WITH APPLICATIONS

Laser applications are now so varied and widespread that it is difficult to describe the field as a single subject. Topics such as surgery, welding, surveying, communications, printing, pollution detection, isotope enrichment, heat treatment of metals, eye treatment, drilling, and laser art involve many different disciplines. Before these individual topics are reviewed, the various characteristics or features of lasers that make them so versatile will be summarized. These properties include beam quality, focusing capabilities, laser energy, wavelength properties, mode quality, brightness properties, and pulse duration.

1. Beam Quality

When a laser beam emerges from a laser cavity, after having evolved while reflecting back and forth between carefully aligned mirrors, it is highly directional, which means that all of the rays are nearly parallel to each other. This directional property, if considered as originating from a single longitudinal mode emerging from a laser cavity, would have such a low beam divergence that a laser aimed at the moon would produce a spot on the moon only 5 miles in diameter after having travelled a distance of 239,000 miles! That is a very small divergence when compared to a flashlight beam which expands to five times its original size while traveling across a room.

2. Focusing Capabilities

Because of the parallel nature of the beam when it passes through a lens (of good quality) all of its rays are concentrated to a point at the focus of the lens. Since diffraction properties of the light must be taken into account, the beam cannot be focused to an infinitely small point but instead to a spot of a dimension comparable to the wavelength of the light. Thus a green light beam could be focused to a spot 5.0×10^{-5} cm in diameter, a size significantly smaller than that produced by focusing other light sources.

3. Laser Energy

The primary limitations on laser energy and power are the restrictions on the size of the amplifier and the damage thresholds when the beam arrives at the laser mirrors and windows. A 1 W continuous carbon dioxide laser beam, focused on a fire brick, will cause the brick to glow white hot and begin to disintegrate. A carbon dioxide laser has been made to continuously operate at a level 10^5 times that powerful! Similarly, a pulsed laser of 5.0×10^{-4} J, when focused into the eye, can cause severe retinal damage. A pulsed neodymium laser has been made that produces 10^8 times that energy!

4. Wavelength Properties

The capability of generating a very pure wavelength or frequency leads to many uses. It allows specific chemical reactions to be activated, wavelength sensitive reflective and transmissive effects of materials to be exploited, and certain atomic processes to be preferentially selected. Some of these processes can use the fixed frequencies of gas and solid-state lasers. Others require the broad tunability of dye lasers.

5. Mode Quality

The ultrapure frequency available from a single-mode output of a laser not only makes possible the concentrated focusing capabilities mentioned but it provides a very pure frequency that is capable of being modulated in a controlled way to carry information. Since the amount of information that is carried over an electromagnetic (light or radio) wave is proportional to the frequency of the radiation, a single-mode optical wave can carry over 10^4 times the information of a microwave and 10^9 times that of a radio wave. This realization is possible whether the optical wave is carried through space, through the atmosphere, or through a tiny glass fiber the size of a human hair.

6. Brightness Properties

The brightness properties are represented by the amount of light concentrated in a specific area during a definite amount of time in a specific wavelength or frequency interval. Such properties are applicable for use in holography and other coherent processes. The effective brightness temperature of a laser can be 10^{20} to 10^{30}°C, much higher than any other man-made light source.

7. Pulse Duration

The duration of a laser beam can vary anywhere from a continuous (cw) beam that will last as long as the power is supplied to the amplifier and as long as the amplifier keeps producing gain (50,000 h for some gas lasers and potentially many tens of years for some semiconductor lasers) to pulses as short as 6.0×10^{-15} sec for the shortest pulses from a pulse-compressed, mode-locked dye laser. Many applications require very short pulses, including high-speed digital signal transmission for communications and also surgical applications that would result in damage to surrounding tissue via heat conduction if longer pulses were used. Other applications require very long or continuous light fluxes to produce effects in materials that are relatively insensitive to the light beam but would be destroyed by shorter duration, higher-intensity beams.

B. COMMUNICATIONS

One of the earliest recognized applications of the laser, because of its capability of producing a very pure frequency, was in the field of communications. Earlier developments using radio waves and then microwaves naturally led to thoughts of using optical frequencies in a similar manner to take advantage of the increased information carrying bandwidth that would be essential for the information age. This concept has now progressed to the point where the semiconductor laser with its small size, low power consumption, and high reliability, in conjunction with optical fibers as transmission media and rapidly developing product lines of optical connectors, couplers, modulating devices, etc., is a major component in communication systems. Using very short, discrete pulses of light to transmit digitally encoded signals is a more recent advancement that will most likely become the ultimate technique for both long and short range communications.

C. MEDICAL

Medical applications are often the most publicized uses of lasers primarily because they affect the general public more directly than other applications. The first applications were in the field of ophthalmology where the laser was used to weld detached retinas and photocoagulate blood vessels that had grown into the region in front of the retina, thereby blocking vision. In such instances the laser beam easily passes through the transparent portions of the eye, including the cornea and lens, to the region of its intended use where the laser energy is absorbed for treatment. Lasers were also developed as diagnostic tools in cell sorting devices for blood treatment and in fluorescent analysis of Pap smears. The more recent applications in laser surgery are perhaps the most far reaching applications of lasers in medicine. The ability of a laser beam to perform very localized cutting along with cauterization of the incision has produced a wide range of surgical procedures including eye operations (see Fig. 8), gynecologi-

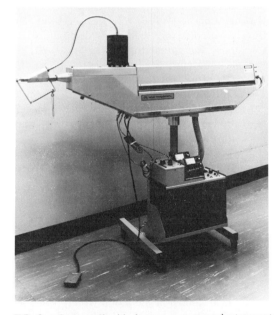

FIG. 8. Carbon dioxide laser eye surgery instrument showing articulated laser beam delivery arm on the left. [Courtesy AT&T Bell Laboratories.]

cal operations, throat and ear surgery, removal of birthmarks, and the most recent prospect of reaming out clogged arteries.

D. Materials Processing

The accurate focusing capability of the laser, allowing the concentration of high power into a small area, makes it a natural device for working with materials. Drilling accurate, tiny holes using a computerized control is now routine in many manufacturing centers, ranging from drilling holes in nipples of baby bottles to drilling holes through high-strength steel. Cutting and welding of high-temperature materials with a laser beam are also effective procedures due to the laser's ability to concentrate energy and power. Another use of lasers is in the heat treatment of metallic surfaces. Pistons of automobile engines, for example, were previously treated in ovens thereby requiring the heating of the entire piston, whereas now a high-power laser can be quickly scanned over specific locations of the piston, locally melting and resolidifying the surface in a short time without having to treat the entire piston.

E. Construction

The use of the laser in the construction industry primarily involves its effectiveness as a straight line reference beam. It is used in the construction of buildings by using a device that has three mutually perpendicular laser beams emerging to provide the reference lines for the sides and vertical alignment of the buildings. It is used as a reference beam for leveling and grading. In this procedure, a laser beam is arranged to continually scan a region under development so that the beam regularly provides a reference level for a grading tractor. Lasers are also used in surveying, as a horizontal reference level.

F. Information Processing

Information processing includes many applications involving reading information in one form and transmitting and converting it to another form. Supermarket bar code readers are probably the most well-known application of this type in which a laser beam scans the digitally encoded "bar code" on grocery items. The reflected, digitally encoded beam is then transmitted to a computer where the product is identified and the information is sent to the checkout

register printout. In compact disc players, a diode laser is focused into the grooves on the disc and the reflected, modulated light is detected, converted to an electrical signal, and sent to an amplifier and speaker system. In holograms, the light that is reflected from an object is recorded on film in a way that denotes not only the intensity of the light from various portions of the object but also the relative phase of each of those portions. Reconstruction of the image then provides a three-dimensional image of the original object. Use of the laser in projection is accomplished by scanning various colors of lasers across a screen much as an electron beam is scanned in a television set. The light of the various laser colors is adjusted in intensity at each point on the screen to produce a realistic image of the original object.

G. Remote Sensing

A laser beam provides a unique opportunity to access a distant region without having to install sensing devices in that region. The region could be a hazardous or polluted area, a mine where poisonous gases are potentially located, or an ozone layer in the upper atmosphere. By various techniques the laser can scatter a small portion of its intense beam off of the impurity species, giving off characteristic radiation that is detected back at the source of the laser without requiring human access to the remote region. This technique makes it possible to do quantitative measurements in real time to provide valuable information that is often not available by other techniques.

H. Military

Military applications of lasers most often fall in the category of radar or ranging devices which determine accurate distances to specific targets or map out regions for future access. Lasers also can be directed onto specific targets (usually using an invisible infrared laser) to serve as an illuminating source that will guide infrared sensitive bombs or provide aiming of gunfire. More recently they have begun to be developed as a directed energy weapons; however, the size of such devices prevents them from becoming relatively portable field weapons.

I. Laser Fusion and Isotope Enrichment

Both of these processes involve the use of lasers in developing alternative energy sources for the future when oil supplies begin to dwindle. The laser fusion program is attempting to produce miniature hydrogen fusion reactions in which the fusion energy will be captured as heat that can be used to drive electric power generators. In a conceptual power plant, powerful lasers (1 MJ of energy) will be focused on tiny pellets of special hydrogen isotopes, compressing and heating them to temperatures as high as 10^8 degrees, densities 10^3 times that of solid densities and durations of the order of one billionth of a second (the conditions under which the fusion reactions occur). Laser isotope enrichment has already been shown to be an effective process for enriching concentrations of special isotopes of uranium for use in atomic fission reactors, in a way that is much less costly than the gaseous diffusion and centrifuge processes that have been used since World War II. This enrichment process is accomplished by using selective laser wavelengths that react with the desired rare isotope but not with the more common isotope of uranium.

J. LITHOGRAPHY AND PRINTING

Lasers can be used as sources to expose paper in printing systems, to expose printed circuits for electronic circuit design, or to expose photoresist material to make electronic microchips. They can also be used to deposit circuit material in specific locations on microchips. This procedure is done by focusing the laser onto locations where deposition is required, thereby producing a localized chemical reaction that precipitates the solid material from a molecular gas or vapor containing that material.

K. LASER ART

Lasers are used in many artistic media. The laser light show is probably the most well known. In such a show, laser beams of various colors are directed around a dark room or on a screen in synchronization with a musical presentation. The complicated but artistic images provide a spectacular visual display. Laser are also used in etching or burning artistic patterns on various media.

L. MISCELLANEOUS APPLICATIONS

Several other applications that do not fall into the other categories are worth mentioning. Laser marking devices are used to put product labels, serial numbers, and other information on items that are difficult to permanently mark by other means, such as high-strength metal and ceramic materials. The laser beam actually melts or ablates the surface in the region where the marking is desired. In a similar way a laser is used to remove material from weights on automobile wheels as determined by computerized balancing instruments, while the wheels are still rotating, thus saving time and improving balancing precision. Lasers also serve as ultrasensitive reference beams in earthquake detection instruments.

BIBLIOGRAPHY

Ready, J. F. (1978). "Industrial Applications of Lasers." Academic Press, New York.
Thyagarajan, K., and Ghatak, A. K. (1981). "Lasers—Theory and Applications." Plenum, New York.
Yariv, A. (1985). "Optical Electronics," 3rd ed. Holt, New York.

MOLECULAR OPTICAL SPECTROSCOPY

Thomas M. Dunn   *University of Michigan*

I. Introduction
II. Theory of Molecular Optical Spectroscopy
III. Polyatomic Molecules

GLOSSARY

Internal conversion: Process whereby energy in one electronic state passes to another electronic state of the same spin multiplicity.

Intersystem crossing: Process whereby energy in one electronic state passes to another electronic state of a different spin multiplicity.

Potential function: Plot of the interaction energy of two atoms as a function of their distance apart. If there is no minimum in the function, the potential function is said to be repulsive. If there is a minimum in the function, it is said to be a bonding potential function. Dissociation is the energy required to separate the two atoms in a diatomic molecule to an infinite distance from one another, each of them being in its ground electronic state.

Ro-vibronic level: Refers to a specific rotational level of a single vibrational level in a specific electronic state, hence ro-vibronic.

Vibrational progression: Series of electronic transitions that either arise or terminate on a single vibrational level of an electronic state and terminate or originate on a succession of different vibrational levels (e.g., 0-0, 0-1, 0-2, ... 0-v″ is a ground state progression and 0-0, 1-0, 2-0, ... v′-0 excited state progression).

Vibrational sequence: Closely grouped series of bands which differ by the same number of vibrational quanta in the ground and excited states (i.e., 0-0, 1-1, 2-2, ... v′-v″ where $\Delta v = 0$).

Vibronic level: Refers to any particular vibrational level of a specific electronic state, hence vibr-onic.

Spectroscopy is the science of analyzing either emitted or absorbed electromagnetic radiation (light) in terms of its wavelength or frequency distribution and relating these to some physical model of the emitter or absorber. Each of these distributions, which may originate from either atoms or molecules, is called a spectrum. It may be continuous or discontinuous over any particular range of wavelengths. Traditionally, analysis in terms of wavelength, rather than frequency, has been more common because of the precision with which standard wavelengths can be compared experimentally with the distributions obtained in practice.

Optical spectroscopy refers to spectra obtained using the kind of optical equipment that has been conventionally used in the study of visible light—glass lenses, prisms, etc. More fundamentally, and as the definition appropriate to this article, optical spectra are those generated by the excitation of a valence electron in a molecule or molecular solid, or when the excited electron returns to a lower state emitting light in the process. Electrons are excited either electrically or by an incident light beam of the appropriate wavelength, as well as by other specialized methods. Molecular optical spectra are optical spectra generated exclusively by molecules, rather than by atoms.

I. Introduction

A. ENERGY CONSIDERATIONS

The energy necessary to excite a valence electron in gaseous or solid materials is usually in the range 1–12 eV, which is between the wavelength limits of about 12,000 and 1000 Å (1 Å = 10^{-10} m). This range includes the visible part of the electromagnetic spectrum (7000–4000 Å), and most light (color) in this wavelength range results from electronic excitation or emission processes.

Optical devices, such as lenses and prisms, made of glass are not transparent to radiation of wavelengths 3500–1000 Å, but lenses and windows made from fused silica or crystalline materials such as fluorite (calcium fluoride), lithium fluoride, etc. are, and have traditionally been used instead of glass. Because of the commonality of the origin of optical spectroscopy over this range, including visible, of wavelengths, the term "optical" is used even though some of the techniques are ones that will not necessarily be familiar to the reader.

Electron volts and reciprocal centimeters (cm^{-1}) are not Système Internationale units and are related to joules and to each other as follows:

$$1 \quad eV = 1.602189 \times 10^{19} \quad J$$

$$1 \quad cm^{-1} = 1.9865 \times 10^{-23} \quad J$$

$$1 \quad eV = 8065.5 \quad cm^{-1}$$

B. Description

The primary distinction between optical spectra generated by atoms and molecules is that while, according to quantum theory, the energy levels of both are quantized (can have only certain restricted values), the energy levels of the atom are uniquely concerned with *electronic* excitation. Molecules, however, possess internal degrees of freedom involving vibrational and rotational energy whose (quantized) states coexist with the electronic. Thus, changes in electronic states of atoms yield only relatively widely spaced lines [see Fig. 1(a)], whereas when there are changes in the electronic states of *molecules*, there are also simultaneous changes in the vibrational and rotational states. This adds greatly to the complexity in the appearance of the spectrum and the density of the lines observed, but, because of the greater regularity of the quantized levels of vibrational and rotational states, the molecular optical spectra are also more regular in appearance—that is, they display an obvious repetitive pattern. Under relatively low dispersion (as in Fig. 1), the coarse or gross structure is quite obvious, while the much finer (rotational) structure that gives each of the coarser features its "banded" appearance is not fully resolved. [Some resolution of the rotational structure can be seen in Fig. 1(b) in the feature that extends from ~3050 to ~3250 Å.] It was this blurred appearance that was first recognized as distinguishing molecular (band) spectra from atomic (line) spectra, and it is from an analysis of this structure that information about bond lengths, force constants, moments of inertia, etc. is obtained.

C. Generation of Radiation

There is no simple method to generate radiation over all the wavelengths (energies) of interest simultaneously, that is, over the range 12,000–1000 Å. White light covers the range

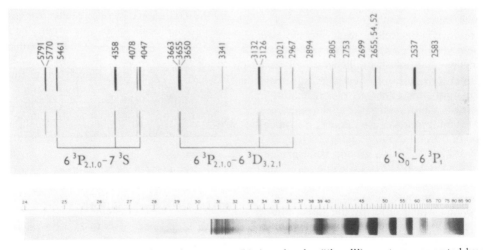

FIG. 1.   (a) The line spectrum of atomic mercury. (b) A molecular ("band") spectrum generated by an oxyacetylene flame. The spectrum extends from about 9000 to about 2800 Å.

7000–4000 Å and is usually generated by passing electric current through a fine resistance wire when it attains "white heat," such that all wavelengths in the visible are emitted. Heated filaments of this kind also emit in the range 12,000–7000 Å, where the radiation is perceived as heat, rather than light. Higher-energy and higher-intensity light sources must be generated by specialized methods, but, for the most part, they employ electrically excited plasmas, which consist of gas streams of very high kinetic temperature. The wavelengths emitted by these plasmas depend on both the current density and the nature of the gas, but there is usually some combination of the two that will generate an essentially continuous range of wavelengths over any particular spectral region. Figure 2 indicates the light sources and their effective wavelength range that are common in optical spectroscopy. More recently, wavelengths shorter than ~1500 Å have been very effectively generated using synchrotrons.

D. DISPERSION OF RADIATION

In order to analyze, that is, to find out what wavelengths of light are actually present in any light source, the radiation must be separated into its constituent wavelengths. This dispersion process was traditionally accomplished using glass, fused silica, or fluorite prisms, which spread, or dispersed, the wavelengths so that they could be individually examined. Modern technology, however, enables us to make much more efficient use of diffraction gratings for this purpose, and the majority of optical spectroscopes, spectrographs, etc. depend on gratings for their dispersive element. The gratings are usually reflection gratings: they are coated with aluminum or some other reflecting coating, and the light reflects off the grating, rather than being transmitted through it, as well as being dis-

persed. Transmission gratings in the visible are used, however, for cheap instruments and are quite effective. Even more recently, when the highest possible resolution of the wavelengths is necessary, modern interferometers are used and, in general, all of the techniques that are now associated with lasers. [*See* OPTICAL DIFFRACTION.]

E. DETECTION OF RADIATION

The eye is capable of detecting radiation over the visible range of the electromagnetic spectrum, but it has the property of perceiving a particular mixture of the visible colors as "white." On the other hand, white light can be dispersed through a prism or a grating, after which the eye has considerable discrimination in its ability to distinguish between the different colors, ranging from deep red at about 6500 Å down to the violet at 4000 Å. The very simplest spectroscopes employ direct visual examination of the spectrum, usually through a telescope or eye piece. More often, however, a photographic emulsion is the detector, and emulsions can be prepared with such fine particles of silver as their active ingredient that they can distinguish between lines as close together as one two-thousandth of a millimeter.

More frequently, radiation in the wavelength range 12,000–1000 Å is detected by a photomultiplier, which converts the light into an electrical signal. This signal can be recorded in a variety of ways but is usually converted into a plot of light intensity (signal) versus wavelength (Fig. 3). The same kinds of plots are also obtained from photographic plates by using a microdensitometer to convert plate blackening to an electrical signal. These records constitute the basic data to be analyzed. The precise variety of photographic emulsion or photomultiplier to be used in any particular experiment depends on the wavelength range, the intensity of the light, the wavelength resolution necessary, the acceptable "background noise," etc. and will not be further discussed here.

The number of angstroms (or other wavelength unit) per unit length on the photographic plate or recording chart is the reciprocal dispersion ($Å\ mm^{-1}$ or $Å\ cm^{-1}$), while the capability of the spectrograph or spectrophotometer to distin-

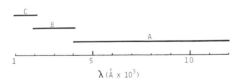

FIG. 2. Wavelengths that may be generated by (A) a hot filament, (B) a molecular hydrogen discharge, or (C) an electrical discharge through Xe and Kr.

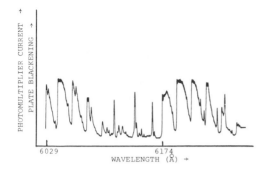

FIG. 3. Microdensitometer record of part of the emission system in the red region of the spectrum of the diatomic molecule NbN. The width of the features is due to the presence of unresolved rotational structure in the molecular bands. A trace of the band at 6029 Å under high resolution where the rotational structure can be clearly seen is shown in Fig. 8(d).

guish two adjacent lines separated by a wavelength difference $\Delta\lambda$ (at the wavelength λ) is defined as the resolving power and is measured by $\Delta\lambda/\lambda$. Conventional optical spectroscopy is usually done with spectrographs (or spectrophotometers) having reciprocal dispersions in the range ~ 10 Å mm^{-1} to ~ 0.1 Å mm^{-1} and resolving powers of ~ 5000–$500,000$. Resolving powers using lasers can, in special circumstances, attain values of 1×10^6–2×10^6.

F. Techniques in Optical Spectroscopy

The precise details of the methods of obtaining molecular optical spectra are now so varied as to make it impossible to give an exhaustive treatment. However, there are some general categories, and a few of the more specialized techniques will be discussed at the appropriate place.

In general, molecular spectra are divided into *emission spectra* and *absorption spectra*. Since most molecules are less robust than atoms, molecular emission spectra are seldom generated by high-energy plasmas etc., although the spectra of some diatomic—and even a few small polyatomic—molecules are observed to emit from the cooler parts of such plasmas. On the other hand, there are also low-temperature plasmas—that is, those in which the free electron energies are not in equilibrium with the temperature of the gases themselves—and these kinds of *glow discharges* are often useful in obtaining emission spectra of even quite large polyatomic molecules. Most emission spectra result when a reservoir of the molecular species is irradiated with a high-intensity light source that excites the molecules into electronically excited states from which they reemit. The most useful light source in this category is now the laser. Emission spectra generated by light sources of these kinds are known as *fluorescence spectra*. If the reservoir of molecules being excited is at such a low pressure that there is little probability of collisions occurring before reemission of the excitation, the spectra are further termed resonance fluorescence spectra.

The spectra of most polyatomic molecules are obtained as absorption spectra. The molecular reservoir, containing windows transparent to the radiation being used, is interposed between a light source containing a continuous distribution of wavelengths (blackbody lamp, hydrogen continuum, etc.) and the spectrograph or spectrophotometer that records the intensity of the transmitted radiation as a function of wavelength. The molecules under study may be dissolved in a solvent that is transparent to the radiation concerned, they may be in the gas phase, or they may be condensed as a solid. In the particular case of solids, the spectrum may be recorded over a considerable temperature range—as low as 2 K.

Finally, it is also possible to study unstable molecular species (radicals, ions, molecular fragments, etc.) by a number of methods. The reservoir containing the molecules in the gas phase may be subjected to an extremely intense short-duration flash of light that photolyzes the original molecule into photofragments (radicals etc.), which are then observed in absorption or emission, or else streams of different molecules may be mixed and, in the ensuing reaction between them, light may be emitted as chemiluminescence.

In these various ways, optical spectra of stable and unstable molecules have been intensively studied since 1925, and the characteristics of molecules as small as H_2 to molecules with molecular weights of thousands are known in great detail. In the following theoretical discussion, the spectra will be considered to have been obtained in absorption: that is, it is assumed that all molecules are initially in their ground states and are excited to higher excited states by absorbed radiation. It is the nature of the internal molecular motions resulting from the absorption of this radiation that is usually the object of study.

II. Theory of Molecular Optical Spectroscopy

Even though the lines obtained from a molecular spectrum are usually measured as wavelengths, all theoretical interpretations are performed in terms of energy. The relationship between wavelength and energy is provided by the Planck equation

$$E = h\nu$$

and since, for electromagnetic wave motion, $c = \lambda\nu$, then

$$E = hc/\lambda$$

Since E is inversely proportional to λ, it is more convenient to define a unit such that E is directly proportional to it; that is, we define $\bar{\nu} = 1/\lambda$. This is called the *wave number* and represents the number of waves per unit length (m^{-1} or cm^{-1}); thus $E = hc\bar{\nu}$, and since E is in joules, then $E/hc = \bar{\nu}$. Thus, it is usual to express "energies" in terms of cm^{-1} or m^{-1} (the SI unit is clearly m^{-1} but the cm^{-1} is well established in practical usage, just as the angstrom, 10^{-10} m, is also well established). Spectroscopic constants and energies are usually expressed in cm^{-1}.

A. DIATOMIC MOLECULES

Molecules that consist of only two atoms bonded together are the simplest molecular systems, and, by analogy with macroscopic analogs, we consider the diatomic molecule to be able to vibrate and to rotate. These two extra degrees of freedom are what distinguish molecular systems from atomic, where the only internal degree of freedom is electronic. As in atoms, so for molecular systems, the electronic energy is quantized. There is no generalized solution of the Schrodinger equation for the electronic energy levels of diatomic molecules, since the potential under which the electrons move is not simply defined and varies from molecule to molecule. The Coulomb potential for a hydrogen atom results in an inverse quadratic disposition of electronic energy levels, while a uniform potential results in a quadratic distribution. The outer electrons in a diatomic or polyatomic molecule move in a potential field that is somewhat intermediate between these two extremes, but in practice it is closer to that of the Coulomb potential: the energy levels converge toward the ionization potential of the molecule.

Since diatomic molecules have only one degree of vibrational freedom, it is possible to plot the potential energy $V(r)$ as a function of the distance r between the two atoms as they approach one another from an infinite distance ($V = 0$). The atoms X and Y are imagined to start at the point P where both of them are in their lowest electronic states. There are two possibilities: there is a minimum in the potential function (curve A, Fig. 4), or there is increasing repulsion at all internuclear distances (curve B, Fig. 4). Both energies must increase steeply as $r \to 0$ because of the large internuclear coulombic repulsion but, in the first case, a low energy equilibrium is reached between attractive and repulsive interactions at the equilibrium internuclear distance r_e. The reduction in total energy, below that of the atoms X and Y at an infinite distance from one another, is the dissociation energy D_{X-Y} of the bond; that is, it is the energy required to reverse the bond-forming process and remove the atoms X and Y to $r = \infty$.

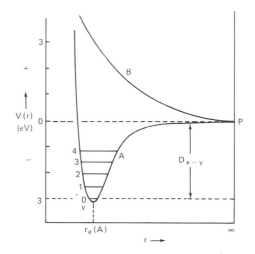

FIG. 4.   Plot of potential energy $V(r)$ as a function of the internuclear distance r. Curve A illustrates the situation where binding occurs, and curve B illustrates the case where the two atoms repel at all distances. (The distance—energy—between successive vibrational levels v has been exaggerated for diagrammatic clarity.)

Since potential curves of this kind, $V(r) : r$, are known to be almost quadratic in the region of the minimum, it is possible to solve the Schrödinger equation for the vibrational motion using a quadratic potential function, simulating the simple harmonic oscillator where $V(r) \propto r^2$. The solutions are $E_v = (v + \frac{1}{2})hc\bar{v}$ or $E_v/hc = (v + \frac{1}{2})\bar{v}$ where $v = 0, 1, 2, \ldots$ and is the vibrational quantum number for the state and where $\bar{v}$ is the "frequency" (cm^{-1}) of the oscillator. Classically, the angular frequency is $\omega = 2\pi\nu$ and ν $(sec^{-1}) = c\bar{v}$.

Whereas, in general, the spacing between *electronic* energy levels is $10–25 \times 10^3$ cm^{-1}, the spacing between successive vibrational levels in any individual potential function is ~200–2000 cm^{-1}. Thus, each separate electronic state is imagined to contain a series of quantized vibrational levels that merge together as the internuclear distance increases and the molecule dissociates into atoms (Fig. 4). Theoretically the nonlinear separation of these levels can be simulated using anharmonic oscillators instead of harmonic ones, that is, ones for which $V(r) = kr^2 + pr^3 + \cdots$ where $k \gg p$.

The second internal degree of freedom is that of molecular rotation. Since the potential energy for the motion is zero, the Schrödinger equation yields solutions that are quadratically spaced according to the relationship $E_J = BJ(J + 1)$ where $J = 0, 1, 2, 3, \ldots$ and $B = h^2/8\pi^2 I$ ($I = \mu r^2$ and $\mu = m_1 m_2 / m_1 + m_2$), where B is called the *reciprocal moment of inertia* and is usually expressed in cm^{-1} as follows: $E_J/hc = \bar{B}J(J + 1)$, where $\bar{B} = h/8\pi^2 Ic$.

The energy level formula for E_J neglects the interaction of the rotation and vibration; that is, it ignores the fact that at high rotational speeds the molecule stretches due to centrifugal force and the separation between rotational levels is no longer purely quadratic in J. Taking account of this interaction, the eigenvalues become $E_J = BJ(J + 1) - DJ^2(J + 1)^2$, where D is the centrifugal distortion constant ($\bar{D}$ in cm^{-1}). It is relatively small compared with $\bar{B}$ and, generally, $\bar{D} \approx 10^{-5}\bar{B}$. Thus, information regarding the internuclear distance is contained in the constants $\bar{B}$ and $\bar{D}$, which are obtainable from the spectrum. A calculation of the magnitude of $\bar{B}$ for diatomic molecules, where m_1 and m_2 are atomic masses appropriate to the first or second row elements, shows that the spacing between $J = 0$ and $J = 1$ is of the order of $0.1–5$ cm^{-1}, that is, much smaller than the vibrational spacings. Thus we may well imagine the energy levels of a diatomic molecule—and, in fact, any molecule—to be made up of an electronic potential function with a series of roughly linearly spaced vibrational energy levels, each of which has its own much finer spaced set of rotation levels. The rotational levels are, of course, not representable upon a $V(r) : r$ diagram, since they do not depend on r, but the sets of vibration–rotation states may be imagined as in Fig. 5.

The molecule thus has a whole set of potential functions, each with its vibrational and rotational sublevels in the case of bound states. The set of states—somewhat simplified—for the molecule C_2 is given in Fig. 6 where, now, the zero of energy has been shifted to the $v = 0$ level of the lowest level, since all electronic transitions in absorption start from there.

There is no reason why the lowest state of a molecule should be bound, and, in general, this is not so. Thus, the lowest electronic state of the molecule He_2 is dissociative (repulsive), and the molecule is only stable in higher excited states. For convenience, however, the examples given can be assumed to have stable ground electronic states unless otherwise specified.

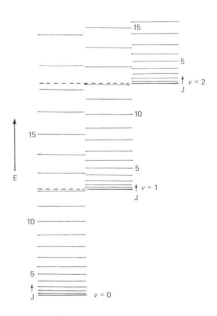

FIG. 5.   The rotational energy levels J of a diatomic molecule. The diagram illustrates their quadratic energy distribution and indicates that each separate vibration level v can be regarded as having its own rotational stack.

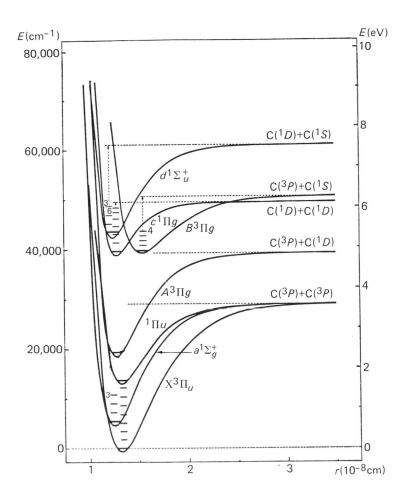

FIG. 6.   The set of $V(r):r$ potential functions for the C_2 molecule. (This diagram corresponds to the energy level scheme for which the $a\,^1\Sigma_g^+$ state was assumed to be above the $X\,^3\Pi_u$ state. Experimentally, in fact, $^1\Sigma_g^+$ is known to be the lower.)

B. OPTICAL TRANSITIONS IN DIATOMIC MOLECULES

When a molecule in its lowest electronic state E'' is placed in a radiation bath such as a light beam, it is able to absorb only certain highly selective frequencies from the bath and become electronically excited to the state E'. The transition energy ΔE is dictated by the precise nature of the states E' and E'', and these must now be precisely specified.

Each different electronic state of a diatomic molecule may be represented by potential curves of the type illustrated in Fig. 4. As for atoms, the electronic angular momentum is quantized, but, since an internal axis is defined by the line joining the two nuclei, it is the component of the electronic angular momentum along this internuclear axis that is quantized in the diatomic molecule. This component is designated Λ (cf. L for atoms) and Σ, Π, Δ, Φ, ... are used when the magnitudes of this angular momentum are 0, 1, 2, 3, ... etc. (cf. s, p, d, f, ... for atoms). In addition, a superscript is used to indicate the *spin multiplicity*, $(2S + 1)$, which is identical to the usage in atomic spectra. For those molecules that have identical nuclei (homonuclear), the electronic wave function may be symmetric or antisymmetric with respect to inversion through the center of symmetry (i.e., the midpoint of the internuclear axis). This property is denoted by subscripts g and u, which are abbreviations for the German words for even and odd (*gerade* and *ungerade*). Each group of these three symbols is defined as a *molecular term* (e.g., $^3\Pi_g$) for the diatomic molecule and, in general, electronic transitions take place between different terms.

Not all transitions are allowed between the states designated by different terms and, for the absorption or emission of electric dipole radiation, the transitions are restricted to $\Delta\Lambda = 0, \pm 1$ and $\Delta S = 0$. The component of the electron spin along the internuclear axis is quantized and designated Σ ($\Sigma = S, S - 1, ..., -S$), and the total angular momentum along the internuclear axis is designated Ω ($\Omega = \Lambda + \Sigma$). There is also an interaction between the electronic angular momentum Λ and the spin Σ such that for $\Lambda \neq 0$ the values of Ω no longer have equal energies. This phenomenon is called *spin–orbit coupling,* and the coupling increases rapidly in magnitude ($\sim Z_{\text{eff}}^2$) with the atomic number of the atoms from which the molecule is constructed. This separation of, for example, a $^2\Pi$ term into two components $^2\Pi_{1/2}$ and $^2\Pi_{3/2}$ ($\Lambda + \Sigma \equiv 1 + \frac{1}{2}$ and $1 - \frac{1}{2}$) is not always represented on potential curves of the type given in Figs. 4 and 6 since, for molecules in the first row of the periodic table, the splitting distance is small compared with the vibrational intervals. Taking due account of this coupling, the overall selection rule for the absorption or emission of electric dipole radiation is $\Delta\Omega = 0, \pm 1$.

The rotational motion of the diatomic molecule is also subject to selection rules, since total angular momentum must be conserved during a transition. Accordingly, $\Delta J = 0, \pm 1$ (except for $\Sigma \leftrightarrow \Sigma$ transitions where $\Delta J = \pm 1$ only). Vibrational motion has no angular component, so that it does not have a selection rule of this kind but, instead, there is an important propensity rule known as the Franck–Condon principle. Somewhat simplistically, it is assumed that the kinetic energy of the nuclei in a vibrating molecule cannot be changed during a transition. Since the kinetic energy depends directly on r, transitions can be represented by vertical lines on a $V(r):r$ diagram. It is usual to regard transitions as occurring from the two points of intersection of the vibrational level with its potential curve, since it is here, where the kinetic energy is zero (i.e., at the turning points of the motion) that the mole-

cule spends most of its time. Since the $v = 0$ level has a maximum probability at the center of the line representing the vibrational level (because of the unnoded probability function of this state), transitions from $v = 0$ are represented by a line drawn from the center of the level.

All these considerations are taken into account by the three vertical lines drawn from the lower potential function to the higher in Fig. 7(a). It will be noted that because of the larger r_e value of the upper state, transitions arising from $v = 0$ of the ground state terminate on higher vibrational levels of the excited electronic state. Vertical lines represent the maximum in transition probability; that is, transitions to other vibrational states will occur but with lower intensities. It is this circumstance that leads to the coarse (vibrational) structure that was indicated by the arrows in Fig. 1(b), and it is, of course, a feature unique to molecular, as distinct from atomic, spectroscopy. The succession of features separated by quanta of the upper-state vibrational frequency $\bar{\nu}'$ are referred to as *progressions:* upper-state progressions for differences in the upper-state quantum level in absorption, and [see Fig. 7(b)] ground-state progressions for a system in emission where the features are separated by lower-state vibrational intervals $\bar{\nu}''$. In the diagram for the emission spectrum, it has been assumed that only the levels $v' = 0$ and 1 are populated and capable of emitting to the ground state.

The final internal degree of freedom that gives rise to the unresolved part of the spectrum given in Fig. 1(b) is that due to the changes in rotational energy upon transition. In the simplest possible case where the reciprocal moments of inertia ($\bar{B}$) of both the upper and lower states are identical, such that the minima of the two potential functions are directly above one another, the selection rule $\Delta J = 0$ simply results in an extremely strong line at a single frequency [ν_0, say—see Fig. 8(a)]. Figure 8(a) also shows that the transitions with $\Delta J = +1$ form a group to the high-frequency side of ν_0, while those with $\Delta J = -1$ form a similar group on the low-frequency side of ν_0. Thus, the spectrum occurs in three parts that are due to $\Delta J = +1$, $\Delta J = 0$, and $\Delta J = -1$, and these are called R, Q, and P *branches,* respectively. For the more frequent case where the potential minima are not directly above one another and where, therefore, $I' > I''$ or $I' < I''$, since the level distributions for both the lower and upper states are quadratic in character, the difference is also quadratic, and the absorption or emission rotational structure will form a

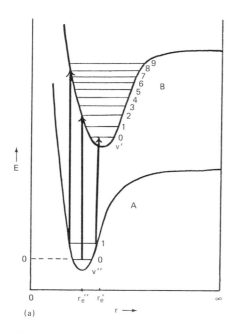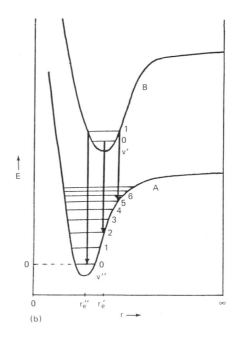

FIG. 7. The process of transitions between different electronic states of a diatomic molecule. (a) The process of absorption of energy from the incident light beam, resulting in population of a distribution of excited state B and vibrational levels v'. (b) The emission of radiation (light) from an excited electronic state to an energetically lower one. Both parts illustrate that vibrational-level populations change during transitions according to the Franck–Condon propensity rule.

head, in the R branch if $I' < I''$ and in the P branch if $I' > I''$. These two cases, together with the "undegraded" example for $I' = I''$, are illustrated in Figs. 8(b), 8(c), and 8(d), respectively. It is the unresolved quadratic distribution of these lines that gives the molecular spectra such a characteristic "band" appearance as was shown in Fig. 1(b). Thus, immediate inspection of the appearance of an optical transition of a diatomic molecule shows whether the equilibrium internuclear distance is larger or smaller in the excited state than in the ground state. It is more usual for the internuclear distance to increase upon excitation, but many examples are also known where the opposite is true. It is a relatively straightforward matter [see for example Herzberg (1950), pp. 168 *et seq.*] to obtain both $\bar{B}'$ and $\bar{B}''$ from the line positions of the rotational structure, provided that they are adequately resolved at high reciprocal dispersion and resolution. From this information, the internuclear distances for both electronic states may be obtained and, from other features of the spectrum (for example, the absence of the Q branch which would indicate a $\Sigma \leftrightarrow \Sigma$ transition), the

nature of the terms for both electronic states may be fully determined. When the upper and lower electronic potential functions are displaced from one another so that the transitions terminate on a wide variety of vibrational levels, it is also possible to obtain the internuclear distances for those vibrational levels and, in this way, to build up a knowledge of the shape of the potential function. Potential curves of the kind given in Figs. 4, 6, and 7 are often referred to as *Morse curves,* after P. M. Morse, who first constructed an approximate potential function to describe them. Table I gives a brief collection of some common diatomic molecules together with their internuclear distances for the ground states, the vibrational frequency interval for that state, and the force constant obtained from regarding the oscillator as approximately harmonic in character. An immense bibliography of these same constants—as well as others—for many different electronic states of these molecules, as well as hundreds of other diatomic molecules, may be found in the appropriate literature [see for example Huber and Herzberg (1979)].

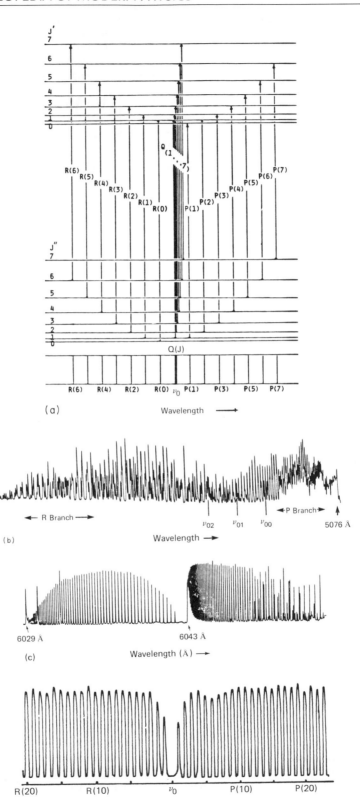

(a)

Wavelength ⟶

(b)

ν_{02} ν_{01} ν_{00}

◄─ R Branch ─►

◄─P Branch─►   5076 Å

Wavelength ⟶

(c)

6029 Å   6043 Å

Wavelength (Å) ⟶

R(20)   R(10) ν_0 P(10)   P(20)

◄─ P branch ─►   WAVELENGTH ⟶   ◄─R branch─►

III. Polyatomic Molecules

For diatomic molecules, even though there are no simple formulas from which to calculate the electronic energy levels, the spectrum is an-

alyzed from the vibrational and rotational structure accompanying the electronic transition. From the theoretical model equations, which are relatively straightforward, the electronic state is then characterized. As already illustrated, this

TABLE I.   Characterizing Constants of Some Diatomic Molecules in Their Electronic Ground States[a]

Molecule	Electronic ground state	$\bar{B}_e$	$\bar{D}_e$	R_e (Å)	$\bar{\nu}_e$	k (N m^{-1})[c]
H$_2$	$^1\Sigma_g$	20.335	—[b]	0.7414	4401	574.8
H^{35}Cl	$^1\Sigma$	5.448	1.39 (4)	1.2746	2991	516.1
AlO	$^2\Sigma$	0.6414	1.08 (6)	1.6179	979	566.9
^{11}BH	$^1\Sigma$	12.021	1.24 (3)	1.242	2367	2032.1
C$_2$	$^1\Sigma_g$	1.8198	6.92 (6)	1.2425	1855	1216.0
C^{35}Cl	$^2\Pi$	0.6936	1.9 (6)	1.6430	867	395.5
CO	$^1\Sigma$	1.9313	6.12 (6)	1.1283	2170	1901.5
^{133}CsF	$^1\Sigma$	0.1844	0.20 (6)	2.3454	353	122.0
I$_2$	$^1\Sigma_g$	0.03737	4.3 (9)	2.6663	215	172.7
CN	$^2\Sigma$	1.8997	6.40 (6)	1.1718	2069	1629.3
CN$^+$	$^1\Sigma$	1.8964	7.0 (6)	1.1729	2033	1573.1
MgO	$^1\Sigma$	0.5743	1.276(6)	1.7490	785	350.1
N$_2$	$^1\Sigma_g$	1.9982	5.76 (6)	1.0977	2359	2294.8
NO	$^2\Pi$	1.6720	0.54 (6)	1.1508	1904	1653.0
O$_2$	$^3\Sigma_g$	1.4377	4.84 (6)	1.2072	1580	1175.9
OH	$^2\Pi$	18.9108	19.38 (4)	0.9697	3738	780.2
^{28}SiO	$^1\Sigma$	0.7267	0.98 (6)	1.5097	1242	926.0
^{48}TiO	$^3\Delta$	0.5354	6.03 (7)	1.6202	1009	719.0

[a] There are similar sets of constants for all of the stable excited states available in the literature. The units are all in cm^{-1} unless otherwise indicated, and the numbers in parentheses represent the power of the exponents by which the given numbers must be multiplied to get the absolute quantity. Thus, 1.39(4) = 1.39 × 10^{-4}.
[b] Not accurately known.
[c] In the older literature, force constants are given in dynes per centimeter: 1 N m^{-1} = 10^3 dynes cm^{-1}.

◀ FIG. 8.   (a) A schematic spectrum of a diatomic molecule in which the excited electronic state potential function is vertically above that of the ground state. Since the rotational levels are (quadratically) spaced the same for both states, all the (Q) $\Delta J = 0$ lines lie on top of each other and the (R and P) $\Delta J = \pm 1$ lines are linearly distributed with a spacing of $2\bar{B}$. Under these circumstances, the band is "undegraded," that is, it has no tendency to form a head. (b) The 0, 0 band of the 5076-Å system of the diatomic molecule AlN. The head forms on the long-wavelength side of the band, in the P branch, indicating that $I' < I''$ ($r_e' < r_e''$). The transition is a complex one, so that it is not possible to follow the individual lines in each branch. The spectrum contains P, Q, and R branches, but even these groups are not easy to discern. [See, however, (c) and (d).] (c) A high-resolution trace of the spectrum of the 6029-Å band of the red system of the diatomic molecule NbN. The band is not strongly degraded, so that the head, in the R branch, occurs only at the very-high-frequency end (at exactly 6029 Å). The low-J lines in the Q branch are unresolved and form the feature at 6043 Å, and the P-branch lines can be seen emerging from under the stronger Q lines, at the long-wavelength end of the spectrum. The spacing between them increases as J increases. (d) A band in the electronic spectrum of the diatomic molecule ion CN$^+$. The band is the 0, 0 (i.e., $v'' = 0$ and $v' = 0$) of the $f^1\Sigma \leftrightarrow a^1\Sigma$, and its origin, indicated by ν_0, is at 2180.6 Å. There is no line at the origin because the selection rule for $\Sigma \leftrightarrow \Sigma$ is $\Delta J = \pm 1$ so that the Q branch ($\Delta J = 0$) does not occur. It should be noted that there is almost no change in the spacing of the lines in either the R or P branches, even though the value of J has reached 20 in both of them. This band is the classical case of a transition where the upper and lower potential functions are vertically disposed, so that r_e' and r_e'' have essentially identical values (r_e for the f state is 1.171 Å, and r_e for the a state is 1.1729 Å). The plate was overexposed to show the first lines in the P and R branches.

characterization is in terms of the equilibrium internuclear distance for the two atoms (i.e., the bond length) in the electronic state, its force constant(s) (anharmonic as well as harmonic), and the interaction between vibration and rotation that appears as a centrifugal distortion parameter ($\bar{D}$).

This characterization is possible because of the relatively few degrees of freedom of the diatomic molecule—apart from the electronic motions, which are expressed empirically only in terms of a single difference in energy between the states giving rise to the transitions. Thus there is only one vibration ("stretching"), and over its whole amplitude, the symmetry of the molecule cannot change—the bond is longer or shorter, but the molecule is still "linear." Also, there is only a single moment of inertia for the rotating system, since the two moments of inertia perpendicular to each other and to the internuclear axis are equal. This simplicity allows the molecule to be characterized by very few parameters for each electronic state. As soon as the molecule has more than two atoms, there is a large increase in the complexity of representing each electronic state and, in fact (see below), the actual distinction between "different" electronic states is no longer obvious or even well defined in some cases.

The primary difference in complexity is caused by the increase in the number of vibrational degrees of freedom. A molecule containing N atoms possesses $3N - 6$ vibrational degrees of freedom if nonlinear and $3N - 5$ if linear [since, if nonlinear, there are $3N$ degrees of freedom altogether of which three are rotations (two for linear molecules) and three are translations]. Thus, a relatively small molecule such as benzene (C_6H_6) has 30 vibrational degrees of freedom, and the representation of the variation of the potential energy with the (30) different vibrations must be viewed as a problem in 30 dimensions! At the most simplistic level it is possible to conceive of 30 different Morse curves, but the problem is that the 30 different vibrations are not perfectly harmonic, and they therefore interact with one another.

The second source of complexity arises from the rotational motion. Except for the linear polyatomic molecule (e.g., CO_2), there is now more than a single moment of inertia—in fact, there are three, I_x, I_y, and I_z, with respect to rotation about the x, y, and z axes. As is well known from classical dynamics, it is possible to select principal axes of inertia with respect to which all

(cross) products of inertia terms vanish. In molecular spectroscopy, these are denoted I_A, I_B, and I_C, where $I_A < I_B < I_C$. Four cases arise:

Spherical top (e.g., CH_4, SF_6)

$$I_A = I_B = I_C$$

Prolate symmetric top (e.g., CH_3F, $CH_3—C{\equiv}CH$)

$$I_A < I_B = I_C$$

Oblate symmetric top (e.g., C_6H_6, NH_3)

$$I_A = I_B < I_C$$

Asymmetric top (e.g., H_2O, $CH_2{=}CH_2$)

$$I_A \neq I_B \neq I_C$$

The optical spectra of polyatomic molecules have very different rotational structure, dependent to which of the above classes the molecule belongs. It should be noted that all linear polyatomic molecules (e.g., CO_2, C_2H_2) are *prolate* symmetric tops, since $I_A \equiv 0$.

The analysis of the spectra of both spherical and asymmetric tops is beyond the scope of this article, but since many asymmetric tops are very nearly symmetric tops (e.g., N_3H, CH_2O, etc.), they can often be dealt with using symmetric-top formalism (see below).

The spectra of polyatomic molecules have one further complicating factor, which arises simply because of the high density of states resulting from the much larger numbers of electronic, vibrational, and rotational energy levels. This high density allows frequent interactions between the rotational and vibrational levels of any particular electronic state (*rovibronic* levels) and results in much greater complexity than for diatomic molecules.

This "mixing" (interaction) between "different" electronic states has a further significant result for polyatomic molecules. Thus, with few exceptions, they emit *fluorescence* only from the lowest excited electronic state having the same spin multiplicity as the ground state. Many polyatomic molecules also exhibit *phosphorescence spectra* when emission to the ground state is observed from the lowest electronic state differing in spin multiplicity from the ground state by 2 (i.e., singlet ↔ triplet, doublet ↔ quartet, etc.). These emissions are usually long-lived (since they are "spin forbidden") and have half-lives greater than or equal to milliseconds or microseconds, compared with nanoseconds for the spin-allowed transitions. The density of states in polyatomic molecules is usually so large that there is always a pathway in which high-energy excitation of the molecule can lead to changes in the vibrational distribution and the electronic state of the molecule, consistent with the conservation of energy. It will be appreciated that even the zeropoint vibrational energy of most polyatomic molecules is $\gtrsim 10^4$ cm^{-1} ($\gtrsim 1$ eV) so that "excess electronic energy" can easily be accommodated in the vibrational modes, since these do not usually have a problem of angular momentum conservation (unlike the rotational energies). Radiationless transitions between electronic states of the same spin multiplicity are called *internal conversions,* while those states with different (usually by ± 2) spin multiplicities are known as *intersystem crossings.*

The theoretically ideal situation would be one in which interactions between electronic and vibrational motions or vibrational and rotational motions do not occur, and this is referred to as the *Born–Oppenheimer approximation.*

With this approximation, the molecular wavefunction can be written as a simple product, that is, $\Psi = \Phi_{el}\Phi_{vib}\Phi_{rot} \cdots$, such that the total energy (E_t) is

$$E_t = E_{el} + E_{vib} + E_{rot} + \cdots$$

This approximation has great simplifying features in that it provides a zeroth-order vocabulary and simple operational equations for each different degree of freedom, but, essentially, it ignores the realities of most of the properties of polyatomic molecules, that is, the interactions between the different degrees of freedom.

From a practical point of view, what this means is that in the optical spectra of polyatomic molecules there is a great emphasis on the magnitudes of those parameters that measure the cross interactions, and these are usually much larger than their equivalents for diatomic molecules. The largest effects are those interactions between electronic and vibrational motions, and these are categorized under two principal "effects": the Renner (Renner–Teller) and the Jahn–Teller Effects.

A. The Renner (Renner–Teller) Effect

This relates to the effect of bending modes on the vibronic levels of degenerate electronic states of *linear* polyatomic molecules—that is, the effect on the vibronic energy levels of those vibrations that destroy the linearity of the atoms making up the molecule. Since Π, Δ, Φ, ... electronic states of any linear (diatomic) molecule (i.e., those states having $\Lambda = \pm 1$, ± 2, ± 3, etc.) are doubly degenerate due to the axial symmetry of the molecule, it follows that if the molecule is bent, the states no longer have equal energies and the degeneracy is removed. Figure 9(a) shows the unperturbed potential function, which is assumed to have a perfectly quadratic form ($V^\pm = kr^2$) but where there is no interaction between the electronic and vibrational motions (ideal). Figure 9(b) shows the splitting of $V^\pm$ into two unequal components for a very small vibronic interaction, and Fig. 9(c) shows the form of the potential functions for larger interactions. In the last example, the potential function has broken up into two more or less distinct parts, V^+ and V^-, and theoretical considerations show that whereas *some* of the original vibronic levels

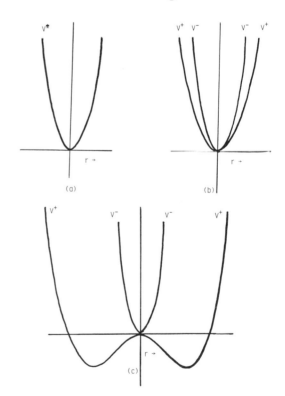

FIG. 9. (a) The potential function for a (doubly degenerate) Π state of a linear molecule as a function of a bending mode (amplitude r) for a zero interaction between the electronic and vibrational degrees of freedom, that is, there are two *coincident* potential functions. (b) The potential function for a Π state of a linear molecule with a very small interaction between the electronic and vibrational modes. (c) The potential function for a Π state of a linear molecule with a large vibronic interaction: that is, the molecule no longer has its lowest energy at the linear configuration.

can be assigned to either V^+ or V^- (the nondegenerate vibronic ones), there are others that split but belong to *both* states (the degenerate vibronic ones), even though we can, with some imprecision, assign the high split component to V^- and the lower to V^+. This ambiguity arises because the two electronic states coincide and are in contact (i.e., do not cross at right angles) at $r = 0$, so that the molecule can change from one "state" to the other unless there are other (symmetry) reasons that forbid it. The result is that many of the *vibronic* states are well defined but are no longer resolvable into electronic and vibrational parts; that is, there is a significant breakdown of the Born–Oppenheimer approximation, and the spectra of linear molecules involving degenerate electronic states must be interpreted taking these considerations into account.

B. THE JAHN–TELLER EFFECT

The second major "theorem" is due to Jahn and Teller, who proved in an *ad moleculam* argument that all degenerate electronic states of polyatomic (nonlinear) molecules are unstable with respect to distortions from the symmetrical nuclear configuration. (Since degeneracy accompanies symmetry, the presence of orbital degeneracy implies a symmetrical nuclear shape for its nondegenerate states.) Thus, for any symmetrical molecule (e.g., CH_4 in a regular tetrahedral shape or C_6H_6 as a regular hexagon) excited to a degenerate electronic state, there will always be a vibration that will distort the molecule from that regular shape, leading to a lower potential energy. The distortions are mandatory theoretically, but the extent of the distortion depends on the nature of the degenerate excited state (i.e., the precise electronic distribution) and the force constant(s) of the distorting vibration(s) (there may be more than one). Figure 10(a) shows the unsplit degenerate potential function and Figs. 10(b) and 10(c) show, diagramatically, the effects of a small and a large Jahn–Teller distortion, respectively.

From the point of view of optical spectroscopy, both the Renner (Renner–Teller) and the Jahn–Teller effects greatly complicate analysis of the spectrum, but for the Renner (Renner–

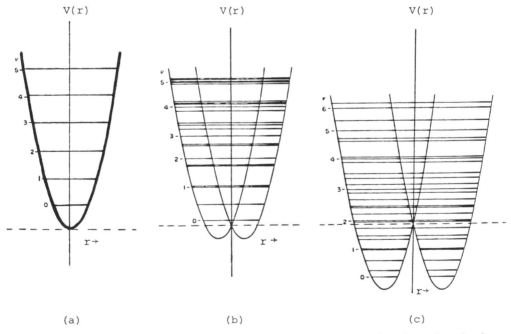

(a)   (b)   (c)

FIG. 10.   (a) The potential function of a doubly degenerate electronic state of a polyatomic molecule as a function of one of the non-totally symmetric vibrational modes for the case of zero vibronic interaction. (b) The potential function of a doubly degenerate electronic state of a polyatomic molecule as a function of one of the Jahn–Teller active vibrational modes with a small vibronic interaction, less than the zero-point energy of the molecule. (c) The potential function of a doubly degenerate electronic state of a polyatomic molecule as a function of one of the strongly interacting Jahn–Teller vibrational modes.

Teller) effect and particularly for small triatomic molecules such as BO_2 and NCO, the theoretical description of the dynamics has been verified. The Jahn–Teller effect is much more complex, and while there are, as yet, few carefully analyzed examples, there is no doubt that the much greater complexity and differences of many optical spectra of polyatomic molecules are due to its operation.

Finally, in terms of the generality of phenomena associated with molecular optical spectroscopy, is the issue of localized and delocalized spectra. It was relatively early recognized that visible color (in particular) and optical absorption in general, are due to the presence of functional groups (e.g., $-N{=}N-$, $\searrow C{=}O$, $\searrow C{=}C\diagdown$, etc.) and that the intensity and wavelength maximum of the absorption are strongly dependent upon the degree to which these groups (*chromophores*) are conjugated with other unsaturated parts of the molecule or with each other.

In molecular orbital terms, each electronic state of the molecule can be written as a linear combination of the wave functions of the outer (valence) electrons of the atoms that make up the molecule. That is,

$$\phi_{el} = (1/\sqrt{N}) \sum_i c_i \chi_i$$

where i runs over all N atoms, χ_i is the atomic wavefunction of the ith atom, and c_i are the coefficients that give the amplitude of the atomic orbital on atom i in the molecular orbital ϕ_{el}. Since only one electron can be assigned to each orbital (exclusive of spin); $\sum_i c_i^2 = 1$. Thus $N^{-1/2}$ is the normalizing factor in the molecular wave function.

If, because of symmetry—or accidentally—all $c_i = 0$ except, say, c_j, then the electron is localized on a single atom. Further, if in the excited state ϕ'_{el}, once again all $c'_i = 0$ except for c'_j (i.e., $c_j^2 = c_j'^2 = 1$), then the transition is said to be *localized* on atom j. This type of optical transition is exemplified by transitions in lanthanide elements (ions) in oxide glasses, in lanthanide ions in solution, and in the gas-phase spectra of lanthanide oxides, etc. This occurs because, at least in the lowest-energy region, the transitions involve excitation between the various (split) terms of the $4f^n$ configurations that are located exclusively on the lanthanide ion. This situation also occurs to a lower degree of approximation in the analogous transition-metal ion complexes, where the spectra originate and terminate among the (split) atomic terms arising from the various d^n configurations. Spectra of this kind are referred to as crystal field or ligand field spectra.

If the molecular orbitals of the ground and excited state are such that *most* of the c_i and $c'_i \neq 0$, then the excitation is spread out over the molecule as a whole, that is, delocalized. The spectra of the aromatic hydrocarbons and all highly conjugated molecules ranging in type from the polyenes to the cyanine dyes carotenes and retinals are of this type. Since the excitation is delocalized, there is usually only a very small change in any (all) of the bond lengths and bond angles, and the transitions are either "vertical" or very nearly so, in the Frank–Condon sense.

In between the localized atom (ion) and the completely delocalized systems are those molecules for which all c_i and $c'_i \sim 0$ except for, say, c_j and c_{j+1} or c_{j-1}, i.e., the transition is localized to a small group (pair) of atoms which, in the older literature, was called a chromophore. Thus, for example, the only optical transition at moderate energies ($\gtrsim 2000$ Å) for a molecule

$$CH_3{-}CH_2{-}\underset{\underset{O}{\|}}{C}{-}CH_2CH_3 ,$$

such as diethyl ketone, is localized almost entirely on the carbonyl group and occurs at ~ 3100 Å. The spectrum shows vibrational activity in only the C=O group, all other modes being almost absent, that is, unaffected by the excitation.

From a qualitative point of view, then, it is always desirable to assess the character of the polyatomic molecule whose spectrum is under study in order to see if the problem can be partitioned and thereby reduced in complexity before undertaking a more holistic approach to the analysis.

C. ROTATIONAL LEVELS OF SYMMETRIC TOPS

In the symmetric top, two of the moments of inertia are different from the third. Just as in the case of atoms and diatomic molecules, the total angular momentum is quantized, but now there are two different components. The quantum number for the total angular momentum is symbolized by J ($= 0, 1, 2, ...$), while its component along the unique axis of the symmetric top (i.e., the largest inertial axis for an oblate top and the smallest for a prolate top) is symbolized by K. Since K represents the *component* of the *total* angular momentum in a molecule fixed direction, then $|K| = 0, 1, \ldots, J$, where the modulus sign indicates that the momentum can be in either the clockwise or counterclockwise directions with respect to the unique axis.

Relatively straightforward (quantum) dynamical arguments show that (ignoring centrifugal distortion)

$$E_{J,K} = BJ(J + 1) + (A - B)K^2$$
for prolate tops

$$= BJ(J + 1) + (C - B)K^2$$
for oblate tops

where E_{JK} is the total energy of a molecule with a total of J quanta of angular momentum, K of which are with respect to the top axis, and where $\bar{A}, \bar{B}$, and $\bar{C}$ are the reciprocal moments of inertia such that $\bar{A} = h/8\pi^2 I_A c$, $\bar{B} = h/8\pi^2 I_B c$, and $\bar{C} = h/8\pi^2 I_C c$. Whereas for the diatomic molecule E_J has a simple quadratic dependence on J, E_{JK} is more complex, and it is desirable to divide it into rotational stacks with specified K each having values of $J = K, K + 1, \ldots$ (since $K \leq J$). Thus Fig. 11(a) shows the E_{JK} levels for a prolate top and Fig. 11(b) those for an oblate top (remembering that $A > B > C$, since $I_A < I_B < I_C$).

D. VIBRATIONAL LEVELS OF POLYATOMIC MOLECULES

The set of vibrations of a polyatomic molecule are usually classified according to their symme-

try point group irreducible representations, but it is sufficient to recognize that there are two distinct subsets—those vibrations that, during their motion, preserve all the symmetry characteristics of the molecule, and those that do not. The simple stretching modes of diatomic molecules clearly leave its symmetry unchanged and are, therefore totally symmetric, while, for example, the bending modes of a linear molecule or the out-of-plane modes of a planar molecule clearly destroy at least some elements of the symmetry of the nonvibrating molecule and are denoted non-totally symmetric modes.

If all the atoms of a complex polyatomic molecule are set vibrating at random (and *vibrating* means that there is no net translation or rotation of the molecule as a whole) and the molecule is viewed with a stroboscope that is capable of having its frequency varied at will, it will be observed that there are particular stroboscopic frequencies at which *all* of the atoms are moving at the same (stroboscopic) frequency and with the same phase as each other. These particular motions are called normal coordinate motions, and there are $3N - 6$ of them for each N-atomic nonlinear polyatomic molecule. In some of these modes it can be seen that, for example, only a single bond, such as C=O, is stretching and compressing, while the rest of the atoms are essentially motionless. A normal mode of this kind that can be almost localized to a small region of the molecule is called a local mode, and

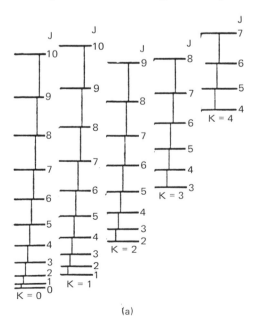

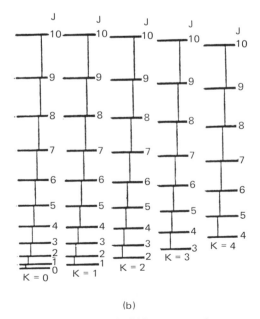

FIG. 11.   Energy-level diagram for symmetric-top molecules. (a) Prolate. (b) Oblate symmetric.

the description is often used as an approximation even when the motion is not completely of this type. The complex motion observed in the absence of the stroboscope is known as Lissajou motion. It can always be resolved into a set of normal coordinate forms.

E. Optical Spectra of Polyatomic Molecules

1. Vibrational Structure

Each electronic state of a polyatomic molecule therefore possesses a set of $3N - 6$ vibrations divisible into the totally and non-totally symmetric subsets, and the molecule may have any or all of these excited and, additionally, be rotating. Since the moments of inertia of a polyatomic molecule depend on the distances of each of the constituent atoms from the principal axis of inertia, it follows that the molecule has slightly different moments of inertia for each vibrational state, so that the rotational structure "superimposed" on the vibrational level is slightly different for each vibrational mode, and the reciprocal moments of inertia derived from analysis of the spectrum are given subscripts to indicate this—that is, A_{v_i}, B_{v_i}, C_{v_i}, where v_i is the ith vibrational mode and i runs over the complete set of $3N - 6$ modes.

When an electron in a polyatomic molecule is excited from a low-lying molecular orbital to a higher (unoccupied) molecular orbital, the potential function is changed—sometimes significantly and sometimes only trivially, dependent on the precise nature of the two orbitals. In general, there are three types of possibilities:

1. That the force constants change by only trivial amounts and the $(3N - 6)$-dimensional potential surface of the excited electronic state is virtually unchanged. (This might happen for the excitation of a nonbonding electron into a nonbonding excited orbital.)

2. One or two (or three …) of the force constants change significantly, but the molecule retains the same *shape* in its excited state and merely becomes slightly smaller or larger: that is, the equilibrium internuclear distances (potential minima) of the totally symmetrical modes change slightly but *none* of the non-totally symmetrical minima change (even though the force constants change).

3. One or two (or three …) of the force constants change significantly and the molecule changes its shape in the excited state. Under these circumstances, one can analyze the situa-

tion in terms of there having been changes in the positions of the minima of the non-totally symmetric vibrational potential functions from their ground-state values.

The optical spectrum can be analyzed in these three cases more or less in terms of the Franck–Condon principle discussed above.

1. Since there are no changes in the minima of any of the $3N - 6$ potential functions, the spectrum contains no progressions but, instead, consists simply of a set of *sequences* ($v_i'' \rightarrow v_i'$, i.e., $\Delta v_i = 0$) clustered around the electronic origin of the transition (the O—O of the diatomic molecule and the O_0^0 of the polyatomic, where the sub- and superscripts refer to the number of modes excited in the ground and excited states, respectively). For sequences to be seen in the spectrum, the particular vibration, v_i'' say, must be populated in the ground state, and if the force constant of v_i'' changes (increases *or* decreases) so that the frequency of the vibration changes from v_i'' in the ground state to v_i' in the excited state, then the sequence $v_{i1}'^1$ will appear displaced by $v_i'' - v_i'$ (cm^{-1}) from O_0^0. (Every vibration in a polyatomic molecule is assigned an identity number 1, …, $3N - 6$.)

This produces a spectrum with most of its intensity in the O_0^0 band (i.e., the band whose intensity arises from all those normal coordinates with unchanged force constants) and with additional (weak) bands v_{i1}^1, v_{j1}^1, v_{k2}^2, etc. This situation is illustrated in Fig. 12.

2. Upon the presumption that the molecule does not change its shape but only "expands" or "contracts" slightly, *progressions* ($v_i'' \rightarrow v_{i+1}'$, v_{i+2}', …, i.e., $\Delta v_i \neq 0$) can be observed only in those modes v_i that are known to be totally symmetric. Conversely, and from the point of view of analysis, *if* progressions are observed in vibrations that are known from the infrared and Raman spectrum of the molecule to be *all* totally symmetric, then the molecule retains the same shape in both ground and excited states. This case is illustrated in Fig. 13 for the 2713-Å system of p-difluorobenzene, which has a total of six totally symmetric vibrations.

3. Following on from the arguments in (1) and (2), *if* progressions are observed in vibrations that are known to be non-totally symmetric, then the molecule has a different *shape* (and perhaps *size* as well) in the two electronic states. The most famous and well-documented example of this behavior (as well as it being the first case unambiguously analyzed) is that of the mole-

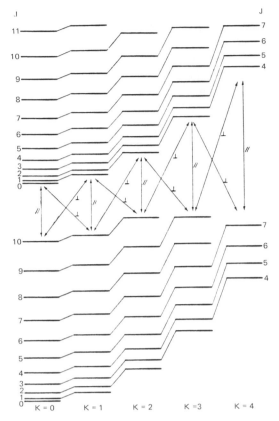

FIG. 12. Diagrammatic representation of the electric dipole-allowed transitions of symmetric top molecules. The J selection rule is $\Delta J = 0, \pm 1$ for both parallel and perpendicular transitions.

cule acetylene, $H-C\equiv C-H$, which is linear in its electronic ground state but which is trans bent as

$$
\begin{array}{c}
H \\
\quad \diagdown \\
\qquad C = C \\
\qquad\qquad \diagdown \\
\qquad\qquad\quad H
\end{array}
$$

in its lowest excited electronic state, ~ 2300 Å. This example is illustrated in Fig. 14. Each of the vibrational bands observed in the spectrum has its own rotational fine structure whose pattern depends on the type of electronic and vibrational transition involved and that obeys the selection rules for the type of molecule (spherical symmetric or asymmetric top).

2. Rotational Structure

Each vibrational level of each electronic level of a symmetric top consists of a set of rotational levels of one of these two kinds and, for electric dipole transitions between two sets of rotational levels, the selection rules are

$\Delta J = 0, \pm 1$ $\Delta K = 0$ for parallel transitions

$\Delta J = 0, \pm 1$ $\Delta K = 1$ for perpendicular transitions

where parallel and perpendicular refer to the di-

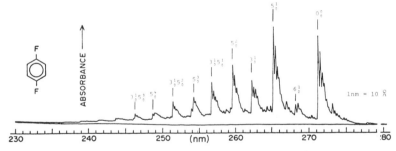

FIG. 13. Low-resolution vapor-phase spectrum of the 2713-Å system of p-difluorobenzene. The O_0^0 (origin) band is very strong (the second strongest band in the system), and the only other bands that appear with any strength are those involving vibrations 3, 5, and 6, all of which are known to be totally symmetric. In approximate terms, 3 involves a change in length of the C–F bond, 5 is a mode that increases the size of the benzene ring slightly, and 6 changes the benzene ring angles slightly. From the relative intensities of the various bands in the spectrum, it is immediately clear that the excited state differs from the ground-state shape by relatively small amounts in the directions implied by the presence of these particular modes. It should be understood that casual inspection alone cannot decide between, for example, the C–F bond increasing *or* decreasing in length. The only thing that can be decided is that there are *changes* of the kinds that correspond to the modes of the specific vibrations. All other bands in the spectrum are sequences, that is, various combinations of the v_{in}^n.

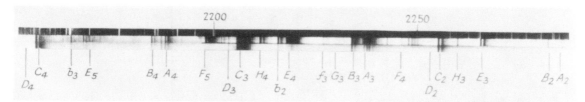

C_4 b_3 E_5 B_4 A_4 F_5 C_3 H_4 E_4 f_3 G_3 B_3 A_3 F_4 C_2 H_3 E_3 B_2 A_2
D_4 D_3 b_2 D_2

FIG. 14. The low-resolution vapor-phase spectrum of acetylene (C_2H_2). The top, black, strip is a spectrum of the iron arc used to obtain accurate wavelength measurements. The second strip is the spectrum of acetylene taken at room temperature (15°C). The third strip is the spectrum of acetylene taken at -80°C in order to depopulate as many vibrational levels in the ground state as possible, thereby simplifying the spectrum greatly. Under these conditions, the most intense features are indicated by C_i, where i indicates the successive members of the progression due to the upper-state vibration having the frequency C_3–C_2 (or C_2–C_1). It turns out that this vibration is the one that bends the linear acetylene molecule into a trans-bent shape. The spectrum shows that the (Franck–Condon) maximum in intensity is at C_3, and analysis shows that the band is one in which the molecule has six or seven quanta of this bending mode. The conclusion is that the excited state of acetylene *is* bent in a trans mode.

rection of the oscillating electric transition moment relative to the top axis.

Detailed discussion is beyond the scope of this article, but, for a typical prolate top, the parallel transitions have the form shown in Fig. 15(a), while the perpendicular have that shown in Fig. 15(b). (The legend for the figure should be carefully noted.)

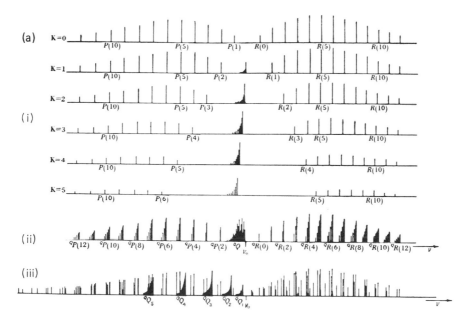

FIG. 15. (a) Subbands of a ∥ band and complete ∥ band of a symmetric top. The subbands in (i) are directly superimposed in (ii). In both (i) and (ii) only a slight difference between $A'-B'$ and $A''-B''$ is assumed. In (iii) the same subbands are superimposed but with shifts corresponding to a much larger difference between $A'-B'$ and $A''-B''$. Here also the lines of the Q branches have not been drawn separately. The heights of the lines indicate the intensities calculated on the basis of the assumption that $A'' = 5.25$, $B'' = 1.70$ cm^{-1}, and $T = 144$ K. (b) Subbands of a ⊥ band and complete ⊥ band of a symmetric top. The complete band is shown in the bottom strip. The spectrum is drawn under the assumption that $A' = 5.18$, $A'' = 5.25$, $B' = 0.84$, $B'' = 0.85$ cm^{-1} and $\zeta_i = 0$. The intensities were calculated for a temperature of 144 K. It should be realized that if the lines of an individual Q branch are not resolved, the resulting "line" would stand out much more prominently than might appear from the spectrum given. (*Figure continues.*)

(b)

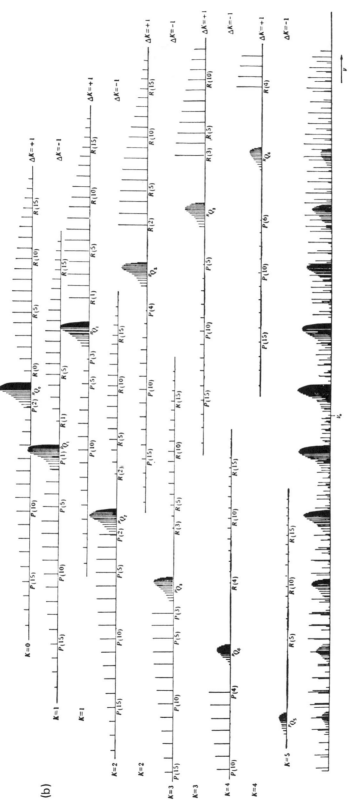

FIG. 15. *(continued)*

In principle, then, the structure of a perpendicular transition and that for a parallel transition are distinctly different and can be visually analyzed in favorable cases. In general, the perpendicular or parallel character of the selection rules is decided by the electronic selection rules, so that *each* vibrational band in the optical spectrum of a symmetric top will be of one type *or* the other.

The individual vibrational features in the spectrum consist of the superposition of a number of sets of rotational structure very similar to the simple pattern found for diatomic molecules but with each set specified by a particular value of K. Each of these is called a *subband*. In the case of a parallel transition, reference to Fig. 15(a) shows that the subbands lie almost on top of one another (they are *exactly* coincident if the upper and lower states have the same moments of inertia—which they almost never do), while the perpendicular are spread out evenly on either side of the $K = 0$ subband [Fig. 15(b)]. Differences in the upper- and lower-state inertial constants cause heads to form in each of the subbands and also cause the set of subbands to have a quadratic frequency distribution, that is, also tend toward forming heads in special cases.

Based upon the analysis of these bands, it is possible to assign the electronic symmetry of the two states involved in the transition and, frequently, to relate them to the molecular orbital structure of the molecule.

F. TYPES OF MOLECULAR OPTICAL SPECTRA

The concepts of optical spectra have been discussed above in general terms, without specific reference to any special molecular types or classifications. The general terms *absorption* and *emission* spectra each subsume a cluster of more specialized types, and these will now be briefly outlined. Most of the optical spectra have been ones in which an outer- (valence-) shell electron is excited to (or drops from) an antibonding valence-shell orbital of the molecule, giving rise to so-called valence-shell transitions. On the other hand, some transitions involve orbitals outside the valence shell, and the spectra are known as *Rydberg* spectra. These can—and do—occur in the optical region, and the excited states are analogous to those in atoms having principal quantum numbers greater than the valence shell. They form the series of levels leading to ionization of the atom, the *Rydberg series*. In general, they have electron probability distributions that

lie outside of the dimensions of the molecule and, except for the lowest of them, are effectively nonbonding. In transitions from the ground state to high Rydberg states, the Franck–Condon nature of the transition is, therefore, an excellent indication of the nature of the electron in the ground state that was excited, that is, bonding, nonbonding, or antibonding.

Polyatomic molecules that possess neither a center of symmetry nor alternating axes of symmetry exhibit *optical activity*. From a phenomenological point of view, the molecule (gas phase or solution) is able to rotate the plane of polarization of a light beam (polarized) that passes through it. The magnitude of the angular rotation depends on the wavelength as well as the specific nature of the molecule, and in this respect *optical rotatory dispersion* (ORD) is the optical-activity analog of the ordinary refractive index. In the same way as the refractive index depends on the optical absorption spectrum, the ORD spectrum depends on the *circular dichroism* (CD) spectrum. Thus, if circularly polarized (left *and* right) beams of light are passed through an optically active medium, there is a difference between the two spectra. The difference is the CD spectrum, just as the ordinary (unpolarized) absorption spectrum is the difference between the beam that passes through the vapor or solution and the one that does not.

From a molecular point of view, the absence of the symmetry elements (center and alternating axes) is what is required for the molecular transition to *simultaneously* have nonvanishing electric-dipole and magnetic-dipole transition moments, since the rotation strength is proportional to the scalar product of the electric and magnetic moments.

Not all molecules are optically active, but any molecule that absorbs electric-dipole radiation can be given an *induced magnetic moment* by placing it in a very strong magnetic field ($\sim 10^5$ G = 10 T). The spectrum that is obtained from the difference of absorption of beams of left and right circularly polarized light is called the *magnetic circular dichroism* (MCD) spectrum. Associated with this spectrum is the dispersion phenomenon, such that the magnitude of rotation of the beam of polarized light depends on the wavelength, and the spectrum is called the *magnetic optical rotation* (MOR) spectrum.

All four of these spectra—CD, MCD, ORD, and MOR—are invaluable in assigning the precise nature of the ground and excited states—

more specifically, the nature of the highest occupied and lowest unoccupied molecular orbitals—although the derivative spectra (ORD and MOR) are more often used purely for diagnostic purposes, particularly for proteins, as was the refractive index in the early days of optical spectroscopy.

G. Laser-Related Optical Spectra

The advent of the laser has widened and changed many aspects of molecular optical spectroscopy with respect both to methodology and to new types of spectroscopy. The changes in methodology are due to the very high powers of monochromatic radiation now available from both single-frequency and tunable lasers. Thus, fluorescence spectra have been observed in many molecules where the quantum yields are so low as to be undetectable using conventional light sources. This also applies to the whole field of *Raman spectroscopy,* where the availability of high-power excitation sources has enabled high resolution to be obtained in the gas phase very close in energy to the exciting line. In solids this has been essential.

Somewhat more esoteric, but extremely important in terms of the observation of new electronic states, is the use made of the laser power for *multiphoton laser excitation* spectroscopy. Thus, the molecule is now able to absorb two (or more) photons *simultaneously,* and the selection rules that apply to the one-photon transitions can be overcome. Additionally, the absorption of two or more photons is capable of ionizing many molecules, so that studies of the electronics states of ions are now possible. It is also possible to reduce the laser power and allow for the *consecutive* absorption of two (or more) photons. The first photon excites the molecule to the standard one-photon excited states, and the second ionizes the molecule from that state. This allows the primary absorption profile to be followed by ion counting (as well as by standard ion-mass spectrometry techniques), and high resolution is obtained in some cases. The particular *two-photon* absorption is known as *reso-*nant two-photon ionization (R2PI) spectroscopy and is finding wide analytical and spectroscopic applications. [*See* MULTIPHOTON SPECTROSCOPY.]

Somewhat connected with R2PI spectroscopy but quite different is *laser optico-galvanic* (LOG) spectroscopy. It has been known for over a century that a condenser can be discharged because of the presence of ions in a flame placed between the plates, and that the presence of easily ionized species such as alkali metals (Na, K, etc.) accentuates this effect. In LOG spectroscopy, a laser beam is passed through a plasma discharge, which usually consists of an inert carrier gas such as neon or argon and traces of the compound under examination. The current flowing between two probes inserted in the plasma is measured as a function of the laser frequency, and sharp discontinuities are observed when species are excited, which then either increases or decreases the ion current in the plasma. The technique is difficult but holds some promise in the investigation of plasma phenomena generally.

There are many other laser-related applications to molecular optical spectroscopy that are too specialized for this article.

Bibliography

Herzberg, G. (1950). ''Molecular Spectra and Molecular Structure,'' Vol. 1. van Nostrand Company Inc., Princeton, New Jersey.

Herzberg, G. (1966). ''Molecular Spectra and Molecular Structure,'' Vol. 3. van Nostrand Company Inc., Princeton, New Jersey.

Herzberg, G. (1971). ''The Spectra and Structures of Simple Free Radicals. An Introduction to Molecular Spectroscopy.'' Cornell University Press, Ithaca and London.

Huber, K. P., and Herzberg, G. (1979). ''Molecular Spectra and Molecular Structure,'' Vol. 4. Constants of Diatomic Molecules. Van Nostrand-Reinhold, New York.

King, G. W. (1964). ''Spectroscopy and Molecular Structure.'' Holt, Rinehart and Winston, Inc. New York.

MULTIPHOTON SPECTROSCOPY

Y. Fujimura *Tohoku University*
S. H. Lin *Arizona State University*

I. Introduction
II. Theory
III. Experimental Methods
IV. Characteristics and Spectral Properties
V. Applications
VI. Prospects

GLOSSARY

Doppler-free multiphoton spectroscopy: Multiphoton spectroscopy aims at eliminating inhomogeneous Doppler broadening.

Ion-dip spectroscopy: A high-resolution multiphoton spectroscopy based on competition between ionization and stimulated emission (or stimulated absorption).

Multiphoton absorption: Simultaneous absorption of multiple numbers of photons by materials under the irradiation of laser light with high intensities.

Multiphoton ionization mass spectroscopy: Multiphoton ionization method combined with mass detection. This spectroscopy allows us to identify molecules by the optical spectrum related to the resonant intermediate state as well as by their mass.

Polarization dependence: Dependence of the cross sections of multiphoton transitions on photon polarizations of laser used.

Resonance enhancement: A drastic increase in the ability of the multiphoton process observed when laser is tuned and the energy of the laser approaches that of a real intermediate state. See multiphoton absorption.

Rydberg states: States of a valence electron orbiting about a positively charged core consisting of the nucleous and inner electrons in atoms or molecules. Jumps of an electron between Rydberg states are called Rydberg transitions. The transition frequency ω is given by the simple expression

$$\omega_n = I_p - R/(n - \delta)^2$$

where I_p is the ionization potential, R the Rydberg constant, n the principal quantum number of the Rydberg electron, δ the quantum defect, and $R/(n - \delta)^2$ the term values. The classical orbit radius of the Rydberg electron increases as n^2, while the orbital velocity decreases as n^{-1}.

Multiphoton spectroscopy consists of the simultaneous interaction of atoms or molecules with two or much more number of photons. That is, it is spectroscopy with the use of multiphoton transitions, or, more generally, the spectroscopic research field of the interaction between matter and two or more photons.

I. Introduction

In Fig. 1, several typical multiphoton processes are shown. The result of the material–multiphoton interaction is usually detected through direct absorption, fluorescence, ionization current, or a photoelectron detection system. The excited-state structures of these materials in gases, liquids, or solids, such as electronic, vibrational, or rotational states or fine structure, which are not found in ordinary single-photon spectroscopy because of their difference selection rules and low transition intensity, can be seen in a wide frequency range from lower electronic excited states to ionized continua.

Multiphoton spectroscopy requires an intense light source. The first experimental observation of the simplest multiphoton transition, two-photon absorption of an Eu^{2+}-doped CaF_2 crystal in

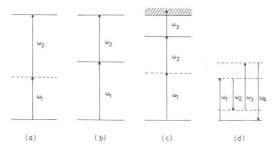

(a)   (b)   (c)   (d)

FIG. 1.   Several multiphoton processes seen in atoms and molecules: (a) a nonresonant two-photon absorption process; (b) a resonant two-photon absorption process; (c) a two-photon resonant three photon ionization; (d) a four-wave mixing process. Solid lines and broken lines represent real and virtual states, respectively; ω_i denotes photon frequencies.

the optical region, by Kaiser and Garrett (1961), was made possible only after a high-power monochromatic ruby laser was developed as the intense incident light source, although the possibility of simultaneous two-photon absorption or stimulated emission was pointed out in 1931 by Goeppert-Mayer. [*See* MOLECULAR OPTICAL SPECTROSCOPY.]

The main reasons for wide interest in the multiphoton spectroscopy are due to the advent of dye lasers for tunability and of multiphoton ionization technique for detecting information from the excited state created by the multiphoton excitation.

The tunability of dye lasers is particularly important for multiphoton excitation, because one can obtain an excitation source by using only a single-frequency laser beam rather than the two or more lasers of different frequencies. The multiphoton ionization technique consists of collecting free electrons produced by the multiphoton ionization process after irradiation by a tunable laser pulse, amplifying ion currents, and recording the signal as a function of the laser frequency. In general, even ion currents of only a few charges per second can be detectable. Therefore, by using this method, one can detect and characterize extremely small amounts of atoms or molecules, even in a rarefied gas. The sensitivity exceeds that of fluorescence and other detections. The multiphoton ionization technique is also important in practical applications such as isotope separation, laser-induced fusion, and the dry etching process.

Multiphoton transitions related to the

multiphoton spectroscopy have several characteristic features: laser intensity dependence, resonance enhancement, polarization dependence, etc. For example, the transition probability of the nonresonant two-photon absorption process shown in Fig. 1(a) with $\omega_1 = \omega_2$, $W^{(2)}_{i \to f}$, can be written as

$$W^{(2)}_{i \to f} = I^2 \sigma^{(2)}/(\hbar \omega_R)^2 \qquad (1)$$

where $\sigma^{(2)}$, I, and ω_R denote the cross section for the two-photon absorption, the laser intensity, and the laser frequency, respectively. Equation (1) indicates that the two-photon transition probability is proportional to the square of the laser intensity applied. This is called the formal intensity law. If no saturation of photon absorption takes place during the multiphoton processes, the order or the multiphoton transition can be experimentally determined from measuring the slope of log–log plots of the transition probability as a function of the laser intensity as shown in Fig. 2:

$$\ln W^{(2)}_{i \to f} = 2 \ln I + C \qquad (2)$$

for the nonresonant two-photon process.

Multiphoton spectroscopy usually utilizes resonance enhancement: that is, a dramatic increase in the multiphoton transition ability can be seen when the exciting laser is tuned and its frequency approaches a real intermediate electronic state called a resonant state. In this case, the level width of the resonant state plays a significant role in determining the transition ability. It is well known that photons can be regarded as particles of mass 0 and spin 1. Polarization de-

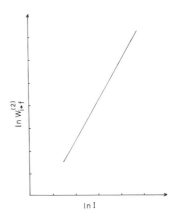

FIG. 2.   The formal intensity law for a nonresonant two-photon process. Here I and $W^{(2)}_{i \to f}$ denote the laser intensity and two-photon transition probability from the i to the f state, respectively.

pendence of multiphoton processes is associated with the spin angular momentum. Polarization dependence and symmetry selection rules of multiphoton transitions are of great importance in characterizing the multiphoton transition process and in determining the symmetry of the states relevant to the transitions. For example, for a two-photon transition of a molecule with a center of symmetry, the initial and final states have the same parity, which is in contrast to the parity selection rule of one-photon spectroscopy governed by the opposite parity. Therefore, one- and two-photon spectra are complementary for measuring vibronic states of the molecule. This relationship between one- and two-photon spectroscopic techniques is similar to that between the infrared (IR) absorption governed by the opposite parity and Raman spectroscopy by the same parity. These characteristic features mentioned briefly are described in detail in Section IV after the theoretical treatment and experimental techniques for the multiphoton spectroscopy are introduced.

II. Theory

In order to analyze multiphoton spectra, various theoretical treatments based on the ordinary time-dependent perturbation method, the Green's function method, the density matrix method, the susceptibility method, etc. have been applied to deriving the expressions for the multiphoton transition probability. In this section these theoretical treatments will be considered, focusing on their advantages and on the restrictions on their application to the multiphoton process.

A. General Considerations

The theory of multiphoton processes can be developed based on the semiclassical or quantum-mechanical formalisms. Here the quantum-mechanical theory in which the photon field is described in the second quantization is presented. In some cases the semiclassical formalism in which the photon field is treated classically and matter–radiation field interaction is characterized by a time-dependent behavior is convenient.

The total Hamiltonian of the total system consisting of atoms or molecules plus radiation field is written as

$$H = H_S + H_R + V = H_0 + V \qquad (3)$$

where H_S is the Hamiltonian of the particles, H_R the Hamiltonian of the free radiation field, and V the interaction between them. The radiation Hamiltonian can be written in terms of photon creation and annihilation operators as

$$H_R = \sum_{\mathbf{k}} \hbar\omega_k \left(b_{\mathbf{k}}^{\dagger} b_{\mathbf{k}} + \frac{1}{2} \right) \qquad (4)$$

where $\mathbf{k}$ specifies both the wave vector and polarization, $\omega_k = c|\mathbf{k}|$ the angular frequency of the $\mathbf{k}$th photon mode, and $b_{\mathbf{k}}^{\dagger}$ and $b_{\mathbf{k}}$ are the creation and annihilation operators of the photon, respectively. The unperturbed Hamiltonian of the total system satisfies

$$H_0 |I\rangle = \varepsilon_I |I\rangle \qquad (5)$$

where

$$\varepsilon_I = \varepsilon_i + \sum_{\mathbf{k}} \left(n_{\mathbf{k}} + \frac{1}{2} \right) \hbar\omega_k \qquad (6)$$

in which ε_i is an eigenvalue of H_S, and $n_{\mathbf{k}}$ is the number of photons with frequency ω_k. The eigenstate of the unperturbed Hamiltonian H_0, $|I\rangle$, can be written as the product of that of the system $|i\rangle$ and that of the radiation field $|n(k_1)n(k_2) \cdots\rangle$.

The interaction Hamiltonian V originates from the coupling of the vector potential $\mathbf{A}(\mathbf{r})$ with moving charged particles with mass m_j and electric charge e_j, and is given by

$$V = \sum_j \left(-\frac{e_j}{m_j c} \right) \mathbf{P}_j \cdot \mathbf{A}(\mathbf{r}_j)$$
$$+ \sum_j \left(\frac{e_j^2}{2m_j c^2} \right) \mathbf{A}^2(\mathbf{r}_j) \qquad (7)$$

where the vector potential is written as

$$\mathbf{A}(\mathbf{r}) = \sum_l \left(\frac{\hbar}{2\varepsilon_0 L^3 \omega_l} \right)^{1/2} \mathbf{e}_l \{ b_l \exp(i\mathbf{k}_l \cdot \mathbf{r})$$
$$+ b_l^{\dagger} \exp(-i\mathbf{k}_l \cdot \mathbf{r}) \} \qquad (8)$$

where L^3 is a cubic box of the photon field and $\mathbf{e}$ the polarization unit vector of the photon.

The interaction Hamiltonian, Eq. (7), can be written in terms of the multipole expansion as

$$V = V_d + V_q + \cdots \qquad (9)$$

where V_d and V_q denote the electric dipole and the electric quadrupole interactions, respectively, and they are given by

$$V_d = -e\mathbf{r} \cdot \mathbf{E} \qquad (10)$$

and

$$V_q = -\frac{e}{2} \sum_{ij} \mathbf{Q}_{ij} \nabla_j E_i \qquad (11)$$

in which $\mathbf{Q}$, the quadrupole dyadic, is defined by

$$\mathbf{Q}_{ij} = \mathbf{r}_i \mathbf{r}_j - \tfrac{1}{3}\mathbf{r}^2 \delta_{ij} \quad (\mathbf{r}_i = x, y, z) \qquad (12)$$

Since the wavelength of photons of optical frequencies is much larger than atomic or molecule dimension, the spatial dependence of the photon field can be neglected in calculating the transition probabilities (dipole approximation). The first term of the interaction Hamiltonian V_d makes a significant contribution to ordinary multiphoton transitions, although effects of the quadrupole interaction term have been observed in some cases. In the dipole approximation the interaction Hamiltonian can be expressed as

$$V = -ie \sum_l (\hbar\omega_l/2\varepsilon_0 L^3)^{1/2} \mathbf{r} \cdot \mathbf{e}_l (b_l - b_l^\dagger) \qquad (13)$$

B. Ordinary Time-Dependent Perturbation Theory

In the ordinary time-dependent perturbation theory, the first-order transition probability per unit time from I to F states, $W_{I \to F}^{(1)}$, is given by

$$W_{I \to F}^{(1)} = \frac{2\pi}{\hbar} |V_{FI}|^2 \, \delta(E_F - E_I), \qquad (14)$$

and the mth-order transition probability per unit time, $W_{I \to F}^{(m)}$, is given by

$$W_{I \to F}^{(m)} = \frac{2\pi}{\hbar} \left| \sum_{M_1} \sum_{M_2} \cdots \sum_{M_m} \frac{V_{FM_m} \cdots V_{M_2 M_1} V_{M_1 I}}{\hbar\omega_{M_m I} \cdots \hbar\omega_{M_1 I}} \right|^2$$
$$\times \delta(E_F - E_I), \qquad (15)$$

where $M_1 \cdots M_m$ specify the intermediate states for the multiquantum transition. The Dirac δ function $\delta(E_F - E_I)$ expresses the energy conservation between initial and final states of the transition. It should be noted that Eqs. (14) and (15) have been derived in the long time limit $t \to \infty$. These expressions are usually called Fermi's golden rule for the transition probability.

The expression for the multiphoton transition probability can be derived by using Fermi's golden rule. The expression for the m-photon (quantum) transition probability comes from that of the mth-order transition probability in Fermi's golden rule. For example, the expression for a two-photon absorption probability from initial state i to the final state f of the system of interest, $W_{i \to f}^{(2)}$, induced by irradiation of

two types of the laser light with frequencies ω_l and $\omega_{l'}$ and polarization unit vectors $\mathbf{e}_l$ and $\mathbf{e}_{l'}$, takes the form

$$W_{i \to f}^{(2)} = \frac{2\pi}{\hbar^2} \sum_l \sum_{l'} \left(\frac{e^2 n_l \omega_l}{2\varepsilon_0 L^3} \right) \left(\frac{e^2 n_{l'} \omega_{l'}}{2\varepsilon_0 L^3} \right)$$
$$\times |S_{fi}(\omega_l \mathbf{e}_l, \omega_{l'} \mathbf{e}_{l'})^2|$$
$$\times \delta(\omega_{fi} - \omega_l - \omega_{l'}) \qquad (16)$$

where

$$S_{fi}(\omega_l \mathbf{e}_l, \omega_{l'} \mathbf{e}_{l'})$$
$$= \sum_m \left[\frac{\mathbf{e}_{l'} \cdot \mathbf{R}_{fm} \mathbf{e}_l \cdot \mathbf{R}_{mi}}{\omega_{mi} - \omega_l} + \frac{\mathbf{e}_l \cdot \mathbf{R}_{fm} \mathbf{e}_{l'} \cdot \mathbf{R}_{mi}}{\omega_{mi} - \omega_{l'}} \right] \qquad (17)$$

in which $\mathbf{R}_{fm}$ is the transition matrix element defined by $\langle f|\mathbf{r}|m \rangle$, and ω_{mi} the frequency difference between m and i levels.

The summations over the photon modes in Eq. (16) can be replaced by integrations after taking $L \to \infty$ according to

$$\sum_l \to \frac{L^3}{8\pi^3 c^3} \int_0^\infty d\omega_l \, \omega_l^2 \int_{\Omega_l} d\Omega_l \qquad (18)$$

together with $n_l \to n(\omega_l)$, where Ω_l is a solid angle about the polarization vector $\mathbf{e}_l$. The resulting expression for the two-photon transition probability is given by

$$W_{i \to f}^{(2)} = 2\pi (2\pi\alpha)^2 I_1 I_2 \omega_1 \omega_2 |S(\omega_1 \mathbf{e}_1, \omega_2 \mathbf{e}_2)|^2$$
$$\times \delta(\omega_{fi} - \omega_1 - \omega_2) \qquad (19)$$

where I_1 and I_2 are the photon fluxes in units of numbers of photon per second per area, and I_i is given by

$$I_i = \frac{1}{8\pi^3 c^2} \int d\omega_i \, \omega_i^2 n(\omega_i) \int_{\Omega_i} d\Omega_i \qquad (20)$$

In Eq. (19), α denotes the fine-structure constant and is given by $\alpha = e^2/(4\pi\varepsilon_0 \hbar c) \simeq 1/137.04$.

An expression for an m-photon transition probability can be obtained from the Fermi's golden rule in a similar manner. The resulting expression explains the formal intensity law and is valid for nonresonant multiphoton transitions. In order to study resonance effects and saturation phenomena, one has to utilize other theoretical treatments taken into account an infinite perturbation procedure.

C. Green's Function (Resolvent) Method

The Green's function method can be applied to the interpretation of optical Stark effects and also resonance effects. The Green's function (resolvent) is defined by

$$(E - H)G(E) = 1 \qquad (21)$$

The transition amplitude from I to F states, $U_{FI}(t)$, is expressed in terms of the time-independent Green's function,

$$U_{FI}(t) = \langle F | \exp(-itH/\hbar) | I \rangle$$

$$= \frac{1}{2\pi i} \int dE \exp\left(\frac{-iEt}{\hbar}\right) G_{FI}(E) \qquad (22)$$

where $G_{FI}(E)$ is the matrix element of the Green's function.

The transition probability per unit time is given by

$$W_{I \to F} = \lim_{t \to \infty} \frac{d}{dt} |U_{FI}(t)|^2 \qquad (23)$$

The matrix elements of the Green's function are evaluated by using the Dyson equation,

$$G(E) = G^0(E) + G^0(E)VG(E) \qquad (24)$$

where the zero-order Green's function $G^0(E)$ satisfies

$$(E - H_0 + i\eta)G^0(E) = 1 \qquad (25)$$

with $\eta \to 0^+$.

One of the merits of using the Green's function method is that the effect of level width Γ_M and level shift D_M of the intermediate states of M in the expression for the multiphoton transition probability can easily be taken into account. The width and shift originate from the interaction of atoms or molecules with the photon field and/or the heat bath. For example, for two-photon processes such as two-photon absorption, and Raman scattering, after utilizing the Dyson equation in evaluating the relevant matrix elements of the Green's function, the matrix element $G_{FI}(E)$ can be expressed as

$$G_{FI}(E) = \sum_M \frac{G_{FF}^0(E) V_{FM} V_{MI} G_{II}^0(E)}{E - E_M^0 - \Lambda_{MM}(E)} \qquad (26)$$

where energy-dependent self-energy $\Lambda_{MM}(E)$ can be written as

$$\Lambda_{MM}(E) = D_M(E) - (i/2)\Gamma_M(E) \qquad (27)$$

in which the level shift $D_M(E)$ and the width $\Gamma_M(E)$ are given by

$$D_M(E) = P \sum_B \frac{|\langle B | V | M \rangle|^2}{E - E_B^0} \qquad (28)$$

where P denotes the principal part, and

$$\Gamma_M(E) = 2\pi \sum_B |\langle B | V | M \rangle|^2 \, \delta(E - E_B^0), \quad (29)$$

respectively. The term B appearing in Eqs. (28) and (29) excludes the initial, final, and intermediate states, and denotes the states combined with the intermediate states through the system–photon field interaction and/or the system–perturber interactions. From Eqs. (26), (22) and (23), an expression for the two-photon transition probability $W_{I \to F}^{(2)}$ can be derived. The resulting expression is identical to that derived in the ordinary perturbation approach, except for the inclusion of the level shifts and the widths of the intermediate states.

The Green's function method just described can be applied to the multiphoton processes in the low-temperature case. In order to take into account temperature effects on the multiphoton processes, other methods, such as the temperature-dependent Green's function or the density matrix method, are commonly used.

D. DENSITY MATRIX METHOD

The density matrix method is widely applied to investigation of mechanisms of multiphoton transitions and to derivation of an expression for the nonlinear susceptibility for nonlinear optical processes of atoms and molecules in the presence of the heat bath. Collision-induced multiphoton transitions that are induced by an elastic interaction between the system and the heat bath during the photon absorption, sometimes referred to as optical collisions, can be explained by using the density matrix method.

The density matrix for the total system, including the heat bath and the photon field $\rho(t)$, is defined by

$$\rho(t) = \sum_i N_i |\psi_i(t)\rangle \langle \psi_i(t)| \qquad (30)$$

where $\psi_i(t)$ is the wave function for the ith quantum system of the total system at time t and N_i the weighting factor. The time evolution of the total system is determined by the Liouville equation for the density matrix

$$i\hbar \frac{\partial}{\partial t} \rho(t) = [H, \rho(t)] \qquad (31)$$

where the square-bracketed term denotes the commutator.

Let us consider a contribution of a collision-induced two-photon absorption by irradiation of two kinds of lasers with a near-resonance frequency ω_1 between initial and resonant states and with frequency ω_2 to the rate constants. In describing the near-resonant two-photon absorption processes, it is necessary to obtain the fourth-order solution of the Liouville equation of Eq. (31). The perturbative solution can be written as

$$\rho(t) = (i\hbar)^{-4} \int_{-\infty}^{t} dt_1 \int_{-\infty}^{t_1} dt_2 \int_{-\infty}^{t_2} dt_3$$
$$\times \int_{-\infty}^{t_3} dt_4 \left[V(t_1), \left(V(t_2), \{ V(t_3), \right. \right.$$
$$\left. \left. [V(t_4), \rho(-\infty)] \} \right) \right] \quad (32)$$

where $V(t)$ denotes the system–photon field interaction Hamiltonian in the interaction picture, and $\rho(-\infty)$ is the density matrix in the initial state. After tracing out over the photon-field variables and the heat-bath variables, the diagonal matrix element representing the final state density of the system $\rho_{ff}^{(S)}(t)$ in a three-level model shown in Fig. 3 can be expressed as

$$\rho_{ff}^{(S)}(t) = 2 \, \text{Re}\{A_{ff}(t) + B_{ff}(t) + C_{ff}(t)\} \quad (33)$$

where

$$A_{ff}(t) = \frac{1}{\hbar^4} \int_{-\infty}^{t} dt_1 \int_{-\infty}^{t_1} dt_2 \int_{-\infty}^{t_2} dt_3$$
$$\times \int_{-\infty}^{t_3} dt_4 \, \langle E_1^{(+)}(t_2) E_1^{(-)}(t_4) \rangle$$
$$\times \langle E_2^{(+)}(t_1) E_2^{(-)}(t_3) \rangle$$
$$\times g_{fm}(t_1 - t_2) g_{fi}(t_2 - t_3) g_{mi}(t_3 - t_4)$$
$$(34a)$$

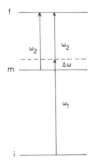

FIG. 3. Three-level model for a near-resonant two-photon absorption. The detuning frequency is defined by $\Delta\omega = \omega_1 - \omega_{mi}$, where ω_{mi} is the frequency difference between the resonant and initial levels.

$$B_{ff}(t) = \frac{1}{\hbar^4} \int_{-\infty}^{t} dt_1 \int_{-\infty}^{t_1} dt_2 \int_{-\infty}^{t_2} dt_3$$
$$\times \int_{-\infty}^{t_3} dt_4 \, \langle E_1^{(+)}(t_3) E_1^{(-)}(t_4) \rangle$$
$$\times \langle E_2^{(+)}(t_1) E_2^{(-)}(t_2) \rangle$$
$$\times g_{fm}(t_1 - t_2) g_{mm}(t_2 - t_3) g_{mi}(t_3 - t_4)$$
$$(34b)$$

and

$$C_{ff}(t) = \frac{1}{\hbar^4} \int_{-\infty}^{t} dt_1 \int_{-\infty}^{t_1} dt_2 \int_{-\infty}^{t_2} dt_3$$
$$\times \int_{-\infty}^{t_3} dt_4 \, \langle E_1^{(+)}(t_3) E_1^{(-)}(t_4) \rangle$$
$$\times \langle E_2^{(+)}(t_2) E_2^{(-)}(t_1) \rangle$$
$$\times g_{fm}(t_1 - t_2) g_{mm}(t_2 - t_3) g_{mi}(t_3 - t_4)$$
$$(34c)$$

In Eq. (34), $\langle \cdots \rangle$ denotes the photon-field correlation function. The matrix element g, for example $g_{mi}(t_\alpha - t_\beta)$, is that of the time evolution operator for the density matrix representing the system and heat bath, and can be expressed phenomenologically as

$$g_{mi}(t_\alpha - t_\beta) = \exp[-i(t_\alpha - t_\beta)\omega_{mi} - |t_\alpha - t_\beta|\Gamma_{mi}] \quad (35)$$

where ω_{mi} is the frequency difference of the system between m and i states, and Γ_{mi} is the dephasing constant relevant to these states. The structure of the dephasing constant can be clarified by using the density matrix method combined with the projection operator or the cumulant expansion technique. In the Markoff approximation, the dephasing constant is given by

$$\Gamma_{mi} = \tfrac{1}{2}(\Gamma_{mm} + \Gamma_{ii}) + \Gamma_{mi}^{(d)} \quad (36)$$

where Γ_{mm} (Γ_{ii}) represents the population decay constants of the m (i) states, with $\Gamma_{mi}^{(d)}$, in which $m \neq i$ is the pure dephasing constant originating from the elastic interaction between the system and the heat bath.

In qualitatively understanding the mechanism of multiphoton processes it is convenient to use a diagrammatic representation of time evolution of the ket and bra vectors. The diagrammatic representation in the case of the two-photon absorption is shown in Fig. 4. Figure 4a–c correspond to the time evolution in Eqs. (34a), (34b),

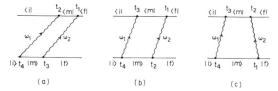

(a)   (b)   (c)

FIG. 4. Diagramatic representations of the ket and bra vectors for a near-resonant two-photon absorption: (a), (b), and (c) correspond to the time evolution of the density matrix in Eqs. (34a), (34b), and (34c), respectively.

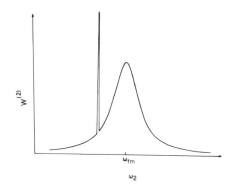

FIG. 5. A line shape of a two-photon absorption as a function of the second laser frequency ω_2. The broad band centered at $\omega_2 = \omega_{fm}$ originates from the collision-induced sequential mechanism and the sharp band at $\omega_2 = \omega_{fm} - \Delta\omega$ from the coherent two-photon mechanism. The band width of the former is mainly characterized by the dephasing constant due to the collision between the system and the perturbers and that of the latter by the laser band width.

and (34c), respectively. The lower and upper lines represent the time evolution of the ket $|\,\rangle$ and bra $\langle\,|$ vectors of the system, respectively. Time develops from the left-hand to right-hand sides. The wavy lines represent the system–photon interaction. The dotted points indicate the interaction points. In the diagram in Fig. 4a, two wavy lines with ω_1 and ω_2 overlap each other during time $t_2 - t_3$. This corresponds to a simultaneous two-photon process. Figure 4b,c describes sequential two-photon processes in which interactions of the system with the photon fields 1 and 2 are independent of each other.

The two-photon transition probability per unit time, $W^{(2)}_{i \to f}$, is defined as

$$W^{(2)}_{i \to f} = \lim_{t \to \infty} \frac{d}{dt} \rho_{ff}(t) \qquad (37)$$

For an idealized steady-state laser excitation characterized by a negligibly small band width for both lasers, the transition probability per unit time takes the form

$$W^{(2)}_{i \to f}$$

$$\propto \frac{1}{(\omega_{mi} - \omega_1)^2 + \Gamma^2} \{\pi\delta(\omega_f - \omega_i - \omega_1 - \omega_2)$$

$$+ \frac{\Gamma^{(d)}_{mi}}{\Gamma_{mm}} \frac{\Gamma}{[(\omega_f - \omega_m - \omega_2)^2 + \Gamma^2]} \qquad (38)$$

In deriving Eq. (38), $\Gamma_{fm} = \Gamma_{mi} = \Gamma$ and $\Gamma_{fi} = 0$ have been assumed for simplicity. The first term in Eq. (38) represents the coherent two-photon transition, and the second term the collision-induced (sequential) two-photon transition due to the system–heat bath elastic interaction. The latter transition rate constant is proportional to pressure of the perturbers added. A line-shape function of the resonant two-photon absorption as a function of the frequency of the second laser ω_2 is drawn schematically in Fig. 5 to demonstrate contribution of the collision-induced two-photon transition to the total line shape. The

broad line shape represents that for the collision-induced sequential two-photon transition and the sharp line shape that of the simultaneous process.

E. Susceptibility Method

The susceptibility method is widely applied to the explanation of nonlinear optical phenomena, such as harmonic generation, sum- and difference-frequency generation, stimulated scattering, and multiphoton absorption, that originate from nonlinear interaction between a coherent laser field and material. The macroscopic polarization of the material induced by the incident radiation field P can be expanded as

$$\mathbf{P} = \chi^{(1)} \cdot \mathbf{E} + \chi^{(2)} : \mathbf{EE} + \chi^{(3)} \vdots \mathbf{EEE} + \cdots \qquad (39)$$

where $\mathbf{E}$, the incident radiation field, is expressed as

$$\dot{\ } = \frac{1}{2} \sum_i \{\mathbf{e}_i E_i \exp[i(\mathbf{k}_i \cdot \mathbf{r} - \omega_i t)] + \text{c.c.}\} \qquad (40)$$

and $\chi^{(n)}$ is called the nth-order susceptibility. For isotropic materials characterized by inversion symmetry, the lowest nonlinear susceptibility $\chi^{(2)}$ vanishes, and the nonlinearity in these materials is usually described in terms of the third-order nonlinear susceptibility $\chi^{(3)}$, neglecting higher order ones. The electric field $\mathbf{E}$ originating from the nonlinear polarization satisfies the Maxwell wave equation

$$\nabla \times \nabla \times \mathbf{E} + \frac{1}{c^2}\frac{\partial^2 \mathbf{E}}{\partial t^2} = -\frac{1}{\varepsilon_0 c^2}\frac{\partial^2 \mathbf{P}}{\partial t^2} \quad (41)$$

For a polarization at the sum frequency $\omega_4 = \omega_1 + \omega_2 + \omega_3$, in the presence of the three kinds of lasers with amplitudes E_1, E_2, and E_3, the electric field amplitude E_4 at distance l can be expressed as

$$E_4 = \chi^{(3)}E_1 E_2 E_3 l \, [\exp(i\,\Delta k\,l) - 1/\Delta k] \quad (42)$$

where Δk, called the phase match parameter, is given by $\Delta k = k_1 + k_2 + k_3 - k_4$. The intensity of the net field I_4 can be written as

$$I_4 = \frac{256\pi^4 \omega_1 \omega_2 \omega_3 \omega_4^3}{c^8 k_1 k_2 k_3 k_4}\,|\chi^{(3)}|^2\,I_1 I_2 I_3 l^2$$

$$\times\,[\sin(\Delta k\,l/2)/\Delta k\,l/2]^2 \quad (43)$$

in which $k_i = \varepsilon_i^{1/2}\omega_i/c$. Here the dielectric constant $\varepsilon_i = 1 + 4\pi\chi^{(1)}$ is equal to the square of the index of the refraction n_i. So far the nonlinear wave mixing processes have been treated in the classical method. It has been shown that the signal is proportional to the products of the intensity of each incident laser field, and for the signal to be observed the phase matching condition $\Delta k = 0$ has to be satisfied.

In order to investigate the frequency dependence of the material, especially resonance enhancement of nonlinear process, it is necessary to clarify the structure of the third-order nonlinear susceptibility. This can be carried out by using the semiclassical method, in which the radiation field is treated classically and the material system quantum-mechanically in solving equation of motion for the density matrix, Eq. (31). The resulting polarization is given in terms of the expectation value of the dipole moment $\boldsymbol{\mu} = e\mathbf{r}$ as

$$\mathbf{P} = N\,\mathrm{Tr}(\rho\boldsymbol{\mu}) \quad (44)$$

where ρ is a solution of Eq. (31) in the steady state condition. The third-order susceptibility consists of 24 terms as

$$\chi^{(3)} = \frac{Ne^4}{4\hbar^3}\sum_m \sum_{m'} \sum_{m''} \frac{\mathbf{e}_4 \cdot \mathbf{R}_{im''3} \cdot \mathbf{R}_{im''m'}}{(\omega_{m''i} - \omega_1 - \frac{1}{2}i\Gamma_{m''i})}$$

$$\times \frac{\mathbf{e}_2 \cdot \mathbf{R}_{m'm}}{(\omega_{m'i} - \omega_1 - \omega_2 - \frac{1}{2}i\Gamma_{m'i})}$$

$$\times \frac{\mathbf{e}_1 \cdot \mathbf{R}_{mi}}{(\omega_{mi} - \omega_4 - \frac{1}{2}i\Gamma_{mi})}$$

$$+\ \text{permutations of indices } 1\text{–}4 \quad (45)$$

where i denotes the initial state of the material.

The 23 other permutations of indices 1–4 arise from assuming different sequences in which the photon absorption at ω_1, ω_2, and ω_3 and photon emission at ω_4 take place. Each of the terms in Eq. (45) makes a contribution when either one of the applied laser frequencies or some linear combinations of these correspond to the difference in energy between two levels of atoms or molecules, $\hbar\omega_{mi} = \varepsilon_m - \varepsilon_i$.

The real physical fields contain both positive and negative frequency components. The frequencies ω_1, ω_2, and ω_3 may be chosen positive or negative, or equal to each other. Depending on these choices, specific optical nonlinear processes are described. Typical third-order optical mixing processes, third-harmonic generation, sum-frequency generation, and coherent anti-Stokes Raman scattering (CARS) are schematically shown in Fig. 6.

Information of the nonresonant two-photon absorption of a material can be obtained by taking $\omega_3 = -\omega_2$ and $\omega_4 = \omega_1$, and neglecting dephasing constants in the intermediate state, and in this case the resonance condition between initial and final states is satisfied, $\omega_{m'i} = \omega_1 + \omega_2$ with $m' = f$ (final state). The imaginary part of the third-order nonlinear susceptibility $\chi''^{(3)}$ is written as

$$\chi''^{(3)}(-\omega_1,\,\omega_1,\,\omega_2,\,-\omega_2)$$

$$\simeq \frac{Ne^4}{2\hbar^3\Gamma_{fi}}\left|\sum_m \left\{\frac{\mathbf{e}_2 \cdot \mathbf{R}_{fm}\mathbf{e}_1 \cdot \mathbf{R}_{mi}}{\omega_{mi} - \omega_1}\right.\right.$$

$$\left.\left.+\ \frac{\mathbf{e}_1 \cdot \mathbf{R}_{fm}\mathbf{e}_2 \cdot \mathbf{R}_{mi}}{\omega_{mi} - \omega_2}\right\}\right|^2 \quad (46)$$

This corresponds to the expression for the two-photon absorption, Eq. (19), derived in ordinary time-dependent perturbation theory.

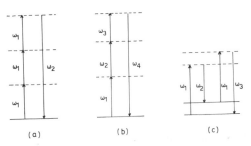

FIG. 6. Schematic energy level diagrams for third-order optical mixing processes: (a) third-harmonic generation, $\omega_2 = 3\omega_1$; (b) sum-frequency generation, $\omega_4 = \omega_1 + \omega_2 + \omega_3$; (c) coherent anti-Stokes Raman scattering (CARS), $\omega_3 = 2\omega_1 - \omega_2$.

III. Experimental Methods

Since two-photon absorption is a second-order process, it is rather weak at the moderate light intensities available from tunable dye lasers. This demands a sensitive technique for detection of only a few two-photon absorption events in the sample. In principle, higher light intensities could be used, but then higher-order processes become more probable and the measured spectrum may be a superposition of two- and three-photon spectra. Even ionization and fragmentation of the molecules is possible at high light intensities, and in this case also the multiphoton spectrum of a fragment may be superposed. For this reason in most cases one should refer to highly sensitive detection techniques and moderate light intensities, rather than high light intensities and nonsensitive detection technique. First, a series of sensitive detection techniques will be discussed in this section.

A. Measurement of the Photon Absorption Due to a Two-Photon Absorption Process

The measurement of absorption contains two steps: first the light power in front of the sample has to be measured, and second the light power after passing the sample. The detection limit is strongly dependent on the special technique used in the experiment, on the integration time, etc. Usually, one is not able to detect differences in light intensities less than 0.1%. Qualitatively, it is clear that real absorption measurements in two-photon spectroscopy are only possible for samples of high density (e.g., liquids and solids). It is not a feasible method for gas-phase spectroscopy, since it is impossible to get the laser light highly focused over a long absorption path length. The first absorption measurement in two-photon spectroscopy was done using the combination of a ruby laser and continuum flash lamp. The two-photon absorption was monitored on an oscilloscope as a short dip in the transmitted flash-lamp light intensity that coincides with the laser pulse. For this experiment, an accurate overlap of both light beams over a long distance is necessary in order to get a high level of absorption.

A special setup was developed by Hopfield *et al.*, who used a crystal that acts as a light guide for both light beams.

In a more recent work, a combination of a high-power pulsed dye laser and a fixed-fre-

quency continuous-wave (cw) Kr⁺ ion laser was used to detect absorption differences as small as 0.1%. Even though in this way a spectrum is hard to measure, very accurate absolute values of the two-photon absorption cross section were obtained for diphenylbutadiene at a special wavelength. In addition to the spectrum, the absolute two-photon cross section yields further arguments whether the two-photon absorption is pure electronic or is vibrationally induced.

A very sensitive technique for detection of weak absorption is the intracavity absorption technique. Here the weakly absorbing sample is placed in the cavity of a dye laser. As a consequence, the dye-laser emission is quenched at those wavelengths where the material absorbs. In this way an absorption as small as 10^{-4} can be detected in a time interval. This technique is appropriate to detect small amounts of material with sharply structured spectra, rather than to measure a complete spectrum. So far there is only one application to two-photon spectroscopy. Two-photon absorption in anthracene solution was detected by intracavity absorption.

B. Detection of Fluorescence after Two-Photon Excitation of Molecules

The most convenient and sensitive method for the measurement of two-photon spectra is the detection of the fluorescence of the excited molecules. As a consequence of a two-photon absorption, most molecules emit a photon at about twice the energy of the absorbed photons, as shown in Fig. 7. These emitted photons are detected with high sensitivity. A typical set-up is shown in Fig. 8. The light from a tunable dye laser is focused within the fluorescence cell con-

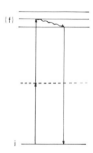

FIG. 7. Principle of a nonresonant two-photon absorption measurement by proving the photon emission. The straight lines with arrow denote the photon absorption and emission processes, and a wavy line represents a relaxation process from the level excited to the lowest one in the final electronic state.

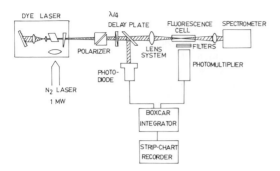

FIG. 8. Experimental setup of the fluorescence detection after a two-photon excitation of a gaseous sample.

taining the molecular gas or the solution. In the small focus, simultaneous absorption of two (visible) photons takes place and ultraviolet (UV) photons are emitted with a quantum yield typical for the molecule under investigation. The emitted UV photons are observed with a high-gain photomultiplier. Special filters are used in order to discriminate them from the intense exciting visible light. This can easily be accomplished since there usually is a large frequency shift between the exciting visible and the emitted UV light in two-photon spectroscopy. This greatly improves the signal-to-noise ratios relative to one-photon excitation and represents a side benefit of this detection technique. In the gas and in solution of low concentration there is no reabsorption of the emitted photons; however, this can be a problem in pure liquids and in crystals with a large absorption cross section. The signal of the photomultiplier is then fed into a boxcar integrator and integrated there within a short time interval of 10 nsec to some picoseconds, depending on the lifetime of the fluorescence. The integrated signal then is recorded on a strip-chart recorder. Finally, when continuously scanning the wavelength of the dye laser, one obtains the two-photon excitation spectrum of the molecule under investigation. For very weak signals and for time-resolved fluorescence measurements, the boxcar integrator may be replaced by a transient digitizer and a data-processing system.

The solid angle for observation of fluorescence is about 6×10^{-1} sr in this setup. The detection limit of the setup is reached when the signal to noise ratio in the measured spectrum is better than 1 : 1. For a conventional photomultiplier, the recording of two-photon spectra is possible when 200 photons are transmitted from the focus for a single exciting laser pulse.

For molecules with a fluorescence quantum yield of unity, this means that some 200 two-photon absorption events should have taken place during the laser pulse in order to be able to measure a two-photon excitation spectrum. Unfortunately, the fluorescence quantum yield of most molecules is smaller than 1, this being a fundamental disadvantage of the fluorescence detection method. In this case the detection by resonance-enhanced multiphoton ionization might be useful, as discussed in the next section. It is also possible to detect phosphorescence if the molecules undergo a fast intersystem crossing process into the triplet system. For detection of vibronic states with high excess energies above S_1, it might be useful to use a high-pressure buffer gas producing a fast collisional deactivation of the excited levels down to the vibrationless ground state. At high pressures this collisional deactivation might compete with the internal radiationless process, and fluorescence from the thermalized S_1 with laser fluorescence quantum yields is observed. By this method, vibronic states with excess energy as large as 6000 cm^{-1} above the vibrationless electronic S_1 state have been observed.

C. Detection of Two- or Three-Photon Absorption by Multiphoton Ionization of Molecules

After a molecule has been excited in an intense light field by two- or three-photon absorption, there is a high probability of absorbing further photons, which finally result in an ionization of the molecule. Another sensitive detection technique for two- or three-photon absorption processes in molecular gases is based on the subsequent ionization of the molecule after the two- or three-photon excitation.

The interesting feature of the multiphoton ionization from the spectroscopic point of view is the resonance enhancement by resonant intermediate states. Since the ionization efficiency is strongly enhanced when the photon energy comes to resonance with real intermediate states, a wavelength scan of the laser leads to a modulation of the ion current, which reflects the spectrum of the intermediate states. Thus it is possible to measure the intermediate-state (two- or three-photon) spectrum by measuring the ion currents. This method is used to measure two-photon spectra of polyatomic molecules by a three- and four-photon ionization.

1. The Ionization Cell

In Fig. 9, the ionization cell is shown in detail. The laser is focused into the cell containing the molecular gas at a typical pressure of a few torr. Very common is a device with a thin wire that is axially positioned in a cylindrical metal plate biased with a positive voltage of some 100 V. The potential drives the free electrons produced by the multiphoton ionization process to the electrode with positive charge. If enough voltage is applied between the electrodes, and if the particle density is sufficient, charge amplification by collisions can take place, increasing the detectability of the electrons. If the gas pressure is very low, the addition of a buffer gas is necessary in order to get charge multiplications in the ionization cell. The voltage produced by the current at a 1-MΩ impedance is then amplified in a preamplifier by one order of magnitude fed into a boxcar integrator and then integrated with a gate width of some 10 μsec. The modulation of the current as a function of wavelength is then recorded on a strip-chart recorder and reflects the intermediate-state resonance. Without a charge amplification in the ionization cell, the empirically found detection limit is about 1000 ions produced within one laser pulse.

2. Ion Detection in a Mass Spectrometer

Ion detection in an ionization cell is the simplest method and is very sensitive. There is, however, no information about the type of ions produced in the multiphoton ionization process. As discussed in Section VI,A, it has been shown that there often is a strong fragmentation of the ions in the multiphoton process.

In order to shed light on the ionization process, it is useful to detect the ions in a mass spectrometer, which allows one to determine the mass of the ions. The scheme of the setup for mass-selective ion detection is shown in Fig. 10. The tunable laser light or the frequency-doubled light is focused into an effusive molecular beam close to the aperture of the nozzle. The ions produced by the multiphoton ionization process are withdrawn through an ion lens system into the mass filter, mass-analyzed, and finally recorded with an ion multiplier. Ion detection in a mass spectrometer clearly reveals that in a typical two-photon experiment, molecules are not only excited to the two-photon state but molecules are ionized, and a rich pattern of fragment ions is formed. Therefore, in a two- or, especially, three-photon experiment with highly intense laser light, we must check whether these fragments may influence the measured two- or three-photon spectrum. This is certainly of importance for the three- and four-photon ionization experiments.

The ions produced in a multiphoton ionization process have particular features: they are produced within a short time interval of some nanoseconds during the laser pulse in the small volume of about 10^{-5} cm^3 given by the focus of the laser light. This spotlike type of pulsed ion sources makes feasible a special type of mass spectrometer. Mass detection for these types of ion sources can be achieved in a more appropriate manner by a time-of-flight mass analyzer.

D. PHOTOACOUSTIC DETECTION METHOD

In this section, the photoacoustic detection method will be briefly discussed. This method is principally different from the most convenient methods discussed in Sections III,B and III,C. Fluorescence detection and, to some degree, ionization detection decrease in sensitivity when a fast competing intra- or intermolecular relaxa-

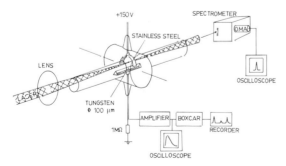

FIG. 9. Ionization cell and the setup for recording multiphoton intermediate-state spectra of gaseous samples.

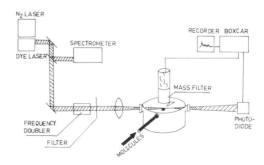

FIG. 10. Experimental setup for multiphoton ionization mass spectroscopy.

tion takes place from the multiphoton excited level. In this case there is a dissipation of the energy selectively released in the excited level into thermal energy. After pulsed excitation there is a rapid conversion of absorbed energy into pressure fluctuations, which then can be detected by a microphone. This means that photoacoustic spectroscopy is based on the detection of those effects that are loss channels in fluorescence and ionization detection. Apparently, photoacoustic detection technique should be principally suitable for observing two-photon spectra in weakly fluorescing materials.

For the time being, there have been only a few successful attempts to measure multiphoton spectra of molecules by the photoacoustic methods. The two-photon spectrum of liquid benzene has been published with the main vibronic bands and a value for the two-photon absorption cross section.

E. DETECTION OF PHOTOELECTRONS

This method is a combination of a photoelectron spectroscopic technique and a multiphoton spectroscopic technique with a multiphoton ionization laser system. The photoelectron intensity is measured as a function of the kinetic energy of electrons released. This is called multiphoton ionization photoelectron spectroscopy. The observed spectra contain information about the final states of the ions produced by multiphoton excitation when the resonant intermediate states are well identified or about the resonant states when the final ionic states are well defined. A schematic model is shown in Fig. 11 for observing the multiphoton ionization photoelectron spectra. The photoelectron energy is denoted by K. The photoelectrons from a sample irradiated by a pulsed tunable dye laser

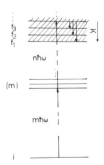

FIG. 11. Principle of a multiphoton ionization photoelectron spectroscopy; K denotes the photoelectron energy.

are detected by a time-of-flight energy analyzer. The photoelectron kinetic energy $K(\text{eV})$ is calculated by the formula $v = (5.93 \times 10^7)K^{1/2}$, where v (cm/sec) is the velocity of the observed photoelectron.

F. MISCELLANEOUS DETECTION METHODS

Some other methods have been proven to be sensitive enough for detection of a two-photon process.

One of these methods is thermal blooming, which is also based on the energy-loss mechanisms in the excited states. The transfer of energy from the excited states into heat causes a change of refractive index of the material under investigation. This is then detected by changes in the optical behavior of the material, that is, by a weak focusing or defocusing of the exciting laser beam. The thermal blooming technique has been applied to the test molecule benzene. Two-photon spectra of liquid benzene were measured point by point in the range between 360 and 530 nm. More recently a spectrum generated by this technique yielded additional information about the position of the $^1E_{1g}$ state of benzene. Another detection method for the two-photon absorption is based on changes of the susceptibility of the material under investigation in the presence of the light field. In a strong light field there are higher-order contributions $\chi^{(n)}$ to the susceptibility: the real part of the nonlinear susceptibility Re $\chi_{ijkl}^{(3)}$ produces an intensity-dependent index of refraction that may turn the polarization vector of the incoming light wave, and the imaginary part of the third-order susceptibility Im $\chi_{ijkl}^{(3)}$ produces an intensity-dependent absorption coefficient, which increases if there is a resonance at the two-photon energy. These two-photon resonances in the third-order susceptibility $\chi^{(3)}$ may be detected in several ways.

Two-photon resonances of gases such as SO_2 and NO can be detected in a four-wave mixing experiment. Since this method creates no real population of the resonant state, the detection of the resonances does not directly depend on the dynamical pathway followed by the excited state. The dynamics of the resonant states enters only through a damping parameter, thereby limiting the magnitude of the resonance term. It has been shown that this method is suitable for obtaining absolute two-photon cross section by comparison of the two-photon resonances with coherent anti-stokes Raman resonances of $\chi_{ijkl}^{(3)}$. Two-photon cross sections are then given in terms of the accurately known Raman cross sections. As demonstrated for Na vapors, the change of polarization produced by the imaginary part of the third-order susceptibility $\chi^{(3)}_{ijkl}$ can be detected for observation of the two-photon spectrum. An extension of this method of the case of molecules seems possible even though the sensitivity is not expected to be better than that of the other methods discussed above. The general virtue of two-photon absorption detection via $\chi^{(3)}_{ijkl}$ is the calibration of two-photon cross sections on the basis of Raman cross sections.

G. DOPPLER-FREE MULTIPHOTON SPECTROSCOPY

One of the advantages of multiphoton spectroscopy is elimination of the inhomogeneous Doppler broadening in the spectra. For an ambient atomic or molecular gas, there is an isotropic velocity distribution that brings about different shifts for the atoms or molecules with different velocity components in the direction of light propagation. The average of these shifts results in a Doppler broadening in the optical transition. For a Maxwell–Boltzmann velocity distribution, a Gaussian line profile in the spectra is characterized by a full-width-at-half-maximum (FWHM)

$$\Delta\omega_D = (2\omega_0/c)[2 \ln(2kT/m)]^{1/2}$$
$$= (2.163 \times 10^{-7})\omega_0(T/M)^{1/2} \qquad (47)$$

where ω_0 is the optical transition angular frequency, T the temperature (K), m the mass in kilograms, and M the molecular weight of the particles in atomic mass units. Typically, the Doppler width in angular frequency $\Delta\omega_D$ for a polyatomic molecule such as benzene C_6H_6 with $M = 78$ at room temperature is $\Delta\omega_D = 1.67$ GHz for $\omega_0 = 40,000$ cm^{-1}. This value of the Doppler width is several times larger than the average spacing of rovibronic transitions in polyatomic molecules. Therefore, in this case it is not possible to observe single Doppler-broadened lines, but the envelope of the line produces a typical rotational contour of the vibronic band.

The frequency shifts of Doppler-limited and Doppler-free two-photon absorption are shown in Fig. 12, in which the interaction of the particle with two monochromatic light beams with frequencies ω_1 and ω_2 is presented. In general, the Doppler broadening in angular frequency is given by $\Delta\omega = \Delta\mathbf{k} \cdot \mathbf{v}$, where $\Delta\mathbf{k}$ is the change in momentum of the laser light and $\mathbf{v}$ is the atomic or molecular velocity. For each particle

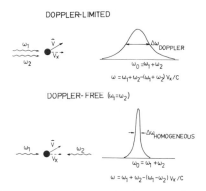

FIG. 12. Principle of Doppler-limited and Doppler-free two-photon absorptions.

with velocity $\mathbf{v}$ whose propagation component in the propagation direction of the laser beams is v_x, the optical frequency ω_0 is shifted by $(\omega_1 + \omega_2)v_x/c$. When an ensemble of the particles in thermal equilibrium is investigated, this yields a broadening of the transition line according to Eq. (47). If two light beams with frequencies ω_1 and ω_2 propagate in opposite direction, as shown in the lower part of Fig. 12, and the particle absorbs one photon from each beam, then the corresponding Doppler shifts have opposite signs, and the residual Doppler shift is given as $(\omega_1 - \omega_2)v_x/c$. The shift cancels exactly to zero if the frequencies of both laser beams are equal to each other.

A typical experimental setup for observing the Doppler-free two-photon absorption is shown in Fig. 13. There are optics for polarization of photon and focusing laser beam, and there is a sample cell between the laser and detection system. Applications of the Doppler-free two-photon absorption of atoms and molecules will be presented in Section V.

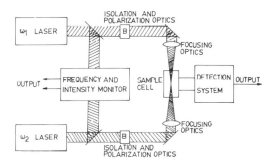

FIG. 13. Experimental setup for observing the Doppler-free two-photon absorption.

IV. Characteristics and Spectral Properties

Laser-intensity dependence and polarization behavior in multiphoton transitions, which are the main characteristics of multiphoton spectroscopy, are presented first. An application of the polarization behavior to a two-photon transition of a molecular system is described. Spectral properties of atomic and molecular multiphoton transitions are finally described, focusing on their difference between nonresonant and resonant multiphoton transitions.

A. LASER-INTENSITY DEPENDENCE

Measurement of the laser-intensity dependence is very important to understand the mechanism of multiphoton processes. The laser-intensity dependence observed in the multiphoton transitions can be classified into two types. One type originates from an intrinsic laser intensity dependence, called the formal intensity law, and the other type from geometrical effects of the focused laser beam applied in the region of material–photon interaction. The latter is sometimes called the $\frac{3}{2}$-power law in multiphoton ionization dissociation experiments, because product yields are proportional to $I^{3/2}$, the $\frac{3}{2}$-power of laser intensity, or in some cases to noninteger powers of the laser intensity irrelevant to the intrinsic intensity dependence. The $I^{3/2}$ dependence has been explained by the laser intensity change due to conical focusing in the material–photon interaction region.

The formal intensity law means that the n-photon transition rate constant is proportional to the nth order of laser intensity, I^n. This can be understood from the nonresonant multiphoton rate expression derived in the ordinary time-dependent perturbation theory; for an n-photon absorption experiment in which a single laser beam with frequency ω_R is irradiated, the transition probability from i to f states, $W_{i \to f}^{(n)}$, is written as

$$W_{i \to f}^{(n)} = I^n \sigma_{i \to f}^{(n)} \qquad (48a)$$

or

$$W_{i \to f}^{(n)} = \frac{I^n \sigma_{i \to f}^{(n)}}{(\hbar \omega_R)^n} \qquad (48b)$$

where I is the photon flux in units of photons per area per time and the photon intensity in units of energy per area per time in Eq. (48a) and (48b), respectively, and $\sigma^{(n)}$ is the cross section in units of (area $)^n$ (time$)^{n-1}$, written as

$$\sigma_{i \to f}^{(n)} = 2\pi (2\pi\alpha)^2 \omega_R^n \left| \sum_{m_1} \sum_{m_2} \right.$$

$$\times \cdots \sum_{m_{n-1}}$$

$$\times \left. \frac{\langle f | \mathbf{r} \cdot \mathbf{e} | m_{n-1} \rangle \cdots \langle m_1 | \mathbf{r} \cdot \mathbf{e} | i \rangle}{[\omega_{m_{n-1}} - \omega_i - (n-1)\omega_R]} \cdots (\omega_{m_1} - \omega_i - \omega_R) \right|^2$$

$$\times \delta(\omega_f - \omega_i - n\omega_R) \qquad (49)$$

The formal intensity law has been utilized to determine orders of multiphoton processes. It should be noted that this law holds for nonresonant multiphoton transitions in low-intensity-laser experiments. In Fig. 14, log–log plots of ionization of aniline ($C_6H_5NH_2$) are shown, in which the formal intensity law for $n = 2$ in the case of unfocused laser excitation and the $\frac{3}{2}$-power laser can be seen. The geometric effect of the focused laser beam can be eliminated by setting up the laser in the crossed atomic or molecular beam. Use of a high-intensity laser for resonant multiphoton transitions may result in a deviation from I^n dependence even after elimination of the geometric effect. A qualitative interpretation of the deviation can be made by using the rate equation approach. For simplicity, let us consider a resonant two-photon ionization

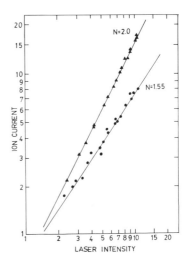

FIG. 14. Intensity dependence of the ion yield produced by resonant two-photon ionization via the $^1B_2$ state of aniline: (a) with a focused high-power laser, (b) with an unfocused laser. [From J. H. Brophy and C. T. Rettner (1971). *Chem. Phys. Lett.* **67,** 351.]

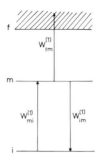

FIG. 15. A simple model for a resonant two-photon ionization.

shown in Fig. 15 and see the conditions under which the deviation from the formal intensity law takes place. The rate equations associated with a sequential two-photon ionization for state i to f through a resonant state m can be expressed as

$$d\rho_i(t)/dt = -W_{ii}^{(1)}\rho_i(t) + W_{im}^{(1)}\rho_m(t) \quad (50a)$$

$$d\rho_m(t)/dt = -W_{mm}^{(1)}\rho_m(t) + W_{mi}^{(1)}\rho_i(t) \quad (50b)$$

$$d\rho_f(t)/dt = W_{fm}^{(1)}\rho_m(t) \quad (50c)$$

where $\rho_i(t) = \rho_{ii}(t)$ is the density matrix element for the initial state with the initial condition $\rho_i(t = 0) = N_0$. The rate constants satisfy $W_{mm}^{(1)} = W_{im}^{(1)} + W_{fm}^{(1)}$ and $W_{ii}^{(1)} = W_{mi}^{(1)}$. The ionization rate is given by

$$\frac{d\rho_f(t)}{dt} = \frac{W_{fm}^{(1)}W_{mi}^{(1)}N_0}{\alpha_1 - \alpha_2}$$
$$\times [\exp(\alpha_1 t) - \exp(\alpha_2 t)] \quad (51)$$

where α_1 and α_2 are the solution for the equation

$$[\alpha + W_{ii}^{(1)}][\alpha + W_{mm}^{(1)}] + W_{im}^{(1)}W_{mi}^{(1)} = 0 \quad (52)$$

The ion number in the case of a square laser pulse with duration t_p can be expressed as

$$\rho_f(t) = \frac{W_{fm}^{(1)}W_{mi}^{(1)}N_0}{\alpha_1 - \alpha_2}$$
$$\times \left[\frac{\exp(\alpha_1 t_p) - 1}{\alpha_1} - \frac{\exp(\alpha_2 t_p) - 1}{\alpha_2}\right]$$
$$(53)$$

In the case of the weak laser field in which the effect of stimulated emission is negligible, $\alpha_{1,2} = -W_{fm}^{(1)}$ and $-W_{mi}^{(1)}$, and, furthermore, in the time regions of $W_{mi}^{(1)}t \leq 1$, the ionization rate is approximately given by

$$d\rho_f(t)/dt \simeq W_{fm}^{(1)}W_{mi}^{(1)}N_0 t \quad (54)$$

which indicates a quadratic intensity dependence. The ion number is also given by the quadratic intensity dependence. On the other hand, in the case in which a strong laser intensity is applied and the stimulated emission process cannot be neglected, the spontaneous emission can safely be omitted $W_{im}^{(1)} = W_{mi}^{(1)}$ and

$$\alpha_{1,2} = \tfrac{1}{2}\{-[2W_{mi}^{(1)} + W_{fm}^{(1)}]$$
$$\pm \sqrt{4[W_{mi}^{(1)}]^2 + [W_{fm}^{(1)}]^2} \quad \} \quad (55)$$

For the time scale $t < (\alpha_1 - \alpha_2)^{-1}$, the ionization rate is given by

$$d\rho_f(t)/dt \simeq W_{fm}^{(1)}W_{mi}^{(1)}N_0 t \quad (56)$$

and for the time scale $t > (\alpha_1 - \alpha_2)^{-1}$,

$$\frac{d\rho_f(t)}{dt} \simeq \frac{W_{fm}^{(1)}W_{mi}^{(1)}N_0}{\{4[W_{mi}^{(1)}]^2 + [W_{fm}^{(1)}]^2\}^{1/2}} \quad (57)$$

From the above expression, the deviation of the ionization rate from the quadratic dependence is found to take place at $t > (\alpha_1 - \alpha_2)^{-1}$ under the strong laser irradiation, and

$$\frac{d\rho_f(t)}{dt} \simeq \frac{W_{fm}^{(1)}N_0}{2} \quad \text{for} \quad W_{mi}^{(1)}$$
$$\simeq W_{im}^{(1)} > W_{fm}^{(1)} \quad (58)$$

and

$$\frac{d\rho_f(t)}{dt} \simeq W_{mi}^{(1)}N_0 \quad \text{for} \quad W_{im}^{(1)} < W_{fm}^{(1)} \quad (59)$$

For the former case, an equilibrium between the initial and resonant states has been achieved, that is, a saturation has been reached, and the excitation process from the resonant to the final state can be observed as the apparent transition. In the latter case, in which the resonant and ionized states are strongly coupled in terms of the dipole transition, the initial excitation process $i \rightarrow m$ can be observed as the apparent transition. This case is commonly seen in resonant multiphoton ionization spectra of atoms and molecules such as shown in Section IV,C.

B. POLARIZATION BEHAVIOR

Cross sections of multiphoton transitions of atoms or molecules in solids, liquids, and gases depend on whether linearly or circularly polarized laser light is applied. This is called the polarization dependence, one of the important characteristics of multiphoton spectroscopy. Measurement of the polarization dependence makes it possible to assign the excited-state symmetry of the material and to obtain information about the mechanism of the transition. In order to see the origin of the polarization behaviors, the concept of photon angular momentum will be briefly mentioned first. Polarization behaviors in multiphoton ionization of atoms and those in two-photon absorptions of both nonrotating and rotating molecules will then be described.

1. Photon Angular Momentum and Polarization Effects

A photon state is characterized not only by its linear momentum but also by the polarization vector $\mathbf{e}_\lambda$, which transforms like a vector and is considered to be the intrinsic angular momentum (spin). Circular polarization vectors constructed by the linear polarization vectors $\mathbf{e}_x$ and $\mathbf{e}_y$,

$$\mathbf{e}_\pm = \mp(1/\sqrt{2})(\mathbf{e}_x \pm i\mathbf{e}_y) \qquad (60)$$

are associated with the spin component $m = \pm 1$ of spin 1 where the quantization axis has been chosen in the photon propagation (z) direction. On the other hand, linear polarization vectors $\mathbf{e}_x$ and $\mathbf{e}_y$ are eigenstates of the intrinsic angular momentum but are not eigenstates of the projection of the spin on the polarization direction. Therefore, in an electronic transition induced by one-photon absorption or emission, one unit of angular momentum is transferred between the electromagnetic field and the electron in the dipole approximation. Because of a different selection rule for the spin component between linearly and circularly polarized photons—that is, the selection rule of the magnetic quantum number $\Delta m = 0$ for linearly polarized photons along the z axis and $\Delta m = \pm 1$ for photons circularly polarized and propagating along the z axis, in which (+) and (−) correspond to right and left polarization, respectively—transition rates of multiphoton processes in which more than one unit of angular momentum are transferred depend on the polarization applied. A simple example (Fig. 16) in a hydrogenic, one-electron model shows the angular-momentum channels available for four-photon transition from an S initial state. As a result of the difference in the number of available channels with different cross sections, the total transition rates of the multiphoton process depend on the polarization.

One of the polarization effects can be seen in multiphoton ionization of atoms and molecules, which is analyzed by using a method of multiphoton ionization photoelectron spectroscopy. The final ionized state is in the continuum

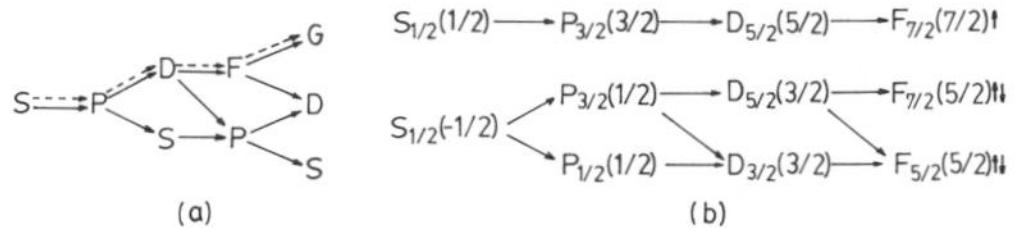

FIG. 16. Angular momentum channels available for 4-photon ionization of an atomic S state. (a) The linearly (———) and circularly (———) polarized cases are illustrated in the case of a negligibly small spin–orbit coupling. (b) The case for atoms with spin–orbit split levels and right circular polarization. The numbers in parentheses are the spin projection quantum numbers, while the arrows illustrate the possible orientations of the free electron spin. [From P. Lambropoulos (1976). *Adv. Atomic Mol. Phys.* **12**, 87.]

and can adequately be expressed as a superposition of partial waves with well-defined angular momentum l; its electronic part is written as

$$|f(r)\rangle = 4\pi \sum_{l=0}^{\infty} i^l \exp(-i\delta_l) G_l(\mathbf{k}, \mathbf{r})$$
$$\times \sum_{m=-l}^{l} Y_{lm}^*(\Theta, \Phi) Y_{lm}(\theta, \phi) \qquad (61)$$

where δ_l is the phase shift, G_l the radial part of the partial wave; and Y_{lm}^* and Y_{lm} are spherical harmonics. The spherical coordinates of the wave vector/radius vector ($\mathbf{k}$) and ($\mathbf{r}$) are denoted by (k, Θ, Φ) and (r, θ, ϕ), respectively. All angular momenta in the final state are available for the multiphoton transition; that is, a photon of any arbitrary polarization leads to ionization, in contrast to a bound–bound transition. The total ionization rate again depends on the photon polarization. For example, a three-photon ionization with a circularly polarized photon leads to a photoelectron of orbital angular momentum $l = 3$ (F wave); on the other hand, in the case of a linearly polarized photon, it leads to a photoelectron whose state is a superposition of $l = 1$ (P wave) and $l = 3$, as shown in Fig. 16.

Another interesting example of polarization behavior in multiphoton processes can be seen in bound–bound transitions of molecules.

2. Polarization Behavior of Nonrotating Molecules

Let us consider a nonresonant two-photon absorption of randomly oriented nonrotating molecules excited by two lasers with polarization vectors $\mathbf{e}_1$, $\mathbf{e}_2$, that is, ($\mathbf{e}_x$, $\mathbf{e}_y$) or ($\mathbf{e}_+$, $\mathbf{e}_-$), and angular frequencies ω_1 and ω_2. The two-photon transition probability from i to f states, $W_{i \to f}^{(2)}$, is given by

$$W_{i \to f}^{(2)} = 2\pi(2\pi\alpha)^2 I_1 I_2 \omega_1 \omega_2 |S_{fi}(\omega_1 \mathbf{e}_1, \omega_2 \mathbf{e}_2)|^2$$
$$\times \delta(\omega_{fi} - \omega_1 - \omega_2) \quad (62)$$

where $S_{fi}(\omega_1 \mathbf{e}_1, \omega_2 \mathbf{e}_2)$, the two-photon transition amplitude, takes the form

$$S_{fi}(\omega_1 \mathbf{e}_1, \omega_2 \mathbf{e}_2) = \sum_m \left[\frac{\mathbf{e}_2 \cdot \mathbf{R}_{fm} \mathbf{e}_1 \cdot \mathbf{R}_{mi}}{\omega_{mi} - \omega_1} \right.$$
$$\left. + \frac{\mathbf{e}_1 \cdot \mathbf{R}_{fm} \mathbf{e}_2 \cdot \mathbf{R}_{mi}}{\omega_{mi} - \omega_2} \right]$$
$$(63)$$

The polarization vectors are usually expressed in a laboratory coordinate system, while the electronic transition moment operator $\mathbf{r}$ is expressed in the molecular system, and therefore a coordinate transformation has to be taken into account in order to evaluate the two-photon transition amplitude. This can be carried out by introducing the Euler angles, which specify the molecular coordinate system with respect to the laboratory one. After the transformation is accomplished, the two-photon transition probability average over all the molecular orientations, $\langle W_{i \to f}^{(2)} \rangle$, can be written as

$$\langle W_{i \to f}^{(2)} \rangle = \delta_F F + \delta_G G + \delta_H H \quad (64)$$

where F, G, and H, which are experimentally controllable polarization variables, are given by

$$F = 4|\mathbf{e}_1 \cdot \mathbf{e}_2|^2 - 1 - |\mathbf{e}_1 \cdot \mathbf{e}_2^*| \quad (65a)$$
$$G = -|\mathbf{e}_1 \cdot \mathbf{e}_2|^2 + 4 - |\mathbf{e}_1 \cdot \mathbf{e}_2^*| \quad (65b)$$
$$H = -|\mathbf{e}_1 \cdot \mathbf{e}_2|^2 - 1 + 4|\mathbf{e}_1 \cdot \mathbf{e}_2^*|^2 \quad (65c)$$

In Eq. (64), δ_F, δ_G, and δ_H, which are characterized by the molecular quantities, laser flux, and laser detuning, take the form

$$\delta_F = \frac{1}{15} (2\pi\alpha)^2 I_1 I_2 \omega_1 \omega_2$$

$$\times \sum_a \sum_b S_{fi}^{aa} S_{fi}^{bb*} \delta(\omega_{fi} - \omega_1 - \omega_2) \quad (66a)$$

$$\delta_G = \frac{1}{15} (2\pi\alpha)^2 I_1 I_2 \omega_1 \omega_2$$

$$\times \sum_a \sum_b S_{fi}^{ba} S_{fi}^{ba*} \delta(\omega_{fi} - \omega_1 - \omega_2) \quad (66b)$$

$$\delta_H = \frac{1}{15} (2\pi\alpha)^2 I_1 I_2 \omega_1 \omega_2$$

$$\times \sum_a \sum_b S_{fi}^{ba} S_{fi}^{ab*} \delta(\omega_{fi} - \omega_1 - \omega_2) \quad (66c)$$

where S_{fi}^{ba}, the component of the two-photon transition tensor in the molecular coordinate system, is defined by

$$S_{fi}^{ba} = \sum_m \left[\frac{\langle f|r_b|m \rangle \langle m|r_a|i \rangle}{\omega_{mi} - \omega_2} + \frac{\langle f|r_a|m \rangle \langle m|r_b|i \rangle}{\omega_{mi} - \omega_1} \right]$$
$$(67)$$

It should be noted that since Eq. (66a) is expressed in terms of the absolute square of the trace of the two-photon transition tensor, $\delta_F \neq 0$ is satisfied only for the case of a transition to a totally symmetric state when the initial state is of a totally symmetric.

Depending on the combination of polarizations used, some experimental cases can be considered:

(1) two linearly polarized photons with parallel polarization;
(2) two linearly polarized photons with perpendicular polarization;
(3) one linear and one circular with linear polarization perpendicular to the plane of the circular polarization;
(4) both circular, in either the same or opposite sense, with perpendicular propagation;
(5) both linear, with $\theta = 45°$ between the two polarization vectors;
(6) one linear and one circular, with linear polarization in the plane of the circular polarization;
(7) both circular in the same sense, with parallel propagation; and
(8) both circular in the opposite sense, with parallel propagation.

In Table I, the values of F, G, and H that correspond to these cases are presented.

As a simple example of the polarization dependence, let us consider two cases of a two-photon transition from a totally symmetric ground state to a nontotally symmetric state: in one case the transition is excited by the laser beam with two linearly polarized photons [case (1), in which the transition probability is denoted by $\langle W_{i \to f}^{(2)} \rangle^{\uparrow\uparrow}$], and in the other case the transition is induced by lasers of two circularly polarized photons with parallel propagation [case (7), in which the transition probability is denoted by $\langle W_{i \to f}^{(2)} \rangle^{CC}$]. From Table I in this case, the ratio is given by

$$\langle W_{i \to f}^{(2)} \rangle^{CC} / \langle W_{i \to f}^{(2)} \rangle^{\uparrow\uparrow} = \tfrac{3}{2}$$

When one assigns a two-photon absorption of molecules, it is important to know the tensor

TABLE I.   Values of the Polarization Variables F, G, and H for Eight Two-Photon Transitions[a]

Polarization variable	Case							
	1	2	3	4	5	6	7	8
F	2	-1	-1	$-\frac{1}{4}$	$\frac{1}{2}$	$\frac{1}{2}$	-2	3
G	2	4	4	$\frac{7}{2}$	3	3	3	3
H	2	-1	-1	$-\frac{1}{4}$	$\frac{1}{2}$	$\frac{1}{2}$	3	-2

[a] From P. R. Monson and W. M. McClain (1970). *J. Chem. Phys.* **53**, 29.

patterns, which depend only on the symmetry of the molecular states relevant to the transition. The tensor patterns are tabulated in Table II. Here the initial state is assumed to belong to the totally symmetric representation A. The tabulated quantity is

$$S_{fi}^{ba} = \langle f^J | {}^A S^{ba} + {}^B S^{ba} + \cdots | i^A \rangle$$

where A, B, ..., J represent the names of the symmetry species (A, B, E, T, etc.).

3. Polarization Behavior of Rotating Molecules

Rotational structures in multiphoton spectra of molecules in gases are well resolved by using a narrow-band tunable dye laser. In these experiments, the different rotational branches of the same band of two-photon excitation spectra of gaseous molecules differ in their polarization behavior. In this subsection, the polarization be-

TABLE II.   Cartesian Tensor Patterns for Two-Photon Processes[a]

1. Groups C_1 and C_i

$$A = \begin{bmatrix} s_1 & s_2 & s_3 \\ s_4 & s_5 & s_6 \\ s_7 & s_8 & s_9 \end{bmatrix}$$

2. Groups C_2, C_3, and C_{2h}

$$A = \begin{bmatrix} s_1 & s_4 & 0 \\ s_5 & s_2 & 0 \\ 0 & 0 & s_3 \end{bmatrix}, \qquad B = \begin{bmatrix} 0 & 0 & s_6 \\ 0 & 0 & s_7 \\ s_8 & s_9 & 0 \end{bmatrix}$$

3. Groups C_{2v}, D_2, and D_{2h}

$$A_1 = A = \begin{bmatrix} s_1 & 0 & 0 \\ 0 & s_2 & 0 \\ 0 & 0 & s_3 \end{bmatrix}, \qquad A_2 = B_1 = \begin{bmatrix} 0 & s_4 & 0 \\ s_5 & 0 & 0 \\ 0 & 0 & 0 \end{bmatrix}$$

$$B_1(C_{2v}) = B_2 = \begin{bmatrix} 0 & 0 & s_6 \\ 0 & 0 & 0 \\ s_7 & 0 & 0 \end{bmatrix}, \qquad B_2(C_{2v}) = B_3 = \begin{bmatrix} 0 & 0 & 0 \\ 0 & 0 & s_8 \\ 0 & s_9 & 0 \end{bmatrix}$$

4. Groups C_4, C_{4h}, and S_4

$$A = \begin{bmatrix} s_1 & s_3 & 0 \\ -s_3 & s_1 & 0 \\ 0 & 0 & s_2 \end{bmatrix}, \qquad B = \begin{bmatrix} s_4 & s_3 & 0 \\ s_5 & -s_4 & 0 \\ 0 & 0 & 0 \end{bmatrix}$$

TABLE II (*Continued*)

$$E = \begin{bmatrix} 0 & 0 & s_6 \\ 0 & 0 & -is_6 \\ s_7 & -is_7 & 0 \end{bmatrix} \quad \text{and} \quad \begin{bmatrix} 0 & 0 & s_6^* \\ 0 & 0 & is_6^* \\ s_7^* & is_7^* & 0 \end{bmatrix}$$

5. Groups C_{4v}, D_4, D_{2d}, and D_{4h}

$$A_1 = \begin{bmatrix} s_1 & 0 & 0 \\ 0 & s_1 & 0 \\ 0 & 0 & s_2 \end{bmatrix}, \quad A_2 = \begin{bmatrix} 0 & s_3 & 0 \\ -s_3 & 0 & 0 \\ 0 & 0 & 0 \end{bmatrix}$$

$$B_1 = \begin{bmatrix} s_4 & 0 & 0 \\ 0 & -s_4 & 0 \\ 0 & 0 & 0 \end{bmatrix}, \quad B_2 = \begin{bmatrix} 0 & s_5 & 0 \\ s_5 & 0 & 0 \\ 0 & 0 & 0 \end{bmatrix}$$

$$E = \begin{bmatrix} 0 & 0 & s_6 \\ 0 & 0 & -is_6 \\ s_7 & -is_7 & 0 \end{bmatrix} \quad \text{and} \quad \begin{bmatrix} 0 & 0 & s_6^* \\ 0 & 0 & is_6^* \\ s_7^* & is_7^* & 0 \end{bmatrix}$$

6. Groups C_3 and $S_6 = C_{3h}$

$$A = \begin{bmatrix} s_1 & s_3 & 0 \\ -s_3 & s_1 & 0 \\ 0 & 0 & s_2 \end{bmatrix}$$

$$E = \begin{bmatrix} s_4 & is_4 & s_5 \\ is_4 & -s_4 & -is_5 \\ s_6 & -is_6 & 0 \end{bmatrix} \quad \text{and} \quad \begin{bmatrix} s_4^* & -is_4^* & s_5^* \\ -is_4^* & -s_4^* & is_5^* \\ s_6^* & is_6^* & 0 \end{bmatrix}$$

7. Groups C_{3v}, D_3, and D_{3d}

$$A_1 = \begin{bmatrix} s_1 & 0 & 0 \\ 0 & s_1 & 0 \\ 0 & 0 & s_2 \end{bmatrix}, \quad A_2 = \begin{bmatrix} 0 & s_3 & 0 \\ -s_3 & 0 & 0 \\ 0 & 0 & 0 \end{bmatrix}$$

$$E = \begin{bmatrix} s_4 & is_4 & s_5 \\ is_4 & -s_4 & -is_3 \\ s_6 & -is_6 & 0 \end{bmatrix} \quad \text{and} \quad \begin{bmatrix} s_4^* & -is_4^* & s_5^* \\ -is_4^* & -s_4^* & is_5^* \\ s_6^* & is_6^* & 0 \end{bmatrix}$$

8. Groups C_{3h}, C_6, and C_{6h}

$$A = \begin{bmatrix} s_1 & s_3 & 0 \\ -s_3 & s_1 & 0 \\ 0 & 0 & s_2 \end{bmatrix}$$

$$E_1 = \begin{bmatrix} 0 & 0 & s_4 \\ 0 & 0 & -is_4 \\ s_5 & -is_5 & 0 \end{bmatrix} \quad \text{and} \quad \begin{bmatrix} 0 & 0 & s_4^* \\ 0 & 0 & is_4^* \\ s_5^* & is_5^* & 0 \end{bmatrix}$$

$$E_2 = \begin{bmatrix} s_6 & -is_6 & 0 \\ -is_6 & -s_6 & 0 \\ 0 & 0 & 0 \end{bmatrix} \quad \text{and} \quad \begin{bmatrix} s_6^* & is_6^* & 0 \\ is_6^* & -s_6^* & 0 \\ 0 & 0 & 0 \end{bmatrix}$$

9. Groups C_{6v}, D_{3h}, D_6, and D_{6h}; groups $C_{\infty v}$ and $D_{\infty v}$

$$A_1 = \Sigma^+ = \begin{bmatrix} s_1 & 0 & 0 \\ 0 & s_1 & 0 \\ 0 & 0 & s_2 \end{bmatrix}, \quad A_2 = \Sigma^- = \begin{bmatrix} 0 & s_3 & 0 \\ -s_3 & 0 & 0 \\ 0 & 0 & 0 \end{bmatrix}$$

(*continues*)

TABLE II   (*Continued*)

$$E_1 = \Pi = \begin{bmatrix} 0 & 0 & s_4 \\ 0 & 0 & -is_4 \\ s_5 & -is_5 & 0 \end{bmatrix} \quad \text{and} \quad \begin{bmatrix} 0 & 0 & s_4^* \\ 0 & 0 & is_4^* \\ s_5^* & is_5^* & 0 \end{bmatrix}$$

$$E_2 = \Delta = \begin{bmatrix} s_6 & -is_6 & 0 \\ -is_6 & -s_6 & 0 \\ 0 & 0 & 0 \end{bmatrix} \quad \text{and} \quad \begin{bmatrix} s_6^* & is_6^* & 0 \\ is_6^* & -s_6^* & 0 \\ 0 & 0 & 0 \end{bmatrix}$$

10. Groups T and T_h, $\omega = \exp(2\pi i/3)$

$$A = \begin{bmatrix} s_1 & 0 & 0 \\ 0 & s_1 & 0 \\ 0 & 0 & s_1 \end{bmatrix}$$

$$E = \begin{bmatrix} s_2 & 0 & 0 \\ 0 & \omega s_2 & 0 \\ 0 & 0 & \omega^* s_2 \end{bmatrix} \quad \text{and} \quad \begin{bmatrix} s_2^* & 0 & 0 \\ 0 & \omega^* s_2^* & 0 \\ 0 & 0 & \omega s_2^* \end{bmatrix}$$

$$T = \begin{bmatrix} 0 & 0 & 0 \\ 0 & 0 & s_3 \\ 0 & s_4 & 0 \end{bmatrix} \quad \text{and} \quad \begin{bmatrix} 0 & 0 & s_4 \\ 0 & 0 & 0 \\ s_3 & 0 & 0 \end{bmatrix} \quad \text{and} \quad \begin{bmatrix} 0 & s_3 & 0 \\ s_4 & 0 & 0 \\ 0 & 0 & 0 \end{bmatrix}$$

11. Groups O, O_h, and T_d, $\omega = \exp(2\pi i/3)$

$$A = \begin{bmatrix} s_1 & 0 & 0 \\ 0 & s_1 & 0 \\ 0 & 0 & s_1 \end{bmatrix}$$

$$E = \begin{bmatrix} s_2 & 0 & 0 \\ 0 & \omega s_2 & 0 \\ 0 & 0 & \omega^* s_2 \end{bmatrix} \quad \text{and} \quad \begin{bmatrix} s_2^* & 0 & 0 \\ 0 & \omega^* s_2^* & 0 \\ 0 & 0 & \omega s_2^* \end{bmatrix}$$

$$T_1 = \begin{bmatrix} 0 & s_2 & 0 \\ -s_2 & 0 & 0 \\ 0 & 0 & 0 \end{bmatrix} \quad \text{and} \quad \begin{bmatrix} 0 & 0 & -s_3 \\ 0 & 0 & 0 \\ s_3 & 0 & 0 \end{bmatrix} \quad \text{and} \quad \begin{bmatrix} 0 & 0 & 0 \\ 0 & 0 & s_3 \\ 0 & -s_3 & 0 \end{bmatrix}$$

$$T_2 = \begin{bmatrix} 0 & s_4 & 0 \\ s_4 & 0 & 0 \\ 0 & 0 & 0 \end{bmatrix} \quad \text{and} \quad \begin{bmatrix} 0 & 0 & s_4 \\ 0 & 0 & 0 \\ s_4 & 0 & 0 \end{bmatrix} \quad \text{and} \quad \begin{bmatrix} 0 & 0 & 0 \\ 0 & 0 & s_4 \\ 0 & s_4 & 0 \end{bmatrix}$$

[a] The tabulated quantity is $S^{ba}(A \rightarrow J)$. The table is divided into 11 sets of groups. Within each set, the groups either are isomorphic or differ from each other only by the inclusion of the center of inversion element. The tensors are labeled by the symbol J only, and that symbol is often simplified by dropping primes and subscripts, due to the variability of nomenclature among different groups within the same isomorphic set. The word "and" is written between tensors that belong to the different parts of a degenerate transition; such pairs must always be used together, for example, for a twofold degeneracy, according to $W^{(2)} = |\mathbf{e}_1 \cdot \mathbf{S}^{I} \cdot \mathbf{e}_2|^2 + |\mathbf{e}_1 \cdot \mathbf{S}^{II} \cdot \mathbf{e}_2|^2$. The basis sets for symmetry species A and B are always unambiguous. When the group has one E species, the basis set is $[x + iy, x - iy]$, except in the tetrahedral and octahedral groups (sets 10 and 11), where the basis is $[u + iv, u - iv]$, with $u = 2z^2 - x^2 - y^2$ and $v = 3^{1/2}(x^2 - y^2)$. When the group has two E species, the basis of E_1 is $[x + iy, x - iy]$ and the basis of E_2 is $[(x + iy)^2, (x - iy)^2]$. In groups T and T_h (set 10), the basis of species T is (x, y, z). In groups O, O_h, and T_d, the basis of T_1 is (x, y, z) and the basis of T_2 is (yz, zx, xy). In sets 10 and 11, note that $1 + \omega + \omega^* = 0$. [W. M. McClain and R. A. Harris (1977). *Excited States* **3**, 2]

havior seen in rotational contour in a nonreso-nant two-photon absorption of a rotating symmetric molecule is treated. The initial and final rovibronic states are respectively specified as $|i\rangle = |i, J_i, K_i, M_i\rangle$ and $|f\rangle = |f, J_f, K_f, M_f\rangle$, where i and f in the right-hand side denote the vibronic states of the initial and final states, re-spectively, and J, K, and M refer to the total angular momentum, the component of J along the molecular fixed z-axis, and that of J along the Z axis of the laboratory coordinates.

In the absence of a magnetic field, each JK level is $(2J + 1)$-fold degenerated. After summa-tions over M_i and M_f, the transition probability is

$$W^{(2)}_{J_iK_i \to J_fK_f} = 2\pi(2\pi\alpha)^2 I_1 I_2 \omega_1 \omega_2 \frac{1}{2J_i + 1}$$
$$\times \sum_{M_i} \sum_{M_f} |\langle J_f K_f M_f|$$
$$\times \hat{S}_{fi}(\omega_1\mathbf{e}_1, \omega_2\mathbf{e}_2)|J_i K_i M_i\rangle|^2$$
$$\times \delta(\omega_{fi} - \omega_1 - \omega_2) \quad (68)$$

where the rotational quantum number depen-dence of the two-photon transition operator, $\hat{S}_{fi}(\omega_1\mathbf{e}_1, \omega_2\mathbf{e}_2)$, has been omitted, i.e.,

$$\hat{S}_{fi}(\omega_1\mathbf{e}_1, \omega_2\mathbf{e}_2) = \sum_m \left[\frac{\mathbf{e}_2 \cdot \mathbf{r}|m\rangle\langle m|\mathbf{e}_1 \cdot \mathbf{r}}{\omega_{mi} - \omega_1} \right.$$
$$\left. + \frac{\mathbf{e}_1 \cdot \mathbf{r}|m\rangle\langle m|\mathbf{e}_2 \cdot \mathbf{r}}{\omega_{mi} - \omega_2} \right] \quad (69)$$

in which m denotes the vibronic states of the intermediate states. This approximation is valid for nonresonant transitions. After performing the transformation between the laboratory and molecular coordinates in the two-photon transi-tion amplitude by using the spherical-coordinate

basis set and evaluating the matrix element, the two-photon transition probability can be written in the product form of the geometrical factor C, the molecular factor M, and the rotational factor R as

$$W^{(2)}_{J_iK_i \to J_fK_f} = \sum_{J=0}^{2} C_J M_J R_J \quad (70)$$

where the geometrical factors C_J, which are functions only of the polarization vectors $\mathbf{e}_1$ and $\mathbf{e}_2$, are given by

$$C_0 = |\mathbf{e}_1 \cdot \mathbf{e}_2|^2/3$$
$$C_1 = |\mathbf{e}_1 \times \mathbf{e}_2|^2/6 \quad (71)$$
$$C_2 = (1 - C_0 - 3C_1)/5$$

The molecular factors M_J are written as

$$M_J = |M^{(J)}_{\Delta K}|^2$$
$$= (2J + 1) \left| \sum_a \sum_b \begin{pmatrix} 1 & 1 & J \\ -a & -b & \Delta K \end{pmatrix} S^{ba}_{fi} \right|^2 \quad (72)$$

with $|\Delta K| \leq J$. The matrix ( ) is called the Wigner $3j$ symbols. Equation (72) can be evalu-ated with the aid of the symmetry property of the Wigner $3j$ symbols. All the possible values of the molecular factors are tabulated in Table III.

The rotational factors R_j are defined as

$$R_J = (2J_f + 1)(2J_i + 1) \begin{pmatrix} J_i & J_f & J \\ -K_i & K_f & \Delta K \end{pmatrix}^2 \quad (73)$$

From this expression the rotational selection rule can be obtained as

$$R_0 \neq 0 \quad \text{for} \quad \Delta J = 0,$$
$$\Delta K = 0; \quad \text{Q branch only}$$

TABLE III.   Molecular Factors M_j of the General Expression in Eq. (72) for Two-Photon Absorption in Rotating Molecules in Spherical and Cartesian Coordinates[a]

J	ΔK	Molecular factors $M_J = |M^{(J)}_{\Delta K}|^2$ with $|\Delta K| \leq J$	
		Spherical coordinates	Cartesian coordinates
2	± 2	$|M_{\pm\pm}|^2$	$\frac{1}{4}(M_{xx} - M_{yy})^2 + \frac{1}{4}(M_{xy} + M_{yx})^2$
	± 1	$\frac{1}{2}|M_{\pm 0} + M_{0\pm}|^2$	$\frac{1}{4}(M_{xz} + M_{zx})^2 + \frac{1}{4}(M_{yz} + M_{zy})^2$
	0	$\frac{1}{6}|M_{+-} + M_{-+} + 2M_{00}|^2$	$\frac{1}{6}(2M_{zz} - M_{xx} - M_{yy})^2$
1	± 1	$\frac{1}{2}|M_{\pm 0} - M_{0\pm}|^2$	$\frac{1}{4}(M_{xz} - M_{zx})^2 + \frac{1}{4}(M_{yz} - M_{zy})^2$
	0	$\frac{1}{2}|M_{+-} - M_{-+}|^2$	$\frac{1}{2}(M_{xy} - M_{yx})^2$
0	0	$\frac{1}{3}|M_{+-} + M_{-+} - M_{00}|^2$	$\frac{1}{3}(M_{xx} + M_{yy} + M_{zz})^2$

[a] Molecular factors are listed for different ΔK and given in the molecular frame system. Spherical coordinates r_+, r_0, and r_- correspond to $-2^{-1/2}(x + iy)$, z, and $2^{-1/2}(x - iy)$, respectively. Matrix elements M_{ba} correspond to S^{ba}_{fi} in Eq. (72). [F. Metz, W. E. Wunsch, L. Weusser, and E. W. Schlag (1978). *Proc. R. Soc. London Ser. A.* **363**, 381.]

$R_1 \neq 0$ for $\Delta J = 0, \pm 1,$

             $\Delta K = 0, \pm 1;$ P, Q, R branches

$R_2 \neq 0$ for $\Delta J = 0, \pm 1, \pm 2,$

             $\Delta K = 0, \pm 1, \pm 2;$ O, P, Q, R, S branches

As an application of the theory described above, let us consider the ratio of the linearly to the circularly polarized two-photon absorption probability. Noting that the geometrical factors for the linearly and circularly polarized laser beams are given by $C_0 = \frac{1}{3}$, $C_1 = 0$, and $C_3 = \frac{2}{15}$, and $C_0 = 0$, $C_2 = 0$, and $C_2 = \frac{1}{5}$, respectively; the ratio can be expressed as

$$\frac{W^{(2)\uparrow\uparrow}_{J_iK_i \rightarrow J_fK_f}}{W^{(2)\circlearrowright\circlearrowright}_{J_iK_i \rightarrow J_fK_f}} = \frac{2}{3} + \frac{5M_0R_0}{3M_2R_2} \tag{74}$$

This equation indicates that for nontotally symmetric transitions characterized by $M_0 = 0$, the ratio is independent of the rotational quantum numbers J and K, and is given by $\frac{2}{3}$. The same behavior of the ratio is also expected for the rotational lines of branches except Q branch of a totally symmetric transition, because $M_0 \neq 0$ and $R_0 = 0$. The ratio of the Q branch in which $M_0 \neq 0$ and $R_0 \neq 0$ deviate from $\frac{2}{3}$, and the magnitude of the deviation depends strongly on M_0R_0/M_2R_2, in which usually $M_0 > M_2$ and $R_0 > R_2$, and then on the rotational quantum numbers.

C. Spectral Properties of Multiphoton Transitions

Spectral properties of the multiphoton transitions of atoms and molecules depend strongly on whether the resonance condition is satisfied. In this subsection, resonance effects in atomic multiphoton transitions are first presented, focusing on the role of the intermediate states, and then the appearance of the vibronic structures in molecular multiphoton transitions is presented, focusing on the difference in the spectral intensity distribution between nonresonant and resonant multiphoton transitions.

1. Resonance Effects in Atomic Multiphoton Transitions

A drastic intensity enhancement in the multiphoton transitions can be observed when the energy of the photon is close to that of an intermediate state. An example of the resonance enhancement is shown in Fig. 17, in which two-photon transition rates of ^{23}Na for the 3S (hyperfine level $F = 2$, see Section V,A) → 4D$_{5/2}$ and

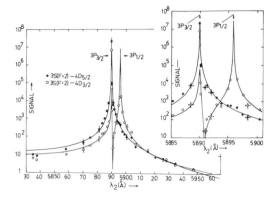

FIG. 17. Resonant enhancement of the two-photon absorption rate of Na; $\hbar(\omega_1 + \omega_2)$ is fixed at the 3S($F = 2$)–4D$_{3/2}$ or 3S($F = 2$)–4D$_{5/2}$ transition, while $\hbar\omega_1$ is tuned through the (one-photon) yellow doublet. The points are experimental and the curves are theoretical. The insert shows the behavior in the region from 5885 to 5900 Å with an expanded horizontal axis. [From J. E. Bjorkholm and P. F. Liao (1974). *Phys. Rev. Lett.* **33**, 128.]

3S ($F = 2$) → 4D$_{3/2}$ transitions are recorded as a function of the wavelength of the second laser λ_2 by detecting the fluorescence from the excited D state. The frequency of the first laser is adjusted to make it close to the 3S–3P doublet. As the frequency approaches to 3S–3P$_{3/2}$ absorption line, the cross sections for the two-photon transitions to the 4D$_{5/2}$ and 4D$_{3/2}$ levels are enhanced by a factor of 10^8. The deep minimum, known as the Fano profile, originates from an interference between 3P$_{3/2}$ → 4D$_{1/2}$ and 3P$_{3/2}$ → 4D$_{5/2}$ transi-

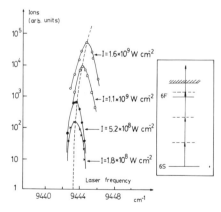

FIG. 18. Variation of the number of ions in the four-photon ionization of Cs as a function of laser frequency in the neighborhood of the resonant three-photon transition 6S → 6F. Dashed line shows the resonance shift for increasing values of laser intensity I. [From G. Mainfray and C. Manus (1980). *Appl. Opt.* **19**, 3934.]

tions, and on the other hand, since $3P_{1/2} \leftrightarrow 4D_{5/2}$, there is no such interference in the two-photon transition via the $3P_{1/2}$ state. Generally, resonance peak positions depend on the laser intensity in a moderate laser intensity range, as shown in Fig. 18. Here variations of the number of ions in four-photon ionization of Cs atoms are drawn as a function of laser frequency in the neighborhood of the resonant three-photon transition $6S \rightarrow 6F$. The dashed line shows the resonance shift linear with respect to the laser intensity I. The origin of the resonance shift has been explained in Section II,C.

2. Vibronic Coupling in Molecular Two-Photon Transitions

From the symmetry argument in a previous section it can be understood that two-photon absorptions between the g and g or u and u electronic states of molecules with inversion symmetry take place with high intensity. However, two-photon transitions between electronic states with different inversion symmetry $u \leftrightarrow g$ are also observed, although their intensity is very weak. The $u \leftrightarrow g$ transition, called the forbidden two-photon transition or vibronically induced two-photon transition, can be allowed by coupling of nuclear vibrations of u inversion symmetry with electrons (vibronic coupling). As an example of the vibronically induced two-photon transition, one can see the two-photon absorption from the ground state $^1A_{1g}$ to the first excited singlet state $^1B_{2u}$ in benzene. For the two-photon transition of benzene, using two photons with identical frequency, the tensor pattern belongs to the A_{1g}, E_{1g}, and E_{2g} irreducible representations of the D_{6h} point group. The species of vibrations inducing the two-photon absorption ($^1B_{2u} \leftarrow {}^1A_{1g}$), Γ_j can be specified from the symmetry consideration, $\Gamma_j \times B_{2u} \times A_{1g} =$

$\{A_{1g}, E_{1g}, E_{2g}\}$, that is, the inducing modes belonging to the b_{2u}, e_{2u}, and e_{1u} species. In Fig. 19, the normalized four-photon ionization spectrum of benzene two-photon resonant with the S_1 state is shown. The vibronic structure consists mainly of the ν_1 totally symmetric mode progression that starts from the vibronically induced 14_0^1 band. The inducing mode ν_{14} is the C—C bond-alternating vibration belonging to the b_{2u} species. The two-photon fluorescence excitation spectrum of the $^1B_{2u} \leftarrow {}^1A_{1g}$ transition of benzene shows very similar vibronic structure.

In order to predict which mode acts as the inducing mode and to evaluate the two-photon transition probability, magnitudes of the transition amplitude have to be calculated. This can be carried out first by expanding the transition moments to the first order of Q_j in the Born–Oppenheimer approximation, for example,

$$\langle m|\mathbf{r}|i\rangle \simeq \langle \Theta_{mv}(Q)|\Theta_{iv}(Q)\rangle$$
$$\times \langle \Phi_m(r, Q)|\Phi_i(r, Q)\rangle|_{Q=Q_0}$$
$$+ \sum_j \langle \Theta_{mv}(Q)|Q_j|\Theta_{iv}(Q)\rangle$$
$$\times \frac{\partial}{\partial Q_j}\langle \Phi_m(r, Q)|\mathbf{r}|\Phi_i(r, Q)\rangle|_{Q=Q_0} \quad (75)$$

where $\Phi(r, Q)$ and $\Theta(Q)$ denote the electronic and vibrational wave functions, respectively, and $\langle \Theta_{mv}(Q)|\Theta_{iv}(Q)\rangle$ is the optical Franck–Condon overlap integral. The vibronically induced transition originates from the second term of Eq. (75). The term Q_j denotes the inducing modes. Summation over the intermediate states in the transition amplitude is then performed neglecting the vibrational quantum number dependence in the energy denominator of the transition amplitude in the case of nonresonant two-photon transitions. In resonant cases, on the other hand, the vibrational quantum number dependence has to be evaluated explicitly. Several methods—the Green's function method, numerical method, path integral method, and so on—can be applied to the evaluation of the resonant two-photon transition probability of molecules.

V. Applications

In this section, application of the multiphoton spectroscopic methods described in Section III to Doppler-free multiphoton transitions in atoms and molecules, to Rydberg transitions and valence-state transitions of molecules, and to above-threshold ionization of atoms is presented.

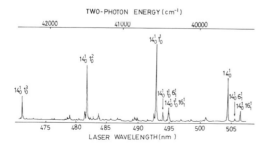

TWO-PHOTON ENERGY (cm⁻¹)

FIG. 19. Normalized multiphoton ionization spectrum of benzene resonant with S_1. [From J. Murakami, K. Kaya, and M. Ito (1980). *J. Chem. Phys.* **72**, 3263.]

A. Doppler-Free Multiphoton Transitions in Atoms and Molecules

In the Doppler-free multiphoton spectroscopy, one can observe many interesting phenomena that cannot be resolvable in a Doppler-limited spectroscopy, such as fine and hyperfine splittings of excited states, isotope shifts, Zeeman splittings, collision-induced broadenings and shifts, and rovibronic structures. Doppler-free two-photon absorption was first demonstrated in sodium vapor by using the fluorescence detection method. A hyperfine splitting of the 3^2S-5^2S transition at 6022 Å of ^{23}Na is shown in Fig. 20. The ^{23}Na nucleus has spin $\frac{3}{2}$, and it interacts with the spin of the unpaired electron by means of a magnetic hyperfine Hamiltonian $H = A\mathbf{I} \cdot \mathbf{S}$. The splitting of the $F = 2$ and $F = 1$ hyperfine levels of an S state is thus $2A_{nS}/h$ in frequency units. The selection rules for the transition result in two absorption lines separated by $\Delta\nu = (A_{3S} - A_{nS})/h$; since the ground-state hyperfine splitting is well-known for sodium, the splitting of the 5S state can be measured.

In the case of the two-photon absorption of

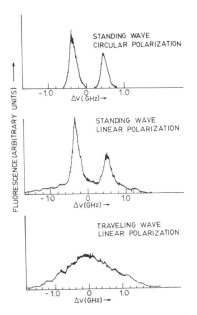

FIG. 20. Two-photon absorption signal on the 3S → 5S transition of atomic ^{23}Na. The experimental traces record the observed resonance fluorescence intensity at 330 nm (4P–3S transition), following the two-photon absorption. [From N. Bloembergen and M. D. Levenson (1976). Doppler-free two-photon absorption spectroscopy. In "High Resolution Laser Spectroscopy" (K. Shimoda, ed.), pp. 315–369. Springer-Verlag, Berlin.]

linearly polarized light traveling in the same direction, the hyperfine splitting cannot be resolved, as shown in the bottom of Fig. 20. The suppression of the Doppler background in an even-quantum absorption process can be carried out based on the angular-momentum selection rules when the initial and final state angular momentum are equal. Since the orbital angular momentum vanishes in the initial and final states of sodium 3S → 5S transition, the selection rule $\Delta L = 0$, $\Delta m = 0$ applies. In the case of using circularly polarized light, the transitions can take place only when the atoms absorb one quantum with angular momentum +1 from one laser beam and a quantum with angular momentum −1 from the oppositely propagating beam, as shown in the upper part of Fig. 20. Absorption of two quanta from a single circularly polarized beam requires $\Delta m = \pm 2$; it is impossible to excite the atoms. On the other hand, when linearly polarized laser beams pointing in opposite directions are used, as shown in the middle part of Fig. 20, effects of the Doppler broadening still exist. This originates from the absorption of the two-quanta photon of the linearly polarized beam, $\Delta m = 0$, traveling in the same direction.

The ratio of the intensity of the $F = 2$ to that of the $F = 1$ line is $5:3$ from the statistical weights of the F states in the 3S ground level. From the separation of the doublet, a value of $A_{5S}/h = 78 \pm 5$ MHz can be obtained for the hyperfine interaction constant in the 5S state. With the success of the initial experiments with sodium vapor, characteristics of alkali atom states (for example, see Fig. 17) and those of other atomic states have been clarified by using Doppler-free multiphoton spectroscopy.

Applications of Doppler-free multiphoton spectroscopy to measurement of rovibronic states of molecules such as Na_2, NO, and C_6H_6 (benzene) have been reported in detail. Especially, spectral lineshapes for rovibronic transitions in polyatomic molecules like benzene overlap due to Doppler broadening, they cannot be resolved in conventional spectroscopy, and the Doppler-free spectroscopy is necessary. Information about geometrical structures of electronically excited states and nonradiative processes such as intramolecular vibrational energy redistribution and electronic energy relaxation can be obtained from the line position and width only after removal of the Doppler broadening. In Fig. 21 is a part of the two-photon spectrum ($S_1 \leftarrow S_0$) of benzene as measured with different spectral resolutions shown. Many lines corresponding to the rotational transitions can be

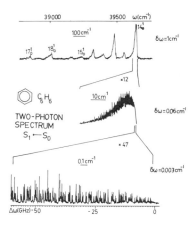

FIG. 21.   Part of the two-photon spectrum of C_6H_6 as measured with different spectral resolution. The middle trace represents the highest resolution possible in Doppler-limited spectroscopy. Only in the Doppler-free spectrum (lower part) are single rotational lines resolved. [From S. H. Lin, Y. Fujimura, H. J. Neusser, and E. W. Schlag (1984). "Multiphoton Spectroscopy of Molecules." Academic Press, Orlando, Florida.]

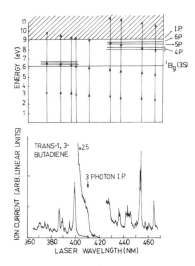

FIG. 22.   Multiphoton ionization spectrum of *trans*-1,3-butadiene, along with an energy-level diagram showing some of the states that appear in the spectrum. [From P. M. Johnson (1980). *Acc. Chem. Res.* **13**, 20.]

seen only in the Doppler-free spectrum shown at the bottom of Fig. 21. This spectrum is the blue edge of the Q branch ($\Delta J = 0$) of the totally symmetric transition 14_0^1 induced by the vibration ν_{14} in the $^1B_{2u}$ electronic state. For this type, two-photon absorption, the Doppler-free spectrum can be observed by using countercircularly polarized light as indicated in atomic S–S two-photon transitions.

B. Multiphoton Ionization of Molecules via Rydberg States

There have been many experimental results on the multiphoton ionization of molecules involving Rydberg states (e.g., iodine, nitric oxide, and aromatic molecules). Mechanisms of the ionization and characteristics of the Rydberg states have been clarified. In Fig. 22, the resonant (2 + 1) and (3 + 1) multiphoton ionization spectrum of *trans*-1,3-butadiene is shown together with the energy level diagram. The spectrum is separated into two regions, one to the blue side of 410 nm and one to the red side. In the former region, the structure of the multiphoton ionization spectrum is characteristic of an allowed two-photon resonance with the B̃ state designated by Herzberg: that is, the rate-determining step of the (2 + 1) multiphoton transition is the initial two-photon transition process. The B̃ state with ^1B_g symmetry is formed by removal

of a π-electron to an S-type Rydberg orbital. Many three-photon resonances with Rydberg states have been measured, from the (3 + 1) four-photon ionization regions to below 410 nm. The structures of the observed spectra reflect those of the initial three-photon absorption process, which is very similar to the vacuum UV spectra because three-photon transitions have the same selection rules as one-photon transitions in a C_{2h} molecule. Quantum defects for the Rydberg series have been identified to clarify the character of the Rydberg orbitals.

C. Low-Lying Electronic Excited States (^1A_g) of Linear Polyenes

Much experimental and theoretical attention has been given to locating low-lying "hidden" electronic excited valence states (^1A_g) of linear polyenes because of their photochemical and biochemical interest. One of the fruitful applications of multiphoton spectroscopy is the direct observation of the excited states of linear polyenes, *trans*-1,3-butadiene, *trans*-1,3,5-hexatriene, *trans,trans*-1,3,5,7-octatetraene, and so on. The ^1A_g-^1A_g transitions that are forbidden for the one-photon process have been observed in the two-photon excitation spectra. The two-photon excitation spectrum of all-*trans*-diphenylhexatriene in ether-isopentane-ethanol (EPA) solvent at 77 K is shown in Fig. 23. The origin of

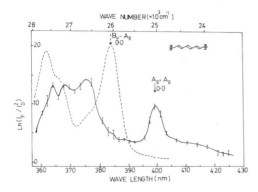

FIG. 23. Two-photon excitation spectrum of all-*trans*-diphenylhexatriene in EPA at 77 K. The one-photon absorption is shown by the dashed curve. [From H. L. B. Fang, R. J. Thrash, and G. E. Leroi (1978). *Chem. Phys.* **57**, 59.]

the lowest excited A_g state is located at 25,050 cm^{-1}, which is at about 900 cm^{-1} below the origin of the first one-photon allowed $^1B_u \leftarrow {}^1A_g$ transition. The ordering of the electronic energy levels of linear polyenes depends, of course, on the substitution groups, as well as on experimental conditions such as the solvent used and the temperature.

D. ABOVE-THRESHOLD IONIZATION OF ATOMS

Multiphoton ionizations of atoms by a strong laser field can lead to the production of electrons at energies corresponding to the absorption of extra photons as well as to the absorption of the minimum number of required photons. This process involving the extra photon absorption is called the above-threshold (multiphoton) ionization or continuum–continuum transition. Developments in measuring the energy spectrum of photoelectrons makes it possible to observe the above-threshold ionization phenomenon. Since measurement of the above-threshold ionization of xenon atoms under the irradiation of a frequency-doubled Q-switched Nd:YAG (yttrium aluminum garnet) laser, experimental and theoretical studies on the mechanism of the above-threshold ionization have both become very active because by analyzing the spectra one can obtain information on the magnitude of the continuum–continuum transition probability and can also understand dynamics taking place above the ionization threshold, where the simple perturbation theory breaks down.

Figure 24 shows electron energy spectra arising from multiphoton ionization of Xe by Nd:YAG laser of 1064 nm wavelength at several

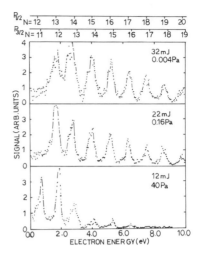

FIG. 24. Electron spectra from multiphoton ionization of xenon at 1064 nm. The vertical scales are normalized. The pulse energy (in units of millijoules) and pressure at which each spectrum is taken are given. The estimated intensity is pulse energy (mJ) $\times 2.10^{12}$ W/cm^2. [From P. Kruit, J. Kimman, H. G. Muller, and M. J. van der Wiel (1983). *Phys. Rev.* **A28**, 248.]

pulse energies indicated. Xenon pressure chosen in such a way that the total electron signal in each spectrum is 25–50 electrons per pulse is also indicated in Fig. 24. From this figure it can be recognized that 12-photon and up to 19-photon ionization processes with higher relative probability take place in addition to 11-photon ionization. A schematic explanation for the origin of the electron energy spectrum of Xe that involves two ionization potentials, at 12.127 eV for the $^2P_{3/2}$ core and at 13.44 eV for the $^2P_{1/2}$ core, is given in Fig. 25. Angular distributions of photoelectrons resulting from the above-thresh-

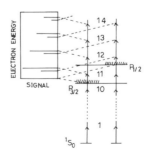

FIG. 25. Energy-level diagram of xenon. A part of the above-threshold ionization of xenon with 1074-nm photons is schematically shown. [From P. Kruit, J. Kimman, H. G. Muller, and M. J. van der Wiel (1983). *Phys. Rev.* **A28**, 248.]

old ionization are measured to study the mechanism. The above-threshold ionization phenomenon is also observed for other atoms, such as cesium atoms.

VI. Prospects

In the previous sections, theoretical foundations and experimental principles of the multiphoton spectroscopy of atoms and molecules have been presented. The field of the multiphoton spectroscopy is still in rapid development. In view of the increasing interest in multiphoton transitions one can expect new experimental methods will be developed in the near future, such as multiphoton circular dichroism and multiphoton magnetic circular dichroism, for example, in the molecular spectroscopic field. In this section, other experimental methods for multiphoton spectroscopy recently developed, multiphoton ionization mass spectroscopy and ionization-dip spectroscopy, are briefly described.

A. MULTIPHOTON IONIZATION MASS SPECTROSCOPY

This method consists of the multiphoton ionization combined with mass detection and allows us to identify atoms and molecules by the optical spectrum related to a resonant intermediate state as well as by their mass. Isotopic species, for example the ^{13}C molecule, can be preferentially ionized in a natural isotopic mixture by shifting the wavelength from the absorption band of the light species if the intermediate state

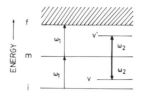

FIG. 27.   Simplified energy level diagram for ionization-dip spectroscopy.

spectrum shows sharp features. The scheme of the setup for mass-selective ion detection has already been shown in Fig. 10.

The fragmentation patterns in the multiphoton ionization mass spectroscopy depend on the laser intensity, and they are different from those obtained by electron impact excitation, charge-exchange excitation, and other methods. In the multiphoton ionization mass spectroscopy, ions with small mass weights can be produced compared with those obtained by other methods. In Fig. 26 is the fragmentation pattern in the mass spectrum of benzene (C_6H_6) shown. One can see that atomic fragment ions C^+ are produced with a high probability. To interpret the fragmentation pattern of polyatomic molecules, several theoretical treatments based on information-theoretical statistical and rate equation approaches have been proposed.

B. IONIZATION-DIP SPECTROSCOPY

This is a new high-resolution multiphoton spectroscopy based on competition between ionization and stimulated emission (or stimulated absorption). Figure 27 shows the principle for the two-photon ionization case. Two photons at ω_1 induce efficient ionization because resonance enhancement occurs through inter-

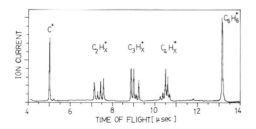

FIG. 26.   Fragmentation pattern in the mass spectrum of benzene (C_6H_6) obtained by two-step photoionization with UV laser light at 2590.1 Å. At high intensities ($>10^7$ W cm^{-2}), smaller molecular ions are observed. The molecular ions are analyzed with the time-of-flight mass spectrometer. [From U. Boesl, H. J. Neusser, and E. W. Schlag (1978). *Z. Naturforsch., Teil.* **A:33**, 1546.]

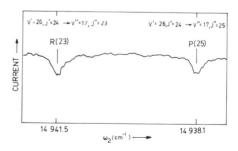

FIG. 28.   Ionization-dip spectrum of I_2 recorded with $\omega_1 = 18393.9$ cm^{-1}. [From D. E. Cooper, C. M. Klimcak, and J. E. Wessel (1981). *Phys. Rev. Lett.* **46**, 324.]

mediate electronic state m. A high-intensity laser probe at frequency ω_2 is introduced. When $\omega_1 - \omega_2$ matches a suitable vibrational frequency ω_{vi} in the electronic ground state, stimulated emission competes with ionization; this decreases the effective population in state m and reduces the ionization signal (i.e., ion dip). Ion dips may also take place if state v is at a higher energy than state m. An application of the ionization dip spectroscopy to iodine (I_2) is shown in Fig. 28. Here, a probing laser induces spectrally sharp dips ($\Delta\omega < 0.2$ cm^{-1}) in the ionization current at $\omega_2 = 14{,}941.5$ cm^{-1} and $\omega_2 = 14{,}938.1$ cm^{-1} in the presence of ω_1 set at the multiphoton ionization peak at $18{,}393.9$ cm^{-1}. These ω_2 transitions are assigned to R(23) and P(25) bands of $B(v' = 26) \rightarrow X(v'' = 17)$ of I_2, respectively.

BIBLIOGRAPHY

Chin, S. L., and Lamboropoulos, P. (eds.) (1984). "Multiphoton Ionization of Atoms." Academic Press, Canada.

Levenson, M. D. (1982). "Introduction to Nonlinear Laser Spectroscopy". Academic Press, New York.

Lin, S. H. (ed.) (1984). "Advances in Multi-photon Processes and Spectroscopy." World Scientific, Singapore.

Lin, S. H., Fujimura, Y., Neusser, H. J., and Schlag, E. W. (1984). "Multiphoton Spectroscopy of Molecules." Academic Press, Orlando.

Rahman, N. K., and Guidotti, C. (eds.) (1984). "Collisions and Half-Collisions with Lasers." Harwood Academic, New York.

Reintjes, J. F. (1984). "Nonlinear Optical Parametric Processes in Liquids and Gases." Academic Press, Orlando.

NEUTRON SCATTERING

William B. Yelon   *University of Missouri Research Reactor*

I. Introduction
II. Neutron Production
III. Instrumentation
IV. Applications of Neutron Scattering

GLOSSARY

Cold source: Moderator at cryogenic temperature, typically H_2 or D_2, used to shift the average neutron energy to ~ 5 meV.

Elastic scattering: Neutron scattering in which the neutron energy is the same before and after the scattering event. Used as a probe of static structure.

Guide tubes: Polished, flat, coated tubes that transport neutrons over long distances with minimal loss through internal reflection. These are especially useful for long ($4-10$ Å) wavelengths.

Incoherent scattering: Scattering in which phase information is not preserved, leading to intensity at all momenta.

Inelastic scattering: Scattering in which the neutron energy after the event is different from the initial energy, with energy transferred to or from the scattering system. Used as a probe of the interatomic and single particle excitations.

Moderator: Medium with a large neutron scattering cross section used to slow fission or spallation neutrons by repeated collisions.

Polarized neutrons: Neutrons prepared either by scattering or absorption in which only one magnetization state is present. Used to enhance the sensitivity to magnetic scattering events relative to nuclear scattering.

Quasielastic scattering: Events in which the neutron energy, on average, is unchanged after scattering but in which some neutrons lose energy and some gain energy in a continuous spectrum centered about $\Delta E = 0$. Used as a probe of diffusive type motions.

Scattering length: Measure of the scattering power of the neutron by a nucleus, defined as

$b = \sqrt{4\pi\sigma_{coh}}$, where σ_{coh} is the coherent cross section.

Spallation source: Source of neutrons produced by bombarding a heavy metal target, such as ^{238}U or W, by high-energy ($500-800$ MeV) protons.

Supermirrors: Neutron reflectors consisting of alternating layers of material with large differences in scattering length, for example, Ni and Ti, with a graded layer spacing to give a large critical angle for total reflection.

Thermal neutron: Neutron that has been slowed to near thermal equilibrium, typically with energy of order 25 meV, wavelength 1.8 Å, and velocity 2200 m/sec.

Time-of-flight: Scattering method in which neutron energies are distinguished by their different velocities, which lead to different flight times over a fixed path.

The thermal neutron, as an uncharged, weakly interacting particle has become one of the most powerful and versatile probes of the structure and dynamics of condensed (solid and liquid) matter. Since the first neutron scattering experiments in the late 1940s, improvements in neutron sources and instrumentation have led to a rapid growth in utilization and increased sophistication in the types of experiments carried out. Today, neutron scattering has become an important tool in physics, chemistry, biology, polymer science, and materials science, to name a few, and major centers have been created to accommodate users from these disciplines.

I. Introduction

A. SPECIAL FEATURES OF NEUTRON–ATOM INTERACTIONS

The power of neutron scattering is a consequence of unique characteristics of the interaction of thermal neutrons ($E \approx 25$ meV, $\lambda \approx 1.8$ Å, and $v \approx 2200$ m/sec) with atoms. These characteristics have made neutron scattering a bulk probe of atomic and magnetic arrangements and of the dynamics of solids and liquids, that is, the interatomic forces in these systems.

1. Coherent Scattering

The cross section for coherent scattering, σ_{coh}, and the related quantity the scattering length,

$b = \sqrt{4\pi\sigma_{coh}}$, are quantities related to the neutron–nuclear interaction and show wide variability from species to species, including between isotopes of the same element (Table I). Unlike the cross section for X-ray scattering (which is proportional to the number of electrons), b does not increase monotonically with atomic number z and, in fact, is as large for many light elements as for heavy elements. Consequently, light elements contribute significantly to the scattering, and adjacent elements are often distin-

TABLE I.   Cross Section Data for Selected Elements[a, b]

Element	b	σ_{coh}	σ_s	σ_a	μ
^{1}H	−0.374	1.76	81.67	0.3326	
^{2}H	0.667	5.60	7.63	0.0005	
He	0.326	1.34	1.34	0.0075	
Li	−0.190	−0.454	1.28	70.5	3.32
^{6}Li	$0.20-0.03i$	−0.51	0.97	940.	43.57
Be	0.779	7.63	7.64	0.0076	0.94
B	$0.53-0.021i$	3.54	5.24	767.	108.9
^{10}B	$-0.01-0.107i$	0.14	3.1	3837.	541.
C	0.665	5.55	5.55	0.0035	0.627 (graphite)
N	0.936	11.01	11.50	1.90	
O	0.580	4.23	4.23	0.0002	
Na	0.363	1.66	3.28	0.530	0.097
Al	0.345	1.495	1.50	0.231	0.104
Si	0.415	2.16	2.18	0.171	0.122
Ti	−0.330	1.37	4.04	6.09	0.573
V	−0.038	0.018	5.21	5.08	0.724
Mn	−0.373	1.75	2.15	13.3	1.26
Fe	0.954	11.44	11.83	2.56	1.22
Co	0.250	0.79	5.6	37.2	3.81
Ni	1.03	13.3	18.5	4.49	2.06
Cu	0.772	7.49	8.01	3.78	0.998
Zr	0.716	6.44	6.60	0.185	0.288
Nb	0.705	6.25	6.26	1.15	0.404
Mo	0.695	6.07	6.35	2.55	0.570
Ag	0.592	4.41	4.99	63.3	4.00
Cd	$0.51-0.07i$	3.3	5.7	2520.	116.9
La	0.824	8.53	9.66	8.97	0.497
Nd	0.769	7.43	18.	50.5	1.99
Sm	$0.42-0.16i$	2.5	52.	5670.	172.8
Gd	$0.95-1.36i$	34.5	192.	48890.	1483.3
Er	0.803	8.10	9.3	159.	5.49
Ta	0.691	6.00	6.02	20.6	1.47
W	0.477	2.86	4.86	18.4	1.46
Pt	0.963	11.65	11.78	10.3	1.46
Au	0.763	7.32	7.75	98.65	6.27
Pb	0.940	11.11	11.12	0.171	0.373
B$_1$	0.853	9.15	9.15	0.034	0.259
U	0.842	8.90	8.91	7.57	0.788

[a] Adapted from Sears, V.F. (1986). *In* "Neutron Scattering," *Methods of Experimental Physics* (K. Skold and D. L. Price, eds.), Vol. 23, Part A, pp. 533–547. Academic Press, Orlando, Florida.
[b] b = Coherent scattering length ($\times 10^{-12}$ cm); σ_{coh} = coherent cross section ($\times 10^{-24}$ cm); σ_a = absorption cross section for 2200 m/sec neutrons ($\times 10^{-24}$ cm); σ_s = total scattering cross section ($\times 10^{-24}$ cm); μ = attenuation length for polycrystalline material (cm^{-1}).

guishable. Furthermore, isotope substitution allows significant changes in scattering without changes in the physics and chemistry of the system. The scattering length is independent of neutron energy except in a few cases where there is a nearby nuclear resonance. The form factor for coherent nuclear scattering is essentially independent of $Q = 4\pi \sin \theta/\lambda$, and significant scattering may be observed even at high scattering angles.

2. Absorption Cross Section

The absorption cross section, σ_a, is small for all but a handful of isotopes (notably ^{10}B, ^{3}He, Cd, Sm, and Gd), and neutron beams will penetrate millimeters, if not centimeters, into samples (Table I). Consequently, large samples are often used, and it is quite easy to develop special sample environments, such as cryostats and furnaces. Low absorption has led to a class of experiments in which a small internal volume of a macroscopic sample is studied.

3. Magnetic Interactions

The neutron has a magnetic moment and is sensitive to the magnetic fields present inside solids, static and fluctuating. Neutron scattering is able to establish the arrangement of spins and the spin dynamics. Since the magnetic moments in solids normally arise from the unpaired electron spins (which have an appreciable spatial extent), the magnetic scattering amplitude p is characterized by a form factor similar to that for X-ray scattering. Typically, magnetic neutron scattering is observed only to $Q \approx 1$ Å^{-1}.

4. Incoherent Scattering

In addition to the coherent scattering cross section, which contributes to discrete peaks in ω–Q (energy–momentum) space, neutrons may be scattered incoherently either because of the different spin states of the nuclei (spin incoherence) or the difference in coherent scattering of different isotopes on the same site (isotopic incoherence). While the effects of interatomic correlations (position and motion) are probed by the coherent scattering, single-particle effects may be studied via incoherent scattering.

5. Inelastic and Quasielastic Scattering

Unlike X-rays, in which a probe with a wavelength comparable to interatomic distances (0.5–5.0 Å) has energy far exceeding the characteristic lattice energies, thermal neutrons simultaneously match energy and wavelength with the probed system. Consequently, creation or annihilation of an excitation in the solid or liquid leads to significant change in the neutron energy and wavelength, which can subsequently be analyzed by diffraction or time of flight. Thus, neutrons are a uniquely powerful probe of the dynamics of solids and can be used to observe excitations at all momenta to $Q \geq 10$ Å^{-1} and energies from microelectron volts to ~ 0.5 eV.

II. Neutron Production

In order to carry out neutron scattering studies, it is necessary to have a substantial neutron flux. Portable sources, such as ^{241}Am, do not provide sufficient flux, nor do laboratory generators. In fact, only large installations (reactor- or accelerator-based sources) have sufficient intensity to support meaningful scientific programs. These facilities are specially designed to provide high power and flux densities.

A. Research Reactors

The only reactors with sufficient neutron flux for scattering are the so-called research reactors. These are typically at 5-MW or greater power levels with compact cores, often of highly enriched ^{235}U. They may be light water (H$_2$O) cooled, beryllium reflected and water moderated, or, heavy water (D$_2$O) cooled and reflected, or, occasionally, some hybrid of these. Neutron fluxes at the edge of the core are typically above 10^{13}/cm^2 sec and may exceed 10^{15}/cm^2 sec for the most advanced designs. The limit for reactors, barring major breakthroughs, appears to be $\sim 5 \times 10^{15}$/cm^2 sec.

B. Pulsed Sources

A high flux of neutrons can also be produced by bombarding a heavy element target with high-energy particles. While electrons have been used to produce neutrons by photoevaporation, the most efficient method is to use high-energy protons (~ 500–800 MeV). This spallation method produces of order 30 neutrons per proton and, when combined with a rapid cycling synchrotron, can give a performance comparable to the best reactors for scattering. Most reactor experiments are performed in a steady-state mode using a crystal monochromator (and analyzer for inelastic scattering); pulsed source experiments use time of flight (ToF) to discriminate between different neutron energies.

C. Moderators

Neutrons produced either by fission or spallation have energies in the megaelectron volt range and must be slowed (moderated) to be useful for scattering research. This is accomplished by repeated collisions of the neutron with the moderating material, thereby transferring most of the neutron energy to the moderator. By adjusting the temperature of the moderator, the neutron spectrum may be shifted in energy to hot ($\sim$ 500 meV), thermal (60 meV), or cold (5 meV), where the energies in parentheses correspond to the peak in the Maxwellian neutron spectrum emerging from the moderator. Moderators are typically hydrogenous materials (H_2O, CH_4, H_2) or the deuterium equivalents for thermal and cold applications, and graphite for hot sources.

D. Beam Tubes and Guides

To transport the neutron beams from the source to the scattering experiment, either active or passive elements are used. Beam tubes carry the beam through the biological shield accepting all neutrons emerging within the solid angle of acceptance. Guide tubes, made of flat, polished reflecting material, can transport neutrons over greater distances by using total internal reflection. The critical angle for total reflection is wavelength dependent and is given by

$$\theta_c = 2\lambda\sqrt{\rho b}/2\pi$$

where ρ is the atom number density. Nickel is the optimum "natural" coating for guides, giving a critical angle $\theta_c \approx 0.1\ \lambda°$ (that is, 0.1° at 1 Å, 1° at 10 Å). The good acceptance for large wavelengths has opened up research opportunities at these wavelengths by transporting beams to guide halls (areas of low background with room to install multiple instruments on each beam line). This has not been possible at short ($\sim$ 1 Å) wavelengths. Presently, much effort is being concentrated on supermirror guides with θ_c of order three times greater than for Ni, which would then confer similar advantages to short wavelength beams.

E. Neutron Sources

In the United States, neutron scattering programs are supported by both steady-state sources, at national laboratories [Oak Ridge, Brookhaven, NIST (formally NBS)], universities (Missouri and MIT), and pulsed sources at Argonne and Los Alamos. In Europe, both national and multinational facilities exist, including the High Flux Reactor in Grenoble (Institut Laue-Langevin). Most nations have at least one research reactor, and pulsed sources have been established in Britain (ISIS) and Switzerland (SIN). Suc-

cessful neutron scattering programs are also maintained in Japan and India.

III. Instrumentation

Although it is impossible to detail neutron scattering instrumentation, a few general comments are worthwhile. Some additional detail will also be given in Section IV. Generally speaking, the experiments are divided between pulsed experiments using choppers and steady-state experiments using crystals, although hybrids of these exist. Pulsed experiments are not restricted to pulsed sources; a steady beam may be pulsed as well.

A. Energy Selection—Monochromation

If a beam of neutrons is pulsed either by pulsing the neutron production or opening a shutter to allow a beam to pass, the different neutron energies will be separated by their ToF to a given point. Thus, timing provides the energy discrimination. For steady-state experiments, energy selection is accomplished by diffraction, using a crystal of known d-spacing and using Bragg's law, $\lambda = 2d \sin \theta$.

B. Polarization

Neutron beams can be prepared in a polarized spin state, which gives great power in the analysis of magnetic structure and spin dynamics. For steady-state sources, polarizing monochromators are used. In these, the nuclear and magnetic scattering for one spin state ($b - p$) cancel, while for the other ($b + p$), they add. Therefore, only one spin state is diffracted. For pulsed beams, most often polarizing filters or polarizing guides are employed. While the latter techniques generally produce lower polarizations than the crystal methods, they allow a wide energy spectrum to be accommodated.

C. Neutron Detection

Since neutrons are uncharged, the only effective method of detection is via nuclear reaction whereby a neutron is absorbed and a charged particle(s) is emitted. For example, $^3\text{He} + \text{n} \rightarrow \text{p}^+ + {}^3\text{H}$. ^{10}B or ^3He are commonly used in gas-filled counters; ^6Li is often used in scintillating material. Because of the relatively low neutron fluxes available for scattering, the largest gains in data collection rates can be accomplished by increasing the detection area through the use of multi- and position-sensitive detectors.

These have made especially large impacts in powder diffraction, small angle scattering, diffuse scattering with steady-state sources, and in all forms of pulsed neutron research.

IV. Applications of Neutron Scattering

Neutron scattering has application to an extremely wide range of disciplines, including physics, chemistry, biology, materials science, and geology. Rather than attempting to draw examples from each discipline, examples will demonstrate neutron scattering techniques of diffraction, inelastic scattering, etc.

A. Neutron Diffraction

Experiments designed to elucidate the interatomic structure of condensed matter fall into this category. In principle, one is interested in only the elastic scattering but, in practice, the experiments accept both elastic and inelastic scattering, and corrections for the inelastic scattering may be necessary. With steady-state sources, diffraction experiments are typically carried out with a fixed incident wavelength and variable scattering angle; with pulsed sources, experiments are performed at fixed angles in a ToF mode of wavelength determination. Examples of diffraction studies include the determination of the structure and oxygen ordering of high-T_c superconductors, determination of the inter-atomic arrangements (pair-correlation functions) of liquids and amorphous materials using isotopic substitution to separate the contributions to the scattering from the different atoms, and accurate location of hydrogen atoms in organometallic compounds (Fig. 1).

Powder diffraction studies have expanded vastly in the past decade because of the availability of Rietveld methods, which is a least-squares fitting of the data giving detailed structural information. Single-crystal diffraction is benefiting from the use of area detectors, which allow rapid data acquisition from crystals with large unit cells such as proteins.

Diffraction studies of magnetic materials provide unique information about the magnetic structure and spin density. While powder diffraction is routinely used for refinement of magnetic structures, studies with polarized neutrons, primarily on single crystals, provide a higher degree of sensitivity. Determining the flipping ratio, the ratio of intensity of scattering from a polarized sample with neutrons having spin-up and spin-down, gives extremely high precision in spin density studies. Among the most important re-

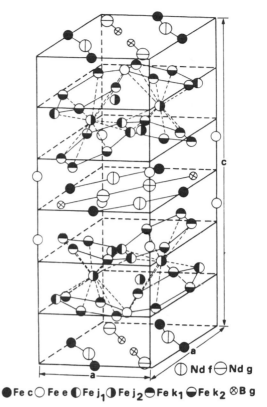

Nd f ⊖ Nd g

● Fe c ○ Fe e ◐ Fe j₁ ◑ Fe j₂ ⊖ Fe k₁ ⊖ Fe k₂ ⊗ B g

FIG. 1.   Crystallographic unit cell of $Nd_2Fe_{14}B$. The structure and stoichiometry of this important, new, hard magnet were solved from neutron powder diffraction data. [From Herbst, J. F., Croat, J. J., Pinkerton, F. E., and Yelon, W. B. (1984). *Phys. Rev.* **B29**, 4176.]

cent works have been studies of magnetic ordering in the high-T_c superconductors and the relationship between magnetism and superconductivity (Fig. 2).

B. Small-Angle Scattering

There is an inverse relation between the momentum $Q = 4\pi \sin \theta / \lambda$ and the length scale probed, that is, studies at high angles and with short wavelengths are sensitive to short distances, while studies at small Q probe large structures. Small angle scattering (more properly small-Q scattering) with neutrons (SANS) typically probes objects with dimensions of 20–1000 Å. These objects may be any fluctuation in scattering density, such as a void or precipitate in metallic systems, biological molecules in solution, or density fluctuations in solids. SANS has been especially valuable in studies of polymers. In homopolymers, it is possible to achieve scattering density fluctuations by replacing some protonated chains by deuterated chains without changing the

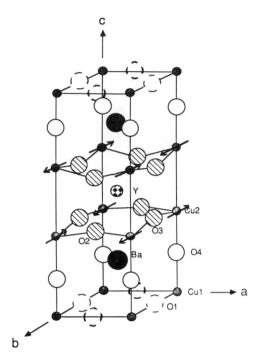

FIG. 2. Crystallographic and magnetic structure of $YBa_2Cu_3O_{7-y}$, a high-T_c superconducting phase. The O_1 oxygen sites are empty for $y = 1$. [From Sinha, S. K. (1988). *Mater. Res. Soc. Bull.* **XIII**(6), 24.]

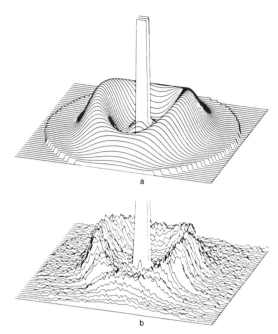

FIG. 3. Small-angle neutron scattering from a block copolymer system under uniaxial stress showing anisotropic elongation. The calculated curve (a) is based on two confocal ellipsoidal models. [From Hammouda, B., Yelon, W. B., Lind, A. C., and Hansen, F. Y. (1989). *Macromolecules* **22**, 418. Copyright © 1989 American Chemical Society.]

chemistry. In this case, the single-chain conformation can be studied, often as a function of temperature, elongation, or other parameter. In polymer alloys and blends, differences in scattering density may exist even when no electron density difference is present (which would prohibit X-ray investigation). In these cases, measurements of phase separation, particle organization, etc., are possible and have yielded important information (Fig. 3).

C. DIFFUSE SCATTERING

Elastic scattering from an infinite perfect crystal will be in the form of delta function Bragg peaks broadened by the instrumental resolution. Loss of perfection may lead to peak broadening (Huang scattering) and to diffuse scattering away from the Bragg peaks. For example, defective carbides, such as $VC_{0.75}$, show short-range ordering of the carbon vacancies, which are manifested by contours of diffuse scattering. Charge density waves often lead to incommensurate satellite peaks away from the Bragg peaks. Close to a phase transition, critical scattering may give broad peaks around Bragg reflections or at other positions in reciprocal (scattering) space asso-

ciated with the symmetry of the phase present below the transition. Diffuse scattering may be either nuclear or magnetic (or mixed), and polarized beam techniques may be useful in determining the nature of the scattering.

D. INELASTIC SCATTERING

Of all probes (X-rays, light, electrons, etc.), only neutrons most nearly match wavevector and energy criteria for inelastic scattering. Experiments are distinguished by the choice of pulsed or steady-state conditions and also by the energy of the incident neutrons, cold, thermal, or hot. Steady-state experiments are most often carried out with a triple-axis spectrometer and investigate excitations that are well defined in ω–Q space, such as acoustic phonons or magnons (Fig. 4). Pulsed sources are more often concerned with excitations that vary slowly in ω-Q space, such as optic phonons or crystal field excitations. Use of hot sources has extended the range of the triple axis to as high as several hundred millielectron volts; while pulsed sources, with a relatively rich supply of epithermal neutrons, offer the opportunity

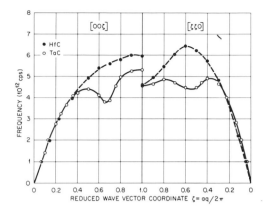

FIG. 4. Phonon dispersion curves in HfC and TaC. The prominent dips in the TaC curves at approximately twice the Fermi momentum are related to the onset of superconductivity, which is absent in the HfC system. [From Smith, H. G. (1970). *Phys. Rev. Lett.* **25**, 1611].

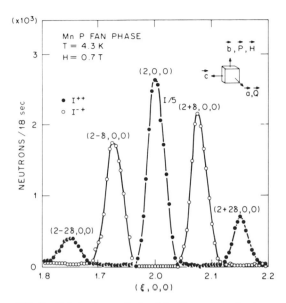

FIG. 5. Polarization analysis of the Bragg scattering in a single crystal of MnP with $+ +$ indicating nonspin flip and $- +$ indicating spin flip scattering. The spin flip peaks at $(2 \pm \delta, 0, 0)$ indicate a sinusoidally modulated moment along **c** with a propagation vector along **a**. The nonspin flip peak at $(2 \pm \delta, 0, 0)$ indicates a modulated component along **b**, indicating a fan magnetic phase. [From Moon, R. M. (1984). Shelter Island Workshop, CONF-8410256. Argonne National Laboratory.]

of studying excitations into the electron volt range. Cold neutrons, combined with special energy analysis modes, such as backscattering or spin echo, have extended the energy down to microelectron volts. Closely related to inelastic scattering are quasielastic scattering methods, which investigate diffusion and critical phenomena by observing the energy width of the scattering centered at $\Delta E = 0$. The high-energy resolution associated with cold neutrons is especially valuable in these studies.

E. POLARIZATION ANALYSIS

One aspect of neutron scattering deserves special mention: the ability first to prepare a beam with a fixed polarization and later to analyze the polarization of the scattered neutrons allows the experimenter to distinguish unambiguously spin flip from nonspin flip scattering, in order to distinguish nuclear from magnetic Bragg peaks or excitations. In many cases of scattering from low-dimensional magnetic systems, as well as other complex systems, such analysis is crucial to any interpretation of the physics of the system (Fig. 5).

BIBLIOGRAPHY

Elcombe, M. M., and Hicks, T. J., eds. (1988). "Neutron Scattering Advances and Applications: Proceedings of a Symposium held at the New South Wales Institute of Technology, Sydney, Australia, August 8–11, 1987." *Materials Science Forum*, Vols. 27 and 28. Trans Tech Publications Ltd., Aedermannsdorf, Switzerland.

Lander, G. H., and Robinson, R. A., eds. (1986). "Neutron Scattering: Proceedings of the International Conference on Neutron Scattering held in Santa Fe, New Mexico, August 19–23, 1985." Elsevier Science, Amsterdam. Reprinted from *Physica* **136B**(1–3).

Lovesey, S. W. (1984). Nuclear scattering. *In* "Theory of Neutron Scattering from Condensed Matter," Vol. 1. Oxford Univ. Press (Clarendon), London and New York.

Lovesey, S. W. (1984). Polarization effects and magnetic scattering. *In* "Theory of Neutron Scattering from Condensed Matter," Vol. 2. Oxford Univ. Press (Clarendon), London and New York.

Skold, K., and Price, D. L. (1986). "Neutron Scattering," *Methods of Experimental Physics*, Vol. 23, Part A. Academic Press, Orlando, Florida.

Skold, K., and Price, D. L. (1987). "Neutron Scattering," *Methods of Experimental Physics*, Vol. 23, Part B. Academic Press, San Diego, California.

Skold, K., and Price, D. L. (1987). "Neutron Scattering," *Methods of Experimental Physics*, Vol. 23, Part C. Academic Press, Orlando, Florida.

Windsor, C. G. (1981). "Pulsed Neutron Scattering." Taylor and Francis, Ltd., London; Halsted Press, New York.

NONLINEAR OPTICAL PROCESSES

John F. Reintjes *Naval Research Laboratory*

I. Optical Nonlinearities—Physical Origins
II. Optical Nonlinearities—Mathematical Description
III. Specific Nonlinear Processes
IV. Applications

GLOSSARY

Anti-Stokes shift: Difference in frequency between the pump wave and the generated wave in a stimulated scattering interaction when the generated wave is at a higher frequency than the pump wave.

Brillouin shift: Difference in frequency between the incident wave and the scattered wave in a stimulated Brillouin interaction. It is equal to the sound wave frequency.

Coherence length: Distance required for the phase of a light wave to change by $\pi/2$ relative to that of its driving polarization. Maximum conversion in parametric processes is obtained at L_{coh} when $\Delta k \neq 0$.

Constitutive relations: Set of equations that specify the dependence of induced magnetizations or polarizations on optical fields.

Depletion length: Distance required for significant pump depletion in a frequency-conversion interaction.

Mode-locked laser: Laser that operates on many longitudinal modes, each of which is constrained to be in phase with the others. Mode-locked lasers produce pulses with durations of the order of 0.5–30 psec.

Nonlinear polarization: Electric dipole moment per unit volume that is induced in a material by a light wave and depends on intensity of the light wave raised to a power greater than unity.

Parametric interactions: Interactions between light waves and matter that do not involve transfer of energy to or from the medium.

Phase conjugation: Production of a light wave with the phase variations of its wave front reversed relative to those of a probe wave.

Phase matching: Act of making the wave vector of an optical wave equal to that of the nonlinear polarization that drives it.

Population inversion: Situation in which a higher-energy level in a medium has more population than a lower-energy one.

Q-switched laser: Laser in which energy is stored while the cavity is constrained to have large loss (low Q) and is emitted in an intense short pulse after the cavity Q is switched to a high value reducing the loss to a low value.

Raman shift: Difference in frequency between the incident wave and the scattered wave in a stimulated Raman interaction. It is equal to the frequency of the material excitation involved in the interaction.

Slowly varying envelope approximation: Approximation in which it is assumed that the amplitude of an optical wave varies slowly in space compared to a wavelength and slowly in time compared to the optical frequency.

Stimulated scattering: Nonlinear frequency conversion interaction in which the generated wave has exponential gain and energy is transferred to or from the nonlinear medium.

Stokes shift: Difference in frequency between the pump wave and the generated wave in a stimulated scattering interaction when the generated wave is at lower frequency than is the pump wave.

Susceptibility, *n*th order: Coefficient in a perturbation expansion that gives the ratio of the induced polarization of order n to the nth power of the electric or magnetic fields.

Wave number (cm^{-1}): Number of optical waves contained in 1 cm. It is equal to the reciprocal of the wavelength in centimeters and has the symbol $\bar{\nu}$. It is commonly used as a measure of the energy between quantum levels of a material system or the energy of the photons in an optical wave and is related to the actual energy by $E = h\bar{\nu}c$, where h is Planck's constant and c is the speed of light.

Wave vector: Vector in the direction of propagation of an optical wave and having magnitude equal to 2π divided by the wavelength.

Wave-vector mismatch: Difference between the wave vector of an optical wave and that of the nonlinear polarization that drives it. It is also called the phase mismatch.

Zero-point radiation: Minimum energy allowed in optical fields due to quantum-mechanical effects.

Nonlinear optics is a field of study that involves a nonlinear response of a medium to intense electromagnetic radiation. Nonlinear optical interactions can change the propagation or polarization characteristics of the incident waves, or they can involve the generation of new electromagnetic waves, either at frequencies different from those contained in the incident fields or at frequencies that are the same but with the waves otherwise distinguishable from the incident waves—for example, by their direction of polarization or propagation. Nonlinear optical interactions can be used for changing or controlling certain properties of laser radiation, such as wavelength, bandwidth, pulse duration, and beam quality, as well as for high-resolution molecular and atomic spectroscopy,

materials studies, modulation of optical beams, information processing, and compensation of distortions caused by imperfect optical materials. Nonlinear optical interactions are generally observed in the spectral range covered by lasers, between the far infrared and the extreme ultraviolet (XUV) (Fig. 1), but some nonlinear interactions have been observed at wavelengths ranging from X rays to microwaves. They generally require very intense optical fields and as a result are usually observed only with radiation from lasers, although some nonlinear interactions, such as saturation of optical transitions in an atomic medium, can occur at lower intensities and were observed before the invention of the laser. Nonlinear optical interactions of various kinds can occur in all types of materials, although some types of interactions are observable only in certain types of materials, and some materials are better suited to specific nonlinear interactions than others.

I. Optical Nonlinearities— Physical Origins

When an electromagnetic wave propagates through a medium, it induces a polarization (electric dipole moment per unit volume) and magnetization (magnetic dipole moment per unit volume) in the medium as a result of the motion of the electrons and nuclei in response to the fields in the incident waves. These induced polarizations and magnetizations oscillate at frequencies determined by a combination of the properties of the material and the frequencies

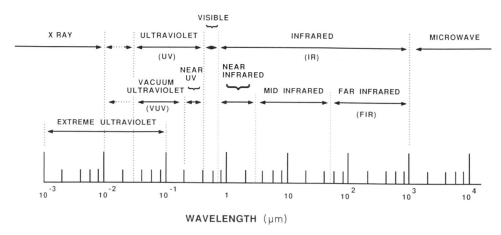

FIG. 1.   Designations of various regions of the electromagnetic spectrum between microwaves and X rays. Dashed arrows indicate inexact boundaries between spectral regions.

contained in the incident light waves. The optical properties of the medium and the characteristics of the radiation that is transmitted through it result from interference among the fields radiated by the induced polarizations or magnetizations and the incident fields.

At low optical intensities, the induced polarizations and magnetizations are proportional to the electric or magnetic fields in the incident wave, and the response of the medium is termed linear. Various linear optical interactions can occur, depending on the specific properties of the induced polarizations. Some of the more familiar linear optical effects that result from induced polarizations that oscillate at the same frequency as the incident radiation are refraction, absorption, and elastic scattering (Rayleigh scattering from static density variations, or Tyndall or Mie scattering). Other linear optical processes involve inelastic scattering, in which part of the energy in the incident wave excites an internal motion of the material and the rest is radiated in a new electromagnetic wave at a different frequency. Examples of inelastic scattering processes are Raman scattering, which involves molecular vibrations or rotations, electronic states, lattice vibrations, or electron plasma oscillations; Brillouin scattering, which involves sound waves or ion-acoustic waves in plasmas; and Rayleigh scattering, involving diffusion, orientation, or density variations of molecules. Although these inelastic scattering processes produce electromagnetic waves at frequencies different from those in the incident waves, they are linear optical processes when the intensity of the scattered wave is proportional to the intensity of the incident wave.

When the intensity of the incident radiation is high enough, the response of the medium changes qualitatively from its behavior at low intensities, giving rise to the nonlinear optical effects. Some nonlinear optical interactions arise from the larger motion of the electrons and ions in response to the stronger optical fields. In most materials, the electrons and ions are bound in potential wells that, for small displacements from equilibrium, are approximately harmonic (i.e., have a potential energy that depends on the square of the displacement from equilibrium), but are anharmonic (i.e., have a potential energy that has terms that vary as the third or higher power of the displacement from equilibrium) for larger displacements, as shown in Fig. 2. As long as the optical intensity is low, the electron or ion moves in the harmonic part of the well. In

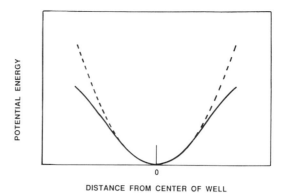

FIG. 2. Schematic illustration of an anharmonic potential well (solid line) that can be responsible for nonlinear optical interactions. A harmonic potential well that does not result in nonlinear interactions is shown for comparison (dashed line).

this regime, the induced polarizations can oscillate only at frequencies that are contained in the incident waves and the response is linear. When the incident intensity is high enough, the charges can be driven into the anharmonic portion of the potential well. The additional terms in the potential introduce terms in the induced polarization that depend on the second, third, or higher powers of the incident fields, giving rise to nonlinear responses. Examples of nonlinear optical processes that occur in this manner are various forms of harmonic generation and parametric frequency mixing.

A second type of nonlinear response results from a change in some property of the medium caused by the optical wave, which in turn affects the propagation of the wave. The optical properties of a material are usually described in the limit of extremely weak optical fields and are therefore considered to be intrinsic properties of the medium. When the optical field is strong enough, however, it can change certain characteristics of the medium, which in turn changes the way the medium affects the optical wave, resulting in a nonlinear optical response. An example of such a response is a change in the refractive index of a medium induced by the optical wave. Such changes can occur, for example, because of orientation of anisotropic molecules along the incident fields, or because of changes in the density of the medium as a result of electrostriction or as a result of a temperature change following absorption of the incident wave. Many of the propagation characteristics of optical waves are determined by the refractive index of the medium. In the situations just described, the refractive index depends on the intensity of the optical wave, and, as a result, many of the propagation characteristics depend on the optical intensity as well.

II. Optical Nonlinearities— Mathematical Description

A. MAXWELL'S EQUATIONS FOR NONLINEAR MATERIALS

The propagation of an optical wave in a medium is described by the Maxwell equation for the optical field, including the effects of the induced polarizations and magnetizations, and a specification of the dependence of the induced polarizations and magnetizations on the optical fields. The Maxwell equation for the electric field of an optical wave in a medium is

$$\nabla^2 E(r, t) - (1/c^2)\, \partial^2 E(r, t)/\partial t^2$$
$$= \mu_0\, \partial^2 P(r, t)/\partial t^2 + \partial[\nabla \times M(r, t)]/\partial t \quad (1)$$

Here $E(r, t)$ is the electric field of the optical wave, and the terms on the right are the induced polarizations (P) and magnetizations (M) that describe the effects of any charge distribution that may be present in the propagation path. In vacuum these terms are zero, and in this situation Eq. (1) reduces to the familiar expression for vacuum propagation of electromagnetic radiation.

B. NONLINEAR POLARIZATIONS

The problem of propagation in a medium is completely specified when the relations between the polarization and the magnetization and the optical fields, termed constitutive relations, are given. In general these relations can be quite complicated. However, for many situations encountered in the laboratory a set of simplifying assumptions can be made that lead to an expansion of the induced polarization in a power series in the electric field of the light wave of the form

$$P = \varepsilon_0 \chi^{(1)} E + \varepsilon_0 \chi^{(2)} E^2 + \varepsilon_0 \chi^{(3)} E^3 + \cdots$$
$$+ \varepsilon_0 Q^{(1)} \nabla E + \varepsilon_0 Q^{(2)} \nabla E^2 + \cdots \quad (2a)$$

For magnetic materials a similar expansion can be made for the magnetization:

$$M = m^{(1)} H + m^{(2)} H^2 + m^{(3)} H^3 + \cdots \quad (2b)$$

The types of expression in Eqs. (2a) and (2b) are generally valid when the optical fields are weak compared to the electric field that binds the electrons in the material and when the coefficients of the various terms in Eqs. (2a) and (2b) are constant over the range of frequencies contained in the individual incident and generated fields. In addition, the wavelength of the radiation must be long compared to the dimension of the scattering centers (the atoms and molecules of the nonlinear medium), so that the charge distributions can be accounted for with a multipole expansion.

When the first of these conditions is not met, as for example with extremely intense radiation, the perturbation series will not converge and all powers of the incident fields must be used. When the second of the conditions is not met, as, for example, can happen when certain resonance conditions are satisfied, each term in the response of the medium will involve convolutions of the optical fields instead of simply powers of the fields multiplied by constants. In these situations the polarizations induced in the medium must be solved for as dynamical variables along with the optical fields.

The coefficients in the various terms in Eqs. (2a) and (2b) are termed the nth-order susceptibilities. The first-order susceptibilities describe the linear optical effects, while the remaining terms describe the nth order nonlinear optical effects. The coefficients $\chi^{(n)}$ are the nth-order electric dipole susceptibilities, the coefficients $Q^{(n)}$ are the nth-order quadrupole susceptibilities, and so on. Similar terminology is used for the various magnetic susceptibilities. For most nonlinear optical interactions, the electric dipole susceptibilities are the dominant terms because the wavelength of the radiation is usually much longer than the scattering centers. These will be the ones considered primarily from now on.

In some situations, however, the electric quadrupole susceptibilities can make a significant contribution to the response. For example, symmetry restrictions in certain classes of materials can prevent some nonlinear processes from occurring by dipole interactions, leaving the quadrupole interactions as the dominant response. This occurs, for example, with second-harmonic generation in media with inversion symmetry. Other situations in which quadrupole interactions are significant are those in which they are enhanced by resonances with appropriate energy levels or for interactions involving sufficiently short-wavelength X radiation.

The nonlinear susceptibilities are tensors and, as such, relate components of the nonlinear polarization vector to various components of the optical field vectors. For example, the second-order polarization is given by

$$P_i = \varepsilon_0 \chi_{ijk}^{(2)} E_j E_k$$

where i, j, and k refer to the spatial directions x, y, and z. The susceptibility tensors χ have the symmetry properties of the nonlinear medium. As a result they can restrict the combinations of vector components of the various optical fields that can be used effectively. In some situations, such as those involving nonlinear optical processes that change the polarization vector of the optical wave, the tensor properties play a central role in the nonlinear interaction. In other situations, however, the tensor properties are important only in determining which combinations of vector components can, or must, be used for the optical fields and the nonlinear polarizations, and beyond that the nonlinear optical susceptibilities can usually be treated as scalars.

The magnitudes of the nonlinear susceptibilities are usually determined by measurement, although in some cases, most notably for third- and higher-order interactions in atomic gases or second-order interactions in some crystals, they can be calculated using various theories. In some cases they can be estimated with varying degrees of accuracy from products of the linear refractive indices at the various wavelengths involved.

C. Calculation of Optical Fields

In order to calculate the optical fields involved in the nonlinear interactions, we assume that they can be expressed in the form

$$E(r, z, t) = \tfrac{1}{2}[A(r, z, t)e^{i(kz-\omega t)}$$

$$+ \text{ complex conjugate}] \qquad (3)$$

This expression describes a wave of amplitude A propagating in the z direction with a frequency ω and a wave vector $k = 2\pi n/\lambda$, where λ is the wavelength of the light in vacuum, and n is the linear refractive index, which takes into account the effects of the linear susceptibilities. Many nonlinear effects involve the interaction of two or more optical fields at different wavelengths. In such situations the optical field is written as

$$E(r, z, t) = \frac{1}{2}\sum[A_i(r, z)e^{i(k_i n_i z - \omega_i \tau)}$$

$$+ \text{ complex conjugate}] \qquad (4)$$

Here the index for the summation extends over all fields in the problem, including those fields that are generated in the nonlinear interaction as well as those that are incident on the medium.

The nonlinear polarization is written as

$$P(r, z, t) = \frac{1}{2}\sum[P_i(r, z)e^{i(k_i^p z - \omega_i t)}$$

$$+ \text{ complex conjugate}] \qquad (5)$$

Here P_i is the amplitude of the nonlinear polarization with frequency ω_i. It is determined by an appropriate combination of optical fields and nonlinear susceptibilities according to the expansion in Eq. (2a). It serves as a source term for the optical field with frequency ω_i. The wave vector of the nonlinear polarization k_i^p is in general different from the wave vector of the optical field at the same frequency. This difference plays a very important role in determining the effectiveness of many nonlinear optical interactions.

The amplitudes of the various waves can be calculated from the expressions given in Eqs. (1) and (2), using the fields and polarizations with the forms given in Eqs. (4) and (5). In deriving equations for the field amplitudes, a further simplifying assumption, the slowly varying envelope approximation, is also usually made. This approximation assumes that the field envelopes $A_i(r, z, t)$ vary slowly in time compared to the optical frequency and slowly in space compared to the optical wavelength. It is generally valid for all of the interactions that are encountered in the laboratory. It enables us to neglect the second derivatives of the field amplitudes with respect to z and t when Eqs. (4) and (5) are substituted into Eq. (1). When this substitution is made, we obtain expressions of the following form for the various fields involved in a nonlinear interaction:

$$\nabla_\perp^2 A_1 + 2ik_1[\partial A_1/\partial_z + (n_1/c)\,\partial A_1/\partial t]$$
$$= -\mu_0\omega_1^2 P_1 e^{-i\Delta k_1 z} \qquad (6a)$$

$$\nabla_\perp^2 A_2 + 2ik_2[\partial A_2/\partial_z + (n_2/c)\,\partial A_2/\partial t]$$
$$= -\mu_0\omega_2^2 P_2 e^{-i\Delta k_2 z} \qquad (6b)$$

$$\vdots$$

$$\nabla_\perp^2 A_n + 2ik_n[\partial A_n/\partial_z + (n_n/c)\,\partial A_n/\partial t]$$
$$= -\mu_0\omega_n^2 P_n e^{-i\Delta k_n z} \qquad (6c)$$

Here A_i is the amplitude of the ith optical field, which may be either an incident field or a field generated in the interaction, and Δk_i is the wave-vector mismatch between the ith optical field and the polarization that drives it and is given by

$$\Delta k_i = k_i - k_i^p \qquad (7)$$

In nonlinear interactions that involve the generation of fields at new frequencies, the nonlinear polarization for the generated field will involve only incident fields. In many cases the interactions will be weak and the incident fields can be taken as constants. In other situations, the generated fields can grow to be sufficiently intense that their nonlinear interaction on the incident fields must be taken into account. In still other nonlinear interactions, such as those involving self action effects, an incident field amplitude can occur in its own nonlinear polarization. The set of Eq. (6) along with the constitutive relations in Eq. (2) can be used to calculate the optical fields in most nonlinear interactions.

D. CLASSIFICATIONS OF NONLINEAR INTERACTIONS

Nonlinear interactions can be classified according to whether they are driven by electric or magnetic fields, whether they involve the generation of fields at new frequencies or change the propagation properties of the incident fields, and whether or not they involve a transfer of energy to or from the nonlinear medium. To a certain extent these classifications are overlapping, and examples of each type of interaction can be found in the other categories.

A list of the more common nonlinear interactions is given in Table I. Most of the commonly observed nonlinear optical interactions are driven by the electric fields. They can proceed through multipole interactions of any order, but the dipole interactions are most common unless certain symmetry or resonance conditions are satisfied in the medium. The interactions are classified first according to whether they involve frequency conversion (that is, the generation of a wave at a new frequency) or self-action (that is, whether they affect the propagation of the incident wave). The frequency-conversion interactions are further classified according to whether they are parametric processes or inelastic processes. Parametric processes do not involve a transfer of energy to or from the nonlinear material. The material merely serves as a medium in which the optical waves can exchange energy among themselves. On the other hand, inelastic frequency conversion processes, generally termed stimulated scattering, do involve a transfer of energy to or from the nonlinear medium.

Self-action effects involve changes in the propagation characteristics of a light wave that are caused by its own intensity. They can involve both elastic and inelastic processes and can affect almost any of the propagation characteristics of the light wave, including both absorption and focusing properties. They can also change the spatial, temporal, and spectral distributions of the incident wave, as well as its state of polarization.

Coherent effects involve interactions that occur before the wave functions that describe the excitations in the nonlinear medium have time to get out of phase with one another. They can involve changes in propagation characteristics as well as the generation of new optical signals.

TABLE I. Common Nonlinear Optical Interactions

Nonlinear process	Description	Incident radiation	Generated radiation
Frequency conversion			
Parametric processes			
qth-Harmonic generation	Conversion of radiation from ω to $q\omega$	ω	$q\omega$
Sum- or difference-frequency mixing	Conversion of radiation at two or more frequencies to radiation at a sum- or difference-frequency combination	Three-wave (second order) ω_1, ω_2	$\omega_g = \omega_1 \pm \omega_2$
		Four-wave (third order)	
		ω_1, ω_2	$\omega_g = 2\omega_1 \pm \omega_2$
		ω_1, ω_2, ω_3	$\omega_g = \omega_1 \pm \omega_2 \pm \omega_3$

TABLE I (*Continued*)

Nonlinear process	Description	Incident radiation	Generated radiation
Parametric down-conversion	Conversion of radiation at frequency ω_1 to two or more lower frequencies	Six-wave (fifth order) ω_1, ω_2	$\omega_g = 4\omega_1 \pm \omega_2$
		Three-wave (second order) ω_1	$\omega_{g1} + \omega_{g2} = \omega_1$
		Four-wave (third order) ω_1	$\omega_{g1} + \omega_{g2} = 2\omega_1$
Optical rectification	Conversion of radiation at frequency ω_1 to an electric voltage	ω_1	$\omega_g = 0 = \omega_1 - \omega_1$
Inelastic processes			
Stimulated scattering	Conversion of radiation at frequency ω_1 to radiation at frequency ω_2 with excitation or deexcitation of an internal mode at frequency ω_0	Stokes scattering ω_1, ω_2	$\omega_2 = \omega_1 - \omega_0$
		Anti-Stokes scattering ω_1, ω_2	$\omega_2 = \omega_1 + \omega_0$
Self-action effects			
Self-focusing	Focusing of an optical beam due to a change in the refractive index caused by its own intensity		
Self-defocusing	Defocusing of an optical beam due to a change in the refractive index caused by its own intensity		
Self-phase modulation	Modulation of the phase of an optical wave due to a time-dependent change in the refractive index caused by its own intensity		
Optical Kerr effect	Birefringence in a medium caused by an intensity-induced change in the refractive index		
Ellipse rotation	Intensity-dependent rotation of the polarization ellipse of a light wave due to the nonlinear refractive index		
Raman-induced Kerr effect	Change of refractive index or birefringence of a light wave at one wavelength due to the intensity of a light wave at another wavelength		
Multiphoton absorption	Increase in the absorption of a material at high intensities		
Saturable absorption	Decrease in the absorption of a material at high intensities		
Coherent effects			
Self-induced transparency	High transmission level of a medium at an absorbing transition for a short-duration light pulse with the proper shape and intensity		
Photon echo	Appearance of a third pulse radiated from an absorbing medium following application of two pulses of proper duration, intensity, and separation		
Adiabatic rapid passage	Inversion of population of a medium by application of a light pulse whose frequency is swept quickly through that of an absorbing transition		
Electrooptic effects			
Pockels effect	Birefringence in a medium caused by an electric field and proportional to the electric field strength		
Kerr effect	Birefringence in a medium caused by an electric field and proportional to the square of the electric field strength		
Magnetooptic effects			
Faraday rotation	Rotation of the direction of linear polarization of a light beam resulting from changes in the relative velocities of oppositely circular polarizations caused by a magnetic field		
Cotton–Mouton effect	Birefringence induced in a material caused by a magnetic field		
Miscellaneous			
Optical breakdown	Rapid ionization of a material by avalanche production of electrons caused by an intense optical field		

Electrooptic and magnetooptic effects involve changes in the refractive index of the medium caused by external electric or magnetic fields resulting in changes in the phase of the optical wave or in its state of polarization.

III. Specific Nonlinear Processes

A. Frequency-Conversion Processes

Frequency-conversion processes are those that involve the generation of radiation at wavelengths other than the ones that are contained in the incident radiation. A typical frequency conversion geometry is illustrated in Fig. 3. Pump radiation, consisting of one or more optical waves at one or more frequencies, is incident on a nonlinear medium and interacts with it to generate the new wave. Depending on the interaction, the generated wave can be at a lower or higher frequency than the waves in the pump radiation, or, in some situations, can be an amplified version of one of the incident waves.

Frequency conversion interactions can either be parametric processes or stimulated scattering interactions. In parametric frequency conversion, the incident wave generates a new wave at a frequency that is a multiple, or harmonic, of the incident frequency, or, if more than one frequency is present in the incident radiation, a sum or difference combination of the frequencies of the incident fields. Parametric frequency conversion can also involve the generation of radiation in two or more longer-wavelength fields whose frequencies add up to the frequency of the incident field. In parametric frequency conversion, energy is conserved among the various optical waves, with the increase in energy of the generated wave being offset by a corresponding decrease in energy of the incident waves.

Parametric frequency-conversion interactions can occur in any order of the perturation expansion of Eq. (2). The most commonly observed processes have involved interactions of second and third order, although interactions up to order 11 have been reported. Parametric interactions are characterized by a growth rate for the intensity of the generated wave that depends on a power, or a product of powers, of the intensities of the incident waves, and they are strongly dependent on difference in wave vectors between the nonlinear polarization and the optical fields.

Stimulated scattering generally involves inelastic nonlinear frequency-conversion processes. They are nonlinear counterparts of the various linear inelastic scattering processes that were mentioned earlier. In these processes, an incident wave at frequency ω_{inc} is scattered into a wave with a different frequency, ω_{scat}, with the difference in energy between the incident and scattered photons taken up by excitation or deexcitation of an internal mode of the material. When the internal mode is excited in the nonlinear process, the scattered wavelength is longer than the incident wavelength, while it is shorter than the incident wavelength when the internal mode is deexcited.

Various types of stimulated scattering processes can occur, including stimulated Raman, stimulated Brillouin, and stimulated Rayleigh scattering, each of which involves a different type of internal mode of the medium, as will be described in Section III,A,2. The optical wave generated in stimulated scattering processes is characterized by exponential growth similar to the exponential growth experienced by laser radiation in the stimulated emission process. [*See* LASERS.]

Frequency-conversion interactions are used primarily to generate coherent radiation at wavelengths other than those that can be obtained directly from lasers, or to amplify existing laser radiation. Various frequency-conversion processes have been used to generate coherent

FIG. 3.   Schematic illustration of a typical geometry used in frequency conversion interactions. Input radiation is supplied at one or more incident frequencies ω_{pump}. The wave at ω_{gen} is generated in the nonlinear interaction.

radiation at wavelengths ranging from the extreme ultraviolet to the millimeter wave region. The radiation generated through nonlinear frequency-conversion processes has all of the special properties usually associated with laser radiation, such as spatial and temporal coherence, collimation, and narrow bandwidths, and can be tunable if the pump radiation is tunable. The generated radiation can have special properties not possessed by existing laser radiation in a given wavelength range, such as tunability, short pulse duration, or narrow bandwidth, or it can be in a spectral range in which there is no direct laser radiation available. An example of the first is the generation of tunable radiation in spectral regions in the vacuum ultraviolet or the infrared where only fixed-frequency lasers exist. An example of the second is the generation of coherent radiation in the extreme ultraviolet at wavelengths less than 100 nm, a spectral region in which lasers of any kind are just beginning to be developed.

Various frequency-conversion interactions can also be used for spectroscopy, image conversion, control of various properties of laser beams, and certain optical information-processing techniques.

1. Parametric Frequency-Conversion Processes

Parametric frequency-conversion processes include harmonic conversion of various order, various forms of frequency mixing, parametric down-conversion, and optical rectification. They are used primarily for generation of radiation at new wavelengths, although some of the interactions can be used for amplifying existing radiation. Various forms of parametric frequency-conversion interactions have been used to generate coherent radiation at wavelengths ranging from almost the X-ray range at 35.5 nm to the microwave range at 2 mm. Parametric frequency-conversion interactions are usually done in the transparent region of the nonlinear medium. Different interactions, nonlinear materials, and lasers are used to generate radiation in the various wavelength ranges. Some of the more commonly used interactions, types of lasers, and types of nonlinear materials in various wavelength ranges are given in Table II.

a. Second-Order Effects. Second-order effects are primarily parametric in nature. They include second-harmonic generation, three-wave sum- and difference-frequency mixing, parametric down-conversion, parametric oscillation, and optical rectification. The strongest second-order processes proceed through electric dipole interactions. Because of symmetry restrictions, the even-order electric dipole susceptibilities are zero in materials with inversion symmetry. As a result, the second-order nonlinear interactions are observed most commonly only in certain classes of crystals that lack a center of inversion. However, some second-order processes due to electric quadrupole inter-

TABLE II.   Wavelength Ranges for Various Nonlinear Parametric Interactions

Nonlinear interaction	Wavelength range	Typical laser	Type of material
Second-harmonic, three-wave sum-frequency mixing	$2-5~\mu m$ Visible Ultraviolet (to 185 nm)	CO_2, CO Nd:YAG, Nd:glass Harmonics of Nd, ruby, dye, argon	Crystal without inversion center Crystal without inversion center Crystal without inversion center
Three-wave difference-frequency mixing	Visible to infrared $(25~\mu m)$	CO, CO_2, Nd, dye, Parametric oscillator, spin-flip Raman	Crystal without inversion center
Parametric oscillation	Visible to infrared	Nd, dye, CO_2	Crystal without inversion center
Third harmonic, four wave sum-frequency mixing, four-wave difference-frequency mixing	Infrared (3.3 μm) to XUV (57 nm)	CO_2, CO, Nd, ruby, dye, excimer	Rare gases, molecular gases, metal vapors, cryogenic liquids
Fifth- and higher-order harmonic and frequency mixing	UV (216 nm) XUV (106–35.5 nm)	Nd, harmonics of Nd, excimer	Rare gases, metal vapors

actions have been observed in solids (e.g., sodium chloride) and gases that have a center of inversion. In addition, some second-order processes, such as second-harmonic generation, have also been observed in gases in which the inversion symmetry has been lifted with electric or magnetic fields. In these situations the interactions are actually third-order ones in which the frequency of one of the waves is zero.

The nonlinear polarization for the second order processes can be written as

$$P_{NL}^{(2)} = 2\varepsilon_0 dE^2 \qquad (8)$$

where d is a nonlinear optical susceptibility related to $\chi^{(2)}$ in Eq. (2) by $d = \chi^{(2)}/2$ and E is the total electric field as given in Eq. (4). The right-hand side of Eq. (8) has terms that oscillate at twice the frequency of the incident waves and at sum- and difference-frequency combinations and terms that do not oscillate. Each of these terms gives rise to a different nonlinear process. In general they are all present in the response of the medium. The term (or terms) that forms the dominant response of the medium is usually determined by phase matching, as will be discussed later.

The nonlinear polarization amplitude as defined in Eq. (5) can be evaluated for each of the specific nonlinear processes by substituting the form of the electric field in Eq. (4) into Eq. (8) and identifying each term by the frequency combination it contains. The resulting form of the nonlinear polarization for the sum frequency process $\omega_3 = \omega_1 + \omega_2$ is

$$P_i^{NL}(\omega_3) = g\varepsilon_0 d_{ijk}(-\omega_3, \omega_1, \omega_2)A_j(\omega_1)A_k(\omega_2)$$

$$\qquad (9)$$

Here i, j, k refer to the crystallographic directions of the nonlinear medium and the vector components of the various fields and polarizations, and g is a degeneracy factor that accounts for the number of distinct permutations of the pump fields.

Similar expressions can be obtained for the other second-order processes with minor changes. For second-harmonic generation, $\omega_1 = \omega_2$, while for difference-frequency mixing processes of the form $\omega_3 = \omega_1 - \omega_2$, ω_2 appears with a minus sign and the complex conjugate of A_2 is used. The degeneracy factor g is equal to 1 for second-harmonic generation and 2 for the second-order mixing processes.

Because the susceptibility tensor d has the same symmetry as the nonlinear medium, certain of its components will be zero, and others will be related to one another, depending on the material involved. As a result, only certain combinations of polarization vectors and propagation directions can be used for the incident and generated waves in a given material. For a specific combination of tensor and field components, the nonlinear polarization amplitude can be written as

$$P^{NL}(\omega_3) = g\varepsilon_0 d_{eff}(-\omega_3, \omega_1, \omega_2)A(\omega_1)A(\omega_2) \qquad (10)$$

where d_{eff} is an effective susceptibility that takes into account the nonzero values of the d tensor and the projections of the optical field components on the crystallographic directions and $A(\omega_i)$ is the total optical field at ω_i. The nonlinear polarization amplitudes for the other second-order processes have similar expressions with the changes noted above. The forms for the polarization amplitude for the various second-order interactions are given in Table III.

The k vector of the nonlinear polarization is given by

$$k_3^p = k_1 \pm k_2 \qquad (11)$$

where the plus sign is used for sum-frequency combinations and the minus sign is used for difference-frequency combinations. The wave-vector mismatch of Eq. (7) is also given in Table III for each nonlinear process. The nonlinear polarization amplitudes and the appropriate wave-vector mismatches as given in Table III can be used in Eq. (6) for the field amplitudes to calculate the intensities of the generated waves in specific interactions.

The second-order frequency-conversion processes are used primarily to produce coherent radiation in wavelength ranges where radiation from direct laser sources is not available or to obtain radiation in a given wavelength range with properties, such as tunability, that are not available with existing direct laser sources. They have been used with various crystals and various types of lasers to generate radiation ranging from 185 nm in the ultraviolet to 2 mm in the microwave region. The properties of some nonlinear crystals used for second-order interactions are given in Table IV, and some specific examples of second-order interactions are given in Table V.

i. SECOND-HARMONIC GENERATION. In second-harmonic generation, radiation at an incident frequency ω_1 is converted to radiation at

TABLE III. Second-Order Nonlinear Polarizations

Nonlinear process	Nonlinear polarization	Wave-vector mismatch	Plane-wave phase-matching condition
Second-harmonic generation	$P(2\omega) = \varepsilon_0 d_{\text{eff}}(-2\omega, \omega, \omega)A^2(\omega)$	$\Delta k = k(2\omega) - 2k(\omega)^a$ $\Delta k = k(2\omega) - k_1(\omega) - k_2(\omega)^b$	$n(2\omega) = n(\omega)^a$ $n(2\omega) = n_1(\omega)/2 + n_2(\omega)/2^b$
Three-wave sum-frequency mixing	$P(\omega_3) = 2\varepsilon_0 d_{\text{eff}}(-\omega_3, \omega_1, \omega_2)A(\omega_1)A(\omega_2)$	$\Delta k = k(\omega_3) - k(\omega_1) - k(\omega_2)$	$n(\omega_3)/\lambda_3 = n(\omega_1)/\lambda_1 + n(\omega_2)/\lambda_2$
Three-wave difference-frequency mixing	$P(\omega_3) = 2\varepsilon_0 d_{\text{eff}}(-\omega_3, \omega_1, -\omega_2)A(\omega_1)A^*(\omega_2)$	$\Delta k = k(\omega_3) - k(\omega_1) + k(\omega_2)$	$n(\omega_3)/\lambda_3 = n(\omega_1)/\lambda_1 - n(\omega_2)/\lambda_2$
Parametric down-conversion	$P(\omega_3) = 2\varepsilon_0 d_{\text{eff}}(-\omega_3, \omega_1, -\omega_2)A(\omega_1)A^*(\omega_2)$ $P(\omega_2) = 2\varepsilon_0 d_{\text{eff}}(-\omega_2, \omega_1, -\omega_3)A(\omega_1)A^*(\omega_3)$	$\Delta k = k(\omega_3) + k(\omega_2) - k(\omega_1)$	$n(\omega_1)/\lambda_1 = n(\omega_2)/\lambda_2 + n(\omega_3)/\lambda_3$
Optical rectification	$P(0) = 2\varepsilon_0 d_{\text{eff}}(0, \omega, -\omega)A(\omega_1)A^*(\omega_1)$	$\Delta k = k(0) - k(\omega) + k(\omega) = 0$	Automatically satisfied

[a] Type I phase matching.
[b] Type II phase matching.

TABLE IV. Properties of Selected Nonlinear Optical Crystals

Nonlinear material	Symmetry point group	Nonlinear susceptibility d $(10^{-12}$ m/V$)$	Transparency range (μm)	Effective nonlinearity, $\lvert d_{\text{eff}} \rvert$	
				Type I phase matching[a]	Type II phase matching
Ag_3AsS_3 (proustite)	$3m$	$d_{22} = 22$; $d_{15} = 13$	0.6–13	$d_{15} \sin\theta - d_{22} \cos\theta \sin 3\phi$	$d_{22} \cos^2\theta \cos 3\phi$
Te (tellurium)	32	$d_{11} = 649$	4–25	$d_{15} \sin\theta + \cos\theta(d_{11} \cos 3\phi - d_{22} \sin 3\phi)$	$d_{11} \cos^2\theta$
Tl_3AsSe_3 (TAS)		$d_+ = 40$	1.2–18	d_+	
$CdGeAs_2$	$\bar{4}2m$	$d_{36} = d_{14} = d_{25} = 236$	2.4–17	$d_{36} \sin\theta$	$d_{36} \sin 2\theta$
$AgGaS_2$	$\bar{4}2m$	$d_{36} = d_{14} = d_{25} = 12$	0.6–13	$d_{36} \sin\theta$	$d_{36} \sin 2\theta$
$AgGaSe_2$	$\bar{4}2m$	$d_{36} = d_{14} = d_{25} = 40$	0.7–17	$d_{36} \sin\theta$	$d_{36} \sin 2\theta$
GaAs	$\bar{4}3m$	$d_{36} = d_{14} = d_{25} = 90.1$	0.9–17	$d_{36} \sin\theta$	$d_{36} \sin 2\theta$
$LiNbO_3$ (lithium niobate)	$3m$	$d_{15} = 6.25$; $d_{22} = 3.3$	0.35–4.5	$d_{15} \sin\theta - d_{22} \cos\theta \sin 3\phi$	$d_{22} \cos^2\theta \cos 3\phi$
$LiIO_3$ (lithium iodate)	6	$d_{31} = 7.5$	0.31–5.5	$d_{31} \sin\theta$	
$NH_4H_2(PO_4)_2$ (ammonium dihydrogen phosphate, ADP)	$\bar{4}2m$	$d_{36} = d_{14} = d_{25} = 0.57$	0.2–1.2	$d_{36} \sin\theta$	$d_{36} \sin 2\theta$
$KH_2(PO_4)_2$ (potassium dihydrogen phosphate, KDP)	$\bar{4}2m$	$d_{36} = d_{14} = d_{25} = 0.5$	0.2–1.5	$d_{36} \sin\theta$	$d_{36} \sin 2\theta$
$KD_2(PO_4)_2$ (potassium dideuterium phosphate, KD*P)	$\bar{4}2m$	$d_{36} = d_{14} = d_{25} = 0.53$	0.2–1.5	$d_{36} \sin\theta$	$d_{36} \sin 2\theta$
$RbH_2(AsO_4)_2$ (rubidium dihydrogen arsenate, RDA)	$\bar{4}2m$	$d_{36} = d_{14} = d_{25} = 0.47$	0.26–1.46	$d_{36} \sin\theta$	$d_{36} \sin 2\theta$
$RbH_2(PO_4)_2$ (rubidium dihydrogen phosphate, RDP)	$\bar{4}2m$	$d_{36} = d_{14} = d_{25} = 0.48$	0.22–1.4	$d_{36} \sin\theta$	$d_{36} \sin 2\theta$
$NH_4H_2(AsO_4)_2$ (ammonium dihydrogen arsenate, ADA)	$\bar{4}2m$			$d_{36} \sin\theta$	$d_{36} \sin 2\theta$
$KD_2(AsO_4)_2$ (potassium dideuterium arsenate, KD*A)	$\bar{4}2m$	$d_{36} = d_{14} = d_{25} = 0.4$	0.22–1.4	$d_{36} \sin\theta$	$d_{36} \sin 2\theta$
$CsH_2(AsO_4)_2$ (cesium dihydrogen arsenate, CDA)	$\bar{4}2m$	$d_{36} = d_{14} = d_{25} = 0.48$	0.26–1.43	$d_{36} \sin\theta$	$d_{36} \sin 2\theta$
$CsD_2(AsO_4)_2$ (cesium dideuterium arsenate, CD*A)	$\bar{4}2m$	$d_{36} = d_{14} = d_{25} = 0.48$	0.27–1.66	$d_{36} \sin\theta$	$d_{36} \sin 2\theta$
$KTiOPO_4$ (potassium titanyl phosphate, KTP)	$mm2$	$d_{31} = 7$; $d_{32} = 5.4$; $d_{33} = 15$; $d_{24} = 8.1$; $d_{15} = 6.6$	0.35–4.5	$d_{31} \cos 2\theta + d_{32} \sin 2\theta$	
$LiCHO_2 \cdot H_2O$ (lithium formate monohydrate, LFM)	$mm2$	$d_{31} = d_{15} = 0.107$ $d_{32} = d_{24} = 1.25$ $d_{33} = 4.36$	0.23–1.2	$d_{31} \cos 2\theta + d_{32} \sin 2\theta$	
$KB_5O_8 \cdot 4H_2O$ (potassium pentaborate, KB5)	$mm2$	$d_{31} = 0.046$ $d_{32} = 0.003$	0.17–>0.76	$d_{31} \cos 2\theta + d_{32} \sin 2\theta$	
Urea	$\bar{4}2m$	$d_{36} = d_{14} = d_{25} = 1.42$	0.2 –1.43	$d_{36} \sin\theta$	$d_{36} \sin 2\theta$

[a] θ is the direction of propagation with respect to the optic axis, ϕ is the direction of propagation with respect to the x axis.

Damage threshold (10^6 W/cm^2)	Uses
12–40	Harmonic generation, frequency mixing, parametric oscillation in mid infrared
40–60 (at 5 μm)	Harmonic generation, frequency mixing, parametric oscillation in mid infrared
32	Harmonic generation, frequency mixing, parametric oscillation in mid infrared
20–40	Harmonic generation, frequency mixing, parametric oscillation in mid infrared
12–25	Harmonic generation, frequency mixing, parametric oscillation in mid infrared
>10	Harmonic generation, frequency mixing, parametric oscillation in mid infrared
	Harmonic generation, frequency mixing, parametric oscillation in mid infrared; difference frequency generation in far infrared
50–140	Harmonic generation, frequency mixing, parametric oscillation in near and mid infrared; harmonic generation and frequency mixing in visible and near ultraviolet
125	Harmonic generation, frequency mixing, parametric oscillation in near and mid infrared; harmonic generation and frequency mixing in visible and near ultraviolet
500 (60 nsec, 1.064 μm)	Harmonic generation, frequency mixing, parametric oscillation in near infrared; harmonic generation and frequency mixing in visible and near ultraviolet
400 (20 nsec, 694.3 nm)	Harmonic generation, frequency mixing, parametric oscillation in near infrared; harmonic generation and frequency mixing in visible and near ultraviolet
23,000 (200 psec, 1.064 μm)	
500 (10 nsec, 1.064 μm)	Harmonic generation, frequency mixing, parametric oscillation in near and mid infrared; harmonic generation and frequency mixing in visible and near ultraviolet
20,000 (30 psec, 1.064 μm)	
350 (10 nsec, 694.3 nm)	Harmonic generation, frequency mixing in near ultraviolet
200 (10 nsec, 694.3 nm)	Harmonic generation, frequency mixing in near ultraviolet
	Harmonic generation, frequency mixing in near ultraviolet
	Harmonic generation, frequency mixing in near ultraviolet
500 (10 nsec, 1.064 μm)	Harmonic generation, frequency mixing and parametric oscillation in visible and near infrared
>260 (12 nsec, 1.064 μm)	Harmonic generation, frequency mixing and parametric oscillation in visible and near infrared
160 (20 nsec, 1.064 μm)	Harmonic generation, frequency mixing and parametric oscillation in visible and near infrared
>1000	Harmonic generation and frequency mixing in the ultraviolet
	Harmonic generation and frequency mixing in the ultraviolet
1.4×10^3	Harmonic generation and frequency mixing in the ultraviolet

TABLE V. Performance for Selected Second-Order Frequency Conversion Interactions

Laser	Incident wavelength (μm)	Generated wavelength (μm)	Nonlinear material	Conversion efficiency		Conditions, comments
				Power (%)	Energy (%)	
Second harmonic generation						
CO_2	10.6	5.3	Ag_3AsS_3 (proustite)	~1		
	9.6	4.8	Te	5		Limited by multiphoton absorption
			Tl_3AsSe_3	40	25	1 J/cm² pump, 100 nsec
			$CdGeAs_2$		27	
			$AgGaSe_2$	60	14	2.1 cm, 75 nsec, 30 mJ, 12 MW/cm²
			GaAs	2.7		Phase matching with stacked plates
Nd:YAG	1.064	0.532	ADP	23		
			KDP	83		346 J output (with Nd: glass amplifiers)
			KD*P		75	30 psec pulses, 10¹⁰ W/cm² pump
			CDA	60		
			CD*A	57		
			$LiNbO_3$	30		
			KTP	42		
Nd:glass (silicate)	1.059	0.5295	KDP	92		5 psec pulses
Nd:glass (phosphate)	1.052	0.526	KDP		80	
Ruby	0.6943	0.347	ADP	40		
			RDA	40		
Nd:YAG	0.532	0.266	ADP		85	30 psec pulses, 10¹⁰ W/cm² pump intensity
			KDP	70		50 J output (with Nd: glass amplifiers)
			KD*P	75		30 psec, 10¹⁰ W/cm² pump intensity
Dye lasers	0.560–0.620	0.280–0.310	ADP,KDP	8–9		
	0.460–0.600	0.230–0.300	LFM	1–2		
	0.434–0.500	0.217–0.250	KB5	0.2–2		
Ar^+	0.5145	0.2572	ADP,KDP	30		300 mW, intracavity, cw

Three-wave sum-frequency mixing ($\omega_3 = \omega_1 + \omega_2$)

Laser	λ (μm)	Output (μm)	Crystal	Efficiency (%)	Comments
CO_2	9.6, 4.8	3.53	Tl_3AsSe_3	1.6	1 J/cm^2 pump, overall efficiency from 9.6 μm
Nd:YAG	1.064, 0.532	0.3547	KDP	55	41 J output (with Nd:glass amplifiers)
			RDP	21	
Nd:YAG	1.064, 0.266	0.2128	ADP	60	Measured with respect to power at 266 nm
			KDP	0.002	1 kW output, 20 nsec pulses
Nd:YAG, Dye (second harmonic)	1.064, 0.258–0.300	0.208–0.234	ADP	10	
Nd:YAG, Dye	0.266, 0.745–0.790	0.1966–0.199	KB5	3	40 kW, 5 nsec at 0.1966
Dye (Second harmonic)	0.622–0.652, 0.310–0.326, 0.740–0.920, 0.232–0.268;	0.185–0.2174	KB5	8–12	

Difference-frequency mixing ($\omega_3 = \omega_1 - \omega_2$)

Laser	λ (μm)	Output (μm)	Crystal	Efficiency (%)	Comments
Nd:YAG, Dye	0.532, 0.575–0.660	2.8–5.65	$LiIO_3$	0.92	Measured with respect to total power
Nd:YAG, Dye	1.064, 0.575–0.640	1.25–1.6	$LiIO_3$	0.7	70 W output
Ruby	0.6943				
Dye	0.84–0.89	3–4	$LiNbO_3$	0.015	6 kW output pulses
	0.80–0.83	4.1–5.2	$LiIO_3$	6×10^{-3}	100 W output pulses
	0.78–0.87	3.2–6.5	Ag_3AsS_3	2.7×10^{-3}	110 W pulses at 5 μm
	0.73–0.75	10–13	Ag_3AsS_3	2×10^{-4}	100 mW at 10 μm
	0.74–0.818	4.6–12	$AgGaS_2$	2×10^{-4}	300 mW at 11 μm
	0.72–0.75	9.5–17	GaSe	0.075	300 W, 20 nsec at 12 μm
Dye	0.81–0.84, 0.81–0.84	50–500	$LiNbO_3$	10^{-5}–10^{-6}	10–100 mW output
Dye	0.440–0.510, 0.570–0.620	1.5–4.8	$LiIO_3$	10^4	400 mW output power

(continues)

TABLE V (*Continued*)

Laser	Incident wavelength (μm)	Generated wavelength (μm)	Nonlinear material	Conversion efficiency		Conditions, comments
				Power (%)	Energy (%)	
Dye	0.586, 0.570–0.620	8.7–11.6	AgGaS$_2$	5×10^{-6}		100 μW output power
Dye		300–2 mm	ZnTe, ZnSe			
Argon, Dye	0.5145, 0.560–0.620	2.2–4.2	LiNbO$_3$	10^{-3}		1 μW continuous
CO$_2$	Various lines between 9.3 and 10.6	70–2 mm	GaAs	10^{-3}		10 W output power
CO$_2$ InSb spin-flip	10.6 11.7–12	90–111	InSb	4×10^{-8}		2 μW output power
Parametric Oscillators ($\omega_1 \rightarrow \omega_2 + \omega_3$)						
Nd:YAG	0.266	0.420–0.730	ADP	25		
Nd:YAG	0.472, 0.532 0.579, 0.635	0.55–3.65	LiNbO$_3$	50	30	100 kW output power, 30 psec pulses
Nd:YAG	1.064	1.4–4.0	AgGaS$_2$	40	16	20 nsec pulses
	1.064	1.4–4.4	LiNbO$_3$	10^{-4}		
Ruby	0.6943	66–200	LiNbO$_3$	1–10		2–100 kW output
Ruby	0.6943	0.77–4	LiIO$_3$	8		10 kW output
Ruby	0.347	0.415–2.1	LiNbO$_3$	10		800 W output, 300 nsec
Hydrogen fluoride	2.87	4.3–4.5 8.1–8.3	CdSe			

twice the frequency, and one-half the wavelength, of the incident radiation:

$$\omega_2 = 2\omega_1 \qquad (12)$$

For this interaction, the incident frequencies in the nonlinear polarization of Eq. (10) are equal, the generated frequency is ω_2, and $g = 1$. The intensity of the harmonic wave can be calculated for specific configurations from the appropriate equation in Eq. (6) using the nonlinear polarization amplitude $P(2\omega) = \varepsilon_0 d_{eff}(-2\omega, \omega, \omega)A(\omega)^2$ from Table III. The simplest situation is one in which the incident radiation is in the form of a plane wave and the conversion efficiency to the harmonic is small enough that the incident intensity can be regarded as constant. The harmonic intensity, related to the field amplitude by

$$I(2\omega) = cn_2\varepsilon_0|A(2\omega)|^2/2, \qquad (13)$$

where n_2 is the refractive index at ω_2, is then given by

$$I(2\omega) = [8\pi^2 d_{eff}^2/n_2 n_1^2 c\varepsilon_0\lambda_1^2]$$
$$\times I_0(\omega)^2 \, \text{sinc}^2(\Delta k \, L/2) \qquad (14)$$

Here $\text{sinc}(x) = (\sin x)/x$, $I_0(\omega)$ is the incident intensity at ω, L is the crystal length, and

$$\Delta k = k(2\omega) - k_1(\omega) - k_2(\omega) \qquad (15)$$

where $k(2\omega)$ is the usual wave vector at 2ω, and $k_1(\omega)$ and $k_2(\omega)$ are the wave vectors of the components of the incident wave at ω. If the incident radiation has only one polarization component relative to the principal directions of the crystal (i.e., is either an ordinary or an extraordinary ray), $k_1(\omega) = k_2(\omega) = k(\omega)$ and $\Delta k = k(2\omega) - 2k(\omega)$, whereas if the incident radiation has both ordinary and extraordinary polarization components, $k_1(\omega)$ is in general not equal to $k_2(\omega)$.

In the low conversion regime, the harmonic intensity grows as the square of the incident intensity. This is one of the reasons effective harmonic conversion requires the intense fields that are present in laser radiation. The harmonic intensity is also very sensitive to the value of the wave-vector mismatch Δk. Its dependence on Δk at fixed L and on L for different values of Δk is shown in Fig. 4a,b, respectively. If $\Delta k \neq 0$, the harmonic wave gradually gets out of phase with the polarization that drives it as it propagates through the crystal. As a result the harmonic intensity oscillates with distance, with the harmonic wave first taking energy from the incident wave, and then, after a phase change of π relative to the nonlinear polarization, returning it to the incident wave. The shortest distance at

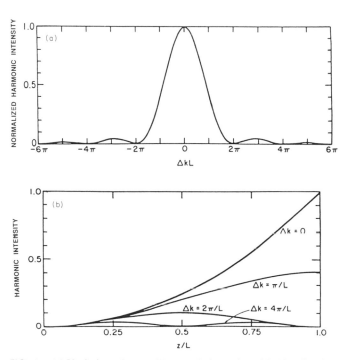

FIG. 4.   (a) Variation of second harmonic intensity with Δk at fixed L. (b) Variation of second harmonic intensity with L for various values of Δk.

which the maximum conversion occurs is termed the coherence length and is given by

$$L_{coh} = \pi/\Delta k \qquad (16)$$

When $\Delta k \neq 0$, maximum conversion is obtained for crystals whose length is an odd multiple of the coherence length, while no conversion is obtained for crystals whose length is an even multiple of the coherence length.

When

$$\Delta k = 0 \qquad (17)$$

the harmonic wave stays in phase with its driving polarization. The process is then said to be phase-matched, and the harmonic intensity grows as the square of the crystal length. Under this condition the harmonic intensity can grow so large that the depletion of the incident radiation must be accounted for. In this case the harmonic and incident intensities are given by

$$I(2\omega_1) = I_0(\omega_1)\,\tanh^2(L/L_{dep}) \qquad (18a)$$

$$I(\omega_1) = I_0(\omega_1)\,\text{sech}^2(L/L_{dep}) \qquad (18b)$$

where L_{dep} is a depletion length given by

$$L_{dep} = [n_2 n_1^2 c\varepsilon_0 \lambda_1^2/8\pi^2 d_{eff}^2 I_0(\omega_1)]^{1/2} \qquad (19)$$

When $L = L_{dep}$, 58% of the fundamental is converted to the harmonic. The harmonic and fundamental intensities are shown as a function of $(L/L_{dep})^2$ in Fig. 5. Although energy is conserved between the optical waves in second-harmonic generation, the number of photons is not, with two photons being lost at the fundamental for every one that is created at the harmonic.

The wave-vector mismatch Δk occurs because of the natural dispersion in the refractive index that is present in all materials. Effective second-harmonic conversion therefore requires that special steps be taken to satisfy the condition of Eq. (17). This process is termed phase matching. For the noncentrosymmetric crystals used for second-order processes, the most common method of phase matching involves use of the birefringence of the crystal to offset the natural dispersion in the refractive indices. In general this means that the harmonic wave will be polarized differently from the incident wave, with the particular combinations being determined by the symmetry properties of the nonlinear crystal. The value of the wave-vector mismatch can be adjusted by varying the direction of propagation of the various waves relative to the optic axis or by varying the temperature of the crystal for a fixed direction of propagation. The angle at which $\Delta k = 0$ is called the phase-matching angle, and the temperature at which $\Delta k = 0$ when $\theta = 90°$ is called the phase-matching temperature.

In the simplest situation, termed type I phase matching, the incident radiation is polarized as an ordinary or an extraordinary ray, depending on the properties of the nonlinear material, and the harmonic radiation is polarized orthogonally to the incident radiation as shown in Fig. 6a. The wave-vector mismatch is then given by

$$\Delta k = k(2\omega) - 2k(\omega) = 4\pi[n(2\omega) - n(\omega)]/\lambda_1 \qquad (20)$$

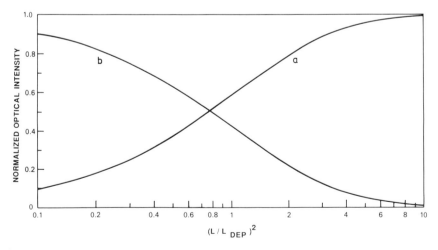

FIG. 5. Variation of the second harmonic (a) and fundamental (b) intensity with $(L/L_{dep})^2$, where L_{dep} is the second-harmonic depletion length, for plane waves at perfect phase matching. [Reproduced from J. Reintjes (1985). Coherent ultraviolet and vacuum ultraviolet sources, in "Laser Handbook," Vol. 5 (M. Bass and M. Stitch, eds.). North-Holland, Amsterdam.]

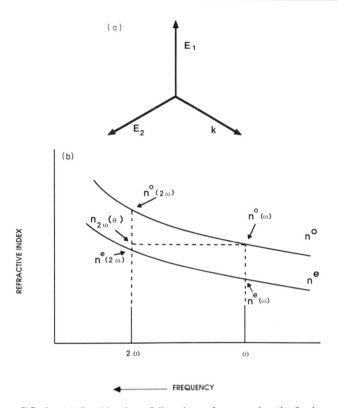

(a)

E_1

E_2 k

(b)

REFRACTIVE INDEX

$n^o(2\omega)$

$n_{2\omega}(\theta)$

$n^e(2\omega)$

$n^o(\omega)$

n^o

n^e

$n^e(\omega)$

2ω ω

FREQUENCY

FIG. 6. (a) Combination of directions of propagation (k), fundamental polarization (E_1), and second harmonic polarization (E_2) for type I second-harmonic generation. (b) Typical variation of the ordinary (n^o) and the extraordinary (n^e) refractive indices with wavelength for a negative uniaxial crystal, showing conditions necessary for type I phase matching of second-harmonic generation. The refractive index of the extraordinary ray at the harmonic varies between n^o and n^e as the angle of propagation relative to the optic axis is varied. Phase matching occurs when $n_{2\omega}(\theta) = n_\omega^o$.

Phase matching entails choosing conditions such that the refractive index for the field at the fundamental wavelength equals the refractive index for the field at the harmonic, that is, $n(2\omega) = n(\omega)$. For example, in a material in which the extraordinary refractive index is less than the ordinary refractive index, the pump wave is polarized as an ordinary ray and the harmonic wave is polarized as an extraordinary ray. The variation of the refractive index of these rays with wavelength is shown in Fig. 6b. The ordinary index is independent of the propagation direction, but the extraordinary index depends on the angle θ between the optic axis and the propagation direction according to the relation

$$[1/n_{2\omega}(\theta)]^2 = (\cos^2\theta)/(n_{2\omega}^0)^2$$
$$+ (\sin^2\theta)/(n_{2\omega}^e)^2 \qquad (21)$$

As θ varies from 0 to 90°, $n_{2\omega}(\theta)$ varies from $n_{2\omega}^o$ to $n_{2\omega}^e$. In this situation, phase matching consists of choosing the propagation direction such that $n_{2\omega}(\theta) = n_\omega^o$.

In type II phase matching, the incident radiation has both ordinary and extraordinary polarization components, while the harmonic ray is an extraordinary ray (or vice versa, if the nonlinear medium has the property $n^o < n^e$). In type II phase matching, the more general relation among the k vectors of Eq. (15) must be used, and the relationship between the various refractive indexes required for phase matching is more complicated than that given in Eq. (20) for type I, but the propagation direction is again chosen such that Eq. (17) is satisfied.

Second-harmonic generation has been used to generate light at wavelengths ranging from 217 nm (by doubling radiation from a dye laser at 434 nm) to 5 μm, by doubling radiation from a CO_2 laser at 10.6 μm. Examples of specific results are given in Table V. The usefulness of a particular material for second-harmonic generation is determined by a combination of the magnitude of its nonlinear susceptibility, its ability to phase match the process and its degree of transparency to the harmonic and fundamental wavelengths. The short wave limit of second-harmonic conversion is currently set by an inability to phase match the harmonic process at shorter wavelengths, while the infrared limit of 5 μm is determined by increasing absorption at the fundamental.

Second-harmonic conversion is commonly used with high-power pulsed lasers such as Nd : YAG, Nd : glass, ruby, or CO_2 to generate high-power fixed-frequency radiation at selected wavelengths in the infrared, visible, and ultraviolet. It is also a versatile and convenient source of tunable ultraviolet radiation when the pump radiation is obtained from tunable dye lasers in the visible and near infrared.

Second-harmonic conversion efficiencies range from the order of 10^{-6} for conversion of continuous wave radiation outside the laser cavity to over 90% for high-power pulsed radiation. The laser power required for high efficiency depends on the configuration and nonlinear material. A typical value for 75% conversion of radiation from an Nd : glass laser at 1.06 μm to 530 nm in a 1-cm-long crystal of potassium dihydrogen phosphate (KDP) is 10^8 W/cm^2. Intensities for conversion in other materials and with other lasers are noted in Table V. Conversion of radiation from high-power pulsed lasers is typically done in collimated-beam geometries with angle-tuned crystals chosen for their high damage thresholds. Conversion of radiation from lower-power pulsed lasers or continuous-wave (cw) lasers is often done in focused geometries in materials with large nonlinear susceptibilities. In many situations the maximum conversion efficiency that can be achieved is not determined by the available laser power but by additional processes that affect the conversion process. Some of these competing processes are linear or nonlinear absorption, imperfect phase matching because of a spread of the laser divergence or bandwidth, incomplete phase matching because of intensity-dependent changes in the refractive index (see Section III,B,1), or damage to the nonlinear crystal caused by the intense laser radiation.

ii. THREE-WAVE SUM-FREQUENCY MIXING. Three-wave sum-frequency mixing is used to generate radiation at higher frequencies, and therefore shorter wavelengths, than those in the pump radiation according to the relation

$$\omega_3 = \omega_1 + \omega_2 \tag{22}$$

where ω_3 is the frequency of the radiation generated in the interaction and ω_1 and ω_2 are the frequencies of the pump radiation. Calculation of the generated intensity is similar to that of second-harmonic generation with the appropriate form of the nonlinear polarization as given in Table III used in Eq. (6). At low conversion efficiencies, the generated intensity has the same $\sin^2(\Delta k\,L/2)/(\Delta k\,L/2)^2$ dependence on the wavevector mismatch as has second-harmonic conversion, but the phase-matching condition is now $\Delta k = k(\omega_3) - k(\omega_1) - k(\omega_2) = 0$, which is equivalent to

$$n(\omega_3)/\lambda_3 = n(\omega_1)/\lambda_1 + n(\omega_2)/\lambda_2 \tag{23}$$

The intensity of the generated radiation grows as the product of the incident pump intensities at low conversion efficiency. In the three-wave sum-frequency mixing interaction, one photon from each of the pump waves is annihilated for each photon created in the generated wave. Complete conversion of the total radiation in both pump waves at perfect phase matching is possible in principle if they start with an equal number of photons. Otherwise the conversion will oscillate with pump intensity or crystal length, even at exact phase matching.

Three-wave sum-frequency mixing is done in the same types of materials as second-harmonic generation. It is used to generate both tunable and fixed-frequency radiation at various wavelengths ranging from the infrared to the ultraviolet. It allows radiation to be generated at shorter wavelengths in the ultraviolet than can be reached with second harmonic conversion if one of the pump wavelengths is also in the ultraviolet.

Examples of specific three-wave sum-frequency mixing interactions are given in Table V. This interaction has been used to produce radiation at wavelengths as short as 185 nm in potassium pentaborate (KB$_5$), with the cutoff being determined by the limits of phase matching. Three-wave sum-frequency mixing can also be used to improve the efficiency of the generation of tunable radiation, as compared to second-harmonic generation, by allowing radiation from a relatively powerful fixed-frequency laser to be

FIG. 7. Schematic illustration of the waves used in parametric down-conversion. Radiation at ω_1 is supplied, and radiation at ω_2 and ω_3 is generated in the nonlinear interaction.

combined with radiation from a relatively weak tunable laser. Another application is in the generation of the third harmonic of certain fixed frequency lasers such as Nd : YAG, Nd : glass, or CO_2 through two second-order processes: second-harmonic conversion of part of the laser fundamental followed by sum-frequency mixing of the unconverted fundamental with the second harmonic. Under certain conditions this process is more efficient than direct third-harmonic conversion (see Section III,A,1,b). Three-wave sum-frequency mixing has also been used for up-conversion of infrared radiation to the visible, where it can be measured by more sensitive photoelectric detectors or photographic film.

iii. THREE - WAVE   DIFFERENCE - FRE - QUENCY MIXING. Three-wave difference-frequency mixing is used to convert radiation from two incident waves at frequencies ω_1 and ω_2 to a third wave at frequency ω_3 according to the relation

$$\omega_3 = \omega_1 - \omega_2 \qquad (24)$$

Just as for sum-frequency mixing, the generated intensity at low conversion efficiency grows as the product of the pump intensities at ω_1 and ω_2. In this situation a photon is created at both ω_3 and ω_2 for every photon annihilated at ω_1. The wave-vector mismatch is $\Delta k = k(\omega_3) - [k(\omega_1) - k(\omega_2)]$, and the phase matching condition is

$$n(\omega_3)/\lambda_3 = n(\omega_1)/\lambda_1 - n(\omega_2)/\lambda_2 \qquad (25)$$

The materials used for difference-frequency mixing are of the same type as those used for second-harmonic and sum-frequency mixing. Difference-frequency mixing is generally used to produce coherent radiation at longer wavelengths than either of the pump wavelengths. It is used most often to produce tunable radiation in the infrared from pump radiation in the visible or near infrared, although it can also be used to generate radiation in the visible. It has been used to produce radiation at wavelengths as long as 2 mm in GaAs, with the limit being set by a

combination of the increasing mismatch in the diffraction of the pump and generated radiation and the increasing absorption of the generated radiation in the nonlinear medium at long wavelengths.

iv. PARAMETRIC   DOWN - CONVERSION. Parametric down-conversion is used to convert radiation in an optical wave at frequency ω_1 into two optical waves at lower frequencies ω_2 and ω_3 according to the relation

$$\omega_1 = \omega_2 + \omega_3 \qquad (26)$$

This process is illustrated schematically in Fig. 7. The wave at ω_1 is termed the pump wave, while one of the waves at ω_2 or ω_3 is termed the signal and the other the idler. When $\omega_2 = \omega_3$ the process is termed degenerate parametric down-conversion and is the opposite of second-harmonic generation, whereas if $\omega_2 \neq \omega_3$, the process is called nondegenerate parametric down-conversion and is the opposite of sum-frequency generation. The individual values of ω_2 and ω_3 are determined by the phase-matching condition, which for plane-wave interactions is given by

$$\Delta k = k(\omega_1) - k(\omega_2) - k(\omega_3) = 0 \qquad (27)$$

and can be varied by changing an appropriate phase-matching parameter of the crystal such as the angle of propagation or the temperature. Off-axis phase matching, as illustrated in Fig. 8, is also possible. The phase-matching condition of Eq. (27) is the same as that for the sum-frequency process $\omega_2 + \omega_3 = \omega_1$. The relative phase of the waves involved determines

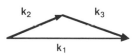

FIG. 8. Off-axis phase-matching diagram for parametric down-conversion. Direction of arrows indicates direction of propagation of the various waves.

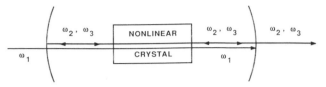

FIG. 9.   Illustration of a doubly resonant cavity for a parametric oscillator.

whether the sum process or the parametric down-conversion process will occur.

In the absence of pump depletion, the amplitudes of the waves generated in a parametric down-conversion process are given by

$$A(\omega_2) = A_0(\omega_2) \cosh \kappa z$$
$$+ iA_0^*(\omega_3)(\omega_2 n_3/\omega_3 n_2)^{1/2} \sinh \kappa z \quad (28a)$$

$$A(\omega_3) = A_0(\omega_3) \cosh \kappa z$$
$$+ iA_0^*(\omega_2)(\omega_3 n_2/\omega_2 n_3)^{1/2} \sinh \kappa z \quad (28b)$$

where

$$\kappa^2 = [2\omega_2\omega_3 d_{\text{eff}}^2/n_2 n_3 c^2] I_0(\omega_1) \quad (29)$$

and $A_0(\omega_i)$ and $I_0(\omega_i)$ are the incident field amplitude and intensity, respectively, at ω_i. At low pump intensities the generated field amplitudes grow in proportion to the square root of the pump intensity, while at high pump intensities the growth of the generated waves is exponential.

If there is no incident intensity supplied at ω_2 or ω_3, the process is termed parametric generation or parametric oscillation, depending on the geometry. For this situation the initial intensity for the generated waves arise from the zero-point radiation field with an energy of $h\nu/2$ per mode for each field. If the interaction involves a single pass through the nonlinear medium, as was shown in Fig. 7, the process is termed parametric generation. This geometry is typically used with picosecond pulses for generation of tunable infrared or visible radiation from pump radiation in the ultraviolet, visible, or near infrared. Amplification factors of the order of 10^{10} are typically required for single-pass parametric generation.

Parametric down-conversion can also be used with a resonant cavity that circulates the radiation at either or both of the generated frequencies, as shown in Fig. 9. In this geometry the process is termed parametric oscillation. If only one of the generated waves is circulated in the cavity, it is termed singly resonant, whereas if both waves are circulated, the cavity is termed doubly resonant. In singly resonant cavities the

wave that is circulated is termed the signal, while the other generated wave is termed the idler. Optical parametric oscillators are typically used with pump radiation from Q-switched lasers with pulses lasting several tens of nanoseconds, allowing several passes of the generated radiation through the cavity while the pump light is present.

One of the primary uses of parametric down-conversion is the generation of tunable radiation at wavelengths ranging from the visible to the far infrared. Its wavelength range is generally the same as that covered by difference-frequency mixing, although it has not been extended to as long a wavelength. Tuning is done by varying one of the phase-matching parameters such as angle or temperature. A typical tuning curve for a parametric oscillator is shown in Fig. 10. As

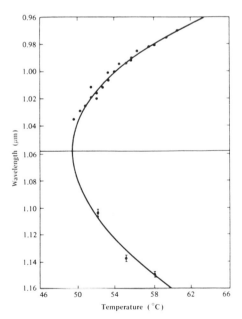

FIG. 10.   Tuning curve for a LiNbO$_3$ parametric oscillator pumped by radiation from the second harmonic of a Nd laser at 529 nm. [Reproduced from J. A. Giordmaine and R. C. Miller (1965). *Phys. Rev. Lett.* **14**, 973.]

the phase-matching condition is changed from the degenerate condition, ω_2 increases and ω_3 decreases in such a way as to maintain the relation in Eq. (26). The extent of the tuning range for a given combination of pump wavelength and nonlinear material is generally set by the limits of phase matching, although absorption can be important if ω_3 is too far in the infrared. Radiation in different wavelength ranges can be produced by using different nonlinear materials and different pump sources. Parametric down-conversion has been used to produce radiation at wavelengths ranging from the visible to 25 μm. Some of the combinations of pump sources, nonlinear materials, and tuning ranges are listed in Table V.

Parametric down-conversion can also be used to amplify radiation at ω_2 or ω_3. In this arrangement radiation is supplied at both the pump wavelength and the lower-frequency wave to be amplified, which is termed the signal. The process is similar to difference-frequency mixing, and differs from the difference-frequency process only in that the incident intensity in the signal wave is considerably less than the pump intensity for parametric amplification, whereas the two incident intensities are comparable in difference-frequency mixing. In principle, very high gains can be obtained from parametric amplification, with values up to 10^{10} being possible in some cases.

Optical rectification is a form of difference-frequency mixing in which the generated signal has no carrier frequency. It is produced through the interaction

$$\omega_3 = 0 = \omega_1 - \omega_1 \qquad (30)$$

It does not produce an optical wave, but rather an electrical voltage signal with a duration corresponding to the pulse duration of the pump radiation. Optical rectification has been used with picosecond laser pulses to produce electrical voltage pulses with durations in the picosecond range, among the shortest voltage pulses yet produced. These pulses do not propagate in the bulk of the nonlinear crystal as do the optical waves, and observation or use of them requires coupling to an appropriate microwave strip line on the nonlinear crystal.

b. Third - and Higher - Order Processes.
Third- and higher-order parametric processes are also used for frequency conversion. They have been used to generate radiation ranging from 35.5 nm, almost in the soft X-ray range, to

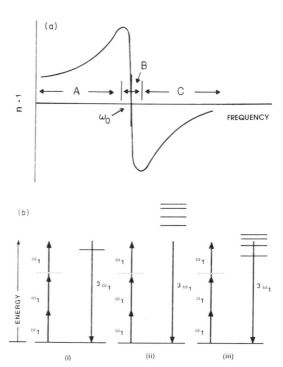

FIG. 11. (a) Variation of the refractive index near an allowed transition showing regions of normal dispersion (A), anomalous dispersion (B), and negative dispersion (C). (b) Energy level diagrams for third-harmonic generation that provide negative dispersion (i) and positive dispersion (ii and iii).

about 25 μm in the infrared. The most commonly used interactions of this type are given in Table VI, along with the form of the nonlinear polarization, the wavevector mismatch, and the plane-wave phase-matching conditions. These interactions include third- and higher-order harmonic conversion, in which an incident wave at frequency ω_1 is converted to a wave at frequency $q\omega_1$, and various forms of four- and six-wave frequency mixing in which radiation in two or more incident waves is converted to radiation in a wave at an appropriate sum- or difference-frequency combination as indicated in Table VI. Four-wave parametric oscillation, in which radiation at frequency ω_1 is converted to radiation at two other frequencies, ω_2 and ω_3, according to the relation

$$2\omega_1 = \omega_2 + \omega_3 \qquad (31)$$

has also been observed.

Third order and higher odd-order processes can be observed with electric dipole interactions in materials with any symmetry. They are used most commonly in materials that have a center

TABLE VI.   Nonlinear Polarizations for Third- and Higher-Order Parametric Frequency Conversion Processes

Nonlinear interaction	Process	Nonlinear polarization
qth Harmonic generation	$\omega_q = q\omega_1$	$P(q\omega) = \varepsilon_0\chi(-q\omega, \omega, \omega, \ldots, \omega)A^q(\omega)/2^{(q-1)}$
Four-wave sum-	$\omega_4 = 2\omega_1 + \omega_2$	$P(\omega_4) = 3\varepsilon_0\chi(-\omega_4, \omega_1, \omega_1, \omega_2)A^2(\omega_1)A(\omega_2)/4$
frequency mixing	$\omega_4 = \omega_1 + \omega_2 + \omega_3$	$P(\omega_4) = 3\varepsilon_0\chi(-\omega_4, \omega_1, \omega_2, \omega_3)A(\omega_1)A(\omega_2)A(\omega_3)/2$
Four-wave difference-	$\omega_4 = 2\omega_1 - \omega_2$	$P(\omega_4) = 3\varepsilon_0\chi(-\omega_4, \omega_1, \omega_1, -\omega_2)A^2(\omega_1)A^*(\omega_2)/4$
frequency mixing	$\omega_4 = \omega_1 + \omega_2 - \omega_3$	$P(\omega_4) = 3\varepsilon_0\chi(-\omega_4, \omega_1, \omega_2, -\omega_3)A(\omega_1)A(\omega_2)A^*(\omega_3)/2$
	$\omega_4 = \omega_1 - \omega_2 - \omega_3$	$P(\omega_4) = 3\varepsilon_0\chi(-\omega_4, \omega_1, -\omega_2, -\omega_3)A(\omega_1)A^*(\omega_2)A^*(\omega_3)/2$
Four-wave parametric	$2\omega_1 \rightarrow \omega_2 + \omega_3$	$P(\omega_2) = 3\varepsilon_0\chi(-\omega_2, \omega_1, \omega_1, -\omega_3)A^2(\omega_1)A^*(\omega_3)/4$
oscillation		$P(\omega_3) = 3\varepsilon_0\chi(-\omega_3, \omega_1, \omega_1, -\omega_2)A^2(\omega_1)A^*(\omega_2)/4$
Six-wave sum-	$\omega_6 = 4\omega_1 + \omega_2$	$P(\omega_6) = 5\varepsilon_0\chi(-\omega_6, \omega_1, \omega_1, \omega_1, \omega_1, \omega_2)A^4(\omega_1)A(\omega_2)/16$
frequency mixing		
Six-wave difference-	$\omega_6 = 4\omega_1 - \omega_2$	$P(\omega_6) = 5\varepsilon_0\chi(-\omega_6, \omega_1, \omega_1, \omega_1, \omega_1, -\omega_2)A^4(\omega_1)A^*(\omega_2)/16$
frequency mixing		

[a] Phase matching in mixtures or by angle.
[b] Phase optimization in single-component media.
[c] Requirement for optimized conversion.
[d] Positive dispersion allowed but not optimal.

of symmetry, such as gases, liquids, and some solids, since in these materials they are the lowest-order nonzero nonlinearities allowed by electric dipole transitions. Fourth-order and higher even-order processes involving electric dipole interactions are allowed only in crystals with no center of symmetry, and, although they have been observed, they are relatively inefficient and are seldom used for frequency conversion.

The intensity generated in these interactions can be calculated from Eq. (6) using the appropriate nonlinear polarization from Table VI. In the absence of pump depletion, the intensity generated in qth harmonic conversion varies as the qth power of the incident pump intensity, while the intensity generated in the frequency-mixing interactions varies as the product of the incident pump intensities with each raised to a power corresponding to its multiple in the appropriate frequency combination. The generated intensity in the plane wave configuration has the same $[(\sin \Delta k\,L/2)/(\Delta k\,L/2)]^2$ dependence on the wave-vector mismatch as do the second-order processes. The plane-wave phase-matching conditions for each of these interactions is the same as for the second-order interactions, namely, $\Delta k = 0$, but the requirements on the individual refractive indexes depend on the particular interaction involved, as indicated in Table VI.

Third- and higher-order frequency conversion are often done with beams that are tightly fo-

cused within the nonlinear medium to increase the peak intensity. In this situation, optimal performance can require either a positive or negative value of Δk, depending on the interaction involved. Phase-matching requirements with focused beams for the various interactions are also noted in Table VI.

The isotropic materials used for third- and higher-order parametric processes are not birefringent, and so alternative phase-matching techniques must be used. In gases, phase matching can be accomplished through use of the negative dispersion that occurs near allowed transitions as shown in Fig. 11. Normal dispersion occurs when the refractive index increases with frequency and is encountered in all materials when the optical frequency falls below, or sufficiently far away from, excited energy levels with allowed transitions to the ground state. Anomalous dispersion occurs in narrow regions about the transition frequencies in which the refractive index decreases with frequency, as shown in Fig. 11a. Negative dispersion occurs in a wavelength range above an allowed transition in which the refractive index, although increasing with frequency, is less than it is below the transition frequency. Regions of negative dispersion occur in restricted wavelength ranges above most, but not all, excited levels in many gases. Examples of the energy-level structures that give positive and negative dispersion for third-harmonic generation are shown in Fig. 11b.

Wave-vector mismatch	Plane-wave-matching condition	Dispersion requirement for focused beams in infinitely long media
$\Delta k = k(q\omega) - qk(\omega)$	$n(q\omega) = n(\omega)$	$\Delta k < 0$
$\Delta k = k(\omega_4) - 2k(\omega_1) - k(\omega_2)$	$n(\omega_4)/\lambda_4 = 2n(\omega_1)/\lambda_1 + n(\omega_2)/\lambda_2$	$\Delta k < 0$
$\Delta k = k(\omega_4) - k(\omega_1) - k(\omega_2) - k(\omega_3)$	$n(\omega_4)/\lambda_4 = n(\omega_1)/\lambda_1 + n(\omega_2)/\lambda_2 + n(\omega_3)/\lambda_3$	$\Delta k < 0$
$\Delta k = k(\omega_4) - 2k(\omega_1) + k(\omega_2)$	$n(\omega_4)/\lambda_4 = 2n(\omega_1)/\lambda_1 - n(\omega_2)/\lambda_2$	$\Delta k = 0^a$, $\Delta k \lesssim 0^b$
$\Delta k = k(\omega_4) - k(\omega_1) - k(\omega_2) + k(\omega_3)$	$n(\omega_4)/\lambda_4 = n(\omega_1)/\lambda_1 + n(\omega_2)/\lambda_2 - n(\omega_3)/\lambda_3$	$\Delta k = 0^a$, $\Delta k \lesssim 0^b$
$\Delta k = k(\omega_4) - k(\omega_1) + k(\omega_2) + k(\omega_3)$	$n(\omega_4)/\lambda_4 = n(\omega_1)/\lambda_1 - n(\omega_2)/\lambda_2 - n(\omega_3)/\lambda_3$	$\Delta k > 0$
$\Delta k = k(\omega_2) + k(\omega_3) - 2k(\omega_1)$	$2n(\omega_1)/\lambda_1 = n(\omega_2)/\lambda_2 + n(\omega_3)/\lambda_3$	$\Delta k = 0^a$, $\Delta k \lesssim 0^b$
$\Delta k = k(\omega_6) - 4k(\omega_1) - k(\omega_2)$	$n(\omega_6)/\lambda_6 = 4n(\omega_1)/\lambda_1 + n(\omega_2)/\lambda_2$	$\Delta k < 0$
$\Delta k = k(\omega_6) - 4k(\omega_1) + k(\omega_2)$	$n(\omega_6)/\lambda_6 = 4n(\omega_1)/\lambda_1 - n(\omega_2)/\lambda_2$	$\Delta k < 0^c$, $\Delta k > 0^d$

Phase matching can be accomplished by using a mixture of gases with different signs of dispersion. In this situation each component makes a contribution to the wave-vector mismatch in proportion to its concentration in the mixture. The value of the wave-vector mismatch can be controlled by adjusting the relative concentration of the two gases until the appropriate phase-matching condition is met for either collimated or focused beams, as shown in Fig. 12.

Phase matching can also be done in single-component media with focused beams, provided that the dispersion of the medium is of the correct sign for the interaction and wavelengths involved. With this technique the wave-vector mismatch depends on the density of the gas, and the pressure is adjusted until the proper wave-vector mismatch is achieved. Alternatively, phase matching in a single-component medium can be done by choosing the pump and gener-

ated frequencies to lie on either side of the transition frequency so that the phase-matching condition is satisfied. This technique is usually used in gases with plane-wave pump beams.

A fourth method for phase matching is the use of noncollinear waves, as shown in Fig. 13. This technique can be used for sum-frequency processes in media with negative dispersion and for difference-frequency processes in media with positive dispersion. It is commonly used, for example, in liquids for the difference frequency process $\omega_4 = 2\omega_1 - \omega_2$.

The conversion efficiency can be increased significantly if resonances are present between certain energy levels of the medium and the incident and generated frequencies or their sum or difference combinations. This increase in conversion efficiency is similar to the increase in linear absorption or scattering that occurs when the incident wavelength approaches an allowed transition of the medium. For the nonlinear interactions, however, a much greater variety of resonances is possible. Single-photon resonances occur between the incident or generated frequencies and allowed transitions just as with linear optical effects. The effectiveness of these resonances in enhancing nonlinear processes is limited, however, because of the absorption and dispersion that accompanies them.

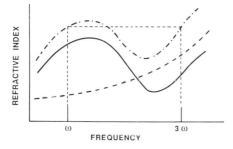

FIG. 12.   Illustration of the use of mixtures for phase matching. The solid curve gives the refractive index variation of a negatively dispersive medium, the dashed curve shows a positively dispersive medium, and the chain curve shows the refractive index of a mixture chosen so that the refractive index at ω is equal to that at 3ω.

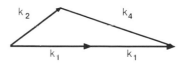

FIG. 13.   Off-axis phase-matching diagram for four-wave mixing of the form $\omega_4 = 2\omega_1 - \omega_2$.

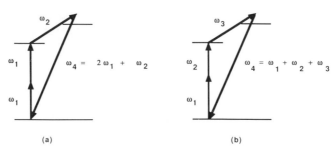

FIG. 14. Level diagrams showing two-photon resonances at (a) $2\omega_1$ and (b) $\omega_1 + \omega_2$ in four-wave sum-frequency mixing.

Resonances in the nonlinear effects also occur when multiples, or sum or difference combinations, of the incident frequencies match the frequency of certain types of transitions. The most commonly used of these resonances is a two-photon resonance involving two successive dipole transitions between levels whose spacing matches twice the value of an incident frequency or a sum or difference of two input frequencies, as indicated in Fig. 14. In single atoms and in centrosymmetric molecules, the energy levels involved in such two-photon resonances are of the same parity, and transitions between them are not observable in linear spectroscopy, which involves single-photon transitions. Near a two-photon resonance, the nonlinear susceptibility can increase by as much as four to eight orders of magnitude, depending on the relative linewidths of the two-photon transition and

the input radiation, resulting in a dramatic increase in the generated power as the input frequency is tuned through the two-photon resonance. An example of the increase in efficiency that is observed as the pump frequency is tuned through a two-photon resonance is shown in Fig. 15. Other higher-order resonances are also possible, but they have not been used as commonly as the two-photon resonances. Resonantly enhanced third-harmonic generation and four-wave mixing have proven very useful in allowing effective conversion of tunable radiation from dye lasers to the vacuum ultraviolet to providing high-brightness, narrow-band sources of radiation for high-resolution spectroscopy and other applications.

Some of the applications of third- and higher-order frequency conversion are given in Table VII. The qth harmonic generation is used to pro-

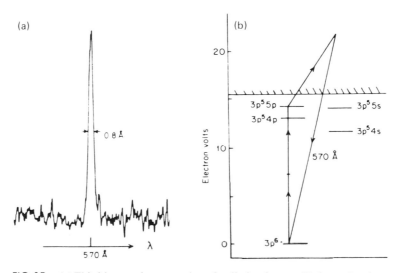

FIG. 15. (a) Third-harmonic conversion of radiation from an Xe laser showing enhancement as the wavelength of the laser is tuned through a two-photon resonance. (b) The resonant-level structure. [Reproduced from M. H. R. Hutchinson *et al.* (1976). *Opt. Commun.* **18,** 203. Copyright North-Holland, Amsterdam.]

TABLE VII. Selected Results for Third- and Higher-Order Frequency Conversion Processes

Interaction	Laser	Pump wavelength (nm)	Generated wavelength (nm)	Nonlinear material	Efficiency (%)
Third harmonic	CO_2	10.6 μm	3.5 μm	CO (liquid), CO (gas), BCl_3, SF_6, NO, DCl	8 (liquid CO)
	Nd:YAG	1.064 μm	354.7	Rb, Na	10
	Nd:YAG	354.7	118.2	Xe	0.2
	Xe_2	170	57	Ar	
	Dye	360–600	120–200	Xe, Kr, Sr, Mg, Zn, Hg	Up to 1
Fifth harmonic	Nd:YAG	266	53.2	He, Ne, Ar, Kr	10^{-3}
	XeCl	308	61.6	He	
	KrF	249	49.8	He	
	ArF	193	38.6	He	
Seventh harmonic	Nd:YAG	266	38	He	10^{-4}
	KrF	249	35.5	He	10^{-9}

duce radiation at a frequency that is q times the incident frequency. The most commonly used interaction of this type is third-harmonic conversion. It has been used to produce radiation at wavelengths ranging from the infrared to the extreme ultraviolet. Third-harmonic conversion of radiation from high power pulsed lasers such as CO_2, Nd:glass, Nd:YAG, ruby, and various rare-gas halide and rare-gas excimer lasers has been used to generate fixed-frequency radiation at various wavelengths ranging from 3.5 μm to 57 nm, as indicated in Table VII. It has also been used with dye lasers to generate radiation tunable in spectral bands between 110 and 200 nm. The extent of the spectral bands generated in this manner is determined by the extent of the negative dispersion region in the nonlinear materials.

Four-wave sum- and difference-frequency mixing interactions of the form

$$\omega_4 = 2\omega_1 \pm \omega_3 \qquad (32a)$$

and

$$\omega_4 = \omega_1 + \omega_2 \pm \omega_3 \qquad (32b)$$

where ω_1, ω_2, and ω_3 are input frequencies, are also commonly used to produce radiation in wavelength ranges that are inaccessible by other means. These processes can be favored over possible simultaneous third-harmonic generation by the use of opposite circular polarization in the two pump waves, since third-harmonic conversion with circularly polarized pump light is not allowed by symmetry. They have been used to generate radiation over a considerable range of wavelengths in the vacuum ultraviolet,

extreme ultraviolet, and the mid infrared. In particular, they have been used to generate tunable radiation over most of the vacuum ultraviolet range from 100 to 200 nm.

These interactions can be used to generate tunable radiation in resonantly enhanced processes, thereby increasing the efficiency of the process. In this situation the pump frequency at ω_1 or the sum combination $\omega_1 + \omega_2$ is adjusted to match a suitable two-photon resonance, while the remaining pump frequency at ω_3 is varied, producing the tunable generated radiation, as illustrated in Fig. 16. The difference frequency processes $\omega_4 = 2\omega_1 - \omega_3$ and $\omega_4 = \omega_1 + \omega_2 - \omega_3$ can be optimized with focused beams in media with either sign of dispersion. As a result, their

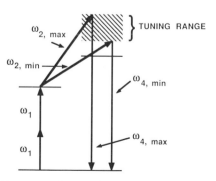

FIG. 16. Level diagram for producing tunable radiation with two-photon resonantly enhanced four-wave sum-frequency mixing. Pump radiation at ω_1 is tuned to the two-photon resonance, pump radiation at ω_2 is tuned over the range $\omega_{2,min}$ to $\omega_{2,max}$, and the frequency of the generated radiation tunes over the range $(2\omega_1 + \omega_{2,min})$ to $(2\omega_1 + \omega_{2,max})$.

usefulness is not restricted to narrow wavelength ranges above dispersive resonances, and they have been used to generate tunable radiation over extensive ranges in the vacuum ultraviolet between 140 and 200 nm in the rare gases Xe and Kr. Tunable radiation generated in this manner in Xe between 160 and 200 nm is illustrated in Fig. 17.

The difference-frequency processes

$$\omega_4 = 2\omega_1 - \omega_3 \tag{33a}$$

and

$$\omega_4 = \omega_1 \pm \omega_2 - \omega_3 \tag{33b}$$

have also been used to generate tunable radiation in the infrared by using pump radiation from visible and near infrared lasers. In some interactions all the frequencies involved in the mixing processes are supplied externally and in others some of them are generated in a stimulated Raman interaction (see Section III,A,2). Because the gases used for these nonlinear interactions are not absorbing at far-infrared wavelengths, it can be expected that they will ultimately allow more efficient generation of tunable far-infrared radiation using pump radiation in the visible and near infrared than can be achieved in second-order interactions in crystals, although they have not yet been extended to as long wavelengths. Ultimately the limitations on conversion can be expected to arise from difficulties

with phase matching and a mismatch between the diffraction of the pump and generated wavelengths. To date, the four-wave difference-frequency mixing interactions have been used to produce coherent radiation at wavelengths out to 25 μm.

Resonances between Raman active molecular vibrations and rotations and the difference frequency combination $\omega_1 - \omega_3$ can also occur. When the four-wave mixing process $2\omega_1 - \omega_3$ or $\omega_1 + \omega_2 - \omega_3$ is used with these resonances it is termed coherent anti-Stokes Raman scattering (CARS). The resonant enhancement that occurs in the generated intensity as the pump frequencies are tuned through the two-photon difference-frequency resonance forms the basis of CARS spectroscopy (see Section IV,A).

Various forms of higher-order interactions are also used for frequency conversion. These consist primarily of harmonic conversion up to order seven and six-wave mixing interactions of the form $\omega_6 = 4\omega_1 \pm \omega_2$, although harmonic generation up to order 11 has been reported. Generally, the conversion efficiency in the higher-order processes is lower than it is in the lower-order ones, and the required pump intensity is higher. As a result, higher-order processes have been used primarily for the generation of radiation in the extreme ultraviolet at wavelengths too short to be reached with lower-order interactions. The pump sources have for

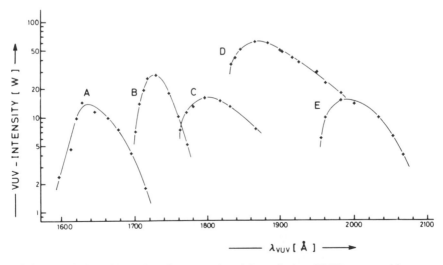

FIG. 17.   Variation of intensity of vacuum ultraviolet radiation (VUV) generated in xenon through the process $2\omega_1 - \omega_2$ in the range 160–200 nm. The different regions A–E correspond to pump radiation obtained from different dye lasers. The radiation is continuously variable within each of the regions. [Reproduced from R. Hilbig and R. Wallenstein (1982). *Appl. Opt.* **21,** 913.]

the most part involved mode-locked lasers with pulse durations under 30 psec and peak power levels in excess of several hundred megawatts.

Fifth-harmonic conversion has been used to generate radiation at wavelengths as short as 38.6 nm using radiation from an ArF laser at 193 nm, seventh-harmonic conversion has been used to generate radiation at wavelengths as short as 35.5 nm with radiation from a KrF laser, and ninth-harmonic conversion has been used to generate radiation at 117.7 nm with radiation from a Nd : glass laser at 1.06 μm. Radiation at various wavelengths in the extreme ultraviolet between 38 and 76 nm using fifth- and seventh-harmonic generation and six-wave mixing of radiation from harmonics of a Nd : YAG laser has also been generated.

Observed conversion efficiencies for many of the third- and higher-order processes are noted in Table VII. They range from about 10% for third-harmonic conversion of Nd : YAG laser radiation in rubidium (1.064 μm $\rightarrow$ 354.7 nm) or CO_2 laser radiation in liquid CO (9.6 μm $\rightarrow$ 3.2 μm) to values of the order of 10^{-11} for some of the higher-order processes. The pump intensities used vary between several hundred kilowatts per square centimeter for resonantly enhanced processes to 10^{15} W/cm^2 for nonresonant processes.

The largest conversion efficiencies that can be achieved with third- and higher-order processes are generally less than those that can be obtained with second-order interactions, because competing processes that limit efficiency are more important for the higher-order interactions. Some of the important limiting processes are listed in Table VIII, along with situations for which they are important. As a result the higher-order processes are most useful in generating radiation in spectral regions, such as the vacuum ultraviolet or far infrared, that are inaccessible by the second-order interactions, or for certain applications such as phase conjugation or spectroscopy.

2. Stimulated Scattering Processes

Stimulated scattering processes are nonlinear interactions in which an incident wave at frequency ω_{inc} is converted to a scattered wave at a different frequency ω_{scat}. The difference in photon energy between the incident and scattered frequencies is taken up or supplied by the nonlinear medium, which undergoes a transition be-

TABLE VIII.   Competing Processes for Third- and Higher-Order Frequency Conversion[a]

Competing process	Effect on conversion efficiency	Conditions under which competing process can be expected to be important
Linear absorption of generated radiation	Loss of generated power Reduction of improvement from phase matching Limitation on product NL	UV or XUV generation in ionizing continuum of nonlinear medium Generated wavelength close to allowed transition
Nonlinear absorption of pump radiation	Loss of pump power Saturation of susceptibility Disturbance of phase matching conditions Self focusing or self defocusing	Two-photon resonant interactions
Stark shift	Saturation of susceptibility Self focusing or self defocusing Disturbance of phase-matching conditions	Resonant or near-resonant interactions, with pump intensity close to or greater than the appropriate saturation intensity
Kerr effect	Disturbance of phase-matching conditions Self-focusing or self defocusing	Nonresonant interactions Near-resonant interactions when the pump intensity is much less than the saturation intensity
Dielectric breakdown, multiphoton ionization	Disturbance of phase-matching conditions Saturation of susceptibility Loss of power at pump or generated wavelength	Conversion in low-pressure gases at high intensities Tightly focused geometries

[a] Reproduced from J. Reintjes (1985). Coherent ultraviolet and vacuum ultraviolet sources, in "Laser Handbook," Vol. 5 (M. Bass and M. L. Stitch, eds.). North-Holland, Amsterdam.

tween two of its internal energy levels, as illustrated in Fig. 18.

If the medium is initially in its ground state, the scattered wave is at a lower frequency (longer wavelength) than the incident wave, and the medium is excited to one of its internal energy levels during the interaction. In this situation the frequency shift is termed a Stokes shift, in analogy to the shift to lower frequencies that is observed in fluorescence (which was explained by Sir George Stokes), and the scattered wave is termed a Stokes wave. The incident (laser) and scattered (Stokes) frequencies are related by

$$\omega_S = \omega_L - \omega_0 \tag{34}$$

where ω_L and ω_S are the frequencies of the laser and Stokes waves and ω_0 is the frequency of the internal energy level of the medium.

If the medium is initially in an excited state, the scattered wave is at a higher frequency (shorter wavelength) than the incident wave and the medium is deexcited during the interaction, with its energy being given to the scattered wave. In this situation the scattered wave is termed an anti-Stokes wave (shifts to higher frequencies are not possible in fluorescence, as explained by Stokes), and the frequency shift is termed the anti-Stokes shift. The laser and anti-Stokes frequencies are related by

$$\omega_{AS} = \omega_L + \omega_0 \tag{35}$$

Various types of stimulated scattering processes are possible, each involving a different type of internal excitation. Some of the more common ones are listed in Table IX, along with the types of internal excitations and the types of materials in which they are commonly observed. Stimulated Brillouin scattering involves interactions with sound waves in solids, liquids, or gases or ion-acoustic waves in plasmas, and stimulated Rayleigh scattering involves interactions with density or orientational fluctuations of molecules. Various forms of stimulated Raman scattering can involve interaction with molecular vibrations, molecular rotations (rotational Raman scattering), electronic levels of atoms or molecules (electronic Raman scattering), lattice vibrations, polaritons, electron plasma waves, or nondegenerate spin levels in certain semiconductors in magnetic fields. The magnitude of the frequency shifts that occur depends on the combination of nonlinear interaction and the particular material that is used. Orders of magnitude for shifts in different types of materials for the various interactions are also given in Table IX.

Stimulated scattering processes that involve Stokes waves arise from third-order nonlinear interactions with nonlinear polarizations of the form

$$P(\omega_S) = \tfrac{3}{2}\varepsilon_0\chi^{(3)}(-\omega_S, \omega_L, -\omega_L, \omega_S)|A_L|^2 A_S \tag{36}$$

The nonlinear susceptibility involves a two-photon resonance with the difference frequency

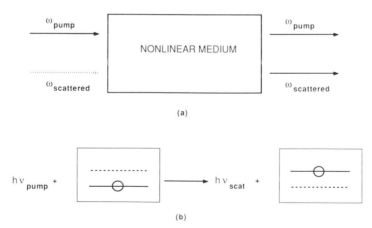

(a)

(b)

FIG. 18.   (a) Schematic illustration of a typical stimulated scattering interaction. The scattered wave can be supplied along with the pump radiation or it can be generated from n ⬤ ⁚ in the interaction. (b) Representation of a stimulated Stokes scattering interaction, illustrating the transitions in the nonlinear medium.

TABLE IX.   Stimulated Scattering Interactions

Interaction	Internal mode	Type of material	Typical range of frequency shift $\Delta\nu$ (cm^{-1})
Stimulated Raman scattering	Molecular vibrations	Molecular gases	600–4150
		Molecular liquids	600–3500
	Molecular rotations	Molecular gases	10–400
	Electronic levels	Atomic and molecular gases, semiconductors	7000–20,000
	Lattice vibrations	Crystals	10–400
	Polaritons	Crystals	10–400
	Electron plasma waves	Plasmas	100–10,000
Stimulated Brillouin scattering	High-frequency sound waves	Solids, liquids, gases	100 MHz–10 GHz
	Ion-acoustic waves	Plasmas	0.1–10 MHz
Stimulated Rayleigh scattering	Density fluctuations, orientational fluctuations	Gases, liquids	0.1–1

combination $\omega_L - \omega_S$. As with other two-photon resonances, the energy levels that are involved are of the same parity as those of the ground state in atoms or molecules with a center of inversion. When the two-photon resonance condition is satisfied, the susceptibility for the stimulated scattering interactions is negative and purely imaginary and can be written as

$$\chi^{(3)}(-\omega_S, \omega_L, -\omega_L, \omega_S)$$
$$= -i\chi''^{(3)}(-\omega_S, \omega_L, -\omega_L, \omega_S) \qquad (37)$$

Stimulated scattering processes can be in the forward direction, involving a scattered wave that propagates in the same direction as the incident laser, as shown in Fig. 18, or in the backward direction, involving a scattered wave that propagates in the opposite direction to the incident laser, as shown in Fig. 19. The field amplitude of the wave generated in the forward direction by a susceptibility of the type in Eq. (37) is described by the equation

$$dA_S/dz = (3\omega_S/4n_Sc)\chi''|A_L|^2 A_S \qquad (38)$$

When pump depletion is negligible, the intensity of the Stokes wave is given by

$$I_S = I_S(0)e^{gIL} \qquad (39)$$

where the quantity g is the gain coefficient of the process given by

$$g = (3\omega_S/n_S n_L c^2 \varepsilon_0)\chi'' \qquad (40)$$

The waves generated in stimulated scattering processes have exponential growth, in contrast with the power-law dependences of the waves generated in the parametric interactions. The exponential gain is proportional to the propagation distance and to the intensity of the pump radiation. Phase matching is generally not required for stimulated scattering processes, since the phase of the material excitation adjusts itself for maximum gain automatically.

Photon number is conserved in stimulated scattering interactions, with one photon being lost in the pump wave for every one created in the Stokes wave. The energy created in the Stokes wave is smaller than that lost in the pump

FIG. 19.   Backward stimulated scattering in which the scattered radiation propagates in the direction opposite to the pump radiation. The scattered wave either can be supplied along with the pump radiation, or can be generated from noise in the interaction.

wave by the ratio ω_S/ω_L, termed the Manly–Rowe ratio, and the difference in photon energy between the pump and Stokes waves represents the energy given to the medium.

When the energy in the Stokes wave becomes comparable to that in the incident laser pump, depletion occurs and the gain is reduced. In principle, every photon in the incident pump wave can be converted to a Stokes photon, giving a maximum theoretical conversion efficiency of

$$\eta_{max} = [I_S(L)/I_L(0)] = \omega_S/\omega_L \qquad (41)$$

In practice, the efficiency is never as high as indicated in Eq. (41) because of the lower conversion efficiency that is present in the low-intensity spatial and temporal wings of most laser beams. Photon conversion efficiencies of over 90% and energy-conversion efficiencies over 80% have, however, been observed in certain stimulated Raman and Brillouin interactions.

a. Stimulated Raman Scattering. Stimulated Raman scattering can occur in solids, liquids, gases, and plasmas. It involves frequency shifts ranging from several tens of reciprocal centimeters for rotational scattering in molecules to tens of thousands of reciprocal centimeters for electronic scattering in gases. Forward stimulated Raman scattering is commonly used for generation of coherent radiation at the Stokes

wavelength, amplification of an incident wave at the Stokes wavelength, reduction of beam aberrations, nonlinear spectroscopy (see Section IV,C), and generation of tunable infrared radiation through polariton scattering. Backward stimulated Raman scattering can be used for wave generation, amplification, pulse compression and phase conjugation (see Section IV,A).

The susceptibility for stimulated Raman scattering in molecules or atoms is given by

$$\chi = -\left[\frac{i}{6\Gamma\hbar^3}\left(1 - \frac{i\Delta}{\Gamma}\right)\right]$$

$$\times \left[\sum \mu_{0i}\mu_{i2}\left\{\frac{1}{\omega_{i0} - \omega_L} + \frac{1}{\omega_{i0} + \omega_S}\right\}\right]^2 \quad (42)$$

where $\Delta = \omega_0 - (\omega_L - \omega_S)$ is the detuning from the Raman resonance, Γ the linewidth of the Raman transition, ω_{i0} the frequency of the transition from level i to level 0, and μ_{0i} the dipole moment for the transition between levels 0 and i.

Gain coefficients and frequency shifts for some materials are given in Table X.

Amplification of an incident Stokes wave generally occurs for exponential gains up to about e^8 to e^{10}, corresponding to small signal amplifications of the order of 3000 to 22,000, although under some conditions stable gains up to e^{19} can be obtained. Raman amplifiers of this type are

TABLE X.   Stimulated Raman Shifts and Gain Coefficients at 694.3 nm

Material	$\Delta\nu_R$ (cm^{-1})	$g \times 10^3$ (cm/MW)	
Liquids			
Carbon disulfide	656	24	
Acetone	2921	0.9	
Methanol	2837	0.4	
Ethanol	2928	4.0	
Toluene	1002	1.3	
Benzene	992	3	
Nitrobenzene	1345	2.1	
N_2	2326	17	
O_2	1555	16	
Carbon tetrachloride	458	1.1	
Water	3290	0.14	
Gases			
Methane	2916	0.66	(10 atm, 500 nm)
Hydrogen	4155 (vibrational)	1.5	(above 10 atm)
	450 (rotational)	0.5	(above 0.5 atm)
Deuterium	2991 (vibrational)	1.1	(above 10 atm)
N_2	2326	0.071	(10 atm, 500 nm)
O_2	1555	0.016	(10 atm, 500 nm)

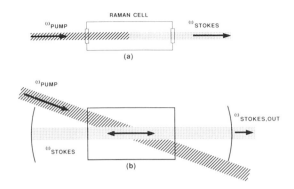

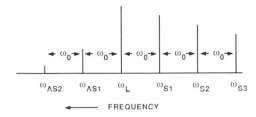

FIG. 21. Schematic illustration of the spectrum produced by multiple Stokes and anti-Stokes scattering.

FIG. 20. (a) Generation of a Stokes wave in single-pass stimulated Raman scattering. (b) Generation of a Raman–Stokes wave in a Raman laser oscillator.

used to increase the power in the Stokes beam. When the pump beam has a poor spatial quality due to phase or amplitude structure, Raman amplification can be used to transfer the energy of the pump beam to the Stokes beam without transferring the aberrations, thereby increasing the effective brightness of the laser system.

Generation of a new wave at the Stokes frequency can be done in a single pass geometry, as shown in Fig. 20a, or in an oscillator configuration at lower gains, as shown in Fig. 20b. The threshold for single-pass generation depends on the degree of focusing that is used but is generally certain for gains above about e^{23} (10^{10}). Once the threshold for single-pass generation is reached, the process generally proceeds to pump depletion very quickly. As a result, the Stokes frequency that is generated in a Raman oscillator usually involves the Raman-active mode with the highest gain.

If sufficient conversion to the Stokes wave takes place, generation of multiple Stokes waves can occur. In this situation the first Stokes wave serves as a pump for a second Stokes wave that is generated in a second stimulated Raman interaction. The second Stokes wave is shifted in frequency from the first Stokes wave by ω_0. If sufficient intensity is present in the original pump wave, multiple Stokes waves can be generated, each shifted from the preceding one by ω_0 as illustrated in Fig. 21. Stimulated Raman scattering can thus be used as a source of coherent radiation at several wavelengths by utilizing multiple Stokes shifts, different materials, and different pump wavelengths.

Continuously tunable Stokes radiation can be generated by using a tunable laser for the pump radiation or in some situations by using materials with internal modes whose energies can be changed. An example of this type of interaction is the generation of tunable narrow-band infrared radiation from spin-flip Raman lasers in semiconductors, as illustrated in Fig. 22. In these lasers the energy levels of electrons with different spin orientations are split by an external magnetic field. The Raman process involves a transition from one spin state to the other. For a fixed pump wavelength—for example, from a CO_2 laser at 10.6 μm—the wavelength of the Stokes wave can be tuned by varying the magnetic field, which determines the separation of the electron energy levels. Multiple Stokes shifts and anti-Stokes shifts can also be obtained. A list of materials, laser sources, and tuning ranges for various spin-flip lasers is given in Table XI. Radiation that is tunable in bands between 5.2 and 16.2 μm has been generated in

TABLE XI. Spin-Flip Raman Lasers

Pump laser (wavelength, μm)	Material	Tuning range (μm)	Raman order
NH_3 (12.8)	InSb	13.9–16.8	I Stokes
CO_2 (10.6)	InSb	9.0–14.6	I, II, III, Stokes I, Anti-Stokes
CO (5.3)	InSb	5.2–6.2	I, II, III, Stokes I, Anti-Stokes
HF	InAs	3–5	

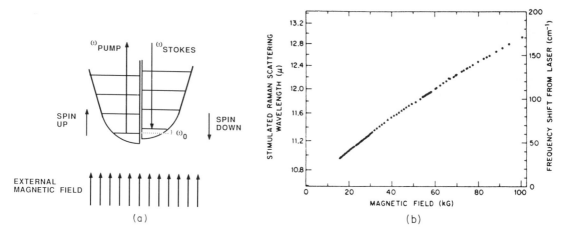

FIG. 22.   (a) Level diagram for a spin-flip Raman laser. (b) Typical tuning curve for an InSb spin-flip Raman laser pumped by a CO_2 laser at 10.6 μm. [Part (b) reproduced from Patel, C. K. N., and Shaw, E. D. (1974). *Phys. Rev. B* **3**, 1279.]

this manner with linewidths as narrow as 100 MHz. This is among the narrowest bandwidth infrared radiation available and has been used for high resolution spectroscopy.

Backward stimulated Raman scattering involves the generation or amplification of a Stokes wave that travels in the opposite direction to the pump wave, as was shown in Fig. 19. Backward stimulated Raman scattering requires radiation with a much narrower bandwidth than does forward Raman scattering and is usually observed for laser bandwidths less than about 10 GHz. The backward traveling Stokes wave encounters a fresh undepleted pump wave as it propagates through the nonlinear medium. As a result, the gain does not saturate when the pump wave is depleted, as it does in the forward direction. The peak Stokes intensity can grow to be many times that of the incident laser intensity, while the duration of the Stokes pulse decreases relative to that of the incident pump pulse, allowing conservation of energy to be maintained. This process, illustrated in Fig. 23,

is termed pulse compression. Compression ratios of the order of 50 : 1, producing pulses as short as a few nanoseconds, have been observed.

Anti-Stokes Raman scattering involves the generation of radiation at shorter wavelengths than those of the pump wave. Anti-Stokes scattering can occur in one of two ways. The more common method involves a four-wave difference frequency mixing process of the form

$$\omega_{AS} = 2\omega_L - \omega_S \qquad (43a)$$

(see Section III,A,1,b) in media without a population inversion. In this interaction the Stokes radiation is generated with exponential gain through the stimulated Raman interaction as described above. The anti-Stokes radiation is generated by the four-wave mixing process, using the Stokes radiation as one of the pump waves. The anti-Stokes radiation that is generated through this interaction grows as part of a mixed mode along with the Stokes radiation. It has the same exponential gain as does the Stokes radia-

FIG. 23.   Illustration of pulse compression with backward stimulated Raman scattering. Backward-traveling Stokes pulse sweeps out the energy of the pump in a short pulse.

tion, but the amplitude of the anti-Stokes radiation depends on the phase mismatch just as for other four-wave mixing interactions. The anti-Stokes generation is strongest for interactions that are nearly phase matched, although neither wave has exponential gain at exact phase matching. For common liquids and gases that have normal dispersion, the anti-Stokes process is not phase matched in the forward direction but is phase matched when the Stokes and anti-Stokes waves propagate at angles to the pump wave, as shown schematically in Fig. 24a. The anti-Stokes radiation is thus generated in cones about the pump radiation in most of these materials. The opening angle of the cone depends on the dispersion in the medium and on the frequency shift. It is of the order of several tens of milliradians for molecular vibrational shifts in liquids and can be considerably smaller for rotational shifts in molecular gases. An example of anti-Stokes emission in H_2 is shown in Fig. 24b. Here the pump radiation was at 532 nm. The anti-Stokes radiation at 435.7 nm was concentrated near the phase-matching direction of

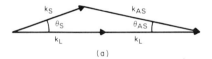

(a)

about 7 mrad, with a dark band appearing at the exact phase-matching direction and bright emission appearing in narrow bands on either side of phase matching.

Multiple anti-Stokes generation can occur through interactions of the form

$$\omega_{AS,n} = \omega_1 + \omega_2 - \omega_3 \quad (43b)$$

where ω_1, ω_2, ω_3 are any of the Stokes, anti-Stokes, or laser frequencies involved in the interaction that satisfy the relations

$$\omega_1 - \omega_3 = \omega_0 \quad (44a)$$

$$\omega_{AS,n} - \omega_2 = \omega_0 \quad (44b)$$

Just as with multiple Stokes generation, the successive anti-Stokes lines are shifted from the preceding one by ω_0 as shown in Fig. 21. Multiple Stokes and anti-Stokes Raman scattering in molecular gases have been used to generate radiation ranging from 138 nm in the ultraviolet to wavelengths in the infrared. Some of the combinations of lasers and materials are listed in Table XII.

In media with a population inversion between the ground state and an excited Raman-active level, radiation at the anti-Stokes wavelength can be produced through a process similar to the one just described for Stokes generation in media that start from the ground state. This combination, illustrated in Fig. 25, is termed an anti-Stokes Raman laser. It has been used to generate radiation at wavelengths ranging from 149 to 378 nm using transitions in atomic gases such as Tl, I, or Br.

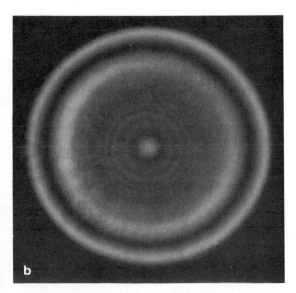

FIG. 24. (a) Off-axis phase-matching diagram for anti-Stokes generation. (b) Anti-Stokes rings produced in stimulated Raman scattering from hydrogen. The dark band in the anti-Stokes ring is caused by suppression of the exponential gain due to the interaction of the Stokes and anti-Stokes waves. [Reproduced from M. O. Duncan *et al.* (1986). *Opt. Lett.* **11**, 803. Copyright © 1986 by the Optical Society of America.]

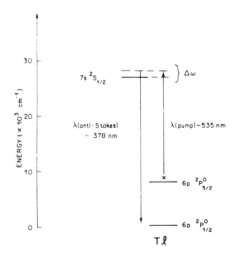

FIG. 25. Level diagram of anti-Stokes Raman laser in Tl. [Reprinted from White, J. C., and Henderson, D. (1982). *Opt. Lett.* **7**, 517. Copyright © 1982 by the Optical Society of America.]

TABLE XII. Wavelength Range (nm) of Tunable UV and VUV Radiation Generated by Stimulated Raman Scattering in H_2 ($\Delta\nu$ = 4155 cm^{-1}) with Dye Lasers[a]

Process (order)	Pump wavelength (nm)					
	600–625 (Rh 101)[b]	570–600 (Rh B)[c]	550–580 (Rh 6G)[b]	300–312.5 (Rh 101, SH)[b]	275–290 (Rh 6G, SH)[b]	548 (Fluorescein 27)[c]
AS (13)						138
AS (12)						146
AS (11)						156
AS (10)						167
AS (9)	185.0–187.3					179
AS (8)	200.4–203.1	196.9–200.4				194
AS (7)	218.6–221.8	214.4–218.6	194.5–198.1			211
AS (6)	240.4–244.3	235.4–240.4	211.6–215.9			231
AS (5)	267.1–271.9	261.0–267.1	256.7–263.0	184.8–189.5		256
AS (4)	300.4–306.6	292.7–300.4	287.3–295.3	200.2–205.7	188.7–195.7	286
AS (3)	343.3–351.3	332.2–343.3	326.3–336.6	218.3–224.9	204.8–213.0	325
AS (2)			377.5–391.4	240.1–248.1	223.8–233.7	376
AS (1)				266.7–276.6	246.8–258.8	
S (1)				342.7–359.1	310.5–329.7	
S (2)					356.5–382.1	

[a] From J. Reintjes (1985). Coherent ultraviolet and vacuum ultraviolet sources, *in* "Laser Handbook," Vol. 5 (M. Bass and M. L. Stitch, eds.), p. 1. North-Holland, Amsterdam.
[b] Data from Wilke and Schmidt (1979). *Appl. Phys.* **18**, 177.
[c] Data from Schomburg *et al.* (1983). *Appl. Phys. B* **30**, 131.

b. Stimulated Brillouin Scattering. Stimulated Brillouin scattering (SBS) involves scattering from high-frequency sound waves. The gain for SBS is usually greatest in the backward direction and is observed most commonly in this geometry, as shown in Fig. 26a. The equations describing backward SBS are

$$dI_B/dz = -g_B I_L I_B \tag{45a}$$

$$dI_L/dz = -g_B(\omega_L/\omega_B)I_L I_B \tag{45b}$$

where the intensity gain coefficient g_B is given by

$$g_B = \frac{\omega_B^2 \rho_0 (\partial\varepsilon/\partial\rho)^2}{4\pi c^3 n\nu\Gamma_B\varepsilon_0} \tag{45c}$$

where ρ is the density, v the velocity of sound, $\partial\varepsilon/\partial\rho$ the change of the dielectric constant with density, and Γ_B is the linewidth of the Brillouin transition. The wave-vector diagram is shown in Fig. 26b. The incident and generated optical waves are in the opposite directions, and the wave vector of the acoustic wave is determined by the momentum matching condition

$$\boldsymbol{k}_v = \boldsymbol{k}_L - \boldsymbol{k}_S \approx 2\boldsymbol{k}_L \tag{46}$$

where $\boldsymbol{k}_v$ is the $\boldsymbol{k}$ vector of the sound wave and $\boldsymbol{k}_L$ and $\boldsymbol{k}_S$ are the wave vectors of the laser and scattered waves. Because the speed of light is so much greater than the speed of sound, the magnitude of the $\boldsymbol{k}$ vector of the incident wave is almost equal to that of the scattered wave. The corresponding frequency shift of the scattered wave, termed the Brillouin shift, is equal to the frequency of the sound wave generated in the interaction and is given by

$$\Delta\omega_B = 2\omega_L v/c \tag{47}$$

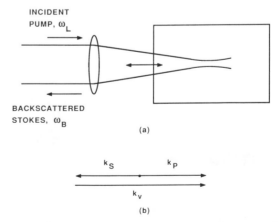

FIG. 26. (a) Schematic diagram of a typical configuration used for stimulated Brillouin scattering. (b) The k-vector diagram for stimulated Brillouin scattering.

where v is the velocity of sound and we have used the approximation that $k_L = -k_S$.

Stimulated Brillouin scattering can be observed in liquids, gases, and solids, and also in plasmas, in which the scattering is from ion-acoustic waves. The SBS shift generally ranges from hundreds of megahertz in gases to several tens of gigahertz in liquids, depending on the particular material and laser wavelength involved. The SBS shifts for some materials are given in Table XIII. The threshold intensity for SBS ranges from 10^8 to 10^{10} W/cm^2, depending on the nonlinear material and the focusing conditions, and maximum reflectivities (the ratio of the intensity of the scattered wave to that of the incident pump wave) commonly exceed 50%. The acoustic waves generated in these interactions are among the most intense high-frequency sound waves generated.

Because the response time of the acoustic phonons is relatively long (of the order of 1 nsec in liquids and up to several hundred nanoseconds in some gases), SBS is observed most commonly with laser radiation with a bandwidth less than 0.1 cm^{-1}. Generally, SBS has a higher gain than stimulated Raman scattering in liquids and usually dominates the interaction for wave generation when the laser radiation consists of narrow-band pulses that are longer than the response time of the acoustic phonon. Stimulated Raman scattering is commonly observed in these materials only for relatively broad-band radiation (for which the SBS interaction is suppressed), for short-duration pulses (for which SBS does not have time to grow), or at the beginning of longer-duration, narrow-band pulses.

In liquids and gases, SBS is used most commonly for phase conjugation (see Section IV,C) and for pulse compression in a manner similar to that described above for stimulated Raman scattering. SBS in solids can also be used for these purposes but is less common because the materials can be easily damaged by the acoustic wave that is generated in the medium.

B. SELF-ACTION EFFECTS

Self-action effects are those that affect the propagation characteristics of the incident light beam. They are due to nonlinear polarizations that are at the same frequency as that of the incident light wave. Depending on the particular effect, they can change the direction of propagation, the degree of focusing, the state of polarization or the bandwidth of the incident radiation, as was indicated in Table I. Self-action effects can also change the amount of absorption of the incident radiation. Sometimes one of these effects can occur alone, but more commonly two or more of them occur simultaneously.

The most common self-action effects arise from third-order interactions. The nonlinear polarization has the form

$$P(\omega) = \tfrac{3}{4}\varepsilon_0 \chi^{(3)}(-\omega, \omega, -\omega, \omega)|A|^2 A \quad (48)$$

The various types of self-action effects depend on whether the susceptibility is real or imaginary and on the temporal and spatial distribution of the incident light. Interactions that change the polarization vector of the radiation depend on the components of the polarization vector present in the incident radiation, as well as on the tensor components of the susceptibility.

The real part of the nonlinear susceptibility in Eq. (48) gives rise to the spatial effects of self-focusing and self-defocusing, spectral broadening, and changes in the polarization vector. The imaginary part of the susceptibility causes nonlinear absorption.

1. Spatial Effects

The real part of the third-order susceptibility in Eq. (48) causes a change in the index of refraction of the material according to the relation

$$n = n^L + n_2 \langle E^2 \rangle = n^L + \tfrac{1}{2} n_2 |A|^2 \quad (49)$$

where

$$n_2 = (3/4n^L)\chi' \quad (50)$$

In these equations, $\langle E^2 \rangle$ is the time average of the square of the total electric field of Eq. (3), which is proportional to the intensity, n^L is the linear refractive index, χ' is the real part of χ, and n_2 is termed the nonlinear refractive index.

TABLE XIII. Stimulated Brillouin Shifts and Gain Coefficients at 1.064 μm

Material	$\Delta\nu_B$ (GHz)	g (cm/MW)
Carbon disulfide	3.84	0.13–0.16
Methanol	2.8	0.014
Ethanol	3	0.012
Toluene	3.9	0.013
Benzene	4.26	0.021
Acetone	3	0.019
n-Hexane	2.8	0.023
Cyclohexane	3.66	0.007
Carbon tetrachloride	1.9	0.007
Water	3.75	0.006

Self-focusing occurs as a result of a combination of a positive value of n_2 and an incident beam that is more intense in the center than at the edge, a common situation that occurs, for example, as a result of the spatial-mode structure of a laser. In this situation the refractive index at the center of the beam is greater than that at its edge and the optical path length for rays at the center is greater than that for rays at the edge. This is the same condition that occurs for propagation through a focusing lens, and as a result the light beam creates its own positive lens in the nonlinear medium. As the beam focuses, the strength of the nonlinear lens increases, causing stronger focusing and increasing the strength of the lens still further. This behavior results in catastrophic focusing, in which the beam collapses to a very intense, small spot, in contrast to the relatively gentle focusing that occurs for normal lenses, as illustrated in Fig. 27.

Self-focusing can occur in any transparent material at sufficiently high intensities and has been observed in a wide range of materials, including glasses, crystals, liquids, gases, and plasmas. The mechanism causing self-focusing varies from one material to another. In solids and some gases the nonlinear index is due to interaction with the electronic energy levels which causes a distortion of the electron cloud, which results in an increase in the refractive index. In materials such as molecular liquids with anisotropic molecules, the nonlinear index arises from orientation of the molecules so that their axis of easy polarization is aligned more closely along the polarization vector of the incident field, as shown in Fig. 28. In such materials the molecules are normally arranged randomly, resulting in an isotropic refractive index. When the molecules line up along the optical field, the polarizability increases in that direction, result-

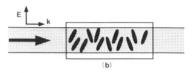

FIG. 28. (a) Random molecular orientation in a liquid with anisotropic molecules produces an isotropic refractive index. (b) Partial alignment with a laser beam produces a nonlinear index and optically induced birefringence.

ing in both an increase in the refractive index for light polarized in the same direction as the incident field and, because the change in the refractive index is less for light polarized perpendicular to the incident radiation, birefringence. This effect is termed the optical Kerr effect, and the materials in which it occurs are termed Kerr-active. Self-focusing is observed most commonly in these materials.

A nonlinear index can also arise from electrostriction, in which the molecules of the medium move into the most intense regions of the electric field. The resulting increase in density causes an increase in the refractive index near the regions of intense fields. Because of the relatively slow response time of moving molecules, electrostriction has a longer time constant than molecular orientation and is typically important only for pulses that last for several tens to several hundreds of nanoseconds or longer.

Self-focusing in plasmas occurs because of a form of electrostriction in which the electrons move away from the most intense regions of the beam. Because the electrons make a negative contribution to the refractive index, the change in their density distribution results in a positive lens.

In order for a beam to self-focus, the self-focusing force must overcome the tendency of the beam to increase in size due to diffraction. This requirement leads to the existence of a critical power defined by

$$P_c = 0.04\varepsilon_0\lambda^2 c/n_2 \qquad (51)$$

For incident powers above the critical power, the self-focusing overcomes diffraction and a beam of radius a focuses at a distance given by

$$z_f = \frac{0.369\,ka^2}{\sqrt{P/P_c} - 0.858} \qquad (52)$$

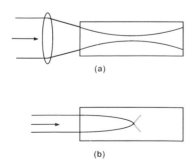

FIG. 27. Schematic of focal region produced by (a) a normal and (b) a self-focusing lens.

For incident powers below the critical power, the self-focusing cannot overcome the spreading due to diffraction and the beam does not focus, although it spreads more slowly than it would in the absence of the nonlinear index. Values of the nonlinear index and the critical powers for some materials are given in Table XIV.

When the power in the incident beam is just above the critical power, the entire beam focuses as described above in a process termed whole-beam self-focusing. When the incident power exceeds the critical power significantly, the beam usually breaks up into a series of focal regions, each one of which contains one or a small number of critical powers. This behavior is termed beam break-up, and the resulting focusing behavior is termed small-scale self-focusing. If the incident beam contains a regular intensity pattern, the distribution of focal spots can be regular. More commonly, however, the pattern is random and is determined by whatever minor intensity variations are on the incident beam due to mode structure, interference patterns, or from imperfections or dust in or on optical materials through which it has propagated. An example of a regular pattern of self-focused spots developed on a beam with diffraction structure is shown in Fig. 29.

Once the self-focusing process has started, it will continue until catastrophic focus is reached at the distance given in Eq. (52). The minimum size of the focal point is not determined by the third-order nonlinear index but can be determined by higher-order terms in the nonlinear index, which saturate the self-focusing effect. Such saturation has been observed in atomic gases. For self-focusing in liquids, it is thought that other mechanisms, such as nonlinear absorption, stimulated Raman scattering, or multiphoton ionization, place a lower limit on the size of the focal spots. Minimum diameters of self-focal spots are of the order of a few micrometers to a few tens of micrometers, depending on the material involved.

If the end of the nonlinear medium is reached before the catastrophic self-focal distance of Eq. (52), the material forms an intensity-dependent variable-focal-length lens. It can be used in conjunction with a pinhole or aperture to form an optical limiter or power stabilizer.

When the incident beam has a constant intensity in time, the focal spot occurs in one place in the medium. When the incident wave is a pulse that varies in time, the beam focuses to a succession of focal points, each corresponding to a dif-

TABLE XIV. Self-Focusing Parameters for Selected Materials

Material	$n_2 \times 10^{-22}$ (MKS units)	Critical power at 1.064 μm (kW)
Carbon disulfide	122	35.9
Methanol	41	106
Ethanol	32	136
Toluene	29	152
Benzene	21	207
Acetone	3.4	1274
n-Hexane	2.6	1645
Cyclohexane	2.3	1880
Carbon tetrachloride	1.8	2194
Water	1.4	3038
Cesium vapor	-2.9×10^{-16} N	

ferent self-focal distance according to Eq. (52). This gives rise to a moving self-focal point, which, when observed from the side and integrated over the pulse duration, can have the appearance of a continuous track. In special cases the beam can be confined in a region of small diameter for many diffraction lengths in an effect termed self-trapping. This happens, for example, when the nonlinear index is heavily saturated, as, for example, in an atomic transition. When the pulse duration is short compared to the response time of the nonlinear index, which can vary from several picoseconds in common liquids to several hundred picoseconds in liquid crystals, the back end of the pulse can be effectively trapped in the index distribution set up by the front of the pulse. This behavior is termed dynamic self-trapping.

Self-focusing in solids is generally accompanied by damage in the catastrophic focal regions. This results in an upper limit on the intensity that can be used in many optical components and also results in a major limitation on the intensity that can be obtained from some solid-state pulsed lasers.

Because of the tensor nature of the nonlinear susceptibility, the intensity-dependent change in the refractive index is different for light polarized parallel and perpendicular to the laser radiation, resulting in optically induced birefringence in the medium. The birefringence can be used to change linearly polarized light into elliptically polarized light. This effect forms the basis of ultrafast light gates with picosecond time resolution that are used in time-resolved spectroscopy and measurements of the duration of short pulses. The birefringence also results in the rota-

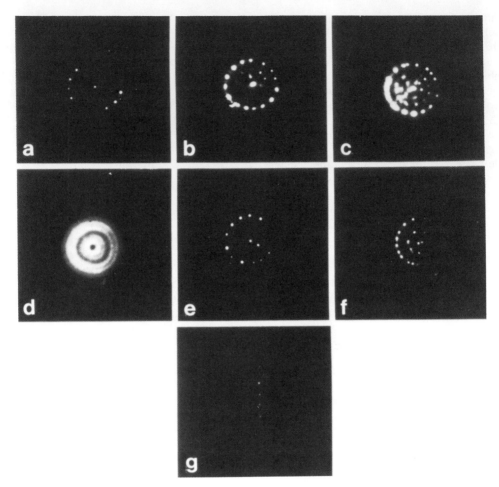

FIG. 29.   Pattern of self-focal spots (a–c, e–f) obtained from a beam with the ring pattern in (d). (g) Pattern produced by a beam with a horizontal straight edge. [Reproduced from A. J. Campillo *et al.* (1973). *Appl. Phys. Lett.* **23**, 628.]

tion of the principal axis of elliptically polarized light, an effect used in nonlinear spectroscopy and the measurement of nonlinear optical susceptibilities.

2. Self-Defocusing

Self-defocusing results from a combination of a negative value of n_2 and a beam profile that is more intense at the center than at the edge. In this situation the refractive index is smaller at the center of the beam than at the edges, resulting in a shorter optical path for rays at the center than for those at the edge. This is the same condition that exists for propagation through a negative-focal-length lens, and the beam defocuses.

Negative values of the nonlinear refractive index can occur because of interaction with electronic energy levels of the medium when the la-

ser frequency is just below a single-photon resonance or just above a two-photon resonance. Generally, self-defocusing due to electronic interactions is observed only for resonant interactions in gases and has been observed in gases for both single- and two-photon resonant conditions. A more common source of self-defocusing is thermal self-defocusing or, as it is commonly called, thermal blooming, which occurs in materials that are weakly absorbing. The energy that is absorbed from the light wave heats the medium, reducing its density, and hence its refractive index, in the most intense regions of the beam. When the beam profile is more intense at the center than at the edge, the medium becomes a negative lens and the beam spreads. Thermal blooming can occur in liquids, solids, and gases. It is commonly observed in the propagation of high-power infrared laser beams through the atmosphere and is one of the major limitations on atmospheric propagation of such beams.

3. Self-Phase Modulation

Self-phase modulation results from a combination of a material with a nonlinear refractive index and an incident field amplitude that varies in time. Because the index of refraction depends on the optical intensity, the optical phase, which is given by

$$\phi = kz - \omega t = \frac{2\pi}{\lambda}\left[n^L + \tfrac{1}{2}n_2|A(t)|^2\right]z - \omega t \quad (53)$$

develops a time dependence that follows the temporal variation of the optical intensity. Just as with other situations involving phase modulation, the laser pulse develops spectral side bands. Typical phase and frequency variations are shown in Fig. 30 for a pulse that has a bell-shaped profile. The phase develops a bell-shaped temporal dependence, and the frequency, which is the time derivative of the phase, undergoes an oscillatory behavior as shown. For a medium with a positive n_2, the down-shifted part of the spectrum is controlled by the leading part of the pulse and the up-

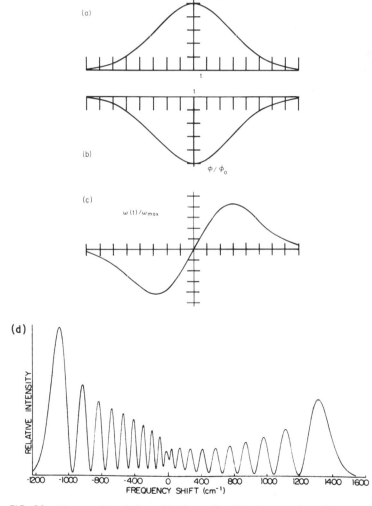

FIG. 30. Temporal variation of (a) intensity, (b) phase, and (c) frequency produced by a medium with a positive nonlinear refractive index. (d) Spectrum of a self phase modulated pulse. The asymmetry results from a rise time that was 1.4 times faster than the full time. [Part (d) reproduced from J. Reintjes (1984). "Nonlinear Optical Parametric Processes in Liquids and Gases." Academic Press, Orlando, Florida.]

shifted part of the spectrum is controlled by the trailing part of the pulse. The number of oscillations in the spectrum is determined by the peak phase modulation, while the extent of the spectrum is determined by the pulse duration and the peak phase excursion. For pulses generated in Q-switched lasers, which last for tens of nanoseconds, the spectral extent is relatively small, usually less than a wave number. For picosecond-duration pulses generated in mode-locked lasers, the spectrum can extend for hundreds of wave numbers and, in some materials such as water, for thousands of wave numbers.

In many instances, self-phase-modulation represents a detrimental effect, such as in applications to high-resolution spectroscopy or in the propagation of a pulse through a dispersive medium when the self-phase-modulation can cause spreading of the pulse envelope. In some instances, however, self-phase-modulation can be desirable. For example, the wide spectra generated from picosecond pulses in materials such as water have been used for time-resolved spectroscopic studies. In other situations, the variation of the frequency in the center of a pulse, as illustrated in Fig. 30c, can be used in conjunction with a dispersive delay line formed by a grating pair to compress phase-modulated pulses in an arrangement as shown in Fig. 31, in

a manner similar to pulse compression in chirped radar.

Pulse compressions of factors of 10 or more have been achieved with self-phase-modulation of pulses propagated freely in nonlinear media resulting in the generation of subpicosecond pulses from pulses in the picosecond time range. This technique has also been very successful in the compression of pulses that have been phase-modulated in glass fibers. Here the mode structure of the fiber prevents the beam break-up due to self-focusing that can occur for large phase modulation in nonguided propagation. Pulse compressions of over 100 have been achieved in this manner and have resulted in the generation of pulses that, at 8 fsec (8×10^{-15} sec), are the shortest-duration optical pulses yet produced. Self-phase-modulation in fibers, coupled with anomalous dispersion that occurs for infrared wavelengths longer than about 1.4 μm, has also been used to produce solitons, which are pulses that can propagate long distances without spreading because the negative dispersion in the fiber actually causes the phase-modulated pulse to narrow in time. Soliton pulses are useful in the generation of picosecond-duration pulses and in long-distance propagation for optical-fiber communication.

4. Nonlinear Absorption

Self-action effects can also change the transmission of light through a material. Nonlinear effects can cause materials that are strongly absorbing at low intensities to become transparent at high intensities in an effect termed saturable absorption or, conversely, they can cause materials that are transparent at low intensities to become absorbing at high intensities in an effect termed multiphoton absorption.

Multiphoton absorption can occur through absorption of two, three, or more photons. The photons can be of the same or different frequencies. When the frequencies are different, the effect is termed sum-frequency absorption. Multiphoton absorption can occur in liquids, gases, or solids. In gases the transitions can occur between the ground state and excited bound states or between the ground state and the continuum. When the transition is to the continuum, the effect is termed multiphoton ionization. Multiphoton absorption in gases with atoms or symmetric molecules follow selection rules for multiple dipole transitions. Thus two-photon absorption occurs in these materials between lev-

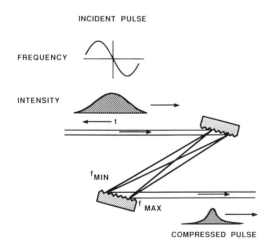

FIG. 31.   Pulse compression produced by a self-phase-modulated pulse and a pair of diffraction gratings. The dispersion of the gratings causes the lower-frequency components of the pulse to travel a longer path than that of the higher-frequency components, allowing the back of the pulse to catch up to the front.

els that have the same parity. These are the same transitions that are allowed in stimulated Raman scattering but are not allowed in a single-photon transitions of linear optics. Multiphoton absorption in solids involves transitions to the conduction band or to discrete states in the band gap. In semiconductors, two- or three-photon absorption can be observed for near-infrared radiation, while for transparent dielectric materials multiphoton absorption is generally observed for visible and ultraviolet radiation. Multiphoton absorption increases with increasing laser intensity and can become quite strong at intensities that can be achieved in pulsed laser beams, often resulting in damage of solids or breakdown in gases. This can be one of the major limitations on the intensity of radiation that can be passed through semiconductors in the infrared or other dielectric materials in the ultraviolet.

The simplest form of multiphoton absorption is two-photon absorption. It is described by an equation of the form

$$dI/dz = -\beta I^2 \tag{54}$$

where β is the two-photon absorption coefficient. This equation has a solution of the form

$$I(L) = I_0/(1 + \beta I_0 L) \tag{55}$$

for the intensity transmitted through a material of length L, where I_0 is the intensity incident on the material at $z = 0$. The form of the solution is indicated graphically in Fig. 32. Note that the transmission as a function of distance is quite different from that encountered for linear absorption, for which the transmitted intensity decreases with distance as $e^{-\alpha L}$. In the limit of large values of $\beta I_0 L$, the transmitted intensity approaches the constant value $1/\beta L$, independent of the incident intensity. Two-photon absorption can thus be used for optical limiting. It can also be used for spectroscopy of atomic and molecular levels that have the same parity as the ground state and are therefore not accessible in linear spectroscopy. Finally, two-photon absorption can be used in Doppler-free spectroscopy in gases to provide spectral resolutions that are less than the Doppler width.

5. Saturable Absorption

Saturable absorption involves a decrease in absorption at high optical intensities. It can occur in atomic or molecular gases and in various types of liquids and solids, and it is usually observed in materials that are strongly absorbing at low light intensities. Saturable absorption occurs when the upper state of the absorbing transition gains enough population to become filled, preventing the transfer of any more population into it. It generally occurs in materials that have a restricted density of states for the upper level, such as atomic or molecular gases, direct bandgap semiconductors, and certain liquids, such as organic dyes. Saturable absorption in dyes that have relaxation times of the order of several picoseconds have been used to mode-lock solid-state and pulsed-dye lasers, producing optical pulses on the order of several tens of picoseconds or less. These dyes have also been used outside of laser cavities to shorten the duration of laser pulses. Saturable absorbers with longer relaxation times have been used to mode-lock cw dye lasers, producing subpicosecond pulses. Saturable absorption can also be used with four-wave mixing interactions to produce optical phase conjugation. Saturation of the gain in laser amplifiers is similar to saturable absorption, but with a change in sign. It is described by the same equations and determines the amount of energy that can be produced by a particular laser. In a linear laser cavity, the gain can be reduced in a narrow spectral region at the center of a Doppler-broadened line, which forms the basis of a spectroscopic technique known as Lamb dip spectroscopy that has a resolution less than the Doppler width.

C. Coherent Optical Effects

Coherent nonlinear effects involve interactions that occur before the wave functions that describe the excitations of the medium have time to relax or dephase. They occur primarily when the nonlinear interaction involves one- or

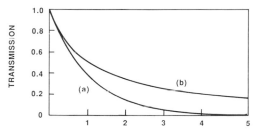

FIG. 32. Dependence of transmitted intensity with distance for (a) linear absorption and (b) two-photon absorption. For (a) $z_0 = \alpha^{-1}$, while for (b) $z_0 = (\beta I_0)^{-1}$.

two-photon resonances, and the duration of the laser pulse is shorter than the dephasing time of the excited state wave functions, a time that is equivalent to the inverse of the linewidth of the appropriate transition. Coherent nonlinear optical interactions generally involve significant population transfer between the states of the medium involved in the resonance. As a result, the nonlinear polarization cannot be described by the simple perturbation expansion given in Eq. (2), which assumed that the population was in the ground state. Rather, it must be solved for as a dynamic variable along with the optical fields.

Virtually any of the nonlinear effects that were described earlier can occur in the regime of coherent interactions. In this regime the saturation of the medium associated with the population transfer generally weakens the nonlinear process involved relative to the strength it would have in the absence of the coherent process.

Coherent interactions can also give rise to new nonlinear optical effects. These are listed in Table XV, along with some of their characteristics and the conditions under which they are likely to occur.

Self-induced transparency is a coherent effect in which a material that is otherwise absorbing becomes transparent to a properly shaped laser

TABLE XV.   Coherent Nonlinear Interactions

Interaction	Conditions for observation
Self-induced transparency	Resonant interaction with inhomogeneously broadened transition; pulse duration less than dephasing time
Photon echoes	Resonant interaction with inhomogeneously broadened transition; pulse duration less than dephasing time; two pulses spaced by echo time τ, with pulse areas of $\pi/2$ and π, respectively
Adiabatic following	Near-resonant interaction; pulse duration less than dephasing time
Adiabatic rapid passage	Near-resonant interaction; pulse duration less than dephasing time; frequency of pulse swept through resonance

pulse. It occurs when laser pulses that are shorter than the dephasing time of the excited-state wave functions propagate through materials with absorbing transitions that are inhomogeneously broadened. In self-induced transparency, the energy at the beginning of the laser pulse is absorbed and is subsequently reemitted at the end of the laser pulse, reproducing the original pulse with a time delay. In order for all the energy that is absorbed from the beginning of the pulse to be reemitted at the end, the pulse field amplitude must have the special temporal profile of a hyperbolic secant. In addition, the pulse must have the correct "area," which is proportional to the product of the transition dipole moment and the time integral of the field amplitude over the pulse duration. The pulse area is represented as an angle that is equivalent to the rotation angle from its initial position of a vector that describes the state of the atom or molecule. In self-induced transparency, the required pulse area is 2π, corresponding to one full rotation of the state vector, indicating that the medium starts and ends in the ground state. If the incident pulse has an area greater than 2π it will break up into multiple pulses, each with area 2π, and any excess energy will eventually be dissipated in the medium. If the initial pulse has an area less than 2π it will eventually be absorbed in the medium.

Self-induced transparency is different from ordinary saturated absorption. In saturated absorption, the energy that is taken from the pulse to maintain the medium in a partial state of excitation is permanently lost to the radiation field. In self-induced transparency, the energy given to the medium is lost from the radiation field only temporarily and is eventually returned to it.

Photon echoes also occur in materials with inhomogeneously broadened transitions. In producing a photon echo, two pulses are used with a spacing of τ, as shown in Fig. 33. The first pulse has an area of $\pi/2$ and produces an excitation in the medium. The second pulse has an area of π and reverses the phase of the excited-state wave functions after they have had time to dephase. Instead of continuing to get further out of phase as time progresses, the wave functions come back into phase. At a time τ after the second pulse, the wave functions are again all in phase and a third pulse, termed the echo, is produced. Photon echoes are observed most easily when the pulsed spacing τ is larger than the inhomogenous dephasing time caused, for example, by Doppler broadening, but smaller than the

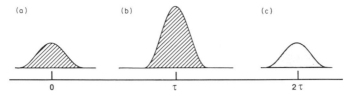

FIG. 33. Arrangement of pulses for generating photon echoes: (a) with an area of $\pi/2$, and (b) with an area of π, are supplied spaced by time τ; and (c) the echo, is generated in the interaction at a time τ after the second pulse.

homogenous dephasing time caused by collisions or population decay.

Other coherent interactions include optical nutation and free induction decay, in which the population oscillates between two levels of the medium, producing oscillations in the optical fields that are radiated, and adiabatic rapid passage, in which a population inversion can be produced between two levels in a medium by sweeping the frequency of a pulse through the absorption frequency in a time short compared to the dephasing time of the upper level.

D. Electrooptic and Magnetooptic Effects

Electrooptic and magnetooptic effects involve changes in the refractive index of a medium caused by an external electric or magnetic field. These are not normally thought of as nonlinear optical effects but are technically nonlinear optical processes in which the frequency of one of the fields is equal to zero. Various electrooptic and magnetooptic effects can occur depending on the situation. Some of these were listed in Table I.

In the Pockels effect, the change in the refractive index is proportional to the external electric field, whereas in the quadratic Kerr effect the change in refractive index is proportional to the square of the electric field. The Pockels effect occurs in solids without inversion centers, the same types that allow second-order nonlinear effects. The quadratic Kerr effect occurs in materials with any symmetry and is commonly used in liquids with anisotropic molecules such as nitrobenzene. The Faraday and Cotton–Mouton effects produce changes in the refractive index that are proportional to the magnetic field strength. Electrooptic and magnetooptic effects generally cause different changes in the refractive indexes in different directions relative to the applied field or the crystal axes, resulting in field-induced birefringence.

Electrooptic and magnetooptic effects are commonly used in light modulators and shutters. The field-dependent changes in the refractive index can be used directly for phase or frequency modulation. This is most commonly done with the Pockels effect, and units with modulation frequencies of up to the order of 1 GHz are commercially available. Field-induced birefringence is used to change the polarization state of a light beam and can be used for both modulation and shuttering. A light shutter can be constructed with either a Pockels cell or a Kerr cell by adjusting the field so that the polarization of the light wave changes by 90°. These are commonly used for producing laser pulses with controlled durations or shapes and for Q-switching of pulsed lasers. The Faraday effect produces a birefringence for circular polarization, resulting in the rotation of the direction of polarization of linearly polarized light. When adjusted for 45° and combined with linear polarizers, it will pass light in only one direction. It is commonly used for isolation of lasers from reflections from optical elements in the beam.

IV. Applications

In the previous sections the basic nonlinear optical interactions have been described, along with some of their properties. In the following sections we shall describe applications of the various nonlinear interactions. Some applications have already been noted in the description given for certain effects. Here we shall describe applications that can be made with a wide variety of interactions.

A. Nonlinear Spectroscopy

Nonlinear spectroscopy involves the use of a nonlinear optical interaction for spectroscopic studies. It makes use of the frequency variation of the nonlinear susceptibility to obtain information about a material in much the same way that variation of the linear susceptibility with frequency provides information in linear spectroscopy. In nonlinear spectroscopy the spectroscopic information is obtained directly from the nonlinear interaction as the frequency of the pump radiation is varied. This can be contrasted with linear spectroscopy that is done with radiation that is generated in a nonlinear interaction and used separately for spectroscopic studies. [*See* MOLECULAR OPTICAL SPECTROSCOPY.]

Nonlinear spectroscopy can provide information different from that available in linear spectroscopy. For example, it can be used to probe transitions that are forbidden in single-photon interactions and to measure the kinetics of excited states. Nonlinear spectroscopy can allow measurements to be made in a spectral region in which radiation needed for linear spectroscopy would be absorbed or perhaps would not be available. It can also provide increased signal levels or spectral resolution.

Many different types of nonlinear effects can be used for nonlinear spectroscopy, including various forms of parametric frequency conversion (harmonic generation, four-wave sum- and difference-frequency mixing), degenerate four-wave mixing, multiphoton absorption, multiphoton ionization, and stimulated scattering. Some of the effects that have been used for nonlinear spectroscopy are given in Table XVI, along with the information that is provided and the quantities that are varied and detected.

An example of improved spectral resolution obtained through nonlinear spectroscopy is the Doppler-free spectroscopy that can be done with two-photon absorption, as illustrated in Figs. 34 and 35. In this interaction, two light waves with the same frequency are incident on the sample from opposite directions. As the frequency of the incident light is swept through one-half of the transition frequency of a two-photon transition, the light is absorbed and the presence of the absorption is detected through fluorescence from the upper state to a lower one through an allowed transition. Normally, the spectral resolution of absorption measurements is limited by the Doppler broadening caused by the random motion of the atoms. Atoms that move in different directions with different speeds absorb light at slightly different frequencies, and the net result is an absorption profile that is wider than the natural width of the transition. In the nonlinear measurement, however, the atom absorbs one photon from each beam coming from opposite directions. A moving atom sees the frequency of one beam shifted in one direction by the same amount as the frequency of the other beam is shifted in the opposite direction. As a result, each atom sees the same sum of the frequencies of the two beams regardless of its speed or direction of motion, and the absorption profile is not broadened by the Doppler effect. This type of spectroscopy can be used to measure the natural width of absorption lines underneath a much wider Doppler profile. An example of a spectrum obtained with

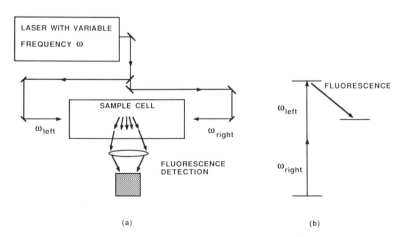

(a)   (b)

FIG. 34.   The use of two-photon absorption with counterpropagating beams for Doppler-free spectroscopy. (a) Experimental arrangement. (b) Level diagram of transitions used.

TABLE XVI.   Application of Nonlinear Optics to Spectroscopy[a]

Nonlinear interaction	Quantity varied	Quantity measured	Information obtained
Multiphoton absorption			
Atom $+ 2\omega_{laser} \to$ atom* [b]	ω_{laser}	Fluorescence from excited level	Energy levels of states with same parity as ground state; sub-Doppler spectral resolution
Multiphoton ionization			
Atom $+ 2\omega_{laser} + \mathcal{E}^a \to$ atom$^+$	ω_{laser}	Ionization current	Rydberg energy levels
Atom (molecule) $+ (\omega_1 + \omega_2) \to$ atom$^+$ (molecule$^+$)[c]	ω_2	Ionization current	Even-parity autoionizing levels
Four-wave mixing			
Sum-frequency mixing			
$\omega_4 = 2\omega_1 + \omega_2$	ω_2	Optical power at ω_4	Energy structure of levels near ω_4, e.g., autoionizing levels, matrix elements, line shapes
Third-harmonic generation			
$\omega_3 = 3\omega_1$	ω_1	Optical power at ω_3	Energy levels near $2\omega_1$ with same parity as ground state
Difference-frequency mixing			
$\omega_4 = 2\omega_1 - \omega_2$ (CARS)	ω_2	Optical power at ω_4	Raman energy levels Solids Polaritons Lattice vibrations Gases Measurement of nonlinear susceptibilities Solids Liquids Concentrations of liquids in solutions Concentrations of gases in mixtures Temperature measurements in gases Measurements in flames, combustion diagnostics Interference of nonlinear susceptibilities Time-resolved measurements Lifetimes, dephasing times CARS background suppression
Four-frequency CARS			
$\omega_4 = \omega_1 + \omega_2 - \omega_3$	$\omega_1 - \omega_3$ $\omega_2 - \omega_3$	Optical power at ω_4	Same as CARS, CARS background suppression
Raman-induced Kerr effect			
$\omega_{2,y} = \omega_1 - \omega_1 + \omega_{2,x}$	ω_2	Polarization changes at ω_2	Same as CARS, CARS background suppression
Higher-order Raman processes			
$\omega_4 = 3\omega_1 - 2\omega_2$	ω_2	Optical power at ω_4	Same as CARS
Coherent Stokes–Raman spectroscopy			
$\omega_4 = 2\omega_1 - \omega_2(\omega_1 < \omega_2)$	ω_2	Optical power at ω_4	Same as CARS
Coherent Stokes scattering			
$\omega_S = \omega_1(t) - \omega_1(t) + \omega_S(t) + \omega_1(t + \Delta t) - \omega_1(t + \Delta t)$	Δt	Optical power at $\omega_S(t + \Delta t)$	Lifetimes, dephasing times, resonant contributions to $\chi^{(3)}$
Raman gain spectroscopy			
$\omega_S = \omega_L - \omega_L + \omega_S$	ω_S	Gain or loss at ω_S	Raman energy levels
Saturation spectroscopy	ω_{laser}	Induced gain or loss	High-resolution spectra

[a] From J. F. Reintjes (1984). "Nonlinear Parametric Processes in Liquids and Gases," pp. 422–423. Academic Press, New York.
[b] An atom or molecule in an excited state is designated by *.
[c] An ionized atom or molecule is designated by $^+$ and $\mathcal{E}^a$ designates ionizing energy supplied by an external electric field.

two-photon Doppler-free spectroscopy is shown in Fig. 35.

Nonlinear spectroscopy can also be used to measure the frequency of states that have the same parity as the ground state and therefore are not accessible through linear spectroscopy. For example, S and D levels in atoms can be probed as two-photon transitions in multiphoton absorption or four-wave mixing spectroscopy. Nonlinear spectroscopy can also be used for spectroscopic studies of energy levels that are in regions of the spectrum in which radiation for

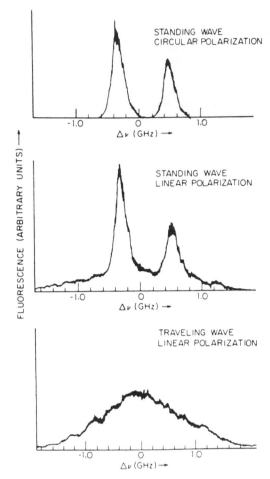

FIG. 35. Example of high-resolution spectrum in sodium vapor obtained with two-photon Doppler-free spectroscopy. [Reproduced from Bloembergen, N., and Levenson, M. D. (1976). Doppler-free two-photon absorption spectroscopy. *In* "High Resolution Laser Spectroscopy" (K. Shimoda, ed.), p. 355. Springer, New York.]

linear spectroscopy either is absorbed or is not available. Examples of such applications are spectroscopy of highly excited levels in the vacuum ultraviolet or extreme ultraviolet or of levels in the far infrared. Nonlinear effects can also be used for spectroscopic studies of excited levels, providing information on their coupling strength to other excited states or to the continuum. Time-resolved measurements can also be made to determine the kinetics of excited states such as lifetimes, dephasing times, and the energy-decay paths.

One of the most extensively used nonlinear processes for spectroscopy is coherent anti-Stokes Raman scattering (CARS). This is a four-

wave mixing process of the form $\omega_{AS} = 2\omega_L - \omega_S$, where ω_L and ω_S are the laser and Stokes frequencies that are provided in the incident radiation and ω_{AS} is the anti-Stokes frequency that is generated in the interaction. The process is resonantly enhanced when the difference frequency $\omega_L - \omega_S$ is tuned near a Raman active mode with frequency $\omega_0 = \omega_L - \omega_S$, and the resulting structure in the spectrum provides the desired spectroscopic information. The Raman-active modes can involve transitions between vibrational or rotational levels in molecules or lattice vibrations in solids. Effective use of the CARS technique requires that phase-matching conditions be met. This is usually done by angle phase matching, as was shown in Fig. 24a. Coherent anti-Stokes Raman scattering offers advantages over spontaneous Raman scattering of increased signal levels and signal-to-noise ratios in the spectroscopy of pure materials, because the intensity of the generated radiation depends on the product $(NL)^2$, where N is the density and L is the length of the interaction region as compared with the (NL) dependence of the radiation generated in spontaneous Raman scattering. As a result, spectra can be obtained more quickly and with higher resolution with CARS. Coherent anti-Stokes Raman scattering has been used for measurements of Raman spectra in molecular gases, Raman spectra in flames for combustion diagnostics and temperature measurements, and in spatially resolved measurements for molecular selective microscopy, an example of which is shown in Fig. 36.

Although CARS offers the advantage of increased signal levels in pure materials for strong Raman resonances, it suffers from the presence of a nonresonant background generated by all the other levels in the medium. In order to overcome these limitations, several alternative techniques have been developed involving interactions, such as the Raman induced Kerr effect, in which the sample develops birefringence, causing changes in the polarization of the incident waves as their frequency difference is tuned through a resonance, and four-frequency CARS, in which the variation of the CARS signal near a resonance with one set of frequencies is used to offset the background in the other set.

B. INFRARED UP-CONVERSION

Three-wave and four-wave sum-frequency mixing has been used for conversion of radiation from the infrared to the visible, where photographic film can be used and photoelectric de-

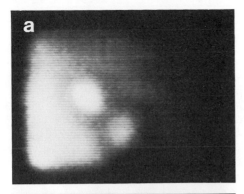

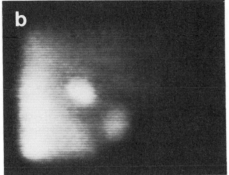

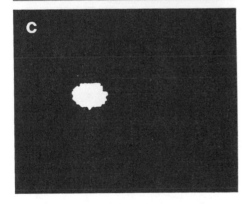

10 μm

FIG. 36. Molecular selective microscopy using coherent anti-Stokes Raman scattering. (A) On-resonant picture of deuterated and nondeuterated liposomes with two liposomes visible are shown. Only the deuterated liposomes have a Raman resonance for the radiation used. (B) The same picture as that in (A), but the two pump waves have been detuned from the Raman resonance. (C) The nonresonant signal has been subtracted from the resonant one, leaving only the deuterated liposome visible. [Reproduced from M. D. Duncan (1984). *Opt. Comm.* **50,** 307. Copyright © North-Holland, Amsterdam.]

tectors offer the advantages of improved sensitivity and increased temporal resolution. Infrared up-conversion has been used for up-conversion of infrared images to the visible and for time-resolved studies of the rotational spectra of molecules formed in explosions, giving information as to the time dependence of the temperature.

C. OPTICAL PHASE CONJUGATION

Optical phase conjugation, also referred to as time reversal or wavefront reversal, is a technique involving the creation of an optical beam that has the variations in its wavefront, or phase, reversed relative to a reference beam. If the optical field is represented as the product of an amplitude and complex exponential phase,

$$E = Ae^{i\phi} \tag{56}$$

then the process of reversing the sign of the phase is equivalent to forming the complex conjugate of the original field, an identification that gives rise to the name phase conjugation. When optical phase conjugation is combined with a reversal of the propagation direction, it allows for compensation of distortions on an optical beam, which develop as a result of propagation through distorting media, for example, the atmosphere, or imperfect or low-quality optical components such as mirrors, lenses, or windows. Such distortions are familiar to people who have looked through old window glass, through the air above a hot radiator, or in the apparent reflections present on the highway on a hot day. Optical phase conjugation can also be used for holographic imaging and can allow images to be transmitted through multimode fibers without degradation due to the difference in phase velocity among the various modes. It can also be used in various forms of optical signal processing such as correlators and for spectroscopy.

The concept of correction of distortions by optical phase conjugation is as follows. When a wave propagates through a distorting medium, its phase contour acquires structure that will eventually diffract, leading to increased spreading, reduced propagation length, reduced focal-spot intensity, and image distortion. The basic idea of compensation of distortions is to prepare at the entrance of the medium a beam whose wave front is distorted in such a way that the distortion introduced by the medium cancels the one that would develop on the beam, resulting in

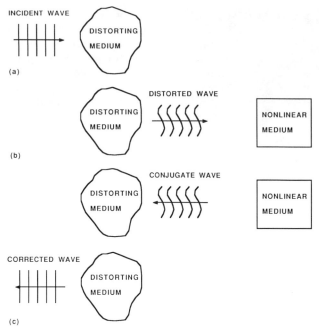

FIG. 37.   The use of optical phase conjugation for compensation of distortions. (a) An initially smooth beam propagates from right to left through a distorting medium. (b) After emerging from the distorting medium, the acquired phase variations are conjugated and the direction of propagation is reversed in a suitable nonlinear interaction. (c) After the second pass through the nonlinear medium, the beam emerges with a smooth wavefront.

an undistorted wave front at the exit of the medium.

The wave front required for this compensation to occur is the conjugate of the wave front obtained by propagation of an initially plane wave through the medium. It is usually obtained by propagating a beam through the distorting medium and then into a nonlinear medium that produces a phase-conjugate beam that propagates in the reverse direction, as illustrated in Fig. 37b. Various nonlinear interactions can be used for optical phase conjugation, including degenerate four-wave mixing in transparent, absorbing, and amplifying media and various forms of stimulated scattering such as Raman, Brillouin, and Rayleigh and stimulated emission. The two most widely used techniques are degenerate four-wave mixing and stimulated Brillouin scattering.

Degenerate four-wave mixing configured for phase conjugation is illustrated in Fig. 38. Here two strong waves, A_1 and A_2, are incident on the

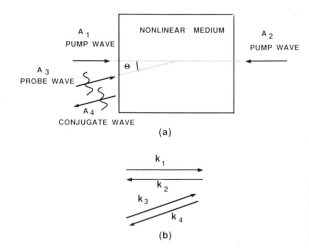

FIG. 38.   (a) Configuration for the use of degenerate four-wave mixing for phase conjugation. The frequencies of all waves are equal; the two pump waves, A_1 and A_2, propagate in opposite directions; and the angle θ between the probe wave and the forward pump wave is arbitrary. (b) The k vector diagram is shown.

nonlinear medium from opposite directions. The wave front carrying the information to be conjugated is the probe wave A_3 and the conjugate wave A_4 that is generated in the interaction $\omega_4 = \omega_1 + \omega_2 - \omega_3$ propagates in the direction opposite to A_3. Because the waves are present in counterpropagating pairs and all the waves have the same frequency, the process is automatically phase matched regardless of the angle between the probe wave and the pump waves. Three-wave difference-frequency mixing of the form $\omega_3 = \omega_1 - \omega_2$ can also be used in crystals for phase conjugation, but the phase-matching requirements in nonlinear crystals restrict the angles that can be used, and hence the magnitude of the distortions that can be corrected.

The generation of conjugate waves in nonlinear interactions involves the creation of volume holograms, or diffraction gratings, in the medium through interference of the incident waves. The interference that occurs, for example, between the waves A_1 and A_3 in degenerate four-wave mixing is illustrated in Fig. 39a. The backward wave is created by the scattering of the pump wave A_2 off the diffraction grating created by the interference of waves A_1 and A_3. The

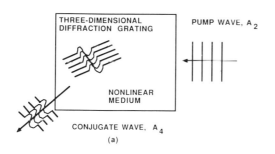

(a)

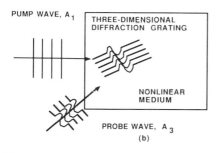

(b)

FIG. 39. Illustration of the formation of phase conjugate waves by scattering from the interference patterns formed by pairs of the incident waves. (a) Interference between the probe wave and the forward pump wave forming a contoured grating in the nonlinear medium. (b) Scattering of the backward pump wave from the contoured grating to form the phase-conjugate wave.

information as to the distortions on the incoming wave appears as bending of the contours of the grating. It is transferred to the phase contour of the backward wave as it is created but with its sense reversed. A similar interference occurs between all pairs of the incident waves, and each interference pattern creates a contribution to the backward-propagating phase-conjugate wave.

In stimulated Brillouin scattering, the incident wave serves as both the pump wave for the nonlinear process and the distorted wave to be conjugated. When the incident wave is highly aberrated, different components of the pump wave interfere in the focus, producing regions of relatively high and low gain. The wave that is ultimately generated in the Brillouin process is the one with the highest average gain. This wave in turn is one that is the phase conjugate of the incident wave, because it is has the most constructive interference in the focus. The interference of the various components of the pump beam in the focus can also be viewed as the creation of diffraction gratings or holograms in a manner similar to that described for degenerate four-wave mixing. The phase-conjugate wave is then produced by scattering of a backward-traveling wave from these diffraction gratings.

Optical phase conjugation has been used for reconstruction of images when viewed through distorting glass windows or other low-quality optical components, for removal of distortions from laser beams, for holographic image reconstruction, and for image up-conversion. An example of image reconstruction using optical phase conjugation is shown in Fig. 40.

D. OPTICAL BISTABILITY

Nonlinear optical effects can also be used to produce bistable devices that are similar in operation to bistable electric circuits. They have potential use in optical memories and the control of optical beams and can perform many of the functions of transistors such as differential gain, limiting, and switching. Bistable optical elements all have a combination of a nonlinear optical component and some form of feedback. A typical all-optical device consisting of a Fabry–Perot (FP) cavity that is filled with a nonlinear medium is illustrated in Fig. 41. The FP cavity has the property that when the medium between its plates is transparent, its transmissivity is high at those optical wavelengths for which an integral number of wavelengths can be contained

CONJUGATOR

CONJUGATOR + DISTORTER

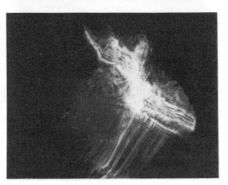

MIRROR

MIRROR + DISTORTER

FIG. 40.   Example of image reconstruction with phase conjugation. The unaberrated image is shown at the lower left using a plane mirror and at the upper left using a conjugate mirror. The image at the lower right shows the effect of an aberrator (a distorting piece of glass) on the image obtained with a normal mirror, while the image at the upper right shows the corrected image obtained with the aberrator and the conjugate mirror. [From J. Feinberg (1982). *Opt. Lett.* **7,** 488. Copyright © 1982 by the Optical Society of America.]

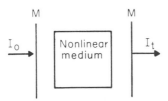

FIG. 41.   Illustration of nonlinear Fabry–Perot cavity for optical bistability. The end mirrors are plane and parallel, and the medium in the center exhibits either a nonlinear refractive index or saturable absorption.

between the plates, and its reflection is high for other wavelengths.

The operation of a nonlinear FP cavity can be illustrated by assuming that the nonlinear medium has a refractive index that is a function of the intensity inside the FP cavity. The wavelength of the incident light is chosen to be off resonance so that at low incident intensities the reflectivity is high, the transmission is low, and not much light gets into the FP cavity. As the incident intensity is increased, more light gets

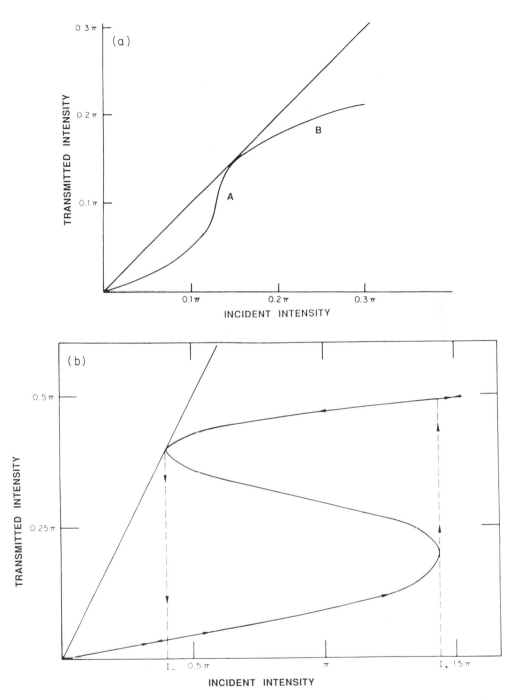

FIG. 42.   Illustration of nonlinear transmission with a nonlinear Fabry–Perot cavity using a medium with a nonlinear refractive index. (a) The transmission curve does not show hysteresis and can be used for differential gain (in region A) or optical limiting (in region B). (b) The transmission shows hysteresis and can be used for optical bistability. The difference between the differential gain and bistable operation is determined by the initial detuning from the resonance. [Reproduced from J. Reintjes (1984). "Nonlinear Optical Parametric Processes in Liquids and Gases." Academic Press, Orlando, Florida.]

into the FP cavity, changing the refractive index of the medium, and bringing the cavity closer to resonance. As a result, the fractional transmission increases and the intensity of the light transmitted through the cavity increases faster than does the intensity of the incident light. When the incident intensity is sufficiently high the cavity is brought into exact resonance, allowing full transmission. For further increases in the incident intensity, the cavity moves beyond the resonance and the fractional transmission decreases.

For conditions in which the transmitted intensity increases faster than the incident intensity, the nonlinear FP cavity can be used for differential gain in a manner similar to transistor amplifiers. An example of the transmission curve under these conditions is shown in Fig. 42a. In this situation, a small modulation on the incident light wave can be converted to a larger modulation on the transmitted wave. Alternatively, the modulation can be introduced on a different wave and can be transferred to the forward wave with a larger value. If the incident intensity is held just below the value at which the transmission increases and the cavity is illuminated by another, weaker wave that changes the transmission level, the nonlinear FP cavity can act as a switch. In the range of intensities for which an increase in intensity moves the FP cavity away from resonance, the nonlinear FP acts as an optical limiter. For certain conditions, the level of light inside the cavity is sufficient to maintain the resonance condition once it has been achieved, resulting in hysteresis of the transmission curve with the incident intensity, and providing the conditions for bistability. An example of the transmission curve under this condition is shown in Fig. 42b.

A medium with saturable absorption can also be used for the nonlinear medium. In this case, the resonance condition is always met as far as the wavelength is concerned, but the feedback from the second mirror, which is required for resonant transmission, is not obtained until sufficient light is present inside the cavity to bleach the nonlinear medium. Devices of this type can show differential gain and optical bistability but not optical limiting.

Bistable optical devices can also be constructed in such a way that the transmitted light signal is converted to an electrical voltage that is used to control the refractive index of the material inside the cavity. These hybrid optical–electrical devices can operate with very low levels of light intensity and as such are especially compatible with integrated optical components.

Optical bistability has also been observed in a wide range of optical configurations, such as laser cavities, degenerate four-wave mixing, and second-harmonmic generation. Under some conditions, bistable devices can exhibit chaotic fluctuations in which the light output oscillates rapidly between two extremes.

BIBLIOGRAPHY

Fisher, R. (ed.) (1983). "Optical Phase Conjugation." Academic Press, New York.

Levenson, M. D. (1982). "Introduction to Nonlinear Spectroscopy." Academic Press, New York.

Pepper, D. M. (1986). Applications of optical phase conjugation. Sci. Am. 254 (January) (6), 74.

Reintjes, J. (1984). "Nonlinear Optical Parametric Processes in Liquids and Gases." Academic, New York.

Reintjes, J. (1985). Coherent ultraviolet sources. In "Laser Handbook" (M. Bass and M. L. Stitch, eds.), Vol. 5, pp 1–202. North-Holland Publ., Amsterdam.

Shen, Y. R. (1984). "The Principles of Nonlinear Optics." Wiley, New York.

Shkunov, V. V., and Zeldovich, B. Y. (1985). Optical phase conjugation. Sci. Am. 253, (December) (6), 54.

NUCLEAR PHYSICS

Lawrence Wilets   *University of Washington*

  I. Twentieth-Century History
  II. The Lightest Nuclei
  III. Gross Properties of Nuclei
  IV. Nuclear Decays
  V. Collective Modes
  VI. Shell Structure of Nuclei
  VII. Nuclear Reactions
  VIII. Energy Sources
  IX. Nuclear Forces

GLOSSARY

Atomic number (Z):   Integer equal to the number of protons in a nucleus; the order of an element in the periodic table.

Atomic weight or nuclear mass number (A): An integer equal to the sum of the number of protons Z and neutrons N.

Isomer: Nuclide in a state of energy excitation.

Isobar: Member of a set of nuclides of the same atomic weight A.

Isospin: Set of quantum numbers relating to the charge of elementary particles. For the nucleon, the total isospin is $T = \frac{1}{2}$, and the third component is $T_3 = +\frac{1}{2}$ for the proton and $T_3 = -\frac{1}{2}$ for the neutron. For the pion, the total isospin is $T = 1$ and the third components are $T_3 = 1, 0, -1$ for the π^+, π^0, π^-.

Isotone: Member of a set of nuclides of the same neutron number N and various Z.

Isotope: Element (atomic number Z) characterized by various N.

Nuclide: Nucleus characterized by Z and N, and by the state of energy excitation.

Parity: Symmetry of wave function under inversion of the coordinate system: $\mathbf{r} \rightarrow -\mathbf{r}$. The wave function either remains unchanged (even, or +, parity) or changes sign (odd, or −, parity).

Spin: Angular momentum in a (nuclear) state, measured in units of $\hbar$. The spin can be either an integer (A even) or a half-odd integer (A odd).

The atom consists of a central, massive core, the nucleus, surrounded by an electron cloud. A nucleus is characteristically 1/10,000 the size of the electron cloud and 4000 times as massive. A nucleus contains Z protons and N neutrons. Protons and neutrons are collectively called nucleons. The number of electrons in a neutral atom is equal to the number of protons, and this is also known as the atomic number, the order in which an atom appears in the periodic table of the elements. The sum of the number of protons and neutrons is denoted by A, the mass number. The isotope of an element of chemical name X is denoted by $^A_Z X_N$, or simply by $^A X$ since Z and N can be deduced from X and A. In contrast to atoms, where the size of the electron cloud changes very slowly with atomic number Z, nuclear density is essentially the same for all nuclei, hence the nuclear radius increases as $A^{1/3}$. Neutrons and protons are bound together in nuclei by strong forces, mediated by intermediate-mass elementary particles called mesons. The protons and neutrons themselves possess a substructure consisting of three quarks bound together by gluons. In nature, all elements up to uranium ($Z = 92$) are to be found, with the exception of technetium ($Z = 43$) and promethium ($Z = 61$), which have been produced in the laboratory. Elements up to $Z = 109$ have been produced artificially and identified, but elements 104–109 have not yet been named. Many naturally occurring and artificially produced nuclides are unstable (metastable) and exhibit radioactivity by decay (transmutation) through the emission of electrons (beta particles), alpha particles (helium nuclei), or fission. Nuclear reactions can be induced in the laboratory by bombardment of targets with a wide variety of projectiles: gamma rays, electrons, elementary particles, and other nuclei. The most stable nuclide is ^{56}Fe. The heaviest nuclei can release energy by fissioning into intermediate-weight fragments; the lightest nuclei can release energy by fusion. The latter process is the source of power production in the sun and stars.

I. Twentieth-Century History

The foundations of both modern nuclear physics and modern atomic physics were established by Ernest (Lord) Rutherford through a series of celebrated experiments first published in 1911. He used alpha particles from naturally radioactive emitters as projectiles to bombard a variety of targets, and he detected the scattered alpha particles by visually observing scintillations on a phosphorescent screen. From the distribution of scattered particles, he was able to demonstrate that the interaction of alpha particles with atoms obeyed Coulomb's inverse-square law down to distances of the order of 10^{-13} m $= 100$ fm (1 fm $= 10^{-15}$ m).

The picture that emerged from Rutherford's experiments was that of an atom consisting of a massive core—the nucleus—of positive electric charge Ze, where $-e$ is the charge on the electron and Z is the atomic number. The nucleus is surrounded by a negatively charged electron gas. Earlier atomic theories fell, most notably J. J. Thomson's model of electrons embedded in a positively charged "jelly." In 1913, Niels Bohr announced his atomic theory of electrons circling the nucleus in quantized planetary orbits. Further studies in atomic physics led to the discovery (invention) of quantum mechanics by Werner Heisenberg (1925) and Erwin Schrödinger (1926).

The discovery of the neutron by James Chadwick in 1932 clarified both the problem of isotopic composition and the connection between atomic weight A and nuclear spin. (Earlier models would have the nucleus consisting of A protons and $A - Z$ electrons. This had the difficulty that the nuclear spin would then be an odd or even multiple of $\frac{1}{2}\hbar$ according to whether Z is odd or even, whereas nuclear spin is odd or even according to whether A is odd or even.) With protons and neutrons now known to be the building blocks of nuclei, the study of nuclear structure was launched.

In 1935, Hideki Yukawa postulated the existence of a new, intermediate-weight elementary particle, which he called the mesotron, to act as the agent to bind neutrons and protons together in the nucleus. Some confusion ensued when Carl Anderson and Seth Neddermeyer discovered a candidate particle in 1938 that did not seem to interact strongly with nuclei. The problem was resolved in 1947 by Cecil Powell and collaborators who identified two particles, the mu and the pi mesons, the latter being the Yukawa mesotron (now called the pion); the mu meson, or muon, is the Anderson–Neddermeyer particle. This was a remarkable triumph of speculative theoretical induction. It also completed the first phase in the microscopic description of nuclear structure. Subsequently, a host of elementary particles has been found, many of which play important roles in nuclear physics.

The discovery of fission by Otto Hahn and Fritz Strassmann in 1939 led to the development of the atomic (more properly nuclear) bomb during World War II and the attendant development of fission reactors for electrical power generation. The fusion process, which is the mechanism by which the sun and stars generate their energy, was the basis for development of the hydrogen bomb in the 1950s, and there has been intense research during the subsequent decades to harness thermonuclear fusion as a power source. At the same time, nuclear physics and chemistry have provided radioactive isotopes, radioactive and stable isotope identification techniques, nuclear magnetic resonance, etc. for medical diagnosis and treatment, geological and archeological dating, tracing of water and atmospheric flow patterns, planetary and solar system histories, and numerous other applications.

At present, much interest is being concentrated on nuclear substructure, namely, the constituents of the protons, neutrons, and other elementary particles. The subparticles are called quarks; the proton and the neutron each contain three quarks.

II. The Lightest Nuclei

A. THE PROTON

The simplest nucleus is $^{1}_{1}H_0$. It has the following characteristics: mass, $m_p = 938.2796$ MeV $= 1836\, m_e$; angular momentum $J = \frac{1}{2}\hbar$; magnetic moment $\mu = 2.7928456\, m_N$ [1 m_N (nuclear magneton) $= e\hbar/2m_p c$]; root-mean-square charge radius $\langle r^2 \rangle^{1/2} = 0.82$ fm.

B. The Neutron

A constituent of nuclei, but not itself a nucleus because it does not form an atom, is the neutron. Its characteristics are $m_n = 939.5731$ MeV; $J = \frac{1}{2}\hbar$; $\mu = -1.91304184\ m_N$; $\langle r^2 \rangle = -0.12$ fm^2.

The neutron is heavier than the proton by 1.29 MeV. It decays into a proton, an electron ($m_e = 0.511$ MeV), and an antineutrino ($m_\nu = 0$) with a half-life of 10.4 min.

C. Light Nuclei

The isotopes of hydrogen have been given special names: deuterium for D $\equiv$ ^{2}H and tritium for T $\equiv$ ^{3}H. The corresponding nuclides are called the deuteron and the triton, respectively. The deuteron is a loosely bound structure (on the nuclear scale), having a binding energy of 2.2 MeV and a root-mean-square charge radius of 4.2 fm. The triton is unstable and decays into ^{3}He, accompanied by the emission of an electron and of an antineutrino with a half-life of 8.5 yr. Although ^{3}H is more tightly bound than ^{3}He, the decay occurs because the neutron is heavier than the proton, and ^{3}H is heavier than ^{3}He.

There is no bound state of two neutrons (the dineutron) or two protons (^{2}He).

Helium ($Z = 2$) also comes in two isotopes, ^{3}He and ^{4}He, both of which are stable. ^{4}He is especially tightly bound, and its central density is the highest of any nucleus.

There are no isotopes of any element with mass number $A = 5$.

III. Gross Properties of Nuclei

A. Nuclear Sizes and Shapes

Nuclear density is remarkably constant with respect to Z and A. This leads to the statement that nuclear volumes are proportional to A and that the radii are proportional to $A^{1/3}$. Measures of the nuclear radii are discussed below.

The electron and muon (which behaves like a heavy electron) are ideal probes for mapping out the charge distributions of nuclei. Both have no measured structure of their own (they are point particles) and interact with nuclei only through the electromagnetic field. (The weak interaction, which is responsible for beta decay, is quite negligible here.) Charge distribution experiments are mainly of two classes: atomic spectra, and scattering.

Common atoms, of course, contain electrons, and atomic isotope shifts have yielded information on the differences in charge distributions among various isotopes. The muon can also be captured in the electric field of a nucleus to form a hydrogen-like atom. Since the muon is 207 times as massive as the electron, its Bohr orbits are 1/207 times the size of the corresponding electron orbits. Thus muons can probe nuclear charge distributions more deeply than electrons.

The most detailed information on charge distributions comes from electron and muon scattering over a large range of energies and scattering angles.

The electric charge distribution deduced from experiments is often characterized by the two-parameter formula

$$\rho(r) = \frac{\rho(0)}{1 + e^{(r-R)/a}} \qquad (1)$$

A graph that summarizes such fits for a number of nuclei is shown in Fig. 1, where some of the fits include a third parameter; the vertical scale is the electric charge density multiplied by A/Z, which gives a measure of the nuclear matter density. Note that He shows the greatest central density of any nuclei. Most heavier nuclei can be characterized by the same central density, and with parameters $R = 1.12A^{1/3}$ fm and $a \approx 0.57$ fm. Here R is (very nearly) the half-density radius and a is a measure of the surface thickness. Another measure of the surface thickness is the distance over which the density falls from 90% to 10% of its central value. This is denoted by $t = 4.39a$ and is approximately equal to 2.5 fm. The approximate constancy of the central density is one aspect of the phenomenon known as nuclear saturation.

More precise analyses of the experiments yield more details of the charge distributions, including granularity beyond the smooth curves shown in Fig. 1, as well as detailed differences between nuclides.

What has been shown above is the angular average of the nuclear charge distribution. Most nuclei are not spherical, but have intrinsic nonspherical shapes, or distortions. Some nuclei execute oscillations about a spherical shape while others exhibit a permanent intrinsic deformation. Prolate (football-shaped) quadrupole deformations are found for a large number of nuclei. Octupole (pear-shaped) and higher-order deformations have also been observed.

B. Nuclear Masses

From the Einstein relationship $E = mc^2$ there is a basic equivalence between mass and energy,

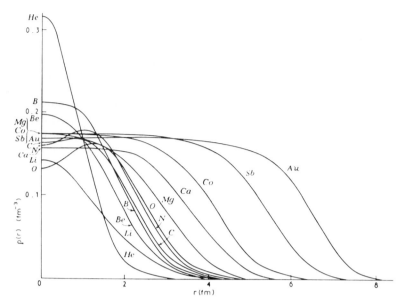

FIG. 1. Nucleon density distributions (A/Z times charge distributions) determined from electron scattering. Most fits have been made using the two-parameter form of Eq. (1), but some involve a three-parameter form which allows for a central depression or peak. [After Elton, L. R. B. (1961). "Nuclear Sizes," Oxford University Press, Oxford.]

and the two terms are used interchangeably; both are frequently measured in the same units, mega-electron-volts. Atomic masses quoted for nuclides are the masses for the corresponding neutral atoms. Nuclear binding energies can be identified by equating the atomic mass to the following:

$$E(Z, N) = M(Z, N)c^2$$
$$= m_n c^2 N + m_p c^2 Z + m_e c^2 Z - B_e(Z)$$
$$- B_N(Z, N) \qquad (2)$$

where the rest-mass energies of the constituent neutrons, protons, and electrons have been explicitly removed. The term B_e is the binding energy associated with the atomic electrons. The nuclear physics is contained in the term $B_N(Z, N)$. The minus sign associated with B_N is a matter of convention. The condition $B_N > 0$ corresponds to binding, a lowering of the total energy of the system. The lower the energy, the more stable is the nuclide.

Figure 2 shows the distribution of values for B_N/A as a function of A. (For any A there are usually several nuclides, or isobars.) The plot shows a maximum at $A = 56$, and ^{56}Fe is the most stable nuclide.

A physically useful parameterization of nuclear binding energies is given by the von Weiz-

säcker semiempirical mass formula, inspired by analogy to a classical liquid drop. It reads as follows:

$$-B_N(Z, N) = C_v A + C_s A^{2/3} + C_{sy} \frac{(N - Z)^2}{A}$$
$$+ C_C \frac{Z^2}{A^{1/3}} + \delta_p(Z, N) + \Delta(Z, N)$$
$$(3)$$

The various terms have the following interpretation and values:

$C_v A$ is the volume energy; $C_v = -15.68$ MeV would be the energy per nucleon in an infinite nucleus with equal numbers of neutrons and protons if there were no electrostatic (Coulomb) repulsion between the protons. The existence of this term is another manifestation of nuclear saturation.

$C_s A^{2/3}$ is called the surface energy. It represents the surface tension constant times the surface area; $C_s = +18.56$ MeV, and the + sign indicates a loss of binding due to the surface.

$C_{sy}(N - Z)^2/A$ is the volume symmetry energy. Nuclear matter (in the absence of Coulomb forces) is most strongly bound for symmetric ($N = Z$) nuclei. Deviation from symmetry results in a loss of binding. $C_{sy} = 28.1$ MeV.

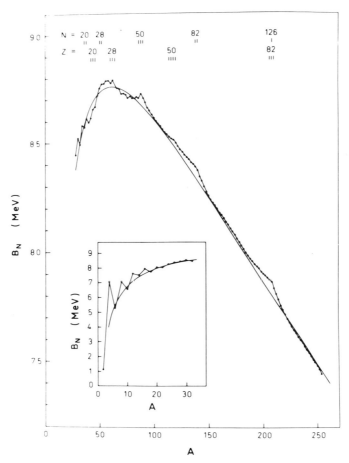

FIG. 2. Binding energy per nucleon, as a function of A. For each A, several isobars may be present. Neutron and proton magic numbers are indicated. [From Bohr, A., and Mottelson, B. R. (1969). "Nuclear Structure," Vol. I, Benjamin, New York.]

$C_C Z^2/A^{1/3}$ is the Coulomb energy corresponding to a uniformly charged sphere of radius $R \propto A^{1/3}$. The numerical value is $C_C = 0.717$ MeV.

$\delta_p(Z, N)$ is called the pairing energy. In their ground states, even–even (even Z, even N) nuclei are more strongly bound than even–odd or odd–even (i.e., odd A) nuclei, and odd–odd nuclei are less strongly bound yet. This can be approximated by the formula

$$\delta_p(Z, N) \approx \frac{34}{A^{3/4}} \quad \text{MeV}$$

$$\times \left\{ \begin{array}{ll} +1 & \text{for odd–odd} \\ 0 & \text{for } A \text{ odd} \\ -1 & \text{for even–even} \end{array} \right\} \quad (4)$$

$\Delta(Z, N)$ contains further details of nuclear structure, especially of what is known as shell structure. In practice, the parameters associated with the preceding terms are fit to the masses of all known nuclides by the method of least-square deviation; then $\Delta(Z, N)$ is the residual. (Some analyses have included explicit shell correction terms in the semiempirical mass formula.) When the residuals $\Delta(Z, N)$ are plotted against either Z (for various N) or against N (for various Z), the result is a sawtooth curve (or band) that decreases monotonically, with breaks at certain "magic" numbers. The breaks at the magic numbers are characteristically a few mega-electron-volts. Nuclei with magic numbers of protons or neutrons are especially stable, and nuclei with both proton and neutron numbers magic are exceptionally stable.

A more representative indicator of magic numbers is to be found in the nucleon separation

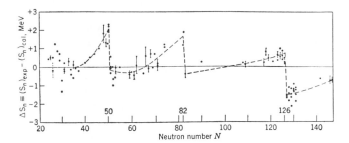

FIG. 3. The difference between observed neutron separation energies $(S_n)_{exp}$ and the values $(S_n)_{cal}$ calculated on the basis of a semiempirical mass formula. Discontinuities of the order of 2 MeV are evident for $N = 50$, 82, and 126; the evidence for a shell closure at $N = 28$ is less conclusive. [From Evans, R. D. (1955). "The Atomic Nucleus," McGraw-Hill, New York.]

energy, the energy required to remove a neutron or a proton [see Eqs. (12) and (13)]. This becomes large as one approaches a magic number from below, and decreases just above a magic number. This is shown for neutron separation energies as a function of neutron number in Fig. 3.

The following numbers are magic for both neutrons and protons: 2, 8, 20, (28), 50, and 82. (The evidence for 28 is less compelling.) The value $N = 126$ is also magic. The next higher magic numbers have not yet been observed, but theoretical calculations indicate that $Z = 114$ and $N = 184$ may be magic.

The terminology "magic" is antiquated but colorful. It arose historically because stability of certain proton numbers (irrespective of N) and certain neutron numbers (irrespective of Z) was unexpected. These numbers are now fully understood in terms of the closing of shells in an independent-particle model, quite analogous to the closing of electron shells in atoms at the noble gases with $Z = 2$, 10, 18, 36, 54, and 86.

IV. Nuclear Decays

A. BETA DECAY

The elementary β^- decay

$$n \rightarrow p + e + \bar{\nu} \qquad (5)$$

can proceed for the free neutron because the neutron mass is greater than the sum of the masses of the proton and electron combined; the neutrino and antineutrino are massless. The same process can proceed in nuclei if it is ener-

getically possible. Furthermore, β^+ (positron) decay

$$p \rightarrow n + e^+ + \nu \qquad (6)$$

or atomic electron capture

$$p + e \rightarrow n + \nu \qquad (7)$$

can also occur in nuclei if energetically possible.

The energy requirement for β^- decay [Eq. (5)] is given in terms of the atomic energies (masses),

$$E(Z, N) - E(Z + 1, N - 1) \equiv Q_{\beta^-} > 0 \qquad (8a)$$

For electron capture it is

$$E(Z, N) - E(Z - 1, N + 1) \equiv Q_{ec} > 0 \qquad (8b)$$

The rest mass of the electron or positron created in the process is included in the atomic energy. For β^+ decay, however, the condition is

$$E(Z, N) - E(Z - 1, N + 1) - 2m_e c^2$$
$$\equiv Q_{\beta^+} > 0 \qquad (8c)$$

where the two electron masses come from keeping track of the number of electrons in the neutral atoms. If β^+ decay is energetically possible, so also is electron capture, but the converse may not be true.

For fixed A, the lowest energy isobar defines the valley of beta stability, the solid curve shown in Fig. 4. Because of the Coulomb energy, the heavier nuclei have $N > Z$. According to the semiempirical mass formula, the equation for the stable valley can be expressed most simply for Z as a function of A:

$$Z = \frac{\frac{1}{2}A}{1 + 0.0064 A^{2/3}} \qquad (9)$$

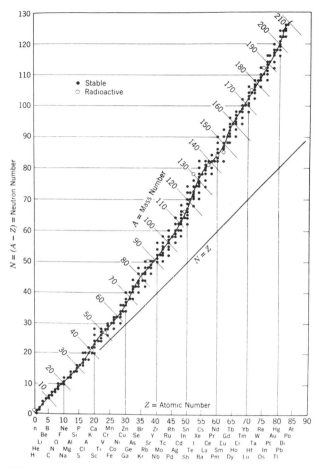

FIG. 4.   Naturally occurring β-stable nuclides plotted against N and Z. [From Evans, R. D. (1955). "The Atomic Nucleus," McGraw-Hill, New York.]

An example of the energetics of beta decay is shown in Fig. 5 for a string of odd-A isobars. Note that there is only one stable member of the string; this is generally true of odd-A isobars. The situation is quite different for even-A isobars, as shown in Fig. 6. The odd–odd nuclides lie higher in energy than the even–even ones by the pairing energy $2\delta_p$. Several even–even isobars can be stable against beta decay. The odd–odd nuclides can almost always beta decay, in some cases by either β^- or β^+ from the same nuclide. For nuclides heavier than $^{17}_7N_7$, there are no cases of stable odd–odd nuclei! (An exception may be $^{180}_{73}Ta_{107}$, which has a half-life greater than 10^{13} yr.) Although it is energetically possible for some even–even nuclei to decay by the emission of two electrons (+ or −), such double beta decay has not been observed, although it has been searched for intensively.

Beta-decay half-lives depend sensitively on the energy release Q, decreasing rapidly with increasing Q, and on details of the nuclear structure. In general, the half-life increases rapidly with increasing change in the nuclear spins between the mother and the daughter. The half-lives tend to decrease as one moves along an isobaric string away from the stable valley.

B. Alpha Decay

Alpha-particle decay is a common phenomenon among heavy nuclei, and all nuclei heavier than ^{209}Bi can decay by α-particle emission (although other modes may dominate). Even when the energy available, Q_α, is positive, the decay is inhibited by the Coulomb barrier

$$2Ze^2/r - Q_\alpha, \qquad r > R \qquad (10)$$

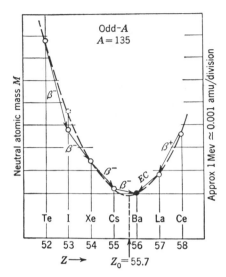

FIG. 5. An example of the mass "parabola" for the odd-A isobars of $A = 135$. The bottom of the mass–energy valley occurs at $Z \equiv Z_0 = 55.7$. The only stable isobar is $_{56}$Ba. [From Evans, R. D. (1955). "The Atomic Nucleus," McGraw-Hill, New York.]

which must be penetrated. Such penetration is forbidden in classical mechanics but is possible quantum-mechanically. The inhibition factor depends very sensitively on Q_α, decreasing rapidly

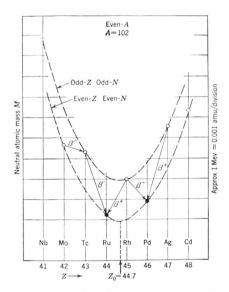

FIG. 6. Cross section of the mass-energy valley for the even-A isobars, $A = 102$, showing the characteristic two "parabolas" separated by twice the pairing energy δ_p. Both $_{44}$Ru and $_{46}$Pd are stable (solid circles). There are no stable odd–odd isobars here. [From Evans, R. D. (1955). "The Atomic Nucleus," McGraw-Hill, New York.]

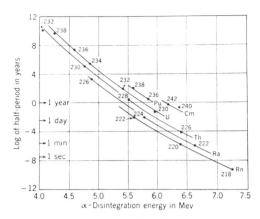

FIG. 7. Logarithm of α-particle lifetimes as a function of the disintegration energy Q_α. The solid curves, which connect isotopes of the same element, are based on calculations incorporating quantum-mechanical barrier penetration. [From Evans, R. D. (1955). "The Atomic Nucleus," McGraw-Hill, New York.]

with decreasing energy. In Fig. 7 are displayed the measured lifetimes for various isotopic sequences plotted as a function of the energy release Q_α. Although the height of the Coulomb potential at the nuclear surface increases with Z, the energy release increases more rapidly, and the lifetimes in general decrease rapidly with Z. This leads to one of the limits of stability discussed in Section IV,D below.

C. SPONTANEOUS FISSION

Within the concept of the liquid drop model of the nucleus, there is competition between the surface tension, which tends to stabilize the droplet in a spherical shape, and the electrostatic (Coulomb) repulsion, which tends to disrupt the system. A measure of this competition is given by the Bohr–Wheeler fissionability parameter,

$$x = \frac{E_{\text{Coulomb}}}{2E_{\text{surface}}} \approx \frac{Z^2/A}{50} \qquad (11)$$

The spherical configuration of a nucleus with $x > 1$ is unstable against deformation along the path which leads path to fission. If $x < 1$, the spherical shape is locally stable, but fission is still energetically allowable for $x > 0.35$ ($x = 0.35$ corresponds to $Z \approx 35$). In the range $0.35 < x < 1$, the process can only proceed by way of quantum-mechanical barrier penetration, so that spontaneous fission half-lives increase rapidly with decreasing x. Odd-A nuclei have a higher barrier against fission than do even–even nuclei.

For ^{238}U, $x \approx 0.71$, spontaneous fission does not play a significant role in its radioactivity, which is dominated by alpha decay. In fact, for almost all known nuclei, alpha decay tends to dominate spontaneous fission. However, spontaneous fission eventually is a limiting factor in how high in Z one can go in producing new elements.

D. LIMITS OF METASTABILITY

For the following discussion, please refer to Fig. 8, where β-stable species are plotted for N and Z.

The stable and long-lived nuclides cluster along the valley of beta stability. As one moves away from the valley on either side, the beta decay rates become faster (i.e., shorter half-lives).

To the lower right side of the valley is the neutron-rich region. The energy required to remove a neutron, called the separation energy, is defined by

$$S_n = B_N(Z, N) - B_N(Z, N - 1) \quad (12)$$

The separation energy decreases as one moves away from the stable valley. When $S_n < 0$, a neutron can be emitted spontaneously in a time comparable with the transit time for a neutron inside the nucleus, which is of the order of 10^{-22}–10^{-21} sec. The line $S_n = 0$ is called the neutron drip line. Beyond this line, nuclides do not live long enough even to be called metastable.

A similar situation occurs on the proton-rich (upper left) side of the stable valley. For protons, as for alpha particles, there is a Coulomb barrier to be surmounted even when the emission is energetically allowable. Barrier penetration is easier for protons than for alpha particles because they have half the charge and one-fourth the mass. The vanishing of the proton separation energy

$$S_p = B_N(Z, N) - B_N(Z - 1, N) \quad (13)$$

represents a practical limit to metastability, although not as severe as for neutron emission.

The corresponding limit for alpha particles occurs when an alpha particle in a nucleus has an energy greater than its Coulomb barrier, as discussed in Section IV,C. The limit imposed by spontaneous fission is $x = Z^2/50A = 1$.

These various limits, as evaluated from the semi-empirical mass formula, are indicated in Fig. 8. These are gross limits, and can be strongly modified by details of nuclear shell structure. There is speculation, backed by theo-

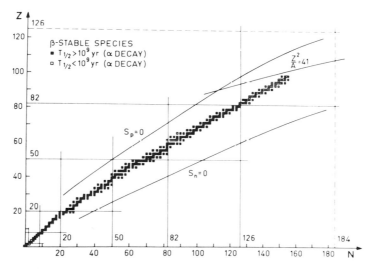

FIG. 8. Qualitative sketch showing some of the limiting conditions for stability that follow from the semiempirical mass formula. The β-stable nuclides are again plotted as a function of N and Z. Two groups of α-decay lifetimes are indicated. Two solid curves represent the limit of stability with respect to neutron and proton emission. The curve labeled $Z^2/A = 41$ corresponds to the limit where the spontaneous fission lifetime is approximately equal to 0.3 sec. [From Bohr, A., and Mottelson, B. R. (1969). "Nuclear Structure," Vol. I, Benjamin, New York.]

retical calculations, that another region of (meta)stability should occur in the region of closed nucleon shells at $Z = 114$ and $N = 184$. (Some recent calculations place the island at $Z = 110$ and $N = 178$ or 180, with half-lives of about 200 days.) The search for such an "island of stability" has been a topic intense investigation. (The terminology "island" is poetic, and requires plotting the negative of the atomic masses in order to implement the metaphor. Then the valley of beta stability becomes the "ridge of stability"; beyond lies the "sea of instability." The challenge is to cross the sea from the ridge to the island.)

V. Collective Modes

Collective motion in nuclei implies participation by many particles, such as is the case in fluid flow. The simplest example of nuclear collective motion is oscillatory expansion and contraction of the whole nucleus, called the breathing mode. Nuclear matter is quite stiff to changes in density. The compression modulus, defined by

$$K \equiv R^2 \frac{\partial^2(E/A)}{\partial R^2} \qquad (14)$$

can be determined by identifying observed monopole excitations with the breathing mode, which has an excitation frequency ω_0 and energy

$$\hbar\omega_0 = \frac{\pi}{3} \left(\frac{\hbar^2 K}{mR^2} \right)^{1/2} \qquad (15)$$

this leads to the value $K \approx 200$ MeV. The implication of this high value for K is that most nuclear collective motion does not involve changes in density. It can involve volume-conserving surface motion, the flow of neutrons with respect to protons (keeping the total density approximately constant), or the flow of nucleon spin. These last two are called isospin and spin waves, respectively.

A. Surface Vibrations

Nuclei near magic proton or neutron numbers have spherical equilibrium shapes. In the context of the liquid drop model, the stiffness of the surface to deformation is dependent on the surface tension, which favors spherical symmetry, and the Coulomb energy, which favors disruption. The normal mode that is softest to deformation is the quadrupole mode ($\lambda = 2$), corresponding to ellipsoidal deformation. The higher modes lie successively higher in frequency. Under the assumption of irrotational fluid flow, the frequencies of the modes ω_λ, or rather the energies of excitation $\hbar\omega_\lambda$, can be approximated by

$$\hbar\omega_\lambda = \left\{ \frac{\lambda}{A} \left[\frac{(\lambda - 1)(\lambda + 2)}{4} - x \frac{5\lambda - 1}{2\lambda + 2} \right] \right\}^{1/2} \qquad (17.4 \quad \text{MeV}) \quad (16)$$

The mode index λ is equal to the number of nodal lines in the surface of deformation. For the lowest, most common deformation, $\lambda = 2$, the formula is

$$\hbar\omega_2 = \left[\frac{2}{A} (1 - x) \right]^{1/2} \qquad (17.4 \quad \text{MeV}) \quad (17)$$

Here x is the fissionability parameter defined in Eq. (11). For ^{208}Pb, as an example, $\hbar\omega_2 \approx 1$ MeV. This is quite low compared with the characteristic particle excitation; see Section VI. The first excited vibrational states of even–even nuclei have spin $I = \lambda$. In the case of quadrupole excitations, the next states have energy $2\hbar\omega_2$ and spins 0, 2, 4.

B. Equilibrium Deformations and Rotations

In regions between closed-shell configurations, nucleonic orbits (Section VI,B) stabilize the nuclei at nonspherical shapes. The overwhelming preference is for a prolate spheroidal (i.e., football) shape. A measure of the distortion is the deformation parameter, defined by

$$\varepsilon = \frac{a - b}{(a + 2b)/3} \qquad (18)$$

where a and b are the major axes of the ellipsoid of revolution, a along the symmetry axis and b perpendicular to the symmetry axis.

Large permanent deformations occur with great regularity in certain regions of Z and N; see Fig. 9. Near the stable valley, these are $19 < A < 25$, $150 < A < 185$, and $220 < A$. (Note that doubly magic nuclei appear at $A = 8 + 8 = 16$, $50 + 82 = 132$, and $82 + 126 = 208$.) The two heavier regions correspond, by coincidence, with the chemical rare earth regions. Large deformations begin abruptly at $N = 90$ (near A of 150) and depend more strongly on N than on Z in the first rare-earth region. Similarly, large deformations begin again at $Z = 88$ (near A of 220) and depend more strongly on Z in the second rare-earth region. Other regions of large deformation occur among the metastable nuclei away from the stable valley.

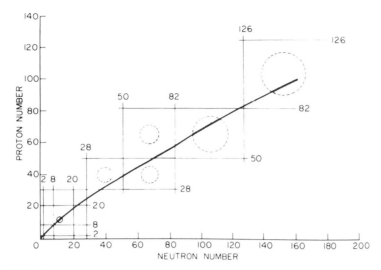

FIG. 9. Regions where deformed nuclei are expected are marked with circles. The solid curve is the line of β stability. Shaded areas mark regions of known deformed nuclei. [From Harvey, B. G. (1969). "Introduction to Nuclear Physics and Chemistry," Prentice-Hall, Englewood Cliffs, N.J.]

Permanent deformations lie in the range $0.1 \leqslant \varepsilon \leqslant 0.4$. The identification of such shapes is based on (1) electrical quadrupole moments, first observed in atomic hyperfine studies, (2) electrical quadrupole transitions induced by electric fields from scattered charged particles (especially α particles), (3) the nature of the rotational spectra, (4) beta and gamma transition rates, (5) atomic isotope shifts, (6) spectra of muonic atoms, etc.

The lowest mode of excitation of a strongly deformed nucleus is rotational. For axially symmetric shapes, the moment of inertia is very small about the axis of symmetry, and rotation occurs about an axis perpendicular to the symmetry axis, analogous to a rotating dumbbell or diatomic molecule (see Fig. 10). The rotational motion is then nearly decoupled from the other collective modes. The excitation energy follows the law

$$E_{\mathrm{Rot}} = \hbar^2/2\mathcal{I}\,[I(I + 1) - I_0(I_0 + 1)] \quad (19)$$

plus corrections for high spin. All band members have the same parity. Here I_0 is the spin of the lowest member of the band. The spin I assumes the values I_0, $I_0 + 1$, $I_0 + 2$, ... except that for even–even nuclei one has $I = 0, 2, 4, \ldots$. (For even–even nuclei, the ground-state spin and parity are always 0^+.) The moment of inertia $\mathcal{I}$ is greater than that of an irrotational fluid, but less than the solid body value.

Rotational energies are small compared with vibrational or particle excitations. The first rotational excitation energy ranges from 50 keV in heavy nuclei to several hundreds of kilo-elec-

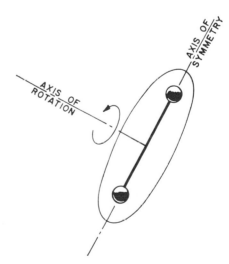

FIG. 10. Rotation of a nucleus, in analogy with a rotation dumbbell (or diatomic molecule). Because the moment of inertia is very small along the axis of symmetry, the object rotates about an axis perpendicular to the symmetry axis. The example is for an even-Z, even-N nucleus, where there is no internal angular momentum to be considered. [From Wilets, L. (1959). *Science* **129**, 361.]

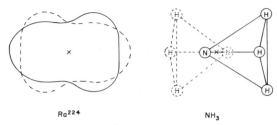

Ra²²⁴   NH₃

FIG. 11. Schematic representation of a pear-shaped nucleus, compared with an equally schematic representation of the ammonia molecule, NH₃. The broken curves indicate the inverted configurations. The crosses are the centers of mass. [From Wilets, L. (1959). *Science* **129**, 361.]

tron-volts for light nuclei. Rotational bands have been observed with spins as high as 60.

Vibrational excitations can also be found on permanently deformed nuclei. Collective excitations for "transition" nuclei (neither good vibrators or rotators) are more complex.

Nuclei also occur with equilibrium octupole deformations. When compounded with a quadrupole deformation, the nuclei assume a pear-shaped surface deformation, as shown in Fig. 11. Such have been identified in the even–even nuclei around $_{88}$Ra. An indicator of such shapes is the existence of low-frequency "tunnelling" vibration corresponding to inversion of the shape. As a result, even–even nuclei exhibit a band of odd spins, $I = 1, 3, 5, \ldots$, with odd parity, displaced from the even-spin band by the tunnelling frequency, which is of the order of 100 keV.

C. Fission

As noted in Section IV,C, nuclei for which the fissionability parameter x is less than unity are stable against small deformation. Nevertheless the energy release Q_f is positive for nuclei down to $x = 0.35$. The fissile nuclei, thorium and heavier, have $Q_f \gtrsim 190$ MeV. This large amount of energy can be identified with the electrostatic energy of the fragments at the point of scission, $Z_1 Z_2 e^2 / (R_1 + R_2)$; during the process of separation the potential energy is converted into kinetic energy of the fragments, neutron emission, and gamma emission.

The potential energy-of-deformation barrier against fission is about 6 MeV for the even–even isotopes of uranium. If this much energy is deposited, the nucleus undergoes rapid fission. This can be achieved by neutron capture on an odd-A isotope (e.g., ^{235}U + n → ^{236}U* → fission,

where the asterisk indicates a state of energy excitation), by gamma absorption, or by any number of nuclear reactions. It is the emission of neutrons ($\gtrsim 2.5$) during the fission process that makes possible a sustained chain reaction.

Mass division during the fission process is usually very asymmetric. The two fragments that emerge have a distribution of masses, and division into equal mass fragments is highly unlikely at low excitation energy. When compared for various fissioning nuclei, the distribution associated with the more massive fragment is seen to be stationary about the doubly magic numbers $Z = 50$, $N = 82$, hence $A = 132$ (see Fig. 12). With increasing excitation energy, the probability of symmetric mass division increases.

A frequent feature of fission is shape isomerism. It plays an important role in understanding the fission process and in testing nuclear models. For uranium and heavier nuclei, the energy of deformation plotted against (say) the quadrupole deformation parameter along the path from spherical to scission generally is characterized by two peaks and two minima. The lowest point corresponds to the most stable (ground-state) configuration, which is usually nonspherical. There is another local minimum at larger deformation that is about 3 MeV higher. During nuclear reactions, some fraction of the events results in populating the second minimum. Because of the smaller barrier to fission (compared with the ground state), either spontaneous fission or decay to the ground state can occur with reduced, but measurable, half-life. The observed half-lives range from 10^{-3} to 10^{-9} sec.

D. Isospin and Spin Modes

The most dramatic example of "isospin" waves is the giant dipole resonance. This collective state corresponds to an oscillation separating neutrons from protons, while still maintaining rather constant total nuclear density. The energy of excitation follows the approximate formula

$$E = \hbar\omega \simeq 78A^{-1/3} \quad \text{MeV} \qquad (20)$$

for $A > 60$. This mode can be excited by various mechanisms, such as gamma ray absorption. The resonance is quite broad, with a full width at half-maximum for gamma-ray excitation cross sections varying from 3 to 10 MeV.

Higher multipoles of various types are possible and have been studied. As an example, nu-

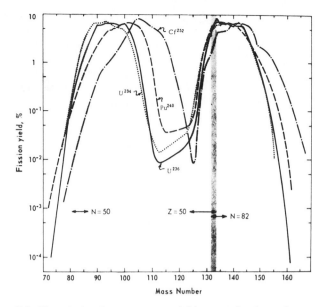

FIG. 12.   Fission-fragment mass yield curves for thermal neu-
tron-induced fission of ^{233}U, ^{235}U, and ^{239}Pu, and spontaneous
fission of ^{252}Cf. Figure is labeled by A of compound nucleus.
Shaded areas indicate approximate positions of nuclear shells.
[After Wahl, A. C. (1965). *Proc. IAEA Symp. Phys. Chem.
Fission, Salzburg,* Vol. I, p. 317, IAEA, Vienna.]

cleons of (say) spin up can oscillate collectively
against nucleons of spin down.

VI. Shell Structure of Nuclei

While the concept of a nuclear fluid is useful,
the analog fluid is somewhere between a liquid
and a gas. The nucleus is self-bound with a well-
defined surface (like a liquid), but the nucleons
move about freely inside (like a gas). This is
possible because nuclear forces, which are classified as "strong," are, when scaled appropriately, actually weaker than interatomic or intermolecular forces—weaker even than the He–He
force, which does lead to a liquid at low temperatures. This is discussed further in Section IX.

Nucleons moving freely inside a nucleus is a
concept similar to that of electrons moving
freely inside an atom. It is called the "shell
model" or "independent-particle model." Each
nucleon experiences the same potential, and the
wave functions, or orbitals, can be calculated by
solving the Schrödinger equation.

A. Spherical Nuclei

For spherical nuclei, the one-body nuclear potential is usually written

$$V(r) = V_c(r) + \gamma \frac{1}{r} \frac{dV_c(r)}{dr} \mathbf{l} \cdot \mathbf{s} \qquad (21)$$

with V_c of the Wood–Saxon form

$$V_c(r) = \frac{V_0}{1 + e^{(r-R_v)/a_v}} \qquad (22)$$

Here $\mathbf{l} \cdot \mathbf{s}$ is the spin–orbit operator; γ is a constant equal to a number between 5 and 10. (In
atomic physics, $\gamma = -1$.) The protons experience a Coulomb potential as well.

The form of the central potential is similar to
that of the charge distribution of Eq. (1). The
strength is $V_0 \approx -50$ MeV. The size parameter
R_v is a few tenths of a fermi (femtometer) greater
than the radius parameter for charge distribution; the surface thickness is about the same.

An orbital is characterized by the quantum
numbers n, l, j, m, where n is the radial quantum
number, l is the orbital angular momentum, j is
the total (spin plus orbital) angular momentum,
and m is the projection of total angular momentum onto (say) some z axis. The Pauli principle
demands that no two identical particles occupy
the same orbital. The lowest state of a nucleus is
generally obtained by filling the lowest-energy
orbitals. The orbitals are degenerate (equal in
energy) for different m. Since m can assume the
values $-j, -j + 1, ..., +j$, there are $2j + 1$ orbitals of the same energy.

The effect of the spin orbit term is to lower
states with $j = l + \frac{1}{2}$ with respect to states with
$j = l - \frac{1}{2}$ (same l).

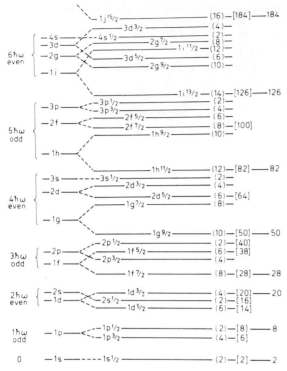

FIG. 13. Schematic representation of single particle energy levels in the spherical shell model. On the left are shown the levels in a harmonic oscillator; in the middle are shown the low levels for a realistic potential; on the right are shown the levels where spin–orbit coupling is included. [From Mayer, M. G., and Jensen, J. H. D. "Elementary Theory of Nuclear Shell Structure," Wiley, New York.]

A crude but useful estimate of single-particle energies and wave functions can be obtained by approximating $V_c(r)$ with a harmonic oscillator potential, $V_c(r) \approx \frac{1}{2}kr^2 + V_0$. The corresponding oscillator frequency would be $\omega = \sqrt{k/m}$, where m is the mass of the neutron or proton. The value of ω is the same functional form as that given in Eq. (20) above, but with numerical coefficient 41; $2\pi/\omega$ is the round-trip transit time of a nucleon in any orbit (in this approximation), which is approximately $A^{1/3} \times 10^{-22}$ sec.

In a schematic, rather than rigorous, way, one can plot the energy levels for the protons and neutrons in any spherical nucleus on a single diagram. This has meaning only for the low-lying states, since the higher levels (i.e., $\varepsilon_{nljm} > 0$) are continuous, rather than discrete. The result is shown in Fig. 13. Normally the orbitals are filled, one per nucleon, in the order of increasing energy, up to the number of protons or neutrons in the nucleus. The numbers at the right denote the number of orbitals up to that position, taking

into account the $2j + 1$ degeneracy. Note that $2j + 1$ is an even integer. These numbers are at gaps in the spectrum, corresponding to shell closures (i.e., the magic numbers: an additional proton or neutron is less tightly bound by the magnitude of the gap than the previous nucleons).

The attractive nature of the nucleon–nucleon force is such that even numbers of protons or of neutrons (separately) couple together to zero total angular momentum. All ground states of even–even nuclei have total angular momentum 0 and positive parity, 0^+ in the notation J^π. Odd-A nuclei have $J^\pi = j^\pi$, where j^π is the angular momentum and parity of the odd nucleon in the unfilled shell.

B. DEFORMED NUCLEI

Non-closed-shell nuclei prefer nonspherical shape. The situation is rather analogous to a marble moving in a circular orbit inside a rubber balloon (see Fig. 14): the centrifugal force of the marble deforms the balloon into an oblate spheroid. When a new shell is begun, the added nucleons prefer equatorial orbits ($m = \pm j$). As the shell is filled, the orbits tend to reinforce the deformation, filling in the order $m = \pm j, \pm(j - 1), ..., \pm\frac{1}{2}$. This leads to a cooperative, or collec-

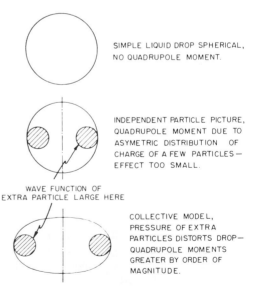

FIG. 14. An illustration of Rainwater's principle for an individual nucleon deforming a nuclear core. The shaded area represents the nucleon's wave function (orbit). Centrifugal force exerted by the nucleon deforms the core into an oblate spheroidal shape, as a marble orbiting rapidly inside a balloon would do. [From Hill, D. L., and Wheeler, J. A. (1953). *Phys. Rev.* **89**, 1106.]

tive, enhancement of the deformation, which is greatest near the middle of the shell filling. Beyond the midpoint, the orbitals reduce deformation. When the shell is completely filled, all $2j + 1$ m states are occupied and the system is again spherical. A different sequence of filling that also favors deformation would begin with the polar orbitals (smallest m) and proceed in the order $m = \pm\frac{1}{2}, \pm\frac{3}{2}, \ldots, \pm j$. These simple arguments would seem to imply that nuclei with shells less than half filled should be oblate, and those with shells more than half filled should be prolate, switching abruptly at half filling. Details of the nucleon–nucleon interaction, however, do favor the prolate shape, and this is clearly the case for all strongly deformed nuclei. The resultant deformed nuclei exhibit large intrinsic electric quadrupole moments. They are larger, by a factor of between 5 and 20, than the moments contributed by a single proton.

In the case of deformed nuclei, the shell-model potential that each nucleon experiences is described by a nonspherical, usually (but not necessarily) axially symmetric potential. In fact, the states can no longer be properly described by the quantum numbers l and j, since the resultant wave functions are linear combinations of states of different l and j. Parity and m, however, remain "good quantum numbers." Furthermore, the single-particle energies are no longer degenerate with respect to m, but (for axial symmetry) there is a degeneracy with respect to the sign of m. Single-particle energy levels as a function of deformation are displayed

in Fig. 15. The ground states of deformed even–even nuclei are still 0^+, but for odd-A nuclei one finds $J^\pi = m^\pi$, where m^π refers to the last odd nucleon.

The structure of transition nuclei is more complex.

VII. Nuclear Reactions

In addition to natural decay processes, nuclear reactions can be induced by any of the elementary particles, by gamma rays, and by other nuclei. Through various reaction mechanisms, nuclei can be excited or transmuted. The resultant products can be analyzed to obtain an understanding of the physics or utilized for practical applications. The following subsections deal with a small part of the wealth of nuclear reactions that have been studied.

A. NUCLEON SCATTERING

Protons and neutrons of energies ranging from a fraction of an electron volt (for neutrons) to many billions of electron volts (for protons) have been used as projectiles to bombard nuclei. The simplest process is that of elastic scattering, where the projectile is deflected but the target nucleus is left undisturbed. The process can be described mathematically by considering the projectile to move in a potential of the form given by Eq. (21). However, some of the events involve inelastic processes, such as energy excitation followed by decay, or some kind of transmutation. These can be included by making the

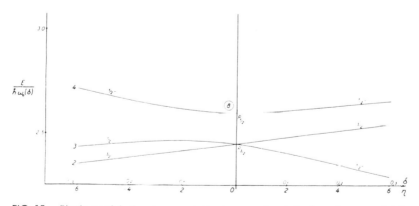

FIG. 15. Single-particle (neutron or proton) energy levels in the nilsson model for quadrupole deformations. Here δ has the same meaning as the ε of Eq. (18). The curves are labeled by the magnetic quantum number and parity m^π. The circled numbers are the magic numbers, corresponding to the number of orbitals below. The first $s_{1/2}$ level is not shown. The figure continues on p. 315. [From Nilsson, S. G. (1955). *Mat. Fys. Medd. Dan. Vid. Selsk.* **29**, No. 16.] *(Figure continues.)*

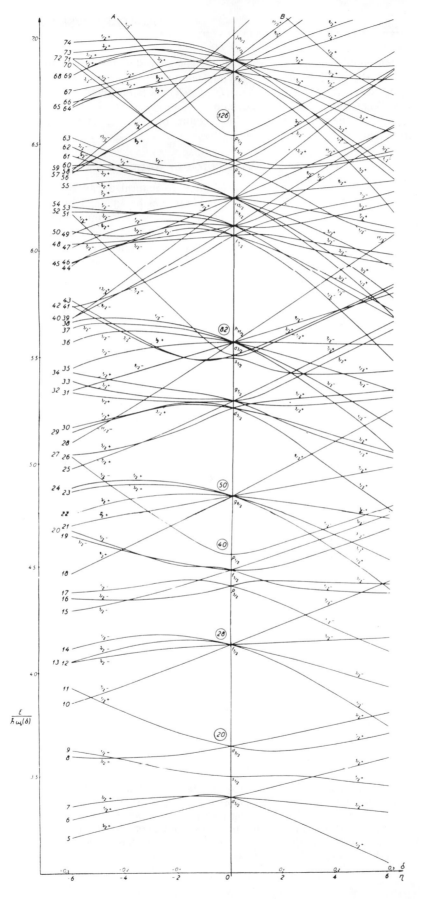

FIG. 15 (*Continued*)

potential complex (i.e., containing an imaginary part as well as a real part). The imaginary part of the potential removes probability from the elastic "channel," and the resultant wave function describes only the elastic process. This description is called the "cloudy crystal ball model," in analogy with the scattering of light from a diffractive, absorptive medium.

Nonelastic scattering refers to any process that leads to a different energy state of the target (or projectile, if it is composite) or any rearrangement of the constituents. A special case is inelastic scattering, which retains the composition of the projectile and target but leaves one or the other in an excited state.

It is frequently useful to separate the reaction process into an initial, "direct" reaction, followed by the subsequent decay of the excited intermediate state. If the nucleus passes through complicated intermediate states, one speaks of "compound nucleus" formation and subsequent decay of the compound nucleus.

The standard notation for nucleus reactions is $T(p, p')T'$ for $p + T \rightarrow T' + p'$ or sometimes simply (p, p'). Here p is the incident projectile and T the target; the primed quantities are the corresponding objects at the end of the reaction. In the above, the projectile p is generic; it can be a proton (p), neutron (n), deuteron (d), etc. Direct nucleon-induced reactions include the following: (a) charge exchange, (n, p) or (p, n); (b) capture, $p + {}_Z X_N \rightarrow {}_{Z+1} X_N^*$ or $n + {}_Z X_N \rightarrow {}_Z X_{N+1}^*$; (c) pickup, (p, d) or (n, d).

The final reaction products can include emitted gamma rays (photons), beta particles, further nucleons, alpha particles, fragments, fission, etc.

B. Other Elementary Particles

Gamma rays, electrons and positrons, muons, pions and other mesons, and antiprotons have been used as projectiles to bombard nuclei. Each is unique in probing different aspects of nuclear structure: Electrons and muons, as discussed in Section III,A, can be used to explore nuclear charge distributions because they interact as point particles and only through the electromagnetic field. Photons, of course, also interact only electromagnetically, and deposit a discrete amount of energy and momentum upon absorption. Pions interact with nuclei through the strong force; they can be absorbed (depositing the large rest mass of the pion as energy of

excitation), scattered (with target excitation), or undergo charge exchange ($\pi^+ + {}_Z X_N \rightarrow \pi^0 + {}_{Z+1} X_{N-1}$, and similarly for the other charged pions) or, more rarely, double charge exchange ($\pi^+ + {}_Z X_N \rightarrow \pi^- + {}_{Z+2} X_{N-2}$, etc.).

Antiprotons interact strongly with nuclei, but also annihilate easily. Therefore, they explore the periphery of nuclei. Like other negatively charged elementary particles, they can also form hydrogen-like atoms with nuclei. From the spectrum of gamma rays produced through atomic transitions, information on their interaction with nuclei can be deduced.

C. Nuclei as Projectiles

Nuclei of all masses have been used as projectiles. Heavy nuclei are especially effective in exciting electric modes or high spin states. They have also been used in the search for new regions of stability—the superheavy nuclei.

Two heavy nuclei (e.g., U + U) can fuse to form a superheavy nucleus, but the compound system is not stable. It decays rapidly by fission into two or more fragments with the emission of other particles.

An interesting and novel process is the collision between two heavy nuclei at energies below the Coulomb barrier, so that the nuclear forces between the objects do not act. Although the two nuclei are close together for only a very short time, it is still long compared with electron transit times and the system behaves as a quasi-molecule or a quasi-atom. The high total electric charge of the nuclei can give rise to exotic atomic processes, including spontaneous positron emission and the formation of a "charged vacuum."

VIII. Energy Sources

Nuclear reactions are an important source of energy in nature and in technology. Radioactive substances of moderate half-lives have been used, especially in space missions, as a source of energy to operate low-power equipment. More important are the large energy yields of fusion and fission reactions, which are outlined briefly below. Both reactions are exothermic because the most stable nuclei are of intermediate mass, which is the region of maximum binding energy.

A. Fusion

In the early stages of stellar evolution, stars are fueled by the fusion of hydrogen nuclei. This is the case with our own sun. The process requires high temperatures in order to bring the electrically charged nuclei close enough together to interact. In fact, fusion in stars is generally a very slow process, fed by the relatively few particles in the high-velocity tail of the thermal distribution. Two main mechanisms are distinguished in solar hydrogen burning: the direct p + p reaction, and the catalytic C–N–O cycle. Both lead to the overall result

$$4^1H \rightarrow {}^4He + 2e + 2\bar{\nu} \qquad (23)$$

The energy release, including the energy carried off by the antineutrinos is 25.7 MeV, or about 0.7% of the rest mass of the four hydrogen atoms.

Fusion is also the source of energy in hydrogen bombs and has been studied as a source of controlled thermonuclear energy generation. A reaction that occurs with high probability and is therefore promising for a power source is

$$^2H + {}^3H \rightarrow {}^4He + n \qquad (24)$$

with an energy release of 17.6 MeV. The specific energy (energy per unit mass of reactant) release in this process is nearly five times that of fission.

The technological problem in controlled fusion is the production of a high-temperature plasma at high density for a sustained period of time. Actually, "high density" here may only be a tiny fraction of 1 atm and confinement times may be only a small fraction of a second. Various techniques are currently under study: magnetic confinement, inertial implosion by laser or particle beams, and muon catalysis. [*See* PLASMA SCIENCE AND ENGINEERING.]

B. Fission Reactors

The energy release during fission is approximately 210–219 MeV. Coulomb energy of the fragments at separation (scission) accounts for about 80% of this energy, and this is primarily converted into the kinetic energy of the fragments. The remainder is released mainly in the form of gamma rays, neutrons, beta rays, and neutrinos. The specific energy release is about 2 million times that of coal or oil.

Even–even heavy, fissile nuclei have a fission threshold of about 6 MeV. This energy can be injected by the capture of a (slow) neutron on the next lighter even–odd isotope. The number of neutrons released per fission event is about 2.5 for slow neutrons on ^{235}U. The multiplication of neutrons provides the mechanism for a sustained fission reaction: the neutrons emitted can be used to produce more fissions. This is called a chain reaction.

Neutron-capture cross sections at low energies are proportional to the inverse of the neutron velocity. Thus low-energy neutrons are more effectively captured than high-energy neutrons. However, natural uranium contains only 0.72% of the isotope ^{235}U, and most neutrons would be captured onto ^{238}U and lost to the chain. A moderating substance, ^2H or ^{12}C, is often introduced to slow the neutrons before they interact and to take advantage of the fact that slow neutrons have a much higher cross section for fission on ^{235}U than for capture on ^{238}U. Nevertheless, such slow, or moderated, fission reactors require a higher enrichment of ^{235}U.

An alternative to the ^{235}U-enriched reactor is the breeder reactor, which can be either slow (moderated) or fast. Since neutron capture on ^{238}U or ^{232}Th leads, after beta decay, to the fissile nuclei ^{239}Pu or ^{233}U, the captured neutrons are not "wasted" but are utilized to breed new fissile nuclei. However, at least two neutrons are required for each breeding: one for the initial fission reaction, and one for each conversion of a nonfissile nucleus into a fissile one. Of the 2.5 neutrons per fission, 0.5 still remain to be wasted by escape from the vessel or nonuseful absorption. The world supply of ^{238}U and ^{232}Th is, for such purposes, essentially inexhaustible.

IX. Nuclear Forces

A. General Features

Nuclear forces are, to a high degree of accuracy, charge independent. That is to say, except for the explicit electromagnetic part, the neutron–neutron, neutron–proton, and proton–proton interactions are equal—when compared in the same state.

The nucleon–nucleon force is primarily central: the force acts along the line joining the nucleons, and the corresponding potential is a function only of the distance of separation of the nucleons. However, noncentral components are significant, especially the spin–orbit and tensor components. The interaction also depends on the relative orbital angular momentum and the relative orientation of the two nucleon spins. In fact, the interaction contains all of the complications allowed by the fundamental symmetries of nature.

The central two-body potential decreases more rapidly then the r^{-1} Coulomb form. The long and intermediate range part of the potential is negative and attractive. This is responsible for nuclear binding. The short range part is positive and strongly repulsive; it is sometimes approximated by a hard (infinite) core of radius 0.4–0.5 fm. The strong repulsion plays a crucial role in nuclear saturation.

B. Meson Theory

The "strong" force between nucleons is mediated by the exchange of various mesons, which are so named because they are (usually) intermediate in mass between electrons and nucleons. In a first approximation, the form of the potential energy of interaction is that given by Yukawa,

$$f^2 e^{-\mu r}/r$$

where $\mu = mc/\hbar$ is the inverse of the range of the interaction and m is the mass of the exchanged meson. At short distances ($\mu r \ll 1$), the potential varies as r^{-1}, just as the Coulomb potential. However, the exponential factor causes a more rapid decrease with distance.

The lightest meson is the pion, originally conjectured by Yukawa. The pion comes in three charges: $+$, 0, $-$. Its mass is about one-seventh that of the nucleon, giving a range of interaction of $\mu_\pi^{-1} = 1.4$ fm. The strength of the interaction is $f^2 \approx 0.08\hbar c$, which is about 11 times that of the Coulomb interaction at short distances. The pion contribution to nuclear forces dominates at large distances, $r \gtrsim 1$ fm, but is relatively weak compared with the shorter range contributions discussed next.

The intermediate range attractive region (0.5 fm $\leqslant r \leqslant$ 1.0 fm) is governed by the exchange of two pions. The short range, repulsive region ($r \leqslant 0.5$ fm) is dominated by the exchange of heavier mesons, especially the rho and omega mesons, which are $5\frac{1}{2}$ times the mass of the pion and hence give a range of 0.25 fm.

It is, however, more appropriate to describe the short-range part of the nucleon–nucleon interaction in the framework of nucleonic substructure. This is an area of current investigation.

C. Comparison of Internucleonic and Interatomic Forces

Both internucleonic and interatomic forces share similar features: long-distance attraction and short-distance repulsion. Nuclear forces are millions of times stronger than atomic forces, but their range is only 10^{-5} of the range of the atomic forces. When the appropriate comparison is made, it is surprising to note that nuclear forces are actually *weaker* than most atomic forces! The comparison made here is in the context of describing the quantum-mechanical many-body system.

As an example, consider the atomic He–He potential. The interaction is extremely weak on the atomic scale. Helium is a noble gas; there is no bound state of the diatomic molecule He_2, although most gases do form diatomic molecules and generally admit many excited bound states. On the nuclear side of the comparison there is the dinucleon, deuterium, which has only one bound state, and there is no stable dineutron or diproton. (Note that this does not violate charge independence of nuclear forces. The deuteron is in a state where the two nucleon spins are parallel; this state is forbidden to the dineutron and diproton systems by the Pauli principle. The spin parallel state has a stronger interaction than the spin antiparallel state.)

In order to effect comparison between nuclear and atomic potentials, it is convenient to multiply the strengths and lengths by dimensioned constants in order to reduce them to dimensionless quantities. In each case, one case one sets $r = \lambda x$ where λ is a constant appropriate to the size of the system in question and x is now the dimensionless length. The strength of the potential is also multiplied by a constant which renders the product dimensionless. This gives the scaled potential

$$v(x) = \frac{2m\lambda^2}{\hbar^2} V(\lambda x) \qquad (25)$$

Here m is the mass of the particle. In Fig. 16 are plotted $v_N(x)$ and $v_{He}(x)$ with the λ's chosen for each so that the potentials intersect the abscissa at the same x, namely, $x = 1$. Note that the potentials thus plotted are quite comparable.

A consequence of the weakness of nuclear forces is that nuclei are not solid structures. They are better described as a fluid intermediate between a gas and a liquid. Nucleons move about freely within a nucleus, like a gas, but form a self-bound system, like a liquid.

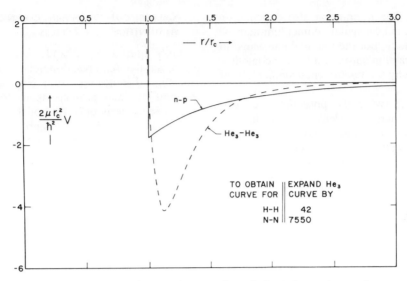

FIG. 16. A comparison of the nucleon–nucleon singlet (spin zero) central potential with the interatomic helium–helium potential, scaled so that the dimensionless radius where each passes through zero is the same. The spin 1 N–N potential is somewhat stronger and can support one bound state, the deuteron.

A nucleus in its ground state is to be compared with a many-atom system at zero absolute temperature. Nearly all materials solidify at zero temperature; an exception is the atomic system made up of isotopic ^{3}He, which liquifies but does not solidify.

B. SUBNUCLEAR STRUCTURE

It is now generally accepted that the fundamental theory of the strong interactions is quantum chromodynamics (QCD). The particles of the theory are quarks, which are spin-$\frac{1}{2}$ fermions and come in six known *flavors*, named *up, down, strange, charm, top,* and *bottom.* (The top quark has not yet been verified experimentally.) Quarks carry electrical charge in multiples of one-third of the electronic charge. They also carry another quantum number called *color,* and associated with color is color charge. There are three colors, conventionally denoted by the three primary colors red, green, and blue. Antiquarks carry anticolor (e.g., anti-red, anti-green and anti-blue). The quarks interact with a force field, the quanta of which are called gluons.

The gluon field is very similar to the electromagnetic field with the important difference that, unlike the photons of electromagnetism, the gluons carry (color) charge and therefore interact with each other as well as with quarks. This renders the theory nonlinear and very diffi-

cult to solve. General features of the theory can be stated, however. These include the following:

1.   Physical particles can exist only in combinations of subparticles that have net color-neutral charge, technically called a color-singlet state. Thus nucleons consist of three quarks, one of each of the primary colors. Mesons contain one quark, and one antiquark that carries the anticolor of the quark. Isolated quarks cannot exist. This property is called color confinement.

2.   The electric charges associated with the various quarks are such that only integral multiples of the electronic charge are allowed for physical particles.

3.   The color analog of the square of the electric charge, called the "strong coupling constant," is not a true constant but depends on the size of the interacting structure (or the magnitude of the momentum transfer). The smaller the size of the interacting structure (the higher the momentum transfer), the weaker is the interaction. This is called asymptotic freedom. Conversely, the larger the separation, the stronger the interaction.

The lightest quarks are the up and down quarks. They have a mass of only a few thousandths of the mass of the nucleons. The up quark has electric charge $+\frac{2}{3}e$; the down quark has electric charge $-\frac{1}{3}e$. A proton consists of two up quarks and one down quark, for a net charge of $+e$; a neutron consists of one up quark and two down quarks, for a net charge of zero. The mass of the nucleons comes from the interaction of the quarks with the gluon field.

The nucleon is described as consisting of a core of three quarks confined within a sphere (sometimes called a "bag") of radius about 0.5–1.0 fm. Surrounding the core is a cloud of pions, which are themselves composed of quark–antiquark pairs, and can carry electric charge. Within the region of the nucleon core, the strong coupling constant has a value of between 1.0 and 2.0. This is to be compared with the corresponding electromagnetic fine structure constant, $\alpha \equiv e^2/\hbar c \approx 1/137$. [See ELEMENTARY PARTICLE PHYSICS.]

BIBLIOGRAPHY

Bohr, A., and Mottelson, B. R. (1969). "Nuclear Structure," Vol. I. (1974). "Nuclear Structure," Vol. II. Benjamin, New York.

Close, F. (1979). "An Introduction to Quarks and Partons." Academic Press, New York.

Cohen, B. L. (1971). "Concepts of Nuclear Physics." McGraw-Hill, New York.

DeBenedetti, S. (1964). "Nuclear Interactions." Wiley, New York.

deShalit, A., and Feshbach, H. (1974). "Theoretical Nuclear Physics," Vol. 1: Nuclear Structure. Wiley, New York.

Eisenberg, J. M., and Greiner, W. (1970). "Nuclear Theory." Vol. 1, Nuclear Models; Vol. 2, Excitation Mechanisms of the Nucleus; Vol. 3, Microscopic Theory of the Nucleus. North Holland, Amsterdam.

Elton, L. R. B. (1966). "Nuclear Theory." Saunders, Philadelphia.

Harvey, B. G. (1969). "Introduction to Nuclear Physics and Chemistry." Prentice-Hall, Englewood Cliffs, N.J.

Roy, R. R., and Nigam, B. P. (1967). "Nuclear Physics." Wiley, New York.

Vandenbosch, R., and Huizenga, J. R. (1973). "Nuclear Fission." Academic Press, New York.

Wilets, L. (1964). "Theories of Nuclear Fission." Oxford University Press, Oxford.

OPTICAL DIFFRACTION

Salvatore Solimeno   *University of Naples*

I. History
II. Mathematical Techniques
III. Helmholtz–Kirchhoff Integral Theorem
IV. Diffraction Integrals
V. Geometric Theory of Diffraction
VI. Coherent and Incoherent Diffraction Optics

GLOSSARY

Airy pattern: Far field diffracted by a circular aperture illuminated by a plane wave or field on the focal plane of an axial symmetric imaging system.

Babinet's principle: The sum of the field diffracted by two complementary screens (corresponding to a situation in which the second screen is obtained from the first one by interchanging apertures and opaque portions) coincides with the field in the absence of any screen.

Boundary diffraction wave: Representation introduced by Rubinowicz of the deviation of the diffracted field from the geometrical optics one as a wave originating from the screen boundary.

Diffraction-limited system: Optical system in which aberrations are negligible with respect to diffraction effects.

Far-field pattern: Diatribution of the field diffracted at an infinite distance from an obstruction.

Fourier methods: Analysis of diffraction and aberration effects introduced by optical imaging systems based on the decomposition of the field in a superposition of sinusoids.

Fraunhofer diffraction: Diffraction at a very large distance from an aperture.

Fresnel diffraction: Diffracted in proximity to an aperture.

Fresnel number of an aperture: Parameter related to the deviation of the field behind an aperture from the geometrical optics one.

Geometric theory of diffraction: Method introduced by J. B. Keller for calculating the field diffracted from an obstacle based on a suitable combination of geometrical optics and diffraction formulas for particular obstacles (wedges, half-planes, cylinders, spheres, etc.).

Green's theorem (Helmholtz–Kirchhoff integral theorem): Representation of the field by means of an integral extending over a closed surface containing the field point.

Huygens–Fresnel principle: Every point on a primary wave front serves as the source of secondary wavelets such that the wave front at some later time is the envelope of these wavelets.

Kirchhoff method: Expression of the field amplitude at any point behind a screen by means of an integral extending only over the open areas in the screen and containing the unobstructed values of the field amplitude and its normal derivative.

Modulation transfer function: Function measuring the reduction in contrast from object to image, that is, the ratio of image-to-object modulation for sinusoids of varying spatial frequencies.

Optical transfer function: Function measuring the complex amplitude of the image transmitted by an optical system illuminated by a unit amplitude sinusoidal pattern, versus the spatial frequency.

Spatial frequency: Representation of an object or an image as a superposition of sinusoids (Fourier components).

Spread function: Image of a point source object; its deviation from an impulse function is a measure of the aberration diffraction effects.

Any deviation of light rays from rectilinear paths which cannot be interpreted as reflection or refraction is called diffraction. As it stands today diffraction optics (DO) deals mainly with the description of fields in proximity to caustics, foci, and shadow boundaries of wave fronts delimited by apertures and with the far fields. Imaging systems, diffraction gratings, optical resonators, and holographic and image processing systems are examples of devices that depend for their ultimate performance on DO. Rudimentary solutions to DO problems can be obtained by using the Huygens–Fresnel principle. More accurate solutions are obtained by solving the wave equation with the Helmholtz–Kirchhoff integral formula. The modern geometric theory of diffraction combines ray optical techniques with the rigorous description of the field diffracted by typical obstacles (wedges, cylinders, spheres) to represent with great accuracy the light field diffracted by a generic obstruction.

I. History

A. THE FIRST IDEAS

The phenomenon of diffraction was discovered by the Jesuit Father Francesco Maria Grimaldi and described in the treatise *Physico Mathesis de Lumine, Coloribus et Iride,* published in 1665, two years after his death. As it stands today, diffraction optics (DO) is the result of a long and interesting evolution originating from the ideas illustrated in 1690 by Christian Huygens in his celebrated "Traité de la Lumière." Thomas Young had the great merit of introducing the ideas of wave propagation to explain, following Newton, the corpuscular theory of optical phenomena. He introduced the principle of interference, illustrated in his three Bakerian lectures read at the Royal Society in 1801–1803. By virtue of this principle, Young was able to compute for the first time the wavelengths of different colors. Later, in 1875, Augustine Jean Fresnel presented to the French Academy the famous treatise "La diffraction de la Lumière," in which he presented a systematic description of the fringes observed on the dark side of an obstacle where he was able to show agreement between the measured spacings of the fringes observed and those calculated by means of the wave theory.

B. THE WAVE THEORY AND MAXWELL EQUATIONS

A remarkable success of the wave theory of light was recorded in 1835 with the publication in the Transactions of the Cambridge Philosophical Society of a fundamental paper by Sir George Biddell Airy, director of the Cambridge Observatory, in which he derived his famous expression for the image of a star seen through a well-corrected telescope. The image consists of a bright nucleus, known since then as *Airy's disk,* surrounded by a number of fainter rings, of which only the first is usually bright enough to be visible to the eye. Successive developments, until Maxwell's publication in 1873 of the "Treatise on Electricity and Magnetism," took advantage of Fresnel's ideas to solve a host of diffraction problems by using propagation through an elastic medium as a physical model. In particular, in 1861 Glebsch obtained the analytic solution for the diffraction of a plane wave by a spherical object.

C. FOURIER OPTICS

Since the initial success of the Airy formula, the theory of diffraction has enjoyed increasing popularity, providing the fundamental tools for quantitatively assessing the quality of images and measuring the ability of optical systems to provide well-resolved images. To deal with this complex situation, Duffieux proposed in 1946 that the imaging of sinusoidal intensity patterns be examined as a function of their period. Then the optical system becomes known through an optical transfer function (OTF) that gives the system response versus the number of lines of the object per unit length. More recently, optical systems have been characterized by means of the numerable set of object fields that are faithfully reproduced. This approach, based on the solution of Fredholm's integral equations derived from the standard diffraction integrals, has allowed the ideas of information theory to be applied to optical instruments.

II. Mathematical Techniques

Diffraction optics makes use of a number of analytical tools:

1. Spectral representations of the fields (plane, cylindrical, spherical wave expansions; Hermite–Gaussian and Laguerre–Gaussian beams; prolate spheroidal harmonics).

2. Diffraction integrals.

3. Integral equations.

4. Integral transforms (Lebedev–Kontorovich transform, Watson transform).

5. Separation of variables.

6. Wiener–Hopf–Fock functional method.

7. Wentzel–Kramers–Brillouin (WKB) asymptotic solutions of the wave equations.

8. Variational methods.

In many cases the solutions are expressed by complex integrals and series, which can be evaluated either asymptotically or numerically by resorting to a number of techniques:

1. Stationary-phase and saddle-point methods.

2. Boundary-layer theory.

3. Two-dimensional fast Fourier transform (FFT) algorithm.

III. Helmholtz–Kirchhoff Integral Theorem

A. Green's Theorem

Let $u(\mathbf{r})e^{i\omega t}$ be a scalar function representing a time-harmonic ($e^{i\omega t}$) electromagnetic field which satisfies the Helmholtz wave equation

$$\nabla^2 u + \omega^2 \mu \varepsilon u - 0 \tag{1}$$

where $\nabla^2 = \partial^2/\partial x^2 + \partial^2/\partial y^2 + \partial^2/\partial z^2$ is the Laplacian operator and μ and ε are the magnetic permeability (expressed in henry per meter in mksa units) and the dielectric constant (faradays per meter), respectively, of the medium. It can be shown that the field $u(\mathbf{r}_0)$ at the point $\mathbf{r}_0$ depends linearly on the values taken by $u(\mathbf{r})$ and the gradient ∇u on a generic closed surface S containing $\mathbf{r}_0$ (Green's theorem):

$$u(\mathbf{r}) = \oiint_S \left[G(\mathbf{r}, \mathbf{r}') \frac{\partial u(\mathbf{r}')}{\partial n_0} - u(\mathbf{r}') \frac{\partial G(\mathbf{r}, \mathbf{r}')}{\partial n_0} \right] dS' \tag{2}$$

where $G(\mathbf{r}_0, \mathbf{r}) = \exp(-ikR)/4\pi R$, $R = |\mathbf{R}| = |\mathbf{r} - \mathbf{r}_0|$, and $\partial/\partial n_0$ represents the derivative along the outward normal $\hat{\mathbf{n}}_0$ to S.

For a field admitting the ray optical representation $u(\mathbf{r}) = A(\mathbf{r})e^{-ikS(\mathbf{r})}$ the integral in Eq. (2) reduces to

$$u(\mathbf{r}) = \frac{i}{2\lambda} \oiint_S \frac{A(\mathbf{r}')}{R} e^{-ik(R+S)}(\cos\theta_i + \cos\theta_d)\, dS' \tag{3}$$

where θ_i and θ_d are, respectively, the angle between the ray passing through $\mathbf{r}'$ and the inward

normal $\hat{\mathbf{n}}_0$ to S and the angle between the direction $-\hat{\mathbf{R}}$ along which the diffracted field is calculated and $\hat{\mathbf{n}}_0$. In particular, when the surface S coincides with a wave front, $\theta_i = 0$ and $\cos\theta_i + \cos\theta_d$ reduces to the *obliquity factor* $1 + \cos\theta_d$.

B. Huygens–Fresnel Principle

The integral of Eq. (3) allows us to consider the field as the superposition of many elementary wavelets of the form

$$du(\mathbf{r}) = \frac{e^{-ikR}}{4\pi R}$$

$$\left[\frac{\partial u(\mathbf{r}')}{\partial n_0} + u(\mathbf{r}')\hat{\mathbf{n}}_0 \cdot \mathbf{R}\left(ik + \frac{1}{R}\right) \right] dS' \tag{4}$$

According to the above equations every point on a wave front serves as the source of spherical wavelets. The field amplitude at any point is the superposition of the complex amplitudes of all these wavelets. The representation of a field as a superposition of many elementary wavelets is known as the *Huygens principle*, since it was formulated in his "Traité de la Lumière" in 1690. Fresnel completed the description of this principle with the addition of the concept of interference.

C. Kirchhoff Formulation of the Diffraction by a Plane Screen

Let us consider the situation in which a plane Π, of equation $z' = $ const, separates the region I containing the sources from the homogeneous region II, where the field must be calculated. In this case, replacing the function G in Eq. (2) by

$$G_\pm(\mathbf{r}, \mathbf{r}') = \frac{\exp(-ik|\mathbf{r} - \mathbf{r}'|)}{4\pi|\mathbf{r} - \mathbf{r}'|}$$

$$\pm \frac{\exp(-ik|\mathbf{r}_s - \mathbf{r}'|)}{4\pi|\mathbf{r}_s - \mathbf{r}'|} \tag{5}$$

with $\mathbf{r}_s$ referring to the specular image of $\mathbf{r}$ with respect to Π, it is easy to verify that Eq. (2) reduces to

$$u(\mathbf{r}) = -2 \iint_\Pi \frac{\partial u(x', y', z')}{\partial z'} \frac{e^{-ikR}}{4\pi R} dx'\, dy'$$

$$= 2 \iint_\Pi u(x', y', z') \left(ik + \frac{1}{R}\right) \frac{e^{-ikR}}{4\pi R} \tag{6}$$

$$\times \cos\theta_d dx'\, dy',$$

use having been made of the relation $\partial G_+(\mathbf{r}', \mathbf{t})/\partial n_0 = G_-(\mathbf{r}', \mathbf{r}) = 0$ for $\mathbf{r}'$ on Π. In particular, for an aperture Σ on a plane screen illuminated by the incident field $u_i(\mathbf{r})$, Eq. (6) can be approximated by

$$u(\mathbf{r}) = 2 \iint_{-\infty}^{+\infty} P(x', y')u_i(x', y', z')\left(ik + \frac{1}{R}\right)$$

$$\times \frac{e^{-ikR}}{4\pi R} \cos\theta_d \, dx' \, dy' \qquad (7)$$

where $P(x', y')$ (the *pupil function*) takes on the value one if (x', y') belongs to the aperture and vanishes otherwise. In writing Eq. (7) we have tacitly assumed that the field on the exit aperture coincides with that existing in the absence of the aperture; this approximation, known as the *Kirchhoff principle*, is equivalent to the assumption that a finite exit pupil does not perturb the field on the pupil plane. Since presumably the actual perturbation is significant only near the pupil edge, we expect the error related to the application of Kirchhoff's principle to be neglected provided the aperture is sufficiently large. Exact analysis of the effects produced by some simple apertures (e.g., half-plane) confirms the validity of Kirchhoff's hypothesis for calculating the field near the *shadow boundaries* separating the lit region from the shady side; the error becomes relevent only for field points lying deep in either the lit or the dark regions.

IV. Diffraction Integrals

A. FRESNEL AND FRAUNHOFER DIFFRACTION FORMULAS

Let us consider a field different from zero only on a finite plane aperture Σ. If we indicate by a the radius of the smallest circumference encircling the aperture and assume that $|z - z'| \gg a$, we can approximate the distance R with $|z - z'| + \frac{1}{2}[(x - x')^2 + (y - y')^2]/|z - z'|$, so that the diffraction integral of Eq. (7) reduces to the *Fresnel formula*

$$u(x, y, z) = \frac{i}{\lambda d} e^{-ikd} \iint_{\Sigma} u(x', y', z')\exp\left\{-i\frac{k}{2d}\right.$$

$$\left. \times [(x - x')^2 + (y - y')^2]\right\} dx' \, dy' \qquad (8)$$

where $d = |z - z'|$.

When $d \gg a^2\pi/\lambda$, where a is the radius of the aperture, we can neglect the terms of the exponential in the integrand of Eq. (8) proportional to $x'^2 + y'^2$, so that

$$u(x, y, z) = \frac{i}{\lambda d} e^{-ikR_0} \iint_{\Sigma} u(x', y', z')$$

$$\times \exp\left(ik\frac{xx' + yy'}{d}\right) dx' \, dy' \qquad (9)$$

where $R_0 = d + (x^2 + y^2)/2d$. The above equation, referred to as the *Fraunhofer diffraction formula*, allows us to express the far field in terms of the two-dimensional Fourier transform of u on the aperture plane, evaluated for $k_x = kx/d$ and $k_y = ky/d$. The Fraunhofer fields of some typical apertures illuminated by plane waves are plotted in Fig. 1.

B. ROTATIONALLY INVARIANT FIELDS AND AIRY PATTERN

The far field radiated by a circular aperture can be obtained by assuming A [see Eq. (3)] in-

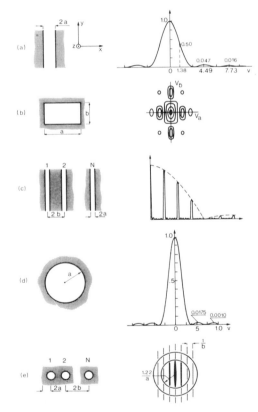

FIG. 1.   Diffraction patterns of typical apertures. The pattern functions $|G(\theta, \phi)|^2$ are respectively proportional to (a) $[\sin(v)/v]^2$, $v = ka \sin\theta \cos\phi$; (b) $\{[\sin(v_a)/v_a][\sin(v_b)/v_b]\}^2$, $v_a = ka \sin\theta \cos\phi$, $v_b = kb \sin\theta \sin\phi$; (c) $\{[\sin(v_a)/v_a][\sin(Nv_b)/\sin v_b]\}^2$, $v_a = ka \sin\theta \cos\phi$, $v_b = kb \sin\theta \sin\phi$; (d) $(2J_1(v)/v)^2$, $v = ka \sin\theta$; (e) $\{[2J_1(v_a)/v_a][\sin(Nv_b)/\sin v_b]\}^2$, $v_a = ka \sin\theta$, $v_b = kb \sin\theta \cos\phi$.

dependent of the angle ϕ and integrating the integral on the right-hand side of Eq. (9) with respect to ϕ, thus obtaining

$$u(\rho, z) = -i\,\mathrm{NA}\,ka\,\exp(-ikd - ikd\theta^2/2)$$
$$\times \int_0^1 A(ax)J_0(ka\theta x)$$
$$\times \exp[-ikS(ax)]x\,dx \qquad (10)$$

where $\rho = \sqrt{x^2 + y^2}$, $\theta = \rho/d$, NA is the numerical aperture of the lens; and $u = A\exp(-ikS)$. In particular, for $A = 1$, $S = 0$, Eq. (10) gives

$$u(\rho, z) \propto 2J_1(ka\theta)/ka\theta \qquad (11)$$

J_1 being the Bessel function of first order. The intensity distribution associated with the above field is known as the *Airy pattern* (see Fig. 1d). Because of the axial symmetry, the central maximum corresponds to a high-irradiance central spot, known as the *Airy disk*. Since $J_1(v) = 0$ for $v = 3.83$, the angular radius of the Airy disk is equal to $\theta = 3.83/ka$.

The central spot is surrounded by a series of rings corresponding to the secondary maxima of the function $J_1(v)/v$, which occur when v equals 5.14, 8.42, 11.6, etc. On integrating the irradiance over a pattern region, one finds that 84% of the light arrives within the Airy disk and 91% within the bounds of the second dark ring.

The Airy pattern can be observed at a finite distance by focusing a uniform spherical wave with a lens delimited by a circular pupil. In this case the quantity v is replaced by $k\,\mathrm{NA}\,\rho$, where $\rho = \sqrt{x^2 + y^2}$.

C. RESOLVING POWER

If we consider the diffraction images of two plane waves, it is customary to assume as a resolution limit the angular separation at which the center of one Airy disk falls on the first dark ring of the other (*Rayleigh's criterion of resolution*). This gives for the angular resolution

$$\theta_{\min} \simeq 1.22\lambda/D \qquad (12)$$

D representing the diameter of the exit pupil.

D. FIELDS IN THE FOCAL REGION OF A LENS

Imaging systems are designed with the aim of converging a finite conical ray congruence, radiated by a point source on the object plane, toward a focal point on the image plane. In most cases, the field relative to the region between the source and the exit pupil can be calculated by geometrical optics methods, that is, by evaluating the trajectories of the rays propagating through the sequence of refracting surfaces. However, downstream from the exit pupil, we are faced with the unphysical result of a field vanishing abruptly across the shadow boundary surface formed by the envelope of the rays passing through the edge of the *exit pupil*. To eliminate this discontinuity it is necessary to resort to the diffraction integral representation. In particular, the field on the exit aperture can be assumed to coincide with that existing in the absence of the aperture and can be calculated by geometrical optics methods.

When the numerical aperture of the beam entering or leaving the lens is quite large, it is necessary to account for the vector character of the electric field $\mathbf{E}$. This occurs, for example, in microscope imaging, where the aperture of the beam entering the objective can be very large. As a consequence, for rotationally symmetric lenses, the focal spot for linearly polarized light is not radially symmetric, a fact that affects the resolving power of the instrument.

If we choose a Cartesian system with the z axis parallel to the optic axis and the plane $z = 0$ coinciding with the Gaussian image of the object plane $z = z_0\,(<0)$ (see Fig. 2), it can be shown for a field point very close to the Gaussian image that Eq. (3) generalizes to the *Luneburg–Debye integral*

$$\mathbf{E}(\mathbf{r}) = i\frac{e^{-ikV_0}}{\lambda} \iint_{\bar{A}} \mathbf{E}'(\hat{\mathbf{n}}_0)\exp[-ik(p(x - \bar{x})$$
$$+ q(y - \bar{y}) + rz)]\,d\Omega \qquad (13)$$

where $\mathbf{E}' = Re^{ikR}\mathbf{E}$, $p = -n_{0x}$, $q = -n_{0y}$, and $r = -n_{0z}$ are the *direction cosines* of the ray passing through (ξ, η, ζ) (see Fig. 3) while $d\Omega =$

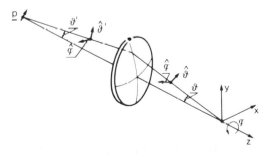

FIG. 2. Mutual orientation of the spherical coordinate systems relative to the source and the image formed by a lens.

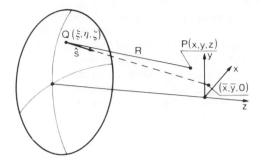

FIG. 3. Schematic representation of field point $P(x, y, z)$, wave front $Q(\xi, \eta, \zeta)$, and Gaussian image $(\bar{x}, \bar{y}, 0)$.

$d\bar{A}/R^2$ is the solid angle that the surface element $d\bar{A}$ of the exit pupil subtends at the focus.

It is now useful to introduce *optical coordinates* v, $\bar{u}$ defined by

$$v = k\rho \, \mathrm{NA}, \qquad \bar{u} = kz \, \mathrm{NA}^2 \qquad (14)$$

where $\mathrm{NA} = \sin \theta_{max}$ represents the *numerical aperture*, θ_{max} indicating the half-aperture in the *image space*. In this way, the Luneburg–Debye integral, for NA sufficiently small and a field

linearly polarized along the x axis, reduces to

$$\mathbf{E}(\mathbf{r}) \propto \hat{\mathbf{x}} \exp(-i\bar{u}/\theta_{max}^2)$$
$$\times \int_0^{2\pi} d\phi \int_0^1 d\Theta \, \Theta \, \mathbf{E}_i(\Theta, \phi)$$
$$\times \exp[iv\Theta \cos(\phi - \psi) + i\bar{u}\Theta^2/2] \qquad (15)$$

A three-dimensional plot of the field amplitude on a focal plane is shown in Fig. 4, and the streamlines of the Poynting vector in a focal region are represented in Fig. 5.

E. Field near a Caustic

When the field point approaches a *caustic*, then two or more rays having almost equal directions pass through P (see Fig. 6). In this case the diffraction integral I for a two-dimensional field takes the form

$$I \propto \frac{1}{\lambda^{1/2}} \int_{-\infty}^{+\infty} \exp[-ik(as + bs^3)] \, ds$$
$$\propto \frac{R^{1/2}}{\lambda^{1/2}} \left(\frac{2}{k\rho_c}\right)^{1/3} \mathrm{Ai}\left[-\left(\frac{k^2\rho'^6}{4\rho_c^4}\right)^{1/3}\right] \qquad (16)$$

$\mathrm{Ai}(\cdot)$ being the Airy function. A plot of the field in proximity to the caustic is shown in Fig. 7.

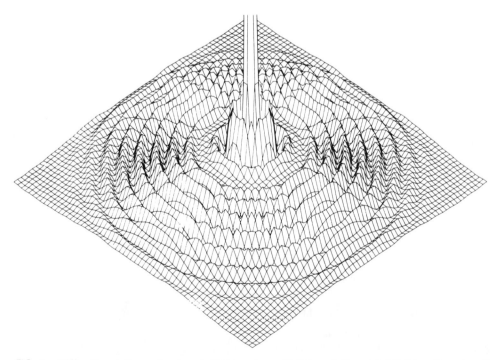

FIG. 4. Diffraction pattern of a ring-shaped aperture with internal diameter 0.8 times the external one, illuminated by a plane wave at a distance corresponding to a Fresnel number of 15. The three-dimensional plot was obtained by using an improved fast Fourier transform algorithm. [From Luchini, P. (1984). *Comp. Phys. Commun.* **31**, 303.]

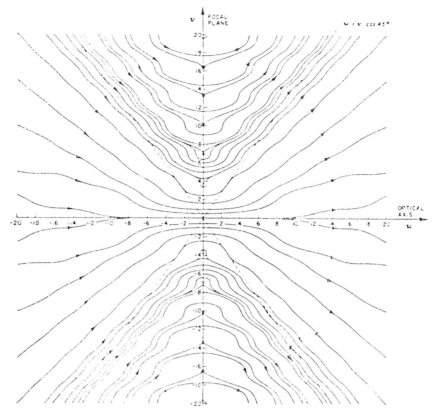

FIG. 5.   Flow lines of the Poynting vector near the focus of an aplanatic system with angular semiaperture of 45. [From Boivin, A., Dow, J., and Wolf, E. (1977). *J. Opt. Soc. Am.* **57,** 1171.]

V. Geometric Theory of Diffraction

As we know from experience, the edges of an illuminated aperture shine when observed from the shadow region. This fact was analyzed by Newton, who explained it in terms of repulsion of light corpuscles by the edges. In 1896 Arnold Sommerfeld obtained the rigorous electromagnetic solution of the half-plane diffraction problem. Using this result, it can be shown that the total field splits into a geometrical optics wave

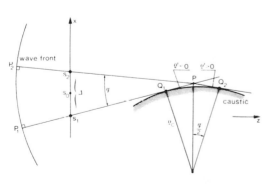

FIG. 6.   Geometry for calculation of the field in proximity to a caustic.

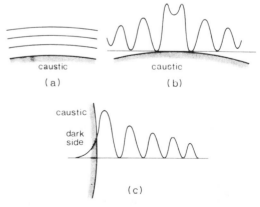

FIG. 7.   Fringes in proximity to a caustic (a) and field amplitude along a ray (b) and a normal to a caustic (c).

and a diffracted wave originating from the edge. Subsequently (1917), Rubinowicz recast the (scalar) diffraction integral for a generic aperture illuminated by a spherical wave in the form of a line integral plus a geometrical optics field. Parallel to this development, J. B. Keller (1957) successfully generalized the concept of ray by including those diffracted by the edges of an aperture.

In order to emphasize the geometric character of his approach, Keller called it the *geometric theory of diffraction* (GTD).

A. DIFFRACTION MATRIX

The asymptotic construction of diffracted fields will be illustrated by using as example the metallic wedge (see Fig. 8) illuminated by a plane wave having the magnetic field parallel to the edge (*s-wave or TM wave*). It can be shown that for $\rho \to \infty$ the diffracted field is given by

$$u(\rho, \phi) \sim D_s(\phi, \phi') \frac{e^{-ik\rho}}{\rho^{1/2}}$$
$$+ \sum_q \exp[-ik\rho \cos(\phi - \phi')]$$
$$\times [U(\phi - \beta_q + 2\pi) - U(\phi - \beta_q)] \quad (18)$$

where $U(x)$ is the unit step function and

$$D_s(\phi, \phi') = -\frac{e^{-i\pi/4}\lambda^{1/2}}{N}$$
$$\left\{ \frac{1}{\cos(\pi/N) - \cos[(\phi - \phi')/N]} \right.$$
$$\left. + \frac{1}{\cos(\pi/N) - \cos[(\phi + \phi')/N]} \right\} \quad (19)$$

where $N = 2\pi/\Phi$ and D_s is the *diffraction coefficient* of the wedge illuminated by an s-wave. For a wave having the electric field parallel to the edge (p-wave) the coefficient D_p can be obtained from D_s by changing to minus the plus sign be-

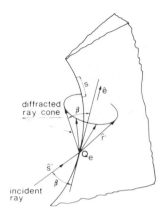

FIG. 9. Cone of rays diffracted by an edge point Q_e.

fore the second term on the right side of Eq. (19).

The above results have been extended to a metallic wedge illuminated by rays forming an angle β with the tangent $\hat{e}$ to the edge at the diffraction point Q_e (see Figs. 9 and 10). It can be shown that the diffraction rays form a half-cone with axis parallel to $\hat{e}$ and aperture equal to the angle β formed by the incident ray with the edge. If we consider the projection of the incident and diffracted rays on a plane perpendicular to the edge at Q_e, the position of the diffracted rays, forming a conical surface, is given by the angle ϕ_e, while the direction of the incident ray is defined by ϕ'_e. The electric component of the edge-diffracted ray can be expressed in the form

$$\mathbf{E}_d(\mathbf{r}) = [\rho_1/[r(\rho_1 + r)]]^{1/2} e^{-ikr}$$
$$\times \mathbf{D}(\phi, \phi'; \beta) \cdot \mathbf{E}_i(Q_e) \quad (20)$$

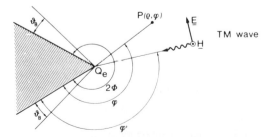

FIG. 8. Geometry of a wedge-shaped region.

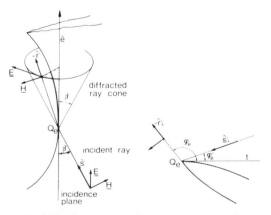

FIG. 10. Geometry for an edge-diffracted field.

where $\mathbf{r}$ represents the distance of the field point P from Q_e, and ρ_1 stands for the distance of P from the focal point along the diffracted ray. The *diffraction matrix* $\mathbf{D}$ has been derived by Kouyoumjian and Pathak and put in the form

$$\mathbf{D}(\phi,\ \phi';\ \beta) = \hat{\boldsymbol{\beta}}_d \hat{\boldsymbol{\beta}}_i D_p + \hat{\boldsymbol{\phi}} \hat{\boldsymbol{\phi}}' D_s \qquad (21)$$

where $\hat{\boldsymbol{\beta}}_d = \hat{\boldsymbol{\phi}} \times \hat{\mathbf{r}}$, $\hat{\boldsymbol{\beta}}_i = \hat{\boldsymbol{\phi}}' \times \hat{\mathbf{s}}'$, and D_p and D_s, which are a generalization of the diffraction coefficient (19) for $\beta \neq \pi/2$, are given by

$$D_{\substack{p \\ s}} = \frac{e^{-i\pi/4} \sin(\pi/N) \lambda^{1/2}}{N \sin \beta}$$

$$\times \left[-\frac{1}{\cos(\pi/N) - \cos[(\phi - \phi')/N]} \right.$$

$$\left. \pm \frac{1}{\cos(\pi/N) - \cos[(\phi + \phi')/N]} \right] \qquad (22)$$

B. DIFFRACTION FROM A SLIT

The GTD formalism can be applied conveniently to the calculation of the field diffracted by a slit of width $2a$ and infinite length. For simplicity, we assume a plane incident wave normal to the edges. As a first approximation we take the field on the aperture coincident with the incident field (Kirchhoff approximation). Then, following J. B. Keller, we can say that the field point P at a finite distance is reached by two different rays departing from the two edges and by a geometrical optics ray, if any (see Fig. 11a). The contribution of the diffracted rays can be expressed in the form

$$\mathbf{E}_d(\theta) = \frac{e^{-ik\rho}}{\lambda^{1/2}} \left[\mathbf{D}\left(\frac{3}{2}\pi + \theta, \frac{\pi}{2}; \frac{\pi}{2}\right) \right.$$

$$\left. + \mathbf{D}\left(\frac{3}{2}\pi - \theta; \frac{\pi}{2}; \frac{\pi}{2}\right) \right] \cdot \mathbf{E}_i \qquad (23)$$

A point that deserves comment is the use of the Kirchhoff approximation, which can be considered valid when the slit width is much larger than the field wavelength. This approximation can be improved, as shown by Keller, by taking into account the multiple diffraction undergone by the rays departing from each edge and diffracted from the opposite one (see dotted line in Fig. 11b).

VI. Coherent and Incoherent Diffraction Optics

An ideal imaging system can be described mathematically as a mapping of the points of the object plane Π_o into those of the image plane Π_i. For a finite wavelength and delimited pupil a unit point source located at (x_0, y_0) produces a field distribution $K(x, y; x_0, y_0)$ called the *impulse response*, which differs from a delta function $\delta^{(2)}(x - \bar{x}, y - \bar{y})$ centered on the Gaussian image of the object having coordinates $(\bar{x}, \bar{y})$. As a consequence, diffraction destroys the one-to-one correspondence between the object and the image.

The departure of K from a delta function introduces an amount of uncertainty in the reconstruction of an object through its image. This is indicated by the fact that two point sources are seen through an optical instrument as clearly separate only if their distance is larger than a quantity W roughly coincident with the dimension of the region on Π_i where K is substantially different from zero. The parameter W, which measures the smallest dimension resolved by an instrument, is proportional to the wavelength. This explains the increasing popularity of UV sources, which have permitted the implementation of imaging systems capable of resolutions better than 0.1 μm, a characteristic exploited in microelectronics for the photolithographic production of VLSI circuits.

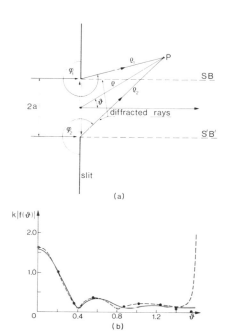

FIG. 11. Diffraction by a slit of width $2a$ (a) and relative far-field pattern (b) for a p-wave and $ka = 8$. The solid curve results from single diffraction; the dashed curve includes the effects of multiple diffraction. The dots represent the exact solution. [From Keller, J. B. (1957). *J. Appl. Phys.* **28**, 426.]

A. IMPULSE RESPONSE AND POINT SPREAD FUNCTION

Let us consider a unit point source in (x_0, y_0, z_0) producing a spherical wave front transformed by a composite lens into a wave converging toward the paraxial image point $(\bar{x}, \bar{y}, z)$. Using the Luneburg–Debye integral, we can express the impulse response $K(x, y; x_0, y_0)$ on Π_i as an integral extended to the wave front of the converging wave:

$$
K(x, y; x_0, y_0) = \exp\left(-ik\,\frac{x_0^2 + y_0^2}{2d_o} - ik\,\frac{\bar{x}^2 + \bar{y}^2}{2d}\right.
$$
$$
\left. + iv_x\bar{p} + iv_y\bar{q}\right) \times \frac{\Omega}{2\pi}
$$
$$
\times \iint P(p, q)A(p, q)
$$
$$
\exp[i(v_x p + v_y q)]\, dp\, dq
$$
$$
\equiv e^{-i\Phi}\bar{K}(x - \bar{x}, y - \bar{y}) \quad (24)
$$

where $v_x = k(x - \bar{x})\,\text{NA}$ and $v_y = k(y - \bar{y})\,\text{NA}$ are the optical coordinates of the point (x, y) referred to the paraxial image $(\bar{x}, \bar{y})$, NA is the numerical aperture of the lens in the image space, P the pupil function, p and q are proportional to the optical direction cosines of the normal to the converging wave front in the image space in such a way that $p^2 + q^2 = 1$ on the largest circle contained in the exit pupil, and d_o and d are the distances of the object and the image from the respective principal planes.

Physically we are interested in the intensity of the field, so it is convenient to define a *point spread function* $t(x, y; x_0, y_0)$ given by

$$
t(x, y; x_0, y_0) = |K(x, y; x_0, y_0)/K(\bar{x}, \bar{y}; x_0, y_0)|^2
$$
$$
(25)
$$

In particular, for *diffraction-limited instruments* with circular and square pupils, respectively, $\bar{K}$ is

$$
\bar{K} \propto 2J_1(v)/v \qquad \bar{K} \propto \frac{\sin v_x}{v_x}\,\frac{\sin v_y}{v_y} \quad (26)
$$

with

$$
v_x = k\text{NA}(x + Mx_0), \qquad v_y = k\text{NA}(y + My_0)
$$
$$
v = (v_x^2 + v_y^2)^{1/2} \quad (27)
$$

M being the *magnification* of the lens and J_1 the Bessel function of first order.

B. COHERENT IMAGING OF EXTENDED SOURCES

We can apply the superposition principle to calculate the image field $i(x, y, z)$ corresponding to the extended object field $o(x_0, y_0, z_0)$ defined on the plane region Σ_0, obtaining

$$
i(x, y, z) = \iint_{\Sigma_0} K(x, y, z; x_0, y_0, z_0)o(x_0, y_0, z_0)
$$
$$
\times dx_0\, dy_0 \quad (28)
$$

If we are interested in the intensity, we have

$$
|i(x, y, z)|^2 = \iint_{\Sigma_0} dx_0\, dy_0 \iint_{\Sigma_0} dx_0'\, dy_0'
$$
$$
\times K(x, y, z; x_0, y_0, z_0)
$$
$$
\times K^*(x, y, z; x_0', y_0', z_0)
$$
$$
\times o(x_0, y_0, z_0) \quad (29)
$$

C. OPTICAL TRANSFER FUNCTION

If we are interested in the intensity distribution $I(x, y) \propto \langle i(x, y)i^*(x, y)\rangle$ Eq. (29) gives for isoplanatic systems

$$
I(x, y) = \frac{1}{M^2} \iint_{\Sigma_0} I_0\left(-\frac{x_0}{M}, -\frac{y_0}{M}\right)
$$
$$
E(x - x_0, y - y_0)\, dx_0\, dy_0 \quad (30)
$$

where M is the magnification. With Fourier transformation, this relation becomes

$$
\bar{I}(\alpha, \beta) \propto T(\alpha, \beta)\bar{I}_0(\alpha, \beta) \quad (31)
$$

where $\bar{I}$ is the two-dimensional Fourier transform of $I(x, y)$ and $T(\alpha, \beta)$ the *optical transfer function* (OTF) of the systems (see Fig. 12). For

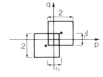
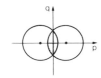

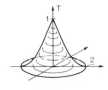

FIG. 12. Optical transfer functions related to incoherent illumination of square (left) and round (right) pupils. T is proportional to the convolution of the pupil function (see top diagram and Eq. 32).

an isoplanatic system, $T(\alpha, \beta)$ is proportional to the Fourier transform of the point-spread function t, so in absence of aberrations

$$T(\alpha, \beta) = \frac{\iint_{-\infty}^{+\infty} \bar{t}(v_x, v_y) \exp(i\alpha v_x + i\beta v_y) \, dv_x \, dv_y}{\iint_{-\infty}^{+\infty} \bar{t}(v_x, v_y) \, dv_x \, dv_y}$$

$$= \frac{\iint_{-\infty}^{+\infty} P(p + \alpha/2, q + \beta/2)}{P(p - \alpha/2, q - \beta/2) \, dp \, dq}{\exp[-ik \iint_{-\infty}^{+\infty} P(p, q) \, dp \, dq}$$

(32)

by virtue of convolution and Parseval's theorems. The modulus of T is called the *modulation transfer function* (MTF).

For a circular pupil, the OTF reads

$$T(\omega) = \begin{cases} (2/\pi)[\arccos(\omega/2) \\ \quad - \omega(1 - \omega^2/4)^{1/2}] & 0 \le \omega \le 2 \\ 0 & \text{otherwise} \end{cases}$$

(33)

where $\omega = (\alpha^2 + \beta^2)^{1/2}$. Then for a circular pupil the spatial frequency ω is limited to the interval $(0, 2)$. The dimensionless frequency ω, conjugate of the optical coordinate v, is related to the

spatial frequency f, expressed in cycles per unit length, by $f = k\omega$ NA. The expression just given for the OTF indicates that the diffraction sets an upper limit to the ability of the system to resolve a bar target with a normalized spatial frequency greater than 2.

BIBLIOGRAPHY

Barakat, R. (1980). *In* "The Computer in Optical Research" (B. R. Frieden, ed.) p. 35–80. Springer-Verlag, Berlin and New York.

Born, M. and Wolf, E. (1970). "Principles of Optics." Pergamon, Oxford, 1970.

Duffieux, P. M. (1983). "The Fourier Transform and its Application to Optics." Wiley, New York.

Gaskill, J. D. (1978). "Linear Systems, Fourier Transforms and Optics," Wiley, New York.

Hansen, R. C., ed. (1981). "Geometric Theory of Diffraction." IEEE Press, New York.

Kouyoumjan, R. G., and Pathak, P. H., *Proc. IEEE* **62**, 1448 (1974).

Northover, F. H., "Applied Diffraction Theory." Amer. Elsevier, New York, 1971.

S. Solimeno, B. Crosignani and P. Di Porto, "Guiding, Diffraction and Confinement of Optical Radiation." Academic Press, Orlando, 1986.

PLASMA SCIENCE AND ENGINEERING

J. L. Shohet   *University of Wisconsin*

I.   Introduction
II. Basic Plasma Properties
III. Plasma Physics
IV. Plasma Diagnostics
V.   Plasma–Surface Interactions
VI. Conclusion

GLOSSARY

Attachment: Phenomenon in which a neutral particle and an electron combine, producing an negatively charged ion.

De-excitation: Inverse of excitation. Radiation is usually emitted.

Diffusion: Phenomenon in which particles diffuse in position space, velocity space, and time.

Discharge machining: In this process, plasmas are used to provide a cutting surface between a thin wire and the work to be cut, usually by passing an arc between them through water.

Excitation: Process in which neutral particles or ions (that are not fully stripped) gain energy, which is evident by orbital electrons moving to higher energy states.

Fusion plasmas: Plasmas with relatively high temperatures and that are composed of light atoms, particularly hydrogen or its isotopes.

Industrial plasmas: Plasmas whose composition is generally of masses above hydrogen (they can have molecular weights of several thousand), and that are usually of two types: thermal (equilibrium) plasmas, in which the electron and ion temperatures are approximately equal, and nonequilibrium (glow-discharge) plasmas, which tend to have relatively high electron temperatures compared to the ion temperatures. Many chemical reactions can only take place in the plasma state.

Ionization: Inverse of recombination.

Ion milling: Beams of ions can be used to cut or "mill" narrow regions of materials to great accuracy.

Plasma: Collection of charged particles, usually of opposite sign, that tends to be electrically neutral. We often describe a plasma as the fourth state of matter. Adding energy to a solid melts it and it becomes a liquid; adding energy to a liquid boils it and it becomes a gas; adding energy to a gas ionizes it and it becomes a plasma.

Plasma-assisted chemical vapor deposition (CVD): Here, plasmas can be used to provide a mechanism to successfully deposit various chemicals on surfaces, either by treating the surface before deposition, or by providing a chemical pathway for successful deposition.

Plasma electronics: Includes applications in which the unique properties of plasmas are used directly in devices, such as arc melters, microwave sources, switchgear, plasma displays, welders, analytical instrumentation, arc lamps, and laser tubes.

Plasma etching: Process in which plasmas etch materials. That is, certain chemical reactions occur on surfaces, creating a volatile compound from the surface material and the plasma, which can then be pumped away.

Plasma polymerization: By ionizing a monomer gas, certain types of polymers can be made, which can be deposited as coatings on various materials. This process often occurs during etching, since by changing the ratio of the various plasma components, etching can be turned into polymerization.

Plasma processing: Encompasses applications in which plasmas or particle beams (charged or neutral) are used to alter an existing material, as in plasma etching, ion milling, ion implantation, or surface modification through plasma cleaning, hardening, or nitriding.

Plasma spray: Coating process that sputters heavy particles (clumps) from an arc system and then directs the spray of these particles to a surface for coating.

Plasma synthesis: Refers to applications in which plasmas are used to drive or assist chemical reactions to synthesize compounds, alloys, polymers, or other complex species starting from simpler starting materials.

Recombination: Phenomenon in which ions and electrons recombine to form neutral particles; radiation is sometimes emitted.

Sheath: Generally a region near a surface in which the plasma is not electrically neutral. Some sheaths can form in the main body of the plasma and are called double layers.

Sputter deposition: In this case, plasmas are used to sputter or knock off from a target electrode particles that are then deposited on a particular material.

Surface modification: Plasmas can be used to modify the properties of materials by interacting on the surface of those materials in several ways. For example, tool steel can be hardened considerably by subjecting the tools to a nitrogen plasma. Turbine blades can be plasma coated for improved mechanical and thermal properties.

Plasmas are composed of mixtures of electrons and positive and/or negatively charged ions as well as neutral particles. They are affected by electric and magnetic fields, which can be used to modify their properties. The temperature of such plasmas can be quite high. As a result, many interactions between particles are substantially different when they are in the plasma state. Thus, new materials can be manufactured that can have improved properties, new chemical compounds may be produced, and surfaces of existing materials can be altered. These aspects will have an increasingly significant role in the future of technology.

I. Introduction

Industrial use of plasma has applications that cover a broad range of activities and has a multibillion-dollar yearly impact in the economy. In order to understand how this occurs, we must first describe what a plasma is, how it behaves under the influence of electric and magnetic fields, and how it is characterized. Normally, we specify the following quantities when we describe a plasma: composition, electron and ion temperatures, and electron and ion densities. We divide plasmas into two general types as follows.

1. Industrial plasmas. The ions in these plasmas are generally composed of masses above hydrogen (the molecular weights can reach several thousand), and are usually of two types: thermal (equilibrium) plasmas, in which the electron and ion temperatures are approximately equal, and nonequilibrium (glow-discharge) plasmas, which tend to have relatively high electron temperatures compared to the ion temperatures.

2. Fusion plasmas. These plasmas have much higher temperatures than industrial plasmas and are composed of light atoms, particularly hydrogen or its isotopes, and are designed to produce energy by means of a thermonuclear reaction.

The graph shown in Fig. 1 indicates some general groupings of industrial plasmas according to their application and how they compare with fusion plasmas. One axis of the graph is proportional to temperature and the other to density.

At present, industrial plasma applications are largely empirical in nature. Further progress will require a much more thorough understanding of the plasma behavior as well as of the interaction of the plasma with solid materials. Design tools, transportable diagnostics, and models are needed.

We characterize industrial applications in three broad and somewhat overlapping areas. They are:

1. *Plasma processing,* which encompasses applications in which plasmas or particle beams (charged or neutral) are used to alter an existing material, as in plasma etching, ion milling, ion implantation, or surface modification through plasma cleaning, hardening, or nitriding.

2. *Plasma synthesis,* which refers to applications in which plasmas are used to drive or assist chemical reactions to synthesize compounds, alloys, polymers, or other complex species starting from simpler starting materials. This could also include the inverse processes of plasma decomposition. *Many chemicals and/or chemical*

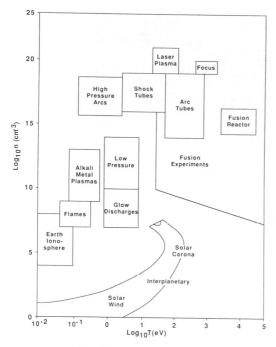

FIG. 1.   Types of plasmas.

reactions can only take place in the plasma state.

3.   *Plasma electronics,* which includes applications in which the unique properties of plasmas are used directly in devices, such as arc melters, microwave sources, switchgear, plasma displays, welders, analytical instrumentation, arc lamps, or laser tubes.

There are four common fundamental requirements needed for progress in industrial applications of plasmas. They are:

1.   Theory, modeling, and systems concepts.
2.   Plasma chemistry and interactions.
3.   Plasma diagnostics and characterization.
4.   Plasma generation and confinement.

It is the purpose of this article to provide a general introduction to plasma science and engineering in order to show both the applications and how advances in this field can be made. The following short description of industrial applications of plasma processing and technology shows how widespread the impact of plasma technology is.

Plasma polymerization. By ionizing a monomer gas, certain types of polymers can be made, which can be deposited as coatings on various materials. There is an important application of this work in the biotechnology field, since biocompatible polymers can be used to coat various implant materials that would otherwise be rejected by the body. Various pharmaceuticals and other "exotic" chemicals can only be made with this process, which is often a result of the combination of ion and free radical generation by the plasma.

Plasma-assisted CVD (chemical vapor deposition). Here, plasmas can be used to provide a mechanism to successfully deposit various chemicals on surfaces, either by treating the surface before deposition or by providing a chemical pathway for successful deposition.

Sputter deposition. In this case, plasmas are used to sputter from a target electrode particles that are then deposited on a particular material.

Plasma etching. The major application of this work is to the semiconductor industry. As the spacing between lines in integrated circuits shrinks to 1 μm and below, conventional "wet" etching using chemicals begins to fail. This is because such processing acts in a spherical direction and undercuts the walls between the etch regions. Appropriately designed plasma etching (dry etching), perhaps combined with electric and magnetic fields or ion beams, offers a dramatic improvement in the etch process, and it is believed that the future of the entire semiconductor fabrication industry will rest with plasma processing.

Ion milling. Beams of ions can be used to cut or "mill" narrow regions of materials to great accuracy.

Surface modification. Plasmas can be used to modify the properties of materials by interacting on the surface of those materials in several ways. For example, tool steel can be hardened considerably by subjecting the tools to a nitrogen plasma. Turbine blades can be plasma coated for improved mechanical and thermal properties.

Welding. The use of plasmas in welding, especially in arc welding, has been known for some time. However, many problems continue to exist with welding, and much of it is due to the lack of understanding of the plasma composition, the plasma temperature and density, and the electric field and current distribution in the welding arc.

Discharge machining. In this process, plasmas are used to provide a cutting surface between a thin wire and the work to be cut, usually by passing an arc between them through water. This process was used to cut a magnet coil that was required to be in a particular twisted shape.

FIG. 2.   Electrical discharge machining (EDM) of a magnet coil.

The tooling and resulting magnet coils are shown in Figs. 2 and 3.

Plasma displays. Many new uses for such displays are being found. A portable computer whose screen is a full plasma display has recently been introduced.

Arc devices. A major component of American industry has involved the use of arc technology in the electrical power system field. Switchgear today is still being designed empirically without the understanding of plasma-surface interactions. The U.S. switchgear industry is suffering greatly from foreign competition as a result. Arc illumination is a major application of this technology as well, with activity above $20 billion per year.

Arc melting. Arc furnaces have been in use for many years. Their applications to the refining and extraction of ores have been many. Yet there can be major improvements made if the proper understanding of the interaction of plasmas with the ores and metals are understood. For example, in the melting of iron ore, 20 lb of graphite electrode are used up per ton of ore. If

FIG. 3.   Three-dimensional twisted magnet coils, made by electrical discharge machining. There is not a single bend in the coils, since they are made by "slicing" a solid block of aluminum in this shape.

just 1 lb graphite were saved, $20 million per year savings would result.

Plasma spray. This is a coating process that sputters heavy particles (clumps) from the cathode of an arc system and then directs the spray of these particles to a surface for coating. It has applications where spray coatings are required.

Much needs to be done to formulate appropriate understanding of the plasma and plasma–surface interactions. Measurements have often found that intuitive understandings of plasma behavior have not been borne out when the actual measurements are made. However, if the measurement system itself results in a perturbation of the plasma, the results may be unclear, and thus noninvasive diagnostics need to be developed.

II. Basic Plasma Properties

A. Density, Temperature, Composition

The mixture of ions, electrons, and neutral particles making up a plasma must be describable in terms of quantities that can be used to describe them so that their properties can be analyzed. There are several of these quantities that provide a useful comparison. They are:

Density, described by n_e, n_{ix}, and n_x, which refer to the electron density, the ion density of species x, and the neutral density of species x, respectively. It is important to note that most plasmas contain ions of several different species (positively and/or negatively charged), and the number density (usually expressed in units of particles per cubic centimeter) is a very important quantity. The density is usually a function of both position and time. It should be measured experimentally and is then often modeled theoretically, depending on the nature of the various processes that act to change them. In particular, we refer to the following processes:

Attachment, in which a neutral particle and an electron combine producing a negatively charged ion.

Diffusion, in which particles diffuse in position space or velocity space. Thermal diffusion is related to particle diffusion, but refers to energy, not particle transport.

Recombination, in which ions and electrons recombine to form neutral particles; radiation is sometimes emitted.

Ionization, the inverse of recombination.

Excitation, in which neutral particles or ions (that are not fully stripped) gain energy, which is evident by orbital electrons moving to higher energy states.

De-excitation, the inverse of excitation. Often, radiation is emitted.

There are many ways in which these processes can occur, such as ionization by electron impact, chemical ionization, or radiation absorption.

It is often surmised that a plasma is electrically neutral, but such a condition usually does not occur when a plasma is in contact with a surface. Under these circumstances, a "sheath" is developed in which either electrons or ions are the dominant species. Usually, this results in

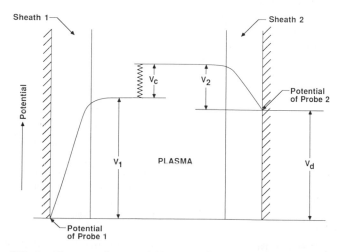

FIG. 4.   Electrostatic potential between the plasma and an electrode in the sheath region.

a net electric field in the sheath. Sheaths have considerably different properties than do the neutral plasma, and care must be taken to understand them. Figure 4 shows the electrostatic potential between a plasma and a conducting material in the sheath region. The conductivity of a plasma may actually be quite high, often greater than metals.

Another important quantity that is needed to characterize a plasma is the temperature of the individual components, i.e., T_e, T_{ix}, and T_x. Temperature is also a quantity that is not usually constant in space or time.

The composition of the plasma is of paramount importance. Here, one needs to know the mass numbers of all of the ions and neutral particles in the plasma. In many cases, it is desirable to know this as a function of both time and position, and the nature of the diagnostic devices needed to determine this ranges from very simple to very sophisticated, indeed.

B. PLASMA PRODUCTION

Plasmas may be generated by passing an electric current through a gas. Normally, gases are electrical insulators, but there are always a few charge carriers present that can be accelerated by the electric field and can then collide with neutral particles, producing an avalanche breakdown, thus making the plasma. The electric field needed for breakdown can be made with a potential set up between a pair of electrodes, with an "electrodeless" rf (radio-frequency) induction coil, with shock waves, with lasers, or with charged or neutral particle beams. The latter processes can also produce gaseous plasmas if they impinge on a solid target. In addition, heating various materials (usually alkali metals) in ovens or furnaces will cause not only evaporation of neutral particles but also ionization, and plasmas may be made in this way. Many chemical processes can also cause ionization.

III. Plasma Physics

A. PLASMA DYNAMICS

The dynamics of the motion of the charged particles in a plasma are governed by the fundamental equation of motion:

$$\mathbf{F} = q(\mathbf{E} + \mathbf{v} \times \mathbf{B}) \qquad (1)$$

where $\mathbf{F}$ is the force on the charged particle, $\mathbf{E}$ is

the electric field, $\mathbf{B}$ is the magnetic field, and $\mathbf{v}$ is the charged particle's velocity. If one knew all of the values of the electric and magnetic fields, including those produced by all of the particles (self-consistent fields) at the location of each particle, and if collisions between particles could be neglected, then the trajectories of all the particles could be obtained simultaneously with a computer. However, as can be seen from Fig. 1, such a computer cannot exist, since literally billions of equations would need to be solved simultaneously. In addition, collisions between particles, including charged particle–neutral particle collisions, must occur, so such a method cannot be totally practicable. However, much understanding can be made with numerical simulation of "clumps" or "clouds" of plasmas using modern supercomputers, where upward of 1 million simultaneous equations can be solved over a reasonable time period.

Thus, we need to develop a means to consider collisional interactions between particles making up the plasmas. Such interactions are classified into two types—elastic and inelastic collisions. In elastic collisions, kinetic energy, linear momentum, and angular momentum of the two colliding particles are conserved. In inelastic collisions, some of the energy and momentum is changed into or from internal vibration energy, chemical energy (such as chemical bonds), or to conduct the processes of ionization, excitation, recombination or de-excitation and potentially generate electromagnetic radiation during the process as well.

The most common representation for such interactions is called the collision cross section. We develop this formulation as follows. Let us assume that a beam of particles of density n particles/cm^3 travelling with a velocity v and of cross section A passes through the plasmas a distance dx. Let N be the number of plasma particles/cubic centimeter. The number of beam particles colliding with plasma particles per unit time may then be written as

$$\frac{dn}{dt} = -[(N\sigma A\ dx)/(A\ dx)]nv \qquad (2)$$

The term $N\sigma A\ dx$ is the probability of collision in the volume $A\ dx$, and nv is the particle current density of incoming beam particles; σ is the collision cross section for this particular process. We may also write $v\ dt = dx$ and $N\sigma = p_0 p_c$, and we can rewrite Eq. (2) to be

$$n = n_0 \exp(-p_0 P_{cx}) \qquad (3)$$

It is also convenient to write

$$n = n_0 \exp(-p_0 P_c vt) = n_0 \exp(-\nu_c t) \quad (4)$$

introducing an average collision frequency $\nu_c = p_0 P_c v$. In the above equations, P_c is the *probability of collision* for a particular process and p_0 is the "reduced" pressure $= 273 p/T$, which expresses a concentration $N/V = 3.54 \times 10^{16} p_0$ molecules/cm^3. The term $p_0 P_c$ has units of $1/$ length, or $1/(p_0 P_c) = \lambda$, the mean free path.

Each process has its own cross section, which in many cases can only be determined experimentally.

1. Particle Diffusion

a. Free Diffusion. To examine this condition, we first consider particles as *free* to move in a plasma of similar particles. For simplicity, we assume particle flow in one dimension along which a density gradient dn/dx has been established. By using Newton's second law, we can write, for the change in momentum due to *elastic* collisions between the particles, each of mass m

$$\frac{d(nmv_x)}{dt} = -mv_x n\nu_m \quad (5)$$

The velocity v_x is acquired by the particle due to random collisions with the other particles and is therefore independent of the coordinates, as long as the particles are all assumed to be in energy equilibrium. Thus, we may write

$$\frac{d(nv_x)}{dt} = v_x \frac{d(nv_x)}{dx} = v_x^2 \frac{dn}{dx} = -nv_x \nu_m \quad (6)$$

When we average over v_x and define the *particle current* to be $\Gamma_x = nv_x$ and make the approximation that $v_x^2 = v_T^2/3$, where v_T is the average thermal speed, we obtain

$$\Gamma_x = -(v_T^2/3\nu_m)\frac{dn}{dx} = -D\, dn/dx \quad (7)$$

which is the free diffusion equation, and D is the free diffusion coefficient,

$$D = (v_T^2/3\nu_m) = (v\lambda/3) \quad (8)$$

The value of D is not normally equal for electrons and ions, since, if they have the same temperature, the average thermal speed v_T is not equal.

Written in three dimensions, if the plasma is isotropic, we obtain $\Gamma = -D\,\nabla n$, where ∇ is the gradient operator. In this case Γ is a vector. If there are spatial variations of the diffusion coeffi-

cient, we should write:

$$\Gamma = -\nabla(Dn) \quad (9)$$

The diffusion coefficient for a given species of particles could have components that are due to several processes, each of which give spatial diffusion. For example, electrons are simultaneously colliding with other electrons, ions, and neutral particles. Each process contributes a "mean free path" and a collision frequency, which must be added together, reciprocally in the case of the former and directly in the case of the latter, to obtain the combined mean free path and free diffusion coefficient for all of the combined processes.

If we consider that the diffusion has not reached a steady state, then we can calculate the time variation in concentration in any element of space, if there is no source or sink of particles, from the equation

$$\frac{\partial n}{\partial t} + \nabla \cdot \Gamma = 0 \quad (10)$$

which leads directly to the time-dependent diffusion equation

$$\frac{\partial n}{\partial t} = D\,\nabla^2 n \quad (11)$$

This equation is separable and can often be solved as a boundary value problem. Inclusion of other effects that generate or destroy particles in an element of space requires a modification of Eq. (10) to

$$\frac{\partial n}{\partial t} = D\,\nabla^2 n + \alpha n^2 - \beta n^2 \quad (12)$$

where α and β are the source and sink terms, respectively. These can be ionization (source), recombination (sink), etc.

b. Mobility and Time-Varying Fields. In the presence of time-varying electric fields and collisions, we may approximate Eq. (1) as

$$m(dv/dt) + (mv_m)v = qE_0 \exp(j\omega t) \quad (13)$$

where q is the charge of the particle, and is negative for electrons and either positive or negative for ions, depending on their nature. The steady-state velocity is then

$$v = \left(\frac{q/m}{j\omega + \nu_m}\right) E \quad (14)$$

We may take the quotient of v/E, which is defined as the *mobility*,

$$\mu = v/E = \left(\frac{q/m}{j\omega + \nu_m}\right) \quad (15)$$

If we are interested in the case of a dc electric field where $\omega = 0$, or a case at high pressure where the collision frequency is much larger than the ac frequency of the electric field, then the mobility reduces to

$$\mu = q/(m\nu_\mathrm{m}) \qquad (16)$$

In many cases, the mobilities of the different species of particles in a plasma must be measured. Neutral particles, being uncharged, do not have a mobility.

The ratio of the diffusion coefficient to the mobility is an important quantity because it is a measure of the average particle energy. In a dc electric field, the ratio is

$$\frac{D}{\mu} = \frac{v_\mathrm{T}^2 m}{3q} = \frac{2}{3q}\left(\tfrac{1}{2}mv_\mathrm{T}^2\right) = \frac{2}{3q}\,u_\mathrm{avg} \qquad (17)$$

The quantity u_avg is the average kinetic energy of the particles, and the numerical constant depends upon how the averaging is carried out. The value $\frac{2}{3}$ is correct for a Maxwellian distribution of velocities.

If we write the particle flow equation for the particles discussed previously and add the contribution through the mobility from the electric field, we obtain

$$\Gamma = -D\,\nabla n - n\mu\mathbf{E} \qquad (18)$$

Assuming that the electric field is in the z direction and that a steady-state condition applies ($d/dt = 0$), we may apply Eq. (16) to the continuity equation to obtain

$$\nabla^2 n = -\frac{\mu}{D}\,\mathbf{E}\cdot\nabla n \qquad (19)$$

c. Ambipolar Diffusion.

So far, we have assumed that the ions and electrons have diffused freely with no interaction between them. When the density is high, this is not the case. That is, the free diffusion coefficients, not normally being equal, will result in the enhanced transport of one species of particle. Charge separation will occur, and, as a result, an ambipolar electric field is established that will affect the motion of the particles directly through Eq. (17). We may write, for the positively charged particles,

$$\Gamma_+ = -D_+\nabla n_+ + \mu_+\mathbf{E}_\mathrm{s} n_+ = n_+\mathbf{V}_+ \qquad (20)$$

where $\mathbf{E}_\mathrm{s}$ is the space-charge field. In terms of the velocity,

$$\mathbf{V}_+ = -\frac{D_+}{n_+}\,\nabla n_+ + \mu_+\mathbf{E}_\mathrm{s} \qquad (21)$$

By writing a similar expression for the negatively charged particles,

$$\mathbf{V}_- = -\frac{D_-}{n_-}\,\nabla n_- - \mu_-\mathbf{E}_\mathrm{s} \qquad (22)$$

If we eliminate $\mathbf{E}_\mathrm{s}$ between these equations and set $n_+ = n_- = n$ as well as the gradients and velocities equal, we obtain the result

$$\mathbf{V} = -\left(\frac{D_+\mu_- + D_-\mu_+}{\mu_+ + \mu_-}\right)\nabla n \qquad (23)$$

The quantity in the parentheses is a diffusion coefficient for the two signs of particles interacting on each other *so that they both diffuse together*. This quantity is called the *ambipolar diffusion coefficient* and is defined as

$$D_\mathrm{amb} = \frac{D_+\mu_- + D_-\mu_+}{\mu_+ + \mu_-} \qquad (24)$$

The ambipolar time-dependent diffusion equation (for both species of particles) is thus

$$\frac{\partial n}{\partial t} = D_\mathrm{amb}\,\nabla^2 n \qquad (25)$$

2. Heat Transport

The above discussion centered on particle transport. In addition to the transfer of particles by diffusion, *energy* can also be transported. We write the energy flux, in a way similar to the particle flux, as

$$Q = n(\tfrac{1}{2}mv^2)v \qquad (26)$$

Let E denote the thermal energy of a particle. Then $\bar{E}$, the mean thermal energy of particles at a given point, is a function of the local temperature T at that point, and, as in the particle diffusion case, we shall assume only a one-dimensional variation for this development. The *specific heat* of the plasma is given by the relation

$$c_\mathrm{v} = \frac{d}{dT}\,(\bar{E}/m) \qquad (27)$$

The net *particle* flux passing a given point, for particles passing the origin of the x direction leaving the region between x and $x + dx$ without having made a collision, is

$$d\Gamma_{0x} = +\frac{v}{6\lambda}\,n(x)e^{-(x/\lambda)}\,dx \qquad \text{if } x > 0 \quad (28)$$

Each of the particles passing the origin carries with it a thermal energy equal, on the average, to the value of $\bar{E}$ at the regon between x and $x + dx$. We may write a similar expression for the particle flux from the points where $x < 0$.

We previously expanded the density $n(x)$ in a Taylor series about the origin. We shall do a similar thing to obtain the energy flux, and expand the local energy in terms of the energy at the origin. In order to do this, we need the average velocity in a given coordinate direction, which is in the, say, positive or negative direction of that coordinate. Note that the average velocity otherwise would be zero. The number density of those particles is, on the average, half the actual density. Assuming a Maxwellian distribution of velocities, we may write the total average velocity to be

$$n\bar{V}_+/2 = \int Vf \, d^3V$$
$$= (2kT/\pi m)^{1/2} = \tfrac{1}{2}V_T \qquad (29)$$

where V_T is the thermal velocity $(3kT/m)^{1/2}$. Thus, the total thermal energy that the particles from x and $x + dx$ carry across the origin is

$$nV_T\bar{E}/4 = \tfrac{1}{4}nV_T(\bar{E} + \lambda u \, \partial\bar{E}/\partial x) \qquad (30)$$

and for particles produced in the region to the left of the origin, the energy flux is

$$nV_T\bar{E}/4 = \tfrac{1}{4}nV_T(\bar{E} - \lambda u \, \partial\bar{E}/\partial x) \qquad (31)$$

In these last two equations, the energy is now written in terms of the energy at the origin. In addition, λ is the mean free path, and u is a numerical constant roughly of order unity. We now obtain the net rate of flow of energy from the negative to the positive side as

$$\tfrac{1}{4}nV_T(\bar{E} - \lambda u \, \partial\bar{E}/\partial x) - \tfrac{1}{4}nV_T(\bar{E} + \lambda u \, \partial\bar{E}/\partial x)$$
$$= -\tfrac{1}{2}nV_T\lambda u \, \partial E/\partial x \qquad (32)$$

We can now obtain this result in terms of the specific heat and the temperature to be

$$Q = -\tfrac{1}{2}mnV_T\lambda u c_v \, \partial T/\partial x \qquad (33)$$

Thus, heat flow is proportional to the temperature gradient, rather than the density gradient. The thermal conductivity χ is then

$$\chi = \tfrac{1}{2}mnV_T\lambda u c_v \qquad (34)$$

so the heat flux is then:

$$\mathbf{Q} = -\chi \, \nabla T \qquad (35)$$

3. Effects of AC Electric and Magnetic Fields

In order to determine the response of the plasma to an electromagnetic wave, we must first examine its response to ac electromagnetic fields. We assume that the plasma is "cold" in that the motion of the individual charged particles can be considered to be entirely due to the electric fields they experience, in this case, externally imposed fields only. We do this by reexamining the equation of motion of a charged particle:

$$\mathbf{F} = q\mathbf{E} = m\frac{d\mathbf{v}}{dt} \qquad (36)$$

If we assume that the electric field is driven by an ac source, we may adopt a "phasor" notation for the ac part of the signal. That is, we shall assume all quantities vary as

$$\exp(-j\omega t) \qquad (37)$$

as far as their time variation, where ω is the driving frequency. Thus, the electric field, for example, is:

$$\mathbf{E} = \mathbf{E}_0 \exp(-j\omega t) \qquad (38)$$

We only need to solve for the spatial part of the variation. Thus, Eq. (38) becomes

$$qE_0 = -j\omega v_0 \qquad (39)$$

We use Eq. (38) to find the ratio of velocity to electric field, that is, the mobility. It is simply

$$\mu = |\mathbf{v}_0/\mathbf{E}_0| = \frac{-q}{j\omega m} \qquad (40)$$

The *conductivity* of the plasma is determined by noting that the electric current density may be written as

$$\mathbf{J} = nq\mathbf{v} = nq\mu\mathbf{E} = -\frac{nq^2\mathbf{E}}{j\omega m} = \sigma\mathbf{E} \qquad (41)$$

where σ is the conductivity of the plasma in the cold plasma approximation. If a plane monochromatic wave propagates through a plasma, we find that the wave will cut off (no longer propagate) where the density of the plasma reaches a certain value. To determine this, we examine Maxwell's equation:

$$\nabla \times \mathbf{H} = \mathbf{J} + \frac{\partial \mathbf{D}}{\partial t} \qquad (42)$$

If we continue to use the exponential time notation, then we may write Eq. (42) as

$$\nabla \times \mathbf{H}_0 = \sigma\mathbf{E}_0 - j\omega\varepsilon_0\mathbf{E}_0 \qquad (43)$$

or

$$\nabla \times \mathbf{H}_0 = -j\omega\varepsilon_0 \left(1 - \frac{\sigma}{j\omega\varepsilon_0}\right)\mathbf{E}_0 \qquad (44)$$

Using Eq. (41) we finally obtain

$$\nabla \times \mathbf{H}_0 = -j\omega\varepsilon_0 \left(1 - \frac{n_e q_e^2}{m_e \varepsilon_0 \omega^2}\right)\mathbf{E}_0 \qquad (45)$$

The effective permittivity ε of the plasma is then

$$\varepsilon = \varepsilon_0 \left(1 - \frac{n_e q_e^2}{m_e \varepsilon_0 \omega^2}\right) = \varepsilon_0[1 - (\omega_{pe}^2/\omega^2)] \quad (46)$$

Equation (46) shows that whenever the electron plasma frequency, $\omega_{pe} = [n_e q_e^2/(m_e \varepsilon_0)]^{1/2}$, greater than the driving frequency ω, the effective permittivity becomes negative and the wave no longer propagates. Note that it is always *less* than the permittivity of free space as well. If $\varepsilon/\varepsilon_0 < 0$, then we do not have normal propagation of waves, but evanescence.

B. Types of Plasmas

In examining Fig. 1, we can now consider the properties of those plasmas that are of direct interest to modern industrial problems. We concentrate on two types: glow (nonequilibrium) and arc (thermal) discharges. In general, if one considers a set of electrodes across which a dc potential is applied, one may classify the glow and arc discharges roughly according to the graph shown in Fig. 5.

The vertical axis is the voltage across the discharge, and the horizontal axis is the current. Note that as the current increases, the discharge goes from non-self-sustaining, to a glow discharge, to an abnormal glow, to an arc discharge. The voltage across the electrodes drops as the abnormal glow region is entered, rises, and drops again as the arc region is entered.

1. Glow Discharges

Figure 6 displays a typical picture of a glow discharge plasma between planar electrodes. The appearance of this discharge is complicated. It is maintained by electrons produced at the cathode by positive-ion bombardment. In the

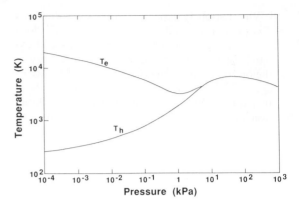

FIG. 6. Transition between thermal and nonequilibrium plasmas.

Aston dark space there is an accumulation of these electrons, which gain energy through the Crookes dark space. The cathode glow results from the decay of excitation energy of the positive ions on neutralization. When the electrons gain sufficient energy in the Crookes dark space (also called the cathode fall, cathode dark space, or Hittdorf dark space) to produce inelastic collisions, the excitation of the gas produces the negative glow. The end of the negative glow corresponds to the range of electrons with sufficient energy to produce excitation, and in the Faraday dark space the electrons once more gain energy as they move to the anode.

The positive column is the ionized region that extends from the Faraday dark space almost to the anode. It is not an essential part of the discharge, and for very short discharge tubes it is absent. In long tubes it serves as a conducting path to connect the Faraday dark space with the anode. This portion of the discharge is nearly electrically neutral, and the main electron and ion loss occurs by ambipolar diffusion. In the last few mean free paths, the electrons may gain energy high enough to excite more freely as the positive ions are forced away from the anode, producing the anode glow.

It is important to note in Fig. 6 the curves for electric field strength and net space charge. In general, most plasma processing tends to be done with the object to be processed placed on the cathode, other electrodes, or a "target" electrode. As a result, knowledge of the electric field and potential distributions near these electrodes is very important for an understanding of the nature and energies of the particles as they traverse the sheath region around the cathode.

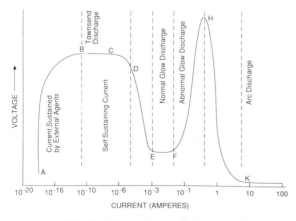

FIG. 5.   Glow and arc discharges.

2. Arc Discharges

A true arc discharge is characterized by a low cathode fall of the order of the ionization potential of the neutral atoms in the plasma. In glow discharges, the cathode fall tends to be much higher, perhaps of the order of 200 V. From Fig. 5, one can see that the current voltage characteristic is falling and the current density of the arc is very high. From this information it is clear that electron emission from the cathode must be governed by some mechanism other than positive-ion bombardment of the cathode. Usually the transition from glow to arc is a rapid and discontinuous process.

Arc discharges may be classified by the emission process that occurs at the cathode. Four types of arcs may be defined.

1. The thermionic arc in which the cathode is heated by the discharge and the arc *is* self-maintaining.
2. The thermionic arc in which the cathode is heated by an external source and the arc *is not* self-maintaining.
3. Field-emission arcs in which the electron current at the cathode is due to a very high electric field at the cathode surface.
4. Metal arcs.

In addition, arcs are frequently classified as high- or low-pressure arcs if the gas pressure is roughly above or below 1 atm. A high-pressure arc is characterized by a small, intensely brilliant core surrounded by a cooler region of flaming gases, sometimes called the aureole. If the arc occurs between highly refractory electrodes, such as carbon or tungsten, both the anode and the cathode are incandescent.

At low pressures, the appearance of the column depends upon the shape of the discharge tube. A constricted tube gives rise to a highly luminous column even at low pressures. The most important difference between low- and high-pressure arcs is in the temperature of the positive column. The high-pressure column is at a very high temperature (5000 K or higher). The ions, electrons, and neutral atoms of the high pressure positive column are in *local thermal equilibrium,* and therefore such plasmas are often called *thermal plasmas.*

The neutral gas temperature of the low-pressure arc is never more than a few hundred kelvins, whereas the electron temperature may be of the order of 40–50,000 K. (1 eV = 11,600 K.)

The difference between glow and arc plasma is shown in Fig. 6.

The current density at the cathode of an arc is very much greater than that of the glow discharge (Fig. 7). In many cases, the cathode current density is practically *independent* of the arc current. As a result, the plasma tends to concentrate in a small area near the cathode and produces *cathode spots.*

Thus low-pressure arc and glow discharges are sometimes called nonequilibrium (nonthermal) discharges.

3. RF Discharges

In most cases, rf and/or microwave radiation can be used to break down and maintain a discharge. The advantage of this process is that it is not necessary to have electrodes in contact with the plasma.

An important use of rf discharges occurs when we might wish to coat an electrode with an electrically insulating material (dielectric), by means of a plasma sputtering or chemical vapor deposition process. Normally, the cathode of the discharge would be the electrode that would receive the positive ions for the coating process.

However, if a dc glow discharge is used to produce the plasma, the dielectric that is depos-

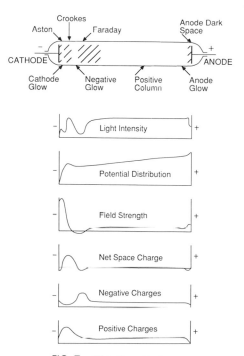

FIG. 7.   The glow discharge.

ited on the cathode will charge up, and the fluxes of both ions and electrons to the surface will become equal, regardless of the potential applied to the electrode itself. The result is that electrons and ions will combine at the dielectric, and no current will be drawn through the electrode to sustain the discharge and it will go out.

An ac discharge, if its frequency were high enough, could be used to maintain the discharge, because as the potential on the electrode backing the dielectric is reversed the charge on the dielectric will leak off. Then when the cycle is reversed again, the deposition process will reoccur. Typically, rf frequencies greater than about 1 MHz are required to maintain a discharge. Below this frequency, the discharge will be extinguished before the potential is reversed.

4. Breakdown

As the electric field in a discharge tube increases from zero, a small dark current (Townsend discharge) is drawn, but at some point there is a sudden transition to one of the several forms of self-sustaining discharge. Figure 8 shows the critical breakdown voltage as a function of the pressure–distance product, where the distance is measured between the electrodes.

From Fig. 8, it is seen that in all cases shown a minimum value of the voltage required for breakdown appears at a critical value of the pressure–distance product. At low pressures, below the critical value of the pressure–distance product (*pd*), the discharge will take place in the *longer* of two possible paths. This means that bringing electrodes closer together at the lower pressures will provide better insulation.

However, in bringing metal parts closer together in order to provide a higher breakdown voltage, care must be taken in the disposition of solid insulating material because the electrostatic field on the surface of the electrode will tend to increase. Above a certain value, cold or "field" emission of electrons will take place, and a vacuum arc (metal arc) will begin.

At high pressures, the minimum breakdown distance is so short that it can be virtually impossible to make any measurements. The minimum breakdown voltage is nearly equal to the normal cathode fall potential in the glow discharge.

C. Magnetic Fields

If a dc magnetic field is superimposed on the plasma, the basic equation of motion [Eq. (1)] now includes the magnetic field vector **B**, which will greatly affect the motion of the particles. First, the equation of motion will now be different depending on whether the component equation will be parallel to or perpendicular to the magnetic field. In the direction parallel to the field, the motion is nearly as though there were no magnetic field. This is not totally the case, since if the energy and magnetic moment of the charged particles are conserved, the orbit in the spatially varying magnetic field will result in a coupling between the parallel and perpendicular components of motion.

The basic effect in a uniform field is that the particles tend to follow the magnetic field lines and orbit around them. The result is a collimation along the field lines, which has important consequences. The plasma conductivity becomes a tensor, and hence the plasma in a dc magnetic field is anisotropic.

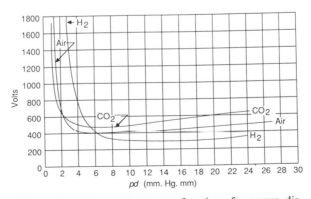

FIG. 8.   Breakdown voltage as a function of pressure–distance product.

D. Plasma Potential

In a nonthermal glow-discharge plasma, we can make several conclusions about the temperatures and densities of the three components, electrons, ions, and neutrals. The electron and ion densities will tend to be equal. Usually, glow-discharge plasmas are not highly ionized, so the neutral density tends to be much larger than the plasma (ion plus electron) density. The thermal speed of the electrons tends to be much greater than the ions or the neutrals.

Suppose we suspend a small, electrically isolated dielectric material into the plasma. Initially it will be struck by electrons and ions with current densities

$$J_e = q_e n_e v_e \qquad J_i = q_i n_i v_i \qquad (47)$$

However, if $v_i \ll v_e$, then the dielectric tends to build up a negative charge and its potential becomes negative with respect to the plasma. The resulting electric field surrounding this material will have a marked effect on the motions of the particles near the material. Since the potential is negative, electrons will tend to be repelled from and ions attracted to the dielectric.

The dielectric tends to keep charging up negatively until the electron flux is reduced and the ion flux increased sufficiently so that the electron and ion fluxes balance. We call the potential of the material at this point V_p or the plasma potential.

E. Debye Length

The Debye length is a measure of how far into the plasma the potential of an electrode or probe is observed. It can be expressed as

$$\lambda_{de} = (\varepsilon_0 k T_e / n_e q_e^2)^{1/2} \qquad (48)$$

for the electrons, where ε_0 is the permittivity of free space, k is Boltzmann's constant, and q_e is the charge of the electron. A similar expression for the Debye length of the ions can be written with the appropriate ion quantities. Normally, electrons tend to congregate around a positive potential, so the electron Debye length will be important under these conditions, and vice versa for the ions. Often $\lambda_{di} = \lambda_{de}$.

IV. Plasma Diagnostics

In order to understand what is happening in a plasma, it is necessary to "diagnose" it. We may break down the various diagnostic measurement techniques into two parts: invasive

TABLE I. Diagnostic Techniques

Invasive	Noninvasive
Langmuir probes	Radiation spectroscopy
Magnetic probes	Optical
Current probes	Microwave
Beam probes	X Rays
Radiation probes	Far infrared
Optical, micro-	Ultraviolet
wave, etc.	Particle Collectors
	Energy analysis
	Mass analysis

and noninvasive techniques. Table I lists some of these.

A noninvasive technique only "listens" or collects what comes out of the plasma. In this case, either radiation or particles are expelled, and we design instrumentation that can detect and analyze them. Invasive diagnostics require either the insertion of a probe or the injection of a particle or radiation beam into the plasma to have the diagnostic work. All other things being equal, noninvasive diagnostics are the most desirable, since they will perturb the plasma the least. However, some of the beam and radiation probes usually perturb the plasma so slightly that for many practical purposes they can be considered noninvasive diagnostics.

A. Probes

Figure 9 shows a sketch of a Langmuir probe and its associated circuit. The battery supplies a

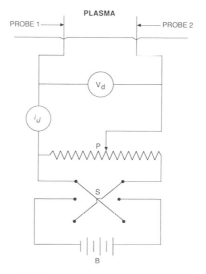

FIG. 9.   A Langmuir probe circuit.

potential to the probe, and the return circuit to the battery must come from the plasma. The method by which this happens is not always obvious. Such probes are usually very simple devices, consisting only of an insulated wire. The problems with such probes are that sheaths can form around the wire and the resulting measurements will be significantly changed by their presence. A "typical" plot of the probe current versus probe voltage characteristic is shown in Fig. 10.

The characteristic is taken either continuously, by varying the battery voltage, or point-by-point in a pulsed discharge, if the probe bias is changed from pulse to pulse. In addition, the probe voltage can be rapidly swept to obtain a characteristic dynamically.

At the point V_s, the probe is at the same potential as the plasma. The electric field across the sheath is zero at this point, and charged particles travel to the probe based entirely on their own thermal velocities. Since electrons move much faster than ions, because of their small mass, if the temperatures of the electrons and ions are about the same then what is collected is primarily electron current.

If the probe voltage is made positive with respect to the plasma, electrons are accelerated toward the probe and the ions are repelled. Thus, the small ion current present at potential V_s now vanishes, and an electron-rich sheath builds up until the total net negative charge in the sheath is equal to the positive charge on the probe. The thickness of the sheath is of the order of the Debye length, and outside of it there is very little electric field, so that the bulk of the

plasma is undisturbed. The electron current is the current that enters the sheath through random thermal motions, and since the area of the sheath is relatively constant as the probe voltage is increased, we have the fairly flat portion A of the characteristic. This is called the electron saturation current region.

If the probe is now made negative relative to V_s, we begin to repel electrons and accelerate ions. The electron current falls as V_p decreases in region B, which we call the transition region. If the electron velocity distribution were Maxwellian, the shape of the curve here after the contribution from ions is subtracted would be exponential.

At large negative values of probe potential, almost all of the electrons are repelled. We then have an *ion sheath* and *ion saturation current*, as shown in region C. This is similar to region A, but there are two differences between the ion and electron saturation currents caused by the mass difference. The first is that often the ion and electron temperatures are not equal, and as a result the Debye lengths are unequal, and the sheath widths are considerably different. The second problem is that if a magnetic field is present, the motion of the electrons is affected much more than the motion of the ions.

The shape of part B of the characteristic is related to the distribution of electron energies and can be used to determine T_e if the distribution is Maxwellian. The magnitude of the electron saturation current is proportional to $n(kT_e/m_e)^{1/2}$, from which n can be obtained (if T_e is previously found). The magnitude of the ion saturation current depends on n and kT_e, but only slightly on kT_i if $T_i \ll T_e$, so ion temperature is not easily measured with probes.

The space potential is found by locating the "knee" or junction between regions A and B of the curve, or by locating the point V_p (the zero current floating potential), and calculating V_s.

Experimental complications for the successful use of probes are many. A partial list of them is as follows:

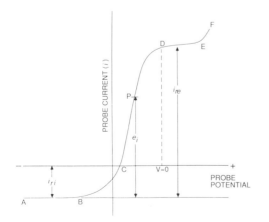

FIG. 10. Current–voltage characteristic of a Langmuir probe.

Surface Layers
Perturbation of the Plasma
Change of Probe Area   Effect of the Probe
Reflections   Shield
Macroscopic Gradients   Oscillations
Metastable Atoms   Photoemission
Secondary Emission   Negative Ions
   and Arcing   Ion Trapping

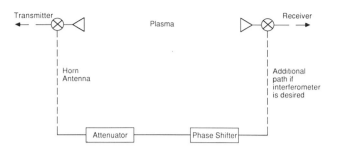

FIG. 11.   A microwave interferometer.

B. Radiation Probes

An example of a radiation probe of a plasma is a microwave interferometer. Figure 11 shows a drawing of such an interferometer.

The plasma introduces a phase shift of the microwave signal that passes through it. By comparison with the reference arm of the interferometer, it yields a signal propositional to the phase difference, which can then be calibrated to yield density directly.

The reason for the phase shift is easily seen if we adopt an exponential notation for the time and spatial variation of a plane monochromatic wave. That is, a wave will propagate as

$$\exp j(kz - \omega t) \tag{49}$$

where k is the wave number. The wave equation

$$\frac{\partial^2 E}{\partial z^2} = \frac{1}{c^2} \frac{\partial^2 E}{\partial t^2} \tag{50}$$

then becomes

$$-k^2 E = -(\omega^2/c^2)E \tag{51}$$

and thus the relationship between k and ω can be determined to be

$$k = \omega/c = \omega(\varepsilon\,\mu)^{1/2} \tag{52}$$

or

$$k = \omega(\varepsilon_0\mu_0)^{1/2}\,[1 - (\omega_p/\omega)^2]^{1/2} \tag{53}$$

The phase shift is the difference in the product of the wave number k in the presence of the plasma, as defined above, times a fixed path length L with the free-space wave number $k_0 = \omega(\varepsilon_0\mu_0)^{1/2}$ times that same path length. That is,

$$\Delta\phi = (k - k_0)L \tag{54}$$

which is directly related to density. The interferometer will detect density as it increases until cutoff is obtained, whereupon no further information is available. The higher the density required to be measured, the higher the frequency of the interferometer.

There are other restrictions with such a system. For one thing, we are assuming plane monochromatic waves. If the plasma has a dimension that is comparable to the wavelength, the plane-monochromatic wave approximation may not be valid. The plasma is also assumed to be cold. Hot plasmas can significantly affect the results. Finally, the interferometer measures *integrated* line density along its path, so the extent of spatial resolution is severely limited. Nevertheless, a microwave interferometer is one of the more important diagnostics.

C. Radiation Spectroscopy

The ultimate object in using spectroscopy in a plasma is the interpretation of all spectroscopic observations in terms of a fully consistent theoretical plasma model. Because spectra arise from atomic processes, the model must emphasize the particulate nature of the plasma, rather than the fluid aspects that characterize many of the properties of the plasmas.

Development of an appropriate model must take account of numerous atomic collision processes, so that it depends on atomic physics for its basic data. The step from atomic physics to a theoretical model for the plasma consists of discovering ways of taking into account the numerous possible atomic processes to give a composite picture of the spectrum. In many ways, the problems of laboratory spectroscopy are similar to those encountered in interpreting astrophysical spectra. A major difference might be in the possibility that laboratory plasmas have properties that change rapidly with time, so that time-dependent, rather than steady-state, solutions are required.

It is convenient to divide the treatment into

two parts: (1) considerations of the intensities of lines and continuum, and (2) the *shape* of the lines. We begin with the first part.

1. Plasma Models

The spectroscopic radiation that we shall be concerned with is emitted when an electron makes a transition in the field of an atom or ion. The observed intensity of the radiation thus emitted depends on three processes:

1. The probability of there being an electron in the upper level of the transition.
2. The atomic probability of the transition in question.
3. The probability of the photons thus produced escaping from the volume of the plasma without being reabsorbed.

Considerable simplification is achieved if the effect of the interaction of radiation (process 3) with the plasma is considered separately. There are, in fact, physically realizable circumstances where this effect may be neglected (optically thin plasmas).

2. The LTE Model

In the local thermodynamic equilibrium (LTE) model, it is assumed that the distribution of electrons is determined exclusively by particle collision processes and that the latter take place sufficiently rapidly so that the distribution responds *instantaneously* to any change in the plasma conditions. In such circumstances, each process is accompanied by its inverse, and these pairs of processes occur at equal rates by the principle of detailed balance. Thus, the distribution of energy levels of the electrons is the same as it would be in a system in complete thermodynamic equilibrium. The population of energy levels is therefore determined by the law of equipartition of energy and does not require knowledge of atomic cross sections for its calculation.

Thus, although the plasma temperature and density may vary in space and time, the distribution of population densities at any instant and point in space depends entirely on *local* values of temperature, density, and chemical composition of the plasma. The uncertainties in predictions of spectral line intensities from an LTE-model plasma depend on the uncertainties in the values of the plasma parameters and of the atomic transition probabilities.

If the free electrons are distributed among the energy levels available to them, then, according

to statistical mechanics, their velocities have a Maxwellian distribution. The number of electrons of mass m and with velocities between v and $v + dv$ is

$$dn_e = n_e 4\pi (m/2\pi kT_e)^{3/2} \exp(-mv^2/2kT_e)v^2 \, dv$$

(55)

where n_e is the total density of free electrons and T_e is the electron temperature. For the **bound levels,** the distributions of electrons are given by the Boltzmann and Saha equations, which are, respectively,

$$\frac{n(p)}{n(q)} = \frac{\omega(p)}{\omega(q)} \exp[\chi(p, q)/kT_e] \qquad (56)$$

$$\frac{n(z + 1)n_e}{n(z, g)} = \frac{\omega(z + 1, g)}{\omega(z, g)} 2(2\pi kT_e/h^2)^{3/2}$$
$$\times \exp[\chi(z, g)/kT_e] \qquad (57)$$

where $n(p)$, $n(z + 1, g)$, and $n(z, g)$ are the population densities of various levels designated by their quantum numbers p, q, and g (the last for the bound level) and ionic charge $z + 1$ and z. The term $\omega(z, p)$ is the statistical weight of the designated level, $\chi(p, q)$ is the energy difference between levels p and q, and $\chi(z, g)$ is the ionization potential of the ion of charge z in its ground level g. Equations (55)–(57) describe the state of the electrons in a LTE model plasma.

If the plasma is optically thin, then the intensity $I(p, q)$ of the spectral line emerging from a transition between bound levels p and q is given by

$$I(p, q) = \frac{1}{4\pi} \int n(p)A(p, q)h\nu(p, q) \, ds \quad (58)$$

where $A(p, q)$ is the atomic transition probability and $h\nu(p, q)$ is the photon energy. The integration is made over that depth of the plasma that is viewed by the detector, and the intensity of radiation $I(p, q)$ is measured in units of power per unit area per unit solid angle.

D. Particle Analysis

Determination of the composition of plasma and neutral particles is important in plasma science and engineering, since the composition determines the nature of the physical and chemical processes that can occur. In addition, analysis of the products of the reaction is important for a determination of the effectiveness of a particular process.

Several different methods are currently in use for this purpose. We discuss one particular

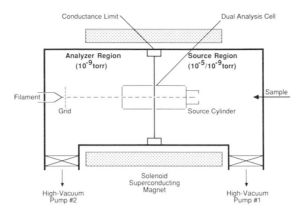

FIG. 12. Fourier transform mass spectrometer.

method, which is a procedure for analyzing the masses of different ions based on the fact that the cyclotron frequency in a dc magnetic field is proportional to the charge-to-mass ratio of that ion. In this case, the ions can be either single ionized atoms or complex molecules.

Figure 12 shows the diagram of the analysis device, called a Fourier transform mass spectrometer (FTMS). A sample of material is placed at the right-hand side of the cell and an electron beam or a laser is directed at the material. Ions produced by this reaction can be electrostatically confined between the "trap" plates shown in the figure if a small voltage is applied (approximately 1–2 V).

The ions orbit at their appropriate cyclotron frequencies. An rf excitation pulse that is "swept" over a large range of possible cyclotron frequencies is applied between the "excitation" plates in the figure. This will excite the ions so that their orbits increase in radius and *ions of equal charge-to-mass ratio will move coherently,* so that a "tube" of ions moves around the cell. This tube induces an alternating charge pulse between the "detection" plates in the figure, and the output signal from these plates is Fourier analyzed and the resulting spectrum is displayed on an oscilloscope.

These devices are capable of extremely high resolution, which means that particles of very high mass numbers that are similar (such as mass numbers 1000 and 1001) can be displayed and analyzed. Other mass spectrometers, using different principles, can be used in other applications.

V. Plasma–Surface Interactions

Surfaces in contact with plasmas are bombarded by electrons, ions, neutral particles, and photons. Electron and ion bombardment is particularly important, especially because the particles are so energetic compared with neutral particles.

Several effects occur at once during this process: nondissociative chemisorption, physical adsorption, surface diffusion, dissociative chemisorption, and formation of product molecules. Figure 13 shows a schematic representation of these processes.

A. PLASMA-ASSISTED CHEMICAL VAPOR DEPOSITION (PACVD)

In this case, we desire to deposit a thin film of some material on the surface of a material, which we call a substrate. Previously, vaporizing the material to be deposited and allowing the vapor to come in contact with the substrate under vacuum resulted in the deposition of a material film on the substrate. However, in order for good deposition to occur, it was often necessary to heat the substrate to elevated temperatures.

PACVD allows lower substrate temperatures. This is particularly important in electronics applications where coatings are deposited onto device structures. In this process, a plasma is placed in contact with the substrate, and either the plasma particles themselves or neutral particles may form the coating. In general, the method of deposition is far from being well understood.

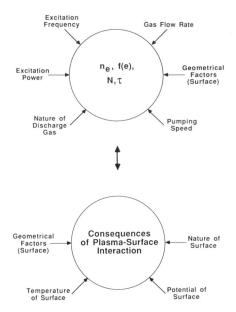

FIG. 13.　Processes occurring in plasma etch.

B. ETCHING

1. Physical Aspects

Figure 14 shows cross sections of films etched with liquid or plasma etchants. The isotropic profile represents no *overetch* and can be generated with either wet or dry (plasma) etching techniques. The anisotropic profile requires plasma etching.

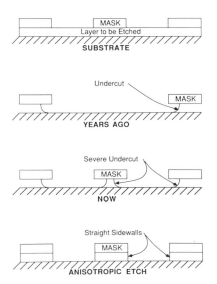

FIG. 14.　Variation of etching conditions.

Until recently, liquid (wet) etching techniques have been the main method of pattern delineation. This is because of two reasons: (1) the technology involved in liquid etching is well established, and (2) the selectivity (ratio of the etch rate of the film being etched to the etch rate of the underlying film or substrate) is generally infinite with typical liquid etchant systems.

Unfortunately, wet etching presents several problems for micrometer and submicrometer geometries. Because of the acid environments of most etchant solutions, photoresists can lose adhesion, thereby altering patterning dimensions and preventing linewidth control. In addition, as etching proceeds downward, it proceeds laterally at essentially an equal rate. This undercuts the mask and generates an isotropic etch profile, as shown in Fig. 14. Because the film thickness and etch rate are often nonuniform, a certain degree of *overetching* is required. If the film thickness is small, relative to the minimum pattern dimension, undercutting is inconsequential. However, if the film thickness is comparable to the lateral film dimension, as is the case for current and future devices, undercutting can be intolerable.

In addition, as device geometries decrease, spacings between stripes of resist also decrease. With micrometer and submicrometer patterns, the surface tension of etch solutions can cause the liquid to *bridge* the space between two resist stripes, and etching of the underlying film is eliminated.

Plasma etching has demonstrated viable solutions to essentially all the problems encountered with liquid etching. Adhesion does not seem to be critical with dry etching techniques. Undercutting appears to be controllable by varying the plasma composition, the gas pressure, and the electrode potentials.

Two additional considerations are in favor of dry etching. First, wet etching requires the use of relatively large volumes of dangerous acids and solvents, which must be handled and ultimately recycled or disposed of. Dry etching uses relatively small amounts of chemicals, although many of the gases used in these processes are also toxic.

In making plasmas for dry etching, a fill gas is broken down by means of application of an external electric field. As the electric field increases, free electrons, whose velocities increase by the action of the field, gain energy. However, since they lose this energy by collisional processes, an increase in pressure, which decreases the mean free path, then decreases the electron energy. What is important therefore, is a measurement of the velocity of an electron or ion as a function of the ratio of E/p (electric field divided by pressure). Figure 15 is a graph of the drift velocity (in an electric field) as a function of this ratio E/p.

The glow-discharge plasmas currently used for microelectronic applications can be characterized by the following parameters:

$$\text{Pressure} = 50 \text{ mtorr to 5 torr}$$

$$n_e = 10^9 - 10^{12} \quad \text{cm}^{-3}$$

$$T_e = 1\text{--}10 \quad \text{eV}$$

Usually the electron temperature is greater than the ion temperature by a factor of about 10. The plasmas are usually very weakly ionized, well below 1%.

These characteristics give the plasma special properties. The electron temperatures are high enough so that chemical bonds can be broken by electron-neutral collisions. As a result, highly reactive chemical species can be produced for etching or deposition. In addition, the *surface chemistry* occurring in glow discharges is generally modified by the impingement of ions and electrons (and photons) onto the film being etched. The combination of both of these processes results in etch rates and etch profiles unattainable with either process individually.

The energy of ions and electrons striking surfaces in a glow discharge is determined by the potentials established within the reaction chamber. Etching and deposition are generally carried out in a plasma produced by rf, which is often capacitively coupled to the plasma. The important potentials are the *plasma potential,* the *floating potential,* and the potential of the *externally biased (powered) electrode.*

Usually, under these circumstances, the electrode surfaces are at a negative potential with respect to the plasma. The result of this is that positive ions bombard the surfaces. The energy of the bombarding ions is established by the difference in potential between the plasma and the surface which the ion strikes. Because these potentials may range from a few volts to a few thousand volts, surface bonds can be broken, and, in certain instances, sputtering of film or electrode material can occur.

In addition, exposure of materials to energetic radiation can result in *radiation damage*. Positive ions can cause *implantation* or displacement damage, while electrons, X rays, or ultraviolet photons can result in ionization. Defects thus created can serve as trapping sites for electrons or holes, resulting in an alteration of the electrical properties of the materials. Such alterations may be very beneficial and can improve the surface properties of many materials considerably.

Figure 16 shows a capacitively coupled configuration used for etching. Consider an rf field established between the two plates. On the first half cycle of the field, one electrode is negative and attracts positive ions; the other is positive and attracts electrons. At the frequencies used (50 kHz to 40 MHz), and because the mobility of electrons is greater than that of the ions, the electron current is much larger than the ion current. This causes a depletion of electrons in the plasma and thus a positive plasma potential. On

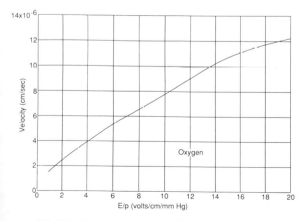

FIG. 15. Electron drift velocity as a function of E/p.

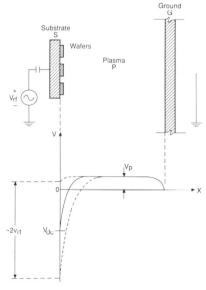

FIG. 16. Capacitively coupled plasma etching reactor.

the second half cycle, a large flux of electrons flows to the electrode that received the small flux of ions. Further, since plasma etching systems generally have a dielectric coating on the electrodes, and/or have a series (blocking) capacitor between the power supply and the electrode, no net charge can be passed. Therefore, on each subsequent half cycle, negative charge continues to build on the electrodes (and on other surfaces in contact with the plasma), so that electrons are repelled and positive ions are attracted to the surface.

When a sufficiently large negative bias is achieved on the electrodes, such that the fluxes of electrons and positive ions striking these surfaces are equal, then the transient situation ceases. At this point, the time-averaged (positive) plasma and (negative) electrode potentials are established.

The plasma potential is essentially uniform throughout the observed glow volume in an rf discharge. Between the glow and the electrode is a narrow region wherein a change from the plasma potential to the electrode (or surface) potential occurs. This is the sheath or dark space, and ions that reach the edge of the glow region are accelerated across the potential drop and strike the electrode or substrate surface.

Because of the series capacitor and/or the dielectric coating of the electrodes, the negative potentials established on the two electrodes in the system may not be the same. For example, the ratio of the voltages on the electrodes has been shown to be inversely proportional to the fourth power of the ratio of the relative electrode areas:

$$V_1/V_2 = (A_2/A_1)^4 \qquad (59)$$

If V_1 is the voltage on the powdered electrode and V_2 is the voltage on the grounded electrode, the voltage ratio is the inverse ratio of the electrode ares to the fourth power. However, for typical etch systems, the exponent is generally less than 4. Although the actual electrodes in a plasma reactor often have the same area, in Eq. (59) A_2 is the *grounded* electrode area, that is, the area of *all grounded* surfaces in contact with the plasma. Because of this, the average potential distribution is similar to that shown in Eq. (59).

In this case, the energy of the ions striking the powered electrode will be higher than that of ions reaching the grounded electrode.

Other plasma parameters can also affect the electrical characteristics. For example, rf power levels and frequency can radically change things. As the frequency is raised, there will be a point at which the ions can no longer follow the alternating voltage, so that the ion cannot traverse the sheath in one half cycle. Above this frequency, ions experience an accelerating field that is an average over a number of half cycles. Such an average motion is described with the oscillation center approximation. The drift for such a motion is given by

$$\mathbf{r}_0 = -\nabla\Phi \qquad (60)$$

where Φ is the *ponderomotive* potential,

$$\Phi = \frac{q^2}{m^2\omega^2}\{E_0^2/2\} \qquad (61)$$

Thus, the net drift, which is *independent* of the sign of the charge, is toward lower average electric fields. The oscillation center drifts in the high-frequency field as if subjected to this potential.

At lower frequencies, the ions are accelerated by instantaneous fields and can attain the maximum energy corresponding to the maximum instantaneous field across the sheath. As a result, for a constant sheath potential, ion bombardment energies are higher at lower frequencies.

2. Chemical Aspects

Figure 13 showed the primary processes occurring during a plasma etch. There are six required steps, and if any one of them does not occur, the entire processing stops. They are (1) generation of reactive species, (2) diffusion to surface, (3) adsorption, (4) reaction, both chemical and physical (such as sputtering), (5) desorption, and (6) diffusion into bulk gas.

The reactive species must be generated by electron–molecule collisions. This is a vital step, because many of the gases used to etch thin-film materials do not react spontaneously with the film. For instance, carbon tetrafluoride, CF_4, does not etch silicon. However, when CF_4 is dissociated via electron collisions to form fluorine atoms, etching of silicon occurs rapidly.

The etchant species diffuse to the surface of the material and adsorb onto a surface site. It has been suggested that free radicals have fairly large *sticking coefficients* compared with relatively inert molecules such as CF_4, so adsorption occurs easily. In addition, it is generally assumed that a free radical will *chemisorb* and react with a solid surface. *Surface diffusion* of the adsorbed species or the produce molecule can also occur.

Product desorption is a crucial step in the etch process. A free radical can react rapidly with a solid surface, but *unless the product species has a reasonable vapor pressure so that desorption occurs, no etching takes place.* For instance, when an aluminum surface is exposed to fluorine atoms, the atoms adsorb and react to form AlF_3. However, the vapor pressure of AlF_3 is approximately 21 torr at 1240°C, and thus etching is precluded at room temperatures.

The chemical reactions taking place in glow discharges are often extraordinarily complex. However, two general types of chemical processes can be categorized. They are (1) homogeneous gas-phase collisions and (2) heterogeneous surface interactions. In order to completely understand and characterize plasma etching, one must understand the fundamental principles of both processes.

Figure 17 shows how two etch processes may result in a synergism in which the resulting etch rate is greater than the sum of the two. In this case a X_eF_2 plasma and an argon ion beam are used together.

a. Homogeneous Gas-Phase Collisions. These represent the manner in which reactive free radicals, metastable species, and/or ions are generated. As shown in Table II, electron impact can result in a number of different reactions.

Due to the electronegative character of many of the etch gases currently used (O_2, CF_4, CHF_3, CCl_4, BCl_3, etc.), electron attachment often takes place, thereby generating negative as well as positive ions in the plasma. *Although these negative ions affect the plasma energetics, they probably have little if any effect on surface reactions, because they are repelled by the negative electrode potential.*

C. PLASMA POLYMERIZATION

In the plasma etching process, a competing process that can dominate over etching can occur, which is called *polymerization*. A polymer is defined as a high-molecular-weight compound made up from a small repeating organic unit called a monomer. The magnitude of the molecular weight ranges from 1000 to several million atomic mass units (amu) and, depending on conditions, the reaction product could have a statistical distribution of molecular weights.

In order for monomers to form polymers one of two reactions must occur: condensation and addition. The condensation reaction usually results in the loss of a small portion of the original starting molecules. Addition polymers are those that result from the reaction of an unsaturated monomer with an initiator that begins a chain reaction at an actiated site to start the growing polymer chain.

As the ratio of fluorine to carbon is increased, polymerization ceases and etching begins at a critical value, which depends on the potential applied to the surface.

Two general schemes have been proposed for organization of chemical and physical information on plasma etching and polymerization. Both have dealt primarily with carbon-containing gases, but with slight modifications can be easily applied to other gases. Figure 18 is a schematic of the influence of the fluorine-to-carbon ratio and electrode bias on etching and polymerization.

This model does not consider the specific chemistry occurring in a glow discharge, but rather views the plasma as a ratio of fluorine to carbon species which can react with a silicon

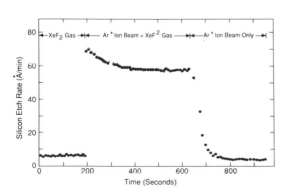

FIG. 17.   Etching synergism of two processes.

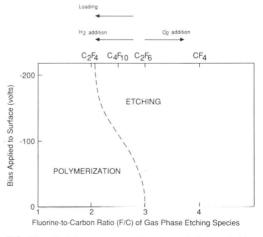

FIG. 18.   Etching rate as a function of electrode bias.

TABLE II. Results of Electron Impact

Excitation (rotational, vibrational, electronic)	$e + A_2 \rightarrow A_2 + e$
Ionization	$e + A_2 \rightarrow A_2^+ + 2e$
Dissociative ionization	$e + A_2 \rightarrow A^+ + A + 2e$
Dissociation	$e + A_2 \rightarrow 2A + e$ or $A^- + A^+ + e$
Dissociative attachment	$e + A_2 \rightarrow A^- + A$

surface. The generation or elimination of these active species by various processes or gas additions then modifies the initial fluorine-to-carbon ratio of the inlet gas.

The F/C ratio model accounts for the fact that in carbon-containing gases, *etching and polymerization* occur simultaneously. The process that dominates depends on etch-gas stoichiometry, reactive-gas additions, amount of material to be etched, and electrode potential, and on how these factors affect the F/C ratio. For instance, the F/C ratio as seen in Fig. 18 determines whether etching or polymerization is favored. If the primary etchant species for silicon (F atoms) is consumed either by a loading effect or by reaction with hydrogen to form HF, the F/C ratio decreases, thereby enhancing polymerization. Such effects are caused primarily by enhanced energies of the ions striking these surfaces.

In the *etchant-unsaturate* model described by Eqs. (62)–(65), specific chemical species derived

Saturated   Unsaturated

$$e + \text{Halocarbon} \rightarrow \text{saturated radicals} \atop + \text{ unsaturated radicals} + \text{atoms} \tag{62}$$

$$\left.\begin{array}{l}\text{Reactive atoms}\\\text{Reactive molecules}\end{array}\right\} + \text{unsaturates} \rightarrow \text{saturates} \tag{63}$$

$$\text{Atoms} + \text{surfaces} \rightarrow \left[\begin{array}{l}\text{chemisorbed layer}\\\text{volatile products}\end{array}\right. \tag{64}$$

$$\text{Unsaturates} + \text{surfaces} \rightarrow \text{films} \tag{65}$$

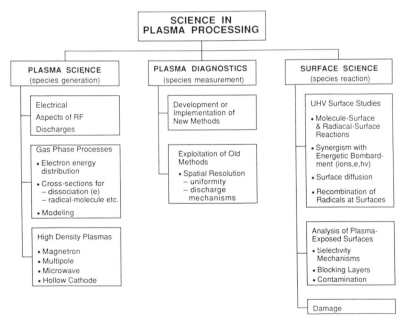

FIG. 19. Needed skills for plasma processing of semiconductors.

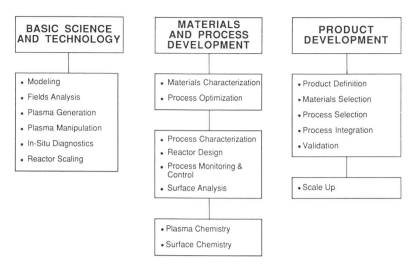

FIG. 20.   Needed skills for plasma processing in the pharmaceutical industry.

from electron collisions with etchant gases are considered.

Application of this model to a CF_4 plasma results in the chemical scheme described by

$$2e + 2CF_4 \rightarrow CF_3 + CF_2$$
$$+ 3F + 2e \qquad (66)$$

$$F + CF_2 \rightarrow CF_3 \qquad (67)$$

$$4F + Si \rightarrow SiF_4 \qquad (68)$$

$$nCF_2 + surface \rightarrow (CF_2)_n \qquad (69)$$

Depending on the particular precursors generated in the gas phase, etching, recombination, or film formation (i.e., polymerization) can occur. Also, gas-phase oxidant additives (O_2, F_2, Cl_2, etc.) can dissociate and react with unsaturate species. As an example, O_2 can undergo the following reactions in a CF_4 plasma:

$$e + O_2 \rightarrow 2O + e \qquad (70)$$

$$O + CF_2 \rightarrow COF_{2-x} + xF \qquad (71)$$

Mass spectrometer studies of oxidant additions in fluorocarbon and chlorocarbon gases have demonstrated that the relative reactivity of atoms with unsaturate species in a glow discharge follows the sequence F = O > Cl > Br. Of course, the most reactive species present will preferentially undergo saturation reactions that reduce polymer formation and that may increase halogen atom concentration. Ultimately, determination of the relative reactivity of the plasma species allows prediction of the primary atomic ethants in a plasma of specific composition.

VI. Conclusion

There are many areas of importance for successful progress in these fields. Two views from industry (from the semiconductor and pharmaceutical industries) have provided the charts of Figs. 19 and 20, which show the various problems needed to successfully advance the field.

BIBLIOGRAPHY

Blaustein, B. C., ed. (1969). "Chemical Reactions in Electrical Discharges." American Chemical Society, New York, Advances in Chemistry Series no. 80.

Brown, S. C. (1959). "Basic Data of Plasma Physics." Wiley, New York.

Bunshah, R. F., ed. (1982). "Deposition Techniques for Films and Coatings." Noyes, Park Ridge, N.J.

Chen, F. F. (1974). "Introduction to Plasma Physics." Plenum, New York.

Coburn, J. W. (1980). *Plasma Chem. Plasma Process.* **2**, 2.

Holloban, J. R., and Bell, A. T., eds. (1974). "Techniques and Applications of Plasma Chemistry." Wiley, New York.

Knights, J. C., ed. (1984). "The Physics of VLSI." American Institute of Physics, New York.

Mucha, J. A., and Hess, D. W. (1983). "History of Dry Etching." American Chemical Society, New York.

Mucha, J. A., and Hess, D. W. (1983). "Plasma Etching." American Chemical Society, New York.

National Materials Advisory Board. (1985). "Plasma Processing," Report of the National Materials Advisory Board. National Academy Press, Washington, D.C.

Nowogrozki, M. (1984). "Advanced III–V Semiconductor Materials Technology Assessment." Noyes, Park Ridge, N.J.

Shohet, J. L. (1971). "The Plasma State." Academic Press, New York.

Venugopalan, M., ed. (1971). "Reactions under Plasma Conditions," Wiley, New York.

POSITRON MICROSCOPY

Arthur Rich and James Van House *The University of Michigan*

I. Production of Low-Energy Positron Beams
II. Overview of Positron Microscopy
III. Theoretical Basis
IV. The Transmission Positron Microscope
V. Positron Reemission Microscopy

GLOSSARY

Brightness: Particle intensity of a beam per unit solid angle per unit energy.

Brightness enhancement: Increase in the brightness of a positron beam by use of one or more stages of moderation. The nonconservative moderation process circumvents the restrictions described by Liouville's theorem.

Liouville's theorem: States that an ensemble of particles occupies a constant volume in phase space when acted on by conservative forces.

Moderator: Material with a negative surface affinity for positrons that is used to stop a flux of high-energy positrons. Some fraction of the incident flux is reemitted at an energy, typically 1 eV, determined by the negative surface work function.

Positronium: Hydrogenlike bound state of a positron and an electron.

The possibility of obtaining images of the microscopic world with resolutions of the order of atomic dimensions (~ 1 Å) was made possible by the introduction of the transmission electron microscope (TEM) over 50 years ago. Since that time, a variety of instruments have been constructed to improve on this resolution or to obtain different contrasts by using a particle other than the electron as a microscopic probe of the target being investigated. During 1988, a variety of microscopes have been demonstrated for the first time that use the electron's antimatter counterpart, the positron. Because of the positron's positive charge and other antimatter properties, such as annihilation with an electron to produce γ-rays, positron imaging promises to reveal features of materials not previously seen with electrons.

I. Production of Low-Energy Positron Beams

All the positron microscopes exploit the now well-developed technology of slow positron beams. These beams are typically generated from a radioactive source (usually ^{22}Na or ^{58}Co) by a process called moderation. In the moderation process, the initially high-energy positrons generated by the decay of radioactive nuclei are slowed from their initial energy of several hundred kiloelectron volts to thermal energies ($\sim \frac{1}{40}$ eV) in a material, called a moderator, whose surface has a negative work function for positron emission. About 1 in 10^3 of the incident high-energy positrons diffuse to the surface where the negative work function for positrons causes the spontaneous emission of the positrons into the vacuum. The emission kinetic energy is usually about 1 eV, and the emission energy spread is about 0.1 eV. These reemitted slow positrons are then accelerated and focused into a beam. The moderation process has the virtue that the initial energy spread (typically 500 keV) of the positrons emitted by the radioactive source is concentrated into an energy band of 0.1 eV in the reemitted positron beam, resulting in a 5000-fold increase in the brightness of the positron flux emitted from the radioactive source. Brightness, which is defined as the intensity of particles per unit solid angle per unit energy, is a measure not only of the beam intensity, but of its angular divergence as well. Increases in beam brightness for systems acted on by conservative processes are prohibited by Liouville's theorem, which states that an ensemble of particles occupies a constant phase space volume when acted on by conservative forces. The moderation of positrons, however, is not a conservative process, thus Liouville's restriction does not apply. Furthermore, as originally proposed by A. P. Mills, Jr., one could generate a positron beam with enhanced brightness by accelerating and focusing the beam onto a small spot on a crystal. That crystal, the remoderator, then reemits low-energy positrons from a smaller area than the original moderator diameter, and thus has enhanced brightness. Through successive stages of this brightness enhancement, one loses intensity (typically less than half the incident positrons are reemitted from the remoderator for 1000-V incident energy), but one gains a factor of 10–100 in brightness per stage.

II. Overview of Positron Microscopy

The first positron microscope image (Fig. 1) was taken at the University of Michigan and appeared in the January 18, 1988, issue of *Physical Review Letters*. It was taken by the positron analogue of the TEM, and is called the transmission positron microscope (TPM). While the initial TPM resolution of 5 μm falls significantly short of TEM standards, improvements in resolution to below 1 nm should be quite feasible, as discussed later. Using the TPM in direct conjunction with a TEM, the TPM and TEM images can be compared. The differences observed in such a comparison, caused by the opposite sign of the coulomb interaction for positrons versus electrons, will contain information on atomic form factors, as well as on the material composition of the sample being studied. In addition, a small fraction of

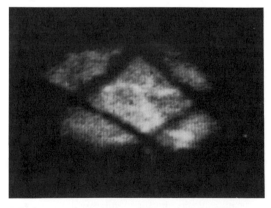

FIG. 1. The first positron microscope image. The image, discussed in detail in Section II, is of a V.Y.N.S. foil. The image magnification is 29× and the resolution is 4.5 μm. [From Van House, J., and Rich, A. (1988). *Phys. Rev. Lett.* **60**, 169–172.]

the positrons will annihilate, during passage through the sample, into two or three γ-rays. Detection of these γ-rays could provide information on the momentum space distribution of the electrons bound in the sample.

The potential for developing the positron counterpart to the scanning electron microscope (SEM) was recently demonstrated at Brandeis University. As discussed in the February 1988 edition of the *Review of Scientific Instruments*, using a brightness enhanced beam, the Brandeis group, in collaboration with Bell Laboratories, have developed a scanning positron microprobe that utilizes the annihilation of a positron with an electron to give γ-rays as a signal. This instrument was used to produce a one-dimensional γ-ray scan of a target with a resolution of 20 μm. Such γ-ray images will be sensitive to large scale defects, because positrons are readily trapped in sample defects. While the demonstrated resolution is not as good as the resolution of the prototype TPM, and falls far short of electron SEM standards, resolutions of 100 nm (limited by positron diffusion) should eventually be possible using higher intensity radioactive sources and a sufficient number of brightness enhancement stages.

The potentially most useful positron microscope exploits the phenomenon of spontaneous positron reemission by the negative surface work function. The first image from such a positron reemission microscope (PRM) was taken by the University of Michigan group. This first PRM operated in a reflection geometry and had a resolution of 2 μm. Independently, the Brandeis group, together with Bell Labs, operated a PRM using a transmission geometry with a resolution of 0.3 μm—the limit of optical microscopes—and with a magnification of 1150. Both groups feel that further improvements can give the PRM a resolution below 1 nm.

These new positron reemission microscopes are variants of a proposal first suggested by L. Hulett, J. M. Dale, and S. Pendyala of Oak Ridge National Laboratory. In their original conception, a small sample was to be placed on a 100-nm-thick substrate, and the positron beam was to illuminate the back side of the substrate. The positrons would slow to thermal energies by inelastic scattering in the substrate and would be reemitted from the front side. They would image the sample lying on the substrate by a shadowing effect. The versions of the PRM actually built differ from the original proposal in that the sample serves as the reemission surface, greatly enhancing the prospects of observing positron specific contrasts, as discussed later.

III. Theoretical Basis

Since the number of low-energy (~ 1 eV) positrons directly emitted by a radioactive source that may be used to form a positron beam of the quality needed for imaging is essentially zero, the recent successful demonstration of positron microscopes is partially based on the fact that the effective brightness of a radioactive source for positron emission at low energies may be increased enormously by the moderation process discussed previously. The e^+ beam so formed is then focused onto a target. The e^+ transmitted through, or reemitted from the target, are then imaged by an objective lens through a contrast aperture. They are then imaged by a projector lens onto a channel electron multiplier array (CEMA) with a phosphor screen anode. The CEMA/phosphor combination converts each individual positron into a spot of light that can be detected by a camera. The camera output is then fed into a computer, which digitizes the image and adds it to events previously stored in computer memory, in order to eventually form a digitized image. This signal averaging is crucial to successful positron microscopy since it allows an image to be built up from the signal events registered by the CEMA at rates as low as $20-200$ Hz. Such rates are a factor of 10^4 lower than the lowest intensities typically used in an electron microscope.

The optimum resolution of a positron microscope occurs when the detector resolution and lens resolution are matched. The detector resolution is determined by the image magnification, as well as the single particle image size produced by the detector. The magnification is, in turn, determined by the desired particle current density at the detector. This detector current density is determined by the initial current density of the beam, the mechanical construction of the instrument, and the acceptance parameters of the focusing elements.

The optimum resolution determined by this matching of lens and detector resolutions is given by

$$R = \left\{ \frac{C_c}{6} \left(\frac{\Delta E}{E} \right) \left(\frac{E_m}{E} \right)^{1/2} \sin \theta_m \left(\frac{\rho_d}{\rho_m} \right)^{1/2} r_d \right.$$

$$\left. \times \left[1 + \sqrt{1 + \frac{9 C_s r_d \sin \theta_m ((\rho_d/\rho_m)(E_m/E))^{1/2}}{(C_c(\Delta E/E))^2}} \right]^{1/2} \right\} \tag{1}$$

Here, C_s and C_c are the spherical and chromatic aberration coefficients of the imaging lenses (primarily the objective lens), E is the energy of the beam at the anode of the objective lens, and ΔE is the variation

in E ($\Delta E \ll E$). The quantities ρ_d and ρ_m are the beam current densities at the detector ($\rho_d = I_d/A_d$, where A_d and I_d are the beam area and the current at the detector, respectively) and the current density at the moderator ($\rho_m = I_m/A_m$, where A_m is the moderator area and I_m the positron current at the moderator). The angle θ_m is the spread in emission angle of the positrons at the moderator, E_m the emission energy, and r_d is the size of the single particle image produced by the detector.

In the case where spherical aberrations dominate R, which commonly occurs in transmission microscopy, Eq. (1) simplifies to

$$R = \left(\frac{C_s}{4} \right)^{1/4} \left(\frac{\rho_d}{\rho_m} \frac{E_m}{E} \right)^{3/8} (r_d \sin \theta_m)^{3/4} \tag{2}$$

We note that these expressions are valid for both the TPM and the PRM, as well as other possible types of microscopes that are based on acceleration and focusing of charged particles using electromagnetic lenses to produce a final image. These expressions are also valid for positron sensitive detectors other than the CEMA/phosphor detector, for example, a charge coupled device (CCD). Using Eqs. (1) and (2), developed to describe the optimum resolution, we now discuss two of the positron microscopes that have been demonstrated within the last year.

IV. The Transmission Positron Microscope

A. Description and Operation

Since the positron microscope is so new, we will discuss the operation of the TPM in some detail in order to illustrate some of the features unique to positrons. The first transmission positron microscope is shown in Fig. 2. Positrons from a 40-mCi, 5-mm

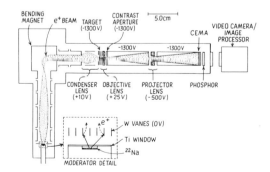

FIG. 2. The transmission positron microscope. [From Van House, J., and Rich, A. (1988). *Phys. Rev. Lett.* **60,** 169–172.]

diam, 15-mg/cm²-thick ²²Na source are incident on a set of W vanes annealed at 2200°C. After losses from source self-absorption, approximately 4×10^{-4} of the positrons emitted by the source are reemitted at an energy of 2 V, accelerated and then focused into a beam of 7×10^5 positrons/sec. This beam is transported to a bending magnet that turns it through 90° and is subsequently focused into a parallel beam that is then incident on a low aberration condenser lens. The condenser lens focuses the beam down to a 1.7-mm-diam spot on the target at an energy of 1.3 keV. The e^+ transmitted through the target are imaged by an objective lens and then by a projector lens onto a three-plate CEMA with a phosphor screen anode. The CEMA/phosphor combination converts each e^+ into a 200-μm-diam spot of light that is detected by an image analysis system (Fig. 2). The system adds the event to the appropriate memory location in a 384×384 array, resulting in a digital signal averaging that allows an image to be built up at rates as low as 20–200 Hz. The actual rate used at the detector depends upon the detector noise level and the desired signal averaging time, with shorter signal averaging times requiring higher rates. The magnification and resolution predicted by Eq. (1) for this prototype TPM are $M = 125 \times$, and $R = 2.2\ \mu$m. In practice, transmission losses in the target must be taken into account. These losses result in lower magnifications, and thus resolutions other than given by Eq. (1). In the prototype TPM, it was found that the typical magnifications were around $M = 55 \times$, where the predicted resolution is $R = 4\ \mu$m.

The images that were obtained using the TPM were of polyvinylchlorylacetate copolymer (V.Y.N.S.) foils less than 800 Å thick supported on a 100 line, 82% transmitting copper mesh. The first TPM image is shown in Fig. 1. It required 4 hr of signal averaging to accumulate. The magnification was calibrated from the known 250-μm grid wire spacing. Gaussian fits to a histogram of one of the grid wires in Fig. 1, both with and without a V.Y.N.S. film, yielded, respectively (errors are statistical), $R_m = (9 \pm 1)\ \mu$m and $R_m = 4.2 \pm 0.5\ \mu$m. The first measurement shows that the effects of chromatic aberrations because of inelastic scattering in the target degrades the resolution by only a factor or two. Such an effect is not serious enough to prevent the use of the TPM at energies down to 1 keV, where the largest new contrasts are expected to occur, as discussed later. The second measurement shows that the predictions of Eq. (1) are verified to within experimental error.

The potential values of resolution that can be achieved with the TPM as a function of the current

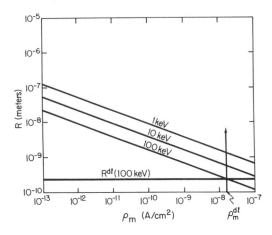

FIG. 3. The resolution of the TPM as a function of the positron current density at the moderator. The diffraction limit on resolution is indicated by the arrow. [From Van House, J., and Rich, A. (1988). *Phys. Rev. Lett.* **60,** 169–172.]

density of positrons emitted from the moderator are shown in Fig. 3. The diffraction limit on resolution is reached at current densities of 1.6×10^{-8} A/cm². Such current densities are well within the reach of the current technological development of slow positron beams, if such beams make use of the technique of brightness enhancement discussed in Section I.

B. Applications of the TPM

Because of the opposite sign of the coulomb interaction, a number of well-known differences exist between positrons and electrons in their scattering interactions with matter. Comparison of the resulting contrast differences between TEM and TPM images taken under otherwise identical conditions could help isolate the effects of particular terms in the scattering cross section on image formation in a manner that cannot be done using the TEM alone. In particular, the screening of the nucleus is more effective for positrons, resulting in substantially reduced small-angle scattering. Because of this, a strongly Z-dependent (Z is the atomic number) fractional difference in the amplitude contrast, ranging at an energy of 50 keV, from 10% for $Z = 8$ to 130% for $Z = 80$, exists between the TPM and TEM. These differences should be 2–3 times larger at lower energies. Comparison of the contrast differences (once they are calibrated) between TPM and TEM images may provide information on atomic form factors because of the screening differences, and could eventually, because of the strong Z dependence, prove a sensitive microanalysis technique. The repulsion of the positrons by the nuclear charge will also alter the phase shifts of the partial waves compared to those of electrons. This will result in differences in the phase contrast that could be as large as those calculated previously for the amplitude contrast.

A number of smaller signals should also prove to yield interesting information. Among these are several new signals that have no analog in the TEM, but which will appear in the TPM because of positron annihilation in flight. These signals contain information on the binding energies and momentum space distribution of the target electrons. Exploitation of these signals will require a scanning technique in order to establish a spatial correlation between the signal and the target.

V. Positron Reemission Microscopy

A. DESCRIPTION AND ANALYSIS OF RESOLUTION

The first PRM is shown in Fig. 4, and its operation is described in the caption. The PRM differs from the TPM in that the nature of the slow positron emission process and the short instrumental depth of information combine to make the PRM an extremely surface sensitive instrument, sampling positron interactions that occur at an energy of 1–3 eV. By comparison, in the TPM, contrast formation occurs for positron interactions at energies of 1 keV and above, sampling primarily bulk properties of thin targets. The areas of application of the PRM are thus radically different than the TPM.

The Michigan PRM was used to conclusively demonstrate the unique defect sensitivity of the PRM, as well as the feasibility of directly imaging biological samples and semiconductor devices. An image demonstrating the defect sensitivity is shown in Fig. 5. For this image, a W foil target was masked with a 100 line, 82% transmitting Cu grid and sputtered at normal incidence with 2-keV argon ions to a fluence of $\simeq 10^{16}/cm^2$. The resulting image, after removal of the masks, unambiguously demonstrates a defect map of the surface. This defect sensitivity will be of particular interest in the study of surface atom diffusion and surface diffusion of defects in a variety of materials, including semiconductors.

Working independently, a group at Brandeis University has developed a transmission geometry PRM, which closely follows the original Oak Ridge concept. In an image taken at $0.3\text{-}\mu m$ resolution with their positron microscope, the Brandeis group identified what they describe as a finger of greatly at-

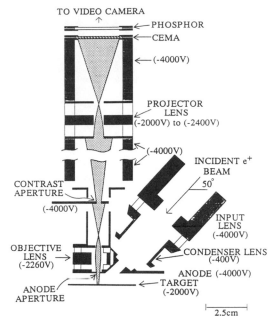

FIG. 4. The reflection positron reemission microscope. The incident slow positron beam of 7×10^5 positrons/sec is generated by a 40-mCi ^{22}Na source and W vane combination and transported by standard electron optics to the entrance of the PRM. A low aberration condenser lens focuses the beam into a 3000-V/cm electric field in a geometry designed to deflect the beam to the target (positioned on the axis of the imaging optics) with minimum distortion of the incident beam shape. The reemitted positrons are accelerated by the electric field and are focused by the anode aperture and then by the objective lens through a variable diameter contrast aperture to the projector lens. The projector lens focuses the positrons that constitute the final image onto a channel electron multiplier array with a phosphor screen anode that generates a $100\text{-}\mu m$-diam single particle image for each incident positron. The single particle image is detected and signal averaged by an image analysis system. [From Van House, J., and Rich, A. (1988). *Phys. Rev. Lett.* **61**, 488–491.]

tenuated emissivity in the middle of a region of relatively high positron counts. They suggest that the positrons in the finger region may have been trapped by boundaries at large tilt angles.

The optimum resolution of either the transmission or reflection PRM can be deduced from Eq. (1), with two important modifications:

1. In the TPM, the resolution was determined based upon the properties of the moderator used to generate the slow positron beam. In the PRM, the target serves as the moderator for the subsequent image beam. Consequently, all quantities that relate to

FIG. 5. A demonstration of the near-surface defect sensitivity of the PRM. The preparation of the damaged W foil is described in Section III. The dark sputtered regions have low positron reemission because of positron trapping in defects. The bright, high reemission areas were masked from sputter damage. The image is an argon sputtered W foil and was taken at a magnification of 17×, with a resolution of 4.6 μm. [From Van House, J., and Rich, A. (1988). *Phys. Rev. Lett.* **61**, 488–491.]

the moderator in Eq. (1) must now be replaced with the corresponding target properties.

2. The acceleration of the reemitted slow positrons by the electric field between the target and anode results in the formation of a virtual target at a distance $l^* \approx 2l$ behind the real specimen. The dependence of l^* upon the emission energy V_e gives rise to nonnegligible spherical and chromatic aberrations during the acceleration of the reemitted positrons from the target to the anode. The complete spherical and chromatic aberration coefficients, C_s and C_c, consist of a linear sum of the spherical and chromatic aberration coefficients of the objective lens referred back to the object space of the virtual specimen, C_s^0 and C_c^0, to the spherical and chromatic aberrations of the accelerating field, yielding

$$C_s = \frac{2l}{27}\sqrt{\frac{V_a}{V_e}} + C_s^0$$

$$C_c = \frac{l\sqrt{V_e V_a}}{\Delta V_e}\left(1 - \sqrt{1 + \frac{\Delta V_e}{V_e}}\right) + C_c^0 \qquad (3)$$

Here, l is the target–anode separation, V_a is the accelerating potential of the anode (all voltages are given with respect to the target potential), V_e is the emission energy, and ΔV_e is the variation in V_e.

Using parameters similar to those discussed in Section II for the TPM, it can be deduced from Eqs. (1) and (3) that the diffraction limit on the resolution for the transmission PRM operating at 100

keV is 6 Å, whereas for the reflecton PRM operating at 5 keV (limited by the geometry), the diffraction limit on the resolution is about 15 Å. Obtaining such resolutions will require positron current densities at the target of greater than 10^{-5}A/cm^2, about 3 orders of magnitude higher than the 10^{-8}A/cm^2 diffraction limit current density required for the TPM. This higher current density is required primarily to compensate for higher spherical and chromatic aberration coefficients given by Eq. (3).

B. CONTRAST FORMATION IN THE PRM AND PRM APPLICATIONS

The mechanisms responsible for the contrast expected to occur in PRM images can be loosely grouped into two categories: topographical contrast and materials contrast. It is in the second category that new types of contrast, and thus new applications, unique to positrons are expected to appear in the PRM.

The topographical contrast in the PRM should be identical to that which appears in the photoelectron emission microscope (PEM), since contrast caused by the effects of the specimen relief on the accelerating field are identical for both positrons and electrons. Heights of 30 Å are detectable experimentally in the PEM, and comparable heights should ultimately be detectable in the PRM. The analysis of Hulett indicates that resolutions approaching atomic dimensions may be detectable in the transmission PRM under conditions involving tip and edge effects of objects placed on a single crystal source of reemitted positrons. Additional sources of topographical contrast include shadowing caused by the incident beam striking a protrusion at an angle. In addition, as discussed by the Bell–Brandeis group, positron reemission microscopy may lend itself to a form of holographic imaging, in which one records the interference between the positrons emitted from a substrate and those scattered from atoms within the molecule lying on the surface. The multiple scattering would make this technique extremely complex, but it has the possibility of circumventing the 0.6-nm limit on resolution imposed by diffraction.

The material contrast is related to the positron reemission yield of a substance, $Y(E)$, which is defined as the number of slow positrons reemitted per incident positron. The yield decreases as the incident positron energy E increases and has a maximum value Y_0 at $E = 0$. The magnitude of $Y(E)$ depends on the various interactions that occur as the incident positrons slow down to near thermal energies and then diffuse through the bulk to the emission surface, as well as interactions that occur at the point of emission.

The manner in which this information will appear in the PRM image is dependent on the nature of the imaging system and on positron diffusion in the bulk of the target. The electron optics form a Gaussian image that is sharply defined at the emitting surface (hence the high topographical resolution discussed previously). Consequently, the surface information will display the approximately 5-Å resolution given by Eq. (1), whereas information at depths below the instrumental depth of field (given by similar relations for the PEM) will appear as a diffuse background. The long positron diffusion length will also result in a randomization of information at depths below 100 Å. Consideration of these two factors leads to the conclusion that, in the PRM, the most important new contrasts will be due to phenomena that occur near the surface and near the point of emission of the slow positron, that is, the PRM will be an extremely surface sensitive instrument.

Slow positron emission has a demonstrated sensitivity to the density of surface and near surface defects in metals (Fig. 5). Areas of application include studies of defect formation at material interfaces and around the edges of islands and terraces of submonolayer thin film overlayers on metal substrates. If a similar defect sensitivity is obtained in insulators and semiconductors, applications to studies of high-density integrated chips and color centers in phosphors are indicated. Assuming that atomic resolutions at edges can be achieved, imaging of single atom defects may be possible. The defect sensitivity of the PRM will be of particular interest in the study of surface atom diffusion and surface diffusion of defects, and may also be useful in understanding the role of surface defects in crack formation.

The sensitivity of $Y(E)$ to the surface work function will give rise to an orientational contrast that should allow studies of polycrystalline materials, such as metal alloys and geological samples. The demonstrated sensitivity of this work function to surface conditions on metals will allow determination of surface cleanliness and should aid in the study of the effects of monolayer and submonolayer coverages of adsorbates on single crystal substrates.

Possible biological applications of the PRM may include elimination of the need for the staining agents required to give sufficient contrast. Such staining agents often produce sample distortion—the sample is no longer in its natural state. The Michigan group has already shown that the PRM can directly image the surface of biological specimens without the use of staining agents, and hence, distortion can be prevented. Another application of the PRM may be in the study of polar surface sites on biological cells. Special labeling compounds attached to site specific antibodies (which then attach to the polar site on the cell) are typically used in light microscopy to link the cell's biological function to its observable topography. Unfortunately, only polar sites specific to the particular antibodies used can be indirectly studied through the labeling compound. The PRM may, however, be directly sensitive to broad classes of polar surface phenomena through two possible mechanisms: positron trapping at negatively charged sites and positron emission after positronium breakup at the polar site, that is, positronium chemistry. Both mechanisms would directly affect the PRM image contrast at the polar site without the use of any labeling or attachment compounds.

BIBLIOGRAPHY

Brandeis, G. R., Canter, K. F., Horsky, T. N., Lippel, P. H., and Mills, A. P., Jr. (1988). *Rev. Sci. Instrum.* **59**, 228–232.

Brandeis, G. R., Canter, K. F., Horsky, T. N., and Mills, A. P., Jr. (1988). *Appl. Phys.* **A46**, 335–337.

Brandeis, G. R., Canter, K. F., Horsky, T. N., and Mills, A. P., Jr. (1988). *Phys. Rev. Lett.* **61**, 492–495.

Rich, A., and Van House, J. (1988). *J. Electron Miscrosc. Tech.* **9**, 209.

Van House, J., and Rich, A. (1988). *Phys. Rev. Lett.* **60**, 169–172.

Van House, J., and Rich, A. (1988). *Phys. Rev. Lett.* **61**, 488–491.

QUANTUM MECHANICS

Albert Thomas Fromhold, Jr.   *Auburn University*

I. Classical Trajectory Motion
II. Classical Model of the Atom
III. Special Relativity
IV. Quantum Concepts
V. Wave Motion
VI. Wave–Particle Duality
VII. Wave Equations for Particles
VIII. Electron Transport in Solids
IX. Summary

GLOSSARY

Eigenfunction: Wavefunction for a specific stationary state of a physical system.

Eigenvalue: Specific value of a physical quantity corresponding to a specific eigenfunction.

Eigenvalue equation: Particular mathematical relation in which a mathematical form known as an operator acts on an eigenfunction to produce the product of the corresponding eigenvalue with the eigenfunction.

Excited states: States of a system having energies above the ground state.

Exclusion principle: Statement that no two particles of a particular statistical type that includes electrons can simultaneously occupy the same state.

Ground state: Lowest energy state of a system.

Lifetime: Mean time before an excited state spontaneously decays to another state such as the ground state.

Particlelike: Localized and acting in an individual manner as entities.

Probability density: Relative probability of particle being at a specified position in space; alternately, the relative probability of some other physical quantity, such as momentum.

Quantization: Discrete value or set of values for a physical quantity such as energy or angular momentum.

Stationary state: Specific state of a physical system characterized by fixed value of some physical quantity such as energy.

Time-dependent Schrödinger equation: Equation for determining the time dependence and space dependence of the eigenfunctions of a system.

Time-independent Schrödinger equation: Eigenvalue equation used to obtain energy eigenvalues and energy eigenfunctions for a system.

Uncertainty relation: Mathematical form relating the maximum precision of measurement of some physical quantity such as position or energy to the precision of some related quantity such as momentum or time.

Wavefunction: Mathematical form, usually complex, used to deduce the probability density.

Wavelike: Nonlocalized and periodic with capability of interacting constructively or destructively.

Wave–particle duality: Coexistence of wavelike and particlelike aspects in a physical entity such as an electron.

Quantum mechanics is a theory that correlates and predicts the behavior of atomic and subatomic systems. It is the only theory to date that is adequate for the microscopic domain in nature. Quantum mechanics correctly predicts the results of physical observations both in the microscopic world where classical physics is unsuccessful and in the macroscopic world which we experience with our senses.

I. Classical Trajectory Motion

A. Historical Perspective

To appreciate the discipline of quantum mechanics, one must first have an understanding of the definition and meaning of mechanics and a historical perspective on its development, successes, and failures.

Mechanics is concerned with the location of material objects and with the motion of these objects. Although the beginning of the study of motion was probably in the aiming and throwing of stones by prehistoric people for self-defense and to attain meat to satisfy hunger, quantitative considerations arose during observation of the seemingly ever-changing positions of the heavenly bodies: namely, the sun, moon, planets, and stars. From the viewpoint of philosophical considerations, the discipline of mechanics dates to Aristotle (384–322 B.C.) who considered an object to be in its most natural state when at rest. Copernicus (1473–1543 A.D.) reflected in a more scientific way on the motions of the earth and the planets relative to the sun, while Galileo (1564–1642 A.D.) reached a conclusion seemingly in conflict with Aristotle, namely, that an object is in a natural state even when in constant-velocity motion, the state of rest being only a special case. This leads to the concept of Galilean reference frames (known also as inertial reference frames), which are nothing more than coordinate systems moving at constant velocity with respect to one another. These play an important role in the theory of special relativity conceived by Einstein (1879–1955).

Galileo developed the telescope and used it to make experimental observations. The results supported the theory of Copernicus that the motions of the earth and planets are most easily explained on the basis of a heliocentric (sun-centered) model. Brahe (1546–1601), without the advantage of a telescope but with the use of an instrument known as a quadrant, mapped out carefully over a 20-yr period the angular positions of the planets and stars, with an experimental precision of 0.01 deg. These data enabled Kepler (1571–1630) to formulate three empirical laws describing the periods and orbits of the planets.

The genesis of the quantitative development of the laws of mechanics, however, is credited to the pondering of Newton (1642–1727), who as legend would lead us to believe, was moved to his understanding of forces and the effect of such on motion by the falling of an apple on his head while he rested under an apple tree. We accept as fact the notion that an apple will fall from the tree toward the earth, as contrasted with its falling the opposite way into the sky; nearly everyone accepts the premise that a mysterious force known as gravity is responsible for this behavior. It took the genius of Sir Isaac Newton, however, to develop concrete scientific statements, known as laws, to describe these observations precisely and quantitatively.

B. Newton's Laws of Motion

The underlying concepts of Newton's laws of motion are that the motion of an object is unchanged until the object is acted upon by a force (e.g., a push or a pull) and that a given force acts with different degrees of effectiveness on different objects, the key point being the difference in the so-called masses of the different objects. These concepts are the basis of Newton's first and second laws. [Newton's first law: An object travels in a straight line at a constant velocity in the absence of a net force. Newton's second law: A net force acting on an object produces a change in velocity (an acceleration) of the object, the change in velocity being proportional to the force and inversely proportional to the mass of the object.] Yet a third law was required to explain the way in which the push or pull of a force is transmitted between the agent of the force and the object, that is, the interaction between agent and object necessary for execution of the force on the object. [Newton's third law: To each action force there is an equal magnitude but oppositely directed reaction force.] Forces between objects thus always occur in pairs, there being a reaction force on the agent equal in value but opposite in direction to the force that the agent exerts on the object.

The tale of the falling apple evokes the mental image of an initially stationary object (the apple hanging from the tree) being acted upon by a zero net force achieved by a delicate balance of counteracting nonzero forces, that is, the force of gravity pulling the apple toward the earth and the opposing force provided by the stem holding the apple to the tree. These two forces are exactly balanced to give a net force of zero, and hence no acceleration, so long as the stem does not snap. Force is thus seen to be a vector quantity having both value and a direction, so mutual cancellation of forces can occur. If suddenly the apple stem snaps under the stress, the balance of forces is destroyed. The force resisting gravity is lost, so the net force becomes the force of gravity, and the apple then accelerates toward the earth. The apple becomes transformed from a state of rest (zero velocity) to a continuum of states of ever increasing velocity with respect to the earth, until finally the earth (or else the head of Sir Isaac Newton) is struck by the apple.

We speak of the moving apple (or other moving object) as having a momentum p of value mv, where m is its mass and v its speed, with the direction of the vector $\mathbf{p}$ being parallel to the motion at any given time. An object having momentum p in a specified direction has associated with this linear momentum a scalar kinetic energy $\mathscr{E}_K = p^2/2m$. The concept of potential energy (energy of position) is then introduced quite naturally, and we thereby arrive at the concept of a transformation of one type of energy into another, with the constraint of conservation of total energy. Being scalar in nature, the total energy is the algebraic sum of its parts, which in the present example consists of kinetic energy (energy of motion) and potential energy (energy of position).

C. ACCELERATION DUE TO EARTH GRAVITY

If we neglect the countering effects of the air to motion, falling apples and other objects of various sizes and masses are found to be given the same value for the acceleration near the surface of the earth by the action of gravity. If Newton's second law is to be obeyed, which maintains that the acceleration $\mathbf{a}$ produced by a force $\mathbf{F}$ acting on an object is inversely proportional to the mass inertia M of the object, namely $\mathbf{a} = \mathbf{F}/M$, then the fact that earth gravity produces the same value g for the acceleration of different masses requires us to accept the fact that the force of gravity on different objects must increase or decrease directly in proportion to the inertial mass of the object,

$$F_j = M_j g \qquad (1)$$

The same g is experienced by all objects (e.g., falling apples, peaches, and balls thrown into the air), provided that the various objects are located at the same distance from the center of the earth and the only force acting on each of the objects is the gravitational force between the object and the earth. These conditions are often closely approximated for dense objects falling from distances above the earth even as high as

tall buildings. Noticeable deviations that can be attributed to differences in relative separation distances from the center of the earth, however, occur at higher altitudes, such as, for example, those reached by the space shuttle. The distance from the center of the earth to the earth's surface also varies a great deal with different locations on the earth's surface, a manifestation of the fact that the earth is not exactly spherical in shape. The poles are closer to the center of the earth than points on the equator, for example. The value of g is found to be larger at the poles than at the equator.

D. NEWTON'S LAW OF UNIVERSAL GRAVITATION

Newton's law of universal gravitation states that an attractive force exists between two gravitational masses m_1 and m_2 that is directly proportional to the product $m_1 m_2$ of the two mass values and inversely proportional to the square of the separation distance r_{12} between the centers of the masses. To return to the example of apples on a tree, the individual distances r_j from the center of the earth to each of the apples labeled j on a tree are almost the same as the distance r_E from the center of the earth to the earth's surface at the position of the tree, the percentage difference between the r_j and r_E being extremely tiny. Thus we can write

$$F_j = G m_j m_E / r_E^2 \qquad (2)$$

with G being a proportionality factor.

E. EQUIVALENCE OF INERTIAL AND GRAVITATIONAL MASS

From Eqs. (1) and (2),

$$g = G(m_j/M_j)(m_E/r_E^2) \qquad (3)$$

For g to be a constant, the ratio m_j/M_j in this equation cannot depend on the mass j, so

$$m_j/M_j = \text{constant} \qquad (4)$$

This means that the gravitational mass of an object is directly proportional to its inertial mass. The choice of unity for this constant gives

$$g = G m_E / r_E^2 \qquad (5)$$

with the usual values for G and g, namely, $G = 6.673 \times 10^{-11}$ N m^2/kg^2 and $g = 9.8$ m/sec^2. Choice of a different value for the ratio in Eq. (4) would merely redefine the values of the constants G and g in Eqs. (1) and (2). Thus we can say that the gravitational mass of an object is equal to its inertial mass. Therefore, values deduced for the inertial masses of two objects by experiments involving applied forces and measured accelerations can be used to predict the gravitational attraction of the two objects.

Although the equivalence of inertial and gravitational masses arrived at above is of the nature of a conclusion, the premise is that Newton's second law and Newton's law of universal gravitation are each valid, independently of one another. Newton probably arrived at the law of universal gravitation from the second law by implicitly assuming the equivalence.

Thus there is either the implicit assumption or else the conclusion that the inertial masses M_j determining the acceleration of objects under a given force in Newton's second law have the same values as the gravitational masses m_j needed for computing the forces between any pair of masses in Newton's law of universal gravitation. This profound result later led Einstein to formulate the principle of equivalence, namely, the effects of a gravitational force cannot be distinguished from those of an acceleration, which underlies the general theory of relativity. [See RELATIVITY, GENERAL.]

F. MOTION OF NATURAL AND ARTIFICIAL SATELLITES

A quantitative analysis of satellite motion about the earth illustrates the predictions of classical mechanics and provides the basis for the classical planetary model of the atom. The gravitational attraction between the mass m_E of the earth and the mass m_S of a satellite has the magnitude

$$F = \Gamma/R^2 \tag{6}$$

where

$$\Gamma = Gm_E m_S \tag{7}$$

and R is the distance from the center of one mass to the center of the other. The gravitational force given by Eq. (6) provides the necessary centripetal acceleration to cause the satellite to travel in a circular orbit, so

$$mv^2/R = \Gamma/R^2 \tag{8}$$

This relation leads to the expression

$$\mathscr{E}_K = \frac{1}{2}mv^2 = \frac{1}{2}\frac{\Gamma}{R} \tag{9}$$

for the kinetic energy.

The potential energy is that energy stored internally in a conservative system by the action of external forces. It is reasonable to consider the potential energy to be zero when the masses are infinitely separated, where the force between them is zero. Then the potential energy for separation distance R is equal to the external work expended in bringing m_S from a distance $r = \infty$ to the position $r = R$, which moreover is the negative of the energy expended in traveling in the reverse direction from R to ∞,

$$\mathscr{E}_P = -\int_R^\infty \mathbf{F}_{applied} \cdot d\mathbf{r} = -\int_R^\infty -\mathbf{F}_{internal} \cdot d\mathbf{r}$$

$$= -\int_R^\infty -\left(-\frac{\Gamma}{r^2}\right) dr = \frac{\Gamma}{r}\Big|_R^\infty = -\frac{\Gamma}{R} \tag{10}$$

Thus by comparison of Eqs. (9) and (10), we see that

$$\mathscr{E}_K = -\tfrac{1}{2}\mathscr{E}_P \tag{11}$$

The total energy $\mathscr{E}_T$ is the sum of the kinetic and potential energies,

$$\mathscr{E}_T = \mathscr{E}_P + \mathscr{E}_K = \mathscr{E}_P - \tfrac{1}{2}\mathscr{E}_P = \tfrac{1}{2}\mathscr{E}_P = -\mathscr{E}_K \tag{12}$$

Equations (9)–(12) yield the radius of the orbit in terms of the total energy,

$$R = -\Gamma/\mathscr{E}_P = -\Gamma/2\mathscr{E}_T = \Gamma/(2|\mathscr{E}_T|) \tag{13}$$

The angular frequency ω is given by

$$\omega = \frac{v}{R} = \frac{(2\mathscr{E}_K/m_S)^{1/2}}{-\Gamma/2\mathscr{E}_T} = \left(-\frac{8\mathscr{E}_T^3}{m_S\Gamma^2}\right)^{1/2}$$

$$= \frac{m}{\Gamma}\left(\frac{2|\mathscr{E}_T|}{m_S}\right)^{3/2} \tag{14}$$

The speed can also be expressed readily in terms of the total energy,

$$v = \left(\frac{2\mathscr{E}_K}{m_S}\right)^{1/2} = \left(-\frac{2\mathscr{E}_T}{m_S}\right)^{1/2} = \left(\frac{2|\mathscr{E}_T|}{m_S}\right)^{1/2} \tag{15}$$

We shall use analogous formulas for the motion of the electron about the nucleus in the planetary model of the single-electron atom.

G. SUCCESS OF CLASSICAL PHYSICS

Mechanics within the framework of the above-described laws, being the first discovered and formulated, are in some sense "classical," so this branch of physics is commonly referred to as classical mechanics. Classical mechanics actually encompasses additional but equivalent formulations, such as those of LaGrange (1736–1813) and Hamilton (1805–1865). These lead formally to more abstract systems of equations of motion, but the important point is that different classical mechanical formulations yield the same predictions for motion. The beauty of these abstract formulations of mechanics may be compared to the beauty of the music and opera created by Mozart (1756–1791) and Verdi (1813–1901) who lived in the same time period as LaGrange and Hamilton.

Classical mechanics has proved adequate for describing and predicting the motions and trajectories of objects that are large enough for us to see directly with our eyes. This includes everyday experiences of balls thrown, kicked, or batted into trajectories, such as a forward pass or a field goal in football, or a tennis ball slammed toward the net by a hand-held racket in a tennis game. It likewise includes the motion of the planets about the sun, the motion of the moon about the earth, the trajectories of the comets, and the motion of artificial satellites.

Historically, the success of Newton's theory in giving an understanding of the motion of the planets constituted a major intellectual breakthrough. A large body of work, including the detailed, laborious calculations for the trajectories of the planets, led to the area of classical mechanics known as celestial mechanics. In fact, the deviation from such predictions for the trajectory of the planet Mercury provided some of the first experimental evidence for the validity of the later general relativity theory (1916) of Einstein.

Another area of classical mechanics involves the application of Newton's laws to the motions of fluids; this is the discipline known as fluid mechanics, which makes extensive use of the concepts of energy and momentum and in addition uses extensions of mechanics provided by thermodynamics. Yet another discipline, statistical mechanics, describes and predicts the average behavior of enormous ensembles of particles, each particle undergoing a motion predictable in principle by classical mechanics. [See STATISTICAL MECHANICS.]

II. Classical Model of the Atom

A. STRUCTURE OF THE ATOM BY RUTHERFORD SCATTERING

Rutherford's interpretation in 1911 of his extensive scattering experiments gave overpowering evidence that an atom consists of a dense, positively charged nucleus surrounded by a cloud of electrons. The electron had been discovered a few years earlier by J. J. Thomson in 1887, who attributed a definite charge-to-mass ratio to the particle. Before the discovery of the

wave properties of the electron in 1927 by G. P. Thomson, the son of J. J. Thomson, it was expected that the mechanical properties of atoms could readily be explained by applying classical mechanics in a straightforward way to this model. In fact, the successful model of planetary motion around the more massive sun under the action of gravitational forces between the planets and the sun, and the perturbations to such motion due to the gravitational forces between planets, provide the closely related classical analog model for describing the motion of electrons about a single massive nucleus, the electrical forces between charged particles replacing the gravitational forces in the solar system. The major differences in the two cases are the following: The forces between electrons are repulsive instead of attractive; and these repulsive forces are of the same order of magnitude as the attractive forces between the electrons and the nucleus, not merely perturbations as is the case of the gravitational forces among the planets orbiting the sun. [See ATOMIC PHYSICS.]

It was expected in particular that the hydrogen atom, containing but a single electron and thus being free of the more complicating repulsive Coulomb force between electrons, which must be considered in two-electron and many-electron atoms, should be easily amenable to treatment by classical mechanics. The picture would be a simple model in which the single electron orbits the far more massive proton nucleus under the action of a centripetal force provided by the attractive electrical force between electron and nucleus. In the planetary analogy of the moon orbiting the earth, the centripetal force provided by the attractive gravitational force between moon and earth depends upon the constant Γ in Eq. (7). Equations (6)–(15) were expected to hold for the electron–proton case, provided that Γ is replaced by

$$\Gamma = -q_1 q_2 / 4\pi\varepsilon \qquad (16)$$

B. CLASSICAL RADIATION BY AN ORBITING ELECTRON

Let us explore the classical viewpoint a bit further by means of the planetary model of the one-electron atom. One electronic charge $-e$ is considered in motion about an equal magnitude charge of opposite sign (e.g., one proton charge $+e$) placed at the center of the orbit. Thus we have a rotating electric dipole, though admittedly the center of rotation is located essentially at one end of the dipole. A time-varying electric field is produced by this rotating dipole. At a distant point in space, the field is essentially that which would be produced by a corresponding dipole rotating about its center. Furthermore, the electric field direction is cyclic, being the same as the rotation frequency of the dipole. An oscillating electric field is thus produced at the observation point, the frequency ν of the field so produced being equal to the frequency of rotation of the dipole. According to classical electrodynamics the energy associated with this oscillating electric field can propagate outward in space or be absorbed by some process at the field point, as for example, by accelerating the conduction electrons in a metal.

The energy radiated away or absorbed must come at the expense of the potential energy of the configuration of the two charges if it is not supplied by some source connected with the rotation of the dipole. For a hydrogen atom there is no such source. The consequent decrease in total energy of the electron in the one-electron atom means that the total energy becomes more negative, and this corresponds to an increase in the kinetic energy of the electron in accordance with Eqs. (10) and (11). This corresponds to a decrease in radius of the electron orbit and an increase in the rotation frequency. On the basis of the classical model of radiation, the frequency of the radiated electromagnetic wave would then increase. The changes that occur should be continuous, on the basis of this classical physics approach, and there would be theoretically no limit to the process even as the total energy of the electron–proton system approached negative infinity, with the consequence that an infinite amount of energy would be radiated away.

Clearly the results deduced on the basis of this classical model are unreasonable. In addition, the predictions do not correspond in any way to the experimental observations of optical spectra, which is not continuous as predicted but contains many sharp spectral lines. Hence this classical model cannot be used to describe the mechanical properties of an atom.

The true state of affairs is that all atoms have a lowest energy state, labeled the ground state by Bohr, for which no further energy emission is possible. Furthermore, even while in a given higher-energy configuration, the atom does not emit energy continuously; excited states emit energy only sporadically, each such event being accompanied by a sudden jump to a lower-energy configuration. Bohr called the various discrete energy states of an atom stationary states,

since such states are stable until the time when a transition to a lower energy state actually takes place.

C. Failure of Classical Physics

The logical deduction from the above discussion can only be that the classical approach, which is so successful for describing the motion of the planets, fails completely for the hydrogen atom. So much for arguing by analogy. Moreover, the failure of classical mechanics for the hydrogen atom is manifested not in our lack of ability to see and follow the electron because it happens to be too tiny to be seen with the eye or even with a microscope, but rather in experimental measurements yielding data such as the light spectrum emitted by excited gases of atoms, which could in no way be explained without additional assumptions quite foreign to classical mechanics. We shall return later to the interesting topic of the relationship of the range and intensity of the various colors emitted and absorbed by a gas to the mechanics of the motion and trajectories of the electrons in the atoms making up the gas. This topic played an extremely important role in the discovery of quantum theory and in the subsequent development of quantum mechanics.

D. Logical Limit of Classical Mechanics

An attractive inverse-square force leads to a negative potential energy that varies inversely with the separation distance between charges or masses, as the case may be, for electrical or gravitational forces. The potential energy is conservative, which means that a decrease in separation distance yields energy that can be extracted by an external agency or converted into heat. If the separation distance can be imagined to decrease to zero, then the potential energy approaches negative infinity. This implies that simultaneously an infinite amount of energy is either extracted or else converted into heat. Conceptually all machines in the world could be run for a day or a month or a year, or as long as one wishes, by allowing an electron to come closer and closer to a proton in a controlled manner. This means that in the original creation of the opposite charges an infinite amount of energy was expended. Alternately, we could imagine causing an explosion that would dwarf even that of the most powerful fission or fusion weapon available today by merely allowing or triggering the collapse of a configuration of two point charges of opposite sign. These incomprehensible conclusions, derived solely from logic, are sufficient in themselves to force us to the realization that nature must actually behave under more constraints than those contained in the formalism of classical mechanics. To bring in other evidence bearing on this matter, electron–positron pairs can be produced from γ rays, and it is known from these experiments on pair production that an infinite amount of energy is not required to separate the oppositely charged particles so produced. (The positron has mainly the same values for its physical properties as the electron; its charge, however, is opposite in sign to that of the electron.) In fact, the γ-ray energy required is only of the order of the rest-mass energies of the electron and the positron. Likewise, pair annihilation does not lead to the emission of an infinite quantity of γ-ray energy, so again we must conclude that the negative Coulomb energy must be restricted by nature to have a finite magnitude. This end could be accomplished by restricting the separation distance between opposite charges to some minimum nonzero value, corresponding perhaps to the electron–proton separation distance in the so-called ground state configuration of the hydrogen atom. There will be more discussion of this state later. [*See* ELEMENTARY PARTICLE PHYSICS.]

E. NEW FOUNDATIONS OF MECHANICS

It is important, however, to note that classical mechanics does provide a generally adequate theoretical description of motion for all objects that can be seen in an ordinary way and that are traveling at ordinary speeds, *ordinary* generally meaning observable by means of the human eye and traveling at speeds low enough for the eye to follow the position of the object. In fact, the theory is found to apply even in the domain of far smaller particles, such as those that can be seen only with the aid of an optical microscope, so long as the speed is well below the speed of light propagation. However, the extremely versatile framework provided for our use by classical mechanics gives inaccurate results in two domains that can be distinctly different:

1. Particles traveling at speeds approaching the speed of light. (Light travels in free space with a speed c of approximately 3×10^8 m/sec. Objects traveling at speeds v of the order of $0.1c$ or larger manifest marked departures from Newton's law predictions based on a fixed mass.)

2. Particles having very tiny masses, such as the smaller atoms and nuclei (e.g., the hydrogen

and helium atoms, the proton, neutron, and the α particle) and especially the even less massive electron. [The ratio of the mass of the smallest nucleus (the proton) to the mass of the electron is 1836.]

In both of these domains, namely, the domain of very fast particles and the domain of very low mass particles, entirely new forms of mechanics were erected. These new forms had essentially different premises from those inherent in classical mechanics. This was necessary because classical mechanics simply fails to describe and predict correctly the motions and trajectories of objects under such conditions. The first domain, very high speed motion, requires the use of the theory of special relativity (1905) of Einstein. The second domain, very small mass particles, requires the use of the theory of quantum mechanics. Special relativity and quantum mechanics constitute a pair of theories that enable us to make accurate calculations in the separate domains listed above. The relativistic mechanics provided by Einstein's theory of special relativity accurately describes and predicts motions of particles and bodies moving with speeds approaching the speed of light, and quantum mechanics describes and predicts experimental observations for very low-mass particles, atoms, and the properties of ensembles of atoms. Einstein's theory of special relativity alone, however, is unable to account for the properties of atoms and the discreteness of the emitted radiation constituting the optically observed atomic spectra. [*See* RELATIVITY, SPECIAL.]

For very small mass particles traveling at relativistic speeds, there have been formulated more complicated theories (relativistic quantum mechanics), that have had their successes, one of the most prominent being Dirac's theory predicting the existence of the positron. Ideally one would like to have a very general framework that not only would predict correctly the motion of particles for each of these domains but also would give the correct results under more ordinary conditions, where classical mechanics already does a completely satisfactory job. It must be admitted, however, that the basis for a perfectly general theory of mechanics, if such can be conceived, still remains to be developed. Even the general theory of relativity of Einstein does not contain the power to interpret the microscopic world; instead it gives answers to cosmological questions relating to the macroscopic world. Simply put, no completely general theory exists today. Let us throw down the gauntlet to those readers of this encyclopedia who have an unbridled imagination and the tenacity to persevere.

III. Special Relativity

A. Essential Relations for Quantum Mechanics

Because the topic of special relativity is covered adequately in another part of this encyclopedia, only the relationships essential for the present development are presented here. It is vital to point out that the profound prediction by Einstein of the variation of the measured mass of an object with its speed has immediate implications for motion that are quite contrary to those predicted by Newton's second law applied in the restricted sense of a fixed, unvarying mass for any given object.

Defining the velocity $\mathbf{v}$ as a vector that has magnitude equal to the speed and pointing in the direction of motion, the momentum $\mathbf{p}$ is then simply

$$\mathbf{p} = m\mathbf{v} \tag{17}$$

Newton considered m to have a fixed value m_0. The acceleration $\mathbf{a}$ is defined as the time rate of change of the velocity,

$$\mathbf{a} = d\mathbf{v}/dt \tag{18}$$

Newton's second law $\mathbf{F} = m_0\mathbf{a}$ can thus be written in terms of the time rate of change of the momentum,

$$\mathbf{F} = d\mathbf{p}/dt \tag{19}$$

This form is valid even in special relativity, as also are Newton's first and third laws. Time t, however, is not absolute in special relativity as it is in the framework of Newton's formalism. Also we know from special relativity that mass m depends upon speed in accordance with

$$m = \gamma m_0 \tag{20}$$

where m_0 is the rest mass and the parameter γ is determined by v/c, the ratio of the speed of the particle to the speed of light, in accordance with

$$\gamma = [1 - (v/c)^2]^{-1} \tag{21}$$

Energy is related to mass m in special relativity,

$$\mathscr{E} = mc^2 \tag{22}$$

This relation holds true for any speed v. When $v = 0$, this reduces to the rest mass energy

$$\mathscr{E}_0 = m_0c^2 \tag{23}$$

The energy versus momentum relation for a free particle in special relativity is

$$\mathscr{E}^2 = \mathscr{E}_0^2 + p^2c^2 \tag{24}$$

This differs from the energy–momentum relation in Newtonian mechanics given by the kinetic energy expression. However, $\mathscr{E}$ reduces to the sum of the rest-mass energy $\mathscr{E}_0$ and the kinetic energy in the limit of low velocities.

The above results are used in our later development of the de Broglie relation, which underlies the Schrödinger equation of quantum mechanics.

B. Deterministic Basis

It is unlikely that Einstein could have foreseen in 1905, the year of publication of his works on special relativity and the photoelectric effect, the great philosophical differences that mark the quantum way of thinking and the relativity way of thinking. Relativistic mechanics, like classical mechanics, leads to a world view of an absolutely predictable (in principle, at least) future, based on the specified state of the universe at any given time. Quantum mechanics, on the contrary, contains an intrinsic margin of uncertainty regarding the future evolution of the system, even if the state of the universe is known as completely as possible at any given time. For example, descriptions of the motion of an electron by Newtonian and Einsteinian mechanics are precise, though not necessarily in agreement with each other or correct, in predicting a specific position and velocity at a time t, given the exact position and velocity at any other time t' and the precise specification of all forces that act on the particle in the time interval between t' and t; quantum mechanics, on the other hand, gives only relative probabilities for a continuous range of possibilities.

C. Important Implications for Quantum Mechanics

Contained in the theory of special relativity is not only the experimentally observed mass variation with speed, but also the concept of the equivalence of mass and energy. This latter concept, which is directly verified in electron–positron pair production, has, of course, far-reaching consequences with regard to present-day energy production in nuclear fission reactors and possible future energy production by nuclear fusion reactors. The knowledge of how enormous amounts of energy can be quickly released by nuclear fission and fusion likewise provides the basis for weapons of exceedingly great destructive power. In his later years, Einstein made great efforts to point out the potentially devastating consequences for the future of all peoples of the weapons developed and used in World War II.

There is unquestionably a great philosophical difference between Einstein's special relativity and modern quantum mechanics. Superficially it is difficult to conceive that such a successful theory as special relativity might be intrinsically flawed on either philosophical or scientific grounds. However, the same statement could likewise be made for Newton's theory in light of its success in explaining and predicting planetary motion. Newton's theory has a limited sphere for predicting results in exact agreement with experiment. Although the theory of special relativity gives predictions that agree much better with experimental observations for cases of rapidly moving particles, both theories have the same deterministic basis. Both are highly successful in certain domains, but neither is completely successful in all domains.

Special relativity gives predictions that are essentially the same as the predictions of Newton's equations at low speeds but are different and, incidentally, in much better agreement with experimental measurements, in the domain of high speeds. Special relativity includes some of those properties of the physical world, such as the velocity dependence of mass and length, that are not included in ordinary classical mechanics. It is thus a more complete theory. Ordinary classical mechanics is still used to describe low-speed phenomena because for most specific applications the equations are easier to solve.

A scientist must be flexible enough to accept new experimental facts. One would be most foolish, for example, to reject relativistic mechanics simply because one dislikes the concept that our perception of the mass of an object varies with speed of the object. In the same way, experimental demonstration of the wave properties of matter requires that our theoretical formulations be broadened to include these properties. If this leads naturally to the introduction of some indeterminism into our predictions of an event, then this must be accepted. In fact, it would seem to be scientifically inappropriate to decide the merits or demerits of any given theory solely on the basis of whether or not the theory is deterministic, since the preference (or lack of such) for a deterministic theory merely reflects a peculiarity in one's philosophical outlook. We know, however, that the very lack of determinism inherent in quantum mechanics was the Achilles heel that eventually led Einstein to the conclusion that quantum mechanics provides, at best, an incomplete description of the universe. His powerful intuition, which should not be lightly discounted, led him to the belief that quantum mechanics would more than likely be superseded eventually by a more complete theory, probably completely deterministic in form. This would allow exact predictions of the complete future evolution of a system, given only the initial conditions at some prior point in time. This viewpoint, which means that nothing is left to chance, was hotly disputed by prominent contemporaries of Einstein. It would seem to belie even the possibility of free will in man. Planck and Dirac, on the contrary, believed that the uncertainty inherent in quantum mechanics was in fact an intrinsic reality in nature, never to be overcome by any better theory that purports to be more comprehensive or more general. For these philosophically minded physical theorists, the important uncertainty principle is elevated from its role as an integral part of quantum mechanics theory to an even more lofty philosophical principle.

It is of course one's prerogative to entertain by philosophical speculation if one so wishes, but nevertheless it is important to keep in mind that such speculation is not physics. Thinking people do not confine their thoughts and or even comments to the extremely limited domain of their greatest expertise and should not be criticized for speaking out in other spheres of interest. However, if only physics is to be done on some given day, then on that day one should ask different questions, such as how best to describe and predict experimental observations in a logical manner from models developed from a combination of observation and intuition. To go beyond this end by inquiring why nature behaves in the fashion it does and no other fashion (e.g., why classical mechanics, which describes so well the motions of visible bodies, fails in the microscopic world of electrons and atoms), often proves interesting and stimulating but otherwise does not in itself constitute the furthering of the subject proper of physics. With this as a basic premise, it seems that the best approach is to develop formulations that represent a synthesis of all experimentally verified properties of matter, including variation of mass with velocity and wavelike interference of particles. We adopt this approach in the present treatise, not hesitating to use special relativity when appropriate. In fact, the roots of the most fundamental relation (i.e., the de Broglie relation) in the wavelike description of particles are to be found most naturally in Einstein's explanation of the photoelectric effect and in the equations of special relativity.

Although we have laid the foundation for exploration and development of the field of quantum mechanics by delineating the domains in which classical mechanics provides erroneous predictions and have inferred the condition of very small mass for which quantum mechanics is unique in yielding correct predictions, it remains for us to explain the implications and appropriateness of the adjective *quantum* in the term *quantum mechanics*. A proper understanding requires that we examine conclusions arrived at by Einstein during the period preceding the development of quantum mechanics. This is done in the following section. It is an enigma that the seeds of quantum theory lie in, indeed constitute the entirety of, one of Einstein's creative explanations of a very puzzling experimental observation, namely, the photoelectric effect. The enigma arises from the remarkable fact that Einstein's best-known work, the theory of special relativity, is an entirely separate and indeed philosophically remarkably different theory from that of quantum mechanics.

IV. Quantum Concepts

A. Early Quantum Theory

We distinguish between what is known as quantum mechanics and what is usually termed old quantum theory. Old quantum theory (sometimes referred to simply as quantum theory) is essentially the patchwork of classical and quantum ideas that were pieced together to yield Planck's theory of blackbody radiation, Einstein's explanation of the photoelectric effect, and Bohr's theory of the one-electron atom. Old quantum theory included the concept of the particle nature of radiation (i.e., the photon) and the quantization of the radiation energy in elemental units $h\nu$. However, it did not include the concept of the wave nature of matter.

B. Einstein Concept of Light as an Energy Quantum

The origin and meaning of the adjective *quantum* can be understood from its use in the explanation of some of the properties of light. It was Einstein, in his explanation of the interaction of light with metal surfaces, who gave impetus to the concept that electromagnetic radiation is made up of discrete increments of energy. At that time (1905), light was generally considered to be an electromagnetic wave, a wave somewhat similar to water waves or sound waves, with an accompanying energy density that could lead to energy exchange with another body. A light wave in free space may be viewed classically as coupled, orthogonal, time-dependent electric and magnetic fields that are in phase, each varying periodically in time and space. This leads to signal propagation through free space and through material media. The electric field of the wave leads to an electric force on charged particles; charged particles can thus be accelerated and thereby gain kinetic energy at the expense of the energy density of the electromagnetic wave. This picture leads to the prediction of electrons in a solid being accelerated to various energies, depending upon the acceleration time, with no sharp cutoff for the maximum energy that can be attained and with no indication that there should be a wavelength (or color) dependence of the effect. It also leads naturally to the conclusion that a more intense light wave of the same wavelength should exert a larger electric force and hence produce more acceleration and consequently higher-energy electrons. In addition it might be expected that if the light is of extremely low intensity, there would be a measurable time delay before the electron could gain sufficient energy to overcome the surface energy barrier (given by the work function) and escape from the metal. These so-called classical predictions were not in accord with the experimental data. No time delay was observed: Electron emission from the metal surface was either observed or not observed, depending upon whether the wavelength was less than or greater than some critical wavelength, that critical wavelength being characteristic of the metal used for the experiment. Furthermore, all emitted electrons had a kinetic energy equal to or less than a given maximum energy that increased as the wave frequency. (The wave frequency ν is given by c/λ, where c is the speed of light and λ the wavelength; it is also the reciprocal of the time period of the oscillation, thus having units typically of cycles per second.)

Einstein pointed out that all these observations could be explained by postulating that the energy transfer between light wave and electron occurs only in well-defined discrete quantities of size $h\nu$, with h being the fundamental constant introduced by Planck in the year 1900 to resolve the blackbody radiation spectrum problem. According to this postulate, light is quantized in elemental units of energy, these elemental units being incapable of further decomposition. From this viewpoint the properties of light are similar to the properties of elemental particles such as the electron and the proton, which have basic units of mass and charge of specific values that are ordinarily incapable of further decomposition.

The idea that light has particlelike properties was in one sense a resurrection of the corpuscular theory of light espoused by Newton. Since that theory had been abandoned, in view of the wavelike properties of diffraction and interference discovered by Huygens (1625–1695), the particlelike picture invoked by Einstein to explain the photoelectric effect was again revolutionary. Even Planck, who had chanced upon the explanation of the blackbody radiation spectrum by introducing the concept that the radiation in a cavity could only have discrete values for the energy, did not believe that the discreteness was associated with light in any fundamental way. Instead, Planck believed initially that the discreteness proceeded only from constraints on the absorption and emission of radiation by the cavity walls themselves. Light itself was generally viewed as a wave phenomenon, with periodic variations in the electric and magnetic fields but no sharp discontinuities in space and time. To postulate, as Einstein did, that the energy could be localized in space so that it could be exchanged with the electron in a metal instantaneously and in discrete quantities, when already the wave nature of light had been experimentally confirmed, was quite revolutionary. Indeed, it was nearly akin to postulating that light had dual properties.

C. Transitions between Atomic States and Photon Emission

Looking into the phenomenon of light production in a light source, that is, the origin of optical spectra, we are led to the concept of excited states of the atoms in the source. An atom becomes excited when it absorbs energy in some way, as from the heat associated with a rise in temperature. Such an excited atom can give off energy in the form of a light wave, or more generally, by emitting some form of electromagnetic radiation. From the viewpoint of classical me-

chanics the atom can have a continuous range of energies in the excited state and can therefore emit a continuous range of electromagnetic wave energies. The optical absorption and emission spectra would from this classical viewpoint be quite nondescript and moreover would vary little from element to element in the periodic table. Such simplicity, however, is belied by the experimental optical spectra: The optical spectrum obtained from a gas discharge is usually very complex, and it is so characteristic of the atoms making up the gas that it can serve as a fingerprint identifier of the element making up the gas. From a classical view, the most unexpected characteristic of the experimental optical spectrum are the sharp peaks in intensity at certain wavelengths (or frequencies); such are difficult or impossible to explain on the basis of a classical mechanics picture of the atom.

If Einstein's view of light as discrete quantities of energy is to be sustained, these discrete quantities being known today as photons, then it is natural (if not absolutely necessary) to conclude that these entities are produced by the atoms in the light source. The relationship between light-wave frequency and the light-quantum energy existent in the photoelectric effect leads one to speculate that the sharp emission spectrum at certain wavelengths in optical spectra, so characteristic of the atoms of the source, indicates that the changes in the atom configuration leading to light emission must occur in a manner such that individual quanta of energy $h\nu$ are emitted. The experimentally unconfirmed expectation on the basis of classical mechanics that the optical spectrum should be rather more or less continuous, with no sudden sharp changes in intensity with wavelength, had therefore to be superseded by the view that the changes in the mechanical configuration are catastrophic and instantaneous. Classically, a sudden change in the total energy of an inverse square force system (see Section I,F) must be accompanied by a sudden change in the radius of the orbit and also a corresponding change in the rotation frequency of the particle in the orbit. The accompanying change in the energy must therefore occur suddenly, the difference in energy between the two characteristic configurations determining the energy of the emitted photons. The mechanical states of the physical system are thus specific and characterized by fixed energies, with the transition between any two such fixed energy states occurring suddenly with the emission of a fixed quantum of energy. Collapse must occur from one characteristic configuration to another characteristic configuration of lower energy when the photon is created. Each configuration may individually be compared to the intricate workings of a fine watch, with the change in configurations occurring as a sudden transition between two well-defined and self-regulated states.

D. Intensity Peaks in Optical Spectra

Einstein's realization of the quantization of light, derived from the interaction of light with the conduction electrons in metals (viz., the photoelectric effect), led to a rejection of the simple classical picture of an atom derived from a planetary model of one or more electrons revolving about the nucleus. In the classical planetary model, there could be arbitrary values for the radius of each electron orbit, and the electron would be capable of having any one of a continuous range of values of kinetic energy and angular momentum. Instead, we are led to the viewpoint that only certain characteristic mechanical configurations of the atom are allowed.

Consider for a moment the analogous classical problem of an artificial earth satellite held in orbit by the gravitational attraction of the earth. From the simple view of classical mechanics, the speed of the artificial satellite can be changed continuously by adding or extracting arbitrarily small quantities of energy. As the speed and frequency of rotation change, corresponding continuous changes occur in the radius of the orbit. If this classical picture were applied to an electron in orbit about a proton, the constant Γ in Eq. (6) would have the value $\Gamma = |q_1 q_2 / 4\pi\varepsilon_0|$ given by Eq. (16), where q_1 and q_2 are the charges of the proton and the electron, and ε_0 is the dielectric constant of free space. According to classical electrodynamics an accelerated charge radiates energy, so the potential energy of the electron would steadily decrease. This ultimately leads to the picture of the catastrophe described in Section II,D.

From a quantum viewpoint, however, the radiation energy can be abstracted only in units of quanta $h\nu$ of radiation energy, where ν is the frequency of the electromagnetic wave produced. The total energy therefore always changes by discrete photon energy increments $h\nu$. Thus, to the extent that the electron can be viewed as changing speed in its orbital motion to yield the emission of energy, then that change in speed must occur in a single-step process.

Clearly an entirely new type of mechanics is required for the description of atomic systems having these properties. Because of the discreteness of the stationary states, as evidenced by the discrete frequency of the electromagnetic wave emitted upon transition between such stationary states, and the quantum nature of the energy emitted as photons, this new type of mechanics was called quantum mechanics.

E. Ideas of de Broglie, Heisenberg, and Schrödinger

Matter was discovered to have wavelike properties by means of the electron diffraction experiments of Davisson and Germer and G. P. Thomson (1927). This was preceded (1923) by de Broglie's deduction from special relativity that a particle of energy $h\nu$ and momentum p had associated with it a wavelength λ. Furthermore de Broglie postulated that there exists a wave–particle duality of all of nature. In the same way that electromagnetic radiation can be observed to behave in a wavelike manner under certain conditions and in a particlelike manner under other conditions (witness radio wave interference on the one hand and the photoelectric effect on the other), wave–particle duality asserts that nonzero-mass particles such as electrons, protons, neutrons, and atoms can be observed to behave in a wavelike manner under some conditions, and in a particlelike manner under other conditions.

This very important concept of particles possessing an intrinsic wave character underlies the Schrödinger equation, that cornerstone of present-day quantum mechanics dating from 1926. The birth of quantum mechanics, in fact, dates to the developments of Heisenberg and Schrödinger in 1925 and 1926, which contain intrinsically the wave properties of matter, though somewhat differently. Whereas Schrödinger directly developed a wave equation to describe the behavior of matter, Heisenberg incorporated the wave properties into a theory in a somewhat different way, setting up a noncommuting matrix operator formulation. Thus he was able to show that certain pairs of physical observables represented by noncommuting operators could not in principle be measured to arbitrary precision, but rather that the more precise that one of the pair was measured, the less precise would be the knowledge of the other. Thus was born the Heisenberg uncertainty principle. This type of uncertainty follows naturally from a wavelike description of matter such as that represented by the Schrödinger equation. The Schrödinger formulation can be used to deduce a matrix formulation of quantum mechanics analogous to that of Heisenberg, so in this sense the two theories are equivalent.

F. Bohr Quantized Energy Levels for Hydrogen Atom

It is remarkable that Bohr was able to put together a successful theory of the hydrogen atom, since its formulation in 1913 predated by more than a decade the discovery of the wave diffraction of particles, and even predated by a decade the postulate of wave–particle duality by de Broglie and the deduction of de Broglie that a particle of momentum p has associated with it a wavelength λ. Fundamentally, Bohr utilized the data given by experimental optical spectra, together with heuristic arguments based on the classical limit, to put together his model of the hydrogen atom. We shall follow a somewhat different line of reasoning, based on the experimentally confirmed de Broglie relation, in deducing the quantized energy levels of Bohr.

As a preliminary to our development of the Schrödinger equation utilizing the experimentally verified de Broglie relation

$$\lambda = h/p \qquad (25)$$

between wavelength λ associated with a particle and the momentum p of that particle, where h is Planck's constant having the value 6.6262×10^{-34} J sec, let us use that same relation as a selection device for choosing a series of discrete orbits for an electron imagined to circle a proton from the continuous range of orbits allowed in the purely classical planetary model of the one-electron atom. The key addition is the requirement that a wave associated with a trajectory must be a single-valued function of position on the trajectory. For a circular orbit, this requires that the wavelength be commensurate with the circumference of the orbit, viz.,

$$2\pi r/\lambda = n_B \qquad (26)$$

where r is the radius of the circular orbit, λ the wavelength, and n_B any integer equal or greater than one. (The subscript B is affixed to the integer n to denote that we are considering essentially the Bohr planetary model of the one-electron atom.) Note that we make no attempt to interpret the meaning of the wave that we assume to exist around the circumference of the orbit, but instead we rely only on one of the

most general properties of waves (viz., single valuedness) for writing the condition given by Eq. (26) above. The logic of this approach is merely that if the properties of matter are wave-like, and if a planetary orbit exists, then condition (26) should be met. In the treatment of the hydrogen atom by means of the Schrödinger equation, it is found that the concept of a well-defined planetary orbit is too naïve except in what we designate the classical limit, due to the fact that a wave does not usually have a precise localization; but nevertheless there do exist well-defined values of the most likely position of the electron relative to the nucleus of the one-electron atom.

Substituting Eq. (25) into (26) gives

$$2\pi r p/h = n_B \qquad (27)$$

which in terms of the new constant $\hbar$ defined as

$$\hbar = h/2\pi \qquad (28)$$

takes the form

$$rp = n_B \hbar \qquad (29)$$

Because, for a circular orbit, the product rp is the angular momentum L, the last equation constitutes a restrictive condition on the angular momentum. That is, the values of the angular momentum L are restricted to the discrete set of values L_{n_B} given by

$$L_{n_B} = n_B \hbar \qquad (n_B = 1, 2, 3, \ldots) \qquad (30)$$

This is a statement of the quantization of angular momentum. Thus we arrive at the startling conclusion that not only is angular momentum conserved for the one-electron atom, as can be readily deduced from classical mechanics, but in addition we have the new quantum condition that the angular momentum must be quantized. The elemental unit for the mechanical angular momentum is thus $\hbar$. Because angular momentum quantization proceeds entirely from the wave description, it is a direct consequence of the wave nature of particles.

Now we show that the quantization of the angular momentum leads to the quantization of the energy values for the one-electron atom. The electrical force between a nucleus of charge Ze and an electron of charge $-e$ is

$$F_E = \frac{KZe^2}{r^2} \qquad (31)$$

where K is a constant that for SI units has the value

$$K = 1/4\pi\varepsilon_0 \qquad (32)$$

where ε_0 is the electric permittivity, having in free space the value (in SI units) of 8.854×10^{-12} F/m, and r the separation distance between electron and nucleus. As in our treatment of the gravitational force between satellite and earth in Section I,F, we make the assumption here that the nucleus is stationary and the electron travels around the nucleus; although this view is absolutely correct in classical mechanics only if the nucleus is infinitely massive, it can be shown to be approximately correct whenever the nucleus is much more massive than the electron, and this is satisfied for all one-electron atoms. In the present derivation we consider the radius of the electron orbit to be equal to the separation distance between electron and nucleus; corrections for the deviations caused by the finite nuclear mass can easily be considered in the treatment of the hydrogen atom by means of the Schrödinger equation. The electrical force provides the centripetal force mv^2/r that causes the electron to travel in the hypothesized circular orbit, so

$$mv^2/r = KZe^2/r^2 \qquad (33)$$

This gives at once an expression for the kinetic energy $\mathcal{E}_K$ of the electron,

$$\mathcal{E}_K = \tfrac{1}{2} mv^2 = KZe^2/2r \qquad (34)$$

The potential energy $\mathcal{E}_P$ associated with the Coulomb force is, in analogy with Eq. (10) for the gravitational earth–satellite problem, given by

$$\mathcal{E}_P = -KZe^2/r \qquad (35)$$

so that

$$\mathcal{E}_K = -\tfrac{1}{2} \mathcal{E}_P \qquad (36)$$

The sum of potential and kinetic energies gives the total energy $\mathcal{E}_T$

$$\mathcal{E}_T = \mathcal{E}_P + \mathcal{E}_K = -2\mathcal{E}_K + \mathcal{E}_K = -\mathcal{E}_K \qquad (37)$$

The kinetic energy is intrinsically positive, as can be noted from Eq. (34), so the total energy is negative. This means that the electron is bound to the nucleus. To emphasize this point, consider a motion of the electron in the radial direction. As the electron separation from the nucleus increases, the potential energy increases toward zero, so the potential energy becomes the total energy, and the corresponding kinetic energy will be zero before the electron can separate to infinity. If the electron cannot separate to an arbitrarily large distance from the nucleus, we say that it is bound to the nucleus.

The centripetal force relation given by Eq. (33) relates the electron speed v to the radius r of the classical orbit,

$$r = KZe^2/mv^2 \qquad (38)$$

or equivalently,

$$v = (KZe^2/mr)^{1/2} \qquad (39)$$

This not only allows the kinetic energy to be expressed in terms of the radius, as given by Eq. (34), but also allows the angular momentum $L = pr = mvr$ for an electron in a circular orbit to be expressed in terms of the radius,

$$L = pr = mvr = mr \, (KZe^2/mr)^{1/2}$$
$$= (KZe^2mr)^{1/2} \qquad (40)$$

Equivalently, this gives the radius in terms of the angular momentum,

$$r = L^2/KZe^2m \qquad (41)$$

which can then be substituted into Eq. (39) to give the speed in terms of the angular momentum,

$$v = \left(\frac{KZe^2}{m} \frac{KZe^2m}{L^2}\right)^{1/2} = \frac{KZe^2}{L} \qquad (42)$$

The kinetic energy in terms of the angular momentum is then given by

$$\mathcal{E}_K = \frac{1}{2} mv^2 = \frac{mK^2Z^2e^4}{2L^2} \qquad (43)$$

Now it should be kept in mind that Eqs. (31)–(40) are based on classical mechanics. The quantum condition given by Eq. (30) is now introduced into Eqs. (34)–(41) to give discrete values for the radius

$$r_{n_B} = \frac{(n_B\hbar)^2}{KZe^2m} \qquad (44)$$

and the following quantized values for the total energy of the Bohr atom,

$$\mathcal{E}_T = -\mathcal{E}_K = \frac{-mK^2Z^2e^4}{2n_B^2\hbar^2} \qquad (n_B = 1, 2, 3, \ldots)$$
$$\qquad (45)$$

Using the SI value for K, this expression for the total energy can be written as

$$\mathcal{E}_{n_B} = \frac{-mZ^2e^4}{32\pi^2\varepsilon_0^2 n_B^2\hbar^2}$$
$$= \frac{-mZ^2e^4}{8\varepsilon_0^2 n_B^2 h^2} \qquad (n_B = 1, 2, 3, \ldots) \quad (46)$$

Thus the average value of the orbit radius is predicted to increase with increase in the integer

n_B while the energy increases algebraically from negative values to approach the asymptotic limit of zero corresponding to an unbound state. At the other extreme of small values for n_B the negative total energy becomes algebraically smaller corresponding to tighter binding of the electron and more localization of the electron in the neighborhood of the nucleus. The lowest energy state is given by $n_B = 1$, so this is the ground state of the one-electron atom. Denoting the total energy in this state by $\mathcal{E}_0$ we have

$$\mathcal{E}_0 = \frac{-mZ^2e^4}{32\pi^2\varepsilon_0^2\hbar^2} = \frac{-mZ^2e^4}{8\varepsilon_0^2h^2} \tag{47}$$

Substituting the value $m = 9.1096 \times 10^{-31}$ kg, $e = -1.6022 \times 10^{-19}$ C, $\hbar = h/2\pi = 1.0546 \times 10^{-34}$ J sec, and $\varepsilon_0 = 8.854 \times 10^{-12}$ F/m gives

$$\mathcal{E}_0 = Z^2 \times 2.180 \times 10^{-18} \quad \text{J}$$
$$= Z^2 \times 13.60 \quad \text{eV} \tag{48}$$

as the ground-state energy for the one-electron atom. For hydrogen, $Z = 1$, so the ground-state energy of the hydrogen atom is predicted to be 13.6 eV. As we shall deduce shortly by using the above results to examine the predictions for optical spectra, the theory gives results that are in good agreement with experiment. Thus this amalgam of classical mechanics and the assumption of a wavelike character for the electron in orbit as given by the de Broglie relation leads to a new picture for electrons in atoms.

Needless to say, the electron is too small to be observed directly in its orbit, even if such a picture were tenable from a fundamental standpoint, and so the theory cannot be proved or disproved in that way. On the other hand, optical spectra are capable of experimental measurement, and thus optical measurements can serve as a point of contact between microscopic mechanical models of the atom and the world of observation. The predictions of the model can therefore be tested in this manner. We shall show in the next section that the Bohr theory leads to a reasonably accurate explanation for the optical spectra for the one-electron atom.

Despite the success of the Bohr theory for one-electron atoms, it is incapable of giving realistic predictions for atoms having two or more electrons, and a fortiori, it fails to give a general explanation of the periodic table for the great variety of elements to be found in nature. Therefore we conclude that a stronger wave theory is needed for understanding and predicting nature. The planetary model as modified in the simplest possible way by the concept of the wave nature of matter is sufficiently successful to indicate a promising way to proceed, since it gives us some insight into the types of thought processes and concepts required to begin a better treatment. That treatment is presented herein in the form of the Schrödinger equation. Before we enter into the development of that topic, however, it is appropriate to examine the predictions of the Bohr theory for the optical spectrum of hydrogen.

G. Optical Spectrum of Hydrogen

Because we cannot observe directly an electron in orbit about a proton, it is impossible to say exactly what the separation distance is between electron and proton at any given instant. That aspect of the Bohr theory would be difficult to check experimentally, even if it were measurable in principle. However, the energy absorption and emission due to transitions of the hydrogen atom between various states of excitation can indeed be measured. This is the experimental point of contact with the theory. If the electron in the hydrogen atom with energy corresponding to the integer n_B suddenly undergoes a transition to an energy corresponding to a different integer n_B', with the energy difference being positive so as to create a photon of energy $h\nu$, then energy conservation gives the relation

$$h\nu = \mathcal{E}_n - \mathcal{E}_{n'}$$
$$= -\mathcal{E}_0 \left[\left(\frac{1}{n'}\right)^2 - \left(\frac{1}{n}\right)^2 \right] \tag{49}$$

where $\mathcal{E}_0$ is given by Eqs. (47)–(48), with $Z = 1$. Since for the photon

$$h\nu = h\left(\frac{c}{\lambda}\right) = \frac{2\pi\hbar c}{\lambda} \tag{50}$$

the above relation can be used to write

$$\frac{1}{\lambda} = \frac{h\nu}{2\pi\hbar c} = \frac{-\mathcal{E}_0}{2\pi\hbar c} \left[\left(\frac{1}{n'}\right)^2 - \left(\frac{1}{n}\right)^2 \right] \tag{51}$$

This is usually written in the form

$$\frac{1}{\lambda} = R \left[\left(\frac{1}{n'}\right)^2 - \left(\frac{1}{n}\right)^2 \right] \tag{52}$$

where

$$R = -\frac{\mathcal{E}_0}{2\pi\hbar c} = \frac{me^4}{64\pi^3\hbar^3\varepsilon_0^2 c} \tag{53}$$

is known as the Rydberg constant. Thus transitions from the various excited states ($n > 1$) to the ground state ($n' = 1$) leads to a series of spectral lines with frequencies $\nu = c/\lambda$ given by

$$\nu_{n' \to 1} = cR \left[1 - \left(\frac{1}{n} \right)^2 \right] \quad (n = 2, 3, 4, \ldots)$$

(54)

This series of lines is known as the Lyman series, which is in the ultraviolet region of the spectrum. (The corresponding optical wavelengths λ are of course given in every case by c/ν, where c is the speed of light.) Yet a second series of spectral frequencies can be generated by transitions from excited states with $n > 2$ to the state $n' = 2$. It can be noted that these frequencies are given by

$$\nu_{n' \to 2} = cR \left[\left(\frac{1}{2} \right)^2 - \left(\frac{1}{n} \right)^2 \right]$$

$$(n = 3, 4, 5, \ldots) \quad (55)$$

This series of spectral lines, known as the Balmer series, has wavelengths in the near ultraviolet and visible region of the spectrum. Similar series of lines determined from $n' = 3, 4$, and 5 can be written using the above prescription, these three series being labeled Paschen, Brackett, and Pfund, with wavelengths in the infrared region of the spectrum.

These various series of spectral lines were known from experimental observation long before the invention of the Bohr model of the atom and indeed constituted some of the major evidence that classical mechanics was inadequate for the treatment of atomic systems. The frequency relationships deduced above from the Bohr theory agree quite well with the experimental spectra. Even better agreement is obtained when the finite mass of the nucleus is included in the theory, the correction coming about because electron and proton revolve about the center of mass, thus giving a slightly smaller orbit radius than that obtained above assuming that the radius is the same as the separation distance between electron and proton. This is the so-called reduced mass effect, which is purely classical in nature. [See MOLECULAR OPTICAL SPECTROSCOPY.]

H. The Laser

The absorption and emission of light due to electron transitions between quantized energy levels provides the basis for one of the most powerful research tools developed over the past 50 years, namely, the laser. Apart from basic research applications, the laser is used in a host of practical applications in fields as diverse as metallurgy and medicine. Although nearly everyone is familiar with the term *laser,* not everyone remembers that the word is an acronym for *light amplification by stimulated emission of radiation.* A similar device, the *maser* (an acronym for *microwave amplification by stimulated emission of radiation*), predated the laser and is based on analogous physical principles.

The key property of a laser beam is its phase coherence. This phase coherence of the beam is to be contrasted with the random phase relationships between the light emitted from various regions of an ordinary light source such as an incandescent filament. The phase-coherent beam in a laser is produced by reflecting the emitted light back and forth between parallel mirror surfaces that bound the emitting medium, so that any light already emitted triggers additional emission and influences the phase of such additional light emission from the excited atoms of the medium. The phase of the newly emitted light turns out to be that of the already emitted light. It is not obvious that this should necessarily occur, but in fact it is basic to the nature of the process referred to as stimulated emission. Stimulated emission is to be distinguished from spontaneous emission, which prevails in ordinary light sources which lack phase coherence.

The emission of light in a laser can initiate because of prior electronic transitions in atoms in a gas (e.g., a helium–neon mixture) or in a solid (e.g., ruby) that have been pre-excited by some process such as the flash of an ordinary discharge lamp. As the phase-coherent beam builds in intensity in the laser, a portion of it is allowed to emerge continuously from the active lasing medium by a partial transmission through one of the mirrored surfaces.

With an intense phase-coherent beam such as can be obtained from the laser, many experiments involving light interference are possible. The historical Michaelson–Morley experiment for determining the relative speed of light in different directions relative to the earth's instantaneous velocity vector in its orbit about the sun is based on the concept of a phase shift between two beams, the two beams being produced by a partial reflection and partial transmission by a mirror beam splitter. Although the Michaelson–Morley experiment was first carried out before the invention of the laser, that experiment can be performed with greater precision with the laser. The laser has enabled the measurement of the speed of light to much greater precision than ever before, making it possible to use the wavelength of a laser beam as an accurate standard for the measurement of length. The angular divergence of a laser beam can be held to quite small values, so that reflection of a measurable signal from very great distances (e.g., from earth to moon and return) becomes possible. By measuring the time for round trip travel, we can measure large distances very accurately.

The intensity of a laser beam in watts per square meter can be extremely high so that localized melting and welding become possible with this tool. In medicine the laser beam can be used as a scalpel, with minimum bleeding resulting because the intense beam actually cauterizes the blood vessels as it cuts through the flesh. The detached retina of an eye can also be repaired by means of another medical laser instrument.

It is clear that the above-described features of a laser hinge critically on the concept of quantized energy levels; otherwise, the light emission would occur in a continuum of energies (and hence a continuum of wavelengths), and phase coherence in the sense described would be impossible. The quantum mechanical computation of energy levels and the attendant probabilities for transitions between levels are thus essential for the design of new lasers.

V. Wave Motion

A. Development of Wave Equation for Transverse Vibrations

The essential equations for classical wave motion constitute the foundation for developing intuitive insight into the fundamentals of quantum mechanical wave equations for particles. It is well known that wave motion can describe phenomena as varied as the transverse displacement of a wire in tension, the propagation of sound in gases, and the longitudinal displacements associated with elastic waves in a solid. All of these classical phenomena can be treated theoretically by means of the same differential equation technique, which moreover proves to be adequate for describing the wave properties of particles.

Consider, for example, the basic classical problem of the time and position dependence of the transverse displacement of a vibrating wire. The transverse displacement $y(x, t)$ of a thin wire under tension $\mathcal{T}$, with mass per unit length along the x-direction given by ρ, is determined as a function of time t by a differential equation that can be obtained from a straightforward application of Newton's second law of motion, $F = ma$, where F is the force and a the acceleration. A net force F_y acting in the y-direction on a length Δx produces an acceleration a_y that is inversely proportional to the mass $\rho \Delta x$,

$$F_y = (\rho \, \Delta x)a_y = (\rho \, \Delta x)(d^2y/dt^2) \quad (56)$$

The net force F_y is given by the difference between the y-directed components of the uniform axial tension force $\mathcal{T}$ at the two ends of the element Δx. Considering θ to measure the angle in radians between the wire and the horizontal line representing the wire when it has no transverse displacement, we can visualize θ as varying with position x along the wire and with time t as the wire vibrates. The tensile force $\mathcal{T}$ is considered to have a line of action that is always parallel to the wire. This leads to a projected component in the y-direction given by ($\mathcal{T} \sin \theta$). At the ends of the segment Δx the tensile force acts in opposite directions, as required of a tensile force. If the segment is not to be accelerated in the x-direction, the x-components of these end forces must essentially cancel. The component of $\mathcal{T}$ projected in the x-direction is given at any point by ($\mathcal{T} \cos \theta$), so this cancellation condition is met adequately if θ is small enough. In the small θ limit, the net force in the y-direction on the segment Δx, namely,

$$F_y = \mathcal{T} \sin \theta_{x+\Delta x} - \mathcal{T} \sin \theta_x \quad (57)$$

reduces to the approximation

$$F_y \simeq \mathcal{T}\theta_{x+\Delta x} - \mathcal{T}\theta_x \quad (58)$$

which in turn can be approximated by

$$F_y = \mathcal{T} \tan \theta_{x+\Delta x} - \mathcal{T} \tan \theta_x \quad (59)$$

Because $\tan \theta$ is the slope $\partial y/\partial x$, this last equation is equivalent to

$$F_y = \mathcal{T}\left[\left(\frac{\partial y}{\partial x}\right)_{x+\Delta x} - \left(\frac{\partial y}{\partial x}\right)_x\right] \quad (60)$$

Equating this expression for F_y to the partial derivative equivalent of Eq. (56) gives

$$(\rho \, \Delta x)\left(\frac{\partial^2 y}{\partial t^2}\right) = \mathcal{T}\left[\left(\frac{\partial y}{\partial x}\right)_{x+\Delta x} - \left(\frac{\partial y}{\partial x}\right)_x\right] \quad (61)$$

Dividing through by Δx and taking the limit $\Delta x \to 0$ gives

$$\rho \left(\frac{\partial^2 y}{\partial t^2} \right) = \mathcal{T} \left(\frac{\partial^2 y}{\partial x^2} \right) \qquad (62)$$

The unit of tension $\mathcal{T}$ in SI units is N, and the units of ρ are kg/m, so $\mathcal{T}/\rho$ has the dimensions of m²/sec², the square of a speed. Denoting the square root of this quantity by v_p,

$$v_p = (\mathcal{T}/\rho)^{1/2} \qquad (63)$$

where v_p has the units of velocity, the above equation for a vibrating wire takes the form

$$\frac{\partial^2 y}{\partial x^2} = \frac{1}{v_p^2} \left(\frac{\partial^2 y}{\partial t^2} \right) \qquad (64)$$

This is the well-known classical wave equation. It is a linear second-order differential equation that governs in the present example the transverse displacement $y(x, t)$ as a function of position along the wire as time proceeds. Both standing-wave modes and running-wave modes are possible, depending upon the boundary conditions imposed on the wire.

B. Solutions to the Classical Wave Equation

Let us attempt a trial solution for the wave equation (64) of the form

$$y = C \exp[i(kx - \omega t)] \qquad (65)$$

where C, k, and ω are at the moment unspecified constants, perhaps even complex numbers. Substituting this expression into the wave equation gives a condition on the constants ω and k in terms of v_p,

$$k^2 = \omega^2/v_p^2 \qquad (66)$$

The complex form of the trial solution is troublesome for individuals concerned with more than the mathematics, and we must limit our consideration in the present situation to the subset of possibilities for which the displacement is a real quantity. That is, the mathematical solution does not in itself have associated with it any physical interpretation. To be able to clothe the mathematical solution with physical content, it is often necessary to restrict somewhat the range of solutions that can be accepted. The displacement being a real quantity, it is necessary to limit our consideration to those mathematical forms that can yield real numbers in the complex domain. One approach would be to work only

with real solutions, but this unnecessarily restrictive approach would eliminate one of the finest ways to solve the problem in a simple way. As an alternative, the complex solutions can be used provided the stipulation is made that only the real part is considered as the actual solution to the physical problem. This may seem to constitute a sleight-of-hand of questionable validity, but it can be justified mathematically. The easiest way to see this is to recognize that there is no way that the real and imaginary parts of a complex trial solution can become mixed [either by squaring of the factor $i = (-1)^{1/2}$ or by the multiplication of real and imaginary parts] merely by substituting into a linear equation such as Eq. (64). If substitution of the complex variable gives a solution to the equation, then, of necessity, both the real and the imaginary parts of the complex variable individually satisfy the equation. Thus if physical reality requires us to restrict our consideration to real solutions only, then we obtain a range of possibilities by taking only the real parts of the complex solutions. The reason for developing the complex variable approach for this problem is that mathematically it is often simpler to work with exponential functions than it is to manipulate sine and cosine functions.

The physically meaningful solutions for the present problem of a vibrating wire are thus given by

$$y \Big|_{\substack{\text{physically} \\ \text{meaningful} \\ \text{subset}}} = \text{Re}\left[C \exp[i(kx - \omega t)] \right] \qquad (67)$$

where Re means take the real part. It is expedient at this point to restrict the constants k and ω to real values. However, it proves very useful, as will be shown, to maintain the constant C in its most general complex form,

$$C = A + iB = D \exp(i\Theta)$$
$$= D \cos \Theta + iD \sin \Theta \qquad (68)$$

where D and Θ are real numbers. Since this equation denotes that $A = (D \cos \Theta)$ and $B = (D \sin \Theta)$, it is evident from the above relations that

$$A^2 + B^2 = D^2 = C^*C \qquad (69)$$

$$\frac{B}{A} = \sin \Theta/\cos \Theta = \tan \Theta \qquad (70)$$

The trial solution can therefore be written in the form

$$y = D \exp[i(kx - \omega t + \Theta)] \qquad (71)$$

thus giving real solutions of the form

$$y = D \cos(kx - \omega t + \Theta) \qquad (72)$$

The right-hand side has the sort of space dependence and time dependence that we see in the laboratory for vibrating wires. Larger values of ω yield shorter repetition periods in time, and larger values of k yield shorter repetition periods in space. Considering that a cosine function repeats itself when its phase is increased by 2π, it can be concluded that $k\lambda = 2\pi$ and $\omega\tau = 2\pi$ give the basic spatial unit λ and temporal unit τ for repetition. These parameters are called wavelength and period. Note that

$$k = 2\pi/\lambda \qquad (73)$$

and

$$\omega = 2\pi/\tau \qquad (74)$$

The temporal frequency ν is the reciprocal of the period τ,

$$\nu = 1/\tau \qquad (75)$$

so

$$\omega = 2\pi\nu \qquad (76)$$

The explicit assumption was made above that the constants ω and k are real. For wave motion this is consistent with the above physical interpretations relating these parameters to the temporal frequency and the reciprocal of the spatial periodicity.

C. Phase Velocity of Waves

It is a universal property of wave motion that

$$\omega/k = 2\pi\nu/(2\pi/\lambda) = \lambda\nu = \lambda/\tau \qquad (77)$$

But this ratio gives the speed v_{phase} of the sinusoidal wave, where τ is the period for one temporal oscillation. This is readily understood by visualizing a moving sinusoidal spatial wave to pass by a given point in space, the length λ passing in time τ. The speed of an individual sinusoidal wave is called phase velocity,

$$v_{\text{phase}} = +\omega/k \qquad (78)$$

Equation (63) therefore tells us that the quantity $v_p = (T/\rho)^{1/2}$ is the speed with which the sinusoidal wave moves along the wire. The phase velocity may depend upon frequency or wavelength in some instances. Note that in the present example, however, the phase velocity depends directly upon the square root of the tension in the wire and inversely upon the square root of the mass per unit length of the wire, independent of the frequency or the wavelength of the wave.

D. Dispersion Relations for Waves

The ω versus k relation for any wave is called the dispersion relation for the wave. The dispersion relation always gives the magnitude of the phase velocity v_{phase} of the wave in accordance with $v_{\text{phase}} = \omega/k$. With this generalization, Eq. (64) can be viewed as a more general equation for one-dimensional classical wave motion. The same form of the equation holds for electromagnetic wave propagation in free space, for example. The phase velocity is then the speed of light, and the appropriate dispersion relation is

$$\omega = ck \qquad (79)$$

Light propagates in a dielectric essentially according to the same wave equation as for free space except that the phase velocity now depends upon the refractive index n of the medium,

$$v_{\text{phase}} = c/n \qquad (80)$$

where n usually varies with the wavelength. The action of a prism in dispersing the colors from incident white light, for example, is due primarily to the fact that the phase velocity is color dependent. Red light travels approximately 0.5% faster than blue light in fused quartz. Both colors travel in fused quartz at approximately 2/3 the speed of light in free space, but since the speed of the blue is decreased slightly more than that of the red as the white light enters the quartz from the vacuum, the blue light is bent more toward the normal direction than the red. The colors thus become dispersed as the white light enters the prism at an angle with respect to the normal to the surface of the prism. The dispersion relation for this situation is

$$\omega = v_{\text{phase}}k = \frac{ck}{n(k)} \qquad (81)$$

where n explicitly depends on k, denoting a wavelength dependence since $\lambda = 2\pi/k$. This example will also be useful to us in understanding group velocity in the development given below.

E. Boundary Conditions

The next consideration is the type of boundary conditions that are imposed by the physical situation. Often, for a wire, the ends are fixed. If the endpoints of a wire of length L are fixed at positions $x = 0$ and $x = L$, then the fact that the displacements must be zero at these points leads to the requirement that the wave must have stationary nodes at these points. A similar type of boundary condition holds for electromagnetic waves trapped in a highly conducting metal box, since the metal walls cannot support an electric field. Equation (72) cannot satisfy such boundary conditions as time progresses, but two such solutions having the same magnitude k but differing in sign of k can be superimposed to yield a solution that can satisfy the condition. For example, the superposition solution

$$y = D \left[\cos \left(kx - \omega t - \frac{\pi}{2} \right) \right.$$
$$\left. + \cos \left(-kx - \omega t + \frac{\pi}{2} \right) \right]$$
$$= D[\sin(kx - \omega t) + \sin(kx + \omega t)]$$
$$= 2D \sin(kx) \cos(\omega t) \qquad (82)$$

satisfies the boundary condition at $x = 0$, and the boundary condition at $x = L$ is satisfied if the wavelength is chosen to have any value $\lambda = L/n$, where n can be any positive integer. The amplitude of this wave is $2D$. The nodes ($y = 0$) and extremum points ($y = \pm 2D$) on the wave do not change position x in time t, so this solution is called a standing wave.

On the other hand, for an infinitely long wire, there may be no constraints on the displacement at any particular position, so the solution described by Eq. (72) may be quite acceptable. The particular values of position x for which the wave has nodes and extremum values changes with time t, so this wave is not stationary. It is thus called a running wave. For positive values of ω and k, we see from Eq. (72) that a given value of the phase is maintained as time increases if one focuses on an observation point x' that moves along the x-axis linearly with t. Thus we say the phase velocity ω/k is constant, and the wave moves in the positive x-direction. If k is negative, but ω is again chosen to be positive, as is conventional, then the phase velocity is negative, and the wave moves in the negative x-direction. We see that the standing wave, obtained in Eq. (82) to satisfy the fixed boundary conditions, can be viewed simply as the superposition of two equal-amplitude running waves having the same frequency and wavelength but moving in opposite directions. More complicated superpositions are considered below.

F. SUPERPOSITION SOLUTIONS

A superposition of solutions of the type given in Eq. (82), each of which is a solution to the given linear differential equation, also satisfies the same equation. That is, there is no way for the terms to mix as products if the equation is linear, so individually the terms for each individual solution add to zero. It is not even necessary that every component have the same value for the phase velocity. In that case v_{phase} would be a function of the frequency (or wavelength). The superposition could be written as a sum of terms having different weighting coefficients D_j,

$$y(x, t) = \sum_j D_j \cos(k_j x - \omega_j t + \Theta_j) \quad (83)$$

where each ratio ω_j/k_j has the appropriate value of v_{phase}. The superposition could yield a shape for $y(x)$ at a given time t that differs markedly from any of the sinusoidal components. If each component moves in the same direction with the same speed, corresponding to a value of v_{p} that is a fixed constant, then we would anticipate motion but not distortion of shape. Variations in the phase velocity can lead to a change in shape, also, since the phase relationships between the individual components continually change if they are traveling at different speeds.

G. GROUP VELOCITY OF WAVES

It is known from the theory of Fourier series and Fourier integrals that the superposition of sinusoidal waves can yield almost any physically reasonable periodic or nonperiodic function. Thus by superposing solutions of different wavelength, nearly any function shape can be generated at any given time. The superposition of waves bearing a harmonic relationship to one another yields a spatially periodic function, whereas a superposition of waves having amplitudes only over a narrow band of frequencies yields a spatially localized function. The waves in the superposition may or may not have the same phase velocity v_{p}. Consider, for example, the case of electromagnetic wave motion in dielectrics where the physical properties of the medium determining the wave speed are frequency (or wavelength) dependent. Whether or not the shape changes in time, it generally moves. The problem we now address is the evaluation of the speed of motion of the shape.

To simplify the problem, let us confine our attention to a narrow band of wavelengths, corresponding to a narrow range of k values centered about some central value k_0. For example, a pulse of light having a certain shape could have a range of wavelengths closely centered about the green 5461 Å line of mercury. If the amplitude is a continuous function of the wavelength over the band, as contrasted with the superposition of a finite number of discrete wavelength components, then an amplitude function $\chi(k)$ can be defined that peaks at k_0 and that has very low values outside the narrow band of interest. It is known from the theory of Fourier integrals that the shape of $\chi(k)$ as a function of k depends directly on the position-dependent shape of the pulse. If $\psi(x, t)$ denotes the pulse in space at time t, then a Fourier inversion yields directly $\chi(k)$. It can be shown that the widths of the two functions are related inversely. That is to say, the broader is $\chi(k)$ over k, the more narrow is $\psi(x, t)$ over x, and vice versa. This is a manifestation of a relationship between the spread (or uncertainty) in wavelength and the spread (or uncertainty) in position of the wave packet. This point is essential for understanding the Heisenberg uncertainty principle but is not directly relevant to the present development for pulse velocity.

For any reasonable shape for $\chi(k)$, the superposition can be written

$$\psi(x, t) = \int_{-\infty}^{\infty} \chi(k) \exp[i(kx - \omega t)] \, dk \quad (84)$$

This complex form has the character of a complex Fourier integral. Alternately the same problem can be treated in terms of integrals involving the real functions, sine and cosine, with the real argument $(kx - \omega t)$ replacing the imaginary argument $[i(kx - \omega t)]$ of the exponential form. To obtain general properties of the superposition, no specific choice is made regarding the functional form of $\chi(k)$, except that it be chosen to be moderate in functional behavior and smooth. It is in general complex, with the role of the real and imaginary parts being exactly the same as that of the real and imaginary parts of $C = A + iB$ utilized in Section V,B, namely, to supply both amplitude and phase information. In this situation both are supplied as a function of k. Let us assume that the dispersion relation ω versus k is known for the wave in question in the neighborhood of k_0. These conditions are then sufficient to use a Taylor series expansion of $\chi(k)$ with k about the value k_0,

$$\omega(k) = \omega(k_0) + \left(\frac{d\omega(k)}{dk}\right)_{k=k_0} (k - k_0)$$

$$+ \frac{1}{2!} \left(\frac{d^2\omega(k)}{dk^2}\right)_{k=k_0} (k - k_0)^2 + \cdots \quad (85)$$

Substituting the first two terms of this expansion yields the following result for the wave packet under consideration,

$$\psi(x, t) = \int_{-\infty}^{\infty} \chi(k) \exp\left[i\left(kx - \left[\omega(k_0)\right.\right.\right.$$

$$\left.\left.\left. + \left(\frac{d\omega(k)}{dk}\right)_{k=k_0} (k - k_0)\right]t\right)\right] dk \quad (86)$$

In terms of the following defined quantities

$$v_{group} = \left(\frac{d\omega(k)}{dk}\right)_{k=k_0} \quad (87)$$

$$\beta_0 = \omega(k_0) - k_0 v_{group} \quad (88)$$

the above expression becomes

$$\psi(x, t) = \int_{-\infty}^{\infty} \chi(k) \exp[i(k[x - v_{group}t] - \beta_0 t)] \, dk \quad (89)$$

To find out where ψ is peaked in space at any given t, one can consider individually the real and imaginary parts of this expression. Writing

$$\chi(k) = \chi_r(k) + i\chi_i(k) \quad (90)$$

then the real part of the above expression is

$$\psi(x, t)\bigg|_{\substack{real \\ part}} = \int_{-\infty}^{\infty} \chi_r(k) \cos(k[x - v_{group}t] - \beta_0 t) \, dk$$

$$- \int_{-\infty}^{\infty} \chi_i(k) \sin(k[x - v_{group}t] - \beta_0 t) \, dk \quad (91)$$

The imaginary part can be obtained similarly. Focusing our attention first on the cosine integral, we note that the coefficient of k, namely, $[x - v_{group}t]$, determines how rapidly the cosine function oscillates as k is varied during the integration over k. Rapid oscillations lead to much cancellation between positive and negative contributions to the integral, with the consequence that the integral is not large. On the other hand, very small values of the argument mean far fewer oscillations over the band of wavelengths where $\chi_r(k)$ is significant, and the smaller amount of cancellation means that the integral is larger. The same arguments can be employed for the sine function integral. Thus for a given t, the maximum value of ψ occurs at the position where the argument is zero, namely, at $x = v_{group}t$. Since this position of the peak moves with the velocity v_{group}, the reason for calling this quantity the group velocity is apparent. The identical state of affairs holds for the imaginary part of ψ, as the reader can quickly see by referring to Eqs. (89) and (90). Thus we arrive at a very general expression for the one-dimensional group velocity, namely, that given by Eq. (87).

The additional term $\beta_0 t$ in Eq. (91) can be written as a phase factor $\Theta(t)$,

$$\Theta(t) = \beta_0 t \qquad (92)$$

which increases linearly with the time. Because it does not depend upon the variable k of integration and does not depend upon the position x, it is not a very important quantity for determining how relatively large ψ will be as a function of x. Its primary effect is to shift the phase of all superimposed waves by the same constant factor Θ at any given time t. This leads to a modulation of the spatial function with time, the frequency ν_{mod} of such modulation being given by $\beta_0 t = 2\pi\nu_{\text{mod}}t$, or simply $\nu_{\text{mod}} = \beta_0/2\pi$. The modulation period is $\tau_{\text{mod}} = 1/\nu_{\text{mod}} = 2\pi/\beta_0$, where β_0 is given by Eq. (88).

More effects come into play if the third term in the Taylor series expansion of $\omega(k)$ is brought into the wave packet integral. The packet can then spread and possibly change shape as time evolves. That, however, is not of immediate concern to us.

Let us now apply our results for the group velocity to the situation of electromagnetic wave propagation through a dielectric. We presume that there is a narrow band of wavelengths that are superimposed to form the packet, and we attempt to obtain the group velocity in terms of the refractive index $n(k)$ of the dielectric. Consider a narrow band of wavelengths of green light centered about the 5461-Å line of mercury as those waves travel in fused quartz. The phase velocity being $c/n(k)$, the dispersion relation is $\omega(k) = ck/n(k)$, so the group velocity is

$$v_{\text{group}} = \left(\frac{d\omega(k)}{dk} \right)_{k=k_0}$$

$$= \left[\frac{c}{n(k)} \left(1 - \frac{k}{n(k)} \frac{dn(k)}{dk} \right) \right]_{k=k_0} \qquad (93)$$

Since $k = 2\pi/\lambda$, it follows that

$$\frac{dn}{dk} = \frac{(dn/d\lambda)}{(dk/d\lambda)} = -\left(\frac{\lambda^2}{2\pi} \right)\left(\frac{dn}{d\lambda} \right) \qquad (94)$$

Thus an alternate form of the group velocity for this situation is

$$v_{\text{group}} = \left[v_{\text{phase}} \left(1 + \frac{\lambda}{n(k)} \frac{dn(k)}{d\lambda} \right) \right]_{k=k_0} \qquad (95)$$

Consider $n(k)$ in this expression in terms of its λ-dependence. The phase velocity v_{phase} is that for the central component wave in the packet,

namely, at wavelength 5461 Å. Note from this result that for situations in which the refractive index is independent of λ, the group velocity is equal to the phase velocity. Estimates for fused quartz, however, give $n = 1.46$ and $dn/d\lambda = -4 \times 10^{-6}$ Å^{-1} at $\lambda = 5461$ Å. Using these numbers in the above relation gives $v_{\text{group}}/v_{\text{phase}} = 1 - 0.015 = 0.985$. Thus the group velocity for the wave packet of green light in fused quartz is approximately 1.5% less than the phase velocity. This may seem to be an unimpressive figure until it is realized that the mere 0.5% difference in phase velocities between 4500 and 6500 Å yields the dispersion in a fused quartz prism that enables the splitting of incident white light into its spectrum of colors.

Although it is interesting to consider that the group and phase velocities differ, as shown above, there is actually nothing so mysterious about it. The individual sinusoidal waves extend throughout space and in this sense are nonlocalized. Any given point on any one of the waves, such as one of the extremum points or one of the nodes, moves with the phase velocity. The superposition of a group of such waves to form a localized packet does so by a subtle phase interference that is on the whole constructive only over a localized region in space and is, on the whole, destructive throughout the remainder of space. The phase interference changes with time if the component waves move with different speeds, so the spatial peak in the localized shape can move in time, even relative to a given point on the moving component wave that has a k-value exactly equal to that of the center of the band of superimposed waves.

H. Wave Motion in Three Dimensions

The classical wave equation given by Eq. (64) likewise has a three-dimensional form to describe motion in arbitrary directions in space. There are various ways to generalize Eq. (64), but it is simplest first of all to think of wave motion along the z-axis in contrast to the x-axis. The simple change in dependent variable from x to z of course yields the relevant equation. It is clear that, to include the three independent directions in space, there must be three terms involving spatial derivatives in place of the single term in Eq. (64). If we denote the three Cartesian coordinates by x, y, and z, then it is necessary to use some alternate notation for the wave displacement. Denoting the wave displacement by $\psi(x, y, z, t)$, then the classical three-dimensional wave equation can be written as

$$\nabla^2 \psi = \left(\frac{1}{v_{\text{phase}}}\right)^2 \frac{\partial^2 \psi}{\partial t^2} \qquad (96)$$

where ∇^2 symbolizes the differential operator $[(\partial^2/\partial x^2) + (\partial^2/\partial y^2) + (\partial^2/\partial z^2)]$.

Next, suppose that a packet is moving in some direction in space that is not parallel to the x-axis of our coordinate system. Then the group velocity has components along the three axes of the coordinate system. The dispersion relation takes the form $\omega = \omega(\mathbf{k})$ where $\mathbf{k}$ is a vector pointing in a given direction. Since ω is a scalar quantity, $\omega(\mathbf{k})$ maps out a type of three-dimensional surface for the dependence of frequency on direction and magnitude of the $\mathbf{k}$-vector. The $\mathbf{k}$-vector represents the direction of propagation of one component wave of the packet, and the magnitude of the $\mathbf{k}$-vector still has the interpretation of being $2\pi/\lambda$, where λ is the wavelength of the component wave in its propagation direction. The group velocity components in the x, y, and z directions in space can be obtained by generalizing Eq. (87), which was derived on the assumption that there was only x-motion. A proper way to carry out the derivation is to use the Taylor series expansion of $\omega(\mathbf{k})$ in three dimensions. We obtain three first derivative terms in place of one, and the multipliers of k_x, k_y, and k_z are respectively $[x - v_{\text{group}}^{(x)} t]$, $[y - v_{\text{group}}^{(y)} t]$, and $[z - v_{\text{group}}^{(z)} t]$, where

$$v_{\text{group}}^{(x)} = \left(\frac{\partial \omega(\mathbf{k})}{\partial k_x}\right)_{\mathbf{k}=\mathbf{k}_0} \qquad (97)$$

$$v_{\text{group}}^{(y)} = \left(\frac{\partial \omega(\mathbf{k})}{\partial k_y}\right)_{\mathbf{k}=\mathbf{k}_0} \qquad (98)$$

$$v_{\text{group}}^{(z)} = \left(\frac{\partial \omega(\mathbf{k})}{\partial k_z}\right)_{\mathbf{k}=\mathbf{k}_0} \qquad (99)$$

The conclusion to be reached is that the group velocity is a vector with components given by Eqs. (97)–(99). These results can be abbreviated by writing

$$\mathbf{v}_{\text{group}} = \nabla_{\mathbf{k}}\omega(\mathbf{k})|_{\mathbf{k}=\mathbf{k}_0} \qquad (100)$$

as the general expression for the group velocity for wave motion in three-dimensional space.

VI. Wave–Particle Duality

A. Ideas of de Broglie

The discovery that particles diffract like waves was one of the most important in physics,

since the entire discipline of quantum mechanics is based on a wave description of matter. Preceding that discovery, de Broglie carried out an analysis of the hypothetical wave properties of matter by employing special relativity theory with an insightful hunch that matter is not so different in its fundamental wave behavior from electromagnetic radiation. His idea was the following: Electromagnetic radiation had been shown to have both wavelike and particlelike properties, so perhaps nonzero rest mass "particles" could also manifest both wavelike and particlelike properties. However, different experimental conditions may be required for matter to manifest one property or the other. De Broglie termed his idea wave–particle duality.

Elsasser suggested in 1926 that the de Broglie hypothesis could be tested by aiming an electron beam at the surface of a crystalline solid to see if diffraction spots were obtained in the same way as were observed in X-ray diffraction. Experiments were carried out shortly thereafter by Davisson and Germer and also by G. P. Thomson, so that by 1927 the de Broglie hypothesis had been experimentally verified. Thus the electron discovered by J. J. Thomson in 1897 enjoyed its status as a classical particle for a period of less than 30 years before its dual nature surfaced. Amazingly, it was later confirmed that not only electrons have this dual nature, but in addition, other particles including neutrons and atoms also can be diffracted. Particles somehow have the inherent property of being able to experience mutual interference even while maintaining such an extremely high degree of localization in space that they appear not to have any physical overlap. This boggles the mind.

Because electromagnetic radiation can manifest discreteness properties, with the photon energy $\mathscr{E}$ related to frequency according to

$$\mathscr{E} = h\nu = \hbar\omega \qquad (101)$$

de Broglie reasoned in analogy that any wave properties of particles would likewise have some associated frequency. Moreover, in accordance with the concept of wave–particle duality and the similarity of particles and electromagnetic radiation in nature, he assumed that the energy–frequency relation for particles would be the same as the energy–frequency relation for photons given above.

Particle is presently the term we use to refer to those fundamental entities for which the rest mass m_0 is nonzero. We freely admit that we do not know exactly what the frequency of a particle refers to, but it could be imagined that there is some internal mode of oscillation. If this is the case, it is not too unreasonable to believe that the measured frequency might change if the particle were set into motion. After all, it is well known from the classical physics treatment of sound waves that relative motion gives rise to Doppler shifts in the frequency. Light also experiences Doppler shifts, but these must be treated quantitatively by means of special relativity.

B. DEVELOPMENT OF THE DE BROGLIE RELATION

Now de Broglie developed his ideas in terms of equations involving the well-known time-dilation phenomenon in special relativity, where moving clocks have longer periods than stationary clocks. Because that approach is somewhat abstruse, we shall develop an alternate derivation of the de Broglie relation.

If a particle can ever be localized to any extent in space, as we certainly have every right to expect, and simultaneously have its experimentally observed wavelike character described in some fashion by waves, then it is quite logical to consider a wavepacket of the sort treated in the last section. The waves forming the packet would necessarily be matter waves, but the diffraction results lead us to believe that these waves have the same linear superposition properties as all other types of waves with which we are familiar. The waves must in some way be associated with the presence of the particle, so it is reasonable to expect that the group velocity $\mathbf{v}_{\text{group}}$ of the packet will be equal to the particle velocity $\mathbf{v}_{\text{particle}}$,

$$\mathbf{v}_{\text{group}} = \mathbf{v}_{\text{particle}} = \mathbf{p}/m \qquad (102)$$

where $\mathbf{p}$ is the momentum of the particle of mass m. For simplicity, in much of this section we confine our attention to a single direction in space, but we generalize to three dimensions whenever there are important implications.

We now use the group velocity expression $v_{\text{group}} = d\omega/dk$ given by Eq. (87) which was derived in Section V,G, together with the derivative of the energy–frequency relation (101) postulated by de Broglie, to obtain

$$d\mathscr{E} = \hbar \, d\omega = \hbar v_{\text{group}} \, dk \qquad (103)$$

Next we differentiate the energy–momentum relation

$$\mathscr{E} = (\mathscr{E}_0^2 + p^2 c^2)^{1/2} \qquad (104)$$

given by Eq. (24) and the mass–energy relation given by Eq. (22) in Section III to obtain a second independent expression for $d\mathscr{E}$,

$$d\mathscr{E} = \frac{c^2 p \, dp}{(\mathscr{E}_0^2 + p^2 c^2)^{1/2}} = \frac{c^2 p \, dp}{mc^2}$$

$$= \frac{p}{m} \, dp = v_{\text{particle}} \, dp \qquad (105)$$

Equating the two independent expressions given for $d\mathscr{E}$ by Eqs. (103) and (105) gives

$$v_{\text{particle}} \, dp = v_{\text{group}} \hbar \, dk \qquad (106)$$

Identifying the group velocity with the particle velocity, according to Eq. (102), and dividing Eq. (106) by this quantity, gives

$$dp = \hbar \, dk \qquad (107)$$

This remarkable result relates the change in the reciprocal of the wavelength of a particle matter-wave component to a change in the momentum of the particle. That is, $k = 2\pi/\lambda$, so the relation can also be written in the more physically meaningful form,

$$dp = \frac{h}{2\pi} d\left(\frac{2\pi}{\lambda}\right) = -\frac{h}{\lambda^2} d\lambda \qquad (108)$$

Integrating Eqs. (107) and (108) from any arbitrary reference point denoted by the subscript 0 gives

$$p - p_0 = \hbar k - \hbar k_0 = h\left(\frac{1}{\lambda} - \frac{1}{\lambda_0}\right) \qquad (109)$$

Alternately, indefinite integrals can be used with a constant of integration. For example,

$$p = \hbar k + Y_0 \qquad (110)$$

with Y_0 being the constant. Evaluation of Y_0 is equivalent to specifying the reference state relationship between p and k in Eq. (109). The result given by Eq. (110) is applicable to particles of any mass and speed.

To proceed, it is helpful to recognize that the derivation is thus far quite general, holding for any value of the rest mass and being valid even in the relativistic limit. Of course, we have built into our development the fundamental hypothesis of de Broglie that matter has associated with it a frequency that is related to the total energy in the same way that the frequency of electromagnetic radiation is related to the individual photon energy. This, in itself, gives us a clue as to how the constant of integration in Eq. (111) may be evaluated. Applying the energy–momentum relation (24) to the case of photons, which have zero rest mass and hence zero rest-mass energy in accordance with $\mathscr{E}_0 = m_0 c^2$, gives

$$\mathscr{E}_{\text{photon}} = \pm p_{\text{photon}} c \qquad (111)$$

In what follows we choose the sign of the root so as to give a positive value for the photon energy, since at the present moment we do not have a physical interpretation for negative photon energy.

Next let us use the energy–frequency relation given by Eq. (101) to obtain an independent relation for the photon energy,

$$\mathscr{E}_{photon} = \hbar\omega_{photon} = \hbar(2\pi\nu_{photon})$$
$$= h\nu_{photon} = hc/\lambda_{photon} \quad (112)$$

Equating the two independent results of Eqs. (111) and (112) for the photon energy and dividing through by the light velocity c gives

$$p_{photon} = h/\lambda_{photon} \quad (113)$$

This relation is consistent with the experimentally proven hypothesis that light has associated with it a momentum. This fact is also a part of classical electromagnetic theory, but the form is somewhat different from the quantum result given above.

Considering the photon to be a quantum particle in every respect, except for being a limiting case from the standpoint of having zero rest mass, then we can use Eq. (110). That result, applied to photons, gives

$$p_{photon} = \hbar k_{photon} + Y_0 = h/\lambda_{photon} + Y_0 \quad (114)$$

Subtracting Eq. (113) from Eq. (114) gives

$$Y_0 = 0 \quad (115)$$

Since the integration constant is independent of the rest mass of the particle, the result deduced, albeit admittedly for a special case, can be employed quite generally. Substituting the result $Y_0 = 0$ into Eq. (110) gives

$$p = \hbar k = h/\lambda \quad (116)$$

Thus we have a relationship between momentum of a particle and the wavelength of the matter wave associated with a particle having this precisely defined momentum. Therefore we have derived the de Broglie relation,

$$\lambda = h/p \quad (117)$$

for the effective wavelength of a nonzero rest-mass particle in terms of the momentum p of the particle. As is evident from the development, the relation is valid for particles of any rest mass, including the limiting case of zero rest mass, and it is valid at any velocity, including low velocity and relativistic velocity.

It seems a bit strange that such a fundamental relation for quantum mechanics is deduced from the theory of special relativity, considering our previous discussion in Section III,C of the dif-

ference in philosophical bases for the two theories. To be specific, we have used the fundamental equations of a deterministic theory to deduce a wavelength relationship for particles, the precision of this wavelength determining to a large extent the inherent uncertainty in the specification of the location of the particle. Do not be misled, however, into believing that relativity theory has predicted the wave nature of particles. Instead, the wave nature of particles is the hypothesis we started with, and that in itself is supported by experimental observation that particles diffract from crystal lattices in a wavelike manner.

C. PHASE VELOCITY FOR FREE PARTICLES

The relations $\omega = \mathscr{E}/\hbar$ and $k = p/\hbar$ can be used to evaluate the phase velocity v_{phase} of matter waves,

$$v_{phase} = \frac{\omega}{k} = \frac{\hbar\omega}{\hbar k} = \frac{\mathscr{E}}{p} = \frac{mc^2}{mv_{particle}}$$
$$= \frac{c^2}{v_{particle}} \quad (118)$$

The phase velocity of matter waves thus depends upon the particle velocity: As the particle speed increases, the phase velocity decreases.

Although the phase velocity takes its most natural form when expressed in terms of the particle velocity, it can also be expressed in terms of the momentum,

$$v_{phase} = \omega/k = \mathscr{E}/p = p^{-1}(\mathscr{E}_0^2 + p^2c^2)^{1/2}$$
$$= c[1 + (p_C/p)^2]^{1/2} \quad (119)$$

where $p_C = m_0c$. This can also be expressed as

$$v_{phase} = c[1 + (k_C/k)^2]^{1/2}$$
$$= c[1 + (\lambda/\lambda_C)^2]^{1/2} \quad (120)$$

where $k_C = p_C/\hbar = m_0c/\hbar$ and $\lambda_C = 2\pi/k_C = h/m_0c$. The parameter λ_C is called the Compton wavelength whenever m_0 is the electron rest mass. (It is not the wavelength of an electron at rest, of course, since the de Broglie relation yields infinity as the wavelength of a particle with zero momentum.)

For the special case of photons and other particles that have zero rest mass, the phase velocity is given by

$$v_{phase}\Big|_{m_0=0} = \frac{\mathscr{E}}{p} = \frac{1}{p}[m_0c^2 + p^2c^2]^{1/2}\Big|_{m_0=0}$$
$$= \frac{pc}{p} = c \quad (121)$$

Matter waves have phase velocities ranging upward from c, with the phase velocity approaching infinity as the particle speed approaches zero. The phase velocity represents the speed of motion of a single isolated and completely extended component wave having no position modulation of its shape, so by itself it can carry no information. Thus there is no conflict with the tenet of special relativity that a signal cannot travel with a velocity exceeding the speed of light.

To summarize for matter waves, the group velocity must be equal to the particle velocity, and the phase velocity varies inversely with the particle velocity. The critical quantity for evaluating the wavelength is the momentum, whereas the critical quantity for evaluating the phase velocity is the particle velocity. Particles having different rest-mass values have the same wavelength whenever the velocities differ such as to give the same values for the momentum. Particles having different rest-mass values, however, have the same phase velocity whenever the momentum values differ by the amount needed to give the same values for the particle velocity.

D. DISPERSION RELATION FOR FREE PARTICLES

It has been shown (see Section V,D) that the dispersion relation for light is $\omega = ck$. This gives immediately the phase velocity $\omega/k = c$ and the group velocity $d\omega/dk = c$.

The dispersion relation for free particles is deduced by utilizing the energy–momentum relation from special relativity and substituting $\mathscr{E} = \hbar\omega$ and $p = \hbar k$,

$$\hbar\omega = (\mathscr{E}_0^2 + \hbar^2 k^2 c^2)^{1/2} \tag{122}$$

This free-particle dispersion relation can be written also in the form

$$\omega = \omega_0[1 + (\hbar/m_0 c)^2 k^2]^{1/2} \tag{123}$$

where $\omega_0 = (m_0 c^2/\hbar)$, the frequency corresponding to the rest mass m_0. In terms of parameters $\lambda_C = h/m_0 c$ and $k_C = 2\pi/\lambda_C$ this becomes

$$\omega = ck_C[1 + (k/k_C)^2]^{1/2} \tag{124}$$

Binomial expansion illustrates a quadratic dependence of ω on k with a cutoff in frequency below the rest mass frequency ω_0,

$$\omega = \omega_0[1 + (k^2/2k_C^2) + \cdots] \tag{125}$$

For low values of the particle momentum, k is small, and the frequency does not depart very much from the rest-mass frequency. This is merely a reflection of the fact that $\mathscr{E} = mc^2 =$

$\gamma m_0 c^2 = \hbar\omega$ predicts that

$$\omega/\omega_0 = \gamma \tag{126}$$

where γ is the parameter defined by Eq. (21) in Section III.

E. ENERGY–MOMENTUM RELATIONS FOR PARTICLES WITH POTENTIAL ENERGY

It may be appreciated that developing the dispersion relation $\omega(k)$ versus k can be sufficient preparation for evaluating the phase velocity $v_{\text{phase}} = \omega(k)/k$, according to Eq. (78). Because most applications of quantum mechanics involve particles with some type of potential energy, it is necessary to consider this situation.

A force $\mathbf{F}$ changes the momentum $\mathbf{p}$ of a free particle in accordance with $\mathbf{F} = d\mathbf{p}/dt$, or equivalently, the change in particle momentum is given by $\mathbf{F}\,dt$. Since the work done by force in moving a particle through a vector distance $d\mathbf{r}$ is $dW = \mathbf{F} \cdot d\mathbf{r}$, the power input $\mathscr{P} = dW/dt$ from the source is $\mathbf{F} \cdot d\mathbf{r}/dt = \mathbf{F} \cdot \mathbf{v}$. In one dimension, $\mathscr{P} = Fv$. In our case the power to change the motion energy of the particle comes from the internal potential energy. For a conservative system, the change in potential energy with position leads to an internal force on the particle,

$$\mathbf{F} = -\nabla U(\mathbf{r}) \tag{127}$$

which for one dimension is simply

$$F = -dU(r)/dr \tag{128}$$

Thus,

$$dp/dt = F = -dU(r)/dr \tag{129}$$

or equivalently,

$$dU(r) = -(dp/dt)\,dr \tag{130}$$

But $v_{\text{particle}} = dr/dt$, so $dr = v_{\text{particle}}\,dt$, and we therefore obtain

$$dU(r) = -(dp/dt)v_{\text{particle}}\,dt = -v_{\text{particle}}\,dp$$
$$= -(p/m)\,dp \tag{131}$$

This relation cannot be integrated directly because m depends upon particle velocity and hence upon p. However, substituting $m = \mathscr{E}/c^2$, with $\mathscr{E}$ given by the usual energy–momentum relation of special relativity [Eq. (24)], leads to a perfect differential,

$$dU(r) = -(c^2 p\,dp)[\mathscr{E}_0^2 + p^2 c^2]^{-1/2}$$
$$= -d(\mathscr{E}_0^2 + p^2 c^2)^{1/2} \tag{132}$$

Integrating from a classical turning point, defined as a position where $p = 0$ so that the potential energy at the point is the total energy $\mathscr{E}_T^{(Nt)}$ in classical Newtonian physics, we obtain

$$U(r) - \mathscr{E}_T^{(Nt)} = m_0 c^2 - [\mathscr{E}_0^2 + p^2 c^2]^{1/2} \quad (133)$$

The classical total energy is conserved as long as no rest mass is created or annihilated. A new constant $\mathscr{E}_T^{(Re\ell)}$ can thus be defined,

$$\mathscr{E}_T^{(Re\ell)} = \mathscr{E}_T^{(Nt)} + m_0 c^2 \quad (134)$$

and Eq. (133) can be written as

$$\mathscr{E}_T^{(Re\ell)} - U(r) = [\mathscr{E}_0^2 + p^2 c^2]^{1/2} \quad (135)$$

This is the energy–momentum relation for a particle having a potential energy. Note from this relation that when the potential energy is zero, $\mathscr{E}_T^{(Re\ell)}$ becomes the total relativistic energy $\mathscr{E}$ of a free particle. The constant $\mathscr{E}_T^{(Re\ell)}$ in lieu of $\mathscr{E}$ is the conserved quantity when a potential energy is included.

Let us also define a quantity $\mathscr{E}_K^{(Re\ell)}$,

$$\mathscr{E}_K^{(Re\ell)} = \mathscr{E}_K^{(Nt)} - U(r) \quad (136)$$

or alternatively, $\mathscr{E}_T^{(Nt)} = \mathscr{E}_K^{(Re\ell)} + U(r)$. By employing Eq. (134), this can be written in the form

$$\mathscr{E}_K^{(Re\ell)} = [\mathscr{E}_T^{(Re\ell)} - m_0 c^2] - U(r)$$
$$= [\mathscr{E}_T^{(Re\ell)} - U(r)] - m_0 c^2 \quad (137)$$

and by employing Eq. (135) it can then be written in the form

$$\mathscr{E}_K^{(Re\ell)} = (\mathscr{E}_0^2 + p^2 c^2)^{1/2} - m_0 c^2 \quad (138)$$

Equation (137) in the form

$$\mathscr{E}_T^{(Re\ell)} = \mathscr{E}_K^{(Re\ell)} + U(r) + m_0 c^2 \quad (139)$$

seems intuitive from a scalar energy viewpoint. This form has its primary usefulness in the low-velocity limit. In the relativistic limit, the best form for $\mathscr{E}_T^{(Re\ell)}$ is that obtained from Eq. (135),

$$\mathscr{E}_T^{(Re\ell)} = (\mathscr{E}_0^2 + p^2 c^2)^{1/2} + U(r) \quad (140)$$

since this form is intuitive from a free-particle viewpoint.

Squaring Eq. (135) and solving for p^2 gives

$$p^2 = (1/c^2)[(\mathscr{E}_T^{(Re\ell)} - U(r))^2 - \mathscr{E}_0^2] \quad (141)$$

Substituting Eq. (137) then gives

$$p^2 = (1/c^2)[(\mathscr{E}_K^{(Re\ell)} + m_0 c^2)^2 - (m_0 c^2)^2] \quad (142)$$

which relates the momentum to the quantity $\mathscr{E}_K^{(Re\ell)}$. Rearranging Eq. (142) and using the quadratic formula to obtain $\mathscr{E}_K^{(Re\ell)}$ explicitly in terms of p gives

$$\mathscr{E}_K^{(Re\ell)} = -m_0 c^2 + m_0 c^2 \left| \left(1 + \frac{p^2 c^2}{m_0^2 c^4}\right)^{1/2} \right| \quad (143)$$

Employing the binomial expansion then gives the series approximation

$$\mathscr{E}_K^{(Re\ell)} \simeq \left(\frac{p^2}{2m_0}\right)\left[1 - \frac{p^2}{4m_0^2 c^2} + \cdots\right] \quad (144)$$

In the low-velocity limit defined by $\gamma^2 \simeq 1$, which requires $(v/c)^2 \ll 1$ [see Eq. (21)], then $p \simeq m_0 v$, and the quantity $\mathscr{E}_K^{(Re\ell)}$ thus reduces in lowest order to the Newtonian expression for the kinetic energy,

$$\mathscr{E}_K^{(Nt)} = \tfrac{1}{2} m_0 v^2 \quad (145)$$

F. Dispersion Relations for Particle with Potential Energy

To develop a wave dispersion relation applicable to particles having a potential energy, we again associate the wavelength with the particle momentum by means of the de Broglie relation $\lambda = h/p$. Equivalently, $p = h/\lambda = (h/2\pi)(2\pi/\lambda) = \hbar k$, in accordance with Eq. (116).

According to the de Broglie hypothesis, a frequency ω can be associated with the particle energy. Although we do not yet understand the nature of the frequency of a particle, we can be guided by familiar results from the particular limiting case of a zero rest-mass quantum particle, namely, the photon. We must remain alert, however, to avoid introducing errors in using the analogy.

The photon has a wavelength λ and a frequency ν. As the photon travels in free space, it has a speed $c = \lambda\nu$. When the photon enters a dielectric material medium of refractive index n, its speed decreases to c/n, but its frequency ν does not change. The photon energy $\mathscr{E} = h\nu$ thus does not change. The wavelength decreases in accordance with $\lambda_{pho}^{med} = \lambda_{pho}^{vac}/n$, where λ_{pho}^{med} is the wavelength in the medium and λ_{pho}^{vac} the wavelength in free space. The photon velocity is still given by the product of photon frequency and the appropriate photon wavelength, whether the medium be a dielectric or free space. The momentum $p_{pho}^{vac} = h/\lambda_{pho}^{vac}$ of the photon in free space changes to $p_{pho}^{med} = h/\lambda_{pho}^{med}$ as it enters the dielectric medium. Thus the constant of motion for the photon is the frequency $\nu = \omega/2\pi$, with the wavelength λ and momentum p being medium dependent. It is an interesting observation that from the viewpoint of Eq. (138) all energy of the photon is kinetic.

We might be tempted to assume, in analogy with the photon, that the frequency of a particle located in a purely conservative potential energy region is a constant of motion. This would give rise to the picture of a particle entering a region of varying potential energy where it could be accelerated. Its speed would change as it accelerates. Its momentum would change, so its wavelength would change also. However, the total energy of the particle would not change. Both remaining unchanged, the frequency would then maintain a direct relationship to the total energy. The analogy with the behavior of a photon as it enters a region of varying refractive index would in this way be exploited to the fullest. However, this would do great violence to our earlier conclusion that the frequency of a free particle depends upon its speed. In the present case we would be considering the particle changing speed due to the change in internal potential energy as it moves through the potential energy region, yet we would be proposing to hold its frequency ω to a constant. Moreover, the constant frequency would not be equal to the "proper" frequency ω_0 for the particle in the relativistic sense but would have some value that would necessarily be related to the choice of $U(r)$. Even the choice of the zero point for measuring $U(r)$ would affect the value of ω. This strange state of affairs would be of course quite unacceptable.

A better approach is to consider the particle as initially created in a zero potential energy region with a proper frequency, which it then maintains always during its lifetime. Any perceived difference in frequency due to motion is then merely due to the relativistic shift in energy with Galilean reference frame. This is analogous to the increase in mass with the particle speed. That is, $m = \gamma m_0$, so $\mathcal{E} = mc^2 = \gamma m_0 c^2$; thus $\mathcal{E} = \hbar\omega = \gamma\hbar\omega_0$, thereby giving $\omega = \gamma\omega_0$ in accordance with Eq. (126). This does maintain the de Broglie premise that a particle has a frequency associated with it even when stationary, that frequency being dependent upon the rest mass m_0 in accordance with the relation $\mathcal{E}_0 = m_0 c^2 = \hbar\omega_0$. As the particle enters into, or moves around in, a region of varying potential energy, its frequency changes relative to an observer fixed in the laboratory frame. This is in contrast to the total energy, which for a conservative system is a constant of the motion. It is suggested, therefore, that if the frequency of the particle is to be associated with the energy of the particle, according to the suggestion of de Broglie, then the particular energy chosen should be the relativistic energy most closely analogous to that of a free particle, namely, $\mathcal{E} = (\mathcal{E}_0^2 + p^2 c^2)^{1/2}$. The

dispersion relation for the particle is therefore reasoned to be given by

$$\omega = \hbar^{-1}(\mathcal{E}_0^2 + p^2 c^2)^{1/2} = \hbar^{-1}(\mathcal{E}_0^2 + \hbar^2 c^2 k^2)^{1/2}.$$

$$(146)$$

This is the same dispersion relation given by Eq. (122).

G. Phase Velocity of Particle with Potential Energy

Taking the frequency of the particle wave from Eq. (146) and using the de Broglie relation $k = p/\hbar$ given by Eq. (117), leads to an evaluation of the phase velocity,

$$v_{\text{phase}} = \omega/k = \hbar\omega/\hbar k = \mathcal{E}/p = p^{-1}(\mathcal{E}_0^2 + p^2 c^2)^{1/2}$$
$$= mc^2/p = mc^2/mv_{\text{particle}} = c^2/v_{\text{particle}}$$

$$(147)$$

This relation is the same as that obtained for a free particle [see Eq. (118)], but in the present case the particle speed changes as it moves in the potential energy region.

VII. Wave Equations for Particles

A. Master Equations for Particle with Potential Energy

It is assumed at the outset in this section that the reader has already studied carefully the earlier material presented, especially the fundamentals of classical wave motion in Section V and the ideas of wave–particle duality in Section VI. The present section relies heavily on the concepts of dispersion relations and phase velocity in wave motion covered in those sections.

Using the three-dimensional classical wave equation [Eq. (96)] and substituting Eq. (147) for the phase velocity gives the following wave equation for particles,

$$\nabla^2\psi = \left[\frac{v_{\text{particle}}^2}{c^4}\right]\frac{\partial^2\psi}{\partial t^2} \qquad (148)$$

For a particle having potential energy $U(\mathbf{r})$, the speed v_{particle} must be expressed in terms of that energy. Generally we cannot expect v_{particle} to be a constant of motion, although it can be when the potential energy has a constant value. Equation (141) can be used, along with $p = mv$, to write

$$v_{\text{particle}}^2 = \frac{p^2}{m^2} = \frac{1}{m^2 c^2}[(\mathcal{E}_T^{(\text{Rel})} - U(\mathbf{r}))^2 - \mathcal{E}_0^2]$$

$$(149)$$

In general, the mass m is velocity dependent. In accordance with Eqs. (22) and (24),

$$\mathcal{E} = mc^2 = (p^2c^2 + \mathcal{E}_0^2)^{1/2} \quad (150)$$

which can be used with Eq. (140) to obtain

$$m = (1/c^2)[\mathcal{E}_T^{(Re\ell)} - U(\mathbf{r})] \quad (151)$$

Substituting Eq. (151) into Eq. (149) gives

$$v_{particle}^2 = \frac{c^2[(\mathcal{E}_T^{(Re\ell)} - U(\mathbf{r}))^2 - \mathcal{E}_0^2]}{(\mathcal{E}_T^{(Re\ell)} - U(\mathbf{r}))^2} \quad (152)$$

It is convenient to define a parameter $\mathcal{R}$,

$$\mathcal{R} = v_{particle}^2/c^4 \quad (153)$$

which denotes the quantity $(1/v_{phase})^2$. Then Eq. (152) can be used to obtain

$$\mathcal{R} = \left(\frac{1}{c^2}\right)\left[1 - \frac{\mathcal{E}_0^2}{(\mathcal{E}_T^{(Re\ell)} - U(\mathbf{r}))^2}\right] \quad (154)$$

It should be noted that $\mathcal{R}$ depends explicitly upon position $\mathbf{r}$ but does not depend explicitly upon the time t. Therefore the variables can be easily separated in Eq. (148) and a solution carried out by the separation-of-variables technique. Writing Eq. (148) in the form

$$\left(\frac{1}{\mathcal{R}}\right)\nabla^2\psi = \frac{\partial^2\psi}{\partial t^2} \quad (155)$$

then substituting the product trial solution

$$\psi(\mathbf{r}, t) = \phi(\mathbf{r})T(t) \quad (156)$$

and dividing through by ψ, we obtain

$$\frac{1}{\mathcal{R}}\frac{1}{\phi}\nabla^2\phi = \frac{1}{T}\frac{\partial^2 T}{\partial t^2} \quad (157)$$

The usual argument for the separation of variables holds. The right-hand side is a function (at most) of the variable t, while the left-hand side is a function (at most) of the variable $\mathbf{r}$. Since the two sides are required to be equal, then neither can vary with t or $\mathbf{r}$. Setting the right-hand side to the constant Γ and employing the trial solution

$$T(t) = T_0 \exp(i\Omega t) \quad (158)$$

yields a solution, provided that

$$\Gamma = -\Omega^2 \quad (159)$$

Since T_0 is unspecified, it is convenient at times to choose it to be unity. The right-hand side of Eq. (157) equals Γ, so the left-hand side must equal Γ also. Thus we obtain a differential equation for $\phi(\mathbf{r})$,

$$\frac{1}{\mathcal{R}}\frac{1}{\phi}\nabla^2\phi = -\Omega^2 \quad (160)$$

or equivalently,

$$\nabla^2\phi(\mathbf{r}) + \Omega^2\mathcal{R}\phi(\mathbf{r}) = 0 \quad (161)$$

There still remains the task of evaluating the constant Ω. This is most easily done in the nonrelativistic limit, which moreover is the limit of most interest at this point. Since the equations are valid in general, both for relativistic speeds and for nonrelativistic speeds of the particle, the evaluation of the constant can be carried out in either limit. The result can then be used quite generally in application of the master equations.

B. Approximations to Master Equations for Nonrelativistic Limit

The key to obtaining an expression valid in the low-velocity limit is to write $\mathcal{R}$, given by Eq. (154), in a form suitable for expansion. Develop the right-hand side of Eq. (154) by noting that the numerator in Eq. (153) is a perfect square and thus can be factored,

$$\frac{(1/c^2)[(\mathcal{E}_T^{(Re\ell)} - U(\mathbf{r}) - \mathcal{E}_0)(\mathcal{E}_T^{(Re\ell)} - U(\mathbf{r}) + \mathcal{E}_0)]}{(\mathcal{E}_T^{(Re\ell)} - U(\mathbf{r}))^2} \quad (162)$$

Then Eq. (137) can be used to convert this quantity into the form

$$\frac{\mathcal{E}_K^{(Re\ell)}(\mathcal{E}_K^{(Re\ell)} + 2\mathcal{E}_0)}{c^2(\mathcal{E}_K^{(Re\ell)} + \mathcal{E}_0)^2} \quad (163)$$

The kinetic energy is small compared with the rest mass energy in the nonrelativistic limit, so the ratio of the two can be considered an expansion parameter $\delta \ll 1$, where

$$\delta = \frac{\mathcal{E}_K^{(Re\ell)}}{\mathcal{E}_0} \quad (164)$$

Utilizing the expansion

$$1/(1 + \delta)^2 \simeq 1 - 2\delta + 3\delta^2 - \mathcal{O}(\delta^3) \quad (165)$$

where $\mathcal{O}(\delta^3)$ means terms of the order of δ^3 or higher, leads to

$$\mathcal{R} \simeq (1/c^2)[\delta(\delta + 2)(1 - 2\delta + 3\delta^2 - \cdots)]$$
$$= (1/c^2)(2\delta - 3\delta^2 + \cdots) \quad (166)$$

where the numerator of Eq. (163) is converted to terms involving the ratio δ. In the nonrelativistic limit, Eq. (144) yields $p^2/2m_0$ as the lowest-order approximation to $\mathcal{E}_K^{(Re\ell)}$. Inserting this value, together with the value $\mathcal{E}_0 = m_0c^2$, into the result given by Eq. (166) yields

$$\mathcal{R} = p^2/m_0^2c^4 - \mathcal{O}(p^4) \quad (167)$$

Equation (161) then becomes, to lowest order in p,

$$\nabla^2\phi(\mathbf{r}) + \Omega^2(p^2/m_0^2c^4)\phi(\mathbf{r}) = 0 \qquad (168)$$

However, for the nonrelativistic limit,

$$p^2 = 2m_0 [\mathscr{E}_T^{(Nt)} - U(\mathbf{r})] \qquad (169)$$

where $\mathscr{E}_T^{(Nt)}$ is the Newtonian total energy discussed in Section VI,E. Thus Eq. (168) can be written in the form

$$\nabla^2\phi(\mathbf{r}) + \Omega^2(2/m_0c^4) [\mathscr{E}_T^{(Nt)} - U(\mathbf{r})] \phi(\mathbf{r}) = 0 \qquad (170)$$

To evaluate the constant Ω^2, consider the particular case in which the potential energy is a constant, viz., $U = U_0$. Equation (170) then becomes very simple because there is no position dependence of the bracketed term. The trial solution

$$\phi = \phi_0 \exp(i\mathbf{k}_0 \cdot \mathbf{r})$$
$$= \phi_0 \exp[i(k_x x + k_y y + k_z z)] \qquad (171)$$

then satisfies the equation. This solution represents a wave having wavelength $2\pi/k$, with $k_0 = (k_x^2 + k_y^2 + k_z^2)^{1/2}$, which is traveling in the direction of the vector $\mathbf{k}_0$. In one dimension, for example, a particle moving along the x-direction, the exponential function reduces simply to $\exp(ik_0x)$. The k-vector involves the wavelength for the particle wave, and therefore it is associated with the particle momentum p in accordance with the de Broglie relation $\mathbf{p}_0 = \hbar\mathbf{k}_0$. The momentum in this case of a constant potential energy is a constant of motion. Substituting Eq. (171) into Eq. (170) with $U = U_0$ thus leads to the requirement

$$-k_0^2 = -\Omega^2(2/m_0c^4)(p_0^2/2m_0) \qquad (172)$$

which yields the needed evaluation of Ω,

$$\Omega = \pm(m_0c^2/\hbar) \qquad (173)$$

Thus $\Omega = \pm\omega_0$, where ω_0 is the frequency associated with the rest mass m_0 of the particle. Substituting this evaluation into Eq. (170) yields the general nonrelativistic equation for the motion of a particle in a potential energy region,

$$\nabla^2\phi(\mathbf{r}) + (2m_0/\hbar^2) [\mathscr{E}_T^{(Nt)} - U(\mathbf{r})]\phi(\mathbf{r}) = 0 \qquad (174)$$

C. THE SCHRÖDINGER EQUATION

Equation (174), known as the time-independent Schrödinger equation, is usually written in the form

$$-(\hbar^2/2m_0) \nabla^2\phi_j + U(\mathbf{r})\phi_j = \mathscr{E}_j\phi_j \qquad (175)$$

where the subscript j denotes a set of possible solutions ϕ_j with attendant discrete values for the Newtonian total energy $\mathscr{E}_T^{(Nt)}$. This equation is an energy eigenvalue equation, since it can be written in the form

$$\mathscr{E}_T^{(op)}\phi_j(\mathbf{r}) = \mathscr{E}_j\phi_j(\mathbf{r}) \qquad (176)$$

with the total energy operator $\mathscr{E}_T^{(op)}$ defined by

$$\mathscr{E}_T^{(op)} = -(\hbar^2/2m_0) \nabla^2 + U(\mathbf{r}) \qquad (177)$$

It is of the nature of an eigenvalue equation that an operator representing some physical quantity operates on a function known as the eigenfunction [in this case $\phi_j(\mathbf{r})$] to produce the product of a constant with the eigenfunction. The constant is the eigenvalue. It represents the constant of motion associated with the physical state represented by the eigenfunction.

Suppose that we define an oscillatory time-dependent function involving the Newtonian total energy $\mathscr{E}_T^{(Nt)} = \mathscr{E}_j = \hbar\omega_j$ for the particle,

$$\theta_j(t) = \theta_0 \exp[-(i/\hbar)\mathscr{E}_j t] = \theta_0 \exp(-i\omega_j t) \qquad (178)$$

Multiplying Eq. (175) through by this factor leads to the following equation

$$-(\hbar^2/2m_0) \nabla^2\psi_j^{(Sch)} + U(\mathbf{r})\psi_j^{(Sch)} = \mathscr{E}_j\psi_j^{(Sch)} \qquad (179)$$

where

$$\psi_j^{(Sch)} = \phi_j(\mathbf{r})\theta_j(t) \qquad (180)$$

The superscript (Sch) denotes the Schrödinger wave function. Note that

$$i\hbar(\partial/\partial t)\psi_j^{(Sch)} = \mathscr{E}_j\psi_j^{(Sch)} \qquad (181)$$

so that Eq. (179) can be written in the form

$$-(\hbar^2/2m_0) \nabla^2\psi_j^{(Sch)} + U(\mathbf{r})\psi_j^{(Sch)} = i\hbar(\partial/\partial t)\psi_j^{(Sch)} \qquad (182)$$

This is called the time-dependent Schrödinger equation. It can be noted from Eq. (156) that with $T_0 = 1$, $\psi = \phi(\mathbf{r}) \exp(\pm i\omega_0 t)$. In Eq. (180), with the choice $\theta_0 = 1$, $\psi^{(Sch)} = \phi(\mathbf{r}) \exp(-i\omega_S t)$. Whereas ω_0 is the frequency associated with the rest mass energy m_0, ω_j is a parameter having the dimensions of frequency that is obtained from the total Newtonian energy $\mathscr{E}_T^{(Nt)}$ for the state j. The ratio of the two wave functions is

$$\psi/\psi^{(Sch)} = \exp[(i/\hbar)(\pm\mathscr{E}_0 + \mathscr{E}_T^{(Nt)})t] \qquad (183)$$

Choice of the $+$ sign in Eq. (183) converts the exponential to $\exp[(i/\hbar)\mathscr{E}_T^{(Rel)}t]$, in accordance with Eq. (134). Because the function $\psi^{(Sch)}$ has physical meaning only from the standpoint of the square of its magnitude, and the magnitude is the same for both functions, ψ and $\psi^{(Sch)}$ give identical physical information within the framework of the Schrödinger representation.

D. INTERPRETATION OF SCHRÖDINGER EQUATION RESULTS

Solution of the Schrödinger equation gives functions $\psi^{(Sch)}$ that are usually complex. The square of the magnitude of $\psi^{(Sch)}$ evaluated at position $\mathbf{r}$ has been interpreted by Born (1882–1970) as the relative probability that the particle is located at that position. By judicious choice of values for the otherwise unspecified constants, such as T_0 and θ_0 in Eqs. (158) and (178), the integral of the probability over all space can be set equal to unity, a process called normalization. This is physically meaningful because the particle is presumed to be somewhere, but not at more than one place at a given time. The normalization process requires that the wavefunction decrease sufficiently rapidly with distance away from the general location of the particle to give a bounded value for the integral of the square of the function. This restriction severely limits the physical set of solutions from the great number of mathematical solutions that formally satisfy the Schrödinger equation.

For the hydrogen atom, for example, where the relevant potential energy for the time-independent Schrödinger equation [Eq. (175)] is given as $U(\mathbf{r}) = -e^2/4\pi\varepsilon_0 r$ by Eqs. (32) and (35) with $Z = 1$, the various functions ϕ_j satisfying the equation and meeting the physically reasonable boundary conditions constitute a discrete set. The attendant energies $\mathscr{E}_j$ are essentially those given by Eq. (46), which we deduced by a simpler technique.

A natural extension of the potential energy to include the energy of the electron spin magnetic moment in an external magnetic field leads to a close correlation of the quantum mechanical predictions for the energy levels of the one-electron atom and detailed experimental optical spectral data. There are many excellent treatises available on the hydrogen atom problem, especially in the older works on quantum mechanics. The reader is referred to those books for details. The mathematical treatment of the hydrogen atom indeed provides an outstanding problem for the do-it-yourself student of quantum mechanics, since it covers virtually every aspect of single-particle problems and also provides a good application of the applied mathematics of vector operations in spherical polar coordinates.

There are other standard problems that the reader can attack with the knowledge developed to the present point. One such problem is that of the one-dimensional simple harmonic oscillator, characterized by the potential energy $U(x) = \frac{1}{2}Kx^2$, where K is a constant. Other revealing problems are those of a particle trapped in one-dimensional and three-dimensional boxes. These problems have application in a number of fields. For example, the harmonic oscillator is important for quantized lattice vibrations in solids. It is also important for the analysis of electromagnetic radiation at thermal equilibrium with the walls of a cavity. The density of allowed modes is computed essentially the same way for radiation in a rectangular cavity with conducting walls as it is for a particle trapped in a potential well or a box. The particle-in-a-box model is quite important for the free-electron theory of metals.

In the following section we examine the predictions of the Schrödinger equation for an electron in the presence of a periodic potential energy. Periodicity in the potential is an attribute of the atom array in a crystalline solid. This problem involves the population of many energy levels at one time, so the quantum statistics governing the behavior of many electrons in the same system plays an important role. This leads into the interesting and important topic of energy bands which are so successful in explaining the difference between metals, semiconductors, and insulators.

VIII. Electron Transport in Solids

A. FAILURE OF CLASSICAL PHYSICS

One of the primary difficulties encountered by classical mechanics is in the area of electron transport in solids. If we view condensed matter as merely an agglomeration of hard-sphere atoms packed so close together that they are in contact, it seems intuitively clear that any particle, however small, while moving through the agglomerate in a straight line would rebound from one or another of the atoms before traveling very far. Even allowing for the fact that the atoms in the solid are usually in a lattice configuration, there still are very few directions through the ordered array where a particle could travel unimpeded. Despite this, it can be deduced from experimental measurements that under conditions of very low strain, very high purity, and quite low temperatures the conduction electrons can travel distances involving hundreds of atoms without being scattered. Devising an acceptable explanation for such easy flow of electrons in metals thus constituted a problem that could not be resolved by means of a mechanics based on a purely classical viewpoint.

Before delving into the quantum mechanical explanation of easy electron transport in metals, let us first ask how we know that atoms in a solid are actually in contact. Next, let us ask how we know that electrons have such long mean free paths in metals in which the atoms are in contact. Classical radii for atoms can be deduced in a variety of ways. Scattering experiments initiated by Rutherford in the early 1900s give direct evidence that atoms have a tiny dense nucleus that is surrounded by a cloud of electrons extending for distances of the order of angstroms $(1 \text{ Å} = 10^{-10} \text{ m})$. Viscosity of fluids and the molecular flow of gases also provide some data on atom and molecule sizes. The X-ray diffraction of crystalline solids yields lattice distances that are of the same order as the atom sizes. Compressibility data for solids lend credence to the view that forces between atoms increase as a high power of the separation distance, as might be expected from a hard-sphere picture of atoms in contact. These indications, together with a variety of other types of data, lead us to picture a solid as an array of hard-sphere atoms in contact.

The second point, namely, the existence of the long mean free path of electrons in metals, can be deduced from the simple picture that resistance is due to electron scattering, coupled with experimental data on the temperature dependence of the resistivity of metals and the dependence of resistivity on purity and crystal preparation techniques. Perhaps the most salient point is that the resistivity decreases by many orders of magnitude when the purity of the metal is increased and the metal is grown as a strain-free single crystal.

The limiting factor on the electron mean free path in metals thus actually appears to depend upon the residual imperfections, impurities, and grain boundaries in the sample instead of the atom density. There is hardly any way a classical picture can explain the fact that the ordinary atoms making up the ordered solid themselves provide so little resistance to electron flow. The classical picture of pointlike electrons scattering in a billiard-ball manner from a hard-sphere atom fails completely.

B. QUANTUM MECHANICS APPROACH

Quantum mechanics enables us to rationalize these classically unexplainable observations. Even neglecting the ordinary Coulomb repulsion between electrons, there remains a quantum mechanical tendency for electrons to remain separated. This tendency can be treated in the framework of what we call the Pauli exclusion principle, which states that no two electrons in a system can have the same set of quantum numbers. Practically speaking, this requires higher and higher average kinetic energies for the electrons as the electron density increases. This explains why adjacent atoms resist electron cloud overlap even though the electron cloud would otherwise be expected to be rather soft and easily deformable under compression, and so accounts for the hard-sphere view of atoms in a crystal lattice.

The unimpeded motion of electrons moving through a lattice of such hard-sphere atoms in a solid can be understood from the wavelike properties of the electron. Even classically it can be shown that the collective scattering of waves from a periodic array of scattering centers differs quite dramatically from the scattering of waves from a random array of scattering centers. The difference between these two situations is that a random array leads to random phases between the scattered wavefronts, whereas phase coherence between the scattered wavefronts is possible if the scattering centers are located in a periodic array. (Indeed, X-ray diffraction of crystalline solids hinges on phase coherence.) In the random array case, movement of an incident wave through the array is grossly impeded because of the partial cancellation of wavefronts having random phase with respect to one another, whereas in the periodic array case, propagation of the wavefront becomes quite possible.

Even in the periodic case, however, there are situations in which propagation is retarded, as, for example, when a portion of the wavefront reflected from one plane of the crystalline array is superimposed upon and has a 180° phase difference with respect to another portion of the wavefront reflected from a different plane of the array. Such waves interfere destructively. Propagation, on the other hand, is enabled by a constructive interference of the scattered waves in the direction of propagation.

These facts of classical wave propagation are applicable immediately to electron propagation in solids once it is admitted that electrons have a wavelike character. Thus we say that, due to the wavelike properties of electrons, the perfectly periodic array of atoms in a solid may not scatter electrons out of their straight-line path. In this sense, the periodic array may be considered to offer no resistance whatsoever to electron motion, thus rationalizing the long mean free paths for electrons in strain-free metal single crystals of high purity held at low temperatures.

The emergent picture is that electrical resistance is not due to the scattering of electrons by the atoms of the periodic array per se but by the departures from periodicity in the crystalline array. Such departures from periodicity are provided by impurities, vacancies, strained regions, dislocations, and grain boundaries, and also by thermal fluctuations of the atom array. Increased scattering at higher temperatures due to temperature dependent thermal fluctuations in the lattice can be shown to lead to the linear temperature dependence of the resistivity of metals. The residual resistance at extremely low temperatures is due to scattering from the impurity atoms and structural defects. A quantum mechanical approach involving the Schrödinger equation, based as it is on the wavelike behavior of particles, provides a suitable framework for rationalizing and treating these varied contributions to the electron resistivity of metals.

C. PERIODIC POTENTIAL FOR CRYSTALLINE SOLIDS

The Schrödinger equation can be applied to describe conduction electrons in metals, each conduction electron being considered to be under the influence of a potential energy function that has the same periodicity as the lattice. If we denote the lattice positions by

$$\mathbf{R_j} = j_1\mathbf{d}_1 + j_2\mathbf{d}_2 + j_3\mathbf{d}_3 \qquad (184)$$

where j_1, j_2, and j_3 are integers and $\mathbf{d}_1, \mathbf{d}_2$, and $\mathbf{d}_3$ are elemental vectors denoting the basic three units of periodicity in a three-dimensional crystalline solid, then a satisfactory potential energy $U(\mathbf{r})$, denoted by $V(\mathbf{r})$ for this special case, has the periodicity requirement

$$V(\mathbf{r} + \mathbf{R_j}) = V(\mathbf{r}) \qquad (185)$$

The time-independent Schrödinger equation [Eq. (175)] then takes the form

$$-\frac{\hbar^2}{2m}\nabla^2\phi_\ell + V(\mathbf{r})\phi_\ell = \mathscr{E}_\ell\phi_\ell \qquad (186)$$

The next step is to use some mathematical func-

tion for $V(\mathbf{r})$. This can be assumed to be a simple form, as for example, a one-dimensional periodic step function (Krönig–Penney model), or it may be quite complex, as for example, when one attempts to simulate mathematically the actual potential energy that would be sensed by an electron that probes different positions in each atom and between atoms in the periodic array. One very general method that lends great insight into the general problem of the motion of an electron in a periodic solid is to express the potential energy as a type of Fourier series,

$$V(\mathbf{r}) = \sum_{\mathbf{n}} V_{\mathbf{n}}e^{i\mathbf{G_n}\cdot\mathbf{r}} \qquad (187)$$

with the amplitudes $V_{\mathbf{n}}$ for the various harmonics chosen so that the function $V(\mathbf{r})$ reproduces any periodic potential energy of interest. For a lattice in which the basic spatial periodicity vectors $\mathbf{d}_1, \mathbf{d}_2$, and $\mathbf{d}_3$ are orthogonal, the vectors $\mathbf{G_n}$ turn out to be especially simple in form, but for the more general case of a nonorthogonal triad $\mathbf{d}_1, \mathbf{d}_2$, and $\mathbf{d}_3$, it greatly facilitates the problem to define a triad $\mathbf{b}_1, \mathbf{b}_2$, and $\mathbf{b}_3$,

$$\mathbf{b}_1 = (\mathscr{V}_{\text{cell}})^{-1}\,\mathbf{d}_2 \times \mathbf{d}_3 \qquad (188)$$

$$\mathbf{b}_2 = (\mathscr{V}_{\text{cell}})^{-1}\,\mathbf{d}_3 \times \mathbf{d}_1 \qquad (189)$$

$$\mathbf{b}_3 = (\mathscr{V}_{\text{cell}})^{-1}\,\mathbf{d}_1 \times \mathbf{d}_2 \qquad (190)$$

where

$$\mathscr{V}_{\text{cell}} = \mathbf{d}_1 \cdot (\mathbf{d}_2 \times \mathbf{d}_3) \qquad (191)$$

is the volume of a unit cell in the real lattice. In terms of these vectors, the vectors $\mathbf{G_n}$ appearing above are given by

$$\mathbf{G_n} = 2\pi(n_1\mathbf{b}_1 + n_2\mathbf{b}_2 + n_3\mathbf{b}_3) \qquad (192)$$

where n_1, n_2, and n_3 are integers. The symbol $\mathbf{n}$ is used to represent the integer triplet (n_1, n_2, n_3). The set $\mathbf{G_n}$ maps out a lattice of points in the same manner that the set $\mathbf{R_j}$ maps out a lattice, but the two lattices are usually quite different. The vectors $\mathbf{R_j}$ are said to map out a real or direct lattice, whereas the vectors $\mathbf{G_n}$ are said to map out the reciprocal lattice. The vectors $\mathbf{G_n}$ are referred to as reciprocal lattice vectors. The functions $\exp(i\mathbf{G_n} \cdot \mathbf{r})$ can be shown to have the properties required of basis functions for a Fourier series representation of arbitrary functions having the lattice periodicity.

Once the periodic potential energy is defined, then the Schrödinger equation [Eq. (175)] can be solved by various methods. One way, leading to great insight into this problem, is to assume a general form for the eigenfunctions ϕ_ℓ by using a Fourier series description with the periodicity of the entire solid, namely,

$$\phi_{\mathbf{m}} = \sum_{\mathbf{m}} B_{\mathbf{m}} e^{i\mathbf{k}_{\mathbf{m}} \cdot \mathbf{r}} \qquad (193)$$

This leads to a coupled set of algebraic equations for the unknown coefficients $B_{\mathbf{m}}$ which contain the energy eigenvalues $\mathscr{E}\ell$. The vectors $\mathbf{k}_{\mathbf{m}}$ can be obtained similar to the way the $\mathbf{G_n}$ were constructed. The algebraic equations so obtained constitute a homogeneous set. Self-consistency then requires the determinant of the coefficients of the unknowns $B_{\mathbf{m}}$ to be zero. This determinant is called the secular determinant. This leads directly to an algebraic equation, known as the secular equation, for the energy eigenvalues $\mathscr{E}\ell$. Choice of a specific eigenvalue $\mathscr{E}\ell'$ for solution of the set of algebraic equations containing the lattice potential energy coefficients $V_{\mathbf{n}}$ yields the eigenfunction $\phi\ell'$ corresponding to that energy. Repeating the procedure for each energy eigenvalue in principle gives the complete set of energy eigenfunctions for that periodic potential energy.

The energy eigenfunctions obtained for a periodic potential energy are known as Bloch functions, after the physicist Felix Bloch (1905–1983). Bloch functions have the general form

$$\phi_{\mathbf{m}'}(\mathbf{r}) = u_{\mathbf{m}'}(\mathbf{r}) e^{i\mathbf{k}_{\mathbf{m}} \cdot \mathbf{r}} \qquad (194)$$

where the $u_{\mathbf{m}'}(\mathbf{r})$ are functions having the periodicity of the real lattice. This periodicity condition is

$$u_{\mathbf{m}'}(\mathbf{r}) = u_{\mathbf{m}'}(\mathbf{r} + \mathbf{R_j}) \qquad (195)$$

The vectors $\mathbf{k}_{\mathbf{m}'}$ are propagation-type vectors for plane wave functions having wavelengths greater than the basic unit of lattice periodicity but less than the length of the crystal in the propagation direction.

A Bloch function for an electron in a solid may be likened to an individual harmonic of sound in a musical cabinet. Bloch functions can be shown to have a number of very interesting properties, as, for example, completeness of the set, linear independence, and orthogonality. The reader should consult the bibliography to become more enlightened concerning these properties.

D. Energy Bands and Energy Gaps

The picture that thereby emerges is that of groups of closely spaced discrete allowed energies that can be populated by electrons, with the groups of allowed levels being separated by energy ranges called gaps that contain no allowed energy values for conduction electrons. Each group of closely spaced, allowed discrete energies is called an energy band. Each allowed energy value within a band is characterized by a set of quantum numbers. With the additional consideration of electron spin, these are four in number. An electron in one of the allowed levels characterized by specified values of these quantum numbers travels unscattered by the atoms of the crystal lattice, the straight-line motion being allowed because of a wavelike propagation through the spatially periodic lattice potential energy. This provides the quantum mechanical explanation of the long mean free path for conduction electrons in metals.

The electrons in a solid are as a rule characterized by the vector $\mathbf{k}$ which denotes the propagation direction. The $\mathbf{k}$-vector is the analog of the momentum for a particle in a solid. (In free space, $\mathbf{p} = \hbar\mathbf{k}$, according to the de Broglie relation derived in Section VI.) The energy of the electron state is denoted by $\mathscr{E}(\mathbf{k})$. The goal of solid-state band structure calculations is to evaluate $\mathscr{E}(\mathbf{k})$ for a specified periodic potential energy. This periodic potential may be considered to be available to us at the outset, although the problem is best approached from the standpoint of computing the potential self-consistently with the electron states deduced in the calculation. Since $\mathscr{E} = \hbar\omega$, the function $\mathscr{E}(\mathbf{k})$ obtained by means of a band structure calculation represents the dispersion relation for electrons in the solid. As recalled from Sections V and VI, the dispersion relation provides the basis for determining the group velocity of the particle. Thus from Eq. (100) we can write

$$\mathbf{v}_{\text{group}} = \nabla_{\mathbf{k}}\omega(\mathbf{k}) = (1/\hbar) \nabla_{\mathbf{k}}\mathscr{E}(\mathbf{k}) \qquad (196)$$

This relation is very useful for obtaining the conduction electron velocity as a function of energy for any particular direction in the crystal. In fact, this approach must be used in lieu of the free space relation $\mathbf{v}_{\text{particle}} = \mathbf{p}/m$ because the inertia of an electron in a crystal is governed by an effective mass m^* instead of its actual mass m. The difference in value between m^* and m is a measure of the average conduction electron interaction with the periodic potential of the lattice. Although a perfectly periodic lattice does not offer a resistance by scattering the conduction electrons, it does offer a resistance to the acceleration of the electron under an applied force (such as an electric field) by affecting its inertial response to the force.

In considering the population of the various

energy levels, we must add in as an essential component the Pauli exclusion principle, that very soul of quantum mechanics which disallows any two electrons to occupy the same state, a state being denoted by the specification of values for the complete set of quantum numbers, including electron spin. Once we weave this into the fabric, we must consider how the energy eigenstates are occupied by the available electrons for conduction. Under thermal equilibrium conditions at temperatures near absolute zero, the lowest energy states are certainly occupied. The lower energy states in any of the energy bands represent states having low kinetic energies. Because the Pauli exclusion principle does not allow more than one conduction electron to crowd into any low-energy state, higher-energy states are populated to the degree required for all conduction electrons to be accommodated.

The requirements by quantum mechanics that all electrons be in different states has no analog in classical mechanics. Classically there is therefore no lower limit to the energy of the conduction electrons. In fact, in a classical description, the kinetic energy of the conduction electrons decreases to zero as the absolute temperature approaches zero, but in a quantum mechanical description, the average kinetic energy of the conduction electrons in a metal decreases asymptotically to a still relatively high value as the absolute temperature approaches zero.

E. Metals

The average kinetic energy for the highest populated energy band can be estimated by invoking a particle-in-a-box model of a metal holding its conduction electrons within the boundary walls, with no consideration given to the actual periodic potential energy of the lattice. This simple approach, known as the Free-Electron Model, often yields surprisingly accurate quantitative values for a number of physical properties of metals associated with the conduction electrons. In such cases a full solution of the Schrödinger equation for the actual periodic potential may not be required.

The kinetic energy of the highest filled state in a given energy band at 0 K is designated the Fermi energy. A computation of how the average energy changes with increases in thermodynamic temperature of the system yields the specific heat of the conduction electrons. The

accurate predictions obtained by quantum mechanics for the specific heat of metals at low temperatures was another gigantic success for the theory, to be sharply contrasted to the total failure on the part of the classical approach to provide an adequate quantitative estimate of this physical property of metals.

Quantum mechanics gives great insight into the scattering of conduction electrons by imperfections in a metal. The quantum nature of the scattering of conduction electrons places the following restriction on the process: Scattering can take place only to vacant quantum states of the system. This means that at 0 K an elastic scattering event can occur only for a conduction electron having an energy equal to the Fermi energy, since that is the only energy for which there simultaneously exist both filled and empty states. The situation is not quite so restrictive at higher temperatures, because there is then a statistical probability that nearby states are occupied or unoccupied over a range of energy at least $k_B T$ in width in the neighborhood of the Fermi energy $\mathscr{E}_F$. [Boltzmann constant $k_B = 1.38 \times 10^{-23}$ J/K; T = absolute (Kelvin) temperature.] Nevertheless, electron scattering (and hence the electrical resistivity) is still severely restricted in metals by the requirements of the Pauli exclusion principle.

F. Insulators

The reader who accepts these reasons for the success of quantum mechanics in describing the properties of the long mean free path in metals must then feel quite puzzled when confronted with the experimental fact that some crystals, even in the limit of high purity, low temperature, and perfect periodicity, do not ordinarily manifest the free and easy motion of electrons. These materials we call electrical insulators.

What is it about insulators that causes them to differ from metals in ability to conduct electrons? The answer is again quantum mechanical in origin. It is hardly more abstruse than the answer to the question of the existence of the long mean free path in metals. The solution of the Schrödinger equation for a periodic potential energy in the way previously outlined yields a division of the energy scale into interspersed allowable and forbidden regions of energy. Over the allowable regions (the energy bands) we find very closely spaced, discrete energy eigenvalues, but in the forbidden regions (the energy gaps) there are no such energy eigenvalues. This property of the energy eigenvalue spectrum characteristic of the periodic potential, however, is of itself insufficient to explain the basic nature of insulators. As in the situation for electron scattering by impurities, we must allow for the consequences of the all-pervading Pauli exclusion principle.

A force for directed electron motion invariably leads to the prediction of a nonzero electrical current from a purely classical viewpoint; however, it does not necessarily lead to such a consequence in quantum mechanics. The reason is that acceleration of electrons by a force leads to a change in electron momentum and generally to an accompanying change in the electron energy. A change in electron momentum is synonymous with a change in the quantum numbers characterizing the occupied electronic state. That is, the acceleration of an electron in quantum mechanics is described by the electron vacating the state it initially occupied, as it simultaneously enters a different allowed state, which by the requisites of the Pauli exclusion principle must necessarily be vacant before any occupation can occur. Quantum mechanically we view the electron as being induced to enter a succession of adjacent allowed states by electric-field-induced transitions. This view can be contrasted sharply with the classical picture of a continuous acceleration of the electron through a continuous range of momentum vectors. For a metal having a partly filled energy band, there indeed exists the requisite sequence of nearby unoccupied states adjacent to the filled states so that transitions can be induced by the electric force acting on the conduction electrons.

For the specific case of electrical insulators, consider the seemingly unlikely situation that there are precisely enough conduction electrons to fill every energy state up to the start of a given energy gap. Both the scattering of the conduction electrons and the electric-field excitation of electrons to nearby empty states are then impossible at 0 K, and for all practical purposes, nearly impossible even at nonzero temperatures for which $k_B T$ is far smaller than the energy gap $\mathscr{E}_{gap}$ extending to the next-higher empty energy level. Although zero scattering might seem to constitute the ideal situation leading to a low (or even zero) resistivity, there nevertheless is no existing electrical current in the presently described situation, nor can there be any induced electric current. There is no existing current because in such a situation all electrons occupy pairs of occupied states representing equal mag-

nitude but oppositely directed momentum, so there is no net charge transport. There can be no induced current since the electric force cannot change the momentum of the electrons in any of the filled states because there are no nearby unoccupied states for the transitions. This is the required situation for an insulator.

In actuality, the seemingly unlikely situation of there being exactly enough electrons to fill a band, with none left over for the next-higher empty band, is not too unlikely. The reason for this is again a bit abstruse, but in simplest terms an energy band is found to contain one allowed state per atom in the solid for each degree of freedom of the electron spin. There being two degrees of freedom for the electron spin (viz., two spin states, designated up and down), then the requirement for a filled energy band is simply that there be two conduction electrons furnished by each atom in the crystal. Since the valence electrons become the conduction electrons in a solid, this is not an unlikely condition.

G. Semiconductors

The energy level distribution for a solid can be quite a bit more complex than described above, since there is the possibility of overlapping energy bands (i.e., energy bands unseparated by the usual energy gap). In addition, there can be energy gaps that are quite small relative to laboratory values of $k_B T$. It can be readily appreciated that overlapping bands promote metallic conduction or else can lead to what is known as a semimetal, whereas narrow energy gaps can lead to what we call a semiconductor. In semiconductors, an increase in temperature leads to more excitation of electrons from the highest-energy, filled band across the gap to the adjacent empty band, thus giving electrons that are capable of conducting in what would otherwise be an empty band. Those electrons excited across the gap leave behind empty states (called electron holes) in the otherwise filled band. These empty states can also promote conduction in the following sense: Filled states near in momentum and energy to the newly provided empty states can undergo transitions to the empty states by means of electric-field excitation, all such transitions being impossible in the 0-K equilibrium situation where all states in the band are filled. In this way two distinct carrier types are simultaneously provided, namely, electrons in an almost empty band and electron holes in an almost filled band.

The number of electrons excited across the energy gap increases nearly exponentially with increasing temperature. To the extent that electron transport increases as the number of carriers, the conductivity of the semiconductor increases almost exponentially with the temperature. A material having the properties just described is called an intrinsic semiconductor.

The prediction of the experimentally observed exponential increase of the conductivity with increasing temperatures in intrinsic semiconductors represents another triumph of quantum mechanics. In a classical description, there are no energy gaps and hence no parallel to the predictions of an exponentially increasing conductivity due to excitation across an energy gap.

The exponential increase of conductivity with temperature in semiconductors also contrasts markedly with the temperature dependence of the conductivity of metals. In that case, the conductivity decreases (instead of increasing) with increasing temperature. This decrease of conductivity in metals, which is more or less linear with increasing temperature, is due to the increase in the thermal vibrations of the atoms of the lattice at higher temperatures. Thermal vibrations yield greater departures of the atom array from perfect periodicity, thus leading to more random scattering of the electrons and a consequent decrease in the electron current.

The exponential increase in conductivity with temperature described for excitation of electrons across an energy gap in an intrinsic semiconductor is paralleled in another type of solid, called an extrinsic semiconductor. In extrinsic semiconductors, however, there is no band-to-band excitation. Instead, the source of electrons for the empty band is a doping concentration of impurities that have outer electrons at energies just below the empty band (so-called donor impurities, or simply donors) or alternately have empty levels at energies just above the filled band (so-called acceptor impurities, or simply acceptors). Semiconduction then takes place by means of electrons donated to the empty band by donor impurities, or alternately by electron holes remaining in the filled band because of electrons accepted from that band by the acceptor impurities.

H. SUPERCONDUCTORS

It is another well-known experimental fact that a number of metals go into a state of zero resistance at very low temperatures, below the so-called transition temperature characteristic of the material in question. The concept of energy gaps likewise turns out to play an important role in understanding this incredible phenomenon, although in a different way from that just described for semiconductors. The energy gap in the case of superconductors is attributed to a condensation of the electrons carrying the charge into so-called Cooper pairs, the binding energy of the pairs being attributed to indirect Coulomb-force-induced interaction between electrons as mediated by the intervening ion cores on the lattice sites in the metal.

Consider, for example, pairs of electrons passing one another while traveling in opposite directions through the lattice of ion cores surrounded by the attendant electron clouds. One electron exerts a force on the nearby ionic lattice, which responds to that force. The resulting disturbance in the periodic potential sensed by the second electron of the pair can lead to an effective lowering in the total energy, relative to the situation of a rigid nonresponding lattice. The energy lowering is the greatest for pairs of electrons having equal magnitude but oppositely directed momentum, so Cooper pairs are characterized by two electrons having this property. The lowering of energy constitutes a superconducting energy gap. The energy gap so produced is quite small, so very low temperatures are required for the pairs to remain unbroken by thermal fluctuations.

As the temperature is reduced through the superconducting transition, we speak of the condensation of the electron system into the paired state. It is evident that electrons with oppositely directed momentum values will in a short time become separated spatially, so the pairing process must be statistical in nature. Pairs must continually exchange partners, as required, for example, in some forms of folk dancing.

The dance of the conduction electrons, while maintaining this property of electron pairing, is a many-body problem of some complexity. The wavefunction for the entire system of electrons as a unit must be considered instead of merely considering individually the single-particle wave functions. The establishment of a net electric current requires a suitable modification in the zero current wave function for the system.

The importance of electron pairing for electrical resistance is this: The scattering of conduction electrons in the paired state is ineffective in randomizing the net electron momentum, so there can be a zero-resistivity state so long as Cooper pairs exist in the system.

Thermal fluctuations can break Cooper pairs to yield electrons in the normal state. The breaking of electron pairs by this means is tantamount to an excitation across the energy gap. In contrast to the case of semiconductors, in the present instance the excitation across the energy gap leads to an increase in the resistivity. Raising the temperature of a superconductor through the superconducting transition temperature means that the thermal fluctuations become so large that essentially no Cooper pairs remain in the metal to provide superconductivity.

I. SUCCESS OF QUANTUM MECHANICS

Thus quantum mechanics provides a framework for understanding the widely different electrical conduction properties of superconductors, normal metals, semiconductors, and insulators. This remarkable success, coupled with the failure of classical physics to lend understanding to these areas, has led to the nearly universal acceptance of quantum mechanics for most calculations in solid-state physics.

IX. Summary

Quantum mechanics is a theory that evolved in the 1920s to correlate and predict the behavior of atomic and subatomic systems. Werner Heisenberg (1901–1976) and Erwin Schrödinger (1887–1961) played prominent roles in the development of this theory, which proves to be the only formulation adequate for the microscopic domain of nature. Heisenberg stressed the importance of including physical observables and experimental observations of optical spectral lines in his formulation, while Schrödinger based his work on a differential equation for the wavelike behavior of small mass particles that includes the possibility of constructive and destructive interference of waves presumably associated with the presence of a particle. Inherent in quantum mechanics is the germ concept that accurate predictions of future trajectories of particles and the time evolution of a system involving one or more particles are at best statisti-

cal, involving a range of possibilities specified exactly only in terms of precise values for the relative probabilities. This precludes the deterministic prediction of an exactly specified future path, regardless of how accurately the initial conditions of the system are specified. Also inherent in the theory is the impossibility of measuring exactly even the initial conditions, such as initial position and initial linear momentum. That is due to some inherent uncertainty in the value of one of these variables following a determination of the value of the other of the variables to some specified degree of precision. The philosophical implications of an inherent uncertainty in quantum mechanical predictions, as contrasted with the absolute determinism inherent in classical mechanical predictions, initially led many (including Albert Einstein) to doubt whether the discipline had any fundamental merit beyond that of being an elegant and elaborate computation tool for obtaining predictions of a statistical nature. To date, we have no better theory for the microscopic world. [*See* ELECTRODYNAMICS, QUANTUM; QUANTUM OPTICS.]

BIBLIOGRAPHY

Ashby, N., and Miller, S. C. (1970). "Principles of Modern Physics." Holden-Day, San Francisco.

Born, M. (1969). "Physics in My Generation." Springer-Verlag, New York.

Eisberg, R., and Resnick, R. (1974). "Quantum Physics." Wiley, New York.

Feynman, R. P., Leighton, R. B., and Sands, M. (1965). "Feynman Lectures on Physics—Quantum Mechanics." Addison-Wesley, Reading, Massachusetts.

Fromhold, A. T., Jr. (1981). "Quantum Mechanics for Applied Physics and Engineering." Academic Press, New York.

Ikenberry, E. (1962). "Quantum Mechanics for Mathematicians and Physicists." Oxford University Press, New York.

McGervey, J. D. (1983). "Introduction to Modern Physics." Academic Press, New York.

Pauli, W. (1973). "Pauli Lectures on Physics," Vol. 5. Wave Mechanics. MIT Press, Cambridge, Massachusetts.

QUANTUM OPTICS

J. H. Eberly   *University of Rochester*
P. W. Milonni   *Los Alamos National Laboratory*

I.   Introduction
II.   Induced Atomic Coherence Effects
III. Radiation Coherence and Statistics
IV.   Quantum Interactions and Correlations

GLOSSARY

AC Stark effect: Effective increase of the Bohr transition frequency of a two-level atom which is being excited by a strong laser beam, the amount of increase being the Rabi frequency.

Bloch vector: Fictitious vector whose rotations are equivalent to the time dependence of the wave function or quantum mechanical density matrix associated with a two-level atom.

Coherent state: Quantized state of a light field whose fluctuation properties are Poissonian; it is considered the most classical quantized field state.

Coherence time: Limiting time interval between two segments of a light beam beyond which the superposition of the segments will no longer lead to interference fringes.

Degree of coherence: Normalized measure of the ability of a light beam to form interference fringes.

Optical bistability: Existence of two stable output intensities for a given input intensity of a steady light beam transmitted through a nonlinear optical material.

Optical Bloch equations: Dynamical equations that determine the motion of the Bloch vector; they are a special type of quantum Liouville equation.

Photon echo: Burst of light emitted by a collection of two-level atoms signaling the realignment of their Bloch vectors after initial dephasing; similar to the spin echo of nuclear magnetic resonance.

Rabi frequency: Steady frequency of rotation of the Bloch vector of an atom exposed to a constant laser beam, proportional to the atom's transition dipole moment and the laser's electric field strength.

Superradiance: Spontaneous emission from many atoms exhibiting collective phase-coherence properties, such as radiation intensity proportional to the square of the number of participating atoms.

Two-level atom: Fictitious atom having only two energy levels which is used as a model in theoretical studies of near-resonance interactions of atoms and light, particularly laser light.

Quantum optics is the study of the statistical and dynamical aspects of the interaction of matter and light. It is concerned with phenomena ranging from spontaneous emission and single photon absorption to the highly nonlinear processes induced by laser fields and has connections with laser physics, nonlinear optics, quantum electronics, quantum statistics, and quantum electrodynamics.

I. Introduction

A. CENTRAL ISSUES OF QUANTUM OPTICS

Planck's quantum, announced to the Prussian Academy on October 19, 1900, as a solution to the blackbody puzzle re-opened the wave--particle question in optics, a question that Fresnel and Young had settled in favor of waves almost two centuries earlier. Planck's quantum could not be confined to light fields. Within three decades, all of particle mechanics had been quantized and rewritten in wave mechanical form, and wave–particle duality was understood to be both universal and probabilistic.

Quantum optics is fundamentally concerned with coherence and interference of both photons and atomic probability amplitudes. For example, it provides one of the main avenues at the present time for detailed study of wave–particle duality. The central issues of quantum optics deal with light itself, with quantum mechanical states of matter excited by light, and with the process of interaction of light and matter.

Questions arising in the description of a single atom and its associated radiation field, as the atom makes a transition between two energy states and either emits or absorbs a photon, are among the most central questions in quantum optics. Observations of individual optical emission and absorption events are possible, and the interpretation of such observations is at the heart of quantum theory.

Various elements of these central considerations are to a degree independent of each other and are understood separately. Among these are (1) the probability that an atomic electron occupies one or another state and the rate at which these occupation probabilities change, (2) the statistical nature of the photons emitted during transitions, (3) correlations between atom and photon states, (4) the characteristic parameters that control the light–matter interaction, and (5) the intrinsically quantum mechanical features of the atom's response to the radiation.

Quantum optics also concerns itself with problems that grow out of these central considerations and whose answers can be expressed within the conceptual framework established by the central problem. Areas related in this way to the core of quantum optics deal, for example, with correlated many-atom light–matter interactions; near-resonant transitions among three and more states of an atom or molecule or solid; optical tests of quantum electrodynamics and measurement theory; multiphoton processes; quantum limits to noise and linewidth; quantum theory of light amplification and laser action; and manifestations of nonlinearity, bistability, and chaos in optical contexts. A wide variety of quantum optical phenomena that bear on one or another of these issues are now known and

widely studied. [*See* NONLINEAR OPTICAL PROCESSES.]

B. THE *A* AND *B* COEFFICIENTS OF EINSTEIN

The second half of the twentieth century has seen remarkable advances in our understanding of light, of its generation, propagation, and detection. The laser is one manifestation of these advances. Lasers generally depend on the quantum mechanical properties of atoms, molecules, and solids because quantum properties determine the ways that matter absorbs light and emits light. Conversely, the properties of laser beams have made optical studies of quantum mechanics possible in a variety of new ways. It is this interplay that has created the field of quantum optics since about 1960. [*See* LASERS.]

From a different historical perspective, however, quantum optics is much older than the laser and even older than quantum mechanics. The quantum concept first entered physics in 1900 when Planck invented the light quantum to help understand black body radiation. The understanding of other quantum optical phenomena, such as the photoelectric effect, first explained by Einstein in 1905, was well underway almost two decades before a quantum theory of mechanics was properly formulated in 1925 and 1926 by Heisenberg and Schrödinger. Indeed, these early developments in quantum optics played an essential role in the first quantum pictures of atomic matter given by Bohr and others in the period from 1913 to 1923.

Only two parameters are needed to understand the interaction of light with atomic (and molecular) matter, according to Einstein. These two parameters are called *A* and *B* coefficients. These coefficients are important because they control the rates of photon emission and absorption processes in atoms, as follows. Let the probability that a given atom is in its *n*th energy level be written P_n. Suppose there are photons present in the form of radiation with spectral energy density (J/m³ Hz) denoted by $u(\omega)$. Then the rate at which the probability P_n changes is due to three fundamental processes:

$$(dP_n/dt)_{\text{absorption of light}} = +Bu(\omega)P_m \quad \text{(1a)}$$

$$(dP_n/dt)_{\text{spontaneous emission of light}} = -AP_n \quad \text{(1b)}$$

$$(dP_n/dt)_{\text{stimulated emission of light}} = -Bu(\omega)P_n \quad \text{(1c)}$$

Here P_m is the probability that the atom is in a lower level, the *m*th, which is related to the *n*th through the energy relation $E_n - E_m = \hbar\omega$, where $\hbar = h/2\pi$ and h is Planck's famous quantum constant. Einstein's great insight was to include stimulated emission [Eq. (1c)] among the three elementary processes, in effect, to recognize that an atom in an upper energy state could be encouraged by the presence of photons [the existence of $u(\omega)$] to hasten the rate at which it would drop down to a lower state.

The three contributions to the rate of change of P_n shown in Eq. (1a–c) can be added to make an overall single equation for the total rate of change of P_n:

$$dP_n/dt = +Bu(\omega)P_m - AP_n - Bu(\omega)P_n \quad (2a)$$

Einstein applied this equation to an examination of blackbody light. He showed that the steady-state solution

$$P_m/P_n = 1 + A/Bu(\omega) \quad (2b)$$

implies the validity of Planck's formula for $u(\omega)$:

$$u(\omega, T) = \frac{\hbar\omega^3}{\pi^2 c^3} \frac{1}{\exp\{\hbar\omega/kT\} - 1} \quad (3)$$

and the value of the prefactor is just the ratio A/B:

$$\frac{A}{B} = \frac{\hbar\omega^3}{\pi^2 c^3} \quad (4)$$

where k is Boltzmann's constant and T the temperature in degrees Kelvin. Table I contains the values of physical constants used in evaluating various radiation formulas. For typical optical radiation, the value of the fundamental ratio A/B is approximately 10^{-14} J/m³ Hz. The corresponding intensity, namely cA/B, is approximately 3×10^{-6} J m⁻², or $6\pi \times 10^{-6}$ W m⁻² per Hz of bandwidth. The value of the spectral intensity of thermal radiation is usually many orders of magnitude lower than this because of the second factor in Eq. (3). At optical wavelengths, the sec-ond factor is much smaller than 1 for all temperatures less than about 5000 K.

After the development of a fully quantum mechanical theory of light by Dirac in 1927, it was possible to give expressions for A and B separately:

$$A = \frac{1}{4\pi\varepsilon_0} \frac{4D^2\omega^3}{3\hbar c^3} \quad (5)$$

$$B = \frac{1}{4\pi\varepsilon_0} \frac{4\pi^2 D^2}{3\hbar^2} \quad (6)$$

In these formulas we have separated the factor $1/4\pi\varepsilon_0 = 8.9874 \times 10^9$ N m²/C² to display A and B in atomic units as well as SI units, and D denotes the quantum mechanical "dipole matrix element" associated with the $m \to n$ transition under consideration.

The values of these important coefficients can be obtained for transitions of interest in quantum optics by assuming that the dipole matrix element is approximately equal to the product of the electron's charge and a "typical" electron displacement from the nucleus. Thus, we take $\omega = 2\pi\nu$ and e and h from Table I and $D = er$, with r equal to about 1 to 3 Å ($1–3 \times 10^{-10}$ m). In this case the values are $A \approx 10^8$ sec⁻¹ and $B \approx 10^{22}$ m² J⁻¹ sec⁻², respectively.

The advantage of Einstein's approach, and the reason it still provides one basis for understanding light–matter interactions, is that it breaks the interaction process into its separate elements, as identified above in Eqs. (1a–c). To repeat this important identification, these processes are (a) absorption, (b) spontaneous emission, and (c) stimulated emission.

However, it must be pointed out that Einstein's formulas are not universally valid, and Eqs. (1) and (2) can be seriously misleading in some cases, particularly for laser light. Laser light typically has a very high spectral energy density $u(\omega)$. In this case different formulas and

TABLE I. Physical Constants Used in Evaluating Radiation Formulas

Constant	Value
h (Planck's constant)	6.6×10^{-34} J sec⁻¹
k (Boltzmann constant)	1.38×10^{-23} J K⁻¹
c (Speed of light)	3×10^8 m sec⁻¹
e (Electric charge)	1.6×10^{-19} C
λ (Typical optical wavelength, yellow)	600×10^{-9} m
ν (Typical optical frequency)	5×10^{14} Hz

equations, and even entirely different concepts with their origins in wave mechanics, may be required.

A large body of experimental evidence has accumulated since 1960 showing that many aspects of the interaction between light and matter depend on electric radiation field strength E directly, not only on energy density $u \approx E^2$. Just those aspects of the light–matter interaction that depend directly on E also depend directly on quantum mechanical state amplitudes ψ, not only on their associated probabilities $|\psi|^2$. Issues of coherence and interference of both radiation fields and probability amplitudes are fundamental to these studies. It is principally the experiments and theories that deal with light and matter in this domain that make up the field of quantum optics.

C. Two-State Atom and Maxwell–Bloch Equations

As Einstein's arguments suggest, in quantum optics it is often sufficient to focus attention on just two energy levels of an atom—the two levels that are closest to resonance with the radiation, satisfying the energy condition

$$E_2 - E_1 \approx \hbar\omega$$

where $\omega = 2\pi\nu$ is the angular frequency of the radiation field. This is shown schematically in Fig. 1.

Under these circumstances the wave function of the atom is a sum of the wave functions for the two states

$$\psi(\mathbf{r}, t) = C_1\phi_1(\mathbf{r}) + C_2\phi_2(\mathbf{r}) \qquad (7a)$$

For simplicity of description we will assume each level corresponds to a single quantum state and will usually use "level" and "state" synonymously.

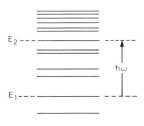

FIG. 1. Schematic energy level diagram showing a two-level subsystem.

The assumption that the electron is certainly in one or the other or a combination of these two levels is expressed mathematically by the equality

$$|C_1|^2 + |C_2|^2 = 1 \qquad (7b)$$

Each term in this equation is called a level probability, and all probabilities must add to 1, of course. These probabilities are the quantities labeled by the letter P in the Einstein equations. The C's themselves are not probabilities, and they have no counterpart in classical probability theory. They are called probability amplitudes. It is the remarkable nature of quantum mechanics that the fundamental equation (the Schrödinger equation) governs these amplitudes C, not the probabilities $|C|^2$.

The Schrödinger equation for either one of the amplitudes is

$$i\hbar \, dC_m/dt = E_mC_m + V_{mn}C_n \qquad (8)$$

where m and n take either the value 1 or 2, but $m \neq n$. Here $\hbar$ is the usual abbreviation for $h/2\pi$, and V_{mn} is called the interaction matrix element between the atom and the radiation field. In almost all cases of interest in quantum optics, this interaction comes from the potential energy $-\mathbf{d} \cdot \mathbf{E}(\mathbf{r}, t)$ of the atomic dipole in the electric radiation field (a dipole $\mathbf{d} = e\mathbf{r}$ exists because of the separation of the negative electronic charge in its planetary orbit from the position of the positively charged nucleus).

In principle $\mathbf{E}$ is a quantum operator field (see Section III,C). The so-called semiclassical theory of radiation uses its average (or "expectation") value instead. This approximation is usually justified when the field is intense, because quantum fluctuations, which almost always occur at the single photon level, are then negligible. In this section we describe the semiclassical theory and ignore quantum field effects entirely.

The dipole interaction is distributed over the atomic orbitals involved, with the result that

$$V_{mn} = \langle \phi_m | - e\mathbf{r} \cdot \mathbf{E}(\mathbf{r}, t) | \phi_n \rangle$$

$$\approx \int d^3r \phi_m^*(\mathbf{r})[-e\mathbf{r} \cdot \hat{\mathbf{e}}E(\mathbf{r}_N, t)]\phi_n(\mathbf{r}) \qquad (9a)$$

$$= -\mathbf{d}_{mn} \cdot \hat{\mathbf{e}}E(\mathbf{r}_N, t)$$

In Eq. (9a) we have written

$$\mathbf{E}(\mathbf{r}, t) = \hat{\mathbf{e}}E(\mathbf{r}, t) \approx \hat{\mathbf{e}}E(\mathbf{r}_N, t)$$

where $\hat{\mathbf{e}}$ is the polarization and $E(\mathbf{r}_N, t)$ is the amplitude of the electric field at the position of the atom (i.e., of the nucleus) r_N instead of at the position of the electron. This approximation is usually well justified because the electron's orbit, and thus the range of the integral, extends mainly over a region much smaller than an optical wavelength. Thus, over the whole range of the integral $E(\mathbf{r}, t) \approx E(\mathbf{r}_N, t)$ and the only $\mathbf{r}$ dependence comes from the dipole moment $e\mathbf{r}$ itself. This is called the dipole approximation. The integral in Eq. (9a) is called the dipole matrix element, d,

$$d = \hat{\mathbf{e}} \cdot \mathbf{d}_{mn} = \hat{\mathbf{e}} \cdot \int d^3r\phi_m^*(\mathbf{r})e\mathbf{r}\phi_n(\mathbf{r}) \quad (9b)$$

If only one atom is under consideration it is common to put it at the origin of coordinates and write $E(0, t)$ or simply $E(t)$. If several or many atoms are under consideration this is generally not possible [see, e.g., Eq. (20)]. Equation (8) can be written in a simpler form by anticipating an interaction with a quasi-monochromatic radiation field

$$E(t) = \mathcal{E}_0(t)e^{-i\omega t} + \text{c.c.} \quad (10)$$

that is nearly resonant with the atom, where c.c. means complex conjugate. This means that the angular frequency $\omega = 2\pi\nu$ of the radiation field is approximately equal to the angular transition frequency of the atom: $\omega_{21} = (E_2 - E_1)/\hbar$. In this case there is a strong synchronous response of the atom to the radiation, and the equations simplify if one removes the synchronously "driven" component of the atomic response by defining new variables a_n:

$$a_1 = C_1 \quad (11a)$$

$$a_2 = C_2 e^{i\omega t} \quad (11b)$$

Note that $|a_1|^2 + |a_2|^2 = |C_1|^2 + |C_2|^2 = 1$, thus, the a's are also probability amplitudes.

If $E(t)$ is sufficiently monochromatic (the field amplitude, i.e., $\mathcal{E}_0$ is practically constant in time; which means $|d\mathcal{E}_0/dt| \ll \omega|\mathcal{E}_0|$), then a rotating wave approximation (RWA) is valid if the anticipated atom field resonance is sufficiently sharp, that is, if $|\omega_{21} - \omega| \ll \omega$. The most important consequence of the RWA is that factors such as $1 + e^{\pm 2i\omega t}$, which appear in the exact equations for the new amplitudes a_1 and a_2, may be replaced by 1 to an excellent approximation. The result is that the frequencies ω_{21} and ω individually pay no further role in the Schrödinger equation, and Eq. (8) takes the extremely compact RWA form:

$$i\, da_1/dt = -\tfrac{1}{2}\chi a_2 \quad (12a)$$

$$i\, da_2/dt = \Delta a_2 - \tfrac{1}{2}\chi a_1 \quad (12b)$$

where

$$\Delta = \omega_{21} - \omega \quad (12c)$$

is called the detuning of the atomic transition frequency from the radiation frequency, and

$$\chi = 2d\mathcal{E}_0/\hbar \quad (12d)$$

is called the Rabi frequency of the interaction. For simplicity, the Rabi frequency χ will be assumed here to be a real number.

For a strictly monochromatic field (time-independent $\mathcal{E}_0$) the solution to Eq. (12a and b) is easily found in terms of $\Omega = \sqrt{\chi^2 + \Delta^2}$, where Ω is called the generalized or detuning-dependent Rabi frequency. In the most important single case the atom is in the lower state at the time the interaction begins, which means that $a_1(0) = 1$ and $a_2(0) = 0$. In this case the solution is

$$a_1(t) = [\cos \Omega t/2 + (i\,\Delta/\Omega)\sin \Omega t/2]e^{-i\Delta t/2}$$
$$(13a)$$

$$a_2(t) = i[(\chi/\Omega)\sin \Omega t/2]e^{-i\Delta t/2} \quad (13b)$$

and the corresponding probabilities are

$$P_1 = \cos^2(\Omega t/2) + (\Delta/\Omega)^2\sin^2(\Omega t/2) \quad (14a)$$

$$P_2 = (\chi/\Omega)^2\sin^2(\Omega t/2) \quad (14b)$$

These solutions describe continuing oscillation of two-level probability between levels 1 and 2. They have no steady state. Figure 2 shows graphs of $P_2(t)$. One already sees, therefore, that the quantum amplitude equations [Eqs. (12a–d)] make strikingly different predictions from the two-level equations of Einstein [recall the steady-state solution Eq. (2b)].

Within the RWA, the dynamics remain unitary or probability conserving: $P_1 + P_2 = 1$ for

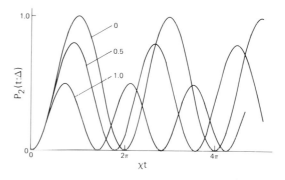

FIG. 2.   Rabi oscillations for different values of Δ/χ.

all values of t. As a result, the compact version [Eq. (12)] of Schrödinger's equation remains valid even for very strong interactions between the atom and the radiation. This is important when considering the effects on atoms of very intense laser fields. The limits of validity of Eq. (12) are determined to a large extent by a generalized statement of the limits of the RWA:

$$\omega_{21} \text{ and } \dot{\omega} \gg \sqrt{\Delta^2 + \chi^2} \qquad (15)$$

There are two other real quantities associated with the amplitudes a_1 and a_2 of the radiation–atom interaction. They belong to the atomic dipole's expectation value $\langle \mathbf{d} \rangle$ which, according to Eqs. (7a), (9b), and (11), is

$$\langle \psi(t)|e\mathbf{r}|\psi(t) \rangle = a_1(t)a_2^*(t)e^{i\omega t}\mathbf{d}_{21} + \text{c.c.} \quad (16)$$

The new quantities are designated by u and v:

$$u = a_1^* a_2 + a_1 a_2^* \qquad (17a)$$

$$v = i(a_1^* a_2 - a_1 a_2^*) \qquad (17b)$$

Along with u and v, a third variable, the atomic inversion $w = P_2 - P_1$ plays an important role. The solutions for u, v, and w corresponding to Eqs. (13) and (14) above are

$$u = (\Delta\chi/\Omega^2)(1 - \cos \Omega t) \qquad (18a)$$

$$v = (\chi/\Omega)\sin \Omega t \qquad (18b)$$

$$w = -(1/\Omega^2)(\Delta^2 + \chi^2 \cos \Omega t) \qquad (18c)$$

which share the oscillatory properties of the probabilities. They also obey the important conservation law:

$$u^2 + v^2 + w^2 = 1 \qquad (18d)$$

which is the same as $|a_1|^2 + |a_2|^2 = 1$.

For many purposes in quantum optics the semiclassical dipole variables u and v, and the atomic inversion w, are the primary atomic variables. They obey equations which are equivalent to Eq. (8) and take the place of Schrödinger's equation for the level's probability amplitudes a_1 and a_2 (or C_1 and C_2):

$$du/dt = -\Delta v \qquad (19a)$$

$$dv/dt = \Delta u + \chi w \qquad (19b)$$

$$dw/dt = -\chi v \qquad (19c)$$

All of these considerations assume that the field is monochromatic or nearly so.

Although the field $\mathbf{E}(\mathbf{r}, t)$ is not considered an operator in the semiclassical formulation, it can still have a dynamical character, which means it is not prescribed in advance, but obeys its own equation of motion. It is naturally taken to obey the Maxwell wave equation, with the usual source term $\mu_0 d^2/dt^2\mathbf{P}(z, t)$. In the semiclassical theory $\mathbf{P}(z, t) = N\langle \mathbf{d} \rangle$ where N is the density of two-level atoms in the source volume and $\langle \mathbf{d} \rangle$ the average dipole moment of a single atom, already calculated in Eq. (16).

As Eq. (16) indicates, $\langle \mathbf{d} \rangle$ is quasimonochromatic, essentially because a two-level atom is characterized by a single transition frequency. Therefore the $\mathbf{E}$ that it generates may also be regarded as monochromatic or nearly so, as Eq. (10) assumed. This internal consistency is an important consideration in the semiclassical theory. In many cases it is also suitable to regard the field as having a definite direction of propagation, say z, and to neglect its dependence on x and y (plane wave approximation):

$$\mathbf{E}(z, t) = \hat{\mathbf{e}}\mathcal{E}(z, t)e^{-i(\omega t - kz)} + \text{c.c.} \qquad (20)$$

where $k = \omega/c$ and the amplitude or envelope function $\mathcal{E}(z, t)$ is a complex generalization of $\mathcal{E}_0(t)$ in Eq. (10). It obeys a "reduced" wave equation in variables z and t:

$$[\partial/\partial z + \partial/\partial ct]\mathcal{E}(z, t) = i(\pi N d\omega/4\pi\varepsilon_0 c)[u - iv]$$
$$(21)$$

The reduced wave equation Eq. (21) and Eqs. (19) for u, v, and w are called the semiclassical coupled *Maxwell–Bloch equations*.

Inspection of the semiclassical Eqs. (19) and (21) reveals their major flaw. One easily sees that the semiclassical approach to radiation theory does not include the process of spontaneous emission. A completely excited atom will not emit a photon in this theory. That is, if the atom is in its excited state, then $a_2 = 1$ and $a_1 = 0$, so $w = 1$ and $u = v = 0$. Thus, according to Eq. (21) no field can be generated. By the same token, if $\mathcal{E} = 0$ then $\chi = 0$ and $dw/dt = 0$, and no evolution toward the ground state can occur. The flaw in the semiclassical Maxwell–Bloch approach arises from the assumption that all of the dynamics can be reduced to a consideration of average values, that is, averages of dipole moment and inversion and field strength. In reality, fluctuations about average values are an important ingredient of quantum theory and essential for spontaneous emission.

Spontaneous emission does not play a dominant role in many quantum optical processes, particularly those involving strong radiation fields. The semiclassical Maxwell–Bloch equations give an entirely satisfactory explanation of these effects, and the next sections describe some of them.

In nature there are no actual two-level atoms, of course. However, the selection rules for allowed optical dipole transitions are sufficiently restrictive, and optical resonances can be sufficiently sharp that very good approximations to two-level atoms can be found in nature. A good example is found in a pair of levels in atomic sodium, and they have been used in recent quantum optical experiments.

In sodium the nucleus has spin $I = 3/2$, and the lowest electronic energy level is $3S_{1/2}$ with hyperfine splitting ($F = 1$ and $F = 2$). The first excited levels $3P_{1/2}$ and $3P_{3/2}$ are responsible for the well-known strong sodium D lines of Fraunhofer in the yellow region of the optical spectrum at wavelengths 589.0 nm and 589.6 nm. The "two-level" transition is between the $m_F = +2$ magnetic sublevel of $F = 2$ of the ground state and the $m_F = +3$ magnetic sublevel of $F = 3$ in the $3P_{3/2}$ excited state. Circularly polarized dye laser light can be tuned within the 15-MHz natural linewidth of the upper level, and the $\Delta m = +1$ selection rule for circular polarization prevents excitation of the other magnetic sublevels.

Spontaneous decay from the upper state, which obeys no resonance condition, is also restricted in this example. The final state of spontaneous decay could, in principle, have $m_F = 4$, 3, 2. However, in sodium there are no $m_F = 4$ or 3 states below the $3P_{3/2}$ state, and only one $m_F = 2$ state, namely the one from which the excitation process began. Thus, the state of the sodium atom is very effectively constrained to this two-state subset out of the infinitely many quantum states of the atom.

II. Induced Atomic Coherence Effects

The ability of an atomic system to have a coherent dipole moment during an extended interaction with a radiation field is a necessary condition for a wide variety of effects associated with quantum optical resonance. The dipole moment should be coherent in the sense that it retains a stable phase relationship with the radiation field. Long-term phase memory may be difficult to achieve, for example because spontaneous emission and collisions destroy phase memory at a rate that is typically in the range 10^8 sec^{-1} or

much greater. Coherence is also lost if the radiation bandwidth is too broad. The importance of dipole coherence effects shows that light–matter interactions do depend most fundamentally on the dipole moment and electric field strength, not on radiation intensity and the B coefficient.

A. π Pulses and Pulse Area

The solution for the level probabilities given in Eq. (14) is the single most important example in quantum optics of the coherent response of an atom to a monochromatic radiation field. "Coherence" in this context has several connected meanings, all associated with the well-phased steady oscillation of the probabilities considered as a function of time. This time dependence was shown in Fig. 2 for several values of the parameters Δ and χ. The significance of the Rabi frequency χ is clear—it is the frequency at which the inversion oscillates when the atom and the radiation are at exact resonance, when $\Delta = 0$, since

$$w(t) = -\cos \chi t \qquad (22)$$

Fruitful connections to the spin vector formalism of magnetic resonance physics are obtained by regarding the triplet $[u, v, w]$ as a vector $\mathbf{S}$. Equations (19) for u, v, and w can then be written in compact vector form:

$$d\mathbf{S}/dt = \mathbf{Q} \times \mathbf{S}, \qquad (23)$$

where $\mathbf{Q}$ is the vector of length Ω with components $[-\chi, 0, \Delta]$. The vector $\mathbf{Q}$ can be called the torque vector for $\mathbf{S}$, which is variously called the pseudo-spin vector, the atomic coherence vector, and the optical Bloch vector.

This vector formulation in Eq. (23) of Eqs. (19) shows that the evolution of the two-level atom in the presence of radiation is simply a rotation in a three-dimensional space. The space is only mathematical in quantum optics because the components of the optical Bloch vector are not the components of a single real physical vector, whereas in magnetic resonance they are the components of a real magnetic moment. In both cases the nature of the torque equation leads to a useful conservation law: $d/dt(\mathbf{S} \cdot \mathbf{S}) = 0$. That is, the length of the Bloch vector is constant. In the u, v, w notation this means $u^2 + v^2 + w^2 = 1$, and it implies that the vector $[u, v, w]$ traces out a path on a unit sphere as the two-level atom changes its quantum state.

Further consideration of the on-resonant atoms ($\Delta \to 0$ and $\Omega \to \chi$) gives information about the interaction of atoms with (nonmonochromatic) pulsed fields. According to Eq. (19a) $u(t)$ can be neglected if $\Delta = 0$ and a new form of solutions to Eqs. (19b and c) follows immediately:

$$v(t) = -\sin \phi(t) \quad (24a)$$

$$w(t) = -\cos \phi(t) \quad (24b)$$

where $\phi(t)$ is called the "area" of the electric field because it is related to the time integral of the electric field envelope:

$$\phi(t) = \int_0^t \chi(t') \, dt' = (d/\hbar) \int_0^t \mathcal{E}_0(t') \, dt' \quad (25)$$

In the monochromatic limit when $\mathcal{E}_0 =$ constant, then $\phi(t) \rightarrow \chi t$.

Recall that the rotating wave approximation (RWA) will not permit $\mathcal{E}_0$ to vary too rapidly. If τ_p is the pulse length, then $\mathcal{E}_0$ is not too rapidly varying if τ_p is long enough, namely if $1/\tau_p \ll \omega$. Pulse lengths in the range 1 nsec $\geq \tau_p \geq 10$ fsec are of interest and are compatible with the RWA [1 fsec (femtosecond) $= 10^{-15}$ sec].

The significance of $\phi(t)$ is evident in Eq. (24). The term ϕ is just the angle of rotation of the on-resonance Bloch vector $\mathbf{S} = [u, v, w]$ during the passage of the light pulse. A light pulse with $\phi = \pi$ is called a π pulse, that is, a pulse that rotates the initial vector $[0, 0, -1]$, which points down, through 180° to the final vector $[0, 0, +1]$, which points up. Thus, a π pulse completely inverts the atomic probability, taking the atom from the ground state to the excited state. A 2π pulse is one that returns the atom via a 360° rotation of its Bloch vector to its initial state, after passing through the excited state. The remarkable nature of this rotation is not so much that an inversion of the atomic state is possible, but that it can be done fully coherently and without regard for the pulse shape. Only the total integral of $\mathcal{E}_0(t)$ is significant.

The physics behind Bloch vector rotation is essentially the same in magnetic resonance, but perhaps less remarkable since the Bloch vector in that case is a physically "real" magnetic moment. In optical resonance there is no "real" electric moment vector whose Cartesian components can be identified with $[u, v, w]$. Early evidence of the response of the optical Bloch vector to coherent pulses was obtained in experiments of Tang, Gibbs, Slusher, Brewer, and others (see Sections II,B and II,C), and further experiments probing these properties continue to be of interest.

B. Photon Echoes

Photon echoes are an example of spontaneous recovery of a physical property that has been dephased after many relaxation times have elapsed. In the case of echoes the physical property is the macroscopic polarization of a sample of two-level atoms. Recovery of the polarization means recovery of the ability to emit radiation, and the signature of a photon echo is the appearance of a burst of radiation from a long-quiescent sample of atoms. The burst occurs at a precisely predictable time, not randomly, and is due to a hidden long-term memory. The echo principle was discovered and spin echoes were observed by Hahn in 1950 in magnetic resonance experiments. Photon echoes were first observed by Hartmann and co-workers in 1965.

Photon echoes are possible when a sample of atoms is characterized by a broad distribution $g(\Delta)$ of detunings. This may occur in a gas, for example, because of Doppler broadening or in a solid because of crystalline inhomogeneities. For the latter reason it is said that $g(\Delta)$ indicates the presence of *inhomogeneous broadening*. The Maxwellian distribution of velocities in a gas is equivalent to a Maxwellian distribution of detunings since each atom's Doppler shift is proportional to its velocity. If the number of atoms in the sample is N, then the fraction with detuning Δ is given by $Ng(\Delta) \, d\Delta$, where

$$g(\Delta) = \frac{1}{\sqrt{(2\pi)}\delta\omega_D} e^{(-1/2)[(\Delta - \bar{\Delta})^2/(\delta\omega_D)^2]} \quad (26)$$

Here $\bar{\Delta}$ is the average detuning and $\delta\omega_D$ is the Doppler linewidth, $\delta\omega_D = \omega_L(kT/mc^2)^{1/2}$.

The Bloch vector picture is well suited for describing photon echoes. Assume that the Bloch vectors for a collection of N atoms all lie in the equatorial $(u - v)$ plane of the unit sphere along the negative v axis, that is, $\mathbf{S} = [0, -1, 0]$. This arrangement can be accomplished by excitation from the ground state $[0, 0, -1]$ with a strong $\pi/2$ pulse for which $\mathbf{Q} = [-\chi, 0, \Delta] \approx [-\chi, 0, 0]$ if $\chi \gg \Delta$. The total Bloch vector is then $\mathbf{S}_N = [U, V, W] = [0, -N, 0]$, which corresponds to a macroscopic dipole moment of magnitude Nd. After this excitation pulse the sample begins to radiate coherently at a rate appropriate to the dipole moment Nd. However, following the excitation pulse we again have $\chi = 0$ and $\mathbf{Q} = [0, 0, \Delta]$, and according to Eq. (23) the individual Bloch vectors immediately begin to precess freely about the w axis (in the $u - v$ plane) at rates depending on their individual detunings Δ. Specifically, $(u - iv)_t = (u - iv)_0 e^{-i\Delta t}$ for an atom with detuning Δ, if $\chi = 0$.

As a consequence, the total Bloch vector will rapidly shrink to zero in a time $\delta t \approx 1/\delta\omega$, where $\delta\omega$ is the spread in angular velocities in the N atom collection. That is, the coherent sum of N dipoles rapidly dephases and $[0, N, 0] \to [0, 0, 0]$, with the result that the sample quickly stops radiating. This is called *free precession decay*, or free induction decay after the similar effect in magnetic resonance, because the decay is due only to the fact that the individual dipole components $u - iv$ get out of phase with each other due to their different precession speeds, not because any individual dipole moment is decaying.

If the distribution of Δ's is determined by the Doppler effect, as in Eq. (26), then since $(u - iv)_0 = i$, one finds

$$U - iV = iN \int d\Delta \, g(\Delta) e^{-i\Delta t}$$
$$= iN e^{-i\Delta t} \exp[-\tfrac{1}{2}(\delta\omega_D^2)^2 t] \quad (27)$$

The decay is very rapid if the Doppler width is large. Typically $\delta\omega_D \approx 10^9$ to 10^{10} sec^{-1}, thus, within a few tenths of a nanosecond $U - iV \to 0$ and radiation ceases.

The echo method consists in applying a second pulse to the collection of atoms after U and V have vanished, that is, at some time $T \gg (\delta\omega_D)^{-1}$. Each single atom with its detuning Δ, after the time interval T, still has $(u - iv)_T = ie^{-i\Delta T}$. Only the *sum* of these u and v values is zero due to their different Δ values. The torque vector describing the second pulse is $\mathbf{Q} = [-\chi, 0, \Delta]$, which can again be approximated by $[-\chi, 0, 0]$ if $\chi \gg \Delta$. The effect of the second pulse is again to rotate the Bloch vectors about the u axis. The ideal second pulse is a π pulse, in which case $u \to u$, $v \to -v$ and $w \to -w$, that is, a rotation by 180°. Thus, for times $t \geq T$ after the π pulse the $u - v$ components are

$$(u - iv)_t = (u - iv)_{T'} e^{-i\Delta(t-T)}$$

where T' signifies the rotated coherence vector immediately following the π pulse at time T:

$$(u - iv)_{T'} = (u + iv)_T = -ie^{i\Delta T}$$

The remarkable feature of a π pulse is that it accomplishes an effective reversal of time. Following it the Bloch vectors do not continue to dephase, but begin to rephase:

$$(u - iv)_t = -ie^{i\Delta T} e^{-i\Delta(t-T)}$$
$$= -ie^{-i\Delta(t-2T)} \quad (28)$$

Thus, at the exact time $t = 2T$, the individual u's and v's all rephase perfectly: $[u, v, w] = [0, 1, 0]$. Their Bloch vectors are merely rotated 180° from their positions after the original $\pi/2$ pulse, and they again constitute a macroscopic dipole moment $\mathbf{S}_N = [0, N, 0]$, and therefore the collection will begin to radiate again.

Because of the timing of this radiation burst, exactly as long (T) after the π pulse as the π pulse was after the original $\pi/2$ pulse, it is natural to call the signal a *photon echo*. Because of the separation by the intervals T and $2T$ from the π and $\pi/2$ excitation pulses, the observation of an echo can be in practice an observation that is very noise free. Following the echo pulse, the coherence vectors again immediately begin to dephase, but they can again be rephased using the same method, and a sequence of echoes can be arranged.

Because of collisions with other atoms, the individual u and v values will actually get smaller during the course of an echo experiment, independent of their Δ values. Thus, the rephased Bloch vector is not quite as large as the original one. One of the possible uses of an echo experiment is to measure the rate of collisional decay of u and v, say as a function of gas pressure, by measuring the echo intensity in a sequence of experiments with different values of T since the echo intensities will get smaller as collisions reduce the length of the rephased Bloch vector. The way in which Eq. (23) is rewritten to account for collisions is taken up in Section II,D.

C. SELF-INDUCED TRANSPARENCY AND SHORT PULSE PROPAGATION

The polarization of a dielectric medium of two-level atoms, such as any atomic vapor excited near to resonance, is linearly related to the incident electric field strength $\mathscr{E}_0$ at low-light intensities but becomes nonlinear in the strong-field, short-pulse regime. This is most evident in Eq. (18), where the sine and cosine functions contain all powers of $\chi = 2d\mathscr{E}_0/\hbar$. These nonlinearities have striking consequences for optical pulse propagation.

If a 2π pulse is injected into a collection of on-resonant two level atoms, it cannot give any energy to them because after its passage the atoms have been dynamically forced back into their initial state. Thus, a 2π pulse has a certain energy stability and so does a 4π pulse and every $2n\pi$ pulse for the same reason. However, all other pulses are obviously not stable since they must give up some energy to the atoms if they do not rotate the atomic coherence vectors all the way back to their initial positions.

The effect on the injected pulse due to atomic absorptions is given by the Maxwell equation (21). When there is a broad distribution $g(\Delta)$ of detunings among the atoms, then Eq. (21) leads to a so-called area theorem, a nonlinear propagation equation for a pulse of area ϕ:

$$d\phi(z)/dz = -\tfrac{1}{2}N\sigma \sin \phi(z) \qquad (29)$$

Here $\phi(z)$ means the total pulse area $\int \chi(z, t')dt'$, where the integral extends over the duration of the pulse. The solution is $\tan \phi(z)/2 = e^{-\alpha z/2}$, and the attenuation coefficient is $\alpha = N\sigma$, where N is the density of atoms and σ is the inhomogeneous absorption cross section:

$$\sigma = \int g(\Delta')\sigma_a(\Delta')d\Delta' \qquad (30)$$

where $\sigma_a(\Delta')$ is the single-atom cross section (see Section II,E). For very weak pulses with $\phi \ll \pi$, one can replace $\sin \phi(z)$ by $\phi(z)$ and recover from Eq. (29) the usual linear law for pulse propagation: $d\phi(z)/dz = -(\alpha/2)\phi(z)$, which predicts exponential attenuation of the pulse: $\phi(z) = \phi(0) \exp[-\alpha z/2]$. The factor of $\tfrac{1}{2}$ arises because ϕ is proportional to the electric field amplitude, not the intensity.

Remarkably, one of the "magic" pulses with $\phi = 2\pi n$, which do not lose energy while propagating, also preserves its shape. This is the 2π pulse, for which there is a constant-shape solution of the reduced Maxwell–Bloch equations:

$$\chi(z, t) = (2/\tau_p)\mathrm{sech}[(t - z/V)/\tau_p] \qquad (31)$$

That is, the entire pulse moves at the constant velocity V, which can be several orders of magnitude slower than the normal light velocity in the medium. In ordinary light propagation this would correspond to an index of refraction $n \approx 1000$. All of these remarkable features were discovered by McCall and Hahn in the 1960s and labeled by the term *self-induced transparency* to indicate that a light pulse could manipulate the atoms in a dielectric in such a way that the atoms cannot absorb any of the light.

Self-induced transparency is an example of soliton behavior. The nonlinearity of the coupled Maxwell–Bloch equations opposes the dispersive character of normal light transmission in a polarizable medium to permit a steady nondispersing solitary wave (or soliton) [Eq. (31)] to propagate unchanged. This happens only for fields sufficiently strong that the 2π pulse condition can be met. The Maxwell–Bloch equations can in many cases be shown to be equivalent to the sine–Gordon soliton equation or generalizations of it.

D. RELAXATION

Both photon echoes and self-induced transparency demonstrate the existence of optical phenomena depending on $\chi \approx \mathscr{E}$, and not on $\mathscr{E}^2$, that is, on the Maxwell–Bloch equations and not on the Einstein rate equations. How are these two approaches to light–matter interactions connected? To answer this question it is necessary to extend the scope of the Bloch equations and include the effects of line-broadening and relaxation processes.

The upper levels of any system have a finite lifetime, and so $|a_2|^2$ cannot oscillate indefinitely as Eq. (14b) implies, but must relax to zero. This is most fundamentally due to the possibility of spontaneous emission of a photon, accompanied by a transition in the system to the lower level. Such transitions occur at the rate A, as in Einstein's equation (1b).

Other relaxation processes also occur. For example, collisions with other atoms cause unpredictable changes in the state of a given two-level atom. These collisional changes typically affect the dipole coherence of the two-level atom instead of the level probabilities, that is, they affect u and v instead of w. We suppose that the rate of such processes is γ. Although γ does not appear in Einstein's rate equations, its existence is implied. This will be clarified below.

The fundamental equations of optical resonance, Eqs. (19), can be rewritten to include these relaxations as follows:

$$du/dt = -\Delta v - (\gamma + A/2)u \qquad (32a)$$

$$dv/dt = +\Delta u + \chi w - (\gamma + A/2)v \qquad (32b)$$

$$dw/dt = -\chi v - A(w + 1) \qquad (32c)$$

In the absence of relaxation ($\gamma = A = 0$), the solutions of Eqs. (32) are purely oscillatory [recall Eq. (18)] and are said to be *coherent*. In the absence of the radiation field ($\chi = 0$), the on-resonance solutions are completely nonoscillatory:

$$u = u_0 e^{-(\gamma + A/2)t}$$

$$v = v_0 e^{-(\gamma + A/2)t}$$

$$w = -1 + (w_0 + 1)e^{-At} \qquad \text{all for } \chi = 0$$

These solutions are said to be *incoherent*. In each case "coherence" refers to the existence of oscillations with a well-defined period and phase. In Bloch's notation, the relaxation rates are written $\gamma + A/2 = 1/T_2$, where $A = 1/T_1$, and T_1 and T_2 are called the "longitudinal" and "transverse" rates of relaxation.

Relaxation theory is a part of statistical physics, and in quantum theory statistical properties of atoms and fields are usually discussed with the aid of the quantum mechanical density matrix ρ. The density matrix for a two-level atom has four elements, ρ_{11}, ρ_{12}, ρ_{21}, and ρ_{22}. These are related to u, v, and w by the equations $u = \rho_{12} + \rho_{21}$, $v = -i(\rho_{12} - \rho_{21})$, and $w = \rho_{22} - \rho_{11}$, which have the inverse forms:

$$\rho_{12} = \tfrac{1}{2}(u + iv) = \langle a_1 a_2^* \rangle \tag{33a}$$

$$\rho_{21} = \tfrac{1}{2}(u - iv) = \langle a_2 a_1^* \rangle \tag{33b}$$

$$\rho_{11} = \tfrac{1}{2}(1 - w) = \langle a_1 a_1^* \rangle \tag{33c}$$

$$\rho_{22} = \tfrac{1}{2}(1 + w) = \langle a_2 a_2^* \rangle \tag{33d}$$

Here the brackets $\langle \ldots \rangle$ are understood to refer to an average over an ensemble of parameters and variables inaccessible to direct and deterministic evaluation, such as the inital positions and velocities of all the atoms in a collection that may collide with and disturb a typical two-level atom.

The equations given in Eqs. (19) for u, v, and w can also be obtained from ρ via the Liouville equation of quantum statistical mechanics: $i\hbar \, d\rho/dt = [H, \rho]$. The equations for the density matrix elements (Eqs. (33)] are

$$d\rho_{21}/dt = -(\gamma + A/2 + i\Delta)\rho_{21}$$
$$- (i\chi/2)(\rho_{22} - \rho_{11}) \tag{34a}$$

$$d\rho_{12}/dt = -(\gamma + A/2 - i\Delta)\rho_{12}$$
$$+ (i\chi/2)(\rho_{22} - \rho_{11}) \tag{34b}$$

$$d\rho_{11}/dt = A\rho_{22} - (i\chi/2)(\rho_{12} - \rho_{21}) \tag{34c}$$

$$d\rho_{22}/dt = -A\rho_{22} + (i\chi/2)(\rho_{12} - \rho_{21}) \tag{34d}$$

We now demonstrate the connection between Einstein's equations and the quantum optical Eqs. (34). Consider the weak-field limit $\chi \ll |\gamma + A/2 + i\Delta|$. In this limit the rate of change of the "off-diagonal" density matrix elements ρ_{21} and ρ_{12} is dominated by the first factor $-(\gamma + A/2 \pm i\Delta)$, and both ρ_{21} and ρ_{12} decay rapidly. If in addition $A \ll |\gamma + A/2 \pm i\Delta|$, then ρ_{22} and ρ_{11} change relatively slowly, and so ρ_{21} and ρ_{12} rapidly adjust themselves to the small quasi-steady values:

$$\rho_{21} \approx -(i/2)\chi[\gamma + A/2 + i\Delta]^{-1}(\rho_{22} - \rho_{11}) \tag{35a}$$

$$\rho_{12} \approx (i/2)\chi[\gamma + A/2 - i\Delta]^{-1}(\rho_{22} - \rho_{11}) \tag{35b}$$

These solutions show that the off-diagonal elements of the density matrix can be determined from constant numerical factors and combinations of the diagonal density matrix elements. They can then be eliminated from Eqs. (34c and d).

This procedure is referred to as *adiabatic elimination* of off-diagonal coherence because the remaining equations for ρ_{11} and ρ_{22} no longer exhibit coherence. That is, they no longer have oscillatory solutions. The term *adiabatic* is appropriate in the sense that ρ_{21} and ρ_{12} are entrained by the slower ρ_{11} and ρ_{22}. The reverse procedure, the elimination of ρ_{11} and ρ_{22} in favor of ρ_{21} and ρ_{12}, is not possible because the reverse inequality $|\gamma + A/2 \pm i\Delta| \ll A$ is not possible.

These adiabatic off-diagonal solutions, once inserted into Eq. (34) lead to an equation for the slowly changing ρ_{22} as follows:

$$\frac{d\rho_{22}}{dt} = -A\rho_{22}$$
$$- \left[\frac{1}{2}\chi^2 \frac{\gamma + A/2}{\Delta^2 + (\gamma + A/2)^2} \right](\rho_{22} - \rho_{11}) \tag{36}$$

Recall $\rho_{22} = |a_2|^2$ is the probability that the atom is in its upper state, and thus plays the same role as P_n in the Einstein equation (2a). Similarly ρ_{11} plays the role here of P_m there. By comparing Eqs. (2a) and (36) one sees that they are identical in form and content if one identifies the coefficient of ρ_{11} in Eq. (36) with the coefficient of P_m in Eq. (2a). In other words, the density matrix equations [Eqs. (34a–d)] of quantum optics contain Einstein's equation in the weak field and adiabatic limits $\chi \ll |\gamma + A/2 \pm \Delta|$ and $A \ll |\gamma + A/2 \pm i\Delta|$. The B coefficient can be derived in this limit (see Section II,E) if one properly interprets the factor $\tfrac{1}{2}\chi^2(\gamma + A/2)/[\Delta^2 + (\gamma + A/2)^2]$.

Relaxation processes affect the Maxwell field as well as the Bloch variables. To determine the form of relaxation to assign to Maxwell equation (21) it is sufficient to consider the conservation of energy. From Eq. (21) and (19c) it follows that

$$[\partial/\partial z + \partial/\partial ct]|\mathcal{E}|^2 = -(N\hbar\omega/4\varepsilon_0 c)[\partial w/\partial t]$$

which is equivalent to an equation for photon flux and level probability

$$[\partial/\partial z + \partial/\partial ct]\Phi(z, t) = -N \, \partial P_2/\partial t \tag{37a}$$

since $\hbar\omega\Phi \equiv I = 2c\varepsilon_0|\mathcal{E}|^2$ and $w = 2P_2 - 1$.

Equation (37a) is Poynting's theorem for a "one-dimensional" medium. It expresses the conservation of photon flux in terms of atomic excitations. However, in the absence of the resonant two-level atoms ($N = 0$), Eqs. (37) predicts $\Phi(z, t) = \Phi_0(z - ct)$, which means that the photon flux has the constant value $\Phi(z_0)$ at every point $z = z_0 + ct$ that travels with the pulse. This is contradictory to ordinary experience in two respects. The medium that is host to the two-level atoms (e.g., other gas atoms, a solvent, or a crystal lattice) always causes both dispersion and absorption. They can be taken into account by modifying Eq. (37) slightly:

$$[\partial/\partial z + \kappa + \partial/\partial v_g t]\Phi(z, t) = -N \,\partial P_2/\partial t \quad (37b)$$

where v_g is the group velocity for light pulses in the medium and κ its linear attenuation coefficient.

This form of the flux equation implies a similar alteration of Maxwell's equation (21):

$$[\partial/\partial z + \kappa/2 + \partial/\partial v_g t]\mathcal{E} = i(Nd\omega/4\varepsilon_0 c)[u - iv] \quad (38)$$

This form of Maxwell's equation is useful in describing the elements of laser theory (see Section II,G).

E. Cross Section and the *B* Coefficient

How does one use quantum optical expressions to obtain basic spectroscopic formulas, such as for the absorption cross section and B coefficient? That is, given expressions derived from Eqs. (34), which are based on the Rabi frequency χ instead of the more familiar radiation intensity I or spectral energy density $u(\omega)$, how does one recover a cross section, for example? Consider the quantum optical derivation of the Einstein formula in Eq. (36). The transition rate (absorption rate or stimulated emission rate) can be identified readily. With the abbreviation $\beta = \gamma + A/2$, one obtains:

$$\text{abs. rate} = \frac{1}{2}\chi^2 \frac{\beta}{\Delta^2 + \beta^2} \quad (39a)$$

The absorption rate is a peaked function of Δ whose value drops to $\frac{1}{2}$ of the maximum value at $\Delta = \pm\beta$. Thus, β is called the *half width at half maximum* (HWHM) of the absorption lineshape. This shows that relaxation leads to *line broadening*, and since β applies equally and individually to every atom, it is an example of a *homogeneous linewidth*. Recall (Section I,B) that inhomogeneous broadening is not a characteristic of

individual atoms but of a collection of them. From expression (39a) at exact resonance one obtains the relationship:

reasonant transition rate $= \chi^2/$full linewidth

This is the single most concise relationship between the parameters of incoherent optical physics (transition rate and linewidth) and the central parameter of coherence (Rabi frequency). It holds in situations much more general than the present example and allows rapid and accurate translation of formulas from one domain to the other.

With the use of $\chi = 2d\mathcal{E}/\hbar$ and $I = 2c\varepsilon_0\mathcal{E}^2$, Eq. (39a) becomes

$$\text{abs. rate} = \frac{D^2 I}{3\varepsilon_0 c\hbar^2} \frac{\beta}{\Delta^2 + \beta^2} \quad (39b)$$

The introduction of the new dipole parameter D here is based on the assumption that all orientations of the atomic dipole matrix element $\mathbf{d}_{21}$ are possible (in case, e.g., all magnetic sublevels of the main levels 1 and 2 are degenerate). Then $d^2 \equiv |\mathbf{e} \cdot \mathbf{d}_{21}|^2$ must be replaced by its spherical average, that is, by $D^2/3$, where $D^2 = |\mathbf{d}_{21} \cdot \mathbf{d}_{12}|$. We assume this is appropriate in the remainder of Section II.

The atomic absorption cross section σ_a is, by definition, the ratio of the rate of energy absorption, $\hbar\omega_{21} \times$ (abs. rate), to the energy flux (intensity) I of the photons being absorbed. From Eq. (39b) this ratio is

$$\sigma_a(\omega; \omega_{21}) = \frac{D^2\omega}{3\varepsilon_0\hbar c} \frac{\beta}{(\omega_{21} - \omega)^2 + \beta^2} \quad (40)$$

where $\gamma = \beta - A/2$ is the specifically collisional contribution to the halfwidth. Formula (40) shows that a quantum optical approach to light absorption through the weak field limit of Eqs. (34) gives conventional results of atomic spectroscopy. If Doppler broadening is present, then Eq. (40) must be integrated over the Doppler distribution of detunings Eq. (26) as was done in Eq. (30).

Lineshape plays a key role in understanding the relationship of Eqs. (39) to the Einstein expression (1a) relating absorption rate to the B coefficient. The absorption cross section can be written $\sigma_a = \sigma_t S_a(\omega; \omega_{21})$, where σ_t is the total frequency-integrated cross section:

$$\sigma_t = \pi D^2\omega_{21}/3\varepsilon_0\hbar c \quad (41a)$$

and S_a is the atomic lineshape

$$S_a(\omega; \omega_{21}) = \frac{\beta/\pi}{(\omega_{21} - \omega)^2 + \beta^2} \quad (41b)$$

which is normalized according to $\int d\omega\, S_a = 1$, and in this case has a Lorentzian shape. A lineshape also exists for the radiation field and is expressed by $u(\omega)$, the spectral energy density function. One connects $u(\omega)$ with I by the frequency integral $c \int u(\omega')d\omega' = I$. In the monochromatic case $cu(\omega')$ takes the idealized singular form $cu(\omega') = I\delta(\omega - \omega')$, and Eq. (39b) is the result for the absorption rate.

In the general nonmonochomatic case, the expression for absorption rate involves the integrated overlap of $u(\omega)$ and the atomic lineshape function:

$$\text{abs. rate} = \frac{c\sigma_t}{\hbar\omega_{21}} \int S_a(\omega'; \omega_{21})u(\omega')d\omega' \quad (42)$$

This has the desired limiting form, involving $u(\omega)$, and not $I = c \int u(\omega)d\omega$, if the spectral width $\delta\omega_L$ of $u(\omega)$ is very broad in comparison to the width β of the absorption lineshape S_a. In this limit, which is implicit in Einstein's discussion, S_a acts in Eq. (42) like a δ-function peaked at $\omega' = \omega_{21}$, and $u(\omega)$ is evaluated at $\omega = \omega_{21}$. From Eq. (42) one can then extract the Einstein B coefficient:

$$B = \frac{\pi D^2}{3\varepsilon_0\hbar^2} \quad (43)$$

A simple relation obviously exists between B and σ_t, namely $\hbar\omega_{21}B = c\sigma_t$.

Another quantity of interest is $\sigma_a(0)$, the on-resonance or peak cross section:

$$\sigma_a(0) = \frac{D^2\omega_{21}}{3\varepsilon_0\hbar c\beta} \quad (44)$$

By definition, $\sigma_a(0) = \sigma_a(\omega = \omega_{21})$; or, conversely, $\sigma_a(\omega; \omega_{21}) = \pi\beta\sigma_a(0)S_a(\omega; \omega_{21})$. Representative values of $\sigma_a(0)$ for an optical resonance transition lie in the range $\sigma_a(0) \approx 10^{-13}$ to 10^{-17} cm^2 for absorption linewidths in the range $\beta \approx 10^8$ to 10^{11} sec^{-1}.

F. Strong Field Criterion and Saturation

The inequality $\chi \ll |\gamma + A/2 \pm i\Delta|$, on which the absorption rate formula (39) and thus the Einstein B coefficient is based, is important in quantum optics and radiation physics generally because it provides a criterion for distinguishing weak radiation fields from strong radiation fields. The inequality implies that there is a critical value $\mathcal{E}_{cr}$ for field strength that gives a universal meaning to the terms *weak field* and *strong field*, namely $\mathcal{E}_0 \ll \mathcal{E}_{cr}$ and $\mathcal{E}_0 \gg \mathcal{E}_{cr}$, respectively, where:

$$\mathcal{E}_{cr} = (\hbar/d)|\gamma + A/2 \pm i\Delta| = (\hbar/d)|\beta \pm i\Delta| \quad (45)$$

However, since the parameters γ, A, d, and Δ may vary by many orders of magnitude from case to case, the numerical value of $\mathcal{E}_{cr}$ may fall anywhere in an extremely wide range. Thus, it is possible that in one experiment a laser with the power level 10^{20} W m^{-2} must be designated "weak," while another laser in a different experiment with the power level 1 W m^{-2} must be considered "strong." This factor of 10^{20} is one indication of the great extent of the domain of quantum optics.

In conventional spectroscopy one sometimes encounters saturation effects. These are of course strongest in the strong field regime and are of interest in quantum optics.

There are two distinct time regimes of saturation phenomena. If $\chi \gg \beta$, there is a range of times $t \ll \delta T \approx \beta^{-1}$ that can still contain many Rabi oscillations since $\delta T \gg \chi^{-1}$. During the time $0 \leq t \ll \delta T$, the fully coherent undamped formula (14b) can be used for the upper state probability. On average, during this time the probability that the atom is in its upper level is

$$P_2 = \tfrac{1}{2}\chi^2/[\Delta^2 + \chi^2] \quad \text{(short time average)} \quad (46a)$$

which is a Lorentzian function of $\Delta = \omega_{21} - \omega$ with the power-broadening halfwidth $\delta\omega_p = \chi$. Power broadening is a saturation effect, because if $\chi \gg \Delta$, then $P_2 \to \frac{1}{2}$ on average, which is obviously saturated, that is, unchanged if χ is made still larger.

Another saturation regime exists for long times, $t \gg \delta T \approx \beta^{-1}$. The solution of Eqs. (24) for $P_2 = \rho_{22}$ in this limit is

$$P_2 = \frac{1}{2}\chi^2 \frac{\beta/A}{\Delta^2 + \beta^2 + \chi^2\beta/A} \quad (46b)$$

In this case χ begins to dominate the width when $\chi^2 > \beta A$, which defines the saturation value of χ:

$$\chi_{sat} = \sqrt{(\beta A)} \quad (47)$$

The power-broadening part of the width of Eq. (46b) is different than in Eq. (46a), namely $\delta\omega_p = \chi\sqrt{(\beta/A)}$. Depending on the value of $\gamma \equiv \beta - A/2$, the power width here can be anything between a minimum of $\chi/\sqrt{2}$, if $\gamma = 0$, and a maximum of $\chi\sqrt{(\gamma/A)}$, if $\gamma \gg A$. This distinction between the saturated power-broadened linewidth $\delta\omega_p$ predicted by Eq. (46a) for short times and by Eq. (46b) for asymptotically long times has caused some confusion in the past. Further study indicates that Eq. (46b) breaks down for sufficiently large χ, basically because Bloch-type relaxation, such as assumed in Eqs.

(32), becomes invalid. The first experimental reports of this regime of optical resonance were made by Brewer, DeVoe, Mossberg, and others in the early 1980s.

In the case of asymptotically long times the expression for P_2 can be written in several ways, using Eq. (47) or (40):

$$P_2 = \frac{\frac{1}{2}(\chi/\chi_{sat})^2}{1 + (\Delta/\chi_{sat})^2 + (\chi/\chi_{sat})^2}$$

$$= \frac{\sigma_a I}{\hbar\omega_{21} A + 2\sigma_a I}$$

$$= \frac{\Phi/\Phi_{sat}}{1 + 2\Phi/\Phi_{sat}} \qquad (48)$$

where $\Phi = I/\hbar\omega_{21}$, and we have introduced the saturation flux required for saturation of the transition $\Phi_{sat} = A/\sigma_a$ (or $= 1/T_1\sigma_a$ in Bloch's notation, which is more appropriate if there are other contributions than spontaneous emission rate A to the level lifetimes). Figure 3 shows the effect of both power broadening and saturation on the steady-state probability P_2.

In common with all two-level saturation formulas, Eq. (48) and Fig. 3 predict $P_2 = \frac{1}{2}$ at most. However, this prediction is valid only for weak fields or for long times. As the solutions in Eq. (14) and in Fig. 2 show, strong monochromatic resonance radiation can repeatedly transfer the electron to the upper level with $P_2 \approx 1$ for times $\ll \beta^{-1}$. Experiments that show $P_2 > \frac{1}{2}$ have been practical only with lasers. Laser pulses are both short and intense, allowing $\chi \gg \beta$ as well as $t \ll \beta^{-1}$.

G. SEMICLASSICAL LASER THEORY

The coupled Maxwell–Bloch equations can be used as the basis for laser theory. It is a semiclassical theory, but still adequate to illustrate the most important results, such as the roles of

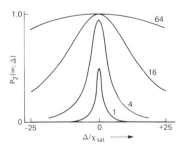

FIG. 3.   Population saturation for different values of χ/χ_{sat}.

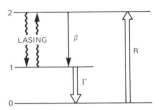

FIG. 4.   Schematic diagram of three-level laser system.

inversion and feedback, the existence of threshold, and the presence of frequency pulling in steady-state operation. A fully detailed semiclassical theory of laser operation was already given by Lamb in 1964.

The density matrix equations (24) must be modified to allow for pumping of the upper laser level and to allow the lower level to decay to a still lower level labeled 0 and not previously needed. The rates of these new processes are denoted R and Γ, respectively. This is basically the scheme of a so-called three-level laser, as sketched in Figure 4.

The diagonal equations change to

$$dp_{11}/dt = -\Gamma\rho_{11} + A\rho_{22}$$
$$+ (i/2)\chi_m(\rho_{12} - \rho_{21}) \qquad (49)$$

$$dp_{22}/dt = -A\rho_{22} - (i/2)\chi_m(\rho_{12} - \rho_{21}) + R \qquad (50)$$

Here the index m or χ_m shows that we are dealing with the electric field of the mth mode of the laser cavity. The laser is easy to operate only if $\Gamma \gg A$ because only then can a large positive inversion be maintained between the two levels with a value of R that is not too high and without seriously depleting the population of level 0. Under other conditions, the equation for the density matrix element ρ_{00} would also have to be included. The off-diagonal density matrix equations (24a and b) are basically unchanged if we interpret γ as including another contribution $\Gamma/2$, where Γ is the decay rate of the lower level probability to level 0 in Fig. 4.

The reduced Maxwell equation (28) is now useful. We will ignore the difference between v_g and c, and write Eq. (38) in terms of χ instead of $\mathscr{E}$, and use $\rho_{21} = \frac{1}{2}(u - iv)$ to find

$$[\partial/\partial z + \kappa/2 + \partial/\partial ct]\chi_m = i(Nd^2\omega/\varepsilon_0\hbar c)\rho_{21} \qquad (51)$$

Note that in a cavity κ can arise principally from mirror losses and not from absorption in the cavity volume. The same adiabatic elimination of dipole coherence undertaken in Eqs. (35) provides the value for ρ_{21} to insert into Eq. (51), which in steady state $(\partial \chi_m / \partial t = 0)$ becomes:

$$[\partial / \partial z + \kappa / 2 - \tfrac{1}{2}(g + i\delta k)]\chi_m = 0 \quad (52)$$

where

$$g + i\delta k = (ND^2\omega / 3\varepsilon_0 \hbar c)[w_{ss}/(\beta + i\Delta)] \quad (53)$$

and now $\beta = \gamma + (A + \Gamma)/2$. By using Eq. (40) one can obtain

$$g = N\sigma_a(\Delta)w_{ss} \quad (54a)$$

$$c\delta k = -(gc/\beta)(\omega_{21} - \omega) \quad (54b)$$

where Eq. (54a) has been used to simplify Eq. (54b). Here w_{ss} denotes the inversion $\rho_{22} - \rho_{11}$ in steady state.

It is clear that g is the intensity gain coefficient, since if $g > \kappa$, Eq. (52) would predict exponential growth, $|\chi_m|^2 \approx \exp[(g - \kappa)z]$, in an open-ended medium such as a laser amplifier. It is thus clear that $g = \kappa$ is the threshold condition for amplification or laser operation. Also obviously, g is not positive unless w_{ss} is positive. Recall that in an ordinary noninverted medium $w = -1$, and in this case $g = -N\sigma_a = -\alpha$, where α is the ordinary absorption coefficient.

Rather than growing indefinitely, the field in a laser cavity must conform to the spatial period determined by the mirrors. Thus, at steady state $\chi_m(z) \approx \chi_0 \exp[i\Delta k_m z]$, where the phase $\Delta k_m z$ is the difference between the actual phase of steady-state laser operation $k_m z$ and the phase $kz = \omega z / c$ that was assumed initially in defining the field carrier wave and envelope functions in Eq. (20). Since χ_m does not depend on the transverse coordinates x and y, this theory cannot describe transverse mode structure, and $k_m = m\pi / L$, where L is the cavity length and m is the longitudinal mode number. Operating values could be $m \approx 10^6$ and $L \approx 10$ cm, in which case the laser would run at a frequency near to $m\pi c / L \approx 3 \times 10^{15}$ sec$^{-1} \approx 5 \times 10^{14}$ Hz.

In laser operation $\partial / \partial z$ can be replaced in Eq. (52) by $i\Delta k_m$ and the imaginary part of Eq. (52) becomes $\Delta k_m = \tfrac{1}{2}\delta k$, which leads directly to a condition for the operating frequency ω:

$$\omega_m - \omega = -(\kappa c / 2\beta)(\omega_{21} - \omega) \quad (55)$$

This requires ω to lie somewhere between the empty cavity frequency $\omega_m = m\pi c / L$ and the natural transition frequency ω_{21} of the atom, and it is said that the two-level laser medium ''pulls'' the operating frequency away from the cavity frequency $m\pi c / L$. The solution of Eq. (55) for ω is:

$$\omega = [\beta \omega_m + \tfrac{1}{2}\kappa c \omega_{21}]/[\beta + \tfrac{1}{2}\kappa c] \quad (56)$$

H. OPTICAL BISTABILITY

The input–output relationships between light beams injected into and transmitted through an empty optical cavity are linear relationships. However, the situation changes dramatically if the cavity contains atoms. This is obvious in the case of laser action. However, even if the atoms are not pumped, their nonlinearities can be significant.

Consider a laser beam injected into an optical cavity filled with two-level atoms. For simplicity we consider the case where the frequencies of the atoms, the cavity, and the injected laser light are all equal: $\omega = \omega_c = \omega_{21}$. Only the $v-w$ Bloch equation are needed then:

$$dv/dt = -\beta v + (\chi_m + \chi_0)w \quad (57a)$$

$$dw/dt = -A(1 + w) - (\chi_m + \chi_0)v \quad (57b)$$

Here χ_0 and χ_m are the Rabi frequencies associated with the injected field strength and the cavity mode field generated by the atoms, and $\beta = \gamma + A/2$. The Maxwell equation for the internally generated χ_m is

$$(\kappa / 2 + \partial / \partial ct)\chi_m = (Nd^2\omega / 2\varepsilon_0 \hbar c)v \quad (58)$$

Note that there is no term $i\Delta k_m$ from $\partial / \partial z$, as there is in the discussion of laser operation, only because of the three-way resonance assumption.

The question is, how does the presence of the atoms in the cavity affect the transmitted signal? Since the transmitted signal differs only by a factor of mirror transmissivity from the total field in the cavity $\chi_t = \chi_m + \chi_0$, we ask for the relation between χ_0 and χ_t. The dynamical evolution of the system is complicated. Early attention was given to this situation in the 1970s by Szöke, Bonifacio, Lugiato, McCall, and others.

As with the laser, steady state is sufficiently interesting, so we put $dv/dt = dw/dt = d\chi/dt = 0$ and solve for χ_0 or χ_t. They obey a simple but nonlinear relation:

$$\chi_0 = \chi_t + (\alpha / \kappa)(\beta A)[\chi_t/(\beta A + \chi_t^2)] \quad (59)$$

where α is the (on-resonance) two-level medium's absorption coefficient, $\alpha = ND^2\omega / 3\varepsilon_0 \hbar c\beta$, and $\beta A = \chi_{sat}^2$, the saturation parameter identified in Eq. (47). If both χ_0 and χ_t are normalized with respect to χ_{sat} by defining $\xi = \chi_0/\chi_{sat}$ and $\eta = \chi_t/\chi_{sat}$ then one finds the dimensionless relation:

$$\xi = \eta + (\alpha / \kappa)[\eta/(1 + \eta^2)] \quad (60)$$

Of course the inverse relation $\eta = \eta(\xi)$, that is, the total field strength as a function of the input

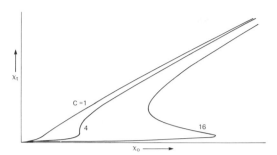

FIG. 5.   Bistable input–output curves for different values of $C = \alpha/2\kappa$.

field strength, is more interesting. Figure 5 shows the important features of this relation. It demonstrates that the total field η is double-valued as a function of input field ξ if α/κ is larger than a certain critical value. In this simple model the critical value is $\alpha/\kappa = 8$, and the vertical segment in the central curve is an indication of this. The double-valued nature of the curves for $\alpha/\kappa > 8$ is termed *optical bistability*. In the bistable region of the third curve hysteresis can occur, and a hysteresis loop is shown.

The elements of a primitive optical switch are evident in the bistable behavior shown here. If the input field is held near to the lower turning point, then a very small increase in ξ can lead to a very large jump in η, the transmitted field. The possibility of optical logic circuits and eventually optical computers is clearly suggested even by the simple model described here, and efforts being made around the world to realize these possibilities in a practical way have already achieved limited success.

III. Radiation Coherence and Statistics

A. COHERENCE OF LIGHT

In quantum optics the coherence of light is treated statistically. The need for a statistical description of light, whether quantized or classical, is practically universal since all light beams, even those from well-stabilized lasers, have certain residual random properties that are not uniquely determined by known parameters. These random properties lead to fluctuations of light. A satisfactory description can be based on a scalar electric field:

$$E(\mathbf{r},\, t) = E^{(+)}(\mathbf{r},\, t) + E^{(-)}(\mathbf{r},\, t) \qquad (61a)$$

where $E^{(+)}$ is the *positive frequency part* of E. That is, $E^{(+)}$ is the inverse Fourier transform of the positive frequency half of the Fourier transform of E. From this definition, one has $[E^{(-)}(\mathbf{r},\, t)]^* \equiv E^{(+)}(\mathbf{r},\, t)$. The split-up into positive and negative frequency parts $E^{(\pm)}(\mathbf{r},\, t)$ is motivated by the great significance of quasi-monochromatic fields, for which one can write

$$E^{(+)}(\mathbf{r},\, t) = \mathscr{E}(\mathbf{r},\, t)e^{-i\omega t} \qquad (61b)$$

$$E^{(-)}(\mathbf{r},\, t) = \mathscr{E}(\mathbf{r},\, t)^* e^{i\omega t} \qquad (61c)$$

The term *quasi-monochromatic* means that $\mathscr{E}(\mathbf{r},\, t)$ is only slowly time dependent, that is, $|d\mathscr{E}/dt| \ll \omega|\mathscr{E}|$.

The intensity of the light beam associated with this electric field is given by:

$$\begin{aligned} I(\mathbf{r},\, t) &= c\varepsilon_0 E(\mathbf{r},\, t)^2 \\ &= c\varepsilon_0[2|\mathscr{E}(\mathbf{r},\, t)|^2 + \mathscr{E}^2(\mathbf{r},\, t)e^{-2i\omega t} + \text{c.c.}] \end{aligned}$$
$$(62)$$

where c.c. means complex conjugate. In principle I is rapidly time dependent, but the factors $e^{\pm 2i\omega t}$ oscillate too rapidly to be observed by any realistic detector. That is, a photodetector can respond only to the average of Eq. (62) over a finite interval, say of length T beginning at t, where $T \gg 2\pi/\omega$. The $e^{\pm 2i\omega t}$ terms average to zero and one obtains:

$$\bar{I}_T(t) = (1/T)\int_0^T I(t + t')dt' = 2c\varepsilon_0|\mathscr{E}(t)|^2 \quad (63)$$

The overbar thus denotes a coarse-grained average value that is not sensitive to variations on the scale of a few optical periods. We have dropped the $\mathbf{r}$ dependence as a simplification. If the beam is steady and the averaging interval T is long enough, then both the length T and the beginning value $I(t)$ are unimportant, and $\bar{I}_T(t)$ is also independent of t. This is the property of a class of fields called *stationary*. Obviously even nonstationary fields can be considered stationary in the sense of a long T average, and we will adopt stationarity as a simplification of our discussion and write $\bar{I}_T(t)$ simply as $\bar{I}$.

Consider now the operation of a *Michelson interferometer* (Fig. 6). An incident beam is split into two beams a and b, and after traveling different path lengths, say ℓ and $\ell + \delta$, the beams are recombined and the beam intensity $I_{a+b} = c\varepsilon_0[E_a(t) + E_b(t)]^2$ is measured. If the beam splitter sends equal beams with field strength $E(t)$ and intensity $\bar{I}$ into each path and there is no absorption during the propagation, then one

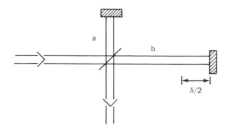

FIG. 6.   Sketch of Michelson interferometer.

measures

$$\bar{I}_{a+b} = (T)^{-1} \int_0^T c\varepsilon_0[E_a(t + t')$$

$$+ E_b(t + t')]^2 dt'$$

$$= (T)^{-1} \int_0^T c\varepsilon_0[E(t + t')$$

$$+ E(t + \delta/c + t')]^2 dt'$$

$$= \bar{I} + \bar{I} + 4c\varepsilon_0$$

$$\times \operatorname{Re}[\langle \varepsilon^*(t)\varepsilon(t + \delta/c)\rangle e^{-i\omega\delta/c}] \quad (64)$$

where Re means "real part" and the angular brackets indicate the time average. If $\delta = 0$, then $\bar{I}_{a+b} = 4\bar{I}$, because the two beams interfere fully constructively. The quantity $\Gamma(\delta/c) = \langle \varepsilon^*(t)\varepsilon(t + \delta/c)\rangle e^{-i\omega\delta/c}$ is called the *mutual coherence function* of the electric field, and the appearance of fringes at the output plane of the interferometer is due to the variation of Γ with δ. From the factor $e^{-i\omega\delta/c}$ it is clear that a fringe shift (a shift from one maximum of $\bar{I}_{a+b}$ to the next) corresponds to a shift of δ by $2\pi c/\omega = \lambda$.

The mutual coherence function is conveniently normalized to its maximum value which occurs when $\delta = 0$, and the normalized function $\gamma(\delta/c)$ is called the *complex degree of coherence*. That is,

$$\gamma(\delta/c) = \langle \varepsilon(t)\varepsilon^*(t + \delta/c)\rangle e^{+i\omega\delta/c}/\langle \varepsilon(t)\varepsilon^*(t)\rangle \quad (65)$$

and the output intensity can be written:

$$\bar{I}_{a+b} = 2\bar{I}[1 + \operatorname{Re} \gamma(\delta/c)] \quad (66)$$

where Re γ must satisfy $-1 \leq \operatorname{Re} \gamma \leq 1$.

It is common experience that if the path difference δ is made too great in the interferometer, the fringes are lost and the output intensity is simply the sum of the intensities in the two beams. This can be accounted for by introducing a *coherence time* τ for the light by writing

$$\gamma(\delta/c) = \gamma(0)e^{-|\delta|/c\tau}e^{+i\omega\delta/c} \quad (67)$$

This representation for γ has the correct behavior since it vanishes whenever δ is large enough,

specifically whenever $\delta \gg c\tau$. For obvious reasons, $c\tau$ is called the *coherence length* of the light. This does not explain the fundamental origin of the coherence time τ, but the lack of such a deep understanding of the light beam can be one reason that a statistical description is necessary in the first place. Typically one adopts Eq. (67) as a convenient empirical relation and interprets τ from it.

The fringe visibility is usually the important quantity that describes an interference pattern, not the absolute level of intensity. The *visibility* is defined by $V = (\bar{I}_{\max} - \bar{I}_{\min})/(\bar{I}_{\max} + \bar{I}_{\min})$, and this is directly related to the complex degree of coherence. Since

$$\operatorname{Re} \gamma = |\gamma(0)|e^{-|\delta|/c\tau} \cos(\omega\delta/c + \phi)$$

one has

$$V = \frac{4\bar{I}|\gamma(0)|e^{-|\delta|/c\tau}}{4\bar{I}} = |\gamma(\delta/c)| \quad (68)$$

Thus, the magnitude of the complex degree of coherence is a directly measurable quantity.

One of the foundations of quantum optics was established by Wolf and others in classical coherence theory when it became understood in the 1950s how to describe optical interference effects in terms of measurable autocorrelation functions such as γ. In a sense, the first example of this was provided much earlier by Wiener in 1930 when he showed that the spectrum $S(\omega)$ of a stationary light field is given essentially by the Fourier transform of γ, considered as a function of a time difference τ rather than a path difference δ:

$$S(\omega) = 2\bar{I} \operatorname{Re} \int d\tau' e^{-i\omega\tau'}\gamma(\tau') \quad (69)$$

As a specific example, suppose a light beam with carrier frequency ω_L has a Gaussian degree of coherence, with coherence time τ, that is, $\gamma(\tau') = |\gamma(0)|e^{i\omega_L\tau'} \exp[-\frac{1}{2}(\tau'/\tau)^2]$. This is another example, similar to the exponential $\gamma(\delta/c)$ given above, in which a simple analytic function is used to model the normal fact that correlation functions have finite coherence, that is, that $\gamma(\tau) \to 0$ as $\tau \to \infty$. In this example the spectrum is Gaussian:

$$S(\omega) = 2\bar{I}|\gamma(0)|\sqrt{(2\pi\tau^2)} \exp[-\frac{1}{2}(\omega - \omega_L)^2\tau^2]$$

$$(70)$$

The effective bandwidth is the frequency range over which $S(\omega)$ is an appreciable fraction of its peak value. In this case the spectrum is centered at $\omega = \omega_L$, and the bandwidth is given by $\Delta\omega = 2\pi\Delta\nu = 1/\tau$.

This is an example of the general rule that the bandwidth is the inverse of the coherence time. A laser beam with bandwidth $\Delta\nu = 100$ MHz has a coherence time $\tau = 1.6$ nsec and a coherence length $\Delta\delta = c\tau = 0.5$ m, while sunlight with a bandwidth six orders of magnitude broader ($\Delta\nu = 10^{14}$ Hz) has a coherence time $\tau = 1.6 \times 10^{-3}$ psec $= 1.6$ fsec and a coherence length $c\tau = 0.5$ μm that are six orders of magnitude smaller. In the latter case, $\Delta\nu \approx \nu$, and $c\tau \approx \lambda$, so sunlight cannot be called quasi-monochromatic in any sense.

A hierarchy of correlation functions can be defined for statistical fields. One denotes by the *degree of first-order coherence* the normalized first-order correlation of positive and negative frequency parts of the field:

$$g^{(1)}(\mathbf{x}_1, \mathbf{x}_2) = \frac{\langle E^{(+)}(\mathbf{x}_1)E^{(-)}(\mathbf{x}_2)\rangle}{[\langle|E^{(+)}(\mathbf{x}_1)|^2\rangle\langle|E^{(-)}(\mathbf{x}_2)|^2\rangle]^{1/2}} \quad (71)$$

where $\mathbf{x} = (\mathbf{r}, t)$ for short. In the case of a quasi-monochromatic field $g^{(1)}$ is just the same as γ defined above. The terms "first-order coherent," "partially coherent," and "incoherent" refer to light with $|g^{(1)}|$ equal to 1, between 1 and 0, and 0, respectively.

Definition (71) and other average values can be interpreted in an ensemble sense, as well as in a time average sense. In the ensemble interpretation the angular brackets $\langle \ldots \rangle$ are associated with an average over "realizations," that is, over possible values of the fields at given points $\mathbf{x}$, weighted by appropriate probabilities. Two important examples are (i) so-called chaotic fields, fields that are stationary complex Gaussian random processes, and (ii) constant intensity fields. The chaotic distribution is characteristic of ordinary (thermal) light such as omitted by the sun, flames, light bulbs, etc. The constant intensity distribution is an idealization associated with highly stabilized single-mode laser (coherent) light. The ergodic assumption that time and ensemble averages are equal is usually made. We will write $\langle I \rangle$ and $\bar{I}$ interchangeably.

If $p[\mathscr{E}, t]d^2\mathscr{E}$ is the probability that the complex field amplitude has the value $\mathscr{E}$ within $d^2\mathscr{E} = \mathscr{E}d|\mathscr{E}|d\phi$ in the complex $\mathscr{E}$ plane, then the thermal (chaotic) distribution for the complex amplitude $\mathscr{E}(t)$ is a Gaussian function:

$$p^{\text{th}}[\mathscr{E}] = (2\pi\mathscr{E}_0^2)^{-1/2}\exp[-\tfrac{1}{2}(|\mathscr{E}|^2/\mathscr{E}_0^2)] \quad \text{(thermal)}$$
$$(72a)$$

and the coherent distribution is a delta function:

$$p^{\text{coh}}[\mathscr{E}] = (|\mathscr{E}_0|/\pi)\delta^2(|\mathscr{E}|^2 - |\mathscr{E}_0|^2) \quad \text{(coherent)}$$
$$(72b)$$

The spatial and temporal coherence properties of radiation, measured via γ or $g^{(1)}$ or $S(\omega)$, determine the degree to which the fields at two points in space and time are able to interfere. By the use of pinholes, lenses and filters any sample of ordinary light can be made equally monochromatic and directional as laser light. Its spatial and temporal coherence properties can be made equal to those of any laser beam. If the laser light is then suitably attenuated so that it has the same !ow intensity as the ordinary light, one may ask whether any important differences can remain between the ordinary light and the laser light.

Surprisingly, the answer to this fundamental question is yes. The differences can be found in higher order correlation functions, involving $\mathscr{E}^*\mathscr{E}$ to powers higher than the first, such as $\langle E^2(t)E^2(t + \tau)\rangle$. These can be referred to as intensity correlations. The first measurements of optical intensity correlation functions were made on thermal light by Brown and Twiss in the 1950s before lasers were available.

Discussions of intensity correlations are typically made with the aid of a quantity called the *degree of second-order coherence* and denoted $g^{(2)}$. Higher order degrees of coherence are also defined. If the fields are stationary only the time delays between the measurement points are significant, and one has:

$$g^{(2)}(\tau_1) = \langle I(t)I(t + \tau_1)\rangle/\langle I\rangle^2 \quad (73a)$$
$$g^{(3)}(\tau_1, \tau_2) = \langle I(t)I(t + \tau_1)I(t + \tau_2)\rangle/\langle I\rangle^3 \quad (73b)$$

and so on. Because of stationarity we can put $t = 0$ everywhere. We will deal mostly now with $g^{(2)}$ and no confusion should occur if we omit the index (2) whenever we have a single delay time τ. The assumption that the light is stationary gives $g(\tau) = g(-\tau)$. Since I is an intrinsically positive quantity, it is clear that $g(\tau)$ is positive. There is no upper limit, so g satisfies $\infty > g(\tau) \geq 0$. In addition, for any distribution of I one has $\langle I^2\rangle \geq \langle I\rangle^2$, so obviously $g(0) \geq 1$; and one can also show that $g(0) \geq g(\tau)$.

With the aid of the thermal and coherent probability distributions given above, several higher order coherence functions can be calculated immediately. For example, if $\tau_1 = \tau_2 = \ldots = 0$, the nth order moments of the intensity are simply $\langle I^n\rangle \approx \int |\mathscr{E}|^{2n}p[\mathscr{E}]d^2\mathscr{E}$:

$$\langle I^n \rangle = n! \langle I \rangle^n = n!\bar{I}^n \quad \text{(thermal)} \quad (74a)$$

$$\langle I^n \rangle = \langle I \rangle^n = \bar{I}^n \quad \text{(coherent)} \quad (74b)$$

For large values of n these are obviously quite different, even if $\bar{I}$ is the same for two light beams, one thermal and the other coherent. The difference plays a role in multiphoton ionization experiments with $n = 2$ and larger. This is one example of a fundamental difference between ordinary light and ideal single-mode laser light.

It should be clear that in addition to thermal light and laser light there may be still other forms of light, characterized by distributions other than those in Eq. (72). Quantum theory also predicts that there can even be kinds of light beams for which the underlying probability distribution does not exist in a classical sense, for example, it may be negative over portions of the range of definition. Generalized phase space functions that play the quantum role of classical probability densities were developed by Glauber and Sudarshan in the 1960s. These are still a principal theoretical tool in studies of photon counting.

B. PHOTON COUNTING

In photon counting experiments the arrival of an individual photon is registered at a photodetector, which is essentially just a specially designed phototube that gives a signal when an arriving photon ionizes an atom at the phototube surface. A typical experiment consists of many runs of the same length T in each of which the number of photons registered by the photodetector is counted. The counts can be organized into a histogram (Fig. 7), which is interpreted as giving the probability $P_n(T)$ for counting n photons during an interval of length T.

An expression for $P_n(T)$ follows from a consideration of the photodetection process, which begins with the ionization of a single atom by a single photon at the surface of the phototube. In any event, the rate of counting is proportional to the intensity of the light beam, so one writes $\alpha I(t)dt$ for the probability of counting a photon at the time t in the interval dt. The factor α takes account of the atomic variables governing the ionization process as well as the geometry of the phototube. It was first shown by Mandel in the 1950s that $P_n(T)$ is then given by

$$P_n(T) = \langle e^{-\alpha \int_0^T dt' I(t')} [\alpha \int_0^T dt' I(t')]^n / n! \rangle \quad (75)$$

where the average is over the variations in intensity during the (relatively long) counting intervals. Alternatively, it can be considered an average over an ensemble of identically prepared runs in which the value of $\bar{I}$ is statistically distributed in some way.

The simplest example occurs if the light intensity does not fluctuate at all, which is characteristic of an ideal single-mode laser (coherent) light beam. In this case Eq. (75) is independent of the t' average and, with $\bar{n} = \alpha \bar{I} T$,

$$P_n = e^{-\bar{n}}(\bar{n})^n / n! \quad \text{(coherent)} \quad (76)$$

which is the well-known Poisson distribution. It is easily verified that $\bar{n} = \Sigma \, n P_n(T)$. That is, as its form indicates, $\bar{n}$ is the average number of photons counted in time T. It is a feature of the Poisson distribution that its dispersion is equal to its mean:

$$\langle (\Delta n)^2 \rangle = \langle n \rangle^2 2 - \langle n \rangle^2 = \bar{n} \quad \text{(Poisson)} \quad (77)$$

A plot of an ideal Poisson photocount distribution is shown in Fig. 7. The Poisson distribution is also called the coherent or coherent-state distribution because it is predicted by the quantum theory of light to be applicable to a radiation field in a so-called coherent state (see Section III,C). A well-stabilized single-mode laser gives the best realization of a coherent state in practice.

It should be obvious that the same Poisson law will be found even if $\langle I(t') \rangle$ is not constant, so long as T is made great enough that all fluctuations associated with a particular interval are averaged out. The counting fluctuations associated with steady $\langle I \rangle$ are due to the discrete single-photon character of the atomic ionization event that initiates the count and are called particle fluctuations.

Although the Poisson distribution is the result for $P_n(T)$ in the simplest case, constant $\langle I(t') \rangle$, it is not the correct result for ordinary thermal light unless T is very long, in fact $2\pi\Delta\nu T \gg 1$ is

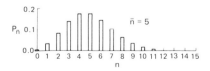

FIG. 7. Ideal Poisson photocount distribution. [Adapted with permission from Loudon, R. (1983). "The Quantum Theory of Light," 2nd ed. Oxford Univ. Press, Oxford. Copyright 1983 Oxford University Press.]

necessary. We have posed a question at the end of the last section about differences between laser light and thermal light with equal (very narrow) bandwidths. In that case $\Delta\nu$ is very small, so we cannot automatically assume $2\pi\Delta\nu T \gg 1$. In fact it illustrates the point to assume the reverse. If $2\pi\Delta\nu T \ll 1$, we can assume $\langle I(t')\rangle$ is constant over such a short time T. However, $\langle I(t')\rangle$ can still fluctuate with t', that is, from run to run. In order to evaluate the average over t', which is now essentially an average over runs, we use the thermal distribution Eq. (72a) for $\mathscr{E}$, which is equivalent to the normalized exponential distribution for I

$$p(I) = (1/\bar{I})e^{-I/\bar{I}} \qquad \text{(thermal)} \qquad (78)$$

since $I \approx |\mathscr{E}|^2$. Then Eq. (75) gives

$$P_n(T) = (\bar{n})^n/(1 + \bar{n})^{1+n} \qquad \text{(thermal)} \qquad (79)$$

which is variously called the thermal, chaotic, or Bose–Einstein distribution, and $\bar{n}$ is defined as before, $\bar{n} = \alpha\bar{I}T$, with the understanding that here $\bar{I}$ means the average over many runs, all shorter than $1/2\pi\Delta\nu$. The difference between the photocount distributions for thermal light under the two extreme conditions $2\pi\Delta\nu T \ll 1$ and $2\pi\Delta\nu T \gg 1$ is shown in Fig. 8. The nature of the difference between the thermal and coherent probability distributions (76) and (77), and thus of the fundamental difference between natural light and single-mode laser (coherent) light, can show up directly in the record of photocount measurements, as is clear by inspecting Figs. 7 and 8. Other photon count distributions than these two correspond to light that is somehow different from both laser light and thermal light. In quantum optics the most interesting examples are examples of purely quantum mechanical light beams.

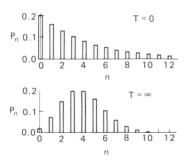

FIG. 8. Thermal photocount records for $T\Delta\nu \gg 1$ and $T\Delta\nu \ll 1$. [Adapted with permission from Loudon, R. (1983). "The Quantum Theory of Light," 2nd ed. Oxford Univ. Press, Oxford. Copyright 1983 Oxford University Press.]

C. QUANTUM MECHANICAL STATES OF LIGHT

The quantum theory of light assigns operators in a Hilbert space to the fields of electromagnetism. Any field $F(\mathbf{r})$, quantum or classical, can be written as a sum of plane waves (as a three-dimensional Fourier series or integral):

$$\begin{aligned} F(\mathbf{r}) &= \sum_k f_k \exp[i\mathbf{k} \cdot \mathbf{r}] \\ &= \sum_k \{f_k \exp[i\mathbf{k} \cdot \mathbf{r}] \\ &\quad + f_{-k} \exp[-i\mathbf{k} \cdot \mathbf{r}]\} \qquad k \gtreqqless 0 \quad (80) \end{aligned}$$

and $f_{-k} = f_k^*$ if $F(\mathbf{r})$ is a real function. If F is operator-valued real (i.e., hermitean), then the expansion coefficients f_k are operators and one writes $f_{-k} = f_k^\dagger$, where the dagger denotes hermitean adjoint in the usual way. In the case of electromagnetic radiation a given plane wave mode has the transverse electric field

$$\mathbf{E}_k(\mathbf{r}) = \hat{\mathbf{e}}N_k[a_k e^{i\mathbf{k}\cdot\mathbf{r}} + a_k^\dagger e^{-i\mathbf{k}\cdot\mathbf{r}}] \qquad (81)$$

where N_k is an appropriate normalization constant. The full field is obtained by summing over k: $\mathbf{E}(\mathbf{r}) = \sum_k \mathbf{E}_k(\mathbf{r})$.

Because a_k and $a_k^\dagger$ are operators rather than numbers, $a_k^\dagger a_k \neq a_k a_k^\dagger$. This is expressed by saying that they obey the "canonical" commutation relations:

$$[a_k, a_k^\dagger] = 1 \qquad (82)$$

where the square bracket means the difference $a_k a_k^\dagger - a_k^\dagger a_k$, as is usual in quantum mechanics. All other commutators among a's and $a^\dagger$'s for this mode, and all commutators of a_k or $a_k^\dagger$ with operators of all other modes vanish, for example, $[a_k, a_\ell] = 0$. The time dependences of these operators, and thus of the electric and magnetic fields, are determined dynamically from Maxwell's equations, which remain exactly valid in quantum theory. In the absence of charges and currents the time dependences are

$$a_k(t) = a_k e^{-i\omega_k t} \quad \text{and} \quad a_k^\dagger = a_k^\dagger e^{i\omega_k t} \quad (83)$$

where $\omega_k = kc$.

The hermitean operator $a^\dagger a$ (we drop the mode index k temporarily) has integer eigenvalues and is called the photon number operator:

$$a^\dagger a|n\rangle = n|n\rangle, \qquad n = 0, 1, 2, \dots \quad (84)$$

and the eigenstate $|n\rangle$ is called a Fock state or *photon number state*. The operators $a^\dagger$ and a are called the photon *creation operator* and *destruc-*

tion operator because of their effect on photon number states:

$$a^\dagger|n\rangle = \sqrt{(n + 1)}|n + 1\rangle, \quad \text{and}$$

$$a|n\rangle = \sqrt{n}|n - 1\rangle \tag{85}$$

That is, $a^\dagger$ and a respectively increase and decrease by 1 the number of photons in the field state. The photon number states are mutually orthonormal: $\langle m|n\rangle = \delta_{mn}$, and they form a complete set for the mode, and all other states can be expressed in terms of them. They have very attractive simple properties. For example, the photon number is exactly determined, that is, the dispersion of the photon number operator $a^\dagger a$ is zero in the state $|n\rangle$ for any n:

$$\langle(\Delta n)^2\rangle = \langle n|(a^\dagger a)^2|n\rangle - \langle n|\,a^\dagger a|\rangle^2$$
$$= n^2 - n^2 = 0.$$

However, the nice properties of the $|n\rangle$ states are in some respects not well suited to the most common situations in quantum optics. For example, loosely speaking, their phase is completely undefined because their photon number is exactly determined. As a consequence, the expected value of the mode field is exactly zero in a photon number state. That is $\langle n|\mathbf{E}(\mathbf{r})|n\rangle = 0$, because Eq. (85) and the orthogonality property together give $\langle n|a|n\rangle = \langle n|a^\dagger|n\rangle = 0$.

In most laser fields there are so many photons per mode that it is difficult to imagine an experiment to count the exact number of them. Thus, a different kind of state, called a *coherent state*, is usually more appropriate to describe laser fields than the photon number states. These states $|\alpha\rangle$ are right eigenstates of the photon destruction operator a:

$$a|\alpha\rangle = \alpha|\alpha\rangle \tag{86}$$

They are normalized so that $\langle\alpha|\alpha\rangle = 1$. Since a is not hermitean, its eigenvalue α is generally complex. The expected number of photons in the mode in a coherent state is $\langle n\rangle = \langle a^\dagger a\rangle = \langle\alpha|a^\dagger a|\alpha\rangle = |\alpha|^2$. Thus, α can be interpreted loosely as $\sqrt{\langle n\rangle}$, the square root of the mean number of photons in the mode. The number of photons in the mode is not exactly determined in the state $|\alpha\rangle$. This can be seen by computing the dispersion in the number:

$$\langle(\Delta n)^2\rangle = \langle\alpha|(a^\dagger a)^2|\alpha\rangle - \langle\alpha|a^\dagger a|\alpha\rangle^2$$
$$= \langle\alpha|a^\dagger(a^\dagger a - 1)a|\alpha\rangle - \langle\alpha|a^\dagger a|\rangle^2$$
$$= \langle\alpha|a^\dagger a|\alpha\rangle = |\alpha|^2 = \langle n\rangle \tag{87}$$

Thus, the dispersion is equal to the mean number, a property already noticed for the Poisson distribution (77). If $\langle n\rangle \gg 1$, then the relative dispersion $\langle(\Delta n)^2\rangle/\langle n\rangle^2$ is very small, and the number of photons in the state is well determined in a relative sense. At the same time, a relatively well-defined phase can also be associated with the state.

The clearest interpretation of the amplitude and phase associated with a coherent state $|\alpha\rangle$ is obtained by computing the expected value of $\mathbf{E}$ in the state $|\alpha\rangle$. Since $\langle\alpha|a^\dagger|\alpha\rangle = \langle\alpha|a|\alpha\rangle^* = \alpha^*$, one easily determines that

$$N_k = \sqrt{(\hbar\omega_k/2\varepsilon_0 V)} \tag{88}$$

where V is the mode volume, and then N_k can be interpreted loosely as the electric field amplitude associated with one photon. This shows that $N_k\alpha = \sqrt{(\hbar\omega_k/2\varepsilon_0 V)}\alpha$ is the mean amplitude of the expected electric field, or what is sometimes called the classical field amplitude. The phase of α thus determines the phase of the field described by the state $|\alpha\rangle$.

The coherent states $|\alpha\rangle$ can be expressed in terms of the photon number states:

$$|\alpha\rangle = \exp[-|\alpha|^2/2]\sum_m\left[\frac{a^m}{\sqrt{m!}}\right]|m\rangle \tag{89}$$

where the sum runs over all integers $m \geq 0$. From Eq. (89) one can compute the probability $p_n(\alpha)$ that a given coherent state $|\alpha\rangle$ contains exactly n photons. According to the principles of quantum theory this is given by $|\langle n|\alpha\rangle|^2$, and we find:

$$p_n(\alpha) = e^{-\langle n\rangle}\langle n\rangle^n/n! \tag{90}$$

which is exactly the "purely random" Poisson probability distribution shown in Fig. 7. This is interpreted as meaning that photon-counting measurements of a radiation field in the coherent state $|\alpha\rangle$ would find exactly n photons with probability (90).

The photon creation and destruction operators are not hermitean and are generally considered not observable, but their real and imaginary parts (essentially the electric and magnetic field strengths, or in other terms, the in-phase and quadrature components of the optical signal) are in principle observable. Thus, one can introduce the definitions

$$a = \tfrac{1}{2}(a_1 - ia_2), \quad a^\dagger = \tfrac{1}{2}(a_1 + ia_2) \tag{91}$$

and their inverses

$$a_1 = a + a^\dagger, \quad a_2 = i(a - a^\dagger) \tag{92}$$

What are the quantum limitations on measurement of the hermitean operators a_1 and a_2? They must, of course, obey the Heisenberg *uncertainty relation* $\langle(\Delta a_1)^2\rangle\langle(\Delta a_2)^2\rangle \geq |\frac{1}{2}\langle[a_1, a_2]\rangle|^2$, where $\langle...\rangle$ indicates expectation value in any given quantum state, and $\Delta a_1 \equiv a_1 - \langle a_1\rangle$. Since $[a, a^\dagger] = 1$, it follows from Eq. (91) that $[a_1, a_2] = -2i$, so the Heisenberg relation for a_1 and a_2 is:

$$\langle(\Delta a_1)^2\rangle\langle(\Delta a_2)^2\rangle \geq 1 \qquad (93)$$

A coherent state $|\alpha\rangle$ can be shown to produce the minimum simultaneous uncertainty in a_1 and a_2. That is, $\langle\alpha|(\Delta a_1)^2|\alpha\rangle = 1$ and $\langle\alpha|(\Delta a_2)^2|\alpha\rangle = 1$. Both a_1 and a_2 are therefore said to reach the *quantum limit* of uncertainty in a coherent state.

However, it is only the product of the operator dispersions that is constrained by the Heisenberg uncertainty relation, and there is no fundamental reason why either a_1 and a_2 could not have a dispersion equal to $\mu \ll 1$, so long as the other had a dispersion at least as large as $1/\mu \gg 1$. A quantum state of the radiation field that permits one of the components of the destruction operator to have a dispersion smaller than the quantum limit is said to be a *squeezed state*. Squeezed states could in principle provide the ability to make ultraprecise measurements such as are projected for gravity wave detection. A squeezed state of radiation was first generated and measured by Slusher and others in 1985.

IV. Quantum Interactions and Correlations

It should be remembered that the highly successful semiclassical version of the quantum theory of light (Sections I and II) does not ignore quantum principles, or put $\hbar \rightarrow 0$, but it does ignore quantum fluctuations and correlations. It is a theory of coupled quantum expectation values.

In the following sections a number of phenomena depending directly on quantum fluctuations and correlations are described. None of them have been successfully treated by the semiclassical theory or by any other theory than the generally accepted and fully quantized version of the quantum theory of light. For this reason the observation of these and similar quantum optical effects can offer a means of testing the accepted theory.

Such tests are of great interest for two related reasons. Because the quantum theory of light (or quantum electrodynamics) is the most carefully studied quantum theory, and because it serves as a fundamental guide to all field theories, it plays a key role in our present understanding of quantum principles and should be tested as rigorously as possible. Moreover, tests of the effects described below play a special role because they bear on the theory in a different way compared with traditional tests, such as high precision measurements of the Lamb shift, the fine structure constant, the Rydberg, and the electron's anomalous moment.

A. Fully Quantized Interactions

The electric field is mainly responsible for optical interactions of light with matter, and the magnetic field plays a subsidiary role, becoming significant only in situations involving magnetic moments or relativistic velocities. The most important light–matter interaction is the direct coupling of electric dipoles to the radiation field through the interaction energy $-\mathbf{d} \cdot \mathbf{E}$.

A systematic description of the fully quantized interactions of quantum optics begins with the total energy of the atom H_A and the radiation H_R and their energy of interaction:

$$H = H_A + H_R - \mathbf{d} \cdot \mathbf{E} \qquad (94)$$

In the quantum theory the atomic, radiation, and interaction energies are given by the Hamiltonian operators

$$H_A = \sum_j E_j|j\rangle\langle j| \qquad (95a)$$

$$H_R = \sum_k \hbar\omega_k a_k^\dagger a_k \qquad (95b)$$

$$-\mathbf{d} \cdot \mathbf{E} = -\sum_i \sum_j \sum_k \hbar f_{ij}^k\{a_k^\dagger + a_k\}|i\rangle\langle j| \qquad (95c)$$

Here E_j and $|j\rangle$ are the quantized energies and eigenstates of the atom including level-shifting and level-splitting due to static fields that give rise to Zeeman and Stark effects, etc. The dipole coupling constant is $\hbar f_{ij}^k = N_k\hat{\mathbf{e}} \cdot \mathbf{d}_{ij}$, in the notation of Eqs. (9) and (39). Also, $a_k^\dagger$ and a_k are the photon creation and destruction operators introduced in Section III,C.

The two-level version of this Hamiltonian is the most used in quantum optics. It is obtained by restricting the sums over i and j to the values 1 and 2. It is not necessary that $|1\rangle$ and $|2\rangle$ be the two lowest energy levels. In photoionization, which is the physical process underlying the operation of photon counters, the upper state is not even a discrete state but lies in the continuum of energies above the ionization threshold (see Section IV,B). In its two-level version H becomes

$$H = E_1|1\rangle\langle 1| + E_2|2\rangle\langle 2| + \sum_k \hbar\omega_k a_k^\dagger a_k$$
$$- \sum_k \hbar f_{12}^k \{a_k^\dagger + a_k\}\{\sigma + \sigma^\dagger\} \quad (96)$$

where $\sigma = |1\rangle\langle 2|$ and $\sigma^\dagger = |2\rangle\langle 1|$, and f_{12}^k has been taken to be real for simplicity. Note that σ has the effect of a lowering transition when it acts on the two-level atomic state $|\Psi\rangle = C_1|1\rangle + C_2|2\rangle$. That is, $\sigma|\Psi\rangle = C_2|1\rangle$, so σ takes the amplitude C_2 of the upper state $|2\rangle$ and assigns it to the lower state $|1\rangle$. By the same argument $\sigma^\dagger$ causes a raising transition.

The term $a_k^\dagger \sigma^\dagger$ in Eq. (96) is difficult to interpret because it has $\sigma^\dagger$ raising the atom into its upper state together with $a_k^\dagger$ creating a photon. One expects photon creation to be associated only with a lowering of the atomic state. The term $a_k \sigma$ presents similar difficulties. It can be shown, however, that these two terms are the source of the very rapid oscillations $\exp[\pm 2i\omega t]$ which were discussed above Eq. (12) and were eliminated by the rotating wave approximation (RWA). The adoption of the RWA eliminates them here also. With the RWA and the convenient convention that $E_1 = 0$ (and therefore that $E_2 = \hbar\omega_{21}$), the working two-level Hamiltonian is

$$H = \tfrac{1}{2}\hbar\omega_{21}(\sigma_z + 1) + \sum_k \hbar\omega_k a_k a_k^\dagger$$
$$- \sum_k \hbar f_{12}^k \{a_k^\dagger \sigma + \sigma^\dagger a_k\} \quad (97)$$

The operators σ, $\sigma^\dagger$, and σ_z are closely related to the 2×2 Pauli spin matrices:

$$|1\rangle\langle 2| = \sigma \rightarrow \tfrac{1}{2}(\sigma_x - i\sigma_y) = \begin{bmatrix} 0 & 0 \\ 1 & 0 \end{bmatrix} \quad (98a)$$

$$|2\rangle\langle 1| = \sigma^\dagger \rightarrow \tfrac{1}{2}(\sigma_x - i\sigma_y) = \begin{bmatrix} 0 & 1 \\ 0 & 0 \end{bmatrix} \quad (98b)$$

$$|2\rangle\langle 2| - |1\rangle\langle 1| \rightarrow \sigma_z = \begin{bmatrix} 1 & 0 \\ 0 & -1 \end{bmatrix} \quad (98c)$$

One can easily confirm that the matrix representation of σ is a "lowering" operator in the two-dimensional space with basis vectors

$$|2\rangle \rightarrow \begin{bmatrix} 1 \\ 0 \end{bmatrix} \quad \text{and} \quad |1\rangle \rightarrow \begin{bmatrix} 0 \\ 1 \end{bmatrix} \quad (98d)$$

and $\sigma^\dagger$ is represented by the corresponding "raising" operator. The σ's obey the commutator relations:

$$[\sigma, \sigma_z] = 2\sigma, \qquad [\sigma^\dagger, \sigma_z] = -2\sigma^\dagger,$$
$$[\sigma^\dagger, \sigma] = \sigma_z \quad (99)$$

Changes in the radiation field occur as a result of emission and absorption of photons during transitions in the atom. One consequence is that a_k obeys an equation obtained from the Heisenberg equation $i\hbar \partial O/\partial t = [O, H]$ that is valid for all operators O:

$$\partial a_k/\partial t = -i\omega_k a_k + if_k\sigma \quad (100)$$

Here we have simplified the notation, rewriting f_{12}^k as f_k. The solution of Eq. (100) is:

$$a_k(t) = a_k(0)e^{-i\omega_k t} + if_k \int_0^t dt' \, e^{-i\omega_k(t-t')}\sigma(t') \quad (101)$$

The first term represents photons that are present in the mode k but not associated with the two-level atoms (e.g., from a distant laser), and the second term represents the photons associated with a transition in the two-level atom.

The σ operators also change in time in the course of emission and absorption processes. Their time dependence is also determined by Heisenberg's equation and one finds the equations:

$$d\sigma/dt = -i\omega_{21}\sigma - i\sum_k f_k a_k \sigma_z \quad (102a)$$

$$d\sigma^\dagger/dt = i\omega_{21}\sigma^\dagger + i\sum_k f_k a_k^\dagger \sigma_z \quad (102b)$$

$$d\sigma_z/dt = 2i\sum_k f_k(\sigma^\dagger a_k - a_k^\dagger \sigma) \quad (102c)$$

These are the operator equations underlying the semiclassical equations for the Bloch vector components given in Eq. (19).

The correspondence between the quantum and semiclassical sets of variables is

$$\sigma \rightarrow \tfrac{1}{2}(u - iv)e^{-i\omega t} \quad (103a)$$
$$\sigma^\dagger \rightarrow \tfrac{1}{2}(u + iv)e^{i\omega t} \quad (103b)$$
$$\sigma_z \rightarrow w \quad (103c)$$

This identification is precise if the radiation field is both intense and classical enough. This means that one retains only one mode and replaces a_k and $a_k^\dagger$ by their coherent state expectation values $\alpha = \alpha_0 e^{-i\phi}$ and $\alpha_0^* e^{i\phi}$. Then Eqs. (102) are identical with Eq. (19) under the previous assumptions, that is, the field is quasi-monochromatic so $\phi = \omega t$, and $N_k\alpha$ is the slowly varying electric field expectation value, which is interpreted as the classical field amplitude.

Thus, one has

$$2f_k\alpha_0 \to 2d^c\mathscr{E}_0/\hbar = \chi \tag{104}$$

The part of a_k that depends on σ in Eq. (100) acts as a radiation reaction field when substituted into Eq. (102). It causes damping and a small frequency shift in the atomic operator equations even if there are no external photons present and thus is associated with spontaneous emission. The damping constant is exactly the correct Einstein A coefficient for the transition, because the A coefficient is a two-level parameter.

The frequency shift is only a primitive two-level version of a more general many level radiative correction such as the Lamb shift. One observes here a natural limitation of any two-level model. It is intrinsically incapable of dealing with any precision with effects, such as radiative level shifts, that depend strongly on the contributions of many levels. However, in the cases of interest described in the remainder of Section IV, the effects of other levels are negligible and the numerical value of the frequency shift is irrelevant. It can be assumed to be included in the definition of ω_{21}.

B. QUANTUM LIGHT DETECTION AND STATISTICS

The quantum theory of light detection is based on the quantum theory of photoionization because photon counters are triggered by an ionizing absorption of a photon. Photoionization is a weak field phenomenon because the effective γ is so large [recall Eq. (45)]. Thus, perturbative methods are adequate and one computes the absolute square of the ionization matrix element, in this case given by $\langle F| -e\mathbf{r} \cdot \mathbf{E}(\mathbf{r})|I\rangle$, where $|I\rangle$ and $|F\rangle$ are the initial and final states of the photoionization. The initial state consists of an atom in its ground state, described by the electronic orbital function $\phi_0(\mathbf{r})$, and the initial state of the radiation field $|\Psi\rangle$. The final state consists of the atom in an ionized state described by an electronic orbital function $\phi_f(\mathbf{r})$ appropriate to a free electron with energy the ionization threshold, and another state of the radiation field $|\Psi'\rangle$. Then the matrix element becomes

$$\langle F| -e\mathbf{r} \cdot \mathbf{E}|I\rangle$$
$$= -\int \phi_f^*(\mathbf{r})e\mathbf{r} \cdot \langle \Psi'| \mathbf{E}(\mathbf{r})|\Psi\rangle\phi_0(\mathbf{r})d^3r$$
$$= -\mathbf{d}_{f0} \cdot \Sigma_k\hat{\mathbf{e}}_k N_k\langle\Psi'| a_k |\Psi\rangle \tag{105}$$

where $\mathbf{d}_{f0}$ is the so-called dipole matrix element for the $0 \to f$ transition in the atom. We have

also made the "dipole" approximation, in which $\exp[i\mathbf{k} \cdot \mathbf{r}] \approx \exp[i0] \approx 1$ over the entire effective range of the matrix element integral [recall the discussion following Eq. (9)].

Only the part of $\mathbf{E}$ that lowers the photon number of the field, namely the "a" part, is effective in ionization, essentially because ionization is a photon absorption process. In addition we can take a single k value if the incident light is monochromatic. Thus, the ionization rate depends on $f_k^2|\langle\Psi'|a_k|\Psi\rangle|^2$, where $f_k = |\mathbf{d}_{f0} \cdot \hat{\mathbf{e}}_k|N_k$. The actual final states of the atom and of the field are never completely observed, and all the unobserved features must be allowed for, that is, included by summation:

$$\text{rate} \approx f_k^2 \sum_{\Psi'} \langle\Psi| a_k^\dagger |\Psi'\rangle\langle\Psi'| a_k |\Psi\rangle$$
$$\approx f_k^2\langle\Psi| a_k^\dagger a_k |\Psi\rangle \approx \langle\Psi| a_k^\dagger a_k |\Psi\rangle$$

since $\Sigma_{\Psi'} |\Psi'\rangle\langle\Psi'| = 1$ for a complete set of final states. In this expression for the ionization rate we write "$\approx$" instead of "$=$" because we are really interested here only in the effects of field quantization on the rate, not the exact numerical value of the rate. If the radiation field is quantum mechanical we do not know perfectly the properties of the incident light, and these properties must be averaged. This average over the properties of the incident light is the same average discussed from a classical standpoint in Sections III,A and III,B. Thus, we finally have

$$\text{rate} \approx \langle a_k^\dagger a_k\rangle \tag{106}$$

where the angular brackets now mean an average over the initial field, that is, the quantum mechanical expectation value in state $|\Psi\rangle$.

The significance of Eq. (106) is in the ordering of the field operators. The nature of the photoionization process mandates that they be in the given order and not the reverse. Since $a^\dagger a$ is not equal to $aa^\dagger$ for a quantum field, the order makes a difference. The order given, in which the destruction operator is to the right, is called *normal order*. Photoionization is a normally ordered process by its nature, and therefore so is photodetection and photon counting. This has fundamental consequences for quantum statistical measurements, as we now explain.

Let us consider the degree of second-order coherence $g^{(2)}$ in quantum theory. This was written in Eq. (73a) as an intensity correlation: $g^{(2)}(\tau) = \langle I(t)I(t + \tau)\rangle/\langle I\rangle^2$. Because of the normally ordered character of photoionization, if $g^{(2)}$ is measured with photodetectors as usual, its correct definition according to the quantum theory of ionization is normally ordered:

$$g^{(2)}(\tau)$$

$$= \frac{\langle E^{(-)}(t)E^{(-)}(t+\tau)E^{(+)}(t+\tau)E^{(+)}(t)\rangle}{\langle E^{(-)}E^{(+)}\rangle\langle E^{(-)}E^{(+)}\rangle} \quad (107)$$

If photon fields were really classical, and these quantum mechanical fine points were unnecessary, then the ordering would make no difference, since the fields $E^{(\pm)}$ would be numbers, not operators, and the original expression for $g^{(2)}$ would be recovered. However, we now exhibit the effects of these quantum differences in a few specific cases.

In the case of a single-mode field, there are no time dependences and $g^{(2)}(\tau) = g^{(2)}(0)$ simplifies to

$$g^{(2)}(0) = \langle a^\dagger a^\dagger a a\rangle/\langle a^\dagger a\rangle^2 \quad (108)$$

which we evaluate in Table II. Among these examples the Fock state is special because its $g^{(2)}$ violates the condition $g^{(2)}(0) \geq 1$, which is one of the classical inequalities given below Eqs. (73). The Fock state is therefore an example of a state of the radiation field for which the quantum and classical theories make strikingly different predictions. It has not yet been possible to study a pure Fock state of more than one photon in the laboratory.

Photon bunching is a term that refers to the fact that photon beams exist in which photons are counted with statistical fluctuations greater than would be expected on the basis of purely random (that is, Poisson) statistics. In fact, almost any ordinary beam (thermal light) will have this property, and this is reflected in that $g^{(2)} > 1$ for thermal light. Photon bunching therefore arises from the Bose–Einstein distribution [Eq. (39)]. A coherent state with its Poisson statistics is purely random and does not exhibit bunching.

A qualitative classical explanation of photon bunching is sometimes made by saying that light from any natural source arises from broadband multimode photon emission by many independent atoms. There are naturally random periods of constructive and destructive interference among the modes, giving rise to large intensity "spikes," or "bunches" of photons, in the light beam. Unbunched light comes from a coherently regulated collection of atoms, such as from a well-stabilized single-mode laser. From this point of view, unbunched coherent light is optimally ordered.

However, *photon antibunching* can also occur. There are "antibunched" light beams, in which photons arrive with lower statistical fluctuations than predicted from a purely coherent beam with Poisson statistics. Antibunched light beams have values of $g^{(2)} < 1$, in common with a pure Fock state beam, and are therefore automatically nonclassical light beams.

The first observation of an antibunched beam with $g^{(2)} < 1$ was accomplished by Mandel and others in 1977 in an experiment with two-level atoms undergoing resonance fluorescence. Antibunching occurs in such light for a very simple reason. A two-level atom "regulates" the occurrence of pairs of emitted photons very severely, even more so than the photons are regulated in a single-mode laser. A second fluorescent photon cannot be emitted by the same two-level atom until it has been re-excited to its upper level by the absorption of a photon from the main radiation mode. Thus, a high Rabi frequency χ permits the degree of second-order coherence $g^{(2)}(\tau) = \langle a^\dagger a^\dagger(\tau)a(\tau)a\rangle/\langle a^\dagger a\rangle^2$ to be nonzero after a relatively short value of the time delay τ, but $g^{(2)}$ is strictly zero for $\tau = 0$. A graph showing the experimental observation is given in Fig. 9.

The significance of photon statistics and photon counting techniques in quantum optics and in physics is clear. They permit a direct examination of some of the fundamental distinctions between the quantum mechanical and classical concepts of radiation.

C. Superradiance

Nonclassical photon counting statistics arise from the multiphoton correlations inherent in specific states of the light field. Similarly, multi-atom quantum correlations can give rise to unusual behavior by systems of radiating atoms, as was pointed out by Dicke in 1954. The most dramatic behavior of this kind is called *Dicke superradiance* or *superfluorescence*.

Multi-atom correlations can exist in N-atom systems even if N is as small as $N = 2$. A pair of two-level atoms labeled a and b can have quantum states made of linear combinations of the

TABLE II. Second-Order Degree of Coherence for Single-Mode Quantum Mechanical Fields in Different States

Quantum field state	Value of $g^{(2)}(0)$	
Vacuum state $|0\rangle$	0	
Fock state $|n\rangle$	0	if $n < 2$
	$1 - 1/n$	if $n > 2$
Coherent state $|\alpha\rangle$	1	
Thermal state	2	

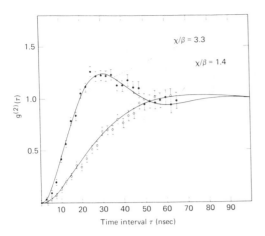

elementary two-atom states

$$|a+\rangle * |b+\rangle \tag{109a}$$

$$|a+\rangle * |b-\rangle \tag{109b}$$

$$|a-\rangle * |b+\rangle \tag{109c}$$

$$|a-\rangle * |b-\rangle \tag{109d}$$

Here the sign $*$ indicates a tensor product of the vector spaces of the two atoms, and $+$ and $-$ designate the upper and lower energy levels, with energies E_2 and E_1, in each of the identical atoms. The two middle states Eq. (109b and c) are degenerate, with total energy $E_1 + E_2 = E_2 + E_1$. It is useful to define the "singlet" and "triplet" states $\|S\rangle$ and $\|T\rangle$ that are linear combinations of the degenerate states as follows:

$$\|S\rangle = (1/\sqrt{2})\{|a\rangle * |b-\rangle - |a-\rangle * |b+\rangle\} \tag{110a}$$

$$\|T\rangle = (1/\sqrt{2})\{|a\rangle * |b-\rangle + |a-\rangle * |b+\rangle\} \tag{110b}$$

in analogy to singlet and triplet combinations of two spin $-\frac{1}{2}$ states.

The two-atom interaction with the radiation field is through $\mathbf{d} \cdot \mathbf{E}$ as in the one-atom case [recall Eq. (95)], and here both atoms contribute a dipole moment:

$$\hat{\mathbf{e}} \cdot \mathbf{d} = \hat{\mathbf{e}} \cdot [\mathbf{d}(a) + \mathbf{d}(b)]$$

$$= \hat{\mathbf{e}} \cdot (\mathbf{d}_{21})_a\{|a+\rangle\langle a-| + |a-\rangle\langle a+|\}$$

$$+ \hat{\mathbf{e}} \cdot (\mathbf{d}_{21})_b\{|b+\rangle\langle b-| + |b-\rangle\langle b+|\}$$

$$= d\{|a+\rangle\langle a-| + |a-\rangle\langle a+|$$

$$+ |b+\rangle\langle b-| + |b-\rangle\langle b+|\} \tag{111}$$

where we have taken equal matrix elements: $\hat{\mathbf{e}} \cdot (\mathbf{d}_{21})_a = \hat{\mathbf{e}} \cdot (\mathbf{d}_{21})_b = d$.

The main features of superradiance lie in the fact that the two-atom dipole interaction causes transitions between the various states of the system at different rates, and the reason for the difference is the existence of greater internal two-atom coherence in the case of the triplet state. Suppose that the two-atom system is fully excited into the state $\|+2\rangle = |a+\rangle * |b+\rangle$, which has energy $2E_2$, and then emits one photon. The system must drop into a state with energy $E_2 + E_1$. From this state it can decay further by emission of a second photon to the ground state $\|-2\rangle = |a-\rangle * |b-\rangle$ which has energy $2E_1$.

According to the Fermi Golden Rule, the rate of these transitions depends on the square of the interaction matrix element between initial and final states, $\langle F|\mathbf{d} \cdot \mathbf{E}|I\rangle$, summed over all possible final states. We consider the second transition, so there is only one possible final atomic state, namely $\|-2\rangle$. The matrix elements factor into an atomic part and a radiation part. We can use Eq. (81) in the dipole approximation to write

$$\langle F|\mathbf{d} \cdot \mathbf{E}|I\rangle$$

$$= \sum_k \langle F_A|\mathbf{d} \cdot \hat{\mathbf{e}}_k|I_A\rangle\langle F_R|N_k(a_k + a_k^\dagger)|I_R\rangle$$

Since superradiance deals only with spontaneous emission, the initial radiation state is the empty or vacuum state. Thus, the field contribution to the matrix element comes just from $a_k^\dagger$ and is the same in all cases and not interesting.

The various possible atomic matrix elements are

$$\langle -2\|\mathbf{d} \cdot \hat{\mathbf{e}}_k|a+\rangle * |b-\rangle = d \tag{112a}$$

$$\langle -2\|\mathbf{d} \cdot \hat{\mathbf{e}}_k|a-\rangle * |b+\rangle = d \tag{112b}$$

$$\langle -2\|\mathbf{d} \cdot \hat{\mathbf{e}}_k\|0, S\rangle = 0 \tag{112c}$$

$$\langle -2\|\mathbf{d} \cdot \hat{\mathbf{e}}_k\|0, T\rangle = (\sqrt{2})d \tag{112d}$$

so their squares have the relative values $1 : 1 : 0 : 2$. That is, the triplet initial state can radiate twice as strongly as either of the original degenerate states, and the singlet state cannot radiate at all. Both the triplet and singlet states are said to be two-body "cooperative" states because they cannot be factored into one-atom states.

It is tempting to interpret these results by saying that the triplet state radiates more rapidly because it has a larger dipole moment than the others and that the singlet state has no dipole moment. Such an interpretation is in the spirit of semiclassical radiation theory, as described in Section I,C, where the expectation values of quantum operators are treated as if they were classical variables. This interpretation has a number of useful features, but must also contain serious flaws, because the observation that it is based on is not true. A calculation of the dipole expectation value shows $\langle \Psi \| \mathbf{d} \| \Psi \rangle = 0$, where $\| \Psi \rangle$ can be any of the two-atom states above, including the rapidly radiating triplet state.

A state with a large dipole moment expectation does exist, namely the factored state

$$\| \Psi_d \rangle = \tfrac{1}{2} \{ |a+\rangle + |a-\rangle \} * \{ |b+\rangle + |b-\rangle \}$$

(113a)

This state is actually an eigenstate of the total dipole operator:

$\hat{\mathbf{e}} \cdot \mathbf{d} \| \Psi_d \rangle$

$$= [\hat{\mathbf{e}} \cdot (\mathbf{d}_{21})_a \{ |a+\rangle\langle a-| + |a-\rangle\langle a+| \}$$

$$+ \hat{\mathbf{e}} \cdot (\mathbf{d}_{21})_b \{ |b+\rangle\langle b-| + |b-\rangle\langle b+| \}] \| \Psi \rangle$$

$$= 2d \| \Psi \rangle$$

(113b)

so one has $\langle \Psi_d \| \mathbf{d} \cdot \hat{\mathbf{e}} \| \Psi_d \rangle = 2d$. This state is also predicted to radiate strongly.

The extrapolation of these predictions to an N-atom system leads to the prediction of a very large N-atom emission intensity I_N. Related predictions are that the N-atom cooperative process begins with a relatively slow buildup. After a delay of average length $\delta\tau_N$, an ultrashort burst of radiation of duration τ_N occurs. In the ideal case one predicts

$$I_N \approx N^2 \hbar \omega_{21} / \tau_1$$

(114a)

$$\tau_N \approx \tau_1 / N$$

(114b)

$$\delta\tau_N \approx \tau_1 \ln(N)$$

(114c)

where τ_1 is the single-atom radiative lifetime: $1/\tau_1 \equiv A$.

If one imagines even small collections of atoms with $N \approx 10^{12}$, then N^2 is impressively very much bigger than N and the term superradiance is indeed apt. If $A \approx 10^8 \text{ sec}^{-1}$ is taken as typical for 2 eV optical transitions, then $N \approx 10^{12}$ suggests that 2×10^{12} eV of energy can be expected at the rate 10^{20} sec^{-1} for a purely spontaneous power output of $2 \times 10^{32} \text{ eV sec}^{-1} \approx 3 \times 10^{11}$ W.

Other aspects of superradiance are equally interesting on fundamental grounds. For example, which of the two large dipole states, $\| T \rangle$ or $\| \Psi_d \rangle$, is actually responsible for superradiance? They both predict $I_N \approx N^2 \hbar \omega_{21}$, but their correlation properties are completely different, in somewhat the same way that a photon number state $|n\rangle$ and a coherent state $|\alpha\rangle$ have very different correlation properties even if they predict the same mode energy $|\alpha|^2 \hbar\omega = n\hbar\omega$. Consider only the fluctuations in $\mathbf{d}$ itself for the two states.

If one calculates the expectation of the dispersion of $\Delta \mathbf{d}^2 \equiv \langle [\mathbf{d} \cdot \hat{\mathbf{e}} - \langle \mathbf{d} \cdot \hat{\mathbf{e}} \rangle]^2 \rangle$, one finds:

$$\langle \Psi_d \| \Delta \mathbf{d}^2 \| \Psi_d \rangle = 0$$

(115)

$$\langle T \| \Delta \mathbf{d}^2 \| T \rangle = d^2$$

(116)

One can infer that radiation from the state $\| T \rangle$ can be expected to exhibit strong fluctuations of a kind completely absent in radiation from the state $\| \Psi_d \rangle$.

The fluctuations predicted from the state $\| T \rangle$ are consistent with the fact that it is exactly the state connected directly to the initial fully excited state $\| +2 \rangle$ by the total dipole operator $\mathbf{d}$. That is, $\hat{\mathbf{e}} \cdot \mathbf{d} \| +2 \rangle = (\sqrt{2})d \| T \rangle$. The fluctuations can be associated with the quantum uncertainty in the emission time of the first photon. Such fluctuations will influence all subsequent evolution, and if $N \gg 1$, they can be regarded as an example of a *macroscopic quantum fluctuation*, that is, a fluctuation with quantum mechanical origins that achieves direct macroscopic observability.

For a period of years, superradiance was a controversial and unobserved phenomenon. The intense and highly directional light beam predicted for the effect suggests that each emitted photon contributes to a spontaneous radiation field, which helps to stimulate the emission of further photons. Such a self-reactive process would provide a feedback analogous to that provided by mirror reflections in a laser cavity. It has been suggested that the physical origins of superradiance and laser emission are in fact the same thing. Important differences exist, however. During laser action, the dipole coherence of an individual atom is interrupted by collisions extremely frequently. The incoherent adiabatic solution for ρ_{21} is a quite satisfactory element of all laser theories (recall Section II,G). By contrast, in ideal Dicke superradiance all N-atom dipole coherence is fully preserved during the entire radiation process.

The experimental observation of superradiance was first achieved in the 1970s by Feld, Haroche, Gibbs, and others. Agreement has been found with the correlated state predictions, particularly with the statistical nature of the delay time fluctuations, and there is no longer any controversy over its existence. However, important questions about quantum and propagation effects on the spatial coherence properties of superradiance remain open.

D. Two-Level Single-Mode Interaction

We have emphasized that beginning with Einstein's reconsideration of Planck's radiation law, the most fundamental interacting system in quantum optics is a single two-level atom coupled to a single mode of the radiation field. This interaction was described semiclassically in Section II,A, and the quantum mechanical origin of the semiclassical equations was explained in Section IV,A. Fully quantum mechanical studies of the quantum coherence properties of this simplest interacting system were initiated by Jaynes in the 1950s. Some of the differences between general quantum and semiclassical theories have been clarified in this context.

When only one mode is significant the Hamiltonian [Eq. (97)] can be reduced to:

$$H_{JC} = \tfrac{1}{2}\hbar\omega_{21}(\sigma_z + 1) + \hbar\omega a^\dagger a$$
$$+ \hbar\lambda(a^\dagger\sigma + \sigma^\dagger a) \qquad (117)$$

Here $\lambda = f_{12}^k$, which is assumed real for simplicity. Remarkably, the effective Hamiltonian (Eq. (117)] for such a truncated version of quantum electrodynamics (the so-called Jaynes–Cummings model) has a number of important properties, and experimental studies by Haroche and Walther and others were begun in the early 1980s. Exact expressions are known for the eigenvalues and eigenvectors of H_{JC}. With $\hbar = 1$ for simplicity and $\Delta = \omega_{21} - \omega$, the eigenvalues are

$$E_{n,\pm} = E_1 + n\omega + \tfrac{1}{2}[\Delta \pm \Omega_n] \qquad (118a)$$

where we have re-inserted $E_1 \neq 0$ for the lower level energy, and the corresponding eigenvectors are given by

$$\|n, +\rangle = \cos\Phi\,|n - 1\rangle * |+\rangle$$
$$+ \sin\Phi\,|n\rangle * |-\rangle \qquad (118b)$$
$$\|n, -\rangle = -\sin\Phi\,|n - 1\rangle * |+\rangle$$
$$+ \cos\Phi\,|n\rangle * |-\rangle \qquad (118c)$$

where $\cos\Phi = \sqrt{(\Omega_n + \Delta)/2\Omega_n}$ and $\sin\Phi = \sqrt{(\Omega_n - \Delta)/2\Omega_n}$. Here the use of Ω_n for $\sqrt{4\lambda^2 n + \delta^2}$ is a deliberate reminder of Ω defined following Eqs. (12) because they play the same roles in their respective quantum and semiclassical theories. Similarly, one writes $\chi(n)$ as a reminder of χ for the same reason, that is, $\chi(n) = 2\lambda\sqrt{n}$ is the QED equivalent of the Rabi frequency $2d\mathscr{E}_0/\hbar$ [and not to be confused with the χ_m of Eqs. (49) and (50)].

One of the observable quantities of the Jaynes–Cummings model is the atomic energy. Its expectation value can be calculated exactly, without approximation:

$$\langle\sigma_z(t)\rangle = -\Sigma_n\, p_n \cos\chi(n)t \qquad \text{(quantum)}$$
$$(119a)$$

Here p_n is the probability that the single mode has exactly n photons. This result can be contrasted with Eq. (18c) in the same limit $\Delta = 0$ and under the same circumstances, namely, with the field intensity $I \approx \chi^2$ distributed with some probability $p(I)$ over a range of values:

$$w(t) = -\int p(I) \cos\chi(I)t\, dI \qquad \text{(semiclassical)}$$
$$(119b)$$

The apparent similarity of these results disguises fundamental differences in dynamical behavior between Eqs. (119a and b). These differences arise from the discreteness of the allowed photon numbers in Eq. (119a), which are discrete precisely because the field is quantized, that is, because the mode contains only whole photons and never fractional units of the energy $\hbar\omega$. By contrast, in the semiclassical theory (recall Section I,C) the intensity I and Rabi frequency χ can have any values.

For the quantum photon distribution associated with a coherent state, for which p_n is given by Eq. (90), a plot of $\langle\sigma_z(t)\rangle$ is shown in Fig. 10. The nearly immediate disappearance or "collapse" of the signal shortly after $t = 0$ can be explained on the basis of the interference of many frequencies $\chi(n)$ in the quantum sum of Eq. (119a). Such a collapse can be expected for any broad distribution p_n and would be predicted by the semiclassical Eq. (119b) as well, if $p(I)$ is a broad distribution function. However, the predicted reappearances or "revivals" of the signal are a sign that the field is quantized. They occur, and at regular intervals, only because p_n is a discrete distribution. The semiclassical expression (119b) leads inevitably to an irreversible collapse. Only quantum theory can provide the step-wise discontinuous photon number distribution that is the basis for the revivals.

The revivals and other quantum mechanical predictions implied by the truncated Hamiltonian [Eq. (117)] are of interest because this Hamiltonian is simple enough to permit exact calculations, without further approximations of the kind familiar in most of radiation theory. For

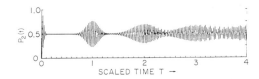

FIG. 10. Quantum collapse and revival of atomic inversion.

example, the expression for quantum inversion in Eq. (119a) has the following unusual properties:

(a) it is not restricted to any finite range of t values;

(b) it holds for all values of the coupling constant λ, which is contained in $\chi(n) = 2\lambda\sqrt{n}$;

(c) it is completely free of decorrelations, such as the commonly used approximation $\langle a^\dagger a\sigma_z\rangle \approx \langle a^\dagger a\rangle\langle\sigma_z\rangle$;

(d) it is finite even at exact resonance ($\omega_{21} = \omega$) without the aid of ad hoc complex energies;

(e) it is fully quantum mechanical with nontrivial commutators preserved: $[a, a^\dagger] = 1$, $[\sigma, \sigma^\dagger] = 2\sigma_z$, etc.; and

(f) it is realistically nonlinear (it saturates because the atomic energy cannot exceed E_2).

This combination of properties is unique in atomic radiation theory. They indicate, for example, that a system obeying Hamiltonian [Eq. (117)] would permit some fundamental questions in quantum electrodynamics to be studied independently of the restrictions of the usual perturbation methods that are based on short-time expansions and a small coupling constant. Experimental realization of the model is unlikely in the optical frequency range because of the restriction to a single radiation mode. However, quantum optical techniques, including the detection of single photons, are rapidly being extended to much lower frequencies, where single-mode cavities can be built. Observations of the Jaynes–Cummings model is expected to play a guiding role in microwave single-mode experiments with Rydberg atoms.

The energy spectrum of the Jaynes–Cummings Hamiltonian makes it clear how this can be done. In Fig. 11 the RWA energy spectrum is shown in the absence of a strong resonant interaction (i.e., with $\lambda = 0$), and also with $\lambda \neq 0$. The spectrum shows that the state $|n\rangle * |1\rangle$, which corresponds to the atom in its lower level and n photons in the mode when $\lambda = 0$ is pushed down to become the state $\|n, -\rangle$ when $\lambda \neq 0$. That is, for Eq. (118a) we obtain

$$E_{n,-} = E_1 + n\omega - \tfrac{1}{2}[\Omega_n - \Delta] \qquad (120)$$

Since $\Omega_n \gtrsim \Delta$, the lower level is pushed down. Similarly, the corresponding upper state $|n\rangle * |2\rangle$ is pushed up by the same amount. This shift is called the *AC Stark shift* because it is due to the interaction with an oscillating (alternating) electric field. The size of the AC Stark shift δ_{AC} varies as a function of Δ in the range

$$\tfrac{1}{2}\chi_n \geq \delta_{AC} \geq \chi_n^2/4\Delta \qquad (121)$$

depending on whether the atom and the field mode are near to or far from resonance.

An external probe of the coupled two-level plus single-mode system can reveal these details of its spectrum. For example, the two nearly degenerate states $\|n, +\rangle$ and $\|n, -\rangle$ are split by twice the AC Stark shift. This splitting can be observed by absorption spectroscopy if a weak second radiation field is allowed to induce transitions to a third level in the atom. This was first described and observed in 1955 by Autler and Townes.

A different kind of probe is provided by *resonance fluorescence*, that is, by spontaneous emission into modes other than the main mode.

E. AC STARK EFFECT AND RESONANCE FLUORESCENCE

Just as the Bloch vector provides a powerful descriptive framework for a wide variety of quantum optical phenomena, so does the Jaynes–Cummings model. The Hamiltonian [Eq. (117)] can be regarded as a zero-order approximation to a "true" Hamiltonian, in which the atom is allowed more than two levels or the field has more than one mode. It is an unusual zero-order approximation because it includes the interaction Hamiltonian as well as the noninteracting atomic and radiation Hamiltonians.

If the atom interacts resonantly with a single strong mode of the field, then its interactions with other modes, perhaps involving other levels of the atom, can be treated approximately.

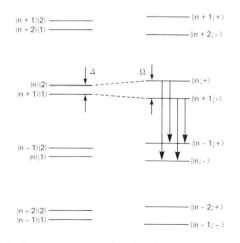

FIG. 11. Jaynes–Cummings RWA energy spectrum.

In this case a strong laser field provides the main mode radiation. Dipole selection rules determine that from the same nearly degenerate states $\|n, +\rangle$ and $\|n, -\rangle$ spontaneous transitions can be made only to the next lower pair of nearly degenerate states, $\|n-1, +\rangle$ and $\|n-1, -\rangle$. There are four separate emission lines predicted, as shown in Fig. 11. The strengths of the four lines are all equal on resonance, but since two of them have the same transition frequency only three lines are actually expected. They have the intensity ratio $1:2:1$, but the side peaks have different widths than the center peak, and the peak height ratio is $1:3:1$. This fluorescence triplet was predicted in the late 1960s, and after a period of controversy about the exact line structure, the predictions mentioned here were verified experimentally in the mid-1970s by Stroud, Walther, Ezekiel, and others. It should be clear from Fig. 11 that the resonance fluorescence peak separation is equal to the Autler–Townes splitting, which is just the quantum Rabi frequency $\chi(n)$ at resonance. The sequence of spectra shown in Fig. 12 illustrates the increased peak separation that accompanies an increased Rabi frequency when the main mode intensity is increased.

F. Tests of Quantum Theory

It has long been recognized that quantum theory stands in conflict with naive notions that "physical reality" can be independent of observation. As is well known, the *Heisenberg uncertainty principle* mandates a limit on the mutual precision with which two noncommuting variables may be observed. This curious feature was put into sharp focus by Einstein, Podolsky, and Rosen in 1935, but for nearly half a century it remained mainly of philosophical interest, as experimental tests were difficult to conceive or implement. The situation changed dramatically during the 1970s when it was realized that quantum optical experiments could test different conceptions of "reality."

Einstein, Podolsky, and Rosen (EPR) gave a precise meaning to the concept of reality in this context: "If, without in any way disturbing a system, we can predict with certainty [i.e., with probability equal to unity] the value of a physical quantity, then there exists an element of physical reality corresponding to this physical quantity." It was of primary concern to EPR whether quantum theory can be considered to be

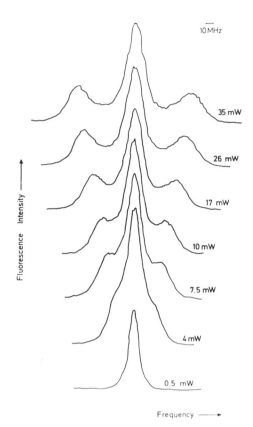

FIG. 12.   Resonance fluorescence spectra in the presence of AC Stark splitting. [Reprinted with permission from Hartig, W., Rasmussen, W., Schieder, R., and Walther, H. (1976). Study of the frequency distribution of the fluorescent light induced by monochromatic radiation, *Zeitschrift für Physik,* **A278**, 205–210.]

a "complete" theory. A necessary condition for completeness of a theory, according to EPR, is that "every element of the physical reality must have a counterpart in the physical theory." Using these definitions of reality and completeness, and the properties of correlated quantum states, EPR concluded that quantum theory does not provide a complete description of physical reality.

An illuminating example due to Bohm is provided by the singlet state of two spin-$\frac{1}{2}$ particles:

$$\|S\rangle = (1/\sqrt{2})[\,|a+, \hat{\mathbf{n}}\rangle * |b-, \hat{\mathbf{n}}\rangle$$
$$- |a-, \hat{\mathbf{n}}\rangle * |b+, \hat{\mathbf{n}}\rangle] \qquad (122)$$

where $|a\pm, \hat{\mathbf{n}}\rangle$ is the state for which particle a has spin up (+) or down (−) in the direction $\hat{\mathbf{n}}$. The unit vector $\hat{\mathbf{n}}$ can point in any direction. In light of the EPR argument, one notes that the spin of particle b in, say, the $\hat{\mathbf{x}}$ direction, can be predicted with certainty from a measurement of the spin of particle a in that direction. If the spin of particle a is found to be up, then the spin of particle b must, for the singlet state, be down, and vice versa. Thus, the spin of particle b in the $\hat{x}$ direction can be predicted with certainty "without in any way disturbing" that particle. According to EPR, therefore, the $\hat{x}$ component of the spin of particle b is an element of physical reality.

Of course, one may choose instead to measure the $\hat{y}$ component of the spin of particle a, in which case the $\hat{y}$ component of particle b can be predicted with certainty. It follows therefore that both the $\hat{x}$ and $\hat{y}$ components of the spin of particle b (and, of course, particle a) are elements of physical reality. However, according to quantum mechanics, the $\hat{x}$ and $\hat{y}$ components of spin cannot have simultaneously predetermined values, because the associated spin operators do not commute. Therefore, quantum theory does not account for these elements of physical reality, and so, according to EPR, it is not a complete theory.

The EPR experiment can be criticized on the ground that "the system" must be understood in its totality and cannot refer to just one of the particles, for instance, of a correlated two-particle system. In the Bohm gedanken experiment just considered, a measurement on particle a does disturb the complete (two-particle) system because the quantum mechanical description of the system is changed by the measurement. One cannot consistently associate an element of physical reality with each spin component of particle b, even though the particles may be arbitrarily far apart and not interacting in any way.

Motivated by the EPR argument, one may ask whether it is possible to formulate a theory in which physical quantities do have objectively real values "out there," independently of whether any measurements are made. These objectively real values may be imagined to be determined by certain *hidden variables* which may themselves be stochastic. One can ask, is it possible, in principle, to construct a hidden variable theory in full agreement with the statistical predictions of quantum mechanics, but which allows for an objective reality in the EPR sense?

In the early 1960s Bell considered the most palatable class of hidden variable theories, the so-called local theories. He demonstrated that such theories cannot fully agree with quantum mechanics. In particular, certain *Bell inequalities* distinguish any local hidden variable theory

from quantum mechanics. Another way of stating "Bell's theorem" is that no local realistic theory can be in full agreement with the predictions of quantum theory.

Bell inequalities are not difficult to derive for the Bohm thought experiment. A local hidden variable theory is postulated to give $\pm\frac{1}{2}$ for each spin component, as determined by the hidden variables. The theory is also supposed to account for the spin correlations: if one is up the other must be down. The difference between such a theory and quantum mechanics is that the spins are predetermined (by the hidden variables) before any measurement, that is, they are objectively real. The condition of locality enters through the additional assumption that a measurement of the spin of each particle is not affected by the direction in which the spin of the other particle is measured. This is certainly reasonable if all the spin components are predetermined, since the two particles may be very far apart when a measurement is made.

The question now is whether any measurements can distinguish such a theory from quantum mechanics. Bell considered $E(\hat{m}, \hat{n})$, the expectation value of the product of the spin components of particles a and b in the $\hat{m}$ and $\hat{n}$ directions, respectively. He obtained the inequality

$$|E(\hat{m}, \hat{n}) - E(\hat{m}, \hat{p})| \leq \tfrac{1}{4} + E(\hat{n}, \hat{p}) \quad (123)$$

which must be satisfied by the entire class of local hidden variable theories. This inequality is violated by quantum theory, as can be seen from the quantum mechanical prediction $E(\hat{m}, \hat{n}) = -\tfrac{1}{4}\hat{m} \cdot \hat{n}$. It is therefore possible to test experimentally the predictions of quantum theory *vis-a-vis* the whole class of plausible "realistic" theories.

Although Bell's theorem promoted philosophical questions about hidden variables to the level of experimental verifiability, it remained difficult to conceive of specific experiments that could be undertaken. In 1969, however, Clauser and co-workers suggested that Bell inequalities could be tested by measuring photon polarization correlations if certain additional but reasonable assumptions about the measurement process were made. The spin considered by Bell is replaced by photon polarization, another two-state phenomenon. Correlated two-photon polarization states are produced in atomic cascade emissions, and efficient polarizers and detectors are available for optical photons.

Consider a $J = 0 \rightarrow 1 \rightarrow 0$ atomic cascade decay, with polarizer–detector systems on the $\pm z$ axes. Linear polarization filters may be employed to distinguish the photons by their energy so that each polarizer–detector system records photons of one frequency but not the other. It may be shown that the two-photon polarization state has the form

$$\| \Psi \rangle = (1/\sqrt{2}[\,| a, \hat{x} \rangle * | b, \hat{y} \rangle$$
$$+ | a, \hat{y} \rangle * | b, \hat{x} \rangle] \qquad (124)$$

where $| a, \hat{x} \rangle$ is the single-photon state in which photon a is linearly polarized along the $\hat{x}$ direction, etc. A similar form applies if a circular polarization basis is used.

The correlated photon state of Eq. (124) is obviously analogous to the spin-$\frac{1}{2}$ correlated state of Eq. (122). A hidden variable theory of such polarization correlations lead to a Bell inequality analogous to Eq. (123):

$$| E(\alpha, \beta) - E(\alpha, \gamma) | \leq 1 - E(\beta, \gamma) \quad (125)$$

where $E(\alpha, \beta)$ now refers to photon polarization components and α and β are the filter orientations with respect to some reference axis. The differences between Eqs. (125) and (123) arise because we are now dealing with spin-one particles and because Eq. (124) describes a positive correlation. The quantum mechanical prediction for $E(\alpha, \beta)$ is simply $\cos 2(\alpha - \beta)$, and Eq. (125) becomes

$$| \cos 2(\alpha - \beta) - \cos 2(\alpha - \gamma) |$$
$$\leq 1 - \cos 2(\beta - \gamma) \qquad (126)$$

which, in fact, is not satisfied for all angles α, β, and γ. Such violations of Bell inequalities have been observed in independent experiments led by Clauser, Fry, and most recently, Aspect. The results of such experiments are in agreement with quantum theory, and appear to rule out any local hidden variable theory. There are possible loopholes in the interpretation of the experiments, but at the present time none of them seem very plausible. According to Clauser and Shimony, "The conclusions are philosophically

unsettling: Either one must totally abandon the realistic philosophy of most working scientists, or dramatically revise our present concept of space–time."

From the viewpoint of quantum optics, the photon polarization correlation experiments measure a second-order field correlation function. Such a correlation function not only distinguishes between classical and quantum radiation theories, but also between quantum theory and local realistic theories. In quantum optics it is often possible to address such questions from essentially first principles and to carry out accurate tests of theory in the laboratory.

BIBLIOGRAPHY

Allen, L., and Eberly, J. H. (1975). "Optical Resonance and Two-Level Atoms." Wiley, New York, reprinted by Dover (1987).

Delone, N. B., and Krainov, V. P. (1985). "Atoms In Strong Light Fields." Springer-Verlag, Berlin.

Fontana, P. (1982). "Atomic Radiative Processes." Academic Press, San Diego.

Knight, P. L., and Allen, L. (1983). "Concepts of Quantum Optics." Pergamon, Oxford.

Knight, P. L., and Milonni, P. W. (1980). The Rabi frequency in optical spectra, In "Physics Reports," Vol. 66, pp. 21–107. North-Holland, Amsterdam.

Loudon, R. (1983). "The Quantum Theory of Light," 2nd ed. Oxford Univ. Press, Oxford.

Mandel, L. (1976). The case for and against semiclassical radiation theory, In "Progress in Optics," (E. Wolf, ed.), Vol. XIII. North-Holland, Amsterdam.

Milonni, P. W. (1976). Semiclassical and quantum-electrodynamical approaches in nonrelativistic radiation theory, In "Physics Reports," Vol. 25, pp. 1–81. North-Holland, Amsterdam.

Perina, J. (1984). "Quantum Statistics of Linear and Nonlinear Phenomena." D. Reidel, Dordrecht.

Stenholm, S. (1984). "Foundations of Laser Spectroscopy." John Wiley, New York.

Yoo, H. I., and Eberly, J. H. (1985). Dynamical theory of an atom with two or three levels interacting with quantized cavity fields, In "Physics Reports," Vol. 118, pp. 239–337. North-Holland, Amsterdam.

QUASICRYSTALS

Marko V. Jarić   *Texas A & M University*

I.   Introduction
II.   Why Are Quasicrystals Puzzling?
III. Crystallography
IV.   Atomic Structure
V.   Physical Properties
VI.   Conclusion

GLOSSARY

Ammann tiling: Icosahedral quasiperiodic tiling of space by two types of rhombohedra that can be obtained by a certain cut through a six-dimensional hypercubic tiling.

Delaunay structure: Set of points such that no two points are closer than a given fixed distance.

Fibonacci sequence: Sequence of small (S) and large (L) objects generated by the substitution $S \rightarrow L, L \rightarrow LS$.

Hypercrystal: Fictitious, higher dimensional crystal used to represent a real incommensurate crystal by a particular cut.

Icosahedral glass: Structure with long-range orientational order and nonzero shear modulus, but without long-range positional order.

Incommensurate: Having irrational ratio.

Incommensurate crystal: Crystal with long-range positional order whose diffraction pattern requires more than three integer indices.

Inflation symmetry: Discrete scale change that leaves a quasilattice invariant.

Long-range orientational order: Order characterized by broken orientational symmetry of the diffraction pattern.

Long-range positional order: Order characterized by discreteness of the Fourier transform, that is, by perfect Bragg diffraction.

Penrose tiling: Pentagonal quasiperiodic tiling of plane, for example, by two types of rhombi, which can be obtained by a certain cut through a five-dimensional hypercubic tiling.

Phason: Degree of freedom associated with the relative phase of two incommensurate waves.

Phason strain: Gradient of a phason field.

Quasilattice: Dense projection of a higher dimensional lattice.

Quasiperiodic: Having a finitely generated discrete but dense Fourier transform.

Shechtmanites: Icosahedral materials sharing the structure of icosahedral aluminum transition–metal alloys.

Quasicrystals are solids with structures that exhibit long-range positional order and noncrystallographic symmetry. An experimental signature of long-range positional order is the sharpness of the diffraction peaks. Quasicrystalline materials with icosahedral, octagonal, decagonal, and dodecagonal symmetry have been discovered. Physical properties of quasicrystals often interpolate between those of crystals and amorphous structures, and some unusual features with potential technological applications are expected to be seen.

I. Introduction

A group of researchers at the National Institute of Standards and Technology (NIST) recently reported the discovery of a new material whose electron diffraction pattern [Fig. 1(a)] reveals a novel structure hitherto assumed impossible. While the diffraction patterns that are commonly seen consist either of diffuse rings, as in amorphous materials [Fig. 1(b)], or sharp peaks occupying a crystallographic lattice, as in ordinary crystals [Fig. 1(c)], the new materials yield diffraction patterns with both sharp peaks and noncrystallographic symmetry, such as the fivefold symmetry shown in Fig. 1(a). This discovery challenged a century-old belief that periodicity and structural order, measured by the sharpness of diffraction peaks, are essentially equivalent.

A. GENERAL

Unlike the incommensurately modulated crystals, which are also not periodic but have sharp diffraction

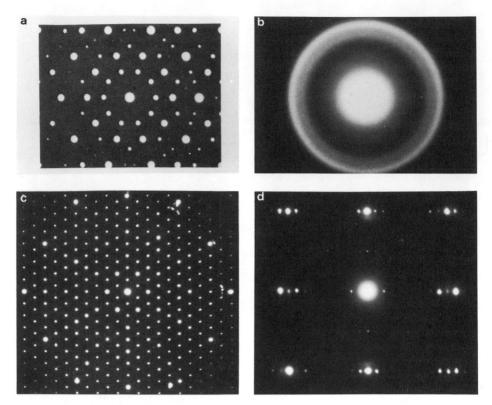

FIG. 1. Typical electron diffraction patterns for (a) an icosahedral quasicrystal, (b) a metallic glass, (c) a crystal, and (d) an incommensurate crystal. The quasicrystal is exemplified by the Al–Mn alloy shechtmanite. [From Shechtman, D., and Blech, I. (1985). *Metall. Trans.* **16A,** 1005.] The metallic glass is exemplified by an Al–Fe–Si alloy, which can also form a quasicrystalline phase. [From Legresy, J. M., *et al.* (1986). *Acta Metall.* **34,** 1759.] The crystal is the μ phase of $Al_{26}Cr_{14}Si_{24}$, which is often found in relationship with a quasicrystalline phase. [From Garçon, S., *et al.* (1987). *Rev. Metall.* **9,** 469 (in French).] The incommensurate crystal is a Ca–Bi–Cu–Sr–O high-temperature superconductor. [Courtesy of M. Audier.]

peaks [Fig. 1(d)], quasicrystals cannot be obtained by small systematic distortions of an underlying crystal structure and, consequently, are much more difficult to model. In fact, most of the initial skepticism about the existence of quasicrystals was based on the lack of appropriate models. However, long before the discovery of quasicrystals, Roger Penrose discovered the aperiodic, highly ordered, pentagonal tilings of a plane (Fig. 2). These tilings were generalized by Robert Ammann to icosahedral, aperiodic tilings of the space. Soon after Penrose's discovery, crystallographer Alan Mackay, fascinated by these tilings, suggested that similar structures might occur in nature. He even presented a demonstration that the diffraction pattern of such structures would consist of numerous sharp peaks. His prophecy was fulfilled by the NIST discovery of shechtmanite, the icosahedral quasicrystalline alloy with the approximate composition Al_6Mn.

As if to echo Mackay's words that quasicrystals "might go unnoticed if not expected," a flood of research that followed the initial discovery uncovered

not only numerous other icosahedral quasicrystals but also octagonal, decagonal, and dodecagonal quasicrystals. For example, the quasicrystalline phase of Al–Li–Cu, although observed over thirty years ago, went virtually unnoticed and was soon forgotten

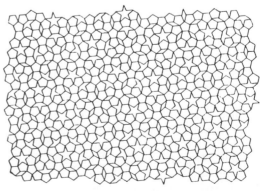

FIG. 2. The Penrose tiling. [From Penrose, R. (1974). *Bull. Inst. Math. Appl.* **10,** 266.]

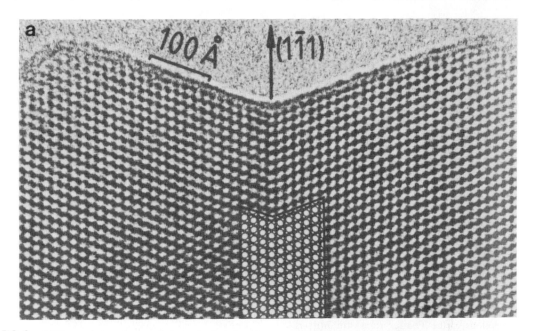

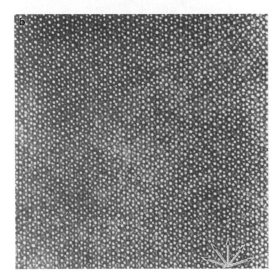

FIG. 3. (a) Transmission electron photograph of a twinned crystal. [Reprinted with permission from Thomas, J. M., *et al.* (1983). *In* "Interzeolite Chemistry" (G. D. Stucky and F. G. Dwyer, eds.), p. 181. Copyright © 1983 American Chemical Society, Washington, D.C.] (b) Transmission electron photograph of an icosahedral quasicrystal. [From Hiraga, K., *et al.* (1985). *Sci. Rep. Res. Inst.* **A32**, 309.]

since it did not fit into any of the known crystallographic schemes.

In this overview, we shall focus on icosahedral quasicrystals. At the present, octagonal, decagonal, and dodecagonal quasicrystals have also been reported. Our emphasis will be on ideal quasiperiodic structures. However, it may be possible to describe quasicrystals in terms of other structures, such as orientationally ordered glasses. In particular, some scientists do not believe quasicrystals can be realized in nature and prefer to interpret the experiments in terms of icosahedral twinning of ordinary crystals.

In Fig. 3, we show transmission electron micrographs of a well-understood twin and a quasicrystal. While the twin boundary is clearly visible in Fig. 3(a), the structure in Fig. 3(b) appears rather uniform. Crystals with unit cells containing thousands of atoms would be necessary to account for the observed quasicrystalline pattern. Moreover, because of limited experimental accuracy, a crystal with a sufficiently large unit cell could always be used to describe any structure. However, such a description may be impractical, leaving the quasicrystal description as a possible simple, albeit perhaps approximate, alternative.

More detailed information can be obtained from the original research papers, books, collections of

papers, and review articles. A limited selection is listed in the bibliography.

B. Diffraction

A typical diffraction experiment is illustrated in Fig. 4. A sample S, characterized by a density of scatterers $n(\mathbf{r})$, is exposed to an X-ray (or electron or neutron) beam with an initial wave vector (or momentum) $\mathbf{k}$. Passing through the sample, the incident beam is scattered on inhomogeneities in the density $n(\mathbf{r})$. A scattered beam with momentum $\mathbf{k} + \mathbf{q}$, where $\mathbf{q}$ is the so-called scattering vector, is detected by D, an X-ray sensitive film. Assuming that no energy is lost in the scattering process and that no double scattering occurs, the measured intensity $I(\mathbf{q})$

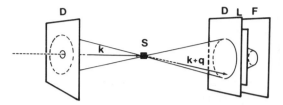

FIG. 4.   Sketch of a typical scattering experiment.

is proportional to the squared magnitude of the Fourier transform $n(\mathbf{q})$ of the density:

$$I(\mathbf{q}) \sim |n(\mathbf{q})|^2 \qquad (1)$$

By turning the sample and by changing the energy of the incident beam, one can measure $I(\mathbf{q})$ for all scattering vectors $\mathbf{q}$. The three-dimensional space of vectors $\mathbf{q}$ is often called the reciprocal space.

Although $I(\mathbf{q})$ can be measured in this way, such measurement is insufficient to determine the density $n(\mathbf{r})$. Obviously, the Fourier transform $n(\mathbf{q})$ is a complex number with an amplitude and a phase $n(\mathbf{q}) = |n(\mathbf{q})|e^{i\theta(\mathbf{q})}$. Only the amplitude can be determined by measuring the diffraction intensity, and consequently, $n(\mathbf{r})$, the inverse Fourier transform of $n(\mathbf{q})$, cannot be determined. Nevertheless, $I(\mathbf{r})$, the inverse transform of the intensity $I(\mathbf{q})$, the so-called Patterson function, contains useful information about $n(\mathbf{r})$ and the structure of the sample. Indeed, $I(\mathbf{r})$ is precisely the pair distribution function, or the auto-correlation function:

$$I(\mathbf{r}) = \int n(\mathbf{r} + \mathbf{r}')n(\mathbf{r}') \, d^3r' \qquad (2)$$

In Fig. 5, we show typical Patterson functions of a quasicrystal and a related crystal.

The transmission electron micrographs are obtained by using a set of lenses (L in Fig. 4) to refocus a part of the diffracted electron beam onto a film (F in Fig. 4). In this manner, an image is obtained that approximates the projection of the density $n(\mathbf{r})$ onto the plane of the film. This can be easily understood by recalling the following theorem and its corollary.

Theorem: The Fourier transform of a product is equal to convolution of the Fourier transforms, and conversely, the Fourier transform of a convolution is equal to the product of the Fourier transforms.

Corollary: The Fourier transform of a projection is equal to a cut of the Fourier transform, and conversely, the Fourier transform of a cut is equal to a projection of the Fourier transform.

For example, in order to explain the electron transmission microscopy, we observe that the Fourier transform of the projection

$$n_p(x, y) = \int n(x, y, z) \, dz \qquad (3)$$

equals the cut $q_z = 0$ of the Fourier transform

$$n_p(q_x, q_y) = n(q_x, q_y, q_z = 0) \qquad (4)$$

The relationship between a cut and a projection is very useful in understanding the structure of and diffraction from quasicrystals.

The previous theorem can also help us understand some general features of the experimentally observed diffraction patterns, such as those shown in Fig. 1. For example, consider a periodic lattice of unit point scatterers. The Fourier transform of this density and, consequently, its diffraction pattern consist of infinitely sharp, delta-function peaks located at the vertices of the reciprocal lattice. If the total number of scatterers is N, then the infinite value of the delta

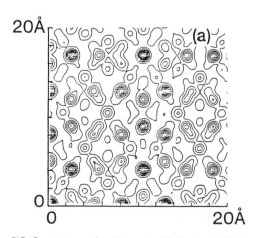

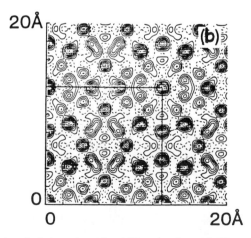

FIG. 5.   Patterson function perpendicular to a twofold axis of (a) an icosahedral quasicrystal and (b) a related crystal. [From Cahn, J. W., *et al.* (1988). *Phys. Rev.* **B38**, 1638.]

function must be replaced by O(N) and the Bragg peak intensity must scale like N^2. In contrast, the peak intensities in a glass scale like N. In a real crystal, the Bragg peaks are always broadened because the scale L over which the crystal is perfectly ordered is finite. This scale is the average spacing between the planar structural defects. In a perfect crystal, it equals the crystal size, $L \sim N^{1/3}$. The acquired peak width is of the order $O(1/L)$.

Conversely, if the point scatterers are smeared, for example, by the thermal motion or static disordering over a scale l, then the Fourier transform and, thus, the intensity are cut off at the scattering vectors of the order of 1/l. This is expressed in the well-known Debye–Waller contribution to the scattering.

In summary, the narrowness of the features in the diffraction pattern is a measure of the long-range order in the structure while the falloff of the intensity at large scattering vectors is a measure of the local disorder. This is consistent with the fact that diffraction from liquids or amorphous materials, which can be considered disordered, shows only broad features.

II. Why Are Quasicrystals Puzzling?

Obviously, the existence of quasicrystals is already puzzling by the mere fact that they have never been seen before. In addition, many researchers are puzzled because of several false but widely held beliefs. First, it is believed that a geometric structure, a set of point scatterers, with sharp diffraction peaks, that is, long-range positional order, and noncrystallographic symmetry cannot satisfy the requirement, the Delaunay condition, that no two points in the structure should be closer than a certain minimal distance. Second, it is a common assumption that in the ground state of a solid only a small number of distinct atomic environments can be present. In a quasicrystal structure, infinitely many different environments are necessarily present. Third, even admitting a possibility of quasicrystalline ground states, it is often doubted that they could grow and be equilibrated on physically reasonable time scales. These beliefs, which nature seems to discount, will be addressed in the following subsections.

A. GEOMETRY

For almost a century, the only known positionally ordered structures were the periodic, crystal structures. A periodic structure can be completely described by a unit cell that viewed as a tile or a brick can be stacked to fill the entire space and by the positions of the atoms inside this unit cell. Typically, a unit cell will contain no more than 10 atoms, al-

though crystals with unit cells accommodating on the order of 1000 atoms are also known.

A crystal structure is characterized by its translation symmetry and by its rotation symmetry. The crystal lattice, which defines the crystal translation symmetry, is generated by three linearly independent basis vectors. As a consequence, the lattice points satisfy the Delaunay condition. However, not all rotation symmetries are compatible with a crystal lattice. As an illustration, consider Fig. 6 in which we assume R_1 and T_1 to be two lattice points and that the lattice has fivefold symmetry. By choosing R_1 on an axis of a fivefold symmetry, we conclude that T_1 must also lie on an axis of fivefold symmetry. Therefore, the points $R_1', R_1'', \ldots$ and $T_1', T_1'', \ldots$ generated by the fivefold rotations must also be lattice points. We can now repeat the same reasoning, starting from the points $R_2 = R_1'$ and $T_2 = T_1'$ to arrive at the lattice points R_3 and T_3. This procedure can be carried out *ad infinitum* leading to arbitrarily close lattice points, and consequently arbitrarily close atoms. Therefore, a periodic lattice cannot exhibit a fivefold symmetry.

The same reasoning can be used to prove that a Delaunay structure cannot have more than one axis of rotation symmetry unless the rotations are of order 1,2,3,4, or 6. These rotations are called crystallographic. Note, however, that a Delaunay structure could possess a unique rotation axis of any order.

Consequently, it was very surprising to discover a material that showed a diffraction pattern with sharp peaks and, at the same time, fivefold symmetry. It is a simple fact that one could construct many functions (densities) $n(\mathbf{r})$ whose Fourier transform would have sharp peaks and fivefold symmetry. However, it seemed impossible that any such density would consist of sharp but well-separated peaks that could be associated with actual atomic positions.

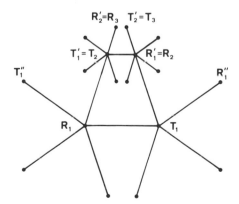

FIG. 6. Incompatibility between the fivefold symmetry and periodicity.

This is in contrast to modulated incommensurate crystals that result from small modulations of ordinary crystals: the lattice vectors (atomic positions) $\mathbf{R}$ of a reference crystal are displaced by a small amount $\mathbf{u}(\mathbf{R})$; $\mathbf{u}(\mathbf{R})$ is a bounded periodic vector function, but with periodicity $\mathbf{a}'$, which is incommensurate with the lattice periodicity. That is, $\mathbf{a}' \cdot \mathbf{b}_i 2\pi$ is irrational for at least one basis vector $\mathbf{b}_i$ of the crystal's reciprocal lattice. In Fig. 1(d), we showed the diffraction pattern of a crystal with an incommensurate modulation. As can be seen, the diffraction pattern of a modulated crystal still consists of sharp peaks, but each original peak acquires an infinite set of satellites. In most cases, only a small number of satellites are detectable. A one-dimensional incommensurate crystal that cannot be described in terms of a weakly modulated crystal is shown in Fig. 7. The short (S) and long (L) segments form the Fibonacci sequence. The two incommensurate basis vectors ($\mathbf{b}_1$ and $\mathbf{b}_2$) are needed in the reciprocal space. The construction of this example will be described later in the text (see Fig. 12). [*See* Encyclopedia article PHASE TRANS-

Since $\mathbf{u}$ is small, a modulated incommensurate structure will possess a minimal distance inherited from the underlying periodic structure. For the same reason, its rotation symmetry will be crystallographic, also inherited from the periodic lattice. In order to obtain a noncrystallographic symmetry, one would need to treat the modulation $\mathbf{u}$ on the same footing with the periodic lattice. That is, $|\mathbf{u}| \sim |\mathbf{a}_i|$ for a basis vector $\mathbf{a}_i$ of the crystal lattice. Consequently, the Delaunay condition could not be assured.

Fortunately, following Penrose's 1974 discovery of an aperiodic tiling of the plane with an underlying pentagonal symmetry and the subsequent work by a number of mathematicians, crystallographers, and physicists, it is now known that Delaunay structures with sharp diffraction peaks and any rotation symmetry can be constructed. A general method for such construction will be described in Section III.

B. Ground State

Even after accepting the existence of quasicrystalline geometric structures that satisfy the Delaunay

condition, many researchers remain skeptical about the possibility that such structures could correspond to the lowest energy states. Common wisdom asserts that in the ground state only a very small number of distinct infinite environments should occur. Therefore, quasicrystals, in which each atom has necessarily a distinct environment, could not be the lowest energy states. However, this intuition already fails in the case of crystals with large unit cells, for example, in tetrahedrally close-packed or Frank–Kasper crystals. It also obviously fails in the case of ordinary, incommensurate crystals that are often experimentally observed.

It is important to emphasize that even if only finite environments are considered, for example, on the scale of a finite-range interparticle potential, a finite set of environments does not guarantee periodicity. This can be clearly seen in aperiodic tilings, such as the Penrose tiling shown in Fig. 2.

No exact proofs exist that a physically relevant interparticle potential, such as the Lennard–Jones potential, should lead to a periodic, crystalline ground state. For example, in the case of the one-component Lennard–Jones system, there is plenty of numerical evidence that the ground state is the primitive face-centered cubic crystal. On the other hand, hard-tile models can be constructed so that a particular periodic or aperiodic tiling is the closest packing, that is, the ground state of the model. A two-dimensional example leading to the Penrose tiling is shown in Fig. 8.

While such models are somewhat unphysical, it is possible to construct two-dimensional models that are more physical and appear to have a quasicrystalline ground state. For example, a four-species, hard-disc system with nonadditive diameters can be constructed so that the Penrose tiling is its ground state. In Fig. 9(a), the equilibrium positions of the discs are shown. The structure is stabilized for the stochiometry $N_1/N_4 = N_2/N_3 = 1$, $N_2/N_1 = t^2$, where $t = (1 + \sqrt{5})/2$ is the golden mean. The phase separa-

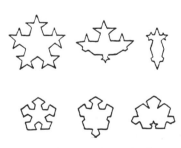

FIG. 8. A set of tiles that force the Penrose tiling of Fig. 2. [From Penrose, R. (1974). *Bull. Inst. Math. Appl.* **10**, 266.]

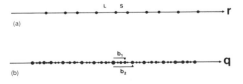

FIG. 7. (a) One-dimensional incommensurate crystal and (b) its diffraction pattern.

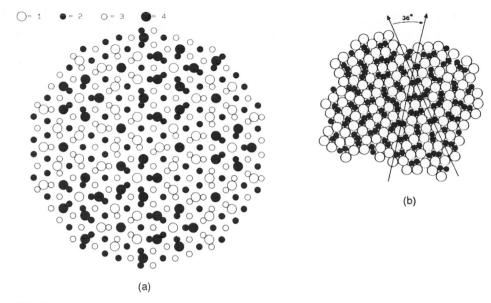

FIG. 9. (a) Two-dimensional quasicrystal ground state for a hard disk. [From Johnson, S. L., and Jarić, M. V. (1989). *Phys. Rev.* (in press).] (b) Two-dimensional quasicrystal ground state for Lennard–Jones interactions. [From Widom, M., *et al.* (1987). *Phys. Rev. Lett.* **58,** 706.]

tion is discouraged since the hard core diameters, which can be identified with the shortest separations in Fig. 9(a), satisfy $d_{ij} < (d_{ii} + d_{jj})/2$. Similarly, computer simulations of a two-dimensional, two-component Lennard–Jones system in which phase separation is discouraged by a favorable interspecies binding energy, rather than by favorable bond lengths, also exhibits ground states, such as the one shown in Fig. 9(b), which could be called quasicrystalline in the sense that the diffraction pattern has algebraically singular peaks and fivefold symmetry.

A finite temperature thermodynamic equilibrium of icosahedral quasicrystal structures has been investigated using the Ginzburg–Landau theory, the Ramkrishnan–Yusouff density functional theory, and a mean-field theory, which couples positional and orientational order parameters. In all of these theories, locally stable quasicrystalline states can be found. Particularly interesting is the theory that shows that a tendency for orientational order, which favors icosahedral orientational ordering, could also promote long-range icosahedral positional order.

Experimentally, most of the quasicrystals related to icosahedral (i) Al–Mn–Si seem to be in a metastable state: upon annealing at a sufficiently high temperature, they transform into a more stable crystalline phase. However, quasicrystals such as i(Al–Li–Cu), i(Al–Fe–Cu), and i(Ga–Mg–Zn), which can grow into millimeter-size faceted single grains, are possibly stable (Fig. 10).

C. GROWTH

Although the ground state for some interparticle potentials can be quasicrystalline, it is puzzling how such structures could actually form. As will be seen in Section III, the known algorithms for constructing quasicrystalline structures are nonlocal. For example, if a construction of the tiling in Fig. 2 is attempted by simply locally attaching individual tiles of Fig. 8, a defect would be created after only several tiles. However, whether a particular attachment of a tile is bound to create a defect cannot be generally determined without growing the tiling to arbitrary distances, possibly including the entire tiling.

This is the most challenging puzzle presented by quasicrystals. Obviously, nature knows how to grow quasicrystals. As mentioned above, some quasicrystalline materials, such as i(Al–Li–Cu) and i(Ga–Mg–Zn), grow nicely faceted quasicrystals that display macroscopic icosahedral symmetry, as shown in Fig. 10.

An approach to resolve this puzzle is based on the experimental finding that even the equilibrium icosahedral quasicrystals show a considerable degree of disorder. This residual disordering could not be removed even after very long equilibration and annealing times. It suggests that a degree of disorder might be intrinsic to quasicrystals, in agreement with their inability to grow locally. Indeed, several numerical simulations of quasicrystal formation in two dimen-

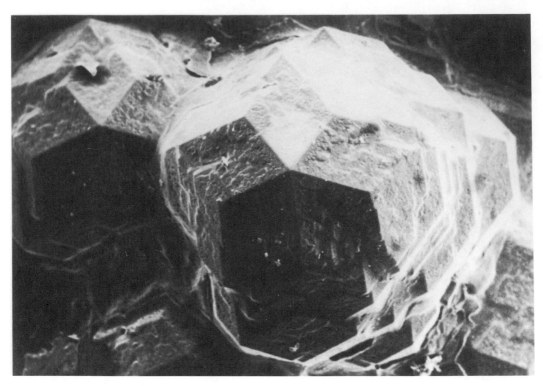

FIG. 10.   A faceted quasicrystal of i(Al$_{6.4}$Li$_{2.7}$Cu$_{0.9}$). [Reprinted by permission from Dubost, B., *et al.* (1986). *Nature (London)*, **324,** 49. Copyright © 1986 Macmillan Magazines Ltd.]

sions indicate formation of disordered structures with almost perfect orientational order and positional order comparable with experiments. The degree of positional ordering depends on the velocity of the solidification front and the temperature gradient perpendicular to the growth front. While several groups still seek local growth algorithms that enforce a perfect quasicrystal structure, their absence seems consistent with the experimental findings.

III. Crystallography

Quasicrystals cannot be described within the framework of the ordinary, three-dimensional crystallography. Although quasicrystals can be related to higher dimensional crystals (hypercrystals), it is also not sufficient to describe them using a higher dimensional crystallography. Consequently, it is necessary to expand the concepts of the traditional crystallography.

A. Indexing

The first quasicrystalline material discovered at the National Bureau of Standards exhibited the icosahe-

dral diffraction pattern reproduced in Fig. 1(a). The presence of two-, three-, and fivefold icosahedral symmetry axes was verified by rotating the sample relative to the incident beam, as shown in Fig. 11. Within the experimental accuracy, the scattering vector $\mathbf{Q}$ of any particular diffraction peak could be written as $\mathbf{Q} = m_i\mathbf{b}_i$, an integral linear combination of six scattering vectors $\mathbf{b}_i$ equal in magnitude and aligned with the six fivefold symmetry axes. In analogy with ordinary crystals, the $D = 6$ integers m_i can be used to index the diffraction peaks.

This method of indexing is quite general and has been used in the past for incommensurate crystals. In fact, a structure is called incommensurate if, by definition, its diffraction pattern can be indexed only by finitely many, but necessarily more than three, integers. Quasicrystals can be viewed as a special case of incommensurate crystals. They are distinguished by having diffraction patterns with noncrystallographic symmetry. As we shall see, the fact that the symmetry is noncrystallographic has certain important consequences related to the physical properties of quasicrystals.

As with ordinary crystals, indexing for incommensurate crystals is not unique since the choice of the

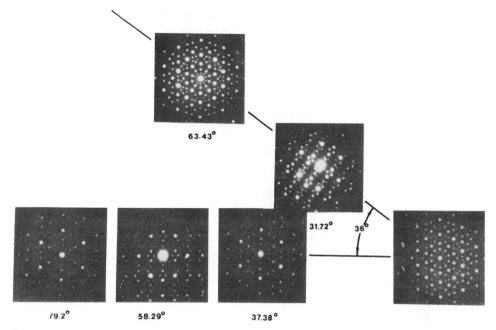

FIG. 11.   The first reported diffraction patterns of an icosahedral quasicrystal. [From Shechtman, D., *et al.* (1984). *Phys. Rev. Lett.* **53**, 1951.]

basis $\{\mathbf{b}_i\}$ is not unique. An important difference, however, is that the vectors in the basis of an incommensurate crystal can be arbitrarily small. This reflects the fact that the diffraction pattern of a perfect incommensurate crystal consists of a dense set of Bragg peaks. In particular, for icosahedral quasicrystals, the entire diffraction pattern and the six basis vectors $\mathbf{b}_i$ can be scaled by a factor t^{3m}, where m is any integer.

B. HYPERCRYSTALS

Before the discovery of quasicrystals, it was well known that the density of an incommensurate crystal can always be represented as a cut through a hypercrystal density. This connection can be formally established by introducing a complementary, $(D - 3)$-dimensional space perpendicular to the physical, three-dimensional space.

Let us denote vectors in the hyperspace by an overbar and their projection into the perpendicular, complementary space by the superscript $\perp$. Then, a basis of the reciprocal hyperlattice can be introduced:

$$\overline{\mathbf{b}}_i = \mathbf{b}_i + \mathbf{b}_i^\perp, \qquad i = 1, \ldots, D \qquad (5)$$

The projections $\mathbf{b}_i^\perp$ are arbitrary except for the condition that the vectors $\overline{\mathbf{b}}_i$, $i = 1, \ldots, D$ should be linearly independent. Therefore, a one-to-one correspondence between the reciprocal quasilattice vectors $\mathbf{Q}$ and the reciprocal hyperlattice vectors $\overline{\mathbf{Q}}$ can be established:

$$\mathbf{Q} = m_i\mathbf{b}_i \Leftrightarrow \overline{\mathbf{Q}} = m_i\overline{\mathbf{b}}_i \qquad (6)$$

Clearly, each $\mathbf{Q}$ is a projection of a $\overline{\mathbf{Q}}$. Therefore, a hypercrystal density

$$\overline{n}(\overline{\mathbf{r}}) = \sum_{\overline{\mathbf{Q}}} n(\overline{\mathbf{Q}})e^{i\overline{\mathbf{Q}}\cdot\overline{\mathbf{r}}} \qquad (7)$$

can be associated with a quasicrystal density

$$n(\mathbf{r}) = \sum_{\overline{\mathbf{Q}}} n(\overline{\mathbf{Q}})e^{i\mathbf{Q}\cdot\mathbf{r}} \qquad (8)$$

so that, indeed,

$$n(\mathbf{r}) = \overline{n}(\overline{\mathbf{r}})\,\big|_{\,\mathbf{r}^\perp = 0} \qquad (9)$$

In order to obtain a quasicrystal of point scatterers, it is necessary to decorate the hypercrystal by $(D - 3)$-dimensional hypersurfaces. If the equation of the hypersurface for the atomic species β is

$$\mathbf{r} = \mathbf{s}_\beta(\mathbf{r}^\perp), \qquad \mathbf{r}^\perp \in v_\beta^\perp \qquad (10)$$

where $v_\beta^\perp$ is the domain of the hypersurface, then the quasicrystal density is given by

$$n(\mathbf{r}) = \sum_{\mathbf{R}}\sum_\beta \delta[\mathbf{r} - \mathbf{R} - \mathbf{s}_\beta(-\mathbf{R}^\perp)]$$
$$\times W_\beta(-\mathbf{R}^\perp) \qquad (11)$$

where

$$W_\beta(\mathbf{r}^\perp) = \begin{cases} 1 & \text{if } \mathbf{r}^\perp \in v_\beta^\perp \\ 0 \end{cases} \qquad (12)$$

and the sum is over the direct hyperlattice whose reciprocal lattice is generated by the basis $\{\mathbf{b}_i\}$. The average density of a species β, easily calculated from this formula, is $v_\beta^\perp / v_c$, where v_c is the volume of the unit hypercell.

An example of a one-dimensional quasicrystal with $D = 2$ is shown in Fig. 12(a). Other examples in two and three dimensions include the Penrose lattice and its three-dimensional generalization, the Ammann lattice. The Penrose lattice requires $D = 4$, but a more symmetrical description can be obtained using $D = 5$. In this case the hyperlattice is simple hypercubic while the atomic surfaces are flat with $v^\perp$ being a rhombic icosahedron, the projection of the hypercubic unit cell into the $\perp$ space.

The Ammann lattice is an important example of the icosahedral quasicrystalline structure. Starting from a $D = 6$ hypercubic lattice aligned with the six-dimensional Cartesian axes, the physical space is spanned by the three vectors,

$$\begin{pmatrix} \hat{\mathbf{e}}_1 \\ \hat{\mathbf{e}}_2 \\ \hat{\mathbf{e}}_3 \end{pmatrix} = [2(t+2)]^{-1/2} \begin{pmatrix} t & t & 1 & 0 & 0 & 1 \\ 0 & 0 & t & 1 & -1 & -t \\ 1 & -1 & 0 & t & t & 0 \end{pmatrix}$$

$$(13)$$

while the $\perp$ space is spanned by the three vectors,

$$\begin{pmatrix} \hat{\mathbf{e}}_1^\perp \\ \hat{\mathbf{e}}_2^\perp \\ \hat{\mathbf{e}}_3^\perp \end{pmatrix} = [2(t+2)]^{-1/2} \begin{pmatrix} 1 & 1 & -t & 0 & 0 & -t \\ 0 & 0 & 1 & -t & t & -1 \\ -t & t & 0 & 1 & 1 & 0 \end{pmatrix}$$

$$(14)$$

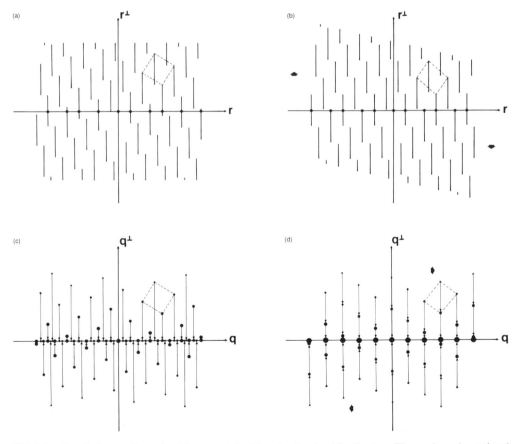

FIG. 12.   Cuts from two-dimensional hypercrystals: (a) an irrational cut leading to a Fibonacci quasicrystal and (b) a rational cut leading to a periodic crystal. Corresponding projections of the reciprocal lattice are shown in (c) and (d), respectively. Thick arrows indicate direction of the shear.

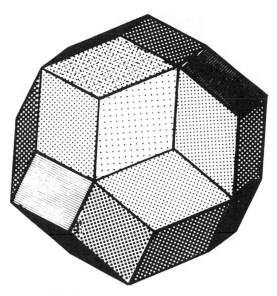

FIG. 13. The rhombic triacontahedron.

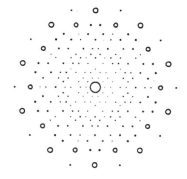

FIG. 14. Calculated diffraction pattern for the Penrose tiling. [From Jarić, M. V. (1986). *Phys. Rev.* **B34**, 4685.]

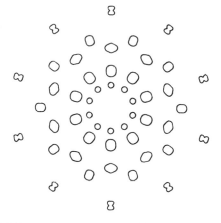

FIG. 15. A contour of constant diffuse scattering for the fivefold plane of an icosahedral quasicrystal. [From Jarić, M. V., and Nelson, D. R. (1988). *Phys. Rev.* **B37**, 4458.]

The atomic surface is again flat. Its domain is a rhombic triacontahedron, as shown in Fig. 13.

C. Diffraction

In order to calculate the diffraction intensities of a quasicrystal, it is necessary to recall Eq. (1) and the corollary stated in the introduction. Therefore, it is first required to Fourier transform the hypercrystal density. Then, the transform must be projected into the physical space to obtain the transform of the quasicrystal density. Finally, the projected transform is squared to obtain the intensity. In this way, using Eqs. (11–14), it follows that the diffraction pattern of a quasicrystal consists of Bragg peaks located at the wave vectors $\mathbf{Q}$ given by Eq. (6), whose intensity is

$$I(\mathbf{Q}) \sim |\sum_{\beta} f_{\beta} W_{\beta}(\overline{\mathbf{Q}})|^2 \qquad (15)$$

where

$$W_{\beta}(\overline{\mathbf{Q}}) = \int_{v_{\beta}^{\perp}} e^{-i\mathbf{Q}\cdot\mathbf{s}(\mathbf{r}^{\perp})-i\mathbf{Q}^{\perp}\cdot\mathbf{r}^{\perp}} \, d^{D-3}r^{\perp} \qquad (16)$$

and f_{β} is the atomic scattering amplitude. As an illustration, the diffraction pattern calculated for Penrose tiling is shown in Fig. 14.

In a more realistic case, Eq. (15) must be generalized to take into account thermal or static disordering. The effect of disordering is, to lowest order, to multiply the intensities by a Debye Waller factor:

$$f_{DW}(\overline{\mathbf{Q}}) = e^{-\overline{\mathbf{Q}}\cdot\overline{\mathbf{c}}\cdot\overline{\mathbf{Q}}} \qquad (17)$$

which depends both on $\mathbf{Q}$ and $\mathbf{Q}^{\perp}$. The $D \times D$ matrix $\overline{\mathbf{c}}$ can be related to the elastic moduli of the quasicrystal. Similar to ordinary crystals, the intensity lost in the Debye–Waller factor reappears in the background as diffuse scattering that exhibits algebraic, $|\mathbf{q} - \mathbf{Q}|^{-2}$, singularities at the Bragg wave vectors. An important difference is that these algebraic peaks show large anisotropy mainly caused by the $\mathbf{Q}^{\perp}$ dependence, which is not present in ordinary crystals. This is illustrated in Fig. 15.

D. Symmetry

It is most convenient to discuss symmetries of quasicrystals in terms of the associated hypercrystal symmetries. However, it is necessary to keep in mind the fact that a quasicrystal is a cut through a hypercrystal so that some hypercrystal symmetries might be redundant. It is also necessary to distinguish between the symmetries of the structure, reciprocal quasilattice, and diffraction pattern.

A hypercrystal is invariant under D-dimensional lattice translations, which define its Bravais class. It is also invariant under rotations, which form its point group. Only rotations isomorphic to the rotations in the physical space need to be considered. The rotations and translations combine to form the space group of the hypercrystal.

For icosahedral quasicrystals, it is convenient to consider only the hypercubic lattices. The icosahedral point group can be lifted from the physical space into the hyperspace. It acts on the $\perp$ space as the algebraic conjugate representation to the representation acting on the physical space. There are five six-dimensional Bravais classes consistent with this symmetry. However, when the associated hyperlattices are projected onto the physical space, pairs of quasilattices related only by a scale change appear, leaving only three distinguishable icosahedral reciprocal quasilattices. These three Bravais classes can be identified with simple, face-centered, and body-centered hypercubic lattices. They are also referred to as simple or vertex, face-centered or edge, and body-centered or face icosahedral quasilattices, respectively.

When combined with the possible rotations and possible fractional translations, the six-dimensional analogs of the ordinary screw rotations and glide planes, these Bravais classes give rise to 16 and 11 icosahedral space groups in 6 and 3 dimensions, respectively. For each of these space groups, a specific relationship exists between the phases of the symmetry related Fourier amplitudes $n(\mathbf{Q})$. Symmetries of all points in the unit hypercell, the Wickoff positions, can also be determined for each of these space groups.

Two new types of symmetry can generally be identified in incommensurate crystals. They can be illustrated with icosahedral quasicrystals. The first one is generated by an icosahedral rotation in the complementary space. In the physical space, this produces a permutation of the reciprocal quasilattice. Although this permutation preserves the length of quasilattice vectors, it cannot be extended to a linear transformation over the entire physical space. Only in special cases will a quasicrystal's microscopic point group symmetry, which is the symmetry of its diffraction pattern, also be its macroscopic point group.

The second type of a new symmetry is the so-called inflation symmetry, invariance of the reciprocal quasilattice under a discrete scale transformation $\mathbf{q} \rightarrow t^{3m}\,\mathbf{q}$, where m is any integer. This symmetry is generated by changing the shape of the unit hyper-

cell. Note that the diffraction pattern is not invariant under this symmetry, only the set of the positions of the Bragg spots is invariant. An incommensurate crystal also cannot have the inflation symmetry in direct space.

The diffraction pattern of an incommensurate crystal is invariant under an arbitrary translation of the hypercrystal. Translations parallel to the physical space correspond to ordinary translations of the quasicrystal. Translations perpendicular to the physical space will generally give rise either to an ordinary translation of the quasicrystal or to a new quasicrystal structure. Since the new quasicrystal structure so generated will have the same diffraction pattern as the original one, it follows that it will also have the same pair distribution function. Consequently, it will also have the same energy.

IV. Atomic Structure

As mentioned in the introduction, there are three basic approaches to describing real quasicrystals: the large unit cell crystal description (with or without icosahedral twinning), the bond–orientationally ordered glass model, and the ideal (quasiperiodic) quasicrystal model. In the case of icosahedral quasicrystals, most of the current structural models are based on the assumption that the real structure can be viewed as a packing of icosahedrally symmetric clusters.

Several classes of icosahedral quasicrystals have been identified. The first quasicrystal, i(Al–Mn) is a representative of the i(Al–Tm) class, the shechtmanites, where i($\cdots$) denotes the icosahedral phase and the transition metal (TM) could be Mn, Fe, Cr, Pd, etc. This class also includes i($Al_{74}Mn_{20}Si_6$), i($Al_{79}Cr_{17}Ru_4$), and many other compositions. As can be seen from Fig. 16, the icosahedral phase often occurs over a wide range of stochiometry. Consequently, the precise stochiometry is often not indicated.

The i(Al–Mg–Zn) class includes i($Al_{25}Mg_{37}Zn_{38}$), i($Al_{44}Mg_{36}Zn_{15}Cu_5$), i($Al_{60}Li_{30}Cu_{10}$), and several other compositions. Two additional classes that have been suggested are represented by i($U_{20.5}Pd_{59}Si_{20.5}$) and i($Ti_{1.8}V_{0.2}Ni$).

The clusters relevant to i(Al–Mn–Si) and to i(Al–Li–Cu) quasicrystals are shown in Fig. 17. These clusters can be identified in crystal structures closely related to the quasicrystalline ones. We shall focus here on the models of ideal quasicrystals. However, since no detailed quantitative structure models are available at present, we shall only present the main ideas on which they are based.

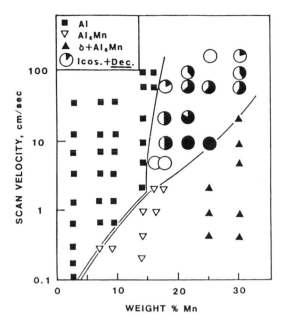

FIG. 16.   The phases dominating resolidified surface melts in Al–Mn alloys. [From Schaefer, R. J., and Bendersky, L. A. (1988). *In* "Aperiodicity and Order": Vol. 1, "Introduction to Quasicrystals" (M. V. Jarić, ed.), p. 124. Academic Press, Boston.]

A. GENERAL

In order to model the structure of real quasicrystals it is sufficient to determine the atomic hypersurfaces decorating the associated hypercrystal. However, the only meaningful description of a structure is one that involves a finite, preferably small, number of parameters. Consequently, an implicit assumption in the hypercrystal approach is that the atomic hypersurfaces can be specified using a small number of parameters, which can be fitted to the experimental data.

Current attempts to model quasicrystals follow several approaches. In the first approach, a rigid geometric network, or a tiling, is established and the icosahedral clusters are placed on its vertices (this approach is also employed in the icosahedral glass models). For example, the Ammann lattice can be used to provide the ideal quasicrystal network. The structure factor splits in this case into a product of the structure factor of the network and that of the cluster. A variant of this approach is to start from a tiling, such as the Ammann tiling, which is based on only two types of rhombohedral tiles, and then to decorate the basic tiles. Unlike the previous case, the structure factor does not split into a product but rather into a sum of products, one term for each tile type.

Typically, placing the basic clusters in a geometrical network does not completely specify the atomic structure of the quasicrystal. Large gaps that are,

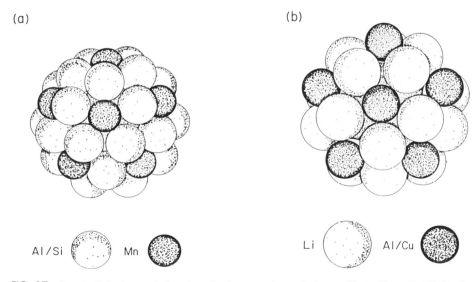

FIG. 17.   Icosahedral clusters believed to dominate quasicrystal phases. [From Elser, V. (1989). *In* "Aperiodicity and Order": Vol. 3, "Extended Icosahedral Structures" (M. V. Jarić and D. Gratias, eds.). Academic Press, Boston.]

however, too small to accommodate the entire clusters remain. They must be filled with the so-called glue atoms. The precise structure inside these gaps can not be determined in this approach, although from the point of view of physical properties of quasicrystals, the glue is as important as the cluster structure.

A quasicrystalline structure obtained by these approaches can always be lifted into six dimensions to give the atomic hypersurfaces. Conversely, it has been attempted to model i(Al–TM) and i(Al–Mg–Zn) by directly postulating certain simple atomic hypersurfaces. A disadvantage of such an approach is that it offers little control over the details of the real-space quasicrystal structure. For example, unphysically short interatomic distances are often generated in this way.

B. CRYSTAL APPROXIMANTS

An alternative approach is to relate the quasicrystal structure to the structure of a related crystal by considering the crystal as a rational (commensurate) approximant of the quasicrystal. That is, the crystal is viewed as a cut through a uniformly strained (sheared) hypercrystal. This is illustrated in Fig. 12(b) for a one-dimensional case. As the distortion brings the hyperlattice into a rational relationship with the cut, for each $\mathbf{Q}_R$, there exists an infinite number of reciprocal hyperlattice vectors $\overline{\mathbf{Q}}_R$ with the identical $\mathbf{Q}_R$ but different $\perp$ projection $\mathbf{Q}^\perp$. Since the atomic hypersurfaces associated with a quasicrystal are bounded, two crystal approximants resulting from two different uniform shifts of the hyperlattice will not be equivalent in general.

The two main steps in the crystal approximant approach are (1) to determine the crystal structure using the conventional crystallography and (2) to lift the known crystal structure into the six-dimensional hyperlattice to give a discrete set of points that are then used to interpolate the atomic hypersurfaces.

This approach is partially motivated by the suggestions that several crystal structures in the i(Al–TM) class appear as successively better rational approximants to the icosahedral quasicrystalline structure. Each approximant would give an independent set of points that could be used in interpolating the atomic surfaces. Presently, the structure of only one of these approximants is known for each of the two main classes of icosahedral quasicrystals, the cubic alpha(Al–Mn–Si) and R(Al–Li–Cu).

This approach is also partially motivated by the systematics it offers in the quasiperiodic repacking of

the atomic environments (clusters) found in the related crystalline structure. It equally systematically describes the departures from these clusters and, consequently, explicitly describes the structure of the glue.

How good a candidate for this approach is a particular crystal structure can be determined by directly using the experimental data. The direct comparison between the structure functions, or their inverse Fourier transforms (the Patterson functions), cannot answer much more than to indicate to what extent the local structures in the crystal and the quasicrystal are similar (see Fig. 5). However, as we learned from studies of ordinary incommensurate crystals, it is often useful to lift the incommensurate Patterson function into the hyperspace. Indeed, the remarkable simplicity of the six-dimensional Patterson functions of i(Al–Mn–Si) and i(Al–Li–Cu) is suggestive of a simple hypercrystal structure. As seen in Fig. 18, these Patterson functions have only two pronounced peaks, at the vertex and the body center of the hypercubic unit cell.

A further test is obtained by comparing the cut through the Patterson function of an appropriately strained hypercrystal with that of the crystal. It is important to note that the Patterson function of a rational cut is not generally equal to the rational cut of

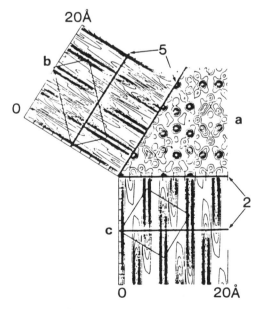

FIG. 18. Hypercrystal Patterson function of i(Al–Mn–Si) along two rational planes and the physical plane perpendicular to the twofold symmetry axis. [From Gratias, D., *et al.* (1988). *Phys. Rev.* **B38**, 1643.]

the Patterson function. The approximate equality only holds when the intensity $I(\mathbf{Q})$ is decaying sufficiently rapidly in the $\perp$ directions, as in the case of i(Al–Mn–Si). However, the decay is not sufficiently rapid for i(Al–Li–Cu). The intensity for the rational cut is given by

$$I_R(\mathbf{Q}_R) = |\sum_{\mathbf{Q}^\perp} \sqrt{I(\mathbf{Q}_R)} e^{i\theta(\mathbf{Q}_R) - i\mathbf{Q}^\perp \cdot \mathbf{u}^\perp}|^2 \quad (18)$$

where $\mathbf{u}^\perp$, the shift of the hyperlattice, is to be chosen to maximize the agreement. The sum over all $\mathbf{Q}^\perp$ with a given $\mathbf{Q}_R$ is infinite.

Therefore, it is generally necessary to test phases $\theta(\mathbf{Q})$ consistent with each of the allowed quasicrystal space groups. If a particular choice leads to a good agreement between the Patterson functions of the rational cut and the actual crystal, then, and only then, is it meaningful to try to use the known crystal structure to interpolate the atomic surfaces of the hypercrystal. Indeed, for the symmorphic, centrosymmetric space group $P\bar{5}32/m$ and an appropriate choice of $\theta(\mathbf{Q})$, the agreement between the rational cut through i(Al–Li–Cu) and the real crystal is considerably improved over that obtained by a simple rational cut through the Patterson function. Therefore, it should be possible to use the crystal structure of R(Al–Li–Cu) to interpolate the atomic hypersurfaces of the i(Al–Li–Cu) class.

However, this approach should not be pushed too far. The discrepancies between real crystals and rational approximants to quasicrystals can occur for several reasons. First, the atomic form factors depend on the local environments and are therefore different between crystals and quasicrystals, and by implication between crystals and rational approximants. Second, there are no phason fluctuations in crystals, whereas rational approximants carry over the phason fluctuations of quasicrystals. Third, without knowing the precise mechanism of quasicrystal disordering that causes finite peak widths, lifting of the diffraction intensities from the physical space into the hyperspace is not unique. Consequently, it is essential that the nature of quasicrystal disordering be understood.

V. Physical Properties

A. ELASTICITY

Like all incommensurate crystals, quasicrystals have low-energy degrees of freedom that are not present in ordinary crystals. They are associated with the relative change of the phase of the density waves and they are called phasons. In terms of the associated hypercrystal, the low-energy degrees of freedom have a simple description: phonons are associated with displacements of the hypersurfaces parallel to the physical space, while phasons are associated with displacements perpendicular to the physical subspace. The energy of a quasicrystal is unchanged as a hypercrystal is uniformly translated, although the structure of the quasicrystal might change.

Elastic distortion of a quasicrystal can be defined in analogy with ordinary crystals: a uniform strain is imposed on the hyperlattice, and then, the atomic surfaces are allowed to relax to minimize the energy. In this way an equilibrium strained quasicrystal is obtained. The difference between the energies of such equilibrium strained and unstrained quasicrystals defines the elastic energy associated with the strain. This strain matrix can be directly measured from the shift of the diffraction spots:

$$\mathbf{Q} \rightarrow \mathbf{Q} - \mathbf{e} \cdot \mathbf{Q} - \mathbf{e}_s \cdot \mathbf{Q}^\perp \quad (19)$$

where $\mathbf{e}$ is the usual symmetric strain, while $\mathbf{e}_s$ is the phason strain. Clearly, the other blocks of the hyperstrain matrix $\bar{\mathbf{e}}$ will not contribute to any change of the quasicrystal structure, beyond a change produced by a uniform phonon or phason translation.

For ordinary incommensurate crystals, for which the atomic hypersurfaces connect into unbounded hypersurfaces, the elastic energy is quadratic, as in ordinary crystals:

$$E = (\tfrac{1}{2}) \, \bar{\mathbf{e}} : \bar{\mathbf{C}} : \bar{\mathbf{e}} \quad (20)$$

where $\bar{\mathbf{C}}$ is the generalized elastic modulus tensor. However, if the atomic hypersurfaces are bounded, as in quasicrystals, the situation is more complex. At zero temperature the phason elastic energy is singular, e.g., $\sim |\mathbf{e}_s|$. The form of the elastic free energy at nonzero temperatures is not known. However, density functional calculations indicate that at a sufficiently high temperature it could be quadratic, as in Eq. (20). This is currently a field of intense investigations.

A related and equally poorly understood question is that of phason dynamics. As mentioned, a phase displacement relates two equal-energy quasicrystal structures. However, it involves atomic rearrangements that cause the two equivalent structures to be separated by an energy barrier. Therefore, the phason dynamics is expected to be diffusive and extremely slow. Results of preliminary ultrasound attenuation experiments are puzzling and seem to indicate unusual relaxation processes seen in neither glasses nor in large unit-cell crystals. They might reflect some unusual phason dynamics. Another consequence of the slow phason dynamics is the low mobility of dislocations and other structural defects. This conclusion is supported by the experimentally observed brittleness of quasicrystals.

Because of the very slow phason dynamics, the response of the quasicrystal will generally depend on the time scale of observation. On the short time scales, the phasons will be frozen and the quasicrystal will respond like an ordinary isotropic solid with the elastic modulus tensor $\mathbf{C}$. However, on a longer time scale, on which the phasons can relax, the quasicrystal will respond to the external stress with the effective elastic modulus tensor:

$$\mathbf{C}_{\text{eff}} = \mathbf{C} - {}^{t}\mathbf{C}_{i} : \mathbf{C}_{s}^{-1} : \mathbf{C}_{i} \qquad (21)$$

where $\mathbf{C}_i$ and $\mathbf{C}_s$ are the phonon–phason and the phason–phason elastic modulus tensors, respectively. Experiments at short time scales and model calculations indicate that these elastic moduli are not dramatically different from the elastic modulus tensor of a crystal.

The low-energy, long-wavelength phason excitations will be present in a quasicrystal at any finite temperature. If the usual hydrodynamic energy density is assumed for phasons, then thermal or quenched static disordering will be responsible for a Debye–Waller factor, as in Eq. (16). The next higher order contribution to the diffraction intensity will be manifested in the characteristic $|\mathbf{q} - \mathbf{Q}|^{-2}$ diffuse scattering around each Bragg peak. Both of these contributions show a $\mathbf{Q}^{\perp}$ dependence in addition to the usual $\mathbf{Q}$ dependence (see Fig. 15).

B. Electronic Structure

Electronic properties of quasicrystals are poorly understood theoretically as well as experimentally. They are supposed to lie somewhere between crystals with only extended states and random systems in which Anderson localizations might occur. Experimental investigations are seriously hampered by the unavailability of good, single quasicrystal samples. However, several preliminary experiments indicate the low-temperature behavior suggestive of weak localization effects.

From a theoretical point of view, electronic properties of quasicrystals are extremely difficult to understand. Not even a simple, exactly solvable quasicrystal model is known in two- or three-dimensions. Naive perturbation theory, the nearly free electron approach, suggests the presence of a Cantor-set spectrum. However, a more careful examination of the perturbation approach reveals convergence difficulties caused by the small denominator problem.

The tight binding approach also runs into difficul-

ties because of the complicated quasicrystal geometry. Not only are the analytical results unavailable, but it is also very difficult to obtain reliable numerical results. Current numerical simulations are carried out on insufficiently large three-dimensional systems.

C. Magnetism

A small number of experiments investigated magnetic properties of quasicrystals. Susceptibility measurements in i(Al–Mn–Si) and i(U–Pd–Si) show several interesting features. At high temperatures, the susceptibilities follow the Curie law indicating that Mn and U have a small local moment. However, Mn has no moment in the related crystalline structure α(Al–Mn–Si).

On the other hand, Mössbauer experiments on i(Al–Mn–Si) can be interpreted by assuming a distribution of local Mn environments, in agreement with the quasicrystalline structure. Presumably, only Mn, on certain sites different from the sites in α(Al–Mn–Si), carries a magnetic moment. It will be difficult to develop a theoretical understanding of the appearance of local moments before the structure of the quasicrystal can be determined.

With the presence of magnetic moments, a question of possible magnetic ordering arises. Indeed, at temperatures of several degrees Kelvin, the susceptibility shows a broad maximum that could indicate some kind of antiferromagnetic or spin-glass ordering. No theories of magnetic ordering on three-dimensional quasilattices exist at present. Even the simplest, mean-field theories are nontrivial in this case.

VI. Conclusion

Four years after the discovery of quasicrystals, it is apparent that a description of their structure in terms of ideal quasicrystals is appropriate. A description in terms of large unit-cell crystals whether twinned or not, although possible in principle, does not seem effective since the experimental measurements set the lower limit on the unit-cell size at 50–100 Å. Therefore, even if the real quasicrystals were truly large unit-cell crystals, their structures, as well as physical properties, would be best described as rational approximants to a quasicrystalline structure.

Another competing description, the icosahedral glass model, requires the introduction of a large degree of order that is evident in experiments but lacking in the simplest version of the model. Therefore, such a model could also be described in terms of disordering of an otherwise ideal, quasicrystalline structure. The difference between the disordered quasicrystal and the ordered glass models is not purely semantic: an important advantage of the quasicrystal approach is the existence of a reference structure, presumably characterized by a small number of parameters.

It should be clear from this overview that much experimental and theoretical work on quasicrystals is yet to be completed. In particular, although the initial effort should go toward the determination of the atomic structure of quasicrystals, their electronic and magnetic properties also need further investigation.

Because of their unique structure, neither crystalline nor amorphous, quasicrystals will necessarily exhibit unique physical properties that will earn them a specific technological niche. Some of the more obvious applications might include the usage of quasicrystals as soft magnets, solid lubricants, catalysts, etc. It is certain that we have just witnessed the beginning of a new and fertile field.

BIBLIOGRAPHY

Currat, R., and Janssen, T. (1988). *In* "Solid State Physics" (H. Erenreich and D. Turnbull, eds.), Vol. 41. Academic Press, Boston.

Gratias, D. (1986). *La Recherche* **17,** 788 (in French), [*Contemp. Phys.* **28,** 219 (1987)].

Gratias, D., and Michel, L., eds. (1986). *Proc. Int. Workshop Aperiodic Cryst. J. Phys. (Paris) Colloq.* **47**(C3).

Grunbaum, B., and Shephard, G. C. (1987). "Tilings and Patterns." Freeman, San Francisco.

Henley, C. L. (1987). Quasicrystal order, its origins and its consequences: A survey of current models, *Comments Cond. Mat. Phys.* **13,** 59–117.

Janot, Ch., and Dubois, J. M. (1988). *J. Phys. F.* **18,** 2303.

Jarić, M. V., ed. (1988, 1989). "Aperiodicity and Order:" Vol. 1, "Introduction to Quasicrystals;" Vol. 2, "Introduction to the Mathematics of Quasicrystals." Academic Press, Boston.

Jarić, M. V., and Gratias, D., eds. (1989). "Aperiodicity and Order:" Vol. 3, "Extended Icosahedral Structures." Academic Press, Boston.

Nelson, D. R. (1986). *Sci. Am.* **255**(2), 42.

Steinhardt, P. J., and Ostlund, S., eds. (1987). "The Physics of Quasicrystals." World Scientific, Singapore.

RADIATION PHYSICS

Andrew Holmes-Siedle   *Fulmer Research Laboratories*

I.   Description of Radiation Physics
II. Radiation Sources
III. Quantifying Radiation
IV. Interaction with Matter
V.   Radiation Effects in Systems and Technology
VI. Useful Radiation Data

GLOSSARY

Absorber: Matter intervening in the path of a radiation beam that absorbs energy from that beam.

Absorption coefficient (μ): Coefficient in Lambert's law, $I = I_0 e^{-\mu d}$, which describes the absorption of the radiation penetrating a material.

Bremsstrahlung: X ray produced when particles (electrons mainly, but also protons) are slowed by interaction with matter. Bremsstrahlung (German, braking radiation) is sometimes called white radiation to distinguish it from the characteristic line spectra produced by elements.

Bulk damage: Disturbance of the lattice structure of matter by the displacement of atoms. Produced by energetic particles and often used in contrast to the terms ionization effect or surface damage.

Dose (D): Used broadly for energy from radiation accumulated in matter. Used in dosimetry for the energy absorbed per unit mass of material, usually by ionization processes. Units are the rad and Gray, which are equivalent, respectively, to 100 ergs gm^{-1} and 1 J kg^{-1}. Therefore, 1 rad = 1/100 Gray or 1 cGy.

Dose rate: Rate at which energy is transferred to a material by a radiation beam, for example, rad per sec (or Gray per sec).

Energy flux: Passage of energy in the form of penetrating particles, not necessarily stopped. Typical units are J cm^{-2} and W m^{-1}.

Exposure: Important term in dosimetry that expresses a fluence in terms of its effect on a dosimetric medium, nearly always air at STP, and the number of air ions produced per unit mass. An exposure of 1 Roentgen is exposure to the fluence of a given radiation that, in air at STP, generates 2.58×10^{-4} coulomb (C) of ionic charge per kilogram. Used in contrast to dose, which, for a given fluence, varies from absorber to absorber.

Fluence: Time-integrated flux of particles or photons. Unit: cm^{-2}. It is useful to add the symbol for the particle, for example e cm^{-2}, or even 1 MeV e cm^{-2}.

Flux: Number of particles passing through some defined zone per unit time. For parallel beams, this is a unit area; for omnidirectional radiation, the zone chosen is usually a sphere with cross section of 1 cm^2. In both cases, the unit is $cm^{-2} sec^{-1}$.

Gray: Radiation absorbed dose unit of the Systeme Internationale, of value 1 J kg^{-1} and equal to 100 rad.

Hardening, hardened: Used to describe improvement in the tolerance of a device or a system to a radiation environment. Originates from the military term for a site that is invulnerable to attack. Secondly, denotes increase of average energy of a beam of radiation due to the selective removal of a lower-energy component of the beam. Thirdly, denotes the changes in mechanical properties of some metals, induced by high fluences of particle radiation.

Integrated circuit: Semiconductor chip on which a large number of interconnecting device functions have been formed.

Ionization effects: Large class of radiation effects that involve the removal of an electron from the ground energy level in an atom and its escape from that atom. For semiconductor devices, the term is used in contrast to bulk damage.

Logic upset: Change in logic state resulting when pulses of radiation of high dose rate generate photocurrents of significant magnitude in semiconductor junctions. These can so alter the voltages at circuit nodes that logic circuits misinterpret the disturbance as a logic signal and change logic state.

Range: Distance into an absorber to which a particle is likely to penetrate. Since stopping is a statistical process, several ranges (practical, maximum, extrapolated, etc.) are defined.

Roentgen: Unit of exposure equivalent to the generation of 2.58×10^{-4} C of air ions per kilogram of air.

Shield: Absorber structure giving protection from radiation.

Single-event upset: Logic upset produced by a single energetic ion.

Surface effect: In metal–oxide semiconductor (MOS) devices and bipolar transistors, the various effects that occur in the passivating surface layers.

Radiation physics deals with the wide range of effects observed when high-energy radiation, in the form of particles or photons, interacts with matter. The field of radiation physics is distinct from nuclear physics. Radiation physics covers the whole range of chemical and physical effects that follows the above interaction, while nuclear physics covers the internal structure of the atomic nucleus. The radiation of interest includes X ray, α-, β-, and γ-rays, cosmic rays, and many forms of artificially accelerated particles, such as beams of electrons, protons, or ions. Important sources of radiation are electrostatic generators including X-ray machines, geomagnetically trapped charged particles, and the neutron and gamma radiation given off during the decay, fission or fusion of radioisotopes, such as in nuclear reactor materials.

The use of radiation sources in medicine requires the use of radiation physics techniques in the planning and control of treatment. Many details of the interaction of radiation with living organisms fall under a separate subject heading, namely, radiobiology. Some machines must survive in high radiation environments (e.g., in space). The field of radiation effects is a branch of applied radiation physics concerned with achieving tolerance to radiation in such systems.

I. Description of Radiation Physics

Radiation physics covers the effects that occur when high-energy radiation interacts with matter. This field is distinct from nuclear physics, even though many of the particles involved are common to both. The radiation types of interest include photons having high energy, such as X rays and γ rays, and a multitude of particles, charged and uncharged. Radiation energy is expressed in kilo- or mega-electron volts (keV or MeV). [*See* NUCLEAR PHYSICS.]

The radiation in question arises from both natural and artificial sources. Natural sources include space radiation and radium and uranium in rocks. Artificial sources include concentrated radioisotopes such as cobalt-60 irradiators, accelerators such as X-ray and electron-beam machines, and nuclear chain reactions in reactors or weapons.

The techniques of radiation physics are used in academic research, industrial technology, aerospace technology, and medicine. In some machines, materials and devices have to withstand high doses of radiation. In the treatment of tumors, beams of radiation have to be generated and controlled with great accuracy. Thus, while the effect of radiation on living tissue falls into the field of radiobiology, the task of irradiating tissue is often carried out by experts in radiation physics.

The field of radiation effects is an important part of radiation physics. Radiation deposits energy in matter. This energy is then distributed among the atoms and molecules in a great variety of ways. The energy excites atoms to higher energy states or moves them about in the material. Figure 1 shows a flow chart containing a list of effects that can be observed when radiation interacts with a solid. A serious disruption in the existing order in the solid can be produced, and this disruption is often called radiation damage. In liquids and gases, the final effects of radiation are very different from those observed in solids. This is because, in fluids, the excited atoms and molecules can move long distances by diffusion and thereby interact chemically with other species. (The field of radiation chemistry and living tissues are not discussed specifically here.)

The scientific disciplines employed in radiation physics center around the interactions of photons or particles with solids, liquids, and gases. The primary transfer of energy is complex but well understood; but the energy trans-

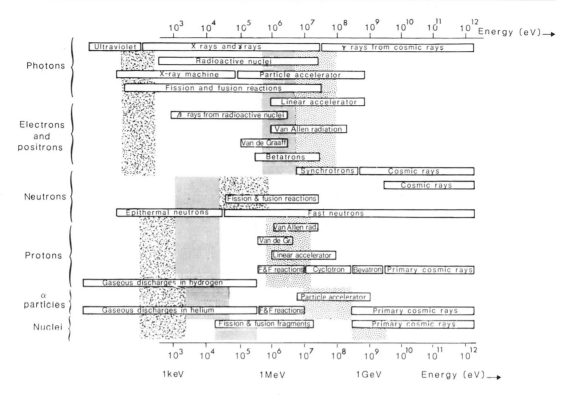

FIG. 1. Energy values for various radiation environments and threshold energies. Approximate threshold for producing ionization ▓, for atomic displacements in solids ▒, and for nuclear transmutations ░. Note: F & F = fission and fusion. (Courtesy of Hughes Aircraft Co.)

ferred then dissipates via secondary interactions with the target material. As Fig. 1 shows, secondary processes are complex and not well understood except in a few cases, such as the various results of the displacement of atoms in silicon by an electron beam. Certain classes of material, important in technology, show very strong responses to radiation. The disciplines of solid-state and plasma physics are important in understanding these responses.

The technological aspects of radiation physics are very wide. Instruments for measuring radiation (dosimeters and counters) are widely used. The prevention of degradation in electronic and optical materials is important. Furthermore, in some cases, controlled irradiation can improve commercial products. For example, food preservation and the toughening of plastics are sometimes most economically achieved by irradiation with noncontaminating radiation (electrons or γ rays). Infectious materials such as sewage can be sterilized by radiation. Radioisotopes that are useful in irradiation experiments are listed in Table I.

II. Radiation Sources

A. OVERVIEW

Figure 1 shows a survey of the broad range of energy values that have to be considered under the term radiation. Most of the radiation of interest has an energy above 10^3 electron volts (eV), but neutrons with much lower energies are still described as particle radiation. Ultraviolet photons also have sufficient energy to cause the same chemical effects as we see with X rays and very high-energy photons. The shaded areas in Fig. 1 show that different fundamental effects in matter occur at different radiation energies.

B. PHOTONS

Photons that have energies in the kilovolt and megavolt range (X rays and γ rays) have the ability to penetrate matter deeply and, when absorbed, to produce strong effects. X rays are generated when an electron beam strikes matter; an X-ray generator consists of a powerful electron gun and a metal target in which the photons are generated. Energies from 30 keV to 3 MeV

TABLE I. Radioisotopes Useful in Irradiation Experiments: Main Emission Energies

Nuclide	Half-life (yr)	Type of decay	Photons Energy (MeV)	Photons Percent emitted (%)	Particles Energy (MeV)	Particles Transition probability (%)
^{60}Co	5.27		1.173	99.86	0.318	99.9
		β^-	1.333	99.98	1.491	0.1
			(av 1.25)			
^{192}Ir	0.526		0.296	29.6		
	(192 d)		0.308	30.7		
			0.316	82.7	0.530	42.6
		β^-	0.468	47.0	0.670	47.2
			0.604	8.2		
			0.612	5.3		
^{137}Cs			0.662	85.1		
					0.512	94.6
	30		0.032 ⎫		1.174	5.4
		β^-	−0.038 ⎬ 8			
			(^{137}Ba) ⎭		Plus internal conver-	
			K X rays		sion electrons	
					0.65 MeV	
^{90}Sr + daughter	28	β^-	0.54	100	—	—
^{90}Y	0.176	β^-	2.27	100	—	—
^{85}Kr	10.6	β^-	0.15	0.7	0.51	0.7
		β^-	0.67	99.3	—	—
^{252}Cf	2.65	Spontaneous fission	—	—	Neutrons, 2 MeV γs, 5.9–6.1 MeV Fission fragments, 80 and 104 MeV	

are common. Small fluxes of X rays are also generated naturally in radioisotope samples by the collision of β rays with the sample itself or with the capsule in which the sample is contained. In electron-beam devices (cathode-ray tubes and electron accelerators), a hazard can be created by the unintentional generation of X rays if the beam collides with the wall of the chamber.

γ rays are high-energy monoenergetic photons. The term is used specifically for photons created during the disintegration of atomic nuclei. A well-known example is the pair of photons created in the spontaneous disintegration of the cobalt-60 atom. These photons have energies of 1.173226 and 1.332483 MeV. γ rays are created by nuclear chain reactions such as those that occur in a nuclear reactor core or a nuclear explosion. The isotopes contained in nuclear fuels represent concentrated sources of γ rays. These can be used for experimental irradiation in spent-fuel ponds. When certain equipment used in reprocessing nuclear fuel is directly ex-

posed to the isotope sample, there is danger that exposed optical and electronic parts may degrade in performance. Thus, the equipment has to be radiation hardened (see later).

C. ELECTRONS

Electron accelerators range in energy from 0.1 keV to 50 MeV; the beam currents range from microamperes to kiloamperes. To reach energies above 50 MeV, electrons are accelerated by radio-frequency energy in a machine called a linear accelerator. Natural sources of high-energy electrons include β rays from isotopes and electrons in space which have been accelerated in magnetic fields. There is a particularly high concentration of electrons trapped around those planets that have high magnetic fields, such as Earth and Jupiter. It is inconvenient for space vehicle designers that these trapped radiation belts cover the altitudes used by unmanned operational satellites. The radiation dose acquired in these regions is a significant source of radiation damage to components. Note that the damage problem applies mainly to *unmanned* satellites. Manned vehicles at present avoid the trapped radiation belts for the sake of the men on board.

D. PROTONS

Proton beams can also be produced by acceleration. One especially well-known form of proton accelerator is the cyclotron, which uses radio-frequency energy. Nuclear reactions also produce protons in a material sample. Proton beams are emitted by the sun, and protons are also found trapped in planetary magnetic fields. High-energy protons are emitted from the sun in bursts associated with solar flares. A less energetic, steady stream of protons is emitted by the sun and is called the solar wind.

E. IONS

α particles (which are ions) consist of high-energy helium nuclei; these too are emitted by radioisotopes and stars and thus are found in interplanetary space. Ion beams can be generated in accelerators. One use for ion beams is the ion implantation of solids to modify their properties. High-current ion implanters are available for industrial use. These beams produce large amounts of radiation damage in the solid so treated. Very energetic ions of all known atomic masses are found in space. These are called cosmic rays.

F. NEUTRONS

The main sources of neutrons are nuclear fission and nuclear fusion reactions. Of these, the fission of uranium is the most common. The primary product of fission is fast neutrons having an energy distribution described as a fission spectrum. This spectrum has a large content with energies above 1 MeV. This is the spectrum that would be observed near a nuclear explosion. In a nuclear reactor, interaction with dense surrounding materials reduces neutron energy. Thermal neutrons or cold neutrons of much lower energy are produced. On collision with matter, fast neutrons produce much damage while thermal neutrons produce radioactivation. Atomic fusion reactions produce neutrons having a much higher energy than fission. For example, one commercial generator of fusion neutrons produces a beam of D–T neutrons at a single energy of 14 MeV.

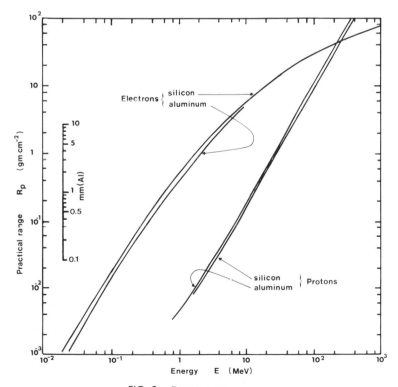

FIG. 2.   Range–energy curves.

III. Quantifying Radiation

A. Flux and Fluence

A parallel beam of radiation passing through free space can be quantified by quoting the number of particles passing through a unit area. See flux, fluence, and energy flux in this article's glossary. If the particles come from a variety of directions (as in outer space), then we quote the number intersecting a sphere of a given cross-sectional area.

B. Energy Spectrum

A curve plotting the population of particles in a given energy range is called the energy spectrum of the particle.

C. Exposure

In radiation physics, we are often concerned with the end-result of the absorption of radiation-borne energy in solids. For this, we first need to quantify the fluence and energy of the particles that impinge on the solid. Even if we cannot do this, we can at least specify an observable effect in a familiar medium such as air. These forms of statement of a quantity of radiation do not describe the exact energy absorbed in a given material. Therefore, we are only specifying an exposure value.

We express exposure in two ways:

1. We note the fluxes impinging on the solid of interest and the energies of the particles or photons concerned, such as 10^{15} cm^{-2} of 1 MeV electrons of 10^{12} cm^{-2} of fission-spectrum neutrons.
2. We measure the quantity of ionization produced by such a flux in a standard medium, usually air.

Method 2 arises from the fact that air ionization chambers are routinely used for quantifying ionizing radiation. Exposure in air due to a given radiation source is expressed in roentgen units, or coulombs per kilogram, based on the number of ions created in a given volume of air. However, for the radiation testing of electronics, it is better to express exposures by method 1.

D. Dose and Kerma

The term dose is a useful general description of the energy per unit mass that has been "dumped" in a material by a high-energy particle on its way through. Given a value for dose, we can calculate biological and nonbiological radiation effects. For crystalline solids such as silicon and metals, some further complications arise. It may be necessary to divide energy deposition into two fractions: ionization and atomic displacement. A new word for energy deposition has been introduced to assist with these distinctions. The word is *kerma*, meaning kinetic energy released in a material. To distinguish between the two forms of energy deposition, we qualify this word and speak of ionization kerma when referring to the fraction of energy going into ionization, and so on.

Radiation dose and kerma are both measures of energy deposited. The new international unit is the Gray (Gy), which represents energy deposition per unit mass of one joule per kilogram. Many authoritative publications still employ the older practical unit, the rad, representing 100 erg per gram. One roentgen represents 86.9 rads in air. A dose of 1 Gy thus equals 100 rad.

E. Range–Energy Relations

Photons and particles, when passing through a slab of material, are attenuated by collisions and other interactions. We can plot the intensity of the emergent radiation as a function of slab thickness d. In any cases, the attenuation can be expressed by an exponential law, of the form

$$\frac{I}{I_0} = e^{-\mu d}$$

where I_0 is the incident radiation intensity, I the emergent intensity, μ the absorption coefficient, and d the thickness of the slab.

For γ rays and X rays, this law is followed exactly for some geometries. For electrons, the law is followed over the early part of the curve but, after a certain distance, the practical range, no electrons emerge. For neutrons, the law is followed approximately, although the interactions with the atoms of the material are very different from those for electrons and γ rays.

For electrons and other charged particles there is a minimum slab thickness W that stops all the particles. This is called the "stopping range". We can plot curves of W versus energy. These are useful for calculations of shielding. Some of these are shown in Fig. 2. For X rays and γ rays, we plot attenuation curves, which

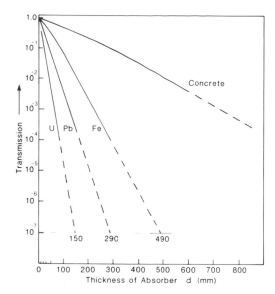

FIG. 3.   Attenuation curves for γ rays in various absorbers.

follow the mathematical law given above. Fig. 3 shows examples of these.

IV. Interaction with Matter

A. PHENOMENA OBSERVED DURING ABSORPTION OF RADIATION

1. General

The laws by which radiation is absorbed by matter are derived from multiple interactions of photons or particles with the atoms of the material. This section describes the primary processes, which are common to all matter. Later sections describe the mechanisms by which these primary processes are linked to the end effects such as atomic displacement and charge build-up.

2. γ Rays and X Rays

γ-ray photons carry little momentum and no charge. The initial interactions of γ photons with a solid are with the electrons. The probability of an interaction is low, so γ rays and X rays are regarded as penetrating radiation, compared with electrons and protons. For each element the probability of interaction with a photon is a function of the energy of the photon and the atomic number Z of the material. Interaction is by three processes: the photoelectric process,

the Compton scattering process, and pair production.

1. Photoelectric process. In this interaction a photon is completely absorbed by an atom, and an electron is ejected. Photon energy $h\nu$ must be greater than the electron binding energy B. The ejected electron has a kinetic energy of $h\nu - B$. The energy absorption spectrum for a particular material exhibits sharp discontinuities, showing that photoelectric processes are influenced by the electron shells of the atom.

2. Compton scattering. In this interaction, only part of the energy of the photon is transferred to an electron. Both the electron and photon are scattered from the collision in a forward direction. Scattering of high-energy photons tends to be sharply concentrated in the original direction. The electrons emitted may have energies as high as 1 MeV.

3. Pair production. In this interaction, high-energy photons act directly on the nucleus. The interaction occurs only if the photon has an energy greater than 1.022 MeV. The photon is annihilated, and an electron–position pair is produced. Any photon energy in excess of 1.022 MeV appears as kinetic energy, divided between the electron and positron. The positron is later annihilated by an electron to produce a pair of secondary photons traveling in opposite directions at a well-defined, single energy.

The absorption coefficient in a given material a arising from photoelectric (τ), compton (σ), and pair production (K) processes can be expressed as follows:

$$\mu_a = \tau_a + \sigma_a + K_a$$

3. Neutron Absorption Mechanisms

Neutrons are uncharged particles that can interact easily with a positively charged atomic nucleus where various capture processes can occur. The probability of such an interaction is low, and hence neutrons are regarded as highly penetrating radiation. In elastic scattering the neutron loses part of its energy by displacing atoms. It can be shown that energy transferable during collisions is greatest for atoms of low atomic weight, especially hydrogen. Therefore, hydrogenous materials are useful for neutron shielding and detection. Energetic protons are created during elastic collisions with hydrogenous materials such as polymers or water. After a number of collisions in a material at room temperature, the neutrons become thermalized, that is, their kinetic energy is of the order of 0.025 eV, the average vibrational energy of the atoms in a solid at room temperature. If the solid is cooled below room temperature, then neutrons of even lower energies, are produced. These are called cold neutrons and have uses in materials physics.

Thermal neutrons can be captured by atomic nuclei. This usually leads to the emission of a γ photon from the excited nucleus and the creation of a new isotope, a process known as transmutation. The new isotope may or may not be radioactive. If it is, then the process is termed radioactivation.

4. Electron (β-Ray) Absorption

The two mechanisms by which electrons shed their kinetic energy are by collision, which yields ionized or excited atoms, and the production of bremsstrahlung radiation.

1. Collisions. As they pass through matter, electrons gradually lose energy by means of collisions with atoms, along a wandering path. The collision energy is then usually taken up by other electrons; these ejected electrons are termed secondary electrons. Occasionally, atoms are displaced from their lattice positions by the collision, and a defect is left in the solid. Secondary electrons, after further collisions, generate electron–hole pairs or electron–ion pairs, depending of the medium in question. In a liquid or a gas, a complex track of electron–ion pairs is produced, which has a core of intense ionization and small spurs where secondary electrons have set off at an obtuse angle to the original track.

2. Electromagnetic radiation. High-energy electrons, when deflected by the electric field around a nucleus, produce X rays, also called bremsstrahlung radiation.

5. Charged Nuclei

Atoms moving at high energy can cause a trail of intense ionization. Particles of interest here include recoil protons, α particles, cosmic rays, ions in accelerator beams, and fission fragments. Light atoms produce a track of significant length. For example, a neon ion at 40 MeV has a track length approaching 100 μm. Over a core region having a diameter of a few nanometers, the dose imparted to the material in the track is 10^6 rads. Over a penumbra region having a diameter of about 10 μm the dose falls off to about 100 rads. Thus, the details of the effects of different energetic atoms on a solid vary strongly according to the atomic weight. This is important in radiation damage calculations and in single-event upsets.

Heavy atoms in motion are found in cosmic rays and fission fragments. During the fission of an atomic nucleus, a large amount of momentum is given to the two fragments of the nucleus. If the fission occurs in a solid, then the fragments are stopped rapidly. The resulting bulk damage to the solid is heavy and is localized in a small volume. Fission fragments from ^{252}Cf can be used in the study of single-event upsets (see glossary).

B. MECHANISMS OF RADIATION EFFECTS

1. Introduction

Radiation changes the performance of some engineering materials. Semiconductors, glasses, plastics, particularly susceptible, and some other materials are mentioned in Table II. The chemical mechanisms that lead to these changes can be divided into three stages:

1. Primary transfer of energy from the moving particle or photon to the stationary network of atoms. This was dealt with in a previous section.

TABLE II. Materials and Devices that Generally Have Poor Tolerance to Radiation

---

Semiconductors
Optical lenses
Optical fibers
Optical windows (e.g., encoder plates)
Elastomers (e.g., plastic bellows)
Plastic bearings
Lubricants
Adhesives
Hydraulic fluids
Paints
Reflective coatings
Wood, paper, string, and cloth
Thin insulators
Photosensitive materials
Gas sensors
Liquid-ion sensors
Surface-active reagents
Piezoelectric transducers
Micropositioners

2. Secondary stages during which that energy is spread to other sites in the solid.

3. Large variety of end effects—electrical, optical, and chemical—that occur when the energy, widely dissipated, takes part in a final, low-energy excitation mechanism.

Although the above effects are obviously complex, we can usually place them in one of two broad classes of effect, namely:

bulk damage, which proceeds through the displacement of atoms, and

ionization, which proceeds through charge generation and bond breaking.

Bulk damage is very important in materials that have to maintain a high degree of perfection to perform their function, such as crystalline semiconductors and optical media. For most other technological materials, ionization effects have a greater impact on performance.

The prediction of radiation effects in electronics is difficult, and the case of microelectronic devices is the most difficult. This is because, in semiconductor microcircuits, two forms of complexity are compounded. An integrated circuit is structurally complex and small; manufacturing technique is constantly changing. Add to this the complexity of the radiation mechanisms described above, and it is not surprising that the interpretation and prediction of effects in such components is a field for specialists. Thus the fields of radiation effects engineering and radiation hardening can be regarded as a combination of radiation physics and advanced solid-state engineering.

2. Ionization Effects

Ionization covers a broad range of phenomena in which electrons are excited and leave their parent atom. In inorganic solids, the positive charge or hole remaining on the atom is also mobile. The electron–hole pair created may recombine, or if a field is present, the pair may drift apart, producing electrical conduction and possibly a space-charging effect. Much of the research on ionization effects in the literature covers these types of behavior. However, in organic materials, another sequel to ionization can occur. In this case, the positive species created is not stable. The molecular fragments combine in new chemical forms (e.g., crosslinking in plastics; see later). These radiolytic effects are also part of the class of ionization effects. For solid-state electronics, charging is an important effect. The trapping of space charge occurs in thin-film oxide dielectrics, particularly when they are operated under a high applied field. This effect has a strong impact on the performance of metal-oxide-semiconductor devices. The effect is shown schematically in Fig. 4. The unwanted charge in the oxide film can change the function of a logic circuit so that the device becomes unusable. Because the oxide films used are insulators, the space charge may persist for many years.

Electrons, protons, and α particles produce intense ionization effects in solids. For electrons, the form of that ionization is a fairly diffuse cloud. For protons, α particles, and fission fragments, the cloud of ionization around the particle track is more intense and structured. Neutrons can also produce ionization in an electronic or optical solid in the form of intense local ionization.

3. Displacement Effects

The momentum imparted to an atom by collision with a high-energy particle may cause the atom to leave its position in the lattice. The knock-on atom leaves a trail or "spike" of damage, as shown in Fig. 5. For any high-energy particle, the maximum energy that can be transferred to the knock-on atom after an elastic collision is a fraction of the energy of the particle. The fraction can be worked out by considerations of momentum exchange. For neutrons, electrons, and most other particles, the recoil energy E_R has a wide distribution; primary recoil atoms usually contain only a small fraction of

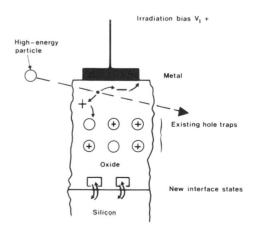

FIG. 4.   Ionization effects in a thin oxide film.

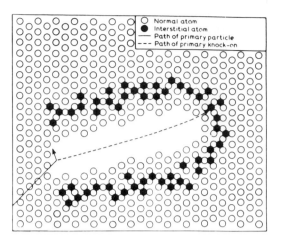

FIG. 5. Displacement effects in a crystal lattice, showing a spike of damage.

the energy E_0 of the incident particles that have struck them. For neutrons, the maximum possible energy that can be imparted to a silicon atom is about one-third of E_0, while for electrons, the value is more like one-tenth of E_0, the difference being due to the different masses of the particles. As a result, the pattern of damage left by neutrons and electrons is different. Defects are produced provided $E_R > E_D$, where E_D is the displacement energy threshold for the silicon crystal lattice. The value of this threshold depends on the energy and type of the radiation. The threshold energy value in silicon is between 15 and 30 eV.

Gaps in the lattice (vacancies) created by collisions may coalesce to form clusters of defects. The displacements due to particles may be distributed throughout the material as a set of isolated damaged regions, sometimes called clusters or spikes. These are composed of a complex arrangement of displaced atoms and other lattice defects.

C. DEGRADATION PROCESSES INTRODUCED BY RADIATION

1. Introduction

We have discussed the modes in which radiation imparts energy to solids in general. We now describe briefly the way in which particular solids react to that liberation of energy. The reactions depend strongly on the atomic arrangement of the material in question. In the case of electronic devices, differences such as chemical bond energy, the presence of $p–n$ junctions, or the presence of an interface give differences in the way the devices respond. Again, the electrical stress applied to an electronic device during irradiation determines how radiation-induced ionization behaves. Figure 6 illustrates the great variety of physical effects that can occur.

2. Effects of Ionization

a. Electrical. The primary interactions between energetic radiation and the electrons surrounding atoms is more complex and varied than the simple transfer of momentum to nuclei of atoms, which we describe later. Ionization produces very different phenomena in various materials. However, despite the initial variety of interaction, much of the energy ultimately converts to electron–hole pairs. In this process, valence-band electrons in the solid are excited to the conduction band and move if an electric field is applied. Under irradiation, any solid (even an insulator) conducts current. The positively charged holes are also mobile, but hole mobility is usually lower than electron mobility.

The net energy required to create an electron–hole pair is relatively small (e.g., 18 eV for SiO_2). Therefore, since momentum transfer is not involved, the energy of the radiation causing the ionization is not so critically important as in an atomic displacement effect. (Of course, the number of pairs created is dependent on the energy of the incident radiation.) Thus the ionization effects produced by neutrons and γ rays may often be simulated by radiation of much lower energy. We can, for example, use X rays, kilovolt electron beams, or even ultraviolet light. The amount of energy deposited in a material by means of ionization is conventionally termed the dose, measured in rads or Grays.

There are three important effects of ionization in semiconductor devices:

1. Photocurrents are produced when the electron–hole pairs are swept apart by a field, particularly the built-in field at a $p–n$ junction. Currents of this type produce malfunctions in integrated circuits, called whiteout in imaging devices.

2. Build-up of a net positive charge in oxide films. In oxides used as insulators or passivation in modern integrated circuits, the charge build-up may in turn produce inversion layers in the silicon. These give rise to severe leakage and the turn-off of some devices that are meant to be "on."

3. Disruption of chemical bonds at inter-

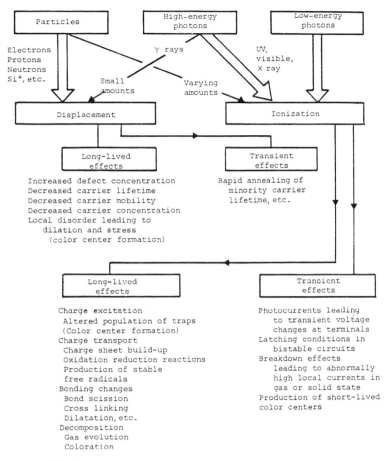

FIG. 6. Summary of radiation-induced responses in solids.

faces between semiconductors and insulators. The resultant interface states again interfere with semiconductor action in the surface region.

Effects 2 and 3 are the ones that probably produce the majority of the engineering problems associated with operating electronics in space and in nuclear facilities. Photocurrent effects are important mainly in intense pulsed-radiation environments, such as that produced from a nuclear weapon or a linear accelerator. Figure 7 shows the responses induced in various sizes of silicon junction.

b. Ionization in Optical Materials. Glasses and many colorless inorganic salts are colored by exposure to radiation. Electrons are transferred into lattice-defect states by excitation with high-energy radiation. The post-excitation state is colored. The reverse process, bleaching, is often possible using light. Most glasses contain color centers fortuitously. The alkali sili-

cates, alkali borates, and phosphates all contain nonbridging oxygen bonds that can give up an unshared electron. The resulting "trapped hole" configuration can absorb light strongly both in the near UV and blue wavelength ranges. Thus, irradiated multicomponent glasses are often deep red-brown after irradiation. It is difficult to avoid this effect, but partial suppression has been achieved by alterations in glass formulation and the addition of dopants: for example, cerium oxide in optical glass. In optical fiber manufacture, pure fused silica is doped to modify the refractive index and melting point. Doping in this case increases the sensitivity of the fiber to color-center formation under irradiation.

Table III gives radiation-induced optical losses for various glasses and optical fibers.

c. Ionization in Polymers. Plastics are used in a variety of ways in devices and electronic

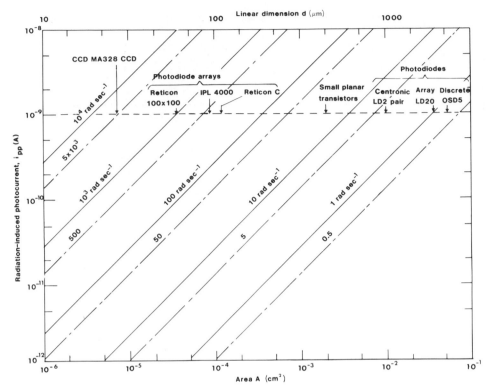

FIG. 7. Radiation-induced photocurrent response versus area in silicon diodes and transistors. Element areas for some available devices are given. Collection width $(W + L) = 20$ μm, 2.76×10^{-8} A rad^{-1} sec cm^{-2}.

systems. The more demanding applications include thin-film insulators (e.g., capacitors), stand-off insulators (e.g., cables), coatings and encapsulations, membranes, adhesives, optical lenses, pyroelectric sensors and electrets. In a plastic, ionization leads to the excitation and breaking of chemical bonds. The bonds subsequently rearrange leading to a large variety of possible products. At an absorbed dose value of 10^7 rad (10^5 Gy), only a few of the more sensitive plastics are altered mechanically or electrically. Even so, many plastics develop a deep coloration at this dose level.

3. Atomic Displacement Effects

a. Semiconductors. A small but significant fraction of the energy of megavolt particles passing through an absorber is dissipated in the transfer of momentum to the atoms of the absorber. If the atom experiencing such a collision receives sufficient kinetic energy (displacement energy E_d), it is removed from its position in the lattice leaving a vacancy, or defect. The atom

removed may meet another such vacancy and recombine, or it may lodge in the lattice in an interstitial position. Being reactive and often mobile at room temperature, a displaced atom may migrate to the surface or to an impurity and combine. The vacancy may also be mobile and often either combines with an impurity atom or forms a cluster with other vacancies. In semiconductors, the resulting vacancy–impurity complexes are usually electronically active. The interstitial atoms are less active. These consequences of displacement in the solid are seen to be numerous in type and difficult to characterize. Research on the processes and mechanisms involved in the production of defects in semiconductors is continuing. The process is conventionally termed displacement or bulk damage.

One of the best-understood examples of bulk damage is that produced by the electron irradiation of silicon. The effect of irradiation on the lifetime τ of minority carrier in silicon is shown in Fig. 8. It can be seen that, over the upper range of electron fluence ϕ, lifetime varies with

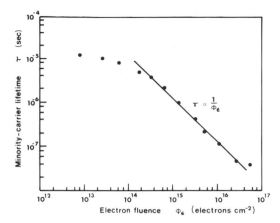

FIG. 8. Atomic displacement damage in silicon: reduction in minority-carrier lifetime. [From Wertheim, G. K. (1957). *Phys. Rev.* **105**, 1730.]

where τ_0 and τ are the values before and after irradiation. The damage constant K of the minority-carrier lifetime expresses the damage (change in lifetime) per unit electron fluence. The value of K for electrons of energy 1 MeV is often taken as the standard against which the effect of all other particles is compared. An electron of higher energy displaces more atoms per unit volume. A proton possessing the same kinetic energy displaces over a thousand times more atoms. The effects of neutrons are intermediate between these two cases.

The damage caused to semiconductors by particles, while not identical for all particles at all energies, is of the same general kind in all cases. Therefore, it is possible to express the effect in silicon of all particles in terms of a damage-equivalent fluence of 1-MeV electrons or neutrons.

Since the angle of incidence of the particle beam may have an effect on the profile of damage versus depth in silicon, the stipulation is usually also made that the particles should be of normal incidence. The unit used is the damage-equivalent normally incident 1-MeV electrons

fluence as follows:

$$\tau \propto \frac{1}{\phi} = \frac{1}{K\phi}$$

$$\frac{1}{\tau} - \frac{1}{\tau_0} = K\phi$$

TABLE III.   Optical Loss Induced by Radiation in Selected Glasses and Optical Fibers[a]

Source	Code	Type	Core material	Form of glass	Response[b] [dB(km rad(Si))$^{-1}$]		
					$\lambda = 0.8\ \mu m$	$0.9\ \mu m$	$1.05\ \mu m$
Corning	(CGW)	5010	Pb flint	Fiber	5.4	2.5	0.50
Pilkington	(PBL)	HYTRAN	Pb flint	Fiber	4.5	1.9	0.50
Galileo	(G)	0001AA	Zn crown	Fiber	1.5	0.49	0.25
Schott	(S)	F2	Pb flint	Bulk	1.3	0.69	0.21
Dividing line for values above and below 1 dB/km at $\lambda = 0.8\ \mu m$.							
NRL		GL2382	BaLa crown	Bulk	0.65	0.35	0.16
Galileo	(G)	0001AB	Zn (0.3% Ce) crown	Fiber	0.27	0.0062	0.0026
NRL[c]	—	GL2364	BaLa (1% Ce) crown	Bulk	0.21	<0.18	—
Owens Corning	(OCF)	X-4147A	Zn crown	Bulk	0.040	0.020	<0.016
Schott[d]	(S)	F2G12	Pb (1.2% Ce) flint	Bulk	<0.1	—	—
Schott	(S)	R1	Pb(~1% Ce)	Fiber	0.0031	0.0015	0.0010
Corning	—	—	SiO_2(Ti)	Fiber	8×10^{-1}	—	—
Corning	—	—	SiO_2(Ge)	Fiber	1.4×10^{-2}	—	—
NRL	—	—	Soda-lime Silicate glass	Bulk	1×10^{-2}	—	—
NRL	—	—	Suprasil I	Bulk	$<1 \times 10^{-5}$	—	—

[a] Reprinted with permission from Evans, B. D., and Sigel, G. E. Jr. (1975). *IEEE Trans. Nucl. Sci.* **NS-22**(6), 2462. Copyright © 1975 IEEE.]

[b] Except where mentioned, readings made 1 hr after γ-irradiation.

[c] 30 min after γ-irradiation.

[d] 9 min after γ-irradiation.

cm^{-2}. Other workers have suggested 3-MeV electrons, 20-MeV protons, and 1-MeV neutrons as the standard particles. It is convenient that, despite the differences in defects produced by neutrons, electrons, protons, and so on, their effect on the minority-carrier lifetime in silicon follows the same functional form in all cases. Variations do, however, occur in the magnitude of the neutron damage, even for the same particle, if variations occur in the impurity concentrations in the semiconductor. In a working device, one also has to know the injection level, namely, the excess minority carriers $\Delta n/N$ that are present.

A reduction in the carrier concentration or minority-carrier lifetime has quite a strong effect, even in the simplest device. For example, in a single $p-n$ junction, these effects include the following:

1. Reduction in the photovoltaic response of the junction
2. Increase in the reverse leakage current
3. Increase in noise effects
4. Reduction in the storage time for excess carriers at reverse-biased junctions
5. Change in the reverse breakdown voltage of junctions
6. Increase in the forward voltage drop of junctions for a given current

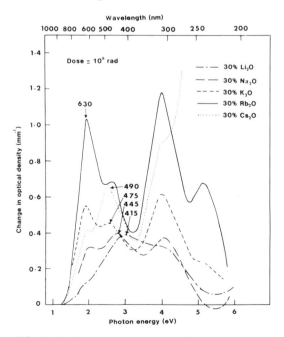

FIG. 9. Radiation-induced absorption of various alkali silicate glasses. (Courtesy of Phillips Research Laboratories.)

b. Optical. When atoms are displaced in an optical medium such as silica or an inorganic salt, the defect produced may constitute an absorber of light, termed a color center. One common type of color center is an anion vacancy containing a trapped electron. Unlike other electrons in the crystal lattice, the trapped electron can be excited to a higher state by visible or near-infrared light. From this cause, potassium chloride becomes purple under X-ray or particle irradiation due to a color-center absorption peak at a wavelength of 550 nm. Pure silica or sapphire develop optical absorptions in the UV under neutron, electron, or other particle irradiation but only become colored if certain impurities are present. A variety of radiation-induced optical absorption effects in glass is shown in Fig. 9.

V. Radiation Effects in Systems and Technology

A. GENERAL

In the foregoing sections, it has been noted repeatedly that practical reasons exist for the study of radiation physics. Advanced electronic and optical systems sometimes have to operate in radiation environments. Humans have to be protected from radiation or exposed to it in controlled amounts as a part of treatment.

High-technology machines are used for the production and control of radiation, and other high-technology devices have to be made to survive the radiation effects we describe. Therefore, it might be said that, at the leading edge of technology, the most sophisticated materials are being exposed to the most complicated effects. It is in this field that radiation physics makes its contribution to modern technology. This section brings together some of the important technological aspects of radiation physics. Table II gives a list of technological materials and devices that generally have poor tolerance to radiation.

B. RADIATION EFFECTS IN SOLID-STATE ELECTRONICS

1. The Problem

In solid-state devices, electronic functions are carried out in microscopic regions of highly structured materials. It is not surprising that, when particles or photons deposit large amounts of energy in such a structure, effects occur that temporarily or permanently interfere with the operation of that device. For example, the *npn* transistor depends for its amplifying actions on the passage of current through a perfectly ordered crystal lattice. This includes current transport across a thin layer known as the base region. If neutrons or high-energy electrons disrupt the crystal lattice, this base transport is disrupted, and the transistor loses its power to amplify electrical signals. A similar effect occurs in single-crystal silicon solar cells. In fact, it was the practical problem of designing solar panels to survive in space that gave rise to one of the first radiation-effects engineering projects.

The degradation of an electronic material in radiation is illustrated in Fig. 10. Here we see the degradation of the output of a silicon solar cell under electron radiation. We see that the curve bears similarities to the curve in Fig. 8, which shows degradation of minority carrier lifetime under electron radiation. This is because the performance of the solar cells is, in fact, controlled by the lifetime parameter.

Ionization severely affects computer circuits that are made using metal-oxide-semiconductor (MOS) technology. This effect is caused by both particles and photons, since a high mass is not needed. The energy imparted to the solid appears in the form of electrons and holes (positive and negative carriers of electricity). If a voltage is present on the metal electrode, the electrons and holes move apart. This leads to the formation of charge sheets in the oxide layer (see Fig. 4). Positive charge sheets near the semiconductor have a profound effect on the ability of the semiconductor to conduct electrical signals. As a result of this type of effect, MOS integrated circuits are often badly affected during space flights. Special design measures are needed to prolong their survival. This may involve either a major adjustment in the processing used to make the devices or the addition of shadow shields to reduce the space radiation dose reaching the component.

For equipment exposed to pulsed radiation such as nuclear weapons, photocurrent effects can upset most integrated circuits inducing photovoltages that act as spurious signals in digital circuits and produce errors in microcomputers, memories, and so forth.

A problem that has arisen with the more recent large-scale integrated circuits is the single-event upset. As the memory and logic cells on integrated circuits become smaller it is possible for a single heavy ion to produce enough current to discharge a logic mode, producing an error. Cosmic rays and radioactive emanations have both been found to produce soft-error upsets in large-scale integrated circuit memory chips. A problem found at one stage in development was

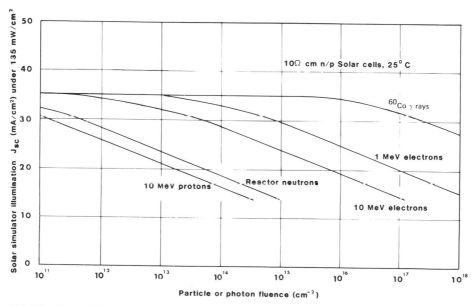

FIG. 10. Degradation of solar cell current density with fluence for various radiations. (Provided through the courtesy of the Jet Propulsion Laboratory, California Institute of Technology, Pasadena, California.)

that the minerals used in the ceramic packaging contained radioactivity that caused soft errors.

2. Data Banks

A large variety of semiconductor device designs is available on the commercial market. Each type has its own make-up, hence, each may have a differing response to radiation. While the laws of radiation physics help us to make an approximate prediction of this response, a firm prediction is obtained only by testing a reasonable sample of a given commercial type in a laboratory radiation source. Such testing has been proceeding for a number of years. It has been sponsored mainly by space and military projects, which have accumulated extensive banks of engineering data. Since new device designs are constantly being produced, the process of testing and putting the results in data banks is continuous. Efficient use of such data banks by the community of aerospace and nuclear engineers is needed to reduce repetitions in the already costly business of analyzing and testing electronic devices under radiation.

3. Radiation Hardening of Devices

Research has gone into ways in which the design of semiconductor structures can be altered to reduce the effects of radiation. The improved devices are sometimes known as radiation-hardened devices. We shall illustrate the approach used in device hardening by the case of the very sensitive MOS integrated circuit. The long-term effects of radiation are localized in the silicon dioxide layers. It has been found that these effects can be alleviated by altering the rate of growth of the oxide layers and controlling the composition of the gases used to promote the growth and annealing of the oxide. Finding the best method of process modification has entailed a large amount of research. The mechanisms of degradation of the oxide films first had to be understood. It is now possible to make megarad-hard microprocessor circuits. The same circuit made by commercial methods would fail at a dose of 10 krads. Only a few hardened production lines exist in the world.

C. OPTICAL SYSTEMS

In the early days of solid-state physics research, it was noticed that crystals of potassium chloride, normally colorless, become purple when exposed to X-rays. No detectable chemical decomposition occurred, so an internal rearrangement must have taken place. It was found that the radiation causes the removal of a chlorine atom from its normal crystal lattice site. The resulting atomic vacancy traps an electron. This acts as an absorber of a certain band of visible light imparting the purple color. Similar color-center effects have since been found in most transparent solids. Most common glasses become dark brown when irradiated.

Physical studies of color centers in transparent solids has contributed greatly to our knowledge of the physics of binding between atoms in crystals and the excitations that occur when radiation and a solid interact. The study of vacancies in lattices is called defect physics. Important tools used in defect physics are optical spectroscopy and electron spin resonance.

The degradation of optical devices under radiation is a serious engineering problem. The lenses of cameras used to inspect the interiors of nuclear reactors quickly lose transmission because of radiation-induced browning. The same may happen with television cameras in space. A particularly vulnerable form of optical device is the optical fiber. If radiation causes an optical medium to darken, then the longer the optical path, the more severe the optical losses produced. In communications, the optical paths in fibers may be several miles long, and the loss of a few decibels per kilometer may cause the failure of the system. Table III lists some results for a variety of commercial glasses.

D. STRUCTURAL MATERIALS

1. Metals

The metal and ceramic parts that go to make up nuclear reactors often receive very high exposure to neutrons and γ rays from the nuclear reactions in the fuels. In the future, similarly intense exposures will occur in the first wall of a fusion power reactor. The radiation damage caused by the neutrons is manifested as the production of voids. The net result is that the metal sample swells significantly. This may lead to the spontaneous rupture of a nuclear fuel—and can and will certainly lead to the loss of mechanical strength. The selection of the alloys for constructing reactors is thus a vital field of research.

2. Polymers

The mechanical strength of polymers depends on the continuity of the carbon backbones of the polymer molecules. Energy deposition from any form of radiation (including UV light) can cause the rupture or scission of the carbon–carbon bonds, leading to loss of strength. However, before this occurs, more complex chemical reactions may actually lead to the toughening of the polymer, due to cross-linking, which leads to a three-dimensional network of molecules. Thus, in several commercial processes, polymers are treated with radiation after molding or extrusion. Radiation also "cures" layers of polymers. For example, uncured polymeric coatings on wires or dipped articles can be hardened in place.

E. STERILIZATION

A beam of electrons or γ rays can kill bacteria and parasites in food, surgical instruments, or sewage without leaving any radioactivity. At the doses required, no serious degradation occurs in the products irradiated, although great caution is exercised before the public is allowed to eat irradiated foods. Radiation sterilization plants will, in future constitute one of largest commercial uses of γ rays in human service. Radiation physics is used in the design of the processing plant and in the accurate control of radiation doses given to the product.

F. MEDICAL APPARATUS

Radiation physicists are widely employed in health care, especially in the areas of instrumenting and running radiation machines and in the planning and performance of radiotherapy. Treatment planning is a highly skilled procedure in which the radiation physicist calculates with precision the radiation doses that can reach a tumor during a treatment and minimizes the doses that reach healthy tissue. As part of the planning, computers are used to calculate and plot the anticipated radiation doses around an organ for any chosen radiation source. The sources may be implanted radioisotopes or external high-energy beams. During exposure, radiation instruments are used to monitor the doses that the patients receive and avoid overdoses.

G. SHIELDING

A radiation shield is a mass of material placed between a source of radiation and an object that requires protection so as to reduce the radiation dose received. The design of shielding is a combination of techniques such as the calculation of radiation transport, the selection of a structure, and the choice of materials to suit the situation of the source and object. For large shields, cost analysis is necessary. For special vehicles such as satellites, of course, weight is of major importance and cost is less important. However, for a large reactor, the aim is to use earth, water, or concrete whenever possible to reduce cost.

The range–energy curves and attenuation curves discussed earlier are useful for a rough evaluation of the effectiveness of a given shield, but for the full design of a radiation shield, detailed calculations of radiation transport are required, using computer programs that trace the progress of individual particles through the various layers of a shield. For power reactors the same programs also calculate the heat produced in shields, the cooling necessary, and the complicating effects of the large ducts that penetrate the shields. In power reactors, the shields become radioactive, remaining so after machine shutdown, thus hampering the maintenance procedure. Research is carried out on low-activation shielding for reactors.

In future, magnetic-confinement fusion reactors that are built for power generation are likely to have a very heavy toroidal shield surrounding the reaction vessel. Extensive ducting will be necessary. In the U.S., the computers used for shield and structure analysis in fusion reactors are some of the most powerful in existence.

In spacecraft, the components of interest are circuits located in electronics boxes. They are surrounded by arrays of other parts, satellite structures, and so on, which act as built-in shielding. Some boxes are thus more protected than others. Computer programs are used to calculate the built-in shielding. The programs used perform ray tracing to determine the directions from which radiation penetrates to the component of interest. A summary of our calculation for a typical spacecraft is presented in Fig. 11.

H. SYSTEMS ENGINEERING

On some occasions, electronically controlled machines have to operate in radiation environments. The prime example is unmanned satellites, which often have to operate in the trapped radiation belts that exist around the earth. The electronics in geostationary communications satellites are designed to last for many years, and internal doses are in the region of 10 krad

Total mission dose,
D, krad (Si)

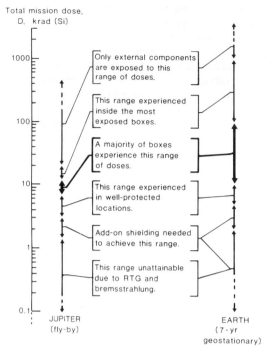

FIG. 11. Summary of the shielding analysis for a spacecraft.

per year. Other examples are military vehicles, which may be exposed to nuclear explosions; remotely handled tools operating in a nuclear reactor facility; and the control and cooling systems in a fusion reactor. The task of designing equipment to survive is a problem in systems engineering. Given the usual constraints on size, weight, cost, and so on, the system has to be optimized. The designer must achieve a balance between the methods available for reducing the effects of radiation. These include (1) changing circuit design to increase tolerance, (2) adding shielding, (3) procuring less sensitive components, (4) making alternative mechanical layouts. Radiation physics is used in the development of the system approach. The first contribution of radiation physics is to make predictions of the degradation or other responses of the electronic or optical components that incur most of the damage. This involves solid-state physics. Calculations must be made of the individual response of each component of the circuit, and then the combined effect of these must be calculated. This involves electrical engineering. Furthermore, methods to reduce the responses where necessary again may involve solid-state physics, electrical and mechanical engineering, and extensive systems engineering.

Finally, a method is developed to compare the merits of the various steps taken.

VI. Useful Radiation Data

A. RADIATION UNITS

$$
\begin{aligned}
1 \text{ rad} &= 100 \text{ erg g}^{-1} \\
&= 6.25 \times 10^{13} \text{ eV g}^{-1} = 10^{-2} \text{ Gy} \\
1 \text{ Mrad} &= 6.25 \times 10^{19} \text{ eV g}^{-1} = 10 \text{ kGy} \\
1 \text{ Gy} &= 1 \text{ J kg}^{-1} = 100 \text{ rad} \\
1 \text{ kGy} &= 10^{5} \text{ rad} = 100 \text{ krad} \\
1 \text{ MGy} &= 10^{8} \text{ rad} = 100 \text{ Mrad} \\
10^{20} \text{ eV g}^{-1} &= 1.6 \text{ Mrad} = 16 \text{ kGy}
\end{aligned}
$$

1 roentgen (R) = 86.9 erg g^{-1}(air) = 2.58×10^{-4} C kg^{-1}(air)

1 R of 1-MeV photons $\equiv 1.95 \times 10^{9}$ photons cm^{-2}.

This fluence deposits	0.869 rad in air
	0.965 rad in water
	0.865 rad in silicon
	0.995 rad in polyethylene
	0.804 rad in LiF
	0.862 rad in Pyrex glass
	(80% SiO_2)

1 curie (Ci) of radioactive material produces 3.700×10^{10} disintegrations per second.

A 1-Ci point source emitting one 1-MeV photon per disintegration gives an exposure of 0.54 R hr^{-1} at 1 m.

A 1-Ci ^{60}Co source gives 1.29 R hr^{-1} at 1 m.

Photon flux at 1 m from a 1-Ci point source = 1.059×10^{9} cm^{-2} hr^{-1} (assuming one γ-ray photon per disintegration).

SI Units recommended by the International Commission on Radiation Units and Measurements (I.C.R.U.) [*Brit. J. Radiology* **49**, 476(1976)] are the following:

— Absorbed dose: the Gray (Gy) = 100 rad = 1 J/kg
— Exposure: the coulomb per kilogram (no name given) = 1 C/kg
— Quantity activity: the Becquerel (Bq) = 1 sec^{-1} = 2.703×10^{-11} Ci, with the old units to be abandoned over 10 yr.

Basic units of biological dose from radiation are the rem and the sievert (Sv). 1 rem (radiation equivalent man) is the absorbed dose to the body of 1 rad weighted by a quality factor Q.F. that is dependent on the type of radiation involved. This is because the energy absorption from radiation is insufficient as a measure of biological damage.

$$1 \text{ rem} = \text{Q.F.} \times 1 \text{ rad}$$
$$1 \text{ Sv} = \text{Q.F.} \times 1 \text{ Gy}$$
$$\text{Q.F.} = 1 \text{ for X rays, } \gamma, \text{ or } \beta$$

The values of Q.F. for α, neutrons, and heavy particles are greater than 1.

B. USEFUL GENERAL CONSTANTS

$$7 \text{ yr} = 3.682 \times 10^6 \text{ min} = 2.209032 \times 10^8 \text{ sec}$$
$$1 \text{ yr} = 5.25960 \times 10^5 \text{ min} = 3.155760 \times 10^7 \text{ sec}$$
$$1 \text{ day} = 1.440 \times 10^3 \text{ min} = 8.6400 \times 10^4 \text{ sec}$$
$$1000 \text{ Å} = 0.1 \text{ } \mu\text{m} = 100 \text{ nm}$$
$$1 \text{ mm} = 0.03937 \text{ in.}$$
$$0.001 \text{ in.} = 25.4 \text{ } \mu\text{m}$$
$$1 \text{ m}^3 = 10^6 \text{ cm}^3$$
$$1000 \text{ cm}^3 = 10^{-3} \text{ m}^3$$
$$1 \text{ liter} = 1000.028 \text{ cm}^{-3} = 0.219 \, 976 \text{ gal}$$
$$1 \text{ gm cm}^{-3} = 1000 \text{ kg m}^{-3}$$
$$1 \text{ eV} = 1.602 \, 192 \times 10^{-19} \text{ J}$$
$$1 \text{ MeV} = 1.602 \, 192 \times 10^{-13} \text{ J}$$
$$1 \text{ J} = 10^7 \text{ erg}$$
$$1 \text{ cal} = 4.187 \text{ J}$$
$$1 \text{ eV/molecule} = 23.1 \text{ kcal/mol}$$
$$\text{Permittivity of free space } \varepsilon_0 = 8.86 \times 10^{-14} \text{ F cm}^{-1}$$
$$= 8.86 \times 10^{-12} \text{ F m}^{-1}$$
$$= 55.4 \text{ electronic charges V}^{-1} \text{ } \mu\text{m}^{-1}$$
$$\text{Permeability of free space } \mu_0 = 1.26 \times 10^{-6} \text{ H m}^{-1}$$
$$\text{Electronic charge } q = 1.602 \, 192 \times 10^{-19}$$
$$1 \text{ C cm}^{-2} = 6.241 \, 459 \times 10^{18} \text{ electrons cm}^{-2}$$
$$1 \text{ } \mu\text{A cm}^{-2} = 6.24 \times 10^{12} \text{ electrons cm}^{-2} \text{ sec}^{-1}$$
$$\text{Velocity of light} = 2.997 \, 925 \times 10^8 \text{ m sec}^{-1}$$
$$1 \text{ N} = 10^5 \text{ dyn}$$
$$1 \text{ mm Hg} = 133.3224 \text{ N m}^{-2}$$
$$\text{Boltzmann's constant } k = 1.380 \, 62 \times 10^{-23} \text{ J K}^{-1}, 6.62 \times 10^{-5} \text{ eV K}^{-1}$$
$$kT \text{ at room temperature} = 0.0259 \text{eV}$$
$$\text{Planck's constant } h = 6.63 \times 10^{-34} \text{ J sec}$$
$$\text{Avogadro's number} = 6.022 \times 10^{23} \text{ mol}^{-1}$$
$$\text{Electron rest mass } m_e = 9.11 \times 10^{-31} \text{ kg}$$
$$\text{Proton rest mass } m_p = 1.67 \times 10^{-27} \text{ kg}$$

C. Dose Rates from γ Emitters[a]

Nuclide	Half-life	Principal γ energies (MeV)	Dose rate at 1 m from 1 Ci (rads/hr in tissue)
Antimony-124	60 day	0.60; 0.72; 1.69; 2.09	0.94
Arsenic-72	26 hr	0.51;[b] 0.63; 0.835	0.97
Arsenic-74	18 day	0.51;[b] 0.596; 0.635	0.42
Arsenic-76	26.5 hr	0.56; 0.66; 1.21; 2.08	0.23
Barium-140[c]	12.8 day	0.16; 0.33; 0.49; 0.54; 0.82; 0.92; 1.60; 2.54	1.19
Bromine-82	35.4 hr	0.55; 0.62; 0.70; 0.78; 0.83; 1.04; 1.32; 1.48	1.40
Caesium-137	30 yr	0.662	0.32
Cobalt-58	71 day	0.51;[b] 0.81; 1.62	0.53
Cobalt-60	5.26 yr	1.17; 1.33	1.27
Gold-198	2.70 day	0.412; 0.68; 1.09	0.22
Iodine-131	8.04 day	0.28; 0.36; 0.64; 0.72	0.21
Iodine-132	2.3 hr	0.52; 0.65; 0.67; 0.78; 0.95; 1.39	1.13
Iridium-192	74 day	0.296; 0.308; 0.316; 0.468; 0.605; 0.613	0.46
Iron-59	45 day	0.19; 1.10; 1.29	0.61
Manganese-52	5.7 day	0.51;[b] 0.74; 0.94; 1.43	1.79
Manganese-54	314 day	0.84	0.45
Potassium-42	12.4 hr	1.52	0.13
Radium-226[d]	1620 yr	0.05–2.43	0.79
Sodium-22	2.6 yr	0.51;[b] 1.28	1.15
Sodium-24	15.0 hr	1.37; 2.75	1.77
Tantalum-182	115 day	0.068; 0.100; 0.222; 1.12; 1.19; 1.22; 1.23	0.64
Thulium-170	127 day	0.052; 0.084	0.002
Zinc-65	245 day	0.51;[b] 1.11	0.26

[a] Reprinted with permission from "The Radiochemical Manual." The Radiochemical Centre, Amersham, England, 1966.

[b] 0.51-MeV γ rays from positron annihilation.

[c] Barium-140 in equilibrium with lanthanum-140.

[d] Radium-226 in equilibrium with daughter products; radiation filtered through 0.5 mm platinum; dose rate from 1 g.

Bibliography

Chilton, A. B., Shultis, J. K., and Faw, R. E. (1984). "Principles of Radiation Shielding." Prentice-Hall, Englewood Cliffs, New Jersey.

Greening, J. R. (1981). "Fundamentals of Radiation Dosimetry." Adam Hilger Ltd., Bristol, UK

Johns, H. E., and Cunningham, J. R. (1969). "The Physics of Radiology," 3rd ed. Charles C. Thomas, Springfield, Illinois

Lamarsh, J. R. (1983). "Introduction to Nuclear Engineering," 2nd ed. Addison-Wesley, Reading, Massachusetts.

Srour, J. R., Long, D. M., Fitzwilson, R. L., Millward, D. G., and Chadsey, W. L. (1984). "Radiation Effects on and Dose Enhancement of Electronic Materials." Noyes Publications, Park Ridge, New Jersey.

Van Lint, V. A. J., Flanagan, T. M., Leadon, R. E., Naber, J. A., and Rogers, V. C. (1980). "Mechanisms of Radiation Effects in Electronic Materials," Vols. I and II. Wiley, New York.

RELATIVITY, GENERAL

James L. Anderson   *Stevens Institute of Technology*

I. Space–Time Theories of Physics
II. Newtonian Mechanics
III. Special Relativity
IV. General Relativity
V. Gravitational Fields
VI. Observational Tests of General Relativity
VII. Gravity and Quantum Mechanics

GLOSSARY

Binary pulsar: Double star system, one of whose components is a neutron star.

Doppler shift: Fractional change in frequency of light due to relative motion between source and observer.

Electrodynamics: Theory of electric and magnetic fields and of the interactions of the charged particles that produce them.

Hubble constant: Ratio of velocity of recession to distance of galaxies.

Mapping: Association of points of a space–time manifold with other points of this manifold.

Perihelion: Point in the orbit of a planet when it is closest to the sun.

Riemannian geometry: Geometry in which the distance between neighboring points is defined by a metric and is quadratic in the coordinate differences between the points.

The general theory of relativity is first and foremost a theory of gravity. At the same time it is also a description of the geometry of space and time that differs profoundly from all such previous descriptions in that this geometry is identified with the gravitational field. The predictions of the general theory differ both quantitatively and qualitatively from those of the Newtonian theory of gravity. Although the quantitative differences between the two theories are usually small, they have been extensively tested in the solar system and other astro-physical systems. Today the agreement between observation and theory is better than 0.5%. However, the qualitative predictions of the theory are its most exciting and challenging feature. Among others, the theory predicts the existence of gravitational radiation. Although this radiation has not been directly observed, the effects of its emission have been observed in the binary pulsar PSR 1913 + 16 and agree with the predictions of the theory to within 3%. The theory also predicts the phenomenon of gravitational collapse leading to the creation of black holes. There is strong observational evidence that such objects exist in the universe. And finally, the general theory serves as the basis for our best description of the universe as a whole, the so-called hot big-bang cosmology.

I. Space–Time Theories of Physics

A. THE SPACE-TIME MANIFOLD

To understand the revolution wrought by the general theory it is useful to set it in a framework that encompasses it as well as the other two major space–time structures of physics, Newtonian mechanics and special relativity. Basic to each of these structures is the notion of the space–time manifold consisting of a four-dimensional continuum of points. It is assumed only that any finite piece of this manifold can be mapped in a one-to-one manner onto a connected region of the four-dimensional Euclidean plane. Otherwise, these points are featureless and indistinguishable from each other, and the manifold as a whole is characterized only by its topological properties. While this manifold is not itself associated with any physical entity, it serves as the basis for the construction of the geometrical structures that are to be associated with such objects.

Since the points of the space–time manifold can be mapped onto the four-dimensional Euclidean plane, one can coordinatize the manifold by assigning to each point the coordinates of its image point in the Euclidean plane, x^μ, where the index μ takes on the values 0, 1, 2, 3. Because the points of the manifold are assumed to be indistinguishable, this mapping is to a large extent arbitrary and hence the coordinatization is also arbitrary. Depending on the topological structure of the manifold, it may be necessary to cover it with several overlapping coordinate "patches" to avoid singularities in the coordinates. If, for example, the manifold has the topology of the surface of a ball, it is necessary to employ two such patches to avoid the coordinate singularity one encounters at the pole when using the customary polar coordinates.

If the manifold is coordinatized in two different ways, for example, by using Cartesian or spherical coordinates, the coordinates used for one such coordinatization must be functions of those used for the other and vice versa. This relation is called a coordinate transformation. In order to preserve the continuity and differentiability of the manifold it must be continuous, nonsingular, and differentiable.

B. Geometrical Structures

In space–time theories, physical entities are associated with geometrical objects that are constructed on the space–time manifold. These objects can be of many different types. A curve can be associated with the trajectory of a particle and is specified by designating the points of the manifold through which it passes. This can be done by giving the coordinates of these points as functions $x^\mu(\lambda)$ of a monotonically varying parameter λ along the curve. Likewise, a two-dimensional surface could be specified by giving the coordinates of the surface as functions of two monotonic parameters, and similarly for three- and four-dimensional regions.

In addition to collections of points, one can introduce geometrical objects that consist of a set of numbers assigned to a point. These numbers are said to constitute the components of the geometrical object. The components of the velocity of a particle at a particular point in its trajectory would constitute such a collection. If the components are specified along a trajectory, a surface, or any other part of the space–time manifold, they are said to constitute a field. The temperature in a room can, for example, be as-sociated with a one-component field. Likewise, the electromagnetic field surrounding a moving charge can be associated with a field consisting of six components.

The basic requirement that must be met in order that an object be geometrical is a consequence of the indistinguishability of the points of the space–time manifold. It is that under a coordinate transformation, the transformed components of the object must be functions solely of its original components and the coordinate transformation. This requirement is simply met in the case of curves, surfaces, and so forth; for example, given a curve, the transformed curve can be immediately calculated given the coordinate transformation.

An especially useful group of geometrical objects for associating with physical entities are those whose transformed components are linear, homogeneous functions of the original components. The simplest example is the single-component object called a scalar $\phi(x)$. Under a coordinate transformation, its transformed value is just equal to its original value. The other linear, homogeneous objects constitute the vectors, tensors, and pseudoscalars, pseudovectors, and pseudotensors. Vectors and pseudovectors are four-component objects (they come in two varieties called covectors and contravectors), while tensors and pseudotensors have larger numbers of components. There are also objects whose transformed components are linear but not homogeneous functions of the original components. Finally, there are objects whose transformed components are nonlinear functions of the original components, although such objects have not been used to any great extent. In all cases, however, the nature of the object is characterized by its transformation law.

It should be pointed out that not all objects one can construct are geometrical. The gradient of a scalar is a geometrical object while the gradients of vectors and tensors in general are not.

C. Laws of Motion

The basis for associating geometrical objects with physical entities is purely utilitarian—there is no general procedure for making this association. The numerical values that these objects can assume are taken to correspond to the observed values of the physical entities with which they are associated. Since not all such values can in general be observed, it is necessary to formulate a set of rules, called here laws of motion, that select from the totality of values a given set of geometrical objects can have, a subset that corresponds to possible observed values. Thus, if it is decided to associate a curve with the trajectory of a planet, one would have to discover a system of equations such as those obtained from Newton's laws of motion to select from the totality of all curves the subset that would correspond to actual planetary trajectories.

One requirement that one would like to be fulfilled by all laws of motion is that of completeness—every set of values allowed by them must, at least in principle, be observable. It is, after all, the purpose of the laws of motion to rule out unobservable sets of values. Nevertheless, there are problems associated with the imposition of such a requirement. There may, for example, be practical limitations on our ability to observe all the values allowed by a given set of laws. It is unlikely that we will ever be able to attain the energies needed to verify some of the predictions of the grand unified theories that are being considered today. However, if one of these theories correctly described all that we can observe about elementary particle interactions, we would not discard it because we could not directly observe its other predictions. More troubling, however, are limitations in principle on what we can observe. When applied to the universe as a whole, the general theory of relativity allows for many different possibilities, yet by its very nature we can observe only the universe in which we live. In the strict sense, then, the general theory should be considered incomplete. Nevertheless, it does correctly describe a vast range of phenomena, and so far there does not exist a more restrictive theory that does so. Therefore, probably, the best we can do is to require that a law not admit values that could be observed if they existed but do not.

D. Principle of General Covariance

However the laws of motion are formulated, they must be such as to be independent of a particular coordinatization of the space–time manifold. This requirement is called the principle of general covariance and was one of the basic principles employed by Einstein when he formulated the general theory. For the principle to hold, the laws of motion for a given set of geometrical objects must be such that all of the transforms of a set of values of these objects that

satisfy the laws of motion must also satisfy these laws.

The principle of general covariance is not, as has sometimes been suggested, an empty principle that can be satisfied by any set of physical laws. If, for example, the geometrical object chosen to be associated with a given physical entity is a scalar field $\phi(x)$, then the only generally covariant law that can be formulated involving only this object is the trivial equation

$$\phi(x) = \text{const} \qquad (1)$$

In order to formulate a nontrivial law of motion for ϕ it is necessary to introduce other geometrical objects in addition to the scalar field. One possibility is to introduce a symmetric tensor field $g_{\mu\nu}(x)$ and its inverse $g^{\mu\nu}(x)$, which are related by the equation

$$g^{\mu\rho}g_{\rho\nu} = \delta^{\mu}_{\nu} \qquad (2)$$

where δ^{μ}_{ν} is the Kronecker delta, with values given by

$$\begin{aligned} \delta^{\mu}_{\nu} &= 1 \qquad \mu = \nu \\ &= 0 \qquad \mu \neq \nu \end{aligned} \qquad (3)$$

and where the appearance of a double index such as ρ implies a summation over its range of values. One can then take as the law of motion for ϕ the equation

$$(\sqrt{-g}\, g^{\mu\nu}\phi_{,\mu})_{,\nu} = 0 \qquad (4)$$

where g is the determinant of $g_{\mu\nu}$ and $,\mu := \partial/\partial x^{\mu}$. The tensor field $g_{\mu\nu}$ cannot itself be given as a function of the coordinates directly since in that case Eq. (4) would not be generally covariant. Rather, it must in turn satisfy a law of motion that is itself generally covariant. If one requires that this law involve no higher than second derivatives of $g_{\mu\nu}$, it can be shown that there are in fact only three essentially different such laws for this object. One can form laws of motion for ϕ other than Eq. (4), but in each case it is necessary to introduce other geometrical objects for the purpose and to formulate laws of motion for them.

E. Absolute and Dynamical Objects

To understand the revolutionary nature of the general theory it is necessary to distinguish between two essentially different types of objects that appear in the various space–time theories. We call them absolute and dynamical, respectively. If the totality of values allowed by the laws of motion for some geometrical object, such as the tensor $g_{\mu\nu}$ introduced above are such that they can all be transformed into each other by coordinate transformations, we say that that object is an absolute object in the theory. This can occur if the law of motion for the object does not involve any of the other objects in the theory. The remaining objects in the theory are called dynamical objects. The electric fields associated with different charge distributions, for example, cannot in general be transformed into one another and hence must be associated with a dynamical object.

Given a theory with absolute objects, it is possible to coordinatize the space–time manifold so that they take on a specific set of values. In the case of the tensor $g_{\mu\nu}$, one of the three possible laws of motion mentioned above is such that every set of values allowed by it can be transformed so that, for every point of the space–time manifold, $g_{\mu\nu} = \text{diag}(1, -1, -1, -1)$. If these values are substituted into the other laws of motion they will no longer be generally covariant, but rather they will be covariant with respect to some subgroup of coordinate transformations. This subgroup will leave invariant the chosen values of the geometrical object (or objects) and will be called the invariance group of the theory. The structure of this group will be independent of which particular set of values allowed by the laws of motion is chosen for the absolute objects. If there are no absolute objects then the invariance group is just the group of all allowed coordinate transformations.

Absolute objects are seen to play a preferred role in a theory—their values are independent of the values of the dynamical objects of the theory while the converse is in general not the case. (If it is, the absolute objects become superfluous and can be ignored.) A theory with absolute objects thus violates a kind of general law of action and reaction. We will see that both Newtonian mechanics and special relativity contain absolute objects while the general theory does not.

II. Newtonian Mechanics

A. Absolute Time and Space

In his formulation of the laws of motion, Newton introduced a number of absolute objects, chief of which were his absolute space and absolute time. Absolute time corresponds to the foliation of the space–time manifold by a one-parameter family of nonintersecting three-dimensional hypersurfaces, which we call planes of absolute simultaneity. All of the points in a given plane are taken to be simultaneous with respect to each other. Furthermore, these planes are such that the curves associated with the trajectories of particles intersect each plane once and only once. The ''time'' at which such an intersection takes place is characterized by the value of the parameter associated with the plane being intersected. These planes are absolute in that their existence and structure are assumed to be independent of the existence or behavior of any other physical system in the space–time.

In Newtonian mechanics the interaction of particles is assumed to be instantaneous as in Newton's action-at-a-distance theory of gravity. Consequently, such interactions take place between the points on the trajectories that lie in the same plane of absolute simultaneity. As a consequence, these planes can be observed by giving an impulse to one of a number of interacting particles and noting where, on the trajectories of the other particles, the transmitted impulse acts.

Newton's absolute space corresponds to a unique three-parameter congruence of nonintersecting curves that fill the space–time manifold; that is, through each point of the manifold there passes one and only one such curve. Furthermore, each curve passes through one and only one point of each plane of absolute simultaneity. The existence of such a congruence would therefore imply that there exists a unique one-to-one relation between the points in any two planes of absolute simultaneity. The ''location'' of a space–time point would be characterized by the parameters associated with the curve of the congruence passing through it.

The notion of absolute space brings with it the notion of absolute rest: a particle is absolutely at rest if its trajectory can be associated with one of the curves of the congruence. However, unlike the planes of absolute simultaneity that are needed in the formulation of the laws of motion of material particles, these laws do not require the existence of the space–time congruence of curves that constitute Newton's absolute space, nor do they afford any way of detecting a state of absolute rest. This property of the Newtonian laws of motion is known as the principle of Galilean relativity. Furthermore, since the congruence is not needed in the formulation of these laws, we can dispense with it and hence with

Newton's absolute space altogether as an unobservable element of the theory.

B. FREE BODIES

In his setting down of the three laws of motion, Newton was careful to give the first law, "Every body continues in its state of rest, or of uniform motion in a right line, unless it is compelled to change that state by forces impressed upon it," as separate and distinct from the second law. He clearly did not consider it, as it is sometimes taken to be, a special case of the second law. In effect, the first law supposes a class of curves, the straight (right) lines, to exist in the space–time manifold. Furthermore, these curves correspond to the trajectories of a class of objects on which no forces act, namely free bodies. As a consequence, these curves are absolute objects of the theory. Furthermore, they, like the planes of absolute simultaneity, are needed to formulate the laws of motion for bodies on which forces act.

C. GALILEAN INVARIANCE

One can always coordinatize the space–time manifold in such a way that the parameter t, which labels the different planes of absolute simultaneity, is taken to be one of these coordinates. When this is done, the equation that defines these planes is simply

$$t = \text{const} \tag{5}$$

Furthermore, the remaining coordinates can be chosen so that the equations of the curves associated with the trajectories of the free bodies are linear in t; that is, they are of the form

$$x_i = v_i t + x_{0i} \tag{6}$$

where the index i takes on the values 1, 2, 3 and v_i and x_{0i} are constants. The constants v_i are the components of the "velocity" of the free body whose trajectory is associated with this curve and the x_{0i} are its initial positions. When expressed in terms of these coordinates, the laws of motion of Newtonian mechanics take on their usual form.

Since the planes of absolute simultaneity and the straight lines constitute the absolute objects of Newtonian mechanics and enter into the formulation of the laws of motion of all Newtonian systems, the subgroup of coordinate transformations that leave them invariant as a whole constitutes the invariance group of Newtonian mechanics. In addition to the group of spatial rotations and translations and time translations, this group consists of the Galilean transformations given by

$$x_i' = x_i + V_i t \tag{7a}$$

and

$$t' = t \tag{7b}$$

where the V_i are the components of the velocity that characterize a particular transformation of the group.

In terms of the primed coordinates, we see that the equations of a straight line (6) take the form

$$x_i' = v_i' t' + x_{0i} \tag{8}$$

where the transformed velocity components v_i' are given by the Galilean law of addition for velocities:

$$v_i' = v_i + V_i \tag{9}$$

III. Special Relativity

A. LIGHT CONES

The transition from Newtonian mechanics to special relativity in the early part of this century involved the abandonment of the Newtonian planes of absolute simultaneity and their replacement by a new set of absolute objects, the light cones. With the completion of the laws of electrodynamics by Maxwell in the middle of the last century it became evident that electromagnetic interactions between charged particles were not instantaneous but rather were transmitted with a finite velocity, the speed of light. This fact, coupled with the Galilean velocity addition law, made it appear possible that some electromechanical experiment could be devised for the detection of a state of absolute rest and thus reinstate Newton's absolute space. However, all attempts to do so, such as those of Michelson and Morley and Trouton and Noble, proved fruitless. In one way or another, these experiments sought to measure the absolute velocity of the earth with respect to this absolute space. Even though they were sensitive enough to detect a velocity as small as 30 km/sec, which is much less than the known velocity of the earth with respect to the galaxy, no such motion was ever detected. [See RELATIVITY, SPECIAL.]

Einstein realized that if all interactions were

transmitted with a finite velocity there was no way objectively to observe the Newtonian planes of absolute simultaneity and that they, like Newton's absolute space, should be eliminated from the theory. It was his analysis of the meaning of absolute simultaneity and its rejection by him that distinguished his approach to special relativity from those of Lorentz and Poincaré. Since, however, unlike absolute space, the planes of absolute simultaneity were needed in the formulation of the laws of motion for material bodies, it was necessary to replace them by some other structure. The key to doing this lay in Einstein's postulate that the velocity of light is independent of the motion of the source. If this is the case, and to date all experimental evidence supports this postulate, the totality of all light ray trajectories form an invariant structure and can be associated with a corresponding family of three-dimensional surfaces in the space–time manifold, the light cones. Just as in Newtonian mechanics, where through each point there passes a unique plane of absolute simultaneity, in special relativity through each point there passes a light cone that consists of all of the points on the curves passing through this point that correspond to the trajectories of light rays.

In Newtonian mechanics the interaction of particles was assumed to take place between the points on the curves associated with their trajectories that lay in the same plane of absolute simultaneity. In special relativity this interaction is assumed to take place, depending on the type of interaction that exists between the particles, either between points that lie in the same light cone or between one such point and points in the interior of the light cone associated with this point. The electromagnetic interaction, for example, takes place between points lying in the same light cone. Consequently, these light cones can be observed by giving an impulse to one member of a group of charged particles and noting where, on the trajectories of the other particles, the transmitted impulse acts.

B. Free Bodies

In addition to the absolute light cones, special relativity assumes, like Newtonian mechanics, a family of curves, the straight lines, that are associated with the trajectories of free bodies. There is, however, an important difference between the two theories. In Newtonian mechanics any straight line that intersects all of the planes of

absolute simultaneity is assumed to correspond to the trajectory of a free body. In special relativity, on the other hand, only the straight lines that correspond to free bodies with velocities less than or equal to the speed of light are assumed to correspond to observable free bodies. These straight lines are such that, given a point lying on one of them, the other points lying on it are either interior to or lie on the light cone associated with that point.

C. Lorentz Invariance

Together, the light cones and straight lines constitute the absolute objects of special relativity. It can be shown that one can coordinatize the space–time manifold in such a way that the points with coordinates x^μ lying on a straight line are given by the equations

$$x^\mu = v^\mu \lambda + x_0^\mu \tag{10}$$

where the v^μ and x_0^μ are constants and λ is a monotone increasing parameter along the line. For the straight lines that correspond to the trajectories of free bodies, the v^μ are constrained by the condition that

$$\eta_{\mu\nu} v^\mu v^\nu \geqq 0 \tag{11}$$

where $\eta_{\mu\nu} = \text{diag}(1, -1, -1, -1)$. Provided that $\eta_{\mu\nu} v^\mu v^\nu > 0$, it is always possible to choose the parameter λ such that $\eta_{\mu\nu} v^\mu v^\nu = 1$. In this case the v^μ are said to constitute the components of the four-velocity of the particle and $\tau = \lambda$ is the proper time along the line.

In addition to the form (10) for the straight lines, coordinates can be chosen so that the points x^μ lying on the light cone associated with the point x_0^μ satisfy the equation

$$\eta_{\mu\nu}(x^\mu - x_0^\mu)(x^\nu - x_0^\nu) = 0 \tag{12}$$

For points interior to this light cone the quantity on the left side of this equation is greater than zero, while for points exterior to it is less than zero. In what follows, coordinates in which the straight lines and light cones are described by Eq. (10) and (12) will be called inertial coordinates.

Since the light cones and straight lines are absolute objects in special relativity, the coordinate transformations that leave these structures invariant constitute the invariance group of special relativity. In an inertial coordinate system, these transformations have the form

$$x'^\mu = \alpha_\nu^\mu x^\nu + b^\mu \tag{13}$$

where α_ν^μ and b^ν are constants. The b^μ are arbitrary while the α_ν^μ are constrained to satisfy the conditions

$$\eta_{\mu\nu}\alpha_\rho^\mu\alpha_\sigma^\nu = \eta_{\rho\sigma} \qquad (14)$$

These transformations form a group, the inhomogeneous Lorentz group, each member of which is characterized by the 10 arbitrary values one can assign to the α_ν^μ and b^μ. This group contains, as subgroups, the three-dimensional rotation group and the group of spatial and temporal translations. It also includes the group of Lorentz transformations, now called Lorentz boosts. A boost along the x axis takes the form

$$\begin{aligned} x'^0 &= \gamma(x^0 + \beta x^1) \\ x'^1 &= \gamma(\beta x^0 + x^1) \\ x'^2 &= x^2 \\ x'^3 &= x^3 \end{aligned} \qquad (15)$$

where $\gamma = (1 - \beta^2)^{-1/2}$ and β is a parameter that characterizes the boost. These transformation equations take their more familiar form if we set $x^\mu = (ct, x, y, z)$ and similarly for x'^μ and $\beta = v/c$, where c is the velocity of light, in which case v is the velocity associated with the transformation. In special relativity, the Lorentz boosts replace the Galilean transformations of Newtonian mechanics just as the light cones replace the planes of absolute simultaneity. Also, the Galilean law of addition for velocities, Eq. (9), is no longer valid. For a boost in the x direction, the transformed components v_i' of the velocity of a body are related to its original components v_i by the equations

$$v_i' = \delta(v_1 + v) \qquad v_2' = \gamma^{-1}\,\delta v_2 \qquad v_3' = \gamma^{-1}\,\delta v_3 \qquad (16)$$

where $\delta = (1 + v_1 v/c^2)^{-1}$.

D. The Space–Time Metric

Equations (10) and (12) for straight lines and light cones are given in a special coordinate system in which they assume these simple forms. It is possible to write generally covariant equations for these objects by introducing a symmetric second rank tensor $g_{\mu\nu}$ of signature -2 together with its inverse $g^{\mu\nu}$. With its help, the light cones can be characterized by the surfaces $\phi(x) = 0$, where ϕ satisfies the covariant equation

$$g^{\mu\nu}\phi_{,\mu}\phi_{,\nu} = 0 \qquad (17)$$

To construct an equation for a straight line we first introduce the Christoffel symbols $\{{}^\mu_{\rho\sigma}\}$ de-

fined by

$$\{{}^\mu_{\rho\sigma}\} = \tfrac{1}{2}g^{\mu\nu}(g_{\rho\nu,\sigma} + g_{\nu\rho,\sigma} - g_{\rho\sigma,\nu}) \qquad (18)$$

These quantities constitute the components of a geometrical object that is linear but not homogeneous. With their help, the equations for the coordinates $x^\mu(\lambda)$ of the points lying on a straight line can now be written as

$$d^2x^\mu/d\lambda^2 + \{{}^\mu_{\rho\sigma}\}(dx^\rho/d\lambda)(dx^\sigma/d\lambda) = 0 \quad (19)$$

where again λ is a monotone increasing parameter along the curve. One can choose λ so that $g_{\mu\nu}\,dx^\mu/d\lambda\,dx^\nu/d\lambda = 1$, in which case it is the proper time along the line. One can also use Eq. (19) to characterize the trajectories of light rays if one adds the condition that $g_{\mu\nu}\,dx^\mu/d\lambda\,dx^\nu/d\lambda = 0$. Such rays have the property that they serve as the generators of the light cones. Equation (19) is usually referred to as the geodesic equation since it has the same form as the equation for a geodesic curve, that is, a curve of minimum length connecting two points in a Riemannian space with a metric $g_{\mu\nu}$.

Having introduced the tensor $g_{\mu\nu}$, it now becomes necessary to construct a law of motion for it. In special relativity this law is taken to be

$$R_{\mu\nu\rho\sigma} = 0 \qquad (20)$$

where the tensor $R_{\mu\nu\rho\sigma}$, called the Riemann–Christoffel tensor, is constructed from the tensor $g_{\mu\nu}$ according to

$$\begin{aligned} R_{\mu\nu\rho\sigma} &= \tfrac{1}{2}(g_{\mu\rho,\nu\sigma} + g_{\nu\sigma,\mu\rho} - g_{\mu\sigma,\nu\rho} - g_{\nu\rho,\mu\sigma}) \\ &\quad + g_{\alpha\beta}(\{{}^\alpha_{\mu\rho}\}\{{}^\beta_{\nu\sigma}\} - \{{}^\alpha_{\mu\sigma}\}\{{}^\beta_{\nu\rho}\}) \qquad (21) \end{aligned}$$

It appears in the equation of geodesic deviation that governs the separation between two neighboring, freely falling bodies. When all of the components of $R_{\mu\nu\rho\sigma}$ vanish, this separation remains constant.

It can be shown that every solution of Eq. (20) can be transformed so that $g_{\mu\nu} = \eta_{\mu\nu}$ everywhere on the space–time manifold, in which case $g_{\mu\nu}$ is said to take on its Minkowski values. In a coordinate system in which $g_{\mu\nu}$ has this form, Eq. (17) becomes

$$\eta^{\mu\nu}\phi_{,\mu}\phi_{,\nu} = 0 \qquad (17a)$$

It is seen that the surface defined by Eq. (12) satisfies this equation. Likewise, in this coordinate system, Eq. (19) reduces to

$$d^2x^\mu/d\lambda^2 = 0 \qquad (19a)$$

and has, as its solution, the curves defined by Eq. (16).

Since $g_{\mu\nu}$ can always be transformed to its Minkowski values, it is seen to be an absolute object in the theory and the group of transformations that leave it invariant is again the inhomogeneous Lorentz group. In all respects $g_{\mu\nu}$ is equivalent to the straight lines and light cones of the theory. Furthermore, one can either construct laws of motion that employ inertial coordinates and that are covariant with respect to the inhomogeneous Lorentz group or construct generally covariant laws with the help of the $g_{\mu\nu}$. When $g_{\mu\nu}$ is transformed to take on the values $\eta_{\mu\nu}$, the latter equations reduce to the former.

The Riemann–Christoffel tensor arose in the study of the geometry of manifolds with Riemannian metrics. In such a geometry one defines a distance ds between neighboring points of the manifold with coordinates x^μ and $x^\mu + dx^\mu$ to be

$$ds^2 = g_{\mu\nu}(x)\, dx^\mu\, dx^\nu \qquad (22)$$

where $g_{\mu\nu}$, a symmetric tensor field, is the metric of the manifold. The vanishing of the Riemann–Christoffel tensor can be shown to be the necessary and sufficient condition for the geometry to be flat; that is, the metric can always be transformed to a constant tensor everywhere on the manifold. Since the tensor $g_{\mu\nu}$ introduced above is an absolute object it is sometimes referred to as the metric of the flat space–time of special relativity. Although we will not need to make use of this geometrical interpretation, it will sometimes prove convenient to give an expression for ds as a way of specifying the components of $g_{\mu\nu}$.

IV. General Relativity

A. The Principle of Equivalence

After Einstein formulated the special theory of relativity he turned his attention to, among other things, the problem of constructing a Lorentz-invariant theory of gravity. Newton thought of gravity as an action-at-a-distance force between massive bodies and as transmitted instantaneously between them. Since special relativity required a finite speed of transmission, Einstein sought to construct a relativistic field theory of gravity. The simplest object to associate with the gravitational field was a scalar field. However, a difficulty presented itself when he came to construct a source for this field. In the Newtonian theory, the gravitational attraction between bodies was proportional to their

masses. In special relativity, however, energy has associated with it an equivalent mass through the relation $E = mc^2$. Consequently, Einstein argued, mass density by itself could not be the sole source of the gravitational field. At the same time, energy density could not be used since it is not, by itself, associated with a geometrical object in special relativity but rather with one component of a tensor field.

While thinking about the problem of gravity, Einstein was struck by a peculiarity of the gravitational interaction between bodies, namely the constancy of the ratio of the inertial to the gravitational mass of all material bodies. In Newtonian mechanics, mass enters in two essentially different ways—as inertial mass in the second law of motion and as gravitational mass in the law of gravitational interaction. Logically, these two masses have nothing to do with one another. Inertial mass measures the resistance of a body to forces imposed on it while gravitational mass determines, in the same way as electric charge determines the strength of the electrical force between charged bodies, the strength of the gravitational force between massive bodies. Galileo was the first to demonstrate this constancy by observing that the acceleration experienced by objects in the earth's gravitational field was independent of their mass. In 1891, Eötvös demonstrated it to an accuracy of one part in 10^8. More recent determinations by groups in Princeton and Moscow have established this constancy to better than one part in 10^{11}.

While there was no explanation for this constancy, Einstein realized that it called into question the existence of one of the absolute objects of both Newtonian mechanics and special relativity, the free bodies. If indeed this ratio was a universal constant, then there could be no such thing as a gravitationally uncharged body since zero gravitational mass would then imply zero inertial mass. Einstein also realized that this constancy meant that it would be impossible to distinguish locally, that is, in a sufficiently small region of space–time, between inertial and gravitational effects through their action on material bodies. An observer in an elevator being accelerated upward with an acceleration equal to that produced by the earth's gravity would see objects fall to the floor of the elevator in exactly the same way that they fall on earth, that is, with an acceleration that is independent of their mass.

After this realization, Einstein made a characteristic leap of imagination. He postulated that it is impossible to distinguish locally between inertial and gravitational effects by any means. One of the consequences of this postulate, called by him the principle of equivalence, is that light should be bent in a gravitational field just as it would appear to be to an observer in an accelerating elevator. But if this is the case, the light cones of special relativity would no longer be absolute objects either, and this in turn would mean that the metric of special relativistic space–time would not be an absolute object.

The principle of equivalence however, implied even more. If inertial and gravitational effects are indistinguishable from each other locally, then one and the same object could be used to characterize both effects. Since it is the metric $g_{\mu\nu}$ that is responsible for the inertial effects one observes in special relativity, $g_{\mu\nu}$ should also be associated with the gravitational field. In effect, geometry and gravity became simply different aspects of the same thing. Actually, one never needs to interpret $g_{\mu\nu}$ as a metric. One can identify it solely with the gravitational field. This identification has, as a consequence, that $g_{\mu\nu}$ must be a dynamical object since the gravitational field clearly must be such. Having recognized this fact, Einstein then turned his attention to the problem of constructing a law of motion for this object.

B. The Principle of General Invariance

In his attempts to construct a law of motion for $g_{\mu\nu}$, Einstein proposed that these laws should be generally covariant. However, we have already seen that the laws of motion of special relativity could be cast in generally covariant form with the introduction of a metric satisfying Eq. (20). But such a metric was absolute and Einstein wanted a law of motion for a dynamical $g_{\mu\nu}$. Consequently, what Einstein was really requiring was not general covariance but rather general invariance, that is, that the invariance group of the laws of motion should be the same as their covariance group, namely the group of all arbitrary coordinate transformations. As we have argued above, this can be the case only if there are no absolute objects in the theory. The absence of absolute objects in the theory satisfies a version of Mach's principle which states that there should be no absolute objects in any physical theory.

Although Einstein did not use precisely the reasoning outlined above, it was his recognition of the preferred role played by the inhomogeneous Lorentz group in special relativity that was crucial to the development of the general theory of relativity. And although he formulated his argument in terms of the relativity of motion, it is clear that he was referring to the invariance properties of the laws of motion. His argument that all motion should be relative—hence the term general relativity—was really a requirement, in modern terms, that these laws should be generally invariant. This is not an empty requirement, as some authors have suggested, but rather severely limits the possible laws of motion one can formulate for $g_{\mu\nu}$.

C. Laws of Motion

The search for a generally invariant law of motion for $g_{\mu\nu}$ occupied a considerable portion of Einstein's time prior to the year 1915. At one point he even argued that such a law could not exist. However, he did succeed in that year in finally formulating this law. If one requires that this law contain no higher than second derivatives of the $g_{\mu\nu}$ and furthermore that it be derivable from a variational principle, then there is, in fact, essentially only one law that fills these requirements. This is in marked contrast to the situation in electrodynamics, where there are an infinite number of laws of motion for the vector potential A_μ that satisfy these requirements.

To formulate the law of motion for $g_{\mu\nu}$, we first construct from the Riemann–Christoffel tensor (21) the Ricci tensor $R_{\mu\nu}$, where

$$R_{\mu\nu} = g^{\rho\sigma}R_{\sigma\mu\rho\nu} \qquad (23)$$

and the curvature scalar R, where

$$R = g^{\mu\nu}R_{\mu\nu} \qquad (24)$$

In terms of these quantities this law can be written as

$$R_{\mu\nu} - \tfrac{1}{2}g_{\mu\nu}R + \Lambda g_{\mu\nu} = \kappa T_{\mu\nu} \qquad (25)$$

where Λ and κ are constants and $T_{\mu\nu}$ is the energy–momentum tensor associated with the sources of the gravitational field.

In general, the components of the Riemann–Christoffel tensor will not vanish even when all of the components of the Ricci tensor do. As a consequence, it follows from the equation of geodesic deviation that the separation between neighboring freely falling masses will change with time. Since such changes appear due to tidal forces in many-practicle systems, the Riemann–Christoffel tensor is thus a measure of such forces and vice versa.

The so-called cosmological term $\Lambda g_{\mu\nu}$ was originally not present in the Einstein field equations. It was later added by him to obtain a static cosmological model with matter. When it was later realized that the universe was expanding and that there were solutions of the field equations without the cosmological term that fit the current observations, the motivation for the inclusion of this term disappeared and it is now not usually included in the equations. Also, measurements made on distant galaxies place an upper limit of 10^{-66} cm^{-2} on $|\Lambda|$.

In addition to the law of motion for $g_{\mu\nu}$ it is necessary to formulate generally invariant laws of motion for the other geometrical objects that are to be associated with the physical quantities being observed. One way to do this is simply to take over the generally covariant form of the laws formulated for these objects in special relativity. The law of motion for the electromagnetic field, when this field is associated with a vector A_μ, can, for example, be written in the form

$$(\sqrt{-g}\ g^{\mu\rho}g^{\nu\sigma}F_{\rho\sigma})_{,\mu} = 4\pi j^\nu \qquad (26)$$

where g is the determinant of $g_{\mu\nu}$, j^μ the current density associated with the sources of the electromagnetic field, and

$$F_{\mu\nu} = A_{\nu,\mu} - A_{\mu,\nu} \qquad (27)$$

Likewise, the equation of motion for a body on which no other forces act can be taken to be given by Eq. (19). (In fact, it can be shown that this law of motion is a consequence of the field equations for the gravitational field $g_{\mu\nu}$ and hence need not appear as a separate postulate in the theory.) Such laws of motion are said to involve minimal coupling to the gravitational field. It is also possible to construct laws of motion that do not couple minimally to the gravitational field. In the case of the electromagnetic field, for example, one could include a factor of $1 + R$, where R is the curvature scalar, inside the parentheses in Eq. (26). The only requirement these laws of motion should satisfy is that they reduce to their special relativistic form when the Riemann–Christoffel tensor vanishes.

D. Clocks, Rods, and Coordinates

It has been argued that some kind of postulate concerning the behavior of clocks and measuring rods is required in general relativity. For example, it has been suggested that a class of objects, ideal clocks, measure proper time along their trajectories, where the proper time along a trajectory is defined to be the integral of the distance ds, given by Eq. (22), along this trajectory. In this view, clocks, and also measuring rods, are assumed to be primitive objects in the theory.

Actually, all such postulates are unnecessary, in both the special and general theories. Clocks, and similarly measuring rods, are, in fact, composite physical systems with laws of motion governing their behavior. Once these laws have been established, there is no need to add additional postulates governing their behavior. It can be shown, for example, that if one takes, as a model for a clock, a classical hydrogen atom, then as long as the forces acting on this clock are small compared to the internal forces acting on its constituents and its dimensions are small compared to the curvature of its trajectory, it will indeed measure approximately the proper time along this trajectory. However, if the forces acting on it are sufficiently strong, the atom will be ionized and cease to measure any kind of time along its trajectory. Thus, the behavior of clocks is seen to be a dynamical question that cannot be decided *a priori* from any kinematic postulate.

In this view, then, clocks and measuring rods, and indeed all measuring devices, are considered to be physical systems with the geometrical objects associated with them obeying their own laws of motion. Furthermore, a physical description would have to be considered incomplete if it did not supply these laws of motion. To avoid the necessity of having to formulate and solve the laws of motion for a particular kind of clock, one may assume that it does satisfy the conditions for measuring proper time, with the proviso that if these conditions are violated it will no longer do so. If this assumption results in inconsistencies it does not mean that a principle of general relativity has been violated but only that these conditions have not been met.

While clocks and rods can be used to measure times and distances, it should be emphasized that these measurements bear no direct relation to the coordinates employed in the formulation of the laws of motion. Since, in all space–time descriptions, these laws are generally covariant, there are no preferred coordinate systems. Consequently, it follows that the predictions of a theory cannot depend on a particular coordinatization. In effect, the coordinates play the same role in space–time theories as do the indices that characterize the various components of a geometrical object and hence, like these indices, are not associated with any physical objects.

It is, however, often convenient to choose a particular coordinatization. Thus in Newtonian mechanics one usually chooses coordinates such that one of them is the parameter that characterizes the planes of simultaneity, and likewise in special relativity one usually employs inertial coordinates. In general relativity one also can employ a coordinatization that is particularly convenient for some purpose. One can, for example, choose coordinates in such a way that one of them corresponds to the time and distance intervals measured by a particular family of clocks and rods. Alternatively, one can choose coordinates so that certain components of the gravitational field have simple values. For example, one can choose coordinates so that $g_{00} = 1$ and $g_{01} = g_{02} = g_{03} = 0$. But in all cases such a choice is arbitrary and devoid of physical content.

V. Gravitational Fields

A. Newtonian Fields

Since Newtonian theory describes, to a high degree of accuracy, the phenomena associated with weak gravitational fields, it is essential that this theory be an approximation to the general theory. Although originally formulated as an action-at-a-distance theory, the Newtonian theory of gravity can also be formulated as a field theory analogous to electrostatics. The gravitational field is characterized, in this version of the theory, by a single scalar field ϕ that satisfies, in suitable coordinates, the field equation

$$\nabla^2 \phi = 4\pi\rho \tag{28}$$

where ρ is the mass density of the sources of the field.

In the general theory we assume that, in the case of weak fields, there exists a coordinate system such that $g_{\mu\nu} = \eta_{\mu\nu} + h_{\mu\nu}$, where $h_{\mu\nu} \ll 1$. We also assume that the velocities of the sources of the gravitational field are all vanishingly small compared to the velocity of light. In this case, the only nonvanishing component of $T_{\mu\nu}$ is $T_{00} = \rho c^2$ and Eq. (25) can be shown to reduce to Eq. (28) if we set $\Lambda = 0$, $\kappa = -8\pi G/c^4$, where G is the Newtonian gravitational constant, and take

$$\phi = (c^2/2)h_{00} \tag{29}$$

Furthermore, one can show that the law of motion (19) reduces to the Newtonian form

$$d^2\mathbf{x}/dt^2 = -\nabla\phi \tag{30}$$

B. The Schwarzschild Field, Event Horizons, and Black Holes

In spite of their enormous complexity, the Einstein field equations (25) possess many exact solutions. One of the first and perhaps still the most important of these solutions was obtained by Schwarzschild in 1916 for the case $\Lambda = 0$ and $T_{\mu\nu} = 0$ by imposing the condition of spherical symmetry on $g_{\mu\nu}$. The nonvanishing components of $g_{\mu\nu}$ are given in spherical coordinates by

$$g_{00} = 1 - 2M/r \qquad g_{11} = -1/(1 - 2M/r)$$
$$g_{22} = -r^2 \qquad g_{33} = -r^2 \sin^2\theta \tag{31}$$

where M is a constant of integration. This solution is seen to be independent of the coordinate x^0 and hence is a static field. The condition that the field be independent of x^0 was originally imposed by Schwarzschild in obtaining his solution of the field equations but has since been shown to be a consequence of the condition of spherical symmetry.

The importance of the Schwarzschild solution lies in the fact that it is the general relativistic analog of the Newtonian field of a point mass. The solution to Eq. (28) in this case is $\phi = Gm/r$, where m is the mass of the point. By making use of Eq. (29) and Eq. (31) for g_{00} we see that $M = Gm/c^2$. The constant $2M$ is referred to as the Schwarzschild radius of the mass m. The Schwarzschild radius of the sun is 2.9 km and of the earth is 0.88 cm. For comparison, the Schwarzschild radius of a proton is 2.4×10^{-52} cm and that of a typical galaxy of mass $\sim 10^{45}$ gm is $\sim 10^{17}$ cm.

The Schwarzschild field has a property that distinguishes it from the corresponding Newtonian field: at $r = 2M$ it becomes singular. Indeed, at this radius g_{11} is infinite! However, this is not a physical singularity, as Eddington first showed in 1924, but rather what is called a coordinate singularity. A final clarification of the structure of the Schwarzschild field came in 1960 with the work of M. Kruskal. He found a coordinate transformation from the Schwarzschild coordinates (x^0, r, θ, ϕ) to the set (u, v, θ, ϕ), where

$$u = a \cosh(x^0/4M) \qquad v = a \sinh(x^0/4M) \tag{32}$$

with $a = [(r/2M) - 1]^{1/2} \exp(r/4M)$, such that the transformed components of the Schwarzschild

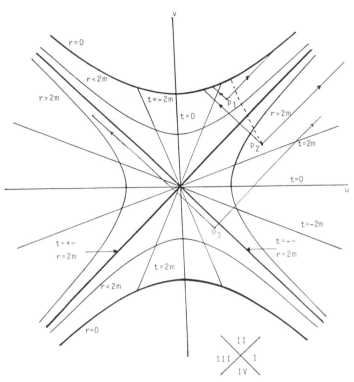

FIG. 1. Kruskal diagram for the Schwarzschild field. The hyperbola $r = 0$ represents a real singularity of the field.

field were given by

$$g_{00} = f \qquad g_{11} = -f \qquad g_{22} = -r^2$$
$$g_{33} = -r^2 \sin^2 \phi \qquad\qquad (33)$$

where $f = (8M/r) \exp(-r/2M)$. In these latter expressions, r is a function of u and v obtained by solving Eq. (32) for this quantity. In this form the field is seen to be singular only at $r = 0$, which is a true physical singularity.

Figure 1 depicts a number of the features of the Kruskal transformation in what is called a Kruskal diagram. In this diagram, curves of constant Schwarzschild r correspond to the right hyperbolas $u^2 - v^2 = \text{const}$, while curves of constant x^0 correspond to straight lines passing through the origin. The region I to the right of the two $r = 2M$ lines corresponds to the entire region $r > 2M$, $-\infty < x^0 < \infty$. Hence it follows that the transformation from Kruskal to Schwarzschild coordinates must be a singular transformation, and indeed it is the singular nature of this transformation that is responsible for the singularity of $g_{\mu\nu}$ at $r = 2M$ in Schwarzschild coordinates.

Many of the properties of the Schwarzschild field can be understood by referring to the Kruskal diagram. In Kruskal coordinates, light rays propagate along straight lines at 45° with respect to the u, v axes. As a consequence, it is seen that all of the rays emitted at a point P_1 in region II above the two lines $r = 2M$ will ultimately reach the singular curve $r = 0$, so no information can be transmitted from this point into region I. Likewise, inward-directed rays emitted at a point P_2 in region I will also reach the $r = 0$ curve while outward-directed rays will continue their outward propagation forever. And finally, some of the rays emitted from a point P_3 in region IV below the two lines $r = 2M$ will ultimately reach points in region I while others will reach the region III to the left of the two lines $r = 2M$.

Because of these features, the surface $r = 2M$ is said to form an event horizon: an observer in region I can never receive information about events taking place in regions II and III of the diagram. In addition, it is seen to be a light cone.

Finally, we note that a material body starting from rest at P_2 will fall into the singularity at $r = 0$ along the dashed path indicated in Fig. 1. Even though it reaches the event horizon at $x^0 = \infty$, the proper time along the curve from P_2 to the point where it reaches the singularity is finite. A local observer falling with the particle would notice nothing peculiar as he passed through the horizon. As long as the particle is outside this horizon it is possible to reverse its motion so that it does not cross it. However, once it does its fate is sealed; its motion can no longer be reversed and it will ultimately reach the $r = 0$ singularity in a finite amount of proper time. For this reason the source of a Schwarzschild field is said to be a black hole—nothing that falls into it can ever get out again and any radiation generated inside the horizon can never be seen from the outside.

C. Kerr–Newman Fields, Naked Singularities

In addition to the Schwarzschild family of solutions, each member of which is characterized by a value of the parameter M, other stationary solutions of the empty-space Einstein field equations have been found by Kerr and Newman. Like the Schwarzschild solution, the Kerr–Newman solutions are asymptotically flat; that is, the value of the Riemann–Christoffel tensor goes to zero as the coordinate r approaches infinity and also the physical singularity in the solution is surrounded by an event horizon. This family of solutions is characterized by three continuously variable parameters, which, because of the asymptotic form of these solutions, can be interpreted as the mass, angular momentum, and electric charge of the black hole.

After the discovery of the Kerr–Newman solutions it was shown by the work of Israel, Carter, Hawking, and Robinson that the Kerr–Newman solutions are the only asymptotically flat, stationary black hole solutions, that is, solutions with event horizons that depend continuously on a finite set of parameters. This result is a version of a conjecture of Wheeler to the effect that "a black hole has no hair." What is still lacking is a proof of uniqueness of the Kerr–Newman solutions, that is, that not more than three parameters is sufficient to characterize completely all stationary, asymptotically flat black hole solutions. To date, the best one can do in this respect is to show that the only such static solutions and the only such neutral solutions belong to the family of Kerr–Newman solutions.

In addition to the Kerr–Newman family of solutions there are solutions of the empty-space field equations that do not have event horizons surrounding the physical singularity in the solution. One such solution is obtained by letting the parameter M in the Schwarzschild solution become negative. Also, a charged Schwarzschild field has no horizon if the charge $Q < M$. In both cases the only singularity occurs at $r = 0$. Because of the absence of a horizon surrounding this point, it is referred to as a naked singularity.

D. Fields with Matter—Gravitational Collapse

In addition to the empty-space solutions discussed above there are numerous solutions of the Einstein equations with nonvanishing energy–momentum tensors. One can, for example, construct a spherically symmetric, nonsingular interior solution that joins on smoothly to an exterior Schwarzschild solution. The combined solution would then correspond to the field of a normal star, a white dwarf, or a neutron star. What distinguishes these objects is the equation of state for the matter comprising them. It is surprising that no interior solutions that could be joined to a Kerr–Newman field have been found.

Normal stars, of course, are not eternal objects. They are supported against gravitational collapse by pressure forces whose source is the thermonuclear burning that takes place at the center of the star. Once such burning ceases due to the depletion of its nuclear fuel, a star will begin to contract. If its total mass is less than approximately two solar masses it will ultimately become a stable white dwarf or a neutron star, supported by either electron or neutron degeneracy pressure. If, however, its total mass exceeds this limit, the star will continue to contract under its own gravity down to a point, a result first demonstrated by Oppenheimer and Snyder in 1939. Although they treated only cold matter, that is, matter without pressure, their result would still hold once the star has passed a certain critical stage even if the repulsion of the nuclei comprising the star were infinite. This critical stage is reached once the radius of the star becomes less than its Schwarzschild radius. When this happens, the surface $r = 2M$ becomes an event horizon and the matter is trapped inside, forming a black hole. Even an infinite pressure would then be unable to halt the continued contraction to a point because such a pressure would contribute an infinite amount to the energy–momentum tensor of the matter, which in turn would result in an even stronger gravitational attraction.

Because of the disquieting features of black hole formation (neither Eddington nor Einstein was prepared to accept their existence), theorists looked for ways to avoid their formation. However, theorems of Penrose and Hawking show that collapse to a singularity is inevitable once the gravitational field becomes strong enough to drag back any light emitted by the star, that is, when the escape velocity at the surface of the star exceeds the velocity of light.

What has not been proved to date is what Penrose calls the hypothesis of "cosmic censorship." This hypothesis asserts that matter will never collapse to a naked singularity, but rather that the singularity will always be surrounded by an event horizon and hence not be visible to an external observer. While the hypothesis is supported by both numerical and perturbation calculations, it has so far not been shown to be a rigorous consequence of the laws of motion of the general theory.

The detection of black holes is complicated by the fact that they are black—by themselves they can emit no radiation. If they exist at all then, they can be detected only through the effects of their gravitational field on nearby matter. If a black hole were a member of a double star system, it would become the source of intense X rays when its companion expanded during the later stages of its own evolution. As matter from the companion fell onto the black hole it would become compressed and thus heated to temperatures high enough for it to emit such high-energy radiation. Of course, it is not enough to find an x-ray-emitting binary system in order to prove the existence of a black hole. It is also necessary that the mass of the x-ray-emitting component be larger than the upper limit on the mass of a stable neutron star, that is, larger than two solar masses.

Until quite recently the best candidate for a black hole was the binary system Cygnus X-1. However, the mass of the x-ray source in this system is difficult to determine from orbital measurements, so possibilities other than its being a black hole cannot be entirely ruled out. In 1983 a more convincing candidate for a black hole was found in an intense x-ray source in the Large Magellanic Cloud. Orbital parameters show that this source has a mass of at least seven solar masses. Also, a second object similar to Cygnus X-1 massive enough to be a black hole has been found in this cloud. Finally, evidence is accumulating that the most likely source of energy in a quasar is a massive black hole. The size and luminosity of quasars are consistent with their being black holes of about 10^8 solar masses.

Also, up to 50% of the mass of matter falling into such a hole can be converted into radiation that escapes. If a few solar masses per year were so converted it would be enough to power a typical quasar.

E. Gravitational Radiation, the Quadrupole Formula

Shortly after he formulated the general theory, Einstein showed that, when linearized around the flat field $g_{\mu\nu} = \eta_{\mu\nu}$, the empty-space field equations with zero cosmological constant possessed plane wave solutions similar to the plane wave solutions of electromagnetism. These waves propagate along the light cones defined by $\eta_{\mu\nu}$, that is, at the speed of light. Like their electromagnetic counterparts, they are transverse waves and possess two independent states of polarization. However, unlike the dipole structure of plane electromagnetic waves, gravitational waves have a quadrupole structure, as would be expected from the tensor nature of the gravitational field. Using coordinates such that $g_{\mu\nu} = \eta_{\mu\nu} + h_{\mu\nu}$ and the direction of propagation as the z axis, the two states of polarization are characterized by $h_{xx}^+ = h_{yy}^+$ and $h_{xy}^\times = h_{yx}^\times$ with, in the two cases, the other components having the value zero.

Although the empty-space field equations have been shown to possess exact solutions with wavelike properties, the so-called plane fronted waves, neither they nor the approximate wave solutions found by Einstein are associated with sources. In electromagnetic theory it is possible to find exact solutions of the field equations associated with sources with arbitrary motions. So far, no such solutions have been found for the highly nonlinear equations of the general theory. In order to construct radiative solutions associated with sources it is necessary to employ approximation methods.

In the past, two such methods have been used, the so-called slow- and fast-motion approximations. As its name implies, the slow-motion approximation assumes that the motion of the sources is small compared to the velocity of light and that the gravitational fields produced by these sources are weak. Unfortunately, this method can be shown to lead to inconsistencies in higher orders of approximation. On the other hand, the fast-motion approximation, which assumes only that the fields are weak, while free of these problems is extremely difficult to apply in practice due to the nonlinearities in the field equations. For gravitationally bound systems such as a binary star, the nonlinear terms can be shown to be of the same order of magnitude as the linear terms in the field equations and hence cannot be ignored.

By combining the best features of these two approximations with the method of matched asymptotic expansions, the work of Burke, Kates, Kegeles, Madonna, and Anderson has led to a satisfactory derivation of the so-called quadrupole radiation formula

$$dE/dt = -\tfrac{1}{5}(G/c^5)\langle d^3Q_{ij}/dt^3 \; d^3Q_{ij}/dt^3\rangle \quad (34)$$

where the angle brackets denote an average over a period of oscillation of the source and the indices i and j take on the values 1 through 3. The quantities Q_{ij} are the components of the mass quadrupole moment:

$$Q_{ij} - \int \rho(x)(x_i x_j - \tfrac{1}{3}\delta_{ij}\delta^{kl}x_k x_l) \; dv \quad (35)$$

where $\rho(x)$ is the mass density of the source. Finally, the quantity E appearing on the left side of this equation is the total Newtonian energy, kinetic plus potential, of the source. As a consequence, the quadrupole formula can be interpreted as an expression of energy conservation, in which case the quantity on its right side becomes the energy carried away by gravitational waves.

The quadrupole formula gives the main contribution to the energy loss of a system due to gravitational radiation. Changes in higher multipoles of the mass will also contribute to this loss, although in general their contribution will be much smaller than that of the quadrupole. However, unlike the electromagnetic case, there is no dipole contribution, since both the total momentum and angular momentum of the radiating system are conserved. And in both theories there is no monopole radiation because of conservation of charge in the one case and conservation of mass in the other.

It should be pointed out that the quadrupole formula is an approximation and is valid only for slow-motion sources, that is, for sources whose velocities are all less than the velocity of light. As a consequence it must be applied with caution to emitters of strong gravitational waves.

The amount of energy emitted by slow-motion sources is in most cases very small. A beryllium rod of length 170 m and weighing 6×10^7 kg, spinning on an axis through its center as fast as it can without disintegrating ($\omega \sim 10^3$ sec^{-1}), would radiate energy at the rate of approximately 10^{-7} erg/sec. For the earth–sun system

Eq. (34) yields a rate of energy loss of about 200 W. Only in the case of extremely massive objects moving at high speeds such as in the binary pulsar will the amount of energy radiated be significant.

VI. Observational Tests of General Relativity

A. GRAVITATIONAL RED SHIFT

In general relativity the apparent rate at which clocks run is affected by the presence of a gravitational field. Like its counterpart in special relativity, this is a kinematic effect and hence is independent of any direct effect of the gravitational field on the internal dynamics of the clock. Only if gradients of this field result in tidal forces that are comparable to the nongravitational forces responsible for the functioning of the clock will this internal dynamics be altered.

The amount of red shift is most easily calculated in the case of a static gravitational field and for two clocks that are at rest in this field. We suppose that one clock, the emitter, sends out light waves whose frequency ν_{em} is the same as its own frequency. A receiving clock, which is identical in construction to the emitting clock, that is, has the same internal dynamics, is used to measure the frequency ν_{rec} of the received radiation. Then it can be shown that

$$Z := (\nu_{rec} - \nu_{em})/\nu_{em} = (1/c^2)(\phi_{em} - \phi_{rec})$$

$$(36)$$

where ϕ_{em} and ϕ_{rec} are the gravitational potentials at the locations of the emitter and receiver, respectively. If ϕ_{em} is less than ϕ_{rec} then the quantity Z is negative, hence the emitted light appears to be red shifted. This effect can be understood by noting that, in going from the emitter to the receiver, a photon will, in this case, gain potential energy. Since its total energy is conserved, its kinetic energy, which is proportional to its frequency, will decrease and hence so will its frequency. If either the emitter or receiver is moving with respect to the background field then Eq. (36) must be amended to take account of the Doppler shift produced by such motion.

The first attempts to observe the gravitational red shift were made on the spectral lines of the sun and known white dwarfs. For the sun, $Z = -2.12 \times 10^{-6}$ while for white dwarfs it would have values 10–100 times as large. In the case of the sun the shift was masked by the Doppler broadening of the spectral lines due to thermal motion. However, observations near its edge are consistent with a red shift of the magnitude predicted by Eq. (36). In the case of the white dwarfs, it was not possible to measure their masses and radii with sufficient accuracy to determine ϕ_{em}, although again red shifts were observed whose magnitudes were consistent with estimates of these quantities.

The first accurate test of the red shift prediction was carried out in a series of terrestrial experiments by Pound and Rebka in 1960, using the Mössbauer effect. The emitter and detector used by them were separated by a vertical height of 74 ft. In this case the gravitational potential can be taken equal to gz, where g is the local acceleration due to gravity and z is the height above ground level. Equation (36) then yields the value $Z = 2.5 \times 10^{-15}$. In spite of its small value, Pound and Rebka were able to observe a shift equal to 1.05 ± 0.10 times the predicted Z. Later experiments with cesium beam and rubidium clocks on jet aircraft yielded similar results.

The possibility of testing the red shift prediction has improved dramatically with the development of high-precision clocks. In 1976 Vessot and Levine used a rocket to carry a hydrogen-maser clock to an altitude of about 10,000 km. Their result verified the theoretical value to within 2 parts in 10^{-4}.

It has been argued that red shift observations do not bear on the question of the validity of general relativity but rather only on the validity of the principle of equivalence. This is, in fact, only partially true. If one assumes that the gravitational field of the earth is uniform over the 74 ft that separated the emitter and receiver of the Pound–Rebka experiment, then indeed their result can be calculated using only this principle. However, one cannot use it alone to determine the result of the Vessot–Levine experiment. In this case it is necessary to make direct use of Eq. (36). The derivation of Eq. (36), however, makes use of Eq. (19) for a light ray, which, in turn, is a consequence of the field equations (25) of general relativity. Also, although the present red shift measurements are not sufficiently accurate to distinguish between different possible equations for the gravitational field, there is nothing in principle that would preclude such a test.

B. Solar System Tests—The PPN Formalism

The first arena used for testing general relativity was the solar system and it remains so to this date. What has changed dramatically over the years, due to the rapid growth of technology, is the degree of accuracy with which the theory can be tested. It is in the solar system that the gravitational field of the sun is of sufficient strength that deviations from the Newtonian theory are observable, but just barely.

In calculating the size of these effects one assumes that the trajectories of planets and light rays obey Eqs. (19). However, rather than take the sun's gravitational field to be the Schwarzschild field in evaluating the Christoffel symbols appearing in these equations, it is useful to use what has become known as the parametrized post-Newtonian (PPN) formalism. In this formalism, first developed by Eddington and extended by Robertson, Schiff, Will, and others, one assumes a more general form for the post-Newtonian corrections to the gravitational field of the sun than those given by general relativity. These corrections are allowed to depend on a number of unknown parameters that one hopes to determine by solar system and other observations. The reason for proceeding in this manner is that it allows one to test the validity of other competing theories of gravity in which these parameters have different values from those they would have if general relativity were valid.

In the most extreme versions of this formalism as many as 10 parameters are employed. However, a number of these parameters can be eliminated if one requires that the equations of motion (19) are a consequence of the field equations for the gravitational field, as they are in general relativity. Also, some of these parameters are known to be small from other experiments. In what follows we will use an abbreviated version of the PPN formalism employed by Hellings in his analysis of the solar system data. In this version the components of the gravitational field have the form

$$g_{00} = 1 - 2U[1 - J_2(R_\odot/r)^2 P_2(\theta)]$$
$$+ 2\beta U^2 + \alpha_1 U(w/c) \quad \text{(37a)}$$

$$g_{0i} = \alpha_1 U w^i/c \quad \text{(37b)}$$

$$g_{ij} = -(1 + 2\gamma U)\delta_{ij} \quad \text{(37c)}$$

where β, γ, and α_1 are PPN parameters $U = GM_\odot/rc^2$ is the Newtonian gravitational potential of the sun, and $R_\odot$ and $M_\odot$ are the radius and mass of the sun. Included in Eq. (37a) is a term proportional to J_2, a dimensionless measure of the quadrupole moment of the sun. In this term P_2 is the Legendre polynomial of order 2 and θ is the angle between the radius vector from the sun's center and the normal to the sun's equator. The so-called preferred frame velocity w^i is taken to be the average of the determinations of the solar system velocity relative to the cosmic blackbody background. In general relativity $\beta = \gamma = 1$ and $\alpha_1 = 0$.

1. Bending of Light

One of the more spectacular confirmations of the general theory came in 1919, when the solar eclipse expedition headed by Eddington announced that they had observed bending of light from stars as they passed near the edge of the sun that was in agreement with the prediction of the theory. Derivations of the bending that use only the principle of equivalence or the corpuscular theory of light predict just half of the bending predicted by the general theory.

The angle of bending for light passing at a distance d from the center of the sun can be computed by using the equation of motion (19) with $g_{\mu\nu} dx^\mu/d\lambda\ dx^\nu/d\lambda = 0$. Using the form for the gravitational field given by Eq. (37), the angle of bending is given by

$$\Theta = (1 + \gamma)2GM_\odot/c^2d \qquad (38)$$

For light that just grazes the edge of the sun $\Theta = 1''.75$ when $\gamma = 1$. Because of this small value it is necessary to observe stars whose light passes very close to the edge of the sun, and this can be done only during a total eclipse. The apparent positions of these stars during the eclipse are then compared to their positions when the sun is no longer in the field of view in order to measure the amount of bending. Unfortunately, such measurements are beset with a number of uncertainties. Thus the measurements made by Eddington and his co-workers had only 30% accuracy. The most recent such measurements were made during the solar eclipse of June 30, 1973 and yielded the value

$$\tfrac{1}{2}(1 + \gamma) = 0.95 \pm 0.11 \qquad (39)$$

The use of long-baseline and very-long-baseline interferometry, which is capable in principle of measuring angular separations and changes in angle as small as 3×10^{-4}, has made possible much more accurate tests of the bending of light. These techniques have been used to observe a number of quasars such as 3C273 that pass very close to the sun in the course of a year. Beginning in 1970, these observations have yielded increasingly accurate determinations, and the most recent, in 1984, agrees with the general relativistic prediction to within 1%.

2. Time Delay

In passing through a strong gravitational field, light not only will be red shifted but also will take longer to traverse a given distance than it would if Newtonian theory were valid. The reason for this delay is that the gravitational field acts like a variable index of refraction, so the velocity of light will vary as it passes through such a field. This effect was first proposed by Shapiro in 1964 as a means of testing general relativity. It can be observed by bouncing a radar signal off a planet or artificial satellite and measuring its round-trip travel time. At superior conjunction, when the planet or satellite is on the far side of the sun from the earth, the effect is a maximum, in which case the amount of delay is given by

$$\delta t = 2(1 + \gamma)(GM_\odot/c^3)\ \ln(4r_e r_p/d^2) \qquad (40)$$

where r_e, r_p, and d are, respectively, the distance from the sun to the earth, the distance from the sun to the target, and the distance of closest approach of the signal to the center of the sun. Since one does not have a Newtonian value for the round-trip travel time with which to compare the measured time it is necessary to monitor the travel time as the target passes through superior conjunction and look for a logarithmic dependence.

The use of a planet such as Mercury or Venus as a target is complicated by the fact that its topography is largely unknown. As a consequence, a signal could be reflected from a valley or a mountaintop without our being able to detect the difference. Such differences can introduce errors of as much as 5 μsec in the round-trip travel time. Artificial satellites such as Mariners 6 and 7 have been used to overcome this difficulty. Furthermore, since they are active retransmitters of the radar signal they permit an accurate determination of their true range. Unfortunately, fluctuations in the solar wind and solar radiation pressure produce random accelerations that can lead to uncertainties of up to 0.1 μsec in the travel time. Finally, spacecraft such as the Mariner 9 Mars orbiter and the Viking Mars landers and orbiters have been used as targets. Since they are anchored to the planet they will not suffer such accelerations. The most recent measurements by Reasenberg et al. in 1979 have yielded a value

$$(1 + \gamma)/2 = 1.000 \pm 0.001 \quad (41)$$

3. Planetary Motion

Long before the general theory was proposed, it was known that there was an anomalous precession of the perihelion (distance of closest approach to the sun) of the planet Mercury that could not be accounted for on the basis of Newtonian theory by taking into consideration the perturbations on Mercury's orbit due to the other planets. At the end of the last century, Newcomb calculated this residual advance to have a value of $41''.24 \pm 2''.09$ of arc per century.

The field values given by Eq. (37) and the equation of motion (19) together yield an expression for the perihelion advance per period that is given, to an accuracy commensurate with the accuracy of the observations, by

$$\delta\bar\omega = (6\pi GM_\odot/c2p)[\tfrac{1}{3}(2 + 2\gamma - \beta)$$
$$+ J_2(R_\odot^2 c^2/12GM_\odot p)] \quad (42)$$

where $p = a(1 - e^2)$ is the semi-latus rectum of the orbit, a its semimajor axis, and e its eccentricity. Using the best current values for the orbital elements and physical constants for Mercury and the sun, one obtains from Eq. (38) a perihelion advance of $42''.95\lambda_p$ of arc per century, where $\lambda_p = [\tfrac{1}{3}(2 + 2\gamma - \beta) + 3 \times 10^3 J_2]$.

The measured value of the perihelion advance of Mercury is known to a precision of about 1% from optical measurements made over the past three centuries and of about 0.5% from radar observations made over the past two decades. If one assumes that J_2 has the value $\sim 1 \times 10^{-7}$, which it would have if it were the consequence of centrifugal flattening due to a uniform rotation of the sun equal to its observed surface rate of rotation, then, using this value, Shapiro gives

$$\tfrac{1}{3}(2 + 2\gamma - \beta) = 1.003 \pm 0.005 \quad (43)$$

which is in excellent agreement with the prediction of general relativity.

This agreement has been called into question by some researchers, notably Dicke and Hill. Observations of the solar oblateness by Dicke and Goldenberg in 1966 led them to conclude that J_2 actually has a value of $(2.47 \pm 0.23) \times 10^{-5}$, leading to a contribution of about $4''$ per century to the overall perihelion advance. If true, this would put the prediction of general

relativity into serious disagreement with the observations. On the other hand, it would agree with the prediction of the Brans–Dicke scalar tensor theory of gravity if an adjustable parameter in that theory were suitably chosen. However, a number of authors have disagreed with the interpretation of their observations by Dicke and Goldenberg. These authors argue that the observations could equally well be explained by assuming a standard solar model with $J_2 \sim 10^{-7}$ and a surface temperature difference of about $1°$ between the pole and the equator. More recently Hill has given a value of $J_2 = 6 \times 10^{-6}$, based on his measurements of normal mode oscillations of the sun. If true, the general relativistic prediction for Mercury would be inconsistent with the observed value by about two standard deviations. Unfortunately, the present measurements of the orbit of Mercury are not sufficiently accurate to separate the post-Newtonian and quadrupole effects.

A resolution of this difficulty has come from an analysis of the ranging data for the planet Mars. Since the quadrupole contribution to the perihelion advance has a different dependence on the semimajor axis from the gravitational effect, it is in principle possible to separate the two by observing the advance for different planets. In spite of the smallness of these effects on the orbit of Mars, the accuracy of the Viking data from Mars, which are accurate to within 7 km, combined with the radar data from Mercury allows such a determination. Using a solar system model that includes 200 of the largest asteroids, Hellings has found, with $J_2 = 0$, that

$$\beta - 1 = (-0.2 \pm 1.0) \times 10^{-3} \quad (44a)$$
$$\gamma - 1 = (-1.2 \pm 1.6) \times 10^{-3} \quad (44b)$$
$$\alpha_1 = (2.2 \pm 1.8) \times 10^{-4} \quad (44c)$$

When J_2 was allowed to have a finite value, he found that

$$J_2 = (-1.4 \pm 1.5) \times 10^{-6} \quad (45a)$$

and

$$\beta - 1 = (-2.9 \pm 3.1) \times 10^{-3} \quad (45b)$$
$$\gamma - 1 = (-0.7 \pm 1.7) \times 10^{-3} \quad (45c)$$
$$\alpha_1 = (2.1 \pm 1.9) \times 10^{-4} \quad (45d)$$

Hellings also used these data to analyze the non-symmetric gravitational theory of Moffat, which was consistent with the Mercury data and Hill's value for J_2. The result was that

$$J_2 = (1.7 \pm 2.4) \times 10^{-7} \quad (46)$$

From the above results it appears that the predictions of general relativity are confirmed to about 0.1%. However, by a suitable adjustment of parameters, several competing theories also share this property. What distinguishes general relativity from these other theories is that, aside from the value for the gravitational constant G, it contains no other adjustable parameters.

4. Time Varying G

In addition to the tests discussed above, the solar system data can be used to test the possibility that the gravitational constant varies with time. Such a possibility was first suggested by Dirac in 1937 on the basis of his large number hypothesis. He observed that one could form, from the atomic and cosmological constants, several dimensionless numbers whose values were all of the order of 10^{10}. Rather than being a coincidence that was valid only at the present time, Dirac proposed that the equality of these numbers was the manifestation of some underlying physical principle and that they held at all times. Since one of these numbers involves the present age of the universe through its dependence on the Hubble "constant" and hence decreases as one moves back in time, the other constants must also change with time in order to maintain the equality between the large numbers. One of these numbers, however, involves only atomic constants, being the ratio of the electrical to the gravitational force between an electron and a proton. Hence the Dirac hypothesis requires that one of these atomic constants must be changing on a cosmic time scale. The constant that is usually taken to vary with time in theoretical implementations of the large number hypothesis is the gravitational constant.

There are several ways of constructing a theory with an effective time-varying gravitational constant. In the Brans–Dicke theory, the effective gravitational constant itself varies with time:

$$G_{\text{eff}} = G[1 + (\dot{G}/G)(t - t_0)] \qquad (47)$$

An alternative proposal by Dirac assumed that cosmic effects couple to local atomic physics so that the ratio of atomic to gravitational time is not constant. The rate of change of gravitational time τ_G with respect to atomic time τ_A is then given as

$$d\tau_G/d\tau_A = 1 + \dot{\phi}(t - t_0) \qquad (48)$$

where ϕ is some cosmological field that is supposed to be responsible for the effect. In both cases, the net effect is to produce an anomalous acceleration in the equations of motion for material bodies.

Since the change in atomic constants is tied to cosmic evolution in the large number hypothesis, the expected rate of change in G should be proportional to the inverse Hubble time:

$$\dot{G}/G - H_0 \cong 5 \times 10^{-11} \text{ yr}^{-1} \qquad (49)$$

On the basis of the Viking lander data, Hellings concludes that

$$\dot{G}/G = (0.2 \pm 0.4) \times 10^{-11} \text{ yr}^{-1} \qquad (50a)$$

$$\dot{\phi} = (0.1 \pm 0.8) \times 10^{-11} \text{ yr}^{-1} \qquad (50b)$$

Since these limits are an order of magnitude smaller than what one would expect from simple cosmic scale arguments, they cast serious doubt on the large number hypothesis.

C. The Binary Pulsar

A new, and essentially unique, opportunity for testing general relativity came with the discovery of the binary pulsar PSR 1913 + 16 by Hulse and Taylor in 1974. It consists of a pulsar in orbital motion about an unseen companion with a period of 7.75 hr. Its relevance for general relativity is twofold: because $v^2/c^2 \sim 5 \times 10^{-7}$ is a factor 10 larger than for Mercury, relativistic effects are considerably larger than any that have been observed in the solar system. Also, the short period amplifies secular changes in the orbit. Thus the observed periastron advance amounts to $4°.2261 \pm 0.0007$ of arc per year compared to the $43''$ of arc per century for Mercury. Furthermore, the pulsar carries its own clock with a period that is accurate to better than one part in 10^{12}. As a consequence, measurements of post-Newtonian effects can be made with unprecedented accuracy. If this were all, the binary pulsar would still be an invaluable tool for testing general relativistic orbit effects. However, it also provides us for the first time with a means for testing an essentially different kind of prediction of general relativity, namely the existence of gravitational radiation.

Considerable effort has gone into identifying the pulsar companion. It was soon found that the pulsar radio signals were never eclipsed by the companion. Also, the dispersion of the pulsed signal showed little change over an orbit, implying the absence of a dense plasma in the system. These two facts together ruled out the possibility of the companion being a main sequence star. Another possibility is that it is a helium star. However, since the pulsar is at a distance of only about 5 kpc from us, such a star would have been seen. In spite of intense efforts, no such star has been observed in the neighborhood of the pulsar. The remaining possibility is that it is a compact object, either a white dwarf, another neutron star, or a black hole.

In the case of conventional spectroscopic binaries, it is usually possible to measure only two parameters of the system, the so-called mass function of the two masses M_1 and M_2 of the components and the product of the semimajor axis a_1 and the sine of the angle i of inclination of the plane of the orbit to the line of sight. However, in the case of the binary pulsar one can use general relativity to determine all four of these parameters from measurements of the periastron advance and the combined second-order Doppler shift and gravitational red shift of the emitted signals. One finds from these combined measurements that $M_1 = M_2 = (1.41 \pm 0.06)M_\odot$. From the fact that the Chandrasekhar limit on the mass of a nonrotating white dwarf is about $1.4M_\odot$, it appears likely that the unseen companion is either a neutron star or a black hole.

In addition to the measurements discussed above, it was discovered that the orbital period P was decreasing with time. Later measurements gave a value for $\dot{P} = (-2.30 \pm 0.22) \times 10^{-12}$ sec/sec^{-1} or about 7×10^{-5} sec per year. If one computes the period change due to loss of energy by the emission of gravitational radiation using the quadrupole formula (34) one obtains a value for $\dot{P} = -2.40 \times 10^{-12}$ sec/sec^{-1}, in excellent agreement with the observed value.

Of course, there are other effects that could change the orbital period, such as tidal dissipation, mass loss or accretion onto the system, or acceleration relative to the solar system. Furthermore, there could be other contributions to the periastron advance such as rotational or tidal deformation of the companion. Only in the case of a helium-star companion would any of these effects contribute significantly to the calculated or observed period change. Furthermore, it would be truly remarkable if some combination of these effects should conspire to give a value for the period change equal to that predicted by the quadrupole formula. It therefore appears safe to say that for the first time we have evidence of a qualitatively new prediction of general relativity, namely gravitational radiation. Finally, the data from the binary pulsar seem to

rule out a number of competing theories of gravity such as the Rosen bimetric theory. In such theories this system can radiate dipole gravitational waves that result in a period increase. Only a very artificial mechanism could then give rise to the observed period decrease.

D. Gravitational Wave Detection

The first comprehensive attempt to detect gravitational waves impinging on the earth was begun by J. Weber in 1961. His antenna consisted of a large aluminum cylinder ~1.5 m long with a resonant frequency of ~1660 Hz. In the early 1970s Weber announced the detection of coincident pulses on two of these antennae separated by a distance of ~1000 mile. However, attempts to duplicate these results by a number of other groups, using somewhat more sensitive detectors than those used by Weber, proved fruitless, and it is now generally agreed that the events recorded by Weber were not caused by gravitational waves.

Since Weber's pioneering efforts, about 15 different groups from around the world have undertaken the construction of gravitational wave detectors. The sensitivity of a detector can be expressed in terms of the smallest strain $\Delta L/L$, where ΔL is a change in the length L of the detector, that can just be measured. This change in length is produced by tidal forces associated with the incident gravitational wave and hence its measurement leads to a determination of the Riemann–Christoffel tensor of the wave. Since this strain is approximately equal to the dimensionless amplitude h of a gravitational wave incident on the detector, the sensitivity of a detector is usually given as the minimum value of h that can be detected.

The original Weber bars had a sensitivity $h \sim 10^{-16}$. At present one of the main limitations on the sensitivity of Weber bars is thermal noise. As a consequence, second-generation Weber bars are being constructed that will be cooled to liquid helium temperatures. Such bars are estimated to have sensitivities of $h \sim 10^{-19}$. It is technically feasible to construct bars for which $h \sim 10^{-21}$, although that would require cooling to the millidegree level. The latter value appears to be a lower limit to what can be attained with presently available technology.

One of the drawbacks of Weber bar detectors is that they are only sensitive to the Fourier component of the incoming signal whose frequency is equal to the resonant frequency of the bar. Furthermore, most bars have resonant frequencies in the kilohertz range with a smallest reported frequency of 60.2 Hz. Unfortunately, most continuous wave sources such as binary star systems have much lower frequencies. In an attempt to overcome this difficulty and to increase sensitivity, a number of groups have undertaken the construction of laser interferometer detectors. In these devices, a gravitational wave would change the lengths of the interferometer arms and one would measure the resulting fringe shifts. Such detectors can, in principle, record the entire waveform of an incoming wave rather than a single Fourier component. It is possible that sensitivities as low as $h \sim 10^{-22}$ might be achieved. It has also been suggested that gravitational radiation could be detected by the accurate Doppler tracking of spacecraft. Such a scheme would, in principle, be able to detect waves with frequencies in the 1 to 10^{-4} range. Present technology is within one or two orders of magnitude of the sensitivity needed to detect possible signals in this frequency range.

Possible sources of gravitational waves can be divided into two groups, those that emit continuously and those that emit in bursts. Possible continuous wave sources are binary stars and vibrating or rotating stars. In the case of binary stars, the strongest emitter known is μ Scorpii, for which $h = 2.1 \times 10^{-20}$. However, its frequency is 1.6×10^{-5} Hz. The largest binary frequency known is 1.9×10^{-3} Hz. However, for this system $h \sim 5 \times 10^{-22}$. Other possible continuous wave sources have values for h that are this small or smaller. Thus, estimates of h for waves from the Crab and Vela pulsars are of order 10^{-24} to 10^{-27}, at frequencies between 10 and 100 Hz. In spite of these low amplitudes, signal integration over an extended time can effectively increase the sensitivity of a detector by an order of magnitude or more, so the detection of such signals is not totally out of the question.

Bursts of gravitational radiation can be expected to accompany cataclysmic events such as supernova explosions, stellar collapse to form neutron stars or black holes, or coalescence of the neutron stars or black holes in a binary system at the end stage of its evolution. One of the problems in dealing with such systems is the determination of the efficiency with which other forms of energy can be converted into gravitational radiation. Estimates range from a maximum of 0.5 to as low as 0.001. Such events would have characteristic frequencies in the range 10^2 to 10^5 Hz and those occurring in our galaxy would have amplitudes estimated to be in the range $h \sim 10^{-18}$ to 10^{-17}. Here the problem for detection is not so much the frequency or the intensity, as it is in the case of continuous emitters, but rather the scarcity of such events. Thus, the supernova rate in our galaxy has been estimated to be 0.03 per year. If one includes such events in other galaxies the rate increases. For example, at a distance of 10 Mpc the estimated supernova rate is one per year. However, the corresponding amplitude would be $h \approx 3 \times 10^{-21}$ to 3×10^{-20}.

By combining the sensitivity estimates for gravitational wave detectors now under construction and the expected amplitudes and frequencies of possible sources we see that the possibility for detection in the near future is good. Furthermore, as the technology improves, an era of gravitational wave astronomy may soon be possible. Since gravitational waves are not absorbed by intervening matter as is electromagnetic radiation, such an astronomy may allow us to explore regions of the universe, such as the centers of galaxies, that are now blocked to our view.

E. Gravitational Lenses

In many of its effects, a gravitational field acts like a medium with a variable index of refraction. Thus, two of the observed effects discussed above, the bending of light and the time delay of signals as they pass through a gravitational field, can be understood on this basis. A further consequence of this notion is that there should exist gravitational lenses with properties similar to those of ordinary optical lenses. The most likely candidates for such lenses are galaxies. If placed between us and a distant point source such as a quasar, a galaxy can, in effect, provide more than one path along which light from the source may reach the observer. As a consequence, one would see multiple images of the source. Applied to a galaxy, the bending formula (38) (with $\gamma = 1$) gives typical bending angles of about 1 arcsec.

As of this date, two "lensed" quasars have been observed. The first of these consists of two quasar images of comparable brightness separated by about 6 arcsec, with identical red shifts of 1.41 within the measuring errors. Both images are also radio sources with a flux ratio that is the same as in the visible. In this case the lens has also been identified as a (probable) elliptic galaxy with a red shift of 0.36. It is in fact a member of a cluster, which must be taken into account in determining the image structure. Shortly after the discovery of the first lensed quasar a second candidate was discovered. It consists of three quasars with identical (to within 100 km/sec) red shifts of 1.722. Although in this case no lensing galaxy has been observed as yet, the configuration can be explained by a nearly edge-on spiral galaxy with a red shift of about 0.8, which would make it hard to detect. In addition to confirming the gravitational lens effect, these observations prove that at least three quasars are at cosmological distances from us.

VII. Gravity and Quantum Mechanics

A. Hawking Radiation and Black Hole Thermodynamics

For most of its history, general relativity has stood apart from quantum mechanics. Early attempts to quantize the gravitational field proved to be largely unsuccessful and quantum theory usually neglected the presence of gravitational fields. For most problems one could justify this neglect. The radius of the first Bohr orbit of a hydrogen atom held together by gravitational rather than electrical forces, for example, would be about 5×10^{30} m, which is almost four orders of magnitude larger than the radius of the visible universe! However, one is not justified in ignoring the effects of strong gravitational fields, such as those that occur near the Schwarzschild radius of a black hole, on the behavior of quantum systems since gravity couples universally to all physical systems. When one takes account of the gravitational field in the quantum description of a system, qualitatively new features emerge. One such feature is the phenomenon of Hawking radiation.

Even in the vacuum, where there are no real quanta, pairs of virtual quanta of the various matter fields observed in nature are being continually created and destroyed in equal numbers. Their presence is manifested in such phenomena as the Lamb shift in hydrogen and the Casimir effect. According to Hawking, a black hole can absorb one member of such a virtual pair, leaving its partner to propagate as a real quantum of the field. The energy needed for this process to occur is supplied by the gravitational

energy of the black hole. Hawking was able to show that, as a black hole formed, such a flux of real quanta should be produced and that it would be equal to the flux produced by a hot body of temperature T given by

$$kT = \hbar g/2\pi c \qquad (51)$$

where k is Boltzmann's constant, $\hbar$ is Planck's constant, and g is the gravitational acceleration at the Schwarzschild radius of the black hole and is equal to $c^4/4GM$, where M is its mass.

For a solar mass black hole this temperature would be 2.5×10^{-6} K. However, for a 10^{12} kg mass black hole it would be 5×10^{12} K. Such a black hole would emit energy at a rate of about 6000 MW, mainly in gamma rays, neutrinos, and electron–positron pairs. Hawking has suggested that "primordial black holes" with such masses might have been formed by the collapse of inhomogeneities in the very early stages of the universe and that some of them might have survived to the present day. If so, they probably represent our only hope of observing Hawking radiation. However, measurements of the cosmic-ray background around 100 MeV place an upper limit for black holes with masses around 10^{15} kg of about 200 per cubic light-year.

The fact that a black hole can radiate real quanta might seem to contradict the fact that no radiation can escape from a black hole. However, that restriction is only true classically. One can think of the emitted radiation as having come from inside the event horizon surrounding the black hole by quantum mechanically tunneling through the potential barrier created by its gravitational field. Actually, it is possible for a black hole to emit almost any configuration of quanta, including macroscopic objects. Since we cannot have direct knowledge of the interior of a black hole, all we can determine are the probabilities for the emission of such configurations. The overwhelming probability is that the emitted radiation is thermal with a temperature given by Eq. (50).

That a black hole should have associated with it a temperature fits in with some analogies between black holes and thermodynamics discovered by Bardeen, Carter, Hawking, and Bekenstein. If the energy density of the matter that went to make up the black hole is nonnegative, it can be shown that, classically, the surface area of the event horizon surrounding it can never decrease with time. Moreover, if two black holes coalesce to form a single black hole, the area of its event horizon is greater than the sum of the areas of the event horizons surrounding the two original black holes. These properties are very similar to those of ordinary entropy. Furthermore, when a black hole forms, all information concerning its structure except its mass, charge, and angular momentum is lost, and when ordered energy is absorbed by a black hole it too is forever lost to the outside world. These considerations led Bekenstein to associate with a black hole an entropy S given by

$$S = ckA/4\hbar \qquad (52)$$

where A is the surface area of the event horizon surrounding the hole.

The existence of Hawking radiation raises the question of the ultimate fate of a black hole. As it radiates, its mass decreases. Its temperature therefore increases, and hence so does the rate of emission of radiation. What is left is, at this point, speculation. It might disappear completely, it might cease radiating when its mass reaches some critical value, or it might continue radiating indefinitely, creating a negative-mass naked singularity. While the latter two possibilities seem unlikely, the first one implies that whatever matter went into making the black hole initially would simply cease to exist. In deriving the emission from black holes, the gravitational field was treated classically while the matter fields were treated quantum mechanically. It has been suggested that when the mass of the black hole becomes comparable to the Planck mass, that is, the mass one can form from the constants c, $\hbar$, and G, namely $(\hbar c/G)^{1/2} \sim 5 \times 10^{-15}$ g, the gravitational field can no longer be treated as a classical field. If so, the fate of a black hole will be decided only when we have a consistent theory of quantum gravity.

B. Quantum Gravity

There seems to be little doubt that, in the final analysis, the gravitational field, like all matter fields, must be quantized. The most telling argument in favor of this assertion is that, if the gravitational field were a classical field, it would be possible to determine both the position and the momentum of its sources by measuring all of its components simultaneously with arbitrary accuracy and thus violate the uncertainty principle. The quantization of the gravitational field is, however, beset with many technical difficulties and there is still considerable debate on how best to accomplish this task.

One of the most troubling problems attendant on the quantization of any field theory is the existence of divergent integrals, which arise in various calculations. In all successful quantum field theories such as quantum electrodynamics, these divergent integrals can be dealt with by a renormalization process whereby they are absorbed into the masses and coupling constants that appear in the theories. The "renormalized" values of these quantities are then taken to be their observed values. A conventional quantization of the gravitational field in the absence of other fields does not lead to divergences, at least in the lowest order of approximation. However, once gravity is coupled to matter, divergences arise that cannot be renormalized. A number of attempts to overcome this difficulty are underway.

One such approach is supergravity. It is an extension of supersymmetric quantum field theory in which every boson in the theory has associated with it a fermion and vice versa. Furthermore, the theory is invariant under the interchange of bosons and fermions. Such theories have been used to construct a unified theory of the strong and electroweak interactions between elementary particles. In supergravity, one associates a spin 3/2 particle called the gravitino with the quantum of the gravitational field. This theory has been shown to have no logarithmic divergences in the lowest two orders of perturbation theory even when gravity is coupled to matter. Whether supergravity or one of its extensions proves to be a viable theory is a matter for future investigation.

Bibliography

Anderson, J. L. (1967). "Principles of Relativity Physics," Academic Press, New York.

Bergmann, P. G. (1968). "The Riddle of Gravitation," Scribner's, New York.

Bertotti, B., de Felice, F., and Pascolini, A. (Eds.). (1984). "General Relativity and Gravitation," Reidel, Dordrecht, Netherlands.

Hawking, S. W., and Israel, W. (Eds.). (1979). "General Relativity," Cambridge Univ. Press, Cambridge, U.K.

Kaufman, W. J., III. (1977). "The Cosmic Frontiers of General Relativity," Little, Brown, Boston.

Misner, C. W., Thorne, K. S., and Wheeler, J. A. (1971). "Gravitation," Freeman, San Francisco.

Rindler, W. (1977). "Essential Relativity," 2nd ed., Springer Verlag, New York.

Sexl, R., and Sexl, H. (1979). "White Dwarfs, Black Holes," Academic Press, New York.

Smarr, L. (Ed.). (1979). "Sources of Gravitational Radiation," Cambridge Univ. Press, Cambridge, U.K.

Wald, R. M. (1984). "General Relativity," Univ. of Chicago Press, Chicago.

Weinberg, S. (1972). "Gravitation and Cosmology," Wiley, New York.

Will, C. (1981). "Theory and Experiment in Gravitational Physics," Cambridge Univ. Press, Cambridge, U.K.

RELATIVITY, SPECIAL

John D. McGervey *Case Western Reserve University*

I. Experimental Background
II. Postulates of Special Relativity
III. Concepts of Space, Time, and Motion
IV. Implications of the Lorentz Transformation
V. Momentum and Energy
VI. Electromagnetism
VII. Consequences of Special Relativity

GLOSSARY

Electric charge: Basic property of particles of matter such that each particle possessing a charge is surrounded by an electric field.

Electric field: Condition in space that causes a force to be exerted on a charged particle. The force is proportional to the product of the magnitude of the charge and the strength of the field; it acts in the direction of the field on a positive charge and in the opposite direction on a negative charge.

Electron volt (eV): Unit of energy equal to 1.6×10^{-19} J, or the kinetic energy given to an electron that has been accelerated by an electric potential difference of 1 V.

Half-life: Time interval during which there is a probability of one-half that a particular radioactive particle or atomic state will decay.

Inertia: Ability of a body to resist an attempt to accelerate it. The greater the inertia, the smaller will be the acceleration produced by a given force.

Magnetic field: Condition in space that causes a force to be exerted on a moving charged particle. The direction of the force is perpendicular to the direction of the field and to the direction of the velocity of the particle.

Mass: Measure of inertia. Masses may be compared by applying equal forces to two bodies and finding their accelerations.

Vector: A directed line segment, representing a quantity that has a magnitude and a direction, such as velocity or force. Vectors may be added or subtracted graphically like displacements (e.g., a northward displacement of 5 miles added to an eastward displacement of 5 miles yields a northeastward displacement of about 7 miles). The northward and eastward displacements are called *components* of the displacement vector.

The special theory of relativity displays the logical consequences of assuming the nonexistence of absolute space or absolute time. This assumption was made by Albert Einstein on the basis of experimental evidence that the speed of light, relative to an observer, is independent of the state of motion of that observer. The theory that proceeded from this assumption explained all the observations known at that time, and it also predicted other phenomena whose existence had not even been suspected. Among other things, the theory showed that, no matter how much energy one gave to a particle, one could not accelerate it to a speed greater than the speed of light. It also showed that the mass of a particle was a measure of its internal energy, thereby giving a clue to the possibility of obtaining nuclear energy; it provided the basic equations for describing the motion of highly energetic particles; and it led to the prediction of the existence of antimatter.

I. Experimental Background

A. ABERRATION OF STARLIGHT

The first well-verified connection between the motion of an observer and the velocity of light was made by the astronomer James Bradley in 1727. He observed a seasonal change in the apparent position of the star Gamma Draconis, relative to other stars. This effect, now known as stellar aberration, is illustrated in Fig. 1.

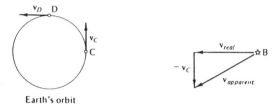

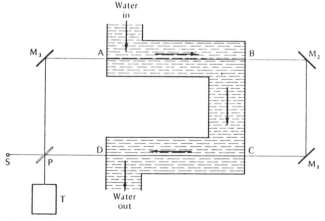

Earth's orbit

FIG. 1. Aberration of starlight. Stars are located at A and B. Light travels in direction $\mathbf{v}_{real}$ to reach earth. When earth is at C, moving with velocity $\mathbf{v}_C$, earth observers see light from B moving in the direction of $\mathbf{v}_{apparent} = \mathbf{v}_{real} - \mathbf{v}_C$. When earth is at D, moving with velocity $\mathbf{v}_D$, light from A is seen to move in direction $\mathbf{v}_{real} - \mathbf{v}_D$. [Reproduced, with permission, from J. McGervey, (1983). "Introduction to Modern Physics," 2nd ed., p. 29, Academic Press, New York.]

Stars A and B lie in the plane of the earth's orbit. In order to reach the solar system from each star, rays of light must travel in the direction of the vector $\mathbf{v}_{real}$, shown at A and B, respectively. When the earth is at point C, with velocity $\mathbf{v}_C$, the ray from A is traveling in the direction $\mathbf{v}_{real}$, relative to earth, when it arrives. But the ray from B appears to be moving in the direction of the vector labeled $\mathbf{v}_{apparent}$ at B. Thus the measured angle between the two stars as

they appear in the sky is less than 90°. But when the earth is at D, three months later, the same two stars are more than 90° apart in the sky.

Recent observations have confirmed, to a high degree of accuracy, that the angle between $\mathbf{v}_{real}$ and $\mathbf{v}_{apparent}$ equals the ratio of the earth's orbital speed to the speed of light, or about 10^{-4} (in radians). A constant aberration, which might result from the motion of the whole solar system through space, cannot be detected. Only the change in aberration angle is seen, as earth changes direction in its orbit.

B. Fizeau Experiment

With the development of the wave theory of light, it was believed that all of space was filled with some substance, called ether, that served as a medium to transmit light waves. It was then suggested that the focal length of a lens would depend on the motion of the earth through this ether, because the speed of light relative to earth would vary as the earth moved in different directions relative to the ether. However, when this idea was tested on starlight, it was found that the light was always bent by a lens as it would be if the earth were at rest, and the apparent direction of the light ray could be considered its "true" direction.

To explain why the motion of the earth does not affect the properties of a lens moving with the earth, Fresnel postulated that, in a medium whose index of refraction is n, the ether density is proportional to n^2, and the excess ether in the

FIG. 2. Fizeau's experiment. Mirrors M_1, M_2, and M_3 cause light beams to pass through moving water along the rectangle ABCD in both directions. [Reproduced, with permission, from J. McGervey (1983). "Introduction to Modern Physics," 2nd ed., p. 33. Academic Press, New York.]

body is carried with it when it moves. Fresnel showed that his "ether drag" always had the consequence that no observer can ever tell, by any method based on the refraction of light, that he is moving through the ether, as long as the observer's speed is much less than c, the speed of light in vacuum.

Fresnel's postulate was put to a direct test by Fizeau in 1851, using the apparatus illustrated in Fig. 2. Water flows through a bent, transparent tube. Light from the source S shines through the water in two directions after passing through plate P, which partially reflects it and partially transmits it. If the water drags the ether with it, the light ray that goes with the water flow will move faster than the ray that moves against the flow. When the rays come together again at the telescope T, the phase difference between them will vary as the water speed varies. The experiment showed that the changes in phase were in agreement with what one would expect from Fizeau's postulate.

The Fizeau experiment involves the speed of light in water, so it is essentially a refraction experiment. If the entire Fizeau apparatus were to move in any direction whatsoever, it would have no effect on the observed phase dif-ference between the two light rays, as long as the ether is dragged according to Fresnel's postulate.

C. MICHELSON–MORLEY EXPERIMENT

Attempts to find small effects of motion through the ether culminated in the Michelson–Morley experiment (Fig. 3). Like Fizeau's experiment, the Michelson–Morley experiment is based on the interference between two light rays that combine after traveling on different paths. But no refraction is involved.

Light from source S is partly reflected and partly transmitted by mirror M to give two rays. The transmitted ray (ray 1) travels to mirror M_1 and back to M; the reflected ray (ray 2) travels to mirror M_2 and back to M. The rays then combine and enter telescope T. If the total travel time of ray 1 is the same as the travel time of ray 2 (between S and a spot on the telescope), then the rays produce a bright region at that spot. This will happen if both rays travel at the same speed, and the total lengths of the two paths are equal.

But if light is carried by an ether, the speed of light will be the same for the two rays only if the apparatus is at rest in the ether. If light travels at

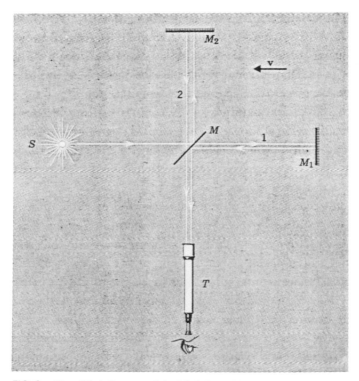

FIG. 3. Simplified diagram of the Michelson–Morley experiment. [Reproduced, with permission, from R. Resnick (1968). "Introduction to Special Relativity," p. 20. John Wiley, New York.]

TABLE I.   Experimental Basis for the Theory of Special Relativity[a][b]

Theory		Light propagation experiments			
	Aberration	Fizeau convection coefficient	Michelson–Morley	Kennedy–Thorndike	Moving sources and mirrors
Ether theories					
Stationary ether, no contraction	A	A	D	D	A
Stationary ether, Lorentz contraction	A	A	A	D	A
Ether attached to ponderable bodies	D	D	A	A	A
Emission theories					
Original source	A	A	A	A	A
Ballistic	A	N	A	A	D
New source	A	N	A	A	D
Special theory of relativity	A	A	A	A	A

[a] From W. K. H. Panofsky and M. Phillips, "Classical Electricity and Magnetism," 2nd ed., © 1962, Addison-Wesley, Reading, Massachusetts. Page 282, Table 15-2. Reprinted with permission.

[b] A, the theory agrees with experimental results; D, the theory disagrees with experimental results; and N, the theory is not applicable to the experiment.

speed c in the ether, and the apparatus moves through the ether, the light will have an apparent speed that differs from c. This apparent speed will depend on the direction in which the light travels. If the apparatus travels through the ether with velocity **v** in the direction shown (from M_1 to P), then ray 1 will take longer than ray 2; the time difference will be approximately Lv^2/c^3, where L is the distance from M to either M_1 or M_2. But if the apparatus is traveling at right angles to this direction, ray 2 will take the longer time. For the length used by Michelson and Morley, with v equal to the orbital speed of the earth, the time difference is about 4×10^{-16} sec.

By rotating the apparatus through a 90° angle, Michelson and Morley were able to reverse the roles of the two rays. If the apparatus were moving through the ether, this should have the effect of changing the locations of bright regions in the field of view of the telescope as the telescope rotated.

The experiment was repeated several times, at different times of the year, to allow for the possibility that the earth happened to be stationary in the ether when the experiment was first performed, but no such effect was ever observed. Michelson had estimated that a speed through the ether of about 2% of the earth's orbital speed should have led to a detectable effect when the apparatus was rotated.

D. PRERELATIVITY EXPLANATIONS

As a result of these and other experiments, it seemed that nature would never permit us to detect motion of the earth through the hypothetical ether. Various explanations were advanced for this situation. It was suggested by Lorentz that bodies moving through the ether were contracted in the direction of motion by just enough to cancel the additional travel time required by ray 1 in the Michelson–Morley experiment. Another possibility was that the earth carried the ether along with it, so the apparatus was always at rest in the ether. It was also suggested that light, unlike other waves, had a speed that depended on the speed of the source. One by one, each theory was tested by additional experiments and found to fail. What was needed was not a series of *ad hoc* theories, but rather, a comprehensive theory based on fundamental principles. Einstein developed that theory—special relativity. It has survived millions of experimental tests.

E. SUMMARY OF EXPERIMENTAL FACTS AND THEORIES

Experiments testing various theories are listed in Table I, showing that each theory, with the sole exception of special relativity, disagrees with one or more experimental results.

Light propagation experiments		Experiments from other fields					
De Sitter spectroscopic binaries	Michelson–Morley using sunlight	Variation of mass with velocity	General mass–energy equivalence	Radiation from moving charges	Meson decay at high velocity	Trouton–Noble	Unipolar induction using permanent magnet
A	D	D	N	A	N	D	D
A	A	A	N	A	N	A	D
A	A	D	N	N	N	A	N
D	D	N	N	D	N	N	N
D	D	N	N	D	N	N	N
D	A	N	N	D	N	N	N
A	A	A	A	A	A	A	A

II. Postulates of Special Relativity

A. THE PRINCIPLE OF RELATIVITY

Galileo referred to the principle of relativity in his "Dialogue on the Great World Systems." The principle may be stated as

No experiment can tell whether one is at rest or moving uniformly

An equivalent way of stating this principle is

The laws of nature are the same in all uniformly moving laboratories

It is understood that uniformly moving laboratories include those considered to be at rest. Another name for a uniformly moving laboratory is inertial frame of reference. The name comes from the fact that Newton's first law of motion, the law of inertia, is valid in such a frame of reference. The law of inertia would not be valid in an accelerated frame of reference.

According to this principle, absolute motion cannot be detected; anyone who is moving uniformly with respect to some object is entitled to consider himself to be at rest and the object to be moving. He can do no experiment to disprove his assumption that he is at rest.

The principle of relativity was considered to be obvious until the concept of the ether was developed. The inability to detect the ether led Einstein to adopt the principle of relativity as the first postulate of his special theory of relativity. This postulate gets rid of the problems associated with the ether by simply stating that there is no ether. The presence of an ether would allow one to detect absolute motion, if one were to assume the ether to be at rest.

B. THE SPEED OF LIGHT

Light is a wave, and waves travel in a material medium with a characteristic speed relative to that medium. Light waves apparently travel in empty space, where some other rule must govern their speed. Einstein, considering the experimental facts, stated that this rule should be the second postulate of special relativity:

Light in empty space always travels with the same definite speed c with respect to any inertial frame of reference, regardless of the state of motion of the body emitting the light.

This postulate explicitly rules out emission theories, which say that the speed of light depends on the speed of the source. Such theories have been disproven experimentally by observation of radiation from binary star systems (Table I). Einstein's postulate, on the other hand, is in agreement with all known observations of the speed of light, and it has led to many other conclusions that have stood the test of experiment.

The fundamental concern of the postulate is *speed*, not light. The existence of the speed c has consequences that can be stated without reference to the behavior of light. One could rephrase the postulate to say:

There is a characteristic speed in the universe, such that anything traveling at that speed with respect to one inertial frame of reference must also travel at that speed with respect to any other inertial frame.

The experiments on light indicate that light travels at that characteristic speed c.

III. Concepts of Space, Time, and Motion

A. Absolute Time

It is natural to think of time in absolute terms, as something that must be the same for all frames of reference. The postulate of an absolute speed conflicts with the notion of absolute time. To appreciate this fact, consider a light source in the middle of a boxcar that is moving uniformly from south to north.

A flash of light emitted in all directions from the source will reach the south end of the car before it reaches the north end, because the south end is moving toward the source and the north end is moving away from it. The light that travels northward (moving at the same speed as the light that travels southward) has a greater distance to travel before it hits its end. But this will be true only in a frame of reference in which the boxcar is moving northward.

An observer inside the boxcar is entitled to assume that the car is at rest. No measurement made inside the boxcar can contradict that assumption. Therefore, for this observer, the same light flash reaches north and south ends simultaneously. In general, events that are simultaneous in one frame and not simultaneous in other frames.

Thus time is not absolute. It is a measure that places all events in a sequence, such that one event may be before, after, or simultaneous with another event. This measure is different in different frames of reference, as we have just seen. The two events, namely, light reaching the north end of the boxcar and light reaching the south end, are simultaneous in only one frame of reference. According to any frame in which the car is moving northward, the same light flash will reach the south end first. There are also frames of reference in which the car is moving southward. Measurements made in any such frame will show that the light will reach the north end first.

B. Transformation of Coordinates

A frame of reference may be thought of as a set of perfect clocks, which can be placed at any point, plus a set of rigid measuring rods to measure the space coordinates of each point. An event is then specified by using these tools to

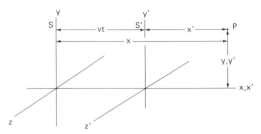

FIG. 4. Two inertial frames with a common x–x' axis and with axis y parallel to y', axis z parallel to z'. Relative to frame S, frame S' is moving in the positive x direction at speed v. Point P gives the location of an event whose space and time coordinates may be determined in each frame of reference. The diagram suggests that the x' coordinate is related to the x coordinate by the equation $x' = x - vt$. The origin of frame S' is chosen so that $x' = x$ at $t = 0$.

obtain the four space–time coordinates x, y, z, and t. The same event in a different (primed) frame of reference would be specified by the four coordinates x', y', z', and t'. If we know x, y, z, and t for a given event, and we know how the primed frame is moving relative to the unprimed frame, we should be able to deduce the values of x', y', z', and t' for that event. The formula for finding these values is called a coordinate transformation.

1. Galilean Transformation

Before relativity, it was simply assumed that time is absolute, so for any event, $t' = t$. We can always choose the directions of our coordinate axes so that the velocity $\mathbf{v}$ of the primed frame relative to the unprimed frame is parallel to the x axis (Fig. 4). From these initial assumptions we obtain the Galilean transformation

$$x' = x - vt, \qquad y' = y,$$
$$z' = z, \qquad\qquad t' = t \tag{1}$$

From Fig. 4, the Galilean transformation seems so obvious that one feels it cannot be wrong. But it conflicts with the second postulate, as we now demonstrate.

A flash of light, starting at $x = 0$ when $t = 0$, will reach coordinate x at time t such that $x/t = c$. The same flash of light will start at $x' = 0$ and $t' = 0$, and its speed in the primed frame will be x'/t'. According to the Galilean transformation, $x'/t' = x'/t = c - v$, the speed of light in the primed frame is not c, and the postulate is violated. Therefore we must rule out the Galilean transformation and find a transformation that gives the same speed of light in both frames of reference.

TABLE II. Coordinates of Events in Two Reference Frames[a]

	Event						
	A	B	C	D	E	F	G
$x =$	0	-3	-1	$+1$	$+3$	$+5$	0
$t =$	0	$+5$	$+5$	$+5$	$+5$	$+5$	$+8$
$x' =$	0	0	$+2.5$	$+5$	$+7.5$	$+10$	$+6$
$t' =$	0	$+4$	$+5.5$	$+7$	$+8.5$	$+10$	$+10$

[a] In this table, the speed of light is one unit, and the relative speed of the two reference frames is $\frac{3}{5}$ unit.

2. Lorentz Transformation

To obtain the correct transformation, one must abandon the assumption that $t' = t$, relying instead on the postulate concerning the speed of light. Therefore the coordinates of the light flash described in the preceding paragraph must always satisfy the equations $x/t = c$ and $x'/t' = c$. This objective is met by the Lorentz transformation

$$x' = (x - vt)\gamma, \qquad y' = y, \qquad z' = z$$
$$t' = [t - (vx/c^2)]\gamma \tag{2}$$

where γ is defined by

$$\gamma = [1 - (v/c)^2]^{-1/2}$$

If we divide the expression for x' by the expression for t', and set x/t equal to c, we find that x'/t' also equals c, as required.

We illustrate the Lorentz transformation in Table II, which lists the coordinates of a set of events A, B, C, etc. This table will permit us to compare measurements made in different frames of reference, without repeated substitution into Eqs. (?). The events in the table are described as follows.

1. Event A is chosen as the event for which all coordinates are zero in both frames.

2. Events B–F are simultaneous in the unprimed frame (at $t = 5$), occurring at equally spaced values of x. Their coordinates in the primed frame are obtained from Eqs. (2) with $v/c = -0.6$ and $c = 1$. (Use of the light-second as the unit of length makes c equal to 1 in light-seconds per second.)

3. Event B occurs at the origin of the primed frame, 4 sec after event A, according to a clock fixed at that origin ($x' = 0$).

4. Event G occurs at the origin of the unprimed frame, 8 sec after event A, according to a clock fixed at that origin ($x = 0$).

The reader may verify for each event in Table II that the x' and t' coordinates in the column for that event are correctly obtained from the x and t coordinates by the use of Eqs. (2) with v equal to -0.6.

If the unprimed frame is considered to be at rest, we verify that the velocity of the primed frame along the x axis is $v = -\frac{3}{5}$ by comparing event A with event B. In a time t of 5 sec, point $x' = 0$ moves from $x = 0$ to $x = -3$, so $v = x/t = -\frac{3}{5}$. (The point $x' = 0$ is a fixed point—a mark on a stationary meter stick—in the primed frame. So its speed in the unprimed frame is the same as the speed of the whole primed frame of reference.)

If the primed frame is considered to be at rest, we find the speed of the unprimed frame by comparing events A and G. In a time (t') of 10 sec, the point $x = 0$ moves from $x' = 0$ to $x' = 6$, so the velocity of the unprimed frame along the x' axis is equal to $\frac{6}{10}$, or $\frac{3}{5}$. Thus both observers agree on the relative speed of the two frames.

From events A and F we see that an object moving with speed $u = c = 1$ in the unprimed frame also moves with speed $u = c = 1$ in the primed frame. A body moving at speed 1 in the unprimed frame, starting from event A, reaches $x = 5$, $t = 5$, which are the coordinates of event F. Event F is described in the primed coordinates by $x' = 10$ and $t' = 10$, so the object moves at speed $x'/t' = \frac{10}{10} = 1$ in that frame also.

IV. Implications of the Lorentz Transformation

A. Length and Time Measurements in a Moving Frame of Reference

1. Addition of Velocities

If an object moves at a speed less than c in one frame of reference, then its speed is different in another frame of reference. For example, refer to Table II and consider an object that moves with constant speed from event A to E. (The fact that this object is present at event E must be clear to an observer in any frame of reference. The general principle is that all observers agree on whether an object is present at a given event. Thus, if a second object were also at event E, the two objects would collide, and the fact that a collision had occurred would be indisputable. The collision could leave marks on each object, and the marks would be visible to all observers, no matter what their speed.)

In the unprimed frame, event E has coordinates $x = 3$ and $t = 5$, so the speed of an object moving from event A to E is $\frac{3}{5}$. In the primed frame, event E has coordinates $x' = 7.5$ and $t' = 8.5$, so the speed of the same object is $7.5/8.5$, or $\frac{15}{17}$. (If we used the Galilean transformation [Eqs. (1)], x' would be 6, t' would be 5, and the speed would be $\frac{6}{5}$—a speed greater than that of light. We see below that the Lorentz transformations cannot give a speed greater than the speed of light, because when v is greater than c, the factor γ in Eqs. (2) is imaginary.)

A general formula giving the velocity of an object in the primed frame in terms of the velocity in the unprimed frame may be found by simple division, using Eq. (2). The components of the velocity along the x', y', and z' axes are, respectively,

$$u'_x = \frac{x'}{t'} = \frac{(x - vt)\gamma}{[t - vx/c^2)]\gamma} = \frac{(x/t) - v}{1 - (vx/c^2 t)}$$

$$= \frac{u_x - v}{1 - (u_x v/c^2)}$$

$$u'_y = \frac{y'}{t'} = \frac{y}{(t - (vx/c^2)} = \frac{u_y}{[1 - (u_x v/c^2)]\gamma} \qquad (3)$$

$$u'_z = \frac{z'}{t'} = \frac{z}{(t - (vx/c^2)} = \frac{u_y}{[1 - (u_x v/c^2)]\gamma}$$

where u_x, u_y, and u_z are the components of the velocity in the unprimed frame, and v is the velocity of the primed frame relative to the unprimed frame. (This velocity is directed along the x axis, as mentioned previously.)

Equations (3) show directly that an object moving at speed c in one frame has the same speed in the other frame, regardless of the value of v. For example, if $u_x = c$, $u_y = 0$, and $u_z = 0$, then direct substitution shows that $u'_x = c$ and the other components are zero.

For the example discussed in connection with Table II, with $u_x = \frac{3}{5}$, $c = 1$, and $v = -\frac{3}{5}$, the first part of Eqs. (3) gives

$$u'_x = [\tfrac{3}{5} - (-\tfrac{3}{5})]/[1 - (\tfrac{3}{5})(-\tfrac{3}{5})] = \tfrac{15}{17}$$

the same speed deduced from Table II.

Equations (3) can also be applied to the Fizeau experiment described in Section I. In the (unprimed) frame of reference in which the water is at rest, the speed of light in the water is $u_x = c/n$, where n is the index of refraction of water. If the laboratory is taken to be the primed frame, and the water moves with speed v in this frame, then the primed frame must be moving with speed v relative to the unprimed frame, and the speed of light in the laboratory should be,

according to Eqs. (3),

$$u'_x = [(c/n) - v]/[1 - (v/cn)] \qquad (4)$$

when the light and the water are moving in opposite directions. Fizeau's result agrees with Eq. (4). (However, it must be admitted that the Fresnel ether-drag formula gives almost the same result. The difference between the predictions of the two formulas is too small to have been observed by the Fizeau experiment.) There are now much better grounds for reliance on Eqs. (3).

2. Synchronization of Clocks

When we speak of the times t and t', we refer to an abstraction that can only be realized by reading a clock. The times shown in Table II are known from clock readings, and these readings can be photographed. Observers in all frames of reference will agree that an unprimed clock that reads 5.00 and a primed clock that reads 4.00 are both present at event B. They also agree that an unprimed clock that reads 5.00 and a primed clock that reads 5.5 are both present at event C.

If these readings are to give what we consider to be the actual time, clocks at different places have to be synchronized. One way to synchronize two clocks at different locations is to start each one with a flash of light that originates midway between them. If both clocks are at rest, the light flash will arrive at each clock and start the clocks simultaneously. Being identical, the clocks will remain synchronized as time passes.

An observer O who is at rest in the unprimed frame of reference will conclude that events B and C are simultaneous, because the clocks in his rest frame at those events both read $t = 5.00$. He will say that clocks in the primed frame are not synchronized, because they read different values at the same time t.

The clocks in the primed frame have indeed been synchronized, but the synchronization was done in the primed frame, that is, by an observer O' who was moving along with the clocks in that frame. Observer O says that the procedure of O' did not work, because the light flash that started the primed clocks did not arrive simultaneously at the two clocks. The primed clock at event C was moving toward the origin of the light flash, and the primed clock at event B was moving away from it; therefore, the clock at event C received the signal first, and it started sooner.

On the other hand, observer O' at rest in the primed frame will say that the primed clocks are indeed synchronized, because they are at rest, relative to O'. Observer O' will say that events B and C are therefore not simultaneous, because the primed clocks at those events read $t' = 4$ and $t' = 5.5$, respectively. Assuming observer O' to be at rest, that observer will say that the clocks of observer O are the ones that are moving and are not synchronized. The general principle here is, if all clocks are synchronized in their own rest frames, then

> Moving clocks at different locations are not synchronized

That is, at any given time, two moving clocks at different places have different readings, no matter how the clocks are constructed (unless the clocks are faulty). This conclusion also follows directly from the Lorentz transformation, or it can be shown from any other method one might use to synchronize two clocks, given the second postulate.

If a starting signal originates halfway between two clocks, the clock that is moving toward the source of the signal will start first. Thus in Table II, giving readings of clocks that are observed at the same time t, the reading of time t' at event C is greater than the reading of t' at event B, because the primed clock at event C, moving toward the signal, started at an earlier time (t) than did the primed clock at event B. The difference between the readings t' of the two clocks is Lv/c^2, where L is the distance between the two clocks (measured in the frame in which they are at rest) and v is the speed of each clock. Table II shows that, for the clocks present at events B and C, $L = 2.5$, $v = \frac{3}{5}$, and $c = 1$, making the difference between the clock readings equal to 1.5, as shown.

From the point of view of observer O', the unprimed clocks are moving in the $+x$ direction. Thus if two unprimed clocks were observed at the same time t', the clock at the larger value of x would give a smaller reading than the other clock (e.g., see events F and G).

3. Lorentz Contraction

Again refer to Table II. Consider a stick that is 5 units long and is at rest in the primed system. Let one end of the stick be fixed at $x' = 0$, and the other end fixed at $x' = 5$. As the primed frame moves past the unprimed frame, observer O measures the position of both ends of the stick simultaneously (at time $t = 5$). One end is taken at $x = 3$ (event B) and the other end is at $x = +1$ (event D). Thus O would logically say that the stick is 4 units long, because that is the dis-

tance between the two ends at a given time. The general principle is

> A moving stick is contracted along the direction in which it moves

The amount of contraction is given by

$$L' = L/\gamma \qquad (5)$$

where L is the length of the stick when it is at rest, L' is its length when it is moving, and γ is defined in Eqs. (2).

Again, either O or O' can be considered to be at rest. If a stick is at rest in the unprimed system, O' measures its length by measuring the position of both ends at the same time t'. Assume that the left end of the stick is at $x = 0$ and the right end at $x = 5$. If the measurement by O' is made at time $t' = 10$, O' will find the right end to be at $x' = 10$. At this time ($t' = 10$) we see from event G that the point $x = 0$ has the coordinate x' equal to 6. Thus, when they are measured at the same time t', the ends of the stick are at $x' = 6$ and $x' = 10$. As before, the stick is found to have a length of 4 units when it is moving at a speed of $\frac{3}{5}$. It is contracted to four-fifths of its rest length, in agreement with the factor γ when $v = \frac{3}{5}$.

4. Time Dilation

Notice events A and B in Table II. Both events occur at the same place ($x' = 0$) in the primed frame of reference. Therefore the times t' of these two events are measured by a single clock, which is at rest in the primed frame. Similarly, events A and G occur at the same place ($x = 0$) in the unprimed frame of reference, and the times t of those events can also be measured by a single clock, which is at rest in the unprimed frame of reference.

In each case, the time interval read by the single clock is smaller than the time interval that is measured between the two events in the other frame of reference, in which that clock is moving. This is a further implication of the Lorentz transformation:

> Moving clocks run slowly relative to identical clocks at rest

The Lorentz transformation can be used to derive a general formula relating the rates of clocks. Consider two events that occur at different places and a clock that is present at both events. If that clock is moving at speed v in any given frame of reference, and the time interval between the events is t_0 in that frame, then the time interval τ measured by the moving clock is given by

$$\tau = t_0/\gamma \qquad (6)$$

where $\gamma = [1 - (v/c)^2]^{-1/2}$ as in Eq. (2).

For example, the clock located at $x' = 0$ is present at events A and B. In the unprimed frame of reference, this clock is moving, with $v/c = -\frac{3}{5}$, and the time interval t_0 between A and B is 5 sec. Inserting these values in Eq. (6) gives $\tau = 4$, which is just the time interval recorded by the clock at $x' = 0$.

Equation (6) can also be applied to events A and G. According to observer O', these events occur at different places (at $x' = 0$ and $x' = 6$, respectively), and the time interval between these events is 10. Thus in Eq. (6) we can put t_0 equal to 10, and a clock that is present at both events must move at a speed of $\frac{6}{10}$, or $\frac{3}{5}$. If we put v/c equal to $\frac{3}{5}$, we again find $\gamma = \frac{5}{4}$, and Eq. (6) then gives $\tau = 8$. This is just the time interval recorded by the clock located at $x = 0$—the clock that is present at both events A and G.

5. Proper Time

Although observers in different frames of reference may disagree on the time interval between two events, and they may disagree on the speed of the clock that is present at both events, they must agree on the times that are recorded by that clock at each event. They must agree for the simple reason that the clock face can be photographed as it appears at each event. The time interval measured by such a clock is called the proper time interval.

A formula that is equivalent to Eq. (6) gives the proper time interval τ in terms of the space and time coordinates of two events as

$$\begin{aligned} c\tau = \{c^2(t_2 - t_1)^2 - (x_2 - x_1)^2 \\ - (y_2 - y_1)^2 - (z_2 - z_1)^2\}^{1/2} \end{aligned} \qquad (7)$$

where the coordinates x, y, z, and t apply to any frame of reference whatsoever, and the subscripts 1 and 2 refer to their values at event 1 and event 2, respectively. Notice that in the frame of reference for which both events occur at the same place,

$$x_2 = x_1, \qquad y_2 = y_1$$
$$z_2 = z_1, \qquad \tau = t_2 - t_1$$

Thus we define a proper time interval by the statement

The proper time interval between any two events is the time interval recorded by a uniformly moving clock that is present at both events

This statement is equivalent to

The proper time interval between any two events is the time interval measured in the frame of reference in which the two events occur at the same place

It can be shown directly from the Lorentz transformation that τ is a quantity that is the same for all observers. When this quantity is zero, a clock would have to move with the speed of light to be present at both events. Such a clock would register zero time interval between any two events; it would be a clock that does not run at all. We can say that, for something traveling at the speed of light, time simply does not flow.

6. Space–Time Interval between Events

There exist many pairs of events for which the quantity τ^2 is negative, so that the proper time interval τ is imaginary. These events are so far apart that a clock would have to travel with a speed greater than c in order to be present at both events. Thus there is no clock that reads this interval directly.

In such a case we can still define $c\tau$ to be the space–time interval between the events. If τ is real, the interval is timelike, and τ is the proper time interval. If τ is imaginary, the interval is spacelike. We have seen that, when the interval is timelike, there is a frame of reference in which the two events occur at the same place. Similarly, when the interval is spacelike, there is a frame of reference in which the two events occur at the same time. In that frame of reference, the square of the distance between the two events is equal in magnitude to the square of the space–time interval. All observers agree on the space–time interval between two events. We say that this is an invariant quantity.

An invariant quantity is one that has the same value in all inertial frames of reference

Thus, although the distance between two events is different in different frames, the space–time interval, as calculated by Eq. (7), is the same in all frames.

Events F and G of Table II are separated by a spacelike interval. In the unprimed frame, F and G are separated by a time of 3 units and a distance of 5 units, so $c^2\tau^2$ is equal to 9 minus 25, or -16. Thus there is a frame of reference in which these two events occur at the same time; in that frame, the distance between the events is equal to 4. The primed frame of Table II is in fact the frame in which these two events occur at the same time, and in that frame the distance between these events is $10 - 6$, or 4. To be present at both events, a clock would have to travel at a speed of $\frac{2}{3}$ relative to the unprimed frame, or at an infinite speed relative to the primed frame. In both the primed frame and the unprimed frame, the square of the space–time interval as defined by Eq. (7) is equal to -4. All observers agree on the value of $c^2\tau^2$. (Test this for other pairs of events; for example, for C and G the value of $c^2\tau^2$ is 8).

B. The Doppler Effect

The Doppler effect is well known in wave phenomena. A sound source has a higher pitch when it is approaching than when it is receding. The Doppler effect for light is similar to the effect for sound, but the formula must be modified for the effect of time dilation—the fact that a moving light source emits light of lower frequency than does an identical source at rest, just as a moving clock runs more slowly than an identical clock at rest. In fact, a light source itself can be used as a standard clock.

For a sound wave, the observed frequency f' may be related to the source's vibration frequency f by the nonrelativistic formula

$$f'/f = (v_o + c_s)/c_s \qquad (8)$$

where v_o is the component of the velocity of the source in the direction toward the observer, c_s is the speed of sound, and the observer is at rest in the medium that transmits the sound.

For light, we can always assume the observer to be at rest, so we can use the same formula, provided we replace c_s by c, and replace f by f/γ. The latter substitution is required by the time dilation effect on the moving source; f is the source's frequency in its rest frame, and the moving source has a lower frequency than it would have if it were at rest. The result is

$$f'/f = (v_o + c)/c\gamma \qquad (9)$$

If the velocity vector $\mathbf{v}$ is directed toward the observer, then v_o becomes equal to the relative speed v, and Eq. (9) can be simplified to

$$f'/f = [(c + v)/(c - v)]^{1/2} \qquad (10)$$

If the source is moving directly away from the observer, Eq. (10) is still valid, but v becomes negative. Finally, if $\mathbf{v}$ is at right angles to the line between source and observer, then v_o equals zero, and Eq. (9) becomes $f' = f/\gamma$. In this case we have what is called the transverse Doppler effect, which results purely from time dilation.

For a sound wave, which travels in a definite medium, the observed frequency f'

is given by a different formula when the source is at rest in the medium and the observer is moving. But for light in empty space, it is meaningless to distinguish between moving source and moving observer, so the same formula, Eq. (9), must hold in all situations, with v_o being the relative speed of approach of source and observer.

To illustrate the use of Eq. (10), let us refer again to Table II. Suppose that a source of light is fixed at $x = 0$, and it emits 100 light flashes per second. Further suppose that an observer O', at rest in the primed frame of reference, is located at $x' = 6$. According to Eq. (9), with $c = 1$ and $v = \frac{3}{5}$, O' will see $f' = 2f$, or 200 flashes/sec as the source approaches.

At time $t' = 0$, O' says that the source is 6 light-sec away, and is moving toward him/her at speed $v = \frac{3}{5}$ light-sec/sec. At time $t' = 10$, the source has arrived at O'. (This is event G.) During this 10-sec time interval, O' has received flashes at the rate of 200/sec, for a total of 2000 light flashes. Some of these flashes were already en route to O' at time $t' = 0$, and the rest were emitted by the source during the following 10 sec. Let us try to account for all of these flashes.

Because the source is moving at $v/c - \frac{3}{5}$, it emits light flashes at a frequency of only four-fifths of the frequency seen in its rest frame [according to the time dilation formula of Eq. (6)]. Therefore O' says that the source emits 80 flashes/sec, and that it emits 80×10, or 800 flashes during those 10 sec. The other 1200 flashes must have already been en route to O' at time $t' = 0$. We confirm this by noting that the flashes are arriving at a rate of 200/sec, so the distance between one flash and the next must be 1/200 light-sec. Since the total distance between O' and the source (according to O') was 6 light-sec, there must have been 6×200, or 1200 flashes that were on their way to O' at time $t' = 0$.

C. The Twin Paradox

Time dilation governs all processes in the moving frame of reference. Biological processes proceed at the same rates as measured by the clock in any given frame (otherwise one could tell if one is in a moving frame). Therefore, if the clocks in a moving frame run slowly (as seen by us in our rest frame), then people moving with those clocks will (as seen by us) age more slowly than will people at rest. However, those people have no way of determining that they are living longer than they would if they were at rest; as far as they are concerned, they are at rest.

These facts suggest an apparent paradox for space travelers moving at high speeds. Consider fraternal twins Jay and Kay. On their 20th birthday, Kay leaves on a space trip while her brother Jay remains on earth. Her journey, at a speed of $\frac{3}{5}c$, is to star system X, which is 15 lt-yr from earth. Upon her arrival there, she will very quickly turn around and return to earth at the same speed.

We can analyze this trip by means of three inertial reference frames: frame J is Jay's rest frame; frame K' is Kay's rest frame while the twins are separating; frame K" is Kay's rest frame while the twins are coming closer together again. The round trip, which takes 50 yr in frame J, takes only 40 yr according to Kay's clocks as she spends 20 yr in frame K' and 20 yr in frame K". She will reach earth again on her 60th birthday, when Jay is observing his 70th birthday.

This result need not be a surprise to either twin, because they can communicate with each other during the entire trip. Let us assume that each twin, counting time by his or her own clocks, sends the other twin a "happy birthday" message—a light signal—on each anniversary of their separation. Kay will send Jay 40 messages, and Jay will receive all of them (including the last one, which will be delivered in person). Jay will send Kay 50 messages, and she will receive all of these. Both twins will agree that Jay has had 50 birthdays and Kay has had 40 birthdays.

The sequence of events is shown graphically in Fig. 5. Both Jay and Kay move on paths in space–time called "world lines." Figure 5a shows these paths in frame J. In frame J, Jay goes nowhere, remaining at $x = 0$; thus his world line is along the t axis (i.e., the line $x = 0$). (He moves in time and is motionless in space.) In this frame, Kay's path is two straight lines: one from the origin to the point $x = 15$, $t = 25$ and another from that point to $x = 0$, $t = 50$.

The light signals that each sends follow straight lines with a slope of 1 (because their speed is 1 lt-yr/yr). The ones sent by Jay at $t = 0, 5, 10, \ldots$ are shown in Fig. 5a. Signal 10 arrives at Kay's ship just as she reaches planet X. Kay is then 40 years old, receiving the message Jay sent when he was 30-yr old. She has received one signal every two years (in agreement with the Doppler formula (10), which gives $f' = \frac{1}{2}$ when $v = -\frac{3}{5}$ and $f = 1$).

During Kay's 20-yr return trip to earth, she receives Jay's birthday messages at the rate of 2/yr. [The Doppler formula [Eq. (10)], with $v = \frac{3}{5}$, $c = 1$, and $f = 1$, yields $f' = 2$]. She receives 40 messages during the return trip, so she is not surprised to find that her brother is $30 + 40$, or 70 years old when she arrives.

Figure 5b shows the trip in Kay's two inertial frames of reference (K' for the first 20 years, K" for the next 20 years). Kay receives signal 10 at $t' = 20$, but to deduce how rapidly Jay has been aging, she must make an inference regarding the time when the signal was sent. From the Doppler effect she can infer that Jay has been reced-

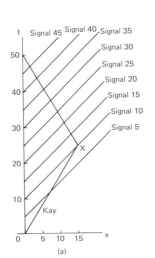

(a)

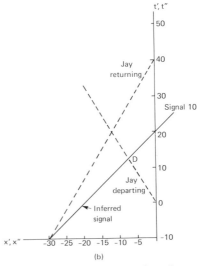

(b)

FIG. 5. World lines for twins Jay and Kay and for the signals sent from Jay to Kay. (a) Seen in Jay's rest frame and (b) seen in Kay's rest frames.

ing from her at a speed of $\frac{3}{5}$. This tells her that his world line must be the dashed line. The intersection of this line with the world line of the light signal (which has a slope of 1 in all frames) shows that the signal was sent at time $t' = 12.5$, from the event marked D. (To verify this conclusion, note that Jay would reach $x' = -7.5$ at time $t' = 12.5$, and that the signal would take 7.5 yr to reach Kay from that point, making the arrival time t' equal to 12.5 + 7.5, or 20 yr, as observed.)

Thus each twin can deduce that the other twin is aging less rapidly. Kay sees that Jay's clock has only recorded 10 yr when $t' = 12.5$ (event D in the figure), and Jay can see that Kay's clock has only recorded 20 yr when his time t is equal to 25 yr (event X). The paradox is in the fact that Jay does indeed age 50 yr (and Kay only 40 yr), because the twins are not in equivalent situations. Jay remains in the same inertial frame for the entire time, but Kay switches from one inertial frame to another. At this moment a strange thing happens: Kay's inference regarding the position of her twin must suddenly change!

In Kay's new frame K″, Jay is judged to be approaching rather than receding. Therefore, Jay's inferred world line is different. From Fig. 5b we see that Jay would have had to be 30 lt-yr away when signal 10 was sent. At a speed of $\frac{3}{5}$, it will take Jay 50 yr to reach Kay, but he will send 40 more signals, showing that he will have aged by another 40 yr when he reaches Kay. Again, Jay's clocks are slow, registering 40 yr instead of 50; but he will be 70 yr old when the twins meet.

If Kay's sudden change of perception when she changes reference frames seems unnatural, the reader must realize that the whole situation is unnatural. Any traveler who attempted to change velocity from $\frac{3}{5}c$ to $-\frac{3}{5}c$ in such a short time would be crushed by the enormous force required. In any actual space trip, the traveler would be subjected to acceleration over extended periods of time, and the special theory of relativity would not be able to show what would happen during the periods of acceleration. The general theory, which does deal with acceleration, leads to the same result as the special theory for examples like the one just described. The traveler is indeed younger than her twin when she returns home. This conclusion obviously has not been tested for human beings, but it has been well verified for short-lived elementary particles, to be discussed in Section VII. [See RELATIVITY, GENERAL.]

V. Momentum and Energy

A. CONSERVATION LAWS

Among the basic laws of physics are conservation laws stating that certain quantities such as energy and momentum are always conserved in any physical process. An observer in one frame of reference may disagree with an observer in another frame of reference in calculating how much energy or momentum is present, but each observer must find that this calculated quantity does not change as time goes by. That is,

The laws of conservation of energy and conservation of momentum are valid in all inertial frames of reference.

in agreement with the principle of relativity.

In prerelativity physics, the momentum of an object of mass m is defined as a vector $\mathbf{p}$ whose components are

$$p_x = mu_x, \qquad p_y = mu_y, \qquad p_z = mu_z \quad (11)$$

where u_x, u_y, and u_z are the three components of the object's velocity.

For a collection of objects, the total momentum $\mathbf{P}$ is defined as the vector sum of the momenta of the individual objects. That is, each component of $\mathbf{P}$ is defined by the equations

$$P_x = m_1 u_{x1} + m_2 u_{x2} + \cdots$$
$$P_y = m_1 u_{y1} + m_2 u_{y2} + \cdots \quad (12)$$
$$P_z = m_1 u_{z1} + m_1 u_{z1} + \cdots$$

where the subscripts 1, 2, etc., refer to the values for objects 1, 2, etc.

When two particles interact, momentum may be transferred from one to the other, but the law of conservation of momentum states that the total momentum vector $\mathbf{P}$ of an isolated collection of particles remains constant, no matter how the particles interact with each other.

1. Effect of Change of Reference Frame

The Galilean transformation equations [Eq. (1)] ensure that $\mathbf{P}$, as defined in Eqs. (12), is conserved in all frames of reference, as long as it is conserved in one frame. This can be illustrated by an interaction (e.g., a collision) between two particles, of mass m and m_2, respectively, and respective speeds u_{x1} and u_{x2} along the x axis. The total momentum before the collision is

$$P_x = m_1 u_{x1} + m_2 u_{x2}$$

In a second frame of reference, moving with velocity v in the $+x$ direction, the respective velocity components are $u_{x1} - v$ and $u_{x2} - v$ according to the Galilean transformation. Then the total momentum in this frame is

$$P_x' = m_1(u_{x1} - v) + m_2(u_{x2} - v)$$

which is easily shown to be

$$P_x' = P_x - (m_1 + m_2)v \tag{13}$$

Thus if P_x does not change in a collision, and the sum of the masses is unchanged (law of conservation of mass), then P_x' does not change, and the law of conservation of momentum is maintained in both frames by use of the Galilean transformation.

2. THE NEED FOR A RELATIVISTIC DEFINITION OF MOMENTUM

We now know that the Galilean transformation is wrong, the Lorentz transformation is the correct one to use, and Eqs. (3) must be used to transform velocities from one frame to another. Suppose that Eqs. (12) were retained as the definition of momentum, but Lorentz velocity transformation [Eqs. (3)] were used instead of the Galilean transformation to derive an expression for P_x' analogous to Eq. (13). This expression could not be written so simply in terms of P_x, as it is in Eq. (13). The result would be that conservation of P_x' would not require that P_x be conserved. In that case the law of conservation of momentum would violate the principle of relativity.

We cannot escape the need for the Lorentz transformation, and the law of conservation of momentum has proven to be very useful. We can preserve both, without violating the principle of relativity, by adopting a new definition of momentum. The new definition must meet two conditions:

1. It must be conserved in all frames of reference, to satisfy the principle of relativity.
2. It must reduce to the older definition in the limit $u \rightarrow 0$.

B. RELATIVISTIC EXPRESSIONS FOR MOMENTUM AND ENERGY

1. Momentum

The clue to the form of the new definition of momentum comes from the definition of proper time. The expression for P_x' in terms of P_x becomes complicated when one uses the old definition [Eq. (12)], because the velocity in each frame is given by the ratio of displacement to time interval, using the time coordinates peculiar to one frame of reference (e.g., $u_x = x/t$). (To simplify the form of the equations, we again assume that the particle starts at $x = y = z = 0$.)

Instead of finding the velocity components by dividing each component of displacement (x, y, and z) by t, let us divide the displacement by the proper time τ (that is, the time in the frame of reference in which the particle is at rest). We then multiply the result by m to obtain each component of momentum. The result is that Eq. (11) is replaced by

$$p_x = mx/\tau = mu_x\gamma, \qquad p_y = my/\tau = mu_y\gamma$$
$$p_z = mz/\tau = mu_z\gamma \tag{14}$$

As $u \rightarrow 0$, $\gamma \rightarrow 1$. In that limit, Eqs. (14) can be approximated by Eqs. (11), so that requirement 2 is satisfied.

The velocity that results from using proper time instead of one's own time is called the four-velocity, because one can define a fourth component by dividing t, the displacement in time, by the proper time γ. Since $t = \tau\gamma$, this fourth component is simply equal to γ. Multiplying this by the mass gives us a fourth component for the momentum vector:

$$p_t = mt/\tau = m\gamma \tag{15}$$

It is logical to be concerned with this fourth component, because when the Lorentz transformation transforms the space components of displacement, the result involves the time. Let us find the components of momentum in another frame of reference, by applying the Lorentz transformation to the coordinates x, y, z, and t that appear in the four components of momentum defined here. We assume that the mass m is the same in all frames of reference, and so is the proper time τ, and we obtain

$$p_x' = (p_x - vp_t)\gamma, \qquad p_y' = p_y$$
$$p_z' = p_z, \qquad p_t' = (p_t - vp_x/c^2)\gamma \tag{16}$$

Equations (16) are identical in form to Eqs. (2); we have simply replaced x by p_x, t by p_t, etc. The four components of momentum are transformed by a Lorentz transformation in exactly the same way as the four components of space (displacement) and time.

2. Energy

It would not be legitimate to have a law that conserved only the first three components of momentum. Therefore, the fourth component must also be conserved. The fourth component of momentum appears to be unrelated to previous theories, but examination of the velocity dependence of this component shows that it is related to the total energy of the particle. The law of conservation of energy, in relativity theory, becomes a part of the law of conservation of momentum.

Let us state the result and then justify it. The energy of an object of mass m and speed u is equal to $c^2 p_t$, or

$$E = mc^2\gamma \tag{17}$$

where $\gamma = (1 - u^2/c^2)^{-1/2}$. We require that this definition of energy, like the definition of momentum, reduce to the classical definition in the limit $u \to 0$. Let us find this limit by expanding Eq. (17) in a power series in u/c. The result is given by the binomial theorem as

$$E = mc^2(1 + u^2/2c^2 + 3u^4/8c^4 + \cdots) \tag{18}$$

When $u^2 \ll c^2$, the third term and all succeeding terms in Eq. (18) may be neglected, and the energy becomes

$$E = mc^2 + mu^2/2 \tag{19}$$

The identification of $c^2 p_t$ with the energy is now justified in the limit $u \to 0$, because the second term in Eq. (19) is the familiar expression for the kinetic energy of an object of mass m and speed u. The first term is the energy possessed by an object when its speed is zero—the rest energy. This term was unknown in classical physics, but since the zero level of energy is arbitrary, its presence does not contradict any previously known law that is valid for small values of u. It does, however, lead to consequences that were quite unexpected when the theory was first proposed.

3. Definition of Kinetic Energy in Relativity Theory

As in classical theory, the kinetic energy of a body is defined as the energy associated with the motion of the body as a whole. The kinetic energy E_k of a body of mass m is thus equal to the total energy minus the rest energy, or

$$E_k = E - mc^2 = mc^2(\gamma - 1) \tag{20}$$

We see from Eqs. (17) and (18) that the kinetic energy of an object of speed u is always greater than the value given by the classical expression. As u increases, the difference between the classical value and the value given by Eq. (17) increases. If u were to equal the speed of light c, the value of γ would be infinite. This demon-strates how the speed of light acts as a limiting speed.

4. Definition of Force

In relativity, force is defined as the rate of change of momentum (in agreement with the classical definition). As the speed of a body approaches the speed of light, the momentum approaches infinity, which means that the increase in speed associated with a given change in momentum becomes smaller and smaller.

Classically, with momentum equal to mass times velocity, force may be defined as mass times acceleration. If we used that definition of force, we would find that as the speed of a body increases a greater force is required to provide a given acceleration. This would lead to the conclusion that the mass increases with increasing speed. The mass that increases is the so-called relativistic mass $m_r = m\gamma$, whose introduction is an artifice that permits one to use equations of the same form as classical ones (e.g., $F = m_r a$ and $p = m_r v$).

On the other hand, the definition of force as rate of change of momentum, with momentum given by Eqs. (14), lets us express everything in terms of one mass, the rest mass, which is independent of speed. Because the introduction of the term relativistic mass is unnecessary and often confusing, the word mass in this article will always mean rest mass.

5. Invariance of Rest Mass

By a simple extension of the principle of relativity, we can say that the mass, like proper time, is invariant. This does not mean that it cannot change as the result of an interaction; it means that it has the same value in all frames of reference. If we assume that all objects are made up of identifiable elementary particles whose masses are determined by laws of physics, and the laws are the same in all frames of reference, then the mass of each particle is the same in all frames of reference.

C. Conservation of Energy and Momentum in Collisions

1. Inelastic Collisions

An inelastic collision is defined classically as one in which kinetic energy disappears, to be replaced by some other form of energy (e.g., heat). In relativity theory, one can use the same definition, with the understanding that all possible forms of energy are included in the total en-

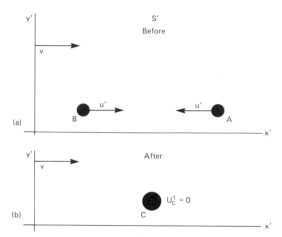

(a)

(b)

FIG. 6. Inelastic collision of equal-mass objects viewed in frame S'. (a) Before collision and (b) after collision.

ergy E, or the fourth component of the momentum vector. Study of an inelastic collision shows how the conservation of all four components of momentum is applied.

Figures 6 and 7 show, in two different frames of reference, a perfectly inelastic collision between two objects of equal mass M. In a perfectly inelastic collision, the colliding particles stick together, and there is a frame of reference S' in which the composite object formed by the collision is at rest, the kinetic energy having disappeared completely in that frame.

Let us apply the law of conservation of energy to the collision in frame S'. In this frame (Fig. 6), bodies A and B each have speed u', so their energies are equal, and the total energy is twice the energy of either one. Therefore, using Eq. (17) with γ equal to $(1 - u'^2/c^2)^{-1/2}$, we have

$$E' = 2Mc^2(1 - u'^2/c^2)^{-1/2}$$

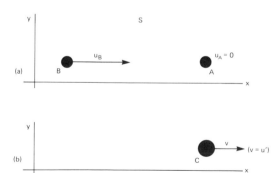

(a)

(b)

FIG. 7. Same collision as in Fig. 6, viewed in frame S: (a) before collision and (b) after collision. S' has velocity $\mathbf{v}$ relative to S.

After the collision, the total energy must remain equal to E'. But now there is only one body, and it is at rest, so E' equals the rest energy of that body:

$$E' = M_f c^2$$

where M_f is the mass of the composite body after the collision.

Equating the two expressions for E', we conclude that the mass of that composite body must be

$$M_f = 2M(1 - u'^2/c^2)^{-1/2} \qquad (21)$$

Thus $M_f \neq 2M$, and we have a violation of the law of conservation of mass; the total mass of the particles in the system is changed by the collision. The law of conservation of mass is no law at all. We invoked it to make the Galilean transformation obey the principle of relativity [see the discussion following Eq. (13)], but we have seen that the Galilean transformation must be abandoned. There exists no reason to retain the law of conservation of mass. That law is replaced by the law of conservation of total energy:

> The total energy of a collection of particles is unchanged by interactions between the particles

In applying this law, it is understood that the energy of each particle is given by Eq. (17), the value of γ being computed for each particle by using the speed of that particle.

In any inelastic collision, the rest mass must increase, because the kinetic energy is decreased, and the sum of the two is the total energy, which is conserved. In such a collision there is also, in any frame of reference, an increase in the temperature of the colliding objects. The objects have acquired internal energy that may in principle be detected by measuring either the change in temperature or the change in the total rest mass of the objects. If the same change in temperature had been produced by other means, such as by building a fire under these objects, the rest mass of the objects would necessarily increase by the same amount.

When the mass changes in one frame, it must change in all frames. Thus in Fig. 7, which shows the collision of Fig. 6 in a different frame of reference S, the total mass before the collision is $2M$. After the collision, the total mass is $2M(1 - u'^2/c^2)^{-1/2}$, as it was in frame S' [Eq. (21)], with the same value of u'.

The law of conservation of energy is just a

special case of the law of conservation of momentum, which may now be stated in similar fashion:

> The value of any component of the total momentum of a collection of particles is unchanged by interactions between the particles

2. A Numerical Example

Let us suppose that, in the collision shown in Figs. 6 and 7, u' and v are both equal to $0.6c$. In that case, $M_f = 2.5 M$. In frame S' it is clear that the total momentum is zero at all times, and the total energy is $M_f c^2$. Let us show that momentum, as defined by Eqs. (14), is also conserved in frame S.

We can find the speed of each particle in frame S by using the velocity-addition formula [Eqs. (3)]. Frame S is moving in the $-x$ direction with speed $v = u' = 0.6c$, relative to frame S'. Thus object A has speed zero in this frame, and the formula gives, for the speed of object B,

$$u_B = (0.6 + 0.6)c/(1 + 0.36) = (15/17)c \quad (22)$$

The momentum of object B is then, according to Eq. (14),

$$p_B = Mu_B(1 - u_B^2/c^2)^{-1/2} = \tfrac{15}{17}Mc/\tfrac{8}{17} = \tfrac{15}{8}Mc \quad (23)$$

The momentum of A being zero, we find that the total momentum is $\tfrac{15}{8}Mc$ before the collision. After the collision, the composite object C has a mass of $2.5M$, a speed of $0.6c$, and hence a momentum of

$$P_C = 2.5Mu_C(1 - u_C^2/c^2)^{-1/2} - 1.5Mc/0.8$$

$$= \tfrac{15}{8}Mc \quad (24)$$

We see that momentum is indeed conserved in frame S. We can also verify that energy is conserved in S. Object A, at rest, has an energy of Mc^2. The energy of object B is, according to Eq. (17),

$$E_B = Mc^2(1 - u_B^2/c^2)^{-1/2} - Mc^2/\tfrac{8}{17} - \tfrac{17}{8}Mc^2$$

so the total energy before the collision is

$$E = E_A + E_B = Mc^2 + \tfrac{17}{8}Mc^2 = \tfrac{25}{8}Mc^2 \quad (25)$$

in frame S. After the collision, object C has a mass of $2.5M$ and a speed of $0.6c$. Consequently, its energy is

$$E_C = 2.5Mc^2(1 - u_C^2/c^2)^{-1/2} = 2.5Mc^2/0.8$$

$$= \tfrac{25}{8}Mc^2$$

C is the only object remaining after the collision,

and its energy equals the original energy E. Therefore, we have verified that energy is conserved in this frame of reference.

3. Energy–Momentum Relation

From Eqs. (14) and (17) it follows that

$$mc^2 = [E^2 - c^2(p_x^2 - p_y^2 - p_z^2)]^{1/2}$$

or

$$(mc^2)^2 = E^2 - c^2p^2 \quad (26)$$

The values for object C of Fig. 7 are [Eqs. (24) and (25)]:

$$p_C = \tfrac{15}{8}Mc, \qquad E_C = \tfrac{25}{8}Mc^2,$$

$$M_C = 2.5M$$

Insertion of these values into Eq. (26) yields

$$(2.5Mc^2)^2 = (\tfrac{25}{8}Mc^2)^2 - (\tfrac{15}{8}Mc^2)^2$$

or

$$6.25 = \tfrac{625}{64} - \tfrac{225}{64}$$

which balances.

Equation (26) can also be deduced from Eq. (7), through multiplication of both sides by mc/τ. Just as we can find the proper time interval between two events by measuring the coordinates of the events in any frame of reference, we can find the mass of a particle by measuring its energy and momentum in any frame of reference, because mass, like proper time, is an invariant quantity.

4. Conversion of Rest Mass into Kinetic Energy

The collision shown in Figs. 6 and 7 could be reversed. Particle C could split up into particles A and B. In that case, the total rest mass of the system would decrease as the kinetic energy increased. It is well known that this process occurs in a nuclear reactor or a nuclear weapon. It is not so widely recognized that this conversion of rest mass into kinetic energy occurs in chemical processes as well. Any process that yields useful energy does so at the expense of rest mass.

D. Particles with Zero Rest Mass

1. The Photon

Maxwell's classical theory of electromagnetism shows that light is an electromagnetic wave. Such a wave carries momentum as well as energy. The theory shows that the energy density in the wave equals c times the momentum density, and this is verified by independent measurements of energy and momentum. For example, the energy density in solar radiation is well known, and the momentum density can be determined from the effect of the resulting radiation pressure on the orbits of low-density artificial satellites (e.g., balloons).

Einstein's theory of the photoelectric effect states that the energy and momentum carried by a light wave cannot be divided into arbitrarily small amounts. Rather, a light wave is made up of particles called photons, whose energy (and hence momentum) is proportional to the frequency of the wave. This theory is consistent with Maxwell's theory provided that the energy E and momentum p of each photon obey the relation $E = pc$.

This relation is consistent with Eq. (26), provided that $m = 0$. We conclude that

the photon is a particle with a rest mass of zero

Furthermore, if we place $v = c$ in Eqs. (14) and (17), we obtain an impossible result—infinite energy or momentum—unless $m = 0$. Thus the photon must have zero mass if it is to travel with the speed of light.

Placing $m = 0$ and $v = c$, in Eq. (14) or (17), yields zero divided by zero, which of course is indeterminate, so these equations do not give us any further information about an individual photon. But Eq. (26), with $m = 0$, is used routinely in studying interactions involving photons.

2. The Neutrino

The neutrino, a neutral particle produced in radioactive decay or in nuclear reactions, is also believed to have a mass that is zero, or so close to zero that the difference is exceedingly hard to detect. Therefore it, too, travels with the speed of light, and its energy E and momentum p are related by the equation $E = pc$.

In spite of its zero mass, the neutrino has many properties that it does not share with the photon. The existence of the neutrino underscores our discussion of the second postulate of relativity. In the theory of relativity, the significant thing about the speed c is not that it is the speed of light; rather, light is of interest because its speed is c.

E. THE REST MASS OF A COMPOSITE BODY

The energy of a moving body can be divided into two parts: (1) the external energy, known as kinetic energy and (2) the internal energy,

known as rest energy. The external energy is easily observed. Changes in internal energy had been detected before relativity theory by means of temperature measurements, but the relativistic equation [Eq. (17)] showed the connection between internal energy and rest mass.

[The factor c^2 in Eq. (17) is simply a conversion factor between our unit of mass and our customary unit of energy. If we used units such that $c = 1$, it would be easier to realize that mass and energy are basically the same thing. What is often called the conversion of mass into energy could be more accurately called the conversion of internal or rest energy into kinetic energy.]

Internal energy may be visualized by considering the body to be a box containing a gas. When the atoms of the gas move about randomly, the box remains at rest, but the kinetic energy of these atoms may be transferred to the outside. When this happens, the box becomes colder and we say that heat is flowing out of it. The rest mass of the box also decreases, because there is less energy in the box. A sufficiently sensitive measurement would show that the box has less inertia when it is colder; it is easier to accelerate it, because the slow atoms inside can be accelerated more easily than faster atoms could be. (It follows from the momentum–energy relation [Eq. (14)] and the ensuing discussion that a given force produces a smaller change in velocity as the speed increases.)

If a box containing a gas is moving uniformly, its external energy includes some of the kinetic energy of the gas atoms inside, because these atoms must move along with the box while they are bouncing around inside the box. If the box is suddenly stopped in an inelastic collision, the motion of these atoms becomes more disordered, as they no longer have a net motion. Their total kinetic energy may remain the same, but the externally observed energy, resulting from the collective motion of the atoms with the box, disappears, being converted into random motions of the atoms (i.e., heat, or increased rest mass of the box).

The connection between rest mass and temperature was not observed before Einstein because any temperature increase that does not vaporize a body is accompanied by a minuscule increase in mass—perhaps one part in 10^{11}, or 1 μg in 100 kg. But the change in mass always is associated with any kind of change in internal energy. For example, it is known that an energy of 13.6 eV is needed to separate the electron from the proton in a hydrogen atom when this atom is in its normal state. Therefore, the mass of the proton plus the mass of the electron exceeds the mass of the normal hydrogen atom by an amount equivalent to this additional energy of 13.6 eV (often written as a mass of 13.6 eV/c^2, because the energy–mass conversion factor is c^2). The total rest energy of the hydrogen atom is 9.4×10^8 eV; the total mass of its constituents exceeds the mass of the atom by only about one part in 100 million.

Rest mass is changed in any reaction—chemical or nuclear—that causes a change in kinetic energy. For example, consider the chemical reaction

$$2H_2 + O_2 \rightarrow 2H_2O \qquad (27)$$

It is well known that this reaction produces kinetic energy; hydrogen and oxygen form an explosive mixture. Since total energy is always conserved, the total rest mass of the two H_2O molecules that are produced must be less than the total rest mass of the two H_2 molecules and the O_2 molecule.

Now let us consider an explosion-proof, perfectly insulated combustion chamber containing hydrogen and oxygen. Let the total mass M of the system—the chamber and its contents—be determined by a precise measurement. The system is at rest, so that its rest energy is all internal energy, equal to Mc^2. Reaction (27) then takes place, and the mass is determined again. Since the chamber is insulated, no energy could escape; therefore the internal energy has not changed, and the mass of the system remains M. However, the rest mass of the constituents has changed, because some of it has been converted to kinetic energy by reaction (27). This kinetic energy is internal energy in this system, and thus it contributes to the rest mass of the system just as did the molecular rest mass from which it was created.

In this example, the system's temperature has been raised by the reaction, because the temperature is related only to the internal kinetic energy, rather than to the total energy, which is unchanged. If we were to remove the insulation, permitting the chamber to cool down, heat would be lost to the outside. This loss of energy would be reflected in a loss of rest mass by the system. If the entire operation were carried out in a system that was in contact with a constant-temperature bath, heat would flow out continuously and the system would lose mass continuously. The difference between the final mass and the initial mass would be simply equal to the difference between the total rest mass of the reactants (hydrogen and oxygen) and the total rest

mass of the resulting individual H_2O molecules. (If the temperature were fixed below the boiling point of water, some additional rest mass would be lost as the H_2O molecules condensed into water.)

It follows that whenever a body loses energy by radiation, the body's rest mass decreases. The photon that is radiated carries away this mass in the form of energy, even though the photon itself has no rest mass. Now consider a photon that is radiated within a body (e.g., inside an oven). For example, suppose that a photon is emitted from an inside wall of an oven and travels to the opposite wall, where it is absorbed. The energy of that photon, while it exists, is part of the internal energy of the oven, and it contributes to the rest mass of the oven. When the photon is emitted from one wall, the mass of that wall decreases because of the loss of energy; when the photon is absorbed by the other wall, the mass of that wall increases as it gains energy. But the overall rest mass of the oven—the mass m that appears in Eqs. (17)–(20)—remains constant throughout the whole process. In summary:

> The rest mass M of a composite object is NOT equal to the sum of the rest masses of its constituent parts. Rather, M is determined by the equation $E = Mc^2$, where E is the total energy that these constituents have, in the frame of reference in which the object is at rest.

VI. Electromagnetism

A. Moving Charges and Magnetism

When the electric and magnetic forces were first studied, it was thought that they were two distinct entities. In the nineteenth century it was found that an electric current could exert a force on a magnet, and Michael Faraday showed that a changing magnetic field could produce an electric current. The connection between magnetic and electric fields became more strongly established when Maxwell's equations of electromagnetism showed that a changing electric field produces a magnetic field, and vice versa.

Unlike the Newtonian equations of mechanics, the Maxwell equations are completely consistent with the special theory of relativity. That means that Maxwell's equations do not change form when a Lorentz transformation is applied to the coordinate system. This should not be unexpected, because the speed of light may be derived from these equations, and the Lorentz transformation is designed to make speed of light independent of coordinate system.

This is not the place for a discussion of Maxwell's equations. Rather, without considering the form of these equations, we simply show how electric and magnetic fields are affected by a Lorentz transfusion. We begin by considering a coordinate system in which there is no magnetic field, but only an electric field. A purely electric field is produced by stationary (or static) electric charges. In general, if any charges are moving, there will be a magnetic field as well as an electric field.

A charge that is at rest at a single point in space in a given frame of reference produces an electric field like that shown in Fig. 8a. The field is represented by lines that show the direction of the field at each other point in space (i.e., the direction of the force that would be exerted on a different charge located at that field point). The density of the lines in any region of space is proportional to the strength of the field (i.e., the magnitude of the force on a charge of a given size) in that region. One could do a set of experiments to measure the field at each point and thus construct all of these lines.

If the charge were moving with velocity **u**, as in Fig. 8b, the results would be different. However, we could find another frame of reference O in which the charge would be at rest, and in that frame the results would necessarily be the same as shown in Fig. 8a. The difference between Fig. 8b and Fig. 8a can be found from a Lorentz transformation.

B. LORENTZ TRANSFORMATION OF ELECTRIC AND MAGNETIC FIELDS

When the observer in frame O sees the electric field lines of Fig. 8a, the observer in O' sees

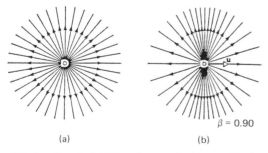

(a)   (b)

FIG. 8. Electric field lines from a point charge: (a) when the charge is at rest; (b) when the charge is moving with velocity **u,** for the case in which u/c is equal to 0.9. [Reproduced, with permission, from R. Resnick (1968). "Introduction to Special Relativity," p. 168. Wiley, New York.]

the entire frame of O contracted in the direction of the velocity **u.** If one takes Fig. 8a and contracts it along the horizontal axis, the result is Fig. 8b. (The effect can be seen by tilting the page sideways and looking from the right or left.)

In any frame of reference in which the charge is moving, there is also a magnetic field. Long before special relativity was developed, it was known, (from the studies of André-Marie Ampère and others) that an electric current (and hence a moving electric charge) produces a magnetic field. In general, no matter what the source of the fields, we can show that, when frame O' is moving relative to O, and there is a purely electric field in O, there will be both electric and magnetic fields in O'.

Without going into the details of electromagnetic fields, we can understand the origin of the magnetic field by considering a simple example. Two particles, each of mass m and charge q, are at rest in frame of reference O—one at the origin, the other directly above it, at $x = 0$, $y = 0$, and $z = d$. According to Coulomb's law, each repels the other with a force of $F = kq^2/d^2$, where k is a constant that depends on the units used for q and d. Let us suppose that this is the only force acting on these charges. Since they are at rest at this time, Newton's law gives the acceleration as $a = F/m = kq^2/d^2m$. This can in principle be confirmed by a measurement of the velocity of each particle as a function of time t, as the particles move apart in the vertical (z) direction.

Next, consider the motion of each charge in reference frame O', which is moving horizontally (in the x direction) with speed v relative to O. In this frame of reference, the acceleration of each charge can be determined in the same way, by measuring the velocity of the charge as a function of the time t'. When the observer in O' measures the vertical component of the velocity, he/she must divide z' by t', just as the observer in O divides z by t. To find the vertical acceleration, he must divide by t' again. According to the Lorentz transformation [Eq. (2)], $z' = z$ and $t' = t\gamma$ (because $x = 0$). Dividing by t' is equivalent to dividing by $t\gamma$, or dividing by t and multiplying by $\sqrt{1 - (v/c)^2}$. When this is done twice, the acceleration a' in O' is found to be the same as in frame O except for a factor of $1 - (v/c)^2$.

An observer in O' who knew nothing of relativity or other frames of reference would therefore say that charges moving in this manner with speed v are subject to a mutual force of

$$F' = ma' = (kq^2/d^2)[1 - (v/c)^2] \quad (28)$$

This reduces to Coulomb's law when $v = 0$. Such an observer would say that the force given by Coulomb's law is augmented by an additional force of $(kq^2/d^2)(v/c)^2$ when the charges are moving parallel to each other with speed v, and that this force is opposite in direction to the electric force. He could then call this additional force the magnetic force, and say that the total force on each charge is given by the vector sum of the electric and magnetic forces, to arrive at the result of Eq. (28).

Magnetic forces were discovered before relativity, but relativity could be used to derive all magnetic forces between moving charges, as we have done in this example. In the general case, with many charges moving in various directions, it is not possible to find a frame of reference like O in the previous example, in which there is no magnetic field. But one can use the Lorentz transformation to find the electric field components E_x', E_y', E_z' and the magnetic field components B_x', B_y', B_z' in O' in terms of the field components in O. If it is assumed that frame O' is moving in the $+x$ direction with speed v relative to frame O, then the result is

$$E_x' = E_x, \qquad\qquad B_x' = B_x$$
$$E_y' = \gamma[E_y - (v/c)B_z], \quad B_y' = \gamma[B_y + (v/c)E_z] \quad (29)$$
$$E_z' = \gamma[E_z + (v/c)B_y], \quad B_z' = \gamma[B_z - (v/c)E_y]$$

Notice that Fig. 7a shows a special case in which the magnetic field is zero in frame 0. In that case, Eq. (29) show that

$$E_x' = E_x, \qquad E_y' = \gamma E_y,$$
$$E_z' = \gamma E_z$$

Thus the components of the vector **E** in the direction perpendicular to the velocity **v** are simply increased by the factor γ resulting from Lorentz contraction of the frame O, in agreement with the configuration of field lines shown in Fig. 7b.

Notice the symmetry between E and B. The equations are basically the same if we replace E by B and B by E everywhere. The components of the vector **E** and the vector **B**, taken together, are part of a single entity, the electromagnetic field. The essential unity of this field only became apparent in the light of special relativity.

VII. Consequences of Special Relativity

A. Atomic Masses

More than 99.9% of the mass of any atom resides in its nucleus, which is composed of neutrons and protons. Atomic nuclei are thus composite bodies. Their masses provide the most well-documented illustrations of the principles governing the rest masses of composite bodies.

Masses of atoms can be determined with great precision by direct observation of their inertia in a mass spectrometer. This device measures the inertia of atoms by deflecting them in a magnetic field. [A neutral atom is not deflected in a magnetic field. Therefore, the atoms must first be ionized by removing (or adding) one or more electrons. Consequently, the direct measurement is that of the mass of an ion, but the very small mass of the missing (or added) electron(s) is accurately known.]

As a result of such measurements, it is known that the mass of any atom (except hydrogen) is almost 1% less than the total mass of its constituents. For example, a helium atom (^{4}He) containing two protons, two neutrons, and two electrons has a mass of (3728.43 ± 0.01) MeV/c^2 (which means that its rest energy is 3728.43 ± 0.01 MeV). The rest energy of an individual proton is 938.280 MeV, that of a neutron is 939.573 MeV, and of an electron 0.511 MeV, so the sum of the rest energies of the six particles in a helium atom is about 3756.73 MeV, or about 28.3 MeV more than the rest energy of the atom.

Thus, to split up a ^{4}He atom completely into its six constituent particles requires that 28.3 MeV be added to the atom. (The energy required to completely split up an atom is called its binding energy. Almost all of this energy is actually the binding energy of the nucleus, because the binding energies of the electrons are minuscule in comparison. For helium, the total energy required to separate the two electrons is only about 80 eV, or 0.0008 MeV.) Binding energies of nuclei or of individual particles in the nucleus (the energy required to separate one particle from the rest) have been measured repeatedly for thousands of atomic species. Measurements have also been made of the energy absorbed or released in nuclear reactions, in which the protons or neutrons are transferred from one atom to another. When the results are compared with mass spectrometer measurements, it is found that the atomic masses are always in exact agreement with the energy measurements and the Einstein formula $E = mc^2$.

B. Nuclear Energy

The fact that energy is equivalent to mass makes it possible to deduce the energy content of an atom by measuring its mass. Unlike chemical reactions, nuclear reactions release an amount of energy that is easily accounted for in a measurement of mass. Thus, even before a given reaction is observed, it can often be recognized from tabulated values of atomic masses that the reaction will be able to release energy—to transform rest energy into kinetic energy.

1. Nuclear Fission

In the 1930s Albert Einstein and others realized the significance of the fact that the uranium atom has about 200 MeV more rest energy than the total mass of lighter bodies into which it can split. One of many possibilities for the splitting of uranium is the induced fission reaction

$$^1n + {}^{235}U \rightarrow {}^{95}Y + {}^{139}I + 2{}^1n \qquad (30)$$

in which a neutron (^1n) causes a uranium (^{235}U) nucleus to split. In this example, the total mass of the four resulting particles [an atom of yttrium (^{95}Y), an atom of iodine (^{139}I), and two neutrons] is less than the rest mass of the uranium and the original neutron by an amount equivalent to an energy of 172 MeV. In this and other induced fission reactions involving uranium, the total energy released averages about 0.1% of the original rest mass of the uranium.

Most of the energy released in reaction (30) is given to the two neutrons, which are able to react with other uranium atoms in the same way to release more energy and additional neutrons, thus causing a chain reaction. The kinetic energy of the resulting swarm of neutrons is transferred to the surroundings, raising the temperature to the point where it can produce an explosion (in a bomb) or boil water (in a fission reactor).

2. Nuclear Fusion

Further study of atomic masses reveals that fusion of the lightest elements into heavier ones also leads to a net loss in rest mass, and thus to a gain of kinetic energy (as in the combination of hydrogen and neutrons to form helium, previously discussed). Such fusion processes are the main source of energy for the sun and other stars.

These processes involve numerous light elements besides hydrogen and helium. For example, three ^4He atoms can fuse to form one atom of carbon (^{12}C), with a decrease of 7.27 MeV in rest energy. The energy thus released is less than 0.1% of the rest energy, but because of the abundance of helium in the sun, it makes a significant contribution to the sun's energy output. The total power radiated by the sun is about 3×10^{26} W. This means that the sun loses a mass of about 3 million tons/sec.

It is tempting to try to build a fusion reactor that would convert the hydrogen in water to helium. The energy that could thus be obtained from a single cup of water would be sufficient to power a thousand all-electric homes for a month. Unfortunately, it has not yet proven possible to produce the combination of high density and high temperature needed to overcome the mutual repulsion of hydrogen atoms and sustain a reaction that would generate useful power.

C. Decay of Subatomic Particles

Modern physics has discovered a huge number of subatomic particles with variety of properties. Most of these particles are unstable: they decay, being spontaneously transformed into other particles. For example, the muon is a particle like the electron but with a rest mass that is 207 times that of the electron. The negatively charged muon decays into an electron and two neutrinos, with a half-life of about 1.5 μsec in the frame of reference in which the muon is at rest. In any frame of reference in which the muon is moving, the observed half-life is larger by the time dilation factor γ. Thus, for muons moving at a speed of 0.8 c, $\gamma = \frac{5}{3}$, and the half-life is 2.5 μ sec.

Half-lives of moving particles have been observed millions of times, for every known unstable particle, in laboratories all over the world, with results that are in exact agreement with the time dilation formula [Eq (6)]. The moving particle acts just like a moving clock.

It is significant that the decay of the muon is not the result of an electromagnetic force, yet special relativity still governs this decay. Relativity is a theory about space and time, not specifically about electromagnetism, even though it was developed by considering the speed of electromagnetic waves.

D. Antimatter: Pair Production and Annihilation

The development of a theory combining special relativity with quantum mechanics led to the prediction, soon verified, of the existence of antimatter. Antimatter behaves in almost all respects like matter. Each particle of matter has an antiparticle with identical mass and opposite charge. An entire universe made up of antimatter would presumably behave like a mirror image of our own universe, with negatively charged atomic nuclei and positively charged electrons (positrons).

In a striking demonstration of the equivalence of rest mass and energy, a particle of matter and its antiparticle can be simultaneously created from the pure energy of a photon. For example, any photon whose energy exceeds twice the rest energy of an electron is capable of converting that energy into rest mass by creating a positron and an electron. Many radioactive materials emit photons of the requisite energy, and this pair production process is routinely observed when these photons pass through matter. (The process cannot occur in empty space, because the momentum of the pair that is created cannot equal that of the photon. Another body must participate in the process, to absorb the remaining momentum as the photon disappears.)

When a particle meets its antiparticle (e.g., when any electron encounters any positron), the pair can be annihilated, producing pure energy—a photon or photons. This process is also governed by the relativistic laws of conservation of energy and momentum. Thus, unless a third particle participates in the process, annihilation of a pair must produce at least two photons. If the total momentum of the pair is zero (as it must be in some frame of reference), creation of a single photon would violate the law of conservation of momentum. A second photon, emitted in a direction exactly opposite to that of the first photon, permits the total momentum of the two photons to be zero.

Many radioactive materials emit positrons, and the resulting pair annihilation process, with the production of two equal-energy photons going in opposite directions, has been observed countless times, always with results in agreement with the equations of special relativity. Observations of this sort are now routinely made by undergraduate students in laboratory physics courses.

E. Red Shifts and Cosmology

Application of relativity formulas is not confined to the domain of subatomic particles. The frequencies of characteristic radiation from ele

ments in distant galaxies can be analyzed with the aid of the relativistic Doppler shift formula [Eq. (10)]. It is found that the radiation from each element in a distant galaxy is shifted toward the red by an amount that can be used, with the aid of Eq. (10), to compute a speed of recession of the galaxy from the earth.

One cannot use this information to verify the correctness of Eq. (10) on this scale, because there is no independent way to verify that the galaxy is actually receding at the calculated speed. However, the distances of the nearer galaxies can be measured independently, because certain stars. (Cepheid variables) can be seen in these galaxies. These stars oscillate in brightness, with a period that is related to the intrinsic average brightness of the star. Knowing the intrinsic brightness of a star, one can calculate its distance from the observed brightness of that star in the sky. Other methods permit estimates of the distances of more remote galaxies. The estimated distances, in conjunction with the redshift measurements, indicate that the speed of recession of a galaxy is directly proportional to its distance from earth.

From this proportionality it has been concluded that the entire universe is expanding. The concept of an expanding universe has led to theories of cosmology that explain numerous details of the known universe. These include the relative abundances of the elements and the presence of low-energy cosmic background radiation that comes uniformly from all directions in space.

Further discussion of this topic would take us from the special theory of relativity to the realm of the general theory, which deals with the relationship between gravitation and space–time.

Bibliography

Einstein, A. (1961). "Relativity: The Special and the General Theory." Crown Publ., New York.
Jackson, J. D. (1975). "Classical Electrodynamics," 2nd ed., Chapter 11. Wiley, New York.
McGervey, J. D. (1983). "Introduction to Modern Physics," 2nd ed., Chapters 2 and 14. Academic Press, New York.
Purcell, E. M. (1965). "Electricity and Magnetism," Chapters 5 and 6. McGraw-Hill, New York.
Resnick, R., and Halliday, D. (1985). "Basic Concepts in Relativity and Early Quantum Theory," 2nd ed. Wiley, New York.

STATISTICAL MECHANICS

W. A. Wassam, Jr. *Centro de Investigación y de Estudios Avanzados del IPN, Mexico*

I. Concepts of Equilibrium Statistical Mechanics
II. Gibbsian Ensembles and Thermodynamics
III. Information Theory and Statistical Mechanics
IV. Liouville Description of Time Evolution
V. Phenomenological Descriptions of Time Evolution
VI. Irreversible Processes and Microscopic Dynamics

GLOSSARY

Density distribution function: Probability density used to describe the macroscopic state of classical systems.

Density operator: Quantum analog of the density distribution function.

Ensemble: Representation of the macroscopic state of a system that corresponds to a virtual collection of systems, each of which possesses identical thermodynamic properties.

Ensemble average: Average value of a quantity over all members of an ensemble.

Entropy: Measure of the statistical uncertainty in the macroscopic state of a system.

Equilibrium: Macroscopic state of a system for which the properties are constant over the time scale of measurements and independent of the starting time for the measurements.

Equilibrium time correlation function: Time-dependent quantity representing the correlation between dynamical events occurring at different times in a system at equilibrium.

Ergodic hypothesis: Idea that an isolated system of fixed volume, energy, and particle numbers spends an equal amount of time, over a long time, in each of the accessible states of the system.

Fluctuation–dissipation relations: Mathematical relations connecting the dissipative behavior of systems to spontaneous equilibrium fluctuations.

Generalized susceptibility: Fourier–Laplace transform of a response function.

Maximum entropy principle: Principle requiring the density distribution function or density operator to be the maximum entropy distribution consistent with all available information defining the macroscopic state of a system.

Principle of equal a priori probabilities: Idea that all accessible states are equally probable for an isolated system of fixed volume, energy, and particle numbers.

Response function: Equilibrium time correlation function representing the response of a system to an instantaneous force exerted on that system.

Statistical mechanics is a branch of physics that utilizes statistical methods in conjunction with the basic principles of classical and quantum mechanics to provide a theoretical framework for understanding the equilibrium and nonequilibrium properties of macroscopic systems. The principal goals of statistical mechanics include (1) constructing mathematical relationships connecting the different properties of a system, (2) relating the properties of a system to the interactions between the particles constituting that system, and (3) developing a sound theory of irreversible processes that automatically includes conventional equilibrium thermodynamics as a limiting case. Usually statistical mechanics is divided into two parts, namely, equilibrium and nonequilibrium statistical mechanics. Equilibrium statistical mechanics is concerned with the time-independent properties of systems, whereas nonequilibrium statistical mechanics is concerned with the time-dependent properties of systems and their time evolution between stationary states.

I. Concepts of Equilibrium Statistical Mechanics

A. Notion of Equilibrium

The macroscopic state of a system can be characterized in terms of its properties, that is, the external parameters defining the system and the values of experimentally measurable attributes of that system. These properties may include volume, pressure, stress, magnetization, polarization, number of particles, and so on.

In performing measurements of the properties of systems, experimentalists often find that the values of the measured properties are constant over the time scale $\Delta t = t - t_0$ of their measurements and independent of the starting time t_0 for the measurements. When such results are obtained, the system of interest is said to be in the state of equilibrium. Of course, we know from microscopic considerations that the properties of a system that are not fixed by external restraints must be a function of time. Thus, the appearance of equilibrium must be due to the smoothing out of fluctuations in the properties of a system on the time scale of macroscopic measurements.

B. Concept of Ensembles

In order to describe the macroscopic state of a system, we introduce the idea of a representative ensemble, first introduced by Gibbs. By an ensemble we mean a virtual collection of a large number $\mathcal{N}$ of independent systems, each of which possesses identical thermodynamic properties. Although all the members of the ensemble are identical from a thermodynamic point of view, they can differ on a microscopic scale. The ensemble corresponding to a given system is said to be a representation of the macroscopic state of that system.

The specification of the representative ensemble of a given system requires us to enumerate the external parameters defining the system and the integrals of motion characterizing that system. According to Gibbs, the macroscopic state corresponding to equilibrium depends only on single-valued additive integrals of motion. By additive, we mean that the integrals of motion are additively composed of the integrals of motion of the subsystems making up the system. As an example, consider a system of fixed energy E, number of particles N, and volume V. This system is energetically isolated and closed to

particle transfer across its boundaries. For such a system, there are four integrals of motion, namely, the total energy E, the total linear momentum $\vec{P}$, the total angular momentum $\vec{L}$, and the total number of particles N. Thus, we expect the macroscopic state of the system to depend on E, $\vec{P}$, $\vec{L}$, N, and V. If the overall system is motionless, $|\vec{P}| = |\vec{L}| = 0$. For this case, the macroscopic state depends only on E, N, and V. This macroscopic state is said to be represented by the microcanonical ensemble (see Section II,A).

The ensemble average of a property of a system is defined as the average value of this property over all members of the ensemble. In usual formulations of equilibrium statistical mechanics, it is postulated that the ensemble average of a property of a system is equivalent to the corresponding thermodynamic property. (For example, the ensemble average of the energy is equivalent to the thermodynamic or internal energy of a system.) In adopting this postulate, one is asserting that the long time average of a property of a system in equilibrium is equal to the ensemble average of this property in the limit $\mathcal{N} \to \infty$, provided that the members of the ensemble replicate the thermodynamic state of the actual system.

C. Macroscopic State of Classical Systems

The dynamical state of a classical particle l is defined by its coordinate (position) $\vec{q}_l$ and momentum $\vec{p}_l$. The state of a collection of N particles is defined by specifying the collective coordinate $\mathbf{q}^N = (\vec{q}_1, \ldots, \vec{q}_N)$ and collective momentum $\mathbf{p}^N = (\vec{p}_1, \ldots, \vec{p}_N)$. Alternatively, we can define the state of a collection of N particles by specifying the phase point $\Gamma^N = (\mathbf{p}^N, \mathbf{q}^N)$ of the system in a $6N$-dimensional space called phase space.

The ensemble representing the macroscopic state of a classical system is described by an object called the density distribution function $\rho(\mathbf{p}^N, \mathbf{q}^N)$, which is defined in such a way that

$$dP\,(\mathbf{p}^N, \mathbf{q}^N) = d\mathbf{p}^N\,d\mathbf{q}^N\,\rho(\mathbf{p}^N, \mathbf{q}^N) \qquad (1)$$

represents the probability of finding the system in the phase space volume element $d\mathbf{p}^N\,d\mathbf{q}^N$ in the neighborhood of the phase point $(\mathbf{p}^N, \mathbf{q}^N)$. Since the sum of the probabilities must equal 1, the density distribution function is normalized to unity, that is,

$$\int d\mathbf{p}^N \int d\mathbf{q}^N\,\rho(\mathbf{p}^N, \mathbf{q}^N) = 1 \qquad (2)$$

Since classical mechanics corresponds to a limit of quantum mechanics, another normalization of the density distribution function is usually adopted. For a system of N identical particles, this normalization is written

$$\int d\Gamma^N \rho(\mathbf{p}^N, \mathbf{q}^N) = 1 \qquad (3)$$

where

$$d\Gamma^N = d\mathbf{p}^N \, d\mathbf{q}^N / N! h^{3N} \qquad (4)$$

is a dimensionless volume element in phase space, with h denoting Planck's constant. The motivation for introducing this normalization is based on the idea that the normalization should reflect Heisenberg's uncertainty principle and the indistinguishability of identical particles in quantum mechanics. The rationale for adopting the explicit form for $d\Gamma^N$ given by Eq. (4) is discussed below.

In view of Heisenberg's uncertainty principle, there is a minimum volume h^3 of phase space that may be associated with a single particle. This minimum volume is called a phase cell. In the phase space of N particles, the volume of a phase cell is h^{3N}. Thus, h^{3N} is a natural unit of volume in phase space for a system of N particles.

The indistinguishability of identical particles is accounted for in quantum mechanics by requiring the state of a system to be invariant to the permutation of particle coordinates. This invariance must be preserved in the classical limit. For a quantum system of N identical particles, there are $N!$ configurations (permutations) corresponding to a given state. Thus, there are $N!$ configurations for a given classical state. (Actually, the situation is more complicated than this. Our discussion corresponds to a limiting form of both Bose–Einstein and Fermi–Dirac statistics called classical or Boltzmann statistics, which is valid only when the number of states available to a system is very large compared with N.)

Introducing the volume unit h^{3N} and the indistinguishability of identical particles, we arrive at the normalization given by Eq. (3). The integration over Γ^N is just a summation over the different states of the classical system. For a system composed of m different kinds of particles, we replace $d\Gamma^N$ by

$$d\Gamma^N = \prod_{j=1}^{m} d\Gamma_j^{N_j} \qquad (5)$$

where

$$d\Gamma_j^{N_j} = d\mathbf{p}_j^{N_j} \, d\mathbf{q}_j^{N_j} / N_j! h^{3N_j} \qquad (6)$$

with $\mathbf{p}_j^{N_j}$ and $\mathbf{q}_j^{N_j}$, respectively, denoting the col-

lective momentum and collective coordinate for type j particles of which there are N_j.

Given the density distribution function $\rho(\mathbf{p}^N, \mathbf{q}^N)$, we can compute the ensemble average or equivalently the average value of any classical dynamical variable $O(\mathbf{p}^N, \mathbf{q}^N)$ associated with a classical system by using the relation

$$\langle O \rangle = \int d\Gamma^N O(\mathbf{p}^N, \mathbf{q}^N) \rho(\mathbf{p}^N, \mathbf{q}^N) \qquad (7)$$

where the angular brackets denote an ensemble average. For example, the average energy of a classical system is given by

$$\langle \mathcal{H} \rangle = \int d\Gamma^N \mathcal{H}(\mathbf{p}^N, \mathbf{q}^N) \rho(\mathbf{p}^N, \mathbf{q}^N) \qquad (8)$$

where $\mathcal{H}(\mathbf{p}^N, \mathbf{q}^N)$ is the classical Hamiltonian of that system.

The Gibbs entropy associated with the ensemble described by $\rho(\mathbf{p}^N, \mathbf{q}^N)$ is defined by

$$S = -k_B \int d\Gamma^N \rho(\mathbf{p}^N, \mathbf{q}^N) \ln \rho(\mathbf{p}^N, \mathbf{q}^N) \qquad (9)$$

where k_B is Boltzmann's constant. This definition may be viewed as another postulate of equilibrium statistical mechanics for the case of classical systems.

D. MACROSCOPIC STATE OF QUANTUM SYSTEMS

In quantum mechanics the state of an N-particle system is described by a state vector $|\psi\rangle$ in Hilbert space. The wave function $\psi(\mathbf{q}^N)$ corresponding to the state $|\psi\rangle$ is defined by the inner product $\psi(\mathbf{q}^N) \equiv \langle \mathbf{q}^N | \psi \rangle$, where $\langle \mathbf{q}^N |$ is the adjoint of the collective coordinate state $|\mathbf{q}^N\rangle$, with $\mathbf{q}^N = (\vec{q}_1, \dots, \vec{q}_N)$ denoting the collective coordinate for the N particles.

The eigenstates $\{|\psi_m\rangle\}$ and eigenvalues $\{E_m\}$ of the quantum Hamiltonian $\mathcal{H}$ of an N-particle system are obtained by solving the equation

$$\mathcal{H}|\psi_m\rangle = E_m|\psi_m\rangle \qquad (10)$$

For a system of N identical particles of mass m in which the interaction between the particles is pairwise and centrally symmetric, the Hamiltonian $\mathcal{H}$ can be written

$$\mathcal{H} = \sum_{i=1}^{N} \frac{\hat{\vec{p}}_i \cdot \hat{\vec{p}}_i}{2m} + \frac{1}{2} \sum_{i,j=1}^{N} {}' \hat{U}_{ij}(|\hat{\vec{q}}_i - \hat{\vec{q}}_j|) \qquad (11)$$

where $\hat{\vec{p}}_i$ and $\hat{\vec{q}}_i$, respectively, denote the momentum and coordinate vector operators for particle i, and $\hat{U}_{ij}$ is the interaction between particles i and j.

Equation (10) assumes the following form in the coordinate representation,

$$\mathcal{H}(\mathbf{q}^N)\psi_m(\mathbf{q}^N) = E_m\psi_m(\mathbf{q}^N) \qquad (12)$$

where $\psi_m(\mathbf{q}^N) = \langle \mathbf{q}^N | \psi_m \rangle$ is an eigenfunction of the differential operator $\mathcal{H}(\mathbf{q}^N)$ corresponding to the quantum Hamiltonian $\mathcal{H}$ in the coordinate representation. The explicit form of $\mathcal{H}(\mathbf{q}^N)$ for the case of Eq. (11) is given by

$$\mathcal{H}(\mathbf{q}^N) = -\sum_{i=1}^{N} \frac{\hbar^2}{2m} \nabla^2_{q_i} + \frac{1}{2} \sum_{i,j=1}^{N} {}' \, U_{ij}(|\vec{q}_i - \vec{q}_j|) \tag{13}$$

where $\nabla^2_{q_i}$ is the Laplacian operator for particle i.

In order to provide a proper description of the N-particle system, it is necessary to specify its spin state as well as energy state. Assuming that $\mathcal{H}(\mathbf{q}^N)$ does not mix the spin states, we should write Eq. (12) as

$$\mathcal{H}(\mathbf{q}^N)\psi_{m,\sigma}(\mathbf{q}^N) = E_{m,\sigma}\psi_{m,\sigma}(\mathbf{q}^N) \tag{14}$$

where the index σ has been introduced to specify the spin. Not all the eigenfunctions of Eq. (14) are permissible. Only those eigenfunctions satisfying certain symmetry requirements are allowed. More specifically, the eigenfunctions for a system of particles with zero or integral/half-integral spin (in multiples of $\hbar = h/2\pi$) are required to be symmetric/antisymmetric with respect to the interchange of particle coordinates. The particles are said to obey Bose–Einstein statistics for the case of zero or integral spin. Otherwise, the particles obey Fermi–Dirac statistics.

One can partition the Hamiltonian $\mathcal{H}(\mathbf{q}^N)$ as follows,

$$\mathcal{H}(\mathbf{q}^N) = \mathcal{H}^{(0)}(\mathbf{q}^N) + U(\mathbf{q}^N) \tag{15}$$

where

$$\mathcal{H}^{(0)}(\mathbf{q}^N) = -\sum_{i=1}^{N} \frac{\hbar^2}{2m} \nabla^2_{q_i} \tag{16}$$

describes the free motion of the particles and

$$U(\mathbf{q}^N) = \frac{1}{2} \sum_{i,j=1}^{N} {}' \, U_{ij}(|\vec{q}_i - \vec{q}_j|) \tag{17}$$

represents the interaction between the particles.

The eigenfunctions $\{\psi_{m,\sigma}(\mathbf{q}^N)\}$ of the Hamiltonian $\mathcal{H}(\mathbf{q}^N)$ can be constructed by taking linear combinations of the eigenfunctions $\{\psi^{(0)}_{l,\sigma}(\mathbf{q}^N)\}$ of $\mathcal{H}^{(0)}(\mathbf{q}^N)$:

$$\psi_{m,\sigma}(\mathbf{q}^N) = \sum_{l} C_{l,\sigma;m,\sigma} \phi^{(0)}_{l,\sigma}(\mathbf{q}^N) \tag{18}$$

From this equation, we see that the eigenfunctions $\{\psi_{m,\sigma}(\mathbf{q}^N)\}$ of $\mathcal{H}(\mathbf{q}^N)$ will possess the proper symmetry when the eigenfunctions $\{\phi^{(0)}_{l,\sigma}(\mathbf{q}^N)\}$ of $\mathcal{H}^{(0)}(\mathbf{q}^N)$ possess the proper symmetry.

For the case of Bose–Einstein statistics, the eigenfunctions $\{\psi_{m,\sigma}(\mathbf{q}^N)\}$ are required to be symmetric with respect to particle interchange. Thus, the wave functions $\{\phi^{(0)}_{l,\sigma}(\mathbf{q}^N)\}$ must be symmetric. These symmetric wave functions can be written

$$\phi^{(0)}_{l,\sigma}(\mathbf{q}^N) = N!^{-1/2} \sum_{\nu} \mathcal{P}_\nu$$
$$\times \, [\chi_{l_1,\sigma_1}(\vec{q}_1) \cdots \chi_{l_N,\sigma_N}(\vec{q}_N)] \tag{19}$$

where $\mathcal{P}_\nu$ is a permutation operator that permutes particle coordinates, the index ν runs over the $N!$ possible permutations of particle coordinates, and $\chi_{l_l,\sigma_l}(\vec{q}_l)$ denotes a one-particle state occupied by particle l. The one-particle states $\{\chi_{l_l,\sigma_l}\}$ are solutions of

$$-(\hbar^2/2m) \, \nabla^2_{q_l} \, \chi_{l_l,\sigma_l}(\vec{q}_l) = E^{(0)}_{l_l,\sigma_l}\chi_{l_l,\sigma_l}) \tag{20}$$

For the case of Fermi–Dirac statistics, the eigenfunctions $\{\psi_{m,\sigma}(\mathbf{q}^N)\}$ are required to be antisymmetric with respect to particle exchange. Thus, the wave functions $\{\phi^{(0)}_{l,\sigma}(\mathbf{q}^N)\}$ must be antisymmetric. These antisymmetric wave functions can be written

$$\phi^{(0)}_{l,\sigma}(\mathbf{q}^N) = N!^{-1/2} \sum_{\nu} \varepsilon_\nu \mathcal{P}_\nu$$
$$\times \, [\chi_{l_1,\sigma_1}(\vec{q}_1) \cdots \chi_{l_N,\sigma_N}(\vec{q}_N)] \tag{21}$$

where $\varepsilon_\nu = +1/-1$ for an even/odd permutation of particle coordinates.

In quantum mechanics the dynamical variables are represented by linear Hermitian operators $\hat{A}$ that operate on the state vectors in Hilbert space. The spectra of these operators determine the possible values of the physical quantities that they represent. Unlike classical systems, specifying the state $|\psi\rangle$ of a quantum system does not necessarily imply the exact knowledge of the value of a dynamical variable. Only for cases in which the system is in an eigenstate of a dynamical variable will knowledge of that state $|\psi\rangle$ provide an exact value of the dynamical variable. Otherwise, only the quantum average of the dynamical variable can be determined. The quantum average of the dynamical variable $\hat{A}$ is given by

$$\langle \hat{A} \rangle \equiv \langle \psi | \hat{A} | \psi \rangle \tag{22a}$$

$$= \int d\mathbf{q}^{\mathbf{N}} \, \psi^*(\mathbf{q}^{\mathbf{N}}) A(\mathbf{q}^{\mathbf{N}}) \psi(\mathbf{q}^{\mathbf{N}}) \tag{22b}$$

where $A(\mathbf{q}^N)$ is defined by

$$\langle \mathbf{q}^N | \hat{A} | \mathbf{q}'^N \rangle = A(\mathbf{q}^N)\, \delta(\mathbf{q}^N - \mathbf{q}'^N) \qquad (23)$$

with $\delta(\mathbf{q}^N - \mathbf{q}'^N)$ denoting a Dirac delta function. For the sake of simplicity we have neglected any possible spin dependence of the quantum average.

The quantum average $\langle \hat{A} \rangle$ of the dynamical variable $\hat{A}$ is actually a conditional average; that is, it is conditional on the fact that the system of interest has been prepared in the state $|\psi\rangle$. In general, we lack sufficient information to specify the state of a system. Thus, additional probabilistic concepts must be introduced that are not inherent in the formal structure of quantum mechanics.

Suppose we know the set of probabilities $\{P(\psi)\}$ for observing a system in the different states $\{|\psi\rangle\}$. Then the average value of the dynamic variable $\hat{A}$ is given by the weighted quantum average

$$\langle \hat{A} \rangle = \sum_{\psi} \langle \hat{A} \rangle_{\psi} P(\psi) \qquad (24)$$

In order to compute the weighted quantum average, we need something to provide us with information about the probabilities $\{P(\psi)\}$. This task is fulfilled by a Hermitian operator $\hat{\rho}$ called the density operator. The density operator $\hat{\rho}$ is the quantum analog of the density distribution function ρ used to describe the macroscopic state of classical systems. Thus, we say that the macroscopic state of quantum systems is described by the density operator $\hat{\rho}$.

The density operator $\hat{\rho}$ possesses the following properties. (1) It is a Hermitian operator, that is, $\hat{\rho} = \hat{\rho}^\dagger$; (2) the trace of $\hat{\rho}$ is unity, that is, $\text{Tr}\,\hat{\rho} = 1$; (3) the diagonal matrix elements of $\hat{\rho}$ in any representation are positive definite, that is, $\langle \alpha | \hat{\rho} | \alpha \rangle \geq 0$; and (4) the probability of finding a system in some state $|\alpha\rangle$ is given by the diagonal matrix element $\rho(\alpha, \alpha) = \langle \alpha | \hat{\rho} | \alpha \rangle$.

The eigenvectors $\{|\psi\rangle\}$ and eigenvalues $\{P(\psi)\}$ of $\hat{\rho}$ are determined by solving the equation

$$\hat{\rho} |\psi\rangle = P(\psi) |\psi\rangle \qquad (25)$$

where

$$P(\psi) = \langle \psi | \hat{\rho} | \psi \rangle \qquad (26)$$

is the probability of finding the system in the state $|\psi\rangle$. In the ψ representation, we write $\hat{\rho}$ as follows:

$$\hat{\rho} = \sum_{\psi'} |\psi'\rangle P(\psi') \langle \psi'| \qquad (27)$$

Although the density operator $\hat{\rho}$ is diagonal in the ψ representation, this may not be the case in any other representation. In other representations there may be nonvanishing off-diagonal matrix elements. For example, we have in the coordinate representation

$$\rho(\mathbf{q}^N, \mathbf{q}'^N) = \sum_{\psi} \psi(\mathbf{q}^N) P(\psi) \psi^*(\mathbf{q}'^N) \qquad (28)$$

where the wave function $\psi(\mathbf{q}^N) = \langle q^N | \psi \rangle$ is the probability amplitude for finding the system with the collective coordinate $\mathbf{q}^N$ given that the system is in the state $|\psi\rangle$. Clearly, $\rho(\mathbf{q}^N, \mathbf{q}^N)$ tells us the probability of finding the system with the collective coordinate $\mathbf{q}^N$. The off-diagonal matrix element $\rho(\mathbf{q}^N, \mathbf{q}'^N)$ tells us something about the phase coherence between the collective coordinate states $|\mathbf{q}^N\rangle$ and $|\mathbf{q}'^N\rangle$, that is, the interference between the probability amplitudes $\{\psi(\mathbf{q}^N)\}$ and $\{\psi^*(\mathbf{q}'^N)\}$.

The macroscopic state described by the density operator $\hat{\rho}$ may be either a pure state or a mixed state. This classification scheme is based on the character of the trace $\text{Tr}\,\hat{\rho}^2$. For pure states

$$\text{Tr}\,\hat{\rho}^2 = 1 \qquad (29)$$

whereas for mixed states

$$\text{Tr}\,\hat{\rho}^2 < 1 \qquad (30)$$

The criterion for a pure state can be satisfied only if the eigenvalues of $\hat{\rho}$ satisfy the relation

$$P(\psi') = \delta_{\psi', \psi} \qquad (31)$$

which implies that the system of interest has been prepared in some quantum state $|\psi\rangle$. Thus, the density operator $\hat{\rho}$ corresponding to a pure state is of the form

$$\hat{\rho} = |\psi\rangle\langle\psi| \qquad (32)$$

If the density operator for the system is not of this form, the criterion for a mixed state will be fulfilled, and thus the density operator will be of the form given by Eq. (27) with a spread in the eigenvalues $\{P(\psi')\}$.

Given the density operator $\hat{\rho}$, we can compute the ensemble average or equivalently the average value of any quantum dynamical variable $\hat{O}$ by using the relations

$$\langle \hat{O} \rangle = \text{Tr}\,\hat{O}\hat{\rho} \qquad (33a)$$

$$= \sum_{\psi} \langle \hat{O} \rangle_{\psi} P(\psi) \qquad (33b)$$

Equation (33a) is more useful than Eq. (33b) for computing the ensemble average $\langle \hat{O} \rangle$. The latter equation requires the determination of the eigenstates of $\hat{\rho}$, which is by no means a trivial task. Since the trace in Eq. (33a) is canonically invariant, we can evaluate this equation by employing any representation satisfying the proper symmetry requirements and any other restraint placed on the system. Note that Eq. (33b) assumes the form of a quantum average when the system has been prepared in a pure state.

The Gibbs entropy associated with the ensemble described by $\hat{\rho}$ is defined by

$$S = -k_B \, \text{Tr} \, \hat{\rho} \ln \hat{\rho} \tag{34}$$

This definition of the entropy is the quantum analog of the definition given for classical systems. Both the quantum and classical forms for the entropy S involve summations over the accessible states of a system. In the representation for which $\hat{\rho}$ is diagonal, the Gibbs entropy S can be written

$$S = -k_B \sum_\psi P(\psi) \ln P(\psi) \tag{35}$$

From Eq. (35), we find that the entropy S associated with a pure state vanishes. This is due to the lack of any statistical indeterminancy in the state of a system prepared in a pure state. For the case of a mixed state, the entropy is nonzero. This reflects the fact that there is some statistical indeterminancy in the description of a system prepared in a mixed state.

II. Gibbsian Ensembles and Thermodynamics

A. Microcanonical Ensemble

1. Classical Systems

Consider an energetically isolated (adiabatic) system composed of a fixed number of particles **N** (closed) and enclosed in a motionless container of fixed volume V. (For the case of a multicomponent system, we use the symbol **N** to represent the collection of particle numbers $\{N_j\}$ for the components of the system. Hereafter, this should be understood.) The energy of the system, denoted by $\mathcal{H}(\mathbf{p}^N, \mathbf{q}^N)$ (classical Hamiltonian), lies in an energy shell of width ΔE, that is, $E \leq \mathcal{H}(\mathbf{p}^N, \mathbf{q}^N) \leq E + \Delta E$. The macroscopic state of this system is called a microcanonical ensemble.

The density distribution function associated with a classical microcanonical ensemble is given by

$$\rho(\mathbf{p}^N, \mathbf{q}^N) = \Omega^{-1}(E, V, \mathbf{N}; \Delta E) \tag{36}$$

for the phase points lying inside the energy shell $E \leq \mathcal{H}(\mathbf{p}^N, \mathbf{q}^N) \leq E + \Delta E$. Outside the energy shell, $\rho(\mathbf{p}^N, \mathbf{q}^N)$ vanishes. In the above, $\Omega(E, V, \mathbf{N}; \Delta E)$ is the number of states inside the energy shell, that is,

$$\Omega(E, V, \mathbf{N}; \Delta E) = \int d\Gamma^N \tag{37}$$
$$E \leq \mathcal{H}(\mathbf{p}^N, \mathbf{q}^N) \leq E + \Delta E$$

According to Eq. (36), all states inside the energy shell are equally probable. In other words, the microcanonical ensemble represents a uniform distribution over the phase points belonging to the energy shell. Such a distribution is often referred to as a mathematical statement of the principle of equal *a priori* probabilities. The implication of the hypothesis that the microcanonical distribution represents the macroscopic state of a closed, isolated system is that such a system spends an equal amount of time, over a long time period, in each of the available classical states. This statement is called the ergodic hypothesis.

The Gibbs entropy of the microcanonical ensemble described by Eq. (36) is given by

$$S = k_B \ln \Omega(E, V, \mathbf{N}; \Delta E) \tag{38}$$

which corresponds to Boltzmann's definition of entropy. Note that this form for the entropy increases as the number of accessible states increases. This behavior of the Boltzmann entropy has led to such qualitative definitions of entropy as "measure of randomness" and "measure of disorder." The microcanonical distribution is the distribution of maximum Gibbs entropy consistent with knowing nothing about the system except the specified values of the external parameters E, V, and **N**.

The entropy of the microcanonical ensemble depends on the variables E, **N**, and V. Thus, the differential dS of S is written

$$dS = \left(\frac{\partial S}{\partial E}\right)_{V,\mathbf{N}} dE + \left(\frac{\partial S}{\partial V}\right)_{E,\mathbf{N}} dV$$
$$+ \sum_l \left(\frac{\partial S}{\partial N_l}\right)_{E,V,\mathbf{N}'} dN_l \tag{39}$$

where the index l runs over the components of the system and

$$\left(\frac{\partial S}{\partial E}\right)_{V,\mathbf{N}} = k_B \left[\frac{\partial}{\partial E} \ln \Omega(E, V, \mathbf{N}; \Delta E)\right]_{V,\mathbf{N}} \tag{40}$$

$$\left(\frac{\partial S}{\partial V}\right)_{E,\mathbf{N}} = k_B \left[\frac{\partial}{\partial V} \ln \Omega(E, V, \mathbf{N}; \Delta E)\right]_{E,\mathbf{N}} \tag{41}$$

$$\left(\frac{\partial S}{\partial N_l}\right)_{E,V,\mathbf{N}'} = k_\mathrm{B}\left[\frac{\partial}{\partial N_l} \ln \Omega(E, V, \mathbf{N}; \Delta E)\right]_{E,V,\mathbf{N}'} \tag{42}$$

In Eqs. (39) and (42), we use the symbol $\mathbf{N}'$ to indicate that the particle numbers $\mathbf{N}' = \{N_j, j \neq l\}$ are fixed for all components of the system except for the particle number N_l for the lth component. (Hereafter, this notation will be adopted.)

In thermodynamics the entropy S is the characteristic function for the variables E, V, and $\mathbf{N}$,

$$dS = \frac{1}{T} dE + \frac{P}{T} dV - \sum_l \frac{\mu_l}{T} dN_l \tag{43}$$

where

$$\frac{1}{T} = (\partial S/\partial E)_{V,\mathbf{N}} \tag{44}$$

$$\frac{P}{T} = (\partial S/\partial V)_{E,\mathbf{N}} \tag{45}$$

$$\mu_l/T = -(\partial S/\partial N_l)_{E,V,\mathbf{N}'} \tag{46}$$

with T, P, and μ_l denoting the temperature, pressure, and chemical potential, respectively.

Assuming that the Gibbs entropy for the microcanonical ensemble is the same as the thermodynamic entropy, Eqs. (40)–(42) and (44)–(46) can be utilized to contruct statistical mechanical expressions for the thermodynamic parameters T, P, and μ_l:

$$1/T = k_\mathrm{B}[\partial \ln \Omega(E, V, \mathbf{N}; \Delta E)/\partial E_{V,\mathbf{N}} \tag{47}$$

$$P/T = k_\mathrm{B}[\partial \ln \Omega(E, V, \mathbf{N}; \Delta E)/\partial V]_{E,\mathbf{N}} \tag{48}$$

$$\mu_l/T = -k_\mathrm{B}[\partial \ln \Omega(E, V, \mathbf{N}; \Delta E)/\partial N_l]_{E,V,\mathbf{N}'} \tag{49}$$

In principle, these results can be employed to compute T, P, and μ_l. In practice, however, this is extremely difficult because the number of states $\Omega(E, V, \mathbf{N}; \Delta E)$ as a function of E, V, and $\mathbf{N}$ is usually not available.

2. Quantum Systems

As for classical systems, the microcanonical ensemble is used to characterize the macroscopic state of energetically isolated and closed quantum systems of fixed volume. The quantum microcanonical ensemble is described by the density operator

$$\hat{\rho} = \Omega^{-1}(E, V, \mathbf{N}; \Delta E)\hat{\mathscr{P}} \tag{50}$$

where $\Omega(E, V, \mathbf{N}; \Delta E)$ is the number of eigenstates $\{|\psi_l\rangle\}$ of the system Hamiltonian $\hat{\mathscr{H}}$ lying in the energy shell $(E, E + \Delta E)$ of width ΔE and $\hat{\mathscr{P}}$ is a projection operator defined by

$$\hat{\mathscr{P}} = \sum_{l \in \Omega} |\psi_l\rangle\langle\psi_l| \tag{51}$$

with the index l enumerating the eigenstates belonging to the energy shell.

The probability $P(\psi_m)$ of finding a system described by the microcanonical density operator $\hat{\rho}$ in a given eigenstate $|\psi_m\rangle$ of $\hat{\mathscr{H}}$ is given by the diagonal matrix element $\langle\psi_m|\hat{\rho}|\psi_m\rangle$. For an eigenstate $|\psi_m\rangle$ lying in the energy shell $(E, E + \Delta E)$, $P(\psi_m)$ is given by

$$P(\psi_m) = \Omega^{-1}(E, V, \mathbf{N}; \Delta E) \tag{52}$$

Otherwise, $P(\psi_m)$ vanishes. From these results, we see that the quantum microcanonical ensemble describes a uniform distribution over the eigenstates of the system Hamiltonian $\hat{\mathscr{H}}$ lying inside the energy shell $(E, E + \Delta E)$. Outside the energy shell, the distribution vanishes.

The Gibbs entropy of the quantum microcanonical ensemble is given by Eq. (38), except that $\Omega(E, V, \mathbf{N}; \Delta E)$ must now be interpreted as the number of eigenstates of the system Hamiltonian lying in the energy shell. Aside from this, the Gibbs entropies of the quantum and classical microcanonical ensembles are identical in form. As with the microcanonical distribution function ρ for classical systems, the microcanonical density operator $\hat{\rho}$ for quantum systems represents the distribution of maximum Gibbs entropy consistent with knowing nothing about the system except the specified values of the external parameters E, V, and $\mathbf{N}$.

In view of the formal identity of the expressions for the Gibbs entropy of the quantum and classical microcanonical ensembles, the statistical mechanical expressions given by Eqs. (47)–(49) for the thermodynamic parameters T, P, and μ_l associated with classical systems also apply to quantum systems. One need only reinterpret the quantity $\Omega(E, V, \mathbf{N}; \Delta E)$ appearing in Eqs. (47)–(49) as the number of eigenstates of the system Hamiltonian $\hat{\mathscr{H}}$ lying in the energy shell $(E, E + \Delta E)$.

B. Canonical Ensemble

1. Classical Systems

Consider a system composed of a fixed number of particles $\mathbf{N}$ enclosed in a motionless container of fixed volume V. The temperature of the system is maintained at temperature T by keeping the system in thermal contact with a large heat bath (at temperature T) with which it is able to exchange energy. Such a system is called a closed, isothermal (fixed-T) system. The macroscopic state of this system is called the canonical ensemble.

The density distribution function associated with a canonical ensemble is given by

$$\rho(\mathbf{p^N}, \mathbf{q^N})$$
$$= Z^{-1} (V, \mathbf{N}, \beta) \exp[-\beta \mathscr{H}(\mathbf{p^N}, \mathbf{q^N})] \qquad (53)$$

where β is a macroscopic parameter to be determined and Z is the canonical partition function

$$Z(V, \mathbf{N}, \beta) = \int d\mathbf{\Gamma^N} \exp[-\beta \mathscr{H}(\mathbf{p^N}, \mathbf{q^N})] \quad (54)$$

The canonical partition function Z depends on the three macroscopic parameters V, $\mathbf{N}$, and β.

The Gibbs entropy of the canonical distribution given in Eq. (53) is written

$$S = k_B \beta \langle \mathscr{H} \rangle + k_B \ln Z(V, \mathbf{N}, \beta) \qquad (55)$$

where $\langle \mathscr{H} \rangle$ denotes the average energy of the system. The canonical distribution is the distribution of maximum Gibbs entropy consistent with the given average energy $\langle \mathscr{H} \rangle$ and the specified values of the external parameters $\mathbf{N}$ and V.

The formal expression given by Eq. (55) for the Gibbs entropy of a canonical ensemble can be rearranged to read

$$\langle \mathscr{H} \rangle = -\beta^{-1} \ln Z(V, \mathbf{N}, \beta) + (k_B \beta)^{-1} S \quad (56)$$

This result resembles the thermodynamic relation

$$E = A + TS \qquad (57)$$

where E is the thermodynamic energy (internal energy), A the Helmholtz free energy, T the temperature, and S the thermodynamic entropy.

Assuming that the Gibbs entropy of the canonical ensemble is equivalent to the thermodynamic entropy and that the average energy $\langle \mathscr{H} \rangle$ is equivalent to the internal energy E, the following correspondences can be made:

$$A(V, \mathbf{N}, T) = -\beta^{-1} \ln Z(V, \mathbf{N}, \beta) \qquad (58)$$
$$\beta = 1/k_B T \qquad (59)$$

The latter correspondence [Eq. (59)] provides us with a relationship between the macroscopic parameter β and the temperature T. The result given by Eq. (58) connects the Helmholtz free energy with the underlying microscopic nature of a system through the canonical partition function Z, which depends on the classical Hamiltonian $\mathscr{H}(\mathbf{p^N}, \mathbf{q^N})$.

In thermodynamics the Helmholtz free energy A is the characteristic function for the variables V, $\mathbf{N}$, and T,

$$dA = -S \, dT - P \, dV + \sum_l \mu_l \, dN_l \qquad (60)$$

where

$$S = -(\partial A/\partial T)_{V,\mathbf{N}} \qquad (61)$$
$$P = -(\partial A/\partial V)_{T,\mathbf{N}} \qquad (62)$$
$$\mu_l = (\partial A/\partial N_l)_{V,T,\mathbf{N'}} \qquad (63)$$

Utilizing the expression given by Eq. (58) for the Helmholtz free energy A and the thermodynamic relations given by Eqs. (61)–(63), one can construct the following statistical mechanical relations for the entropy S, pressure P, and chemical potential μ_l:

$$S = k_B T[\partial \ln Z(V, \mathbf{N}, \beta)/\partial T]_{V,\mathbf{N}}$$
$$+ k_B \ln Z(V, \mathbf{N}, \beta) \qquad (64)$$
$$P = k_B T[\partial \ln Z(V, \mathbf{N}, \beta)/\partial V]_{T,\mathbf{N}} \qquad (65)$$
$$\mu_l = -k_B T[\partial \ln Z(V, \mathbf{N}, \beta)/\partial N_l]_{V,T,\mathbf{N'}} \qquad (66)$$

An expression for the average energy $\langle \mathscr{H} \rangle$ of a closed, isothermal system can be obtained by differentiating $\ln Z$ with respect to β. The result is

$$\langle \mathscr{H} \rangle = -[\partial \ln Z(V, \mathbf{N}, \beta)/\partial \beta]_{V,\mathbf{N}} \qquad (67)$$

Differentiating $\ln Z$ a second time with respect to β gives us expressions for the energy fluctuation in the system:

$$\langle (\mathscr{H} - \langle \mathscr{H} \rangle)^2 \rangle = \langle \mathscr{H}^2 \rangle - \langle \mathscr{H} \rangle^2 \qquad (68a)$$
$$= [\partial^2 \ln Z(V, \mathbf{N}, \beta)/\partial \beta^2]_{V,\mathbf{N}} \quad (68b)$$
$$= -(\partial \langle \mathscr{H} \rangle/\partial \beta)_{V,\mathbf{N}} \qquad (68c)$$

This result allows us to connect the heat capacity $C_V = (\partial \langle \mathscr{H} \rangle/\partial T)_{V,\mathbf{N}}$ at constant volume to the energy fluctuation and microscopic interactions through the relations

$$C_V = k_B \beta^2 \langle (\mathscr{H} - \langle \mathscr{H} \rangle)^2 \rangle \qquad (69a)$$
$$= k_B \beta^2 [\partial^2 \ln Z(V, \mathbf{N}, \beta)/\partial \beta^2]_{V,\mathbf{N}} \quad (69b)$$

2. Quantum Systems

As for classical systems, the canonical ensemble is used to characterize the macroscopic state of closed, isothermal quantum systems. The quantum canonical ensemble is described by the density operator

$$\hat{\rho} = Z^{-1}(V, \mathbf{N}, \beta) \exp(-\beta \mathscr{H}) \qquad (70)$$

where V is the system volume, $\mathbf{N} = \{N_j\}$ is the collection of particle numbers for the system, $\beta = (1/k_B T)$, $\mathscr{H}$ is the system Hamiltonian, and Z is the canonical partition function

$$Z(V, \mathbf{N}, \beta) = \text{Tr} \exp(-\beta \mathscr{H}) \qquad (71)$$

The probability $P(\psi_m)$ of finding a system described by the canonical density operator $\hat{\rho}$ in a given eigenstate $|\psi_m\rangle$ of the system Hamiltonian $\mathcal{H}$ is given by the diagonal matrix element $\langle\psi_m|\hat{\rho}|\psi_m\rangle$. We write

$$P(\psi_m) = Z^{-1}(V, \mathbf{N}, \beta) \exp(-\beta E_m) \quad (72)$$

where E_m is the energy of state $|\psi_m\rangle$. The canonical partition function Z can be written in the form

$$Z(V, \mathbf{N}, \beta) = \sum_l \exp(-\beta E_l) \quad (73)$$

where the summation index l is restricted to symmetry-allowed eigenstates $\{|\psi_l\rangle\}$ of $\mathcal{H}$.

The Gibbs entropy of the quantum canonical ensemble is given by

$$S = k_B\beta\langle\mathcal{H}\rangle + k_B \ln Z(V, \mathbf{N}, \beta) \quad (74)$$

where the average energy $\langle\mathcal{H}\rangle$ can be written in the following equivalent forms,

$$\langle\hat{\mathcal{H}}\rangle = \text{Tr } \hat{\rho}\hat{\mathcal{H}} \quad (75a)$$

$$= \sum_m P(\psi_m)E_m \quad (75b)$$

with $P(\psi_m)$ given by Eq. (72). This result for the Gibbs entropy of a quantum canonical ensemble is identical in form to the Gibbs entropy for a classical canonical ensemble [See Eq. (55)], except that the average energy and canonical partition function in Eq. (74) are to be computed using quantum mechanics rather than classical mechanics. As with the canonical distribution function ρ for classical systems, the canonical density operator $\hat{\rho}$ for quantum systems represents the distribution of maximum Gibbs entropy consistent with the given average energy $\langle\mathcal{H}\rangle$ and the specified values of the external parameters $\mathbf{N}$ and V.

In view of the formal identity of the expressions for the Gibbs entropy of quantum and classical canonical ensembles, all the previously made formal connections between the thermodynamics and the statistical mechanics of closed, isothermal classical systems also apply to closed, isothermal quantum systems [see Eqs. (58)–(69b)]. One need only replace the classical ensemble averages with quantum ensemble averages and reinterpret the classical canonical partition function as a quantum canonical partition function.

C. Isobaric–Isothermal Ensemble

1. Classical Systems

Consider a system composed of a fixed number of particles $\mathbf{N}$ enclosed in a motionless container with a variable volume V and a fixed pressure P. The temperature of the system is maintained at temperature T by keeping the system in thermal contact with a large heat bath (at temperature T) with which it is able to exchange energy. The macroscopic state of this system is called an isobaric–isothermal ensemble.

The density distribution function associated with an isobaric–isothermal ensemble is given by

$$\rho(\mathbf{p^N}, \mathbf{q^N}; V) = \Delta^{-1}(P, \mathbf{N}, \beta)$$
$$\exp\{-\beta[\mathcal{H}(\mathbf{p^N}, \mathbf{q^N}; V) + PV]\} \quad (76)$$

where β and P are macroscopic parameters to be determined and Δ is the isobaric–isothermal partition function

$$\Delta(P, \mathbf{N}, \beta) = \int dV \int d\mathbf{l^N}$$
$$\exp\{-\beta[\mathcal{H}(\mathbf{p^N}, \mathbf{q^N}; V) + PV]\} \quad (77)$$

The isobaric–isothermal distribution function $\rho(\mathbf{p^N}, \mathbf{q^N}; V)$ is normalized as follows:

$$\int dV \int d\mathbf{\Gamma^N} \rho(\mathbf{p^N}, \mathbf{q^N}; V) = 1 \quad (78)$$

In writing Eqs. (76)–(78), we have indicated the explicit volume dependence of the density distribution function $\rho(\mathbf{p^N}, \mathbf{q^N}; V)$ and classical Hamiltonian $\mathcal{H}(\mathbf{p^N}, \mathbf{q^N}; V)$ to reflect the fact that the system volume V is variable.

The previously given forms for the ensemble average of classical dynamic variables [see Eq. (7)] and the Gibbs entropy associated with classical ensembles [see Eq. (9)] must be slightly modified to accommodate the isobaric–isothermal ensemble. We write

$$\langle O \rangle = \int dV \int d\mathbf{\Gamma^N} O(\mathbf{p^N}, \mathbf{q^N}; V)\rho(\mathbf{p^N}, \mathbf{q^N}; V)$$

and $\quad (79)$

$$S = -k_B \int dV \int d\mathbf{\Gamma^N}$$
$$\rho(\mathbf{p^N}, \mathbf{q^N}; V) \ln \rho(\mathbf{p^N}, \mathbf{q^N}; V) \quad (80)$$

The Gibbs entropy of the isobaric–isothermal distribution described by Eq. (76) is given by

$$S = k_B\beta\langle\mathcal{H}\rangle + k_B\beta P\langle V\rangle + k_B \ln \Delta(P, \mathbf{N}, \beta) \quad (81)$$

where $\langle\mathcal{H}\rangle$ and $\langle V\rangle$ respectively represent the average energy and average volume of the system. The isobaric–isothermal distribution is the distribution of maximum Gibbs entropy consistent with the given average energy $\langle\mathcal{H}\rangle$ and average volume $\langle V\rangle$ for the specified value of the external parameter $\mathbf{N}$.

The formal expression given by Eq. (81) for

the Gibbs entropy of an isobaric–isothermal ensemble can be rearranged to read

$$\langle \mathcal{H} \rangle + P \langle V \rangle = -\beta^{-1} \ln \Delta(P, \mathbf{N}, \beta)$$
$$+ (k_B \beta)^{-1} S \qquad (82)$$

This result is formally similar to the thermodynamic relations

$$H = E + PV \qquad (83a)$$
$$= G + TS \qquad (83b)$$

where H is the enthalpy, E the internal energy, P the pressure, V the thermodynamic volume, G the Gibbs free energy, T the temperature, and S the thermodynamic entropy.

Assuming that the Gibbs entropy of the isobaric–isothermal ensemble is equivalent to the thermodynamic entropy and that the average energy $\langle \mathcal{H} \rangle$ and average volume $\langle V \rangle$ are equivalent to their thermodynamic analogs, one can make the following correspondences:

$$G = -\beta^{-1} \ln \Delta(P, \mathbf{N}, \beta) \qquad (84)$$
$$T = 1/k_B \beta \qquad (85)$$

It should be noted that these correspondences require us to assume that the macroscopic parameter P in Eq. (82) is the pressure that appears in the thermodynamic relation given by Eq. (83a). A mathematical relationship connecting the Gibbs free energy G to microscopic interactions is given by Eq. (84).

Gibbs free energy G is the characteristic function in thermodynamics for the variables $\mathbf{N}$, P, T,

$$dG = -S \, dT + V \, dP + \sum_l \mu_l \, dN_l \qquad (86)$$

where

$$S = -(\partial G / \partial T)_{P, \mathbf{N}} \qquad (87)$$
$$V = (\partial G / \partial P)_{T, \mathbf{N}} \qquad (88)$$

and

$$\mu_l = (\partial G / \partial N_l)_{T, P, \mathbf{N}'} \qquad (89)$$

Utilizing the expression given by Eq. (84) for the Gibbs free energy G and the thermodynamic relations given by Eq. (87)–(89), we construct the following statistical mechanical relations for S, V, and μ_l:

$$S = k_B T [\partial \ln \Delta(P, \mathbf{N}, \beta) / \partial T]_{P, \mathbf{N}}$$
$$+ k_B \ln \Delta(P, \mathbf{N}, \beta) \qquad (90)$$
$$V = -k_B T [\partial \ln \Delta(P, \mathbf{N}, \beta) / \partial P]_{T, \mathbf{N}} \qquad (91)$$
$$\mu_l = -k_B T [\partial \ln \Delta(P, \mathbf{N}, \beta) / \partial N_l]_{T, P, \mathbf{N}'} \qquad (92)$$

It should be noted that the volume V appearing on the left side of Eq. (91) is actually the thermodynamic volume, which is the ensemble-averaged volume $\langle V \rangle$. The angular brackets around V have been dropped to render the statistical mechanical notation more similar to the thermodynamic notation.

An expression connecting the enthalpy H to microscopic interactions can be obtained by differentiating $\ln \Delta$ with respect to β. The result is

$$H = -[\partial \ln \Delta(p, \mathbf{N}, \beta) / \partial \beta]_{P, \mathbf{N}} \qquad (93)$$

Differentiating $\ln \Delta$ a second time with respect to β gives us expressions for the fluctuation in the variable $(\mathcal{H} + PV)$:

$$\langle [(\mathcal{H} + PV) - \langle (\mathcal{H} + PV) \rangle]^2 \rangle$$
$$= \langle (\mathcal{H} + PV)^2 \rangle - \langle (\mathcal{H} + PV) \rangle^2 \qquad (94a)$$
$$= [\partial^2 \ln \Delta(P, \mathbf{N}, B) / \partial \beta^2]_{P, \mathbf{N}} \qquad (94b)$$
$$= -[\partial \langle (\mathcal{H} + PV) \rangle / \partial \beta]_{P, \mathbf{N}} \qquad (94c)$$

Since the enthalpy H is equivalent to the ensemble average $\langle (\mathcal{H} + PV) \rangle$, the last equation allows us to connect the heat capacity $C_P = (\partial H / \partial T)_P$ at constant pressure to the fluctuation in the variable $(\mathcal{H} + PV)$ and microscopic interactions through the relations

$$C_P = k_B \beta^2 \langle [(\mathcal{H} + PV) - \langle (\mathcal{H} + PV) \rangle]^2 \rangle \qquad (95a)$$
$$= k_B \beta^2 [\partial^2 \ln \Delta(P, \mathbf{N}, \beta) / \partial \beta^2]_{P, \mathbf{N}} \qquad (95b)$$

2. Quantum Systems

As for classical systems, the isobaric–isothermal ensemble is used to characterize the macroscopic state of closed, isothermal quantum systems with a variable volume and fixed pressure. The quantum isobaric–isothermal ensemble is described by the density operator

$$\hat{\rho}(V) = \Delta^{-1}(P, \mathbf{N}, \beta) \exp\{-\beta[\hat{\mathcal{H}}(V) + PV]\} \qquad (96)$$

where P is the system pressure, $\mathbf{N} = \{N_j\}$ is the collection of particle numbers for the system, $\beta = 1/k_B T$, $\hat{\mathcal{H}}(V)$ is the system Hamiltonian, V is the system volume, and Δ is the isobaric–isothermal partition function

$$\Delta(P, \mathbf{N}, B) = \int dV \, \mathrm{Tr} \, \exp\{-\beta[\hat{\mathcal{H}}(V) + PV]\} \qquad (97)$$

The quantity V corresponding to the system volume is a classical variable. The isobaric–isothermal density operator $\hat{\rho}(V)$ is normalized as follows:

$$\int dV \, \text{Tr} \, \hat{\rho}(V) = 1 \qquad (98)$$

The previously given forms for the ensemble average of quantum dynamic variables [Eqs. (33a) and (33b)] and the Gibbs entropy associated with quantum ensembles [Eqs. (34) and (35)] must be slightly modified to accommodate the isobaric–isothermal ensemble. We write

$$\langle \hat{O} \rangle = \int dV \, \text{Tr} \, \hat{O}(V) \hat{\rho}(V) \qquad (99a)$$

$$= \int dV \sum_m O(\psi_m^V) \rho(\psi_m^V) \qquad (99b)$$

and

$$S = -k_B \int dV \, \text{Tr} \, \hat{\rho}(V) \ln \hat{\rho}(V) \qquad (100a)$$

$$= -k_B \int dV \sum_m \rho(\psi_m^V) \ln \rho(\psi_m^V) \qquad (100b)$$

where

$$O(\psi_m^V) = \langle \psi_m^V | \hat{O}(V) | \psi_m^V \rangle$$

$$\rho(\psi_m^V) = \langle \psi_m^V | \hat{\rho}(V) | \psi_m^V \rangle \qquad (101a)$$

$$= \Delta^{-1}(P, \mathbf{N}, \beta) \exp[-\beta(E_m^V + PV)] \qquad (101b)$$

with E_m^V denoting the energy of the eigenstate $|\psi_m^V\rangle$ of the system Hamiltonian $\mathcal{H}(V)$. The isobaric–isothermal partition function Δ can be written in the form

$$\Delta(P, \mathbf{N}, \beta) = \int dV \sum_l \exp[-\beta(E_l^V + PV)] \qquad (102)$$

It should be noted that both the eigenstates $\{|\psi_m^V\rangle\}$ and eigenvalues $\{E_m^V\}$ of the system Hamiltonian are parametrically dependent on the volume V. The quantity $dV \, \rho(\psi_m^V)$ can be interpreted as the joint probability of finding the system in the state $|\psi_m^V\rangle$ and with a volume in the neighborhood of V.

The Gibbs entropy of the quantum isobaric–isothermal ensemble is given by

$$S = [k_B \beta \langle \mathcal{H} \rangle + k_B \beta P \langle V \rangle + k_B \ln \Delta(P, \mathbf{N}, \beta)] \qquad (103)$$

where $\langle \mathcal{H} \rangle$ and $\langle V \rangle$, respectively, are the average energy and average volume for the system. The above result for the Gibbs entropy of a quantum isobaric–isothermal ensemble is identical in form to the Gibbs entropy for a classical isobaric–isothermal ensemble [Eq. (81)], except

that the ensemble averages and partition function in Eq. (103) are to be computed using quantum mechanics rather than classical mechanics. As with the isobaric–isothermal distribution function ρ for classical systems, the isobaric–isothermal density operator $\hat{\rho}$ for quantum systems represents the distribution of maximum Gibbs entropy consistent with the given average energy $\langle \mathcal{H} \rangle$ and average volume $\langle V \rangle$ and the specified value of the external parameter $\mathbf{N}$.

In view of the formal identity of the expressions for the Gibbs entropy of quantum and classical isobaric–isothermal ensembles, all the previously made formal connections between the thermodynamics and the statistical mechanics of closed, isothermal classical systems with variable volume also apply to closed, isothermal quantum systems with variable volume [see Eqs. (84)–(95b)]. One need only replace the quantum ensemble averages by classical ensemble averages and reinterpret the classical isobaric–isothermal partition function as a quantum isobaric–isothermal partition function.

D. Grand Canonical Ensemble

1. Classical Systems

Our previous discussion of ensembles was limited to closed systems, that is, systems in which the number of particles $\mathbf{N}$ remains fixed. Now we consider an open system, that is, a system that can exchange particles with its surroundings. The volume V of the system is fixed, and the temperature is maintained at temperature T by keeping the system in thermal contact with a large heat bath (at temperature T), with which it is able to exchange both energy and particles. Such a system is called an open, isothermal system. The macroscopic state of this system is called a grand canonical ensemble.

The density distribution function associated with a grand canonical ensemble is given by

$$\rho(\mathbf{p}^N, \mathbf{q}^N) = \Xi^{-1}(V, \beta, \boldsymbol{\mu}) \exp \left\{ -\beta \left[\mathcal{H}(\mathbf{p}^N, \mathbf{q}^N) \right. \right.$$
$$\left. \left. - \sum_l \mu_l N_l \right] \right\} \qquad (104)$$

where β and $\boldsymbol{\mu} \equiv \{\mu_l\}$ are macroscopic parameters to be determined and Ξ is the grand canonical partition function

$$\Xi(V, \beta, \boldsymbol{\mu}) = \sum_{\mathbf{N}=0}^{\infty} \int d\Gamma^N \exp \left\{ -\beta \left[\mathcal{H}(\mathbf{p}^N, \mathbf{q}^N) \right. \right.$$
$$\left. \left. - \sum_l \mu_l N_l \right] \right\} \qquad (105)$$

The grand canonical distribution function $\rho(\mathbf{p}^N, \mathbf{q}^N)$ is normalized as follows:

$$\sum_{N=0}^{\infty} \int d\Gamma^N \rho(\mathbf{p}^N, \mathbf{q}^N) = 1 \qquad (106)$$

The summation over $\mathbf{N}$ in Eqs. (105) and (106) is understood to mean a summation over all the particle numbers $\mathbf{N} = \{N_j\}$. We have used a single summation index $\mathbf{N}$ to simplify the notation.

The previously given forms for the ensemble average of classical dynamic variables [Eq. (7)] and the Gibbs entropy associated with classical ensembles [Eq. (9)] must be slightly modified to accommodate the grand canonical ensemble. We write

$$\langle O \rangle = \sum_{N=0}^{\infty} \int d\Gamma^N \rho(\mathbf{p}^N, \mathbf{q}^N) O(\mathbf{p}^N, \mathbf{q}^N) \qquad (107)$$

and

$$S = -k_B \sum_{N=0}^{\infty} \int d\Gamma^N \rho(\mathbf{p}^N, \mathbf{q}^N) \ln \rho(\mathbf{p}^N, \mathbf{q}^N)$$
$$(108)$$

The Gibbs entropy of the grand canonical distribution described by Eq. (104) is given by

$$S = \left[k_B \beta \langle \mathcal{H} \rangle - k_B \beta \sum_l \mu_l \langle N_l \rangle \right.$$
$$\left. + k_B \ln \Xi(V, \beta, \boldsymbol{\mu}) \right] \qquad (109)$$

where $\langle \mathcal{H} \rangle$ denotes the average energy of the system and $\langle N_l \rangle$ is the average number of particles of type l in the system. The grand canonical distribution is the distribution of maximum Gibbs entropy consistent with the given average energy $\langle \mathcal{H} \rangle$ and averge particle numbers $\{\langle N_l \rangle\}$ for the specified value of the external parameter V.

The formal expression given by Eq. (109) for the Gibbs entropy of a grand canonical ensemble can be rearranged to read

$$\langle \mathcal{H} \rangle + \beta^{-1} \ln \Xi(V, \beta, \boldsymbol{\mu}) = (k_B \beta)^{-1} S$$
$$+ \sum_l \mu_l \langle N_l \rangle \qquad (110)$$

This result resembles the thermodynamic relation

$$E + PV = TS + \sum_l \mu_l N_l \qquad (111)$$

where E is the internal energy, P the pressure, V the volume, T the temperature, S the thermodynamic entropy, μ_l the chemical potential of component l, and N_l the thermodynamic number of particles of type l.

Assuming that the Gibbs entropy of the grand canonical ensemble is equivalent to the thermodynamic entropy and that the average energy $\langle \mathcal{H} \rangle$ and average particle numbers $\{\langle N_l \rangle\}$ are equivalent to their thermodynamic analogs, one can make the correspondence

$$PV = \beta^{-1} \ln \Xi(V, \beta, \boldsymbol{\mu}) \qquad (112)$$

In making this correspondence, it is necessary to assume that the macroscopic parameter β is given by $\beta = 1/k_B T$ and that the macroscopic parameter μ_l in Eq. (110) is equal to the chemical potential appearing in Eq. (111). The result given by Eq. (112) provides us with a relationship connecting the product PV to microscopic interactions.

The product PV is the thermodynamic characteristic function for the variables V, T, and $\boldsymbol{\mu}$,

$$d(PV) = S\, dT + \sum_l N_l\, d\mu_l + P\, dV \qquad (113)$$

where

$$S = [\partial(PV)/\partial T]_{\boldsymbol{\mu}, V} \qquad (114)$$
$$N_l = [\partial(PV)/\partial \mu_l]_{V, T, \boldsymbol{\mu}'} \qquad (115)$$
$$P = [\partial(PV)/\partial V]_{T, \boldsymbol{\mu}} \qquad (116)$$

In Eq. (115), we use the symbol $\boldsymbol{\mu}'$ to indicate that the chemical potentials $\boldsymbol{\mu}' = \{\mu_j, j \neq l\}$ are fixed for all components in the system except for the chemical potential μ_l for the lth component.

Utilizing the expression given by Eq. (112) for the product PV and the thermodynamic relations given by Eqs. (114)–(116), one can construct the following statistical mechanical relations for S, N_l, and P:

$$S = k_B T[\partial \ln \Xi(V, \beta, \boldsymbol{\mu})/\partial T]_{V, \boldsymbol{\mu}}$$
$$+ k_B \ln \Xi(V, \beta, \boldsymbol{\mu}) \qquad (117)$$
$$N_l = k_B T[\partial \ln \Xi(V, \beta, \boldsymbol{\mu})/\partial \mu_l]_{V, T, \boldsymbol{\mu}'} \qquad (118)$$
$$P = k_B T[\partial \ln \Xi(V, \beta, \boldsymbol{\mu})/\partial V]_{\boldsymbol{\mu}, T} \qquad (119a)$$
$$= (k_B T/V) \ln \Xi(V, \beta, \boldsymbol{\mu}) \qquad (119b)$$

It should be noted that the particle number N_l appearing on the left side of Eq. (118) is actually the thermodynamic particle number, which is the ensemble-averaged particle number $\langle N_l \rangle$. The angular brackets around N_l have been dropped to render the statistical mechanical notation more similar to the thermodynamic notation.

Now let us consider the fluctuation in the particle number for a single-component system. Differentiation of $\ln \Xi$ twice with respect to μ gives us

$$\langle (N - \langle N \rangle)^2 \rangle = \langle N^2 \rangle - \langle N \rangle^2 \tag{120a}$$

$$= (k_B T)^2 [\partial^2 \ln \Xi(V, \beta, \mu)/\partial \mu^2]_{V,T} \tag{120b}$$

$$= k_B T (\partial \langle N \rangle / \partial \mu)_{V,T} \tag{120c}$$

The last relation coupled with the thermodynamic relation $N(\partial \mu / \partial N)_{V,T} = V(\partial P / \partial N)_{V,T}$ allows us to connect the isothermal compressibility $\chi_T = -(1/V)(\partial V/\partial P)_{N,T}$ with the fluctuation in the particle number and microscopic interactions. This connection is given by

$$\chi_T = (V/k_B T)\langle (N - \langle N \rangle)^2 \rangle / \langle N^2 \rangle \tag{121}$$

2. Quantum Systems

As for classical systems, the grand canonical ensemble is used to characterize the macroscopic state of open, isothermal quantum systems. The quantum grand canonical ensemble is described by the density operator

$$\hat{\rho}(\hat{\mathbf{N}}) = \Xi^{-1}(V, \beta, \mu) \exp \left\{ -\beta \left[\mathcal{H}(\hat{\mathbf{N}}) - \sum_l \mu_l \hat{N}_l \right] \right\} \tag{122}$$

where V is the system volume, $\beta = 1/k_B T$, $\mu = \{\mu_l\}$ is the collection of chemical potentials, $\mathcal{H}(\hat{\mathbf{N}})$ is the system Hamiltonian, $\hat{\mathbf{N}} = \{\hat{N}_l\}$ is the collection of particle number operators for the components of the system, and Ξ is the grand canonical partition function

$$\Xi(V, \beta, \mu) = \text{Tr} \exp \left\{ -\beta \left[\mathcal{H}(\hat{\mathbf{N}}) - \sum_l \mu_l \hat{N}_l \right] \right\} \tag{123}$$

The eigenstates $\{|\psi_k^N\rangle\}$ of the system Hamiltonian $\mathcal{H}(\hat{\mathbf{N}})$ will also be eigenstates of the particle number operators $\hat{\mathbf{N}}$ when the operators $\mathcal{H}(\hat{\mathbf{N}})$ and $\hat{\mathbf{N}}$ commute. For such cases, we write

$$\mathcal{H}(\hat{\mathbf{N}})|\psi_k^N\rangle = E_k^N|\psi_k^N\rangle \tag{124}$$

and

$$\hat{\mathbf{N}}|\psi_k^N\rangle = \mathbf{N}|\psi_k^N\rangle \tag{125}$$

where $\mathbf{N} = \{N_l\}$ is the collection of particle numbers for the system, $|\psi_k^N\rangle$ is an $\mathbf{N}$-particle state vector satisfying the proper symmetry requirements with respect to particle permutation, and E_k^N is the energy of state $|\psi_k^N\rangle$. In general, the particle number operators $\hat{\mathbf{N}}$ will not commute with the Hamiltonian of a system. This will be the case, for example, in multicomponent reactive systems. For such cases, the Hamiltonian

contains interaction terms that give rise to the transformation of one set of particles into another set of particles. Hereafter, we will not consider such complications and restrict our considerations to nonreactive systems.

The previously given forms for the ensemble average of quantum dynamical variables [Eqs. (33a) and (33b)] and the Gibbs entropy associated with quantum ensembles [Eqs. (34) and (35)] must be slightly modified to accommodate the grand canonical ensemble. We write

$$\langle \hat{O} \rangle = \sum_{\mathbf{N}=0}^{\infty} \text{Tr} \, \hat{\rho}(\mathbf{N}) \hat{O}(\mathbf{N}) \tag{126a}$$

$$= \sum_{\mathbf{N}=0}^{\infty} \sum_k P(\psi_k^N) O(\psi_k^N) \tag{126b}$$

and

$$S = -k_B \sum_{\mathbf{N}=0}^{\infty} \text{Tr} \, \hat{\rho}(\mathbf{N}) \ln \hat{\rho}(\mathbf{N}) \tag{127a}$$

$$= -k_B \sum_{\mathbf{N}=0}^{\infty} \sum_k P(\psi_k^N) \ln P(\psi_k^N) \tag{127b}$$

where

$$O(\psi_k^N) = \langle \psi_k^N | \hat{O}(\hat{\mathbf{N}}) | \psi_k^N \rangle$$

$$\hat{\rho}(\mathbf{N}) = \Xi^{-1}(V, \beta, \mu) \exp \left\{ -\beta \left[\mathcal{H}(\mathbf{N}) - \sum_l \mu_l N_l \right] \right\} \tag{128}$$

In Eqs. (126b) and (127b), $P(\psi_k^N)$ is the probability of finding the system with the collection of particle numbers $\mathbf{N}$ and in the $\mathbf{N}$-particle state $|\psi_k^N\rangle$. The probability $P(\psi_k^N)$ is given by the diagonal matrix element $\langle \psi_k^N | \hat{\rho}(\hat{\mathbf{N}}) | \psi_k^N \rangle$, which can be written in the form

$$P(\psi_k^N) = \Xi^{-1}(V, \beta, \mu) \exp \left[-\beta \left(E_k^N - \sum_l \mu_l N_l \right) \right] \tag{129}$$

where

$$\Xi(V, \beta, \mu) = \sum_{\mathbf{N}=0}^{\infty} \sum_k \exp \left[-\beta \left(E_k^N - \sum_l \mu_l N_l \right) \right] \tag{130}$$

The Gibbs entropy of the quantum grand canonical ensemble is given by

$$S = \left[k_B \beta \langle \mathcal{H} \rangle - k_B \beta \sum_l \mu_l \langle \hat{N}_l \rangle + k_B \ln \Xi(V, \beta, \mu) \right] \tag{131}$$

where $\langle \hat{\mathcal{H}} \rangle$ and $\langle \hat{N}_l \rangle$, respectively, are the average energy of the system and the average number of particles of type l in the system. The above result for the Gibbs entropy of a quantum grand canonical ensemble is identical in form to the Gibbs entropy for a classical grand canonical ensemble [Eq. (109)], except that the ensemble averages and partition function in Eq. (131) are to be computed using quantum mechanics rather than classical mechanics. As with the grand canonical distribution function ρ for classical systems, the grand canonical density operator $\hat{\rho}$ for quantum systems represents the distribution of maximum Gibbs entropy consistent with the given average energy $\langle \hat{\mathcal{H}} \rangle$ and average particle numbers $\{\langle \hat{N}_l \rangle\}$ and the specified value of the external parameter V.

In view of the formal identity of the expressions for the Gibbs entropy of quantum and classical grand canonical ensembles, all the previously made formal connections between the thermodynamics and statistical mechanics of open, isothermal classical systems also apply to open, isothermal quantum systems [see Eqs. (112)–(121)]. One need only replace the quantum ensemble averages by classical ensemble averages and reinterpret the classical grand canonical partition function as a quantum grand canonical partition function.

E. Thermodynamic Equivalence of Gibbsian Ensembles

All the Gibbsian ensembles discussed in Sections II,A–D were defined by specifying certain physical conditions for the systems from which the Gibbsian ensembles are composed. For example, the canonical ensemble is defined by fixing the particle numbers and volume of a system and keeping the system in thermal contact with a large heat bath at some temperature. From a theoretical point of view, the results for the different Gibbsian ensembles hold only for the conditions under which the Gibbsian ensembles are defined. Nonetheless, in actual applications of statistical mechanics, investigators often choose a Gibbsian ensemble on the basis of mathematical convenience rather than on the basis of the physical conditions under which a system is found.

The replacing of one ensemble by another is usually rationalized by appealing to the fact that the size of fluctuations in the variables characterizing the macroscopic state of large systems is usually small in most situations. Of course,

there are exceptional cases in which the fluctuations may not be small, for example, when two phases coexist and when a critical point is realized. Barring these cases and assuming that the fluctuations of energy, particle number, and volume are small, we can for most thermodynamic purposes replace the energy, particle number, and volume by their average values in the various Gibbsian ensembles. In making such replacements one is essentially assuming that the microcanonical, canonical, isobaric–isothermal, and grand canonical ensembles are thermodynamically equivalent; that is, they give the same values for the thermodynamic functions. Actually, the thermodynamic equivalence of these Gibbsian ensembles may be rigorously established in the limit $V \to \infty$, $N \to \infty$, and $V/N =$ const (thermodynamic limit). Nonetheless, the Gibbsian ensembles are not equivalent with respect to the calculation of fluctuations.

III. Information Theory and Statistical Mechanics

A. Isomorphism between Information Theory and Statistical Mechanics

Consider a system that can be described in terms of a set of events. We say that an event occurs when some variable(s) used to characterize that event assumes some value(s). Generally, we lack sufficient information to specify the exact probability distribution for the occurrence of the events. Rather, we possess only a limited amount of information, such as the average value(s) of the variable(s) used to characterize the events. Nonetheless, it is desirable to make use of our limited information to make educated guesses about the possible outcome of the events and to make estimates of the properties of our system. This is the basic problem facing us in statistical mechanics.

The problem of determining the probabilities of a set of events requires us to find a least bias distribution that agrees with the given information about a system. The resolution of this problem underwent a great advance with the advent of Shannon's information theory. Shannon showed that there exists a unique measure of the statistical uncertainty in a discrete probability distribution, which is in agreement with the intuitive notion that a broad distribution represents more uncertainty than a narrow distribution. This measure is called the missing information, which is defined by

$$I = -c \sum_l P(l) \ln P(l) \qquad (132)$$

where c is a positive constant, $P(l)$ is the probability of event l, and $\mathbf{P} = \{P(l)\}$ is the distribution of events. The missing information I is positive, increases with increasing uncertainty, and is additive for independent sources of uncertainty. Clearly, the missing information is just the Gibbs entropy for quantum ensembles.

The generalization of the above expression for the missing information to include the case of a continuous set of events can be accomplished by treating the continuous problem as a limiting case of the discrete problem. For a continuous distribution of events, we write the missing information as

$$I = -c \int d\mathbf{x} \, \rho(\mathbf{x}) \ln \rho(\mathbf{x}) \qquad (133)$$

where $\rho(\mathbf{x})$ is a normalized probability density used to describe the continuous distribution of events. This form for the missing information is identical to the expression for the Gibbs entropy associated with classical ensembles.

In the context of information theory, one chooses the probability distribution for a set of events on the basis of the available information and requires the distribution to possess a maximum in the missing information. Actually, this is the only unbiased probability distribution that we can postulate. The use of any other distribution would be equivalent to making arbitrary assumptions about unavailable information. The actual construction of information-theoretic distributions is accomplished by treating the linearly independent pieces of known information as constraints and utilizing the method of Lagrange multipliers. Distributions constructed in this manner are identical in form to the distributions associated with the Gibbsian ensembles discussed in the last section. As with information-theoretic distributions, the distributions associated with these ensembles are maximum entropy distributions consistent with the information defining them.

The isomorphism between information theory and statistical mechanics has suggested to a number of investigators that statistical mechanics be reinterpreted in such a way that information theory can be used to justify statistical mechanics. This point of view has been pioneered by Jaynes. In the context of the information-

theoretic approach to statistical mechanics, entropy and entropy maximization become fundamental concepts rather than mathematical facts not essential for justifying statistical mechanics. The information-theoretic approach to statistical mechanics not only leads to conceptual and mathematical simplification of this subject but also frees us from such hypotheses as ergodicity and equal a priori probabilities. In addition, it enables us to provide a very clear path for the formal development of statistical mechanics and thermodynamics without having to appeal to the phenomenological equations of thermodynamics to render the statistical mechanical formalism meaningful.

B. Maximum Entropy Principle

1. Classical Version

If we interpret Gibbs entropy in the same spirit as the missing information of information theory, it may be viewed as a measure of statistical uncertainty. Adopting this point of view, it seems natural to treat the following principle as a basic postulate of classical equilibrium statistical mechanics.

Maximum Entropy Principle (Classical Version). The density distribution function ρ used to describe the macroscopic state of a system must represent all available information and correspond to the maximum entropy S distribution consistent with this information, where the entropy S is defined by

$$S = -k_B \, \mathrm{Tr} \, \rho \ln \rho \qquad (134)$$

In Eq. (134) the symbol Tr indicates a classical trace, that is, a sum over all accessible classical states of a system. This sum may involve an integration over the accessible phase space of a system, as well as integrations or summations (or both) over any other classical variables required to define the macroscopic state of that system.

The implementation of the maximum entropy principle can be illustrated by considering a system for which we possess information concerning the average values of a set of observables $\{\langle O_j \rangle, j = 0, \ldots, n\}$, where

$$\langle O_j \rangle = \mathrm{Tr} \, \rho O_j \qquad (135)$$

The classical variable $O_0 \equiv 1$. Thus, $\langle O_0 \rangle = 1$ is simply a statement of the requirement that ρ be normalized to unity.

According to the maximum entropy principle,

the density distribution function ρ describing the macroscopic state of our system must represent all available information, $\{\langle O_j \rangle, j = 0, ..., n\}$, and correspond to a maximum in Gibbs entropy S. The actual construction of ρ can be accomplished by adopting a procedure due to Lagrange that allows us to change a constrained extremum problem into an unconstrained extremum problem.

Adopting the method of Lagrange, we seek an unconstrained extremum of an auxilary function $\mathscr{L}$ called the Lagrangian, which is defined by

$$\mathscr{L} \equiv S - k_B \sum_{j=0}^{n} \Lambda_j \langle O_j \rangle \qquad (136)$$

where $\{\Lambda_j, j = 0, ..., n\}$ are Lagrange parameters. The variation $\delta \mathscr{L}$ in $\mathscr{L}$ due to a variation $\delta \rho$ in ρ is given by

$$\delta \mathscr{L} = -k_B \, \mathrm{Tr} \, \delta \rho \left(\ln \rho + \sum_{j=0}^{n} \Lambda_j O_j \right) \quad (137)$$

Since we seek an unconstrained extremum of $\mathscr{L}$, $\delta \mathscr{L}$ must vanish for arbitrary variations $\delta \rho$. It follows that the density distribution function ρ is given by

$$\rho = \exp\left(-\sum_{j=0}^{n} \Lambda_j O_j \right) \qquad (138)$$

Introducing the definition $\Omega \equiv -\Lambda_0$, we rewrite Eq. (138) in the form

$$\rho = \exp\left(\Omega - \sum_{j=1}^{n} \Lambda_j O_j \right) \qquad (139)$$

where

$$\Omega = -\ln Z \qquad (140)$$

with Z denoting the generalized partition function

$$Z = \mathrm{Tr} \, \exp\left(-\sum_{j=1}^{n} \Lambda_j O_j \right) \qquad (141)$$

In writing Eqs. (139)–(141), we have solved for the Lagrange parameter Ω by imposing the normalization constraint $\mathrm{Tr} \, \rho = 1$.

The remaining Lagrange parameters $\{\Lambda_j, j = 1, ..., n\}$ can be determined by solving the set of equations

$$\langle O_i \rangle = \mathrm{Tr} \, \exp\left(\Omega - \sum_{j=1}^{n} \Lambda_j O_j \right) O_i \quad (142a)$$

$$= \partial \Omega / \partial \Lambda_i \qquad (142b)$$

where Ω is given by Eqs. (140) and (141). A necessary condition for Eqs. (140) and (142b) to be solvable for the Lagrange parameters with arbitrary $\{\langle O_j \rangle, j = 1, ..., n\}$ and a sufficient condition for the solution to be unique is that the determinant $|\chi|$ formed from the matrix χ be nonvanishing, where the elements of the matrix χ are given by

$$\chi_{m,k} = -\partial^2 \Omega / \partial \Lambda_m \, \partial \Lambda_k \qquad (143a)$$

$$= \langle \Delta O_m \, \Delta O_k \rangle \qquad (143b)$$

$$= \mathrm{Tr} \, \rho \, \Delta O_m \, \Delta O_k \qquad (143c)$$

with $m, k = 1, ..., n$ and $\Delta O_{m,k} = O_{m,k} - \langle O_{m,k} \rangle$. It should be noted that the statistical fluctuations in the distribution ρ are described by the elements of the matrix χ.

The above necessary and sufficient condition for a unique solution for the Langrange parameters or equivalently the distribution function ρ requires the pieces of information $\{\langle O_j \rangle, j = 0, ..., n\}$ to be linearly independent. Assuming that this is indeed the case, the distribution function ρ is unique and represents a maximum in Gibbs entropy consistent with the available information defining the macroscopic state of the system. The entropy S of this distribution is given by

$$S = -k_B \Omega + k_B \sum_{j=1}^{n} \Lambda_j \langle O_j \rangle \qquad (144)$$

From the above discussion, it should be clear that all the classical distribution functions discussed in Section II can be constructed by starting with the maximum entropy principle. One need only make use of the available information defining the macroscopic state of the system of interest.

2. Quantum Version

The quantum version of the maximum entropy principle can be stated as follows:

Maximum Entropy Principle (Quantum Version). The density operator $\hat{\rho}$ used to describe the macroscopic state of a system must represent all available information and correspond to the maximum entropy S density operator consistent with this information, where the entropy S is defined by

$$S = -k_B \, \mathrm{Tr} \, \hat{\rho} \ln \hat{\rho} \qquad (145)$$

In Eq. (145) the trace, denoted by Tr, can be carried out by employing any complete set of states consistent with symmetry requirements and any other restraints placed on the system. If classical variables are also required to define the macroscopic state of the system, the trace may also involve integrations or summations (or both) over the relevant classical variables.

Aside from being consistent with the maximum entropy principle, we require the density operator $\hat{\rho}$ to be a Hermitian operator with nonnegative eigenvalues.

The implementation of the quantum version of the maximum entropy principle to costruct density operators can be accomplished by following a path similar to that used in constructing maximum entropy distribution functions. Of course, one must be careful to account properly for the noncommutativity of quantum operators. For the quantum case, our information is the set of averages $\{\langle \hat{O}_j \rangle, j = 0, \ldots, n\}$, where $\{\hat{O}_j\}$ are assumed to be Hermitian operators. Making use of this information and the method of Lagrange, we arrive at the following density operator,

$$\hat{\rho} = \exp\left(\Omega \hat{I} - \sum_{j=1}^{n} \Lambda_j \hat{O}_j\right) \quad (146)$$

where

$$\Omega = -\ln Z \quad (147)$$

with Z denoting the generalized partition function

$$Z = \text{Tr} \exp\left(-\sum_{j=1}^{n} \Lambda_j \hat{O}_j\right) \quad (148)$$

In writing Eqs. (146)–(148), we solved for the Lagrange parameter Ω by imposing the normalization constraint $\text{Tr} \hat{\rho} = 1$.

Now the density operator $\hat{\rho}$ is required to be Hermitian with nonnegative eigenvalues. Since the operators $\{\hat{O}_j\}$ are assumed to be Hermitian, the Hermiticity requirement of $\hat{\rho}$ will be fulfilled, provided that the Lagrange parameters $\{\Lambda_j\}$ are real. Assuming this to be the case, we write the eigenvalues of $\hat{\rho}$ in the form

$$P(\psi) = \exp\left(\Omega - \sum_{j=1}^{n} \Lambda_j \langle \hat{O}_j \rangle_\psi\right) \quad (149)$$

where Ω is given by Eq. (147), with

$$Z = \sum_\psi \exp\left(-\sum_{j=1}^{n} \Lambda_j \langle \hat{O}_j \rangle_\psi\right) \quad (150)$$

and $\langle \hat{O}_j \rangle_\psi = \langle \psi | \hat{O}_j | \psi \rangle$ denoting a quantum average of the operator $\hat{O}_j$. In the above, the index ψ runs over the eigenvectors of the density operator $\hat{\rho}$, which are also eigenvectors of the opera-

tor $\hat{R} = \Omega \hat{I} - \sum_j \Lambda_j \hat{O}_j$. It is clear from Eqs. (149) and (150) that the eigenvalues of $\hat{\rho}$ are nonnegative as required.

The Lagrange parameters $\{\Lambda_j, \ j = 1, \ldots, n\}$ can be determined by solving the set of equations

$$\langle \hat{O}_i \rangle = \text{Tr} \exp\left(\Omega \hat{I} - \sum_{j=1}^{n} \Lambda_j \hat{O}_j\right) \hat{O}_i \quad (151a)$$

$$= \partial \Omega / \partial \Lambda_i \quad (151b)$$

The necessary and sufficient condition for Eqs. (147) and (151b) to give a unique solution for the Lagrange parameters (or equivalently the density operator $\hat{\rho}$) is identical to the classical case except that the elements of the matrix χ are now given by

$$\chi_{m,k} = \langle \Delta \hat{O}_m \Delta \hat{O}_k \rangle \quad (152a)$$

$$= \text{Tr} \hat{\rho} \Delta \hat{O}_m \Delta \hat{O}_k \quad (152b)$$

where

$$\Delta \hat{O}_k = \int_0^1 d\lambda \exp\left(-\lambda \sum_{j=1}^{n} \Lambda_j \hat{O}_j\right) \Delta \hat{O}_k$$

$$\times \exp\left(\lambda \sum_{j=1}^{n} \Lambda_j \hat{O}_j\right) \quad (153)$$

with $\Delta \hat{O}_k = \hat{O}_k - \langle \hat{O}_k \rangle$. The elements of the χ matrix for quantum systems provide us with information concerning statistical fluctuations only when the operators $\{\hat{O}_j\}$ commute or when the noncommutativity of these operators is neglected.

As with the classical case, the necessary and sufficient condition for a unique solution for the Lagrange parameters or equivalently the density operator $\hat{\rho}$ requires that the pieces of information $\{\langle O_j \rangle, j = 0, \ldots, n\}$ be linearly independent. Assuming this to be the case, the density operator $\hat{\rho}$ is unique and represents a maximum in Gibbs entropy consistent with the available information defining the macroscopic state of the system. The entropy S of this distribution is given by

$$S = -k_B \Omega + k_B \sum_{j=1}^{n} \Lambda_j \langle \hat{O}_j \rangle \quad (154)$$

which is identical in form to the classical result.

All the density operators discussed in Section II can be constructed by starting from the quantum version of the maximum entropy principle and making use of the available information defining the macroscopic state of the system of interest.

C. RELATION BETWEEN THERMODYNAMICS AND STATISTICAL MECHANICS

1. Classical Systems

In Section III,B,1, the maximum entropy principle was used to construct the density distribution function ρ characterizing the macroscopic state of some general system for which we possess the information $\{\langle O_j \rangle, j = 1, ..., n\}$ [see Eq. (139)]. Now we turn our attention to the relationship between the thermodynamics and statistical mechanics of this system.

In writing the maximum entropy distribution function ρ, we introduced a quantity Ω that is related to the partition function Z by Eq. (140) [also see Eq. (141)]. Since the partial derivatives $\{\partial\Omega/\partial\Lambda_j, j = 1, ..., n\}$ give us the averages $\{\langle O_j \rangle, j = 1, ..., n\}$ [Eq. (142b)] defining the macroscopic (thermodynamic) state of the system, we refer to Ω as a thermodynamic potential and the Lagrange parameters $\{\Lambda_j, j = 1, ..., n\}$ as thermodynamic parameters.

The macroscopic state of the system depends not only on the averages $\{\langle O_j \rangle, j = 1, ..., n\}$ but also on the external parameters $\{a_l, l = 1, ..., m\}$ defining the system. By external parameters, we mean such quantities as the volume of the system, the number of particles in the system, and the intensity of an external field. Clearly, the thermodynamic potential Ω depends on both the thermodynamic and the external parameters. Thus, Ω can be viewed as the thermodynamic characteristic function of $\{\Lambda_j, j = 1, ..., n\}$ and $\{a_l, l = 1, ..., m\}$.

The differential $d\Omega$ of the thermodynamic potential Ω is given by

$$d\Omega = -\sum_{l=1}^{m} f_l \, da_l + \sum_{j=1}^{n} \langle O_j \rangle \, d\Lambda_j \quad (155)$$

where

$$f_l \equiv -\partial\Omega/\partial a_l \quad (156)$$

The quantity f_l can be thought to represent a "force" arising from the variation in the external parameter a_l.

Recall that the Gibbs entropy of the system is given by Eq. (144). With this result and Eq. (155) at our disposal, we find that the entropy change dS associated with the system undergoing an infinitesimal process is given by

$$dS = k_B \sum_{l=1}^{m} f_l \, da_l + k_B \sum_{j=1}^{n} \Lambda_j \, d\langle O_j \rangle \quad (157)$$

This result allows us to identify the thermodynamic parameters and the "forces" with the partial derivatives of the entropy S:

$$\partial S/\partial a_l = -k_B \, (\partial\Omega/\partial a_l) \quad (158a)$$

$$= k_B f_l \quad (158b)$$

$$\partial S/\partial\langle O_j \rangle = k_B \, \Lambda_j \quad (159)$$

The macroscopic state of the system can be defined in terms of either the thermodynamic parameters $\{\Lambda_j, j = 1, ..., n\}$ or the averages $\{\langle O_j \rangle, j = 1, ..., n\}$. Both sets of variables represent the same information. If the thermodynamic parameters are known, the set of equations given by Eq. (159) can be used to determine the averages. On the other hand, if the averages are known, the thermodynamic parameters can be determined by solving the set of equations given by Eq. (142b). Since we have two sets of variables that correspond to the partial derivatives of two different functions connected by a single relation, namely, Eq. (144), we say that Ω and S are Legendre transforms of each other. The relationship connecting Ω and S is called a Legendre transformation.

If the system undergoes an infinitesimal process in which the entropy change dS vanishes, we say that the system has undergone a quasistatic or reversible process. For such a process, the system passes through an ordered sequence of equilibrium states defined in such a way that the averages $\{\langle O_j \rangle, j = 1, ..., n\}$ adiabatically follow the external parameters $\{a_l, l = 1, ..., m\}$. The manner in which the averages adiabatically follow the external parameters can be expressed as

$$\sum_{j=1}^{n} \Lambda_j \, d\langle O_j \rangle = -\sum_{l=1}^{m} f_l \, da_l \quad (160)$$

which follows from Eq. (157). If the average energy $\langle \mathcal{H} \rangle$ of the system is included among the averages, we can recast Eq. (160) in the form

$$dE = -đW \quad (161)$$

where $dE = d\langle \mathcal{H} \rangle$ is the change in the average energy of the system and

$$đW = \Lambda_1^{-1} \left[\sum_{l=1}^{m} f_l \, da_l + \sum_{j=2}^{n} \Lambda_j \, d\langle O_j \rangle \right] \quad (162)$$

is the work (energy expended) done in changing the external parameters and the averages other than the average energy.

In order for us to introduce the concepts of heat and heat transfer, we must include among the $\{\langle O_j \rangle, j = 1, ..., n\}$ the average energy $\langle \mathcal{H} \rangle$ and bring the system in thermodynamic contact with another system with which it can communicate. So we consider a composite system composed of two subsystems, referred to as subsystems 1 and 2. The macroscopic state of subsystem $\alpha = 1, 2$ is defined in terms of the external parameters $\{a_l^{(\alpha)}, l = 1, ..., m(\alpha)\}$ and the averages $\{\langle O_j^{(\alpha)} \rangle, j = 1, ..., n(\alpha)\}$, where $\langle O_1^{(\alpha)} \rangle = \langle \mathcal{H}^{(\alpha)} \rangle$ is the average energy of subsystem α. For the sake of simplicity, we assume that the sets of variables $\{\langle O_j^{(\alpha)} \rangle, j = 1, ..., n(\alpha)\}$ and $\{a_l^{(\alpha)}, l = 1, ..., m(\alpha)\}$ are identical for both subsystems and subject to the following conservation relations,

$$\langle O_j \rangle = \langle O_j^{(1)} \rangle + \langle O_j^{(2)} \rangle \tag{163}$$

and

$$a_l = a_l^{(1)} + a_l^{(2)} \tag{164}$$

where $\langle O_j \rangle$ and a_l are fixed.

Before the two subsystems of the composite system are brought into thermodynamic contact, they are statistically independent and the values of the variables defining the macroscopic state of one subsystem are uncorrelated with the values of the variables defining the macroscopic state of the other subsystem. With these remarks in mind, it should be clear from the preceding discussions that the entropy S of the composite system before thermodynamic contact is given by

$$S - S^{(1)} + S^{(2)} \tag{165}$$

where the entropy $S^{(\alpha)}$ of subsystem α is written

$$S^{(\alpha)} = -k_B \Omega^{(\alpha)} + k_B \Lambda_1^{(\alpha)} \langle \mathcal{H}^{(\alpha)} \rangle$$
$$+ k_B \sum_{j=2}^{n(\alpha)} \Lambda_j^{(\alpha)} \langle O_j^{(\alpha)} \rangle \tag{166}$$

with $\Omega^{(\alpha)}$ denoting the thermodynamic potential of subsystem α:

$$\Omega^{(\alpha)} = -\ln Z^{(\alpha)} \tag{167}$$

Here, $Z^{(\alpha)}$ is the partition function of subsystem α:

$$Z^{(\alpha)} = \mathrm{Tr} \exp\left(-\Lambda_1^{(\alpha)} \mathcal{H}^{(\alpha)} - \sum_{j=2}^{n(\alpha)} \Lambda_j^{(\alpha)} O_j^{(\alpha)} \right) \tag{168}$$

Now we bring the two subsystems of the composite system into thermodynamic contact, so

that they can communicate via infinitesimal transfer processes described by the differentials $da_l^{(\alpha)}$ and $d\langle O_j^{(\alpha)} \rangle$. The entropy change dS accompanying these infinitesimal transfer processes can be written

$$dS = dS^{(1)} + dS^{(2)} \tag{169}$$

where the entropy change $dS^{(\alpha)}$ associated with subsystem α is given by

$$dS^{(\alpha)} = k_B \Lambda_1^{(\alpha)} [dE^{(\alpha)} + đW^{(\alpha)}] \tag{170}$$

Here, $dE^{(\alpha)} = d\langle \mathcal{H}^{(\alpha)} \rangle$ is the change in the average energy of subsystem α and

$$đW^{(\alpha)} = \Lambda_1^{(\alpha)-1} \left[\sum_{l=1}^{m(\alpha)} f_l^{(\alpha)} da_l^{(\alpha)} + \sum_{j=2}^{n(\alpha)} \Lambda_j^{(\alpha)} d\langle O_j^{(\alpha)} \rangle \right] \tag{171}$$

is the work done by subsystem α in changing its external parameters and averages.

Making use of the conservation relations given by Eqs. (163) and (164), we recast Eq. (169) in the following equivalent form:

$$dS = k_B \sum_{l=1}^{m(1)} [f_l^{(1)} - f_l^{(2)}] da_l^{(1)}$$
$$+ k_B [\Lambda_1^{(1)} - \Lambda_1^{(2)}] d\langle \mathcal{H}^{(1)} \rangle$$
$$+ k_B \sum_{j=2}^{n(1)} [\Lambda_j^{(1)} - \Lambda_j^{(2)}] d\langle O_j^{(1)} \rangle \tag{172}$$

If the composite system is in a state of statistical equilibrium, the entropy S must be a maximum. In view of this extremum property of the entropy, the above entropy change must vanish for arbitrary and independent values of $da_l^{(1)}$ and $d\langle O_j^{(1)} \rangle$ when the composite system is in statistical equilibrium. Assuming this to be the case, it follows that

$$f_l^{(1)} = f_l^{(2)} \tag{173}$$

and

$$\Lambda_j^{(1)} = \Lambda_j^{(2)} \tag{174}$$

which represents a set of conditions of equilibrium between subsystems 1 and 2.

It is quite possible for the composite system to be in a state of statistical equilibrium even though some or none of the external forces and thermodynamic parameters satisfy Eqs. (173) and (174). Equilibrium states of this type are realized when restraints imposed on the composite system prevent the occurrence of some or all the possible transfer processes described by the differentials $da_i^{(\alpha)}$ and $d\langle O_j^{(\alpha)}\rangle$. (If a given transfer process is forbidden, the differential representing that transfer process must be set equal to zero.) In light of the possibility of such restrictions, we conclude that equilibrium between the two subsystems requires identical external forces and thermodynamic parameters to be equal when the transfer processes associated with these quantities are allowed by the restraints placed on the composite system. This is a general thermodynamic criterion for statistical equilibrium between two systems.

If the two subsystems are not in equilibrium, we say that they are out of equilibrium or in a state of disequilibrium. When the composite system is out of equilibrium the entropy change dS associated with an infinitesimal process must be positive. (This statement is equivalent to one version of the second law of thermodynamics.) After the composite system, originally in a state of disequilibrium, has exhausted all allowed spontaneous transfer processes (each with $dS > 0$), the entropy will reach a maximum and the conditions of statistical equilibrium consistent with the imposed restraints will be realized. Any further changes in the macroscopic state of the composite system will be reversible, and the equality $dS = 0$ will hold.

Assuming that the composite system has achieved a state of statistical equilibrium for which $\Lambda_1^{(1)} = \Lambda_1^{(2)} = \Lambda_1$ (energy transfer is allowed), we conclude that the following relation must hold for the composite system undergoing an infinitesimal process:

$$[dE^{(1)} + dW^{(1)}] = -[dE^{(2)} + dW^{(2)}] \quad (175)$$

If the energy increase $dE^{(1)}$ of subsystem 1 is not equal to the work done (energy expended) $dW^{(1)}$ by this subsystem, there will be an energy mismatch given by

$$dQ^{(1)} = dE^{(1)} + dW^{(1)} \quad (176)$$

This energy mismatch is accounted for by an energy transfer $dQ^{(1)}$ from subsystem 2 to subsystem 1. The energy transfer $dQ^{(1)}$ is more commonly called heat transfer.

Rearrangement of Eq. (176) leads to the first law of thermodynamics,

$$dE^{(1)} = dQ^{(1)} - dW^{(1)} \quad (177)$$

which tells us that the energy stored in a system during an infinitesimal quasi-static process is the difference between the heat flux into the system and the work performed by the system.

Making use of the fact that $\Lambda_1^{(1)} = \Lambda_1$, and Eqs. (170) and (176), we write the entropy change $dS^{(1)}$ of subsystem 1 associated with an infinitesimal quasi-static process as

$$dS^{(1)} = k_B\Lambda_1 \, dQ^{(1)} \quad (178)$$

where $k_B\Lambda_1$ plays the role of an integrating factor for converting the inexact differential $dQ^{(1)}$ into an exact differential $dS^{(1)}$. Defining the inverse of the integrating factor $k_B\Lambda_1$ as the absolute temperature T ($T = 1/k_B\Lambda_1$), Eq. (178) assumes the form of one version of the second law of thermodynamics:

$$dS^{(1)} = dQ^{(1)}/T \quad (179)$$

This result tells us that the entropy change associated with a system is connected to the quasi-static heat flux into that system.

The formal development given above constitutes a complete thermodynamic description of the system. The application of these results to the development of the thermodynamics of the various systems discussed in Section II is straightforward. With the maximum entropy principle and the notion of quasistatic processes at our disposal, we showed that the thermodynamics of a system emerges from the statistical mechanics of that system. Unlike conventional approaches to equilibrium statistical mechanics, the information-theoretic approach does not require us to appeal to the phenomenological equations of thermodynamics in order to render the statistical mechanical formalism meaningful.

2. Quantum Systems

The formal development of thermodynamics for classical systems and its connection to equilibrium statistical mechanics required the maximum entropy principle along with the formal definition of entropy and the notion of quasistatic processes. Aside from this, we needed the formal expressions for the averages $\langle O_j\rangle$ and the thermodynamic potential Ω. Since the formal expressions for the averages $\langle \hat{O}_j\rangle$, the thermodynamic potential Ω, and the entropy S for quantum systems are all identical in form to the classical expressions except for the presence of quantum operators rather than classical variables, the thermodynamics that emerges from quantum equilibrium statistical mechanics is identical to the thermodynamics obtained from classical equilibrium statistical mechanics. The only difference between the quantum and classical versions of thermodynamics lies at the microscopic level in the statistical mechanical expressions for the thermodynamic quantities.

IV. Liouville Description of Time Evolution

A. Classical Systems

Consider an isolated system composed of N structureless particles. The state of the system at time t can be defined by specifying the system phase point $\Gamma_t^N = (\mathbf{p}_t^N, \mathbf{q}_t^N)$ in the $6N$-dimensional phase space of the system. The state Γ_t^N of the system changes as it evolves along a path (trajectory) in phase space in accordance with Hamilton's equations,

$$d\vec{q}_j(t)/dt = \partial \mathcal{H}[\mathbf{p}^N(t), \mathbf{q}^N(t)]/\partial \vec{p}_j(t)$$

$$(180)$$

and

$$d\vec{p}_j(t)/dt = -\partial \mathcal{H}[\mathbf{p}^N(t), \mathbf{q}^N(t)]/\partial \vec{q}_j(t)$$

$$(181)$$

where $\vec{q}_j(t)$ and $\vec{p}_j(t)$, respectively, are the coordinate and momentum vectors for particle j at time t, and $\mathcal{H}[\mathbf{p}^N(t), \mathbf{q}^N(t)]$ is the classical Hamiltonian for the system. The system is assumed to be conservative. So the Hamiltonian satisfies the condition $\mathcal{H}[\mathbf{p}^N(t), \mathbf{q}^N(t)] = E$, where E is the energy of the system.

In view of the gross nature of macroscopic measurements, we can never specify exactly the state of a real physical system. There will always be some statistical uncertainty in our measurements. So we adopt a statistical description of classical systems by introducing a time-dependent probability density $\rho(\Gamma^N; t)$, which is defined in such a way that $d\Gamma^N \rho(\Gamma^N; t)$ represents the probability at time t of finding the system in the phase space volume element $d\Gamma^N$ in the neighborhood of the phase point Γ^N. Since the phase points of the system must lie in the system phase space, the probability density $\rho(\Gamma^N; t)$ is normalized as follows:

$$\int d\Gamma^N \rho(\Gamma^N; t) = 1 \qquad (182)$$

It is convenient to think of the probability density $\rho(\Gamma^N; t)$ as describing a fluid occupying the phase space of the system. Adopting this picture, the $6N$-dimensional vector $\dot{\vec{\Gamma}}^N$ represents the velocity of fluid motion and $\rho(\Gamma^N; t)$ represents the fluid density at the point $\dot{\vec{\Gamma}}^N$, where the components of $\dot{\vec{\Gamma}}^N$ are the collection of components of the three-dimensional vectors $\vec{p}_j$ and $\vec{q}_j$ for all the particles in the system. Within the context of the fluid picture of $\rho(\Gamma^N; t)$ the con-

servation of probability can be viewed as the conservation of mass. So $\rho(\Gamma^N; t)$ must evolve according to a continuity equation identical in form to the continuity equation for mass density in fluid mechanics. Thus, we write

$$\partial \rho(\Gamma^N; t)/\partial t = -\vec{\nabla}_{\Gamma^N} \cdot [\dot{\vec{\Gamma}}^N \rho(\Gamma^N; t)] \quad (183)$$

where $\vec{\nabla}_{\Gamma^N}$ is a $6N$-dimensional gradient operator.

The components of the vector $\dot{\vec{\Gamma}}^N$ can be determined from Hamilton's equations. Computing these vector components in this manner, one obtains $\vec{\nabla}_{\Gamma^N} \cdot \dot{\vec{\Gamma}}^N = 0$. Thus, the continuity equation assumes the form

$$\partial \rho(\Gamma^N; t)/\partial t = -\dot{\vec{\Gamma}}^N \cdot \vec{\nabla}_{\Gamma^N} \rho(\Gamma^N; t) \quad (184)$$

The partial derivative $\partial \rho(\Gamma^N; t)/\partial t$ is the time rate of change of the probability density $\rho(\Gamma^N; t)$ at a fixed point in phase space. If we want the time rate of change of $\rho(\Gamma^N; t)$ as seen by an observer moving along a trajectory in phase space, it is necessary to consider the total time derivative

$$d\rho(\Gamma^N; t)/dt = \partial \rho(\Gamma^N; t)/\partial t$$
$$+ \dot{\vec{\Gamma}}^N \cdot \vec{\nabla}_{\Gamma^N} \rho(\Gamma^N; t) \quad (185)$$

It is evident from Eqs. (184) and (185) that the total time derivative $d\rho(\Gamma^N; t)/dt$ vanishes. Thus, the probability density $\rho(\Gamma^N; t)$ remains constant along a given trajectory in phase space.

Utilizing Hamilton's equations to compute the components of the vector $\dot{\vec{\Gamma}}^N$ in Eq. (184), we obtain the classical Liouville equation,

$$\partial \rho(\Gamma^N; t)/\partial t = -i\mathcal{L}(\Gamma^N)\rho(\Gamma^N; t) \quad (186)$$

where the differential operator $\mathcal{L}(\Gamma^N)$, called the Liouville operator, can be written in the following equivalent forms,

$$\mathcal{L}(\Gamma^N) = -i[\vec{\nabla}_{p^N} \mathcal{H}(\mathbf{p}^N, \mathbf{q}^N) \cdot \vec{\nabla}_{q^N}$$
$$- \vec{\nabla}_{q^N} \mathcal{H}(\mathbf{p}^N, \mathbf{q}^N) \cdot \vec{\nabla}_{p^N}] \quad (187a)$$

$$= -i \sum_{l=1}^{N} [\vec{\nabla}_{p_l} \mathcal{H}(\mathbf{p}^N, \mathbf{q}^N) \cdot \vec{\nabla}_{q_l}$$
$$- \vec{\nabla}_{q_l} \mathcal{H}(\mathbf{p}^N, \mathbf{q}^N) \cdot \vec{\nabla}_{p_l}] \quad (187b)$$

$$= -i \sum_{l=1}^{N} \left[\frac{\vec{p}_l}{m_l} \cdot \vec{\nabla}_{q_l} + \vec{F}_l(\mathbf{q}^N) \cdot \vec{\nabla}_{p_l} \right]$$

$$(187c)$$

with

$$\vec{F}_l(\mathbf{q}^N) = -\vec{\nabla}_{q_l} U(\mathbf{q}^N) \qquad (188)$$

In the above, $\vec{\nabla}_{p^N}$ and $\vec{\nabla}_{q^N}$, respectively, are $3N$-

dimensional momentum and coordinate gradient operators, whereas $\vec{\nabla}_{p_l}$ and $\vec{\nabla}_{q_l}$ are three-dimensional gradient operators associated with particle l. The symbol m_l in Eq. (187c) denotes the mass of particle l. The vector $\vec{F}_l(\mathbf{q}^N)$, defined by Eq. (188), is the force on particle l arising from the interaction of this particle with the other particles in the system. In writing Eqs. (187c) and (188), we have assumed that the classical Hamiltonian $\mathcal{H}(\mathbf{p}^N, \mathbf{q}^N)$ can be written in the form

$$\mathcal{H}(\mathbf{p}^N, \mathbf{q}^N) = \sum_{l=1}^{N} \frac{\vec{p}_l \cdot \vec{p}_l}{2m_l} + U(\mathbf{q}^N) \quad (189)$$

where $U(\mathbf{q}^N)$ is the interaction between the particles in the system.

The classical Liouville equation is sometimes written

$$\partial \rho(\mathbf{p}^N, \mathbf{q}^N; t)/\partial t = -\{\mathcal{H}(\mathbf{p}^N, \mathbf{q}^N), \rho(\mathbf{p}^N, \mathbf{q}^N; t)\} \quad (190)$$

where the symbol $\{,\}$ denotes the Poisson bracket, which is defined by

$$\{\mathcal{H}(\mathbf{p}^N, \mathbf{q}^N), \rho(\mathbf{p}^N, \mathbf{q}^N; t)\}$$
$$= \mathcal{H}(\mathbf{p}^N, \mathbf{q}^N)\Lambda(\mathbf{p}^N, \mathbf{q}^N)\rho(\mathbf{p}^N, \mathbf{q}^N; t) \quad (191)$$

In the above,

$$\Lambda(\mathbf{p}^N, \mathbf{q}^N) = \vec{\nabla}_{p^N} \cdot \vec{\nabla}_{q^N} - \vec{\nabla}_{q^N} \cdot \vec{\nabla}_{p^N} \quad (192)$$

is called the Poisson bracket operator. The arrows over the gradient operators indicate their direction of operation when the Poisson bracket operator $\Lambda(\mathbf{p}^N, \mathbf{q}^N)$ is inserted in Eq. (191).

If the initial probability density $\rho(\mathbf{\Gamma}^N; 0)$ is known, we can compute the probability density $\rho(\mathbf{\Gamma}^N; t)$ at time t by utilizing the formal solution of the classical Liouville equation:

$$\rho(\mathbf{\Gamma}^N; t) = \exp[-i\mathcal{L}(\mathbf{\Gamma}^N)t]\rho(\mathbf{\Gamma}^N; 0) \quad (193)$$

Here, $\exp[-i\mathcal{L}(\mathbf{\Gamma}^N)t]$ is the propagator for the probability density.

The classical Liouville equation has the following properties. (1) The classical canonical distribution function is stationary with respect to the classical Liouville equation; (2) the classical Liouville operator $\mathcal{L}(\mathbf{\Gamma}^N)$ is Hermitian; (3) the classical Liouville equation is invariant under the time reversal transformation $t \rightarrow -t$, $\vec{p}_j \rightarrow -\vec{p}_j$, and $\vec{q}_j \rightarrow -\vec{q}_j$; and (4) the Gibbs entropy $S(t)$ for systems out of equilibrium is independent of time when $S(t)$ is computed by using the formal solution $\rho(\mathbf{\Gamma}^N; t)$ of the classical Liouville equation.

Of the properties of the classical Liouville

equation enumerated above, properties (2)–(4) are a bit troublesome. The Hermiticity of the classical Liouville operator $\mathcal{L}(\mathbf{\Gamma}^N)$ implies that its eigenvalues are real. So $\rho(\mathbf{\Gamma}^N; t)$ must exhibit oscillatory temporal behavior and appears not to decay to a unique stationary or equilibrium state in the limit $t \rightarrow \infty$. This raises the question of how we describe the irreversible decay of a system out of equilibrium to a unique equilibrium state. Moreover, the time-reversal invariance of the classical Liouville equation leads us to the conclusion that this equation describes reversible systems with no privileged direction in time. The above problems coupled with the time invariance of the Gibbs entropy raise some serious questions about the compatibility of the second law of thermodynamics, the reversibility of the Liouville equation, the use of Gibbs entropy to describe systems out of equilibrium, and the irreversible decay of a system to a unique equilibrium state. This compatibility problem has preoccupied researchers for many years. At this time, no complete and satisfactory answers have generally been agreed upon.

In principle, the time evolution of a classical dynamic variable $O(\mathbf{\Gamma}^N; t) = O[\mathbf{p}^N(t), \mathbf{q}^N(t)]$ can be determined by solving Hamilton's equations with the initial conditions $\mathbf{p}^N(0) = \mathbf{p}^N$ and $\mathbf{q}^N(0) = \mathbf{q}^N$. Alternatively, $O(\mathbf{\Gamma}^N; t)$ can be determined by utilizing the equation

$$O(\mathbf{\Gamma}^N; t) = \exp[+i\mathcal{L}(\mathbf{\Gamma}^N)t]O(\mathbf{\Gamma}^N) \quad (194)$$

where $\exp[+i\mathcal{L}(\mathbf{\Gamma}^N)t]$ is the propagator for classical dynamic variables. Equation (194) is the formal solution of the equation of motion:

$$dO(\mathbf{\Gamma}^N; t)/dt = i\mathcal{L}(\mathbf{\Gamma}^N)O(\mathbf{\Gamma}^N; t) \quad (195)$$

Given the probability density $\rho(\mathbf{\Gamma}^N; t)$, we can compute the average value $\langle O(t) \rangle$ of a classical dynamic variable $O(\mathbf{\Gamma}^N)$ at time t by utilizing either of the following relations,

$$\langle O(t) \rangle = \int d\mathbf{\Gamma}^N\, O(\mathbf{\Gamma}^N; t)\rho(\mathbf{\Gamma}^N; 0) \quad (196a)$$

$$= \int d\mathbf{\Gamma}^N\, O(\mathbf{\Gamma}^N)\rho(\mathbf{\Gamma}^N; t) \quad (196b)$$

where $\rho(\mathbf{\Gamma}^N; t)$ and $O(\mathbf{\Gamma}^N; t)$, respectively, are given by Eqs. (193) and (194).

The solution of time evolution problems associated with classical systems can be greatly facilitated by introducing a classical phase space representation that plays a role in the description of classical systems in a manner that is formally analogous to the coordinate and momentum representations of quantum mechanics. The state vectors of this representation $\{|\Gamma^N\rangle\}$ enumerate all the accessible phase points for a classical system. The phase function $f(\Gamma^N)$ is regarded as the inner product $f(\Gamma^N) = \langle\Gamma^N|f\rangle$, which can be thought to represent a component of the vector $|f\rangle$ in the classical phase space representation. The application of the classical Liouville operator $\mathscr{L}(\Gamma^N)$ to the phase function $f(\Gamma^N)$ is defined as $\mathscr{L}(\Gamma^N)f(\Gamma^N) \equiv \langle\Gamma^N|\hat{\mathscr{L}}|f\rangle$, where $\hat{\mathscr{L}}$ is an abstract operator to be associated with the differential operator $\mathscr{L}(\Gamma^N)$.

With the classical phase space representation at our disposal, we can write the equations of motion for the density distribution function $\rho(\Gamma^N; t)$ and the classical dynamical variable $O(\Gamma^N; t)$ in the following vector forms:

$$\partial|\rho_t\rangle/\partial t = -i\hat{\mathscr{L}}|\rho_t\rangle \tag{197}$$

$$d|O_t\rangle/dt = i\hat{\mathscr{L}}|O_t\rangle \tag{198}$$

The formal solutions to these equations of motion are given by

$$|\rho_t\rangle = \exp(-i\hat{\mathscr{L}}t)|\rho_0\rangle \tag{199}$$

$$|O_t\rangle = \exp(+i\hat{\mathscr{L}}t)|O\rangle \tag{200}$$

The above results enable us to write the average value of a classical dynamical variable as a matrix element of the propagator $\exp(-i\hat{\mathscr{L}}t)$,

$$\langle O(t)\rangle = \langle O^*|\exp(-i\hat{\mathscr{L}}t)|\rho_0\rangle \tag{201}$$

where the asterisk indicates complex conjugation. Alternatively, we can write $\langle O(t)\rangle$ as $\langle O(t)\rangle = \langle O^*|\rho_t\rangle = \langle O_t^*|\rho_0\rangle$, where $\langle O_t^*| = |O_t^*\rangle^\dagger$, with $|O_t^*\rangle$ given by Eq. (200).

B. Quantum Systems

In quantum mechanics the state of an N-particle system at time t is defined by specifying a state vector $|\psi_t\rangle$. The time evolution of $|\psi_t\rangle$ is governed by the Schrödinger equation,

$$\partial|\psi_t\rangle/\partial t = -(i/\hbar)\hat{\mathscr{H}}|\psi_t\rangle \tag{202}$$

where $\hat{\mathscr{H}}$ is the quantum Hamiltonian for the system. In the principle, we can determine the state vector $|\psi_t\rangle$ from the initial state vector $|\psi\rangle = |\psi_0\rangle$ by using the formal solution of the Schrödinger equation,

$$|\psi_t\rangle = \exp[-(i/\hbar)\hat{\mathscr{H}}t]|\psi\rangle \tag{203}$$

where $\exp[-(i/\hbar)\hat{\mathscr{H}}t]$ is the propagator for the system state vector.

The actual application of Eq. (203) to the description of the time evolution of a quantum system requires us to specify the initial state of the system. Generally, there is some statistical indeterminancy in the initial preparation of a system, so we adopt a statistical description that employs a density operator $\hat{\rho}(0)$ to specify the initial state. As with equilibrium density operators, the density operator $\hat{\rho}(0)$ is assumed to have the following properties: (i) Tr $\hat{\rho}(0) = 1$, (ii) $\hat{\rho}(0)$ is Hermitian, and (iii) the diagonal matrix elements of $\hat{\rho}(0)$ in any representation are positive definite and represent the probabilities for the system to be in the states of that representation.

Given the eigenvectors $\{|\psi\rangle\}$ and eigenvalues $\{P(\psi)\}$ of the density operator $\hat{\rho}(0)$, we can compute the probability $P(\alpha; t)$ of finding the system at the time t in the eigenstate $|\alpha\rangle$ of some Hermitian operator $\hat{A}$ by using the relation

$$P(\alpha; t) = \sum_\psi P(\alpha|\psi; t)P(\psi) \tag{204}$$

From a physical point of view, $P(\psi)$ represents the probability of finding the system at time $t = 0$ in the eigenstate $|\psi\rangle$ of $\hat{\rho}(0)$. The quantity $P(\alpha|\psi; t) = |\langle\alpha|\psi_t\rangle|^2$ is the probability for the system to make a transition from the state $|\psi\rangle$ to the state $|\alpha\rangle$ during the time interval t when the system has been prepared in the state $|\psi\rangle$ at time $t = 0$. $P(\psi)$ is due to the lack of initial information, whereas $P(\alpha|\psi; t)$ is due to the statistical nature of quantum mechanics.

The expression given by Eq. (204) for the probability $P(\alpha; t)$ of finding the system at time t in the state $|\alpha\rangle$ can be rewritten as a diagonal matrix element of a time-dependent density operator $\hat{\rho}(t)$, that is, $P(\alpha; t) = \langle\alpha|\hat{\rho}(t)|\alpha\rangle$, where

$$\hat{\rho}(t) = \exp[-(i/\hbar)\hat{\mathscr{H}}t]\hat{\rho}(0)\exp[+(i/\hbar)\hat{\mathscr{H}}t] \tag{205}$$

with

$$\hat{\rho}(0) = \sum_\psi |\psi\rangle P(\psi)\langle\psi| \tag{206}$$

Although the ψ representation was employed to arrive at Eq. (205), the initial density operator $\hat{\rho}(0)$ need not be expressed in the ψ representation. In actual practice, one usually chooses a representation on the basis of mathematical convenience for the problem at hand.

The result discussed in the preceding paragraph implies that the diagonal matrix elements of the density operator $\hat{\rho}(t)$ in the α representation give us the probabilities of finding the system at time t in the states of that representation. The off-diagonal matrix elements of $\hat{\rho}(t)$ in the α representation provide us with information about the phase coherence between the states of this representation at time t. These remarks apply to any representation for which the matrix elements of $\hat{\rho}(t)$ are defined.

The time derivative of Eq. (205) gives us an equation of motion for the density operator $\hat{\rho}(t)$,

$$\partial\hat{\rho}(t)/\partial t = -(i/\hbar)[\mathscr{H}, \hat{\rho}(t)]_- \qquad (207)$$

where the subscript minus sign in $[\mathscr{H}, \hat{\rho}(t)]_-$ indicates that this quantity is a commutator. Equation (207) is the quantum analog of the classical Liouville equation, so we call it the quantum Liouville equation.

The quantum Liouville equation given by Eq. (207) can be brought into a form that more closely resembles the classical Liouville equation by introducing a quantum Liouville operator $\hat{\mathscr{L}}$,

$$\hat{\mathscr{L}} = (1/\hbar)\hat{\mathscr{H}}^- \qquad (208)$$

where the operator $\hat{\mathscr{H}}^-$ is defined in such a way that

$$\hat{\mathscr{H}}^-\hat{A} = [\mathscr{H}, \hat{A}]_- \qquad (209)$$

for any operator $\hat{A}$. With the above definition of the quantum Liouville operator $\hat{\mathscr{L}}$, we cast Eq. (207) in the form

$$\partial\hat{\rho}(t)/\partial t = -i\hat{\mathscr{L}}\hat{\rho}(t) \qquad (210)$$

The symbol $\hat{}$ over the operators $\hat{\mathscr{L}}$ and $\hat{\mathscr{H}}^-$ indicates that these operators are tetradic operators, that is, operators that require four indices in their matrix representation.

If the initial density operator $\hat{\rho}(0)$ is known, the density operator $\hat{\rho}(t)$ at time t can be determined by using Eq. (205), which corresponds to the formal solution of the quantum Liouville equation given by Eq. (207). In view of the equivalence of Eqs. (207) and (210), $\hat{\rho}(t)$ can also be written

$$\hat{\rho}(t) = \exp(-i\hat{\mathscr{L}}t)\hat{\rho}(0) \qquad (211)$$

where $\exp(-i\hat{\mathscr{L}}t)$ is the propagator for the density operator. Clearly,

$$\exp(-i\hat{\mathscr{L}}t)\hat{\rho}(0) = \exp[-(i/\hbar)\mathscr{H}t]\hat{\rho}(0)$$
$$\times \exp[+(i/\hbar)\mathscr{H}t] \qquad (212)$$

The basic properties of the classical Liouville equation and the troublesome questions they raise are also shared by the quantum Liouville equation. For the quantum case, we summarize these properties as follows. (1) The canonical density operator is stationary with respect to the quantum Liouville equation; (2) the quantum Liouville operator $\hat{\mathscr{L}}$ is Hermitian; (3) the quantum Liouville equation is time-reversal invariant; and (4) the Gibbs entropy $S(t)$ for systems

out of equilibrium is independent of time when $S(t)$ is computed by using the formal solution $\hat{\rho}(t)$ of the quantum Liouville equation.

Given the density operator $\hat{\rho}(t)$, we can compute the average value $\langle\hat{O}(t)\rangle$ of a quantum dynamical variable $\hat{O}$ at time t by using the relation

$$\langle\hat{O}(t)\rangle = \text{Tr } \hat{O}\hat{\rho}(t) \qquad (213)$$

When this relation is employed to compute $\langle\hat{O}(t)\rangle$, we say that we are working in the Schrödinger representation. Alternatively, we can compute $\langle\hat{O}(t)\rangle$ by working in the Heisenberg representation,

$$\langle\hat{O}(t)\rangle = \text{Tr } \hat{O}(t)\hat{\rho}(0) \qquad (214)$$

where

$$\hat{O}(t) = \exp[+(i/\hbar)\mathscr{H}t]\hat{O}\exp[-(i/\hbar)\mathscr{H}t]$$
$$(215)$$

If the system is initially prepared in some state $|\psi\rangle$, then $\hat{\rho}(0) = |\psi\rangle\langle\psi|$. For this case, Eqs. (213) and (214) assume the usual forms for quantum averages employed in quantum mechanics:

$$\langle\hat{O}(t)\rangle = \langle\psi_t|\hat{O}|\psi_t\rangle \qquad (216a)$$
$$= \langle\psi|\hat{O}(t)|\psi\rangle \qquad (216b)$$

The time evolution of the quantum dynamical $\hat{O}(t)$ can be generated by means of the unitary transformation given by Eq. (215), which corresponds to a formal solution of Heisenberg's equation of motion:

$$d\hat{O}(t)/dt = (i/\hbar)[\mathscr{H}, \hat{O}(t)]_- \qquad (217)$$

By introducing the formal definition of the quantum Liouville operator $\hat{\mathscr{L}}$, this equation of motion can be cast in a form that resembles the equation of motion given by Eq. (195) for classical dynamical variables. We write

$$d\hat{O}(t)/dt = i\hat{\mathscr{L}}\hat{O}(t) \qquad (218)$$

In view of the formal equivalence of Eqs. (217) and (218), the time evolution of $\hat{O}(t)$ can be generated not only by Eq. (215) but also by

$$\hat{O}(t) = \exp(+i\hat{\mathscr{L}}t)\hat{O} \qquad (219)$$

which corresponds to the formal solution of Eq. (218).

Matrix representations of the quantum Liouville equation and Heisenberg's equation of motion can be obtained by sandwiching these equations of motion between the basis vectors of any complete orthonormal basis $\{|\phi_k\rangle\}$ defined in the same space as $\mathscr{H}$ and $\hat{\rho}(t)$. Forming such matrix elements, we write

$$\frac{\partial \rho(k, l; t)}{\partial t} = -i \sum_{m,n} \mathcal{L}(k, l; m, n)\rho(m, n; t)$$

$$(220)$$

and

$$\frac{dO(k, l; t)}{dt} = i \sum_{m,n} \mathcal{L}(k, l; m, n)O(m, n; t)$$

$$(221)$$

where

$$\rho(k, l; t) = \langle \phi_k | \hat{\rho}(t) | \phi_l \rangle$$

$$O(k, l; t) = \langle \phi_k | \hat{O}(t) | \phi_l \rangle$$

$$\mathcal{L}(k, l; m, n) \equiv (1/\hbar)[\mathcal{H}(k, m) \, \delta_{l,n}$$
$$- \mathcal{H}(n, l) \, \delta_{k,m}] \qquad (222)$$

with

$$\mathcal{H}(k, m) = \langle \phi_k | \hat{\mathcal{H}} | \phi_m \rangle$$

Although these matrix forms of the quantum Liouville equation and Heisenberg's equation of motion are formally correct, the solution of time evolution problems can be more readily accomplished by working in a representation called the superstate representation. The basis vectors of this representation $\{|N_{kl}\rangle\}$ are abstract states that are associated with the set of operators $\{\hat{N}_{kl} = |\phi_k\rangle\langle\phi_l|\}$ formed from the set of basis vectors $\{|\phi_k\rangle\}$ used in the formulation of Eqs. (220) and (221). The matrix element $A(k, l) = \langle\phi_k|\hat{A}|\phi_l\rangle$ of some operator $\hat{A}$ can be regarded as the inner product $A(k, l) = \langle N_{kl}|A\rangle$, which represents a component of the vector $|A\rangle$ in the superstate representation. The matrix elements of the quantum Liouville operator $\hat{\mathcal{L}}$ are defined as $\mathcal{L}(k, l; m, n) = \langle N_{kl}|\hat{\mathcal{L}}|N_{mn}\rangle$, where $\hat{\mathcal{L}}$ is an abstract operator associated with the operator $\hat{\mathcal{L}}$.

With the superstate representation at our disposal, we can cast the quantum Liouville equation and Heisenberg's equation of motion into vector forms identical to the vector equations of motion given by Eqs. (197) and (198) for classical systems. The only difference between the vector equations of motion for $|\rho_t\rangle$ and $|O_t\rangle$ for classical and quantum systems is the manner in which the components of these vectors and the matrix elements of $\hat{\mathcal{L}}$ are computed. Nonetheless, the matrix form for the average of a quantum dynamical variable does differ from the classical results. For the quantum case, we write

$$\langle \hat{O}(t) \rangle = \langle O^\dagger | \exp(-i\hat{\mathcal{L}}t) | \rho_0 \rangle \qquad (223)$$

where the dagger indicates the Hermitian conjugate. Alternatively, we can write $\langle \hat{O}(t) \rangle = \langle O^\dagger | \rho_t \rangle = \langle O_t^\dagger | \rho_0 \rangle$, where $\langle O_t^\dagger | = |O_t^\dagger\rangle^\dagger$, with $|O_t^\dagger\rangle$ given by Eq. (200).

V. Phenomenological Descriptions of Time Evolution

A. Linear Phenomenological Equations of Motion

The quantum and classical Liouville equations are rarely used in the actual characterization of experimental data concerning the temporal and spectral properties of real physical systems. Instead, we usually adopt phenomenological equations with a few adjustable parameters (phenomenological parameters) that are determined by fitting the phenomenological equations to experimental data. Generally, phenomenological equations describe the motion of a few degrees of freedom or gross observables with much of the underlying microscopic information buried in the phenomenological parameters.

The most commonly used phenomenological equations are linear, irreversible, and Markovian (without memory). Such equations can be cast in a vector form that resembles the vector form of the quantum and classical Liouville equations. Nonetheless, the character of linear, irreversible, Markovian equations is quite different from that of the Liouville equations. Usually, the former type of equations have all or some of the following properties: (1) The density operator or density distribution function corresponds to a reduced or contracted description of a system. (2) The Liouville operator is non-Hermitian with both Hermitian and anti-Hermitian components, and thus the system exhibits damped oscillatory behavior and decays to a unique stationary state (usually thermal equilibrium) in the infinite time limit $t \to \infty$. (3) The equations of motion possess broken time-reversal symmetry; that is, they are not invariant with respect to time-reversal transformations. (4) The Gibbs entropy $S(t)$ computed with the solutions of these equations satisfies the H theorem $S(t) \geq S(0)$.

Examples of linear, irreversible, Markovian equations that possess some or all the above-enumerated properties include the Fokker–Planck, Smoluchowski, Bloch, master, nonadiabatic Fokker–Planck, and stochastic

Liouville equations. Many of these equations, as well as other phenomenological equations, were originally developed on the basis of intuitive arguments about the nature of a system. For the most part, phenomenological equations have been quite successful in the codification of large amounts of experimental data on different systems. This success has generated much theoretical work concerned with the construction of phenomenological equations from basic principles, thus affording us with formal expressions that relate phenomenological parameters to microscopic interactions. In Section VI, we discuss some of the approaches that have been employed to obtain such expressions.

Aside from constructing phenomenological equations and formal expressions for phenomenological parameters, researchers have focused much attention on the construction of new phenomenological equations, studying the range of validity of known phenomenological equations and generalizing such equations to include non-Markovian retardation and nonlinear behavior. In addition, considerable effort has been made to develop powerful techniques that enable us to compute quantities of interest without having to obtain solutions to global equations of motion.

An example of a linear, irreversible, Markovian equation is the N-particle classical Fokker–Planck equation,

$$\partial \rho(\Gamma^N; t)/\partial t = [-i\mathscr{L}_S(\Gamma^N) + \mathbf{L}_{FP}(\Gamma^N)]\rho(\Gamma^N; t) \tag{224}$$

where $\rho(\Gamma^N; t)$ is the density distribution function for the N particles,

$$\mathscr{L}_S(\Gamma^N) = -i \sum_{l=1}^{N} \left[\frac{\vec{p}_l}{m_l} \cdot \vec{\nabla}_{q_l} \right.$$
$$\left. + \vec{F}_l(\mathbf{q}^N) \cdot \vec{\nabla}_{p_l} \right] \tag{225}$$

and

$$\mathbf{L}_{FP}(\Gamma^N) = \sum_{l,n=1}^{N} \vec{\nabla}_{p_l} \cdot \vec{\vec{\xi}}_{l,n}(\mathbf{q}^N)$$
$$\cdot [\vec{p}_n + (m_n k_B T) \vec{\nabla}_{p_n}] \tag{226}$$

Here, $\vec{F}_l(\mathbf{q}^N) = -\vec{\nabla}_{q_l} \bar{U}(\mathbf{q}^N)$ denotes the mean force on particle l due to the mean potential $\bar{U}(\mathbf{q}^N)$ obtained by averaging the total interaction potential for the N particles plus bath over the equilibrium distribution of the bath (assumed to be at the temperature T) and $\vec{\vec{\xi}}_{l,n}(\mathbf{q}^N)$ is a friction tensor that is spatially dependent. The streaming operator $\mathscr{L}_S(\Gamma^N)$ describes the reversible motion of the N particles in a bath at thermal equilibrium. Damping of the reversible mo-

tion is brought about by the Fokker–Planck operator $\mathbf{L}_{FP}(\Gamma^N)$, which leads to energy dissipation. In the infinite time limit $t \rightarrow \infty$, the solution of the Fokker–Planck equation assumes the form of the classical canonical distribution function.

The Fokker–Planck operator $\mathbf{L}_{FP}(\Gamma^N)$ appearing in the Fokker–Planck equation leads to the damping of the momentum degrees of freedom, which is communicated to the spatial degrees of freedom through the coupling terms $(\vec{p}_l/m_l) \cdot \vec{\nabla}_{q_l}$ and $\vec{F}_l(\mathbf{q}^N) \cdot \vec{\nabla}_{p_l}$ in the streaming operator $\mathscr{L}_S(\Gamma^N)$. The momentum relaxation drives the relaxation of the spatial degrees of freedom. On time scales that are long compared with the time scale for momentum relaxation, it is thought that the momentum degrees of freedom are essentially in thermal equilibrium, whereas the spatial degrees of freedom are significantly out of equilibrium. For such cases, the Fokker–Planck equation is assumed to reduce to the Smoluchowski equation,

$$\frac{\partial \rho(\mathbf{q}^N; t)}{\partial t} = \sum_{l,n=1}^{N} \vec{\nabla}_{q_l} \cdot \vec{\vec{D}}_{l,n}(\mathbf{q}^N)$$
$$\cdot \left[\vec{\nabla}_{q_n} - \frac{1}{k_B T} \vec{\vec{F}}_n(\mathbf{q}^N) \right] \rho(\mathbf{q}^N; t) \tag{227}$$

where $\rho(\mathbf{q}^N; t)$ is a spatial density distribution function and $\vec{\vec{D}}_{l,n}(\mathbf{q}^N)$ is a spatially dependent diffusion tensor. In the infinite time limit $t \rightarrow \infty$, the solution of the Smoluchowski equation assumes the form of the classical canonical spatial distribution function $\exp[\Omega_S - (1/k_B T)\bar{U}(\mathbf{q}^N)]$. The Smoluchowski and Fokker–Planck equations have been used to describe a diversity of phenomena, including coagulation, dynamics of colloidal systems, electrolytic processes, chemical reactions, ion transport in biological systems, diffusion of particles on surfaces, and sedimentation.

In the areas of photophysics and magnetic resonance, purely quantum irreversible equations of motion are often employed to describe the time evolution of excited states of atoms and molecules in a heat bath. A commonly employed irreversible equation of motion is the multistate Bloch equation, which is sometimes written in the form

$$\frac{\partial \rho(k, l; t)}{\partial t} = \sum_{m,n} [-i\mathscr{L}_S(k, l; m, n)$$
$$- R(k, l; m, n)]\rho(m, n; t) \tag{228}$$

where

$$\mathscr{L}_S(k, l; m, n) = (1/\hbar)[\bar{\mathscr{H}}(k, m)\, \delta_{l,n}$$
$$- \bar{\mathscr{H}}(n, l)\, \delta_{k,m}] \quad (229)$$

and $R(k, l; m, n)$ denotes a matrix element of a non-Hermitian relaxation tetradic $\hat{\hat{R}}$. The matrix elements $\bar{\mathscr{H}}(k, m) = \langle \phi_k | \hat{\bar{\mathscr{H}}} | \phi_m \rangle$ are those of a mean Hamiltonian $\bar{\mathscr{H}}$ obtained by averaging the Hamiltonian for the system of interest plus bath over the equilibrium density operator of the bath (assumed to be at the temperature T). The reversible motion of the quantum system in a bath at thermal equilibrium is described by the matrix elements $\mathscr{L}_S(k, l; m, n)$ of the tetradic operator $\hat{\hat{\mathscr{L}}}_S = (1/\hbar)\hat{\bar{\mathscr{H}}}^-$. Damping of this reversible motion is brought about by processes described by the matrix elements of the relaxation tetradic $\hat{\hat{R}}$. Matrix elements of the type $R(k, k; l, l)$ describe incoherent transitions between quantum states, that is, transitions that preserve the phase coherence of the system. The interruption of phase coherence (dephasing) is described by matrix elements of the type $R(k, l; m, n)$ for $k \neq l$. The matrix elements of the relaxation tetradic are usually defined in such a way that the quantum system achieves thermal equilibrium in the infinite time limit $t \to \infty$.

It is generally believed that the phase relaxation of quantum system occurs on a much shorter time scale than population decay. For time scales much longer than the time scale for phase relaxation (dephasing), the Bloch equations are assumed to take the form of a Pauli master (kinetic) equation,

$$\partial\rho(k, k; t)/\partial t = -\rho(k, k; t)W(k\to)$$
$$+ \sum_{l \neq k} \rho(l, l; t)W(l \to k) \quad (230)$$

where $W(l \to k)$ is a rate constant for the transition from the state $|\phi_l\rangle$ to the state $|\phi_k\rangle$ and $W(k\to) = \Sigma_{l \neq k} W(k \to l)$ is the total rate constant for transitions out of the state $|\phi_k\rangle$. The rate constants are usually required to satisfy the principle of detailed balance, that is, $\rho(k, k; \infty)\,W(k \to l) = \rho(l, l; \infty)W(l \to k)$, and thus the state populations $\{\rho(k, k; t)\}$ decay to stationary values (usually thermal populations) in the infinite time limit.

B. Classical Brownian Motion Theory

Around the turn of the century researchers in the area of Brownian motion theory were preoccupied with the irregular motion exhibited by colloidal-sized particles immersed in a fluid. Since then the mathematical apparatus of Brownian theory has crept into a number of disciplines and has been used to treat a range of problems involving systems from the size of atoms to systems of stellar dimensions. For the sake of clarity, we will not consider such a range of problems, confining our attention to the more traditional problem of describing an N-particle system immersed in a heat bath.

In classical Brownian motion theory one usually assumes that the time evolution of the density distribution function $\rho(\Gamma^N; t)$ for N particles immersed in a heat bath is governed by the Chapman–Kolmogorov equation,

$$\rho(\Gamma^N; t + \Delta t)$$
$$= \int d\Gamma'^N\, P(\Gamma^N|\Gamma'^N; \Delta t)\rho(\Gamma'^N; t) \quad (231)$$

where $P(\Gamma^N|\Gamma'^N; \Delta t)$ is a conditional transition probability representing the probability of the system making a transition from the phase point Γ'^N to the phase point Γ^N during the time interval Δt. Self-consistency requires the conditional transition probabilities to be normalized to unity, that is, $\int d\Gamma^N\, P(\Gamma^N|\Gamma'^N; \Delta t) = 1$, and to satisfy the boundary condition $P(\Gamma^N|\Gamma'^N; \Delta t = 0) = \delta(\Gamma^N - \Gamma'^N)$. Here, Δt is the time scale of macroscopic measurements, that is, the time resolution of the observer. It is assumed that Δt is short on the time scale t_S characterizing the motion of the N particles (the system) and long on the time scale t_B characterizing the response of the bath to the motion of the system and bath reequilibration. This assumption requires the existence of a separation in the time scales t_S and t_B for the evolution of the system and bath, the latter being faster ($t_B \ll t_S$), so that $t_B \ll \Delta t \ll t_S$.

It is convenient to introduce a rate constant $W(\Gamma'^N \to \Gamma^N; \Delta t)$ to characterize the average rate at which the N-particle system passes from the phase point Γ'^N to the phase point Γ^N during the time interval Δt. We define $W(\Gamma'^N \to \Gamma^N; \Delta t)$ by

$$P(\Gamma'^N \to \Gamma^N; \Delta t)$$
$$= \delta(\Gamma^N - \Gamma'^N) + \Delta t\, W(\Gamma'^N \to \Gamma^N; \Delta t)$$
$$(232)$$

With this definition, we can readily convert the Chapman–Kolmogorov equation to a phase space master equation,

$$\frac{\partial \rho(\mathbf{\Gamma}^N; t | \Delta t)}{\partial t}$$

$$= \int d\mathbf{\Gamma}'^N \, W(\mathbf{\Gamma}'^N \rightarrow \mathbf{\Gamma}^N; \Delta t)\rho(\mathbf{\Gamma}'^N; t) \tag{233}$$

where $\partial \rho(\mathbf{\Gamma}^N; t | \Delta t)/\partial t$ is a phenomenological (macroscopic) time derivative defined by

$$\frac{\partial \rho(\mathbf{\Gamma}^N; t | \Delta t)}{\partial t}$$

$$= [\rho(\mathbf{\Gamma}^N; t + \Delta t) - \rho(\mathbf{\Gamma}^N; t)]/\Delta t \tag{234a}$$

$$= \Delta t^{-1} \int_t^{t+\Delta t} dt' \, \frac{\partial \rho(\mathbf{\Gamma}^N; t')}{\partial t'} \tag{234b}$$

The phenomenological time derivative is a coarse-grained time derivative obtained by time averaging the instantaneous time derivative $\partial \rho(\mathbf{\Gamma}^N; t')/\partial t'$ over the time scale Δt of macroscopic measurements. From a macroscopic point of view, the phenomenological time derivative can be treated as an instantaneous time derivative. In adopting this attitude, one is mimicking the kind of time smoothing actually done in the analysis of experiments, which are always coarse-grained in time.

By performing a derivate moment expansion of the rate constants appearing in the phase space master equation, one can convert this integral equation to an equivalent differential equation called the generalized Fokker–Planck equation:

$$\frac{\partial \rho(\mathbf{\Gamma}^N; t | \Delta t)}{\partial t}$$

$$= \sum_{s_1=0}^{\infty} \cdots \sum_{s_{6N}=0}^{\infty} \frac{(-1)^{(s_1 + \cdots + s_{6N})}}{s_1! \cdots s_{6N}!}$$

$$\times \nabla_{\Gamma_1}^{s_1} \cdots \nabla_{\Gamma_{6N}}^{s_{6N}} [\mathbb{K}_{\Gamma_1,\ldots,\Gamma_{6N}}^{(s_1 + \cdots + s_{6N})}(\mathbf{\Gamma}^N)$$

$$\times \rho(\mathbf{\Gamma}^N; t)] \tag{235}$$

where the derivate moments

$$\mathbb{K}_{\Gamma_1,\ldots,\Gamma_{6N}}^{(s_1 + \cdots + s_{6N})}(\mathbf{\Gamma}^N)$$

are defined by

$$\mathbb{K}_{\Gamma_1,\ldots,\Gamma_{6N}}^{(s_1 + \cdots + s_{6N})}(\mathbf{\Gamma}^N)$$

$$= \int d\mathbf{\Gamma}'^N (\Gamma_1' - \Gamma_1)^{s_1} \cdots$$

$$\times (\Gamma_{6N}' - \Gamma_{6N})^{s_{6N}} W(\mathbf{\Gamma}^N \rightarrow \mathbf{\Gamma}'^N; \Delta t) \tag{236}$$

with Γ_l denoting a component of the $6N$-dimensional vector $\mathbf{\Gamma}^N$. The derivate moments charac-

terize the distribution of spatial and momentum transitions in phase space.

If the derivate moments higher than second order are neglected, Eq. (235) assumes the form of a so-called linear Fokker–Planck equation,

$$\partial \rho(\mathbf{\Gamma}^N; t | \Delta t)/\partial t$$

$$= -\vec{\nabla}_{\Gamma^N} \cdot [\vec{\mathbb{K}}_{\Gamma^N}^{(1)}(\mathbf{\Gamma}^N)\rho(\mathbf{\Gamma}^N; t)]$$

$$+ \tfrac{1}{2}\vec{\vec{\nabla}}_{\Gamma^N} : [\vec{\vec{\mathbb{K}}}_{\Gamma^N,\Gamma^N}^{(2)}(\mathbf{\Gamma}^N)\rho(\mathbf{\Gamma}^N; t)] \tag{237}$$

where the components of $\vec{\mathbb{K}}_{\Gamma^N}^{(1)}(\mathbf{\Gamma}^N)$, denoted by $\mathbb{K}_{\Gamma_i}^{(1)}(\mathbf{\Gamma}^N)$, are called "drift" coefficients,

$$\mathbb{K}_{\Gamma_i}^{(1)}(\mathbf{\Gamma}^N) = \int d\mathbf{\Gamma}'^N (\Gamma_i' - \Gamma_i) \frac{P(\mathbf{\Gamma}'^N | \mathbf{\Gamma}^N; \Delta t)}{\Delta t} \tag{238}$$

and the components of $\vec{\vec{\mathbb{K}}}_{\Gamma^N,\Gamma^N}^{(2)}(\mathbf{\Gamma}^N)$, denoted by $\mathbb{K}_{\Gamma_i,\Gamma_j}^{(2)}(\mathbf{\Gamma}^N)$, are called "diffusion" coefficients,

$$\mathbb{K}_{\Gamma_i,\Gamma_j}^{(2)}(\mathbf{\Gamma}^N) = \int d\mathbf{\Gamma}'^N (\Gamma_i' - \Gamma_i)(\Gamma_j' - \Gamma_j)$$

$$\times \frac{P(\mathbf{\Gamma}'^N | \mathbf{\Gamma}^N; \Delta t)}{\Delta t} \tag{239}$$

The neglect of derivative moments higher than second order is equivalent to assuming that the spatial and momentum transitions in phase space are short ranged.

At this point, it is usually assumed that the average of the classical variables over the conditional transition probabilities, so-called stochastic averaging, is equivalent to averaging the classical equations of motion over the equilibrium distribution of the bath. More specifically,

$$\mathbb{K}_{\Gamma_i}^{(1)}(\mathbf{\Gamma}^N) = \langle [\Gamma_i(\Delta t) - \Gamma_i(0)] \rangle_B / \Delta t \tag{240}$$

and

$$\mathbb{K}_{\Gamma_i,\Gamma_j}^{(2)}(\mathbf{\Gamma}^N) = \langle [\Gamma_i(\Delta t) - \Gamma_i(0)]$$

$$\times [\Gamma_j(\Delta t) - \Gamma_j(0)] \rangle_B / \Delta t \tag{241}$$

where $\langle \ \rangle_B$ denotes averaging over the equilibrium distribution of the bath.

The actual evaluation of Eqs. (240) and (241) requires us to solve the classical equations of motion for the N particles plus bath. This many-body problem is often circumvented by introducing a stochastic description of the N-particle motion with the heat bath treated as a source of thermal noise. This is accomplished by introducing Langevin equations, which correspond to a mean field version of Hamilton's equations for the N particles augmented with additional stochastic terms intended to represent the influence of the bath on the N-particle motion.

For the case of a single particle of mass m in a spatially homogeneous bath, we write the Langevin equations as

$$d\vec{p}(t)/dt = -\xi\vec{p}(t) + \vec{f}(t) \quad (242)$$

and

$$d\vec{q}(t)/dt = \vec{p}(t)/m \quad (243)$$

The first term on the right side of Eq. (242) is a frictional force due to the viscous drag exerted on the particle by the bath. The assumption that the frictional force is proportional to the particle momentum $\vec{p}$ is motivated by Stokes's law. (According to Stokes's law, the frictional force on a spherical particle of mass m and radius r in a medium of viscosity η is $-\xi\vec{p}$, where $\xi = 6\pi r\eta/m$ is called the friction coefficient.) As in Stokes's law, we call the parameter ξ the friction coefficient. The second term on the right side of Eq. (242), $\vec{f}(t)$, is a fluctuating force intended to represent the collisions of the particle of interest with the particles of the bath.

The force $\vec{f}(t)$ is assumed to be a random force that fluctuates on time scales much smaller than the time scale characterizing the motion of the momentum vector $\vec{p}(t)$. The random nature of $\vec{f}(t)$ is expressed by

$$\langle f_\alpha(t)\rangle_B = 0 \quad (244)$$

where the index α indicates a Cartesian component of the vector $\vec{f}(t)$. Since the force $\vec{f}(t)$ is assumed to vary on a much shorter time scale than $\vec{p}(t)$, the time evolutions of $\vec{p}(t)$ and $\vec{f}(t)$ are expected to be uncorrelated; that is, $p_{\alpha_1}(t_1)$ is independent of $f_{\alpha_2}(t_2)$ and vice versa. So we write

$$\langle p_{\alpha_1}(t_1)f_{\alpha_2}(t_2)\rangle_B = 0 \quad (245)$$

In addition to the above assumptions, the force $\vec{f}(t)$ is assumed to be δ correlated in time and Gaussian, that is,

$$\langle f_{\alpha_1}(t_1)f_{\alpha_2}(t_2)\rangle_B = 2mk_BT\xi\,\delta_{\alpha_1,\alpha_2}\,\delta(t_1 - t_2) \quad (246)$$

and

$$\langle f_{\alpha_1}(t_1)f_{\alpha_2}(t_2)\cdots f_{\alpha_n}(t_n)\rangle_B = 0 \quad (247)$$

for $n \geq 3$. Note that the right side of Eq. (246) scales as the temperature T of the bath. The motivation for this scaling and endowing the force $\vec{f}(t)$ with the above properties is to bias the dynamics in such a way that the average kinetic energy of the particle $\langle\vec{p}(t)\cdot\vec{p}(t)\rangle/2m$ decays to its equipartition value $(3k_BT/2)$ in the infinite time limit $t \to \infty$. In other words, so that the particle of interest will come to thermal equilibrium with the bath.

The formal solution of the Langevin equations is given by

$$\vec{p}(t) = \psi(\xi; t)\vec{p}(0) + \int_0^t d\tau\,\psi(\xi; t - \tau)\vec{f}(\tau)$$

$$(248)$$

$$\vec{q}(t) = \vec{q}(0) + (1/m\xi)\vec{p}(0)[1 - \psi(\xi; t)]$$

$$+ (1/m\xi)\int_0^t d\tau\,[1 - \psi(\xi; t - \tau)]\vec{f}(\tau)$$

$$(249)$$

where

$$\psi(\xi; t) = \exp(-\xi t) \quad (250)$$

With these formal results and the above assumptions about the nature of $\vec{f}(t)$, we can readily compute the "drift" and "diffusion" coefficients given by Eqs. (240) and (241). Before doing this, it is instructive to make use of the above results to explore some additional features of Langevin dynamics.

Averaging the formal solution for $\vec{p}(t)$, given by Eq. (248), over the initial density distribution function for the particle plus bath, assuming the bath is at thermal equilibrium, we find that the average values of the Cartesian components of the initial average momentum $\langle\vec{p}(0)\rangle$ all decay exponentially to zero with the same relaxation time $\tau = \xi^{-1}$. So we think of the inverse of the friction coefficient ξ as providing a measure of the time scale for momentum relaxation.

After the particle's initial momentum $\langle\vec{p}(0)\rangle$ has come to thermal equilibrium, there will be equilibrium momentum fluctuations. Such spontaneous equilibrium fluctuations are described by the equilibrium time correlation function

$$C_{p,p}(t) = \langle\vec{p}(t)\cdot\vec{p}(0)\rangle_{EQ} \quad (251)$$

where $\langle\ \rangle_{EQ}$ denotes averaging over the equilibrium distribution for the particle of interest plus bath. The time evolution of this time correlation function can be readily determined by utilizing Eq. (248). The result is

$$C_{p,p}(t) = \exp(-\xi t)C_{p,p}(0) \quad (252)$$

The relaxation time $\tau_C = \xi^{-1}$ provides a measure of the time scale for which the momentum fluctuations are correlated. So we call τ_C the correlation time for the equilibrium time correlation function $C_{p,p}(t)$.

We can also examine spontaneous equilibrium fluctuations by working in the frequency domain and considering the spectral density of equilibrium time correlation functions. For the case of momentum fluctuations, we consider the spectral density

$$\bar{C}_{p,p}(i\omega) = \int_0^\infty dt \, \exp(-i\omega t) C_{p,p}(t) \quad (253a)$$

$$= mk_B T/(i\omega + \xi) \quad (253b)$$

Note that the real part of this spectral density is a Lorentzian with a maximum about $\omega = 0$ and a halfwidth of 2ξ, which provides a measure of the correlation time $\tau_C = \xi^{-1}$ for momentum fluctuations. This correlation time τ_C can be written

$$\tau_C = (1/mk_B T) \lim_{\omega \to 0} \bar{C}_{p,p}(i\omega) \quad (254)$$

Later, we shall find that this result provides us with a formal expression for a quantity $D = (k_B T/m\xi)$, called the diffusion coefficient:

$$D = \frac{1}{3m^2} \int_0^\infty dt \, \langle \vec{p}(t) \cdot \vec{p}(0) \rangle_{EQ} \quad (255)$$

Spontaneous equilibrium fluctuations in the force $\vec{f}(t)$ appearing in the Langevin equations are described by the equilibrium time correlation function:

$$C_{f,f}(t) = \langle \vec{f}(t) \cdot \vec{f}(0) \rangle_{EQ} \quad (256a)$$

$$= 6mk_B T\xi \, \delta(t) \quad (256b)$$

Such fluctuations are δ-correlated in time by assumption.

Now the spectral density of the force fluctuations is given by

$$\bar{C}_{f,f}(i\omega) = \int_0^\infty dt \, \exp(-i\omega t) \langle \vec{f}(t) \cdot \vec{f}(0) \rangle_{EQ}$$

$$\quad (257a)$$

$$= 2mk_B T\xi \quad (257b)$$

Since the force $\vec{f}(t)$ is δ-correlated in time, the spectral density $\bar{C}_{f,f}(i\omega)$ is completely flat (frequency independent). Such a spectrum is called a white spectrum, and its source is called white noise.

It is clear from Eqs. (257a) and (257b) that the friction coefficient ξ is given by the zero frequency limit of the spectral density of spontaneous force fluctuations. Alternatively, we can write

$$\xi = (\tfrac{1}{3}mk_B T) \lim_{\omega \to 0} \bar{C}_{\dot{p},\dot{p}}(i\omega) \quad (258a)$$

$$= \left(\frac{1}{9mk_B T}\right) \int_0^\infty dt \, \langle \dot{\vec{p}}(0) \cdot \dot{\vec{p}}(t) \rangle_{EQ}$$

$$\quad (258b)$$

From these results, we see that the frictional damping of the particle momentum arises from spontaneous equilibrium fluctuations in the force acting on the particle. This is sometimes called a fluctuation–dissipation relation.

Now that we have completed our exploration of Langevin dynamics, let us return to the use of Langevin dynamics to compute the "drift" and "diffusion" coefficients appearing in the linear Fokker–Planck equation. Substituting Eqs. (248) and (249) into Eqs. (240) and (241), we can readily demonstrate that the following are the only nonnegligible "drift" and "diffusion" coefficients for time scales Δt much shorter than the time scale ξ^{-1} for momentum relaxation, that is, $\xi \, \Delta t \ll 1$:

$$\mathbb{K}_{p_\alpha}^{(1)}(\Gamma^N) = -\xi p_\alpha \quad (259)$$

$$\mathbb{K}_{q_\alpha}^{(1)}(\Gamma^N) = p_\alpha/m \quad (260)$$

$$\mathbb{K}_{p_\alpha,p_\alpha}^{(1)}(\Gamma^N) = 2mk_B T\xi \quad (261)$$

For this limit, the linear Fokker–Planck equation given by Eq. (237) assumes the form of the Fokker–Planck equation given by Eqs. (224)–(226) for the case of a single particle with an isotropic friction tensor and in the absence of a force field due to other particles undergoing Brownian motion.

If the time scale Δt is much longer than the time scale ξ^{-1} for momentum relaxation (i.e., $\xi \, \Delta t \gg 1$) all the "drift" and "diffusion" coefficients are negligible except for

$$\mathbb{K}_{q_\alpha,q_\alpha}^{(2)}(\Gamma^N) = 2k_B T/m\xi \quad (262)$$

For this limit, the linear Fokker–Planck equation given by Eq. (237) assumes the form of the Smoluchowski equation given by Eq. (227) for the case of a single particle with an isotropic diffusion tensor and in the absence of a force field due to other particles undergoing Brownian motion. From this result, one can establish that the diffusion coefficient $D = k_B T/m\xi$ is given by

$$D = \lim_{\Delta t \to \infty} [\langle |\vec{q}(\Delta t) - \vec{q}(0)|^2 \rangle_B/6 \, \Delta t] \quad (263)$$

Alternatively, the diffusion coefficient can be written in the form of Eq. (255).

C. PHENOMENOLOGICAL THEORY OF IRREVERSIBLE PROCESSES

According to the second law of thermodynamics, any spontaneous process occurring in an isolated system out of equilibrium will lead to an increase in the entropy inside that system. In spite of the general validity of the second law, we have yet to fully understand the problem of irreversible time evolution in systems out of equilibrium. Of course, Brownian motion theory does provide some insight into the direction that might be pursued in seeking answers to this problem. The most important idea to emerge from Brownian motion theory is the notion that dissipative or irreversible behavior arises from spontaneous equilibrium fluctuations.

Although thermodynamics allows us to make statements about the entropy change associated with the passage of a system from one constrained equilibrium state to another, it lacks the appropriate mathematical apparatus for treating the irreversible time evolution of systems out of equilibrium. Thus, the incorporation of irreversible processes into thermodynamics is incomplete. In part, this problem stems from the lack of any explicit reference to time in the formalism of thermodynamics.

There have been many attempts to incorporate irreversible processes into thermodynamics. Early attempts were confined to the treatment of special classes of phenomena, such as thermoelectric effects. In these attempts, the theoretical formulation of a given irreversible process was usually done in a manner that was disjoint from the formulation of other irreversible processes, each formulation requiring different ad hoc assumptions about the nature of a system. This lack of unity in the theory of irreversible processes existed until Onsager presented a unifying conceptual framework for linear Markovian (without memory) systems. The work of Onsager and later work done by others in the same spirit is usually called the phenomenological theory of irreversible processes.

In the phenomenological theory of irreversible processes one usually divides systems into two distinct types, namely, discrete composite systems and continuous systems. Discrete composite systems are systems for which the time-dependent thermodynamic properties are discontinuous across the boundaries of the subsystems making up the composite system. Continuous systems are systems for which thermodynamic properties vary continuously throughout. We first discuss the problem of irreversible processes in discrete composite systems. We then consider processes in continuous systems.

Consider an adiabatically isolated composite system composed of two subsystems. The two subsystems, designated subsystems $\alpha = 1$ and 2, are characterized by the same sets of external parameters $\{a_l^{(\alpha)}(t), l = 1, \ldots, m\}$ and averages $\{\langle O_j^{(\alpha)}(t) \rangle, j = 1, \ldots, n\}$ at each instant of time t. The external parameters and averages are subject to the conservation relations

$$a_l = a_l^{(1)}(t) + a_l^{(2)}(t) \qquad (264)$$

$$\langle O_j \rangle = \langle O_j^{(1)}(t) \rangle + \langle O_j^{(2)}(t) \rangle \qquad (265)$$

where a_l and $\langle O_j \rangle$ are independent of time.

If the composite system is out of equilibrium, there will be irreversible flows across the boundary separating the two subsystems. (Here, we have used the word *flow* in the sense of the rate of change of a local average or external parameter.) These flows are accompanied by an entropy increase of the composite system.

Clearly, the word *flow* requires the concept of time, so time must be introduced into thermodynamics. This can be done by assuming the validity of the Gibbsian relation given by Eq. (172) for all time t. Hence, we write the entropy change $dS(t)$ associated with the system undergoing infinitesimal transfer processes at time t as

$$dS(t) = \left[\sum_{l=1}^{m} X_l^f(t) I_l^f(t) + \sum_{j=1}^{n} X_j^0(t) I_j^0(t) \right] dt \qquad (266)$$

where

$$I_l^f(t) = da_l^{(1)}(t)/dt \qquad (267)$$

$$I_j^0(t) = d\langle O_j^{(1)}(t) \rangle/dt \qquad (268)$$

$$X_l^f(t) = k_B[f_l^{(1)}(t) - f_l^{(2)}(t)] \qquad (269a)$$

$$= [\partial S(t)/\partial a_l^{(1)}(t)] \qquad (269b)$$

$$X_l^0(t) = k_B[\Lambda_j^{(1)}(t) - \Lambda_j^{(2)}(t)] \qquad (270a)$$

$$= [\partial S(t)/\partial \langle O_j^{(1)}(t) \rangle] \qquad (270b)$$

Here, we use the superscripts f and 0 to distinguish between quantities associated with averages and external parameters. The introduction of the explicit time dependence in Eq. (266) is equivalent to assuming the validity of the Gibbsian definition of entropy for systems out of equilibrium. This assumption is sometimes called the assumption of local equilibrium.

All the transfer processes or flows represented in Eq. (266) are assumed to be allowed. Then the state of equilibrium for the composite system requires all the $X_l^f(t)$ and $X_j^0(t)$ to vanish. Otherwise, the system is out of equilibrium. The sets of quantities $\{X_l^f(t)\}$ and $\{X_j^0(t)\}$ provide a measure of the displacement from equilibrium at time t. Actually, these quantities can be thought to represent forces that are created by displacing the system from equilibrium. The forces $\{X_l^f(t)\}$ and $\{X_j^0(t)\}$ manifest by inducing the flows $\{I_j^0(t)\}$ and $\{I_l^f(t)\}$ along the linearly independent coordinates $\{a_l^{(1)}(t)\}$ and $\{\langle O_j^{(1)}(t) \rangle\}$ characterizing the macroscopic state of the system. Since the system eventually settles down to either a state of equilibrium or a steady state (characterized by constant flows), we go farther and label these forces as thermodynamic driving forces that drive the system to its final state.

The rate of entropy production $\mathcal{P}(t) = dS(t)/dt$ in the composite system due to the flows $\{I_j^0(t)\}$ and $\{I_i^f(t)\}$ is given by

$$\mathcal{P}(t) = \mathcal{P}^f(t) + \mathcal{P}^0(t) \qquad (271)$$

where

$$\mathcal{P}^f(t) = \sum_{l=1}^{m} X_l^f(t)I_l^f(t) \qquad (272)$$

$$\mathcal{P}^0(t) = \sum_{j=1}^{n} X_j^0(t)I_j^0(t) \qquad (273)$$

Here, $\mathcal{P}^f(t)$ is the entropy production due to flows induced by external forces and $\mathcal{P}^0(t)$ is the entropy production due to "free flow." Clearly, the second law of thermodynamics requires the entropy production $\mathcal{P}(t)$ to be nonnegative, that is, $\mathcal{P}(t) \geq 0$. This is sometimes referred to as the local entropy production postulate.

To simplify the notation, let us represent the driving forces $\{X_i^f(t)\}$ and $\{X_j^0(t)\}$ by a single set $\{X_j(t)\}$. Similarly, we lump the flows $\{I_i^f(t)\}$ and $\{I_j^0(t)\}$ into the set $\{I_j(t)\}$. Making these notational changes, we rewrite Eqs. (266) and (271) as

$$dS(t) = \sum_j X_j(t)I_j(t)\,dt \qquad (274)$$

and

$$\mathcal{P}(t) = \sum_j X_j(t)I_j(t) \qquad (275)$$

As stated earlier, the driving forces $\mathbf{X}(t) = \{X_j(t)\}$ are regarded as sources for the flows $\mathbf{I}(t) = \{I_j(t)\}$. Thus, the flows in the system are a function of the driving forces, that is, $I_i(t) = I_i[\mathbf{X}(t)]$. In the state of equilibrium, the driving forces vanish and so must the flows. Hence, $\lim_{\mathbf{X}(t)\to 0} I_i(t) = 0$.

It is reasonable to assume that the flow $I_i(t)$ will be linear in the driving forces when the driving forces are sufficiently small and the system is close to equilibrium. So we write

$$I_i(t) = \sum_j L_{i,j}^{(1)}X_j(t) \qquad (276)$$

where

$$L_{i,j}^{(1)} = \partial I_i[\mathbf{X}(t) = 0]/\partial X_j(t) \qquad (277)$$

is the first-order expansion coefficient appearing in the Taylor series expansion of $I_i(t)$ about the state of equilibrium, that is, the state for which

$\mathbf{X}(t) = 0$. The set of equations given by Eq. (276) are called linear phenomenological equations. The phenomenological coefficient $L_{i,j}^{(1)}$ gauges the interaction or coupling between the flow $I_i(t)$ and driving force $X_j(t)$. In the next section, we shall see that $L_{i,j}^{(1)}$ provides a measure of the correlation between the flows i and j when the system is in the state of equilibrium. This correlation is manifested by the appearance of interference effects between simultaneously occurring flows.

Assuming the validity of the linear response equations given by Eq. (276), the rate of entropy production is written

$$\mathcal{P}(t) = \sum_{i,j} X_i(t)L_{i,j}^{(1)}X_j(t) \qquad (278)$$

Now the entropy production $\mathcal{P}(t)$ must be non-negative. A necessary and sufficient condition for this requirement to be fulfilled is that all the principal minors of the determinant $|\mathbf{L}^{(1)}|$ formed by the matrix $\mathbf{L}^{(1)}$ of phenomenological coefficients be positive definite. A priori, we do not know whether this condition is fulfilled for an arbitrary system. Clearly, it will hold when the matrix $\mathbf{L}^{(1)}$ is symmetric, $L_{i,i}^{(1)} > 0$ for all i, and $L_{i,i}^{(1)}L_{j,j}^{(1)} > L_{i,j}^{(1)2}$.

Thus far, we have introduced three principal postulates: (1) The Gibbsian expression for dS is valid for systems out of equilibrium (so-called assumption of local equilibrium); (2) the entropy production, given by a bilinear form in the forces and flows when the system is close to equilibrium, is nonnegative (local entropy production postulate); and (3) the flows can be expressed as a linear combination of the forces (assumption of linear response). Onsager introduced another postulate (called the Onsager reciprocity theorem) stating that the matrix $\mathbf{L}^{(1)}$ formed by the phenomenological coefficients is symmetric, that is, $L_{i,j}^{(1)} = L_{j,i}^{(1)}$. These symmetry relations are commonly called Onsager reciprocity relations.

In the proof of the reciprocity theorem, Onsager wrote the linear phenomenological equations given by Eq. (276) in the form

$$\frac{\overline{dO_i(t + \Delta t|\mathbf{O}')}}{dt} = \sum_j L_{i,j}^{(1)}\frac{\partial S(\mathbf{O}')}{\partial O_j'} \qquad (279)$$

where $\overline{dO_i(t + \Delta t|\mathbf{O}')}/dt$ is the phenomenological time derivative

$$\overline{dO_i(t + \Delta t|\mathbf{O}')}/dt = [\overline{O_i(t + \Delta t|\mathbf{O}')} - \overline{O_i(t|\mathbf{O}')}]/\Delta t \qquad (280)$$

with $\overline{O_i(t + \Delta t|\mathbf{O}')}$ representing the average value of the observable O_i at time $t + \Delta t$ given that the set of observables **O** possess the values **O**′ at time t.

To proceed further, Onsager did not take the Gibbsian path discussed earlier. Rather, he adopted the Boltzmann definition of entropy and Einstein's theory of fluctuations. Nonetheless, Onsager was led to the following expression for the phenomenological coefficient $L_{i,j}^{(1)}$,

$$L_{i,j}^{(1)} = -(1/k_B\Delta t)[\overline{O_i(t + \Delta t)O_j(t)} - \overline{O_i(t)O_j(t)}] \quad (281)$$

where the quantity $\overline{O_i(t + \Delta t)O_j(t)}$ is an averaged quantity intended to represent the correlation between events $O_i(t + \Delta t)$ and $O_j(t)$. In defining the average $\overline{O_i(t + \Delta t)O_j(t)}$, Onsager made an assumption reminiscent of the assumption made in Brownian motion theory that stochastic averaging is equivalent to ensemble averaging. With the above result in hand and arguments based on the principle of microscopic reversibility (time-reversal symmetry of the microscopic equations of motion), Onsager was led to the reciprocity relation $L_{i,j}^{(1)} = L_{j,i}^{(1)}$. Of course, Onsager realized that the reciprocity relation must read $L_{i,j}^{(1)}(\vec{B}) = L_{j,i}^{(1)}(-\vec{B})$ for systems in an external magnetic field $\vec{B}$.

Today, we realize that the reciprocity relations do not hold for a number of important linear phenomenological equations. They hold only when the observables **O** possess definite time-reversal parity and this parity is the same for all of them.

Now that we have completed the discussion of discrete composite systems, let us consider the phenomenology adopted in the treatment of irreversible processes in continuous systems. Since the thermodynamic properties vary continuously throughout continuous systems, the following Gibbsian relation is assumed at each point $\vec{r}$ in a system at time t,

$$dp_S(\vec{r}; t) = k_B \sum_i \Lambda_i(\vec{r}; t) \, dp_i(\vec{r}; t) \quad (282)$$

where $\rho_S(\vec{r}; t)$ is the entropy density (entropy per unit volume), $\Lambda_i(\vec{r}; t)$ is a local thermodynamic parameter, and $\rho_i(\vec{r}; t)$ is a local density associated with a conserved variable (such variables are energy, particle number, charge, etc.). Equation (282) allows us to identify the local thermodynamic parameters with partial derivatives of the entropy density, that is,

$$k_B\Lambda_i(\vec{r}; t) = [\partial\rho_S(\vec{r}; t)/\partial\rho_i(\vec{r}; t)] \quad (283)$$

For the case of continuous systems, one uses the fluxes $\{\vec{\mathcal{J}}_i(\vec{r}; t)\}$ associated with the local

densities to characterize the irreversible time evolution. These fluxes are viewed as the response of the system to the thermodynamic driving forces $\{\vec{X}_i(\vec{r}; t)\}$:

$$\vec{X}_i(\vec{r}; t) = k_B\vec{\nabla}_r \cdot \Lambda_i(\vec{r}; t) \quad (284)$$

When the system is in equilibrium the forces $\{\vec{X}_i(\vec{r}; t)\}$ are assumed to vanish.

The local rate of entropy production $\mathcal{P}_S(\vec{r}; t) = dp_S(\vec{r}; t)/dt$ due to the irreversible fluxes is given by

$$\mathcal{P}_S(\vec{r}; t) = \sum_i \vec{X}_i(\vec{r}; t) \cdot \vec{\mathcal{J}}_i(\vec{r}; t) \quad (285)$$

So the rate of entropy production $\mathcal{P}(t) = dS(t)/dt$ in the system is written

$$\mathcal{P}(t) = \int_\mathcal{D} dV \, \mathcal{P}_S(\vec{r}; t) \quad (286a)$$

$$= \sum_i \int_\mathcal{D} dV \, \dot{X}_i(\vec{r}; t) \cdot \vec{\mathcal{J}}_i(\vec{r}; t) \quad (286b)$$

where the volume integration is over the system domain $\mathcal{D}$.

Since the local densities $\{\rho_i(\vec{r}; t)\}$ correspond to conserved variables, they must be governed by continuity equations of the form

$$\partial\rho_i(\vec{r}; t)/\partial t = -\vec{\nabla}_r \cdot \vec{\mathcal{J}}_i(\vec{r}; t) \quad (287)$$

With this result at our disposal, we can utilize Gauss's theorem to rewrite Eq. (286b) as

$$dS(t) = d_{\text{EXT}}S(t) + d_{\text{IRR}}S(t) \quad (288)$$

where

$$d_{\text{EXT}}S(t) = k_B \sum_i \int_\mathcal{S} d\mathcal{S} \, \Lambda_i(\vec{r}; t)[\hat{n} \cdot \vec{\mathcal{J}}_i(\vec{r}; t)] \, dt \quad (289)$$

$$d_{\text{IRR}}S(t) = k_B \sum_i \int_\mathcal{D} dV \, \Lambda_i(\vec{r}; t) \frac{\partial\rho_i(\vec{r}; t)}{\partial t} \, dt \quad (290)$$

with the surface integration in Eq. (289) over the boundary surface $\mathcal{S}$ separating the system from the external (EXT) world. The unit vector $\hat{n}$ is normal to $\mathcal{S}$ and directed outward. In the above, $d_{\text{EXT}}S(t)$ is the contribution to the entropy change $dS(t)$ in the time interval dt due to exchange of entropy between the system and external world, which arises from fluxes across the boundary surface $\mathcal{S}$. The contribution to the entropy change $dS(t)$ in the time interval dt due to irreversible processes in the system is given by $d_{\text{IRR}}S(t)$. Clearly, the second law of thermodynamics requires that $d_{\text{IRR}}S(t) \geq 0$ or $\mathcal{P}(t) \geq 0$ for closed and adiabatically isolated systems.

As for discrete systems, linear phenomeno-logical equations are expected to hold when the system of interest is sufficiently close to equilibrium. For the case of continuous systems, we write

$$\vec{\mathcal{J}}_i(\vec{r}; t) = \sum_j \vec{\vec{L}}_{i,j}^{(1)}(\vec{r}) \cdot \vec{X}_j(\vec{r}; t) \qquad (291)$$

where the components of the second-rank tensor $\vec{\vec{L}}_{i,j}^{(1)}$ are given by

$$L_{i_\alpha, j_{\alpha'}}^{(1)}(\vec{r}) = \partial \mathcal{J}_{i_\alpha}[\vec{r}; \mathbf{X}(t) = 0]/\partial X_{j_{\alpha'}}(\vec{r}; t)] \qquad (292)$$

which represents the coupling between the αth Cartesian component of the flux $\vec{\mathcal{J}}_i(\vec{r}; t)$ with the α'th cartesian component of the force $\vec{X}_j(\vec{r}; t)$.

VI. Irreversible Processes and Microscopic Dynamics

A. QUANTUM STATISTICAL MECHANICAL THEORY OF RELAXATION PHENOMENA

The quantum statistical mechanical theory of relaxation phenomena (QSM theory) is a maximum entropy approach to time-dependent phenomena that is in the spirit of the information-theoretic approach to equilibrium statistical mechanics and the phenomenological theory of irreversible processes. QSM theory provides us with a firm statistical mechanical basis for the phenomenological theory of irreversible processes and classical Brownian motion theory. Moreover, it gives us a number of formulas that can be employed in the investigation of the role played by microscopic interactions in a diversity of irreversible phenomena.

The roots of QSM theory lie in Mori's statistical mechanical theory of transport processes and Kubo's theory of thermal disturbances. The version of QSM theory presented here with its refinements and modern embellishments was used by the author to develop an irreversible thermodynamic theory for photophysical phenomena and a quantum stochastic Fokker–Planck theory for adiabatic and nonadiabatic processes in condensed phases.

Consider a composite system composed of the system of interest (the system) and its surroundings (the bath). Assuming the validity of the Gibbsian definition of entropy for the entropy of systems out of equilibrium, we write the entropy $S(t)$ of the composite system at time t as

$$S(t) = -k_B \, \mathrm{Tr} \, \hat{\rho}(t) \ln \hat{\rho}(t) \qquad (293)$$

where $\hat{\rho}(t)$ is a statistical density operator used to describe the macroscopic state of the composite system. We take the macroscopic state of the composite system to be defined by specifying its average energy $\langle \mathcal{H} \rangle$, which is assumed to be independent of time, and certain averages $\{\langle \hat{O}_j(t) \rangle\}$ (thermodynamic coordinates) for the system of interest. Included in the set of averages $\{\langle \hat{O}_j(t) \rangle\}$ are any time-dependent external parameters, such as those characterizing applied fields.

Assuming that the maximum entropy principle can be used at any given instant of time, the statistical density operator $\hat{\rho}(t)$ must be the maximum entropy density operator consistent with the information $\langle \mathcal{H} \rangle$ and $\{\langle \hat{O}_j(t) \rangle\}$ defining the macroscopic state of the composite system. As for the case of equilibrium systems, the averages $\langle \mathcal{H} \rangle$ and $\{\langle \hat{O}_j(t) \rangle\}$ are required to be linearly independent.

Adopting the time-dependent version of the maximum entropy principle, we write the statistical density operator $\hat{\rho}(t)$ as

$$\hat{\rho}(t) = \exp\left[\Omega(t)\hat{I} - \beta\mathcal{H} - \sum_j \Lambda_j(t)\hat{O}_j \right] \qquad (294)$$

where $\Omega(t)$ and $\{\Lambda_j(t)\}$ are time-dependent Lagrange parameters. The Lagrange parameter $\Omega(t)$ is given by

$$\Omega(t) = -\ln Z(t) \qquad (295)$$

where

$$Z(t) = \mathrm{Tr} \, \exp\left[-\beta\mathcal{H} - \sum_j \Lambda_j(t)\hat{O}_j \right] \qquad (296)$$

is a time-dependent partition function.

The entropy of the composite system can be readily computed by using Eqs. (293) and (294):

$$S(t) = -k_B \Omega(t) + k_B \beta \langle \mathcal{H} \rangle$$

$$+ k_B \sum_j \Lambda_j(t) \langle \hat{O}_j(t) \rangle \qquad (297)$$

Note the formal similarity of this result to that of equilibrium statistical mechanics. As with equilibrium systems, $\Omega(t)$ is interpreted as a thermodynamic potential and $\{\Lambda_j(t)\}$ are thought to represent thermodynamic parameters. Since some of the $\{\langle \hat{O}_j(t) \rangle\}$ are time-dependent external parameters, $\{\Lambda_j(t)\}$ also include time-dependent external forces.

In view of the formal similarity of the above form for the entropy $S(t)$ to that for equilibrium systems, the standard thermodynamic relations can be extended to include an explicit time dependence. For instance, the differential $dS(t)$ is written

$$dS(t) = k_B \beta \langle \hat{\mathcal{H}} \rangle$$

$$+ k_B \sum_j \Lambda_j(t) \, d\langle \hat{O}_j(t) \rangle \quad (298)$$

which allows us to identify the thermodynamic parameters β and $\{\Lambda_j(t)\}$ with the partial derivatives

$$[\partial S(t)/\partial \langle \hat{\mathcal{H}} \rangle] = k_B \beta \quad (299)$$

$$[\partial S(t)/\partial \langle \hat{O}_l(t) \rangle] = k_B \Lambda_l(t) \quad (300)$$

The partial derivative given by Eq. (300) is precisely the definition employed in the phenomenological theory of irreversible processes to define the thermodynamic driving force $X_l(t) = k_B \Lambda_l(t)$ for the case of discrete systems.

Now let us consider the observed response of the system to displacements $\langle \Delta \hat{O}_j(t) \rangle = \langle \hat{O}_j(t) \rangle - \langle \hat{O}_j(\infty) \rangle$ from thermal equilibrium, where $\langle \hat{O}_j(\infty) \rangle$ is the average of $\hat{O}_j$ over the canonical density operator $\hat{\rho}(\infty) = \exp[\Omega(\infty)\hat{I} - \beta \hat{\mathcal{H}}]$, where $\Omega(\infty) = -\ln Z(\infty)$, with $Z(\infty)$ denoting the canonical partition function $Z(\infty) = \text{Tr} \exp(-\beta \hat{\mathcal{H}})$. The observed response is defined in terms of the phenomenological currents $\{\langle \Delta \hat{\dot{O}}_i(t; \Delta t) \rangle\}$:

$$\langle \Delta \hat{\dot{O}}_i(t; \Delta t) \rangle = \Delta t^{-1} \int_0^{\Delta t} d\tau \, \langle \Delta \hat{\dot{O}}_i(t + \tau) \rangle \quad (301a)$$

$$= [\langle \Delta \hat{O}_i(t + \Delta t) \rangle - \langle \Delta \hat{O}_i(t) \rangle]/\Delta t \quad (301b)$$

which corresponds to a time average of the instantaneous current $\langle \Delta \hat{\dot{O}}_i(t) \rangle$ over the time scale Δt of macroscopic measurements (time resolution of the macroscopic observer). The average $\langle \Delta \hat{O}_i(t + \Delta t) \rangle$ is a conditional average defined by

$$\langle \Delta \hat{O}_i(t + \Delta t) \rangle = \langle \Delta \hat{O}_i(t + \Delta t) | \{\langle \Delta \hat{O}_j(t) \rangle\} \rangle \quad (302a)$$

$$= \text{Tr} \, \hat{\rho}(t) \, \Delta \hat{O}_i(\Delta t) \quad (302b)$$

where $\Delta \hat{O}_i(\Delta t) = \exp(+i\hat{\mathcal{L}} \Delta t) \Delta \hat{O}_i$, with $\Delta \hat{O}_i = \hat{O}_i - \langle \hat{O}_i(\infty) \rangle$. The above conditional average is defined in such a way that $\langle \Delta \hat{O}_i(t + \Delta t) \rangle$ represents the average value of the displacement from equilibrium associated with the thermodynamic coordinate i at time $t + \Delta t$ given that the observed displacements from equilibrium at time t are given by $\{\langle \Delta \hat{O}_j(t) \rangle\}$. The above definition of the phenomenological currents is essentially the same as that employed by Onsager in writing Eq. (279).

The expression given by Eq. (301a) for the phenomenological currents can be recast in the form

$$\langle \Delta \hat{\dot{O}}_i(t; \Delta t) \rangle = \langle \Delta \hat{\dot{O}}_i(t) \rangle$$

$$+ \int_0^{\Delta t} d\tau \left(1 - \frac{\tau}{\Delta t}\right) \langle \Delta \hat{\ddot{O}}_i(t + \tau) \rangle \quad (303)$$

where $\langle \Delta \hat{\dot{O}}_i(t) \rangle$ and $\langle \Delta \hat{\ddot{O}}_i(t + \tau) \rangle$ are the conditional averages $\langle \Delta \hat{\dot{O}}_i(t) | \{\langle \Delta \hat{O}_j(t) \rangle\} \rangle$ and $\langle \Delta \hat{\ddot{O}}_i(t + \tau) | \{\langle \Delta \hat{O}_j(t) \rangle\} \rangle$, respectively. These conditional averages can be evaluated by making use of Eq. (302b), with $\hat{\rho}(t)$ given by Eq. (294).

Before proceeding, we should remark that, in using the maximum entropy density operator $\hat{\rho}(t)$ to evaluate the phenomenological currents, one is making use of the macroscopic information at time t to predict the phenomenological currents $\{\langle \Delta \hat{\dot{O}}_i(t; \Delta t) \rangle\}$. This is equivalent to asserting that the system of interest is Markovian (without memory) on the time scale of macroscopic measurements. Also, it must be stated that in any choice of averages $\{\langle \hat{O}_i(t) \rangle\}$ there is an implicit assumption that we are considering the system of interest on a coarse-grained time scale Δt such that $t_B \ll \Delta t \ll t_S$, where t_S is the time scale for relaxation of the averages $\{\langle \hat{O}_i(t) \rangle\}$ and t_B is the time scale for the bath to respond to the system and reequilibrate.

Since the phenomenological currents $\{\langle \Delta \hat{\dot{O}}_i(t; \Delta t) \rangle\}$ are measured on a coarse-grained time scale Δt, it is deemed appropriate to write the rate of entropy production as follows:

$$\mathcal{P}(t; \Delta t) = k_B \sum_j \Lambda_j(t) \langle \hat{\dot{O}}_j(t; \Delta t) \rangle \quad (304)$$

One can readily demonstrate that $\mathcal{P}(t; \Delta t) \geq 0$ for $\Delta t > 0$ (which defines the future) when the phenomenological currents are computed by the prescription given above.

By expanding the statistical density operator $\hat{\rho}(t)$ about the state $\hat{\rho}(\infty)$ of the thermal equilibrium, one can show that the phenomenological currents can be written in the form

$$\langle \Delta \hat{\dot{O}}_i(t; \Delta t) \rangle = -\exp[\Delta \Omega(t)] \sum_j \Lambda_j(t) \beta^{-1}$$

$$\times \{ \int_0^\beta d\lambda \, \langle \hat{U}(i\hbar\lambda) \hat{\dot{O}}_j(0) \hat{\mathcal{R}}(\lambda; t)$$

$$\Delta \hat{\dot{O}}_i(0) \rangle_{EQ}$$

$$+ \int_0^{\Delta t} d\tau \, [1 - (\tau/\Delta t)]$$

$$\times \int_0^\beta d\lambda \langle \hat{U}(i\hbar\lambda) \hat{\dot{O}}_j(0)$$

$$\hat{\mathcal{R}}(\lambda; t) \Delta \hat{\ddot{O}}_i(\tau) \rangle_{EQ} \} \quad (305)$$

where the averages denoted by $\langle \rangle_{EQ}$ are over the equilibrium density operator $\hat{\rho}(\infty)$, $\Delta\Omega(t) = \Omega(t) - \Omega(\infty)$, $\hat{U}(i\hbar\lambda) = \exp(\lambda\hat{\mathcal{H}})$, and $\hat{\mathcal{R}}(\lambda; t) = \exp\{-\lambda[\hat{\mathcal{H}} + \beta^{-1} \sum_j \Lambda_j(t)\hat{O}_j]\}$. The thermodynamic parameters $\{\Lambda_j(t)\}$ can be determined by solving the nonlinear equations

$$\langle\Delta\hat{O}_i(t)\rangle = -\exp[\Delta\Omega(t)] \sum_j \Lambda_j(t)\beta^{-1}$$

$$\times \int_0^\beta d\lambda \langle \hat{U}(i\hbar\lambda)\hat{O}_j(0)$$

$$\hat{\mathcal{R}}(\lambda; t) \, \Delta\hat{O}_i(0)\rangle_{\text{EQ}} \qquad (306)$$

Once the thermodynamic parameters have been determined from the above set of equations, the observed response is computed using Eq. (305).

It is reasonable to assume that Eqs. (305) and (306) can be linearized when the time-dependent thermodynamic parameters are sufficiently small and the system is close to equilibrium. The linearized forms of Eqs. (305) and (306) are given by

$$\langle\Delta\hat{O}_i(t; \Delta t)\rangle = \sum_j M_{i,j}^{(1)}\Lambda_j(t) \qquad (307)$$

and

$$\langle\Delta\hat{O}_i(t)\rangle = -\sum_j \chi_{i,j}\Lambda_j(t) \qquad (308)$$

where

$$M_{i,j}^{(1)} = M_{i,j}^{(1),\text{S}} + M_{i,j}^{(1),\text{C}} \qquad (309)$$

and

$$\chi_{i,j} = \langle\Delta\hat{O}_j(0) \, \Delta\hat{\tilde{O}}_i(0)\rangle_{\text{EQ}} \qquad (310)$$

with

$$M_{i,j}^{(1),\text{S}} = -\langle\Delta\hat{O}_j(0) \, \Delta\hat{\dot{\tilde{O}}}_i(0)\rangle_{\text{EQ}} \qquad (311)$$

and

$$M_{i,j}^{(1),\text{C}} = \int_0^{\Delta t} d\tau \left(1 - \frac{\tau}{\Delta t}\right)$$

$$\Delta\hat{\dot{O}}_j(-\tau) \, \Delta\hat{\dot{\tilde{O}}}_i(0)\rangle_{\text{EQ}} \qquad (312)$$

In Eqs. (310)–(312), $\hat{\tilde{O}}_j = \beta^{-1} \int_0^\beta d\lambda \, \hat{O}_j(i\hbar\lambda)$. Equation (307) is just the quantum analog of the linear equations of the phenomenological theory of irreversible processes for the case of discrete systems with $L_{i,j}^{(1)} = M_{i,j}^{(1)}/k_{\text{B}}$ and $X_j(t) = k_{\text{B}}\Lambda_j(t)$. Equations (309), (311), and (312) provide us with microscopic expressions for the phenomenological coefficients.

The classical limit of these results are obtained by allowing $\hbar \to 0$. We write

$$\lim_{\hbar\to0} M_{i,j}^{(1)} = -\langle\Delta O_j(0) \, \Delta\dot{O}_i(0)\rangle_{\text{EQ}}$$

$$+ \int_0^{\Delta t} d\tau \left(1 - \frac{\tau}{\Delta t}\right)$$

$$\langle\Delta\dot{O}_j(-\tau) \, \Delta\dot{O}_i(0)\rangle_{\text{EQ}} \qquad (313)$$

and

$$\lim_{\hbar\to0} \chi_{i,j} = \langle\Delta O_j(0) \, \Delta O_i(0)\rangle_{\text{EQ}} \qquad (314)$$

where $\Delta O_j(0)$ and $\Delta O_i(0)$ are classical dynamic variables and $\langle \rangle_{\text{EQ}}$ denotes an average over the classical equilibrium distribution function for the composite system.

If the average $\overline{O_i(t + \Delta t)O_j(t)}$ in Onsager's expression for the phenomenological coefficient $L_{i,j}^{(1)}$ [see Eq. (281)] is interpreted as an average over the classical equilibrium distribution function and this average is stationary in time, that is, $\overline{O_i(t + \Delta t)O_j(t)} = \overline{O_i(\Delta t)O_j(0)}$, one can readily demonstrate that Onsager's expression for $L_{i,j}^{(1)}$ is identical to the classical result for $M_{i,j}^{(1)}/k_{\text{B}}$. Of course, O_i and O_j must be replaced by ΔO_i and ΔO_j.

The phenomenological coefficients $M_{i,j}^{(1),\text{S}}$ and $M_{i,j}^{(1),\text{C}}$ satisfy the following symmetry relations:

$$M_{i,j}^{(1),\text{S}}(-\vec{B}) = -M_{j,i}^{(1),\text{S}}(-\vec{B}) \qquad (315\text{a})$$

$$= -M_{i,j}^{(1),\text{S}}(\vec{B}) \qquad (315\text{b})$$

$$M_{i,j}^{(1),\text{C}}(-\vec{B}) = M_{j,i}^{(1),\text{C}}(\vec{B}) \qquad (316)$$

where the tilde over the indices i and j indicates that the variables $\Delta\hat{O}_i$ and $\Delta\hat{O}_j$ are to be replaced by their corresponding time-reversed variables. If the variables $\Delta\hat{O}_i$ and $\Delta\hat{O}_j$ possess definite time-reversal parity, the symmetry relations assume the forms

$$M_{i,j}^{(1),\text{S}}(-\vec{B}) = -M_{j,i}^{(1),\text{S}}(-\vec{B}) \qquad (317\text{a})$$

$$= -\varepsilon_i\varepsilon_j M_{i,j}^{(1),\text{S}}(\vec{B}) \qquad (317\text{b})$$

$$M_{i,j}^{(1),\text{C}}(-\vec{B}) = \varepsilon_i\varepsilon_j M_{j,i}^{(1),\text{C}}(\vec{B}) \qquad (318)$$

where $\varepsilon_i/\varepsilon_j$ indicates the time-reversal parity of $\Delta\hat{O}_i/\Delta\hat{O}_j$. The symmetry relations given by Eq. (318) are the Onsager reciprocity relations for the case in which all the variables possess the same time-reversal parity.

The results given by Eqs. (307) and (308) enable us to construct the following linear equation of motion for the displacements from equilibrium,

$$\langle\Delta\hat{\dot{O}}_i(t; \Delta t)\rangle = -\sum_j [i\Omega_{i,j} + K_{i,j}^{(1)}]\langle\Delta\hat{O}_j(t)\rangle \qquad (319)$$

where

$$i\Omega_{i,j} = \sum_k M_{i,k}^{(1),\text{S}}\chi_{k,j}^{-1} \qquad (320)$$

$$K_{i,j}^{(1)} = \sum_k M_{i,k}^{(1),\text{C}}\chi_{k,j}^{-1} \qquad (321)$$

with $\chi_{k,j}^{-1}$ denoting an element of the inverse matrix χ^{-1}. From a macroscopic point of view, Eq. (319) can be treated as a differential equation. Treated as such, the displacements from equilibrium $\{\langle\Delta\hat{O}_i(t)\rangle\}$, in general, will exhibit damped oscillatory temporal behavior and decay to zero in the infinite time limit.

As in classical Brownian motion theory, the damping in Eq. (319), characterized by the damping constants $\{K_{i,j}^{(1)}\}$, arises from spontaneous equilibrium fluctuations described by the equilibrium time correlation functions appearing in the formal expressions for $\{M_{i,j}^{(1),C}\}$. The oscillatory motion of the system is characterized by the frequencies $\{\Omega_{i,j}\}$, which are defined by Eq. (320). The influence of microscopic interactions on the overall time evolution of the displacements from equilibrium is buried in the $\{\Omega_{i,j}\}$ and $\{K_{i,j}^{(1)}\}$ or equivalently $\{M_{i,j}^{(1),S}\}$, $\{M_{i,j}^{(1),C}\}$, and $\{\chi_{i,j}\}$.

In the actual evaluation of the phenomenological coefficients $\{M_{i,j}^{(1),C}\}$ the integration limit Δt is usually extended to $+\infty$ and the time correlation function is multiplied by the convergence factor $\exp(-\varepsilon t)$, where $\varepsilon \to 0^+$ at the end of the calculation. Specifically, $M_{i,j}^{(1),C}$ is written as

$$M_{i,j}^{(1),C} = \lim_{\varepsilon\to 0^+} \int_0^\infty dt\, \exp(-\varepsilon t)$$
$$\times \langle\Delta\dot{\hat{O}}_j(-t)\,\Delta\dot{\hat{O}}_i(0)\rangle_{EQ} \quad (322)$$

The formal expressions given by Eqs. (311) and (322) for the phenomenological coefficients $M_{i,j}^{(1),S}$ and $M_{i,j}^{(1),C}$ can be given a matrix representation similar to that given by Eq. (223) for the average $\langle\hat{O}(t)\rangle$. The matrix representations of $M_{i,j}^{(1),S}$ and $M_{i,j}^{(1),C}$ are written as

$$M_{i,j}^{(1),S} = \langle\Delta O_i^\dagger|\Delta\dot{O}_j\,\rho_{EQ}\rangle \quad (323)$$

and

$$M_{i,j}^{(1),C} = \lim_{\varepsilon\to 0^+} \int_0^\infty dt\, \exp(-\varepsilon t)$$
$$\times \langle\Delta\dot{O}_i^\dagger|\exp(-i\hat{\mathscr{L}}t)|\Delta\dot{O}_j\,\rho_{EQ}\rangle \quad (324)$$

where the vectors $\langle\Delta\dot{O}_i^\dagger|$ and $|\Delta\dot{O}_j\,\rho_{EQ}\rangle$ correspond to the quantum operators $\Delta\dot{O}_i$ and $\Delta\dot{O}_j \times \hat{\rho}_{EQ}$, respectively. In the classical limit, Eqs. (323) and (324) assume the forms [see Eq. (201)]

$$M_{i,j}^{(1),S} = \langle\Delta O_i^*|\Delta\dot{O}_j\,\rho_{EQ}\rangle \quad (325)$$

and

$$M_{i,j}^{(1),C} = \lim_{\varepsilon\to 0^+} \int_0^\infty dt\, \exp(-\varepsilon t)$$
$$\times \langle\Delta\dot{O}_i^*|\exp(-i\hat{\mathscr{L}}t)|\Delta\dot{O}_j\,\rho_{EQ}\rangle \quad (326)$$

where the vectors $\langle\Delta\dot{O}_i^*|$ and $|\Delta\dot{O}_j\,\rho_{EQ}\rangle$ correspond to the classical variables $\Delta\dot{O}_i$ and $\Delta\dot{O}_j\,\rho_{EQ}$, respectively.

A simple example to which the above-described linear theory applies is the kinetic description of the excited-state dynamics of optically excited molecules, atoms, and ions in a heat bath. For this problem, we can write the rate constant $W(l\to k)$ for a transition from some state l to some state k in the form [see Eq. (230)]

$$W(l\to k) = -\lim_{\varepsilon\to 0^+} \frac{1}{\langle\hat{N}_{ll}(\infty)\rangle} \int_0^\infty dt\, \exp(-\varepsilon t)$$
$$\times \langle\Delta\hat{N}_{ll}(-t)\,\Delta\hat{N}_{kk}(0)\rangle_{EQ} \quad (327)$$

where $\hat{N}_{ll}$ and $\hat{N}_{kk}$ are number operators for the states l and k, respectively.

Another problem to which the above-described linear theory applies is the classical Fokker–Planck description of single-particle motion in a spatially homogeneous heat bath. For this problem, the friction coefficient ξ is given by Eq. (258b), where the limit $\varepsilon \to 0^+$ with the convergence factor $\exp(-\varepsilon t)$ is implicitly understood (also see Eqs. (224)–(226)].

The results presented thus far provide a firm statistical mechanical basis for the discrete case of the phenomenological theory of irreversible processes. Nonetheless, their applicability is not limited to the type of discrete composite systems discussed in the exposition on the phenomenological theory. Moreover, the presented results are not restricted to a set of averages with a discrete index. For cases in which the averages $\{\langle\Delta\hat{O}(\mathbf{x}; t)\rangle\}$ possess both a discrete index j and a continuous index $\mathbf{x}$, one need only augment the summations over the discrete indices with integrations over the continuous index $\mathbf{x}$.

The averages $\{\langle\Delta\hat{O}_j(\mathbf{x}; t)\rangle\}$ possess both a discrete and a continuous index, for example, when $\hat{O}_j(\mathbf{x}; t)$ represents some local density j at some point $\mathbf{x}$, which may correspond to some coordinate in a system or perhaps a phase point in the phase space associated with certain degrees of freedom. If these local densities correspond to conserved variables, their evolution is given by a microscopic continuity equation of the form $\dot{\hat{O}}_j(\mathbf{x}; t) = -\vec{\nabla}_x \cdot \hat{\vec{\mathscr{J}}}_j(\mathbf{x}; t)$, where $\hat{\vec{\mathscr{J}}}_j(\mathbf{x}; t)$ is a flux operator. For such cases, we can write the linear equations given by Eq. (307) in the form

$$\langle\Delta\hat{\vec{\mathscr{J}}}_i(\mathbf{x}; t)\rangle = \sum_j \int d\mathbf{x}' \,[\vec{\mathcal{M}}_{i,j}^{(1),S}(\mathbf{x}, \mathbf{x}')\Lambda_j(\mathbf{x}'; t)$$
$$+ \vec{\mathcal{M}}_{i,j}^{(1),C}(\mathbf{x}, \mathbf{x}') \cdot \vec{\nabla}_{x'} \Lambda_j(\mathbf{x}'; t)]$$
$$(328)$$

where

$$\vec{\mathcal{M}}_{i,j}^{(1),S}(\mathbf{x}, \mathbf{x}') = <\Delta\hat{O}_j(\mathbf{x}; 0)\,\Delta\hat{\vec{\mathscr{J}}}_i(\mathbf{x}'; 0)>_{EQ}$$

(329)

and

$$\vec{\mathcal{M}}_{i,j}^{(1),C}(\mathbf{x}, \mathbf{x}') = \lim_{\varepsilon \to 0^+} \int_0^\infty dt \, \exp(-\varepsilon t)$$
$$\times <\Delta\hat{\vec{\mathscr{J}}}_j(\mathbf{x}'; -t)\,\Delta\hat{\vec{\mathscr{J}}}_i(\mathbf{x}; 0)>_{EQ}$$

(330)

Equation (328) represents a generalization of the linear phenomenological equations for continuous systems allowing for local densities with no definite time-reversal parity and effects due to nonlocality in $\vec{\mathcal{M}}_{i,j}^{(1),S}(\mathbf{x}, \mathbf{x}')$ and $\vec{\mathcal{M}}_{i,j}^{(1),C}(\mathbf{x}, \mathbf{x}')$ [see Eq. (291)].

B. RESPONSE OF SYSTEMS TO EXTERNAL DISTURBANCES

Theoretical treatments of the response of a quantum system to an external disturbance, such as the application of an electric or magnetic field (or both), usually start with the following form of the quantum Liouville equation,

$$\partial\hat{\rho}(t)/\partial t = -i\hat{\mathscr{L}}(t)\hat{\rho}(t)$$

(331)

where

$$\hat{\mathscr{L}}(t) = \hat{\mathscr{L}} + \hat{\mathscr{L}}_{EXT}(t)$$

(332)

Here, $\hat{\mathscr{L}} = (1/\hbar)\,\hat{\mathscr{H}}^-$ is the Liouville operator of the system, described by the Hamiltonian $\hat{\mathscr{H}}$, and $\hat{\mathscr{L}}_{EXT}(t) = (1/\hbar)\hat{\mathscr{H}}_{EXT}^-(t)$ is the Liouville operator associated with the external disturbance $\mathscr{H}_{EXT}(t)$.

The most common model, due to Kubo, used in the treatment of the external disturbance problem is to assume that the disturbance is switched on adiabatically in the infinite past, $t = -\infty$, when the system was in the state of thermal equilibrium. Since the external disturbance is absent at $t = -\infty$, $\lim_{t \to -\infty} \mathscr{L}_{EXT}(t) = 0$. With this model in mind, the formal solution of Eq. (331) for the density operator $\hat{\rho}(t)$ can be written in the form of an integral equation,

$$\hat{\rho}(t) = \hat{\rho}(-\infty) - i\int_{-\infty}^t dt' \, \exp[-i\hat{\mathscr{L}}(t - t')]$$
$$\hat{\mathscr{L}}_{EXT}(t')\hat{\rho}(t')$$

(333)

where $\hat{\rho}(-\infty)$ is the canonical density operator $\exp(\Omega\hat{I} - \beta\hat{\mathscr{H}})$.

The response of the system to the external disturbance is described in terms of the resultant displacement $\langle\Delta\hat{B}(t)\rangle = \langle\hat{B}(t)\rangle - \langle\hat{B}(-\infty)\rangle$ of some observable from equilibrium:

$$\langle\Delta\hat{B}(t)\rangle = -i\int_{-\infty}^t dt' \, \mathrm{Tr}\,\hat{B}(0)\exp[-i\hat{\mathscr{L}}(t - t')]$$
$$\times \hat{\mathscr{L}}_{EXT}(t')\hat{\rho}(t')$$

(334)

For sufficiently weak external disturbances, departure from thermal equilibrium is small and the response $\langle\Delta\hat{B}(t)\rangle$ is linear in $\hat{\mathscr{L}}_{EXT}(t')$:

$$\langle\Delta\hat{B}(t)\rangle = -i\int_{-\infty}^t dt' \, \mathrm{Tr}\,\hat{B}(0)\exp[-i\hat{\mathscr{L}}(t - t')]$$
$$\times \hat{\mathscr{L}}_{EXT}(t')\hat{\rho}(-\infty)$$

(335)

The Liouville operator $\hat{\mathscr{L}}_{EXT}(t)$ is often written in the form

$$\hat{\mathscr{L}}_{EXT}(t) = (1/\hbar)F(t)\hat{\mathscr{A}}^-$$

(336)

where $\hat{\mathscr{A}}^-$ is a tetradic operator that acts on the system and $F(t)$ is a periodic classical force given by

$$F(t) = \mathrm{Re}(F_0\exp[+(i\omega + \varepsilon)t])$$

(337)

with $\varepsilon \to 0^+$ at the end of the calculation.

For the external disturbance given by Eq. (337), the linear response $\langle\Delta\hat{B}(t)\rangle$ can be written

$$\langle\Delta\hat{B}(t)\rangle = \mathrm{Re}[\chi_{B,A}(\omega)F_0\exp(+i\omega t)]$$

(338)

where $\chi_{B,A}(\omega)$ is a generalized susceptibility defined by

$$\chi_{B,A}(\omega) = \lim_{\varepsilon \to 0^+} \int_0^\infty dt \, \exp[-(i\omega + \varepsilon)t]\phi_{B,A}(t)$$

(339)

with $\phi_{B,A}(t)$ given by the following equivalent forms:

$$\phi_{B,A}(t) = -(i/\hbar)\langle[\hat{A}(0), \hat{B}(t)]_-\rangle_{EQ}$$

(340a)

$$= \beta\langle\hat{\dot{A}}(0)\hat{B}(t)\rangle_{EQ}$$

(340b)

$$= -\beta\langle\hat{A}(0)\hat{\dot{B}}(t)\rangle_{EQ}$$

(340c)

The quantity $\phi_{B,A}(t)$, called the response function, represents the response of the system to an instantaneous force $F(t)\alpha\,\delta(t)$ applied at $t = 0$. For classical systems, the response function $\phi_{B,A}(t)$ assumes the forms

$$\phi_{B,A}(t) = \langle\{A(0), B(t)\}\rangle_{EQ}$$

(341a)

$$= \beta\langle\dot{A}(0)B(t)\rangle_{EQ}$$

(341b)

$$= -\beta\langle A(0)\dot{B}(t)\rangle_{EQ}$$

(341c)

where $\{,\}$ is the Poisson bracket. These results tell us that the linear response of a system to an external disturbance is related to spontaneous equilibrium fluctuations, described by the response function $\phi_{B,A}(t)$, through the generalized susceptibility $\chi_{B,A}(\omega)$. The formal expressions given by Eqs. (340a)–(341c) enable us to connect the response to microscopic interactions.

The relationships between the generalized susceptibility $\chi_{B,A}(\omega)$ and the Fourier components $f_{B,A}(\omega)$ of the response function $\phi_{B,A}(t)$ is usually called the fluctuation–dissipation theorem:

$$\chi_{B,A}(\omega) = \pi f_{B,A}(\omega) + i \int_{-\infty}^{+\infty} d\omega' \frac{f_{B,A}(\omega')}{\omega' - \omega} \quad (342)$$

where

$$f_{B,A}(\omega) = (2\pi)^{-1} \int_{-\infty}^{+\infty} dt \, \exp(i\omega t)\phi_{B,A}(t) \quad (343)$$

Moreover, $\chi_{B,A}(\omega)$ and $f_{B,A}(\omega)$ are connected by the Kramers–Kroning relations:

$$\text{Re } \chi_{B,A}^{s}(\omega) = \pi f_{B,A}(\omega) \quad (344)$$

$$\text{Im } \chi_{B,A}^{s}(\omega) = \int_{-\infty}^{+\infty} d\omega' \frac{f_{B,A}^{s}(\omega')}{\omega' - \omega} \quad (345)$$

$$\text{Re } \chi_{B,A}^{a}(\omega) = i \int_{-\infty}^{+\infty} d\omega' \frac{f_{B,A}^{a}(\omega')}{\omega' - \omega} \quad (346)$$

$$\text{Im } \chi_{B,A}^{\alpha}(\omega) = -i\pi f_{B,A}^{a}(\omega) \quad (347)$$

where the superscripts s and a refer, respectively, to the symmetric and antisymmetric parts of $\boldsymbol{\chi}(\omega)$ and $\mathbf{f}(\omega)$. In the above, $f_{B,A}^{s}(\omega)$ is real and $f_{B,A}^{a}(\omega)$ is pure imaginary.

A simple example to which the above-described linear response theory applies is the problem of electrical conduction. For this problem, the response function is given by

$$\phi_{\mu,\nu}(t) = \beta\langle \hat{J}_{\nu}(0)\dot{\hat{J}}_{\mu}(t)\rangle_{\text{EQ}} \quad (348)$$

where $\hat{J}_{\nu}$ is a current operator. The response function $\phi_{\mu,\nu}(t)$ describes the response of the current in the μ direction to an instantaneous electric field applied in the ν direction at $t = 0$. The susceptibility $\chi_{\mu,\nu}(\omega)$ corresponding to the response function $\phi_{\mu,\nu}(t)$ is the frequency-dependent conductivity

$$\sigma_{\mu,\nu}(\omega) = \lim_{\varepsilon \to 0^+} \beta \int_0^\infty dt \, \exp[-(i\omega + \varepsilon)t]$$

$$\langle \hat{J}_{\nu}(0)\dot{\hat{J}}_{\mu}(t)\rangle_{\text{EQ}} \quad (349)$$

C. PROJECTION OPERATOR THEORIES

Since most problems require only the average values of certain dynamical variables or a contracted description of a system, Zwanzig sug-

gested that we partition the density operator–distribution function appearing in the Liouville equation for a system into a relevant and an irrelevant part, with the relevant part containing the relevant information for the problem at hand. This partitioning and the subsequent formal development are readily accomplished by working in the matrix language introduced in Section IV, which allowed us to write the Liouville equation and the equation of motion for dynamic variables respectively in the vector forms given by Eqs. (197) and (198), which apply to both quantum and classical systems.

The partitioning of the vector $|\rho_t\rangle$ corresponding to a global density operator or distribution function into a relevant part $|\rho_t^{R}\rangle$ and an irrelevant part $|\rho_t^{I}\rangle$ is achieved by introducing a projection operator $\hat{\mathscr{P}}$ and its complement $\hat{Q} = \hat{I} - \hat{\mathscr{P}}$, which are defined in such a way that

$$|\rho_t\rangle = |\rho_t^{R}\rangle + |\rho_t^{I}\rangle \quad (350)$$

where

$$|\rho_t^{R}\rangle = \hat{\mathscr{P}}|\rho_t\rangle \quad (351)$$

and

$$|\rho_t^{I}\rangle = \hat{Q}|\rho_t\rangle \quad (352)$$

The projection operator $\hat{\mathscr{P}}$ is required to be idempotent, that is $\hat{\mathscr{P}}^2 = \hat{\mathscr{P}}$.

Utilizing the vector form of the Liouville equation [Eq. (197)] and the above partitioning of $|\rho_t\rangle$, one can construct the following equation of motion for $|\rho_t^{R}\rangle$:

$$|\dot{\rho}_t^{R}\rangle = -i\hat{\mathscr{P}}\hat{\mathscr{L}}\hat{\mathscr{P}}|\rho_t^{R}\rangle$$

$$- \int_0^t d\tau \, \hat{\mathscr{P}}\hat{\mathscr{L}}\hat{Q}$$

$$\times \exp(-i\hat{Q}\hat{\mathscr{L}}\hat{Q}\tau)\hat{Q}\hat{\mathscr{L}}\hat{\mathscr{P}}|\rho_{t-\tau}^{R}\rangle$$

$$- i\hat{\mathscr{P}}\hat{\mathscr{L}}\hat{Q} \exp(-i\hat{Q}\hat{\mathscr{L}}\hat{Q}t)\hat{Q}|\rho_0\rangle \quad (353)$$

This equation of motion is called the Zwanzig master equation. It is an exact equation of motion for $|\rho_t^{R}\rangle$ and thus is formally equivalent to the Liouville equation. Since $|\rho_t^{R}\rangle$ depends on itself at earlier times through the convolution integral in Eq. (353), we say that Eq. (353) possesses memory. Such equations of motion are labeled non-Markovian.

The actual use of Eq. (353) requires the specification of the projection operator $\hat{\mathscr{P}}$, which in turn depends on the problem at hand and how we want to bias the final results. For example, suppose we want to construct a non-Markovian analog of Eq. (319). The formal structure of Eqs. (319)–(326) suggests that the projection operator

$\hat{\mathscr{P}}$ be defined as

$$\hat{\mathscr{P}} = \sum_{k,j} |\Delta\bar{O}_k \, \rho_{EQ}\rangle \, \chi_{k,j}^{-1} \langle\Delta O_j^{\dagger}| \qquad (354)$$

where $\chi_{k,j}^{-1}$ is an element of the inverse matrix χ^{-1}, with the elements of χ defined by Eq. (310). The vectors $|\bar{O}_k\rho_{EQ}\rangle$ and $\langle\Delta O_j^{\dagger}|$ respectively, correspond to the operators $\Delta\hat{\bar{O}}_k\hat{\rho}_{EQ}$ and $\Delta\hat{O}_j$, where $\hat{\rho}_{EQ} = \exp(\Omega I - \beta\hat{\mathscr{H}})$. The inner product $\langle\Delta O_j^{\dagger}|\Delta\bar{O}_k\,\rho_{EQ}\rangle$ is the element $\chi_{j,k}$ of the matrix χ.

In Section IV,B, we stated that the average $\langle\Delta\hat{O}_i(t)\rangle$ can be written as the inner product $\langle\Delta O_i^{\dagger}|\rho_t\rangle$. One can readily demonstrate that $\langle\Delta\hat{O}_i(t)\rangle$ is also given by $\langle\Delta\hat{O}_i(t)\rangle = \langle\Delta O_i^{\dagger}|\rho_t^R\rangle$, where $|\rho_t^R\rangle = \hat{\mathscr{P}}|\rho_t\rangle$, with $\hat{\mathscr{P}}$ defined by Eq. (354). With the latter result at our disposal, we can utilize Eqs. (353) and (354) to construct the following non-Markovian analog of Eq. (319),

$$\langle\Delta\hat{O}_i(t)\rangle = -\sum_j \left[i\Omega_{i,j}\langle\Delta\hat{O}_j(t)\rangle \right.$$
$$\left. + \int_0^t d\tau \, \mathbb{K}_{i,j}(\tau)\langle\Delta\hat{O}_j(t-\tau)\rangle \right] + \mathbb{I}_i(t)$$
$$(355)$$

where

$$i\Omega_{i,j} = \sum_k M_{i,k}^S \chi_{k,j}^{-1} \qquad (356)$$

$$\mathbb{K}_{i,j}(\tau) = \sum_k \mathbb{M}_{i,k}^C(\tau)\chi_{k,j}^{-1} \qquad (357)$$

and

$$\mathbb{I}_i(t) = \langle\Delta\dot{O}_i^{\dagger}|\hat{Q}\,\exp(-i\hat{Q}\hat{\mathscr{L}}\hat{Q}t)\hat{Q}|\rho_0\rangle \qquad (358)$$

with

$$M_{i,k}^S = \langle\Delta O_i^{\dagger}|\Delta\dot{\bar{O}}_k\,\rho_{EQ}\rangle \qquad (359)$$

and

$$\mathbb{M}_{i,k}^C(\tau) = \langle\Delta\dot{O}_i^{\dagger}|\hat{Q}\,\exp(-i\hat{Q}\hat{\mathscr{L}}\hat{Q}\tau)\hat{Q}|\Delta\dot{\bar{O}}_k\,\rho_{EQ}\rangle$$
$$(360)$$

The first term in Eq. (355) describes the instantaneous response of the system to the displacements $\{\langle\Delta\hat{O}_j(t)\rangle\}$. This response is characterized by the frequencies $\{\Omega_{i,j}\}$. The second term in Eq. (355) describes memory effects; that is, it relates the displacements from equilibrium at time t to earlier values of these displacements through the memory kernels $\{\mathbb{K}_{i,j}(\tau)\}$. The last term in Eq. (355), $\mathbb{I}_i(t)$, is a source term describing effects due to the initial preparation of the system.

An equation of motion formally equivalent to Eq. (355) was obtained by Mori by applying the projection operator $\hat{\mathscr{P}}$ directly to the vector

equation of motion for the dynamic variables. The result of this application is

$$\langle\Delta\dot{O}_{i,t}^{\dagger}| = -\sum_j \left[i\Omega_{i,j}\langle\Delta O_{j,t}^{\dagger}| \right.$$
$$\left. + \int_0^t d\tau \, \mathbb{K}_{i,j}(\tau)\langle\Delta O_{j,t}^{\dagger}| \right] + \langle f_{i,t}| \qquad (361)$$

where

$$\langle f_{i,t}| = \langle\Delta\dot{O}_i^{\dagger}|\hat{Q}\,\exp(-i\hat{Q}\hat{\mathscr{L}}\hat{Q}t)\hat{Q} \qquad (362)$$

Since this equation of motion resembles the Langevin equations of classical Brownian motion theory, Eq. (361) is often called a generalized Langevin equation, with $\langle f_{i,t}|$ representing a "random force."

Forming the inner product of Eq. (361) with the vector $|\rho_0\rangle$ and then making use of the relation $\langle\Delta\hat{O}_j(t)\rangle = \langle\Delta O_{j,t}^{\dagger}|\rho_0\rangle$, one obtains Eq. (355). Since Eq. (355) can be obtained from either Zwanzig's master equation or Mori's generalized Langevin equation, the above projection operator approaches are commonly called the Mori–Zwanzig projection operator formalism.

We should remark that Mori did not actually use the projection operator $\hat{\mathscr{P}}$ defined by Eq. (354). Instead, Mori used the projection operator $\hat{\mathscr{P}} = \sum_{k,j} |\Delta\bar{O}_k\rangle\chi_{k,j}^{-1}\langle\Delta O_j^{\dagger}|$, where the inner product $\langle\Delta O_j^{\dagger}|\Delta\bar{O}_k\rangle = \text{Tr}\,\hat{\rho}_{EQ}\,\Delta\hat{O}_j\,\Delta\hat{\bar{O}}$, with $\hat{\rho}_{EQ}$ playing the role of a metric. The use of such a definition leads to some difficulties when one is attempting to perform actual calculations. Nonetheless, the result given by Eq. (355) is formally equivalent to Mori's results. Equations (359) and (360) can be brought into a form that more closely resembles Mori's results by writing

$$M_{i,k}^S = \langle\Delta O_i^{\dagger}\rho_{EQ}^{1/2}|\Delta\dot{\bar{O}}_k\,\rho_{EQ}^{1/2}\rangle \qquad (363)$$

and

$$\mathbb{M}_{i,k}^C(\tau) = \langle\Delta\dot{O}_i^{\dagger}\,\rho_{EQ}^{1/2}|\hat{Q}$$
$$\exp(-i\hat{Q}\hat{\mathscr{L}}\hat{Q}\tau)\hat{Q}|\,\Delta\dot{\bar{O}}_k\,\rho_{EQ}^{1/2}\rangle \qquad (364)$$

where $\hat{Q} = \hat{I} - \hat{\mathscr{P}}$, with $\hat{\mathscr{P}} = \sum_{k,j} |\Delta\bar{O}_k\,\rho_{EQ}^{1/2}\rangle\chi_{k,j}^{-1}\langle\Delta O_j^{\dagger}\,\rho_{EQ}^{1/2}|$.

It is usually quite difficult to work with non-Markovian equations of the type given by Eq. (355), so a variety of qualitative arguments are usually invoked to rationalize making a Markovian approximation to this equation. The typical argument is based on the notion that the memory kernels in Eq. (355) should decay on a much faster time scale than the displacements from equilibrium $\{\langle\Delta\hat{O}_j(t)\rangle\}$. Nonetheless, the Markovian approximation is made by writing

$$\mathbb{K}_{i,j}^{(\tau)} = K_{i,j}\,\delta(\tau) \qquad (365)$$

where

$$K_{i,j} = \sum_k M_{i,k}^C \chi_{k,j}^{-1} \qquad (366)$$

with

$$M_{i,k}^C = \lim_{\varepsilon \to 0^+} \int_0^\infty dt \, \exp(-\varepsilon t)$$
$$\times \langle \Delta \dot{O}_i^\dagger | \hat{Q} \exp(-i\hat{Q}\hat{\mathscr{L}}\hat{Q}t)\hat{Q} | \Delta \dot{O}_k \, \rho_{EQ} \rangle \qquad (367)$$

In addition to making a Markovian approximation to Eq. (355), the source term $\mathbb{I}_i(t)$ describing effects due to initial preparation is usually neglected. This assumption is based on the idea that such effects are expected to be unimportant when the system is close to equilibrium. A more rigorous justification for invoking this approximation is to use the maximum entropy principle to describe the initial vector $|\rho_0\rangle$, which leads to the form $|\rho_0\rangle = \exp(\Omega) \exp [-\beta|\mathscr{H}\rangle - \sum_j \Lambda_j(0)|O_j\rangle]$. Then, approximating $|\rho_0\rangle$ by $|\rho_0\rangle \cong \sum_{j,k}|\Delta O_j \, \rho_{EQ}\rangle \chi_{j,k}^{-1}\langle \Delta \dot{O}_k(0)\rangle$ (assuming that the system is initially close to equilibrium), one is led to the conclusion that $\mathbb{I}_i(t)$ vanishes.

Invoking the Markovian approximation to Eq. (355) and neglecting effects due to initial preparation of the system, one is led to a Markovian equation identical in form to the linear equation of motion given by Eq. (319) with $K_{i,j}^{(1)}$ replaced by $K_{i,j}$. [The frequencies $\{\Omega_{i,j}\}$ defined by Eq. (356) are identical to those defined by Eq. (320).] The replacement of $K_{i,j}^{(1)}$ by $K_{i,j}$ is required by virtue of the fact that the formal expressions for $M_{i,k}^{(1),C}$ and $M_{i,k}^C$ differ due to the presence of the operator $\hat{Q}$ in the expression for $M_{i,k}^C$ [see Eqs. (324) and (367)]. For some problems, under certain limiting conditions, $M_{i,k}^{(1),C}$ and $M_{i,k}^C$ are equivalent. In general, this is not the case. Nonetheless, it is common practice to drop the $\hat{Q}$ and assume the approximate equivalence of $M_{i,k}^{(1),C}$ and $M_{i,k}^{(1)}$.

The classical limit of the quantum results given by Eqs. (355)–(360) and (363)–(367) is obtained by writing

$$\mathbb{I}_i(t) = \langle \Delta \dot{O}_i^* | \hat{Q} \exp(-i\hat{Q}\hat{\mathscr{L}}\hat{Q}t)\hat{Q} | \rho_0 \rangle \qquad (368)$$

$$M_{i,k}^S = \langle \Delta O_i^* | \Delta \dot{O}_k \, \rho_{EQ} \rangle \qquad (369)$$

$$\mathbb{M}_{i,k}^C(\tau) = \langle \Delta \dot{O}_i^* | \hat{Q}$$
$$\times \exp(-i\hat{Q}\hat{\mathscr{L}}\hat{Q}\tau)\hat{Q} | \Delta \dot{O}_k \, \rho_{EQ} \rangle \qquad (370)$$

$$M_{i,k}^C = \lim_{\varepsilon \to 0^+} \int_0^\infty dt \, \exp(-\varepsilon t)$$
$$\times \langle \Delta \dot{O}_i^* | \hat{Q} \exp(-i\hat{Q}\hat{\mathscr{L}}\hat{Q}t)\hat{Q} | \Delta \dot{O}_k \, \rho_{EQ} \rangle \qquad (371)$$

and replacing $\langle \Delta \hat{O}_i(t) \rangle$ with $\langle \Delta O_i(t) \rangle$. For the classical limit, $\hat{\mathscr{P}} = \sum_{k,j}|\Delta O_k \, \rho_{EQ}\rangle \chi_{k,j}^{-1}\langle \Delta O_j^*|$, where $\chi_{k,j}$ is given by Eq. (314) and the vectors $\langle \Delta O_j^*|$ and $|\Delta O_k \times \rho_{EQ}\rangle$, respectively, correspond to the classical variables ΔO_j and ΔO_k ρ_{EQ}.

As in the previously discussed theories of irreversible processes, the time evolution of the displacements from equilibrium in the above projection operator theory is connected to spontaneous equilibrium fluctuations. Such fluctuations are built into the theory through the projection operator $\hat{\mathscr{P}}$. The formal expressions for the quantities $\mathbb{I}_i(t)$, $M_{i,k}^S$, $\mathbb{M}_{i,k}^C(\tau)$, and $M_{i,k}^C$ enable us to study the influence of microscopic interactions on the time evolution of displacements from equilibrium.

Projection operators of the type given by Eq. (354) possess a global operational character in the sense that they operate on both the system of interest and the bath. An alternative approach to the problem of constructing contracted descriptions is to utilize projection operators that operate only on the bath. Such bath projection operators have often been employed in conjunction with Zwanzig's master equation.

Aside from the use of projection operator techniques to construct contracted descriptions, they serve as powerful analytic and numerical tools for the solution of time evolution and spectral density (susceptibility) problems. Most notable are the memory function formalism of Mori and the dual Lanczos transformation approach of the author. Mori's memory function formalism provides a framework for determining the time evolution of autocorrelation functions (correlation functions describing self-correlations) and computing the associated spectral densities (susceptibilities) for classical systems without having to obtain solutions of global equations of motion. The dual Lanczos transformation approach is a projection operator theory that can handle very general time evolution and spectral density problems. Unlike Mori's memory function formalism, which requires the Liouville operator to be Hermitian, the dual Lanczos transformation approach does not impose symmetry restrictions on the Liouville operator and thus applies to both reversible and irreversible systems. Moreover, the dual Lanczos transformation approach can be used to determine the time evolution or spectral densities of autocorrelation and cross-correlation functions (the latter describe correlations between different variables), as well as displacements from equilibrium, without the need to obtain solutions of global equations of motion. The Mori and Zwanzig projection operator theories represent special limiting cases of the author's dual Lanczos transformation approach.

BIBLIOGRAPHY

Berne, B. J., ed. (1977). "Modern Theoretical Chemistry, Vol. 6." "Statistical Mechanics, Part B: Time Dependent Processes." Plenum, New York.

Callen, H. B. (1962). "Thermodynamics." Wiley, New York.

de Boer, J., and Uhlenbeck, G. E., eds. (1962). "Studies in Statistical Mechanics," Vol. 1. North-Holland, Amsterdam.

Forster, D. (1975). "Fluctuations, Broken Symmetries, Time Correlation Functions." Benjamin, Reading, Massachusetts.

Jancel, R. (1963). "Foundations of Classical and Quantum Statistical Mechanics." Pergamon Press, London.

Katz, A. (1967). "Principles of Statistical Mechanics—The Information Theory Approach." Freeman, San Francisco.

Lim, E. C., ed. (1982). "Excited States," Vol. 5. Academic Press, New York.

Louisell, W. H. (1973). "Quantum Statistical Properties of Radiation." Wiley, New York.

Prigogine, I. (1980). "From Being to Becoming—Time and Complexity in the Physical Sciences." Freeman, San Francisco.

Reichl, L. E. (1980). "A Modern Course in Statistical Physics." Univ. of Texas Press, Austin.

Ter Haar, D., ed. (1961). "Fluctuation, Relaxation, and Resonance in Magnetic Systems." Oliver and Boyd, Edinburgh.

Wax, N., ed. (1954). "Selected Papers on Noise and Stochastic Processes." Dover, New York.

Yourgrau, W., van der Merwe, A., and Raw, G. (1966). "Treatise on Irreversible and Statistical Thermophysics—An Introduction to Nonclassical Thermodynamics." Macmillan, New York.

SUPERCONDUCTIVITY MECHANISMS

J. Spalek   *Purdue University*

I. Introduction
II. The Bardeen–Cooper–Schrieffer Theory
III. Normal and Magnetic States of Correlated Electrons
IV. Novel Mechanisms of Electron Pairing
V. Conclusions

GLOSSARY

Almost-localized Fermi liquid: Metallic system which under a relatively small change of external parameter such as temperature, pressure, or composition undergoes a transition to the Mott insulating state. In such a metal, electrons have large effective mass. At low temperature the system may order antiferromagnetically or undergo a transition to the superconducting state. Both the nonstoichiometric oxides (such as $La_{2-x}Sr_xCuO_4$), and heavy fermion systems are regarded as almost-localized Fermi liquids.

Bardeen, Cooper and Schrieffer (BCS) theory: Theory describing properties of superconductors in terms of the concept of pairing of electrons with opposite spins and momenta. The pairing of electrons is mediated by lattice deformation and the resultant attractive interaction overcomes their mutual repulsion. At a critical temperature, the electron system undergoes a transition to a condensed state of pairs which is characterized by a zero dc electrical resistance and a strong diamagnetism.

Correlated electrons: Electrons with their kinetic (or band) energy comparable to or smaller than the magnitude U of electron–electron repulsion. This situation is described by the condition $U \gtrsim W$, where W is the width of a starting (bare) energy band. Strictly speaking, we distinguish between the limits of almost-localized Fermi liquid for which $U \lesssim W$, and the limit of strongly correlated electrons (the spin-liquid) in which $U \gg W$. The term *correlated electrons* means that the motion of a single electron is correlated with that of others in the system.

Exchange interaction: Part of the Coulomb interaction between electrons which depends on their resultant spin state. If the spin-singlet configuration is favored in the ground state, then the interaction is called antiferromagnetic. The exchange interaction provides a mechanism of magnetic ordering in Mott insulators; it may also correlate electrons into singlet or triplet pairs in a metallic state, particularly when the pair-exchange coupling J of electron pair is comparable to the kinetic energy of each of its constituents.

Fermi liquid: Term describing the state of interacting electrons in a metal. Equilibrium properties of such systems are modeled by a gas of free electrons with renormalized characteristics such as the effective mass. The properties at low temperatures are determined mainly by the electrons near the Fermi surface. The electron–electron interactions lead to specific contributions to the transport properties of such a system.

Hubbard subband: Term describing each of the two parts of an energy band in a solid which splits when the electron–electron repulsion energy is comparable to their kinetic energy. The Hubbard splitting of the original band induced by the interaction explains in a natural way the existence of the Mott insulating state in the case of an odd number of electrons per atom.

Mott insulator: Insulator containing atoms with partially filled $3d$ or $4f$ shells. These systems order magnetically (usually antiferromagnetically) when the temperature is lowered. Thus, they differ from ordinary (Bloch–Wilson) or band insulators which are weakly diamagnetic and are characterized by filled atomic shells, separated from empty states by a gap. In the antiferromagnetic phase of the Mott insulators, each electron with its (frozen) spin oriented up is surrounded by the electrons with spin down, and vice versa.

I. Introduction

Superconductivity remains among the most spectacular manifestations of electron states in a metal. Experimentally, one observes below a characteristic temperature T_c a transition to a phase with nonmeasurable dc resistance (or with a persistent current), perfect diamagnetism of bulk samples in a weak magnetic field, and quantum tunneling between two pieces of a superconductor separated by an insulating layer of semimacroscopic thickness. In the theoretical domain, one studies the quantum-mechanical (nonclassical) behavior of microscopic particles (electrons) at a macroscopic scale. Here, we summarize our present understanding, the Bardeen–Cooper–Schrieffer (BCS) theory of classical superconductors (see Section II), and we review the current theoretical approaches to new superconductors: the heavy fermion materials and the high-T_c magnetic oxides. The latter subject is discussed in Section IV, after we summarize normal-state properties of correlated electrons in Section III.

A brief characterization of the recent studies of superconductivity is in order. From the time of the first discovery (1911) of superconductivity in mercury (at temperature $T_c \simeq 4.2$ K) by Kammerlingh Onnes until 1986, studies were limited to low temperatures, $T < 25$ K. During the last 2 yr, 5 different classes of new superconducting compounds with critical temperatures of $T_c = 30$ K (for $Ba_{1-x}K_xBiO_3$), 40 K (for $La_{2-x}Sr_xCuO_4$), 90 K (for $YBa_2Cu_3O_{7-\delta}$), 110 K (for $Bi_2Sr_2CaCu_2O_8$), and 125 K (for $Tl_2Ca_2Ba_2Cu_3O_{10-y}$) were discovered and thoroughly studied in a number of laboratories. This sequence of rapidly cascading discoveries provides a proper perspective on the pace of recent development and on the current excitement accompanying these discoveries.

The starting point in both classical and new superconducting materials is the electronic structure that determines the metallic properties in the normal phase (that is, that above T_c). In this respect, the classical superconductors are well described by band theory, and, in some cases, starting from the concept of the electron gas. By contrast, the new materials are characterized as those whose electrons are close to localization, that is, those close to the metal–insulator transition of the Mott–Hubbard type. The latter transition may be induced by a relatively small change in compound composition (cf. the behavior of $La_{2-x}Sr_xO_4$ or $YBa_2Cu_3O_{7-x}$ as a function of x). It is quite interesting to note that oxides such as $YBa_2Cu_3O_{7-x}$ may be synthesized in either insulating ($x \gtrsim 0.5$) or metallic states. Additionally, antiferromagnetic ordering of the $3d$ electrons is observed close to the metal–insulator transition; the magnetic insulating state transforms into a supercon-

ducting state when $0 < x \lesssim 0.5$. Therefore, an account of our understanding of the almost-localized metallic state in a normal or magnetic (that is, nonsuperconducting) phase is highly desirable and provided in Section III. The antiferromagnetic insulating, normal metallic, and superconducting states must all be treated on the same footing for a proper characterization of the high-T_c oxides.

Details of the electronic structure in the high-T_c oxides are also important for two additional reasons. First, as discussed later, in these superconductors the coherence length is quite small, that is, comparable to the lattice constant. Hence, the details of the Bloch wave function on the atomic scale become crucial. Second, a whole class of models (discussed in Section IV) relies on the electron pairing induced by short-range electron–electron interactions. These interactions are strong and also present in the normal phase. Hence, one must develop a coherent theoretical picture of the correlated metallic state that undergoes a transformation either to the Mott insulating or to the superconducting state.

In this article, the properties of correlated electrons in normal, insulating, magnetic and superconducting phases are reviewed within parametrized models, starting from either Hubbard or Anderson-lattice Hamiltonians. These are the models that describe the properties of correlated metallic systems in terms of a few parameters, such as the bandwidth W of starting (uncorrelated, bare) electrons and the magnitude U of short-range (intraatomic) Coulomb interaction. Such models provide an overall understanding of both the nature of correlated metallic and insulating ground states and the underlying thermodynamic properties of those systems. However, the guidance of detailed band structure calculations is often needed in choosing appropriate values for the microscopic parameters.

II. The Bardeen–Cooper–Schrieffer Theory

The Bardeen–Cooper–Schrieffer (BCS) theory [1–10] relies on three features of metallic solids: (1) the electron-lattice interaction; (2) the formation of an electron-pair bound state (the so-called *Cooper pair state*) due to the coupling of the electrons to the lattice; and (3) the instability of the normal metallic state with respect to formation of a macroscopic condensed state of all pairs $(\mathbf{k} \uparrow, -\mathbf{k} \downarrow)$ with antiparallel spins in momentum ($\mathbf{k}$) space. The condensed state exhibits the principal properties of superconductors, such as a perfect diamagnetism, zero dc resistance, and so on. We discuss briefly those three features first and then summarize some consequences of the BCS theory. The BCS theory deals not only with one of the possible (phonon-mediated) mechanisms for superconductivity, but also provides an adequate language for the description of such a condensed state in general terms, independent of the particular pairing mechanism.

A. FROM ELECTRON–PHONON COUPLING TO THE EFFECTIVE ATTRACTIVE INTERACTION BETWEEN ELECTRONS: VIRTUAL EXCHANGE OF PHONONS

The electron-lattice interaction can be described by introducing phonons as quasiparticles representing vibrational modes of the lattice. In this picture, an electron moving in a solid and scattering on the lattice vibration absorbs or emits a phonon with energy $\hbar\omega_q$ and quasi-momentum $\hbar q$. If during such processes the energy of incoming electrons (with energy ε_k and momentum $\hbar k$) and scattered electrons (with energy $\varepsilon_{k'}$) is conserved, then a real scattering process has taken place. Such events lead to the nonzero resistivity of metals at temperature $T > 0$. For these processes

$$\varepsilon_{k'} - \varepsilon_k = \pm\hbar\omega_q \qquad (1)$$

where " $-$ " corresponds to the emission and " $+$ " to the absorption of the phonon. However, in the quantum-mechanical description of scattering processes there also exist virtual processes that do not conserve energy. Such events involve the emission and subsequent reabsorption of a phonon in a time interval Δt such that the uncertainty principle $\Delta E \cdot \Delta t \geq \hbar$ is not violated. The uncertainty of particle energies ΔE is related to the magnitude of the electron–phonon interaction. In effect, this leads to the following electron–electron contribution to the total energy involving a pair (k, k') of electrons

$$V_{kk'q} = |W_q|^2 \frac{\hbar\omega_q}{(\varepsilon_{k'} - \varepsilon_k)^2 - (\hbar\omega_q)^2} \qquad (2)$$

where $(k' - k) = q$, and W_q is the electron–phonon matrix element characterizing the process of single emission or absorption of the phonon by the electron

subsystem. In many electron systems, one represents Eq. (2) by an effective electron–electron interaction, which can be written as

$$H' = \sum_{kk'q} V_{kk'q} c^+_{k+q\sigma} c^+_{k'-q\sigma'} c_{k'\sigma'} c_{k\sigma} \qquad (3)$$

This is a phonon-mediated contribution to the interaction between electrons. More precisely, in this expression $c_{k\sigma}$ symbolizes a destruction or annihilation of an electron in the initial single-particle state $|k\sigma\rangle$, whereas $c^+_{k'\sigma'}$ is the creation of an electron in the state $|k'\sigma'\rangle$ after the scattering process has taken place. The processes represented in Eq. (3) of destruction of the electron pair in the states $|k\sigma\rangle$ and $|k'\sigma'\rangle$ and their subsequent reestablishment in the final states $|k + q\sigma\rangle$ and $|k' - q\sigma'\rangle$ are customarily represented by a diagram of the type drawn in Fig. 1(b). It symbolizes the phonon exchange between the two electrons moving through crystal. The virtual processes are composed of the two parts: one describing phonon emission and the subsequent reabsorption process and one describing the reverse process.

One should notice that if in Eq. (2) $|\varepsilon_{k'} - \varepsilon_k| < \hbar\omega_q$, then $V_{kk'q} < 0$, that is, the interaction is attractive. Hence, if this quantity overcomes the magnitude of the Coulomb repulsion between the electrons in a given medium, this leads to a net attraction between the electrons. Such a net attractive interaction results in a stable superconducting state, as we shall see next.

B. INSTABILITY OF ELECTRON GAS STATE IN CASE OF ATTRACTIVE INTERACTION BETWEEN ELECTRONS: COOPER PAIRS

Following Fröhlich's discovery [11] that the electron–electron attraction can be mediated by phonons (cf. the previous discussion), the next step was taken by Cooper [12] who asked what happens when two electrons are added to an electron gas at $T = 0$. Because of the Pauli exclusion, they must occupy the states above the Fermi level, as shown in Fig. 2.

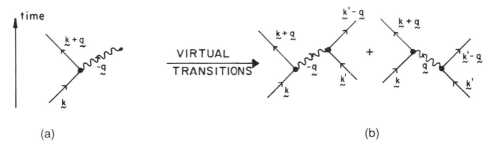

(a)   (b)

FIG. 1. (a) The scattering diagram of electrons with wave vectors $k \rightarrow k + q$ accompanied by the emission of the phonon of wave vector ($-q$). (b) Virtual emission and subsequent reabsorption of the phonon by electrons; the two processes drawn combine into the contribution [2] leading to the effective electron–electron attraction.

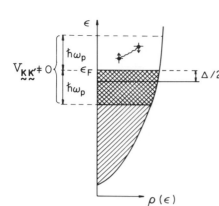

FIG. 2. Schematic representation of the conduction band filled with electrons up to the Fermi level ε_F. The density of states is $\rho(\varepsilon)$ (per one spin direction). The two electrons added to the system attract each other with the potential $V_{kk'} = -V$ if placed within the energy interval $\hbar\omega_p$ counting from ε_F. The attraction leads to a binding energy Δ below ε_F for the pair configuration $(\mathbf{k}\uparrow, -\mathbf{k}\downarrow)$.

Cooper showed that if the attractive potential in Eq. (2) is approximated by a negative nonzero constant $(-V)$ in the energy interval $\hbar\omega_p$ above ε_F (cf. Fig. 2), then such a potential introduces a binding between these two electrons with a binding energy

$$\Delta = -2\hbar\omega_p \left[\exp\left(\frac{2}{\rho V}\right) - 1 \right]^{-1} \quad (4)$$

relative to the energy of those two particles placed at the Fermi energy. In this expression, $\hbar\omega_p$ represents the average phonon energy (related to the Debye temperature θ_D through $\hbar\omega_p = k_B\theta_D$), and ρ is the density of free-particle states at the Fermi energy ε_F for the metal under consideration.

A few important features of the bound state represented by Eq. (4) should be mentioned. First, the binding energy Δ is largest for the state of the pair at rest, that is, with the total pair momentum $\mathbf{k}_1 + \mathbf{k}_2 = 0$. Thus, Δ represents the binding energy of the pair $(\mathbf{k}, -\mathbf{k})$. Second, the spin of the pair is compensated, that is, a singlet state is produced. Finally, the bound state has a lower energy than a pair of free particles placed at the Fermi level. Hence, the electron gas state is unstable with respect to such pair formation. A system of such pairs may condense into a superfluid state if the attractive interaction is stronger than the Coulomb repulsion between the particles constituting a pair. However, the situation is not so simple since the size of the pair is

$$\xi_0 = \langle r^2 \rangle^{1/2} \approx \frac{2\hbar V_F}{\Delta} \approx \frac{\hbar V_F}{k_B T_c} \sim 10^4 \text{ Å} \quad (5)$$

where V_F is the Fermi velocity for electrons. The quantity ξ_0 thus exceeds by far the average classical distance between the electrons, which is comparable

to interatomic distance $a \sim 1-2$ Å. In other words, the wave functions of the different pairs overlap appreciably, forming a condensed and coherent state of pairs in the superconducting phase. The properties of this condensed phase are discussed next. The new length scale ξ appearing in the system when electrons are bound into Cooper pairs is called the coherence length.

C. THE PROPERTIES OF THE SUPERCONDUCTING STATE: THE PAIRING THEORY

The BCS theory [1] provides a method of calculating the ground state, thermodynamic, and electromagnetic properties of a superconductor treated as a condensed state of electron pairs with opposite momenta and spins. The starting microscopic Hamiltonian is

$$\mathcal{H} = \sum_{\mathbf{k}\sigma} \varepsilon_\mathbf{k} n_{\mathbf{k}\sigma} + \sum_{\mathbf{k}\mathbf{k}'} V_{\mathbf{k}\mathbf{k}'} c_{\mathbf{k}\uparrow}^+ c_{\mathbf{k}\downarrow}^+ c_{-\mathbf{k}'\downarrow} c_{\mathbf{k}'\uparrow} \quad (6)$$

The first term describes the single particle (band) energy, $\varepsilon_\mathbf{k}$ being the energy per particle, and $n_{\mathbf{k}\sigma} = c_{\mathbf{k}\sigma}^+ c_{\mathbf{k}\sigma}$, the number of particles in the state $|\mathbf{k}\sigma\rangle$. The second term describes the pairing part [Eq. (3)] for the system of pairs that scatters from the state $(\mathbf{k}', -\mathbf{k}')$ into the state $(\mathbf{k}, -\mathbf{k})$. This term describes the dominant contribution of all processes contained in Eq. (3) (cf. Ref. 10).

In order to obtain eigen energies of the Hamiltonian [Eq. (6)], one can use either the variational method due to Schrieffer [2] or the transformation method developed by Bogoliubov and Valatin [13]. To obtain quasiparticle states in the superconducting phase, one has to combine an electron in the state $|\mathbf{k}\uparrow\rangle$ with one in the state $|-\mathbf{k}\downarrow\rangle$. More precisely, one defines new quasiparticle operators $\lambda_{\mathbf{k}0}^+$ and $\lambda_{\mathbf{k}1}^+$, which are expressed by the operators c^+ and c in the following manner:

$$\lambda_{\mathbf{k}0}^+ = u_\mathbf{k} c_{\mathbf{k}\uparrow}^+ - v_\mathbf{k} c_{-\mathbf{k}\downarrow}$$

and

$$\lambda_{\mathbf{k}1}^+ = v_\mathbf{k} c_{\mathbf{k}\uparrow}^+ + u_\mathbf{k} c_{-\mathbf{k}\downarrow}$$

This transformation expresses the process of creating electrons in the excited state $|\mathbf{k}\uparrow\rangle$ (that is, in the broken-pair configuration) under the restriction that no electron in the state $|-\mathbf{k}\downarrow\rangle$ is present. The coefficients of the transformation fulfill the condition $u_\mathbf{k}^2 + v_\mathbf{k}^2 = 1$.

In effect, the salient features of any superconducting theory with singlet spin pairing are as follows.

The single-particle excitations in the superconducting phase are specified by

$$E_\mathbf{k} = [(\varepsilon_\mathbf{k} - \mu)^2 + |\Delta_\mathbf{k}|^2]^{1/2} \quad (7)$$

where μ is the chemical potential of the system and $|\Delta_\mathbf{k}|$ is the so-called superconducting gap determined from the self-consistent equation

$$\Delta_{\mathbf{k}} = -\sum_{\mathbf{k}'} V_{\mathbf{k}\mathbf{k}'} \frac{\Delta_{\mathbf{k}'}}{2E_{\mathbf{k}'}} \tanh\left(\frac{\beta E_{\mathbf{k}'}}{2}\right) \qquad (8)$$

with $\beta \equiv (k_BT)^{-1}$. One should notice that if $V_{\mathbf{k}\mathbf{k}'}$ is approximated by a negative constant, then $\Delta_{\mathbf{k}} = \Delta$; Eq. (8) then yields as a solution either $\Delta \equiv 0$ or $\Delta \neq 0$, obeying the equation

$$1 = \frac{V}{N} \sum_{\mathbf{k}}{}' \frac{1}{2E_{\mathbf{k}}} \tanh\left(\frac{\beta E_{\mathbf{k}}}{2}\right) \qquad (9)$$

where now

$$E_{\mathbf{k}} = [(\varepsilon_{\mathbf{k}} - \mu)^2 + \Delta^2]^{-1/2} \qquad (10)$$

The primed summation in Eq. (9) is restricted to the regime of $\mathbf{k}$ states where $V \neq 0$. Equations (9) and (10) constitute the simplest BCS solution for an isotropic ($\mathbf{k}$-independent) gap. One sees that $E_{\mathbf{k}}$ is always nonvanishing and reaches a minimum $E_{\mathbf{k}} = \Delta$ for electrons placed on the Fermi surface, where $\varepsilon_{\mathbf{k}} = \mu$. Thus, the meaning of the gap becomes obvious: it is the gap for the single-electron excitations from the superconducting (condensed) phase to a free-particle state. The presence of a gap $\Delta > k_BT$ in the spectrum of single-particle excitations suppresses the scattering of electrons with acoustic phonons. The thermally excited electrons across the gap do not yield the nonzero resistivity because their contribution is short circuited by the presence of the pair condensate that carries a current with no resistance. The same holds true even for the superconducting systems for which the gap vanishes in some directions of $\mathbf{k}$ space.

One should emphasize that all thermodynamic properties are associated with thermal excitations; the energies that are specified by Eq. (7) contain $|\Delta_{\mathbf{k}}|$ or Δ as a parameter to be determined self-consistently from Eq. (8) or (9), respectively. Next, we provide a brief summary of the results that may be obtained within the BCS theory.

D. SUMMARY OF THE PROPERTIES

The solution of Eq. (9) provides these properties:

1. At $T = 0$, Eq. (1) reads

$$1 = (V/2N) \sum_{\mathbf{k}}{}' E_{\mathbf{k}}^{-1} \qquad (11)$$

The value of $\Delta \equiv \Delta(T = 0)$ for $\rho V \ll 1$ is given by

$$\Delta_0 = \frac{\hbar\omega_p}{\sinh(1/\rho V)} \approx 2\hbar\omega_p \exp\left(-\frac{1}{\rho V}\right) \qquad (12)$$

where $\hbar\omega_p \approx k_B\theta_D$[1] and ρ is the density of states at the Fermi energy. One notices a striking similarity between Eqs. (12) and (4), particularly for $\rho V \ll 1$ (this condition represents the so-called weak-coupling limit).

[1] In actual practice, one assumes that $\hbar\omega_p \approx 0.75 \, k_B\theta_D$ (cf. Meservey and Schwartz in Ref. 9).

2. We can choose the origin of energy at μ. Then, Eq. (9) can be transformed into an integral form:

$$1 = V \int_0^{\hbar\omega_p} \frac{\rho(\varepsilon) \, d\varepsilon}{(\varepsilon^2 + \Delta^2)^{1/2}} \tanh\left(\frac{\beta}{2} \sqrt{\varepsilon^2 + \Delta^2}\right) \qquad (13)$$

Since $\hbar\omega_p \ll \mu$, we may take $\rho(\varepsilon) \approx \rho(\varepsilon_F) \equiv \rho$ within the range of integration. This allows for an analytic evaluation of the critical temperature for which $\Delta = 0$:

$$T_c = 1.14\theta_D \exp\left(-\frac{1}{\rho V}\right) \qquad (14)$$

In all these calculations, it is implicitly assumed that $\rho V \ll 1$. Because of the presence of the exponential factor in Eq. (14), the critical temperature T_c is much smaller than the Debye temperature. This is the principal theoretical reason why T_c is so low in the superconductors discovered in the period 1911–1986. The exponential dependence of T_c on the electronic parameter ρV also explains why the parameters pertaining to the electronic structure, which are of the order 1 eV or more, respond to phase transitions on an energy scale that is three orders of magnitude smaller ($k_BT_c \sim 1$ meV). Effects with such a nonanalytic dependence of transition temperature on the coupling constant cannot be obtained in any order of perturbation theory starting with the normal state as an initial state. A similar type of effect is obtained in the studies of the Kondo effect (cf. Section IV).

3. Combining Eqs. (14) and (12), one obtains the universal ratio

$$2\Delta_0/k_BT_c = 3.53 \qquad (15)$$

which is frequently used as a test for the applicability of the BCS model. However, this value can also be obtained in the strong-coupling limit [15] for a particular strength of electron–phonon coupling.

4. By regarding energies $E_{\mathbf{k}}$ as representing electron excitations across the gap, one can write the expression for the entropy of a superconductor in this standard form:

$$S = -2k_B \sum_{\mathbf{k}} [f_{\mathbf{k}} \ln f_{\mathbf{k}} + (1 - f_{\mathbf{k}}) \ln(1 - f_{\mathbf{k}})] \qquad (16)$$

where $f_{\mathbf{k}} \equiv f(E_{\mathbf{k}}) = [1 + \exp(\beta E_{\mathbf{k}})]^{-1}$ is the Fermi–Dirac distribution function. Hence, the free energy of the superconducting state is

$$F_S = 2 \sum_{\mathbf{k}} E_{\mathbf{k}} f_{\mathbf{k}} - TS \qquad (17)$$

One should notice that the thermodynamic properties are determined fully only if the chemical potential $\mu = \mu(T)$ and the temperature dependence of the superconducting gap $\Delta_{\mathbf{k}} = \Delta_{\mathbf{k}}(T)$ are explicitly determined, since only then is the spectrum of single-particle excitations (characterized by the energies $\{E_{\mathbf{k}}\}$) uniquely determined. The quantity $\Delta(T)$ is de-

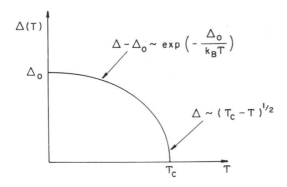

FIG. 3.   Schematic representation of the temperature dependence of the superconducting gap for the isotropic change. T_c is the critical temperature for the transition, and $\Delta_0 \equiv \Delta(T = 0)$.

termined from Eq. (13). The chemical potential is determined from the conservation of the number N_e of particles, that is, from the condition $\Sigma_k f_k = N_e$. The temperature dependence of the gap in the isotropic case is drawn schematically in Fig. 3.

5. By calculating this difference of the free energies $F_S - F_N$ in superconducting (F_S) and normal (F_N) phases and equating the difference with the magnetic free energy $H_c^2 V/8\pi$ (V is the volume of the system), one can obtain an approximate relation of the form

$$H_c(T)/H_c(0) \approx 1 - (T/T_c)^2 \qquad (18)$$

For the applied field $H > H_c$, superconductivity is destroyed because in the thermodynamic critical field H_c the spin singlet bound state is destroyed by the thermal fluctuations. The pair binding energy is then effectively overcome by the magnetic energy, so that the pairs break up into single particles. Strictly speaking, this type of behavior characterizes the so-called superconductors of the first kind.

6. By calculating the specific heat from the standard thermodynamic analysis $[C_S = -T(\partial^2 F_S/\partial T^2)v]$, one obtains at $T = T_c$ a discontinuity of the form

$$\frac{C_S - C_N}{C_N} = 1.43 \qquad (19)$$

where C_N is the specific heat at T_c for the material in its normal state. At low temperatures, the specific heat decreases exponentially:

$$C_S \sim \exp(\Delta_0/k_B T) \qquad (20)$$

for the special case of an isotropic gap. However, if the gap is anisotropic $[\Delta = \Delta_k(T)]$ and has lines of zeros (along which $\Delta_k = 0$), then the low-temperature dependence of C_S does not follow Eq. (20) but rather a power law T^n, with n depending on the details of the gap anisotropy.

The specific heat grows with T because the number

of thermally broken pairs increases with rising temperature; eventually, at $T = T_c$, ($k_B T_c \sim \Delta_0$), all bound pairs dissociate thermally at which point C_S reaches maximum. If the temperature is raised further (above T_c), the excess specific heat drops rapidly to zero since no pairs are left to absorb the energy. This type of behavior is observed in superconductors with an isotropic gap (cf., for example, Hg, Sn). One should notice that this interpretation of the thermal properties is based on the single-particle excitation spectrum [Eq. (10)]; we have disregarded any fluctuation phenomena near T_c, as well as collective excitations of the condensed system. It can be shown that the large coherence length $\xi \sim 10^3$–10^4 Å encountered in classic superconductors [8] is related to the absence of critical behavior near T_c. This is not the case in high-T_c superconductors (discussed in Section IV); hence, the new materials open up the possibility of studies of critical phenomena in superconducting systems.

7. The spin part of the static magnetic susceptibility vanishes as $T \to 0$. This is a direct consequence of the binding of electrons in the condensed state into singlet pairs. Therefore, the Meissner effect (the magnetic flux expulsion from the bulk of the sample) is present at $T = 0$ because of the orbital part of the susceptibility (an electron–pair analog of the Landau diamagnetism of single electrons in a normal electron gas). The expulsion of the magnetic flux from the bulk is measured in terms of the so-called London penetration depth $\lambda = \lambda(T)$, which characterizes the decay of the magnetic induction inside the sample. It decays according to

$$B(z) = H_a \exp(-z/\lambda)$$

where the z direction is perpendicular to the sample surface and the applied magnetic field H_a is parallel to it. The temperature dependence of the penetration depth is given by

$$\frac{\lambda(T)}{\lambda(0)} = \left[\frac{\Delta(T) \tanh(\Delta/2k_B T)}{\Delta_0} \right]$$

$$\approx \left[1 - \left(\frac{T}{T_c} \right)^4 \right]^{-1/2} \qquad (21)$$

This result has been derived under the assumption that the coherence length[2] $\xi \approx \hbar \, V_F/\Delta$ is much larger than λ. One should note that for a bulk sample of dimension $d \gg \lambda$ the induction $B \equiv 0$ almost everywhere. This condition determines the magnetic sus-

_____

[2] The coherence length in a superconductor can be estimated by using the uncertainty relation $\Delta p \xi_0 = \hbar$, where Δp is a change of the electron momentum (at $\varepsilon = \varepsilon_F$) due to the attractive interaction, which can be estimated from the corresponding change of particle kinetic energy $\Delta E = V_F \Delta p$. Taking $\Delta E \approx \Delta_0$, we obtain the desired estimate of ξ.

ceptibility χ of a superconductor regarded as an ideal diamagnet; in cgs units, $\chi \equiv M/H = -1/(4\pi)$.

8. The relative ratio of the two characteristic lengths $\kappa \equiv \lambda/\xi$ determines the type of superconductivity behavior in a magnetic field. From the dependence $\xi \sim \Delta^{-1}$, we infer that as $T \to T_c$, $\xi \sim (T_c - T)^{-1/2}$. The same type of dependence for $\lambda(T)$ can be inferred from Eq. (21) when $T \to T_c$. Within the phenomenological theory of Ginzburg and Landau (which can be derived from the BCS theory as shown by Gorkov [8]), one can show that if $\kappa \lesssim 1/\sqrt{2}$, then the material is a superconductor of the first kind; if $\kappa \gtrsim 1/\sqrt{2}$, then the material is of the second kind. The value of κ is directly related to the penetration depth $\lambda(T)$. The thermodynamic critical magnetic field [Eq. (18)] has the form

$$H_c(T) = \Phi_0 \frac{\sqrt{2}}{\kappa \lambda^2(T)} \qquad (22a)$$

or equivalently

$$H_c(T) = \Phi_0/2\pi\sqrt{2}\xi(T)\lambda(T) \qquad (22b)$$

where $\Phi_0 = \hbar c/2e$ is the magnetic-flux quantum. This value of the field terminates superconductivity of the first kind. For superconductors of the second kind, the corresponding field is given by

$$H_{c2} = \kappa\sqrt{2}H_c = \Phi_0/2\pi\xi(T)^2 \qquad (22c)$$

For the fields $H_{c1} < H_a < H_{c2}$ [with $H_{c1} \equiv H_c(0)$ $\ell n\ \kappa/(\sqrt{2}\kappa)$], the superconducting phase is inhomogeneous, composed of the lattice of vortices, each of the form of a tube containing one flux quantum, penetrating the sample. All of the newly discovered high-T_c superconductors are of the second kind, with very small values of H_{c1} and very large values of H_{c2}. This means that the value of the coherence length ξ is indeed small in those systems.

9. The sound absorption coefficient α_S in the superconducting phase is related to that in the normal phase α_N by

$$\frac{\alpha_S}{\alpha_N} = \frac{2}{1 + \exp(\Delta/k_B T)}$$

This is a very simple result; hence, experimental results for (α_S/α_N) are used to determine the temperature dependence of the gap Δ.

A complete discussion of superconducting states within the BCS theory is provided in Refs. 1–10.

E. STRONG-COUPLING EFFECTS: ELIASHBERG APPROACH

The BCS theory provides a complete though approximate theory of both thermal as well as dynamic properties of superconductors in the weak coupling limit $\rho V \ll 1$. The electron–electron interactions deriving from the electron–lattice interaction are treated in the lowest order and the electron–electron correlations are decoupled in the mean-field type approximation. Generalizations of the BCS treatment concentrate on two main problems: (1) inclusion of the repulsive Coulomb interaction between the electrons [14] and (2) extension of the BCS theory to the situation with arbitrarily large electron–phonon coupling [15], by generalizing the treatment of normal metals, with electron–lattice interactions incorporated in a systematic fashion [16]. Both of these factors have been included in the Eliashberg approach to superconductivity [15].

The Coulomb repulsive interaction reduces the effective attractive interaction between the electrons, so that instead of Eq. (14) one obtains in the BCS approximation

$$T_c = 1.14\theta_D \exp\left(-\frac{1}{\lambda - \mu^*}\right) \qquad (23a)$$

where $\lambda = \rho V$ is the effective electron–phonon coupling and μ^* is the so-called pseudo-Coulomb potential [14].

The Eliashberg correction to the BCS theory must be evaluated numerically. The numerical solution of the Eliashberg equation representing higher order corrections to the BCS theory may be represented by [17]

$$T_c = \frac{\theta_D}{1.45} \exp\left[-\frac{1.04(1 + \lambda)}{\lambda - \mu^*(1 + 0.62\lambda)}\right] \qquad (23b)$$

Figure 4 illustrates the difference between values of T_c obtained by the BCS and Eliashberg theories [18]. We see that the repulsive Coulomb interaction and the higher order electron–phonon effects combine

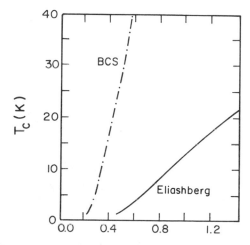

FIG. 4. Numerical solution of T_c versus the electron–phonon coupling constant λ for the Coulomb pseudopotential $\mu^* = 0.1$. The other parameters are taken as for the superconducting element niobium. Note that the Eliashberg theory gives a much slower increase of T_c than does the BCS theory.

to reduce drastically the superconducting transition temperature. This and other results [19] have led to the conclusion that the value of T_c determined within the phonon-mediated mechanism has an upper limit of the order of 30 K.

One should mention a very important feature of the phonon-mediated electron pairing. Namely, the transition temperature is proportional to the Debye temperature θ_D. Hence, T_c given by Eq. (23a) depends on the mass M of the atoms composing the lattice. In the simplest situation we expect that $T_c \sim M^{-1/2}$. A dependence of T_c on the mass M was demonstrated experimentally [19a] by studying the isotope influence on T_c. These observations provided a crucial argument in favor of the lattice involvement in the formation of the superconducting state. If the Coulomb repulsion between electrons is taken into account, then the relation is $T_c \sim M^{-\alpha}$ with [17]

$$\alpha = \frac{1}{2}\left\{1 - \frac{(1 + \lambda)(1 + 0.62\lambda)}{[\lambda - \mu^*(1 + 0.62\lambda)]^2}\right\}$$

In the strong coupling limit ($\lambda \gtrsim 1$) the exponent α is largely reduced from its initial value of $\frac{1}{2}$. Therefore, if the value of α is small, one may interpret this fact as either the evidence for strong electron–phonon coupling or that a new nonphonon mechanism is needed to explain superconductivity.

III. Normal and Magnetic States of Correlated Electrons

A. Narrow-Band Systems

The modern theory of metals derives from the concept of a free electron gas, which obeys the Pauli exclusion principle [20]. The principal influences of the lattice periodic potential on the individual electron states are to renormalize their mass and to change the topology of the Fermi surface. Landau [21] was the first to recognize the applicability of the electron-gas concept to the realistic situation where the repulsive Coulomb interaction between particles is small compared to the kinetic energy of electrons. He incorporated the interaction between electrons into a further (many-body) renormalization of the effective mass and investigated the physical properties, such as specific heat, magnetic susceptibility, sound propagation, and thermal and electric conductivities in the term of quasiparticle contributions.

An important next development was contributed by Mott [22], who pointed out that if the Coulomb interaction between the electrons is sufficiently strong

(that is, comparable to the band energy of the quasiparticles), then electrons in a solid would have to localize on the atoms, with one electron per atom. This qualitative change of the nature of single-electron states from those for a gas to those for atoms is called the metal-insulator or the Mott transition. An empty (unoccupied) state in the Mott insulator (that is, that without electrons available) will act as a mobile hole. In these circumstances, the transport of charge takes place via the correlated hopping of electrons through such hole states. In the Mott-insulator limit, those hole states play a crucial role in establishing the superconductivity of oxides, as discussed in Section IV.

The paramagnetic or magnetically ordered states of electrons comprising the Mott insulator distinguishes this class of materials from ordinary band (Bloch–Wilson) insulators or intrinsic semiconductors; the latter are characterized at $T = 0$ by a filled valence band and an empty conduction band, separated by a gap. The electrons in the filled valence band are paired; hence Bloch–Wilson insulators are diamagnetic.

The basic question now arises whether one can treat Mott insulators and metals within a single microscopic description of electron states by generalizing the band theory of electron states so as to describe Mott insulators within the same microscopic model. The first step in this direction was proposed by Hubbard [24], who showed by use of a relatively simple model that as the interaction strength (characterized by the magnitude U of the intraatomic Coulomb repulsion) increases and becomes comparable to the band energy per particle (characterized by the bare bandwidth W), the original band of single-particle states splits into two halves. Thus, the Mott insulator may be modeled by a lattice of hydrogenic-like atoms with one electron per atom, placed in the lowest $1s$ state. The distinction between the normal metallic and Mott insulating states is shown schematically in Fig. 5(a) & (b), where the metal (a) is depicted as an assembly of electrons represented by the set of plane waves characterized by the wave vector $\mathbf{k}$ and spin quantum number $s = \sigma/2$, where $\sigma \equiv \pm 1$.

The transformation to the Mott localized state may take place only if the number of electrons in the metallic phase is equal to the number of parent atoms, that is, when the starting band of free-electron states is half filled. The collection of such unpaired spin moments will lead to paramagnetic behavior at high temperatures. As the temperature is lowered, the system undergoes a magnetic phase transition; in the case of the Mott insulators, the experimentally observed transition is almost always to antiferromagnetism, as shown in Fig. 5(b), where each electron with

(a)

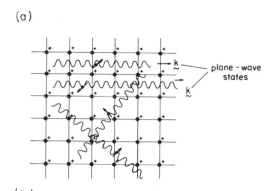

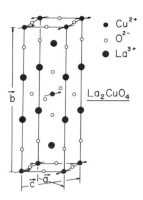

(b)

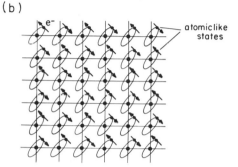

FIG. 5. (a) Schematic representation of a normal metal as a lattice of ions and the plane waves with wave vector **k** representing free electron states. (b) Model of the Mott insulator as a lattice of atoms with electrons localized on them. Note that the ground-state configuration is usually antiferromagnetic (with the spins antiparallel to each other).

FIG. 6. The magnetic structure of La_2CuO_4. The neighboring Cu^{2+} ions in the planes have their spins (each representing $3d^9$ configuration) antiparallel to each other. The antiferromagnetic structure is three-dimensional. (From Y. Endoh *et al.* [23].)

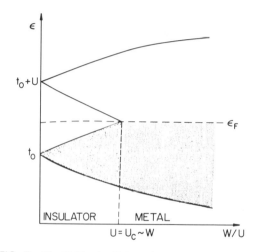

FIG. 7. The Hubbard splitting of the states in a single, half-filled band for the strength of the intraatomic Coulomb interaction $U > U_c$. The state with filled lower Hubbard subband for $U > U_c$ is identified with that of the Mott insulator. (From Ref. 24.)

its spin moment up is surrounded by the electrons on nearest neighboring sites with spins in the opposite direction (down). Such spin configuration reflects a two-sublattice (Néel) antiferromagnetic state. The actual magnetic structure of Cu^{2+} ions in La_2CuO_4, taken from Ref. 23, is shown in Fig. 6.

If the number of electrons in the band of electron states is smaller than the number of available atomic sites, then electron localization cannot be complete because empty atomic sites are available for hopping electrons. However, for the half-filled band case, as the ratio U/W increases, half of the total number of single-particle states in the starting band are gradually pushed above the Fermi level ε_F. An increase in the ratio of U/W may be achieved by lengthening the interatomic distance, thus reducing W, which is directly proportional to the wave-function overlap for the two states located on the nearest neighboring sites. The splitting of the original band into two Hubbard subbands eliminates double (spin-singlet) occupations of the same energy state ε. Effectively, this pattern reflects the situation of electrons being separated from each other as far as possible; however, the correspondence between the Hubbard split-band situation, shown schematically in Fig. 7, and the electron disposition in the spin lattice in real space [cf. Fig. 5(b)] is by no means obvious and requires a more detailed treatment that relates the two descriptions of the Mott insulator. This problem is dealt with in the following section.

1. The Hubbard Model

In discussing narrow band systems, one starts from the model Hamiltonian due to Hubbard [24], which appears to be complicated but can really be interpreted in simple terms, namely,

$$\mathcal{H} = \sum_{\mathbf{k}\sigma} \varepsilon_{\mathbf{k}\sigma} n_{\mathbf{k}\sigma} + U \sum_i n_{i\uparrow} n_{i\downarrow} \qquad (24)$$

where ε_k is the single-particle (band) energy per electron with the wave vector $\mathbf{k}$, U is the magnitude of intraatomic Coulomb repulsion between the two electrons located on the same atomic site i, $n_{k\sigma}$ is the number of electrons in the single–particle state $|k\sigma\rangle$, and $n_{i\sigma}$ is the corresponding quantity for the atomic state $|i\sigma\rangle$. This simple Hamiltonian describes the localization versus delocalization aspect of electron states since the first term provides the gain in energy ($\varepsilon_k < 0$) for electrons in the band state $|k\sigma\rangle$, whereas the second accounts for an energy loss ($U > 0$) connected with the motion of electrons throughout the system that is hindered by encounters with other electrons on the same atomic site. The competitive aspects of the two terms is expressed explicitly if the first term in Eq. (24) is transformed by the so-called Fourier transformation to the site $\{|i\sigma\rangle\}$ representation. Then, Eq. (24) may be rewritten as

$$\mathcal{H} = \sum_{ij\sigma} t_{ij} a_{i\sigma}^+ a_{j\sigma} + U \sum_i n_{i\uparrow} n_{i\downarrow} \qquad (25)$$

where

$$t_{ij} = \frac{1}{N} \sum_{\mathbf{k}} \varepsilon_k \exp[i\mathbf{k} \cdot (\mathbf{R}_j - \mathbf{R}_i)] \qquad (26)$$

is the Fourier transform of the band energy ε_k and $a_{i\sigma}^+$ ($a_{i\sigma}$) is the creation (annihilation of electrons in the atomic (Wannier) state centered on the site $\mathbf{R}_j$. The first term in Eq. (25) represents the motion of an electron through the system by a series of hops $j \rightarrow i$, which are described in terms of destruction of the particle at site j and its subsequent recreation on the neighboring site i. The width of the corresponding band in this representation is given by

$$W = 2 \sum_{j(i)} |t_{ij}| \approx 2z|t| \qquad (27)$$

where z is the number of nearest neighbors (n.n.) and t is the value of t_{ij} for the n.n. pair $\langle ij \rangle$. Thus, the Hamiltonian [Eq. (25)] is parameterized through the bandwidth W and the magnitude U. In actual calculations, it is the ratio U/W that determines the localized versus collective behavior of the electrons in the solid.

2. The Hubbard Subbands and Hole States

The normal-metal case is represented in Eq. (24) by the limit $W/U \gg 1$; the first (band) term then dominates. On the other hand, the complementary limit $W/U \ll 1$ corresponds to the limit of well-separated atoms, since the excitation energy of creating double occupancy on a given atom (with the energy penalty $\varepsilon \sim U$) far exceeds the band energy of individual particles. The transition from the metallic to the atomic type of behavior takes place when

$W \sim U$; this is also the crossover point where the single band in Fig. 7 splits in two. The actual dependence of the density of states for interacting particles is shown in Fig. 8 (taken from Ref. 25). These curves were drawn for the Lorentzian shape of the density of states (DOS), that is, for a starting band with a characteristic width Δ:

$$\rho^0(\varepsilon) = \frac{\Delta}{\pi} \frac{1}{(\varepsilon - t_0)^2 + \Delta^2} \qquad (28a)$$

where t_0 determines the position of the center of the band (usually chosen as $t_0 = 0$). Detailed calculations [27] show that with a growing magnitude of interaction (U/Δ) the DOS [Eq. (28a)] splits into two parts described by the density of states:

$$\rho(\varepsilon) = \frac{\Delta}{\pi} \left[\frac{1 - n/2}{(\varepsilon - t_0)^2 + \Delta^2} \right.$$
$$\left. + \frac{n/2}{(\varepsilon - t_0 - U)^2 + \Delta^2} \right] \qquad (28b)$$

The first term describes the original DOS [Eq. (28a)], with the weighting factor $(1 - n/2)$, whereas the second represents the upper subband (on the energy scale) and with the weighting factor $(n/2)$ and shifted by an amount U. These two terms and the corresponding two parts of DOS in Fig. 8 describe the Hubbard subbands. The dashed line in Fig. 8 represents the inverse lifetime of single-electron states placed in the pseudogap, while the arrows point to

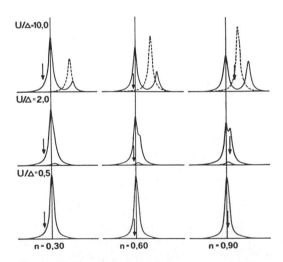

FIG. 8. The Hubbard splitting of the states for different band fillings $n = 0.3, 0.6$, and 0.9, and for different $U/W = 0.5, 2$, and 10, respectively. The x-axis is the axis of particle energy; the y-axis is the value of density of states. The arrow indicates the position of the Fermi energy, whereas the dotted line represents the inverse lifetime of the quasiparticle state in the pseudogap. (From Ref. 25.)

the position of the Fermi energy in each case. For $n = 1$, the Fermi level falls in a pseudogap, where the lifetime of those quasiparticle states is very short. This is reminiscent of the behavior encountered in an ordinary semiconductor, where the states in the band gap are those with a complex wave vector **k**.

To display the similarity between the Mott insulator as a two-band system in which the Hubbard subbands assume a role similar to the valence and conduction bands in an ordinary semiconductor, we have plotted in Fig. 9 the position of the Fermi level as a function of the numbers of electrons, n, per atom in the system. As n moves past unity, a jump in ε_F occurs for $U/\Delta \gg 1$. This is exactly what happens in the ordinary semiconductor when the electrons are added to the conduction band. This feature shows once more that the states near the upper edge of the lower Hubbard subband (that is, the states near ε_F for n close to but less than unity) can be regarded as hole states. We will see that those states are the ones with a high effective mass.

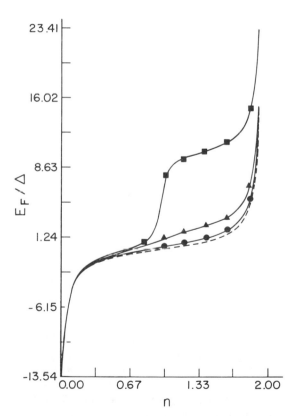

FIG. 9.   The position of the Fermi level ε_F as a function of band filling n, for different values of interaction (from the bottom to the top curve) $U/\Delta = 0$, 0.5, 2, and 10. For $U/\Delta = 10$, the Fermi level jumps between the subbands when $n \approx 1$. (From Ref. 25.)

It should be emphasized that the Hubbard subband structure is characteristic of magnetic insulators and cannot be obtained with a standard band theoretical approach to the electron states in solids. The N states in the lower Hubbard subband are singly occupied; this is directly related to the picture of unpaired spins drawn in Fig. 5(b) and is one of the reasons for calling the electron states for such interacting systems correlated electronic states. The other reason (discussed in detail below) arises because the proper description of electronic states near the localization threshold (the Mott transition) requires that one incorporate two-particle correlations into the quasiparticle states. The Hubbard split-band picture is only the first step in the proper description of the electron states. These correlations will lead to a very heavy mass of quasiparticles near the Mott transition; the heavy mass indicates a strong reduction of the bare bandwidth W as the localization threshold is approached from the metallic side.

3. Localized versus Itinerant Electrons: Metal–Insulator Transitions

The Hubbard split-band picture of unpaired electronic states in a narrow band, shown in Figs. 7 and 8, provides a rationale for the existence of a paramagnetic insulating ground state of the interacting electron system. The corresponding experimentally observed metal–insulator transitions (MITs) at finite temperature are very spectacular, as demonstrated in Fig. 10, where the resistivity (on a logarithmic scale) is plotted as a function of inverse temperature for a canonical system $(V_{1-x}Cr_x)_2O_3$ (the data are from Ref. 26). The number of transitions (1, 2, or 3) depends on the Cr content. Note the presence of an intervening metallic state between the antiferromagnetic insulating (AFI) and paramagnetic insulating (PM) states, as well as the reentrant metallic behavior at high temperatures for $0.005 \lesssim x \lesssim 0.0178$. To rationalize these data, we discuss the physical implications of a model of interacting narrow-band electrons for $U \sim W$ starting from the Hamiltonian [Eq. (25)]. We summarize here the main features of the detailed discussion presented in Refs. 27–29, which provide the main features of the ground state and thermodynamic properties.

In the absence of interactions ($U = 0$), the band energy per particle is $\bar{\varepsilon} = -(W/2)n(1 - n/2)$, where $0 \leq n \leq 2$ is the degree of the band filling; for $n = 1$, this reduces to $\bar{\varepsilon} = -W/4$. When the interactions are present, the band narrows; this is because of a restriction on the electron motion caused by their repulsion, as described earlier. One way of handling this restriction is to adjoin to the bare bandwidth

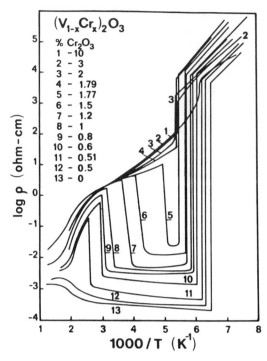

FIG. 10. Experimental measurements [26] pertaining to the variation of resistivity ρ in the logarithmic scale with inverse temperature $1000/T$ for the $(V_{1-x}Cr_x)_2O_3$ system. The atomic content of Cr_2O_3 in V_2O_3 for each curve is specified.

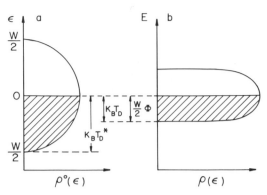

FIG. 11. Schematic representation of the bare (ρ_0) and quasiparticle (ρ) densities of states. The band narrowing factor Φ for interacting electrons (b) is specified. The degeneracy temperature T_D for the interacting electrons and that corresponding to noninteracting electrons (T_D^*) are also indicated. Situation drawn corresponds to the half-filled case ($n = 1$) for which the Fermi energy can be chosen as $\varepsilon_F = 0$.

a multiplying factor Φ. This leads to renormalized density of states for quasiparticles as illustrated in Fig. 11. The factor Φ is a function of the particle–particle correlation function $\eta \equiv \langle n_{i\uparrow} n_{i\downarrow} \rangle$, the expectation value for the double occupancy of a representative lattice site. The quantity η is calculated for $T = 0$ self-consistently by minimizing the total energy E_G (per site), composed of the band energy $E_B = \Phi\bar{\varepsilon}$ and the Coulomb repulsion energy $U\eta$, where the parameter is specified by $\Phi = 8\eta(1 - 2\eta)$ [27, 28]. These two energies represent the expectation values of the two terms in Eq. (25) for the case of a half-filled band. The optimal values of the quantities are given by

$$\eta_0 = \frac{1}{4} (1 - U/U_c) \tag{29a}$$

$$\Phi_0 = 1 - (U/U_c)^2 \tag{29b}$$

and

$$E_G = (1 - U/U_c)^2 \, \bar{\varepsilon} \tag{29c}$$

with $U_c = 8|\bar{\varepsilon}| = 2W$. Thus, as U increases, η_0 decreases from $\frac{1}{4}$ to 0. At the critical value $U = U_c$, $E_B = 0$ and there are no double occupancies for the

same lattice site; this signals the crossover by the system from the itinerant (band) to the localized (atomiclike) state. The point $U = U_c$ corresponds to a true phase transition at $T = 0$; the last statement can be proved by calculating the static magnetic susceptibility, which is [28]

$$\chi = \frac{\chi_0}{\Phi_0 \left[1 - U\rho \dfrac{1 + I/2}{(1 + I)^2} \right]} \tag{30}$$

where $I \equiv U/U_c$, ρ is the density of bare band states at $\varepsilon = \varepsilon_F$, and χ_0 is the magnetic susceptibility of band electrons with energy ε_k at $U = 0$. As $\Phi \to 0$ (that is, $U \to U_c$), the susceptibility diverges. The localized electrons are represented in this picture by noninteracting magnetic moments for which the susceptibility is given by the Curie law $\chi = C/T \to \infty$ as $T \to 0$. Thus, the metal–insulator transition is a true phase transition; η_0 may be regarded as an order parameter, and the point $U = U_c$ as a critical point. We concentrate now on a more detailed description of the metallic phase, which permits a generalization of the previous results to the case $T > 0$. First, as has been said, the increase of magnitude of interaction U reduces the band energy according to $E_B = -W\Phi_0/4$. Eventually, E_B becomes comparable to the interaction part $U\eta$; they exactly compensate each other at $U = U_c$. The resultant electronic configuration (localized versus itinerant) is then determined at $T > 0$ by the very small entropy and the exchange interaction contributions. The entropy of the metallic phase in the low-temperature regime may be estimated by using the linear specific heat expression for electrons in a band narrowed by correlations, namely, $C_v \equiv \gamma T = (\gamma_0 \mid \Phi_0)T$, where $\gamma_0 = 2\pi^2 k_B^2 \rho/3$ is the linear specific heat coefficient (per one atom) for uncorrelated electrons (that is, $U = 0$). Hence, the entropy $S = \gamma T = C_v$. Combining this relation with the resultant energy at $T = 0$, given by Eq. (29c), one can write an explicit expression for the free energy of the metallic phase [29]:

$$\frac{F}{N} = \left(1 - \frac{U}{U_c}\right)^2 \bar{\varepsilon} - \frac{1}{2}\frac{\gamma_0}{\Phi_0}T^2 \qquad (31)$$

This is the free energy per one atomic site. On the other hand, if the exchange interaction between the localized moments is neglected, then each site in the paramagnetic state is randomly occupied by an electron with its spin either up or down. The free energy F_I for such insulating system of N moments is provided by the entropy term for randomly oriented spins, that is,

$$F_I/N = -k_B T \ln 2 \qquad (32)$$

Now, a system in thermodynamic equilibrium assumes the lowest F state. The condition for the transition from the metallic to local-moment phase is specified by $F = F_I$. The phase transition determined by this condition can be seen explicitly when we note that the free energy varies with T either parabolically [Eq. (31)] or linearly [Eq. (32)], depending on whether the system is a paramagnetic metallic (PM) or a paramagnetic simulating (PI) phase. As is illustrated in Fig. 12, several of those curves intersect in one or two points depending on the value U/W. These intersection points determine the stability limits of PM and PI phases. The lowest curve for the PM phase lies below the straight line for the PI state; there is no transition, that is, the metallic Fermi-liquid state with the effective mass enhancement $m^*/m_0 \lesssim 2.5$ is stable at all temperatures. As U/W increases, the parabolas fall higher on the free energy (F/WN) scale and the possibilities for transitions open up. The higher two curves illustrate the case in which the intersections with the straight lines occur at J and K and at L and M, respectively; at low and high temperatures the parabola lies below the straight line for F_I/WN, so that the metallic phase is stable in those T regions. At intermediate temperatures, the PI phase is stable. The loci of the intersections move further apart on the $k_B T/W$ scale as U/W is increased, as may be seen in Fig. 12(b), where the phase boundaries are drawn; this part of the figure represents the temperature of the transitions [the intersection points on Fig. 12(a)] versus the relative magnitude of interaction U/W. We see that the PM

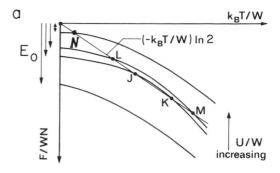

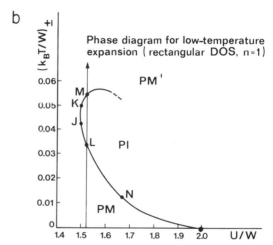

FIG. 12. (a) Plots of the free energies for the paramagnetic Mott insulator (the straight line starting from the origin) and the correlated metal (the parabolas). The parabolic curves' points of crossing at L and J correspond to a discontinuous metal–insulator transition, while those crossing at K and M correspond to the reverse. (b) Schematic representation of the phase diagram between paramagnetic metallic (PM and PM′) and paramagnetic insulating (PI) phases. The points of crossing from (a) are also shown. The vertical line with an arrow represents a sequence of the transitions seen in Fig. 10 for $0.005 \leq x \leq 0.018$ and in the paramagnetic phase.

phase is stable at low temperature; thus, reentrant metallic behavior is encountered at high T. The explicit form of the curve drawn in Fig. 12(b) is obtained from the coexistence condition $F = F_I$, which leads to the following expression for the transition temperatures [29]:

$$\frac{k_B T_\pm}{W} = \frac{3}{2\pi^2}\left[1 - \left(\frac{U}{U_c}\right)^2\right]$$
$$\times \left\{\ln 2 \pm \left[(\ln 2)^2 + \frac{\pi^2}{3}\frac{1 - U/U_c}{1 + U/U_c}\right]^{1/2}\right\} \qquad (33)$$

The root T_- represents the low temperature part, that is, that for $k_B T/W \leq 0.049$; the T_+ part is the one above the point where both curves meet; this takes place at the lower critical value of $U = U_{\ell c}$ such that

$$\frac{U_{\ell c}}{U_c} = 1 - \frac{3\sqrt{2}}{2\pi} \frac{1}{(\rho|\bar{\varepsilon}|)^{1/2}} \approx 0.75 \quad (34)$$

Below the value of $U = U_{\ell c}$, the correlated Fermi liquid is stable at all temperatures. Ultimately, for $1.58 \leq U/W \leq 2.0$, only one intersection (at low T) of the curves remains. This means that in this regime of U/W the reentrant metallic behavior is achieved gradually as the temperature increases.

4. Strongly Correlated Electrons: Kinetic Exchange Interaction and Magnetic Phases

In the limit $W \ll U$, the ground state of the interacting electron system will be metallic only if the number of electrons N_e in the system differs from the number N of atomic sites. Simply, only then can charge transport take place via the hole states in the lower Hubbard subband (for $N_e < N$), that is, when the transport of electrons can be represented via hopping from site to site, avoiding the doubly occupied configurations on the same site. This restriction on the motion of individual electrons is described above in terms of the band narrowing factor Φ, which, in the normal phase, is now equal to [27, 28] $\Phi = (1 - n)/(1 - n/2)$. This shows that the effective quasiparticle bandwidth $W^* \equiv W\Phi$ is nonzero only if the number of holes $\delta \equiv 1 - n > 0$.

For $W \ll U$, there is one class of dynamic processes that is important in determining the magnetic interactions between strongly correlated itinerant electrons, namely, the virtual hopping processes, with the formation of a doubly occupied site configuration in the intermediate state. Such processes have been depicted in Fig. 13, where one electron hops onto the site occupied by an electron with opposite spin, and then hops back to the original site. During such processes, the electrons can exchange positions (and this yields to the spin reversal of the pair with respect to

site i   site j

FIG. 13. Virtual hopping processes between singly and doubly occupied atomic sites that lead to an antiferromagnetic exchange interaction between the neighboring sites. This interaction is responsible for the antiferromagnetism in most of the Mott insulators.

the original configuration) or the same electron hops back and forth. The corresponding effective Hamiltonian, including the virtual-hopping processes in first nontrivial order, has the form

$$\mathcal{H} = \sum_{\mathbf{k}\sigma} \Phi_\sigma \varepsilon_{\mathbf{k}} n_{\mathbf{k}\sigma}$$
$$+ \sum_{ij} \left(2t_{ij}^2/U\right)\left(\mathbf{S}_i \cdot \mathbf{S}_j - \frac{1}{4} n_i n_j\right) \quad (35)$$

where in general the band narrowing factor $\Phi_\sigma = (1 - n)/(1 - n_\sigma)$, $n_\sigma = \langle n_{i\sigma}\rangle$ is the average number of particles per site with the spin quantum number σ, and $n_i \equiv n_{i\uparrow} + n_{i\downarrow}$ is the operator of the number of particles on given site i. Note that in the paramagnetic state $n_\sigma = n_{-\sigma} = n/2$, and Φ_σ reduces to $\Phi = (1 - n)/(1 - n/2)$, the value for the normal state.

One should note that the effective Hamiltonian [Eq. (35)] represents approximately the original Hubbard Hamiltonian for $W \ll U$ (for more precise treatment, see Ref. 30 and Section IV, A). When $n \to 1$, $\Phi \to 0$, and Eq. (35) reduces to the Heisenberg Hamiltonian with antiferromagnetic interaction, which is the reason why most Mott insulators order antiferromagnetically. In the limit of a half-filled band, we also find that the effective bandwidth $W^* \equiv W\Phi = 0$, thus proving that the electrons in that case are localized on atoms. The nature of the wave function for these quasi-atomic states has not yet been satisfactorily analyzed, though some evidence given later shows that they should be treated as soliton states.

For $n < 1$, the normalized band (the first term) and the exchange parts in Eq. (35) do not commute with each other. This means that for the narrow-band system of electrons represented by Eq. (35) the spin dynamics influences the nature of itinerant quasiparticle states of energies $\Phi\varepsilon_{\mathbf{k}}$. What is even more striking, as $n \to 1$, the two terms in Eq. (35) may contribute equally to the total energy. The critical concentration of electrons n_c for which these two terms are comparable is

$$n_c \simeq 1 - \frac{1}{2z}\frac{W}{U} \sim 0.02\text{--}0.05$$

In Fig. 14, we have plotted schematically the commonly accepted phase diagram describing the possible magnetic phases in the plane $n - U/W$. Close to the case of one electron per atom, the antiferromagnetic (AF) phase is stable for any arbitrary strength of interaction. At intermediate filling, the ferromagnetic (F) phase may be stable. On the low-interaction side, the ferromagnetic phase terminates

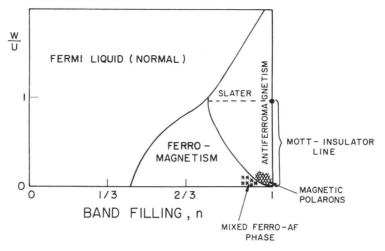

FIG. 14. Commonly accepted magnetic phase diagram for strongly correlated electrons on $n - W/U$ plane.

at points where the Stoner criterion is met, that is, when $\rho^0(\varepsilon_f)U = 1$, where $\rho^0(\varepsilon_f)$ is the value of the bare density of states (per spin) at the Fermi level.

Peculiar features appear in the corner where $n \approx 1$, and $W/U \ll 1$, that is, where the number of holes is small, so that the exchange interaction contribution to the total system energy is either larger than or comparable to the band energy part $\Phi\bar{\varepsilon}$. In such a situation, a mixed ferro-antiferromagnetic phase is possible [31]. When the number of holes is very small, each hole may form a magnetic polaron with a ferromagnetic cloud accompanying it: the hole is self-trapped within the cloud of ferromagnetic polarization it created. We consider those objects next.

5. Magnetic Polarons

It has been proved by Nagaoka [31] that in the limit $W/U \rightarrow 0$ the ground state of the Mott insulator with one hole involves ferromagnetic ordering of spins. This is because in this limit the antiferromagnetic exchange term in Eq. (35) vanishes and the band energy is lowest when $\Phi_{\sigma=\uparrow} = 1$ and $\Phi_{\sigma=\downarrow} = 1 - n$. We can thus choose an equilibrium state with $n_\uparrow = n$, $n_\downarrow = 0$, that is, a state with all spins pointing up.

Mott [22] has pointed out that if W/U is small but finite, a hole may create locally a ferromagnetic polarization of the spins in a sphere of radius R, surrounded by a reservoir of antiferromagnetically ordered spins. The situation is drawn schematically in Fig. 15. The energy of such a hole accompanied by

3d HOLE POLARON

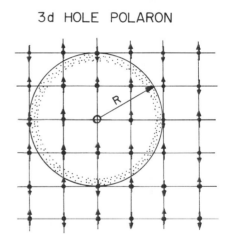

FIG. 15. Representation of the magnetic polaron state, that is, one hole in the antiferromagnetic Mott insulator. This hole produces ferromagnetic polarization around itself and may become self-trapped.

a cloud of saturated polarization can be estimated roughly as [32]

$$E(R) = -\frac{W}{2} + \pi^2|t| \left(\frac{a}{R}\right)^2$$
$$+ \frac{4\pi}{3} \left(\frac{R}{a}\right)^3 \frac{zt^2}{U} \qquad (36)$$

where a is the lattice constant and t is the hopping integral t_{ij} between the z nearest neighbors. In this expression, the first term is the band energy of a free hole in a completely ferromagnetic medium, the second represents the kinetic energy loss due to the hole confinement, whereas the third involves the antiferromagnetic exchange energy penalty paid by polarizing the spins ferromagnetically within a volume $4\pi R^3/3$. Minimizing this equation with respect to R, we obtain the optimal number of spins contained in the cloud:

$$N = \frac{4\pi}{3}\left(\frac{\pi U}{W}\right)^{3/5} \tag{37a}$$

and the polaron energy

$$E_0 = -\frac{W}{2}\left[1 - \frac{5\pi^2}{3z}\left(\frac{W}{U}\right)^{2/5}\right] \tag{37b}$$

Equation (36) holds for a three-dimensional system; for a planar system, the factor $(4/3)\pi(R/a)^3$ in the last term should be replaced by the area $\pi(R/a)^2$. One then obtains the corresponding optimal values

$$N = \pi\left(\frac{2\pi U}{W}\right)^{1/2} \tag{38a}$$

and

$$E_0 = -\frac{W}{2}\left[1 - \frac{2\pi^2}{z}\left(\frac{W}{2\pi U}\right)^{1/2}\right] \tag{38b}$$

These size estimates will be needed later when discussing the hole states at the threshold for the transition from antiferromagnetism to superconductivity in high-T_c oxide materials. One should note that U/W must be appreciably larger than unity in order to satisfy the requirement $N \gg 1$. In other words, the condition $R \gg a$ must be met, so that the spin subsystem (and the hole dynamics) may be treated in the continuous-medium approximation, the condition under which Eq. (36) can be derived.

6. The Spin Liquid

The difference between the electron liquid of strongly correlated electrons (represented, for example, by the holes in the lowest Hubbard subband) and the Fermi liquid can be shown clearly in the limit of relatively high temperatures $W^* \ll k_B T \ll U$, where the quasiparticle band states with energies $(\Phi\varepsilon_k)$ are populated equally, independently of their energy. Namely, if N_e electrons are placed into N available states, the number of configurations for a phase with excluded double occupancies of each state is [33]

$$2^{N_e}\frac{N!}{N_e!(N - N_e)!} \tag{39a}$$

The first factor is the number of spin configurations for the singly occupied sites, while the second specifies the configurational entropy—the number of ways to distribute N_e spinless particles among N states. This leads to molar entropy in the form

$$S_L = R[n\,\ell n\,2 - n\,\ell n\,n$$
$$- (1 - n)\,\ell n(1 - n)] \tag{39b}$$

where $n = N_e/N$ is the degree of the subband filling and R is the gas constant. The above reduces to $S_L = R\ell n\,2$ for $n = 1$, that is, to the entropy of the N spins ($\frac{1}{2}$) on the lattice. By contrast, in a Fermi liquid that obeys the Fermi–Dirac distribution double occupancies are not excluded as illustrated in Fig. 16(a). The corresponding number of configurations is then

$$\left(\frac{N}{N_e/2}\right)^2 = \frac{N!}{(N_e/2)!(N - N_e/2)!} \tag{40a}$$

with the corresponding molar entropy

$$S_F = R[2\,\ell n\,2 - n\,\ell n\,n$$
$$- (2 - n)\,\ell n(2 - n)] \tag{40b}$$

Hence, for $n = 1$, $S_F = 2S_L = 2R\,\ell n\,2$. One should emphasize that only the value for S_L reproduces correctly the entropy of N localized paramagnetic spins

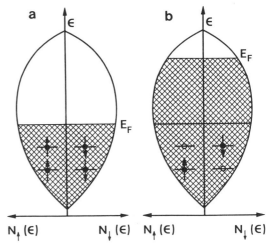

FIG. 16. Schematic representations of the difference in the **k**-space occupation for the ordinary fermions (left) and the strongly correlated electrons (right). Both spin subbands with $\sigma = \uparrow$ and $\downarrow$ are drawn. Note that the holes drawn in (b) do not appear; they are only shown to indicate a single occupancy of each single-particle state. The position of the Fermi level is different for the same number of electrons in the two situations. (From Ref. 33.)

(the electronic part of the entropy for magnetically disordered states of the Mott insulator). Hence, in accord with intuitive reasoning, the Fermi–Dirac distribution that allows for double state occupancy cannot be applied to a strongly correlated electron liquid, which we call the spin liquid. The state of such a liquid reduces to that of the spin system on the lattice if $N = N_e$ (for the Fermi liquid case, the ground state is then a metal with a half-filled band).

One should now ask how these results may be generalized to handle the regime of low temperatures and of arbitrary number of holes. One observes that in Fig. 8 the band states for $U \ll W$ are split for any arbitrary degree of band filling [cf. also Eq. (28b)]. Therefore, in enumerating the distribution of particles in the lower Hubbard subband, one must exclude double occupancies of the same energy (ε) state. Since the quasiparticle energy is labeled by the wave vector $\mathbf{k}$, one can equivalently exclude the double occupancies of given state $|\mathbf{k}\rangle$. Under this assumption, the statistical distribution is given by [33]

$$\bar{n}_{\mathbf{k}\sigma} = (1 - \bar{n}_{\mathbf{k}-\sigma}) \frac{1}{1 + \exp[\beta(E_{\mathbf{k}\sigma} - \mu)]} \quad (41a)$$

where $\beta = (k_B T)^{-1}$, $\bar{n}_{\mathbf{k}\sigma}$ is the average occupancy of the state $|\mathbf{k}\sigma\rangle$, and μ is the chemical potential that is determined from the conservation of the total number of particles

$$N_e = \sum_{\mathbf{k}\sigma}' \bar{n}_{\mathbf{k}\sigma} \quad (41b)$$

The corresponding molar entropy is now given by

$$S_L = -\frac{R}{N} \sum_{\mathbf{k}} [(1 - \bar{n}_{\mathbf{k}}) \ell n(1 - \bar{n}_{\mathbf{k}})$$
$$+ \bar{n}_{\mathbf{k}\uparrow} \ell n \, \bar{n}_{\mathbf{k}\uparrow} + \bar{n}_{\mathbf{k}\downarrow} \ell n \, \bar{n}_{\mathbf{k}\downarrow}] \quad (41c)$$

with $\bar{n}_{\mathbf{k}} = \bar{n}_{\mathbf{k}\uparrow} + \bar{n}_{\mathbf{k}\downarrow}$.

One should notice that the distribution function [Eq. (41c)] differs from the ordinary Fermi–Dirac formula by the factor $(1 - \bar{n}_{\mathbf{k}-\sigma})$, which expresses the conditional probability that there should exist no second particle with the spin quantum number $\mathbf{k}(-\sigma)$ if the state $\mathbf{k}\sigma$ is to be occupied by an electron as shown in Fig. 16(b). If $E_{\mathbf{k}\sigma} = E_{\mathbf{k}}$ (that is, when the particle energy does not depend on its spin direction), then Eq. (41a) reduces to

$$n_{\mathbf{k}} = \frac{1}{1 + (1/2) \exp[\beta(E_{\mathbf{k}} - \mu)]} \quad (41d)$$

This is the same type of formula that applies to the occupation number of simple donors, if the index $\mathbf{k}$ is dropped and ε represents the position of donor level with respect to the bottom edge of the conduction band. At $T = 0$, each state is singly occupied.

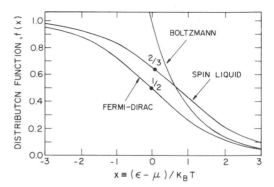

FIG. 17. Comparison of the Fermi–Dirac and Boltzmann distributions for $\bar{n}_{\mathbf{k}\sigma}$ with that for the strongly correlated electrons (the spin-liquid phase); the total occupancy $n_{\mathbf{k}} = n_{\mathbf{k}\uparrow} + n_{\mathbf{k}\downarrow}$ has been taken in the latter case.

This is the principal feature by which the present formula differs from the Fermi–Dirac distribution at $T = 0$, as illustrated in Fig. 17. The distribution [Eq. (41a)] leads to the doubling of the volume enclosed by the Fermi surface in spin-liquid state compared to the Fermi-liquid state. At low temperatures, the application of the distributions [Eq. (41a) or (41c)] yields the Fermi-liquidlike properties: a linear T dependence of the specific heat (of large magnitude if $n \rightarrow 1$) and of the entropy. At high temperatures, the new distribution leads to the entropy of the form of Eq. (39b) and local-moment behavior in the form of the Curie–Weiss law for the susceptibility. Hence, the properties of the spin liquid governed by the distribution [Eq. (41a) or (41d)] interpolate between those of a metal and those of local moments. Such behavior is observed in many correlated systems, for example, in heavy fermions.

One should notice that the entropy expression [Eq. (41c)] can be rewritten for the paramagnetic state in the following form:

$$S_L = -nR \, \ell n \, 2 - k_B \sum_{\mathbf{k}} [n_{\mathbf{k}} \ell n \, n_{\mathbf{k}}$$
$$+ (1 - n_{\mathbf{k}}) \ell n(1 - n_{\mathbf{k}})] \quad (42)$$

The first part represents the entropy of spin moments, while the second represents the entropy of spinless fermions. An alternative decomposition has been put forward [34] in which the dynamics of correlated electrons is decomposed into that of neutral fermions called spinons and the charged bosons called holons. Within this picture, the onset of superconductivity is considered as a combined effect of Bose condensation of the holons with a simultaneous formation of a coherent paired state by the fermion counterpart [35–37]. This problem is discussed in more detail in Section IV.A.

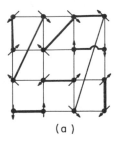
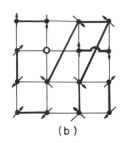

(a)   (b)

FIG. 18. Schematic representation of singlet-spin pairing forming the RVB state. All paired configurations should be taken to calculate the actual ground state. Here, (a) represents the RVB state for the Mott insulator, whereas (b) is that with one hole. The latter case will contain an unpaired spin, as indicated.

The above treatment of the spin liquid deals only with its statistical properties in the $U \to \infty$ limit. The problem now arises as to what happens when the spin part of the form of the second term in Eq. (35) is explicitly included. The problem of the resultant ground state of holes in a Mott insulator is a matter of intensive debate [35–37]. The state called the resonating valence bond (RVB) state has been invoked [35] specifically to deal with this problem; this state is schematically drawn in Fig. 18 for the case without holes (a) and with one hole (b). The connecting lines represent bonds, across which the two electrons form spin-singlet pairs. The resonating nature of bonds is connected with the idea that the RVB ground state is a coherent superposition of all such paired configurations. The dynamic nature of this spin dimerization is connected with the terms $(S_i^+ S_j^- + S_i^- S_j^+)$ in the exchange part of the Hamiltonian [Eq. (35)]. There is the possibility that the RVB state [which, for obvious reasons, differs from the ordinary (Néel) antiferromagnet] is a ground state for the planar CuO_2 planes in high-T_c oxides, such as $La_{2-x}Sr_xCuO_4$, where the long-range magnetic order is destroyed for $x \approx 0.02–0.03$. We return to this problem in Section IV when discussing the boundary line between antiferromagnetism and superconductivity for the high-T_c oxides.

B. THE HYBRIDIZED SYSTEMS

Most of the strongly correlated systems are encountered in oxides and in several classes of organic and inorganic compounds. In oxides, the $3d$ orbitals of cations such as Cu^{2+} and Ni^{2+}, hybridize with the $2p$ orbitals of oxygen, particularly if the atomic ($3d$) states are energetically close to the $2p$ states. The properties of correlated and hybridized states can be

properly discussed in terms of the Anderson lattice model Hamiltonian, which is of the form

$$\mathcal{H} = \varepsilon_f \sum_{i\sigma} N_{i\sigma} + \sum_{k\sigma} \varepsilon_k n_{k\sigma} + U \sum_i N_{i\uparrow} N_{i\downarrow}$$

$$+ \frac{1}{\sqrt{N}} \sum_{k\sigma} (V_k e^{ik\cdot R_i} a_{i\sigma}^+ c_{k\sigma} + \text{H.c.}) \qquad (43)$$

In this Hamiltonian, the first term describes the energy of atomic electrons positioned at ε_f, the second represents the energy of band electrons, the third represents the intraatomic Coulomb repulsion between two electrons of opposite spins, while the last describes the mixing of atomic with band electrons due to the energetic coincidence (degeneracy) of those two sets of states (H.c. refers to a Hermitian conjugate of the hybridization part). In heavy fermions, the atomic states are $4f$ states, whereas they are $3d$ states of Cu^{2+} ions in high-T_c systems; the band states are $5d–6s$ and $2p$ states, respectively. Note that $N_{i\sigma} = a_{i\sigma}^+ a_{i\sigma}$, and $n_{k\sigma} = c_{k\sigma}^+ c_{k\sigma}$ are the number of particles in given atomic (i) or k states, respectively. In this Hamiltonian, the following parameters appear: the atomic-level position ε_f, the width W of a starting band states with energies $\{\varepsilon_k\}$, the magnitude U of the Coulomb repulsion for two electrons located in the same atomic site, and the degree of hybridization (mixing), V_k, characterized by its magnitude V.

Two completely different situations should be distinguished from the outset, namely, (1) $U > W > |\varepsilon_f| \gg |V|$, and (2) $U > W > |V| \gtrsim |\varepsilon_f|$. Case (1) applies when the starting (bare) atomic level is placed deeply below the Fermi level and the atomic states admix weakly to the band states. In case (2), the hybridization is large and is responsible for strong mixing of the two starting sets of states. The band structure corresponding to the hybridized band states in the absence of electron–electron interactions (that is, $U = 0$) is depicted in Fig. 19. We observe a small

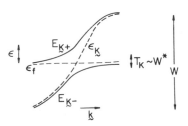

FIG. 19. Schematic representation of the hybridized bands with energies $E_{k\pm}$, which are formed by mixing the band states (with energy ε_k) and atomic states (located at $\varepsilon = \varepsilon_f$). The original band has the width W much wider than the peaked structure of the width W^*.

gap in the hybridized band structure; it occurs around the bare atomic level position ε_f and separates two hybridized bands. Those two bands, which have the energies

$$E_{\mathbf{k}\pm} = \frac{\varepsilon_{\mathbf{k}} + \varepsilon_f}{2} \pm \left[\left(\frac{\varepsilon_{\mathbf{k}} - \varepsilon_f}{2}\right) + |V_{\mathbf{k}}|^2\right]$$

correspond to the bonding and antibonding types of states in molecular systems. The structure of the hybridized bands is demonstrated explicitly in Fig. 20, an example of the density of states for each band. One sees that strongly peaked structures occur in the regions near the gap. If the Fermi level falls within these peaks, a strong enhancement of the effective mass should take place solely because of these peculiarities of the band structure. In some situations, only a pseudogap caused by the hybridization is formed, as shown in Fig. 21. This is so if the hybridization matrix element V depends on the wave vector $\mathbf{k}$ and if along some directions in reciprocal space $V_{\mathbf{k}} = 0$.

The inclusion of the interaction term in Eq. (43) renders the treatment of the Anderson lattice Hamiltonian much more complicated; up to now this problem has not been solved rigorously. A large variety of approximate treatments have been proposed and reviewed recently [38–41], in all of which the principal task was to provide a satisfactory description of heavy fermion materials [42]. In effect, the limiting case of almost-localized strongly correlated electrons was studied, which, among others, provides a quasiparticle electronic structure similar to that shown in Fig. 20, with a very strong enhancement of the DOS near the Fermi surface. This leads to very heavy quasiparticles, which, in some systems, may undergo transitions either to antiferromagnetism or to superconducting states. In this respect, heavy fermion materials are analogous to the high-T_c systems, though with much lower transition temperatures.

C. THE ELECTRONIC STATES OF THE SUPERCONDUCTING OXIDES

The high-T_c superconducting oxides, such as $La_{2-x}Sr_xCuO_4$ (the so-called 214 compounds) and $YBa_2Cu_3O_{7-\delta}$ (the so-called 123 compounds), have one common structural unit: the quasi-two-dimensional structure that is approximated by CuO_2 planes, one of which is drawn schematically in Fig. 22. We discuss mainly the role of these planes since it is widely accepted that the electronic properties of those subsystems are the main factor determining the observed superconductivity, antiferromagnetism, and localization effects in those materials. In stoichio-

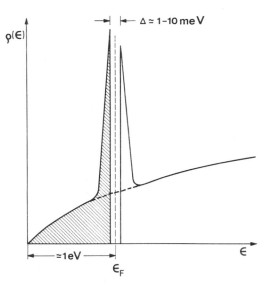

FIG. 20. Density $\rho(\varepsilon)$ of hybridized states versus particle energy ε. Note that the hybridization gap Δ_h may be very small compared to the total width of the band states. The position of Fermi level ε_F corresponds to the filled lower band.

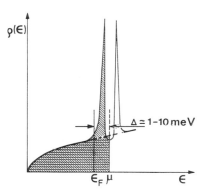

FIG. 21. Same as in Fig. 20 but with the pseudogap among the hybridized bands.

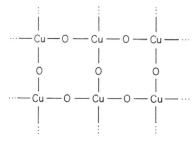

FIG. 22. Schematic representation of the CuO_2 planes in the superconducting oxides in the tetragonal phase. The Cu–Cu distance is $a \approx 1.9$ Å for La_2CuO_4.

metric La_2CuO_4 or $YBa_2Cu_3O_7$, the formal valence of Cu is $2+$, that is, it corresponds to one-hole ($3d^9$) electron configuration. In a strictly cubic structure, with Cu^{2+} surrounded by O^{2-} ions in an octahedral arrangement, the highest band is doubly degenerate and of e_g symmetry, that is, composed of $d_{x^2-y^2}$ and $d_{3z^2-r^2}$ orbitals. However, in high-T_c materials, the octahedra are largely elongated in the direction perpendicular to the CuO_2 planes, so that the bands are further split; it is commonly assumed that the antibonding orbital $d_{x^2-y^2}$ is higher in energy and hence half filled. These d states hybridize with the oxygen $2p_x$ and $2p_y$ orbitals of σ type, as shown schematically in Fig. 23; both the bonding and antibonding configurations are shown; the latter corresponds to the signs of the two p orbitals shown in the brackets.

A simple description of the electronic states for the planar CuO_2 system is obtained by introducing a single band representing Cu d electrons in the tight binding approximation. For the square configuration of the Cu atoms (which reflects the tetragonal structure of La_2CuO_4), such a dispersion of band energies has the form

$$\varepsilon_{\mathbf{k}} = 2t(\cos k_x a + \cos k_y a) \qquad (44)$$

where t is the so-called hopping or Bloch integral $\langle i|V|j \rangle$ between the nearest neighboring ions i and j, and a is the Cu–Cu distance. For La_2CuO_4 and $YBa_2Cu_3O_{6.5}$, this band is half filled, with the Fermi surface determined from the condition $\varepsilon_{\mathbf{k}} = \mu = 0$. As shown in Fig. 24, this leads to a square in reciprocal space connecting the points $(\pi/a)(\pm 1, 0)$ with the points $(\pi/a)(0, \pm 1)$. The oxygen electrons in the $2p$ states are regarded as playing only a passive role of a transmitter of the individual d electrons from one d_{x^2-y} state to its neighbor (note that the O^{2-} valence state has completely filled p shells). If the number of electrons in that band is decreased (for example, by substituting Sr for La in 214 compounds), then the Fermi surface shrinks and gradually transforms into a circle, as shown in Fig. 24 [43]. Within such a model, La_2CuO_4 should be metallic. However, at $T < T_N \simeq 240$ K, this compound orders antiferromagnetically [23], and the ground state is then insulating. The fact that this system remains insulating above the Néel temperature T_N means that the stoichiometric La_2CuO_4 and $YBa_2Cu_3O_{6.5}$ are Mott insulators, not a Slater split-band antiferromagnet; for the latter, the split-band structure for $T < T_N$ should coalesce into one band as $T \to T_N$. The presence of the paramagnetic insulating state for both La_2CuO_4 and $YBa_2CuO_{6+\delta}$ supports the view that those oxides should be regarded as narrow-band systems

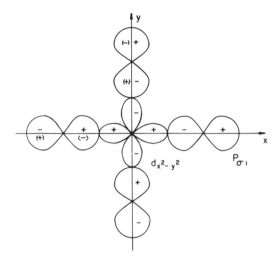

FIG. 23. The configuration of the $3d_{x^2-y^2}$ and p_σ orbitals for bonding configurations. The reverse signs for the two p orbitals (that is, those in the brackets) represent the hybridized configuration for the antibonding state.

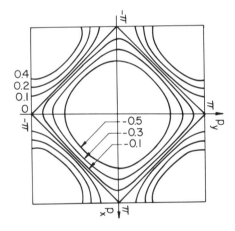

FIG. 24. The shape of the two-dimensional Fermi surface for the band energy of the form of Eq. (44). The values specified represent μ/t as a parameter. The square shape corresponds to $\mu = 0$ or, equivalently, to $n = 1$. (From Ref. 43.)

characterized by strong electron–electron interactions ($U > U_c$), as originally proposed by Anderson [44]. An antiferromagnetic ground state is then expected since the kinetic exchange interaction between the strongly correlated electrons takes place [30].

A principal problem appears when holes occur in the Mott insulator, that is, when we consider $La_{2-x}Sr_xCuO_4$ or $YBa_2Cu_3O_{6.5+x}$. We have already seen at the end of Section III, A that for small x the kinetic energy of the holes and the exchange energy of electrons may become comparable or the latter may become even larger than the former. In such situations, the motion of the holes will be influenced by the setting in of almost instantaneous spin–spin correlations. This means that such metallic states (if formed) cannot be regarded as the Fermi liquid with slowly evolving spin fluctuations; instead, the resonance between various spin configurations must be built into the electron wave function characterizing its itinerant state. The decomposition of the resonating spin configurations into spin pair-singlet configurations constitutes an important characteristic of the RVB theory of the normal state [35, 44]. Some experimental evidence for the quantum spin-liquid state above the Néel temperature has been provided by neutron quasi-elastic scattering [45]; these results were subsequently interpreted [46].

The interpretation of the metallic state in terms of a single, narrow band requires the presence of both $3d^9$ (Cu^{2+}) as well as $3d^8$ (Cu^{3+}) states. Most of the X-ray spectroscopical studies [47] conclude that the satellite peak corresponding to $3d^8$ configuration is actually absent. Therefore, in order to explain both the insulating properties of $La_2Sr_xCuO_4$ and the metallic properties of $La_{2-x}Sr_xCuO_4$ for $x \simeq 0.04$–0.05, one introduces hybridized $2p$–$3d$ states for the holes introduced by the doping. The proper model of such states is then the Anderson lattice model [Eq. (43)]. Band structure calculations by Mattheis [48] for $La_{2-x}Sr_xCuO_4$, shown in Fig. 25, justify a reasonable description within a simple two-dimensional tight binding model with only the Cu $3d_{x^2-y^2}$ and p_σ orbitals on oxygens taken into account. Namely, the structures denoted by A and B in Fig. 25 correspond, respectively, to antibonding and bonding hybridized bands, with respective band energies

$$E_{k\pm} = \frac{\varepsilon_p + \varepsilon_d}{2} \pm \left[\left(\frac{\varepsilon_p - \varepsilon_d}{2} \right)^2 + 4V^2 \left(\sin^2 \frac{k_x a}{2} + \sin^2 \frac{k_y a}{2} \right) \right]^{1/2}$$ (45)

where ε_p and ε_d are atomic level positions for the $3d$ and $2p$ states, respectively. Detailed calculations [48] lead to a nonzero bandwidth of the $2p$ band because of the p–p overlap; then $\varepsilon_p \rightarrow \varepsilon_p + \varepsilon_k$, with

$$\varepsilon_k = 2t_p[\cos(k_x a/\sqrt{2}) + \cos(k_y a/\sqrt{2})]$$

The band structure calculations should be regarded as providing input parameters for the parameterized models, which include electron correlations more accurately. On the basis of various estimates [49] of

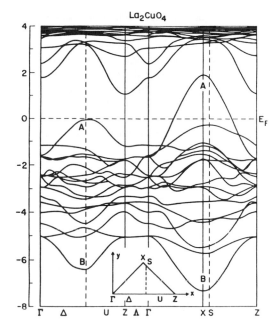

FIG. 25. Energy bands for La_2Cu_4 calculated within a local density approximation for the assumed crystal structure are body-centered tetragonal. A portion of the x–y plane in the extended Brillouin zone scheme in also shown in the inset. The portions B and A correspond to the bonding and antibonding parts of the hybridized band discussed in the text. (From L. F. Mattheis [48].)

those parameters, one can assume that they fall in the range

$$|\varepsilon_p - \varepsilon_d| = 0\text{–}2 \text{ eV}$$

$$|V| \approx 1\text{–}1.5 \text{ eV}$$

$$|t| \leq 0.5 \text{ eV}$$

$$|t_p| \leq 0.5 \text{ eV}$$

and

$$U \leq 8\text{–}10 \text{ eV}$$

From these estimates of the parameters, one sees that $|V| \sim |\varepsilon_p - \varepsilon_d|$. Hence, one may not be able to use the perturbation expansion in $V/(\varepsilon_p - \varepsilon_d)$ of the Anderson lattice Hamiltonian. Such a perturbation expansion was used [50] when transforming the hybridized model represented by the Hamiltonian [Eq. (43)] into an effective narrow-band model represented by Eq. (36).

On the basis of the facts that the present day band calculations do not provide paramagnetic insulating states for the stoichiometric materials, such as La_2CuO_4, and that the antiferromagnetic ground state is difficult to achieve within the local density approximation [48], we conclude that an approach based on the parameterized models discussed in the preceding two sections should be treated in detail. The microscopic parameters obtained from the band-structure calculations should be treated as input parameters in those models. A review of properties obtained within the parameterized models and relevant to high-T_c systems is given in Section IV.

IV. Novel Mechanisms of Electron Pairing

The binding of two fermions into a boson is a prerequisite for condensation of microscopic particles into a coherent macroscopic state. This condensation may take the form of Bose–Einstein condensation if the interaction energy between the pairs is much smaller than the binding energy of a single pair. Such a Bose condensed state of charged particles may exhibit the principal properties of the superconducting state, such as the Meissner–Ochsenfeld effect [51]. In the BCS theory (discussed in Section II), pair condensation occurs under a completely different condition, namely, when the states of different pairs overlap strongly so that the motion of one widely separated pair takes place in the mean field of all other pairs.

The pairing of particles in the BCS theory is described in momentum (reciprocal) space, where it is assumed that the quasiparticle states with a well-defined Fermi surface are formed first; the pairing involves electrons from the opposite points on the Fermi surface $(\mathbf{k}, -\mathbf{k})$ and generates either a spin singlet state (as in the classic superconductors) or a higher angular-momentum state (as in superfluid ^{3}He [52]). Because of a small coherence length ($\xi \sim 10$ Å), the new superconductors offer an opportunity for exploring the possibility of pairing in real (coordinate) space. Moreover, since the carrier concentration determined from the Hall effect measurements [53] for high-T_c oxides is at least one order of magnitude smaller than that for ordinary metals, it is tempting to describe the onset of the superconducting state as a Bose condensation of preexisting pairs. In fact, the situation is not that simple. For example, in $La_{2-x}Sr_xCuO_4$ with $x = 0.04$, the average distance between holes in the normal phase is $\approx 5a$, a magnitude comparable to ξ. These circumstances, when combined with antiferromagnetism and localization effects, render the new superconducting materials unique in the sense that their description requires a

unification of theoretical approaches to phenomena previously regarded as disparate.

The rapidly accumulating evidence for rather strong electron–electron interaction in high-T_c oxides [35, 47] and in heavy fermion systems [38–42] makes it unlikely that electron pairing in these materials is caused by extremely strong electron–phonon interaction. Furthermore, the electron–phonon interaction does not allow for a connection (or strictly speaking, competition) between the observed superconductivity and antiferromagnetism [23]. This is one of the reasons for an intensive search for a purely electronic mechanism of pairing. We now discuss some of the mechanisms that have been proposed. The main emphasis so far has been placed on an exchange–mediated pairing for strongly correlated electrons [44], since for such systems, the pairing, antiferromagnetism, and metal–insulator transitions to the Mott localized phase are derived from a single theoretical scheme. The last two phases have been discussed in Section III; here, we concentrate on the spin-singlet pairing among strongly correlated electrons. Later, we discuss charge-transfer-mediated and phonon-mediated pairings. Finally, we classify the types of correlated metallic states in solids. This classification provides a concise way of characterizing specific properties of these systems by which the almost-localized systems differ from ordinary metals.

A. Exchange Interactions and Real-Space Pairing

1. Narrow-Band Systems

In Section III, A, we have provided an approximate Hamiltonian [Eq. (34)], which includes the antiferromagnetic exchange interactions between the correlated electrons in the limit $U/W \gg 1$. The precise form of this Hamiltonian to second order in W/U is [30, 54]

$$\mathcal{H} = {\sum_{ij\sigma}}' t_{ij} b_{i\sigma}^+ b_{j\sigma}$$
$$+ {\sum_{ij}}' \frac{2t_{ij}^2}{U} \left(\mathbf{S}_i \cdot \mathbf{S}_j - \frac{1}{4} \nu_i \nu_j \right)$$
$$+ \text{(three-site terms)} \qquad (46)$$

where the primed summation means that $i \neq j$. In this Hamiltonian, doubly occupied site i configurations $|i\uparrow\downarrow\rangle$ are excluded. This exclusion is reflected by the presence of creation $(b_{i\sigma}^+)$ and annihilation

$(b_{i\sigma})$ operators for electrons in the state $|i\sigma\rangle$, which are defined as

$$b_{i\sigma}^+ \equiv a_{i\sigma}^+(1 - n_{i-\sigma}) \quad \text{and} \quad b_{i\sigma} \equiv a_{i\sigma}(1 - n_{i-\sigma})$$
(47)

so that

$$\nu_{i\sigma} = b_{i\sigma}^+ b_{i\sigma} \quad \text{and} \quad \nu_i = \sum_\sigma \nu_{i\sigma} \quad (48)$$

The spin operator is defined as

$$\mathbf{S}_i \equiv (S_i^+, S_i^-, S_i^z)$$

$$\equiv [a_{i\uparrow}^+ a_{i\downarrow}, a_{i\downarrow}^+ a_{i\uparrow}, (n_{i\uparrow} - n_{i\downarrow})/2]$$

Note that the same representation at the operator $\mathbf{S}_i$ can be written in terms of projected operators $b_{i\sigma}^+$ and $b_{i\sigma}$. The factor $(1 - n_{i-\sigma})$ in Eq. (47) imposes explicitly the restriction that the creation or the annihilation of electrons in the state $|i\sigma\rangle$ can take place only if there is no second electron already on the same site. Thus, $\nu_i = \Sigma_\sigma n_{i\sigma}(1 - n_{i-\sigma})$ enumerates only the singly occupied sites ($\nu_i = 0$ or 1). In other words, the N states corresponding to the doubly occupied site configurations have been projected out. Thus, Eq. (46) describes the dynamics of strongly correlated electrons for $N_e \leq N$ of electrons. Also, in performing the summations in Eq. (46), one usually considers only the pairs $\langle ij \rangle$ of nearest neighbors; in this approximation the parameters $J_{ij} = J$ and $t_{ij} = t$ can be chosen as constants.

The first term in Eq. (46) describes the single-particle hopping of electrons from the singly occupied to the empty atomic sites; the second describes the exchange interaction induced by virtual hopping between the sites i and j, while the three-site part describes the motion of electrons with spin σ from the singly occupied site located at i to the next nearest neighboring empty site k via the occupied configuration (with electron of opposite spin) located at site j. The various contributions to Eq. (46) are represented graphically in Fig. 26.

If one introduces a new pair of creation and annihilation operators in coordinate space by

$$\widetilde{b}_{ij}^+ = \frac{1}{\sqrt{2}}(b_{i\uparrow}^+ b_{j\downarrow}^+ - b_{i\downarrow}^+ b_{j\uparrow}^+) \quad (49a)$$

and

$$\widetilde{b}_{ij} = \frac{1}{\sqrt{2}}(b_{i\downarrow} b_{j\uparrow} - b_{i\uparrow} b_{j\downarrow}) \quad (49b)$$

then the Hamiltonian [Eq. (46)] with inclusion of the three-site part can be written in the following very suggestive closed form [54]:

$$\mathcal{H} = \sum_{ij\sigma}{}' t_{ij} b_{i\sigma}^+ b_{j\sigma} - \sum_{ijk}(2t_{ij}t_{jk}/U)\,\widetilde{b}_{ij}^+ \widetilde{b}_{kj} \quad (50)$$

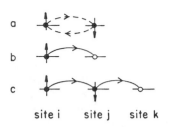

FIG. 26. Various hopping processes in narrow-band systems in a partial band filling case: (a) virtual hopping processes leading to kinetic exchange interaction; (b) single-particle hopping representing the band energy of correlated electrons; (c) contribution to the pair hopping; this process gives the pairing contribution in Eq. (50) with $k \neq i$.

The first term represents, as before, the dynamics of single electrons moving between the empty sites regarded as holes; the second term combines the last two terms in Eq. (46) and expresses the dynamics of the singlet pairs [cf. Eqs. (49a) and (49b)]. The division in Eq. (50) into single-particle and pair parts is in analogy to the BCS Hamiltonian; however, here, the operators are expressed in coordinate space. The term with $i = k$ in the pairing part enumerates the spin-singlet pairs of neighboring spins; the terms with $i \neq k$ represent pair hopping of such singlet pair bonds. Thus, in the language of operators [Eqs. (49)], one adds the bond dynamics to that of single electrons. Moreover, the forms of Eqs. (46) and (50) are completely equivalent; hence, the pairing effect and the antiferromagnetism should be directly linked within this formalism (they are two different expressions of the same part of $\mathcal{H}$).

It is difficult to diagonalize the Hamiltonian [Eq. (50)] to obtain the eigenvalues of the system. Part of the problem arises from the fact that the single-particle operators $b_{i\sigma}$ and $b_{j\sigma}^+$ do not obey the fermion anticommutation relation, and that the pair operators $\widetilde{b}_{ij}^+$ and $\widetilde{b}_{ij}$ do not obey boson commutation relations. Additionally, the two terms in Eq. (50) do not commute, so that the itinerant characteristics of the electrons and the pair-binding effects combine and produce a paired metallic phase, particularly if the two terms are of comparable magnitude. We have seen in Section III that if the number of holes $\delta \equiv 1 - n < \delta_c \sim 0.02$, then the pairing (or exchange) part dominates and antiferromagnetism sets in. Detailed calculations [31] lead to the boundary line between the antiferromagnetic and ferromagnetic phase, as shown in Fig. 27. The energy of the completely saturated ferromagnetic phase (CF) indicated does not depend on the value of exchange integral $J_{ij} \equiv 2t_{ij}^2/U$.

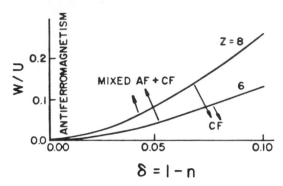

FIG. 27. Phase boundary between mixed ferromagnetic (CF)–antiferromagnetic (AF) phase and pure ferromagnetic phase for the simple cubic ($z = 6$) and bcc cubic ($z = 8$) structures. A similar type of phase boundary can be obtained for other structures. (From Ref. 62.)

We now discuss the superconducting phase for which the pairing part in Eq. (50) plays a crucial role. To make the problem tractable at this point, one replaces the operators [Eqs. (49)] by fermion operators [44, 55], that is,

$$b^+_{i\sigma} \rightarrow a^+_{i\sigma} \qquad b_{i\sigma} \rightarrow a_{i\sigma} \qquad (51a)$$

and introduces the replacement

$$\tilde{b}^+_{ij} \rightarrow b^+_{ij} = \frac{1}{\sqrt{2}}(a^+_{i\downarrow}a^+_{j\uparrow} - a^+_{i\uparrow}a^+_{j\uparrow}) \qquad (51b)$$

and

$$\tilde{b}_{ij} \rightarrow b_{ij} = \frac{1}{\sqrt{2}}(a_{i\downarrow}a_{j\uparrow} - a_{i\uparrow}a_{j\downarrow}) \qquad (51c)$$

Simultaneously, one renormalizes the parameters t_{ij} and J_{ij} in such a manner that they contain the restrictions on particle dynamics due to the projection of doubly occupied site configurations in the expression for the ground-state energy. Within the Gutzwiller–Ansatz approximation [28], Eqs. (49) reduce the starting Hamiltonian to the form

$$\mathcal{H} = \delta \sum_{ij\sigma} t_{ij}a^+_{i\sigma}a_{j\sigma} \sum_{ijk}(2t_{ij}t_{jk}/U)b^+_{ij}b_{kj} \qquad (52)$$

where $\delta = 1 - n$. This Hamiltonian has been solved within the mean-field approximation equivalent to the BCS approximation [55, 56] and with neglect of the pairing terms with $k \neq i$. This leads to the following self-consistent equations for $\Delta_k \neq 0$:

$$\frac{J}{N}\sum_k \frac{\gamma^2_k}{E_k}\tanh\left(\frac{\beta E_k}{2}\right) = 1 \qquad (53)$$

with $J = 2t^2/U$, $E_k = [(\varepsilon_k - \mu)^2 + |\Delta_k|^2]^{1/2}$ and $\gamma_k = \cos(k_x a) + \cos(k_y a)$ for a planar configuration

of the lattice. This equation must be supplemented with the equation for the chemical potential in the superconducting phase of the form

$$\frac{1}{N}\sum_k \left(1 - \frac{\varepsilon_k}{E_k}\right)\tanh\left(\frac{\beta E_k}{2}\right) = n \qquad (54)$$

In solving Eq. (53), solutions of the following type have been considered:

1. extended s wave [55, 57]:
$$\Delta^{(s)}_k = \Delta[\cos(k_x a) + \cos(k_y a)] \qquad (55a)$$

2. d wave [56, 57]:
$$\Delta^{(d)}_k = \Delta[\cos(k_x a) - \cos(k_y a)] \qquad (55b)$$

3. mixed s–d phases [58]:
$$\Delta^{(sd)}_k = s\Delta^{(s)}_k + d\Delta^{(s)}_k \qquad (55c)$$

The mixed phase was found to be the most stable close to the half-filled band case. For the half-filled band case, the ground-state energy for s and d waves states are the same.

The type of solution obtained within the mean-field approximation (cf. Section II for details) is illustrated through Fig. 28, where the temperature dependence of the specific heat is shown for a different number of holes δ and for $|t|/U = 0.1$ and with inclusion of nearest-neighbor repulsive Coulomb interaction V. A discontinuity of $C(T)$ at $T = T_c$ takes place for each δ. For comparison, the dotted lines represent the specific heat for the normal phase.

There is a major problem with the standard mean-field solutions discussed in Refs. 54–58, namely, it

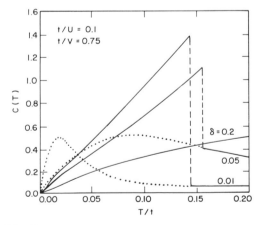

FIG. 28. Temperature dependence of the specific heat $C(T)$ within the mean-field approach to the exchange-mediated pairing in a narrow band. The dotted line represents $C(T)$ for the normal phase, while the discontinuity occurs at the transition to the superconducting phase. (From Ref. 57.)

yields a nonzero (in fact, maximal or almost maximal) value of the superconducting transition temperature T_c for the half-filled band case, which corresponds to the Mott-insulating state. This is a spurious result; it appears because by performing the transformation Eqs. (51) the double site occupancies reappear again for $n < 1$. To remove some of the unphysical features of the mean-field solution, a new formalism has been proposed [57–59] in which auxiliary (slave) bosons are introduced. In this formalism, some of the properties of the projected operators [Eqs. (47) and (49)] are already preserved in the mean-field type approximation involving boson and fermion fields on the same footing. The transition temperature T_c now vanishes, as it should, in the limit $n = 1$. According to Ref. 60, the slave bosons represent holes in the Mott insulators and are regarded as charged, while the fermions are neutral. These entities are called holons and spinons, respectively.

The holon–spinon language is introduced formally by noting that the projected operators [Eq. (47)] are represented as

$$b_{i\sigma}^+ \equiv b_i f_{i\sigma}^+ \qquad \text{and} \qquad b_{i\sigma} = b_i^+ f_{i\sigma} \quad (56)$$

where b_i and b_i^+ are annihilation and creation boson operators located at the atomic site i, while $f_{i\sigma}^+ \equiv a_{i\sigma}^+$ and $f_{i\sigma} \equiv a_{i\sigma}$ are the commonly used fermion operators. Substituting Eq. (56) into Eq. (46), one obtains

$$\mathcal{H} = \sideset{}{'}\sum_{ij\sigma} t_{ij} b_i b_j^+ f_{i\sigma}^+ f_{j\sigma} - \sideset{}{'}\sum_{ij} (2t_{ij}^2/U) b_{ij}^+ b_{ij}$$

$$+ \sum_i \lambda_i \left(b_i^+ b_i + \sum_\sigma f_{i\sigma}^+ f_{i\sigma} - 1 \right)$$

$$- \mu \sum_{i\sigma} n_{i\sigma} + \text{(three-site terms)} \quad (57)$$

The first two terms represent, respectively, the coupled holon–spinon hopping and the binding of spinons into singlet pairs. The third term expresses the fact that the number of holons and spinons is equal to unity on each site; the Lagrange multiplier λ_i thus explicitly provides for the removal of double occupancies. The fourth term represents the conservation of the number of electrons. Now, in a further approximation, one decouples fermions from bosons and then solves the two parts self-consistently. The mean-field treatment discussed earlier corresponds to the approximation in which λ_i is taken as the same at each site ($\lambda_i \to \lambda$) and in which one introduces the replacement $\langle b_i b_i^+ \rangle = \langle b_i \rangle \langle b_i^+ \rangle = |\langle b \rangle|^2 = 1 - n$. The superconducting solution is described in terms of two correlation functions: $\Delta_B \equiv \langle b_i^+ b_j^+ \rangle \approx \langle b^+ \rangle^2$,

characterizing the Bose condensation of holons, and $\Delta_F \equiv \langle f_{i\uparrow} f_{j\downarrow} \rangle$, characterizing the gap in the spectrum of fermion excitations (the site indices i and j denote a pair $\langle ij \rangle$ of nearest neighbors). The nonzero Δ_B occurs only below a temperature T_B, which we call the Bose condensation temperature, whereas the nonzero Δ_F appears only below $T = T_{RVB}$, characterizing the mean-field solution within the RVB theory [55]. The superconducting phase is characterized by nonzero values of both Δ_B and Δ_F simultaneously. This is because in the mean-field approximation $\langle b_{i\uparrow} b_{i\downarrow} \rangle \Delta_B \Delta_F$. Hence, the lower of the two temperatures (T_B and T_{RVB}) determines the superconducting transition temperature. In Fig. 29, taken from Ref. 62, we have plotted these two temperatures as a function $\delta = 1 - n$. One should notice that in order to have $T_B \neq 0$ a small nonzero overlap $t_z = 0.1t$ was taken in the direction perpendicular to the square planar configuration of the atoms. We see that $T_c \to 0$ as $\delta \to 0$, as should be the case.

This treatment is based on several approximations that need further clarification. The most serious of them is that in the limit of $\delta \to 0$ this approach does not properly reproduce the characteristics of the Mott insulator that has the properties of the spin ($\tfrac{1}{2}$) Heisenberg antiferromagnet. Furthermore, the three-site terms that provide the Bose pairing $b_i^+ b_k^+$ have been ignored. Those two problems can be solved (cf. Ref. 64) by (1) replacing the operators in Eq. (47) in **k** space by the operators $b_{\mathbf{k}\sigma}^+$ and $b_{\mathbf{k}\sigma}$, with the statistical mechanical distribution for $\langle b_{\mathbf{k}\sigma}^+ b_{\mathbf{k}\sigma} \rangle$ given by Eq. (41c), that is, by the statistical distribution for

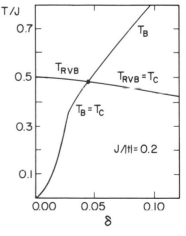

FIG. 29. The critical temperature T_c (the thick line) versus $\delta = 1 - n$. The temperatures T_{RVB} and T_B are those characterizing the onset of coherence for the spinons and the Bose condensation of holons. Note that T_c is determined by the lower of the two temperatures. (From Ref. 62.)

the spin liquid [33]; and (2) by transforming the three-site terms in Eq. (50) to **k** space and introducing the BCS-type approximation by taking only the terms of the type $b_{\mathbf{k}\uparrow}^+ b_{\mathbf{k}\downarrow}^+ b_{-\mathbf{k}'\downarrow} b_{\mathbf{k}'\uparrow}$. In effect, the **k** dependence of the gap $\Delta_{\mathbf{k}} \equiv \langle b_{\mathbf{k}\sigma}^+ b_{\mathbf{k}\sigma} \rangle$ can be derived explicitly; it is of the extended s wave form [Eq. (55a)]. We should see further progress along this line in the near future [64].

An alternative model of the onset of high-T_c superconductivity has been proposed recently [65]. The model involves paired boson condensation (characterized again by the boson $\langle b_i^+ b_i^+ \rangle$ amplitude) because of coherent tunneling of quasiparticle holon pairs between the CuO$_2$ planes. This approach [65] explains the increase of T_c if the number of CuO$_2$ neighboring planes increases; the last result agrees with experimental results for Bi and Tl compounds for which $T_c > 100$ K. Some degree of interplanar coupling is needed in any theory involving the Bose condensation since it is absent in a strictly two-dimensional model.

2. Hybridized Systems

The electron states near the Fermi surface in the high-T_c oxides such as La$_2$CuO$_4$ involve hybridization of electrons of atomiclike $3d_{x^2-y^2}$ states of copper with $2p_\sigma$ of oxygen (cf. Fig. 23 and Section III,C). These electron states can be described by the Anderson lattice Hamiltonian of the type of Eq. (43), with a width of the bare p band $W \approx 4$ eV, the position of the $3d^9$ level at $\varepsilon_f \equiv \varepsilon_d - \varepsilon_p \sim 1$ eV, $U \leq 10$ eV, and hybridization magnitude $|V| \simeq 1.5$ eV [66]. The hybridization is intersite in nature, that is, it involves the $2p$ and $3d$ orbitals located on different sites. Therefore, the effective hybridization energy is $Vz \simeq 6$ eV, where $z = 4$ is the number of nearest neighboring O atoms in the plane for a given Cu atom. We see that $Vz > \varepsilon_f$; hence, the $3d$ and $2p$ states mix strongly, that is, the d electrons can be promoted to $2p$ hole states and vice versa. Additionally, $2p$ electrons can be promoted to form the $3d^{10}$ configurations of the excited states. If $Vz \gtrsim \varepsilon_f$, but $|V|z \ll \varepsilon_f + U$, the above two promotion mixing events are low- and high-energy processes, respectively. The situation is drawn schematically in Fig. 30, where the parameter U is assumed to be by far larger than $|\varepsilon_f|$, W, or $|V|z$. We consider this limiting situation first [67].

The high-energy processes take place only as virtual events, that is, with electron hopping from the p state to the highly excited $3d$ state and back. Such virtual p–d–p processes have been drawn schematically in Fig. 31, where site m labels the $2p_\sigma$ state of the oxygen anion O^{2-} centered at $\mathbf{R}_m$ and site i labels

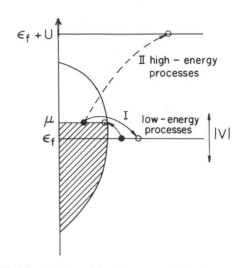

FIG. 30. Division of the charge-transfer (p–d) processes into low- and high-energy parts. The processes II give rise to the Kondo and superexchange interactions when treated perturbationally to second- and fourth-order, respectively.

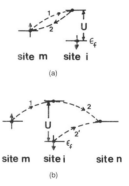

FIG. 31. Schematic representation of the hopping processes induced by the high-energy mixing processes. (a) Virtual hopping and (b) three-site processes. The hoppings labeled 2 and 2' are alternative processes.

$3d_{x^2-y^2}$ due to the Cu^{2+} ion centered at $\mathbf{R}_i$. Then, the effective Hamiltonian can be rewritten in the real-space language, and for large U reads [67]

$$
\begin{aligned}
\mathcal{H} = &\sum_{\mathbf{k}\sigma} \varepsilon_{\mathbf{k}} n_{\mathbf{k}\sigma} + \varepsilon_f \sum_{i\sigma} b_{i\sigma}^+ b_{i\sigma} \\
&+ \sum_{im\sigma} (V_{im} b_{i\sigma}^+ c_{m\sigma} + V_{im}^* c_{m\sigma}^+ b_{i\sigma}) \\
&- \sum_{imn} \frac{2 V_{mi}^* V_{im}}{U + \varepsilon_f} \widetilde{B}_{im}^+ \widetilde{B}_{in}
\end{aligned}
\tag{58}
$$

The first term describes the band energy of itinerant ($2p_\sigma$) electrons, while the third represents the residual mixing since, as in the case of narrow-band electrons, the operators $(b_{i\sigma}^+)$ and $(b_{i\sigma})$ are projected operators [Eq. (47)] for the starting $3d$ states. The last term represents the so-called hybrid (interorbital) pairing with the pairing operators

$$\widetilde{B}_{im}^+ = \frac{1}{\sqrt{2}} (b_{i\uparrow}^+ c_{m\downarrow}^+ - b_{i\downarrow}^+ c_{i\uparrow}^+) \qquad (59a)$$

and

$$\widetilde{B}_{im} = \frac{1}{\sqrt{2}} (b_{i\downarrow} c_{m\downarrow} - b_{i\uparrow} c_{i\uparrow}) \qquad (59b)$$

The meaning of the effective Hamiltonian [Eq. (58)] is as follows. The first three terms provide eigenvalues representing the hybridized quasiparticle states with the structure discussed in Section III,B. The last term provides a singlet pairing for those hybridized states. It expresses (for $m = n$) the Kondo interaction between the p and $3d$ electrons of the form

$$\sum_{im} \frac{2|V_{im}|^2}{U + \varepsilon_f} \left(\mathbf{S}_i \cdot \mathbf{s}_m - \frac{1}{4} v_i n_m \right)$$

It is antiferromagnetic in nature, with the exchange integral

$$J_{im} \equiv \frac{2|V_{im}|^2}{U + \varepsilon_f} \sim 0.5 \text{ eV}$$

hence, the pairing results in a spin singlet state. It must be underlined that Eq. (58) represents hybridized correlated states in the so-called fluctuating valence regime in which $U \gg |V_{im}| \gtrsim \varepsilon_f$. This is the reason why we cannot completely transform out the hybridization. Also, the occupancy n_f of the atomic level is a noninteger because the strong hybridization induces a redistribution of the particles among starting atomic and band states.

When both U and $|\varepsilon_f|$ are much larger than $|V_{im}|$, one can transform out the hybridization completely and obtain instead of Eq. (58) the following effective Hamiltonian:

$$\mathcal{H} = \sum_{\mathbf{k}\sigma} \varepsilon_{\mathbf{k}} n_{\mathbf{k}\sigma} + \varepsilon_f \sum_{i\sigma} b_{i\sigma}^+ b_{i\sigma}$$
$$+ \sum_{ijm\sigma} \frac{V_{mi}^* V_{mj}}{\varepsilon_f} b_{i\sigma}^+ b_{j\sigma} (1 - n_{m\sigma})$$
$$\times \sum_{imn} \frac{2V_{mi}^* V_{in} U}{\varepsilon_f (\varepsilon_f + U)} \widetilde{B}_{im}^+ \widetilde{B}_{in} \qquad (60)$$

We have now a two-band system: the $3d$ electrons acquire a bandwidth $W^* \sim (V^2/\varepsilon_f)(1 - \langle n_{m\sigma}\rangle)$. The spin singlet pairing is again of interband type. The part with $m = n$ in the last term is equivalent to the Kondo interaction derived a long time ago for magnetic impurities [68]. Here, the lattice version of

this Hamiltonian provides both pairing and itinerancy to the bare atomic electrons.

Note that the hybrid pairing introduced in this section expresses both the Kondo interaction (the two-site part) and pair hopping. It is therefore suitable for a discussion of superconductivity of Kondo lattice effects in heavy fermion systems. The pairing part supplements the current discussions of Anderson lattice Hamiltonian in the $U \to \infty$ limit [39–41]. One may state that the Kondo interaction mediated pairing introduced above represents the strong-coupled version of spin fluctuation mediated pairing for almost-localized systems introduced previously [69].

An approach using the slave boson language for hybridized systems has also been formulated [70] and contains a principal feature of the effective Hamiltonian [Eq. (58)]; the solution in the mean-field approximation has also been discussed. Figure 32 illustrates the dependence of the superconducting transition temperature T_c versus hole concentration x_h; this is compared with experimental data recently obtained [71]; dependence of T_c the full concentration range of holes is shown in Fig. 33. The superconductivity appears for $La_{2-x}Sr_xCuO_4$ only for $0.04 \lesssim x_h < 0.34$. The full phase diagram comprising localization, antiferromagnetism, and superconductivity will be considered next.

3. Properties of Correlated Electrons: An Overview

Two alternative models and mechanisms of exchange mediated pairing have been discussed so far: the narrow-band model with d–d kinetic exchange mediated pairing, and the hybridized model with d–p Kondo interaction mediated pairing. The hybridized model should be regarded as a basis of narrow-band

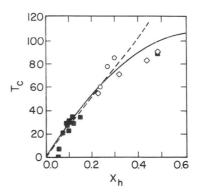

FIG. 32. The superconducting transition temperature T_c versus hole concentration x_h. Squares, experimental data for $La_{2-x}Sr_xCuO_4$; crosses and diamonds, data for $YBa_2Cu_3O_{7-y}$. (From Ref. 70.)

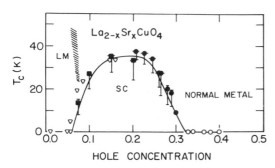

FIG. 33. The superconducting transition temperature T_c versus hole concentration for $La_{2-x}Sr_xCuO_4$ in the full range. LM means regime of local moments (insulating phase). (From Ref. 71.)

behavior in real oxides and in heavy fermion systems since the direct d–d (or f–f) overlap of the neighboring atomic wave functions is extremely small. Next, we give a brief overview of the narrow-band properties of the correlated electrons starting from the hybridized (Anderson lattice) model.

First, we discuss the quasiparticle states in $U \to \infty$ limit. The simplest approximation is to reintroduce ordinary fermion operators $a_{i\sigma}^+$ and $a_{i\sigma}$ in Eq. (58) and readjust the hybridization accordingly [72]. In effect, one obtains the hybridized bands of the form of Eq. (45), that is,

$$E_{\mathbf{k}\pm} = \frac{\varepsilon_f + \varepsilon_{\mathbf{k}}}{2} \pm \left[\left(\frac{\varepsilon_f - \varepsilon_{\mathbf{k}}}{2} \right)^2 + 4|\widetilde{V}_{\mathbf{k}}|^2 \right]^{1/2}$$

(61)

where $\widetilde{V}_{\mathbf{k}} \equiv q^{1/2} V_{\mathbf{k}}$, and $q \equiv (1 - n_f)/(1 - n_f/2)$ for $0 \le n_f \le 1$, while $V_{\mathbf{k}}$ is the space Fourier transform of V_{im}. For the case of the CuO_2 layers [73],

$$|\widetilde{V}_{\mathbf{k}}|^2 = qV^2 \left[\sin^2\left(\frac{k_x a}{2} \right) + \sin^2\left(\frac{k_y a}{2} \right) \right]$$

(62)

If the Fermi level falls into the lower hybridization band and $n_f = 1 - \delta$, with $\delta \ll 1$, then it can be shown that the quasiparticles describing the hybridized states are of mainly quasi-atomic character. In other words, the effective Hamiltonian [Eq. (58)] is approximately of the narrow-band form [Eq. (52)]. The pairing takes place between heavy quasiparticles. This limiting situation describes qualitatively the situation in heavy fermions with Kondo interaction mediating the pairing. By contrast, if the Fermi level falls close to the top of the upper hybridization band (as is the case for high-T_c superconductors, since the p band is almost full and the $3d$ level is almost half filled), then the pairing is mainly due to the band electrons ($2p$ holes in the case of high-T_c

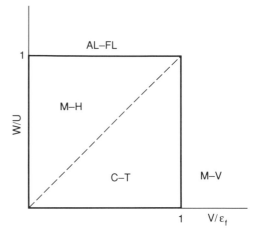

FIG. 34. Schematic representation of the regimes of stability of charge-transfer (C-T) and Mott–Hubbard (M-H) insulating states, as well as of mixed-valent (M-V) and almost-localized Fermi-liquid (AL-FL) metallic states.

oxides). These results are obtained by constructing explicitly the eigenstates corresponding to the eigenvalues [Eq. (61)] and taking the limits corresponding to heavy fermions ($n_f \to 1$) and high-T_c systems ($n = n_d + n_p \approx 3$).

a. Mott–Hubbard Insulators, Charge-Transfer Insulators, and Mixed-Valent Systems. The next problem concerns the Mott localization in systems with hybridized d–p states. The systems such as NiO, CoO, or MnO regarded as classic Mott insulators are, strictly speaking, hybridized $3d$–$2p$ systems. However, these cases are to a good approximation ionic systems in the sense that the electronic configuration in, for example, NiO, is $Ni^{2+}O^{2-}$. Then, the valence $2p$ band is completely full and plays only a passive role in effective d–d charge transfer processes [74], since a $2p \to 3d$ transfer is followed by $3d \to 2p$ transfer from the neighboring $3d$ shell of Ni^{2+}. In effect, the antiferromagnetic exchange interaction in Eq. (46) expresses formally the superexchange interaction that has been known for a long time [75]. In this approach, the kinetic exchange interaction between d electrons (induced by virtual d–d transitions, cf. Section III,A) is expressed as a fourth-order effect in the hybridization V since the virtual d–d transition involves a sequence of d–p and p–d transitions in the fourth order.

The possible macroscopic states of hybridized systems are illustrated in Fig. 34 as a schematic classification of possible states of hybridized systems modeled by the periodic Anderson Hamiltonian [Eq. (43)]. The parameter W/U characterizes the degree of correlation of quasi-atomic electrons that may acquire a nonzero bandwidth mainly due to hybridization; the parameter V/ε_f characterizes the degree of mixing of the states involved. If the d (or f) atomic level lies deeply below the top of the valence band ($V/\varepsilon_f \ll 1$), then we have either Mott–Hubbard (M–H) or charge-transfer (C–T) insulators; for the former the band gap Δ is due to $d^n \rightarrow d^{n+1}$ excitations (that is, $\Delta \sim U$-W), whereas for the latter it is due to $d^n p^2 \rightarrow d^{n+1} p^2$ charge-transfer transitions. The atomic $3d$ (or $4f$) electrons are unpaired in both the C–T and M–H states. If $V/\varepsilon_f \gtrsim 1$ and $W/U \ll 1$, then we enter mixed-valent (M–V) and (close to the border with M–H) heavy fermion regimes. On the other hand, if $W/U \gtrsim 1$, then irrespective of the value of V/ε_f we encounter the correlated metal regime that we call an almost-localized Fermi liquid (AL FL). Both heavy fermion and high-T_c systems are close to the line separating M–H and M–V regimes. A similar classification scheme for transition metal oxides has been proposed in Ref. 76.

The classification shown in Fig. 34 provides only a distinction between insulating and metallic states. A complete phase diagram for the high-T_c system La$_{2-x}$Sr$_x$CuO$_4$ is drawn schematically in Fig. 35 (taken from Ref. 77a). Stoichiometric La$_2$CuO$_4$ or doped with $x \lesssim 0.02$ exhibit antiferromagnetism (AF). In the regime of $0.02 \lesssim x \lesssim 0.04$, the inhomogeneous (SG) magnetic insulating phase sets in, while for $x \gtrsim 0.04$, a transition from insulating (I) to metallic (M) takes place and the system is superconducting until a transition from orthorhombic (O) to tetragonal (T) crystallographic structure occurs. A similar phase diagram was established for

YBa$_2$Cu$_3$O$_{6+x}$ [77b]. Those phase diagrams combine all the features we have separately discussed so far. Main features of this phase diagram are explained next.

b. Magnetic Interactions, Polarons, and Pairing. To rationalize the phase diagram shown in Fig. 35, we note first that antiferromagnetism is stable only close to the half filling of the d band (cf. Fig. 27 and the discussion in Section III,A). In the case of the hybridized model, one has to calculate explicitly the contributions to the d–p and d–d interactions. Within the perturbation expansion for the Anderson lattice model but with only the high-energy mixing processes (cf. Fig. 30) treated in this manner [67, 78], we obtain the magnetic part of the effective Hamiltonian to fourth order as

$$\mathcal{H}_m \simeq J_{pd} \sum_{im} \left(\mathbf{S}_i \cdot \mathbf{s}_m - \frac{1}{4} n_i n_m \right)$$
$$+ J_{dd} \sum_{\langle ij \rangle} \left(\mathbf{S}_i \cdot \mathbf{S}_j - \frac{1}{4} N_i N_j \right)$$
$$+ J_{pp} \sum_{\langle mm' \rangle} \left(\mathbf{s}_m \cdot \mathbf{s}_m' - \frac{1}{4} n_m n_m' \right) \quad (63)$$

where the first term represents the p–d Kondo-type interaction, with the exchange integral

$$J_{pd} \approx \frac{2|V|^2}{U + \varepsilon_f} \left[1 - \frac{|V|^2}{U + \varepsilon_f} (n_d + n_p + 1) \right] \quad (64)$$

The second term expresses the d–d (kinetic exchange) interaction, with $J_{dd} = |V|^4/(U + \varepsilon_f)^3$, and the last term represents the interaction between p holes, with $J_{pp} = |V|^4 n_d/(U + \varepsilon_f)^3 \approx J_{dd}$. The antiferromagnetic p–d and d–d interactions are not compatible; in the hole language, the p hole polarizes its surroundings ferromagnetically, as shown in Fig. 36 (note that the hole may be located in any O^{1-} ion, so its position in the volume of radius R is not fixed). A simple estimate [78] of the canting angle θ between the neighboring $3d$ spins $\mathbf{S}_i$ and $\mathbf{S}_j$ caused by the hole polarization gives

$$\cos \frac{\theta}{2} \approx \frac{J_{pd} - 2J_{dd}}{2J_{dd}} \quad (65)$$

Taking $J_{pd} \approx 0.5$ eV and $J_{dd} \approx 50$ K, we obtain the canting angle θ through the relation $\cos(\theta/2) \approx 25x_p$. The energy E_c of the system with a single hole canting the surrounding spins is

$$E_c = -\frac{1}{2} \frac{(J_{pd} - 2J_{dd})^2}{J_{dd}} z - J_{dd} z \quad (66)$$

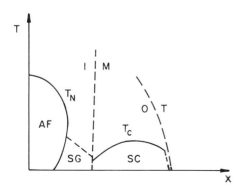

FIG. 35. Schematic phase diagram on the plane T–x for La$_{2-x}$Sr$_x$CuO$_4$. Antiferromagnetic (AF), spin-glass (SG), superconducting (SC), insulating (I), and metallic phases have been drawn, as well as the boundary between orthorhombic (O) and tetragonal (T) crystallographic phases have been specified. (From Ref. 79.)

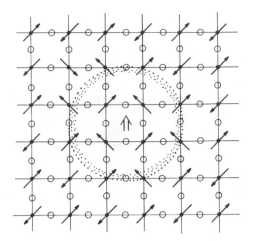

FIG. 36.   Schematic representation of a $2p$ hole polaron in the planar CuO_2 structure. Cu^{2+} ions are indicated by the arrows, while O^{2-} ions are indicated by open circles. The hole creates a canted spin configuration with resultant ferromagnetic polarization and autolocalizes in it. This is the reason why the high-T_c oxides remain insulating when the concentration of the hole does not exceed $x_c \sim 0.04$–0.05.

This energy is lower than the energy $(-J_{dd}z)$ of the antiferromagnetic (Néel) state of antialigned d spins due to Cu^{2+} copper ions. Next, we estimate the radius R of the hole polaron with aligned spins, as depicted in Fig. 36. Applying the same type of reasoning as in Section III,A, we obtain the expression for the energy E_p of a single polaron:

$$E_p = \frac{E_0}{(R/a)^2} - \frac{1}{2} \frac{(J_{pd} - 2J_{dd})^2}{J_{dd}} z \cdot \bar{x}_p^2 \quad (67)$$

where now $\bar{x} = (a/R)^2$ is the probability to finding a p hole on a given oxygen atomic site within the radius R. Minimization with respect to R for the two-dimensional case leads to

$$\frac{R}{a} = \left[\frac{(J_{pd} - 2J_{dd})^2 z}{J_{dd} E_0} \right]^{1/2}$$
$$\approx J_{pd} \sqrt{\frac{z}{J_{dd} E_0}} \approx 4 \quad (68)$$

A metal–insulator transition takes place when the neighboring polarons overlap, that is, when $Rx_{pc}^{-1/2} = 1$; this yields the critical hole concentration $x_c \approx 0.07$. One can also estimate this critical concentration by equating the band energy of holes, which is $-(W/2)x_p(1 - x_p)$, with the magnetic energy gain per hole due to aligning the neighboring d spins $(-J_{pd}^2/2J_{dd})zx_p^2$. This leads again to $x_c \approx 0.068$, in rough agreement with the observed value $x_c \approx 0.04$–0.05. For $x > x_c$, the ground state

of the system is metallic, and the pairing described in Sections IV,A,1 and IV,A,2 can take place. Within the exchange-mediated mechanism, all interactions in Eq. (63) are antiferromagnetic. Hence, in general, one has p–d pairing characterized by the operators of Eqs. (59), d–d pairing characterized by the operators of Eqs. (49), and p–p pairing [79] characterized by the operators

$$p_{mm'}^+ \equiv \frac{1}{\sqrt{2}} (c_{m\uparrow}^+ c_{m'\downarrow}^+ - c_{m\downarrow}^+ c_{m'\uparrow}^+) \quad (69a)$$

and

$$p_{mm'} \equiv \frac{1}{\sqrt{2}} (c_{m\downarrow} c_{m'\uparrow} - c_{m\uparrow} c_{m'\downarrow}) \quad (69b)$$

All three types of pairing may contribute to the superconducting ground state. However, the d–p interaction is much stronger, hence, the d–p hybrid type of pairing is in the limit $U > W > |V| \gtrsim \varepsilon_f$, the dominant one. As stated, this type of pairing may appear effectively as a d–d or p–p type of pairing in the hybridized basis, depending on whether the Fermi level lies close to the top of the lower or upper hybridized bands, respectively. For the sake of completeness, we write down the full effective Hamiltonian with all the pairings specified, namely,

$$\begin{aligned}
\mathcal{H} &= \sum_{\mathbf{k}\sigma} \varepsilon_{\mathbf{k}} n_{\mathbf{k}\sigma} + \varepsilon_f \sum_{i\sigma} b_{i\sigma}^+ b_{i\sigma} \\
&+ \sum_{im\sigma} (V_{im} b_{i\sigma}^+ c_{m\sigma} + V_{im}^* c_{m\sigma}^+ b_{i\sigma}) \\
&+ J_{pd} \sum_{\langle imn \rangle} \widetilde{B}_{im}^+ \widetilde{B}_{in} \\
&+ J_{dd} \sum_{\langle ijk \rangle} \widetilde{b}_{ij}^+ \widetilde{b}_{kj} \\
&+ J_{pp} \sum_{\langle mm'm'' \rangle} p_{mm'}^+ p_{m'm''} \quad (70)
\end{aligned}$$

In deriving this result, one does not assume that $|V| \ll \varepsilon_f$; therefore Eq. (70) is applicable to the situation with fluctuating valence. Next, by introducing a slave boson representation [Eq. (56)], we obtain most general Hamiltonian for treatment of pairing in correlated systems [78]. We should be able to witness a decisive progress in the near future concerning the relative role of hybrid p–d, d–d, and p–p pairings in high-T_c systems using the slave boson technique to Eq. (70). Also, Eq. (70) should serve as a basis for the discussion of antiferromagnetism and superconductivity in heavy fermion systems; in that situation, the role of itinerant $2p$ states is played by hybridized $5d$–$6s$ conduction bands, while the role of $3d$ electrons is played by the $4f$ electrons.

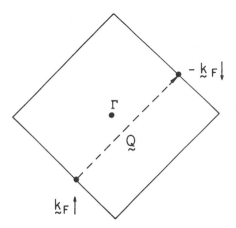

FIG. 37. Two-dimensional Fermi surface for a half-filled band. The opposite points of the surface are related by the wave vector $\mathbf{Q} = (\pi/a)(1,1)$.

c. Coexistence of Antiferromagnetism and Superconductivity. In the previous analysis, we have treated separately antiferromagnetism (AF) and superconductivity (SC). Detailed calculations [42, 61, 80], within the mean-field theory discussed, point to the possibility of coexistence of AF and SC phases. It is possible to visualize this coexistence by considering a narrow-band model with the two-dimensional (almost square) Fermi surface as drawn in Fig. 37. Namely, the band energy of electrons located on the Fermi surface has the property $\varepsilon_{\mathbf{k}+\mathbf{Q}} = -\varepsilon_{\mathbf{k}}$, where $\mathbf{Q} \equiv (\pi/a_1, \pi/a) = 2\mathbf{k}_F$. This is the so-called nesting condition; any system with this property is unstable with respect to formation of the spin density wave (SDW) state with the wave vector $\mathbf{Q}$. The SDW state is identical with a two-sublattice AF state if the magnitude of magnetic moment in the ordered state is $\mu \sim \mu_B$, where μ_B is the Bohr magneton. One should also notice (cf. Fig. 37) that $\mathbf{Q}$ connects two single-particle states on the opposite sides of the Fermi surface since $-\mathbf{k}_F + \mathbf{Q} = \mathbf{k}_F$. Furthermore, both SDW and SC states couple electrons with the opposite spins. This is why AF and SC states are compatible for $n \approx 1$. There is no clear experimental evidence that these two phases coexist in a high-T_c system, though there is some evidence from muon spin rotation that it is so [81]. A clear detection of such coexistence would demonstrate directly the importance of exchange interactions in a superconducting phase. Namely, within the exchange mediated superconductivity, one can show [54] that close to the half-filled narrow-band case, $T_N/T_c \sim 6-8$. The analysis of AF–SC

coexistence conditions within the Anderson lattice Hamiltonian has not yet been performed.

B. PHONONS AND BIPOLARONS

After the discoveries of superconductivity in the 40 K and 90 K ranges [82] the obvious question was posed whether the phonon mediated mechanism of pairing, so successful in the past, can explain the superconductivity with such high value of T_c. It was realized from the outset that one should include specific properties of these compounds, such as the quasi-planar (CuO_2) structure with a logarithmic (Van Hove) singularity in the density of states $\rho(\varepsilon)$ at the middle point of the two-dimensional band [83–85], the polar nature of the CuO bonds rendering applicable the tight binding representation of the electronic states [48, 83], and strong electron lattice coupling [84–86], leading to the local formation of small bipolarons (that is, two electron pairs) [87, 88] that may undergo Bose condensation when the metallic state is reached.

There is no clear evidence for the phonon mediated mechanism of pairing in the classic high-T_c superconductors since the isotope effect in both La systems [89] and Y systems [90] is quite small. However, the recently discovered superconductors $Ba_{1-x}K_xBiO_3$ [91] exhibit large isotope effect [92] and superconductivity with $20\ K \lesssim T_c \lesssim 30\ K$ in the concentration range $0.25 \lesssim x \lesssim 0.4$. Also, the proximity of superconductivity and the charge density wave (CDW) state is observed [93].

The last property, as well as the observed diamagnetism in the insulating phase $x \lesssim 0.25$, is very suggestive [94] that small trapped polarons are formed before the electron subsystem condenses into a superconducting phase. Condensation takes place when the percolation threshold for the insulator–metal transition is reached[3] (at $x \sim 0.2$). The parent compound can be called a bipolaron insulator, which has the formal valency scheme $Ba_2^{2+}Bi^{3+}Bi^{5+}O_6^{2-}$.

Three specific features of $Ba_{1-x}K_xBiO_3$ compounds should be noted. First, the diamagnetic nature of the parent compound $BaBiO_3$ distinguishes the systems from the parent compounds La_2CuO_4 and $YBa_2Cu_3O_7$, which are both antiferromagnetic. Second, the $Ba_{1-x}K_xBiO_3$ systems are copper free

---

[3] The actual percolation threshold for the onset of the metallic phase is $x_c/2 \sim 0.12$ since the bipolarons reside on every alternate Bi lattice site. Also, the holes introduced by K doping must be present in a Bi–O hybridized band for $x > x_c$ to render the bipolarons mobile.

and have truly three-dimensional cubic structure in the SC phase [91]. Third, their main superconducting properties are in accordance with the prediction of the standard BCS theory [95].

The theory of the $Ba_{1-x}K_xBiO_3$ compound must incorporate three additional obvious facts. First, the pairing process $2Bi^{4+} \rightarrow Bi^{3+} + Bi^{5+}$ is possible when the electron lattice coupling leads to an attraction overcoming the e–e repulsion in the Bi^{3+} state relative to the Bi^{5+} state [88]. It involves a relaxation of the O^{2-} octahedra, that is, an optical, almost dispersionless, breathing mode. This can provide a local (on-site) attractive interaction between $6s$ electrons of the type $\lambda n_{i\uparrow} n_{n\downarrow}$, which leads to a scalar (**k**-independent) pairing potential $V_{\mathbf{kk'}} = \lambda$, which, in turn, provides a justification for the observed properties reflecting an isotropic shape of the gap ($\Delta_{\mathbf{k}} \equiv \Delta$), as in the standard BCS theory (cf. Section II).

Second, from the fact that the parent compound $BaBiO_3$ is an insulator, we conclude that either the magnitude V of the Coulomb repulsion between the electrons on nearest neighboring Bi atoms exceeds the width W^* or the bipolaron band [95] of the small bipolarons are self-trapped in the potential created by interaction with nearest neighboring oxygens. The onset of the metallic phase at concentrations near the percolation threshold $x_c \sim 0.1$ for n.n. interaction means that both effects may be important. In either case, the CDW state will set in, so the entropy of the bipolaron lattice vanishes at $T = 0$ (at least, for $x = 0$). The CDW phase plays the same role here as does AF ordering in the La_2CuO_4 and $YBa_2Cu_3O_6$. The properties of $Ba_{1-x}K_xBiO_3$ are similar to those of the $Ba_{1-x}Pb_xBiO_3$ compounds discovered over a decade earlier [97].

Third, the fact that the onset of the superconductivity coincides with the transition from the CDW insulator to an SC metal speaks in favor of preexisting electron pairs already present in the insulating phase. However, the bipolaron concentration is large, and hence, the interpretation of the superconducting transition as Bose condensation of bipolarons may be inapplicable even when the coherence length is small. The overall theoretical situation is nonetheless much clearer for $Ba_{1-x}K_xBiO_3$ compounds than for either the $La_{2-x}Sr_xCuO_4$ or the $YBa_2Cu_3O_{7-\delta}$ series since the accumulated (so far) experimental evidence indicates that (optical?) phonon-mediated pairing takes place [98].

The $Ba_{1-x}K_xBiO_3$ compounds seem to be a natural candidate for a bipolaronic mechanism of electron pairing [99]. This is because the diamagnetic (and charge-ordered) parent system $Ba_2Bi^{3+}Bi^{5+}O_3$ can be regarded as an ordered lattice of locally bound two-electron pairs (bipolarons) located on alternate Bi^{3+} sites; these pair states are stabilized by a strong relaxation of the surrounding oxygen anions. Effectively, the two electrons are attracted to each other. The effect of potassium doping is to make these pairs mobile by diminishing the number of bipolarons per Bi site from the value $\frac{1}{2}$ [100]. In essence, the lattice distortion is responsible for the bipolaron formation in the same manner as in the case of the copper pairs; the difference is due to the circumstance that the bipolarons are locally bound complexes in a direct space that undergo a Bose condensation from an incoherent state of preexisting and moving pairs. The temperature of such condensation is $T_c \sim x^{2/3}$, where x is the dopant (K) concentration [101]. A key feature of the bipolaron theory of superconductivity is that the Bose-condensed state develops from the charge-density-wave (CDW) insulating state, not from the spin-density-wave (antiferromagnetic) state; the latter situation takes place for the cuprates. Further studies are necessary to calculate the physical properties of a bipolaron superconductor and in particular, the differences with an ordinary (phonon-mediated) superconductor.

C. Charge Excitations

In 1964 Little [102] introduced the idea that virtual electron-hole (exciton) excitations may lead to a pairing with a high value of T_c. This idea has been reformulated recently in the context of high-T_c superconductivity by considering the role of charge transfer ($p \rightarrow d$ and $d \rightarrow p$) fluctuations [103], as well as of intraatomic (Cu $d \rightarrow d$) excitations [104]. The charge-transfer fluctuations involve both $Cu^{2+} - O^-$ and $Cu^{3+} - O^{2-}$ low-energy configurations and Cu^+O^- states. The former two configurations are particularly important if the energy difference $|\varepsilon_p - \varepsilon_d|$ is comparable to the magnitude $|V|$ of the $2p - 3d$ hybridization. This is the limit we have considered within the hybridized model in Section IV,A, probably extended to include the $3d-2p$ Coulomb interaction directly. The method of approach is therefore similar to that in Section IV,A in the limit of strongly correlated electrons. In the limit of weakly interacting electrons (that is, for $U \ll W$), the perturbation expansion in the powers of U provides an effective pairing potential in an explicit form. The processes leading then to the pairing are virtual excitations involving charge and antiferromagnetic spin fluctuations [105]. At the moment, it is difficult to see clearly the difference between exchange-mediated and charge-transfer-mediated types of pairing for strongly correlated hybridized systems.

V. Conclusions

In this article, we have concentrated mainly on reviewing the properties of correlated electrons in normal, antiferromagnetic, and superconducting phases in copper-containing systems, in which the last two are phases caused by antiferromagnetic exchange interactions. Two theoretical models have been discussed in detail: the Hubbard model of correlated narrow-band ($3d$) electrons and the Anderson lattice model of correlated and hybridized electrons, involving $2p$ and $3d$ states in the case of high-T_c oxides. The latter model is regarded as more general and applicable to both high-T_c and heavy fermion systems; in some limiting situations discussed previously hybridized bands exhibit a narrow-band behavior.

The principal novel feature of the metallic phase involving either $3d$ (in high-T_c oxides) or $4f$ (in heavy fermion systems) electrons is that for the half-filled band configuration the itinerant electrons states transform into a set of localized states constituting the Mott insulator. The difference between the Fermi liquid (FL) and the liquid of correlated electrons (the spin liquid) is illustrated in Fig. 38, where the high-temperature value of the entropy has been plotted for these two phases as a function of the number n of

electrons per atom [the statistical distribution of Eq. (41d) was used to calculate $S(n)$ for the latter phase]. Only the spin-liquid case correctly reproduces the entropy of localized moments when the Mott-insulator limit is reached for $n \rightarrow 1$. This limiting value of the entropy per mole, $S = R \ln 2$ for $n \rightarrow 1$, represents one of the necessary conditions to be fulfilled by any theory claiming to describe properly the situation near the Mott-insulator limit. Additionally, those systems are characterized by quasiparticles with very heavy effective mass $m^* \sim \delta^{-1}$ or $W^* \sim \delta$. For $\delta \ll 1$, the band energy becomes comparable to the kinetic exchange characterized by $J = W^2/(Uz)$. Itinerant systems for which $J \gtrsim W^*$ are called quantum spin-liquid systems. Although the proper form of the kinetic exchange for the strongly correlated itinerant electrons had been derived in 1976 [30], its connection with the real-space pairing and the quantum spin-liquid state was recognized only in 1987 [44]. The Mott-insulator, the spin-liquid, and the heavy-fermion states are the primary phases of correlated electrons different from the normal metal state. This difference is sketched out in Fig. 39, where the arrows point to both common features for normal and correlated metals, as well as to those specific to the correlated systems.

The correlated systems that interested us here may also be called almost-localized systems. As discussed in Section II, there are two classes of such systems, separated roughly by the Mott–Hubbard boundary

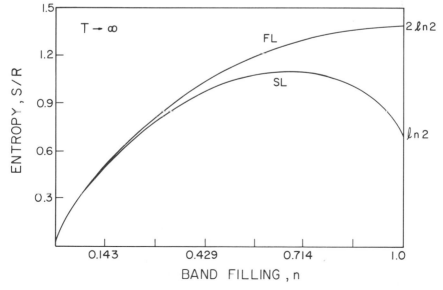

FIG. 38.   The high-temperature limiting value for the entropy (in units of the gas constant R) as a function of n for the cases of the Fermi liquid (FL) and the spin liquid (SL). Note the difference of the values by a factor of two in the limit $n \rightarrow 1$.

$U = U_c \sim W$. Those for which the Coulomb interaction $U < U_c$ are regarded as Fermi liquids and have been treated extensively in Refs. [28] and [29], and those systems for which $U > U_c$ are the spin liquids. This qualitative division is sketched in Fig. 40, where the various thermodynamic phases have been specified for each class (cf. also Fig. 14 for all magnetic phases). The complementary regimes are those with $U/W \ll 1$ and $U/W \gg 1$. Most of the metallic systems can be located between these two limiting situations. It remains to be proven more precisely that the Mott–Hubbard boundary separating Fermi liquid from the Mott insulator for $n = 1$ extends to the part of the diagram with $n \neq 1$. This is a fundamental problem, related in the case of strongly interacting systems to the question of validity of the Luttinger theorem[4] and to the problem of existence of local magnetic moments in the itinerant electron picture, that is, to the problem of validity of the Bloch theorem for a correlated metal. Also, the question of applicability of the Fermi-liquid concept in the limit $U/W \gg 1$ is connected with that concerning the properly defined existence of fermion quasiparticles,[5] interacting only weakly among themselves. One should emphasize that the discussion of the standard mean-field treatment of superconductivity presented in Section IV reduces the whole problem to

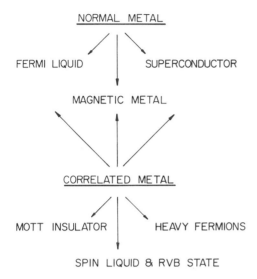

FIG. 39.   Schematic representation of the difference between the normal metal and correlated metal. Only the latter state may lead to the Mott localization, as well as to heavy fermion and spin-liquid metallic phases.

[4] The Luttinger theorem states that as long as the metallic state is stable the volume encircled by the Fermi surface remains independent of the strength of the electron–electron interaction. This theorem is not valid when the Mott transition takes place and the Fermi surface disappears. The volume also doubles when metal is described by a spin liquid discussed in Section II [cf. Figs. 16(a) and (b)].

[5] The holons and spinons are regarded as solitons (cf. Ref. 34).

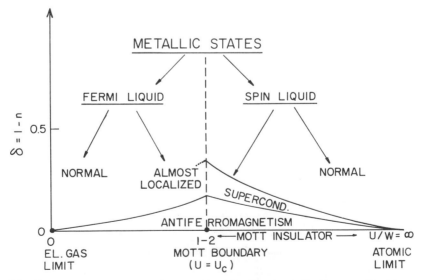

FIG. 40.   Qualitative distinction between Fermi-liquid and spin-liquid states. The Mott boundary $U = U_c$ roughly separates the two limiting phases.

the single-particle approach with a self-consistent field $\sim \Delta_{\mathbf{k}}$. It is not yet completely clear what types of collective excitations (in addition to quasiparticle states) are needed to make the theory complete. The introduction of holons as bosons and spinons as fermions [34] seems to be just one possibility; more natural seems to be a treatment of holons as spinless fermions and of spinons as boson operators that reflect magnonlike properties of local moments.

Recent studies of high-T_c oxides revealed that some of their characteristics are close to those provided by the BCS theory. Namely, the value of $2\Delta_0/k_b T_c \simeq 3.5$ is often indicated [106], the temperature dependence of the London penetration depth, which is close to $[1 - (T/T_c)^4]^{-1/2}$ over a wide temperature range [107], and the electron pairing is in the spin singlet state [108]. Additionally, the shape of the Fermi surface for $YBa_2Cu_3O_7$, as determined by the positron annihilation technique [109], agrees with the predictions of the band structure calculations for an even ($n \approx 4$) number of electrons. These results do not necessarily eliminate the principal features obtained from the theory of strong electron correlations. We think that before discarding the theory based on electron correlations we must show clearly that the stoichiometric La_2CuO_4 or $YBa_2Cu_3O_6$ compounds are not insulating in the paramagnetic phase; actually, they seem to be paramagnetic insulators with well-defined magnetic moments (that is, with the Curie–Weiss law for the magnetic susceptibility obeyed), which supports strongly the view that they are Mott insulators. In this respect, the situation in heavy fermion systems is rather clear since the recent theoretical results [38–41] based on theory of strongly correlated and hybridized states provide a reasonable rationalization of most of the properties of their normal state. The mechanism of pairing in superconducting heavy fermion systems has not yet been determined fully; but in view of the circumstances that some of the superconductors (for example, UPt_3) are antiferromagnetic and exhibit pronounced spin fluctuations in the normal state, the spin-fluctuation mechanism in its version outlined in Section IV,A,2 is a strong candidate. In the next year or so, one should be able to see a clarification of this problem.

Let us end with a methodological remark concerning the analogy of the studies of magnetism and superconductivity. In 1928, Heisenberg introduced the exchange interaction $J_{ij}\mathbf{S}_i \cdot \mathbf{S}_j$ between the magnetic atoms with spins $\{\mathbf{S}_i\}$. The ferromagnetic state was understood in terms of a molecular field $\mathbf{H}_i \sim \langle \mathbf{S}_i \rangle$ which was related to the direct exchange integral J_{ij}. Later, various other exchange interactions have been introduced, such as superexchange, double exchange, RKKY interaction, the Bloembergen–Rowland interaction, Hund's-rule exchange, and kinetic exchange, to explain magnetism in specific systems, such as oxides, rare-earth metals, and transition metals. However, all these new theories provided a description in terms of a single-order parameter—the magnetization $\langle \mathbf{S}_i \rangle$; the particular feature of the electron states in each case (localized states, itinerant states, or a mixture of the two states) is contained only in the way of defining this order parameter or the exchange integral. By analogy, the BCS theory provided a concept of a superconducting order parameter ($\Delta_{\mathbf{k}}$), which is universal for all theories of singlet superconductivity. New mechanisms of pairing should provide a novel interpretation to the coupling constant $V_{\mathbf{kk'}}$ as well as supply some details concerning the specific features of the system under consideration: the gap anisotropy, the role of hybridization, etc. It remains to be seen if some qualitative differences arise if superconductivity should occur as a result of Bose condensation of the preexisting pairs. This question is particularly important in the case when coherence length is small, as in the high-T_c systems.

In the coming years, one should see detailed calculations within the exchange mechanism and comparisons with the experiment concerning the complete phase diagram, as well as the thermodynamic and electromagnetic properties of the new superconductors $La_{2-x}Sr_xCuO_4$ and $YBa_2Cu_3O_{7-\delta}$. It would not be surprising if the final answer for these systems came from a detailed analysis of the model outlined in Section IV,A. The system $Ba_{1-x}K_xBiO_3$ will most probably be described satisfactorily within the standard phonon mediated mechanism. On the other hand, it is too early to say anything definite about Bi and Tl compounds with $T_c > 100$ K, though the suggested influence of the electronic structure near ε_F by the CuO_2 planes seems to be indicating a nontrivial role of the exchange interactions also in those systems when coupled with interlayer pair tunneling. One of the missing links between the properties of the last two classes of compounds with those of $La_{2-x}Sr_xCuO_4$ is the conspicuous lack of evidence for antiferromagnetism in the $Bi_2S4_2CaCu_2O_8$ and the $Tl_2Ca_2Ba_2Cu_3O_{10-y}$ compounds.

ACKNOWLEDGMENTS

I am grateful to J. M. Honig for his critical reading of the manuscript. The financial support of Purdue University through the office of the Dean K. L. Kliewer of the School of Science is acknowledged.

This work was also partially supported by NSF grant DMR 86-16533 and by the Indiana Center for Innovative Superconductor Technology.

REFERENCES

1. J. Bardeen, L. N. Cooper, and J. R. Schrieffer (1957). *Phys. Rev.* **106**, 162; **108**, 1175.

2. J. R. Schrieffer (1964). "Theory of Superconductivity." W. A. Benjamin, Reading, Massachusetts.

3. P. G. De Gennes (1966). "Superconductivity of Metals and Alloys." W. A. Benjamin, Reading, Massachusetts.

4. M. Tinkham (1965). "Superconductivity." Gordon & Breach, New York.

5. G. Rickayzen (1965). "Theory of Superconductivity." Wiley (Interscience), New York.

6. J. M. Blatt (1964). "Theory of Superconductivity." Academic Press, New York.

7. N. N. Bogoliubov, V. V. Tolmachov, and D. V. Sirkov (1958). *A new method in the theory of superconductivity. Akad. Nauk SSSR, Moscow.* Translated and published *in* "The Theory of Superconductivity" (N. N. Bogoliubov, ed.). Gordon & Breach, New York, 1968.

8. L. D. Landau and E. M. Lifshitz (1980). "Statistical Physics," Part 2, Chapter 5, 2nd ed. Pergamon, Oxford; A. A. Abrikosov, L. P. Gorkov, and I. E. Dzyaloshinski (1963). "Methods of Quantum Field Theory in Statistical Physics," Chapter 7. Dover, New York.

9. R. D. Parks, ed. (1969). "Superconductivity" (in two volumes). Dekker, New York.

10. C. G. Kuper (1968). "An Introduction to the Theory of Superconductivity." Oxford Univ. Press (Clarendon), Oxford and New York; A. C. Rose-Innes and E. H. Rhoderick (1969). "Introduction to Superconductivity." Pergamon, Oxford.

11. H. Fröhlich (1952). *Proc. R. Soc. London A* **215**, 291.

12. L. N. Cooper (1956). *Phys. Rev.* **104**, 1189.

13. N. N. Bogoliubov (1958). *Nuovo Cimento* **7**, 794; J. G. Valatin (1958). *Nuovo Cimento* **7**, 843.

14. P. Morel and P. W. Anderson (1962). *Phys. Rev.* **125**, 1263. See also D. J. Scalapino, in Chapter 10 of Ref. 9.

15. G. M. Eliashberg (1966). *Zh. Eksp. Teor. Fiz* **38**, 966 [*Sov. Phys.-JETP* **11**, 696 (1960)]. For review see P. B. Allen and B. Mitrovic (1982). *In* "Solid State Physics" (H. Ehrenreich, F. Seitz, and D. Turnbull, eds.) pp. 2–92. Academic Press, New York.

16. A. B. Migdal (1958). *Zh. Eksp. Teor. Fiz* **34**, 1438 [*Sov. Phys.-JETP* **7**, 996 (1958)].

17. W. L. McMillan (1968). *Phys. Rev.* **167**, 331.

18. F. S. Khan and P. B. Allen (1980). *Solid State Commun.* **36**, 481.

19. M. L. Cohen and P. W. Anderson (1972). *AIP Conf. Proc.* **4**, 17.

19a. E. Maxwell (1950). *Phys. Rev.* **78**, 477; C. A. Reyn-
olds, B. Serin, W. H. Wright, and L. B. Nesbitt (1950). *Phys. Rev.* **78**, 487.

20. A. H. Wilson (1958). "The Theory of Metals." Cambridge Univ. Press, London and New York.

21. For review see G. Baym and C. Pethick (1978). *In* "The Physics of Liquid and Solid Helium" (K. H. Bennemann and J. B. Ketterson, eds.), Chapter 3. Wiley, New York.

22. N. F. Mott (1974). "The Metal-Insulator Transitions." Taylor and Francis, London.

23. D. Vaknin *et al.* (1987). *Phys. Rev. Lett.* **58**, 2802; Y. Endoh *et al.* (1988). *Phys. Rev.* **B37**, 7443.

24. J. Hubbard (1964). *Proc. R. Soc. London A* **281**, 401.

25. M. Acquarone, D. K. Ray, and J. Spalek (1982). *J. Phys.* **C15**, 959.

26. H. Kuwamoto, J. M. Honig, and J. Appel (1980). *Phys. Rev.* **B22**, 2626.

27. J. Spalek, A. M. Oleś, and J. M. Honig (1983). *Phys. Rev.* **B28**, 6802.

28. W. F. Brinkman and T. M. Rice (1970). *Phys. Rev.* **B2**, 4302.

29. J. Spalek, A. Datta, and J. M. Honig (1987). *Phys. Rev. Lett.* **59**, 728. J. Spalek, M. Kokowski, and J. M. Honig (1989). *Phys. Rev. B* **39**, 4175.

30. For the Mott insulators, the kinetic exchange interaction has been introduced in P. W. Anderson (1959). *Phys. Rev.* **115**, 2. This result has been extended to the case of a strongly correlated metal in K. A. Chao, J. Spalek, and A. M. Oleś (1977). *J. Phys.* **C10**, L271; cf. also J. Spalek and A. M. Oleś (1976), Jagellonian University preprint SSPJU-6/76, October.

31. P. B. Vischer (1974). *Phys. Rev.* **B10**, 943; J. Spalek, A. M. Oleś, and K. A. Chao (1981). *Phys. Stat. Sol.(b)* **108**, 329; Y. Nagaoka (1966). *Phys. Rev.* **147**, 392.

32. M. Héritier and P. Lederer (1977). *J. Phys.* **38**, L-209.

33. J. Spalek and W. Wójcik (1988). *Phys. Rev.* **B37**, 1532.

34. S. A. Kivelson, D. S. Rokhsar, and J. P. Sethna (1987). *Phys. Rev.* **B35**, 8865; P. W. Anderson and Z. Zou (1988). *Phys. Rev. Lett.* **60**, 132.

35. P. W. Anderson, Cargese 1988, Lecture Notes, to be published; also in *Proc. Int. School Phys. Enrico Fermi—1987:* "Frontiers and Borderlines in Many-Particle Physics," in press.

36. For a recent review see H. Fukuyama, Y. Hasegawa, and Y. Suzumura (1988). *Physica* **C153–155**, 1630.

37. P. W. Anderson *et al.* (1988). *Physica* **C153–155**, 527.

38. P. A. Lee, T. M. Rice, J. W. Serene, L. J. Sham, and J. W. Wilkins (1986). *Comments Condens. Matter. Phys.* **12**, 99.

39. T. M. Rice (1989). In *Proc. Int. School Phys. Enrico Fermi—1987:* "Frontiers and Borderlines in Many-Particle Physics," in press.

40. D. M. Newns and N. Read (1987). *Adv. Phys.* **36**, 799.

41. P. Fulde, J. Keller, and G. Zwicknagl (1988). *In* "Solid State Physics" (H. Ehrenreich and D. Turn-

bull, eds.), Vol. 41. Academic Press, San Diego, California.

42. For a review of experimental properties of heavy fermions see G. R. Stewart (1984). *Rev. Mod. Phys.* **56**, 755; Z. Fisk *et al.* 1986, *Nature (London)* **320**, 124; F. Steglich (1955) *in* "Theory of Heavy Fermions and Valence Fluctuations" (T. Kasuy and T. Saso, eds.), p. 23ff, Springer-Verlag, Berlin and New York.

43. P. Wróbel and L. Jacak (1988). *Mod. Phys. Lett.* **B2**, 511.

44. P. W. Anderson (1987). *Science* **235**, 1196.

45. G. Shirane *et al.* (1987). *Phys. Rev. Lett* **59**, 1613.

46. S. Chakravarty *et al.* (1988). *Phys. Rev. Lett.* **60**, 1057.

47. M. Nucker *et al.* (1987). *Z. Phys.* **B67**, 9; A. Fujimori *et al.* (1987). *Phys. Rev.* **B35**, 8814; P. Steiner *et al.* (1988). *Z. Phys.* **B69**, 449.

48. L. F. Mattheis (1987). *Phys. Rev. Lett.* **58**, 1028; J. Yu, A. J. Freeman, and J.-H. Xu (1987). *Phys. Rev. Lett.* **58**, 1035; B. Szpunar and V. H. Smith, Jr. (1988). *Phys. Rev.* **B37**, 2338. For review see K. C. Hass (1989). *In* "Solid State Physics" (H. Ehrenreich and D. Turnbull, eds.), Vol. 42, pp. 213–270 Academic Press, San Diego, California.

49. P. Fulde (1988). *Physica* **153–155**, 1769; T. Zaanen *et al.* (1988). *Physica* **153–155**, 1636.

50. F. C. Zhang and T. M. Rice (1988). *Phys. Rev.* **B37**, 3759.

51. M. R. Schafroth (1955). *Phys. Rev.* **100**, 463.

52. For a review see A. Leggett (1975) *Rev. Mod. Phys.* **47**, 331.

53. M. W. Shafer, T. Penney, and B. L. Olson (1987). *Phys. Rev.* **B36**, 4047.

54. J. Spalek (1988). *Phys. Rev.* **B37**, 533; M. Acquarone (1988), *Solid State Commun.* **66**, 937.

55. G. Baskaran, Z. Zou, and P. W. Anderson (1987). *Solid State Commun.* **63**, 973.

56. M. Cyrot (1987). *Solid State Commun.* **62**, 821.

57. A. E. Ruckenstein, P. J. Hirschfeld, and J. Appel (1987). *Phys. Rev.* **B36**, 857.

58. G. Kotliar (1988). *Phys. Rev.* **B37**, 3664.

59. Y. Isawa, S. Maekawa, and H. Ebisawa (1987). *Physica* **148B**, 391.

60. Z. Zou and P. W. Anderson (1988). *Phys. Rev.* **B37**, 627.

61. M. Inui, S. Doniach, P. J. Hirschfeld, and A. E. Ruckenstein (1988). *Phys. Rev.* **B37**, 2320.

62. Y. Suzumura, Y. Hasegawa, and H. Fukuyama (1988). *J. Phys. Soc. Jpn.* **57**, 2768.

63. G. Kotliar and J. Liu (1988). *Phys. Rev.* **B38**, 5142.

64. J. Spalek, *Phys. Rev.* **B40**, in press.

65. J. M. Wheatley, T. C. Hsu, and P. W. Anderson (1988). *Phys. Rev.* **B37**, 5897.

66. For a recent discussion see F. Mila (1988). *Phys. Rev.* **B38**, 11358, and references therein.

67. J. Spalek (1988). *Phys. Rev.* **B38**, 208; J. Spalek (1988). *J. Solid State Chem.* **76**, 224.

68. J. R. Schrieffer and P. A. Wolff (1966). *Phys. Rev.* **149**, 491.

69. P. W. Anderson (1984). *Phys. Rev.* **B30**, 1549; K. Miyake, S. Schmitt-Rink, and C. M. Varma (1986). *Phys. Rev.* **B34**, 6554; D. J. Scalapino, E. Loho, and J. E. Hirsch (1986). *Phys. Rev.* **B34**, 8190; M. R. Norman (1987). *Phys. Rev. Lett.* **59**, 232; M. R. Norman (1988). *Phys. Rev.* **B37**, 4987.

70. D. M. Newns (1987). *Phys. Rev.* **B36**, 5595; D. M. Newns (1988). *Phys. Scripta* **T23**, 113.

71. J. B. Torrance *et al.* (1988). *Phys. Rev. Lett.* **61**, 1127.

72. T. M. Rice and K. Ueda (1985). *Phys. Rev. Lett.* **55**, 995; T. M. Rice and K. Ueda (1986). *Phys. Rev.* **B34**, 6420.

73. K. Miyake, T. Matsuura, K. Sano, and Y. Nagaoka (1988). *J. Phys. Soc. Jpn.* **57**, 722.

74. P. W. Anderson (1959). *Phys. Rev.* **115**, 2. Also *in* "Solid State Physics" (F. Seitz and D. Turnbull, eds.), Vol. 14, pp. 99–213, 1963. Academic Press, New York.

75. For review see, for example, S. V. Vonsovskii (1974), "Magnetism," Wiley, New York.

76. J. Zaanen, G. A. Sawatzky, and J. W. Allen (1985). *Phys. Rev. Lett.* **55**, 418; J. Zaanen, G. A. Sawatzky, and J. W. Allen (1986). *J. Magn. Mat.* **54–57**, 607.

77a. A. Aharony *et al.* (1988). *Phys. Rev. Lett.* **60**, 1330.

77b. J. M. Tranquada *et al.* (1988). *Phys. Rev. Lett.* **60**, 156.

78. J. Spalek, to be published.

79. The p–p pairing has been discussed first in V. J. Emery (1987), *Phys. Rev. Lett.* **58**, 2794, and also in V. J. Emery and G. Reiter (1988), *Phys. Rev.* **B38**, 4547.

80. The coexistence of SDW and SC states has also been discussed within BCS theory: W. Baltensperger and S. Strassler (1963), *Phys. Kondens. Mater.* **1**, 20; M. J. Nass *et al.* (1981), *Phys. Rev. Lett.* **46**, 614; A. W. Overhauser and L. Daemen (1988), *Phys. Rev. Lett.* **61**, 1885. The corresponding problem for exchange-mediated superconductivity has been outlined in R. H. Parmenter (1987). *Phys. Rev. Lett.* **59**, 923.

81. A. Weidinger *et al.* (1989). *Phys. Rev. Lett.* **62**, 102; J. H. Brewer *et al.* (1988). *Phys. Rev. Lett.* **60**, 1073.

82. J. G. Bednorz and K. A. Muller (1986). *Z. Phys.* **B64**, 189; C. W. Chu *et al.* (1987). *Phys. Rev. Lett.* **58**, 405; S. Uchida *et al.* (1987). *Jpn. J. Appl. Phys.* **26**, L1; M. K. Wu *et al.* (1987). *Phys. Rev. Lett.* **58**, 908.

83. J. Labbé and J. Bok (1987). *Europhys. Lett.* **3**, 1225.

84. P. Prelovšek, T. M. Rice, and F. C. Zhang (1987). *J. Phys.* **C20**, L229.

85. J. D. Jorgensen *et al.* (1988). *Phys. Rev. Lett.* **58**, 1024.

86. W. Weber (1987). *Phys. Rev. Lett.* **58**, 1371; S. Barisic, J. Batistic, and J. Friedel (1987). *Europhys. Lett.* **3**, 1231.

87. For a recent review of phonon- and bipolaron-mediated superconductivity see, respectively, A. Oguri (1988), *J. Phys. Soc. Jpn.* **57**, 2133; and L. J. de Jongh (1989), in *Proc. 1st Int. Symp. on Super-*

conductivity, Nagoya—1988, to be published by Springer-Verlag, New York.

88. A. Alexandrov and J. Ranninger (1981). *Phys. Rev.* **B24**, 1164; A. Alexandrov, J. Ranninger, and S. Robaszkiewicz (1986). *Phys. Rev.* **B33**, 4526.

89. B. Batlogg *et al.* (1987). *Phys. Rev. Lett.* **59**, 912; T. A. Faltens *et al.* (1987). *Phys. Rev. Lett.* **59**, 915.

90. K. J. Leary *et al.* (1987). *Phys. Rev. Lett.* **59**, 1236.

91. L. F. Mattheis, E. M. Gyorgy, and D. W. Johnson, Jr. (1988). *Phys. Rev.* **B37**, 3745; R. J. Cava *et al.* (1988). *Nature (London)* **332**, 814; D. G. Hinks *et al.* (1988). *Nature (London)* **333**, 6176.

92. D. G. Hinks *et al.* (1988). *Nature (London)* **335**, 419.

93. S. Pei *et al.* (1989). *Phys. Rev.* **B39**, 811.

94. T. M. Rice (1988). *Nature (London)* **332**, 780.

95. B. Dabrowski, personal communication.

96. C. Varma (1988). *Phys. Rev. Lett.* **61**, 2713.

97. A. W. Sleight, J. J. Gillson, and P. E. Bierstedt (1975). *Solid State Commun.* **17**, 27.

98. For recent critical estimates of isotope shifts of the T_c value, see P. B. Alden (1988), *Nature (London)* **335**, 396.

99. B. K. Chakraverty (1979). *J. Physique Lett.* **40**, L99; A. S. Alexandrov and J. Ranninger (1981). *Phys. Rev.* **B23**, 1796.

100. T. M. Rice (1988). *Nature (London)* **332**, 780.

101. See, for example, L. J. de Jongh (1988). *Proc. NATO Adv. Res. Workshop*, to appear in *J. Chim. Phys.*

102. W. A. Little (1964). *Phys. Rev.* **134A**, 1416; V. L. Ginzburg (1970). *Sov. Phys.-Uspekhi* **13**, 335.

103. C. M. Varma, S. Schmitt-Rink, and E. Abrahams (1987). "Proceedings of the Conference on Novel Mechanisms of Superconductivity" (S. A. Wolff and V. Z. Kresin, eds.), p. 355. Plenum Press, New York; also in *Solid State Commun.* **62**, 681.

104. W. Weber (1988) *Z. Phys.* **B70**, 323.

105. D. J. Scalpino, E. Loh, Jr., and J. E. Hirsch (1986). *Phys. Rev.* **B34**, 8190; D. J. Scalpino, E. Loh, Jr., and J. E. Hirsch (1987). *Phys. Rev.* **B35**, 6694; J. R. Schrieffer, X. G. Wen, and S. C. Zhang (1988). *Phys. Rev. Lett.* **60**, 944.

106. For a recent critical review see W. A. Little (1988). *Science* **242**, 1390.

107. A. T. Fiory, A. F. Hebard, P. M. Mankiewich, and R. E. Howard (1988). *Phys. Rev. Lett.* **61**, 1419.

108. J. Niemeyer, M. R. Dietrich, and C. Politis (1987). *Z. Phys.* **B67**, 155.

109. L. C. Smedskjaer, J. Z. Liu, R. Benedek, D. G. Legnini, D. J. Lam, M. D. Stahulak, and A. Bansil (1988). *Physica* **C156**, 269.

ULTRASONICS

John Szilard   *Loughborough University*

I. Acoustic Waves and Their Properties
II. Generation and Reception of Ultrasound Waves
III. Applications

GLOSSARY

Angular frequency: $\omega = 2\pi f$, where f is frequency.

Particle acceleration: Acceleration of a particle in the wave at time t: $b = -B \sin \omega t = -\omega^2 A \sin \omega t$, where B is the acceleration amplitude.

Particle displacement: Distance to which a particle in the wave has moved out from its equilibrium position at time t: $a = A \sin \omega t$, where A is the displacement amplitude.

Particle velocity: Velocity at which a particle in the wave is moving at time t: $u = U \cos \omega t = \omega A \cos \omega t$, where U is the velocity amplitude.

Radiation pressure: Static pressure exerted by propagating acoustic waves on any reflector or absorber or on a lossy medium itself. On a perfect absorber, radiation pressure = intensity/sound velocity.

Sound intensity: $I = \frac{1}{2}(p^2/\rho c) = \frac{1}{2}\rho c U^2 = \frac{1}{2}\rho c \omega^2 A^2$.

Sound pressure: Pressure in the wave, $p = P \cos \omega t = \rho c U \cos \omega t = \rho c \omega A \sin \omega t$, where P is the pressure amplitude.

Wave number: $2\pi/\lambda$, usual symbol k.

Wavelength: Distance between two nearest points in the wave of equal phase; usual symbol λ.

The term ultrasonics refers to the science and technology dealing with acoustic waves (elastic waves or stress waves, i.e., mechanical waves) the frequency of which is higher than the nominal limit of audibility by the human ear. It must be called nominal limit, since it is not definable in exact terms, only on some statistical basis, because it also depends on sex and age, and it varies considerably from person to person. So as a convention, 20 kHz is usually taken as the lower frequency limit of ultrasound waves.

It should be noted that until the 1940s, occasionally even during the early 1950s, acoustic waves of frequencies above audibility were also called supersonic or hypersonic waves. The term supersonic about that time became reserved for speed exceeding the velocity of sound, while the term hypersonic was simply dropped. However, the latter term is making a comeback in recent years, but with a different meaning. As the use of higher and higher frequencies is becoming feasible, some people are beginning to call acoustic waves from around 1 GHz hypersonic.

In industry definitions are often not adhered to as strictly as in science. Thus, when high-power acoustic energy is used, sometimes more advantageously at frequencies in the audio range, (i.e., below 20 kHz), wrongly they still call it ultrasonics, although simply sonics would be the correct term to use.

I. Acoustic Waves and Their Properties

Wave motion usually means the propagation of a disturbance through a medium in the process of which energy is transmitted, but at the end the medium is in the same place as it would be without the passage of the disturbance. Acoustic waves mean that this disturbance is of an elastic nature; that is, this disturbance consists of some elastic deformation of the medium. According to Hooke's law, in an elastic deformation the restoring force is proportional to the deformation itself.

A. UNDAMPED HARMONIC OSCILLATION

It can be shown that an undamped harmonic oscillation is characterised by

$$a = A \sin \omega t \tag{1}$$

where A is the displacement amplitude. By differentiating the displacement, the velocity is obtained as

$$u = \omega A \cos \omega t \tag{2}$$

where $A\omega$ is the velocity amplitude, while the acceleration is the derivation of Eq. (2):

$$b = -\omega^2 A \sin \omega t \tag{3}$$

$A\omega^2$ being the acceleration amplitude. In Eq. (3) the negative sign indicates that the acceleration always points in the direction opposite to the displacement.

B. FORCED OSCILLATIONS

Consider applying a periodic force $F = F_0 \sin \omega t$ to the oscillating system of the previous case. Now it experiences forced oscillations of frequency $f = \omega/2\pi$, where ω is the angular frequency, and the motion can be described by the equation

$$M \frac{d^2a}{dt^2} + R_m \frac{da}{dt} + \frac{a}{C_m} = F_0 \sin \omega t \tag{4}$$

where M is the oscillating mass, R_m represents the damping in the form of a mechanical resistance proportional to the velocity, and C_m is the compliance of the spring.

Then, under steady-state conditions at a given time, the velocity u, using $j = \sqrt{-1}$, is given by

$$u = \frac{F}{R_m + j\,[\omega M - (1/\omega C_m)]} \tag{5}$$

and the velocity amplitude V is

$$V = \frac{F_0}{\sqrt{R_m^2 + [\omega M - (1/\omega C_m)]^2}} \tag{6}$$

The form of Eqs. (4) through (6) looks identical to those that describe the behavior of an electrical RLC circuit (i.e., a circuit with components resistance R, inductance L, and capacitance C) in response to an alternating voltage V. Thus the mechanical quantities denoted by R_m, M, C_m, and F are seen as analogous to the electrical resistance, inductance, capacitance, and voltage, respectively; likewise a and u are equivalent to the charge Q and current i, respectively.

This analogy can be extended further to the ratio of force and velocity (also changing from lumped to distributed constants), calling it mechanical impedance:

$$Z_m = F/u \tag{7}$$

From Eq. (5) this is seen to be a complex quantity

$$Z_m = R_m + jX_m \tag{8}$$

where $X_m = \omega M - 1/\omega C_m$, which is the mechanical reactance.

C. THE WAVE EQUATION

Consider now a medium consisting of small, elementary masses, or particles coupled together by little springs. If now a group of these particles is made to oscillate, their coupling springs will transmit the oscillation, however, with a time delay, to their neighbors, and so on. Thus a propagating wave motion has been created and sustained as long as vibrational energy is being supplied to the first group of particles.

Assuming wave motion in the form of a plane wave with velocity c along the x direction and ideally without any losses, the particle displacement ξ at any time at distance x along the propagation direction may be given in general terms as

$$\xi = f_1(x - ct) + f_2(x + ct) \tag{9}$$

Here the function $f_1(x - ct)$ represents wave propagation in the positive x direction, while $f_2(x + ct)$ expresses propagation in the opposite direction. Differentiating twice Eq. (9) with respect to x, we get

$$\partial^2\xi/\partial x^2 = f_1''(x - ct) + f_2''(x + ct) \tag{10}$$

and differentiating twice with respect to t,

$$\partial^2\xi/\partial t^2 = c^2[f_1''(x - ct) + f_2''(x + ct)] \tag{11}$$

Equations (10) and (11) can be combined as

$$\partial^2\xi/\partial t^2 = c^2(\partial^2\xi/\partial x^2) \tag{12}$$

which is the one-dimensional wave equation. It can be generalized into three dimensions:

$$\frac{\partial^2\xi}{\partial t^2} = c^2\left(\frac{\partial^2\xi}{\partial x^2} + \frac{\partial^2\xi}{\partial y^2} + \frac{\partial^2\xi}{\partial z^2}\right) \tag{13a}$$

also written as

$$\partial^2\xi/\partial t^2 = c^2 \nabla^2\xi \tag{13b}$$

In general the particle displacement a, being a harmonic oscillatory motion, is

$$a = A \sin[(2\pi/\lambda)(x - ct) + \phi] \tag{14}$$

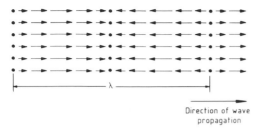

FIG. 1.   Particle motion in a longitudinal wave.

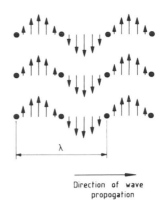

Direction of wave
propogation

FIG. 2.   Particle motion in a transverse wave.

where $2\pi/\lambda = k$, called the wave number, and ϕ denotes the phase of the wave at $t = 0$. If we put $\phi = 180°$, rearranging Eq. (14) and using the angular frequency $\omega = 2\pi f$, Eq. (14) becomes

$$a = A \sin(\omega t - kx) \qquad (15)$$

and it is often more conveniently written in the exponential form as

$$a = A \exp j(\omega t - kx) \qquad (16)$$

D. Modes of Wave Propagation

Ultrasonic waves may propagate in different modes as follows. In an infinite medium (i.e., in a medium whose boundaries are so far away from the site of the wave motion that they have no effect on the waves), if the particles may move backward and forward in line with the axis of propagation (Fig. 1), the density ρ and pressure p of the medium will fluctuate periodically around the ambient value; therefore longitudinal waves are also called pressure waves.

In a transverse wave (Fig. 2) instead of compressional deformation the medium suffers periodical shear deformation and shear stresses; therefore this wave type is also called a shear wave. While pressure waves can propagate in any elastic/compressible medium (solid, fluid, or gas), shear waves can travel only in a medium that has shear elasticity (i.e., in a solid or in a very limited way in some special viscoelastic liquid).

A combination of the preceding two basic wave modes may propagate at or near a boundary of two dissimilar media. Figure 3 shows a surface wave on a solid body, also called a Rayleigh wave. These waves are similar to, but not identical with, waves on water.

In Fig. 4 symmetrical and asymmetrical or antisymmetrical plate waves, also called Lamb waves, are shown.

In rods waves can propagate as bulges and contractions (similar to the symmetrical plate waves, but with a complete axial symmetry), called rod waves, and as flexural waves. Finally, torsional waves (pure shear) are also possible in a rod.

E. Velocity of Different Ultrasound Wave Types

Propagation velocity of the different modes is determined by the density (distributed mass) and material constants of the medium concerned describing its elastic behavior. Formulas for the various more important conditions are given subsequently.

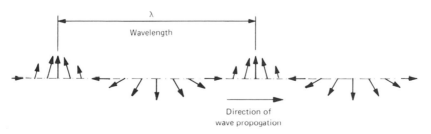

FIG. 3.   Particle motion in a surface wave.

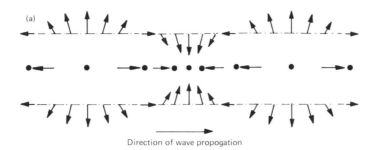

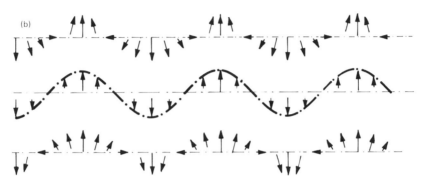

FIG. 4.   Plate waves: (a) symmetrical, (b) asymmetrical or antisymmetrical.

In gases the sound velocity (only pressure waves can travel) is

$$c = \sqrt{\gamma P_0 / \rho_0} \qquad (17)$$

where ρ_0 is the density, γ the specific heat ratio, and P_0 the ambient pressure.

In fluids (again pressure waves only),

$$c = \sqrt{\kappa / \rho} \qquad (18)$$

where κ is the bulk stiffness modulus, which is the reciprocal of the bulk (adiabatic) compressibility. Here the fluctuation of the density is so small that generally it is negligible. This is why the index O has been omitted. Conditions where these fluctuations are not negligible will be discussed later. (See Section I, H.)

The elastic behavior of solids can be described in different ways, using different sets of material constants that determine the velocity of the different modes of propagation. The most commonly used formulas contain the modulus of elasticity E (Young's modulus) and the Poisson's ratio ν. With these

$$c_{\text{pressure}} = \sqrt{\frac{E}{\rho} \cdot \frac{1 - \nu}{(1 + \nu)(1 - 2\nu)}} \qquad (19)$$

and

$$c_{\text{shear}} = \sqrt{\frac{E}{\rho} \cdot \frac{1}{2(1 + \nu)}} = \sqrt{\frac{G}{\nu}} \qquad (20)$$

where G is the modulus of rigidity, also called the shear modulus or torsional modulus. Since for many materials ν is around $\frac{1}{3}$ (for steel it is 0.28 and for aluminium 0.34), one can take as a first approximation as

$$c_{\text{shear}} / c_{\text{pressure}} = 0.5$$

(For steel it is 0.55, for aluminium 0.49.)

For Rayleigh waves

$$c_{\text{surface}} = \frac{0.87 + 1.12\nu}{1 - \nu} \sqrt{\frac{E}{\rho} \cdot \frac{1}{(1 + \nu)}}$$

$$\simeq 0.9 \sqrt{\frac{G}{\rho}} \qquad (21)$$

In a plate or rod the mode and the velocity of propagation can also depend on the ratio of the wavelength to thickness (i.e., on the frequency). For Lamb waves, two transcendental equations can be derived and for the velocities a family of curves worked out for various metals to show their dependence on plate thickness and frequency. More on this subject can be found in textbooks, especially in those dealing with non-destructive testing.

In thin rods (i.e., with diameters less than 0.1λ) the rod wave velocity is not dependent on ν,

$$c_{\text{rod}} = \sqrt{E/\rho} \qquad (22)$$

For other modes of propagation, which are of much less practical significance, the reader should again consult the textbooks.

At boundaries of solid media for waves incident at an angle other than perpendicular, the mode of propagation can change from one to another, in most cases partially but under some conditions wholly. How the propagating energy will be split into the new modes depends on the angle of incidence and the properties of the adjacent media. This phenomenon is called mode conversion.

F. ULTRASONIC FIELD PARAMETERS

It should be noted that the particle displacement amplitude A is always negligibly small compared to λ.

Elastic waves consist of periodic elastic deformations and this means periodic compressional or/and sheer stresses, depending on the particular mode of propagation. For sake of simplicity consider in the following only compressional stress or pressure, called sound pressure, but what will be said about pressure will likewise be applicable to shear stress in shear waves.

Returning now to the electrical analogy and to Eqs. (7) and (8) and taking a step further, the ratio of sound pressure and particle velocity is called specific acoustic impedance Z_{sp}:

Z_{sp} = sound pressure/particle velocity

Dividing this further by the cross-sectional area S across which the acoustic wave is propagating, the acoustic impedance Z_{ac} is obtained:

$$Z_{\text{ac}} = Z_{\text{sp}}/S = R + jX \qquad (23)$$

In general this is a complex expression and its real part R is the acoustic resistance, where the dissipation of energy takes place. The reactive term jX is due to the inertia and stiffness of the medium. The purely resistive term R is also called the characteristic impedance of the medium of propagation, and it is

$$R = \rho c \qquad (24)$$

(i.e., the product of the density ρ and sound velocity c).

In practice, for sake of brevity, the word *characteristic* is usually omitted in connection with the impedance of a medium.

Acoustic impedance Z_{ac} is involved when dealing with acoustic radiation from vibrating surfaces or transmission through lumped acoustic elements. However, when acoustic waves travel in an ordinary medium (with distributed parameters), it is the product ρc of the medium that couples sound pressure with particle velocity. Thus,

$$P = \rho c u = \rho c \omega A \qquad (25)$$

where P is the pressure amplitude.

The intensity of acoustic waves can be expressed in terms of the pressure, the velocity, or the displacement amplitude:

$$I = \tfrac{1}{2}(P^2/\rho c) = \tfrac{1}{2}\rho c U^2 = \tfrac{1}{2}\rho c \omega^2 A^2 \qquad (26)$$

In a wave energy is transported, and with it there will be a net flow of momentum, whether the wave is acoustic, electromagnetic, or any other type. The rate of change of momentum through a unit area is equal to a pressure, here called radiation pressure, and it is the product of instantaneous velocity u and the instantaneous mass flow rate ρu and is given as ρu^2. Although the time average of particle velocity is 0, since

$$U \int_0^{2\pi} \cos(\omega t - kx)\, d(\omega t) = 0 \qquad (27)$$

(i.e., there is no net mass flow), the time average of u^2 is $U^2/2$. Thus the radiation pressure Π is

$$\Pi = \rho(U^2/2) = I/c = E_0 \qquad (28)$$

where E_0 is the energy density (newtons per square meter).

When an acoustic wave hits perpendicularly the boundary of the medium it travels in, one part of it will be reflected and the other will be transmitted, but no mode conversion can take place. The reflection and transmission coefficients T and R, respectively, are determined by the ratio of the impedances of the two media, $m = Z_2/Z_1$. These coefficients are different for the pressure and for the intensity.

For pressure,

$$T_{\text{p}} = \frac{2Z_2}{Z_2 + Z_1} = \frac{2m}{m + 1} \qquad (29)$$

$$R_{\text{p}} = \frac{Z_2 - Z_1}{Z_2 + Z_1} = \frac{m - 1}{m + 1} \qquad (30)$$

and for intensity,

$$T_{\text{i}} = \frac{4Z_2 Z_1}{(Z_2 + Z_1)^2} = \frac{4m}{(m + 1)^2} \qquad (31)$$

$$R_i = \left\{\frac{Z_2 - Z_1}{Z_2 + Z_1}\right\}^2 = \left\{\frac{m - 1}{m + 1}\right\}^2 \qquad (32)$$

It should be noted that when acoustic waves hit a medium of lower impedance (i.e., $Z_2 < Z_1$), R_p turns out to be negative. This means that in this case the reflected wave suffers a phase reversal.

Concerning the geometrical phenomena of reflection, refraction, diffraction, interference, Huyghens' principle, and so on, acoustic waves behave in a way similar to electromagnetic waves. However, one special property of the acoustic waves that distinguishes them from electromagnetic waves must be taken into account: the various modes of propagation, including the possibility of mode conversion. This means two things: (1) longitudinal acoustic waves cannot be polarized, and (2) when hitting the boundary of a solid medium, mode conversion may occur (Fig. 5) with a change of velocity, which must be substituted into Snell's formula.

These two things also mean that ultrasonic lenses, mirrors, prisms, and so on can be made and used similarly as in geometrical optics. Materials, of course, must be chosen appropriately and when the dimensions of these things are considered, the significance of the number of wavelengths across the aperture must not be forgotten.

If in a medium a layer of a different impedance is embedded, transmission and reflection will in-

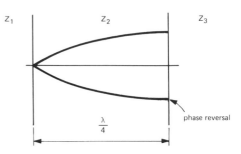

FIG. 6. Quarter wave matching. Here $Z_1 > Z_2 > Z_3$.

volve the phenomena of standing waves, potentially resonance (Fig. 6) and antiresonance.

Transmission between two media of dissimilar impedance can be improved if a material is found the impedance of which is the geometrical mean of that of the two media

$$Z_2 = \sqrt{Z_1 Z_3} \qquad (33)$$

Much greater improvements of transmission is achieved if the intermediate medium is not like a bulk medium, but a layer at best of $\lambda/4$ thickness (Fig. 6). However, it may be an odd multiple of $\lambda/4$. This method is called quarter wave matching.

In accordance with wave motion, in general ultrasonic waves can form a more or less narrow beam in front of their source depending on the ratio D/λ, where D is the diameter of the source.

Close to the source an interference field exists with maxima and minima (Figs. 7 and 8), called near field, or Fresnel zone. It extends from the source to the last interference maximum, and its length N can be expressed as

$$N = (D^2 - \lambda^2)/4\lambda \simeq D^2/4\lambda \qquad (34)$$

In the near field the beam slightly contracts, and then in the far field (Fraunhoffer zone) it is divergent (Fig. 9) and its half-angle θ is given by the theory of diffraction as

$$\sin \theta = 1.22(\lambda/D) \qquad (35)$$

The directivity pattern is better seen in a polar diagram (Fig. 10). Beside the main lobe there are weaker side lobes, and as the main lobe becomes narrower, the side lobes become smaller and more numerous.

G. ATTENUATION

Media in which ultrasonic waves travel are more or less lossy, and so the intensity and pressure decrease with distance. In a plane wave (no

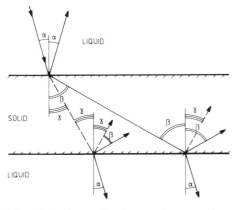

FIG. 5. Reflection, refraction, mode conversion, and its effect on the angle of reflection and refraction, respectively. Solid lines indicate pressure waves and broken lines shear waves. [Reproduced with permission from Szilard, J. (Ed.) (1982). "Ultrasonic Testing: Non-conventional Testing Techniques." Wiley, New York. Copyright © John Wiley & Sons.]

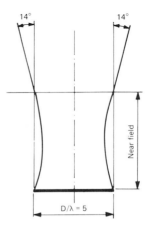

FIG. 9.   Beam divergence in the far field with $D/\lambda = 5$.

geometrical beam spread) the initial amplitude A_0 (either particle displacement or pressure amplitude) will decrease to A_r after traveling a distance r,

$$A_r = A_0 e^{-\alpha r} \tag{36}$$

where α is the amplitude attenuation coefficient. Likewise, since the intensity is proportional to the square of the amplitude,

$$I_r = I_0 e^{-2\alpha r} \tag{37}$$

Taking the natural logarithm of Eq. (36), we can write

$$\alpha = (1/r)\, \ln(A_0/A_r) \tag{38}$$

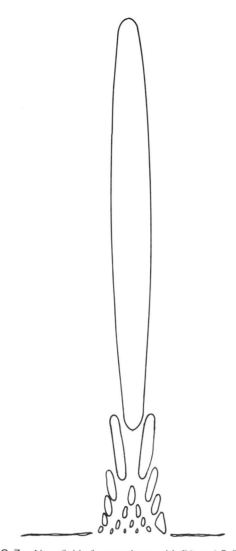

FIG. 7.   Near field of a transducer with $D/\lambda = 6.7$. The closed lines in the beam are of equal intensity, surrounding areas of local maxima. [Reproduced with permission from Szilard, J. (Ed.) (1982). "Ultrasonic Testing: Non-conventional Testing Techniques." Wiley, New York. Copyright © John Wiley & Sons.]

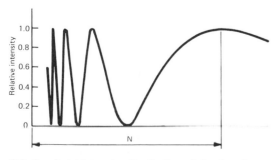

FIG. 8.   Axial intensity distribution of circular piston.

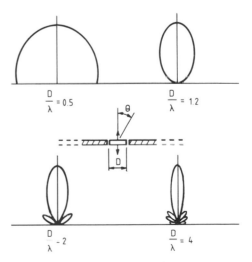

FIG. 10.   Directivity pattern of four transducers. [Reproduced with permission from Hueter and Bolt (1955). "Sonics." Wiley, New York. Copyright © John Wiley & Sons.]

In a homogeneous medium such as a gas, a liquid, an amorphous solid, or a single crystal, the energy lost is simply absorbed and turned into heat. However, in a polycrystalline solid or a basically homogeneous material containing particles of different ρc and having dimensions commensurate with λ, some of the wave energy is lost in the form of scatter. Thus α can be broken down into two components:

$$\alpha = \alpha_{abs} + \alpha_{scat} \qquad (39)$$

In absorption the main roles are played by (1) internal friction (viscosity), (2) elastic hysteresis, (3) heat conduction, and (4) other factors (relaxation, molecular structure, etc.). Owing to the different mechanisms involved in gases and liquids, the absorption normally varies with the square of the frequency, while in solids it is generally a linear function.

The scatter coefficient depends on, among other factors such as the discontinuity of ρc, the wavelength-to-diameter ratio of the inhomogeneities. Three ranges can be distinguished as a function of the average diameter $\bar{D}$ as follows:

1.   Where $\lambda \gg \bar{D}$, $\alpha \propto \bar{D}^3 f^4$, called the Rayleigh scattering range,
2.   Where $\lambda \approx \bar{D}$, $\alpha \propto \bar{D} f^2$, called the random phase (stochastic) scattering range,
3.   Where $\lambda < \bar{D}$, $\alpha \propto 1/\bar{D}$, called the diffusion scattering range.

H. Some Nonlinear Effects

In general in an ultrasound wave the pressure amplitude is so small that in the velocity formulas [Eq. (21), among others] the density ρ and other parameters can be regarded as constant. However, this is an approximation that is not always valid. It is possible to generate ultrasound waves of such intensity that during a cycle the periodic changes of density and similar factors should be taken into account. This means that [e.g., Eq. (18), $c^2 = \kappa_0/\rho_0$] the velocity formula for liquid, being valid up to moderate intensities, will change, with c fluctuating within each cycle:

$$c^2 = \frac{\kappa}{\rho} = \frac{\kappa_0}{\rho_0}(1 + a_1 p + a_2 p^2 + \cdots) \qquad (40)$$

where p is the instantaneous sound pressure,

$$p = P \sin \omega t = \rho c \omega A \sin \omega t \qquad (41)$$

and $a_1, a_2, \cdots$ are constants determined by the properties of the medium. Clearly now at any point in the ultrasonic field, the relationship between the temporal and spatial wave forms of pressure is itself a function of the instantaneous pressure at that point.

At these high intensities the wave equation does not represent a linear system any more, but contains a nonlinear term. As a consequence, a high-intensity wave will be gradually distorted as it travels along. The velocity is pressure dependent and in the higher-pressure crest the wave travels faster and hurries forward, while in the lower-pressure troughs it travels slower and lags behind, gradually changing from a sine wave into a sawtooth shape.

What is happening can be described as the generation of second, third, and higher harmonics and energy is being transferred from the fundamental frequency into the harmonics as the wave propagates. However, absorption being proportional to the square of the frequency, the higher harmonics are attenuated more quickly and the sawtooth wave eventually reverts to a much weaker sine wave of the fundamental frequency.

A further consequence of the nonlinear behavior is that high-intensity sine waves of angular frequencies ω_1 and ω_2 do not simply add together according to the theorem of superposition, but they also interact generating frequencies in the series $k\omega_1 + m\omega_2$. Here k takes all integral values from 0 to ∞ and m from $-\infty$ to $+\infty$, including 0.

The preceding nonlinear effects are exploited mostly in parametric sonar systems, based on the transmission of two interacting high-frequency primary beams. The frequency of the resulting secondary beam is the difference of the two primary beams. This enables the use of low-frequency—thus low-attenuation—beams, which can be made very narrow because the beam width and directivity now depend on the much shorter wavelength, high-frequency primary beam.

II. Generation and Reception of Ultrasound Waves

If airborne ultrasound is wanted and it is generated in a solid electroacoustic transducer, coupling the energy into aid or any other gas is a major problem because their respective impedance ratio is on the order of 10^{-5}. This is why mostly in the low-frequency range up to a few tens of kilohertz it can be cheaper to make use of some nonreversible processes than the reversible ones. For example, to generate airborne ultrasound at a small scale, whistles of some sort are often used, while for larger installations the use of ultrasonic sirens is very efficient. Sirens in the 20–30-kHz range can work at around 30 to 20% efficiency, while the whistles, although being very cheap, can reach only a few percent.

For treating large quantities of liquids, a few special whistles have been developed, although they have become to some extent obsolete.

There are some special cases (e.g., where extremely wide frequency band, high power, or very high frequencies are needed) when short ultrasound pulses are generated by applying a thermal shock to the surface of a medium (solid or even liquid) by very high power laser pulses in a noncontact way and the resultant rapid thermal expansion creates a short ultrasonic pulse. However, this method has its own limitations; in addition, it is very expensive. It is used only in some laboratory experiments.

The use of reversible processes is far more widespread, and naturally, they are also suitable for receiving ultrasound. They are capable of converting electrical energy into acoustical energy and vice versa. They are called electroacoustic transducers or just transducers for short.

Most important of the reversible processes were first the magnetostrictive effect and then the piezoelectric effect.

When ferromagnetic materials (metals, alloys, or ferrites) are subjected to a magnetic field, they change their dimensions, mostly their length. Some contract, others expand. Conversely, when subjected to tensile or compressional stress, their permeability will change (magnetoelastic effect). In some the effect is so pronounced that it can be exploited to make transducers. Figure 11 shows this effect in some metals and alloys. The data are very approximate because the effect depends not only on the exact composition, but also on the processes the metal went through, such work hardening and annealing. It can be seen, for example, that nickel and permendur behave in the opposite way and both are very suitable for making transducers. Interestingly, in iron and cast cobalt the effect changes sign.

Magnetostrictive transducers are normally made out of thin sheets of nickel or a suitable alloy and wound like a transformer (Fig. 12). The length determines the resonant frequency of the transducer, as

$$l = \lambda/2$$

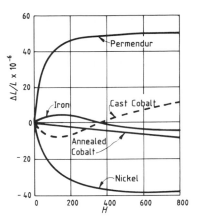

FIG. 11. Magnetostrictive properties of some metals and alloys. [Reproduced with permission from Fredderick, J. R. (1965). "Ultrasonic Engineering." Copyright © John Wiley & Sons.]

where λ is the wavelength in the transducer material.

Magnetostrictive transducers need a biasing magnetic field when transmitting to avoid frequency doubling, and when receiving to produce a signal in the coil as the permeability changes. This biasing field may be produced by a dc coil or by a small permanent magnet inserted into the magnetic circuit.

Magnetostrictive transducers are usable up to around 50 or 60 kHz with increasing losses (eddy currents and hysteresis), and even in the 20–30-kHz range are only moderately efficient. They are expensive but extremely rugged, and they can handle high power. By now they have retreated into the background. They are largely superceded by the more efficient and cheaper piezoelectric sandwich transducers.

Some crystals such as quartz have a lattice structure such that a plate may be cut out of the crystal with certain orientation with respect to

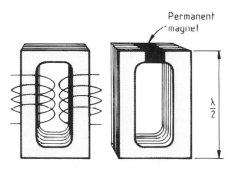

FIG. 12. Simple magnetostrictive transducers.

the crystallographic axes, which shows the piezoelectric effect and its inverse. This means that when subjected to an electric field in the right direction, it will change its dimensions. Conversely, when external forces cause a similar deformation, electrical charges appear on its opposite surfaces. The deformation mentioned may be contraction/expansion or shear deformation, depending on the orientation of the plate and the electric field.

Transducers are normally used in the thickness mode [i.e., when their thickness is equal to $\lambda/2$ in the transducer material (thickness resonance)]. Normally silver electrodes are deposited on the appropriate surfaces.

Natural piezoelectric crystals were known for a long time. These were followed by the development of polycrystalline piezoelectric ceramics, more recently even some piezoelectric plastics. The artificial piezoelectric materials contain small piezoelectric domains (like the magnetic domains in ferromagnetic materials), which are originally of random orientation. Before they can be used, they have to be aligned by a strong electric field (about 25–30 kV/cm), and then they stay aligned for many years. This process is called poling or polarizing. If the poling is done at right angles to the field applied in use, the transducer will vibrate in the shear mode.

A similarity between magnetostrictive and piezoelectric materials is that there is a temperature specific to each particular material at which its crystal structure changes to a different one that does not have the same properties. In both groups this is called the Curie temperature or Curie point. Above the Curie temperature they are not magnetostrictive or piezoelectric, respectively, but when cooled down below it, will regain their original properties. However, ceramics and plastics have to be poled again.

The electrical/mechanical properties of piezoelectric materials are described by a number of constants, but here only the most important ones will be mentioned: sound velocity c (different along different crystallographic axes); density ρ; dielectric constant ε; elastic modulus E; electromechanical coupling coefficient k, such that

$$k^2 = \frac{\text{mechanical energy converted into electrical charge}}{\text{mechanical energy put into transducer}}$$

and also

$$k^2 = \frac{\text{electrical energy converted into mechanical energy}}{\text{electrical energy put into transducer}}$$

Piezoelectric constant d, such that

$$d = \frac{\text{charge (in coulombs)}}{\text{force (in newtons)}}$$

and also

$$d = \frac{\text{deformation (in meters)}}{\text{applied voltage (in volts)}}$$

Piezoelectric constant g, such that

$$g = \frac{\text{field (in volts per meter)}}{\text{applied stress (in newtons per square meter)}}$$

The interrelation of these constants is

$$g = d\varepsilon\varepsilon_0 \tag{42}$$

where ε_0 is the permittivity of vacuum and

$$k = gdE \tag{43}$$

Piezoelectric ceramics most frequently used nowadays belong to the lead zirconate–titanate (abbreviated as PZT) group, especially PZT5 or PZT5A and PZT4. Some manufacturers use various trade names for their products.

Properties of some natural and artificial transducer materials are shown in Table I.

For very high frequencies beginning between 50 and 100 MHz, making good transducers raises new problems, for example, the piezoelectric plates become very thin and fragile. One possibility is using the transducer at high harmonic frequency; another is finding other materials and manufacturing processes. For example, zinc oxide and cadmium sulfide are naturally piezoelectric, and when they are evaporated onto a suitable substrate, their crystallites can be made to grow with such an orientation that the layer may be used as a transducer. In the gigahertz range a quartz rod can act as a transducer when its end is appropriately mounted in a microwave cavity resonator.

For very low frequencies (20–60 kHz) most commonly ceramic disks of a few millimeters thickness are used with a hole in the middle, clamped between two metal cylinders, called sandwich transducers. The resonant frequency is determined mainly by the length of the metal cylinders; the two ceramic disks in the middle play a very small part in it.

For use transducers must be mounted.

To radiate high power a transducer should be chosen with minimum internal losses (i.e., one having a high Q), and it should be mounted in such a way that the amount of vibrational energy lost into the mounting and radiated anywhere else than the intended load is kept to a mini-

TABLE I.   Constants of Various Piezoelectric Materials[a,b]

Physical property	Quartz 0° X-cut	Lithium sulfate 0° Y-cut	Barium titanate type B	Lead zirconate–titanate		Lead meta-niobate	Units
				PZT-4	PZT-5		
Density p	2.65	2.06	5.6	7.6	7.7	5.8	10^3 kg/m^3
Acoustic impedance pc	15.2	11.2	24	30.0	28.0	16	10^6 kg/m^2 sec
Frequency thickness constant ft	2870	2730	2740	2000	1800	1400	kHz mm
Maximum operating temperature	550	75	70–90	250	290	500	°C
Dielectric constant	4.5	10.3	1700	1300	1700	225	—
Electromechanical coupling factor for thickness mode k_{33}	0.1	0.35	0.48	0.64	0.675	0.42	—
Electromechanical coupling factor for radial mode k_p	0.1	—	0.33	0.58	0.60	0.07	—
Elastic quality factor Q	10^6	—	400	500	75	11	—
Piezoelectric modulus for thickness mode d_{33}	2.3	16	149	285	374	85	10^{-12} m/V
Piezoelectric pressure constant g_{33}	58	175	14.0	26.1	24.8	42.5	10^{-3} (V/m)/(N/m^2)
Volume resistivity at 25°C	$>10^{12}$	—	$>10^{11}$	$>10^{12}$	$>10^{13}$	10^9	
Curie temperature	575	—	115	320	365	550	°C
Young's modulus E	8.0	—	11.8	8.15	6.75	2.9	10^{10} N/m^2
Rated dynamic tensile strength	—	—	—	24	27.6	—	10^6 N/m^2

[a] Reproduced with permission from Frederick, J. R. (1965). "Ultrasonic Engineering." Wiley, New York. Copyright © (1965) John Wiley & Sons.

[a] The properties of the ceramic materials can vary with slight changes in composition and processing, and hence the values that are shown should not be taken as exact.

mum. This means that the transducer should be backed by air (Fig. 13) or at best by a material containing as many air-filled pores as possible.

On the other hand, when short pulses are needed (i.e., when a wide frequency response is wanted), the transducer itself should be lossy, having a low Q to reduce ringing. Furthermore, it should be backed by a material with a good enough impedance match and high enough absorption to extract acoustic energy from the transducer after the excitation to reduce its ringing as required. For this purpose mostly epoxy

resins loaded with tungsten powder are used. The transducer, suitably backed, must be enclosed in a protective housing. Then it is called a probe. The probe sometimes also contains a small transformer to match its electrical impedance to that of the instrument with which it is used (Fig. 14).

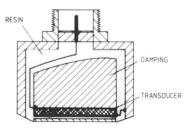

FIG. 14.   Ultrasonic NDT probe: transducer mounted with heavy damping. The transducer is protected from wear by a suitable layer cemented to it. [Reproduced with permission from Szilard, J. (Ed.) (1982). "Ultrasonic Testing: Non-conventional Testing Techniques." Wiley, New York. Copyright © John Wiley & Sons.]

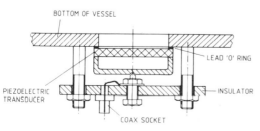

FIG. 13.   Air-backed mounting of transducer for high-power applications.

III. Applications

Applications can be divided into two major groups: passive and active. Passive applications, which can also be called low-power applications, cause no permanent change in the propagation medium but are used to collect information about the presence of defects, obstacles, targets, anatomical structure, and so on about dimensions, boundaries, material properties, flow, and so on (testing, measurements, sonar, medical diagnostics, etc.) or to carry information, such as in underwater navigation and communication, or in delay lines for signal processing.

Active applications, which can also be called high-power applications, include those applications in which ultrasound waves are used as a particular form of energy, active in bringing about some change *in* the medium of propagation or *on* the surface of objects the energy hits, exerting specific, desirable effects. They may improve a technological process or even make a new process possible (e.g., cleaning, machining hard materials, welding dissimilar materials).

A. PASSIVE OR LOW-POWER APPLICATIONS

The following list contains only the more important applications.

1. Nondestructive Testing

Nondestructive testing (NDT) is one of the largest fields of passive applications of ultrasonics. It is eminently suitable for finding internal discontinuities, cracks, laminations, and inclusions because of the following.

1. The impedance of cracks and such is significantly different from that of the parent material and so they strongly reflect ultrasound waves (even extremely thin cracks are very strong reflectors).

2. It is possible to generate short pulses and narrow beams of ultrasound, thereby pinpointing where the echoes come from.

3. The transit time from transmitter to receiver can easily be measured, thereby the distance between the ultrasound source/receiver and the surface producing the echo (either a defect or the far side of the test object) can easily be determined. Thus the location of defects or the thickness of the test object can be found.

By far the most common application of ultrasonics in NDT is in testing steel and other metals, because attenuation in these materials is very low and so high frequencies (in the low-megahertz range) can be used without undue scatter. This in turn means that short wavelength and consequently high resolution is obtainable.

The technique almost exclusively used is based on transmitting very short pulses of mostly 2 to 6 MHz and receiving the echoes. After amplification, for flaw detection the echoes are displayed on an oscilloscope, but for thickness measurement the transit time of the pulse is measured electronically. Using the sound velocity in the metal, the transit time is converted into thickness and is displayed digitally.

A typical application is shown in Fig. 15.

Probes can be coupled to the test object by using a thin layer of oil, grease, or some paste or jelly (contact coupling). Alternatively, the object may be immersed into a water bath and the probe dipped into the water (immersion testing).

It is not necessary to interrogate the test object at normal incidence, and for testing at an angle the probe is mounted on a perspex wedge and coupled to the test object in the usual way. However, at the perspex–metal interface mode conversion takes place and the waves propagate in the shear mode. This is illustrated in Fig. 16, which shows the testing of a weld.

For some testing tasks the signal passing through the object is monitored. Any defect would reduce or completely block the transmission of ultrasound.

Results of the test can be displayed as a simple trace on the oscilloscope screen; this is called A scan. This is the type most frequently

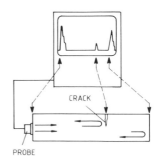

FIG. 15.   Ultrasonic testing at normal incidence (using pressure waves). [Reproduced with permission from Szilard, J. (Ed.) (1982). ''Ultrasonic Testing: Non-conventional Testing Techniques.'' Wiley, New York. Copyright © John Wiley & Sons.]

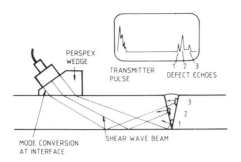

FIG. 16. Weld testing at an angle, using shear waves.

used in NDT, but the echo signals can also be used to build up an image. If the plane of the imaged section of the object is perpendicular to the test surface, it is called B scan; if it is parallel to it, it is called C scan. Ultrasonic images can also be produced by using holographic techniques and/or image converters of one kind or another.

After the basic testing techniques have been worked out, much development is directed toward speeding up the testing process, automating as much as possible, and in the assessment of the results, making use of microprocessors.

Although ultrasonic NDT is used mostly for steel and other metals, it can be applied to many other materials when its limitations are kept in mind. An important example is porcelain isolators for high-voltage overhead lines. Others include composite materials, fiberglass, or carbon-fiber reinforced plastics, and many others.

2. Material Characterization

Ultrasound velocity and attenuation are dependent on material properties, which in turn may be related to other characteristics that can be of special interest to the engineer (e.g., breaking strength of cast iron or porcelain, quality of concrete, grain structure of steel).

3. Flow Measurement

There are various ways to measure flow velocity in a pipe by measuring its effect on the passage of ultrasound waves transmitted and received at the outside of the pipe, without disturbing the flow by inserting anything into the pipe.

4. Level Measurement

Measuring the level in a container, mostly of a liquid, may be done either on the principle of the transit time of a pulse reflected from the liquid–air interface (usually from the bottom upward in the liquid, occasionally from the top downward in the air), powdery material in a container, or in a hopper by cutting off the airborne beam from one side to the other at a certain height. For liquids one can look at the reverberation of ultrasound pulses in the container wall: where there is liquid, the reverberation decays far more quickly than where there is air.

5. Chemical Analysis

Chemical analysis can be done advantageously under certain conditions using ultrasonic measurements. Titration can be automated if a precipitate is formed when equilibrium is just exceeded; namely, a very small amount of precipitate can be detected as it is formed, by its ultrasound scattering effect.

It may be convenient to determine the concentration of a two-component mixture based on the effect the composition has on the ultrasound velocity.

6. Stress under Load and Residual Stress

Stress under load and residual stress can be determined by very accurate (1 in 10^4) shear wave velocity measurements.

7. Temperature of Hot Gases

The temperature of hot gases can be measured by pulse transit time measurement between two ultrasonic probes protected by such delay rods as fused quartz. This technique has been applied to measuring temperature in an internal combustion engine chamber and in a plasma jet.

8. Medical Diagnostics

Medical diagnostics is another major field of application. This field consists mainly, but not exclusively, of examining the anatomical structure of soft tissue on the one hand, or its functioning, while on the other checking the normality/abnormality of blood flow in various vessels. All these examinations are noninvasive.

Examining the anatomical structure is in principle similar to NDT, but differences do exist. The sound velocity in tissue is on average about one-fourth of that in steel or aluminium (the scale of the display is different by this factor). While in metals any discontinuity means a substantial acoustic impedance mismatch and strong reflection, boundaries between different tissues are extremely weak reflectors. Therefore diagnostic instruments use much higher gain than in NDT. In tissues the attenuation is very much higher than in metals, and therefore every instrument must have time-varying gain (TVG) control. Tissue types differ, sometimes quite markedly, in the amount of back scatter they produce. Display is most commonly of the B-scan type. While X rays are most useful for the examination of bones, teeth, and other dense objects such as kidney stones and gall stones, for soft tissues special contrast materials must be ingested or injected into the patient. Ultrasound is in general unsuitable for the examination of bones because of their shape and structure, but very advantageous for soft tissues. It is possible to check for disorders in pregnancy, where an added advantage is the fact that it is quite safe (Fig. 17). So far no harmful effects have been found. Tumors, malignant and benign, cysts, abscesses, and many other disorders can be found and recognized in many organs. Normal or diseased functioning of the heart valves can be visualized. The list of applications grows all the time.

Blood flow measurements are based on the Doppler principle. They can detect constriction of vessels and find their location, and in general they are extremely useful in many diseases that involve various disorders of circulation.

9. Acoustic Microscopy

Ultrasonic imaging at frequencies in the range of hundreds of megahertz, either in the through transmission or the echo mode, are capable of high resolution. Thin, small, opaque objects can be examined, but even transparent objects may be worth examining with an ultrasonic microscope because it shows differences in mechanical properties, not the optical characteristics.

One of the available types uses simple, rectangular mechanical scanning of the object mounted on a mylar film between two fixed sapphire lenses carrying the two transducers. The transmitter may also be used as a receiver if operating in the echo mode (Fig. 18). Going up to 1-GHz resolution in the order of 1 μm is possible.

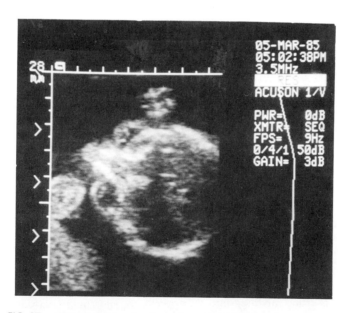

FIG. 17. Ultrasonic scan in obstetrics with high resolution: enlarged from a section through the womb at 19 weeks of pregnancy, scale 1 cm/large division. The plane of imaging runs through the baby's head at the level of the eyes; even the lenses are visible and one of the optic nerves, the nose, and part of the brain. Above the head section through the hand (fingers), beside the nose a section through the lower arm. In the bottom left corner there is the placenta. (Photograph courtesy of Dr. D. O. Cosgrove, Royal Marsden Hospital, London.)

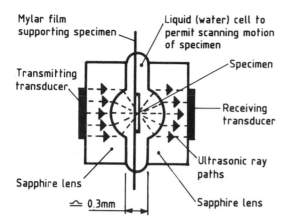

FIG. 18. Ultrasonic microscope in which the beam is stationary and the specimen is scanned.

Another type (tradename Sonoscan) sends plane ultrasonic waves of several 100 MHz through the object and scans the half-silvered surface of a lucite (perspex) plate with a narrow laser beam to optically detect the ultrasound waves passing through the specimen.

10. Surface Acoustic Wave Devices

In the past, ultrasonic waves were used as a fixed value analog delay line by bonding a transmitter and a receiver transducer to a suitable low-loss rod, such as quartz, or to a metal wire. In these devices bulk waves were propagated. Now the most common method employs a piezoelectric plate (either a natural crystal or a prepoled ceramic) with interdigital electrodes formed on it by first silvering it and then photoetching the desired pattern out of it (Fig. 19). The electrode spacing should correspond to the wavelength of surface waves at the desired frequency. Such a set of electrodes will work both forward and backward and waves in the unwanted direction must be stopped by putting some absorber on the substrate.

By suitable configuration of the electrodes, the bandwidth can be controlled. Not only can simple delay lines be made, but the lines can be tapped and other devices made for signal processing (e.g., fast spectrum analyzers, filters, convolvers). At high frequencies they are more economical than digital techniques.

11. Sonar

Sonar is another major field of ultrasonics. The name means *sonic navigation and ranging* and it applies to underwater acoustics, including all the equipment. The relationship of sonar to underwater acoustics is like that of radar to electromagnetics. It is very important in measuring the depth of water, finding underwater obstacles such as rocks and shipwrecks, locating schools of fish and their movements, helping divers and in military applications such as finding submarines, mines, enemy ships, and so on. Targets can be displayed in a pictorial manner showing their positions and distance as on radar screens and in NDT or medical B scans.

12. Acoustic Stress Wave Emission

Acoustic stress wave emission (AE for short) is based on the fact that in a nonhomogeneous material when subjected to increasing stress, the stress field and the material will both vary from point to point. So local unstable conditions can develop long before the material as a whole becomes unstable, giving rise to a local dynamic displacement, be it a dislocation movement, intergranular sliding, or crack propagation. Owing to the rapid deformation, oscillations will be set up at a site of disturbance, which will travel as acoustic waves to the surface of the object. An example of this phenomenon is the familiar "cry of tin," an audible noise attributed to crystal twinning.

Coupling a number of receiver transducers to

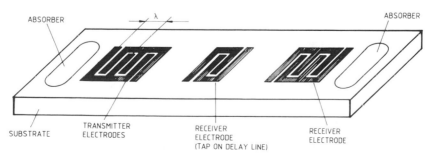

FIG. 19. Delay line in surface acoustic wave devices.

the surface of a test object, loading it, and observing the frequency and energy of the received acoustic (mostly ultrasonic) signals can give valuable information on the strength of the test object.

B. Active or High-Power Applications

1. Effects of High-Power Acoustic Waves

Various mechanisms can be activated by ultrasound energy to bring about or promote certain effects in the medium of propagation. Most of the effects can be related to the following factors.

1. Heat: Losses in the medium of propagation turn the acoustic energy into heat. At certain interfaces losses may be high.

2. Stirring: Intense ultrasound produces violent agitation in a liquid, especially if its viscosity is low.

3. Cavitation: In liquids the low-pressure half-cycle can disrupt the continuity of the liquid to produce small gas- and vapor-filled cavities that collapse during the high-pressure half-cycle. As the cavities collapse and the liquid rushing in from one side meets the opposite side, the local pressure can rise for a short time to the order of 10,000 bars with an accompanying temperature rise to around 10,000 K. At the same time a shock wave is generated at each cavity. In addition, while these cavities exist, their shape is, in contrast to hydrodynamic cavitation, not spherical, but somewhat flattened like a lens. Charged ions can be found on the opposite surfaces, which give rise to electrical discharges.

4. Mechanical effects: Stresses developed can cause ruptures.

5. Diffusion: The rate of diffusion is generally increased.

6. Chemical effects: Chemical activity in general is accelerated, especially oxidation; also catalytic processes are promoted.

For high-intensity applications the ultrasound energy must often be concentrated. At high frequencies (several hundred kilohertz to megahertz range) this can be done by forming the ceramic transducer into a bowl shape. The focal point is the center of curvature.

In the low-frequency (20–100 kHz) range a metal horn of $\lambda/2$ length (often made of titanium) is attached to the transducer, the other end of which has a different area. If the other end is

smaller, the displacement, pressure, and velocity amplitude will increase by a factor dependent on the area ratio and the shape of the horn, whether it is a conical, exponential, Fourier, or stepped horn (Fig. 20). Consequently, horns are often also called velocity transformers.

2. Cleaning

Ultrasonic cavitation can blast off contamination from the surface of many objects very quickly and effectively, especially if the cleaning bath has the right chemical composition (detergents and/or solvents) and temperature. Contamination can effectively be removed from intricate surfaces such as tapped blind holes and assembled instruments. Cleaning tanks are in use from very small size (a few hundred milliliters) driven by an overall power of 50 W or less up to tanks capable of cleaning a complete jet engine requiring power in the order of many kilowatts. The cleanliness achievable is excellent.

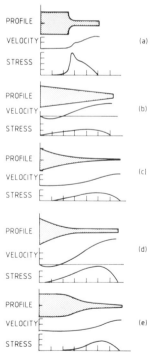

FIG. 20. Ultrasonic horns. Profiles of impedance transformer horns: (a) stepped, (b) conical, (c) exponential, (d) catenoidal, and (e) Fourier. Particle velocity and stress as a function of the length coordinates of the respective cross sections are shown below each profile. [Reproduced with permission from Frederick, J. R. (1965). "Ultrasonic Engineering." Wiley, New York. Copyright © 1965 John Wiley & Sons.]

3. Emulsification and Homogenization

Mixing immiscible liquids like oils and water and homogenization of existing emulsions and mixtures are a popular field of application of high-power ultrasonics in industry, especially the food and cosmetics industry but also in many others. Here again cavitation plays the most significant part.

4. Extraction

Ultrasonic cavitation can break up cell membranes and cell walls and increase the diffusion rate. Consequently, processes requiring extraction from plants or other cellular material can be speeded up and at the same time the yield increased. One example is extracting bitter materials from hops in many breweries.

Ultrasonic extraction is also advantageous in collecting enzymes, antigenes, and other materials from tissues and colonies of bacteria.

5. Making of Aerosols, Atomization of Liquids

If liquid is fed at a suitable rate to the surface of an ultrasonic horn vibrating with high amplitude, so-called capillary waves are set up and the liquid is thrown off into the air in the form of small droplets. Interestingly, the size of these droplets is almost uniform and is determined by the surface tension, the density of the liquid, and the frequency.

This process can be important for production of aerosols of medicines to be inhaled and aerosols of fuel oils (especially heavy oils) in a burner to steady the flame at low loading levels.

6. Impact Grinding

Impact grinding is a special form of machining hard and brittle materials. It means that a suitably shaped tool made of mild steel or brass is mounted on the end of an ultrasonic horn and when it is vibrating, an abrasive slurry is fed onto the object and at the same time the tool is pressed onto the object. However, it is never really in contact with the object, but shoots the abrasive particles at the object at high speed like a miniature machine gun. The abrasive particles do the actual machining by chipping away small bits of the brittle object, while the tool itself is little affected. It cannot be chipped away; rather the abrasive particles become embedded into it. Depending on the shape of the tool, the object may be cut, drilled in any shape, or the negative

of the tool surface formed on it. The machining can be very accurate with fine surface finish.

A typical application is manufacturing a large number of tiny sapphire bearings for watches and instruments, all in one operation with a multiple tool, or making dies of hard metal.

7. Ultrasonic Machining

Ultrasonic machining is adding ultrasonic power to the cutting tool of a lathe. It results in higher cutting speed with less power and smoother, stress-free surface finish than is possible without it. Friction is reduced and the swarf breaks off and is removed more readily. However, it is difficult to mount the tool as firmly as required and yet allow it to vibrate at the ultrasound frequency without much of the ultrasound being lost in the mounting.

8. Forming Metals

Forming metals, especially extrusion and drawing metals, through a die can benefit from high ultrasonic power being fed to the die. Friction is reduced as is the force necessary for the plastic deformation of the metal, since it flows more freely. Reductions in drawing forces of 50% or more have been reported. A further advantage is that greater deformation is possible in one step than without ultrasonic power.

9. Forming Plastics

The forming of plastics, both thermosetting and thermoplastic materials, can be improved by feeding ultrasonic power to the tool or tools. Internal friction is reduced and the material flows more easily, even at lower temperatures than otherwise. Furthermore the elastic recovery of the plastic after the removal of the forming force is reduced by 20% to 50%.

10. Welding Plastics

Welding plastics is based on the fact that if the two pieces of plastic that are to be welded together are pressed down and irradiated, at their boundary, which is a discontinuity reflecting a proportion of ultrasound energy, the intensity and with it the rate of energy absorbed will be higher than elsewhere. As the temperature rises, the plastic softens, but even before melting the ultrasound waves push together and entangle the long chain molecules of the two plastics. Thus a bond is being formed before the temperature is reached where the plastic begins to degrade. The bond is normally stronger than is possible by the application of heat by other means, such as microwaves.

11. Welding Metals

Welding metals is possible by pressing an extension of a horn sideways on the objects to be welded against an anvil. In this way shear vibration takes place in the contact area, in sharp contrast to welding plastics. It results in the asperities (microscopic protrusions) being plastically deformed, the peaks pushed into the valleys, and surface impurities dispersed and atoms of the two metals being brought so close together that a strong bond is formed at room temperature. One of the objects must be in the form of a sheet or wire less than 6 or 7 mm thick.

Since the process does not depend on heat, metals of very dissimilar thickness can be welded together. Dissimilar metals with widely differing melting points cause no problem, and there are no residual stresses. Weldability of metals is contained in Fig. 21. Some metals, like aluminium, can be welded even to certain non-metals.

12. Drying of Heat-Sensitive Materials

Drying of heat-sensitive materials like fine powders or fibers that could easily be blown away by a stream of air can be greatly enhanced by using high-power airborne ultrasound. First, the radiation pressure simply pushes much of the moisture out. Second, the sound waves increase the diffusion rate and also generate microstreaming of the air in the vicinity of the wet particles, which also helps in the drying process.

The time of drying may be reduced by a factor of 2 or 3.

Examples include the manufacturing of powdered sugar and the drying of pulverized coal.

13. Foam Control

Foam control can be important in certain industrial processes. Even moderately high power airborne ultrasound excites the bubbles into vibration sufficiently to break them up almost immediately.

14. Degassing of Liquids and Melts

Degassing of liquids and melts is brought about by ultrasonic cavitation. Dissolved gas tends to migrate more quickly into cavitation bubbles than away from them, and so the bubbles grow and rise to the surface.

15. Metal Casting

Apart from degassing the melt, ultrasound irradiation tends to break up the dendrites as they grow, producing new nuclei and resulting in a

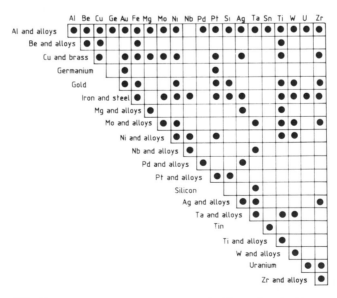

FIG. 21.   Ultrasonic weldability of metals. [Reproduced with permission from Frederick, J. R. (1965). ''Ultrasonic Engineering.'' Wiley, New York. Copyright © 1965 John Wiley & Sons.]

refined grain structure, which means improved mechanical properties such as strength and toughness. However, coupling the ultrasound into the melt can cause problems.

16. Medical Therapy

The main nonspecific effects of ultrasonic irradiation on living tissues are the following.

1. Heat is produced by the energy absorbed.
2. Diffusion rates are increased.
3. Ultrasound waves act like a "microscopic massage."
4. There are some other mechanical effects, still not quite understood.

Many of these effects are interrelated. Heat generated in certain diseased tissues is known to be beneficial and has been in therapeutical use, achieved in other ways. It causes among other things local improvement of blood supply by dilation of the blood vessels. This, together with the increased diffusion rate and acceleration of enzyme activity due to the higher temperature, helps the cells to get fresh nutrients and to get rid of waste products. Ultrasonic irradiation, if correctly applied, has an added specific effect of causing relaxation of muscles and reducing pain. The tissue metabolic rate increases and as a result of all these effects the body's natural healing processes are significantly enhanced, and thus a wide range of conditions and injuries can be treated successfully, considerably speeding up the healing process. Indeed the range is too wide to be treated here.

Excessive dosage can, of course, cause damage by overheating, coagulation, and mechanical destruction, but careful application is quite safe.

The normal range of dosage is around 0.25 to 1.0 W/cm^2, exceptionally 1.5 W/cm^2 (25 to 10 or 15 kW/m^2) at 1 to 3 MHz with a stroking movement of the applicator using a coupling jelly, for about 5 to 20 min, two or three times a week.

17. Surgical Therapy

Surgery can benefit from exploiting the destructive effect of high-intensity ultrasound because by focusing it can be restricted to small areas. A special advantage is that deeper-seated diseased tissues (tumors, etc.) can be accurately destroyed in the brain without damaging important brain tissue between the surface and the lesion. There are many other diverse surgical applications.

BIBLIOGRAPHY

Goldberg, B. B., and Wells, P. N. T. (Eds.) (1983). "Ultrasonics in Clinical Diagnosis," 3rd ed. Churchill Livingstone, Edinburgh.

Hussey, M. (1985). "Basic Physics and Technology of Medical Diagnostic Ultrasound." MacMillan, London.

Krautkrämer, J., and Krautkrämer, H. (1977). "Ultrasonic Testing of Materials," 2nd ed. Springer-Verlag, Berlin, Heidelberg.

Matthews, J. R. (1983). "Acoustic Emission." Gordon and Breach, New York.

Morgan, D. P. (1985). "Surface-Wave Devices for Signal Processing." Elsevier, Amsterdam.

Puškár, A. (1982). "The Use of High Intensity Ultrasound." Elsevier, Amsterdam.

Szilard, J. (Ed.) (1982). "Ultrasonic Testing: Non-conventional Testing Techniques." Wiley, New York.

Wells, P. N. T. (1977). "Biomedical Ultrasonics." Academic Press, New York.

UNIFIED FIELD THEORIES

L. Dolan   *Rockefeller University*

I.   The Fundamental Interactions
II.   The Elementary Particle Spectrum
III. Range of Interactions
IV.   Strength of Interactions
V.   The Renormalization Group
VI.   Kaluza–Klein Theories
VII. Uses of Symmetry to Classify Particles
VIII. Gauge Theories
IX.   Unified String Theories

GLOSSARY

Dual string theory: Model of elementary particles that treats particles with different masses and quantum numbers as different vibrational excitations of a massless relativistic string.

Elementary particle: One of about 60 pointlike particles that by themselves or in combinations account for all the known resonances of particle physics. An elementary particle is an irreducible representation of the Poincaré group, the group of space–time symmetry transformations consisting of space–time translations and the Lorentz transformations of the special theory of relativity.

Field: Function that associates a number with each point in space–time.

Field theory: Mathematical system of equations of motion and an action principle for fields, which determines their dynamics.

Gauge theory: Field theory with a local symmetry that allows transformations to be made independently at each point of space–time and always results in a theory of massless particles.

Massless particle: Elementary particle that travels at the speed of light. Because the speed of light is the same in all reference frames, there is no rest frame for a massless particle, and hence it has no rest mass and is called massless.

Quantum field theory: Field theory according to which the fields associated with each point in space–time are operators instead of classical commuting functions.

Quantum numbers: Numbers that label the elementary particles; quantities such as mass, momentum, electric charge, and color. The quantum numbers are the eigenvalues of the maximal set of commuting generators of the symmetry algebra, that is, the Cartan subalgebra. In general, any such number is called a charge.

Relativistic system: System moving with a velocity near the speed of light, which need not be quantum mechanical. For example, the general theory of gravitation, which describes the bending of light around massive celestial bodies, is a classical not a quantum theory.

Spontaneous symmetry breaking: Situation that occurs when the field theory action and the equations of motion have a symmetry, but the solution is not invariant. This allows renormalizable gauge theories to describe massive gauge bosons.

Tachyon: Unphysical particle that has negative mass squared (imaginary mass). Such a particle travels faster than the speed of light.

A unified field theory describes the physics of elementary particles such as electrons and quarks and what we believe to be the four fundamental interactions of nature: the two familiar forces of electromagnetism and gravity, which act at all distances, both large and small; and the two more anti-intuitive forces, the weak and the strong interactions, which only occur at very short distances and are responsible for radioactivity and nuclear fusion. These four interactions, although very different (in strength, range, and the particles on which they act), when viewed in everyday life and in experiments performed in current-day accelerators, have many underlying mathematical features in common that suggest that at very high energies (not presently available in accelerators) these interactions become the same. Since such questions involve particles moving very fast and coming very close together, a consistent theory necessarily involves both the theory of relativity and quantum mechanics. Quantum mechanics introduces the concept of processes that do not commute with each other, that is, quantities such that $uv \neq vu$. Causality requires that particles at spacelike separations (separated by a distance greater than their time separation) do commute since no information can be transferred faster than the speed of light, a constant given by the theory of relativity. One way to incorporate the fact that whether objects commute depends on their relative position is to describe each particle at a given point in space–time x, t by a field $\phi(x, t)$, such that the fields $\phi(x, t)\phi(y, t') \neq \dot{\phi}(y, t')\phi(x, t)$ for separations $(x - y)^2 \leq (t - t')^2$, but otherwise they do commute. A unified field theory is thus a description of elementary particles in terms of fields, which furthermore unifies all the interactions, that is, treats all particles and interactions on an equal footing both conceptually and technically in such a way as to make a prediction that in some region of energy, measurements will reveal there is just one type of force. As early as the 1920s, Albert Einstein and others suggested the unification of gravity and electromagnetism, the only forces that were well understood at the time; they did this by using classical field theory, which was relativistic but not quantum.

I. The Fundamental Interactions

Electromagnetism is the force between any charged particles. It is proportional to the inverse of the distance squared between the particles. It is carried or mediated by a photon, a massless particle of spin one identified with light.

Gravity acts on the mass, that is, energy of particles, and therefore on all particles. It also is proportional to the inverse separation distance squared and is mediated by a graviton, a massless spin two particle predicted by quantum gravity.

The strong interactions act between particles that are similar to protons and neutrons, that is,

all the strongly interacting particles. These are called hadrons (baryons, mesons, glueballs) and are made of quarks and gluons. In contrast to particles such as electrons, photons, or gravitons, which are pointlike particles and are called elementary, the hadrons are composite. They are of finite size of the order of 10^{-13} cm. The strong interactions do not have a simple force law like $1/r^2$. Inside the hadron the force appears to increase with separation distance. In fact the way the hadrons interact is by the strong interactions of the quarks and gluons, which are elementary, but unlike the other elementary particles they are confined inside the hadrons (i.e., they never appear as free one-particle states). Apart from this phenomenon of quark–gluon confinement, however, the strong interactions are similar to electromagnetism and gravity in that they are also mediated by integer spin particles, the eight non-Abelian color gluons of spin one, which are massless inside the hadron. See Fig. 1.

The weak interactions are responsible for radioactive decay, such as neutron decay into proton, electron, and antineutrino. The weak force is mediated by the three intermediate vector bosons W^+, W^-, and Z^0. They are again of spin one but have mass of the order of 100 GeV. (A giga-electron-volt is a billion electron volts.) At energies much higher than this, of course, the

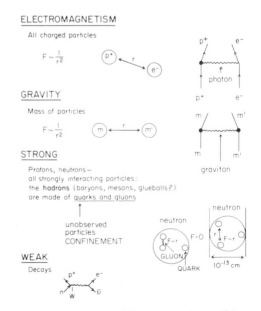

FIG. 1. The fundamental forces and the particles on which they act.

W's and Z appear to be massless, and indeed in this region the weak and electromagnetic interactions do act as the same force. This is the prediction of the Weinberg–Salam unified field theory for the subset of the weak and electromagnetic interactions. It shows that the W's and Z couple in a way similar to the photon and that the electroweak force has just one independent coupling constant.

II. The Elementary Particle Spectrum

A. CLASSIFICATION

The known elementary particles have spin less than or equal to two and mass ranging from zero to the order of 100 GeV. These are the 12 spin one gauge bosons and the one spin two graviton that mediate the interactions together with the three families of spin $\frac{1}{2}$ quarks and leptons made up of 45 left-handed (spin up, i.e., helicity $+\frac{1}{2}$) states representing particles and antiparticles as well as 45 right-handed (spin down, i.e., helicity $-\frac{1}{2}$) states. (The right-handed particles transform as the left-handed antiparticles under the symmetry groups $SU(3)_{color} \times$

$SU(2)_{weak} \times U(1)$ of the strong, weak, and electromagnetic interactions. These are discussed later.) The fact that for a given particle the left- and right-handed states transform differently is a reflection of the parity violation of the weak force. Although in some sense, parity violation is not as fundamental an experimental signal in building the theoretical framework as is, say, the masslessness of the gauge bosons, which requires gauge invariance and is relevant for all the interactions, this left–right asymmetry of the fermions under the weak interactions remains an experimental enigma that must come out in the finer details of the correct unified theory and in this way serves as a guide to its construction. See Table I. Each quark is labeled by its strong interaction color index $i = 1, 2, 3$ (red, white, or blue) and its weak flavor (u, up; d, down; c, charm; s, strange; t, top; and b, bottom). The existence of the top quark has not yet been firmly established experimentally.

One task of a unified theory is to have precisely the particle content of the approximately 58 elementary particles listed in Table I, without a lot of extra freedom to add or subtract particles at will. [*See* ELEMENTARY PARTICLE PHYSICS.]

TABLE I.   Elementary Particle Spectrum

Spin 0   None known yet (the Higgs particles)

Spin $\frac{1}{2}$ (Quarks and leptons)—spin $\uparrow$ and spin $\downarrow$ act differently, for example, no right-handed neutrinos

$$SU(3)\text{triplet}$$

$$SU(2)_{weak}\text{ doublet} \rightarrow \begin{pmatrix} u^i \\ d^i \end{pmatrix}_L \qquad \begin{pmatrix} c^i \\ s^i \end{pmatrix}_L \qquad \begin{pmatrix} t? \\ b^i \end{pmatrix}_L \qquad \text{Quarks}$$

$$\bar{u}^i_L \sim u^i_R \qquad \bar{c}^i_L \sim c^i_R \qquad \qquad ?$$
$$\bar{d}^i_L \sim d^i_R \qquad \bar{s}^i_L \sim s^i_R \qquad \bar{b}^i_L \sim b^i_R$$

$$\begin{pmatrix} \nu^e \\ e^- \end{pmatrix}_L \qquad \begin{pmatrix} \nu^\mu \\ \mu^- \end{pmatrix}_L \qquad \begin{pmatrix} \nu^\tau \\ \tau^- \end{pmatrix}_L \qquad \text{Leptons}$$

$$\bar{\nu}^e_L \sim \nu^e_R$$

$$e^+_L \qquad\qquad \mu^+_L \qquad\qquad \tau^+_L$$

15 left-handed states $\times$ 3 = 45, 3 generations so far

Spin 1   (12 vector bosons)

1 photon	3 weak bosons	8 gluons
γ	W^+, W^-, Z^0	$A^a_\mu(\bar{x})^{1,\dots,8}$

$SU(2)_{weak} \times U(1)$	$SU(3)_{color}$
weak and electromagnetic	strong

Spin $\frac{3}{2}$ None known yet (gravitinos, predicted by supergravity)

Spin 2   1 graviton

B. Recent Key Experimental Discoveries

The experimental evidence for the quark model of hadrons—hadron spectroscopy—has a long history starting with the discovery of the pion in 1947. The $SU(3)_{isospin}$ symmetry introduced by Gell-Mann indicated that members of groups of the low-lying hadrons—(baryons) proton, neutron, lambda, sigmas, cascades and (mesons) pions, kaons—were identical under the strong force and could be envisioned to be each built out of various combinations of just three flavors (u, d, s) of colored quarks. Although surprising and revolutionary itself, this observation became a centerpin of unification when in January 1975 a meson containing a fourth type of quark (c) was discovered. This particle was the J/ψ at 3.1 GeV. It was discovered in two places: at Brookhaven National Laboratory in a fixed target experiment of a proton beam on Beryllium going into J/ψ + anything:

$$p^+ + Be \longrightarrow J/\psi + X$$
$$\downarrow$$
$$\longrightarrow e^+ + e^-$$

and at Stanford Linear Accelerator (SLAC) in an electron–positron colliding beam experiment:

$$e^+ + e^- \longrightarrow hadrons \quad (J/\psi)$$

Then in October of 1977, yet another new meson that could be interpreted as a bound state of the b quark and antiquark, the upsilon (γ) at 9.4 GeV, was found at Fermilab with a 400-GeV proton beam on a fixed proton target:

$$p^+ + p^- \longrightarrow \gamma + X$$
$$\downarrow$$
$$\longrightarrow \mu^+ + \mu^-$$

Thus in the past ten years there has been overwhelming experimental verification for the family structure of quarks. Whether this pattern continues as we build accelerators with higher and higher energy probes so that more and more flavors appear or if only these exist and there really is a "desert" between the mass of the sixth flavored quark and the Planck mass (10^{19} GeV) remains one of the most intriguing questions to be answered by the unified theory.

In January of 1983, the discovery of the three intermediate vector bosons W^+, W^-, Z^0 in the CERN $p\bar{p}$ collider confirmed the unification of the weak and electromagnetic interactions. This was an experiment with 270-GeV protons on 270-GeV antiprotons:

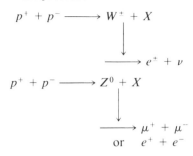

The existence of these massive bosons ($W^\pm \sim$ 81 GeV, $Z^0 \sim$ 91 GeV, the heaviest known elementary particles) was the first experimentally verified prediction of a field theory that unified two of the different forces and put the W's, the Z, and the photon on equal footing. It thus represents a partial unification of all the interactions.

III. Range of Interactions

Particle physicists find it useful to introduce the concept that energy or mass (connected by $E = mc^2$) is related to inverse distance. This is easy to see if we choose units where the speed of light c equals 1 so that 1 sec is approximately 3×10^{10} cm (distance) and Planck's constant h equals 1 so that 1/[1 sec] is approximately 6.6×10^{-25} GeV (energy). In these units, 10^{-13} cm $\sim$ 1/0.1 GeV, which means that in order to see an object the size of a hadron, we must probe with a particle traveling with energy of 0.1 GeV (100 MeV, approximately the mass of the pion). That is to say that the hadron is invisible to probes of energy less than 0.1 GeV, since then its size is smaller than the wavelength of the probe. This further implies that at the highest energies available to us in the latest accelerators ($\sim 10^3$ GeV = 1 TeV), any object of size $<10^{-17}$ cm is unresolvable. A central idea in theories of unification is that space–time has more than four dimensions, with the extra dimensions closed (like a circle or a torus) and small, of the order of the Planck length 10^{-33} cm. Because such small distances cannot be distinguished in today's accelerators, these extra dimensions do not contradict experiment.

With this equivalence of mass and inverse distance, it is simple to describe the range of the four interactions. Because gravity and electromagnetism are mediated by massless particles, they are of infinite range, $1/0 = \infty$. The weak interactions mediated by particles with mass ~100 GeV have range almost zero. The strong interactions occur when the hadrons "overlap" so that the quarks and gluons inside can interact. Thus for this the range is 10^{-13} cm.

IV. Strength of Interactions

The relative strength of the interactions is a measure of how often or how probable is the occurrence of a particular scattering or decay. At energies between 0.0005 and 1 GeV the strengths measured by the coupling constants are

Strong	$g(m_\pi) \sim 1$
Electromagnetic	$\alpha(m_{\text{electron}}) \sim \frac{1}{137}$
Weak	$m_{\text{proton}}^2 G_F \sim 10^{-5}$
Gravity	$m_{\text{proton}}^2 G \sim 10^{-39}$

where α is the fine structure constant, G_F Fermi's constant, and G Newton's constant.

Thus at the pion mass, the strong interactions have a strength or relative frequency of the order of 1, while the electromagnetic coupling measured at the electron mass (0.0005 GeV) is much less strong and is given by the fine structure constant $\sim \frac{1}{137}$. The weak interactions measured at the proton mass (1 GeV) are even weaker still and are scaled by Fermi's constant to give the dimensionless ratio 10^{-5}. Gravity, the weakest force of all at these low energies, is scaled by Newton's constant to be $1/10^{39}$ the strength of the strong force. Unification means that these disparate ranges and coupling constants will become identical at some higher energy. It may well be that this unification does not happen until energies of the order of the Planck mass. If this is the case, then although we have a theory of this physics, the actual direct experimental verification of it may not be possible. This heralds a new era of high-energy physics, where experiment and theory are farther apart and theory must plunge on ahead with the hope that some of the coherent unified theory in 10 dimensions and at the Planck mass will trickle down to accessible four-dimensional phenomenology.

V. The Renormalization Group

The incorporation of quantum mechanics together with relativity is one of the features of field theory. This has been successful at least as far as the special theory of relativity is concerned. The special theory describes fast-moving objects in flat space. This is sufficient for the strong and electroweak interactions since only gravity is responsible for the curvature of space. However, although special relativity (i.e., Lorentz invariance) is a more general symmetry than the nonrelativistic Gallilean invariance, its introduction into quantum mechanics makes the system more and not less complicated. This is because at relativistic energies, even though there is more symmetry, there are also more degrees of freedom since there can now be pair production: because $E = mc^2$, whenever the kinetic energy of a particle gets to be greater than $2mc^2$, it can turn into two particles and there are now twice as many degrees of freedom. This union of special relativity with quantum mechanics in general introduces ultraviolet infinities into the theory. [See QUANTUM MECHANICS; RELATIVITY, SPECIAL.]

The Lorentz invariant field theories that describe the strong and electroweak interactions have ultraviolet divergences order by order in a perturbation expansion in the coupling constant, but these divergences can be subtracted out in a consistent way. Such theories are said to be renormalizable, and they make extremely detailed, accurate predictions for the scattering amplitudes in physical regions where the coupling is small.

The fact that the divergences must be subtracted out results in a mathematical equation called the renormalization group equation, which relates scattering amplitudes at different energies and shows that the coupling constants change with energy. In particular, the strong interactions are asymptotically free, so that g (which is 1 at the pion mass) decreases as the energy increases (i.e., with decreasing distance), while the electromagnetic and weak couplings increase to join each other at ~100 GeV (the electroweak unification scale) with a value still considerably less than 1 so that electroweak perturbation theory is still sufficient to describe physics in the region. This trend can be extrapolated to an even higher energy called the grand unification scale (possibly of the order of 10^{15} GeV), where it is presumed that all three couplings merge with a value less than 1. See Fig. 2.

A feature of the strong interaction color gauge theory is that at low energies it is strongly coupled ($g \sim 1$) so that a perturbative approximation is not useful in describing hadronic physics. A controlled, systematic nonperturbative ap-

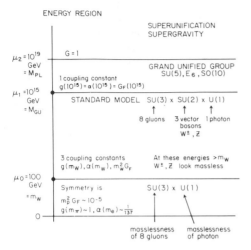

FIG. 2. The different scales of high-energy physics.

proximation for this field theory has remained elusive, although several powerful attempts have been made, namely, the large N approximation (an expansion in the inverse of the number of colors), the instanton semiclassical approximation (an expansion around a solitonlike solution of the classical nonlinear equations of motion), and lattice gauge theories. The absence of analytical nonperturbative techniques has led to a consideration of physics at higher energies such as the grand unified scale, where the coupling is again weak so that perturbation theory can be reliably used. There is a hunch that knowledge of the entire unified theory will introduce enough extra information about the symmetries and other properties of the theory that nonperturbative low-energy effects may then be calculated.

A description of quantum gravity requires the union of quantum mechanics with the general theory of relativity. This invariance of general coordinate transformations introduces more ultraviolet divergence, and there is no way to write down a renormalizable or finite field theory with a finite number of local fields. If instead of a pointlike field, we introduce a one-dimensional object—a string that is a set of points rather than a single point—then it is possible to describe a quantum theory of particles coupling gravitons that is finite (at least in the first nontrivial order of perturbation theory, and presumably to all orders). This is the supersymmetric string theory. It unifies both bosons and fermions (integer and half-integer particles), all the

interactions, and naturally lives in 10 space time dimensions rather than 4. [*See* RELATIVITY, GENERAL.]

VI. Kaluza–Klein Theories

In the 1920s, when Theodore Kaluza, Oskar Klein, Albert Einstein, and others suggested the unification of gravity and electromagnetism, they did so by the ingenious device of postulating the existence of an extra fourth dimension of space–time and making that fifth dimension topologically equivalent to a circle. The Kaluza–Klein miracle is that the extra degrees of freedom of five-dimensional gravity then couple as a Maxwell field (a photon) to gravity in four dimensions. This technique also gives the quantization of electric charge. In addition to the massless gauge fields in four dimensions, there is also an infinite "Kaluza–Klein tower" of massive fields since the field variables in five dimensions are periodic in the fifth coordinate and can be expanded in a Fourier series. The masses of these four dimensional fields are proportional to the inverse of the size of the closed (i.e., "compactified") fifth dimension.

In the 1980s we would like to unify all the interactions in a quantum framework. It is possible to do so with the introduction of a few more extra dimensions, two new symmetries: supergravity, which treats bosons and fermions as partners; and affine Kac–Moody Lie algebras, which put an infinite number of particles in the same multiplet; and the generalized quantum field theory of the dual string model.

VII. Uses of Symmetry to Classify Particles

A. LIE ALGEBRAS

Noether's theorem says that for every continuous symmetry there is a constant of the motion, that is, a conserved quantity. Thus systems that are invariant under time translations conserve energy, theories that predict the same results no matter where you do the experiment (invariant under space translations) have conservation of momentum, and invariance or symmetry under rotations implies the conservation of angular momentum (i.e., spin). These are examples of space–time symmetries. Translations are abelian transformations; this is, they commute with with each other. Space rotations are non-Abelian; they do not commute. They are labeled by the three generators of the Lie algebra $SU(2)$: T_a, $a = 1, 2, 3$, where

$$T_1 T_2 - T_2 T_1 = T_3, \quad \text{etc.}$$

that is to say T_3 is not zero. It is possible to extend this connection between conserved quantum numbers and symmetries to "internal symmetries." The symmetry responsible for the conservation of electric charge is $U(1)$. The symmetry responsible for the fact that the strong interactions conserve isospin is $SU(2)$. Isospin invariance says, for example, that the proton and neutron are identical under the strong force. We can label the two states with the eigenvalues of T_3 in the "spinor" representation. This multiplet is a doublet:

$$T_3 = \begin{cases} +\frac{1}{2} & p \\ -\frac{1}{2} & n \end{cases}$$

The generators $T^{\pm} = T_1 \pm iT_2$ change $p \to n$ and $n \to p$, respectively. (In the quark model, p and n are replaced by the u and d quark.) The values of T_3 are the isospin of the particle. The use of symmetry is that it is now no longer necessary to keep track of the proton and neutron separately, since the strong interactions cannot distinguish between members of a multiplet. The meaning of symmetry is that particles are labeled by a fixed set of numbers—the values of T_3: 0 for a singlet, $+\frac{1}{2}$ and $-\frac{1}{2}$ for a doublet, $-1, 0, 1$ for a triplet representation, and so forth, and the force between particles is invariant under the switch $-\frac{1}{2} \to \frac{1}{2}$ or $-1, 0, 1 \to 1, -1, 0$. Not all symmetry groups have representations of every dimension; $SU(3)$ has eight generators. Two of these commute with each other, $T^3 T^8 = T^8 T^3$, so that the states are now labeled by two quantum numbers, the isospin and the hypercharge. Also $SU(3)$ is the symmetry of both Gell-Mann isospin (u, d, s all in the same multiplet) and the color gauge theory (u^1, u^2, u^3 in a triplet). Its representations have dimension (number of particles in a multiplet) 3, 8, 10, 27, ... For the strong interactions, the color gauge theory has each flavored quark in a colored triplet, the 3, and the eight gluons in the octet, the 8. See Table I.

The weak interaction symmetry group is $SU(2)_{\text{weak}} \times U(1)_{\text{electromagnetism}}$. The u and d left-handed quarks transform as a doublet under this $SU(2)$, while the u and d right-handed quarks transform as two singlets under $SU(2)$. This is the parity violation of the weak interactions. It follows that the u and d left-handed antiquarks transform as two singlets, and the u and d righthanded antiquarks transform as a doublet. This pattern is repeated for the c, d and t, b quarks. The leptons are even more left–right asymmetric in that whereas the left-handed neutrino and electron form an $SU(2)$ doublet and the left-handed positron is a singlet, there simply is no left-handed antineutrino observed at all. It again follows that the right-handed antineutrino and positron form a doublet, the right-handed electron is a singlet, and no right-handed neutrino has been observed. The μ and τ families behave similarly.

Therefore all the known elementary particles are classified by their representations under the gauge symmetry groups $SU(3)_{\text{color}} \times SU(2)_{\text{weak}} \times U(1)_{\text{em}}$ of the strong, weak, and electromagnetic interactions, respectively. For one family of 15 left-handed states, this is written as $(3, 2)_{1/3} + (\bar{3}, 1)_{-4/3} + (\bar{3}, 1)_{2/3} + (1, 2)_{-1} + (1, 1)_2$. The subscripts are the $U(1)$ hypercharge eigenvalue $Y = 2$ (electric charge–weak isospin). The electric charge of the quarks is $\frac{2}{3}$ for u, $-\frac{1}{3}$ for d, $\frac{2}{3}$ for c, $-\frac{1}{3}$ for s, unknown for t, $-\frac{1}{3}$ for b, and of the leptons 0 for ν and 1 for e^-. One says that the fermions of a given helicity form a complex representation of the gauge group. That is to say that the preceding representation is not equivalent to its complex conjugate: $(3, 1)_{4/3} + (3, 1)_{-2/3} + (\bar{3}, 2)_{-1/3} + (1, 1)_{-2} + (1, 2)_1$, which is the representation of the right-handed states. These are called chiral representations. The presence of chiral fermions appears to be fundamental for the weak interactions. Furthermore, such fermions must be massless whenever the $SU(2) \times U(1)$ gauge invariance is not spontaneously broken. Of course, this symmetry is broken below 100 GeV, when the W's and Z no longer appear massless. This implies that the mass of the quarks and leptons cannot be much more than the electroweak scale either, which is consistent with experiment. For the spin one states, the representation is $(8, 1)_0 + (1, 3)_0 + (1, 1)_0$.

B. SUPERGRAVITY

The supersymmetry algebra has generators that have anticommutation relations as well as commutation relations. It changes fermions into bosons and bosons into fermions. An extended ($N = 8$) supergravity multiplet in four dimensions has particles with all spins less than or equal to two: the spin content is $(1(2), 8(\frac{3}{2}), 28(1), 56(\frac{1}{2}), 70(0))$. This multiplet has enough particles to describe all those listed in Table I, so that it is a prototype for a completely unified theory. However, such multiplets do not contain chiral fermions. In unified superstring theory, the major role of supersymmetry is to make the theory finite. This is done in such a way that a realistic theory is contained in the particle spectrum, but a hallmark of such supersymmetric theories is that they always contain more massless particles than those on Table I.

C. Affine Kac–Moody Lie Algebras

In 1981, a new type of symmetry was discovered to be relevant for theories of particle physics. Affine algebras are infinite dimensional, they have an infinite number of generators, and subsequently they have an infinite number of particles in one irreducible multiplet. These algebras were first written down independently by Kac and Moody in 1968. They are derived from a generalization of the Cartan classification of Lie algebras corresponding to allowing the determinant of the Cartan matrix to be equal to zero. The commutation relations of the generators are

$$[T_n^a, T_m^b] = f^{abe} T_{n=m}^e + cn \, \delta_{n,-m} \, \delta^{ab}$$

where n, m are all integers and a, b, e, run over the dimension of some Lie algebra g with structure constants f^{abe}. The generators T_0^a close the algebra g. The second term is the central extension; c is a arbitrary constant. Thus for each finite dimensional Lie algebra g, there is associated an affine Kac–Moody Lie algebra $\hat{g}$. In general, the Cartan subalgebra of $\hat{g}$ has one more generator in it than does that of g, and so there is one new quantum number called affinecharge. In Kaluza–Klein theories, where the infinite number of particles in the affine multiplet describe the infinite Kaluza–Klein tower, the affinecharge coincides with the electric charge. In compactified string theories, the affinecharge measures the contribution to the mass of the particle from the internal (compactified space) degrees of freedom.

These algebras were first seen to be relevant to field theory as symmetries of two-dimensional sigma models and spin models, as well as to the self-dual class of solutions of $SU(N)$ gauge theory. They also have close ties with the transfer matrix solutions and the Kramers–Wannier duality transformations of models of condensed matter physics.

In 1978, the representations of these algebras

were constructed in terms of the vertex operators of the string model, with the restriction that the momentum argument was discrete and took on the values of the root lattice of g. (The root lattice is an infinite set of numbers that are all integer multiples of the roots, that is, structure constants of the algebra.)

That the same (affine) algebra appeared both in the context of the string model, which in 1968 was used as a theory of the hadrons rather than of the elementary particles, and in the context of the $SU(3)$ color gauge theory, which was a completely separate description of hadrons, was startling. It seems that the symmetry is what is fundamental to the physics, more so than the particular model we choose to describe it.

The irreducible representations of $\hat{g}$ are infinite dimensional with the property that they are just towers of the finite dimensional representations of g. In compactified superstring theory, the affine multiplets describe particles at fixed helicity with mass values ranging from zero to infinity, in multiples of the Planck mass. As we shall see, only the massless particles will be used to describe the elementary particles.

VIII. Gauge Theories

A. Massless Particles

Gauge theory is a comprehensive and basic principle of particle physics. It explains and requires the masslessness of the bosons, which mediate the fundamental forces. It does this by requiring a theory to be invariant under *local* instead of *global* transformations. This allows the freedom to make different gauge transformations at different points in space–time. These gauge degrees of freedom can be used to fix some of the degrees of freedom of the gauge field, which results in forcing them to be massless particles. Since this is a symmetry of the entire theory, these particles cannot develop mass from quantum corrections. Thus the presence of massless particles in a theory signals the existence of a gauge symmetry, and the presence of a gauge symmetry requires massless particles. The $U(1)$ invariance of Maxwell's equations predicts the masslessness of the photon; $SU(3)$ color predicts the masslessness of the gluons; $SU(2) \times U(1)$ predicts the masslessness of the W^+, W^-, Z^0; and general coordinate invariance ensures the masslessness of the graviton. Supergravity is the local gauge theory of supersymmetry. Because its generators are fermionic (they have anticommutation relations), supergravity requires a massless gauge fermion: the spin $\frac{3}{2}$ gravitino. If supersymmetry is not broken, then all members of a supergravity multiplet have the same mass. When the supergravity multiplet contains a scalar (spin 0), then it will also be massless, which provides a mechanism for light scalars.

The gauge hierarchy question is why are the two scales of electroweak breaking (100 GeV) and the grand unification or Planck scale (10^{15}–10^{19} GeV) so different, and what sets their difference? An explanation of this can be given by superunification, which (1) sets the bottom scale at zero by treating all the observed elementary particles (these have mass <100 GeV) as members of massless supersymmetric multiplets and (2) sets the top scale at Planck's constant by eliminating the grand unification scale in favor of just one superunification scale. The actual non-zero masses of the quarks, leptons, and intermediate vector bosons are then to be derived from the spontaneous breakdown of $SU(2) \times U(1)$ and supersymmetry.

B. Spontaneous Symmetry Breaking

Spontaneously broken symmetry occurs when the solutions of a theory are not invariant under the symmetry group of the original action. In particular, the physical vacuum or ground state is not invariant. For continuous symmetry groups, the Noether charge Q, which generates the transformation, does not annihilate the vacuum $Q|0\rangle \neq 0$, and there exists an infinite set of degenerate vacua $|0\rangle$, $|0\rangle'$, and so on. In theories with scalar fields ϕ, where $[Q, \phi] \sim \phi$, symmetry breaking occurs when the vacuum expectation value $\langle 0|\phi|0\rangle$ is nonvanishing since then $\langle 0|[Q, \phi]|0\rangle \neq 0$, which implies $Q|0\rangle \neq 0$. In general, whenever a continuous symmetry is spontaneously broken, it gives rise to a massless Goldstone boson. For gauge theories, this becomes the Higgs mechanism, and the scalar degrees of freedom are "eaten" by the massless gauge particles, which then becomes massive (for broken supersymmetry, there is a Goldstone fermion that is eaten by the gravitino; this makes the gravitino massive, which may explain why it has not yet been observed).

Because the intermediate vector bosons are massive, the gauge symmetry $SU(2) \times U(1)$ must be spontaneously broken at a mass scale of about 100 GeV. Although the quarks and leptons are not gauge particles, as long as we are above the electroweak scale they will appear massless because they are chiral. Thus the apparent lightness of the families is tied together with the smallness of the electroweak scale.

Symmetry is an extremely powerful tool in unified theories. Regions where it is broken or not broken are different phases. For the unified groups, the symmetry is broken at low energies (the ordered phase) and restored at high energies (the disordered or symmetric phase.)

This is similar to the confinement of quarks inside hadrons. At low temperatures or densities, the quarks and gluons are in the confined phase, but for very high temperatures there appears to be an unconfined phase of a quark–gluon plasma. These ideas will be tested in the next few years at Brookhaven and Fermilab.

IX. Unified String Theories

Recently there have been some rather interesting developments in unified theories. In particular, it is now possible to describe a quantum theory of gravity (at least in one loop and probably to all orders.) This situation is comparable to Einstein's formulation of the general theory of classical gravitation and to the discovery of quantum mechanics itself.

This superstring theory has a sector that includes the particles of the gauge theories that have already been successful in a description of the strong and electroweak as well as the gravitational interaction. The string thus builds on the earlier understanding of gauge invariance and many of the ideas already incorporated into the ideas of a unified theory.

The string is a very short one-dimensional object in contrast with the pointlike entities of relativistic quantum field theory. Because of this, the string is inherently more ultraviolet convergent. In fact there is no renormalizable or finite description of quantum gravity in terms of a field theory with a finite number of fields. Furthermore, purely bosonic strings have not yet been formulated without tachyon. These particles with negative mass squared signal a wrong choice of the vacuum and subsequently make the scattering amplitudes infinite.

Currently the only candidates for finite string theories are supersymmetric, which implies that the massless gauge sector includes fermions as well as bosons. String theories are just field theories with an infinite number of fields. In the zero-width or tree approximation, these particles lie on Regge trajectories, that is, straight lines on the mass^2 versus spin Chew–Frautchi plot. The mass splitting is given by the Regge slope parameter α, which is equal to $1/m_{\text{Planck}}^2$ for strings whose massless spin two particle couples with Newton's constant. Since the Planck mass is 10^{19} GeV and the highest energy available even at the proposed super collider accelerator (SSC) will be of the order of 10^4 GeV, the elementary particle spectrum of the 12 vector bosons, the graviton, the three quark–lepton families, and possible spin zero Higgs particles will all be contained in the massless supersymmetric multiplets of the superstring.

The actual masses of the fermions and Higgs and the weak vector bosons must be introduced eventually by the spontaneous breakdown of supersymmetry and $SU(2) \times U(1)$.

But even without an understanding of the supersymmetry breaking, we see that the string predicts the particle content uniquely. In this way, the quest for a quantum theory of gravity has led to a unified field theory of the fundamental interactions of nature. This is, of course, not quite what Einstein envisioned, since he was seeking the unification on a classical level of gravitation and electromagnetism.

The original lowest-order amplitudes of the dual string first written down by Veneziano, approximated realistic scattering by zero-width processes. To allow for particle decay, it was subsequently shown that finite widths could be accommodated by a systematic expansion, similar to the loop expansion of field theory, which was constructed to satisfy perturbative unitarity. In the first quantized picture, interactions are included by evaluating the path integral of the two-dimensional sigma model describing the string world sheet. The expansion corresponds to summing over two-dimensional surfaces with different topologies, first the sphere, then for one-loop the torus, and so on. This way of introducing interactions corresponds to the string merely breaking or joining, described by a given vertex operator. So an interacting open string theory necessarily contains closed strings since the two ends of an open string can touch to form a closed one. The massless sector of open strings contains spin one gauge bosons, while for closed strings it contains the graviton. Therefore, whereas quantum field theory eschews gravity, string theory requires gravity.

A very peculiar feature of the strings is that they are 1-dimensional objects that naturally live in 10 dimensions of space–time instead of 4.

This is remiscent of the Kaluza–Klein theories. The extra dimensions are not an obstacle. We assign that 6 of them are small and compact (but mostly flat), and then we can interpret the particle spectrum as that of a 4-dimensional theory where the extra dimensions have become internal degrees of freedom. See Fig. 3. In addition to the infinite Regge tower of particles of the original 10-dimensional formulation there is now an infinite Kaluza–Klein tower as well. It turns out that this particle spectrum carries an irreducible infinite dimensional representation of the infinite parameter affine Kac–Moody algebra, which includes a gauge group whose rank is equal to the number of compactified dimensions. So the original attempt of unification to put all particles in one multiplet has been realized.

With the compactification procedure it is now possible to understand how string theories in different numbers of dimensions are related and that if a consistent theory is found in some dimension how it can be measured in another dimension. We are then naturally led to the question, how many truly different string structures are there? A consistent interacting string model is signaled by the presence of nonanomalous symmetries: general covariance, non-Abelian guage invariance, and 2-dimensional conformal invariance of the sigma model action, and the global reparameterization invariance [for example, the modular symmetry $SL(2, I)$] of the surfaces summed over in the path integral loop expansion. It happens that a model usually has all of these symmetries or none of them, which may indicate that there is one large symmetry group of the string. There is an intimate connection between the vertex operators for compactified strings and elements of the finite sporadic groups. In fact, when the 26-dimensional bosonic string is compactified to 2 dimensions by

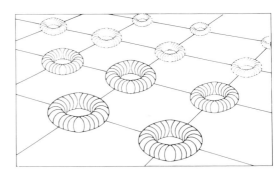

FIG. 3.   Six hidden dimensions.

requiring the internal momenta to take on discrete values given by points on the Leech lattice, the particle spectrum carries irreducible representations of the largest sporadic group, the Monster F_1, It is not unreasonable that the 10-dimensional superstring with its internal fermionic degrees of freedom is related to the higher-dimensional purely bosonic model and that there is just one basic string structure whose very large symmetry group we can call the Kac Cradle, the symmetry group of the string, related to the Monster group and the affine algebras.

Physicists are interested in how much of this coherent theory will trickle down to explain the phenomenology at 100 GeV and below. The fact that the fundamental compactification appears to be from 10 to 4 seems to indicate from the affine arguments that the fundamental gauge group will be a rank six group such as $SU(3) \times SU(2) \times U(1) \times U(1) \times U(1)$ (i.e., first of all there should be a few more Z^0's). The superstring in 10 dimensions has zero cosmological constant to one loop. In order to investigate the status of symmetry breaking, mass ratios for the low-lying particles, Yukawa couplings, proton decay, and the value of the cosmological constant in 4 dimensions, it is reasonable to set up an effective (i.e. approximate) 4-dimensional action and perhaps also and effective 10-dimensional action, which is, of course, a nonrenormalizable and nonfinite field theory of gravity involving a finite number of local fields reflecting the degrees of freedom of the massless sector of the sting. The Feynman graphs computed with this action are equal to the $\alpha' \to 0$ limit of the dual amplitudes in any given order of the loop expansion.

The outstanding problems of quark confinement and dynamical symmetry breaking are still with us. The achievement of a controlled systematic nonperturbative approximation has remained elusive. Many in the physics community have moved on to study interactions in the higher energy regions of the Planck mass where the coupling is weak and perturbation theory is applicable again. Unified string theory does not answer at this point the original questions of hadronic physics; but it most likely does provide

us with a theory of quantum gravity. Whether or not it survives the test of higher loops, and if so, if it is not a theory realized in nature, it has already contributed immeasurably to our intellectual development in particle physics. It is safe to say the old discussion that at Planck's length the Wightman axioms of constructive field theory and causality may break down and that graviton scattering cannot be calculated in a field theoretic context is no longer relevant. The string theory and its supergravity daughters enable us to calculate gravitational quantities on exactly the same footing as the other interactions. And, of course, once something can actually be calculated in an unambiguous fashion, it is usually not too hard to change it slightly to match nature.

BIBLIOGRAPHY

Alvarez-Gaume, L. and Witten, E. (1983). *Nucl. Phys.* B **234**, 269.

Dolan, L. (1984). *Phys. Rep.* **109**, 1.

Dolan, L. (1986). *In* "Proceedings of the Lewes Superstring Workshop" (L. Clavelli and A. Halprin, eds.). World Scientific, Singapore.

Feynman R. and Gell-Mann, M. (1958). *Phys. Rev.* **109**, 193.

Georgi, H. (1982). "Lie Algebras in Particle Physics." Benjamin, Reading, Massachusetts.

Glashow, S. (1961). *Nucl. Phys.* **22**, 579.

Green, M. B. (1983). *Surv. High Energy Phys.* **3**, 127.

Gross, D., Harvey, J., Martinec, E., and Rohm, R., (1985). *Nucl. Phys.* B **256**, 253.

't Hooft, G. (1971). *Nucl. Phys.* B **33**, 173; B **35**, 167 (1971).

Kac, V. G. (1968). *Izv. Akad. Nauk SSSR, Ser. Mat.* **32**, 1323 [Math. USSR Izv. **2**, 1271, *Engl. Transl.*].

Kac, V. G. (1985). "Infinite Dimensional Lie Algebras." Cambridge Univ. Press, London.

Kaluza, Th. (1921). *Sitzungsber. Preuss. Akad. Wiss. Berlin Math. Phys. Kl.* **966.**

Klein, O. (1926). *Z. Phys.* **37**, 895.

Moody, R. (1968). *J. Algebra* **10**, 211.

Salam, A. (1968). *In* "Elementary Particle Theory Relativistic Groups and Analyticity" (Nobel Symposium No. 8), p. 367. (Svartholm, Almqvist and Wiksell, eds.) Stockholm.

Schwarz, J. (1982). *Phys. Rep.* **89**, 223.

Weinberg, S. (1967). *Phys. Rev. Lett.* **19**, 1264.

INDEX

A coefficients, Einstein's, 159–160, 520–522
Aberrations, starlight, 613–614
Above-threshold multiphoton ionization, 350–351
Absolute objects, 591–592
Absolute space, 592–593
Absolute time
  in Newtonian mechanics, 592–593
  special relativity and, 617–618
Absorption
  acoustic waves, 724
  atomic, 58
  gamma rays in matter, 575
  holograms, 268
  light energy, stimulated emission and, 287
  multiphoton, 402–403
  neutrons in matter, 575–576
  nonlinear, in nonlinear optical processes, 402–403
  photons, Einstein's *B* coefficient, 160
  saturable, in nonlinear optical processes, 403
Absorption cross sections
  in quantum optics, 530–531
  selected elements, table, 444
Absorption spectra, molecular optical, 306, 323–324
AC Stark effect, 547–548
Acoustic wave devices, surface, 731
Acoustic waves
  attenuation, 722–724
  definition, 717
  directivity pattern, 722–723
  far field, 722–723
  forced oscillations, 718
  Fraunhoffer zone, 722–723

Fresnel zone, 722–723
  generation, 724–727
  high-power, effects, 732
  Lamb, 719–720
  longitudinal, 719
  mode conversion, 721
  near field, 722–723
  nonlinear effects, 724
  plate, 719–720
  pressure, 719–720
  propagation modes, 719
  propagation velocity, 719–721
  properties, 717–724
  quarter wave matching, 722
  Rayleigh, 719–720
  reception, 724–727
  rod, 719, 721
  scattering, 724
  shear, 719
  stress, emission, 731–732
  surface, 719
  torsional, 719
  transverse, 719
  ultrasonic field parameters, 721–722
  undamped harmonic oscillation, 718
  wave equation, 718–719
Adiabatic approximation, 27–28
Adiabatic rapid passage, 404
Adiabatic systems, 642
Aerosol production, ultrasound applications, 733
Airy discs, 438, 440–441
Airy pattern, 441
Algebra, supersymmetry, 743–744
Alkali halides, structure and bonding, 63
Alloys, magnetostrictive properties, 725

Alpha decay, nuclei, 421–422
Alpha particles
  early detection, 416
  sources, 573
Alternating current, electromagnetic fields, effect on plasmas, 457–458
Ambipolar diffusion
  in plasma physics, 456
  time-dependent diffusion equation, 456
Ambipolar diffusion coefficient, 456
Amman tiling, 552
Amorphous semiconductors
  applications, 13–14
  chalcogenide alloys, 7–9
  chemical modifications, 12
  classification, 2
  Cohen–Fritzche–Ovshinsky (CFO) model, 5
  defects, 9–11
  definition, 1
  doping, 11–12
  electronic structure, 3–9
  mobility edges, 4
  mobility gap, 5
  structure, 2–3
  transient effects, 12–13
Amplitude, attenuation coefficient for acoustic waves, 723–724
Angular momentum, 46–47
  diatomic molecules, 309
  hydrogen atom, 48–49
  photon, polarization effects and, 340
  quantization, 493
  symmetric top molecules, 317–318
Angular selectivity, in holography, 266

Anharmonicity, 219–220
Anisotropy, magnetization and, 255
Annihilation, particles and
    antiparticles, 634–635
Annihilation operators
    for electrons and positrons, 163
    in quantum electrodynamics, 157
Antiferromagnetic materials, 252
Antiferromagnetic order, 256
Antiferromagnetism, coexistence with
    superconductivity, 710
Antimatter, special relativity and,
    634–635
Antiparticles
    charge-conjugation, 175
    leptons, 179
    nature of, 175, 181
    quarks, 181
Antiprotons, nuclear reactions, 431
Arc discharges, 452, 459
Arc melting, plasmas in, 452–453
Argon lasers, 295
Asymptotic freedom
    in quantum chromodynamics,
        191–192
    quarks, 434
Atomic and molecular collisions
    Bethe–Born approximation, 34–36
    binary-encounter approximation,
        33–34
    charge exchange, 29–32
    charge exchange, two-state models,
        36–40
    classical deflection function, 22
    classical oscillator, 34
    cross sections, 16–17
    elastic collision expressions, 20–22
    elastic scattering, 22–24
    energy loss cross sections, 19–22
    impact parameter cross sections,
        17–22
    impulse approximation, 22–24
    inelastic collision expressions, table,
        36–37
    inelastic collisions, 32–42
    interaction potentials, 27–32
    scattering amplitude, 24–25
    semiclassical cross sections, 25–26
    wave mechanics of scattering,
        24–27
Atomic mass
    description, 418
    special relativity and, 633
Atomic number, 415
Atomic physics
    atomic units, 50
    charge-exchange collisions, 60
    complex atoms, 52–56
    constituents of atoms, 47
    cross section, 58–59
    elastic collisions, 59

helium atom, 51–52
history, 44–47
hydrogen atom, 48–50
inelastic collisions, 59
interaction of atoms with radiation,
    56–58
properties of atoms, 47
reactive collisions, 60
Atomic states, photon emission and,
    491–492
Atomization of liquids, ultrasound
    applications, 733
Atoms
    classical model, 485–488
    excited states, 491
    ground states, 486, 495
    Millikan's experiments, 45
    photon–atom interactions,
        158–159
    stationary states, 486
    two-level model, 522–525
Atoms, complex, 52–56
    electronic configuration and
        ionization energies, 54–55
    Hartree–Fock method, 53
    Hartree consistent field method,
        52–53
    j–j coupling, 53–56
    Russell–Saunders coupling, 53–56
Attractors
    Lorenz, 82–84
    strange, 82
Autler–Townes splitting, 547–548
Autocorrelation functions (statistical
    mechanics), 677
Average velocity, for plasmas, 457
Averaging, stochastic (statistical
    mechanics), 664

B coefficients, Einstein's, 160,
    520–522, 530–531
Baker's transformation, 85–87
Balmer formula, 46
Balmer series, 496
Band-gap energy, semiconductor, 232
Band structures, calculation, 514
Bardeen–Cooper–Schrieffer theory,
    680–686
    Cooper pairs, 681–682
    electron–lattice interaction,
        681–682
    Eliashberg correction to, 686–687
    pairing theory, 682–683
    virtual exchange of phonons,
        681–682
Barn cross section unit, 184
Baryonic matter
    composition, 131
    equation of state, 137–138

Baryons, color degree of freedom,
    188–189
Beam techniques, in intramolecular
    dynamics, 229
Becquerel unit, 586
Bell inequalities, 549
Beta decay processes
    double, 421
    nuclear, 420–421
Beta rays
    absorption in matter, 576
    sources, 573
Bethe–Born approximation, 34–36
Binary-encounter approximation,
    33–34
Birefringence, in nonlinear optical
    processes, 399–400
Black holes
    Hawking radiation and, 610–611
    Schwarzschild field and, 601–602
Bloch equations, 662–663
Bohr–Wheeler fissionability
    parameter, 422
Bohr theory, 45–46, 493–495
Bohr's frequency condition, 152
Boltzmann statistics, statistical
    mechanics and, 639
Bonding, in solids
    covalent crystals, 63–64
    crystal structure, 61
    description techniques, 62
    directed valence bonds, 63–64
    electronegativity, 62–63
    groups of structures, 62
    heat of formation, 62–63
    ionic crystals, 62–63
    metallic bonding, 65–66
    mixed covalent and ionic bonding,
        64–65
    molecular crystals, 62
    pseudopotential method for quantum
        structure, 67
    quantum structural analysis, 67
    quantum structural diagrams, 66–67
    radius ratio rules, 63
Boosts, Lorentz, 595
Born approximation, 21, 26, 34–35,
    37
Born–Oppenheimer approximation
    in intramolecular dynamics,
        213–214
    in molecular spectroscopy, 98
    polyatomic molecules and, 315
Born–Oppenheimer separation, 27
Bose–Einstein statistics, 639
Bosons
    experimental discoveries, 740
    in quantum electrodynamics, 154
    weak interactions, 738–739
Boundary conditions, wave motion
    and, 499–500

Brackett series, in quantum mechanics, 496
Bragg angles, in holography, 266
Branches, diatomic molecules, 310
Breakdown, in plasma physics, 460
Brightness, slow positron beams, 473–474
Brillouin scattering, stimulated, 396–397
Brillouin zones, decomposition at center of, 233
Brownian motion, classical, 663–666
Brueckner–Bethe–Goldstone method, for neutron matter, 131

C-systems, in intramolecular dynamics, 218
Carbon dioxide lasers, 295
Casting of metals, ultrasound applications, 734–735
Caustic, fields near, 442
Cavitation, ultrasonic, 732
Ceramics, piezoelectric, 726
Čerenkov radiation detector, 188
Chalcogenide alloys, 7–9
Chaos, classical systems, 206
    definition, 72
    ergodicity, 72
    extreme sensitivity to initial conditions, 73–74
    fractals, 75–76
    mixing, 73
    unpredictability, 74–75
Chaos, dissipative systems
    applications, 84–85
    Hénon map, 81–82
    logistic map, 76–81
    Lorenz attractor, 82–84
    period doubling, 77–80
    strange attractors, 82
    universality, 80
Chaos, Hamiltonian systems
    applications, 91–92
    Baker's transformation, 85–87
    chaotic mixing, 86–87
    Hénon–Heiles model, 89–91
    Kolmogorov–Arnold–Moser theorem, 89
    Poincaré sections, 90–91
    resonance islands, 87–89
    standard map, 87–89
Chaos, quantum, 92, 206, 213
    microwave ionization of highly excited hydrogen atoms, 94–96
    problem of, 92
    standard map, 94
    symptoms of, 92–94
Chaotic energy transfer, 213–214, 227

Chapman–Kolmogorov equation, 663
Characteristic function (statistical mechanics), 643–644
Charge exchange collisions, 29–30, 36–40, 60
Charge-parity violation, see CP violation
Charge transfer fluctuation, role in superconductivity, 711–712
Charge transfer insulating states, 707–708
Charged current interactions, in electroweak theory, 195
Chemical analysis, ultrasound applications, 729
Chemical lasers, 296
Chemical physics
    Born–Oppenheimer approximation, 98
    diffraction methods, 101–102
    dynamics of molecular processes, 102–104
    electric multiple moment, 99
    electronic spectroscopy, 99–100
    equilibrium statistical mechanics, 105–106
    Green–Kubo relations, 106
    linear response theory, 106
    molecular dynamics, 106
    molecular spectroscopy, 98–101
    molecular structures, 101–102
    Monte Carlo simulations, 107
    nonequilibrium statistical mechanics, 106
    numerical statistical mechanics, 106–107
    partition function, 105, 107
    path integral methods, 107
    quantum systems, simulation, 107
    radial distribution function, 105–106
    radiofrequency spectroscopy, 100–101
    statistical mechanics, 105–106
    structure inferred from moments of inertia, 102
    time correlation function, 106
    vibrational spectroscopy, 100
Chemisorption, in plasma etching, 468
Christoffel symbols, 595
Chromophores, polyatomic molecules, 317
Circular dichroism spectrum, 323
Classical mechanics
    logical limit, 486–487
    success of, 484–485
Classical physics
    electron transport in solids and, 511–512
    failure of, 486

Clocks
    in general relativity, 598–599
    Lorentz transformation and synchronization of, 620–621
Coercive force, hysteresis loop and, 255
Cohen–Fritzche–Ovshinsky (CFO) model, 5–6
Coherence, of light, 534–537
Coherence length, 535
Coherence time, 535
Coherent anti-Stokes Raman scattering, 332
    applications, 408
    description, 408
Coherent neutron scattering, 354–355
Collapse, gravitational, 601–602
Collision cross sections, plasma particles, 454–455
Color, quarks, 434
Color centers, radiation effects, 579, 582, 584
Color confinement, quarks, 434
Color degree of freedom, quarks, 188–191
Commutation relations, quantized electromagnetic fields, 158
Composite systems (statistical mechanics), 655
Composition, plasmas, 454
Compressibility, isothermal, 649
Compression modulus, nuclei, 424
Compton scattering, photons, 165
Compton wavelength, 505
Conditional averaging, 671
Conduction bands, exciton, 232
Conductivity, electrical
    dense matter, 138–140
    in plasma physics, 457
Conservation laws
    elementary particle physics, 176–177
    relativistic quantum electrodynamics, 164
Conservation of probability, 657
Conservation relations, 655
Constant of the motion, global, 215
Constrained variational method, for neutron matter, 131
Construction industry, laser applications, 301
Continuity equation (statistical mechanics), 657
Continuum–continuum transitions, in multiphoton ionization of atoms, 350–351
Contracted descriptions (statistical mechanics), 677
Cooper pairs
    excitons and, 248
    formation, 681–682
    superconductors and, 517

Coordinate transformations (special relativity), 618
Copper vapor lasers, 296–297
Corpuscular theory, of light, 150
Correlated electrons
    almost-localized Fermi-liquid metallic states, 707–708
    charge excitations, 711–712
    charge–transfer insulating states, 707–708
    electronic states of superconducting oxides, 698–701
    hole states, 688–689
    Hubbard Hamiltonian, 688–689
    Hubbard subbands, 688–689
    hybridized systems, 697–701, 705–706
    kinetic exchange interaction, 693–694
    magnetic interactions, 708–709
    magnetic phases, 693–694
    metal–insulator transitions, 690–693
    mixed-valent metallic states, 707–708
    Mott–Hubbard insulating states, 707–708
    Mott transition, 687
    narrow-band systems, 687–697, 701–705
    normal and magnetic states, 687–701
    pairings, 708–709
    pairings, phonon-mediated, 710–711
    pairings, real-space spin-singlet, 701–710
    polarons, magnetic, 694–695, 708–709
    properties, 706–710
    spin liquids, 695–697
Cosmic censorship hypothesis, 602
Cosmological term, general relativity and, 598
Cosmology, special relativity and, 635
Cotton–Mouton effect, 405
Coulomb barrier, in alpha decay, 421–422
Coulomb energy, nuclei, 419
Coulomb forces
    interaction potentials in atomic and molecular collisions, 27
    in nonrelativistic quantum electrodynamics, 158
Coupling constants, 183
    charged current interactions and, 195
    in higher-order Feynman diagrams, 184–185
    running, in quantum chromodynamics, 191

    strong, gluons and, 192
    Yukawa, in Higgs mechanism, 196, 198
Coupling term, in intramolecular dynamics, 214
CP violation, 201–202
    CP invariance, 110–111
    CP symmetry, 109–111
    CPT theorem, 110–111
    in kaons, 111–112
    Kobayashi–Maskawa theory, 112
    observation of, 111–112
    research areas, 112–114
    superweak theory, 111–112
    time reversal symmetry, 110
CPT theorem, 110–111
Creation operators
    for electrons and positrons, 163
    in quantum electrodynamics, 157
Critical point (statistical mechanics), 650
Cross-correlation functions (statistical mechanics), 678
Cross sections
    barn, 184
    definition, 16–17
    differential, 58
    diffusion, 20, 22
    electromagnetic force, 184
    impact parameter, 17–22
    neutron scattering, table, 354
    nuclear stopping, 20
    semiclassical, 25–26
    strong force, 184–185
    total, 58
    weak force, 184
Crystals
    covalent, structure and bonding, 63–64
    ionic, structure and bonding, 62–63
    molecular, structure and bonding, 62–63
    nonlinear optical, properties, 372–373
    semiconducting, 231–232
Crystals, structure
    covalent, 63–64
    diamond and zinc-blende, excitons, 234
    ionic, 62–63
    molecular, 62–63
    wurtzite, exciton systems, 233–234
Curie radiation unit, 586
Curie temperature, 253–254

Damping constant, 673
Data banks, radiation effects, 584
Data processing, optical, holography and, 271–272

Data storage, holography and, 272
de Broglie relation, 46, 493, 504–505
Debye length, 461
Decay, subatomic particles, special relativity and, 634
Decoupling, in intramolecular dynamics, 218
Deflection function, center of mass, 22
Deformations, equilibrium, nuclei, 424–426
Degassing, liquids and melts, ultrasound applications, 734
Delaunay structures, quasicrystals, 556–557
Demkov approximation, 37, 39–40
Dense matter
    composition, 126–132
    electrical conductivity, 138–140
    first density domain, equation of state, 132–134
    liquid drop model, 129–130
    neutrino emissivity, 144–146
    neutrino opacity, 146–148
    neutron drip, 129
    neutronization, 126–128
    nuclear semiempirical mass formula, 128–129
    pion condensation, 131–132
    pion condensation domain, equation of state, 136–137
    second density domain, equation of state, 134
    shear viscosity, 142–143
    thermal conductivity, 140–142
    third density domain, equation of state, 134–135
Dense matter physics
    background and scope, 116–117
    basic theoretical method, 117–126
    definition, 116
    electrostatic interaction, estimation, 122–123
    equation of state, 121–122
    Fermi gas method, 119–120
    at high temperatures, 124–126
    Pauli's exclusion principle, 118–119
    pressure equations, 120–121
    relativistic electrons in, 121
    Thomas–Fermi method, 123–124
Density
    local (statistical mechanics), 669, 673
    plasmas, 453–454, 461
Density distribution function
    classical canonical ensembles, 644
    classical grand canonical ensembles, 647
    maximum entropy principle and, 651–652
    microcanonical ensembles, 642
    in statistical mechanics, 638–639

Density matrix method, in multiphoton spectroscopy, 329–331
Density of states, in intramolecular quantum dynamics, 225
Density operators
  maximum entropy principle and, 653
  quantum canonical ensembles, 644
  quantum grand canonical ensembles, 649
  quantum isobaric–isothermal ensembles, 645
  in statistical mechanics, 641
Dephasing, in intramolecular dynamics, 225
Dephasing constant, in multiphoton spectroscopy, 330
Desorption, in plasma etching, 468–469
Deuteron nuclide, 417
Diabatic approximation, 27
Diamagnetism, 251–252
Diamonds, exciton systems in, 234
Diatomic molecules
  angular momentum, 309
  branches, 310
  electronic ground state, 313
  Franck–Condon principle, 310
  heads, 311
  molecular rotation, 308
  molecular term, 310
  Morse curves, 313
  optical spectroscopy, 307–313
  potential functions, 308
  progressions, 310
  reciprocal moment of inertia, 308
  selection rules, 310
  spin–orbit coupling, 310
  spin multiplicity, 309
  vibrational degree of freedom, 307
Dichromated gelatin, in holography, 268
1,2-Dienes, elimination methods in synthesis, 127
Diffraction
  efficiency, holograms, 265, 267
  holographic image reconstruction and, 264
  limit on resolution of positron reemission microscopes, 478
  neutron, 357
  orders, holographic reconstructed beams, 266
Diffuse scattering, neutrons, 358
Diffusion
  ambipolar, 456
  chaotic, Hamiltonian systems, 89
  hydrogen, Mössbauer spectroscopy, 268
  tensors, 662

Diffusion coefficients
  in plasma physics, 455
  in statistical mechanics, 666
Diffusion equation
  ambipolar time-dependent, 456
  time-dependent, 455
Dilation of time, Lorentz transformation and, 621–622
Diode lasers, 298
Dirac delta function, 641
Dirac equation, 47
  in quantum electrodynamics, 154
  relativistic fermions and, 193
  in relativistic quantum electrodynamics, 162
Direct current, magnetic fields, effect on plasmas, 460–461
Discharge machining, 451–452
Disequilibrium, in classical systems, 656
Dispersion, third- and higher-order frequency-conversion processes, 384–385
Dispersion relations
  for free particles, 506
  for particles with potential energy, 507–508
  waves (quantum mechanics), 499
Displacement effects
  mechanism (radiation physics), 577–578
  in optical materials, 582
  in semiconductors, 580–582
Dissipation energy (statistical mechanics), 662
Dissociation, molecular, 207
Disturbances, external, system response to, 674–675
Divergences, in quantum electrodynamics, 154–155
Donor–acceptor pairs, bound-exciton complexes and, 242–243
Doping, amorphous semiconductors, 11–12
Doppler broadening, laser linewidth and, 291
Doppler effect
  laser linewidth and, 292
  transverse, 622
Doppler-free multiphoton spectroscopy, 337, 348–349
Droplets, hole–electron, 247–248
Dry etching, with plasmas, 466
Drying, heat-sensitive materials, ultrasound applications, 734
Dynamical objects, in space–time theories, 591–592
Dynamical symmetry breaking, 203
Dyson's perturbation theory, in quantum electrodynamics, 155

Echoes, photon, 526–527
Effective permittivity (plasma physics), 457–458
Eigenfunctions
  in quantum mechanics, 223
  in statistical mechanics, 640
Eigenstates, 639
Eigenvalues
  in quantum mechanics, 223
  in statistical mechanics, 639
Einstein–Podolsky–Rosen, quantum theory and, 548–549
Einstein's A coefficient, 159–160, 520–522
Einstein's B coefficient, 160, 520–522, 530–531
Elastic collisions, 59
  Born approximation, 21
  impulse estimates, 21
  Lenz–Jensen potential, 22
  Massey–Mohr, 21
  quadratures, 20–21
Elastic scattering
  deflection function, 22
  impulse approximation, 22–24
  wave mechanics, 24–27
Elasticity, quasicrystals, 566–567
Electric dipole radiation, 56
  selection rules, 56–57
Electric fields, Earth
  Lorentz transformation, 632–633
  quantized, electrodynamics, 157
Electrical resistance, 513
Electroacoustic transducers, 724–725
Electrodynamics, classical, 150–151
Electromagnetic fields
  alternating current, effect on plasmas, 457–458
  covariant quantization, 163–164
  direct current, effect on plasmas, 460–461
  quantized, electrodynamics, 157–158
Electromagnetic force
  basic forces and, 177–178
  leptons and, 179
  photons and, 175
  quarks and, 181
  reaction cross sections, 184
  strength relative to weak force, 195
  unification with weak force, 193
Electromagnetism
  Lorentz transformation and, 631–633
  in unified field theory, 738
Electron capture processes, nuclei, 420
Electron diffraction, molecular structures, 102
Electron drift velocity, in plasma etching, 467
Electron g value, bound-exciton complexes, 239

Electron scattering, in quantum mechanics, 515
Electron transport, in solids, 511–518
  energy bands, 514–515
  energy gaps, 514–515
  insulators, 515–516
  metals, 515
  periodic potential for crystalline solids, 513–514
  quantum mechanics approach, 512–513
  semiconductors, 516–517
  superconductors, 517–518
Electronegativities
  ionic crystals and, 62–63
  Pauling's table, 62
Electronic configuration, ionization potentials for elements and, 54–55
Electronic energy levels, diatomic molecules, 307
Electronic excited states, diatomic molecules, 310
Electronic spectroscopy, 99–100
Electronic structure
  amorphous semiconductors, 3–9
  quasicrystals, 567
Electrons
  absorption in matter, 576
  anomalous magnetic moment, 168–169
  Dirac wave function, 193
  discovery, 45
  early models, 416
  in lepton family of elementary particles, 179–180
  magnetic moment, 47
  mass, 174
  nuclear reactions, 431
  in particle colliders and accelerators, 186–187
  properties, 175
  quantum electrodynamics, 154
  relativistic, in dense matter physics, 121
  relativistic quantum electrodynamics, 162–163
  size, 173
  sources, 573
  spin angular momentum, 46–47
Electrostatic interactions, dense matter, 122–123
Electrostriction, in nonlinear optical processes, 363, 398
Electroweak and strong interactions
  charged current interactions, 195
  coupling of quarks, 197–199
  fermions, left- and right-handed, 193
  Higgs mechanism, 196–197

  neutral current interactions, 195–196
  phenomenology, 199–200
  $SU(2) \times U(1)$ gauge theory, 193–196
Elementary particle physics
  affine Kac–Moody Lie algebras, 744
  antiparticles, 175, 181
  basic forces, 177–178
  charge conjugation, 175
  charge conjugation-parity (CP) violation, 201–202
  classification of elementary particles, 739
  conservation laws, 176–177
  dynamical symmetry breaking, 203
  experimental techniques, 185–188
  force-carrying particles, 178–179
  gauge theories, 744–745
  grand unified gauge theories, 202–203
  gravitational force, 177–178
  Kaluza–Klein theories, 742
  particle accelerators, 186–187
  particle colliders, 186–187
  particle detectors, 187–188
  particle interactions, 181–185
  propagator, 183, 199–200
  reaction cross sections, 184–185
  spontaneous symmetry breaking, 745
  standard model, 200–202
  supergravity, 743–744
  supersymmetric gauge field theories, 203–204
  unified string theories, 745–746
Elementary particles
  classification, 739, 742–744
  experimental discoveries, 740
  force-carrying, 178–179
  high-energy particles, see High-energy particles
  massless, gauge theory, 744–745
  properties, 174–175
  range of interactions, 740–741
  size, 173
  spectrum, 739
  strength of interactions, 741
  strong interactions, 738
  tau particles, 179–180
  types of, 174–175
  unified string theories, 745–747
  weak interactions, 738–739
Elements
  electronic configuration, table, 54–55
  ionization energies, table, 54–55
Eliashberg theory, 686–687
Emission
  spontaneous, 56

  spontaneous photon, Einstein's A coefficient for, 159–160
  stimulated photon, Einstein's B coefficient for, 160
Emission spectra, molecular optical, 306, 323–324
Emissivity, neutrino, in dense matter, 144–146
Emulsification, ultrasound applications, 733
Energy, relativistic expressions for, 626
Energy bands, electron transport in solids and, 514–515
Energy conservation, in collisions, 627–629
Energy eigenvalue equation, 510
Energy eigenvalue spectrum, 516
Energy gaps
  electron transport in solids and, 514–515
  superconductors, 517
Energy–momentum relations
  for a free particle, 488
  for particles with potential energy, 506–507
  in special relativity, 629
Energy transfer, intramolecular
  anharmonicity, 219–220
  beam techniques, 229
  Born–Oppenheimer approximation, 214
  C-system, 218
  chaos, quantum and classical, 206, 213–214, 225
  chaotic, NaBrKCl, 207–212, 221
  computer solutions, numerical, 207
  constant of the motion, 215
  coupling term, 214
  decoupling, 218
  density matrix, 224
  density of states, 225
  dephasing, 225
  dissociation, 207
  dynamics, 213–229
  eigenfunctions, 223
  eigenvalues, 223
  energy hypersurface, 217
  ergodic systems, 217
  frequencies of motion, 217
  Hamilton–Jacobi approach, 215
  Hamiltonian, 214
  Hamilton's equations, 214–215
  Hilbert-space vector, 224
  integrable systems, 215–217
  irregular systems, 218
  irreversible energy flow, 206
  irreversible equations, 213
  long-lived intermediates, 208–209
  mixed state, 224

mixing systems, 217–218
molecular systems, 218–219
nodal patterns, 227
nonresonant, 220
normal coordinates, 219
oscillators, 216
phase space, 215, 223
pure state, 224
qualitative dynamics, 207–214
quantization of, 223
quantum beats, 224
quantum/classical correspondence, 206, 213
quantum dynamics, 223–228
quantum trapping states, 212
quasiperiodicity, 216
rate of, 211
regular systems, 215–217
relaxation behavior, 210
resonant, 221
resonant system, 219
reversible energy flow, 206
RRKM theory, 228
scope of, 205
statistical approximations, 228–229
statistical behavior in dynamics, 216–217
statistical dynamics, 206
time-reversal invariant equations, 213
trajectories, 208, 217
trapping effect, 220
tunneling, 223
unimolecular decay, 225, 228
vibrational energy, 210
wavefunctions, 227
Ensembles
concept of (statistical mechanics), 638
Gibbsian, 642–650
Ensembles, canonical
characteristic function, 644
classical systems, 643–644
closed, isothermal systems, 644
Gibbs entropy, 644
heat capacity, 644
Helmholtz free energy, 644
internal energy, 644
partition function, 644
quantum systems, 644–645
Ensembles, grand canonical
classical systems, 647–649
density distribution function, 647
density operator, 649
Gibbs entropy, 649–650
heat bath, 647
isothermal compressibility, 649
number operators, 649
open, isothermal systems, 647
open systems, 647
quantum systems, 649–650

Ensembles, isobaric–isothermal
classical systems, 645–646
density operator, 646
enthalpy, 646
Gibbs entropy, 646
Gibbs free energy, 646
heat bath, 645
quantum systems, 646–647
Ensembles, microcanonical
characteristic function, 643
classical systems, 642–643
density distribution function, 642
energy shell, 642
ergodic hypothesis, 642
Gibbs entropy, 642
principle of equal a priori probabilities, 642
projection operator, 643
quantum systems, 643
Enthalpy, isobaric–isothermal ensembles, 645
Entropy
density, 669
local production postulate, 668
production, local rate of, 669
Equal a priori probabilities, principle of, 642
Equation of state, dense matter, 121–122
first density domain, 132–134
pion condensation domain, 136–137
second density domain, 134
third density domain, 134–135
Equations of state
baryonic matter, 137–138
neutron matter, 135–136
quark matter, 138
Equilibrium
local, 667
statistical, composite systems, 655–656
in statistical mechanics, 638
Equilibrium equations, spontaneous fluctuations, 665–666
Equilibrium time correlation function, 665–666
Equivalence, principle of (general relativity), 596–597
Ergodic hypothesis, 642
Ergodic systems, 217
Ergodicity, classical chaotic systems, 72
Etchant-unsaturate model, in plasma polymerization, 470–471
Event horizons, 600
Exchange forces, helium atom, 51–52
Excitations
charge, role in superconductivity, 711–712
lasers, mechanisms, 287–288

Excited states, 491
Exciton-bound-phonon quasi-particle, 243–245
Excitons, semiconductor
acceptors, 237
band-gap, 232
band symmetries, 233
binding energy determination, 232
Bloch-like functions, 233
bound complexes in degenerate semiconductors, 237–239
bound complexes in nondegenerate semiconductors, 239
Brillouin zone, 233
central-cell corrections, 232
conduction band, 232
Cooper pairs, 248
Coulomb interaction, 232
deformation-potential theory, 236
in degenerate semiconductors, 237–239
density of states, 248
diamond and zinc-blende structures, 234
dipole radiation, 233
in direct degenerate semiconductors, 234
in direct nondegenerate semiconductors, 233–234
donor–acceptor pairs, 242–243
donors, 237
electron g value, 239
exciton-bound-phonon quasi-particles, 243–245
extrinsic (bound-exciton complexes), 237–243
Fabry–Perot etalons, 250
Fermi–Dirac statistics, 248
Fermi surface, 232
Frenkel model, 231
Frohlich electron–phonon interaction, 245
Green's function method, 232
Hamiltonian, 232, 235–238
Hartree–Fock calculations, 232
hole–electron droplets, 247–248
indirect, germanium and silicon, 234 235
indirect transitions, 234–235
intrinsic, 232–237
irreducible representations, 233
k space, 235
Landau-type solutions, 240
laser sources, 245
lasing transitions, 249–250
magnetic fields, 235–236, 239–240
Maxwell–Boltzmann statistics, 248
multibound excitons, 241–242
multiquantum well structure, 246–247

Excitons, semiconductor (*continued*)
  nondegenerate semiconductors, 239
  optical bistability, 250
  oscillator strength, 237
  Pauli principle, 242
  phonons, 234, 243–245
  photons, spatial resonance
      dispersion, 245–246
  semiconducting crystals, 231–232
  shell model, 242
  strain field, 236–237
  stress field, 240–241
  superconductivity, 248–249
  superlattice structures, 246
  two-photon processes, 245–246
  uniaxial strain, 240–241
  uniaxial stress, 236
  unit cell, 237
  valence band, 232
  van der Waals attraction, 243
  Wannier excitons, 232
  wave vectors, 235
  wurtzite structure, 233
Extraction, ultrasound applications,
    733
Extreme sensitivity to initial
    conditions, classical chaotic
    systems, 73–74
Extreme ultraviolet lasers, 294

Fabry–Perot cavity, nonlinear, 411–414
Fabry–Perot etalons, 250
Faraday effect, in nonlinear optical
    processes, 405
Faraday's constant, 44
Fermi–Dirac statistics
  hole–electron droplets, 248
  statistical mechanics and, 639
Fermi coupling constant, in
    electroweak interactions, 199–200
Fermi energy, 515
Fermi gas method, 119–120
Fermions, 174, 188–190, 193–196, 203
  left-handed, 193
  in quantum electrodynamics, 154
  right-handed, 193
Ferrimagnetic materials, 252
Ferrimagnetic order, 256
Ferromagnetic domains, 256–257
Ferromagnetic order, 256
Ferromagnetism
  anisotropy energy, 255
  antiferromagnetic materials,
      251–252
  antiferromagnetic order, 256
  coercive force, 255
  Cooper pairs, 258
  Curie temperature, 253–254

definition, 251–252
diamagnetic materials, 251–252
ferrimagnetic materials, 251–252
ferrimagnetic order, 256
ferromagnetic domains, 256–257
ferromagnetic materials, 258–259
ferromagnetic order, 256
hysteresis loop, 254–255
magnetic change point, 253–254
magnetic susceptibility, 256
magnetization curves, 254
magnetostriction, 257
magnons, 257
Neel temperature, 256
occurrence, 255
paramagnetic materials, 251–252
remanence, 255
saturation moment, 256
superconductivity, 257–258
temperature dependence, 255
Weiss molecular field, 253
Feynman diagrams, 181–185
  coupling constants and, 184–185
  particle decay and, 183
  perturbation calculations, 184–185
  for photon absorption, 160
  photons and, 184
  for quantum electrodynamics, 155
  for relativistic quantum
      electrodynamics, 164–166
  for spontaneous photon emission,
      159
  for Thomas scattering, 161
Fine arts, laser applications, 302
Fine-structure constant, in quantum
    electrodynamics, 154
Fission method, for neutron
    production, 355
Fizeau experiment, 614–615
Floating potential, plasma etching and,
    467
Flow measurements, ultrasound
    applications, 729
Fluctuation–dissipation relation, 666
Fluctuation–dissipation theory, 675
Fluence, radiation, 574
Fluorescence detection, in multiphoton
    spectroscopy, 333–334
Fluorescence spectra
  molecular optical spectroscopy, 306
  polyatomic molecules, 314
Fluorine-to-carbon ratio, in plasma
    polymerization, 469–470
Flux, radiation, 574
Flux operator, 673–674
Foam control, ultrasound applications,
    734
Fokker–Planck equation
  generalized, 664
  in statistical mechanics, 661–662

Force, in relativity theory, 627
Formal intensity law (multiphoton
    spectroscopy), 326, 338–339
Four-wave difference-frequency
    mixing, 387–388
Four-wave sum-frequency mixing,
    387–388
Fourier analysis, optical diffraction
    and, 438
Fourier optics, optical diffraction and,
    438
Fourier transform mass spectrometers,
    particle analysis with, 465
Fractal images, classical chaotic
    systems, 75–76
Franck–Condon principle, 310
Fraunhofer diffraction formula, 440
Free bodies
  in Newtonian mechanics, 593
  in special relativity, 594
Free diffusion equation
  in plasma physics, 455
  time-dependent, 455
Free electron lasers, 298–299
Free electron model, electron transport
    in metals, 515
Frenkel excitons, 231–232
Frequency-conversion processes
  applications, 368–369
  description, 368–369
  nonlinear optical crystals, table of
      properties, 372–373
  parametric, 369–389
  second-order interactions, table,
      374–376
  second-order nonlinear
      polarizations, table, 371
  stimulated scattering, 389–397
  wavelengths for nonlinear
      parametric interactions, table,
      369
Frequency shifts
  in stimulated Brillouin scattering,
      396–397
  in stimulated Raman scattering, 392
Fresnel diffraction formula, 440
Fringes, in holography, 263
Frohlich electron–phonon interaction,
    245
Fusion energy, lasers and, 301–302

Gain coefficients
  in stimulated Brillouin scattering,
      396–397
  in stimulated Raman scattering, 392
Gain media, laser
  bandwidth, 289
  homogeneous linewidth, 291

inhomogeneous linewidth, 291–292
population inversions, 287–289
shape, 289
Galilean invariance, 593
Galilean reference frames, 482
Galilean relativity, 592, 617
Galilean transformation, 618
Gamma rays
absorption in matter, 575
nuclear reactions, 431
sources, 573
Gases
hot, temperature measurement, 729
inversions in, 288
Gauge bosons
classification, 739
in dynamical symmetry breaking,
203
in grand unified gauge theories,
202–203
mass by Higgs mechanism,
196–198
mass prediction by standard model,
200–201
in particle decay, 199
in quantum chromodynamics, 190
$SU(2) \times U(1)$ gauge theory, 194
in supersymmetric gauge field
theories, 203–204
types and properties, 174–175,
178–179
W^+, W^-, and Z^0 particles,
discovery, 740
W^+, W^-, and Z^0 particles, in Higgs
mechanism, 197
W^+, W^-, and Z^0 particles, masses,
200
W^+ and W^- particles in charged
current interactions, 195
Z^0 particle and parity violation, 196
Z^0 particle coupling to fermions, 195
Z^0 particle mass, 174
Gauge field theories, 190
Gauge symmetry, 179, 744
Gauge theories
for massless particles, 744–745
spontaneous symmetry breaking,
745
Gauss' law, 123
General covariance, principle of, 591
Geodesic deviation, 595
Geometric theory, optical diffraction,
443–445
Geometrical objects, in space–time
theories, 590
Geometry, quasicrystals, 556–557
Germanium crystal structures, indirect
excitons, 234–235
Gibbs entropy
canonical ensembles, 644

isobaric–isothermal ensembles,
645
microcanonical ensembles, 642
quantum canonical ensembles, 645
quantum grand canonical ensembles,
649–650
in statistical mechanics, 642
Gibbs free energy, isobaric–isothermal
ensembles, 645
Glow discharges
description, 458
microelectronic applications, 467
molecular optical spectroscopy and,
306
Gluons
classification, 739
description, 434
experimental evidence, 192
properties, 178
in quantum chromodynamics,
190–191
strong coupling constant and, 192
Gradient operators, 657
Grand unified theories
gauge theories, 202–203
supersymmetry, 203–204
Gravitational collapse
black holes, 601–602
dense matter physics, 116
Gravitational constant, 607
Gravitational fields
Newtonian fields, 599
Schwarzschild field, 599–601
Gravitational force, 177–178
Gravitational radiation, 602–603
Gravitational waves, 608–609
Gravity
quantum mechanics and, 610–611
supergravity, 611
in unified field theory, 738
Gray radiation unit, 574
Green–Kubo relations, 106
Green's functions
optical Stark effects in multiphoton
spectroscopy, 328–329
resonance effects in multiphoton
spectroscopy, 328–329
two-body, for excitons, 232
Green's theorem, 439
Grinding, impact, ultrasound
applications, 733
Gross observables, 661
Ground states
atoms, 486, 495
quasicrystals, 557–558
Group velocity
dispersion relations and, 514
waves (quantum mechanics),
500–501
Growth, quasicrystals, 558–559

H_2^+ potentials, 30–31
Hadrons
collision with leptons, 191–192
composition by quarks, 180–181
conservation laws, 176–177
electroweak decay, 199
experimental discoveries, 740
quark model, 188
reactions with elementary particles,
176
strong interactions, 738
Hamilton–Jacobi approach, to classical
dynamics, 215
Hamiltonian
classical, 639
exciton, 235–236
in intramolecular dynamics, 214
Hamiltonian operator
in nonrelativistic quantum
electrodynamics, 158–159
in quantum electrodynamics, 157
Hamilton's equations, in
intramolecular dynamics, 214
Harmonic oscillator, quantum
mechanics, 511
Hartree–Fock methods
for complex atoms, 53
for neutron matter, 131
Hartree consistent field method,
52–53
Hawking radiation, 610–611
Heads, diatomic molecules, 310–311
Heat, concept of, 655
Heat bath, isobaric–isothermal
ensembles, 645
Heat capacity, 644
Heat flow, in plasma physics, 457
Heat of formation, 62–63
Heat transfer, concept of, 655
Heisenberg representation, 660
Heisenberg uncertainty principle
high-energy particles and, 173
in quantum mechanics, 492–493
in statistical mechanics, 639
Helium atom, 51–52
decay of $^3S_1$ state, 51–52
exchange force, 51
ground state, 51
isotopes, 417
orthohelium, 52
parahelium, 52
Pauli exclusion principle, 51
singlet state, 51
triplet state, 51
Helium–cadmium lasers, 296
Helium–neon lasers, 295
Helmholtz energy, 644
Helmholtz–Kirchhoff integral
theorem, 439–440
Hermitian operators, 640

Hidden variables, in quantum optics, 549
Higgs mechanism, in electroweak theory, 196–198
High-energy particles
    elementary particle physics, 173–174
    and Heisenberg uncertainty principle, 173
    in particle detectors, 187–188
Hilbert-space vector, 224
Hénon–Heiles model, 89–91
Hénon map, 81–82
Hole–electron droplets, 247–248
Holograms
    absorption, 268
    amplitude, 266
    classification, 265–267
    definition, 262
    diffraction efficiency, 265
    Fourier transform, 267
    Fraunhofer, 267
    Fresnel, 267
    lensless Fourier transform, 267–269
    multiplex, 273
    phase, 266
    quasi-Fourier-transform, 267–269
    reflection, 267, 270
    thick, 266
    thick-phase, 265
    thin, 265–266
    transmission-type, 266–267
Holographic optical elements, 272
Holography
    angular selectivity, 266
    Bragg angles, 266
    color, 269
    data storage and, 272
    definition, 261
    dichromated gelatin, 268
    diffraction, 264
    diffraction efficiencies of hologram classes, 267
    diffraction orders, 266
    fringe spacing, 263
    fringes, 263
    holographic optical elements, 272
    intensity information, 262
    interference fringe spacing, 264
    interference fringes, 263
    interferometry, 270–271
    nondestructive testing, 269–271
    optical data processing and, 271–272
    optical wavelengths, 265
    phase information, 262
    photochromics, 269
    photoplastics, 269
    photopolymers, 269
    photorefractive materials, 269

polarization of reconstructing beam, 265
reconstruction, 263–265
recording materials, 267–269
refractive index of materials, 266
silver halide photographic emulsions, 268
spatial coherence, 262–263
speckle, 269
stereograms, 270
temporal coherence, 262–263
thick holograms, 266
wavelength sensitivity of recording material, 268
Homogenization, ultrasound applications, 733
Hubbard Hamiltonian, 688–689
Hubbard subbands, 688–689
Huygens–Fresnel principle, 439
Hydrogen atom
    angular momentum, 48–49
    atomic orbitals, 49–50
    atomic units based on, 50
    highly excited, microwave ionization, 94–96
    isotopes, 417
    parity, 49
    problem of, in quantum mechanics, 495, 511
    quantum numbers, 48
    states, 48
Hypercrystals, 560–562
Hysteresis loop, magnetic, 254–255

Immersion testing, ultrasound applications, 728
Implantation damage, in plasma etching, 467
Impulse approximation, 21–24
Impulse response, in optical diffraction, 446
Incoherent neutron scattering, 355
Incommensurate crystals, 559–560
Independent-particle model, nuclei, 427
Index of refraction, holograms, 266
Indexing, quasicrystals, 559–560
Industry, plasma applications, 450–451
Inelastic collisions, 59
    Born approximation, 37
    conservation of energy and momentum in, 627–629
    Demkov (exponential coupling), 37
    impact parameter results, 36
    Landau–Zener–Stueckleberg, 36
    Langevin, 37
    Rosen–Zener (noncrossing), 37
    useful form, 37

Inelastic scattering, neutrons, 355, 358–359
Inflation symmetry, quasicrystals, 563
Information, missing (statistical mechanics), 650–651
Information processing, laser applications, 301
Information theory, and statistical mechanics, 650–651
Infrared lasers, 293–294
Infrared spectroscopy, in chemical physics, 100
Infrared up-conversion, 408–409
Inner product, 639, 659
Insulators, mixed covalent and ionic bonding, 64–65
Integrable systems, in intramolecular dynamics, 215–217
Integrals, motion, 638
Integrating factor, 656
Interaction potentials, 27–28
    elastic scattering, 22–24
    intermediate range potentials, charge exchange and, 29–32
    long-range potentials, 29
    short-range potentials, 28–29
Interface states, in semiconductors, 579
Interferometry
    holographic nondestructive testing and, 270–271
    microwave, plasmas, 463
Internal combustion engines, powertrain, 27–29
Internal conversion, polyatomic molecules, 315
Internal energy (statistical mechanics), 644
Intersystem crossings, polyatomic molecules, 315
Intramolecular processes, rate of energy flow, 211
Invariance
    Galilean, in Newtonian mechanics, 593
    group, in space–time theories, 592
Ion detection, in a mass spectrometer, 335
Ion milling, 451
Ionic crystals, structure and bonding, 62–63
Ionization-dip spectroscopy, 351–352
Ionization effects
    electrical, 578–580
    mechanism, 577
    in optical materials, 579
    in polymers, 579–580
Ionization multiphoton spectroscopy, 334–335

Ionization potential, electronic configuration for elements and, 54–55

Ions
detection in mass spectrometry, 335
sources, 573

Irregular systems (intramolecular dynamics), 218

Irreversible equations, time-reversal invariant equations and, 213

Irreversible processes
microscopic dynamics (statistical mechanics), 670–678
phenomenological theory (statistical mechanics), 666–670

Isomerism, shape, 426

Isospin modes, of nuclei, 426–427

Isothermal compressibility, 649

Isotope enrichment, lasers and, 301–302

Isotopes, notation, 415

$j–j$ coupling, 53–56

Jahn–Teller effect, polyatomic molecules, 316–317

Jaynes–Cummings model, 546

Kaluza–Klein theories, 742

Kaons, CP violation, 111–112

Kerma radiation unit, 574

Kerr effect
optical, 398
quadratic, 405

Kerr–Neuman fields, 601

Kinetic energy
conversion of rest mass into, 629
definition in relativity theory, 627

Kinetic theory of gases, 44–45

Kirchhoff principle, 439–440

Klein–Nishina formula, 165

Kobayashi–Maskawa matrix, 198–199

Kobayashi–Maskawa theory, 112

Kolmogorov–Arnold–Moser theorem, 89

Kramers–Kroning relations, 675

Kruskal diagram, 600

Krypton lasers, 295

Lagrange methods, in statistical mechanics, 652

Lagrange parameters
maximum entropy principle and, 653
in statistical mechanics, 652

Lagrangian, in gauge field theories, 190, 194, 196–199

Lamb shift
in quantum electrodynamics, 155
relativistic treatment, radiative corrections, 167

Landau–Zener–Stueckleberg approximation, 39

Langevin cross section, 37, 40

Langevin dynamics, 665–666

Langevin equations
generalized, 676
in statistical mechanics, 664–665

Langmuir probes, plasmas, 461–462

Laplace operators, in statistical mechanics, 640

Laser beams
gain medium shape, 289
growth, 289
longitudinal modes, 290–291
modes, 290–291
optical cavity, 289–290
optical resonator, 289–290
quality, 299
saturation, 289
transverse modes, 291

Laser cooling
atomic beam slowing, 280
definition, 275
dipole force, 277–278
dipole force fluctuation, 278–279
optical molasses, 280–281
radiation pressure forces, 276
spontaneous force, 276–276
spontaneous force fluctuation, 277

Laser optico-galvanic spectroscopy, 324

Laser trapping
definition, 275
dipole-force trap, single beam, 281–282
optical traps, spontaneous force, 282–283
optodynamic traps, 282

Lasers
amplifiers, 287, 292
argon lasers, 295
in art, 302
brightness properties, 300
carbon dioxide lasers, 295
chemical, 296
coherent properties, 285
color center, 297
communications applications, 300
construction industry applications, 301
copper vapor, 296–297
Doppler broadening, 291
Doppler effect, 292
energy of, 299

excitation mechanisms, 287–288
extreme ultraviolet, 294
fluorescing dyes, 288
focusing capabilities, 299
free electron, 298–299
fusion energy and isotope enrichment, 301–302
gain media, 287–288, 291–292
gas, 294–296
helium–cadmium, 296
helium–neon, 295
history, 286
information processing applications, 301
infrared, 293–294
intensity dependence, in multiphoton spectroscopy, 326, 338
inversions in gas, metal vapors, and plasmas, 288
krypton, 295
line narrowing, cavity effects, 292
linewidth, 291–292
liquid (dye), 288
lithography applications, 302
materials processing applications, 301
metal vapor, 296–297
military applications, 301
mode quality, 299–300
molecular hydrogen, 294
neodymium, 297
optical resonators, 285
photons and, 287
population inversions, 287–289
printing applications, 302
pulse duration, 300
quantum mechanics and, 496–497
rare gas-halide excimers, 295–296
remote sensing applications, 301
ruby, 297
semiconductor, 249–250, 288, 298
soft X-ray, 294
solid-state, 297
stimulated emission, 285
theory, semiclassical, 532–533
ultraviolet, 294
vacuum ultraviolet, 294
visible, 294
wavelengths, 292–293, 299

Lasing transitions, excitons, 249–250

Lattice dynamics, Mössbauer spectroscopy, 286

Lattice gradients, Mössbauer spectroscopy, 287–289

Legendre transformations, 654

Legendre transformtions, 654

Lennard–Jones potential, 22

Lenses
gravitational, 609–610

Lenses (*continued*)
  resolution, positron microscopes, 475
Lenz–Jensen potential, 22, 28, 32
Leptons
  classification, 739
  conservation, 179–181
  definition, 179
  interactions with photons, electrodynamics, 169–170
  muon, decay, 180
  number, 179–180
  properties, 175, 179
  in standard model, 201–203
  tau, decay, 180
  types, 175, 179
Level measurement, ultrasound applications, 729
Lie algebra
  in classification of particles, 742–743
  Kac–Moody, 742, 744
Light
  bending, in PPN formalism, 605
  coherence, 534–537
  cones, special relativity and, 593–594
  early theories, 150
  energy quantum, Einstein's concept, 490–491
  intensity, holographic recording, 262
  quantum mechanical states, 538–540
Light quanta
  Einstein's theory, 490–491
  quantum mechanical states, 538–540
Linear phenomenological equations, 668
Linear response theory, 106
Linewidth, lasers
  in amplifiers, 292
  homogeneous, of gain media, 291
  inhomogeneous, of gain media, 291–292
Liouville equation
  classical, 657–658
  quantum, 660
  stochastic, 661–662
Liouville operator
  classical, 657
  quantum, 660
Liquid drop model
  dense matter, 129–130
  nuclei, 422
Liquid dye lasers, 298
Liquids
  atomization, ultrasound applications, 733

degassing, ultrasound applications, 734
  dyes, inversions in, 288
Lithography, laser applications, 302
Local density, 669, 673
Local entropy production postulate, 668
Local equilibrium assumption, 667
Local thermodynamic parameters, 669
Logistic maps, 76–77
  chaos, 81
  period doubling, 77–80
  universality, 80
Lorentz boosts, 595
Lorentz contraction, 621
Lorentz force, 159
Lorentz invariance, 594–595
Lorentz transformation, 595, 618–619
  of electric and magnetic fields, 632–633
  implications, 619–625
  in relativistic quantum electrodynamics, 162
Lorentzian line profile, 57
Lorenz attractor, 82–84
Luneburg–Debye integral, 441–442
Luttinger theorem, 713
Lyman series, 496

Machining, ultrasonic, 733
Magic numbers
  neutrons, 419–420
  protons, 420
Magnetic change point, 253–254
Magnetic circular dichroism spectrum, 323
Magnetic fields
  bound-exciton complexes, 239–240
  intrinsic excitons, 235–236
  Lorentz transformation, 632–633
  quantized, electrodynamics, 157
Magnetic interactions, correlated electrons, 708–709
Magnetic moments
  anomalous, 168–169
  electron, 47, 168–169
Magnetic neutron scattering, 355
Magnetic optical rotation spectrum, 323
Magnetic order, type of, 256
Magnetic phases, strongly correlated electrons, 693–694
Magnetism
  concept of, 251–252
  origin of, 252–253
  quasicrystals, 567
  Weiss molecular field, 253

Magnetization
  anisotropy energy, 255
  curves, for ferromagnetic materials, 254
  nonlinear, mathematical description, 364
  nonlinear optical processes, 362–363
Magnetostriction, 257
Magnons, 257
Markovian approximations, 676–677
Maser amplifiers, lasers and, 496
Mass, dense matter, 128–129
Mass number, 415
Mass spectrometry, ion detection, 335
Massey–Mohr approximation, 21, 26–27
Materials characterization, ultrasound applications, 729
Materials processing, laser applications, 301
Matter waves, 504
Maximum entropy principle
  classical version, 651–652
  quantum version, 652–653
Maxwell–Bloch equations, 524–525
Maxwell–Boltzmann statistics, exciton gases, 248
Maxwell's equations
  nonlinear materials, 364
  in nonrelativistic quantum electrodynamics, 159
  optical diffraction and, 438
  in quantum electrodynamics, 153
Measuring rods, in general relativity, 598–599
Mechanics, classical, 484–487
Mechanics, Newtonian
  absolute space and time, 592–593
  free bodies, 593
  Galilean invariance, 593
  Galilean relativity and, 592
  simultaneity, 592
Medicine
  laser applications, 300–301
  ultrasound applications, 729–730, 735
Memory effects (statistical mechanics), 675–676
Memory function formalism, 677
Meson theory, for nuclear forces, 433
Mesons
  experimental discoveries, 740
  quark model of hadrons and, 188
Mesotrons, discovery, 416
Metal–insulator transitions, strongly correlated electrons, 693–694
Metal oxide semiconductor transistors, ionization effects, 577, 583

Metal vapor lasers
  copper, 296–297
  helium–cadmium, 296
Metal vapors, inversions in, 288
Metallic oxides, superconducting,
    698–701
Metals
  bonding, 65–66
  casting, ultrasound applications,
    734–735
  electron transport, 515
  formation, ultrasound applications,
    733
  magnetostrictive properties, 725
  radiation effects, 584
  welding, ultrasound applications,
    734
Metastability, nuclear, 421–422
Metric of flat space–time, 595–596
Michelson–Morley experiment,
    615–616
Microelectronics, glow-discharge
    plasmas, 467
Microprobes, scanning positron, 474
Microscopy, ultrasonic, 730–731
Microwave interferometry, plasmas,
    463
Microwave spectroscopy, moment of
    inertia data and, 102
Missing information (statistical
    mechanics), 651
Mixed state, in intramolecular
    quantum dynamics, 224
Mixed-valent metallic states,
    707–708
Mixing, chaotic
  classical systems, 73
  Hamiltonian systems, 86–87
Mixing systems, in intramolecular
    dynamics, 217–218
Mobilities, particles in plasmas,
    455–456
Mobility edges, amorphous
    semiconductors, 4
Mobility gaps, amorphous
    semiconductors, 5
Moderation process, slow positron
    beam production by, 473
Moderators, neutron, 355–356
Modulation transfer function, 447
Molecular dynamics, numerical
    statistical mechanics, 106
Molecular optical spectroscopy
  absorption spectra, 306, 323
  circular dichroism spectrum, 323
  description, 304
  detection of radiation, 305–306
  dispersion of radiation, 305
  emission spectra, 306, 323
  fluorescence spectra, 306

generation of radiation, 304–305
  glow discharges, 306
  induced magnetic moments, 323
  laser-related optical spectra, 324
  magnetic circular dichroism
    spectrum, 323
  magnetic optical rotation spectrum,
    323
  Planck relationship, 307
  Rydberg spectra, 323
  techniques, 306–307
  type of molecular optical spectra,
    323–324
  unstable molecular species, 306
  wave number, 307
Molecular optical spectroscopy,
    diatomic molecules
  angular momentum, 309
  Franck–Condon principle, 310
  heads, 310–311
  Morse curves, 313
  optical transitions, 309–313
  potential functions, 308
  progressions, 310
  reciprocal moment of inertia, 308
  rotation, 308
  Schrödinger equation for vibrational
    motion, 308
  selection rules, 310
  spin–orbit coupling, 310
  spin multiplicity, 309
Molecular optical spectroscopy,
    polyatomic molecules
  Born–Oppenheimer approximation,
    315
  chromophores, 317
  circular dichroism spectrum, 323
  fluorescence, 314
  internal conversions, 315
  intersystem crossings, 315
  Jahn–Teller effect, 316–317
  optical rotatory dispersion, 323
  phosphorescence, 314
  prolate symmetric tops, 314
  Renner effect, 315–316
  rotational levels of symmetric tops,
    317–318
  rotational motion, 314
  rotational structure, 320–323
  rovibronic levels, 314
  vibrational degrees of freedom, 314
  vibrational levels, 318–319
  vibrational structure, 319–320
Molecular orbital diagrams, 31–32
Molecular orbital method, 28
Molecular orbital potentials, 30–31
Molecular processes, dynamics,
    102–104
Molecular species, unstable, molecular
    optical spectroscopy, 306

Molecular spectra, optical, types of,
    323–324
Molecular term, diatomic molecules,
    310
Moliere potential, 32
Momentum
  conservation in collisions, 627–629
  relativistic expressions, 626
Monte Carlo simulations, in chemical
    physics, 107
Mori–Zwanzig projection operator
    formalism, 676
Morse curves, diatomic molecules, 313
Motion
  collective, nuclei, 424
  concepts, in special relativity,
    618–619
  integrals, 638
Motion equations
  charged particles in a plasma, 454
  linear phenomenological, 661–663
  non-Markovian, 675
Mott Hubbard boundary, 712–713
Mott–Hubbard insulating states,
    707–708
Mott transition, 687
Multiphoton ionization mass
    spectroscopy, 351
Multiphoton laser excitation
    spectroscopy, 324
Multiphoton spectroscopy
  above-threshold ionization,
    350–351
  angular momentum, photons, 340
  coherent anti-Stokes Raman
    scattering, 332
  continuum–continuum transition,
    350–351
  density matrix method, 329–331
  dephasing constant, 330
  Doppler-free, 337, 348–349
  Fermi's golden rule, 328
  fluorescence detection after two-
    photon excitation, 333–334
  formal intensity law, 326, 338–339
  Green's function (resolvent) method,
    328–329
  ionization detection, 334–335
  ionization-dip spectroscopy,
    351–352
  laser intensity dependence, 326
  low-lying electronic excited states of
    linear polyenes, 349–350
  multiphoton transitions, spectral
    properties, 346–347
  nonlinear susceptibility, 332
  optical collisions, 329
  optical Stark effects, 328–329
  phase match parameter, 332
  photoacoustic detection, 335–336

Multiphoton spectroscopy (*continued*)
  photoelectron detection, 336
  polarization behavior, 339–346
  polarization dependence, 326–327
  resonance effects, 328–329,
      346–347
  resonance enhancement, 326, 346
  Rydberg states, 349
  susceptibility method, 331–332
  tensor patterns for two-photon
      processes, 342–344
  thermal blooming, 336
  time-dependent perturbation theory,
      328–329
  two-photon spectroscopy, 348–349
  vibronic coupling, 347
Muons
  decay, 199
  discovery, 416
  in lepton family, 179–180
  nuclear reactions, 431
  photon–muon interactions,
      169–170

Naked singularities, 591
Neodymium lasers, 297
Neutral current interactions, in
      electroweak theory, 195–196
Neutrinos
  emissivity in dense matter, 144–146
  opacity in dense matter, 146–148
  zero rest mass, 630
Neutron beams, tubes and guides, 356
Neutron drip, dense matter, 117, 129
Neutron drip line, 423
Neutron matter
  composition, 130–131
  equation of state, 135–136
Neutron scattering
  absorption cross sections, 355
  applications, 357–359
  beam tubes and guides, 356
  coherent scattering, 354–355
  cross section data, table, 354
  diffuse scattering, 358
  incoherent scattering, 355
  inelastic scattering, 355, 358–359
  instrumentation, 356–357
  magnetic scattering, 355
  moderators, 355–356
  neutron–atom interactions, 353–355
  neutron diffraction techniques, 357
  neutron production, 355–356
  polarization analysis, 359
  pulsed sources, 355–356
  quasielastic scattering, 355
  research reactors, 355

small-angle scattering, 357–358
  steady-state sources, 356
Neutronization, dense matter, 117,
      126–128
Neutrons
  absorption in matter, 575–576
  beam tubes and guides, 356
  characteristics, 417
  discovery, 416
  magic numbers, 420
  moderators, 355–356
  production, 355–356
  pulsed sources, 355–356
  research reactors, 355
  separation energy, 423
  single particle energy levels,
      deformation and, 429–430
  sources, 573
  steady-state sources, 356
Newtonian gravitational fields, 599
Newton's laws
  first law of motion, 482
  motion, in quantum mechanics,
      482–483
  second law of motion, 482
  third law of motion, 482
  universal gravitation, in quantum
      mechanics, 483
Nondestructive testing
  holographic, 269–271
  ultrasound applications, 728–729
Nonlinear optical processes
  applications, 405–414
  classification, 366–368
  coherent optical effects, 403–405
  common, 366–367
  description, 362
  electrooptic effects, 405
  frequency-conversion, 368–397
  induced magnetizations, 364
  induced polarizations, 364–365
  infrared up-conversion, 408–409
  magnetooptic effects, 405
  mathematical description, 364–368
  Maxwell's equations, 364
  nonlinear absorption, 402–403
  nonlinear spectroscopy, 405–408
  optical bistability, 411–414
  optical field calculations, 365–366
  optical phase conjugation, 409–411
  optical rectification, 383
  parametric down-conversion,
      381–383
  parametric oscillation, 382
  physical origins, 362–364
  saturable absorption, 403
  second-harmonic generation, 370,
      377–380
  second-order, table, 371–376

second-order effects, 369–370,
      377–383
  self-action effects, 397–403
  self-defocusing, 400–401
  self-focusing, 398–400
  self-phase modulation, 401–402
  spatial effects, 397–400
  three-wave difference-frequency
      mixing, 381
  three-wave sum-frequency mixing,
      380–381
Nonlinear spectroscopy
  applications, 406–408
  description, 405–408
  types, 406–408
Normal coordinates, in intramolecular
      dynamics, 219
Normalization, in quantum mechanics,
      511
Nuclear decay, alpha, 421–422
Nuclear energy, special relativity and,
      633–634
Nuclear fission
  description, 426
  discovery, 416
  fission reactions as source of
      neutrons, 573
  special relativity and, 634
  spontaneous, description, 422–423
Nuclear forces
  general features, 432–433
  internucleonic and interatomic,
      433–434
  meson theory, 433
Nuclear fusion, special relativity and,
      634
Nuclear fusion power
  as energy source, 432
  as neutron source, 573
Nuclear magnetic resonance
      spectroscopy, 100–101
Nuclear masses
  binding energies and, 418
  magic numbers, 419–420
  von Weizsäcker semiempirical
      formula, 418
Nuclear moments, Mössbauer
      spectroscopy, 432
Nuclear physics
  alpha decay, 421–422
  beta decay, 420–421
  collective modes, 424–427
  deuteron, 417
  energy sources, 431
  equilibrium deformations and
      rotations, 424–426
  fission, 426
  fission reactors, 432
  fusion, 432

helium isotopes, 417
history, 416
internucleonic and interatomic
    forces, 433–434
isopin, 426–427
light nuclei, 416–417
meson theory, 433
metastability, 421–422
neutrons, 417
nuclear forces, 432–435
nuclear masses, 417–420
nuclear reactions, 429–431
nuclear sizes and shapes, 417
protons, 416
quantum chromodynamics, 434–435
shell structure of nuclei, 427–429
spin modes, 426–427
spontaneous fission, 422–423
subnuclear structure, 434–435
surface vibrations, 424
triton, 417
Nuclear reactions
    description, 429–431
    as energy source, 431–432
    induction, 415–416
Nuclear reactors
    breeder, 432
    fission, as energy source, 432
Nuclear saturation, 417–418
Nuclei
    alpha decay, 421–422
    charged, absorption in matter, 576
    collective motion, 424
    deformed, 428–429
    electron capture, 420
    equilibrium deformations and
        rotations, 424–426
    fission processes, 426
    gross properties, 417–420
    isospin modes, 426–427
    light, 417
    liquid drop model, 422
    metastability limits, 423–424
    as projectiles, 431
    shapes, 417
    shell structure, 427–429
    sizes, 417
    spherical, description, 427–428
    spin modes, 426–427
    spontaneous fission, 422–423
    subnuclear structure, 434–435
    superheavy, 431
    surface vibrations, 424
Nucleon scattering
    elastic, 429–431
    inelastic, 431
Nucleus, compound, 431
Number operators (statistical
    mechanics), 649

Occupation number, photons, 157
Onsager reciprocity relations, 668
Onsager reciprocity theorem, 668
Open systems, isothermal, 647
Operators (statistical mechanics), 639
Optical bistability
    excitons and, 250
    in nonlinear processes, 411–414
    in quantum optics, 533–534
Optical Bloch vector, quantum optics,
    525–527
Optical cavity, lasers, 289–290, 292
Optical collisions, density matrix
    method, 329–331
Optical coordinates, in diffraction, 442
Optical data processing, holography
    and, 271–272
Optical diffraction
    Airy patterns, 440–441
    coherent and incoherent, 445–447
    coherent imaging of extended
        sources, 446
    description, 438
    field near a caustic, 442
    fields in focal region of lens,
        441–442
    and Fourier optics, 438
    Fraunhofer formula, 440
    Fresnel formula, 440
    geometric theory, 443–445
    and Green's theorem, 439
    and Helmholtz–Kirchhoff integral
        theorem, 439–440
    history, 438
    Huygens–Fresnel principle, 439
    impulse response, 446
    integrals, 440–442
    Kirchhoff formulation, 439–440
    mathematical techniques,
        438–439
    matrix description, 444–445
    and Maxwell equations, 438
    optical transfer function, 446–447
    point spread function, 446
    resolving power, 441
    rotationally invariant fields,
        440–441
    from a slit, 445
    and wave theory, 438
Optical elements, holographic, 272
Optical fields, in nonlinear processes,
    365–366
Optical laser traps, 282–283
Optical materials
    displacement effects, 582
    ionization effects, 579
    radiation effects, 584
Optical molasses technique,
    280–281

Optical phase conjugation
    applications, 411
    techniques, 409–411
Optical rectification, 383
Optical resonators
    in lasers, 289–290
    stable and unstable, in lasers, 290
Optical rotatory dispersion, 323
Optical spectra, quantum mechanics
    and, 491, 495–496
Optical theorem, wave mechanics of
    scattering and, 25–27
Optical transfer function, optical
    diffraction, 438, 446–447
Optical waves, in nonlinear optical
    processes, 363–364
Optodynamic laser traps, 282
Orthohelium, 52
Oscillations
    forces, acoustic waves, 718
    undamped harmonic, acoustic
        waves, 718
Oscillators
    classical, 34
    in intramolecular dynamics, 216
Overetching, 466
Oxide charge, in electrical materials,
    577

Pair annihilation, particles and
    antiparticles, 634–635
Pair production, special relativity and,
    634–635
Pairing energy, nuclei, 419
Pairings
    correlated electrons, 708–709
    phonon-mediated, 710–711
    spin-singlet, correlated electrons,
        701–710
Parahelium, 52
Paramagnetic hyperfine spectroscopy,
    by Mössbauer effect, 26
Paramagnetic materials, 251–252
Parametric down-conversion, optical-
    wave radiation, 381–383
Parametric frequency-conversion
    processes
    applications, 369
    description, 369
    parametric down-conversion,
        381–383
    second-harmonic generation, 370,
        377–380
    second-order, nonlinear
        polarizations, 371
    second-order, performance,
        374–376

Parametric processes (*continued*)
second-order effects, 369–383
third- and higher-order, 383–389
three-wave difference-frequency
mixing, 381
three-wave sum-frequency mixing,
370
wavelength ranges, 369
Parametric oscillation, 382
Parametrized post-Newtonian
formalism
bending of light, 605
gravitational lenses, 609–610
planetary motion, 606–607
time delay, 605–606
time varying G, 607
Parity, hydrogen atom, 49
Parity invariance, 109
Parity violation, in electroweak
interactions, 193, 195–196, 201
Particle accelerators
in elementary particle physics, 186–187
high-energy particles and, 173
Particle analysis, plasmas, 464–465
Particle colliders, in elementary
particle physics, 186–187
Particle decay
charge-parity (CP) violation, 201
Feynman diagrams, 183
muon and tau leptons, 180
neutral current, of quarks, 199
probability, 175
quarks, 181
Particle detectors
bubble chambers, 187
calorimeters, 188
Čerenkov radiation detector, 188
drift chamber, 187
scintillators, 188
semiconductor detectors, 187
Particles
current in plasmas, 455
diffusion in plasmas, 455–456
elementary, *see* Elementary particles
mobilities in plasmas, 455–456
net flux in plasmas, 456
Partition function
in chemical physics, 107
in statistical mechanics, 105, 644
Paschen series, 496
Path integral methods, 107
Patterson functions, quasicrystals,
555, 565–566
Pauli exclusion principle
complex atoms, 52
in dense matter physics, 118–119
in elementary particle physics, 189
helium atom and, 51–52
periodic table classification and, 47
in quantum mechanics, 512

Pauli master (kinetic) equation, 663
Pauling's table of elemental
electronegativities, 62
Penning traps, 168
Penrose tiling, 552–553
Period doubling, 77–80
Permittivity, effective, 457–458
Perturbation calculations, in Feynman
diagrams, 184–185
Perturbation theory, time-dependent,
328–329
Pfund series, 496
Phase cell, 639
Phase function, 659
Phase match parameter, 332
Phase matching
parametric down-conversion,
381–383
second-harmonic generation,
378–379
third- and higher-order frequency-
conversion processes, 384–385
three-wave difference-frequency
mixing, 381
three-wave sum-frequency mixing,
380
Phase point, 638
Phase shifts, wave mechanics of
scattering and, 25
Phase space
classical representation, 658–659
in intramolecular dynamics, 215,
223
master equation, 663–664
in statistical mechanics, 638–639
Phase velocity
free particles, 505–506
laser beams, 496
particles with potential energy, 508
waves (quantum mechanics), 499
Phason dynamics, quasicrystals,
566–567
Phenomenological coefficient, 668
Phenomenological currents, 671
Phenomenological equations
linear, 668
linear motion, 661–663
Phenomenological theory, irreversible
processes, 666–670
Phenomenological time derivative,
664, 668
Phonon-mediated pairings, correlated
electrons, 710–711
Phonons
exciton-bound-phonon quasiparticle, 243–245
exciton interactions, 243–245
Frohlich electron–phonon
interaction, 245
Phonons, virtual exchange, 681–682

Phosphorescence spectra, polyatomic
molecules, 314
Photoacoustic detection method, in
multiphoton spectroscopy,
335–336
Photochromics, for holography, 269
Photocurrent effects, in electrical
materials, 578–579
Photoelectrons, detection in
multiphoton spectroscopy, 336
Photon counting, in quantum optics,
537–538
Photons
absorption, Einstein's B coefficient,
160
antibunching, 543
assembly, in quantum electrodynamics, 156–157
atoms and, interactions, 158–159
bunching, 543
coherent state, 539
Compton scattering, 165
echoes, in nonlinear optical
processes, 404–405
echoes, in quantum optics, 526–527
electromagnetic force and, 175
emission, transitions between atomic
states and, 491–492
exciton–photon interaction, 245–246
in higher-order Feynman diagrams,
184
lepton–photon interactions, 169–170
muon–photon interactions, 169–170
neutral current interactions, 195
nuclear reactions, 431
number state, 538
properties, 175
quantum electrodynamics, 151–152
spontaneous emission, Einstein's A
coefficient, 159–160
squeezed state, 540
statistics, quantum light detection
and, 542–543
stimulated emission, Einstein's B
coefficient, 160
Thomson scattering, 160–161
two-photon processes, 245–246
zero rest mass, 630
Photoplastics, for holography, 269
Photopolymers, for holography, 269
Photorefractive materials, for
holography, 269
Photoresists, for holography, 268–269
π pulses, in quantum optics, 525–526
Piezoelectric materials, properties,
726–727
Piezoelectric transducers, 725–727
Pions
condensation in dense matter,
131–132, 136–137

discovery, 416
nuclear reactions, 431
Planck's constant, 151, 493
Planck's radiation law
  in molecular optical spectroscopy, 307
  in quantum electrodynamics, 151–152
Planetary motion, in PPN formalism, 606–607
Plasma-assisted chemical vapor deposition, 451, 465
Plasma diagnostic techniques
  Fourier transform mass spectrometers, 465
  invasive, 461
  Langmuir probes, 461–462
  microwave interferometry, 463
  noninvasive, 461
  particle analysis, 464–465
  radiation spectroscopy, 463–464
Plasma displays, 452
Plasma electronics, 451
Plasma etching, 451
  chemisorption, 468
  coupled configuration, 467–468
  dry techniques, 466
  externally biased electrode potential, 467
  floating potential, 467
  glow-discharge plasmas for microelectronics, 467
  grounded electrode area, 468
  homogeneous gas-phase collisions, 469
  implantation damage, 467
  net drift, 468
  overetching, 466
  plasma potential, 467–468
  ponderomotive potential, 468
  product desorption, 468–469
  radiation damage, 467
  required steps, 468
  in semiconductor industry, 451
  sticking coefficients, 468
  surface diffusion, 468
  synergism of two processes, 469
  wet techniques, 466
Plasma physics
  ac electromagnetic fields, 457–458
  ambipolar diffusion, 456
  ambipolar diffusion coefficient, 456
  arc discharges, 459
  breakdown, 460
  collision cross sections, 454–455
  composition, 454
  conductivity, 457
  dc magnetic fields, 460–461
  Debye length, 461

density, 453–454
  diffusion coefficient, 455
  dynamics, 454–458
  effective permitivity, 457–458
  elastic collisions, 454–455
  equation of motion, 454
  free diffusion, 455–456
  glow discharges, 458
  heat flux, 457
  heat transport, 456–457
  inelastic collisions, 454
  mobility, 455–456
  net particle flux, 456
  particle current, 455
  particle diffusion, 455–456
  plasma potential, 461, 467–468
  probability of collision, 454–455
  rf discharges, 459–460
  specific heat, 456
  temperature, 454
  thermal conductivity, 457
  thermal plasmas, 459
  time-dependent diffusion coefficient, 455
  time-varying fields, 455–456
  total average velocity, 457
Plasma polymerization, 451, 469–471
  etchant-unsaturate model, 470–471
  fluorine-to-carbon ratio model, 469–470
Plasma potential, 461
  etching and, 467
  glow-discharge plasmas, 467–468
Plasma spray coating, 453
Plasma–surface interactions, 465–471
Plasmas
  fusion, 450
  industrial applications, 450–451
  inversions, 288
  processing, industrial, 450
  production, 454
  properties, 453–454
  synthesis, industrial, 450
Plastics
  formation, ultrasound applications, 733
  piezoelectric, 726
  welding, ultrasound applications, 733–734
Pockels effect, in nonlinear optical processes, 405
Poincaré sections, Hénon–Heiles equations, 90–91
Point spread function, in optical diffraction, 446
Poisson brackets (statistical mechanics), 658

Polarization, nonlinear
  mathematical description, 364–365
  nonlinear optical processes, 362–363
  second-order processes, table, 371
  third- and higher-order frequency-conversion processes, table, 384
Polarization analysis, with scattered neutrons, 359
Polarization dependence, in multiphoton spectroscopy, 326–327, 339–346
Polarization interaction, 38
Polarons, magnetic, 694–695, 708
Polyatomic molecules
  Born–Oppenheimer approximation, 315
  chromophores, 317
  fluorescence spectra, 314
  internal conversions, 315
  intersystem crossings, 315
  Jahn–Teller effect, 316–317
  phosphorescence spectra, 314
  principle axes of inertia, 314
  prolate symmetric tops, 314
  Renner effect, 315–316
  rotational levels of symmetric tops, 317–318
  rotational motion, 314
  rotational structure, 320–323
  vibrational degrees of freedom, 314
  vibrational levels, 318–319
  vibrational structure, 319–320
Polyenes, linear, low-lying electronic excited states, 349–350
Polymers
  ionization effects, 579–580
  radiation effects, 584–585
Ponderomotive potential, 468
Population inversions, laser gain media, 287
Positron microscopy
  beam formation, 475
  beam production, 473
  brightness of beams, 473–474
  channel electron multiplier arrays, 475
  detector resolution, 475
  history, 474
  lens resolution, 475
  moderation process, 473
  optimum resolution, 475
  reemission microscopy, see Positron reemission microscopy
  signal averaging, 475
  spherical aberrations, 475
  transmission positron microscopes, see Transmission positron microscopy

Positron reemission microscopy,
    477–478
  analysis of resolution, 477–478
  applications, 479
  contrast formation, 478–479
  defect sensitivity, 477
  diffraction limit on resolution of,
    478
  history, 474
  material contrast, 478–479
  reflection, 477
  topographical contrast, 478
  transmission geometry, 477
Positrons
  quantum electrodynamics, 154
  relativistic quantum electro-
    dynamics, 162–163
Potential energy, diatomic molecules,
    307
Potential functions
  diatomic molecules, 308
  polyatomic molecules, 315–317
Power-law potentials, 21, 23, 26
Pressure, dense matter, 120–121
Principle of equivalence, quantum
    mechanics and, 483–484
Printing, laser applications, 302
Probability density, propagator, 658
Probability of collision, plasma
    particles, 454–455
Product desorption, in plasma etching,
    468–469
Progressions, diatomic molecules, 310
Projection operators
  in statistical mechanics, 643
  theories, 675–678
Propagator
  elementary particle physics, 183,
    199–200
  for probability density, 658
Proper time
  in general relativity, 598–599
  Lorentz transformation and, 622
Protons
  characteristics, 416
  decay predicted by grand unified
    theories, 203
  magic numbers, 420
  mass, 174
  in particle accelerators, 185–186
  separation energy, 423
  single particle energy levels,
    deformation and, 429–430
  size, 173
  sources, 572–573
Pseudopotential method, quantum
    structure of solids, 67
Pulsars, binary, general relativity and,
    607–608
Pulse area, in quantum optics,
    525–526

Pulsed sources, of neutrons, 355–356
Pupil function, in optical diffraction,
    440
Pure state, in intramolecular quantum
    dynamics, 224

Quadrupole radiation formula,
    602–603
Quantization
  angular momentum, 493
  energy values, 493–494
Quantum average (statistical
    mechanics), 640
Quantum beats, 224
Quantum chromodynamics, 434–435
  description, 434
  gluons, confinement, 178
  gluons, prediction, 192–193
  running coupling constant, 191–192
  theoretical origins, 190
Quantum/classical correspondence,
    206, 213
Quantum electrodynamics
  definition, 149–150
  divergences, 154–155
  early theories, 150
  electrons, 154
  history, 153
  photons, 151–152
  positrons, 154
  renormalization, 154–155
Quantum electrodynamics,
    nonrelativistic
  applications, 159–162
  approximations, 155–156
  photon–atom interactions, 158–159
  photon assembly, 156–157
  quantized electromagnetic fields,
    157–158
Quantum electrodynamics, relativistic
  conservation laws, 164
  electromagnetic field, covariant
    quantization, 163–164
  electrons, 162–163
  Feynman diagrams, 164–166
  photon–lepton interactions,
    169–170
  positrons, 162–163
  radiative corrections for, 166–169
  relativistic theory, 162
  S matrix, 164–166
  symmetries, 164
Quantum energy, Einstein's concept of
    light as, 490–491
Quantum hypothesis, history of, 45
Quantum mechanics
  acceptors, 517
  angular momentum quantization,
    493

Balmer series, 496
band structure calculation, 514
Bloch functions, 514
Bohr theory, 493–495
Brackett series, 496
classical mechanics, 484–486
classical wave equation, 498–499
Compton wavelength, 505
Cooper pairs, 517
covalent crystals, 63–64
de Broglie relation, 493, 504–505
definition, 481
dispersion relations, 499, 506–508
donors, 517
eigenfunctions and eigenvalues, 223
electrical resistance, 513
electron holes, 516
electron orbital wave functions,
    63–64
electron scattering, 515
electron spin, 516
electron transport, 511–512
energy–momentum relation, 488,
    506
energy bands, 514–515
energy eigenvalue equation, 510
energy eigenvalue spectrum, 516
energy gaps, 514–515
energy values, quantization,
    493–494
excited states, 491
Fermi energy, 515
free electron model, 515
Galilean reference frame, 482
gravity and, 610–611
ground states, 486, 495
group velocity, 514
harmonic oscillator problem, 511
Heisenberg uncertainty principle,
    492–493
hydrogen atom problem, 495, 511
insulators, electrical, 515–516
lasers, 496–497
Lyman series, 496
masers, 496
matter waves, 504
Newton's law of universal
    gravitation, 483
Newton's laws of motion, 482–483
normalization, 511
optical spectrum, 491, 495–496
Paschen series, 496
Pauli exclusion principle, 512, 515
Pfund series, 496
phase coherence, 496
phase velocity, 499, 505–506, 508
photons, 491–492
Planck's constant, 493
planetary model, 484–486
principle of equivalence, 483–484
quantum, origin and meaning, 490

quantum structure analysis of
    structure and bonding in solids,
    67
reciprocal lattices, 513
reduced mass effect, 496
Rydberg constant, 496
Schrödinger equation, 510–511
Schrödinger wave function, 510
secular determinant, 514
secular equation, 514
semiconductors, 516–517
spectral lines, 496
stationary states, 486
stimulated emission, 496
structural diagrams, structure and
    bonding in solids, 66–67
superconducting energy gap, 517
superconductor transition
    temperature, 517
superconductors, 517–518
transitions, 491–492, 495–497, 516
unified field theory and, 741–742
wave–particle duality, 492,
    503–508
wave packets, 501
Quantum optics
A coefficients, 159–160, 520–522
absorption cross sections, 530–531
AC Stark effect, 547–548
Autler–Townes effect, 547–548
B coefficients, 520–522, 530–531
Bell inequalities, 549
coherence length, 535
coherence of light, 534–537
coherence time, 535
coherent states, 539
Einstein–Podolsky–Rosen (EPR)
    definitions, 548–549
fully quantized interactions,
    540–542
hidden variables, 549
inhomogeneous broadening, 526
Jaynes–Cummings model, 546
laser theory, 532–533
line broadening, 528, 530–531
Maxwell–Bloch equations,
    524–525
optical bistability, 533–534
optical Bloch equations, 525–527
photon antibunching, 543
photon bunching, 543
photon counting, 537–538
photon echoes, 526–527
photon number states, 538–539
photon statistics, 542–543
π pulses, 525–526
pulse area, 525–526
quantum light detection, 542–543
Rabi frequency, 523
relaxation, 528–530
resonance fluorescence, 547–548

saturation, 531–532
self-induced transparency, 527–528
semiclassical theory of radiation,
    522–525, 546–547
spontaneous emission, 524
squeezed states, 540
strong fields, 531
superradiance, 543–545
two-level atoms, 522–525
two-level single-mode interaction,
    545–547
Quantum statistical mechanical theory,
    relaxation phenomena, 670–674
Quantum systems, simulations in
    chemical physics, 107
Quantum theory, Heisenberg
    uncertainty principle, 492–493
Quantum trapping states, 212
Quantum wells, excitons, 246–247
Quark matter
    composition, 132
    equation of state, 138
Quarks
    antiquarks, 181
    color degree of freedom, 188–190
    conservation laws, 181
    defining properties, 180–181
    description, 415–416, 434–435
    electroweak coupling, 197–198
    model, evidence for, 188
    model of hadrons, 188
    properties, 175–176
    relation to other particles, 175–176
    size, 173
    standard model, 201–203
    types, 175–176
    weak neutral coupling, 198–199
Quasi-elastic neutron scattering, 355
Quasi-periodicity, intramolecular, 216
Quasi-static processes, in statistical
    mechanics, 654
Quasicrystals, 551–552
    Amman tiling, 552
    atomic structure, 563–566
    classes of, 563
    Delaunay structures, 556–557
    diffraction, 553–555, 562
    elasticity, 566–567
    electronic structure, 567
    geometry, 556–557
    ground state, 557–558
    growth, 558–559
    hypercrystals, 560–562
    incommensurate structures,
        559–560
    indexing, 559–560
    inflation symmetry, 563
    magnetism, 567
    modeling, 564–565
    Patterson functions, 555, 565–566
    Penrose tiling, 552–553

phason dynamics, 566–567
rational commensurate
    approximants, 565–566
symmetry, 562–563

Rabi frequency, 523
Rad radiation unit, 574
Radial distribution function, 105–106
Radiation
    absorption process, 58
    angular distribution, 57–58
    detection, in molecular optical
        spectroscopy, 305–306
    dispersion, in molecular optical
        spectroscopy, 305
    electric dipole, 56
    generation, in molecular optical
        spectroscopy, 304–305
    gravitational, 602–603
    Lorentzian life profile, 57
    polarization, 57–58
    spontaneous emission, 56
    stimulated emission, 58
Radiation damage, 577–578
    data banks, 584
    metals, 584
    optical materials, 579, 582, 584
    plasma etching, 467
    polymers, 579–580, 584–585
    radiation hardening and, 584
    semiconductors, 580–582
    shielding, 585
    sterilization, 585
Radiation dose, quantification, 574
Radiation effects
    color centers, 579, 582
    ionization effects, 577–580
    mechanisms, 576–578
    optical systems, 584
    sterilization, 585
    structural materials, 584–585
    systems engineering and, 585–586
Radiation effects, electronic materials,
    578–580
    degradation processes induced by,
        578–582
    displacement, 577–578, 580–582
    solid-state electronics, 582–584
Radiation exposure, 574
Radiation flux, 574
Radiation hardening, solid-state
    devices, 584
Radiation physics
    absorption of radiation in matter,
        575–576
    α particles, 573
    Becquerel, 586
    β rays, 573, 576
    charged nuclei, 576

Radiation physics (*continued*)
  color centers, 579, 582, 584
  constants, table, 587–588
  curie, 586
  defect clusters, 578
  description, 570–572
  dose, 574
  electrons, 573, 576
  exposure, 574
  fission reactions, 573
  flux, 574
  fusion reactions, 573
  γ rays, 572–573, 575
  Gray unit, 574, 578
  ions, 573
  kerma, 574
  neutrons, 573, 575–576
  photocurrent effects, 578–579
  photons, 572–573
  protons, 573
  rad, 574, 578
  radioisotopes useful in irradiation
    experiments, table, 572
  range energy relations, 574–575
  rem, 586–587
  roentgen, 574
  sievert, 586–587
  single-event upsets, 583–584
  X rays, 572–573, 575
Radiation probes, of plasmas, 463
Radiation shielding and protection,
  design of shields, 585
Radiation spectroscopy
  local thermodynamic equilibrium
    model, 464
  plasma models, 464
Radiation theory, semiclassical,
  22–525, 546–547
Radiative corrections, for relativistic
  quantum electrodynamics,
  166–169
Radioactivity
  γ rays, 572–573
  X-ray emission, 572–573, 575
Radiofrequency discharges, plasmas,
  459–460
Radiofrequency spectroscopy, in
  chemical physics, 100–101
Radioisotopes, useful in irradiation
  experiments, table, 572
Radius ratio rules, 63
Rainbow angle, 24
Raman scattering
  anti-Stokes, 394–395
  backward stimulated, 392, 394
  forward stimulated, 392
  stimulated, 392–396
Raman spectroscopy
  in chemical physics, 100
  molecular optical spectroscopy and,
    324

Ramsauer–Townsend effect, 27
Rayleigh scattering, acoustic waves,
  724
Rayleigh's criterion of resolution
  (optical diffraction), 441
Reactive collisions, 60
Reciprocal lattices, crystalline solids,
  513
Reciprocal moment of inertia, 308
Reconstruction, holographic images,
  263–265
Red shifts
  gravitational, 603–604
  special relativity and, 635
Reduced mass effect, 496
Reflection positron reemission
  microscopy, 477
Relativity, general
  absolute objects, 591–592
  binary pulsars and, 607–608
  black holes, 601–602, 610–611
  Christoffel symbols, 595
  cosmic censorship, 602
  cosmological term, 598
  dynamical objects, 591–592
  equivalence, principle of, 596–597
  event horizon, 600
  Galilean invariance, 593
  Galilean relativity, 592
  general covariance, principle of, 591
  general invariance, principle of, 597
  geodesic deviation, 595
  geometrical objects, 590
  gravitational collapse, 601–602
  gravitational fields, 599–603
  gravitational lenses, 609–610
  gravitational radiation, 602–603
  gravitational wave detection,
    608–609
  Hawking radiation, 610–611
  invariance group, 592
  Kerr–Newmann fields, 601
  Kruskal diagram, 600
  Lorentz boosts, 595
  Lorentz invariance, 594–595
  naked singularities, 601
  Newtonian mechanics, 592–593
  PPN formalism, 604–610
  proper time, 598–599
  quadrupole formula, 602–603
  quantum gravity, 611
  red shift, gravitational, 603–604
  Riemann–Christoffel tensor,
    595–596
  Schwarzschild field, 599
  simultaneity, 592
  space–time manifold, 589–590
  space–time metric, 595–596
  supergravity, 611
  tensors, 590
  unified field theory and, 742

Relativity, special
  absolute time and, 617–618
  addition of velocities, 619–620
  antimatter, 634–635
  atomic masses and, 633
  Christoffel symbols, 595
  clocks, synchronization,
    620–621
  conservation laws and, 625–626
  coordinate transformation, 618
  cosmology and, 635
  decay of subatomic particles and,
    634
  deterministic basis, 488
  Doppler effect, 622–623
  electromagnetism and, 631–633
  energy, 626–627
  essential relations for quantum
    mechanics, 488
  experimental background,
    613–617
  force, 627
  free bodies, 594
  Galilean transformation, 618
  geodesic deviation, 595
  gravitational constant, time varying,
    607
  implications for quantum
    mechanics, 488–489
  inelastic collisions, 627–629
  kinetic energy, 627
  light cones, 593–594
  Lorentz boosts, 595
  Lorentz contraction, 621
  Lorentz invariance, 594–595
  Lorentz transformation, 618–619,
    632–633
  momentum, 626
  neutrinos, 630
  nuclear energy and, 633–634
  pair production and annihilation,
    634–635
  photons, 630
  principle of relativity, 617
  proper time, 622
  red shifts and, 635
  rest mass, conversion into kinetic
    energy, 629
  rest mass, invariance, 627
  rest mass, of a composite body,
    630–631
  rest mass, zero, 630
  Riemann–Christoffel tensor,
    595–596
  space–time interval between events,
    622
  space–time metric, 595–596
  time dilation, 621–622
  transverse Doppler effect, 623
  twins paradox, 623–625
  unified field theory and, 741

Relaxation
  behavior, in intramolecular
    dynamics, 210
  phenomena, quantum statistical
    mechanical theory, 670–674
  in quantum optics, 528–530
Rem (roentgen equivalent man),
  586–587
Remanence, hysteresis loop and, 255
Remote-sensing techniques, laser
  applications, 301
Renner effect, polyatomic molecules,
  315–316
Renormalization, in quantum
  electrodynamics, 154–155
Renormalization group, equations in
  unified field theory, 741
Research reactors, for neutron
  scattering, 355
Resolution
  positron microscopes, 475
  positron reemission microscopes,
    477–478
  transmission positron microscopes,
    476
Resonance effects
  in atomic multiphoton transitions,
    346–347
  in multiphoton spectroscopy,
    328–329
Resonance enhancement, in
  multiphoton spectroscopy, 326,
  346
Resonance fluorescence, 547–548
Resonance islands, 87–89
Resonance zones, 219–220
Resonances
  in frequency-conversion processes,
    386
  giant dipole, in nuclear physics, 426
Resonant systems, in intramolecular
  dynamics, 219
Resonant two-photon ionization
  spectroscopy, 324
Response function (statistical
  mechanics), 674
Rest mass
  composite bodies, 630–631
  conversion into kinetic energy, 629
  invariance, 627
  neutrinos, 630
  photons, 630
  zero, particles with, 630
Reversible processes (statistical
  mechanics), 654
Reversible systems (statistical
  mechanics), 658
Riemann–Christoffel tensor, 595–596
Rods, measuring (general relativity),
  598–599
Roentgen radiation unit, 574

Rosen–Zener approximation, 37,
  39–40
Rotation
  internal molecular, diatomic
    molecules, 308
  nuclei, 424–426
  polyatomic molecules, optical
    spectroscopy, 314
Rotational levels, symmetric tops,
  317–318
Rotational structure, polyatomic
  molecules, 320–323
RRKM theory, 228
Ruby lasers, 297
Russell–Saunders coupling, 53–56
Rutherford atom, 45, 416
Rutherford scattering, 485
Rydberg constant, 45
  in quantum mechanics, 496
Rydberg series, 323
Rydberg spectra, 323
Rydberg states
  multiphoton ionization of molecules
    via, 349
  transition to, 323

S matrix, 164–166
Saturation, in quantum optics,
  531–532
Scanning positron microprobes, 474
Scattering, acoustic waves, 724
Schrödinger equation
  for diatomic vibrational motion, 308
  in nonrelativistic quantum
    electrodynamics, 159
  in quantum electrodynamics, 157
  in quantum mechanics, 213, 510–511
  in relativistic quantum electro-
    dynamics, 162
  in statistical mechanics, 659
Schrödinger representation (statistical
  mechanics), 660
Schrödinger wave function, 510
Schwarzschild field, 599–601
Second-harmonic generation
  applications, 379–380
  description, 370, 377–379
  efficiencies, 380
  phase matching in, 378–379
Secular determinant, periodic potential
  for crystalline solids and, 514
Selection rules, rotational motion of
  diatomic molecules, 310
Self-defocusing, in nonlinear optical
  processes, 400–401
Self-focusing, in nonlinear optical
  processes, 398–400
Self-induced transparency, in quantum
  optics, 527–528

Self-phase modulation, in nonlinear
  optical processes, 401–402
Semiclassical approach to collision
  events, 16
Semiconductor lasers, 249–250, 288,
  298
Semiconductors
  acceptors, 517
  amorphous, see Amorphous
    semiconductors
  displacement effects, 580–582
  donors, 517
  electron holes and, 516
  electron transport, 516–517
  excitons, see Excitons,
    semiconductor
  inversions in, 288
  mixed covalent and ionic bonding,
    64–65
  quantum mechanics, 516–517
  transitions, 516
Sensitivity to defects, positron
  reemission microscopes, 477
Separation energy
  neutron, 423
  proton, 423
Shape isomerism (nuclear physics),
  426
Shell model
  multibound excitons, 242
  nuclei, 427
Shell structure, nuclei, 419, 427–429
Sievert radiation unit, 586–587
Signal averaging, in positron
  microscopy, 475
Silica, mixed covalent and ionic
  bonding, 65
Silicon crystal structures, indirect
  excitons, 234–235
Silver halides, in holography, 268
Simultaneity, in Newtonian mechanics,
  592
Single-event upsets, in large-scale
  integrated circuits, 583–584
Singlet states, helium atom, 51
Sirens, ultrasonic, 725
Small-angle scattering with neutrons,
  357–358
Smoluchowski equation, 661–662
Solid-state lasers, 288, 297
  neodymium lasers, 297
  ruby lasers, 297
Solids
  crystalline, periodic potential,
    513–514
  electron transport, 511–518
  inversions, 288–289
Sonar, ultrasound applications, 731
Space, concepts of, 618–619
Space–time intervals, 622
Space–time manifold, 589–590

Space–time metric, 595–596

Spallation method, for neutron production, 355

Spatial coherence, holography and, 262–263

Spatial frequency, in optical diffraction, 447

Specific heats, plasmas, 456

Spectra, of operators (statistical mechanics), 640

Spectral lines, in quantum mechanics, 496

Spectral properties, multiphoton transitions, 346–347

Spectroscopy, Doppler-free, 406–407

Spin
  conservation, elementary particles, 176–177
  in electroweak theory, 193
  elementary particles, 174–175

Spin angular momentum, 46–47

Spin liquids, 695–697

Spin modes, nuclei, 426–427

Spin multiplicity, diatomic molecules, 309

Spin orbit, spherical nuclei, 427

Spin–orbit coupling, diatomic molecules, 310

Spontaneous emission, 56, 524
  Einstein's A coefficient for, 159–160

Sputter deposition, plasmas, 451

Standard maps
  Hamiltonian systems, 87–89
  quantum, 94

Standard model, elementary particle physics, 200–202

Stark effect, optical, 328–329

Starlight, aberration, 613–614

Stationary states, of atoms, 486

Statistical approximations, intramolecular dynamics and, 228–229

Statistical dynamics, intramolecular, 206

Statistical mechanics
  adiabatic systems, 642
  assumption of local equilibrium, 667
  autocorrelation functions, 677
  Bloch equation, 661–663
  Boltzmann statistics, 639
  Bose–Einstein statistics, 639
  broken time-reversal symmetry, 661
  Brownian motion theory, classical, 663–666
  Chapman–Kolmogorov equation, 663
  characteristic function, 643–644
  chemical physics and, 105–106
  closed, isothermal systems, 644

closed systems, 647

commutator quantity, 660

composite systems, 655

conditional average, 671

conservation of probability, 657

conservation relations, 655

constraints, information and, 651

continuity equation, 657

contracted descriptions, 677

critical point, 650

cross-correlation functions, 678

damping constants, 673

definition, 637

density distribution function, 638–639, 647, 651–652

density operator, 641, 644, 646, 649, 653, 659–660

dependent diffusion tensor, 662

diffusion coefficient, 666

Dirac delta function, 641

disequilibrium, 656

dissipation, 662

dual Lanczos transformation, 677

dynamic variables, 639

eigenfunction, 640

eigenstates, 639

eigenvalues, 639

energy shells, 642

enthalpy, 646

entropy production, local rate, 669

equations of motion, non-Markovian, 675

equilibrium, 105–106, 638, 655–656

equilibrium time correlation function, 665

ergodic hypothesis, 642

external disturbances, 674–675

Fermi–Dirac statistics, 639

first law of thermodynamics, 656

fluctuation–dissipation relation, 666

fluctuation–dissipation theory, 675

flux operator, 673

Fokker–Planck equations, 661–662, 664

friction tensor, 662

Gibbs entropy, 639, 642, 644–646, 648–650

Gibbs free energy, 646

Gibbsian ensembles, 642–650

goals of, 637

gradient operator, 657

gross observables, 661

Hamiltonian, 639

heat, 655–656

heat bath, 645, 647, 650

heat capacity, 644

heat flux, 656

heat transfer, 655–656

Heisenberg representation, 660

Heisenberg uncertainty principle, 639

Helmholtz free energy, 644

Hermitian operators, 640

Hilbert space, 639

information theory and, 650–656

inner product, 659

integrals of motion, 638

integrating factor, 656

internal energy, 644

irreversible decay of a system, 658

irreversible processes, microscopic dynamics, 670–678

irreversible processes, phenomenological theory, 666–670

isothermal compressibility, 649

Kramers–Kroning relations, 675

Lagrange method, 652

Lagrange multipliers, 651

Lagrange parameters, 652–653

Langevin dynamics, 665–666

Langevin equations, 664–665, 676

Laplacian operators, 640

Legendre transforms, 654

linear phenomenological equations, 668

Liouville equations, 657–658, 660

Liouville operator, 657, 660

local density, 669, 673

local entropy production postulate, 668

local thermodynamic parameter, 669

Markovian approximations, 676–677

maximum entropy principle, 651–653

mean potential, 662

memory effects, 675–676

memory function formalism, 677

missing information, 650–651

Mori–Zwanzig projection operator formalism, 676

nonequilibrium, 106

nonlocality, 674

number operators, 649

numerical, 106–107

Onsager reciprocity relations, 668

open systems, 647

operators, 639

partition function, 644

Pauli master (kinetic) equation, 663

phase cell, 639

phase function, 659

phase point, 638

phase space, 638–639, 658–659, 663–664

phenomenological coefficient, 668

phenomenological currents, 671

phenomenological equations of motion, 661–663

phenomenological parameters, 661
phenomenological time derivative, 664, 668
Poisson brackets, 658
principle of equal a priori probabilities, 642
projection operator, 643, 675–678
propagator for probability density, 658
quantum average, 640
quantum macroscopic state, 639–642
quasi-static processes, 654
relaxation phenomena, 670–674
response function, 674
reversible processes, 654
reversible systems, 658
Schrödinger equation, 659
Schrödinger representation, 660
second law of thermodynamics, 656
Smoluchowski equation, 661–662
spectra, 640
spontaneous equilibrium fluctuations, 665–666
stochastic averaging, 664
Stoke's law, 665
streaming operator, 662
superstate representation, 661
susceptibility, generalized, 674
tetradic operators, 660
thermodynamic coordinates, 670
thermodynamic equivalence of Gibbsian ensembles, 650
thermodynamic limit, 650
thermodynamic parameters, 654
thermodynamic potential, 654
thermodynamics and, 654–656
time evolution, 657–670
time reversal transformation, 658
transfer processes, 655
transition probability, 663
unitary transformation, 660
wave function, 639
white noise, 666
white spectrum, 666
Zwanzig master equation, 67
Statistical relaxation dynamics, 218
Steady-state sources, of neutrons, 356
Stereograms, in color holography, 270
Sterilization, by radiation, 585
Sticking coefficients, in plasma etching, 468
Stimulated emission, 58
  absorption and, 287
  definition, 285
  lasers, 496
  photon, Einstein's B coefficient for, 160
Stimulated scattering processes
  Brillouin, 396–397

common types, 390–391
conversion efficiency, 392
description, 389–392
origin, 390–391
Raman, 392–396
Stochastic averaging, 664
Stokes Raman scattering, 393–395
Stokes waves
  stimulated Raman scattering, 393–395
  stimulated scattering processes, 390–392
Stokes's Law, 665
Stopping cross sections
  Bethe–Born approximation, 34–36
  binary-encounter approximation, 33–34
  charge exchange, 36–40
  detailed balance, 40
  inelastic collision expressions, 37
  Rosen–Zener approximation, 37
Streaming operator (statistical mechanics), 662
Stress
  under load, 729
  residual, 729
Strong force
  basic forces and, 177–178
  color and, 189–190
  quarks and, 180–181
  reaction cross section, 184–185
Strong interactions, in unified field theory, 738
$SU(2) \times U(1)$ gauge theory, electroweak interactions, 193–199
$SU(3)$ gauge group, in quantum chromodynamics, 190
$SU(3)_{color}$ gauge group, 190, 202
Superconducting oxides, electronic states, 698–701
Superconductivity
  Bardeen–Cooper–Schrieffer theory, see Bardeen–Cooper–Schrieffer theory
  charge transfer fluctuations, 711–712
  coexistence of antiferromagnetism and, 710
  correlated electrons, see Correlated electrons
  Eliashberg theory, 686–687
  exciton mechanism, 248–249
  ferromagnetism and, 257–258
  intraatomic excitations, 711–712
  Luttinger theorem, 713
  Mott–Hubbard boundary, 712–713
Superconductors
  Cooper pairs, 517
  electron transport, 517–518
  energy gap, 517

quantum mechanics, 517–518
transition temperature, 517–518
Supergravity, 611
  in particle classification, 743–744
Superlattices, excitons, 246–247
Superradiance, in quantum optics, 543–545
Supersymmetric gauge field theories, 203–204
Superweak theory, CP violation, 111–112
Surface diffusion, in plasma etching, 468
Surface vibrations, nuclei, 424
Surgery, ultrasound applications, 735
Susceptibility, generalized (statistical mechanics), 674
Susceptibility method, in multiphoton spectroscopy, 331–332
Symmetric top molecules, optical spectroscopy, 317–318
Symmetries, in relativistic quantum electrodynamics, 164
Symmetry
  breaking, in gauge theory, 745
  quasicrystals, 562–563
Symmetry energy, nuclei, 418
Synchronization, clocks, Lorentz transformation and, 620–621

Tau particles
  interactions with photons, 169–170
  in lepton family, 179–180
Technicolor, dynamical symmetry breaking model, 203
Temperature
  high, dense matter physics, 124–126
  plasma components, 454, 461
Temporal coherence, holography and, 262–263
Tensors
  diffusion, 662
  patterns, two-photon processes, 342–344
  Riemann–Christoffel, 595–596
  in space–time theories, 590
Tetradic operators, 660
Thermal blooming
  multiphoton spectroscopy and, 336
  in nonlinear optical processes, 400–401
Thermal conductivity
  dense matter, 140–142
  in plasma physics, 457
Thermal plasmas, 459
Thermodynamic parameters, local, 669
Thermodynamic potential, 654

Thermodynamics, statistical
   mechanics and, 654–656
Third-harmonic conversion
   applications, 387
   description, 387
Third law of thermodynamics, 482
Thomas–Fermi–Dirac method, 124
Thomas–Fermi method, 123–124
Thomson scattering, photons, 160–161
Three-wave difference-frequency
   mixing, 381
Three-wave sum-frequency mixing,
   380–381
Time
   concepts of (special relativity),
      617–618
   dilation, Lorentz transformation
      and, 621–622
   proper (general relativity),
      598–599
Time correlation function, 106
Time delay, in PPN formalism,
   605–606
Time derivative, phenomenological,
   664
Time evolution
   Liouville description, 657–661
   phenomenological descriptions,
      661–670
Time-reversal invariant equations
   in intramolecular dynamics, 213
   irreversible equations and, 213
Time reversal symmetry, 110
   broken, 661
Time reversal transformation, 658
Time reversal violation, 202
Total average velocity, in plasma
   physics, 457
Trajectories, in intramolecular
   dynamics, 208, 217
Transducers, magnetostrictive, 725
Transition probabilities (statistical
   mechanics), 663
Transition temperature, superconductors, 517–518
Transmission electron microscopy,
   quasicrystals, 553–555
Transmission positron microscopy,
   475–477
   applications, 476–477
   history of, 474
   operation of microscope,
      475–477
Transmission positron reemission
   microscopy, 477
Transmutations, dense matter, 127
Transparency, self-induced
   in nonlinear optical processes,
      404
   in quantum optics, 527–528

Trapping effect, of nonlinear
   resonance, 220
Trapping states, quantum, 212
Triplet states, helium atom, 51
Triton nuclide, 417
Tungsten carbide, bonding and
   structure, 65
Tunneling, in intramolecular
   dynamics, 223
Twins paradox, 623–625
Two-photon spectroscopy, one-photon
   spectroscopy and, 327

$U(1)$ phase rotation group, 190, 194
Ultrasonics
   acoustic stress wave emission,
      731–732
   acoustic waves, 717–724
   aerosol production, 733
   applications in microscopy,
      730–731
   chemical analysis, 729
   cleaning, 732
   degassing of liquids and melts, 734
   drying of heat-sensitive materials,
      734
   effects of high-power acoustic
      waves, 732
   emulsification, 733
   extraction, 733
   flow measurement, 729
   foam control, 734
   high-power applications,
      732–735
   homogenization, 732
   impact grinding, 733
   level measurement, 729
   low-power applications,
      728–732
   machining, 733
   material characterization, 729
   medical diagnostics, 729–730
   medical therapy, 735
   metal casting, 734–735
   metal formation, 733
   metal welding, 734
   nondestructive testing, 728–729
   passive applications, 728–732
   plastic formation, 733
   plastic welding, 733–734
   sonar, 731
   stress determination, 729
   surface acoustic wave devices, 731
   surgical therapy, 735
   temperature measurement, 729
   wave generation, 724–727
   wave reception, 724–727
Ultraviolet lasers, 294

Uniaxial stress
   effect on bound-exciton complexes,
      240–241
   exciton splitting, 236–237
Unified field theories
   definition, 737–738
   elementary particle spectrum,
      739–740
   fundamental interactions, 738–739
   gauge, 744–745
   Kaluza–Klein, 742
   Lorentz invariant, 741
   range of interactions, 740–741
   renormalization group, 741–742
   strength of interactions, 741
   string, 745–747
   symmetry uses, 742–744
   Weinberg–Salam, 739
Unimolecular decay, 225–226, 228
Unitary transformation (statistical
   mechanics), 660
Universal gravitation, Newton's law
   of, 483
Universality, nonlinear dynamical
   systems, 80
Unpredictability, classical chaotic
   systems, 74–75

Vacuum expectation value, in Higgs
   mechanism, 196–198
Vacuum state, photons, 156
Vacuum ultraviolet lasers, 294
Valence bands, exciton, 232
Valley of beta stability, 420–421
Van der Waals interaction, 29
Vapor pressure thermometry, below
   4 K, 301
Velocities, Lorentz transformation and,
   619–620
Vibrational degrees of freedom
   diatomic molecules, 307
   polyatomic molecules, 314
Vibrational energy, in intramolecular
   dynamics, 210
Vibrational levels, polyatomic
   molecules, 318–319
Vibrational spectroscopy, 100
Vibrational structure, polyatomic
   molecules, 319–320
Vibrations
   surface, nuclei, 424
   transverse, wave equation, 497–498
Vibronic coupling, in molecular two-
   photon transitions, 347
Viscosity, shear, dense matter, 142–143
Visible lasers, 294
Von Weizscker semiempirical mass
   formula, 418

Wave equation
    for acoustic waves, 718–719
    classical, solutions, 498–499
    for particles, 508–511
    for transverse vibrations, 497–498
Wave functions
    electron orbital, covalent crystals
        and, 63–64
    in statistical mechanics, 639
Wave mechanics of scattering
    Born approximation, 26
    history, 46–47
    Massey–Mohr approximation,
        26–27
    scattering amplitude, 24–25
    semiclassical cross section, 25–26
Wave motion (quantum mechanics)
    boundary conditions, 499–500
    classical wave equation,
        498–499
    dispersion relations, 499
    group velocity, 500–503
    phase velocity, 499
    superposition solutions, 500
    in three dimensions, 502–503
    wave equation for transverse
        vibrations, 497–498
Wave number, molecular optical
    spectroscopy, 307
Wave packets, quantum mechanics,
    501
Wave–particle duality
    in quantum mechanics, 503–508
    quantum optics and, 520

Wave properties, quantum
    electrodynamics, 153
Wave theory
    early, of light, 150
    optical diffraction and, 438
Wavefunctions, erratic nodal patterns,
    227
Wavelength sensitivity, holographic
    recording materials, 268
Wavelengths, laser
    extreme ultraviolet, 294
    infrared, 293–294
    range, 292–293
    ultraviolet, 294
    vacuum ultraviolet, 294
    visible, 294
Waves
    hypersonic, 717
    matter, 504
    supersonic, 717
Weak force
    basic forces and, 177–178
    charge-parity (CP) violation, 201
    leptons and, 179–180
    quarks and, 181
    reaction cross section, 184
    strength relative to electromagnetic
        force, 195
    unification with electromagnetic
        force, 193
Weak hypercharge, in electroweak
    interactions, 194, 198
Weak interactions, in unified field
    theory, 738–739

Weapons, laser applications, 301
Weinberg angle, in electroweak theory,
    195, 199–200
Weinberg–Salam–Glashow theory,
    dense matter, 144, 147–148
Weiss molecular field, 253
Welding
    metals, ultrasound applications, 734
    plasmas, 451
    plastics, ultrasound applications,
        733–734
Wet etching, with plasmas, 466
White noise, 666
White spectrum, 666
Wurtzite crystal structures, excitons,
    233–234

X-ray diffraction
    molecular structures, 101–102
    quasicrystals, 553–555, 562
X-ray small-angle scattering,
    molecular structures, 101
X rays
    absorption in matter, 575
    sources, 573

Zeeman spectra, bound excitons, 237
Zinc-blende crystal structures,
    excitons, 234
Zwanzig master equation, 675